Xenopus

A LABORATORY MANUAL

ALSO FROM COLD SPRING HARBOR LABORATORY PRESS

RELATED TITLES

Mammalian Development: Networks, Switches, and Morphogenetic Processes
Generation and Interpretation of Morphogen Gradients
Calcium Techniques: A Laboratory Manual
Regeneration
Size Control in Biology
Budding Yeast: A Laboratory Manual
Ion Channels: A Laboratory Manual
Neurogenesis
Manipulating the Mouse Embryo

WEBSITE

www.cshprotocols.org

Xenopus

A LABORATORY MANUAL

EDITED BY

Hazel L. Sive
Northeastern University

CSH PRESS COLD SPRING HARBOR LABORATORY PRESS
Cold Spring Harbor, New York • www.cshlpress.org

XENOPUS
A LABORATORY MANUAL

Publisher	John Inglis
Acquisition Editor	Richard Sever
Managing Editor	Maria Smit
Developmental Editors	Jeremy Allen, Heather Cerne, Christin Munkittrick, Martin Winer
Senior Project Manager	Inez Sialiano
Permissions Coordinator	Carol Brown
Production Editor	Kathleen Bubbeo
Production Manager and Cover Designer	Denise Weiss
Director of Product Development & Marketing	Wayne Manos

Cover art: A *Xenopus laevis* transgenic tadpole that is expressing neuronal β-tubulin GFP (green) and that has been stained for actin (red) and DNA (blue) is shown. The image was kindly provided by Helen Rankin Willsey (University of California, San Francisco), who uses *Xenopus* tadpoles like this one to understand the molecular mechanisms underlying psychiatric disorders such as autism by mutating risk genes in half of the animal. For more information, visit willseyfroggers.org.

Library of Congress Cataloging-in-Publication Data

Library of Congress Control Number: 2022949627 |
ISBN 978-1-621821-80-9 (paperback) | ISBN 978-1-621821-77-9 (ePub3)

10   9   8   7   6   5   4   3   2   1

Contents

SECTION 3   Oocytes and Oogenesis
Editor: Kimberly L. Mowry

INTRODUCTION

PROTOCOLS

SECTION 4   Embryology I
Editor: Sally A. Moody

INTRODUCTION

SECTION 5   Embryology II
                 Editor: Sally A. Moody

INTRODUCTION

SECTION 6   **Cell Biology**
Editor: Anna Philpott

INTRODUCTION

PROTOCOLS

SECTION 7   **Xenopus Extract Systems**
Editor: Rebecca Heald

INTRODUCTION

PROTOCOLS

SECTION 8   Chromosome Dynamics In Vitro
Editor: Rebecca Heald

INTRODUCTION

PROTOCOLS

SECTION 11   Genomics
Editor: Michael J. Gilchrist

INTRODUCTION

PROTOCOLS

SECTION 12   Transcriptomics and Proteomics
Editor: Michael J. Gilchrist

INTRODUCTION

PROTOCOLS

SECTION 15 Neurobiology: Physiology and Behavior

Editors: Ben G. Szaro and Hollis T. Cline

INTRODUCTION

PROTOCOLS

SECTION 16 Immunology

Editor: Jacques Robert

INTRODUCTION

PROTOCOLS

SECTION 17   Regeneration

Editor: Juan Larraín

INTRODUCTION

PROTOCOLS

SECTION **18**   Chemical Screening and Toxicity Testing
Editor: André W. Brändli

INTRODUCTION

PROTOCOLS

Foreword: A Brief History of *Xenopus* in Biology

OVERVIEW

Xenopus is one of the premier model systems to study cell and developmental biology in vivo in vertebrates. Here we review how this South African frog came to be favored by a large community of scientists after the explosive growth of molecular biology and examine some of the original discoveries arising from this sturdy frog. Experimental embryology started in *Rana* but developed in newt embryos for historical reasons. A long lineage of mentorship, starting with Boveri, Spemann, Baltzer, Hadorn, and Fishberg, used newt embryos. In Oxford, Fishberg made the transition to *Xenopus laevis* because it was widely available for human pregnancy tests and laid eggs year-round and he fortuitously isolated a 1-nucleolus mutant. This mutant allowed nuclear transfer experiments showing that genetic information is not lost during cell differentiation and the demonstration that the nucleolus is the locus of transcription of the large ribosomal RNAs. With the advent of DNA cloning, the great equalizer in biology, microinjected *Xenopus* oocytes became an indispensable tool providing the first living cell mRNA translation, Polymerase II and III transcription, and coupled transcription–translation systems in eukaryotes. *Xenopus* embryos provide abundant material to study the earliest signaling events during vertebrate development and have been subjected to saturating molecular screens in the genomic era. Many novel principles of development and cell biology owe their origins to this remarkably resilient frog.

INTRODUCTION

We owe a debt of gratitude to Hazel L. Sive for pioneering the Cold Spring Harbor *Xenopus* laboratory courses. The detailed protocols developed there—and compiled in *Early Development of Xenopus laevis* by Sive, Grainger, and Harland published by Cold Spring Harbor Laboratory Press twenty years ago—have guided the *Xenopus* experimental work that has exploded in so many laboratories. The resilient frog is considered the old martyr of science. We can trace fundamental discoveries made with the tenacious *Rana* to Galvani, who in 1791 discovered electricity using isolated frog leg muscles. A few years later, Volta used the same assay to invent the electric battery. Here we relate how a frog from remote South Africa became a powerful model system in biology in the course of a few decades.

EMBRYOLOGY

The beginning of experimental embryology can be traced to work with *Rana*. Wilhelm Roux killed one of the two first two embryonic cells in 1883. In 1895, Thomas H. Morgan further showed that if the dead cell is removed, the surviving cell could self-organize into a tadpole of half the normal size. When biologists realized that development could be interrogated experimentally, research quickly shifted to newt (salamander) embryos, which had advantages for transplantation studies and a long history. This saga starts with Theodore Boveri, the father of European cell biology. He mentored Hans Spemann, who trained Fritz Baltzer, who in turn mentored Ernst Hadorn in Switzerland. A student of Hadorn's, Michail Fishberg, moved to Oxford where instead of newts he started work on the frog *Xenopus* and mentored J.B.G., who in turn mentored many, including E.D.R. and Ron Laskey, who mentored Richard Harland. Much of the current excitement in the *Xenopus* developmental biology field can be traced back to this lineage, culminating in Fishberg and his transition to a frog from distant lands.

FIGURE 1. A clone of *Xenopus laevis* frogs derived by nuclear transplantation of nuclei originating from a single embryo. The donor nuclei were transplanted into the eggs of a wild-type pigmented female (shown) and marked by both the 1-nu mutation and an albino mutation; all members of the clone are male. (Reproduced, with permission, from Gurdon, *Development* **140:** 2449–2456 [2013], © Company of Biologists.)

Xenopus laevis had been previously introduced in the United Kingdom by Lancelot Hogben, who had been teaching in Cape Town, South Africa, for the purposes of pregnancy tests. *Xenopus* females injected into the dorsal lymph sac with human urine containing human chorionic gonadotrophin respond by laying eggs the next day. This proved much more convenient than earlier pregnancy tests using rabbits. Gradually, improved *Xenopus* husbandry allowed the establishment of multiple *Xenopus* colonies throughout Europe and the United States. At the end of World War II, Pieter Nieuwkoop in Holland pioneered embryological studies in *X. laevis*. In the wild, it lives under the paradisiacal blue skies of South Africa, yet it will lay eggs year-round after rains form puddles. This provided a fundamental experimental advantage over *Rana* and newts, which have short springtime breeding seasons.

In Oxford, Fishberg had the wisdom to breed a spontaneously arising mutant that had one nucleolus per diploid cell instead of the normal two. This invaluable genetic marker allowed his graduate student J.B.G. to re-interpret the nuclear transplantation studies started in *Rana* by Briggs and King. This led to the demonstration that the nuclei of differentiated cells did not lose genetic material during development, as they were able to develop into fertile frogs (Fig. 1). Interest in *Xenopus* exploded when Donald Brown in the United States, with J.B.G, showed that 0-nucleolus tadpoles did not synthesize the large ribosomal RNAs. Igor Dawid, a colleague of Brown, used *Xenopus* interspecies hybrids to demonstrate that mitochondria contain DNA, which is always inherited from the mother and not from the sperm. Ron Laskey, a graduate student of J.B.G., showed that components accumulated in the oocyte-initiated DNA replication in the egg.

With the advent of molecular cloning, the great equalizer in biology, research in *Xenopus* was set for extraordinary growth from the 1970s onward.

THE *XENOPUS* OOCYTE AS A LIVING TEST TUBE

The abdomen of an adult *Xenopus* female is filled with thousands of large oocytes of 1.2 mm in diameter. When removed from the mother, oocytes can be cultured for several weeks in a simple saline solution developed by Lester Barth for neural induction studies in salamanders. Whereas the fertilized egg immediately replicates nuclei, the oocyte remains metabolically active but unchanged in simple culture conditions for a few weeks. Microinjected oocytes provided many breakthroughs in the analysis of gene expression in vertebrates.

Microinjected oocytes provided the first translation system for eukaryotic mRNAs. Globin mRNA, purified from immature red blood cells, a gift from Jean Brachet and H. Chantrenne

(Belgium), was shown by J.B.G. and Charles Lane to be efficiently translated into globin protein when injected into *Xenopus* oocyte. A decade later, the use of microinjected synthetic mRNAs transcribed in vitro, introduced by Douglas Melton, opened many doors.

Microinjection of DNA into the large oocyte nucleus, called the germinal vesicle, provided the first transcription system for eukaryotic RNA polymerases II and III. With the advent of DNA cloning, E.D.R. showed that *Xenopus* oocytes offered the first coupled transcription–translation system for the expression of cloned genes. To this day, the stability of transcription complexes is being studied by successive microinjections of DNA.

Proteins translated in oocytes, if they contain the proper signal sequences, can be either secreted or localized to the plasma membrane. The *Xenopus* oocyte provided the premier system to study the properties of cloned receptors and ion channels in electrophysiology. Nucleocytoplasmic transport of proteins and small RNAs microinjected into the cytoplasm, starting with iodinated histones, was demonstrated. Further, gene expression of somatic nuclei from differentiated cells injected into the oocyte were reprogrammed to resemble that of embryos and stem cells.

The simple, but revolutionary, idea that underlies the fertile experimental heritage of the *Xenopus* oocyte is that if one introduces a single purified molecule at a time, the living test tube of the cell will take care of the rest.

EGG CELL-FREE EXTRACTS

Xenopus eggs can be broken by centrifugation to generate concentrated cytoplasmic preparations for the study of many cell biological processes. These extracts were initially developed by Yoshio Masui to study the assembly of the nuclear envelope around DNA. These cell-free extracts have been essential for studies of the cell cycle, mitotic spindle formation, microtubule assembly, and chromatin formation. Modern proteomic techniques have defined the composition of these extracts in detail and, because protein complexes can be depleted or added back, *Xenopus* extracts provide an essential postgenomic biochemical assay system.

GERM LAYER DIFFERENTIATION

A great advantage of the amphibian embryo is that the future tissue allocations can be targeted shortly after fertilization, when a less pigmented dorsal crescent forms as a result of a cortical rotation of the egg. Studies on dorsal–ventral cell lineages, and the effect of microinjected molecules on them, can be followed in great detail. The dorsal crescent becomes the Spemann organizer at gastrula, which is the source of dorsal–ventral differentiation signals (Fig. 2). Explants of animal cap cells at blastula differentiate into epidermis, but if treated with TGF-β growth factors such as activin they will differentiate into mesoderm, and at higher doses into endoderm, as shown by Jim Smith and Makoto Asashima. Within the ectoderm, the inhibition of BMP and TGF-β signaling results in the induction of central nervous system. Studies in *Xenopus* have greatly enhanced our understanding of morphogen gradients in development and provided the guiding principles for the recent protocols used to channel the histotypic differentiation of mammalian embryonic stem cells in culture.

REVERSE GENETICS IN *XENOPUS*

Synthetic mRNAs transcribed from cloned genes allowed many gain- and loss-of-function studies. Dominant-negative constructs can inhibit gene function, and overexpression can test for the function of novel genes. Functional screens to identify novel molecules were carried out in microinjected embryos using pools of mRNAs transcribed from cDNA libraries, followed by sib selection. The

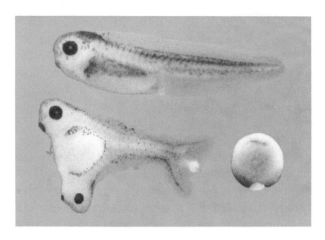

FIGURE 2. The Spemann organizer transplantation experiment in *Xenopus*. A small region of the dorsal blastopore lip transplanted into the ventral side of a gastrula host (*lower right*, albino transplanted tissue) results in a perfectly patterned second body axis. The Spemann organizer region has been subjected to saturating molecular screens that identified many novel secreted growth factor antagonists such as Chordin, Noggin, Frzb, Cerberus, and Dickkopf. (Reproduced, with permission, from De Robertis, *Nat Rev Mol Cell Biol Genet* **7:** 296-302 [2006], © Springer Nature.)

Xenopus Spemann organizer proved a particularly rich fishing ground for novel molecules, many of which were, unexpectedly, found to be secreted inhibitors of growth factors, such as Noggin, Chordin, Frzb, Cerberus, and Dickkopf. Genes transcribed by a common signaling pathway could be identified as synexpression groups in large-scale in situ hybridization screens of embryos. A notable gene isolated in *Xenopus* was the first Hox gene, although in this case E.D.R. used probes previously identified by Walter Gehring in *Drosophila*. (Interestingly Gehring, like Fishberg, was a graduate student of Hadorn's, closing the circle in this lineage of developmental biologists.) A convenient loss-of-function system that depletes both maternal and zygotic mRNAs was provided by antisense morpholino oligonucleotides (MOs), first introduced by Janet Heasman, which have been extremely effective.

One limitation of *X. laevis* is its long generation time of more than a year. This led to the introduction of *Xenopus tropicalis* which reproduces in 4–6 months and is suitable for transgenesis and site-directed mutagenesis and has a diploid genome. *X. laevis* is a subtetraploid frog that originated as a hybrid between two different species. The completion of the *X. laevis* genomic sequence by Richard Harland and colleagues was a crucial landmark that provides an invaluable resource for gene targeting and understanding the role of gene loss and duplication during evolution. With the proliferation of candidate target genes from human genetics, microinjected *Xenopus* embryos provide an in vivo model to rapidly screen their relevance to disease in a vertebrate.

CONCLUSIONS

In a few decades a vibrant scientific fellowship has developed among those who love experimenting with the embryos of this resilient frog with their own hands. This community benefits from the biannual *Xenopus* meetings, a superb Xenbase database, the National *Xenopus* Resource (NXR) supported by the NIH, and last, but not least, by the Cold Spring Harbor Laboratory courses on *Xenopus* development. This new book *Xenopus: A Laboratory Manual*, edited by Hazel L. Sive, provides biologists with an invaluable collection of protocols that ensure that *Xenopus* researchers will continue discoveries on the mysteries of vertebrate development well into the future.

Edward M. De Robertis
Department of Biological Chemistry
University of California, Los Angeles
eddy@mednet.ucla.edu

John B. Gurdon
The Gurdon Institute, University of Cambridge
johngurdon@gmail.com

Acknowledgments

The acknowledgments for this Manual span many years and a multitude of people. Enormous thanks to Cold Spring Harbor Laboratory for a long, wonderful relationship with our community, and to *Cold Spring Harbor Protocols* for their commitment, enthusiasm, and unending patience as we put this large project together. Special thanks to Cold Spring Harbor Laboratory Press, specifically Richard Sever (Assistant Director, Cold Spring Harbor Laboratory Press), Maria Smit (Managing Editor), Inez Sialiano (Senior Project Manager), Denise Weiss (Production Manager), and Kathleen Bubbeo (Production Editor).

Huge gratitude to the editors of this manual who expertly brought their chapters together: Andre Brandli, Daniel Buchholz, Ken Cho, Hollis Cline, Lance Davidson, Mike Gilchrist, Richard Harland, Rebecca Heald, Rob Grainger, Matt Guille, Juan Larrain, Laura Anne Lowery, Sally Moody, Kim Mowry, Anna Philpott, Jaques Robert, Yun-Bo Shi, Ben Szaro, and Gert Jan Veenstra.

Warm thanks and congratulations to the authors of each protocol. Your work will not only assist long-time members of the *Xenopus* community, but will introduce powerful protocols to those exploring the *Xenopus* system for the first time.

We are grateful to Sir John Gurdon and Professor Eddy de Robertis for their excellent Foreword.

Thank you to the NIH, NSF, and all other funding agencies for supporting research that led to development of the protocols included and that contribute to the National *Xenopus* Research (NXR) and the European *Xenopus* Resource (EXRC), so pivotal for effective *Xenopus* research.

Finally, I would like to acknowledge the frogs. The field of vertebrate development has progressed because *Xenopus* embryos are so accessible, so plentiful, and large enough to manipulate, relative to mammalian systems. An underlying principle of initiating the *Xenopus* course at Cold Spring Harbor was to teach colleagues how to handle and look after frogs with kindness and respect. The outcomes of *Xenopus* research have been extraordinary, guiding studies in mammalian models and impacting understanding of human development and disease. Without the frogs, we would be nowhere.

Introduction

Welcome to the World of *Xenopus*! More than 20 years ago, Cold Spring Harbor Press published the first edition of our *Xenopus* Lab Manual with those words at the beginning. I am thrilled that in 2021, together with *Cold Spring Harbor Protocols*, our community has brought together this enormous and seminal collection of *Xenopus* protocols. With 18 Chapters, 18 Topic Introductions, 120 Protocols, and 213 authors, there has never before been a book of this magnitude focusing on the *Xenopus* system.

Congratulations to everyone involved!

I have known *Xenopus* for many decades, starting as an undergraduate at Wits University in Johannesburg, South Africa. On the top floor of our Zoology building was a room full of these frogs. In large concrete tubs, they would hang, weightless, with just nose and eyes in the air. We collected the frogs with nets from local ditches, and in a single sweep scooped up dozens of plump animals. Back in the frog room, if you simply put together a male and female, she would lay hundreds of perfect eggs. Later we put the frogs back in their ditch. For an undergraduate project, I grew thousands of embryos, kept in rows of rectangular dishes. One day as I walked along my rows, I noticed that all embryos of the same age were doing the same thing—when a few hours old, all comprised four cells, next day all were kidney bean–shaped with a patch of black pigment at one end, and the day after all had eyes and long tails. Like magic! Maybe, I thought, one could extract from the embryos chemicals that were making them develop in sync, to understand this magical process. The amazing development of *Xenopus* embryos put me on the career path of developmental biology research.

When I came to the United States to pursue a Ph.D., I felt so grateful for the opportunity, but the transition was difficult and lonely. Happily, it turned out that there was a frog room at the new university. In old bathtubs were the same kind of frogs that had hung in the water at Wits: the South African claw-toed frog, *Xenopus*. I was so pleased to see them! They had such familiar friendly faces, even though they'd been born in the United States. Those frogs made me feel less uneasy in my new home. As a postdoctoral researcher, I began to search for molecules that tell *Xenopus* embryos what to do, developing the cement gland as a positional marker, and focusing on early steps of nervous system and face formation. All the way to the present, with my wonderful research group, *Xenopus* embryos have given significant insights and unusual answers in our research.

Toward the end of my postdoc, I started thinking that we owed the frogs something. In South Africa we had put the frogs back into their ditch after they laid eggs. In the United States, *Xenopus* lived in the laboratory and we had to keep them healthy, and to gently collect eggs from the female. It seemed that new investigators would value instruction into *Xenopus* use and care, and this led me to propose and set up the first Cold Spring Harbor (CSH) *Xenopus* course in 1992. As a new Assistant Professor, I taught the course with Richard Harland and subsequently Robert Grainger, and we three co-edited the first edition of the *Xenopus* Manual. Through leadership from other outstanding *Xenopus* researchers, I am deeply proud that the course runs to this day, almost 30 years on.

Cold Spring Harbor has had a long, excellent relationship with our community. Terri Grodzicker embraced my proposal to set up a *Xenopus* course, and CSH has been pivotal in supporting the course all these years. Our partnership with *CSH Protocols* has made this manual a reality. On behalf of our community, I want to express greatest appreciation to Richard Sever (Assistant Director, Cold Spring Harbor Laboratory Press), Maria Smit (Managing Editor), and Inez Sialiano (Senior Project Manager). Their unending patience, polite prodding, and joy when submissions arrived made working on the manual a positive experience for everyone. Thank you so much!

And now, we have a new celebration of the power of *Xenopus*! In contrast to the first edition of this manual, which focused on the embryo, this second edition includes the broad range of techniques by which *Xenopus* oocytes, eggs, embryos, tadpoles, froglets, or adults can drive fundamental and translational research. The protocols include use of both *Xenopus laevis* and *Xenopus tropicalis*. They include techniques by which *Xenopus* embryos have made unparalleled contributions—defining the genes, signaling pathways, and cellular processes that direct vertebrate development. This new protocol set adds important methods addressing oocytes and oogenesis, cell biology, nuclear and chromosomal dynamics, imaging, genetics and gene editing, genomics, transcriptomics and proteomics, neurobiology, immunology, metamorphosis, regeneration, and chemical or toxicity screening.

A further innovation is that as well as a hard-copy edition, the manual and each protocol will be available online. Comments can be appended to existing protocols and new ones can be added to keep the *Xenopus* collection current and of greatest usefulness.

Our manual includes contributions from eminent senior *Xenopus* investigators, with a Foreword from Professor Eddy de Robertis and Nobel Laureate Sir John Gurdon. But many more junior investigators have contributed protocols, ensuring continuation and innovation for the *Xenopus* system. As noted further in the Acknowledgments, my warmest and most sincere thanks to every chapter editor for planning contents and identifying experts, and my gratitude to every author of a protocol or topic introduction.

Whether you are new to *Xenopus*, and wondering whether this is the system to address your questions, or whether you are a frog veteran wanting to expand your repertoire, please explore the vast range of considerations and technical options laid out in the manual. Congratulations on your choice of system, and welcome back to the world of *Xenopus*!

Hazel L. Sive*
Northeastern University, Boston, Massachusetts
h.sive@northeastern.edu

---

* During preparation of much of this Manual, Hazel Sive was Professor of Biology at the Massachusetts Institute of Technology, and Member, Whitehead Institute for Biomedical Research.

SECTION **1**: Husbandry

Xenopus Husbandry

Richard M. Harland[1] and Hazel L. Sive[2,3]

[1]*Department of Molecular and Cell Biology, University of California, Berkeley 94720-3200, USA;* [2]*Northeastern University, Boston, Massachusetts 02115, USA*

Adult frogs that are well-cared-for will give high-quality eggs and embryos for use in every *Xenopus* protocol. Thoughtful frog husbandry is thus pivotal to successful research using these organisms. Protocols for successfully raising tadpoles, establishing and maintaining water quality, and detecting specific pathogens are key to maintaining healthy frog populations.

A HISTORICAL VIEW

Although frogs have been popular laboratory animals from the beginnings of experimental science, the genus *Xenopus* has become the most-used laboratory amphibian for two reasons. First, ovulation can be induced throughout the year by hormone injection, and, second, the physical attributes of the embryos make them accessible to state-of-the-art technical approaches. Following extensive use in pregnancy tests, embryologists noted that they need not rely on production of embryos from seasonal frogs, and *Xenopus laevis* was popularized for experimental usage by Mikhail Fischberg and his student John Gurdon in Europe, and by Don Brown in the United States (Gurdon and Hopwood 2000).

In the natural state, *Xenopus* live in ponds, ditches, or rivers in Southern Africa. Wild-caught *Xenopus laevis* can yield outstanding numbers and quality of embryos. *X. laevis* can be shipped from South Africa or bred in Europe, Japan, or the United States and supplied to research groups. The frogs can be housed for many years of repeated ovulation and natural fertilization, although males are often harvested for testes and in vitro fertilization. In the case of *Xenopus tropicalis*, animals may be purchased or bred in individual research groups, especially in the case of transgenic animals.

Cursory assessment of the often muddy "natural" habitat of *Xenopus* suggested the animals were tolerant to a range of water conditions, but in the laboratory, optimal health and fecundity require careful attention to water chemistry, including optimal pH, ion levels, and temperature.

Historically, *X. laevis* were kept in bathtubs, artificial ponds or plastic tanks, with a "fill and dump" regimen of dechlorinated water and diets of offal and maggots. These frogs often picked up parasites from slaughterhouse offal, and between cleanings, offensive fouling of the water occurred. Furthermore, the water would spike in pH, causing free ammonia to damage the animals, so veterinarians came to require that the water be kept clean by constant flow of clean water. This brought its own problems, because the incoming water was often cold and saturated with atmospheric gas, leading to "gas bubble disease" (Green 2009). Animals directly from southern Africa were generally overtly

[3]Correspondence: h.sive@northeastern.edu

healthy and vigorous, but brought with them endemic skin and liver parasites, and so importation of wild-sourced animals has declined.

IMPROVING *XENOPUS* HUSBANDRY

Although animals can be held in still tanks, with at least biweekly dump and refill, it is difficult to maintain constant and optimal water quality to ensure health. Opportunistic pathogens may thrive (Feazel et al. 2009) and still-water tanks fed by nonsterile dechlorinated water are therefore not a long-term mechanism for keeping *Xenopus* adults healthy.

The current best solution is to install recirculating systems, which provide optimal water and temperature parameters and can include dozens of tanks and hundreds of frogs (see Protocol 1: Animal Maintenance Systems: *Xenopus laevis* [Shaidani et al. 2020a]; Protocol 2: Animal Maintenance Systems: *Xenopus tropicalis* [Shaidani et al. 2020b]). In these systems, purified water is supplemented with salts, enters the tanks, and on drainage is filtered to remove solid waste. The water then passes through a biological filter, consisting of denitrifying bacteria on a solid support, where breakdown of nitrogenous waste prevents high levels of ammonia that can damage animals. The water is sterilized by UV illumination before reentering the system. In theory, little new water needs to be added to a recirculating system, but in practice a water change of ~10% per day or 50% per week prevents buildup of potential toxins.

A diet of pellet food that includes all needed nutrients can keep frogs well-nourished and prevents fouling of the water that diets such as liver elicit.

Egg production is also affected by temperature fluctuation, and temperatures of >27°C can sterilize males or result in death. Seasonal changes can occur even in a recirculating system due to fluctuations in heating and air conditioning, so a chiller/heater should be installed to maintain the desired temperature.

Although some diseases remain problematic, particularly the chytrid fungus and the opportunistic pathogen *Mycobacteria liflandii*, the introduction of recirculating systems, with scheduled changes of UV sterilizing bulbs, has diminished outbreaks of disease. The incoming water is generally held in a reservoir, where it can degas, and because it only replaces a fraction of the water every day, gas bubble disease is not a problem, even with fairly vigorous recirculation (see Protocol 4: Defining the Specific Pathogen-Free State of *Xenopus* Using TaqMan Assays [Hensley et al. 2020]).

Recirculating systems have different tank sizes. Large tanks allow the frogs to swim, as in the wild, and should be considered optimal. Thus, tanks of several feet in length allow *X. laevis* to swim, and 10 inches of water depth allows them to hang with their nostrils and eyes above the water, as they prefer to do. However, keeping ovulated frogs in groups that can rest for a few months before reovulation can necessitate smaller tanks. For *X. tropicalis*, tanks are often smaller, commensurate with their size (two to three frogs per liter is optimal for adults). Finally, some investigators give their frogs places to hide or rest, providing a richer environment. Wide PVC pipe lengths allow this for both *X. laevis* and *X. tropicalis*, and for *X. tropicalis*, floating items such as plastic lids allow them to retreat from well-lit areas or sit partially submerged. Tanks should have a secure lid, although it is essential to maintain an air gap between the lid and the water surface to prevent the frogs from drowning.

FROM TADPOLES TO FROGS

In addition to maintenance of healthy adults, similar water parameters apply to raising tadpoles (see Sec. 2, Protocol 3: How to Grow *Xenopus laevis* Tadpole Stages to Adult [Ishibashi and Amaya 2021]; Protocol 3: Raising and Maintaining *Xenopus tropicalis* from Tadpole to Adult [Lane et al. 2021]). The tadpoles generally start life as embryos in Petri dishes where they develop into swimming tadpoles. Shortly thereafter, they can be seen pumping water through their gills, and this is when they start feeding on particulate food. The ideal time to transfer them to the aquarium is still a matter of research, but usually they start feeding in Petri dishes, before being gradually introduced to the aquarium water, and then to a gentle flow in the frog system (Ishibashi and Amaya 2021; Lane et al. 2021).

INSTITUTIONAL ANIMAL CARE COMMITTEES AND *XENOPUS* HUSBANDRY

Animal welfare is regulated by the local institutional animal care and use committee (IACUC), but ultimately overseen by federal agencies in the United States. The IACUC has jurisdiction over general issues such as when the animals should be transferred out of the laboratory, to rooms directly supervised by institutional staff or to a recirculating system. Most frequently tadpoles are under the care of the principal investigator's staff. The IACUC has considerable latitude to decide when the tadpoles transition from "prevertebrates," which under law are not considered to be "vertebrate animals," for the purposes of pain management or euthanasia according to the American Veterinary Association Guidelines for the Euthanasia of Animals (National Research Council 2020). Some IACUCs consider the hatching stage or feeding to be the transition to vertebrate animalhood, whereas other IACUCs consider development of a vertebral column at metamorphosis to define a vertebrate. Regardless of definition, it is good and humane practice to ensure that tadpoles are anesthetized before procedures such as dissection or fixation once they develop reflex responses (tailbud stage). It is easy to achieve this by immersion in solutions of tricaine or benzocaine buffered with sodium bicarbonate, which increase adsorption of the drug. Although both these have been shown to be efficacious, an additional criterion has been introduced in the Guide to Lab Animal Care (National Research Council 2011) that administration of anesthetics should use "pharmaceutical-grade" products. This has led to some confusion, because these may not be available other than as ointments or gels, which are impractical to deliver in the water. However, MS-222 from Western Chemical is considered pharmaceutical-grade by at least some veterinarians. Moreover, local IACUCs are empowered to approve conventions based on what is allowed at National Institutes of Health (NIH) laboratories or where U.S. Food and Drug Administration (FDA)-approved fish anesthetics from specific vendors are approved. Investigators should consult their local veterinarians for assistance.

ACKNOWLEDGMENTS

Many thanks to Jacques Robert and Amy Sater for comments on the manuscript. R.M.H. was supported by NIH grant R35GM127069. H.S. was supported by NIH grant 5R01DE021109.

REFERENCES

Feazel LM, Baumgartner LK, Peterson KL, Frank DN, Harris JK, Pace NR. 2009. Opportunistic pathogens enriched in showerhead biofilms. *Proc Natl Acad Sci* **106:** 16393–16399. doi:10.1073/pnas.0908446106

Green SL. 2009. *The laboratory* Xenopus *sp.* CRC Press, Boca Raton, FL.

Gurdon JB, Hopwood N. 2000. The introduction of *Xenopus laevis* into developmental biology: of empire, pregnancy testing and ribosomal genes. *Int J Dev Biol* **44:** 43–50.

Hensley CL, Bowes LM, Feldman H. 2020. Defining the specific pathogen-free state of *Xenopus* using TaqMan assays. *Cold Spring Harb Protoc* doi:10.1101/pdb.prot106179

Ishibashi S, Amaya E. 2021. How to grow *Xenopus laevis* tadpole stages to adult. *Cold Spring Harb Protoc* doi:10.1101/pdb.prot106245

Lane M, Slocum M, Khokha MF. 2021. Raising and maintaining *Xenopus tropicalis* from tadpole to adult. *Cold Spring Harb Protoc* doi:10.1101/pdb.prot106369

National Research Council. 2011. *(US) Committee for the Update of the Guide for the Care and Use of Laboratory Animals. Guide for the care and use of laboratory animals*, 8th ed. National Academies Press, Washington, DC.

National Research Council. 2020. *AVMA guidelines for the euthanasia of animals: 2020 edition*. American Veterinary Medical Association. The National Academies Press. https://www.avma.org/sites/default/files/2020-01/2020-Euthanasia-Final-1-17-20.pdf.

Shaidani N-I, McNamara S, Wlizla M, Horb ME. 2020a. Animal maintenance systems: *Xenopus laevis*. *Cold Spring Harb Protoc* doi:10.1101/pdb.prot106138

Shaidani N-I, McNamara S, Wlizla M, Horb ME. 2020b. Animal maintenance systems: *Xenopus tropicalis*. *Cold Spring Harb Protoc* doi:10.1101/pdb.prot106146

Animal Maintenance Systems: *Xenopus laevis*

Nikko-Ideen Shaidani, Sean McNamara,[1] Marcin Wlizla, and Marko E. Horb[2]

National Xenopus Resource, Marine Biological Laboratory, Woods Hole, Massachusetts 02543, USA

Modular recirculating animal aquaculture systems incorporate UV sterilization and biological, mechanical, and activated carbon filtration, creating a nearly self-contained stable housing environment for *Xenopus laevis*. Nonetheless, minimal water exchange is necessary to mitigate accumulation of metabolic waste, and regular weekly, monthly, and yearly maintenance is needed to ensure accurate and efficient operation. This protocol describes the methods for establishing a new recirculating system and the necessary maintenance, as well as water quality parameters, required for keeping *Xenopus laevis*.

MATERIALS

It is essential that you consult the appropriate Material Safety Data Sheets and your institution's Environmental Health and Safety Office for proper handling of equipment and hazardous materials used in this protocol.

Reagents

Ammonia Freshwater and Saltwater Test Kit (API, Chalfont, PA)
Bleach (6.0% NaClO)

> *Dilute with Type II water to make a 10% bleach solution (final concentration of NaClO is 0.6%).*

Dechlorinator ($Na_2S_2O_3$; ProLine brand Pentair AES, Apopka, FL)

> *This contains both the anhydrous (CAS# 7772-98-7) and pentahydrate salt (CAS# 10102-17-7) molecules. 1.6–2.6 ppm of $Na_2S_2O_3$ per 1 ppm of chlorine is typically sufficient to dechlorinate water.*

Ethanol (190 proof)

> *Dilute to 70% with Type II water.*

Nitrate NO_3 Freshwater and Saltwater Test Kit (API, Chalfont, PA)
Nitrifying bacteria (ProLine brand, Pentair AES, Apopka, FL)
Nitrite NO_2 Freshwater and Saltwater Test Kit (API, Chalfont, PA)
Reef salt (Seachem Laboratories; Madison, GA)
Sodium bicarbonate ($NaHCO_3$; ProLine brand, Pentair AES, Apopka, FL)
Type II water (ASTM International 2018)
Virkon Aquatic (DuPont, Wilmington, DE)
Xenopus laevis frogs

---

[1]Present address: Iwaki Aquatic, Holliston, Massachusetts 01746
[2]Correspondence: mhorb@mbl.edu

Equipment

Air Pump, 40 LPM (Alita Industries, Inc., Baldwin Park, CA)
Orion Star A211 pH Benchtop Meter (Thermo Scientific, Beverly, MA)
PINPOINT Water Hardness Meter (American Marine Inc., Ridgefield, CT)
Recirculating Aquatic System (Iwaki Aquatic Systems and Services, Holliston, MA)

> There are several companies that sell recirculating aquatic systems, including Aquaneering, Tecniplast, and Aquatic Enterprises.

Scrub pads

METHOD

Each section below should be treated independently; these should not be viewed as consecutive procedures as the numbering may imply.

Establishing a New Recirculating System

1. Starting with a new or sterilized system, rinse the biomedia (provided by the system manufacturer) with Type II water and place in the biofilter.

2. Disable the water effluent exchange and UV sterilization, and keep the carbon filter empty.

> It is necessary to initially limit filtration and sterilization functions and allow nitrogenous waste products to build up sufficiently to support healthy growth of the biomedia.

3. Allow flow from a single tank on the end of each rack, and add 5–10 frogs to the system.

> Larger systems require more frogs. This number of frogs is suggested for a system with ~500-L capacity.

4. After 24 h, disable the water flow but keep the biomedia aerated using an external air pump.

5. Add nitrifying bacteria to the biofilter.

> Larger systems require more nitrifying bacteria. Follow the instructions outlined by ProLine.

6. Wait 1–2 h before restarting the water flow.

> During this brief period, the frogs present in the system should be kept in their tanks. The absence of water flow should not be detrimental to their health.

7. Perform daily NH_3/NH_4^+, NO_2^-, and NO_3^- level measurements using the relevant test kits.

> Wait for NH_3/NH_4^+ levels to measure at least 1–2 ppm.

8. Initiate UV sterilization and water exchange, and add carbon to the appropriate filter housing. The levels of NO_2^- and NO_3^- should begin to increase.

> Flow to 1–2 additional tanks can be started. Additional tanks can be filled as needed.

9. Allow for NH_3/NH_4^+ and NO_2^- levels to reduce to 0.5 ppm before adding more frogs to the system.

> The entire process can take 2 wk to 2 mo. Nitrifying bacteria can thrive when water parameters are consistent. If dosing reservoirs are not present, sodium bicarbonate can be used to stabilize pH. Once NH_3/NH_4^+ and NO_2^- levels reach the desired level, gradually add frogs to the system while monitoring water quality parameters. Care should be taken to not open flow to too many tanks at once as this can drain the sump below the set safe level and result in an automatic system shutdown.

Daily Water Assessment and Maintenance Required for Proper Operation

10. Record system readings. See Table 1 for age-appropriate guidelines for water temperature, pH, conductivity, and frog density.

> This allows the user to deduce trends and identify potential issues that may be occurring.

11. Inspect the UV bulb.

> Check system display to confirm the UV bulb is on and functioning correctly. Replace the bulb if necessary.

TABLE 1. Water-quality parameters for *Xenopus laevis*

Age	Water temperature (°C)	pH	Conductivity (μS)	Tank density
Tadpole/froglet	24 (23–25)	7.4 (7.2–7.6)	1000 (900–1100)	2–4 per 1 L
Juvenile	22 (20–24)	7.8 (7.0–8.5)	1600 (1200–1800)	1 per 1–2 L
Adult	20 (18–20)	7.8 (7.0–8.5)	1600 (1200–1800)	1 per 3 L

The water quality parameters and animal stacking density necessary for optimal growth vary, dependent on the life history stage of *X. laevis* (Hilken et al. 1995; Green 2009). These parameters are listed in the table above. The optimal conditions, as well as the allowable tolerance range, are given in parentheses.

12. Regularly scrape internal sides of the tanks with scrub pads to remove any algae buildup.

    Algal buildup will vary depending on several factors including lighting and dissolved nutrients. Excessive buildup should be removed as needed.

13. For systems that use stand pipes with overflows, shake the stand pipes to clear them. Tanks with only an overflow bulkhead will require daily active removal of detritus to keep them clean.

14. Exchange 10% of the system water.

    Modern recirculating aquatic systems can be programmed to perform water exchange automatically. Make sure that the automatic setting is kept at 10% daily exchange.

15. Measure pH and conductivity dosing reservoirs daily.

    Use sodium bicarbonate to buffer and regulate pH and sea salt to control conductivity.

16. Observe carbon and mechanical filters for buildup of waste and replace as needed once passage of water through them is significantly impeded.

17. Observe individual tanks for accumulation of waste and film on the bottom and sides, and sanitize tanks that are particularly dirty.

    To clean tanks, scrub and rinse with Type II water, spray with 10% bleach and let sit for 1 h, rinse again with Type II water, spray with 70% ethanol and let sit for 1 h, rinse once more with Type II water, and allow to air dry.

Weekly Water Assessment and Maintenance Required for Proper Operation

18. Assess NH_3/NH_4^+, NO_2^-, and NO_3^- levels with the relevant API test.

    Acceptable measurements are as follows: NH_3/NH_4^+ =0 ppm (0–0.5 ppm), NO_2^- <1 ppm, NO_3^- <40 ppm, alkalinity >40 ppm. If levels are out of range, perform a 25% water exchange. Repeat water change each day until values reach acceptable levels.

19. Test pH and conductivity with a reliable external probe as a reference for internal system probes.

    If the external and internal probes are not in agreement, clean the system probe and check that it is correctly reading calibration solutions. Recalibrate the probe if necessary.

20. Measure temperature with an external thermometer.

    Heaters and chillers are used to keep the temperature stable. Measuring with an external thermometer confirms that the system is calibrated to correctly engage heaters and chillers as necessary to keep the water temperature stable. Recalibrate or replace the system thermometer if it is not in agreement with the external thermometer.

Annual Water Assessment and Maintenance Required for Proper Operation

21. Replace UV bulbs and check to ensure that the quartz sleeve is intact.

22. Replace pH electrodes.

23. Remove and clean return pipes.

24. Replace all rubber tubing.

Cite this protocol as *Cold Spring Harb Protoc*; doi:10.1101/pdb.prot106138

System Sterilization

System sterilization is necessary to restore full operational capabilities following an outbreak of disease.

25. Disable all system components including the water pump, biomedia agitators, and probes.

26. Discard biomedia or sterilize in a 10% bleach solution.

27. Scrub the inside of the sump and associated parts and remove all detritus.

28. Remove and dispose of all filter pads, cartridges, and carbon.

29. Disassemble and scrub all the tanks and accessories.

30. Rinse the tanks with water, coat with 10% bleach, and let stand for 1 h.

31. Rinse the tanks with water, coat with 70% ethanol, and let stand for 1 h.

32. Rinse the tanks with Type II water and let air dry.

33. Soak accessories in 10% bleach for 1 h.

34. Soak accessories in 10 g/L solution of $Na_2S_2O_3$ in Type II water for 1 h.

35. Soak accessories in 70% ethanol for 10 min, rinse, and air-dry.

36. Drain 75% of the sump volume according to the manufacturer's instructions.

37. Reassemble the system putting back all the tanks, filter housings, and new rubber tubing. Leave the filter housing empty, and do not add any carbon to the system.

38. Start the system.

39. Add bleach to the system.

      Ensure the final concentration of NaClO in the system is 0.06%.

40. Reduce the effluent rate to its lowest setting.

      Do not reduce the daily water exchange rate to <1%.

41. Wait 2–3 d, then increase the daily water exchange rate to ∼50%.

      This will begin removing the bleach from the system.

42. Begin testing the pH of the system after a week.

43. Add Virkon to the system once the system water pH matches the influent water pH.

      The concentration of Virkon should be 3.2 g per 1 m^2 of estimated internal surface area of the entire system.

44. Reduce the effluent rate back to its lowest setting.

      Virkon can cause foam accumulation and in some cases, may cause the system to overflow. To help mitigate this, limit agitation of the water and spray 70% ethanol directly on the foam to help disperse bubbles.

45. Wait 7 d then increase the daily water exchange rate to ∼50%.

      This will wash the Virkon out of the system.

46. Allow the system to run for 2–3 d, replace all rubber tubing, and follow the protocol for establishing a new recirculating system described above to stabilize the system and allow for housing animals.

DISCUSSION

Optimal eggs and embryos are produced by healthy laboratory population of *Xenopus laevis*. Regular monitoring of water-quality parameters, as well as daily monitoring and visual inspection of the animals, is crucial for prevention and early identification of developing conditions that may be detrimental to overall colony health. The use of tap water is discouraged, and starting with a good water source providing Type II water will aid in maintenance of optimal water parameters as well as limit the chances of introducing pathogens and toxic contaminants. Municipal tap water may also be basic and thus may add the additional complication of needing to use acid buffers to aid in stabilizing the pH. The stability of the parameters will also be affected by the animal maintenance system configuration, with recirculating systems being more cost effective and efficient than flowthrough systems and also requiring much less maintenance than static systems. Furthermore, animal health is best when the accumulation of nitrogenous waste is low, thus the initial effort made to establish a stable biofilter in a recirculating system is absolutely necessary. Nitrifying bacteria used in the biofilter aid in establishing an ammonia/nitrogen cycle in which toxic ammonia is oxidized by *Nitrosomonas* bacteria producing toxic nitrite that is further oxidized by *Nitrobacter* bacteria into nontoxic nitrate (Hem et al. 1994). Once this cycle is established, animals can be safely kept in the system and should remain healthy with regular maintenance to keep the system clean and water parameters within the tolerance range of the animals (McNamara et al. 2018). If the systems are operating well and the water parameters are within range yet the animals are not thriving, a potential presence of pathogens should be considered and tested. Presence of highly virulent pathogens can greatly affect animal quality of life and may require euthanasia off all the frogs housed in the system. In such a case, complete system sterilization is necessary before it can be safely used to house animals again.

REFERENCES

ASTM Standard D1193 – 06. 2018. "*Standard specification for reagent water.*" *ASTM International*, West Conshohocken, PA.

Green SL. 2009. *The Laboratory Xenopus sp.* CRC Press, Boca Raton, FL.

Hem LJ, Rusten B, Ødegaard H. 1994. Nitrification in a moving bed biofilm reactor. *Water Res* **28**: 1425–1433. doi:10.1016/0043-1354(94)90310-7

Hilken G, Dimigen J, Iglauer F. 1995. Growth of *Xenopus laevis* under different laboratory rearing conditions. *Lab Anim* **29**: 152–162. doi:10.1258/002367795780740276

McNamara S, Wlizla M, Horb ME. 2018. Husbandry, general care, and transportation of *Xenopus laevis* and *Xenopus tropicalis*. *Methods Mol Biol* **1865**: 1–17. doi:10.1007/978-1-4939-8784-9_1

Cite this protocol as *Cold Spring Harb Protoc*; doi:10.1101/pdb.prot106138

Animal Maintenance Systems: *Xenopus tropicalis*

Nikko-Ideen Shaidani, Sean McNamara,[1] Marcin Wlizla, and Marko E. Horb[2]

National Xenopus Resource, Marine Biological Laboratory, Woods Hole, Massachusetts 02543, USA

Modular recirculating animal aquaculture systems incorporate UV sterilization and biological, mechanical, and activated carbon filtration, creating a nearly self-contained stable housing environment for *Xenopus tropicalis*. Nonetheless, minimal water exchange is necessary to mitigate accumulation of metabolic waste, and regular weekly, monthly, and yearly maintenance is needed to ensure accurate and efficient operation. This protocol describes the methods for establishing a new recirculating system and the necessary maintenance, as well as water quality parameters, required for keeping *Xenopus tropicalis*.

MATERIALS

It is essential that you consult the appropriate Material Safety Data Sheets and your institution's Environmental Health and Safety Office for proper handling of equipment and hazardous materials used in this protocol.

Reagents

Ammonia Freshwater and Saltwater Test Kit (API, Chalfont, PA)
Bleach (6.0% NaClO)

   Dilute with Type II water to make a 10% bleach solution (final concentration of NaClO is 0.6%).

Dechlorinator (Na$_2$S$_2$O$_3$; ProLine brand Pentair AES, Apopka, FL)

   This contains both the anhydrous (CAS# 7772-98-7) and pentahydrate salt (CAS# 10102-17-7) molecules. 1.6–2.6 ppm of Na$_2$S$_2$O$_3$ per 1 ppm of chlorine is typically sufficient to dechlorinate water.

Ethanol (190 proof)

   Dilute to 70% with Type II water.

Nitrate NO$_3$ Freshwater and Saltwater Test Kit (API, Chalfont, PA)
Nitrifying bacteria (ProLine brand, Pentair AES, Apopka, FL)
Nitrite NO$_2$ Freshwater and Saltwater Test Kit (API, Chalfont, PA)
Reef salt (Seachem Laboratories; Madison, GA)
Sodium bicarbonate (NaHCO$_3$; ProLine brand, Pentair AES, Apopka, FL)
Type II water (ASTM International 2018)
Virkon Aquatic (DuPont, Wilmington, DE)
Xenopus tropicalis frogs

---

[1]Present address: Iwaki Aquatic, Holliston, Massachusetts 01746
[2]Correspondence: mhorb@mbl.edu

Equipment

Air Pump, 40 LPM (Alita Industries, Inc., Baldwin Park, CA)
Orion Star A211 pH Benchtop Meter (Thermo Scientific, Beverly, MA)
PINPOINT Water Hardness Meter (American Marine Inc., Ridgefield, CT)
Recirculating Aquatic System (Iwaki Aquatic Systems and Services, Holliston, MA)

> *There are several companies that sell recirculating aquatic systems, including Aquaneering, Tecniplast, and Aquatic Enterprises.*

Scrub pads

METHOD

> *Each section below should be treated independently; these should not be viewed as consecutive procedures as the numbering may imply.*

Establishing a New Recirculating System

1. Starting with a new or sterilized system, rinse the biomedia (provided by the system manufacturer) with Type II water and place in the biofilter.

2. Disable the water effluent exchange and UV sterilization, and keep the carbon filter empty.

   > *It is necessary to initially limit filtration and sterilization functions and allow nitrogenous waste products to build up sufficiently to support healthy growth of the biomedia.*

3. Allow flow from a single tank on the end of each rack and add 5–10 frogs to the system.

   > *Larger systems require more frogs. This number of frogs is suggested for a system with ~500-L capacity.*

4. After 24 h, disable the water flow but keep the biomedia aerated using an external air pump.

5. Add nitrifying bacteria to the biofilter.

   > *Larger systems require more nitrifying bacteria. Follow the instructions outlined by ProLine.*

6. Wait 1–2 h before restarting the water flow.

   > *During this brief period, the frogs present in the system should be kept in their tanks. The absence of water flow should not be detrimental to their health.*

7. Perform daily NH_3/NH_4^+, NO_2^-, and NO_3^- level measurements using the relevant test kits.

   > *Wait for NH_3/NH_4^+ levels to measure at least 1–2 ppm.*

8. Initiate UV sterilization and water exchange, and add carbon to the appropriate filter housing. The levels of NO_2^- and NO_3^- should begin to increase.

   > *Flow to 1–2 additional tanks can be started. Additional tanks can be filled as needed.*

9. Allow for NH_3/NH_4^+ and NO_2^- levels to reduce to 0.5 ppm before adding more frogs to the system.

   > *The entire process can take 2 wk to 2 mo. Nitrifying bacteria can thrive when water parameters are consistent. If dosing reservoirs are not present, sodium bicarbonate can be used to stabilize pH. Once NH_3/NH_4^+ and NO_2^- levels reach the desired level, gradually add frogs to the system while monitoring water quality parameters. Care should be taken to not open flow to too many tanks at once as this can drain the sump below the set safe level and result in an automatic system shutdown.*

Daily Water Assessment and Maintenance Required for Proper Operation

10. Record system readings. See Table 1 for age-appropriate guidelines for water temperature, pH, conductivity, and frog density.

    > *This allows the user to deduce trends and identify potential issues that may be occurring.*

 Cite this protocol as *Cold Spring Harb Protoc*; doi:10.1101/pdb.prot106146

TABLE 1. Water-quality parameters for *Xenopus tropicalis*

Age	Water temperature (°C)	pH	Conductivity (μS)	Tank density
Tadpole/froglet	27 (25–27.5)	7.0 (6.8–7.2)	1000 (900–1100)	2–4 per 1 L
Juvenile	26 (25–27)	7.0 (6.8–7.2)	1200 (1000–1300)	1 per 1–2 L
Adult	25 (24–26)	7.0 (6.8–7.2)	1200 (1000–1300)	1 per 1 L

The water quality parameters and animal stacking density necessary for optimal growth vary, dependent on the life history stage of the animal (Hilken et al. 1995; Green 2009). These parameters are listed in the table above. The optimal conditions, as well as the allowable tolerance range, are given in parentheses.

11. Inspect the UV bulb.

    Check system display to confirm the UV bulb is on and functioning correctly. Replace the bulb if necessary.

12. Regularly scrape internal sides of the tanks with scrub pads to remove any algae buildup.

    Algal buildup will vary depending on several factors including lighting and dissolved nutrients. Excessive buildup should be removed as needed.

13. For systems that use stand pipes with overflows, shake the stand pipes to clear them. Tanks with only an overflow bulkhead will require daily active removal of detritus to keep them clean.

14. Exchange 10% of the system water.

    Modern recirculating aquatic systems can be programmed to perform water exchange automatically. Make sure that the automatic setting is kept at 10% daily exchange.

15. Measure pH and conductivity dosing reservoirs daily.

    Use sodium bicarbonate to buffer and regulate pH and sea salt to control conductivity.

16. Observe carbon and mechanical filters for buildup of waste and replace as needed once passage of water through them is significantly impeded.

17. Observe individual tanks for accumulation of waste and film on the bottom and sides, and sanitize tanks that are particularly dirty.

    To clean tanks, scrub and rinse with Type II water, spray with 10% bleach and let sit for 1 h, rinse again with Type II water, spray with 70% ethanol and let sit for 1 h, rinse once more with Type II water, and allow to air dry.

Weekly Water Assessment and Maintenance Required for Proper Operation

18. Assess NH_3/NH_4^+, NO_2^-, and NO_3^- levels with the relevant API test.

    Acceptable measurements are as follows: NH_3/NH_4^+ =0 ppm (0–0.5 ppm), NO_2^- <1 ppm, NO_3^- <40 ppm, alkalinity >40 ppm. If levels are out of range, perform a 25% water exchange. Repeat water change each day until values reach acceptable levels.

19. Test pH and conductivity with a reliable external probe as a reference for internal system probes.

    If the external and internal probes are not in agreement, clean the system probe and check that it is correctly reading calibration solutions. Recalibrate the probe if necessary.

20. Measure temperature with an external thermometer.

    Heaters and chillers are used to keep the temperature stable. Measuring with an external thermometer confirms that the system is calibrated to correctly engage heaters and chillers as necessary to keep the water temperature stable. Recalibrate or replace the system thermometer if it is not in agreement with the external thermometer.

Annual Water Assessment and Maintenance Required for Proper Operation

21. Replace UV bulbs and check to ensure that the quartz sleeve is intact.

22. Replace pH electrodes.

23. Remove and clean return pipes.

24. Replace all rubber tubing.

System Sterilization

System sterilization is necessary to restore full operational capabilities following an outbreak of disease.

25. Disable all system components including the water pump, biomedia agitators, and probes.

26. Discard biomedia or sterilize in a 10% bleach solution.

27. Scrub the inside of the sump and associated parts and remove all detritus.

28. Remove and dispose of all filter pads, cartridges, and carbon.

29. Disassemble and scrub all the tanks and accessories.

30. Rinse the tanks with water, coat with 10% bleach, and let stand for 1 h.

31. Rinse the tanks with water, coat with 70% ethanol, and let stand for 1 h.

32. Rinse the tanks with Type II water and let air dry.

33. Soak accessories in 10% bleach for 1 h.

34. Soak accessories in 10 g/L solution of $Na_2S_2O_3$ in Type II water for 1 h.

35. Soak accessories in 70% ethanol for 10 min, rinse, and air-dry.

36. Drain 75% of the sump volume according to the manufacturer's instructions.

37. Reassemble the system, putting back all the tanks, filter housings, and new rubber tubing. Leave the filter housing empty, and do not add any carbon to the system.

38. Start the system.

39. Add bleach to the system.

    Ensure the final concentration of NaClO in the system is 0.06%.

40. Reduce the effluent rate to its lowest setting.

    Do not reduce the daily water exchange rate to <1%.

41. Wait 2–3 d, then increase the daily water exchange rate to ~50%.

    This will begin removing the bleach from the system.

42. Begin testing the pH of the system after a week.

43. Add Virkon to the system once the system water pH matches the influent water pH.

    The concentration of Virkon should be 3.2 g per 1 m^2 of estimated internal surface area of the entire system.

44. Reduce the effluent rate back to its lowest setting.

    Virkon can cause foam accumulation and in some cases, may cause the system to overflow. To help mitigate this, limit agitation of the water and spray 70% ethanol directly on the foam to help disperse bubbles.

45. Wait 7 d then increase the daily water exchange rate to ~50%.

    This will wash the Virkon out of the system.

46. Allow the system to run for 2–3 d, replace all rubber tubing, and follow the protocol for establishing a new recirculating system described above to stabilize the system and allow for housing animals.

DISCUSSION

Optimal eggs and embryos are produced by healthy laboratory population of *Xenopus tropicalis*. Regular monitoring of water-quality parameters, as well as daily monitoring and visual inspection of the animals, is crucial for prevention and early identification of developing conditions that may be detrimental to overall colony health. The use of tap water is discouraged, and starting with a good water source providing Type II water will aid in maintenance of optimal water parameters as well as

Cite this protocol as *Cold Spring Harb Protoc*; doi:10.1101/pdb.prot106146

limit the chances of introducing pathogens and toxic contaminants. Municipal tap water may also be basic and thus may add the additional complication of needing to use acid buffers to aid in stabilizing the pH. The stability of the parameters will also be affected by the animal maintenance system configuration, with recirculating systems being more cost effective and efficient than flowthrough systems and also requiring much less maintenance than static systems. Furthermore, animal health is best when the accumulation of nitrogenous waste is low, thus the initial effort made to establish a stable biofilter in a recirculating system is absolutely necessary. The *Nitrosomonas* and *Nitrobacter* bacteria in the biofilter respectively oxidize the toxic ammonia into toxic nitrite and then the nitrite into nontoxic nitrate (Hem et al. 1994). Following establishment of a stable ammonia/nitrogen cycle, regular maintenance and tests of water parameter levels will aid in maintaining the animals in good health (McNamara et al. 2018). Nonetheless, sterilization of the entire system may be necessary if addition of new frogs to the system also introduces pathogens that negatively affect their quality of life. Once sterilized, the system can be put back into operation by again following the steps for establishing a functional and stable biofilter.

REFERENCES

ASTM Standard D1193-06 (2018). "*Standard Specification for Reagent Water.*" *ASTM International*, West Conshohocken, PA.

Green SL. 2009. *The laboratory Xenopus sp.* CRC Press, Boca Raton, FL.

Hem LJ, Rusten B, Ødegaard H. 1994. Nitrification in a moving bed biofilm reactor. *Water Res* **28:** 1425–1433. doi:10.1016/0043-1354(94)90310-7

Hilken G, Dimigen J, Iglauer F. 1995. Growth of *Xenopus laevis* under different laboratory rearing conditions. *Lab Anim* **29:** 152–162. doi:10.1258/002367795780740276

McNamara S, Wlizla M, Horb ME. 2018. Husbandry, general care, and transportation of *Xenopus laevis* and *Xenopus tropicalis*. *Methods Mol Biol* **1865:** 1–17. doi:10.1007/978-1-4939-8784-9_1

Raising and Maintaining *Xenopus tropicalis* from Tadpole to Adult

Maura Lane,[1] Michael Slocum, and Mustafa K. Khokha

Pediatric Genomics Discovery Program, Department of Pediatrics and Genetics, Yale University School of Medicine, New Haven, Connecticut 06510, USA

Xenopus tropicalis is a powerful model organism for cell and developmental biology research. Recently, precise gene-editing methods such as CRISPR–Cas9 have allowed facile creation of mutants. The ability to raise and maintain lines of wild-type and mutant animals through all life stages is thus critical for researchers using this model organism. The long fertile life (>8–10 yr) and relatively hardy nature of *X. tropicalis* makes this a straightforward process. Environmental parameters such as water temperature, pH, and conductivity often vary slightly among husbandry protocols. However, the stability of these variables is essential for rearing success. This protocol describes conditions to optimally raise and maintain *X. tropicalis* from embryos to adulthood.

MATERIALS

It is essential that you consult the appropriate Material Safety Data Sheets and your institution's Environmental Health and Safety Office for proper handling of equipment and hazardous materials used in this protocol.

Reagents

BioVita Fry (1.2-mm and 2.0-mm pellets; BioOregon)

Frog Brittle (Powder) for Tadpole *Xenopus* (Nasco SA05964[LM])

Modified Ringer's (MR; 1/9×) containing 50 µg/mL gentamicin (see Sec. 2, Protocol 5: Obtaining *Xenopus tropicalis* Embryos by In Vitro Fertilization [Lane and Khokha 2021a] or Sec. 2, Protocol 6: Obtaining *Xenopus tropicalis* Embryos by Natural Mating [Lane and Khokha 2021b])

Sera Micron solution
  Prepare a solution that contains 1 tablespoon of sera Micron Growth Food (sera 00720) and 500 mL of system water. Prepare fresh for each feeding.

Sera Vipan Baby (sera 00730) or Tetramin Ciclid Flakes (ground to a powder using a blender)

Xenopus tropicalis embryos, partially dejellied (see Sec. 2, Protocol 5: Obtaining *Xenopus tropicalis* Embryos by In Vitro Fertilization [Lane and Khokha 2021a] or Sec. 2, Protocol 6: Obtaining *Xenopus tropicalis* Embryos by Natural Mating [Lane and Khokha 2021b]

Equipment

Aquatic recirculating system (see Protocol 2: Animal Maintenance Systems: *Xenopus tropicalis* [Shaidani et al. 2020])

---

[1]Correspondence: maura.lane@yale.edu

Cite this protocol as *Cold Spring Harb Protoc*; doi:10.1101/pdb.prot106369

Conical tube (50-mL)
Fish nets (small and medium)
Funnel (small plastic, ~1 mm OD opening at the tip)
Glass Pasteur pipettes (Corning 2239912)
Incubator (22°C–28°C) (VWR 2005)
Petri dishes (150-mm; Falcon 351085)
Pipette pump (VWR 53502-233)
Plastic flotation pads (Bio-Serv Tuff Turf Replacement Pad F9115-R)
Set of measuring spoons (1/8-tsp.–1 Tbsp.)
Squirt bottle (500-mL)
Stereomicroscope
Tank hides (short lengths of PVC pipe cut in half, or commercial, e.g., Bio-Serv K3261)
Turkey baster

METHOD

Although environmental parameters vary slightly among husbandry protocols (Grammer et al. 2005; Showell and Conlon 2009; Jafkins et al. 2012; McNamara et al. 2018), it is important to maintain stable water parameters when raising X. tropicalis. Tadpoles and metamorphs are particularly sensitive to changes in water quality. Adult animals are more tolerant to these changes, but they will become compromised when consistently maintained under undesirable conditions. Table 1 summarizes care in a recirculating system at each life stage—from Stage 43–44 tadpoles to sexually mature adults. Recording the values measured by the system sensors should be done twice daily. Verify these values weekly using a water-quality kit and/or pH/conductivity meter.

The water temperatures provided here are ideal for each life stage. If you are restricted to one temperature, we recommend a range of 23.5°C–24.5°C. Tadpoles will not grow as fast, but they will develop normally. Keeping adults above 25°C can result in poor egg quality, and temperatures above 26°C can cause illness in adults.

The feeding amounts provided here should be viewed as starting points. The amount fed should ultimately depend on consumption.

TABLE 1. Environmental parameters in recirculating system by life stage

Parameter	Stage 44–Wk 1	Wk 2	Wk 3–Stage 48	Stage 49–63	Stage 64–4 mos.	4 mos.–sexual maturity	Sexual maturity
Diet	10 mL sera Micron twice/d	20 mL sera Micron twice/d	30 mL sera Micron twice/d and 1/16 tsp. powdered diet once/d	30 mL sera Micron twice/d and 1/8 tsp powdered diet once/d	3–5 small (1.2-mm) BioVita pellets/froglet twice/d (increase as needed)	1/16 tsp. larger (2.0-mm) BioVita pellets/frog (or as much as eaten in 15 min) twice/d	1/16 tsp. larger pellets/frog once/d for regularly ovulated females or three times/wk for other adults
Flow rate	2–3 drops/sec	0.5 L/min					
Density	100/4–6 L	50/4–6 L	3–4/L		2–3/L		
Monitor	3–4 times/d				Daily		
Temperature	25.0°C–26.0°C				23.5°C–24.5°C[a]		
pH	6.6–7.0						
Conductivity	1000–1300 µS[b]						
Light cycle (h)	12:12						

[a]Not <22°C or >25°C.
[b]Lower conductivity may lead to soft eggs that are more difficult to inject.

Stages 1–44

1. Raise the dejellied embryos, obtained by either IVF or natural mating, in a 150-mm Petri dish to Stage 43–44 (free-swimming) tadpoles at 22°C–28°C.

   Embryos will develop faster at higher temperatures (26°C–28°C); for details, see the table at http://www .xenbase.org/anatomy/static/xenopustimetemp.jsp.

2. Change the solution (1/9× MR containing 50 μg/mL gentamicin) and remove abnormal or dead embryos daily using a stereomicroscope and a pipette pump fitted with a glass Pasteur pipette.

   See Troubleshooting.

Stages 44–63

3. Introduce Stage 43–44 tadpoles to the recirculating system by gently pouring them into a tank containing system water with the following parameters.
   - Use an initial density of 100 tadpoles/4–6 L tank. Thin to 50/4–6 L at week 2; then thin to 3–4 tadpoles/L week 3. Thin to 2–3/L when the tadpoles have reached Stage 64.

     It is not necessary to keep a lid on the tank until stage 63–64. This makes it easier to feed and observe the tadpoles.
   - Maintain the temperature at 25°C–26°C.
   - Keep the pH between 6.6 and 7.0, the conductivity at 1000–1300 μS, and the light:dark cycle at 12 h:12 h.
   - Adjust the flow rate to 2–3 drops/sec, and gradually increase to 0.5 L/min over the first 2 wk.

     The flow rate should be reduced during feeding to prevent removal of food before tadpoles can consume it.

4. Feed as follows.

   i. Add 10 mL of sera Micron solution immediately after adding tadpoles to the tank using a squirt bottle and 50 mL conical tube to measure.

   ii. During week 1, add 10 mL of sera Micron solution/tank twice a day.

   iii. During week 2, add 20 mL of sera Micron solution/tank twice a day.

   iv. During week 3–Stage 48, add 30 mL of sera Micron solution/tank twice a day. Also feed 1/16 teaspoon of sera Vipan Baby (or ground fish flake)/tank once a day.

   v. During Stages 49–63, add 30 mL of sera Micron solution/tank twice a day. Also feed 1/8 teaspoon of sera Vipan Baby or other powdered diet once a day.

      These amounts are starting points; feed more or less based on amount of food consumed. Metamorphs may eat less; this can be a stressful period, as many organs are undergoing remodeling. A tadpole tank with a clean bottom, where animals are actively seeking food, should be fed again.

5. Care as follows.

   i. Monitor health and behavior three to four times daily.

      See Troubleshooting.

   ii. Use a net to remove dead animals daily. Remove excess algae growth (>30% of tank) and accumulated solids as needed using a turkey baster.

      Tadpole tanks develop a thin green layer on the bottom of the tank. This is normal and not generally harmful. Excessive buildup that begins to float should be removed, or tadpoles moved to a clean tank.

   iii. Gently transfer tadpoles to a clean tank weekly. Minimize handling by scooping or gently pouring tadpoles between tanks instead of netting them. Use a net to remove debris from the

Cite this protocol as *Cold Spring Harb Protoc*; doi:10.1101/pdb.prot106369

water during tank changes. Pour the water from the old tank through the net into the new tank, moving the net aside as the tadpoles pass by.

    iv. Move Stage 63–64 and more mature animals to a separate tank to prevent froglets from eating the tails of younger animals.

Stages 64 to ~4 mo (Juveniles)

6. Adjust the system and frogs to the following parameters.
   - Use a density of 2–3 animals/L.

     Keep a lid on the tank from this point on to prevent frogs from escaping.

   - Maintain the temperature at 25.0°C–26°C.
   - Keep the flow rate at 0.5 L/min.

7. Feed a small pelleted diet (e.g., larger chunks sifted from Nasco Frog Brittle [Powder] for Tadpole *Xenopus* or 1.2 mm BioVita Fry pellets). Feed three to five pellets/froglet twice a day initially using a measuring spoon and small funnel inserted through a hole in the lid, and increase as needed.

   Recirculating system tank lids are typically predrilled with small holes. These are convenient for feeding solid food without opening the lid. This reduces feeding labor time and avoids frogs escaping during feeding.

   Froglets and adults should consume pelleted food within 15 min. If animals are actively seeking food after this time, they should be refed. Uneaten pellets will collect on the bottom and grow bacteria if not removed.

8. Care as follows.
   i. Monitor health and behavior at least daily.

   Healthy juvenile (and adult) frogs often float at the surface or rest on the bottom of the tank for prolonged periods when undisturbed. They will seek hiding places; a short length of a half tube of PVC piping or commercially supplied hides or floating pads can provide cover. Healthy frogs become active when disturbed and struggle when handled. There should be vigorous "feeding-frenzy" behavior when feeding.

   See Troubleshooting.

   ii. Change tank as needed, indicated by excess solids in the water and/or growth of algae on the sides of the tank.

4 mo to Sexual Maturity

9. Keep system parameters the same as for juvenile frogs (see Step 6), except for the temperature, which should be decreased to 23.5°C–24.5°C.

10. Feed larger pelleted food (e.g., 2.0-mm BioVita Fry pellets) using a measuring spoon and small funnel. Feed 1/16 tsp./frog, or as much as is eaten in 15 min., twice per day.

11. Care as follows:
    i. Monitor health and behavior daily.

    See Troubleshooting.

    ii. Change tank once/mo or as needed.

12. Distinguish *X. tropicalis* males and females using the following characteristics.
    - Identify sexually mature males as smaller and bullet-shaped (Fig. 1A).

      Sexual maturity in the male is marked by the development of nuptial pads, typically at 4–6 mo of age (Grammer et al. 2005). These are visible as dark areas on the inner forearms (Fig. 1C) that are used to grasp the female during amplexus. Nuptial pads are periodically shed. In the natural habitat, they are most prominent during mating season (Green 2010). In the laboratory setting, males with dark nuptial pads are usually selected for breeding or in vitro fertilization.

    - Identify sexually mature females as larger and pear-shaped (Fig. 1B).

      Sexually mature females (6–8 mo) will produce eggs after ovulation is induced. They have a prominent cloaca (Fig. 1B) that becomes enlarged and reddened during ovulation.

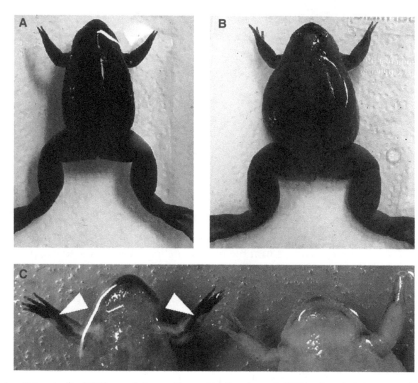

FIGURE 1. Sexing *X. tropicalis*. (*A*) The male is smaller and bullet-shaped. (*B*) The female is larger and pear-shaped. (*C*) Mature males have nuptial pads on inner forearms (white arrows); females lack nuptial pads.

Sexual Maturity (Adulthood)

X. tropicalis males and females can safely be housed together. They will not mate in the laboratory setting unless exogenous ovulation hormone is administered.

13. Keep system parameters the same as in Step 9, with the temperature at 23.5°C–24.5°C.

    Keeping adult frogs above 25°C or below 22°C is not recommended. They may appear healthy and behave normally at temperatures above 25°C, but egg quality will likely suffer. Maintaining temperatures below 22°C or above 26°C may cause lethargy and illness.

14. Feed 1/16 tsp. larger pelleted food per frog (as in Step 10), but only once per day for regularly ovulated females and three times per week for all other adults.

15. Care as follows:

    i. Monitor health and behavior daily.

       See Troubleshooting.

    ii. Change tank once per month or as needed.

TROUBLESHOOTING

Problem (Step 5.i): Tadpoles have a dark overall appearance, display atypical behavior (e.g., floating, swimming in circles, or laying or resting on the tank bottom for extended periods), or die.

Solution: Tadpoles are the first in a colony to be affected by abnormal water parameters. Fluctuations in temperature, pH, or conductivity may result in the sudden die-off of an entire tank of tadpoles. If abnormal tadpole behavior is observed, check and correct water parameters in your system as needed. Animals might be saved by changing out the tank. Carefully pour off about 2/3 of the

Cite this protocol as *Cold Spring Harb Protoc*; doi:10.1101/pdb.prot106369

water in the tank. Gently pour the remaining water containing the tadpoles into a fresh tank. Be sure incoming water has been corrected for any imbalances before turning on the flow.

Note that groups of healthy tadpoles often filter feed in a stationary head down–tail up orientation, especially before the onset of metamorphosis. They will also actively filter material from the bottom of the tank. Healthy metamorphs may also rest on the bottom of the tank; this is not a concern unless most of the animals in the tank are on the bottom.

Problem (Step 8.i, Step 11.i, and Step 15.i): Signs of illness in juveniles and adults include lethargy, skin discoloration and/or lesions, excessive skin sloughing, and bloating. Bloat caused by the accumulation of air within the body will cause frogs to have difficulty diving or staying submerged. Animals affected by fluid bloat can remain submerged but have a bloated appearance.

Solution: While *Xenopus* are relatively hardy, there are diseases that seriously affect them (Parker et al. 2002; Trott et al. 2004; Green 2010). Treatment of a colony is often impractical, so prevention or early intervention is key. Monitoring animals for signs of normal and abnormal behavior is therefore critical to detecting illness or disease. Check and correct water parameters at any sign of abnormal behavior or health. Remove affected animals from the recirculating system immediately and consult your veterinarian.

Problem (Step 15.i): The eggs laid by mature females are soft or of poor quality.

Solution: Soft or poor-quality eggs may be the result of high temperature or inadequate mineral content in the water. If soft eggs are consistently observed during microinjection, be sure the temperature and conductivity in the recirculating system are within the appropriate range (Table 1).

ACKNOWLEDGMENTS

We thank Tim Grammer for helping to develop husbandry protocols, as well as Sarah Kubek for assistance with pictures.

REFERENCES

Grammer TC, Khokha MK, Lane MA, Lam K, Harland RM. 2005. Identification of mutants in inbred *Xenopus tropicalis*. *Mech Dev* **122:** 263–272.

Green SL. 2010. *The laboratory* Xenopus *sp*. CRC, Boca Raton, FL.

Jafkins A, Abu-Daya A, Noble A, Zimmerman LB, Guille M. 2012. Husbandry of *Xenopus tropicalis*. *Methods Mol Biol* **917:** 17–31. doi:10.1007/978-1-61779-992-1_2

Lane M, Khokha MK. 2021a. Obtaining *Xenopus tropicalis* embryos by in vitro fertilization. *Cold Spring Harb Protoc* doi:10.1101/pdb.prot106351

Lane M, Khokha MK. 2021b. Obtaining *Xenopus tropicalis* embryos by natural mating. *Cold Spring Harb Protoc* doi:10.1101/pdb.prot106609

McNamara S, Wlizla M, Horb ME. 2018. Husbandry, general care, and transportation of *Xenopus laevis* and *Xenopus tropicalis*. In Xenopus: *methods and protocols* (ed. Vleminckx K), pp. 1–17. Springer, New York.

Nieuwkoop PD, Faber J. 1994. *Normal table of* Xenopus laevis *(Daudin): a systematical and chronological survey of the development from the fertilized egg till the end of metamorphosis*. Garland, New York.

Parker JM, Mikaelian I, Hahn N, Diggs HE. 2002. Clinical diagnosis and treatment of epidermal chytridiomycosis in African clawed frogs (*Xenopus tropicalis*). *Comp Med* **52:** 265–268.

Shaidani N-I, McNamara S, Wlizla M, Horb ME. 2020. Animal maintenance systems: *Xenopus tropicalis*. *Cold Spring Harbor Protoc* doi:10.1101/pdb.prot106146

Showell C, Conlon FL. 2009. Natural mating and tadpole husbandry in the western clawed frog *Xenopus tropicalis*. *Cold Spring Harb Protoc* doi:10.1101/pdb.prot5292

Trott KA, Stacy BA, Lifland BD, Diggs HE, Harland RM, Khokha MK, Grammer TC, Parker JM. 2004. Characterization of a *Mycobacterium ulcerans*–like infection in a colony of African tropical clawed frogs (*Xenopus tropicalis*). *Comp Med* **54:** 309–317.

Defining the Specific Pathogen-Free State of *Xenopus* Using TaqMan Assays

Casey L. Hensley, Kaitlin M. Bowes, and Sanford H. Feldman[1]

Center for Comparative Medicine, University of Virginia, Charlottesville, Virginia 22908, USA

Colonies of valuable inbred and transgenic laboratory-reared *Xenopus* frogs maintained for research constitute naïve populations of animals susceptible to some opportunistic infectious diseases. Therefore, it is prudent to characterize any new animal acquisitions before introduction into an existing colony as a biosecurity measure to preclude the concurrent introduction of an infectious microorganism associated with the new animal(s). In addition, some pathogens of *Xenopus*, such as *Chlamydia* and *Mycobacterium* spp, are zoonotic diseases, placing frog aquarists at risk for acquiring an infection. Because it is not cost effective to test for all diseases of *Xenopus* frogs, we have defined a subset of prevalent infectious microorganisms and developed TaqMan polymerase chain reaction (PCR) assays to detect these agents. The specific pathogens in our test panel were selected from relatively recent publications where they reportedly caused morbidity and/or mortality in *Xenopus laevis* and/or *X. tropicalis*. The assays herein do not constitute a comprehensive list of infectious diseases of *Xenopus* frogs. Therefore, a frog devoid of the infectious agents in our test panel are characterized as "specific pathogen-free." Three of the described quantitative polymerase chain reaction (qPCR) assays detect many species within their genus (i.e., qPCRs for ranaviruses, *Chlamydia* spp, and *Cryptosporidia* spp).

MATERIALS

It is essential that you consult the appropriate Material Safety Data Sheets and your institution's Environmental Health and Safety Office for proper handling of equipment and hazardous materials used in this protocol.

Reagents

Benzocaine gel (20%; optional; see Step 1)

DNeasy Blood and Tissue Kit (QIAGEN 69504-50) and DNeasy Powerlyzer PowerSoil Kit (QIAGEN 12855-50)

Oligonucleotide primers and TaqMan probes for pathogens of interest (see Table 1)

TaqMan exogenous internal positive control reagents (ThermoFisher 4308321)

TaqMan Universal Master Mix II with UNG (ThermoFisher 4440038)

Target positive control DNA sequences from each of the microorganisms cloned into either pCR4-TOPO or pCR2.1-TOPO (Invitrogen)

> *These are not commercially available but can be obtained from the author under a material transfer agreement with the University of Virginia. Linearize the plasmids before use by cutting at a single locus outside the target with the appropriate restriction enzyme, BglI (pCR4) or BglII (pCR2.1).*

Xenopus laevis and/or *X. tropicalis* frogs

[1]Correspondence: shf2b@virginia.edu

TABLE 1. *Xenopus* specific pathogen-free microorganism panel with corresponding primer and probe sequences

Pathogen	Primer and TaqMan probe sequences[a]		Genomic target[b]	Sample for DNA[c]
Frog virus 3 (Ranavirus)	Sense	TCCTGGGTCAGAGACGTTCAC	DNA pol II	Kidney
	Antisense	AGCCTCTCGTACTCTACCTTCC		
	Probe	ACCACCTGAACTGCGTCCACGACC		
Capillaria xenopi	Sense	GCCACGGTCCTCTAAGTCAAG	18S rRNA	LI
	Antisense	CCAGGACACTCGGTAAAGAGC		
	Probe	ATTGGCTCTGCCGCCGTTGGTCAT		
Batrachochytrium dendrobatidis	Sense	AATGAGTGGACTTGTGCGC	18S rRNA	TWS
	Antisense	CCGAACGTGACTGTCATATCC		
	Probe	CATTGAGTTGGGACATGGTAG		
Mycobacterium chelonae	Sense	TAGGTAGTCGGCAAAACGT	18S rRNA	LI
	Antisense	CAGGGTGACAACACAATGT		
	Probe	CATTGAGTTGGGACATGGTAG		
Mycobacterium xenopi	Sense	CACACTGTTGGGTTTTGA	rRNA ITS	LI
	Antisense	CGCTTACCACAGTCTTTG		
	Probe	TCGCAACCACTATCCAACACTCA		
Mycobacterium liflandii	Sense	GGTTCATCGGTTCGGCATAGC	pMUM2	LI
	Antisense	ACATTGCGCTCGACGTCATC		
	Probe	CCGAGGACCACCGCCACGACACTA		
Mycobacterium marinum	Sense	GCTACG ATTCCCAACATGTCA	pMUM3	LI
	Antisense	CTCCTGTCCATTACCTTCAAGG		
	Probe	CTTCGCCGCCTACACCATTG		
Chlamydia species	Sense	GCGTACCAGGTAAAGAAGCACC	16S rRNA	OS
	Antisense	CCTTTCCGCCTACACGCCCT		
	Probe	AGCACCCTCCGTAATACCGCAGCT		
Cryptosporidium species	Sense	CGCGCGCTACACTGATGCATCCA	18S rRNA	LI
	Antisense	TTCCTCGTTMAAGATCAATAATT[d]		
	Probe	TAGCATCGTGATGGGGATAGATCATTG		
Mycobacterium gordonae	Sense	ATCGTCTGTAGTGGACGAAGAC	rRNA ITS	LI
	Antisense	TTCTCAAACACCACACCCC		
	Probe	GAGGCAACACCCTCGGGTGCTG		

[a]All probes use 5′-FAM as a fluorophore and 3′-TAMRA as a quencher. The reactions are not multiplexed.
[b]DNA target detected (pMUM, mycobacterial plasmid; rRNA ITS, ribosomal internal transcribed spacer).
[c]Tissue from which DNA is isolated (LI, large intestinal content; TWS, toe web skin; OS, oral sample taken on a cotton swab).
[d]Degeneracy at one position (M = A or C) facilitates detection of a broader range of *Cryptosporidium* species.

Equipment

Cotton tip swab (optional; see Step 1)
Microamp Fast 8 Tube Strip (ThermoFisher 4358293)
Microamp Optical 8 Cap Strip (ThermoFisher 4323032)
Microcentrifuge
Microcentrifuge tubes
Micropipettes and aerosol resistant tips
Nanodrop spectrophotometer
Shaker-incubator
Step One Plus quantitative thermal cycler and software (Applied Biosystems; ThermoFisher)
Surgical scissors (sterile; optional; see Step 1)
Thermo Bio 101 Fastprep FP 120 orbital shaker
Tissue forceps (sterile; optional; see Step 1)
Vortex mixer

METHOD

1. Collect samples for qPCR as follows:

   - *For chlamydia testing:* Orally swab the frog with a cotton tip swab. Vortex the cotton tip in 180 μL of QIAGEN ATL buffer and proceed with Step 2.

   - *For ranavirus testing:* Euthanize frogs by application of 20% benzocaine gel to the skin. Confirm euthanization by testing whether frogs respond to a toe pinch by leg withdrawal (~30 min). Obtain a 3 mm^3 kidney sample by dissecting into the coelomic cavity, moving the intestines, egg mass, and fat bodies laterally.

     The kidney is located along the vertebrae in the abdominal area.

   - *For chytrid testing:* Euthanize the frog as for ranavirus testing. After the frog is euthanized, excise a skin sample from between two rear toes with sterile scissors and tissue forceps.

   - *For all other microorganisms:* Obtain a lower gastrointestinal sample by confining one or more frogs to a small aquarium overnight and collecting the fecal sediment the next morning.

2. Isolate DNA from the sample using DNeasy for toe web skin, oral swab, or kidney samples, and Powerlyzer for Soil for large intestine samples according to manufacturer's recommendations.

3. Elute the DNA from the QIAGEN column using AE or TE buffer from the kit heated to 70°C.

   Pigments that co-isolate with DNA will form a pellet and are inhibitory to PCR; therefore, if a pigmented pellet is visibly present after centrifugation during the elution step, transfer the supernatant to a sterile labeled microcentrifuge tube.

4. Measure DNA concentration and purity using nanodrop spectrophotometer.

   A minimal sample DNA concentration of 10 ng/μL is needed, with an $A_{260}/A_{280} \geq 1.4$.

5. Program the Step One Plus quantitative thermal cycler for the "Presence–Absence" assay using thermal cycling parameters listed in Table 2. Do not exceed an annealing temperature of 60°F which is the high annealing temperature limit required by the internal positive control (IPC) reaction.

   The Step One Plus prompts the user to include the following two controls in triplicate: inhibited qPCR non-template control (NTC; the IPC kit contains an inhibitor that is added instead of sample DNA) and NTC (water is included instead of sample DNA). The machine generates a schematic of the reaction plate showing where the blocked reaction and NTC wells are located. You must program into the thermal cycler where your positive control and individual samples are located in the plate, and these will be shown on the schematic.

6. Make the qPCR master mix using (per reaction) 10 μL of 2× TaqMan Universal PCR Master Mix II with UNG, 2 μL 10× IPC Mix, 0.4 μL 50× IPC DNA, 1 μL of 20× solution of primers and probes resulting in primer concentrations of 900 nM of the agent specific primers and 300 nM of

TABLE 2. Thermal-cycling temperatures and times

Assay[a]	Denature	Anneal	Extend	No. cycles
Frog virus 3 (Ranavirus)	95°C–15 sec	55°C–15 sec	72°C–60 sec	40
Capillaria xenopi	95°C–15 sec	60°C–30 sec		40
Batrachochytrium dendrobatidis	95°C–15 sec	55°C–60 sec		40
Mycobacterium chelonae	95°C–15 sec	52°C–15 sec	72°C–30 sec	40
Mycobacterium xenopi	95°C–15 sec	54°C–15 sec	72°C–30 sec	40
Mycobacterium liflandii	95°C–15 sec	60°C–30 sec		40
Mycobacterium marinum	95°C–15 sec	60°C–30 sec		40
Chlamydia species	95°C–15 sec	60°C–30 sec		40
Cryptosporidium species	95°C–15 sec	60°C–30 sec		40
Mycobacterium gordonae	95°C–15 sec	60°C–30 sec		40

[a]Each thermal cycling program starts with a uracil N-glycosylase (UNG) reaction (for 5 min at 50°C) and then a hot-start reaction (for 10 min at 95°C).

Cite this protocol as *Cold Spring Harb Protoc*; doi:10.1101/pdb.prot106179

the agent specific probe, and 4.6 µL water. It is prudent to make one more reaction volume per ten reactions created to compensate for pipetting errors. Mix small volumes of the Master Mix by trituration (pipetting up and down) and larger volumes by vortex.

7. Dispense 18 µL of master mix into each Fast Reaction well according to the Step One Plus-generated microplate schematic on the Step One Plus screen. Add the reagents in the following steps to the appropriate wells according to this schematic.

8. Add 2 µL qPCR blocking agent from the IPC kit to the three inhibited reaction wells.

9. Add 2 µL water to the three NTC reaction wells.

10. Add 2 µL of the appropriate positive cloned control (10–1000) copies to the three positive control reaction wells.

11. Add 2 µL containing 50–100 ng sample DNA to the test wells.

12. Begin the "Presence–Absence" assay on the Step One Plus quantitative thermal cycler.

DISCUSSION

The Step One Plus software will interpret the qPCR results and validate that amplification was not inhibited by confounding substances in the sample DNA. The IPC reaction and qPCR reactions use TaqMan probes labeled with fluorophores VIC and FAM, respectively, in each reaction. These fluorophores have different fluorescence emission wavelengths. The machine compares the VIC fluorescence of the blocked-IPC reactions and the NTC controls to establish an IPC threshold, and quantifies the relative difference in end point fluorescence values (ΔR_f) of these two controls. The pathogen target ΔR_f threshold is calculated from the relative FAM fluorescence difference between the NTC reaction and each DNA sample reaction, and the cloned positive control. When the pathogen FAM ΔR_f of a sample exceeds the ΔR_f IPC threshold, the software interprets the result as the pathogen's DNA being "present" in the sample. A pathogen FAM ΔR_f from a sample reaction that is below the IPC VIC ΔR_f threshold is interpreted by the software as the pathogen DNA being "absent" in the sample. If the sample DNA happens to contain a substance co-isolated with the sample DNA that is inhibitory to the qPCR, the IPC VIC ΔR_f reaction fails to reach the calculated IPC ΔR_f threshold, and the software interprets that assay result as "unconfirmed." Cloned positive controls are not required by the Step One Plus software but are included to demonstrate the ability of the assay to detect a known positive control.

The "presence–absence" assay is used rather than quantifying the number of copies of target DNA sequences in the sample because the density of a pathogen is not as important as whether or not it is present. The "presence" of pathogen DNA sequences in samples from frogs you are considering adding to your colony is a demonstration that the frog harbors the detrimental microorganism(s) and is a threat to the health of your frog colony.

Test samples for most of the described qPCRs can be obtained from live animals. Currently, obtaining toe web skin and kidney samples for chytrid and ranavirus testing, respectively, requires that the frog be euthanized. In the future, it may be possible to test DNA isolated from environmental samples from aquaria, such as water filtrate, for the presence of DNA sequences associated with the presence of pathogenic microorganisms.

REFERENCES

Boyle DG, Boyle DB, Olsen V, Morgan JA, Hyatt AD. 2004. Rapid quantitative detection of chytridiomycosis (*Batrachochytrium dendrobatidis*) in amphibian samples using real-time Taqman PCR assay. *Dis Aquat Organ* **60**: 141–148. doi:10.3354/dao060141

Feldman SH, Ramirez MP. 2014. Molecular phylogeny of *Pseudocapillaroides xenopi* (Moravec et Cosgrov 1982) and development of a quantitative PCR assay for its detection in aquarium sediment. *J Am Assoc Lab Anim Sci* **53**: 668–674.

Fremont-Rahl JJ, Ek C, Williamson HR, Small PLC, Fox JG, Muthupalani S. 2011. *Mycobacterium liflandii* outbreak in a research colony of *Xenopus* (*Silurana*) *tropicalis* frogs. *Vet Pathol* **48**: 856–867. doi:10.1177/0300985810388520

Godfrey D, Williamson H, Silverman J, Small PL. 2007. Newly identified *Mycobacterium* species in a *Xenopus laevis* colony. *Comp Med* **57:** 97–104.

Green SL, Lifland BD, Bouley DM, Brown BA, Wallace RJ Jr, Ferrell JE Jr. 2000. Disease attributed to *Mycobacterium chelonae* in South African clawed frogs (*Xenopus laevis*). *Comp Med* **50:** 675–679.

Green SL, Bouley DM, Josling CA, Fayer R. 2003. Cryptosporidiosis associated with emaciation and proliferative gastritis in a laboratory-reared South African clawed frog (*Xenopus laevis*). *Comp Med* **53:** 81–84.

Iglauer F, Willmann F, Hilken G, Huisinga E, Dimigen J. 1997. Anthelmintic treatment to eradicate cutaneous capillariasis in a colony of South African clawed frogs (*Xenopus laevis*). *Lab Anim Sci* **47:** 477–482.

Longcore JE, Pessier AP, Nichols DK. 1999. *Batrachochytrium dendrobatidis* gen. et sp. nov., a chytrid pathogenic to amphibians. *Mycologia* **91:** 219–227. doi:10.1080/00275514.1999.12061011

Maryam A. 1988. Mycobacterium-induced infectious granuloma in *Xenopus*: histopathology and transmissibility. *Cancer Res* **48:** 958–963.

McNamara S, Wlizla M, Horb ME. 2018. Husbandry, general care, and transportation of *Xenopus laevis* and *Xenopus tropicalis*. *Methods Mol Bio* **1865:** 1–17. doi:10.1007/978-1-4939-8784-9_1

Newcomer CE, Anver MR, Simmons JL, Wilcke BW Jr, Nace GW. 1982. Spontaneous and experimental infections of *Xenopus laevis* with *Chlamydia psittaci*. *Lab Anim Sci* **32:** 680–686.

Ouellet M, Mikaelian I, Pauli BD, Rodriguez J, Green DM. 2005. Historical evidence of widespread chytrid infection in North American amphibian populations. *Conserv Biol* **19:** 1431–1440. doi:10.1111/j.1523-1739.2005.00108.x

Reed KD, Ruth GR, Meyer JA, Shukla SK. 2000. *Chlamydia pneumoniae* infection in a breeding colony of African clawed frogs (*Xenopus tropicalis*). *Emerg Infect Dis* **6:** 196–199. doi:10.3201/eid0602.000216

Robert J, Morales H, Buck W, Cohen N, Marr S, Gantress J. 2005. Adaptive immunity and histopathology in frog virus 3-infected *Xenopus*. *Virology* **332:** 667–675. doi:10.1016/j.virol.2004.12.012

Sánchez-Morgado JM, Gallagher A, Johnson LK. 2009. *Mycobacterium gordonae* infection in a colony of African clawed frogs (*Xenopus tropicalis*). *Lab Anim* **43:** 300–303. doi:10.1258/la.2008.008035

Schwabacher H. 2009. A strain of mycobacterium isolated from skin lesions of a cold-blooded animal, *Xenopus laevis*, and its relation to atypical acid-fast bacilli occurring in man. *J Hygiene* **57:** 57–67. doi:10.1017/S0022172400019896

Trott KA, Stacy BA, Lifland BD, Parker JM, Diggs H. 2004. Characterization of a *Mycobacterium ulcerans*-like infection in a colony of African tropical clawed frogs (*Xenopus tropicalis*). *Comp Med* **54:** 309–317.

Cite this protocol as *Cold Spring Harb Protoc*; doi:10.1101/pdb.prot106179

Obtaining *Xenopus* Eggs and Embryos

Hazel L. Sive[1,3] and Richard M. Harland[2]

[1]*Northeastern University, Boston, Massachusetts 02115, USA;* [2]*Department of Molecular and Cell Biology, University of California, Berkeley, California 94720-3200, USA*

Collecting eggs from adult *Xenopus laevis* and *Xenopus tropicalis* to raise healthy embryos and tadpoles is relatively simple but requires careful handling of the frog. Eggs can be fertilized through natural matings or by in vitro fertilization and examined visually. Here we review how eggs are obtained and how to recognize healthy eggs that will develop into high-quality embryos.

HANDLING FROGS

To collect eggs manually, you will need to pick up frogs. Wear gloves that are powder-free and start by simply putting your hand into a bucket of frogs and leaving it there. The frogs will stay very calm. Gently get used to handling them by first touching the belly of a frog with your finger and then carefully moving one across the bucket with your hand. Learn to pick up *Xenopus laevis* frogs by following the guidance provided by Shaidani et al. (2021a) (see Protocol 1: Obtaining *Xenopus laevis* Eggs) and *Xenopus tropicalis* frogs as Lane et al. (2021) describe (see Protocol 4: Obtaining *Xenopus tropicalis* Eggs). It can be helpful to make a paper frog for practice. Holding the frog in one hand and covering its head with your other hand will help it lie still. If a frog squirms, another strategy is to bring the frog against your body (wearing a clean lab coat) and cover its head that way. Follow the protocol instructions, stay calm, and with time it will become routine. Remember that *X. laevis* are larger than *X. tropicalis* and there are slight variations to handling the smaller species.

OBTAINING FERTILIZED EGGS THROUGH NATURAL MATINGS

There are two ways to collect fertilized eggs for growth into embryos both for *X. laevis* and for *X. tropicalis*. The first is through "natural mating" (see Protocol 1: Obtaining *Xenopus laevis* Eggs [Shaidani et al. 2021a] and Protocol 6: Obtaining *Xenopus tropicalis* Embryos by Natural Mating [Lane and Khokha 2021a]). This refers to placing a male and female frog together, after which they should go into amplexus, in which the male clasps the female around the belly and they deposit eggs and sperm together, leading to fertilization. Note that both male and female frogs will need to be induced to mate by prior hormone injection. Several rounds of fertilized eggs can be produced in this way over the course of a day. The advantage of this method is that male frogs can repeatedly provide sperm rather than being killed and it is simpler than in vitro fertilization (IVF). The significant disadvantage is that

[3]Correspondence: h.sive@northeastern.edu

Cite this introduction as *Cold Spring Harb Protoc*; doi:10.1101/pdb.top106195

fertilized eggs are produced in small numbers and must be diligently collected right after laying to ensure you obtain batches of synchronously developing embryos. Often the adults may not be mature enough or healthy enough to mate successfully. Frogs must be laying and eggs fertilizing well, and one must be well organized to collect embryos early enough for microinjection at the one- or two-cell stage by this method. However, the method can be useful if you want to collect embryos for use at older stages.

OBTAINING FERTILIZED EGGS BY IVF

The second way to obtain fertilized eggs uses IVF, and this is the gold standard for getting plenty of truly synchronously developing embryos (see Protocol 1: Obtaining *Xenopus laevis* Eggs [Shaidani et al. 2021a] and Protocol 5: Obtaining *Xenopus tropicalis* Embryos by In Vitro Fertilization [Lane and Khokha 2021b]). Eggs are manually collected from a female frog induced to ovulate by gentle techniques. When picked up, she will push them out on her own following gentle massage of the region above the cloaca, and you can catch the eggs in a Petri dish. You should never try to squeeze eggs out of a female frog. Squeezing may rupture her internal organs and lead to death. If the female does not lay, she likely has no eggs ready to be released and you should try another frog. Usually, several females are induced for each experiment to cater for this eventuality.

Manually collected eggs may be fertilized from testes harvested from a killed male frog. For *X. laevis*, usually this is done by running a piece of testis over eggs in a dish from which most of the liquid has been removed. For *X. tropicalis*, you can crush the testes in a salt solution, then mix the slurry into dishes of eggs from which excess liquid has been removed (this method may also be used for *X. laevis*). Placing the eggs into a high-salt solution at the right temperature keeps the eggs competent for fertilization for many hours, another advantage of this method over natural matings. IVF can yield hundreds of high-quality embryos, which are immediately available for microinjection of nucleic acid or protein. Indeed, current state-of-the-art techniques using both *X. laevis* and *X. tropicalis* typically require IVF.

RECOGNIZING HEALTHY EGGS

It is easy to recognize "good" eggs that are likely to be successfully fertilized: These are round and firm-looking with distinct dark and cream sides. Eggs should be laid separate from each other; eggs formed in "strings" are often inferior. Completely white eggs are not viable and should be discarded. After fertilization (see Protocol 5: Obtaining *Xenopus tropicalis* Embryos by In Vitro Fertilization [Lane and Khokha 2021b] and Protocol 2: Obtaining *Xenopus tropicalis* Embryos [Shaidani et al. 2021b]), the eggs rotate in their membranes, and in a Petri dish those fertilized will have the darker side facing upward. With good eggs, a fertilization success of >90% can be achieved.

Xenopus eggs are covered by a series of membranes, the outermost being the protective jelly coat. Immediately after laying, eggs have a collapsed jelly coat, which allows fertilization, and this expands soon after the eggs come into contact with low-salt water. Maintaining eggs in a high-salt solution prevents jelly coat expansion. After fertilization, when eggs are placed in low salt, the coat expands. This keeps the eggs at a distance from one another probably to allow optimal oxygenation.

Raising Healthy Embryos and Tadpoles

Once you have fertilized eggs, the next step is to ensure they develop into embryos or tadpoles (see Protocol 5: Obtaining *Xenopus tropicalis* Embryos by In Vitro Fertilization [Lane and Khokha 2021b], Protocol 2: Obtaining *Xenopus tropicalis* Embryos [Shaidani et al. 2021b], and Protocol 6: Obtaining *Xenopus tropicalis* Embryos by Natural Mating [Lane and Khokha 2021a]). Effective fertilization is an excellent harbinger of great embryo quality. But sometimes things go wrong and embryo development

Cite this introduction as *Cold Spring Harb Protoc*; doi:10.1101/pdb.top106195

ceases or looks anomalous. Because the embryos are developing in a Petri dish, you can look at them under a dissecting microscope as often as you like, to follow their progress. *Normal Table of* Xenopus laevis is a useful companion (Nieuwkoop and Faber 1967) and additional guidance on raising *X. tropicalis* embryos is also available (Nakayama and Grainger 2023).

Keeping the embryos at low density and promptly removing those that are not developing optimally will help promote good health of the remaining embryos. Make sure your solutions are fresh and clean. Antibiotics may be added, but these are primarily needed after microdissection. If the embryos all die, do not be discouraged. With practice and patience, you will become an expert in collecting eggs and growing embryos and tadpoles that you can raise into older stages as needed.

ACKNOWLEDGMENTS

R.M.H. was supported by National Institutes of Health (NIH) grant R35GM127069. H.L.S. was supported by NIH grant 5R01DE021109.

REFERENCES

Lane M, Khokha MK. 2021a. Obtaining *Xenopus tropicalis* embryos by natural mating. *Cold Spring Harb Protoc* doi:10.1101/pdb.prot106609

Lane M, Khokha MK. 2021b. Obtaining *Xenopus tropicalis* embryos by in vitro fertilization. *Cold Spring Harb Protoc* doi:10.1101/pdb.prot106351

Lane M, Mis EK, Khokha MK. 2021. Obtaining *Xenopus tropicalis* eggs. *Cold Spring Harb Protoc* doi:10.1101/pdb.prot106344

Nakayama T, Grainger RM. 2023. Best practices for *Xenopus tropicalis* husbandry. *Cold Spring Harb Protoc* doi:10.1101/pdb.prot106252

Nieuwkoop PD, Faber J. 1967. *Normal table of* Xenopus laevis *(Daudin): a systematical and chronological survey of the development from the fertilized egg till the end of morphogenesis.* North-Holland Publishing Company, Amsterdam, reprinted 1994 Garland Publishing, New York. ebook 2020 at https://doi.org/10.1201/9781003064565 with figures at https://www.xenbase.org/anatomy/alldev.do.

Shaidani N-I, McNamara S, Wlizla M, Horb ME. 2021a. Obtaining *Xenopus laevis* eggs. *Cold Spring Harb Protoc* doi:10.1101/pdb.prot106203

Shaidani N-I, McNamara S, Wlizla M, Horb ME. 2021b. Obtaining *Xenopus laevis* embryos. *Cold Spring Harb Protoc* doi:10.1101/pdb.prot106211

Obtaining *Xenopus laevis* Eggs

Nikko-Ideen Shaidani, Sean McNamara,[1] Marcin Wlizla, and Marko E. Horb[2]

National Xenopus *Resource, Marine Biological Laboratory, Woods Hole, Massachusetts 02543, USA*

Nearly a century ago, studies by Lancelot Hogben and others demonstrated that ovulation in female *Xenopus laevis* can be induced via injection of mammalian gonadotropins into the dorsal lymph sac, allowing for egg production throughout the year independent of the normal reproductive cycles. Hormonally induced females are capable of producing thousands of eggs in a single spawning, which can then be fertilized to generate embryos or used as a substrate for generation of egg extracts. The protocol for induction of ovulation and subsequent egg collection is straightforward and robust, yet some of its details may vary among laboratories based on prior training, availability of necessary reagents, or the experimental objectives. As the goal of this protocol is not to describe every single variation possible for acquiring eggs but to provide a simple and clear description that can be easily applied by researchers with no prior working experience with *X. laevis*, we focus on describing the method we use at the National *Xenopus* Resource—that is, inducing ovulation in *X. laevis* via dorsal lymph sac injection of gonadotropic hormones and the stimulation of egg laying through application of gentle pressure to the females.

MATERIALS

It is essential that you consult the appropriate Material Safety Data Sheets and your institution's Environmental Health and Safety Office for proper handling of equipment and hazardous materials used in this protocol.

RECIPES: Please see the end of this protocol for recipes indicated by <R>. Additional recipes can be found online at http://cshprotocols.cshlp.org/site/recipes.

Reagents

Egg-laying solution (optional; see Steps 11–20):
    Marc's modified Ringer solution (MMR) <R>
    Modified Barth's saline for *Xenopus* (MBS) <R>
Hormone solutions for egg laying in *Xenopus* <R>

> *Generally, pregnant mare serum gonadotropin (PMSG) is used for priming (see Steps 1–5), and either human chorionic gonadotropin (hCG) or ovine luteinizing hormone (oLH) is used for boosting (see Steps 6–10).*

Xenopus laevis mature female frogs, preferably no younger than 18 mo old

Equipment

Incubator at 18°C
Petri dish (sterile)

---

[1]Present address: Iwaki Aquatic, Holliston, Massachusetts 01746, USA
[2]Correspondence: mhorb@mbl.edu

Cite this protocol as *Cold Spring Harb Protoc*; doi:10.1101/pdb.prot106203

Standard *X. laevis* maintenance facility (see Sec. 1, Protocol 1: Animal Maintenance Systems: *Xenopus laevis* [Shaidani et al. 2020a])

Syringes (1-mL) and needles (e.g., PrecisionGlide, 27 G × 3/8″, from Becton, Dickinson and Company)

Temporary holding tank (see Step 9)

> Use a plastic mouse cage or a large Tupperware container with an aerated cover that can be secured to prevent the frogs from escaping.

Transfer pipettes

METHOD

Priming

Prime frogs 1–7 d before planned ovulation. The priming injection on its own does not induce egg laying. Instead, it promotes consistent production of a high number of mature eggs following the boosting injection. The use of PMSG for priming has been shown to improve follicular maturity.

1. Remove PMSG stock from −20°C freezer and allow it to thaw at room temperature or 37°C.

2. Fill the 1-mL syringe with 30–50 U of 100 U/mL PMSG.

   > Smaller inbred J-strain frogs require less hormone and can be primed with 30 U of PMSG, whereas larger wild-type frogs can be given 50 U of PMSG.

   > Alternatively, hCG can be used to prime females at 0.1 of the amount used for boosting or 50 U for a wild-type female.

3. Pick up and restrain the female with one hand in such a way that the dorsal surface of the frog is rested against palm of the hand with the head of the frog pointed toward the wrist (Fig. 1A,B). Place the thumb and the middle finger along the posterior sides of the animal abdomen and use the index finger to abduct one of the frog's hind limbs (Fig. 1C).

   > If possible, the palm of the hand should cover the frog's eyes, which will help keep it calm. This hold makes the dorsal lymph sac pocket easily accessible while restraining the animal.

4. With the other hand, insert the needle subcutaneously through the dorsal surface into the dorsal lymph sac in the posterior medial region of the animal proximal to the lateral line "stitch marks" and slowly inject the hormone (Fig. 1C–E).

   > Insert the needle at a shallow angle to prevent penetrating muscle.

5. Return the frog to its permanent tank and do not feed until after the eggs have been collected.

   > Adult frogs can regurgitate food following hormone injection, and the presence of solid waste can reduce the durability of the laid eggs.

Boosting

Perform the boosting injection the evening before egg collection (see Step 10 for timing details). The boosting injection promotes the final steps of oocyte maturation and is necessary for induction of ovulation. Females given a boosting injection are likely to spawn, even if they have not been given a prior priming injection. The priming injection serves to ensure consistent egg-laying response to the boosting injection.

6. Remove oLH or hCG stock from −20°C freezer and allow it to thaw at room temperature or 37°C.

7. Fill a 1-mL syringe with 2 µg of oLH per 1 g of body mass (typically ∼200 µg of 0.4 mg/mL oLH) or with 500 U of 1000 U/mL hCG.

   > Smaller inbred J-strain females require less hormone and can be primed with 120 µg of oLH or 300 U of hCG. Larger wild-type females require 200 µg of oLH or 500 U of hCG.

8. Inject the frog as described in Steps 3 and 4.

9. Prepare a temporary holding tank with system water for the frog.

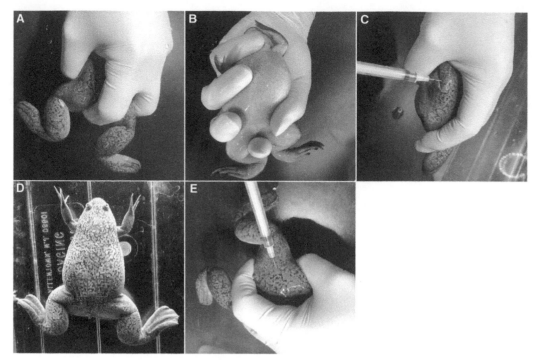

FIGURE 1. Single hand manual frog restraint for gonadotropin injection and egg collection. (*A*) View from the dorsal surface with the index finger in-between both legs, one of the legs being held between the index and the middle fingers, the thumb located around the waist on the other side of the frog. (*B*) View of the hold from the ventral surface with the index finger between the legs, the middle finger and the thumb located on the waist on the opposite sides of the frog, and the tip of the little finger pressed against the throat to secure the head against the base of the palm of the hand, thus aiding to cover the eyes. (*C*) The frog in fully restrained hold with one of the legs pulled back against the abdomen and a needle ready for injection through the dorsal surface. (*D*) *X. laevis* female adult with the general location of the dorsal posterior lymph sacs, the injection target, outlined in red. (*E*) A more relaxed hold on the female with neither leg fully restrained against the abdomen and with a needle ready for injection through the dorsal surface.

10. Place the frog in the holding tank, and place the tank into an 18°C incubator.

    Ovulation will begin ∼8–12 h postinjection and should continue for ∼5 h. Do not feed the frogs while waiting for them to lay eggs.

    The onset of ovulation can be delayed by several hours by keeping the female at a lower temperature of 15°C.

Obtaining Eggs

11. Remove the temporary holding tank with the frog from the 18°C incubator.

    The steps below describe a way of handling a female frog to promote egg laying and subsequent collection of a batch of eggs into a Petri dish. As an alternative, at this point the system water in the frog holding tank can be replaced with an egg-laying solution like 1× MMR or 1× MBS. The viability of eggs laid into system water is lost rapidly but can be maintained for at least 1 h if laid into egg-laying solution. This allows for eggs to be collected throughout the day, without the need to handle the females. The animals tolerate being placed in the egg-laying solution well; however, further handling to promote egg laying should be avoided as it does stress the females. This approach is potentially less reliable at producing a large number of viable eggs at exactly the desired time.

12. Obtain a sterile Petri dish for egg collection.

13. Similar to the hold used for restraining the animal during hormone injection, gently but firmly pick up the female from the dorsal side with the dominant hand, allowing its head to be concealed by the palm (Fig. 1A,B).

    The frog should be in the prone position.

Cite this protocol as *Cold Spring Harb Protoc*; doi:10.1101/pdb.prot106203

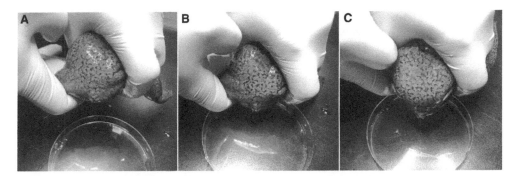

FIGURE 2. Manual two-handed restraint of a female ready for egg expulsion. (A) Initial hold with the female held in the dominant hand. The face of the frog rests against the inside of the base of the palm of the hand. One of the legs is held between the index and the middle finger while the thumb is near the pit of the contralateral arm. The other leg is being restrained between the index and the middle fingers of the other hand. (B) Both hands are used to progressively pull both legs out and forward toward the head. (C) Fully restrained female with both legs held against the abdomen. The hold is loose enough to allow the female to wiggle and flex its leg and abdominal muscles, yet firm enough to prevent it from breaking loose.

14. Place the thumb and middle finger along the posterior side of the female with the little finger against the throat of the animal to restrain it further.

15. Using the index finger, abduct one of the frog's hind limbs to reflect it rostralward. At the same time abduct the second limb using the other hand (Fig. 2A–C).

    This will expose the cloaca/vent and allow for the best control when attempting to position the female. The thumb, ring, and index finger should be available to massage the dorsal and ventral lower trunk.

16. Position the female vertically, with the cloaca over the Petri dish.

    This helps prevent the eggs from running along the body of the female.

17. Use the dominant hand's ring and middle finger tips to apply pressure to the belly and the other hand's thumb to apply pressure to the back of the animal.

18. Gently shift the pressure in an anterior to posterior direction to aid the expulsion of eggs.

    Simply restraining the female in the described hold, without applying additional pressure, is often sufficient for egg expulsion. The hold itself should be tight and stiff enough to prevent the female from breaking loose. Eggs will come out as she flexes her leg and abdominal muscles attempting to escape. If no eggs come out when simply holding the animal, gently increase the applied pressure being sure not to hurt the animal. Monitor the animal for stress indicators such as sudden excess of skin surface mucus. Females should only be squeezed once every hour and no more than three times in a single day.

19. Transfer the frog to a clean tank overnight and allow it 24 h to recover prior to returning it to the system.

    When dealing with albino females in particular, even gentle pressure may result in mild bruising around the eyes and the part of the abdomen handled directly. This bruising will disappear within hours and has no long-term ill effects on the health of the female. X. laevis are hardy, although care should be taken not to apply such pressure that it leads to injury.

20. Immediately after putting the female back, use a transfer pipette to remove any system water that may have dripped into the egg dish during collection.

    Egg viability will decline rapidly if eggs are exposed to a significant amount of system water. If collected eggs are not going to be used immediately, it may be useful to flood the egg collection dish with 1× MMR or 1× MBS, which will prevent them from drying out and will help maintain their viability for at least an hour.

DISCUSSION

The method described here is well-established (Wlizla et al. 2017, 2018) and allows for a simple and efficient way to collect eggs from female *X. laevis* at desired times. These eggs can then be used directly in experiments or fertilized via in vitro fertilization (IVF) to generate embryos (see Protocol 2: Obtaining *Xenopus laevis* Embryos [Shaidani et al. 2020b]). Developing a correct technique for restraining *X. laevis* females for hormone injection and egg collection requires some practice but is essential for decreasing distress caused to the animal and for collecting the best-quality eggs. Restrained frogs appear to struggle less against the hold when their eyes are covered. For those not experienced in holding frogs, a good initial practice may be to place a moist paper towel on a tabletop, put the female on the paper towel, and gently wrap her head to cover her eyes while keeping the posterior dorsal surface still accessible for injection. This way the hard tabletop surface can be used to aid in immobilizing the animal.

Another aspect of this procedure that requires some practice is determining the amount of pressure that should be applied to the animal during egg expulsion. Too little pressure might not produce any eggs, whereas too much may cause the eggs to be damaged or cause injury to the animal. If simply holding the female is not productive, some pressure will be necessary. The pressure should be applied in a way in which it shifts in a rostral to caudal direction so that it is not simply a squeeze, but instead a massage intended to push the eggs toward and out of the cloaca. As an alternative to handling the female, following the boosting hormone injection, she can be placed in a tank containing egg-laying solution. The female will lay eggs at a slower rate but they will maintain their viability for at least 1 h and can be collected directly from the tank; however, it is likely that the total amount of eggs produced will not be as high without the additional manual pressure. Common egg-laying solutions include 1× MMR or 1× MBS, both of which work equally well for maintaining egg viability. The choice of egg-laying solution will depend on prior user experience and experimental design aspects beyond the scope of this protocol.

A final point to address is the use of mammalian gonadotropic hormone injection into the dorsal lymph sac of the *X. laevis* female as a reliable and effective way for induction of ovulation. hCG has been historically used for this purpose; however, during a shortage of commercially available hCG, we determined that other gonadotropins, including PMSG for priming and oLH or human luteinizing hormone for boosting, can stimulate spawning as effectively (Wlizla et al. 2017). Although using hormones not sourced in animals may be the more humane option when considering animal welfare, the awareness of these alternatives is useful when dealing with supply disruptions or when trying to find the most economical option.

RECIPES

Hormone Solutions for Egg Laying in **Xenopus**

Hormone	Concentration in 1× PBS
Human chorionic gonadotropin (hCG; from National Hormone and Peptide Program, Sigma-Aldrich, Merck Animal Health [Chorulon brand], or BioVendor)	1000 U/mL
Ovine luteinizing hormone (oLH; from National Hormone and Peptide Program)	0.4 mg/mL
Pregnant mare serum gonadotropin (PMSG; from BioVendor)	100 U/mL

Dissolve phosphate-buffered saline (PBS) tablets (Sigma-Aldrich P4417) in type 1 ultrapure water, at a concentration of 1 tablet per 200 mL. Sterilize by autoclaving. Prepare individual hormone solutions in 1× PBS at the concentrations listed above. Store all hormone solutions at 20°C until needed for use.

Cite this protocol as *Cold Spring Harb Protoc*; doi:10.1101/pdb.prot106203

Marc's Modified Ringer Solution (MMR)

1 M NaCl
20 mM KCl
10 mM $MgSO_4 \cdot 7H_2O$
20 mM $CaCl_2 \cdot 2H_2O$
50 mM HEPES free acid

Prepare a 10× stock that consists of the reagents listed above, adjusting the pH to 7.4–7.8 with NaOH and sterilizing by autoclaving. Store the stocks indefinitely at room temperature. Dilute stock as necessary with type 1 ultrapure water (ASTM International 2018) without further pH adjustment.

Modified Barth's Saline for Xenopus (MBS)

Make a 10× MBS stock:
    800 mM NaCl
    10 mM KCl
    10 mM $MgSO_4 \cdot 7H_2O$
    50 mM HEPES free acid
    25 mM $NaHCO_3$

Prepare a 10× stock that consists of the reagents listed above, adjusting the pH to 7.8 with NaOH and sterilizing by autoclaving. Prepare a separate 0.1 M stock of $CaCl_2$, and sterilize by autoclaving. Store the stocks indefinitely at room temperature.

    To make 1× MBS, dilute the 10× MBS stock with type 1 ultrapure water (ASTM International 2018), and add $CaCl_2$ stock to a final concentration of 0.7 mM.

REFERENCES

ASTM International. 2018. *ASTM standard D1193-06 (2018).* "Standard specification for reagent water." ASTM, West Conshohocken, PA.

Shaidani N-I, McNamara S, Wlizla M, Horb ME. 2020a. Animal maintenance systems: *Xenopus laevis. Cold Spring Harb Protoc* doi:10.1101/pdb.prot106138

Shaidani N-I, McNamara S, Wlizla M, Horb ME. 2020b. Obtaining *Xenopus laevis* embryos. *Cold Spring Harb Protoc* doi:10.1101/pdb.prot106211

Wlizla M, Falco R, Peshkin L, Parlow AF, Horb ME. 2017. Luteinizing hormone is an effective replacement for hCG to induce ovulation in *Xenopus. Dev Biol* **426:** 442–448. doi:10.1016/j.ydbio.2016.05.028

Wlizla M, McNamara S, Horb ME. 2018. Generation and care of *Xenopus laevis* and *Xenopus tropicalis* embryos. *Methods Mol Biol* **1865:** 19–32. doi:10.1007/978-1-4939-8784-9_2

Obtaining *Xenopus laevis* Embryos

Nikko-Ideen Shaidani, Sean McNamara,[1] Marcin Wlizla, and Marko E. Horb[2]

National Xenopus *Resource, Marine Biological Laboratory, Woods Hole, Massachusetts 02543, USA*

The embryos of the African clawed frog, *Xenopus laevis*, are a powerful substrate for the study of complex fundamental biological and disease mechanisms in neurobiology, physiology, molecular biology, cell biology, and developmental biology. A simple and straightforward technique for generating a large number of developmentally synchronized embryos is in vitro fertilization (IVF). IVF permits simultaneous fertilization of thousands of eggs but requires the death of the parental male, which may not be feasible if the male comes from a stock of precious animals. An alternative to euthanizing a precious male is to use a natural mating, which allows for the collection of many embryos with minimal preparation but with the potential loss of the experimental advantage of developmental synchronization. Here we present both strategies for obtaining *X. laevis* embryos.

MATERIALS

It is essential that you consult the appropriate Material Safety Data Sheets and your institution's Environmental Health and Safety Office for proper handling of equipment and hazardous materials used in this protocol.

RECIPES: Please see the end of this protocol for recipes indicated by <R>. Additional recipes can be found online at http://cshprotocols.cshlp.org/site/recipes.

Reagents

Benzocaine (ethyl 4-aminobenzoate; CAS 94-09-7) (for IVF [Steps 1–13] only)
> *Prepare a 10% solution of benzocaine in 95% ethanol and store at room temperature.*

Culture medium (select one of the following; see Step 7)
   Marc's modified Ringer solution (MMR) <R>
   Modified Barth's saline for *Xenopus* (MBS) <R>
Gentamicin solution (e.g., ThermoFisher Scientific 15710064)
Hormone solutions for egg laying in *Xenopus* <R> (for natural mating [Steps 14–25] only)
Xenopus laevis males and females, sexually mature
> *Females at least 18-mo-old are preferred as they will be more productive; males should be at least 8-mo-old. For IVF (Steps 1–13),* Xenopus laevis *eggs must be obtained in advance (see Protocol 1: Obtaining* Xenopus laevis *Eggs [Shaidani et al. 2020a]).*

Equipment

Items needed for IVF (Steps 1–13) only:
   Forceps

[1]Present address: Iwaki Aquatic, Holliston, Massachusetts 01746, USA

[2]Correspondence: mhorb@mbl.edu

Kimwipes (optional; see Step 6)

Microcentrifuge tubes

Paper towels

Pestle for 1.5-mL microcentrifuge tubes (e.g., USA Scientific, Inc. 1415-5390)

Plastic pipette (fine-tipped) (see Step 11)

Surgical scissors

Items needed for natural mating (Steps 14–25) only:

Incubator at 18°C

Syringes (1-mL) and needles (e.g., PrecisionGlide, 27 G × 3/8″, from Becton, Dickinson and Company)

Temporary holding tank (see Step 22)

> Use a plastic mouse cage or a large Tupperware container with an aerated cover that can be secured to prevent the frogs from escaping.

Items used in both procedures:

Petri dish (sterile)

Standard *X. laevis* maintenance facility (see Sec. 1, Protocol 1: Animal Maintenance Systems: *Xenopus laevis* [Shaidani et al. 2020b])

Transfer pipettes

METHOD

Follow Steps 1–13 for IVF and Steps 14–25 for natural mating.

In Vitro Fertilization

Testes Isolation

1. Anesthetize the male frog via submersion in a bath of 0.15% benzocaine in system water for 15–30 min.

   > The male has reached the proper anesthetic plane when it is no longer responsive to foot pinching and has lost the swallowing reflex when its throat is rubbed.

2. To ensure euthanasia, place the anesthetized male on a moist paper towel, and using surgical scissors perform cervical transection severing the spine at the level of the atlas bone immediately posterior to the skull.

3. With the frog in the supine position (Fig. 1A), cut through the skin (Fig. 1B,C), fascia, and muscle with fine scissors to expose the lower abdominal cavity (Fig. 1D).

   > After cutting through each layer, move it out of the way to allow access to the tissues below and eventually the abdominal cavity. This is a terminal procedure, and because the goal is to gain easy access to the abdominal cavity, the opening should be made large enough to permit that.

4. Locate and isolate the testes along the dorsal midline of the abdominal cavity (Fig. 1E,F).

   > Because of the positioning of the testes on the dorsal side of the cavity, it is often necessary to adjust the position of fat bodies (Fig. 1D) to reveal the testes. In a healthy, fully sexually mature male, each testis resembles in shape and size a grain of puffed rice, ~7-mm-long and 3 mm in diameter. The testis may be closely associated with the fat bodies. One way to distinguish it is that the color of the testis is much paler than that of fat bodies. The testis is white to beige, not yellow, with a diffuse network of fine red vasculature on the surface.

5. Using scissors, cut the testes away from the surrounding organs.

6. Clean the testes on a dry paper towel or a Kimwipe by removing any attached remnants of fat and viscera and blood (Fig. 1G).

   > Removing all tissues besides the testes should aid in fertilization rate and durability during extended storage. Be gentle when handling the testes with forceps, as they can be easily damaged. Keeping them intact before use should also increase their durability.

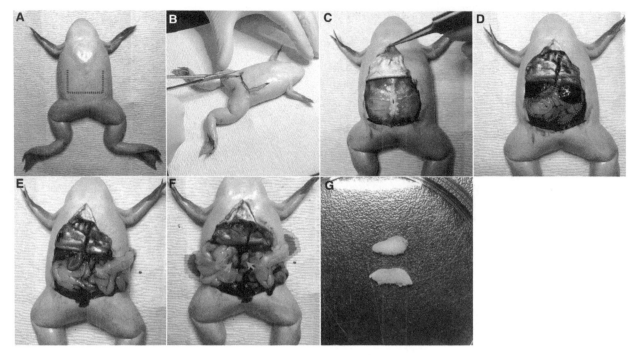

FIGURE 1. Isolation of testes from an *X. laevis* male. (*A*) A ventral view of a dead *X. laevis* male. The dashed line indicates the general location of the opening that needs to be made to give access to the abdominal cavity. (*B*) Incision through the skin. (*C*) The generated flap of skin pulled back to provide access to the layer of ventral abdominal muscle underneath. (*D*) Both the skin and muscle layers pulled back giving access to the open abdominal cavity. The dashed line outlines the area of the abdominal cavity currently obscured by the fat bodies. (*E*) The fat bodies on the male's left side have been pulled aside giving access to the testis indicated by the blue arrow. (*F*) Both testes, indicated by blue arrows, are visible after fat bodies have been moved aside. (*G*) Isolated and clean testes in a dish of 1× MMR and gentamicin.

7. Store the testes in 1× MMR with 10 µg/mL of gentamicin antibiotic at 4°C.

    High salt prevents the sperm from activating, and when stored at 4°C in the presence of the antibiotic, the testes will be viable for at least a week. If preferred, MBS can be substituted for MMR, here and wherever MMR is being used in the following steps. Use the same concentration of MBS as that indicated for MMR.

Applying the Sperm

8. Collect eggs into a sterile Petri dish (see Protocol 1: Obtaining *Xenopus laevis* Eggs [Shaidani et al. 2020a]) and remove system or egg-laying water that is incidentally transferred.

9. Cut approximately 1/8 of a testis, and transfer it into a 1.5-mL microcentrifuge tube with ∼1 mL of 1× MMR (Fig. 2A).

    Alternatively, sperm suspension can be made in 1–2 mL of 0.1× MMR, which activates the sperm immediately. The remaining part of the testis, if not being used immediately, can be returned to storage as in Step 7 above.

10. Use a microcentrifuge pestle to pulverize the testis piece, and then add the sperm solution to the eggs using a transfer pipette (Fig. 2B,C).

    Alternatively, the testis piece can be macerated directly in the dish with forceps.

11. With a fine-tipped plastic pipette, mix the sperm solution with the eggs, spreading them into a monolayer if possible (Fig. 2D,E).

12. After 5 min, flood the dish with 0.1× MMR to activate the sperm.

    Fill the dish when flooding.

13. After 15 min, look for egg rotation and animal pole contraction which are good indicators of a successful fertilization (Fig. 2F).

Cite this protocol as *Cold Spring Harb Protoc*; doi:10.1101/pdb.prot106211

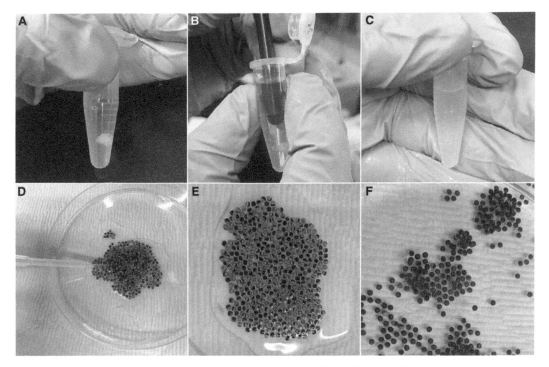

FIGURE 2. In vitro fertilization. (*A*) A 1.5-mL microcentrifuge tube with a small piece of a testis, approximately a quarter of a whole, in 1× MMR. (*B*) A microcentrifuge pestle ready to pulverize the testis piece. (*C*) A cloudy testis solution with not many large tissue pieces left following pulverization with a microcentrifuge tube. (*D*) A fine-tipped plastic pipette is used to mix the testis solution and the eggs together. (*E*) Eggs shortly after having been mixed with the testis solution and spread out into a monolayer. Notice that there is no distinct preference in the orientation of the eggs at this point. (*F*) Approximately 15 min after flooding the dish with 0.1× MMR the successfully fertilized eggs have oriented themselves with the pigmented animal pole facing up.

Natural Mating

Priming

Prime frogs 1–7 d before planned ovulation. The priming injection on its own does not induce egg laying. Instead, it promotes consistent production of a high number of mature eggs following the boosting injection. The use of pregnant mare serum gonadotropin (PMSG) for priming has been shown to improve follicular maturity.

14. Remove PMSG stocks from −20°C freezer and allow to thaw at room temperature or 37°C.

15. Fill a 1-mL syringe with 30–50 U of 100 U/mL PMSG.

    Smaller inbred J-strain females require less hormone and can be primed with 30 U of PMSG, whereas larger wild-type females can be given 50 U of PMSG. hCG can also be used to prime at 30 U for a smaller J-strain female and 50 U for a larger wild-type female.

    Although not necessary, males are also sometimes primed to presumably promote sperm maturation. If priming a male, use half the amount of the hormone as used for the female.

16. Pick up and restrain the frog with one hand in such a way that the dorsal surface of the frog is rested against palm of the hand with the head of the frog pointed toward the wrist. Place the thumb and the middle finger along the posterior sides of the animal abdomen and use the index finger to abduct one of the frog's hindlimbs.

    It is typically easier to handle the animals with one's dominant hand. If possible, the palm of the hand should cover the frog's eyes which will help keep it calm. This hold makes the dorsal lymph sac pocket easily accessible while restraining the animal.

    See Figure 1 in Protocol 1: Obtaining Xenopus laevis *Eggs (Shaidani et al. 2020) for visual guidance on priming female frogs.*

17. With the other hand, insert the needle subcutaneously through the dorsal surface into the dorsal lymph sac in the posterior medial region of the animal proximal to the lateral line "stitch marks" and slowly inject the hormone.

   Insert the needle at a shallow angle so the penetration is superficial to prevent injection into the muscle. Intramuscular injection should have no detrimental effect on animal health but may reduce the desired effect of the hormone.

18. Return the frog to its permanent tank and do not feed it before the boosting injection.

   Adult frogs can regurgitate food following hormone injection and the presence of solid waste can reduce oocyte durability.

Inducing Ovulation and Fertilization

Perform the boosting injection the evening before egg collection (see Step 22 for timing details). The boosting injection promotes the final steps of oocyte maturation and is necessary for induction of ovulation. Females given a boosting injection are likely to spawn, even if they have not been given a prior priming injection. The priming injection serves to ensure consistent egg laying in response to the boosting injection. In males, the boosting injection seems to encourage amplexus and the mating behavior necessary for proper juxtaposition of the male and female cloacas during mating and therefore for rapid and efficient fertilization of eggs as they are being laid by the female.

19. Remove ovine luteinizing hormone (oLH) or human chorionic gonadotropin (hCG) stocks from a −20°C freezer and allow to thaw at room temperature or 37°C.

20. Fill a 1-mL syringe with 2 µg of oLH per 1 g of body mass (typically ∼200 µg of 0.4 mg/mL oLH) or with 500 U of 1000 U/mL hCG.

   Smaller inbred J-strain females require less hormone and can be boosted with 120 µg of oLH or 300 U of hCG, whereas larger wild-type females will require 200 µg of oLH or 500 U of hCG. Males should be injected with half the amount of hormone as the females.

21. Inject both the male and female (see Steps 16 and 17).

22. Place both animals in a temporary holding tank filled with system water, and place the tank into an 18°C incubator.

   Ovulation will begin ∼8 to 12 h postinjection. Do not feed the frogs while waiting for the female to lay eggs.

   When checking on a natural mating in progress, if the male is in amplexus, avoid unnecessarily moving or disturbing the pair, as this may cause the male to release the female, thus aborting the mating.

23. Collect embryos throughout the day, and place them in 0.1× MMR with 10 µg/mL of gentamicin in sterile Petri dishes.

   Survival of early embryonic culture is highly sensitive to microbial contamination. Maintaining the culture in the presence of gentamicin helps prevent contamination. Embryos should be kept in gentamicin containing medium until they are ready to ingest food after stage 45.

24. At the end of the day, transfer the frogs to a clean tank overnight, and allow them 24 h to recover before returning them to the system.

   Males and females should be isolated from each other to prevent prolonged amplexus, which may sometimes cause sores or stress to the female.

25. Embryos kept at 18°C will continue to develop and can be maintained until the desired experimental stage. If they are kept for an extended duration, any dead or dying embryos should be removed daily and the remaining culture should be transferred to a clean Petri dish containing fresh 0.1× MMR with 10 µg/mL of gentamicin.

   X. laevis embryos are most stable at 18°C; however, they can tolerate a range of temperature and can be safely kept between 15°C and 23°C.

Cite this protocol as *Cold Spring Harb Protoc*; doi:10.1101/pdb.prot106211

DISCUSSION

Healthy and mature *X. laevis* females can produce in excess of 1500 eggs per spawning that can be easily fertilized via the strategies described here to generate embryos for use in experiments (Wlizla et al. 2017). The main advantage of using IVF is that it results in synchronized development of the embryos within the batch, which greatly simplifies certain aspects of experimental design and staging as the embryos continue to develop. When using IVF, only a small piece of the testis is required; the rest can be stored in 1× MMR or 1× MBS at 4°C and will maintain viability for at least a week. This permits the same male to be used for several experiments through the week, even after it has been killed.

Nevertheless, the use of natural mating may be necessary if the parental male being used is rare or precious. In such a case, it is useful for two investigators to work as a pair if precisely staged embryos during early development are required for the experiment, such as for microinjection at the one- or two-cell stage. One investigator can collect the embryos from the mating tank at regular intervals, select those at the appropriate stage, and perform any additional necessary treatments such as removal of the jelly coat, while the other investigator performs the experimental manipulations such as microinjection. This permits the maximum use of the embryos generated through natural mating as it may not be trivial for a single person to collect, stage, prepare, and perform experiments on embryos as they are being continuously generated.

Alternatively, users should also consider the extensive transgenic and mutant stocks available at the National *Xenopus* Resource (NXR) and the European *Xenopus* Resource Centre (EXRC) (Horb et al. 2019). Both resource centers can rapidly fulfill requests and ship adult frogs, embryos, isolated testes, and cryopreserved sperm.

Finally, regarding potentially limited resources, it is worth pointing out that several different mammalian gonadotropins can be used for the priming and boosting injections. Historically, hCG has been used for both injections and may be the more humane option when considering effects on animal welfare associated with extraction and purification of gonadotropins from animal sources. However, during a recent commercial shortage of hCG, we showed that the use of PMSG for priming and oLH for boosting is just as effective in inducing ovulation (Wlizla et al. 2017). Awareness of such alternatives is necessary for preventing interruptions in research resulting from supply disruptions.

RECIPES

Hormone Solutions for Egg Laying in Xenopus

Hormone	Concentration in 1× PBS
Human chorionic gonadotropin (hCG; from National Hormone and Peptide Program, Sigma-Aldrich, Merck Animal Health [Chorulon brand], or BioVendor)	1000 U/mL
Ovine luteinizing hormone (oLH; from National Hormone and Peptide Program)	0.4 mg/mL
Pregnant mare serum gonadotropin (PMSG; from BioVendor)	100 U/mL

Dissolve phosphate-buffered saline (PBS) tablets (Sigma-Aldrich P4417) in type 1 ultra-pure water, at a concentration of 1 tablet per 200 mL. Sterilize by autoclaving. Prepare individual hormone solutions in 1× PBS at the concentrations listed above. Store all hormone solutions at 20°C until needed for use.

Marc's Modified Ringer Solution (MMR)

1 M NaCl
20 mM KCl
10 mM $MgSO_4 \cdot 7H_2O$
20 mM $CaCl_2 \cdot 2H_2O$
50 mM HEPES free acid

Prepare a 10× stock that consists of the reagents listed above, adjusting the pH to 7.4–7.8 with NaOH and sterilizing by autoclaving. Store the stocks indefinitely at room temperature. Dilute stock as necessary with type 1 ultrapure water (ASTM International 2018) without further pH adjustment.

Modified Barth's Saline for Xenopus (MBS)

Make a 10× MBS stock:
800 mM NaCl
10 mM KCl
10 mM $MgSO_4 \cdot 7H_2O$
50 mM HEPES free acid
25 mM $NaHCO_3$

Prepare a 10× stock that consists of the reagents listed above, adjusting the pH to 7.8 with NaOH and sterilizing by autoclaving. Prepare a separate 0.1 M stock of $CaCl_2$, and sterilize by autoclaving. Store the stocks indefinitely at room temperature.

To make 1× MBS, dilute the 10× MBS stock with type 1 ultrapure water (ASTM International 2018), and add $CaCl_2$ stock to a final concentration of 0.7 mM.

REFERENCES

Horb M, Wlizla M, Abu-Daya A, McNamara S, Gajdasik D, Igawa T, Suzuki A, Ogino H, Noble A, Centre de Ressource Biologique Xenope team in France, et al. 2019. Xenopus resources: transgenic, inbred and mutant animals, training opportunities, and web-based support. *Front Physiol* 10: 387. doi:10.3389/fphys.2019.00387

Shaidani N-I, McNamara S, Wlizla M, Horb ME. 2020a. Obtaining *Xenopus laevis* eggs. *Cold Spring Harb Protoc* doi:10.1101/pdb.prot106203

Shaidani N-I, McNamara S, Wlizla M, Horb ME. 2020b. Animal maintenance systems: *Xenopus laevis*. *Cold Spring Harb Protoc* doi: 10.1101/pdb.prot106138

Wlizla M, Falco R, Peshkin L, Parlow AF, Horb ME. 2017. Luteinizing hormone is an effective replacement for hCG to induce ovulation in *Xenopus*. *Dev Biol* 426: 442–448. doi:10.1016/j.ydbio.2016.05.028

Cite this protocol as *Cold Spring Harb Protoc*; doi:10.1101/pdb.prot106211

How to Grow *Xenopus laevis* Tadpole Stages to Adult

Shoko Ishibashi and Enrique Amaya[1]

Division of Cell Matrix Biology and Regenerative Medicine, School of Biological Sciences, Faculty of Biology, Medicine and Health, University of Manchester, Manchester M13 9PT, United Kingdom

In *Xenopus laevis*, the tadpole stage is characterized by three forms—those occurring before the initiation of limb development, those covering limb development, and those encompassing metamorphosis. Maximal tadpole growth, especially during the second form, is critically dependent on good husbandry practices. Here we describe a protocol for raising *Xenopus laevis* tadpoles through to adulthood. Each step may need to be modified depending on the aquaria used and local conditions.

MATERIALS

It is essential that you consult the appropriate Material Safety Data Sheets and your institution's Environmental Health and Safety Office for proper handling of equipment and hazardous materials used in this protocol.

Reagents

Gammarus (Tetra; optional as a supplement for young froglets; see Table 1)

ReptoMin (Tetra)

Alternative foods, such as Nasco frog brittle, can be used.

Spirulina powder (available widely, including health food stores)

Water

For tadpoles, allow tap water to sit for at least half a day at room temperature or pass tap water through a carbon filter to dechlorinate before adding to aquaria.

Xenopus laevis embryos (see Protocol 2: Obtaining *Xenopus laevis* Embryos [Shaidani et al. 2020])

Equipment

Air pump for tadpoles (optional)

Aeration will help in keeping water quality, but is unnecessary, if the water is changed regularly.

Aquaria for frogs (2–12 mo old) in recirculating system (50 cm × 30 cm × 20 cm height [30 L] tank, with ~15 cm water depth)

Aquaria for frogs (>1 yr old) in recirculating system (60 cm × 40 cm × 20 cm height [48 L] tank, with ~17 cm water depth)

Aquaria for tadpoles in static water (minimum size of 8 cm height × 23 cm × 15 cm)

We use 15 cm height × 30 cm × 23 cm (about 10 L). It can be bigger but should be a size that permits easy and regular cleaning.

---

[1]Correspondence: enrique.amaya@manchester.ac.uk

Cite this protocol as *Cold Spring Harb Protoc*; doi:10.1101/pdb.prot106245

Conical tube (50 mL)
Nets, large size (25–30 cm) for adult frogs
Nets, medium size (10–13 cm) for tadpoles and juvenile frogs
Petri dish (9 cm)
Scrub brush

METHOD

The Xenopus laevis *tadpole stage encompasses three forms: those occurring before the initiation of limb development, those covering limb development, and those encompassing metamorphosis (Weisz 1945). The first form corresponds to Nieuwkoop and Faber (NF) (Nieuwkoop and Faber 1994) stages 45 through 47 (NF45-47), beginning about 4 d after fertilization. Appearance of the hindlimb bud at NF48 defines the initiation of the second form. This form lasts the longest (two or more months), covers the primary growth phase of the tadpole, and includes the stages when the hindlimbs and forelimbs develop (NF48-66). The third form (NF59-66) features the onset and climax of metamorphosis and takes around 8 d to complete. During metamorphosis, the body undergoes dramatic changes internally and externally, culminating with the tails and gills being resorbed and the froglets starting to swim with their legs and respiring with their lungs. During this phase, the tadpole does not feed, while it changes from an herbivorous to a carnivorous diet.*

Raising *X. laevis* Tadpoles through Metamorphosis

1. Transfer fertilized embryos of any stage cultured in a Petri dish to a container with dechlorinated water at 2–3 cm depth. Keep the tanks at room temperature between 16°C–25°C.

   Thirty to 50 embryos can be accommodated per 10 L tank.

2. When tadpoles reach NF45 stage, transfer them to a tank with dechlorinated water at 5–10 cm depth. Start feeding as below when the tadpoles start swimming actively.

Feeding

3. Suspend about 2 g of spirulina powder in 50 mL of dechlorinated water in a 50-mL conical tube. Mix well by inversion before each use. Add 2–3 mL of spirulina suspension per 30 tadpoles or 4–5 mL per 50 tadpoles daily.

   See Troubleshooting.

   Store the spirulina suspension at 4°C for a few weeks but discard if the suspension changes color from green to blue, as this indicates that the spirulina has become anoxic and is no longer usable.

   Tadpoles tend to feed at the bottom of the tank. Therefore, it is best to add spirulina suspension rather than to adding spirulina powder directly to the tank as the powder tends to remain at the surface of the water. In contrast, the spirulina/water mixture tends to sink to the bottom of the tank where it is more readily consumed by the tadpoles.

4. Feed tadpoles once in the morning and check the water quality and excrement on the bottom of the tank in the evening or the next morning.
   * If the water is not clear, reduce the amount of food per feeding.

     If all the food is eaten, the water will be clear and green strings of excrement will be present on the bottom of the tank.

   * If you do not see much excrement or the tadpoles' growth appears stunted, increase the amount of food per feeding.

     Daily observation and modification of the food intake accordingly is essential for ensuring the best husbandry conditions for optimal growth and raising of healthy tadpoles. See Troubleshooting.

5. Increase the amount of food as tadpoles grow. For example, when tadpoles reach NF50-52 (body length ~1.5–2 cm), feed 5–6 mL of the spirulina suspension per 30 tadpoles per day. At this stage, maintain a maximum of 30 tadpoles in a 10 L tank, and increase the water depth to 10–15 cm. If

Cite this protocol as *Cold Spring Harb Protoc*; doi:10.1101/pdb.prot106245

TABLE 1. *Xenopus laevis* tadpole, froglet density, and adult feeding regiment

Stages	Body length or features	Max tadpole numbers for 10 L/frog numbers for 30 or 48 L	Amount of food per feeding	Frequency of feeding
NF45		30	2–3 mL of spirulina/water mixture	Once/d
tadpoles		50	4–5 mL of spirulina/water mixture	Once/d
NF50	1.5–2 cm	30	5–6 mL of spirulina/water mixture	Once/d
tadpoles				
NF55	Hindlimbs with	20–25	4–5 mL of spirulina/water mixture	Twice/d
tadpoles	digits			
NF58	Developing arms	20	6–7 mL of spirulina/water mixture	Twice/d
tadpoles				
Juvenile	After metamorphosis	30–40/30 L	2–3 small pieces of a ReptoMin stick/frog (optional pinch of crushed Gammarus)	Once/d
froglets				
	3 cm	30–40/30 L	2–4 pieces of 1/3–1/4 cut ReptoMin stick/frog	Once/d
	3–4 cm	30/30 L	1–1.5 of ReptoMin sticks/frog	Once/d
	5–7 cm	20–25/30 L	2–3 of ReptoMin sticks/frog	Once/d
	8–10 cm	10–15/30 L, 20–30/48 L	4–6 of ReptoMin sticks/frog	Once/d
	<10 cm	20/48 L	4–6 of ReptoMin sticks/frog	Once/d
Adult	Females	10/48 L	6–10 of ReptoMin sticks/frog	Every other d
	Males	15–20/48 L	2–4 of ReptoMin sticks/frog	Every other d

necessary, split the tadpoles into two or more tanks or remove the smaller tadpoles to maintain no more than 30 tadpoles per tank.

The amount of food and tadpole density are listed in Table 1; however, please note that this is just a rough guide and adjustment may be required depending on local conditions. See Troubleshooting.

6. When the fingers in the tadpoles' hindlimbs become apparent at around NF55, begin to feed 4–5 mL of spirulina suspension twice a day per 20–25 tadpoles. Make sure that the water is clear at the second feeding. If the water is not clear, skip the second feeding on that day, or reduce the amount of food for each feeding and feed it twice a day. At this stage, split the tadpoles such that 20–25 tadpoles are present per tank.

See Troubleshooting.

7. When the tadpoles' legs are formed completely and arms become visible around NF58, increase the amount of food to 6–7 mL spirulina suspension for 20 tadpoles twice a day. Adjust the amount of food by checking the water clarity and excrement before and after feeding.

8. When the tadpoles' heads start to shrink during metamorphosis, stop feeding them. Only feed the amount required for the remaining premetamorphic growing tadpoles in the same tank. Alternatively, transfer tadpoles undergoing metamorphosis to another tank.

9. When the tadpoles' tails become very short (NF64-65), begin to feed them with frog diet, such as ReptoMin, broken into small pieces. Feed 1–2 pieces of 2–3 mm ReptoMin per frog daily. Spare fertilized early prefeeding stage embryos or unfertilized eggs can also be good food for small froglets just after metamorphosis. Leave the food left in the tank so that they can feed freely, unless the water becomes very dirty.

Cleaning and Changing Water

It is important to keep the bottom of the tank clean for efficient feeding where tadpoles can find food. In order to get rid of excrement and keep the floor clean, the tank should be cleaned at least once per week even though the water may appear clear. The frequency of cleaning should be increased depending on the density of tadpoles in the tank and frequency of feeding.

10. Prepare dechlorinated water in a clean tank.

11. Transfer tadpoles using a net into the tank above.

12. Clean the tank with water and dry the tank.

Cite this protocol as *Cold Spring Harb Protoc*; doi:10.1101/pdb.prot106245

Raising *X. laevis* Tadpoles from Metamorphosis to Adulthood

Feeding

13. Just after metamorphosis, move small froglets to a 30-L tank within the recirculating system with a density of 30–40 frogs/tank.

14. Feed juvenile frogs 2–3 pieces of ReptoMin broken into small pieces per frog daily. Temporarily turn off the water flow to each individual tank before feeding in the morning. In the afternoon, remove any food left over in the tank and restart the flow of water (restart recirculation). If there is a lot of food left over, reduce the amount of food in subsequent feedings.

15. Increase the amount of food as frogs grow according to Table 1. Remember to temporarily turn off water flow to each individual tank during each feed to ensure that food does not flow out of tanks before it can be eaten. Remember to restart flow to each tank once the feeding has ended.

16. Sort frogs by their size regularly, so that all frogs in the tank are able to eat food equally without competing for food.

    Five to seven centimeter juveniles and 8–10 cm juveniles can be accommodated in a 30 L tank with a density of 20–25 frogs and 10–15 frogs, respectively. A 48-L tank can hold twenty to thirty 8–10 cm frogs.

17. When frogs reach around 10 cm, males reach maturity and begin to eat less than females. Therefore, keep monitoring the leftover food after feedings and adjust the amount of food accordingly.

18. When frogs reach adulthood, separate males and females (body length 12–15 cm for females, ~10 cm for males). Place 10 adult female frogs or 15–20 male frogs per 48 L tank.

19. Feed adult frogs every other day according to Table 1.

    Adult frogs can be kept at 16°C–23°C; however, the lower temperature of around 16°C–18°C is preferable for Xenopus laevis, as the risk for outbreak of diseases will be reduced and egg quality tends to be better when the females are maintained at lower temperatures.

    Adult females can be ovulated every 3 mo, and females should not be rested for more than a year between ovulations, as this can result in poorer egg quality.

Cleaning and Changing Water

20. For juveniles fed daily in static water, clean the tank and change water once or twice a week, depending on their density.

21. For juveniles fed daily in the recirculating system with filtered water, clean tanks monthly with clean water and a brush.

22. For adult frogs fed every other day in the recirculating system with filtered water, clean tanks once or twice a year, depending on the system.

TROUBLESHOOTING

Problem (Steps 3–4): Water becomes become putrid and tadpoles are floating or dying.
Solution: Change water immediately. This is likely the result of overfeeding. Therefore, decrease the amount of food per feeding. It may also be caused by high water temperature, which encourages bacterial growth. Temperature should be kept between 16°C–25°C.

Problem (Steps 5–6): Tadpoles appear bent.
Solution: Bending is most likely caused by calcium deficiency or physical damage. Therefore, feed more spirulina or reduce the density of tadpoles in the tank. Take care when you transfer tadpoles with a net.

Cite this protocol as *Cold Spring Harb Protoc*; doi:10.1101/pdb.prot106245

DISCUSSION

The amount of food and frequency of feeding should change depending on the number of tadpoles in a tank and their age. Therefore, daily observations of water quality and tadpole health/growth are the most important things to do when raising tadpoles. One good indication of tadpole health is the color of the blood vessels in their head. The blood vessels should appear bright red if tadpoles have enough food and oxygen. If they are pale, increase the amount of food and consider aerating the tanks. The amount and condition of excrement should also be monitored daily. If tadpoles are fed enough, long green strings will be apparent on the bottom of the tank. However, if there are no strings, especially if they are all fragmented, this suggests that the tadpoles are searching for food and picking their excrement. In this case, increase the amount of food given to the tadpoles.

If tadpoles are fed daily on weekdays, weekend feedings can possibly be skipped, as less food is better than overfeeding, which could lead to deaths of all tadpoles. When feeding on weekends, it may be better to feed less if no one will be able to check water quality later in that day.

REFERENCES

Nieuwkoop PD, Faber J. 1994. *Normal table of* Xenopus laevis *(Daudin). A systematical and chronological survey of the development from the fertilized egg till the end of metamorphosis.* Garland Publishing, Inc, New York.

Shaidani N-I, McNamara S, Wlizla M, Horb ME. 2020. Obtaining *Xenopus laevis* embryos. *Cold Spring Harb Protoc* doi: 10.1101/pdb.prot106211

Weisz PB. 1945. The normal stages in the development of the South African clawed toad, *Xenopus laevis. Anat Rec* **93:** 161–169. doi:10.1002/ar.1090930207

Protocol 4

Obtaining *Xenopus tropicalis* Eggs

Maura Lane,[1,2] Emily K. Mis,[1] and Mustafa K. Khokha[1]

[1]*Pediatric Genomics Discovery Program, Department of Pediatrics and Genetics, Yale University School of Medicine, New Haven, Connecticut 06510, USA*

Xenopus is a powerful model system for cell and developmental biology in part because frogs produce thousands of eggs and embryos year-round. For cell biological studies, egg extracts can mimic many processes in a cell-free system. For developmental biology, *Xenopus* embryos are a premier system, combining cut-and-paste embryology with modern gene manipulation tools. *Xenopus tropicalis* are particularly suited to genetic studies because of their diploid genome, as compared to the tetraploid genome of *Xenopus laevis*. When collecting eggs, there are differences in timing of steps, amounts of hormone administered, and handling of females between these species. In this protocol, *X. tropicalis* females are induced with a hormone that stimulates ovulation, and then eggs are collected. To administer the ovulation hormone and express eggs, it is necessary to be comfortable with handling frogs. Proficient handling of *X. tropicalis* requires practice, as they are relatively small, active, and slippery.

MATERIALS

It is essential that you consult the appropriate Material Safety Data Sheets and your institution's Environmental Health and Safety Office for proper handling of equipment and hazardous materials used in this protocol.

RECIPES: Please see the end of this protocol for recipes indicated by <R>. Additional recipes can be found online at http://cshprotocols.cshlp.org/site/recipes.

Reagents

Human chorionic gonadotrophin (hCG) (Chorulon [Merck Animal Health 057176, 10,000 units])

Xenopus laboratories use a variety of hormones to induce ovulation and mating. Our laboratory achieves the best results using Chorulon hCG, so it is used for both priming and boosting in this protocol. Another commonly used hCG product is made by Sigma-Aldrich (CG10-10VL, 10,000 units). The Chorulon and Sigma-Aldrich brands are diluted to 1 unit/μL. Chorulon hCG comes with a diluent; sterile water (e.g., McKesson Medical 237000, 10 mL) must be purchased separately for Sigma-Aldrich hCG. hCG is a controlled substance in some states. There are nonregulated alternatives used by other laboratories, namely luteinizing hormone–releasing hormone (LHRH; Sigma-Aldrich L4513-1 MG), ovine luteinizing hormone (oLH; National Hormone and Peptide Program, www .humc.edu/hormones), and pregnant mare serum gonadotrophin (PMSG; Accurate Chemicals UCBH119/H). Please refer to Wlizla et al. (2017) and Wlizla et al. (2018) for detailed dilution and use protocols.

Mature *Xenopus tropicalis* females

Ideal females are mature, pear-shaped, and proven to lay well. Using two to four females per experiment increases the likelihood of obtaining high-quality eggs. They can be produced in house (see Sec. 1, Protocol 3: Raising and Maintaining Xenopus tropicalis *from Tadpole to Adult [Lane et al. 2021]) or purchased from an outside vendor or stock center (e.g., Xenopus 1 6020 or Xenopus National Resource NXR_1018). Females reaching maturity (6–8*

[2]Correspondence: maura.lane@yale.edu

Cite this protocol as *Cold Spring Harb Protoc*; doi:10.1101/pdb.prot106344

mo) should be boosted with 50 units of hCG monthly for 2–3 mo to stimulate egg production before being used for egg collection. These females can be boosted in the evening and allowed to lay eggs in a holding bucket (with a maximum of four females per bucket) overnight. These eggs should be discarded the next day, and the females should be returned to the recirculating system.

Modified Barth's saline (MBS) <R> at a concentration of 1×, containing 0.2% (w/v) bovine serum albumin (BSA)

Filter-sterilize. The pH of this solution should be 7.8–8.0. Store at 4°C, and bring to room temperature before use.

Modified Ringer's (MR; 1/9×) or recirculating system water (see Step 4).

Prepare modified Ringer's solution (MR; 1×) <R> with distilled water; the final pH will be 7.8–8.0. Then dilute the 1× MR solution to a 1/9× concentration. If recirculating system water is used instead (see Sec. 1, Protocol 1: Animal Maintenance Systems: Xenopus tropicalis [Shaidani et al. 2020]), the pH should be between 6.8 and 7.4. Use either solution at room temperature.

Equipment

Alcohol pads (sterile)

Holding buckets and lids (e.g., Rubbermaid 6304 and 6509)

Make holes in the lids for air.

Needles (18- to 20-gauge [for Steps 1–2] and 30-gauge × ½″ [for Step 3]; BD)

Petri dish (60 mm; BD 351007)

Stereomicroscope

Syringes (sterile, 1-mL tuberculin [for Steps 2–3] and 10-mL [for Step 1])

METHOD

Prepare Ovulation Hormone

1. Resuspend hCG to 1 unit/µL with sterile diluent (supplied with Chorulon) using a 10-mL syringe with an 18- to 20-gauge needle. Shake until the powder is dissolved.

   Keep hCG refrigerated after resuspension.

2. Wipe the rubber cap with a sterile alcohol pad. Draw the appropriate amount of hCG into a 1-mL tuberculin syringe using a 20-gauge needle.

   Wipe the rubber cap with a sterile alcohol pad before each use to prevent contamination.

3. Switch to a smaller 30-gauge × ½″ needle for injection. Align the beveled opening of the needle with the volume markings on the syringe.

   Follow standard safety guidelines regarding needle use, recapping, and disposal.

Administer Ovulation Hormone

A priming dose of hCG (10 units) is typically given to the female 24–72 h (minimum 12 h to maximum 5 d) before the time eggs are desired to improve ovulation performance. A boosting dose (100–200 units) is later given to induce ovulation. Frogs will begin laying eggs 4–5 h after boosting. Higher hCG doses tend to result in shorter time to ovulation; however, we have observed that higher doses may also result in increased mortality. In general, the lowest dose of hCG that induces ovulation when desired should be used.

For the health of the animal, keep the animal's skin moist while it is out of water, and wear moistened powder-free gloves to avoid removing the frog's protective slime layer.

4. Prepare a holding bucket containing ∼2 L of fresh system water or 1/9× MR at room temperature.

   Do not use another source of water, such as tap water, pure deionized (DI) water, or reverse osmosis (RO) water. Chloramines present in tap water are toxic to the frogs. DI and RO water do not contain salts necessary for the frogs to maintain their osmotic balance, so these can be used only when the conductivity is adjusted using salt.

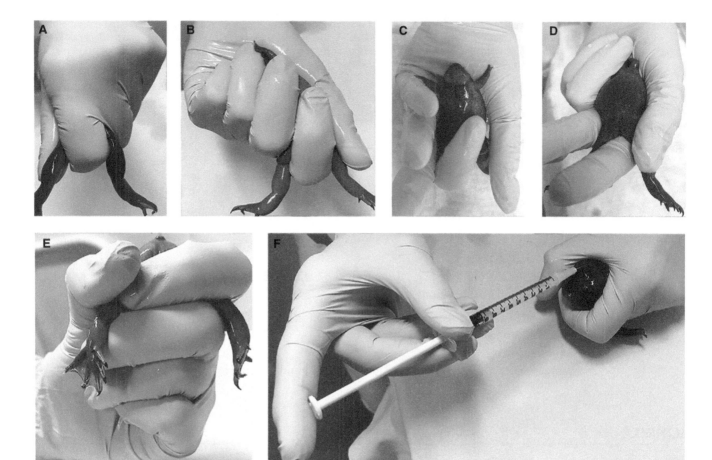

FIGURE 1. Handling *Xenopus tropicalis females* for hormone (hCG) injections. (*A*) Method for general handling, dorsal view. (*B*) Method for general handling, ventral view. (*C*) Ventral view with fingers opened to show position of frog in palm. (*D*) Position of index and pinkie fingers to prevent frog from sliding forward. (*E*) Slide thumb over top of rear leg, securing the legs and exposing the distal portion of the frog's dorsum. (*F*) Inject subcutaneously into the exposed dorsal area. Avoid lateral sense organs running along either side of the body.

5. Using your nondominant hand, grasp the frog with your thumb, middle, ring, and pinkie finger wrapped around the frog's body, and your index finger between its rear legs (Fig. 1A,B). Create a pocket with the palm of your hand and your middle, ring, and pinkie fingers to hold the frog (Fig. 1C). Use your ring finger to apply slight pressure on the underside of the head to prevent the frog from sliding forward out of your grip (Fig. 1D).

    The frog should be able to move slightly in your hand. Although your grasp should be firm enough to prevent escape, excessive squeezing can result in panic and injury. Patience is sometimes required until the frog stops her attempts to escape.

6. When the frog has stopped moving, slide your thumb to join your index finger behind the rear legs. Maintain contact between your thumb and the frog's body during this movement, as she may otherwise take the opportunity to escape. Using your thumb and index finger, gently draw the rear legs toward the frog's head, keeping the rest of your fingers in place (Fig. 1E).

    This results in the posterior of the frog sliding out of your hand slightly, exposing the caudal dorsum (Fig. 1E,F).

    See Troubleshooting.

7. Inject hCG into the subcutaneous dorsal lymph sac, staying within the triangular area formed by your hand and lateral line sense organs. Keep the bevel of the needle up and at a slight angle to gently penetrate the skin. Then reposition the needle parallel to the body, advance 3–5 mm, and inject the hCG (Fig. 1F).

 Cite this protocol as *Cold Spring Harb Protoc*; doi:10.1101/pdb.prot106344

FIGURE 2. Collection of *Xenopus tropicalis* eggs and assessment of egg quality. (*A*) Normal cloaca before hCG treatment. Lateral line sense organs are visible on either side of the dorsal surface. (*B*) Enlarged red cloaca after hCG treatment. (*C*) Position for collecting eggs. (*D*) Good-quality eggs. These are uniform in appearance, with a smooth, round shape. They are discretely contained in individual jelly coats. An evenly pigmented animal pole and a pale vegetal pole is divided by a crisp equatorial band (white arrow). (*E*) Poor-quality eggs. These may include white, dead eggs. Eggs may not be discrete, but instead be laid in tubes of jelly often referred to as "stringy" eggs (dashed arrows). The jelly may be coated in debris (arrowhead).

The lateral line sense organs appear as sutures along either side of the dorsum (Fig. 2A,B).

See Troubleshooting.

8. Withdraw the needle, and apply gentle digital pressure at the injection site for 5 sec.

    See Troubleshooting.

9. Place the frog in the holding tank. Label the tank with the injection time and hCG dose.

    Three to four hours after boosting, the female's normally small cloaca should become swollen and red, indicating a response to the hormone (Fig. 2A,B). The frog should begin producing eggs in another 1–2 h, visible as clumps or individual eggs in the bucket. This signals that the female is now ready for egg collection. See Troubleshooting.

Manual Egg Expression

In this series of steps, eggs are manually expressed into a high-salt solution (1× MBS containing 0.2% BSA), which prevents the eggs from sticking to the dish. As an alternative, actively laying females can be kept in a mating bucket containing 1–2 L of this solution, which, when used in this situation, can extend egg viability by several hours. It also causes the eggs to separate, making collection easier. Eggs can then be collected as needed over several hours for various experiments. Do not leave females in the high-salt solution longer than 5–6 h (S-W Cha, pers. comm).

10. When the female begins to lay eggs, prepare a Petri dish for egg collection. Coat the Petri dish with 1× MBS containing 0.2% BSA.

11. Grasp the female as for injection, and then gently pull the rear legs cranially. Position the cloaca over the Petri dish (Fig. 2C), and apply gentle downward pressure by moving your fingers toward the cloaca.

    The eggs should be readily expelled and fall into the dish. Excessive pressure should not be necessary and can result in stress or injury. A stressed frog will float at the surface and breathe heavily after egg collection. See Troubleshooting.

    See Wlizla et al. (2018) for a slightly different grasp for expressing eggs.

    To fertilize the eggs, proceed to Protocol 5: Obtaining Xenopus tropicalis Embryos by In Vitro Fertilization (Lane and Khokha 2021). Consistent fertilization has been obtained for up to 2 h after eggs are laid into a solution of 1× MBS containing 0.2% BSA. Fertilization becomes variable after 2 h (E Mis, M Lane, unpub. results).

12. Place the female in the holding tank prepared in Step 4.

    Females can be used for egg collection again after 1 h of rest if they do not appear stressed and readily lay eggs. Although egg production lasts for 4–6 h, the first eggs laid are generally the best quality. Do not return the frog to the recirculating system until laying is complete (6–8 h), as eggs can clog the system.

    See Troubleshooting.

13. Examine the eggs under a stereomicroscope.

    Eggs should be uniform in appearance and have a smooth, round shape, an evenly pigmented animal pole, and a pale vegetal pole (Fig. 2D). Stringy eggs, or dishes containing many dead (white) eggs, should not be used (Fig. 2E).

    See Troubleshooting.

14. Return the female to the recirculating system after she has finished laying eggs.

    See Troubleshooting.

TROUBLESHOOTING

Problem (Step 6): It is difficult to properly position the frog for injection.

Solution: Steps 5 and 6 should be practiced until you can reliably position the frog for injection before proceeding to Step 7. With experience, you may develop another method that works best for you. Wrapping the frog in a soaking wet paper towel and exposing the posterior dorsum for injection (i.e., the "burrito" method) is an option for those who have trouble with manual restraint.

Problem (Step 7): The needle does not easily penetrate the frog's skin.

Solution: The needle may be dull. Change the needle to avoid injuring the animal.

Problem (Step 7): Resistance is felt when injecting hormone into the frog (plunger requires more than slight pressure).

Solution: The needle may be in the muscle underlying the subcutaneous space. Withdraw the needle slightly and inject when there is no resistance.

Problem (Step 8): Bleeding occurs at the injection site.

Solution: Bleeding may occur if skin capillaries or underlying muscle tissue are disrupted during injection. Apply pressure to the injection site until bleeding stops.

Problem (Steps 9 and 11): There is a lack of egg production.

Solution: If the female's cloaca is red but no eggs are visible at 4 h, egg production can sometimes be encouraged with gentle squeezing. If no eggs are obtained, wait 15–20 min and check for eggs again. Do not try to force egg production by excessive squeezing. Occasionally, a frog will appear ready to lay but never produce eggs. On the other hand, if the cloaca is not red or swollen (Fig. 2B), the female should not be used. Using two to four females per experiment increases the likelihood of obtaining eggs.

Problem (Steps 12 and 14): An ovulated female appears bloated after 12–24 h. She may appear lethargic and float.

Solution: A female may occasionally become impacted with eggs and septic after ovulation. Euthanize immediately according to your institution's protocol.

Problem (Steps 12 and 14): A female frog has a handling injury and shows signs of physical injury such as red, sloughing skin within the first 24 h after use.

Solution: Improper handling during egg expression can result in serious injury. If the animal is roughly handled or struggles excessively during use, the protective slime layer can be disrupted or the skin can be abraded, leading to infection. Another injury seen is a disjointed rear leg. Euthanize the animal in either case. Provide retraining to handlers to ensure the issues do not recur.

Problem (Step 13): The eggs are of poor quality (Fig. 2E).

Solution: Egg quality is crucial to fertilization success and embryo development. Poor-quality eggs may fertilize but often fail to divide or develop normally. If the eggs are of poor quality, consider the following factors:

- Maintaining water temperatures of >25°C or <22°C can result in delayed egg production and poor-quality eggs.

- Low water conductivity may result in "soft" eggs, which are difficult to microinject. This can be avoided by using a salt higher in calcium and magnesium (e.g., Reef Crystals [Instant Ocean RC1-200]) in your recirculating system and maintaining system conductivity of >900 µS.

- Using two to four females for ovulation increases the likelihood of obtaining high-quality eggs.

- In our experience, a mature female's first one to three ovulations may result in relatively low-yield and low-quality eggs, although subsequent ovulations generally yield much better eggs. A test fertilization (see Protocol 5: Obtaining *Xenopus tropicalis* Embryos by In Vitro Fertilization [Lane and Khokha 2021]) to confirm good egg quality can be performed before routinely using a female.

- Females used routinely for egg collection should be rested for 2–3 mo between uses. Animals rested longer than this may produce poor-quality eggs. Those not being actively used should be boosted every 6 mo with 50 units of hCG.

- *X. tropicalis* may be affected by a reproductive slump during winter months, when quality and number of eggs, as well as fertilization rates, suffer.

RECIPES

$CaCl_2$ (0.1 M)

11.1 g $CaCl_2$

Dissolve 11.1 g $CaCl_2$ in 1 L H_2O. Autoclave and store in aliquots at −20°C or 4°C.

MBS Salts (10×)

NaCl (880 mM)
KCl (10 mM)
$MgSO_4$ (10 mM)
HEPES (50 mM, pH 7.8)

   Omit HEPES if MBS is to be used for oocyte maturation.

$NaHCO_3$ (25 mM)

Adjust pH to 7.8 with NaOH. Autoclave. Store at room temperature.

Modified Barth's Saline (MBS)

$CaCl_2$ (0.1 M)
MBS salts (10×)

For a 1× solution of MBS, mix 100 mL of 10× MBS salts with 7 mL of 0.1 M $CaCl_2$. Adjust the volume up to 1 liter with H_2O. Store at room temperature.

Modified Ringer's Solution (MR; 1×)

0.1 M NaCl
1.8 mM KCl
2.0 mM $CaCl_2$
1.0 mM $MgCl_2$
5.0 mM HEPES (pH 7.6)

Store for up to 1 mo at room temperature.

ACKNOWLEDGMENTS

We thank Michael Slocum for husbandry care and assistance with images. We also thank Sang-Wook Cha for advice on the high-salt procedure.

Cite this protocol as *Cold Spring Harb Protoc*; doi:10.1101/pdb.prot106344

REFERENCES

Lane MA, Khokha MK. 2021. Obtaining *Xenopus tropicalis* embryos by in vitro fertilization. *Cold Spring Harb Protoc* doi:10.1101/pdb.prot106351

Lane MA, Slocum M, Khokha MK. 2021. Raising and maintaining *Xenopus tropicalis* from tadpole to adult. *Cold Spring Harb Protoc* doi:10.1101/pdb.prot106369

Shaidani N-I, McNamara S, Wlizla M, Horb ME. 2020. Animal maintenance systems: *Xenopus tropicalis*. *Cold Spring Harb Protoc* doi:10.1101/pdb.prot106146

Wlizla M, Falco R, Peshkin L, Parlow AF, Horb ME. 2017. Luteinizing hormone is an effective replacement for hCG to induce ovulation in *Xenopus*. *Dev Biol* **426:** 442–448. doi:10.1016/j.ydbio.2016.05.028

Wlizla M, McNamara S, Horb ME. 2018. Generation and care of *Xenopus laevis* and *Xenopus tropicalis* embryos. In Xenopus: *methods and protocols* (ed. Vleminckx K), pp. 19–32. Springer, New York.

Obtaining *Xenopus tropicalis* Embryos by In Vitro Fertilization

Maura Lane[1,2] and Mustafa K. Khokha[1]

[1]*Pediatric Genomics Discovery Program, Department of Pediatrics and Genetics, Yale University School of Medicine, New Haven, Connecticut 06510, USA*

Xenopus is a powerful model system for cell and developmental biology in part because frogs produce thousands of eggs and embryos year-round. In vitro fertilization (IVF) is ideal for obtaining developmentally synchronized embryos for microinjection or when natural mating has failed to produce a fertilization. In IVF, females are induced to ovulate, and then eggs are collected by manual expression. After testes are collected from a euthanized male frog, the eggs are fertilized in vitro. The embryos are then treated with cysteine to remove the sticky protective jelly coat. Dejellied embryos are much easier to manipulate during microinjection or when sorting in a Petri dish. The jelly coat is also very difficult to penetrate with an injection needle. After microinjection, embryos are maintained in Petri dishes until desired stages are reached. Although in vitro fertilization in *X. laevis* and *X. tropicalis* is similar, critical differences in solutions, handling of testis, response of fertilized eggs directly after introduction of sperm, and developmental timing are required for successful fertilization in *X. tropicalis*.

MATERIALS

It is essential that you consult the appropriate Material Safety Data Sheets and your institution's Environmental Health and Safety Office for proper handling of equipment and hazardous materials used in this protocol.

RECIPES: Please see the end of this protocol for recipes indicated by <R>. Additional recipes can be found online at http://cshprotocols.cshlp.org/site/recipes.

Reagents

Anesthetic agents (select one; see Step 2)

Benzocaine base USP (0.05%)

Prepare 5% stock in 100% ethanol in fume hood. Store at −20°C. Dilute 5% stock in distilled water to 0.05%. Adjust pH to 6.8–7.0 with sodium bicarbonate. Store for up to 7 d at room temperature.

Ethyl 4-aminobenzoate hydrochloride (benzocaine HCl USP; ≥250 mg/L)

Prepare fresh in distilled water in fume hood. Adjust pH to 6.8–7.0 with sodium bicarbonate. Use at room temperature.

Tricaine (2 g/L)

Prepare a 5× stock (10 g/L) in distilled water in amber glass bottle in fume hood. Store for up to 1 mo at −4°C or until the solution becomes discolored. Protect from light. Prepare the 2 g/L solution fresh in distilled water from 5× stock. Adjust pH to 7.0–7.4 with sodium bicarbonate. Use at room temperature.

Cysteine (3%, w/v; Sigma-Aldrich W326305) in 1/9× MR

Prepare fresh. Adjust the pH to 7.8–8 with NaOH. Use at room temperature.

[2]Correspondence: maura.lane@yale.edu

Cite this protocol as *Cold Spring Harb Protoc*; doi:10.1101/pdb.prot106351

Cysteine powder will lose potency over time, extending the treatment time required to dejelly embryos. To minimize this issue, purchase smaller amounts (e.g., 100 g vs. 1 kg), and close the lid tightly after use. If possible, use a consistent brand and type of cysteine, and prepare your cysteine solution at the recommended pH. We use cysteine without HCl (see Step 21).

Ficoll PM 400 (3%; GE Healthcare Biosciences 95021-196) in 1/9× MR (for microinjection only; see Steps 27–30)

A Ficoll-containing solution will readily grow contaminants. Autoclave and handle carefully to prevent introducing microorganisms. Use and store at room temperature.

Gentamicin (50 μg/mL) in 1/9× MR

Prepare a 1000× stock (50 mg/mL) of gentamicin (Sigma-Aldrich G3632) in distilled water, filter-sterilize, and store at 4°C. Prepare the dilution in 1/9× MR, and use at room temperature within 1 week.

Materials for egg collection (see Step 11 and Protocol 4: Obtaining *Xenopus tropicalis* Eggs [Lane et al. 2021a])

Materials for euthanizing the male frog (see Step 3)

Mature *X. tropicalis* male with dark, prominent nuptial pads for harvesting testes (e.g., Xenopus 1 6270 or *Xenopus* National Resource NXR_1018)

The male does not need to be primed or boosted but, if desired, can be primed with 10 units of hCG, and/or boosted with 50 units of hCG concurrent with priming and boosting females (see Protocol 4: Obtaining Xenopus tropicalis *Eggs [Lane et al. 2021a]).*

Cryogenically stored sperm can be used in place of freshly harvested testes (Pearl et al. 2017). This can save time, although the fertilization rate may be lower. One advantage of frozen sperm is that stock centers may have stored frozen sperm from genetically modified frog lines, which can be ordered instead of animals, simplifying distribution.

Modified Barth's saline (MBS) <R> at a concentration of 0.1×

Prepare fresh and use at room temperature. The pH should be 7.8–8.0.

Modified Barth's saline (MBS) <R> at a concentration of 1×, containing 0.2% (w/v) bovine serum albumin (BSA)

Filter-sterilize. The pH of this solution should be 7.8–8.0. Store at 4°C, and bring to room temperature before use.

Modified Ringer's solution (MR; 1×) <R>, diluted to a concentration of 1/9×

Prepare the 1× stock with distilled water; the final pH will be 7.8–8.0. To prepare 1/9× MR, dilute the 1× MR with distilled water. The pH should be 7.8–8.0. Use this solution at room temperature within 1 month.

Xenopus tropicalis females (e.g., Xenopus 1 6260 or *Xenopus* National Resource NXR_1018), primed and boosted for egg collection as described in Protocol 4: Obtaining *Xenopus tropicalis* Eggs (Lane et al. 2021a)

Equipment

Absorbent surface liners (Fisherbrand 14-127-47)
Dissecting stereomicroscope
Glass Pasteur pipettes (Corning 2239912)
Holding buckets and lids (e.g., Rubbermaid 6304 and 6509)

Make holes in the lid for air.

Iris forceps, curved, with 1 × 2 teeth (e.g., Miltex V918-790)
Low-temperature incubator (VWR 2005)
Microcentrifuge tube pestle (USA Scientific 1415-5390)
Microcentrifuge tubes (1.5-mL)
Petri dishes (150-mm; Falcon 351085)

For Steps 26 and 30, Petri dishes with a 100-mm diameter (Falcon 351029) can be used instead.

Pipette pump (VWR 53502-233)
Plastic transfer pipettes (Globe Scientific 89209-802)
Sharp scissors, straight (e.g., Miltex 34–80)

METHOD

Confirm Egg Production

1. Immediately before performing Step 2, check the holding bucket containing the boosted females for eggs.

   Even a few eggs can indicate the females are ready to be used. See Troubleshooting.

   Do not actually collect eggs for fertilization until you have isolated the testes as described in Steps 2–10.

Isolate Testes

2. Anesthetize the male before euthanasia by fully immersing in room-temperature benzocaine base (0.05%; USP), ethyl 4-aminobenzoate hydrochloride (benzocaine HCl, USP; ≥250 mg/L), or tricaine (2 g/L) until movement and righting reflex has ceased (typically ∼5 min).

   Lack of withdrawal response to a deep toe pinch indicates adequate anesthesia. Do not proceed until there is no withdrawal response.

3. Euthanize the frog according to your institution's protocol.

   Approved methods of euthanasia may vary by institution but will generally adhere to The AVMA Guidelines for the Euthanasia of Animals (American Veterinary Medical Association 2020) in the United States. Follow your institution's approved methods and drugs for euthanasia.

4. Place the euthanized male on an absorbent surface liner in dorsal recumbency.

5. Using toothed forceps, grasp the skin of the lower abdomen and pull upward, away from the underlying body wall. Use small scissors to make a skin flap extending from the lower abdomen to just below the arms (Fig. 1A). Reflect the flap anteriorly.

6. Repeat with the underlying body wall to expose the abdominal cavity and organs. Blot with an absorbent tissue to remove blood.

   Some bleeding may occur when the midline abdominal vessel is cut.

7. Visualize the testes by reflecting the surrounding organs such as liver, intestines, and fat bodies anteriorly.

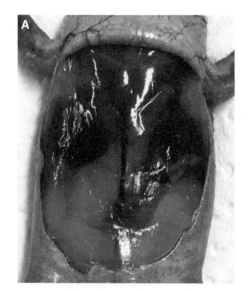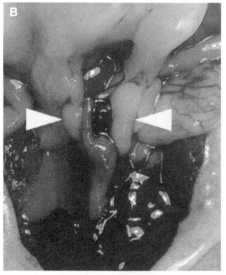

FIGURE 1. Dissecting testes. (*A*) Reflect the skin flap anteriorly to expose the body wall. (*B*) Lift the body wall anteriorly, and move the surrounding organs to reveal testes (white arrows).

Cite this protocol as *Cold Spring Harb Protoc*; doi:10.1101/pdb.prot106351

Testes lie deep in the abdomen, flanking the aorta, and are usually surrounded by yellow fat. The testes appear light pink, slightly translucent, and are typically 2–5 mm in size (Fig. 1B).

8. Carefully remove the testes with fine forceps and small scissors.

   The amount of fat and blood on the testes should be minimized. If desired, use a stereomicroscope to improve visualization.

9. Place the testes in a 1.5-mL microcentrifuge tube containing ∼0.5 mL of 0.2% BSA in 1× MBS.

   Both testes should be used for maximal fertilization rate, especially if there are multiple dishes of eggs. Alternatively, the testes can be collected in separate tubes and used for different experiments. Keep in mind that the fertilization rate will decrease with time and that X. tropicalis sperm have a shorter window of viability than X. laevis sperm. So use both testes within 2 h; the fertilization rate declines substantially after 3–4 h.

10. Dispose of the frog carcass according to your institution's requirements.

Collect Eggs and Sperm

11. Collect eggs as described in Protocol 4: Obtaining *Xenopus tropicalis* Eggs (Lane et al. 2021a).

    If collecting eggs from multiple females, use a separate dish for each to avoid wasting sperm on poor-quality eggs.

12. Release sperm from the testes by crushing them with a micropestle.

    The solution should become cloudy.

Fertilize Eggs

13. Using a plastic pipette, remove excess liquid from the dish(es) of eggs to avoid diluting the sperm solution.

14. Prop the dish up slightly so that the eggs collect at the lowest point.

    The lid of the dish can be used as a prop.

15. Use a plastic pipette to distribute the sperm solution approximately evenly among the dishes of eggs. Mix gently once or twice with a pestle or pipette.

16. Incubate the sperm with the eggs for 3 min at room temperature.

    Sperm will attach to the eggs during this time.

17. After incubation, place the dish flat on the countertop and add enough 0.1× MBS to completely cover the embryos.

18. Wait 10 min.

    The lower salt concentration will activate the sperm to penetrate the eggs.

19. Check for fertilization under a dissecting stereomicroscope at 10× magnification.

    The pigmented animal hemisphere will turn up within 10–15 min of fertilization, whereas unfertilized eggs are impartial. The pigmented hemisphere of the embryo contracts during the 10- to 15-min window, covering less than half of the surface of the embryo. The sperm entry point (SEP) appears as a dot or indentation in the pigmented area. Eggs will remain flaccid and difficult to inject, whereas embryos will be stiff and easily injected.

    See Troubleshooting.

    After confirming fertilization, proceed to Step 20 to dejelly the embryos. Leave any unfertilized eggs in the dish. Embryos will be selected after dejellying.

Dejelly Embryos

20. Use a pipette to remove the medium from the dish containing the fertilized embryos.

21. Carefully fill the dish approximately halfway with the 3% cysteine solution. Gently stir the embryos with a plastic pipette once or twice to distribute the cysteine. Incubate without disturbing the embryos for 5 min at room temperature.

    Cysteine without HCl consistently takes 5–6 min to dejelly embryos and is preferred by our laboratory. Cysteine HCl will also work, though it can take 10 min or more to completely remove the jelly coat. It will not harm embryos.

    Anecdotal reports suggest swirling can result in spinal deformations and/or double axes.

22. After incubation, check for adequate treatment by gently agitating the dish.

    Embryos should separate easily and form a closely packed monolayer.

    See Troubleshooting.

Remove Cysteine

23. Remove most of the cysteine from the dish with a Pasteur or plastic pipette.

24. Carefully fill the dish approximately halfway with 0.1× MBS or 1/9× MR at room temperature. Once the embryos settle, remove the solution with a Pasteur or plastic pipette. Repeat this for a total of three washes to completely remove the cysteine.

    See Troubleshooting.

    If the embryos will be microinjected, proceed to Step 27 to add a Ficoll-containing solution after the final wash. If the embryos will not be microinjected, replace the final wash with 0.1× MBS or 1/9× MR at room temperature, and continue with Step 25.

25. Use a plastic pipette to transfer the embryos to a 150-mm Petri dish filled approximately halfway with 1/9× MR.

    The embryos will spread out and be easier to sort in a larger dish.

26. Under a dissecting scope, use a pipette pump with a glass Pasteur pipette to select and transfer healthy embryos to a new 100- or 150-mm Petri dish filled approximately halfway with 1/9× MR containing 50 µg/mL gentamicin.

    Do not exceed 50 to 60 embryos in a 100-mm dish or 100 to 120 embryos in a 150-mm dish. Overcrowded embryos will quickly use up oxygen and foul the water excessively.

    Proceed to Step 31.

Microinject the Embryos (Optional)

The rate of embryo development is temperature-dependent (Khokha et al. 2002). This is important to consider when performing time-sensitive experiments, such as injections at specific cell stages. Development before microinjection can be controlled by changing the room temperature or by using a cold plate. Temperatures between 16°C and 22°C and between 27°C and 28°C can be tolerated for a short period, to slow or speed early development before injection.

27. Carefully fill the dish approximately halfway with 1/9× MR containing 3% Ficoll. Once the embryos settle, remove the solution with a Pasteur or plastic pipette. Repeat this wash once, and then fill the dish approximately halfway with 1/9× MR containing 3% Ficoll.

28. Under a dissecting scope, use a pipette pump with a glass Pasteur pipette to select and transfer healthy embryos to an injection dish filled approximately halfway with 1/9× MR containing 3% Ficoll.

    Ficoll stabilizes the embryos during injection and helps prevent leakage of injected material from the injection site.

29. Microinject embryos according to Protocol 7: Microinjection of *Xenopus tropicalis* Embryos (Lane and Khokha 2021). Allow the embryos to rest in the injection dish for 1 h at room temperature.

30. Carefully transfer the embryos to a new 100- or 150-mm dish filled approximately halfway with 1/9× MR containing 50 µg/mL gentamicin.

    Do not exceed 50 to 60 embryos in a 100-mm dish or 100 to 120 embryos in a 150-mm dish.

Cite this protocol as *Cold Spring Harb Protoc*; doi:10.1101/pdb.prot106351

Incubate the Embryos

31. Incubate embryos at the desired temperature. Refer to the Development Time/Temperature Chart for *X. tropicalis* at http://www.xenbase.org/anatomy/static/xenopustimetemp.jsp for the approximate time from IVF to the desired stage at different temperatures. Use an incubator to ensure a constant temperature and a predictable rate of embryo development.

    Extreme temperatures should not be used for raising embryos. Embryos can be kept at 28°C for an hour after injection to speed development. Maintain temperatures between 22°C and 26°C during the rest of development, especially during gastrulation.

32. Remove dead or abnormal embryos and replace the solution of 1/9× MR containing 50 µg/mL gentamicin daily.

    See Sec. 1, Protocol 3: Raising and Maintaining Xenopus tropicalis *from Tadpole to Adult (Lane et al. 2021b) for further information on raising* Xenopus *beyond embryo stages.*

TROUBLESHOOTING

Problem (Step 1): There are no eggs in the bucket.

Solution: If the females have enlarged cloacae, give the females a gentle "test squeeze." Eggs should be readily expressed if the frogs are ready for use. Females should be ready to use 4–5 h after boosting. Any female without an enlarged, reddened cloaca should not be used. See Protocol 4: Obtaining *Xenopus tropicalis* Eggs (Lane et al. 2021a) for details on preparing females for egg collection.

Problem (Step 19): There are no signs of fertilization.

Solution: Wait up to an additional 10 min.

Problem (Steps 22 and 24): The embryos continue to stick together or turn white and die.

Solution: There are a number of factors relating to cysteine that affect dejelly time and embryo health. Inadequately treated embryos will stick together after washing, whereas overly treated embryos will begin to turn white and die. It is best to use a consistent brand and type of cysteine, prepare your cysteine solution at the recommended pH, and test dejellying time empirically. Dejellying with cysteine without HCl, as recommended here, generally takes ~2–5 min. If the embryos do not separate after 5 min, continue the cysteine treatment, checking every minute until dejellying is complete. Embryos treated with cysteine with HCl can take 10 or more minutes to dejelly, with no negative effects on the embryos.

RECIPES

CaCl₂ (0.1 M)

11.1 g CaCl₂

Dissolve 11.1 g CaCl₂ in 1 L H₂O. Autoclave and store in aliquots at −20°C or 4°C.

MBS Salts (10×)

NaCl (880 mM)
KCl (10 mM)
MgSO₄ (10 mM)
HEPES (50 mM, pH 7.8)

Omit HEPES if MBS is to be used for oocyte maturation.

NaHCO₃ (25 mM)

Adjust pH to 7.8 with NaOH. Autoclave. Store at room temperature.

Modified Barth's Saline (MBS)

CaCl₂ (0.1 M)
MBS salts (10×)

For a 1× solution of MBS, mix 100 mL of 10× MBS salts with 7 mL of 0.1 M CaCl₂. Adjust the volume up to 1 liter with H₂O. Store at room temperature.

Modified Ringer's Solution (MR; 1×)

0.1 M NaCl
1.8 mM KCl
2.0 mM CaCl₂
1.0 mM MgCl₂
5.0 mM HEPES (pH 7.6)

Store for up to 1 mo at room temperature.

ACKNOWLEDGMENTS

We thank Emily Mis and Helen Willsey for comments on the protocol. We thank Michael Slocum for husbandry advice and assistance with images.

REFERENCES

American Veterinary Medical Association. 2020. *The AVMA guidelines for the euthanasia of animals: 2020 edition.* American Veterinary Medical Association, Schaumburg, IL.

Khokha MK, Chung C, Bustamante EL, Gaw LW, Trott KA, Yeh J, Lim N, Lin JC, Taverner N, Amaya E, et al. 2002. Techniques and probes for the study of *Xenopus tropicalis* development. *Dev Dyn* **225:** 499–510. doi:10.1002/dvdy.10184

Lane MA, Khokha MK. 2021. Microinjection of *Xenopus tropicalis* embryos. *Cold Spring Harb Protoc* doi:10.1101/pdb.prot107644.

Lane MA, Mis EK, Khokha MK. 2021a. Obtaining *Xenopus tropicalis* eggs. *Cold Spring Harb Protoc* doi: 10.1101/pdb.prot106344.

Lane MA, Slocum M, Khokha MK. 2021b. Raising and maintaining *Xenopus tropicalis* from tadpole to adult. *Cold Spring Harb Protoc* doi:10.1101/pdb.prot106369.

Pearl E, Morrow S, Noble A, Lerebours A, Horb M, Guille M. 2017. An optimized method for cryogenic storage of *Xenopus* sperm to maximise the effectiveness of research using genetically altered frogs. *Theriogenology* **92:** 149–155. doi:10.1016/j.theriogenology.2017.01.007

Cite this protocol as *Cold Spring Harb Protoc*; doi:10.1101/pdb.prot106351

Obtaining *Xenopus tropicalis* Embryos by Natural Mating

Maura Lane[1] and Mustafa K. Khokha

Pediatric Genomics Discovery Program, Department of Pediatrics and Genetics, Yale University School of Medicine, New Haven, Connecticut 06510, USA

Xenopus is a powerful model system for cell and developmental biology in part because frogs produce thousands of eggs and embryos year-round. Natural matings are a simple and common method to obtain embryos for injection or other experimental use or to raise to adulthood. This method does not require sacrificing a male as in vitro fertilization (IVF) does. Male and female frogs are injected with an ovulation hormone, placed together in a mating bucket, and left for 4–6 h or overnight to mate. Embryos are then collected, treated with cysteine to remove the sticky jelly coat, and used for injections and/or raised to later stages or adulthood. For embryos raised past free-swimming stages, the cysteine step can optionally be skipped, and tadpoles can be allowed to hatch naturally from the jelly coat. Although there are many similarities between natural mating protocols for *Xenopus laevis* and *Xenopus tropicalis*, there are key differences such as hormone dosage, timing of ovulation, and embryo incubation temperature. Here we provide a specific protocol for inducing natural matings in *X. tropicalis*.

MATERIALS

It is essential that you consult the appropriate Material Safety Data Sheets and your institution's Environmental Health and Safety Office for proper handling of equipment and hazardous materials used in this protocol.

RECIPES: Please see the end of this protocol for recipes indicated by <R>. Additional recipes can be found online at http://cshprotocols.cshlp.org/site/recipes.

Reagents

Cysteine (3%, w/v; Sigma-Aldrich W326305) in 1/9× MR

Prepare fresh. Adjust the pH to 7.8–8 with NaOH. Use at room temperature.

Ficoll PM 400 (3%; GE Healthcare Biosciences 95021-196) in 1/9× MR (for microinjection only; see Step 8)

A Ficoll-containing solution will readily grow contaminants. Autoclave and handle carefully to prevent introducing microorganisms. Use and store at room temperature.

Gentamicin (50 µg/mL) in 1/9× MR

Prepare a 1000× stock (50 mg/mL) of gentamicin (Sigma-Aldrich G3632) in distilled water, filter-sterilize, and store at 4°C. Prepare the dilution in 1/9× MR, and use at room temperature.

Human chorionic gonadotrophin (hCG) (Chorulon [Merck Animal Health 057176, 10,000 units])

See Protocol 4: Obtaining Xenopus tropicalis *Eggs (Lane et al. 2021a) for alternative hormones.*

[1]Correspondence: maura.lane@yale.edu

Mature *Xenopus tropicalis* females (e.g., Xenopus 1 6260 or *Xenopus* National Resource NXR_1018)

Refer to Protocol 4: Obtaining Xenopus tropicalis *Eggs (Lane et al. 2021a) for details on maintaining mature females that lay good-quality eggs.*

Mature *X. tropicalis* male with dark, prominent nuptial pads (e.g., Xenopus 1 6270 or *Xenopus* National Resource NXR_1018)

Modified Barth's saline (MBS) <R> at a concentration of 0.1× (optional; see Step 8)

Prepare fresh and use at room temperature. The pH should be 7.8–8.0.

Modified Ringer's solution (MR; 1×) <R>, diluted to a concentration of 1/9×

Prepare the 1× stock with distilled water; the final pH will be 7.8–8.0. To prepare 1/9× MR, dilute the 1× MR with distilled water. The pH should be 7.8–8.0. Use this solution at room temperature. Use within 1 mo.

Recirculating system water (23°C–24°C; pH 6.8–7.4)

For details on establishing and maintaining a recirculating system for X. tropicalis, *see Sec. 1, Protocol 1: Animal Maintenance Systems:* Xenopus tropicalis *(Shaidani et al. 2020).*

Equipment

Glass Pasteur pipettes (Corning 2239912)

Holding buckets and lids (e.g., Rubbermaid 6304 and 6509)

Make holes in the lid for air.

Low-temperature incubator (VWR 2005) (optional; see Step 13)

Petri dishes (150-mm [Falcon 351085] and 60-mm [Falcon 351007])

For Steps 9, 11, and 13, 100-mm-diameter Petri dishes (Falcon 351029) can be used instead of 150-mm dishes. Maintain a density of 50 to 60 embryos/dish in this case.

Pipette pump (VWR 53502-233)

Plastic transfer pipettes (Globe Scientific 89209-802)

Trim about 2 cm off of the ends of some of the pipettes to make collecting embryos easier (see Step 5).

Syringes, needles, and alcohol pads for injections (see Protocol 4: Obtaining *Xenopus tropicalis* Eggs [Lane et al. 2021a])

METHOD

Set Up the Mating

1. At least 12 h before boosting, prime female and male frogs with 10 units of hCG each as described in Protocol 4: Obtaining *Xenopus tropicalis* Eggs (Lane et al. 2021a).

   Frogs can be primed up to 5 d before boosting; 24–72 h prior is typical.

2. Approximately 4 h before ovulation is desired, boost the female with 100–200 units of hCG and the male with 25–50 units of hCG.

   If producing embryos overnight (see Step 4), a dose as low as 50 units may be adequate for the female.

3. After boosting, place the frogs together in a holding bucket containing ∼2 L of water freshly obtained from the aquatic recirculating system or 1/9× MR at room temperature (23°C–24°C).

4. Place the bucket for 4–6 h or overnight at room temperature in a quiet place.

   Three to four hours after boosting, the pair should assume the typical amplexus mating position (Fig. 1). The male will fertilize the eggs as they are laid over 6–8 h, so the embryos can be collected throughout the day and used (e.g., for injection purposes).

   Alternatively, the frogs can be boosted in the evening and will lay overnight. The embryos can then be collected the next morning; these will be at later stages and can be raised and/or used for experimental purposes. These embryos will be too far developed to use for injection but are often collected for experiments such as in situ hybridization or antibody staining.

   Proceed to Step 5 to dejelly the embryos or to Step 10 to raise the embryos without dejellying them.

FIGURE 1. Amplexus. The male *Xenopus tropicalis* grasps the female.

Collect the Embryos (with or without Dejellying)

Allowing embryos to emerge from their jelly coats naturally (as in Steps 10–12) saves time dejellying. In general, only healthy embryos develop to swimming stages, making collecting and sorting easier as well. This method only works when collecting later stages to raise or to use for later stage experiments. The larva will be well past injection stages when they emerge from the jelly coat on their own. To collect and dejelly the embryos for injection or early-stage experimental uses, follow Steps 5–9.

Collect Embryos and Dejelly with Cysteine

5. Use a trimmed plastic pipette to transfer embryos from the mating tank to a 60-mm Petri dish.

6. Remove excess liquid from the dish using a pipette.

   Embryos will remain adhered in clumps by the jelly coat, so the liquid can easily be removed.

7. To dejelly the embryos, carefully fill the dish approximately halfway with 3% cysteine in 1/9× MR. Gently stir the embryos with a plastic pipette once or twice to distribute the cysteine. Incubate for ~5 min at room temperature.

   Treatment is complete when embryos are no longer clumped together but form a monolayer on the bottom of the dish. They should pack together tightly, appearing to touch each other.

   See Troubleshooting.

8. Carefully fill the dish approximately halfway with 0.1× MBS or 1/9× MR at room temperature. Once the embryos settle, remove the solution using a pipette pump fitted with a glass Pasteur pipette or a plastic pipette. Repeat this for a total of three washes to completely remove the cysteine. Cover the embryos with 0.1× MBS or 1/9× MR after the final wash.

   If embryos are to be used immediately for injection, do not transfer to a larger dish. Instead, perform two additional washes with 3% Ficoll in 1/9× MR. Cover the embryos with 3% Ficoll in 1/9× MR. Proceed to Protocol 7: Microinjection of Xenopus tropicalis Embryos (Lane et al. 2021b).

   See Troubleshooting.

9. Transfer all of the embryos to a 150-mm Petri dish containing 75–100 mL of 1/9× MR. Under a dissecting microscope, use a pipette pump fitted with a glass pipette to select and transfer healthy

embryos from this dish to a 150-mm Petri dish containing 75–100 mL of 1/9× MR containing 50 μg/mL gentamicin.

> *If embryos are to be raised, maintain a density of about 100 to 120 embryos/dish to avoid overcrowding; in this case, proceed to Step 13.*

Collect Embryos without Dejellying

10. Leave the embryos overnight to develop in the mating bucket with the parents at room temperature. If the water is fouled at this point, perform a partial water change.

> *In our experience, removing the parents from the bucket or performing a 100% water change will negatively affect embryo development.*

11. Use an untrimmed plastic pipette to transfer stage 28–33 (Nieuwkoop and Faber 1994) embryos to a 150-mm Petri dish containing 75–100 mL 1/9× MR containing 50 μg/mL gentamicin.

> *Embryos take ~27–32 hr to reach stages 28–33 at 24°C. A maximum density of about 100 to 200 embryos/dish is critical to prevent overcrowding and increased embryo death.*

12. After transfer, remove any debris and dead embryos from the dish using a dissecting microscope and pipette pump fitted with a glass Pasteur pipette.

> *Proceed to Step 13.*

Raise Embryos to Stage 43

13. Raise the embryos to stage 43 in a 150-mm Petri dish at the desired temperature. Maintain a density of 100–120 embryos per dish or less.

> *Keeping the dishes in an incubator will maintain stable temperature and reliably control the rate of development.*

14. Remove dead or abnormal embryos, and change the solution (1/9× MR containing 50 μg/mL gentamicin) daily.

15. Transfer stage 43 embryos to recirculating water system for continued growth (see Sec. 1, Protocol 3: Raising and Maintaining *Xenopus tropicalis* from Tadpole to Adult [Lane et al. 2021c]).

TROUBLESHOOTING

Problem (Step 7): The embryos become white or appear to disintegrate.

Solution: A prolonged cysteine treatment can cause embryo death. Cysteine without HCl, as recommended here, generally takes ~4–5 min to properly dejelly embryos. Cysteine with HCl can take up to 10 or more min to dejelly, with no negative effects on the embryos.

Problem (Step 8): The embryos stick together after washing.

Solution: It is helpful to perform an initial empirical test to determine optimal dejellying time. If embryos stick together after washing, treat with cysteine for another 1–2 min. Keep in mind that cysteine powder will lose potency over time, extending the treatment time required to dejelly embryos. Purchase smaller amounts (i.e., 100 g vs. 1 kg), and close the lid tightly after use to minimize this issue. If possible, use a consistent brand and type of cysteine, and prepare your cysteine solution at the recommended pH.

Cite this protocol as *Cold Spring Harb Protoc*; doi:10.1101/pdb.prot106609

RELATED INFORMATION

For additional discussion on natural matings in *X. tropicalis*, see Showell and Conlon (2009). For IVF in *X. tropicalis*, see Protocol 5: Obtaining *Xenopus tropicalis* Embryos by In Vitro Fertilization (Lane and Khokha 2021). Additionally, del Viso and Khokha (2012) and Wlizla et al. (2018) provide advice on both natural matings and IVF in *X. tropicalis*.

RECIPES

CaCl₂ (0.1 M)

11.1 g CaCl₂

Dissolve 11.1 g CaCl₂ in 1 L H₂O. Autoclave and store in aliquots at −20°C or 4°C.

MBS Salts (10×)

NaCl (880 mM)
KCl (10 mM)
MgSO₄ (10 mM)
HEPES (50 mM, pH 7.8)

> Omit HEPES if MBS is to be used for oocyte maturation.

NaHCO₃ (25 mM)

Adjust pH to 7.8 with NaOH. Autoclave. Store at room temperature.

Modified Barth's Saline (MBS)

CaCl₂ (0.1 M)
MBS salts (10×)

For a 1× solution of MBS, mix 100 mL of 10× MBS salts with 7 mL of 0.1 M CaCl₂. Adjust the volume up to 1 liter with H₂O. Store at room temperature.

Modified Ringer's Solution (MR; 1×)

0.1 M NaCl
1.8 mM KCl
2.0 mM CaCl₂
1.0 mM MgCl₂
5.0 mM HEPES (pH 7.6)

Store for up to 1 mo at room temperature.

ACKNOWLEDGMENTS

We thank Michael Slocum for husbandry care and advice and for assistance with images. We thank Tim Grammer for helping to develop the initial protocols and Emily Mis and Helen Willsey for comments on the text.

REFERENCES

del Viso F, Khokha M. 2012. Generating diploid embryos from *Xenopus tropicalis*. *Methods Mol Biol* **917:** 33–41. doi:10.1007/978-1-61779-992-1_3

Lane MA, Khokha MK. 2021. Obtaining *Xenopus tropicalis* embryos by in vitro fertilization. *Cold Spring Harb Protoc* doi:10.1101/pdb.prot106351

Lane MA, Mis EK, Khokha MK. 2021a. Obtaining *Xenopus tropicalis* eggs. *Cold Spring Harb Protoc* doi:10.1101/pdb.prot106344

Lane MA, Mis EK, Khokha MK. 2021b. Microinjection of *Xenopus tropicalis* embryos. *Cold Spring Harb Protoc* doi:10.1101/pdb.prot107644

Lane MA, Slocum M, Khokha MK. 2021c. Raising and maintaining *Xenopus tropicalis* from tadpole to adult. *Cold Spring Harb Protoc* doi:10.1101/pdb.prot106369

Nieuwkoop PD, Faber J. 1994. *Normal table of* Xenopus laevis *(Daudin): a systematical and chronological survey of the development from the fertilized egg till the end of metamorphosis.* Garland, New York.

Shaidani N-I, McNamara S, Wlizla M, Horb ME. 2020. Animal maintenance systems: *Xenopus tropicalis. Cold Spring Harb Protoc* doi:10.1101/pdb.prot106146

Showell C, Conlon FL. 2009. Natural mating and tadpole husbandry in the western clawed frog *Xenopus tropicalis. Cold Spring Harb Protoc* doi:10.1101/pdb.prot5292

Wlizla M, McNamara S, Horb ME. 2018. Generation and care of *Xenopus laevis* and *Xenopus tropicalis* embryos. In Xenopus: *methods and protocols* (ed. Vleminckx K), pp. 19–32. Springer, New York.

Cite this protocol as *Cold Spring Harb Protoc;* doi:10.1101/pdb.prot106609

Microinjection of *Xenopus tropicalis* Embryos

Maura Lane,[1] Emily K. Mis, and Mustafa K. Khokha

Pediatric Genomics Discovery Program, Department of Pediatrics and Genetics, Yale University School of Medicine, New Haven, Connecticut 06510, USA

Microinjection is an important technique used to study development in the oocyte and early embryo. In *Xenopus*, substances such as DNA, mRNA, and morpholino oligonucleotides have traditionally been injected into *Xenopus laevis*, because of their large embryo size and the relatively long time from their fertilization to first division. In the past few decades, *Xenopus tropicalis* has become an important model in developmental biology; it is particularly useful in genetic studies. The advent and rapid development of CRISPR–Cas9 technology has provided an array of targeted gene manipulations for which *X. tropicalis* is particularly suited. The equipment and protocol for *X. tropicalis* microinjection is broadly transferable from *X. laevis*. There are important differences between the species to consider, however, including the smaller embryo size and faster embryo development time in *X. tropicalis*. There are a number of solutions and reagents that differ in concentration and composition as well. Here we describe a microinjection protocol specifically for studies in *X. tropicalis*.

MATERIALS

It is essential that you consult the appropriate Material Safety Data Sheets and your institution's Environmental Health and Safety Office for proper handling of equipment and hazardous materials used in this protocol.

RECIPES: Please see the end of this protocol for recipes indicated by <R>. Additional recipes can be found online at http://cshprotocols.cshlp.org/site/recipes.

Reagents

Dejellied *Xenopus tropicalis* embryos obtained from Protocol 5: Obtaining *Xenopus tropicalis* Embryos by In Vitro Fertilization (Lane and Khokha 2021a) or from Protocol 6: Obtaining *Xenopus tropicalis* Embryos by Natural Mating (Lane and Khokha 2021b)

Dextran solution for practicing injections (see Step 3)

Prepare a stock solution of a red tracing dye (e.g., dextran, tetramethylrhodamine, 10,000 MW, lysine fixable [fluoro-Ruby]; Invitrogen D1817) at a concentration of 20 µg/µL. Keep sterile and store at −20°C. On the day of the injections, prepare 10 µL of a 1:3 mix of the dextran stock in RNase-free sterile water.

Ethanol (70%)

Ficoll PM 400 (3%; GE Healthcare Biosciences 95021-196) in 1/9× MR

A Ficoll-containing solution will readily grow contaminants. Autoclave and handle carefully to prevent introducing microorganisms. Use and store at room temperature.

Gentamicin (50 µg/mL) in 1/9× MR

[1]Correspondence: maura.lane@yale.edu

TABLE 1. Common reagents for experimental embryo injection

Reagent for microinjection	Preparation	Stock concentration	Dose range[a]	Notes
mRNA encoding gene of interest	Use an in vitro transcription kit such as mMESSAGE mMACHINE T7 Transcription Kit (Ambion AM1344) or SP6 Transcription Kit (Ambion AM1340). Follow kit instructions. Maintain RNase-free conditions.	Varies depending on yield. If yield is too low to achieve desired final concentration in your mix, repeat transcription reaction, or set up two reactions at the outset.	2 pg–500 ng/embryo (or more)	Use a fluorescently tagged gene construct or co-inject an RNase-free fluorescent mRNA tracer (see Table 2).
CRISPR–Cas9	Prepare and use according to Deniz et al. (2018)			
Morpholino oligonucleotides (MOs)	Resuspend MOs in RNase-free water to a 20 ng/µL stock at room temperature. Heat the solution for 5 min to 65°C to fully dissolve MOs. Mix well and store at room temperature in five equal aliquots. Incubate MOs at 37°C–42°C on a heat block for 5–30 min before use.	20 ng/µL	1–30 ng/embryo	MOs are extremely resistant to nucleases, but DEPC treatment can damage the MO. Use nuclease-free water to resuspend your MO and prepare injection mixes. MOs, especially those conjugated with a fluorescein tracer, can precipitate, causing the microinjection needle to clog. Be sure to incubate the MO before use as noted under Preparation.

[a]Appropriate dosage of the above reagents ultimately depends on the gene being targeted and the toxicity of the reagents. These should be empirically tested using a dose titration curve. Three doses are typically injected initially (high, medium, and low, using 20–50 embryos/dose). Controls using co-injection materials (tracers, Cas9, water) alone are also recommended. Embryos are observed for abnormal development after incubating for 24 h at 25°C. Select the highest dose that does not cause abnormal embryo development. Doses may need to be further refined with a revised dose curve.

> Prepare a 1000× stock (50 mg/mL) of gentamicin (Sigma-Aldrich G3632) in distilled water, filter-sterilize, and store at 4°C. Prepare the dilution in 1/9× MR, and use at room temperature.

Mineral oil (heavy, sterile, RNase- and DNase-free)

Modified Ringer's solution (MR; 1×) <R>, diluted to a concentration of 1/9×

> Prepare the 1× stock with distilled water; the final pH will be 7.8–8.0. To prepare 1/9× MR, dilute the 1× MR with distilled water. The pH should be 7.8–8.0. Use this solution at room temperature.

mRNA, CRISPR–Cas9, morpholino oligonucleotides (MO), or other solution for experimental microinjection (see Tables 1 and 2)

Equipment

Borosilicate glass capillaries (World Precision Instruments TW100F-4)

> Capillaries are used to make injection needles. Keep tubes clean and RNase-free.

Dissecting stereomicroscope and associated equipment

Focus mount (Nikon, C-FMCN, MND54000)

LED light column mount (Nikon, 77031107)

LED light source with dual gooseneck

Microscope body (Nikon, SMZ745, MMA36400)

Microscope eyepiece reticle (Klarmann Rulings, KR-271BP, 26-mm)

> This is used to measure the size of air bubble when calibrating the needle. The reticle listed is specific to the diameter of the eyepieces above. Consult your microscope manufacturer and Klarmann to determine the appropriate reticle for your eyepieces.

Microscope stand (Nikon, MND51030)

Wide-field eyepieces (Nikon, C10XB, MMK30102)

Cite this protocol as *Cold Spring Harb Protoc*; doi:10.1101/pdb.prot107644

TABLE 2. Common tracing materials used for experimental co-injection

Reagent for labeling	Preparation	Stock concentration	Working dilution	Notes
Dextran, tetramethylrhodamine, (fluoro-Ruby), 25 mg (Invitrogen D1817)[a]	Dilute in 1250 µL of RNase-free water. Prepare 20 µL aliquots. Store at −20°C.	20 µg/µL	Use 0.5–1.5 µL of stock in a final injection mix volume of 5 µL.	Prepare and maintain in RNase-free conditions. Use with CRISPR–Cas or MO injections to label injected embryos. Excitation/emission: 555/580 (red fluorescence). This dye is visible under the surface of the embryo by eye, so it is also useful to confirm injection when practicing.
Dextran, Alexa Fluor 488, 5 mg (Invitrogen D22910)[a]	Dilute in 600 µL of RNase-free water. Prepare 20 µL aliquots. Store at −20°C, protected from light.	8 µg/µL	Use 0.5–1.5 µL of stock in a final injection mix volume of 5 µL.	Prepare and maintain in RNase-free conditions. Use in CRISPR–Cas or MO injection mixes. Excitation/emission: 495/519 (green fluorescence)
Fluorescent mRNA[b] such as: pCS-H2B-EGFP (Addgene 53744) pCS-H2B-mRFP1 (Addgene 53745) pCS-memb mCherry (Addgene 53750)	Use an in vitro transcription kit such as mMESSAGE mMACHINE T7 Transcription Kit (Ambion AM1344) or SP6 Transcription Kit (Ambion AM1340). Follow kit instructions. Maintain RNase-free conditions. Store at −80°C. Aliquot to minimize freeze–thaw cycles.	Varies depending on yield	Use 2 pg–500 ng/ embryo (or more).	Use when injecting mRNA for overexpression studies to label injected embryos.

[a]Download the Molecular Probes Handbook from the Invitrogen website for information on a wide variety of fluorescent tracers for many applications.
[b]There are a variety of fluorescent mRNA plasmids available through Addgene, in addition to those listed here. Make sure your plasmid has a polymerase promoter site 5′ of your gene and an SV40 polyadenylation site 3′ of your gene.

The microscope and associated equipment listed here are what we use in our laboratory. Similar microscopes from reputable sources such as Zeiss, Olympus, or Leica would work equally well.

Forceps (Dumont #5)

Forceps are used for breaking the injection needle tip. Keep sharp using abrasive paper or sharpening stone under a microscope. Protect tips to preserve the shape.

GELoader tips (0.5–20 µL) (Eppendorf 022351656)

These tips are used for backloading injection needles.

Glass Pasteur pipettes (Corning 7095B-5X)
Hair loop tool (optional; see Step 30)
Injection dishes for *X. tropicalis* embryos <R>
Injection equipment

Electrode holder (Warner Instruments 641278)

The electrode holder is used for mounting injection needles.

Iron plate (IP; Narishige Group)
Magnetic stand (Narishige Group GJ-8)
Micromanipulator (Narishige Group MM3)
Micropipette puller (Sutter Instruments P97)

This puller is used to taper capillary tubes into microinjection needles.

PLI-FS foot switch (Harvard Apparatus 65-0029)

This switch can only be used with Harvard Apparatus microinjectors. Other brands require a model-specific footswitch.

PLI–90A Picoliter microinjector (Harvard Apparatus/Warner Instruments 64-1738)

The microinjector controls the duration of air supplied to the injection needle. Alternates include PV850 (World Precision Instruments), PLI-100 (Harvard Apparatus), and Nanoject II (Drummond Scientific).

Source of pressure-regulated air

This can be an oxygen canister or dedicated air lines with a pressure gauge.

Low-temperature incubator (VWR 2005)
Microcentrifuge tubes (1.5-mL)
Petri dishes (150-mm [Falcon 351058], 100-mm [Falcon 351029], 60-mm [Falcon 351007], and 35-mm [Falcon 351008])
Pipette pump (VWR 53502-233)
Pipettes (P20, P1000)
Plasticine clay or foam strips (see Step 9)
Ruler (small)
Syringe (10-mL) with a Luer tip compatible with the electrode port type (see Step 5)
Toothbrush

METHOD

Standardize Injection Volume

You can greatly simplify the microinjection workflow by using a standard injection volume for all experiments. An amount of 2 nL is an ideal volume to inject into one- to eight-cell stage X. tropicalis embryos (10 nL for X. laevis). Follow Steps 1 and 2 to determine the number of reticle hash marks needed to produce a bubble 2 nL in volume. Repeat Steps 1 and 2 to determine the bubble size (in hash marks) needed for a volume of 10 nL for X. laevis injections. Perform these steps after setting up the microscope and micromanipulator but before starting any experiments. This procedure only needs to be done once during the initial equipment setup.

1. At high magnification, measure the distance (mm) between two adjacent hash marks on the eyepiece reticle (Fig. 1) by focusing on a small ruler placed on the stage.

2. Determine the number of hash marks that will produce a 2-nL bubble by using the equation for the volume of a sphere, $V = 4\pi r^3/3$, where $r =$ diameter/2.

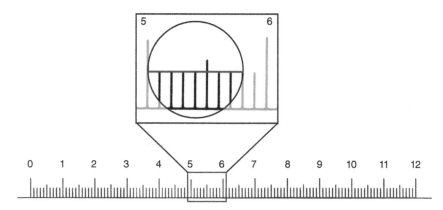

FIGURE 1. Micrometer on the reticle is used to measure injection bubble diameter. The entire micrometer visible through the eyepiece is shown at the *bottom*. Each of the 12 major divisions have 10 smaller divisions, or "hash marks," used to measure the size of the injection bubble. An enlarged section of the micrometer is shown above. The circle represents a bubble of injection solution. In this case, the bubble measures eight hash marks. When measuring the bubble, line one edge up to a major division hash mark (here "5"). This will make it easier to quickly determine the bubble size.

Cite this protocol as *Cold Spring Harb Protoc*; doi:10.1101/pdb.prot107644

Example (X. tropicalis): Assume you have determined that a bubble 8 hash marks in diameter = 2 nL volume. To co-inject 100 pg of mRNA with 25 pg of mRNA tracer, prepare your reaction mix with mRNA at 50 pg/nL and mRNA tracer at 12.5 pg/nL (typical total reaction mix volume of 3–6 µL). You will now deliver the desired dose by simply calibrating your needle (see Steps 18–25) to produce a 2-nL bubble. You can do this for any reaction mix going forward.

Practice Injections

It is important to practice with the equipment and steps involved in microinjection before beginning important experiments. Fertilized embryos advance quickly to the two-cell stage (45 min–1 h), so preparation is key (see Time/Temperature Development Table at http://www.xenbase.org/anatomy/static/xenopustimetemp.jsp).

3. Perform practice injections using a tetramethylrhodamine dextran solution to practice loading and mounting needles and injecting embryos.

    This dye is visible with magnification under the surface of the embryo at the time of injection, making it ideal for practicing and confirming injections. It is also very bright using fluorescence microscopy.

Preinjection Checklist

This preinjection checklist will help save time and frustration. Routinely follow these steps before conducting experiments.

4. Inspect the injection station to be sure the microscope, light source, and micromanipulator are adjusted to your preferences and functioning properly. Ensure that tools and materials such as forceps, pipettes, and oil for breaking needles are available.

5. Visually check the electrode holder (Fig. 2) for broken glass or dried injection material in the bore. If unsure, try inserting a capillary tube. If the tube does not feed smoothly and/or there is a "crunchy" feel, clean the bore. Carefully remove the air tube from the electrode port, loosen the compression cap, and flush with sterile water in a 10-mL syringe until the debris is removed. Use a syringe with a Luer tip compatible with the port type. Dry well with air from an empty syringe.

    Consult Warner Instruments for details (https://www.warneronline.com/product/221).

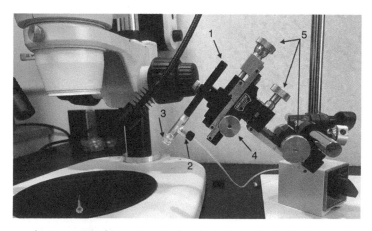

FIGURE 2. Microscope and micromanipulator setup. 1. The electrode holder holds the injection needle used to inject materials into the embryos. 2. Air supply Luer port with attached air supply line. The airline is connected to a source of compressed air through a microinjector (not shown), which allows the operator to control the air burst duration. The airline can be removed at the port to flush debris from the electrode bore using sterile water in a 10-mL syringe fitted with a port adapter. 3. The compression cap and red gasket are used to secure the injection needle (not shown) in the electrode bore. 4. The fine x-axis knob is used to move the electrode holder and needle during microinjection. 5. The coarse x-, z-, and y-axis knobs are used to move the micromanipulator in different planes when making adjustments to the overall position of the unit and electrode holder, ideally before beginning injections. The unit should be positioned so the only knob that needs to be used during injection is the fine x-axis knob to lower and raise the needle in and out of the embryo as you inject.

6. Make sure you have enough injection dishes. Use a toothbrush and ddH₂0 to clean any debris from the dish. Embryos will be pulled through the mesh if it is not attached to the dish, so discard or set aside such dishes for repair.

7. Pull needles as in Step 9. It is advisable to prepare two needles for each solution to be injected, in case the first needle breaks or clogs.

8. Make sure the solutions for your microinjection mix, especially those kept as common use stocks, are available.

Prepare Needles

Prepare needles before beginning microinjection protocol. If obtaining embryos by in vitro fertilization (IVF), prepare needles during the 10-min interval in Step 18 in Protocol 5: Obtaining Xenopus tropicalis *Embryos by In Vitro Fertilization (Lane and Khokha 2021a).*

9. Use a micropipette puller to taper borosilicate glass capillary tubes into injection needles. Use the following settings with TW100F-4 borosilicate glass capillaries: heat 360, pull 180, velocity 80, time 60, and pressure 500.

> *Each capillary tube will make two needles. Refer to the micropipette puller user manual for details on pulling method and adjustments.*
>
> *Handle pulled needles with care, as the tip is very thin and will break easily. Protect unused needles by securing them in a strip of plasticine clay in a 150-mm Petri dish (keep covered) or in foam strips glued in a tray or alternate.*
>
> *The pulled needles are also extremely sharp and can easily penetrate skin and become lodged there. Use caution when handling pulled needles to avoid injury. It is also best practice to make sure you have removed your needle from the electrode holder after use, to avoid others inadvertently being injured. Used needles should be disposed of in glass waste immediately.*

Prepare Injection Solution

10. Refer to Tables 1 and 2 for information on preparing common materials used for microinjection.

11. Prepare the microinjection mix, the final volume of which is generally 3–6 μL.

> *This is enough to reserve some of the mix in case the first needle becomes broken or irreparably clogged.*
>
> *After you are proficient at loading and breaking your needle, you may wish to prepare your microinjection mix after fertilization has been confirmed. This will avoid wasting injection reagents if the fertilization fails but will reduce the time you have to prepare the needle before injecting.*

Load Solution into Injection Needle

12. Draw an aliquot (2–3 μL) of sterile RNase-free injection mix from a microcentrifuge tube using a P10–P20 micropipette and a sterile Seque/Pro capillary tip. Draw the solution up slowly to avoid air bubbles.

> *Reserve some of your mix if possible. If the needle breaks or clogs, you may need to load another needle.*

13. With your nondominant gloved hand, hold the injection needle with the open end facing toward you. Load the injection solution by inserting the capillary tip 15–18 mm into the back of the needle, and then slowly eject the solution while withdrawing the tip from the needle.

> *Loading your solution too fast can force it out of the back of the needle. Twisting the needle while loading can aid in keeping the solution in the tube.*
>
> *You can stabilize the needle by placing your elbow on the benchtop, or by carefully securing the needle in a strip of clean plasticine clay in a 150-mm Petri dish. Load the solution in the same manner as above.*

Cite this protocol as *Cold Spring Harb Protoc*; doi:10.1101/pdb.prot107644

Mount Injection Needle in Electrode Holder

14. Use the fine z-axis knob on the micromanipulator to move the electrode holder away from the microscope platform to provide space between the microscope stage and the electrode holder while mounting the needle (Fig. 2).

15. Loosen the compression cap at the tip of the electrode holder by turning it counterclockwise (Fig. 2). Carefully feed ~15 mm of the back of the needle into the opening in the cap.

    The needle should be visible above the orange gasket in the channel of the electrode holder.

16. Turn the compression cap clockwise to secure the needle in place. Do not overtighten. Test the needle security by pulling down gently before proceeding.

    The orange rubber gasket should appear to expand when tightened sufficiently.

Break the Needle Tip

The unbroken needle is sealed at the tip. It must be snipped to create the proper size opening and needle shape to inject embryos. This requires practice and a steady hand. The ideal needle tip is beveled at ~45° to facilitate penetrating the embryos with minimal damage. If the needle opening is too small, it will be prone to clogging, and may not be stiff enough to penetrate the embryo. If this occurs the needle will need to be rebroken higher up. Conversely, a large bore needle can be difficult to calibrate, and can damage the embryo. If this occurs a new needle must be used.

17. Set the microscope at low to medium power (user preference). Using the fine z-axis micromanipulator knob, slowly lower the needle until it comes into the field of view. Adjust focus on the tip of the needle, then use fine forceps to grasp and break the needle tip.

Adjust Bubble Size

Adjust the bubble size to the diameter previously determined to deliver 2 nL of solution (see Steps 1–2). The diameter can be adjusted by changing the size of the needle opening and/or the duration of air burst. The ideal scenario is to break the needle to a size that allows a bubble of the proper diameter to be ejected using a duration setting of 200–400 msec. The injection dose can then easily be halved or doubled by halving or doubling the duration.

18. Turn on the microinjector and set the air delivery duration to 300 msec. Set the microscope on high magnification.

19. Place a 35-mm Petri dish of heavy mineral oil on the microscope platform and adjust focus to the surface of the oil.

20. Using the fine z-axis knob (Fig. 2), lower the needle until the tip is positioned just below the surface of the oil.

    The needle should be immersed in the oil as soon as possible after breaking it to avoid the solution drying out and clogging the needle.

21. Adjust your focus so the needle tip is sharply in view. Position the eyepiece micrometer over the tip of the needle by rotating the eyepiece.

22. Depress the microinjector foot pedal while observing the needle tip on high power.

    The air pressure will push the solution to the tip and eject a drop of injection solution into the oil.

23. Measure the diameter of the bubble (in "hash marks") at high power.

24. Adjust the size of the bubble by changing the air duration. If the duration goes to >400 msec, rebreak the needle. If the duration goes to <100 msec, start again with a new needle. The needle opening is too large at this point and will injure your embryos.

25. After calibration is complete, move the electrode holder with the needle up to the highest z-axis position to decrease the chance of breaking the needle during Steps 28–31.

    Steps 28–31 should be completed without delay, to avoid the injection solution drying out and clogging the needle while embryos are collected. Alternatively, a basic binocular light microscope can be dedicated to embryo sorting only, in which case the injection needle can be placed back in the oil after Step 25 to prevent clogging.

Select and Position Embryos in Injection Dish

These steps are generally performed at room temperature. If desired, embryo development can be slowed by keeping the dish of embryos on a cold plate or small benchtop incubator at 22°C–23°C. Proceed at room temperature at Step 29.

26. Fill injection dish(es) approximately halfway with 3% Ficoll in 1/9× MR.

27. For each injection dish, label the top and bottom of a 60- or 100-mm Petri dish with the name of the injected material and any other relevant information. Embryos will be transferred into this dish after injecting.

    Steps 26–27 can be done before beginning microinjection protocol to save time.

28. Examine embryos in 3% Ficoll under medium power (2×–4×). Use a pipette pump and glass Pasteur pipette to select healthy embryos.

    Refer to Figure 2D,E in Protocol 4: Obtaining Xenopus tropicalis *Eggs (Lane et al. 2021) for information on distinguishing healthy from sick or dead embryos.*

    Calculate fertilization rate as in Step 19 of Protocol 5: Obtaining Xenopus tropicalis *Embryos by In vitro Fertilization (Lane and Khokha 2021a). If the fertilization rate is ∼70%–100%, it may be faster to draw up embryos indiscriminately, inject all of them, and then remove the unfertilized ones later. Use care to avoid crushing your needle while collecting embryos.*

29. Carefully position 50 to 100 embryos in the injection dish in an orderly fashion (Fig. 3). Arranging the embryos in rows makes it easy to track which ones have been injected. Keep the embryos near the center of the dish to avoid breaking the needle on the dish wall.

30. Using a delicate flexible tool, such as a GELoader tip or a hair loop tool, gently nudge embryos until they are aligned in individual wells of the mesh. If you are targeting a specific cell or area of the embryos, you may also gently position them so the target is tilted toward the needle. Cell fate maps can be found at http://www.xenbase.org/anatomy/static/xenbasefate.jsp.

31. Under medium magnification, use the pipette pump to carefully remove Ficoll until the embryos settle in the wells but are still covered by a small amount of fluid. Keep the tip of the glass pipette close to the edge of the dish to avoid disturbing the embryos.

    If the embryos are stuck in the wells or if the surface of the embryos begin to dry, replace some of the Ficoll. If embryos are dislodged from the wells by the injection needle, remove more Ficoll.

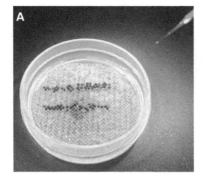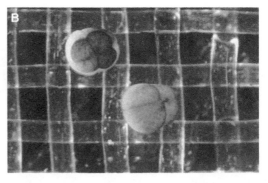

FIGURE 3. Injection dish. (*A*) Embryos arranged in rows in an injection dish. Injection needle with Dextran Alexa-fluor 488 tracer is visible in *upper right*. (*B*) Close-up of embryos in individual wells of injection dish.

Microinjection

32. Under low to medium magnification (depending on your preference), focus on the surface of the embryos and slowly lower the injection needle to the dish. When the needle tip is near the first embryo's surface, depress the foot pedal and look for a bubble of solution at the end of the needle.

    If there is no bubble, the needle may be clogged. Try to remove the clogged material by injecting it into the Ficoll surrounding the embryos and/or temporarily increasing the duration. Gently prodding the tip of the needle against the mesh or bottom of the dish can also dislodge a clog. Ultimately you may need to rebreak and recalibrate your needle or start over with a new needle.

33. Inject the embryo by moving the dish of embryos so the needle is positioned above and slightly off center of the embryo. Advance the needle until it just passes through the surface of the embryo. Do not advance deeply into the embryo, as this may damage it.

    Healthy, properly dejellied embryos will be turgid and easy to inject. The needle should "pop" through the vitelline membrane surrounding the embryo. Unfertilized eggs are flaccid and difficult to inject. Dead embryos are white and will likely explode when injected. These should not be injected and should be removed from the dish during sorting.

34. Depress the foot pedal to eject an aliquot of injection solution.

    If using a bright tracer, you may see color appear below the embryo's surface. You may also see a "bleb" form as the fluid is injected.

35. Wait a few seconds before withdrawing the needle to reduce the chance of solution or embryo contents leaking out.

36. When the needle is fully withdrawn, position it above the next embryo by moving the dish, and repeat the injection process.

37. Continue injecting in an organized manner (i.e., row by row or column by column). At the end of each row or column, make sure the needle is not clogged by expelling a drop of fluid above the embryos.

38. After all embryos are injected, carefully refill the injection dish with 3% Ficoll in 1/9× MR. If many embryos are floating in Ficoll after injection, they can be encouraged to sink into the dish by gently expelling drops of Ficoll from a glass Pasteur pipette in the pipette pump on top of the embryos.

    Embryos that are allowed to float after injection are sometimes damaged and undergo inexact cell divisions.

39. Place the correctly labeled lid from the 60-mm dish described in Step 27 over the smaller injection dish to protect the embryos and keep track of the injection condition.

40. Allow the embryos to heal and recover undisturbed for 1 h at room temperature.

41. After 1 h, use the pipette pump and a glass Pasteur pipette to gently dislodge the embryos from the mesh by slowly pipetting up and down. Make sure not to pick up embryos during this time, or you may shear them. Once the embryos are free of the mesh, carefully select properly dividing embryos by drawing them up into the glass Pasteur pipette and transferring them into the labeled dish containing 1/9× MR with gentamicin. Avoid selecting dead or dying embryos, as these can cause other embryos to die.

42. Incubate embryos at desired temperature (22°C–28°C).

    See Time/Temperature Table at http://www.xenbase.org/anatomy/static/xenopustimetemp.jsp

43. Clean and rinse injection dishes with 70% ethanol and air dry after use. If the dish is used again before it dries, it should be rinsed with ddH$_2$O.

44. The morning following injection, check embryos for the presence of the fluorescent tracer using fluorescence microscopy. Remove all uninjected embryos from the dish.

45. Change 1/9× MR with gentamicin and remove dead embryos in the dish daily.

RECIPES

Injection Dishes for X. tropicalis *Embryos*

MATERIALS

Reagents

Ethanol (95%)
Methylene chloride or chloroform

These reagents are volatile and should be used in a fume hood.

Equipment

Cork (silicone rubber, with a diameter close to that of the Petri dish)
Petri dishes (plastic, 35-mm; Falcon 351008)
Polypropylene mesh (Repligen 146418)

METHOD

Injection dishes for microinjection of Xenopus *oocytes are modified Petri dishes to which a nylon mesh has been fixed to hold oocytes or embryos in place. The mesh also helps to keep track of which specimens have been injected because it holds the specimens in regular rows. Ensure that the grid fits snugly to the edge of the dish and lies flat on the surface, so that the specimens do not become trapped underneath.*

1. Cut the 500-μm polypropylene mesh into circles so that it fits snugly into the Petri dishes.
2. Working in a fume hood, fix the mesh in place by melting the bottom of the plastic Petri dish with five drops of methylene chloride or chloroform. Use the silicone rubber cork to hold the grid flat while the plastic sets (~10 sec). Double-check to be sure that the mesh is completely adhered to the dish.
3. Allow the dishes to sit overnight in a fume hood to ensure complete evaporation.
4. After complete evaporation of the solvent, rinse the dishes thoroughly with 95% ethanol and then with water before use.

Modified Ringer's Solution (MR; 1×)

0.1 M NaCl
1.8 mM KCl
2.0 mM $CaCl_2$
1.0 mM $MgCl_2$
5.0 mM HEPES (pH 7.6)

Store for up to 1 mo at room temperature.

REFERENCES

Deniz E, Mis EK, Lane M, Khokha MK. 2018. CRISPR/Cas9 F_0 screening of congenital heart disease genes in *Xenopus tropicalis*. *Methods Mol Biol* **1865:** 163–174. doi:10.1007/978-1-4939-8784-9_12

Lane MA, Khokha MK. 2021a. Obtaining *Xenopus tropicalis* embryos by in vitro fertilization. *Cold Spring Harb Protoc* doi:10.1101/pdb.prot106351

Lane MA, Khokha MK. 2021b. Obtaining *Xenopus tropicalis* embryos by natural mating. *Cold Spring Harb Protoc* doi:10.1101/pdb.prot106609

Lane MA, Mis EK, Khokha MK. 2021. Obtaining *Xenopus tropicalis* eggs. *Cold Spring Harb Protoc* doi:10.1101/pdb.prot106344

Cite this protocol as *Cold Spring Harb Protoc*; doi:10.1101/pdb.prot107644

Best Practices for *Xenopus tropicalis* Husbandry

Takuya Nakayama and Robert M. Grainger[1]

Department of Biology, University of Virginia, Charlottesville, Virginia 22904, USA

Xenopus tropicalis has been adopted by laboratories as a developmental genetic system because of its diploid genome and short generation time, contrasting with *Xenopus laevis*, which is allotetraploid and takes longer to reach sexual maturity. Because *X. tropicalis* has been introduced more recently to many laboratories, some specific methods more appropriate for handling of eggs and embryos of *X. tropicalis* are still not widely known to researchers who use *X. laevis*. Here we highlight some recommendations and opportunities possible with this model system that complement existing *X. tropicalis* procedures. Of particular importance, because of the value of generating genetically modified lines for researchers using *X. tropicalis*, we describe a procedure for sterilizing embryos, which could be applied to both species of *Xenopus*, but might be particularly useful for raising genetically modified animals in *X. tropicalis*. This protocol will help ensure that a colony will have a high probability of being free of pathogens known to be serious threats to *Xenopus* health.

HISTORICAL OVERVIEW COMPARING EXPERIMENTAL METHODS FOR *XENOPUS LAEVIS* AND *XENOPUS TROPICALIS*

In general, every laboratory has slightly different protocols for handling *Xenopus* embryos and for performing frog husbandry. The differences may stem from many sources: available animal housing and equipment, laboratory procedures handed down from one laboratory member to another, local husbandry regulations, and so on. In many cases, such differences may not be problematic as long as these methods produce healthy embryos and adult animals. However, based on our experience handling both *Xenopus laevis* and *Xenopus tropicalis*, we have identified critical points at which methods that work for *X. laevis* do not necessarily work well for *X. tropicalis* (e.g., Hirsch et al. 2002a). Those who have previously handled *X. laevis* may use methods from this prior experience that may not transfer well to *X. tropicalis*. These researchers may initially find it challenging to raise and handle *X. tropicalis*. In addition, there are procedures specific to *X. tropicalis* for enhancing tadpole raising and, for example, using these embryos in studies with tissue recombinants with mammalian embryos that may be valuable to investigators. These and other protocols were developed to provide standard methods for raising *X. tropicalis* when this model system was first being cultivated as a genetically tractable model to complement *X. laevis* (Hirsch et al. 2002a,b). Finally, a procedure for sterilizing embryos ensures long-term colony health as required for propagation of genetic lines (see below).

---

[1]Correspondence: rmg9p@virginia.edu

ENHANCING NATURAL MATINGS

We use Chorulon (hCG, Intervet Inc) for priming and boosting frogs here and below. To prepare Chorulon, add 5 mL of the supplied diluent to the vial (making 2 units/μL), aliquot in small volumes (e.g., 500 μL), and store at −20°C. Once thawed, store at 4°C and use ideally within 2 weeks to a month. The appropriate amount for aliquoting for freezing depends on the frequency of usage of frogs in each laboratory.

In addition to what is described in Protocol 6: Obtaining *Xenopus tropicalis* Embryos by Natural Mating (Lane and Khokha 2022b), carbon-filtered (i.e., dechlorinated) tap water (approximately pH 7.6–7.8 in our facility) can be used for natural matings, with about 4 L per 6 L bucket (e.g., polycarbonate square bucket with lid; Rubbermaid FG630600CLR and FG650900WHT [lid]). After boosting, we recommend keeping frogs in a quiet place in the dark (e.g., a black cloth to completely cover the bucket is sufficient) at ~25°C so that frogs are less stressed by any stimulation from the surrounding environment. Multiple animals per bucket can be used to increase the success rate. For example, one can mix a female with two to three males (or fewer males and more females) and monitor the animals to find a pair that is amplexed, which usually should happen within 3 h after boosting. Once at least one amplexed pair is confirmed, carefully remove unmated frogs. Because a gravid female often lays many more eggs than are needed for an experiment, males and females can be swapped in the middle of a mating among multiple pairs as needed—for example, when it is desirable to test different combinations of mutant genotypes of a given gene. To avoid contamination of embryos between exchanged animals (depending on the experimental purpose), carefully and patiently remove all (fertilized) eggs attached to animals between exchanges. This can be accomplished by using a transfer pipette (e.g., Fisher 13-711-7M) to remove eggs from the frog epidermis while avoiding any injury to frogs. When frogs are actively mating, even if they are separated, they usually amplex quickly once again. The movement of a mating pair from one bucket to another allows egg collection at intervals, which can be valuable for several purposes. For example, eggs can be fertilized at variable time points, spreading out their development to maximize the number of embryos that can be injected or to stagger embryos for later work at a given developmental stage.

When arranging matings overnight as described in Protocol 6: Obtaining *Xenopus tropicalis* Embryos by Natural Mating (Lane and Khokha 2022b), there is always a risk that frogs may consume eggs or embryos during this extended period. Also, if the fertilization rate is not high, many unfertilized eggs can decay overnight, which can kill fertilized embryos as well. Therefore, unless overnight matings are essential, it may be better to avoid them, or to prepare multiple matings to ensure enough healthy embryos can be collected.

As described in Protocol 4: Obtaining *Xenopus tropicalis* Eggs (Lane et al. 2022a), during matings, females may keep laying eggs for a long time. It is important to confirm that females have stopped laying before moving them back to an animal housing system. Doing so prevents frogs from acquiring an infection and from releasing eggs that may stick to housing tanks. To prevent these problems, after experiments, frogs can be kept overnight in a new bucket filled with a minimum of 1 L per frog of fresh dechlorinated tap water supplemented with sea salts (e.g., Instant Ocean Sea Salt; 2 g/L). As needed, frogs can be isolated for longer with daily water changing (maintaining the salt treatment) because females sometimes keep laying for more than a day.

METHODS FOR FERTILIZING EGGS LAID NATURALLY

Some researchers have historically collected *X. laevis* eggs laid naturally in a high-salt solution (e.g., Leno and Laskey 1991). For example, high-salt modified Barth's saline (MBS) is widely used to collect eggs for in vitro fertilization (e.g., Blow and Laskey 1986; Sive et al. 2007). This method is very useful when frogs do not lay as many eggs as required by manual manipulation, is much less stressful for frogs, and thus has less chance of accidentally injuring frogs. However, this particular method does not work well for collection of fertilizable *X. tropicalis* eggs. Instead, eggs should be collected in 1× MBS (pH 7.5) + 0.5% BSA. See Sec. 10, Protocol 9: Gynogenetic Production of Embryos in *Xenopus tropicalis* Using a Cold Shock Procedure: Rapid Screening Method for Gene Editing Phenotypes (Nakayama et al. 2022) for details including all MBS-containing solutions mentioned herein. Eggs can be stored up to 5 h in this medium for in vitro fertilization, but a longer storage time before fertilization could cause maternal effects leading to abnormal development of embryos (Kosubek et al. 2010). Therefore, we recommend, if possible, collecting eggs every 30 min followed by immediate in vitro fertilization.

ENHANCING *X. TROPICALIS* EGG FERTILIZATION AND RAISING EMBRYOS

Protocol 5: Obtaining *Xenopus tropicalis* Embryos by In Vitro Fertilization (Lane and Khokha 2022a), Protocol 6: Obtaining *Xenopus tropicalis* Embryos by Natural Mating (Lane and Khokha 2022b), Protocol 4: Obtaining *Xenopus tropicalis* Eggs (Lane et al. 2022a), and Protocol 7: Microinjection of *Xenopus tropicalis* Embryos (Lane et al. 2022b) concerning generating and raising *X. tropicalis* embryos thoroughly describe these procedures, but one point to be emphasized is that, although *X. laevis* testes and sperm suspensions are commonly stored on ice, *X. tropicalis* testes and sperm are inactivated under these conditions and must be kept at room temperature. Fertilization is also highly efficient and synchronous if eggs are gently spread as a single layer in a Petri dish before applying the sperm solution. We prepare sperm suspensions in 1× MBS (pH 7.5) + 0.1% BSA.

Although Protocol 5: Obtaining *Xenopus tropicalis* Embryos by In Vitro Fertilization (Lane and Khokha 2022a) uses modified Ringer's solution to raise embryos, other media can be used. We use 0.1× MBS (pH 7.5) for culturing embryos, and the addition of phenol red (e.g., Sigma-Aldrich P0290; 40–50 µL/L) to monitor the pH of the medium during culturing can be very useful. For example, too many embryos maintained in a dish will quickly acidify the medium, at which point one needs to change the medium immediately and reduce the density of animals per dish. During dejellying, monitoring the pH of the cysteine solution is also very helpful. At a lower pH the efficiency of dejellying drops significantly, so addition of phenol red allows the experimenter to judge the time at which the cysteine solution should be refreshed during dejellying.

Because of the nature of their fertilization membrane, young *X. tropicalis* embryos before hatching are much stickier than *X. laevis*, adhering to plastic Petri dishes, which often leads to abnormal development. Sticking can be avoided by coating dishes with a BSA-containing medium at a minimum, but coating dishes with 0.8%–1% agarose (e.g., Thermo Fisher Scientific BP160) dissolved in 0.1× MBS (or H$_2$O) may be more effective (see Sec. 10, Protocol 9: Gynogenetic Production of Embryos in *Xenopus tropicalis* Using a Cold Shock Procedure: Rapid Screening Method for Gene Editing Phenotypes [Nakayama et al. 2022]). Because embryos can be trapped under a loosening agarose coating, this step is only performed during the first day of development, and uncoated dishes are used subsequently. We also coat Pasteur pipettes and plastic transfer pipettes with a BSA-containing solution (e.g., 0.1× MSB + 0.1% BSA) by rinsing pipettes in this solution, to prevent embryos from sticking to pipettes.

Although many laboratories use gentamicin in embryo culture media, we do not always do so, but instead try to keep embryos healthy by maintaining proper density and replacing the medium daily. For some experiments we may use gentamicin (e.g., Corning 61-098-RF, 100 mg/mL in sterile H$_2$O, store at 4°C) at a dose of 50 µg/mL, but only for the first and possibly second day, because gentamicin is a well-known ototoxic antibiotic that damages hair cells in the inner ear (e.g., Lombarte et al. 1993), and *Xenopus* stereociliary bundle formation starts as early as at stage 31 (Quick and Serrano 2005).

MICROINJECTION OF *X. TROPICALIS* EMBRYOS

For microinjection we follow a method established in our laboratory (Ogino et al. 2006) using 6% Ficoll PM400 (e.g., GE Healthcare 17-0300-10) in 0.1× MBS (pH 7.5) with phenol red, which we filter-sterilize. Filtering allows any debris to be removed to avoid clogging injection needles and prevents contamination by microorganisms in the Ficoll solution (as mentioned in Protocol 5: Obtaining *Xenopus tropicalis* Embryos by In Vitro Fertilization [Lane and Khokha 2022a] and Protocol 7: Microinjection of *Xenopus tropicalis* Embryos [Lane et al. 2022b]). We freeze the Ficoll solution at −20°C for long-term storage. The pH can become acidic after storage or after a few days' use, which can be easily determined by monitoring the color change of phenol red. If needed, readjust pH before use. It is important to dejelly *X. tropicalis* eggs completely for microinjection using the nonhydrochloride form of L-cysteine (e.g., Sigma-Aldrich C5360; 2% in 0.1× MBS, pH 7.8–8.0) to perform efficient and fast dejellying.

RAISING AND USING *X. TROPICALIS* EMBRYOS AT DIFFERENT TEMPERATURES

Although it is advisable to raise *X. tropicalis* embryos in the range of 22°C–25°C, it is also possible to increase the temperature for short periods to 28°C to accelerate development for purposes of getting embryos to stages needed for experiments more quickly (e.g., once one batch of embryos reaches a certain stage, it can be split and cultured at a different temperature to obtain embryos of different ages simultaneously). Depending on the experimental purpose, however, one should be careful if comparing embryos that are raised at different temperatures because they may have different metabolic speeds internally, which could affect many biological properties (e.g., rate of cell determination [Ten Cate 1956] or gene expression profiles) even if their outer appearance (developmental stage, i.e., by Nieuwkoop and Faber 1967) appears the same.

    X. tropicalis embryos are tolerant of a wide range of temperatures (up to 32°C) for at least short periods. They can thus be used in recombination experiments with mammalian tissues to assess evolutionary conservation of signaling mechanisms—for example, by determining the ability of *X. tropicalis* embryonic tissue to be induced by mouse tissue toward a particular fate (e.g., Hirsch et al. 2002b).

ENHANCING TADPOLE HEALTH AND ACCELERATING TIME TO SEXUAL MATURITY

Every laboratory may require slightly different methods/protocols depending on the availability or limitations of equipment and systems and the size of the colony being raised and maintained. However, one of the most common needs for monitoring tadpole health is careful observation—not simply following written protocols blindly. Observation of young stage embryos is critical up to several times daily (as described in Sec. 1, Protocol 3: Raising and Maintaining *Xenopus tropicalis* from Tadpole to Adult [Lane et al. 2022c]). Dishes should be checked for overcrowding and pH level. If the medium becomes acidic, it should be changed immediately. Overfeeding of tadpoles should be avoided, as leftover food can be a source of bacteria growth, which quickly clouds the medium. Bacteria are difficult to remove once they become abundant. Additionally, we would like to emphasize a point that many people may not realize, that feces of tadpoles sometimes can be sticky, which can clog their gills, leading to death. It is therefore very important to clean up leftover food and feces as needed, especially for young tadpoles.

    In general, feeding more and cleaning more, accordingly, accelerates the growth of tadpoles, leading to faster maturation of frogs. Our observation suggests that growth conditions of tadpoles will affect the maturation of frogs, and that faster-growing tadpoles will become frogs that mature faster. Also, faster-growing tadpoles tend to become larger froglets. Under ideal circumstances, tadpoles could be froglets (i.e., stage 66 of Nieuwkoop and Faber 1967) as quickly as 6 wk after birth. In our preliminary experiments, we have seen that injection of froglets, which must be large enough to handle, with pregnant mare serum gonadotropin (PMSG; e.g., Prospec HOR-272; 50 units/froglet) accelerates maturation of frogs, especially females (T. Nakayama and R. Grainger, unpubl.).

Cite this protocol as *Cold Spring Harb Protoc*; doi:10.1101/pdb.top106252

We have not observed any harmful effects of PMSG injection, which could be a useful approach to accelerate maturation of frogs, although this approach needs more systematic study in the future.

OPTIMIZING *X. TROPICALIS* COLONY HEALTH FOR LONG-TERM PROPAGATION OF ANIMAL LINES

When raising inbred, transgenic, or mutant lines of *X. tropicalis*, it is especially important to ensure that animals used for experiments are healthy at the outset and remain so after treating or manipulating embryos. Assays for a wide range of *Xenopus* diseases are described in Sec. 1, Protocol 4: Defining the Specific Pathogen-Free State of *Xenopus* Using TaqMan Assays (Hensley et al. 2020).

Animals purchased from commercial sources should be verified to be pathogen-free if they will be used to make animal lines. Although some diseases are treatable (Green 2009), infections with other pathogens, notably *Mycobacterium liflandii* (Trott et al. 2004; Mve-Obiang et al. 2005), are at present impossible to treat effectively, nor can they be detected in asymptomatic animals. Although investigators who are using either *X. laevis* or *X. tropicalis* for transient experiments may not be as concerned about diseases like *M. liflandii*, which may only become evident if animals are stressed (as can occur during mating), it is essential that colonies intended for animal lines be housed separately from those animals that may potentially carry this pathogen.

We have developed a protocol for sterilizing embryos to raise and maintain animal lines, based on a procedure described by Peng et al. (1991) designed to surface sterilize embryos in preparation for subsequent isolation of viable cells to generate cell lines. Seventy percent ethanol, used in this procedure, has been shown to be an effective disinfection agent where it has been tested on mycobacterial species (e.g., Lind et al. 1986). This protocol has been used to generate clean (free of *M. liflandii*) lines in the Grainger laboratory for >15 yr, and for all *X. tropicalis* and *X. laevis* lines at the National *Xenopus* Resource (NXR) at Woods Hole. The goal of the protocol is to eliminate pathogens, like *M. liflandii*, from the entire colony. The procedure involves brief treatment of newly in-vitro-fertilized embryos with ethanol, followed by thimerosol treatment.

Begin with a single-layer sheet of eggs in a noncoated dish followed by in vitro fertilization. Once cleavage is confirmed, wash embryos with 70% ethanol by pouring and decanting for 10 sec. The timing of this step is critical: Embryos should be soaked in 70% ethanol no shorter or longer than 10 sec. A shorter time will result in embryos that may not be sterilized, and a longer time will affect embryo viability. Next, wash embryos with 0.1× MBS (pH 7.5) at least three times as quickly as possible by pouring and decanting the medium. Dejelly the embryos in 2% nonhydrochloride form L-cysteine in 0.1× MBS (pH 7.8–8.0) as quickly as possible (ideally in 5 min). Sort only healthy cleaving embryos into an agarose-coated dish (see above) containing culture medium (0.1× MBS, pH 7.5). Once all healthy embryos are sorted, move the embryos into thimerosal-containing medium (Sigma-Aldrich T8784; 0.1 mg/mL 0.1× MBS) in a coated dish, and rinse embryos by hand-swirling dish for 10 sec. Move the embryos into another fresh thimerosal-containing medium in a coated dish, and gently keep swirling the dish for 10–15 min on a rotator at room temperature. Wash the embryos with 0.1× MBS at least five times and sort only healthy embryos for culture. From this point forward, treat embryos with great care so they are not exposed in any way to potentially infected animals (e.g., via direct contact or indirect contact from laboratory staff that have recently handled potentially infected animals).

DISCUSSION

X. tropicalis husbandry has evolved considerably as the system has grown in its utility for generating and studying mutagenized lines that require large, complex facilities (e.g., the NXR and EXRC [The European *Xenopus* Resource Centre]) for housing stocks. Key steps along the way included the recognition that it was necessary to develop refinements to traditional husbandry protocols for amphibians—for example, the discovery that Sera micron, a high-protein algal food source for raising tadpoles, avoids many of the problems associated with earlier nutritional sources (e.g., Hirsch et al. 2002a; see Showell and Conlon 2009). Early procedures have now evolved into fine-tuned protocols including Sec. 1, Protocol 3: Raising and Maintaining *Xenopus tropicalis* from Tadpole to Adult (Lane et al. 2022c). It is important for investigators to appreciate that variations remain to be found that will continue to increase husbandry efficacy, and careful observation is still required to raise healthy animals, as described herein.

ACKNOWLEDGMENTS

This work was funded by National Institutes of Health (NIH) grants EY06675, EY10283, EY022954, and RR13221 and research awards from the Sharon Stewart Aniridia Trust and Vision for Tomorrow to R.M.G. All of our experimental protocols for *X. tropicalis* are the consequence of multiple modifications for improvement and efficiency of procedures over decades of work in our laboratory. For their contributions to this endeavor, we acknowledge all past members of the Grainger lab, particularly for the initial and seminal work of Lyle B. Zimmerman, as well as notable contributions from Marilyn Fisher, Martin F. Offield, Jessica Gray, Nicolas Hirsch, Hajime Ogino, and Margaret B. Fish. The authors also acknowledge Xenbase (RRID:SCR_003280) for providing essential online resources for *Xenopus* research and NXR (RRID:SCR_013731) and EXRC for providing healthy animals.

REFERENCES

Blow JJ, Laskey RA. 1986. Initiation of DNA replication in nuclei and purified DNA by a cell-free extract of *Xenopus* eggs. *Cell* **47:** 577–587. doi:10.1016/0092-8674(86)90622-7

Green SL. 2009. *The laboratory* Xenopus *sp. (laboratory animal pocket reference)*, 1st ed. CRC Press, Boca Raton, FL.

Hensley CL, Bowes KM, Feldman SH. 2020. Defining the specific pathogen-free state of *Xenopus* using TaqMan assays. *Cold Spring Harb Protoc* doi:10.1101/pdb.prot106179

Hirsch N, Zimmerman LB, Grainger RM. 2002a. *Xenopus*, the next generation: *X. tropicalis* genetics and genomics. *Dev Dyn* **225:** 422–433. doi:10.1002/dvdy.10178

Hirsch N, Zimmerman LB, Gray J, Chae J, Curran KL, Fisher M, Ogino H, Grainger RM. 2002b. *Xenopus tropicalis* transgenic lines and their use in the study of embryonic induction. *Dev Dyn* **225:** 522–535. doi:10.1002/dvdy.10188

Kosubek A, Klein-Hitpass L, Rademacher K, Horsthemke B, Ryffel GU. 2010. Aging of *Xenopus tropicalis* eggs leads to deadenylation of a specific set of maternal mRNAs and loss of developmental potential. *PLoS One* **5:** e13532. doi:10.1371/journal.pone.0013532

Lane M, Khokha MK. 2022a. Obtaining *Xenopus tropicalis* embryos by in vitro fertilization. *Cold Spring Harb Protoc* doi:10.1101/pdb.prot106351

Lane M, Khokha MK. 2022b. Obtaining *Xenopus tropicalis* embryos by natural mating. *Cold Spring Harb Protoc* doi:10.1101/pdb.prot106609

Lane M, Mis EK, Khokha MK. 2022a. Obtaining *Xenopus tropicalis* eggs. *Cold Spring Harb Protoc* doi:10.1101/pdb.prot106344

Lane M, Mis EK, Khokha MK. 2022b. Microinjection of *Xenopus tropicalis* embryos. *Cold Spring Harb Protoc* doi:10.1101/pdb.prot107644

Lane M, Slocum M, Khokha MK. 2022c. Raising and maintaining *Xenopus tropicalis* from tadpole to adult. *Cold Spring Harb Protoc* doi:10.1101/pdb.prot106369

Leno GH, Laskey RA. 1991. DNA replication in cell-free extracts from *Xenopus laevis*. *Method Cell Biol* **36:** 561–579. doi:10.1016/S0091-679X(08)60297-6

Lind A, Lundholm M, Pedersen G, Sundaeus V, Wåhlén P. 1986. A carrier method for the assessment of the effectiveness of disinfectants against *Mycobacterium tuberculosis*. *J Hosp Infect* **7:** 60–67. doi:10.1016/0195-6701(86)90027-7

Lombarte A, Yan HY, Popper AN, Chang JS, Platt C. 1993. Damage and regeneration of hair cell ciliary bundles in a fish ear following treatment with gentamicin. *Hear Res* **64:** 166–174. doi:10.1016/0378-5955(93)90002-I

Mve-Obiang A, Lee RE, Umstot ES, Trott KA, Grammer TC, Parker JM, Ranger BS, Grainger R, Mahrous EA, Small PLC. 2005. A newly discovered mycobacterial pathogen isolated from laboratory colonies of *Xenopus* species with lethal infections produces a novel form of mycolactone, the *Mycobacterium ulcerans* macrolide toxin. *Infection Immun* **73:** 3307–3312. doi:10.1128/IAI.73.6.3307-3312.2005

Nakayama T, Cox A, Howell M, Grainger RM. 2022. Gynogenetic production of embryos in *Xenopus tropicalis* using a cold shock procedure: rapid screening method for gene editing phenotypes. *Cold Spring Harb Protoc* doi:10.1101/pdb.prot107648

Nieuwkoop PD, Faber J. 1967. *Normal table of* Xenopus laevis *(Daudin)*. Garland, New York, NY.

Peng HB, Baker LP, Chen Q. 1991. Tissue culture of *Xenopus* neurons and muscle cells as a model for studying synaptic induction. *Meth Cell Biol* **36:** 511–526. doi:10.1016/S0091-679X(08)60294-0

Ogino H, McConnell WB, Grainger RM. 2006. High-throughput transgenesis in *Xenopus* using I-SceI meganuclease. *Nat Protoc* **1:** 1703–1710. doi:10.1038/nprot.2006.208

Quick QA, Serrano EE. 2005. Inner ear formation during the early larval development of *Xenopus laevis*. *Dev Dyn* **234:** 791–801. doi:10.1002/dvdy.20610

Showell C, Conlon FL. 2009. Natural mating and tadpole husbandry in the western clawed frog *Xenopus tropicalis*. *Cold Spring Harb Protoc* doi:10.1101/pdb.prot5292

Sive HL, Grainger RM, Harland RM. 2007. *Xenopus laevis* egg collection. *Cold Spring Harb Protoc* doi:10.1101/pdb.prot4736

Ten Cate G. 1956. *The intrinsic embryonic development*, pp. 137–176. North Holland, Amsterdam.

Trott KA, Stacy BA, Lifland BD, Diggs HE, Harland RM, Khokha MK, Grammer TC, Parker JM. 2004. Characterization of a *Mycobacterium ulcerans*–like infection in a colony of African tropical clawed frogs (*Xenopus tropicalis*). *Comp Med* **54:** 309–317.

Using the *Xenopus* Oocyte Toolbox

Kimberly L. Mowry[1]

Department of Molecular Biology, Cell Biology, and Biochemistry, Brown University, Box G-L268, Providence, Rhode Island 02912

The *Xenopus* oocyte is a unique model system, allowing both the study of complex biological processes within a cellular context through expression of exogenous mRNAs and proteins, and the study of the cell, molecular, and developmental biology of the oocyte itself. During oogenesis, *Xenopus* oocytes grow dramatically in size, with a mature oocyte having a diameter of ~1.3 mm, and become highly polarized, localizing many mRNAs and proteins. Thus, the mature oocyte is a repository of maternal mRNAs and proteins that will direct early embryogenesis prior to zygotic genome transcription. Importantly, the *Xenopus* oocyte also has the capacity to translate exogenous microinjected RNAs, which has enabled breakthroughs in a wide range of areas including cell biology, developmental biology, molecular biology, and physiology. This introduction outlines how *Xenopus* oocytes can be used to study a variety of important biological questions.

BACKGROUND

The *Xenopus* oocyte has been called a "living test tube" (de Robertis et al. 1977) and has proven to be an invaluable system for investigating a myriad of biological questions. Research using *Xenopus* oocytes has revolutionized the fields of cell and molecular biology, resulting in many "firsts" including, but not limited to, the identification of the nucleolus as the site of ribosomal RNA transcription (Brown and Gurdon 1964), the isolation of a eukaryotic gene (Birnstiel et al. 1968; Brown and Weber 1968), in situ hybridization (Gall and Pardue 1969), in vivo expression of exogenous mRNA (Gurdon et al. 1971), transcription of a cloned eukaryotic gene (Brown and Gurdon 1977), sequencing of a eukaryotic gene (Fedoroff and Brown 1978), and isolation of a eukaryotic transcription factor (Engelke et al. 1980). The large size of the *Xenopus* oocyte and the availability of large quantities of oocytes continues to facilitate microinjection approaches to study cellular functions of exogenous proteins, as well as biochemical approaches to study fundamental cellular processes.

OOGENESIS

As depicted in Figure 1, oogenesis in *Xenopus* is divided into six stages (I–VI) based on size and morphology (Dumont 1972). Stage I oocytes are transparent, lacking yolk, and display the first hallmark of polarity along the animal-vegetal (AV) axis: The Balbiani Body, or mitochondrial cloud (Heasman et al. 1984), is present on the vegetal side of the nucleus or germinal vesicle (GV). Oocytes grow dramatically during oogenesis—from 50 μm at the beginning of stage I to 1.3 mm by the end of

[1]Correspondence: kimberly_mowry@brown.edu

Cite this introduction as *Cold Spring Harb Protoc*; doi:10.1101/pdb.top095844

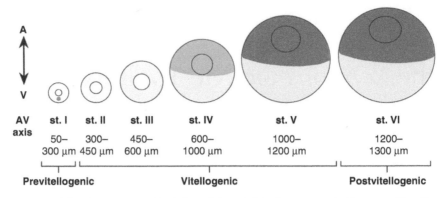

FIGURE 1. Depiction of the six stages (st.; I–VI, from *left* to *right*) of oogenesis in *Xenopus*. The diameters of each oocyte, stages of vitellogenesis (or yolk accumulation, and the animal-vegetal (AV) axis are shown. The nuclei (or germinal vesicles [GVs]) are indicated by circles, and the Balbiani Body in the stage I oocyte is depicted in gray on the vegetal side of the GV, which is in the center. In the stage IV oocyte, the GV is displaced toward the animal hemisphere, where accumulating pigment is indicated by brown shading. The GV is further displaced toward the animal hemisphere in stage V and VI oocytes, where accumulated pigment is indicated by dark brown shading and increased yolk accumulation is indicated by yellow shading.

stage VI—with much of this growth due to the accumulation of yolk during vitellogenesis, which refers to the uptake of the yolk precursor protein vitellogenin (Wallace and Dumont 1968). Vitellogenesis begins in stage II oocytes and continues through stage V of oogenesis (Dumont 1972). By the end of oogenesis (stage VI) the vegetal yolk mass comprises 70% of total oocyte yolk protein (Danilchik and Gerhart 1987) and 80% of total oocyte protein (Wiley and Wallace 1981). Oocyte pigmentation begins in stage III, with polarization of the pigment first evident in stage IV and increasing through stage VI, such that the animal hemisphere is darkly pigmented in fully grown oocytes (Dumont 1972).

In Protocol 1: Isolation of *Xenopus* Oocytes (Newman et al. 2018), procedures for the isolation of various stages of *Xenopus* oocytes are detailed, and each of the other accompanying protocols uses distinct stages of oogenesis. Protocol 7: Whole-Mount In Situ Hybridization of *Xenopus* Oocytes (Bauermeister and Pieler 2018), Protocol 9: Fluorescence In Situ Hybridization of Cryosectioned *Xenopus* Oocytes (Neil and Mowry 2018), and Protocol 8: Whole-Mount Immunofluorescence for Visualizing Endogenous Protein and Injected RNA in *Xenopus* Oocytes (Jeschonek and Mowry 2018), use stage II–IV oocytes. Protocol 2: Isolation and Analysis of *Xenopus* Germinal Vesicles (Morgan 2018) uses stage IV–V oocytes, and Protocol 5: Heterologous Protein Expression in the *Xenopus* Oocyte (Marchant 2018), Protocol 6: Patch-Clamp and Perfusion Techniques to Study Ion Channels Expressed in *Xenopus* Oocytes (Zhang and Cui 2018), and Protocol 4: Oocyte Host-Transfer and Maternal mRNA Depletion Experiments in *Xenopus* (Houston 2018) use stage VI oocytes. Protocol 3: Microinjection of *Xenopus* Oocytes (Aguero et al. 2018) describes procedures for microinjecting DNA or RNA molecules into stage VI *Xenopus* oocytes.

USING *XENOPUS* OOCYTES TO STUDY POLARITY AND EARLY DEVELOPMENT

During oogenesis, *Xenopus* oocytes synthesize the large store of maternal mRNAs and proteins that will be necessary to program early embryogenesis until zygotic transcription begins at the mid-blastula transition (Newport and Kirschner 1982). The highly active transcription required for this results in lampbrush chromosomes, which are a topic of Protocol 2: Isolation and Analysis of *Xenopus* Germinal Vesicles (Morgan 2018). Additionally, during oogenesis, a subset of mRNAs are polarized along the AV axis, providing the basis for developmental polarity in the oocyte and patterning in the embryo (King et al. 2005; Medioni et al. 2012). *Xenopus* oocytes have been an important model system for studying the mechanisms driving mRNA transport and localization (Houston 2013), and several accompanying protocols are relevant to this topic: Protocol 7: Whole-Mount In Situ Hybridization of *Xenopus* Oocytes (Bauermeister and Pieler 2018) presents whole-mount in situ hybridization as a technique for spatial analysis of RNA expression in oocytes. Protocol 8: Whole-Mount Immunofluorescence for Visualizing Endogenous Protein and Injected RNA in *Xenopus* Oocytes (Jeschonek and Mowry 2018) details techniques for analyzing spatial distribution of both endogenous proteins and microinjected RNAs. Protocol 9: Fluorescence In Situ Hybridization of Cryosectioned *Xenopus* Oocytes (Neil and Mowry 2018) describes a FISH approach, combined with cryosectioning, for high-resolution imaging of endogenous RNA distribution in oocytes. Analyzing the contributions of maternally inherited mRNAs to early embryogenesis is of critical importance in *Xenopus* and many systems. Protocol 4: Oocyte Host-Transfer and Maternal mRNA Depletion Experiments in *Xenopus* (Houston 2018) details procedures for using host-transfer techniques to assess effects on embryonic development after manipulating expression of maternal mRNAs in oocytes.

USING *XENOPUS* OOCYTES TO STUDY PHYSIOLOGY, CELL, AND MOLECULAR BIOLOGY

Xenopus oocytes have long been an invaluable tool for expression and analysis of heterologous proteins and, as noted above, inspired characterization as a "living test tube" (de Robertis et al. 1977). Proteins produced from mRNA microinjected into the cytoplasm (Gurdon et al. 1971) or DNA microinjected into the nucleus (Mertz and Gurdon 1977) are robustly expressed and appropriately processed and trafficked. The development of in vitro procedures for production of synthetic RNAs (Krieg and Melton 1984) further expanded the versatility and effectiveness of the oocyte as a system for analyzing the function of expressed proteins. Protocol 3: Microinjection of *Xenopus* Oocytes (Aguero et al. 2018) provides procedures for microinjection of *Xenopus* oocytes and Protocol 5: Heterologous Protein Expression in the *Xenopus* Oocyte (Marchant 2018) details procedures for nuclear injection of cDNA constructs for expression of heterologous proteins in oocytes. Both of these protocols can be adapted to study any gene of interest. The large size of the oocyte is of particular benefit to electrophysiological approaches to study channel proteins, such as acetylcholine receptors (Miledi et al. 1982) and many others (Dascal 1987). In Protocol 6: Patch-Clamp and Perfusion Techniques to Study Ion Channels Expressed in *Xenopus* Oocytes, Zhang and Cui (2018) describe the use of *Xenopus* oocytes to study heterologously expressed ion channels. The protocol can be applied to study a variety of ion channels and neurotransmitter receptors. Together, these protocols provide a toolbox for using *Xenopus* oocytes to carry out functional studies for an investigator's protein of interest.

SUMMARY

For many decades, the *Xenopus* oocyte has proven to be an invaluable system for addressing fundamental biological questions. The protocols introduced here will assist investigators in continuing to use and study this remarkable cell.

ACKNOWLEDGMENTS

I thank S.E. Cabral and L.C. O'Connell for comments on the manuscript. Work in my laboratory using *Xenopus* oocytes is supported by a grant (GM071049) from the National Institutes of Health.

REFERENCES

Aguero T, Newman K, King ML. 2018. Microinjection of *Xenopus* oocytes. *Cold Spring Harb Protoc* doi:10.1101/pdb.prot096974

Bauermeister D, Pieler T. 2018. Whole-mount in situ hybridization of *Xenopus* oocytes. *Cold Spring Harb Protoc* doi:10.1101/pdb.prot097014

Birnstiel M, Speirs J, Purdom I, Jones K, Loening UE. 1968. Properties and composition of the isolated ribosomal DNA satellite of *Xenopus laevis*. *Nature* **219**: 454–463. doi:10.1038/219454a0

Brown DD, Gurdon JB. 1964. Absence of ribosomal RNA synthesis in the anucleolate mutant of *Xenopus laevis*. *Proc Natl Acad Sci* **51**: 139–146. doi:10.1073/pnas.51.1.139

Brown DD, Gurdon JB. 1977. High-fidelity transcription of 5S DNA injected into *Xenopus* oocytes. *Proc Natl Acad Sci* **74**: 2064–2068. doi:10.1073/pnas.74.5.2064

Brown DD, Weber CS. 1968. Gene linkage by RNA–DNA hybridization. II. Arrangement of the redundant gene sequences for 28s and 18s ribosomal RNA. *J Mol Biol* **34**: 681–697. doi:10.1016/0022-2836(68)90189-7

Danilchik MV, Gerhart JC. 1987. Differentiation of the animal-vegetal axis in *Xenopus laevis* oocytes. I. Polarized intracellular translocation of platelets establishes the yolk gradient. *Dev Biol* **122**: 101–112. doi:10.1016/0012-1606(87)90336-8

Dascal N. 1987. The use of *Xenopus* oocytes for the study of ion channels. *CRC Crit Rev Biochem* **22**: 317–387. doi:10.3109/10409238709086960

de Robertis EM, Gurdon JB, Partington GA, Mertz JE, Laskey RA. 1977. Injected amphibian oocytes: a living test tube for the study of eukaryotic gene transcription? *Biochem Soc Symp* **42**: 181–191.

Dumont JN. 1972. Oogenesis in *Xenopus laevis* (Daudin). I. Stages of oocyte development in laboratory maintained animals. *J Morphol* **136**: 153–179. doi:10.1002/jmor.1051360203

Engelke DR, Ng SY, Shastry BS, Roeder RG. 1980. Specific interaction of a purified transcription factor with an internal control region of 5S RNA genes. *Cell* **19**: 717–728. doi:10.1016/S0092-8674(80)80048-1

Fedoroff NV, Brown DD. 1978. The nucleotide sequence of oocyte 5S DNA in *Xenopus laevis*. I. The AT-rich spacer. *Cell* **13**: 701–716. doi:10.1016/0092-8674(78)90220-9

Gall JG, Pardue ML. 1969. Formation and detection of RNA-DNA hybrid molecules in cytological preparations. *Proc Natl Acad Sci* **63**: 378–383. doi:10.1073/pnas.63.2.378

Gurdon JB, Lane CD, Woodland HR, Marbaix G. 1971. Use of frog eggs and oocytes for the study of messenger RNA and its translation in living cells. *Nature* **233**: 177–182. doi:10.1038/233177a0

Heasman J, Quarmby J, Wylie CC. 1984. The mitochondrial cloud of *Xenopus* oocytes: the source of germinal granule material. *Dev Biol* **105**: 458–469. doi:10.1016/0012-1606(84)90303-8

Houston DW. 2013. Regulation of cell polarity and RNA localization in vertebrate oocytes. *Int Rev Cell Mol Biol* **306**: 127–185. doi:10.1016/B978-0-12-407694-5.00004-3

Houston DW. 2018. Oocyte host-transfer and maternal mRNA depletion experiments in *Xenopus*. *Cold Spring Harb Protoc* doi:10.1101/pdb.prot096982

Jeschonek SP, Mowry KL. 2018. Whole-mount immunofluorescence for visualizing endogenous protein and injected RNA in *Xenopus* oocytes. *Cold Spring Harb Protoc* doi:10.1101/pdb.prot097022

King ML, Messitt TJ, Mowry KL. 2005. Putting RNAs in the right place at the right time: RNA localization in the frog oocyte. *Biol Cell* **97**: 19–33. doi:10.1042/BC20040067

Krieg PA, Melton DA. 1984. Functional messenger RNAs are produced by SP6 *in vitro* transcription of cloned cDNAs. *Nucleic Acids Res* **12**: 7057–7070. doi:10.1093/nar/12.18.7057

Marchant JS. 2018. Heterologous protein expression in the *Xenopus* oocyte. *Cold Spring Harb Protoc* doi:10.1101/pdb.prot096990

Medioni C, Mowry K, Besse F. 2012. Principles and roles of mRNA localization in animal development. *Development* **139**: 3263–3276. doi:10.1242/dev.078626

Mertz JE, Gurdon JB. 1977. Purified DNAs are transcribed after microinjection into *Xenopus* oocytes. *Proc Natl Acad Sci* **74**: 1502–1506. doi:10.1073/pnas.74.4.1502

Miledi R, Parker I, Sumikawa K. 1982. Properties of acetylcholine receptors translated by cat muscle mRNA in *Xenopus* oocytes. *EMBO J* **1**: 1307–1312. doi:10.1002/j.1460-2075.1982.tb01315.x

Morgan GT. 2018. Isolation and analysis of *Xenopus* germinal vesicles. *Cold Spring Harb Protoc* doi:10.1101/pdb.prot096958

Neil CR, Mowry K. 2018. Fluorescence in situ hybridization of cryosectioned *Xenopus* oocytes. *Cold Spring Harb Protoc* doi:10.1101/pdb.prot097030

Newman K, Aguero T, King ML. 2018. Isolation of *Xenopus* oocytes. *Cold Spring Harb Protoc* doi:10.1101/pdb.prot095851

Newport J, Kirschner M. 1982. A major developmental transition in early *Xenopus* embryos: I. characterization and timing of cellular changes at the midblastula stage. *Cell* **30**: 675–686. doi:10.1016/0092-8674(82)90272-0

Wallace RA, Dumont JN. 1968. The induced synthesis and transport of yolk proteins and their accumulation by the oocyte in *Xenopus laevis*. *J Cell Physiol* **72**: 73–89. doi:10.1002/jcp.1040720407

Wiley HS, Wallace RA. 1981. The structure of vitellogenin. Multiple vitellogenins in *Xenopus laevis* give rise to multiple forms of the yolk proteins. *J Biol Chem* **256**: 8626–8634.

Zhang G, Cui J. 2018. Patch-clamp and perfusion techniques to study ion channels expressed in *Xenopus* oocytes. *Cold Spring Harb Protoc* doi:10.1101/pdb.prot099051

Cite this introduction as *Cold Spring Harb Protoc*; doi:10.1101/pdb.top095844

Isolation of *Xenopus* Oocytes

Karen Newman, Tristan Aguero, and Mary Lou King[1]

Department of Cell Biology, Miller School of Medicine, University of Miami, Miami, Florida 33136

Xenopus oocytes and oocyte extracts are the starting material for a variety of experimental approaches. Oocytes are obtained by surgical removal of the ovary from anesthetized females. Although oocytes may be used while they remain within their ovarian follicle, it is more practical to work with defolliculated oocytes. Defolliculation can be performed either manually or enzymatically. Here we present a protocol for the isolation and separation of *Xenopus* oocytes at various developmental stages, and guidelines for maintaining oocytes in culture.

MATERIALS

It is essential that you consult the appropriate Material Safety Data Sheets and your institution's Environmental Health and Safety Office for proper handling of equipment and hazardous materials used in this protocol.

RECIPES: Please see the end of this protocol for recipes indicated by <R>. Additional recipes can be found online at http://cshprotocols.cshlp.org/site/recipes.

Reagents

Antibiotics

    Gentamycin (1000× stock solution [10 mg/mL in H_2O])

    Streptomycin (1000× stock solution [100 mg/mL in H_2O])

Collagenase (Type 1; Worthington Biochemical) or Liberase TM (Roche Diagnostics) (see Step 5)

The Type 1 collagenase preparation from Worthington is a crude product known to contain collagenases I and II as well as caseinase, clostripain, and tryptic activity. Liberase TM contains highly purified collagenases I and II blended with Thermolysin, a nonclostridial neutral protease. Liberase is preferred for many applications because it is purified and has significantly higher specific activities with lot-to-lot consistency.

Medium for oocyte culture

    Modified Barth's saline (MBS) for oocytes <R>

    Oocyte culture medium (OCM) <R> (for long-term culture; see Step 11)

A number of formulations exist for OCM (Wallace and Misulovin 1978; Opresko et al. 1980; Wylie et al. 1996). The formulation cited here is used for host transfer studies (Schneider et al. 2010).

    Oocyte reagent 2 (O-R2) <R> (Wallace et al. 1973)

The choice of medium (MBS or OR-2) depends upon what the oocytes will be used for experimentally. For longer-term culture (Step 11), OCM containing antibiotics and bovine serum albumin (BSA) is useful to maintain healthy oocytes.

Xenopus (anesthetized females)

---

[1]Correspondence: mking@med.miami.edu

The selection of the donor female will depend on the desired stage of oocytes. In Xenopus, oogenesis is divided into six stages, each identified by a Roman numeral (Dumont 1972). Sexually mature females will yield higher numbers of stage V–VI oocytes, whereas juvenile females may have predominantly early-stage (I–II) oocytes with few, if any, later staged oocytes. Early stages of oocytes have been ideal to image RNA localization (Heinrich and Deshler 2009; Gagnon and Mowry 2010); extracts from late-stage oocytes have been used to study the cell cycle (Novak and Tyson 1993; Iwabuchi et al. 2000) and signal transduction (Crane and Ruderman 2006).

Equipment

Culture dishes (100-mm [glass] and 150-mm [plastic])
Dissecting microscope and dual gooseneck LED illuminators (low heat)
Forceps (Dumont #4 or #5)

For manual defolliculation (Step 3), two pairs of forceps are used: either two sharpened or one sharpened and one shaped like blunt pliers, according to preference.

Glass bowl for sequential sieving (170 mm in diameter, 90 mm high)
Incubator at 18°C
Sieves (Pro 4" or Low Pro 4") with 475- and 1000-µm mesh (Aquaculture Nursery Farms) (see Step 9. ii)

Nylon mesh filters are also available from Cole-Parmer (U.S. Mesh 18,1000 µm; U.S. Mesh 35,500 µm).

METHOD

Oocyte Harvest and Defolliculation

Defolliculation can be performed manually (Step 3) or by enzymatic digestion (Steps 4–8).

1. Harvest ovarian tissue by surgery from anesthetized female frogs.

2. After surgical harvest, place the ovarian tissue in MBS and tease apart with forceps to expose the oocytes for defolliculation (Fig. 1A,A′). Examine the oocytes before defolliculation to determine the quality, and proceed only if the oocytes are healthy.

   See Step 11 for a description of oocyte quality.

   See Troubleshooting.

Manual Defolliculation

Manual defolliculation is recommended for experiments requiring only a few to a few hundred oocytes and for host transfer experiments (Schneider et al. 2010). Manually defolliculated oocytes can be used immediately for experiments.

Oocytes are surrounded by a vitelline membrane, follicle cells, and connective tissue rich in blood vessels (theca). Manual defolliculation is unlikely to remove the innermost layer of follicle cells. At a protein level, this cellular contamination is likely inconsequential, but may be significant at the DNA level.

3. Perform manual defolliculation using two pairs of forceps.

   This procedure is well-described and shown by Schneider et al. (2010) (see http://www.jove.com/video/ 1864/fertilization-of-xenopus-oocytes-using-the-host-transfer-method) and Miyamoto et al. (2015) (see http://www.jove.com/video/52496/manipulation-vitro-maturation-xenopus-laevis-oocytes-followed).

Enzymatic Defolliculation

4. Cut the teased ovarian tissue into small pieces (0.5–1 cm in diameter) for enzymatic digestion.

5. Transfer ~10 mL of ovarian tissue pieces to 25 mL of enzyme solution (for a total volume of 35 mL). For enzyme solution, use either Type 1 collagenase diluted in medium to 2 mg/mL (~0.23 WU/mL; 9.2 WU/25 mL) or Liberase TM diluted in medium to ~0.26 WU/mL (0.35 mL/25 mL).

   Calcium is required for optimal collagenase activity and is included in the medium.

   Although enzymatic defolliculation is mainly targeting collagen fibers comprising the theca, the most effective products are not pure collagenase but a preparation that contains a mixture of proteases.

Cite this protocol as *Cold Spring Harb Protoc*; doi:10.1101/pdb.prot095851

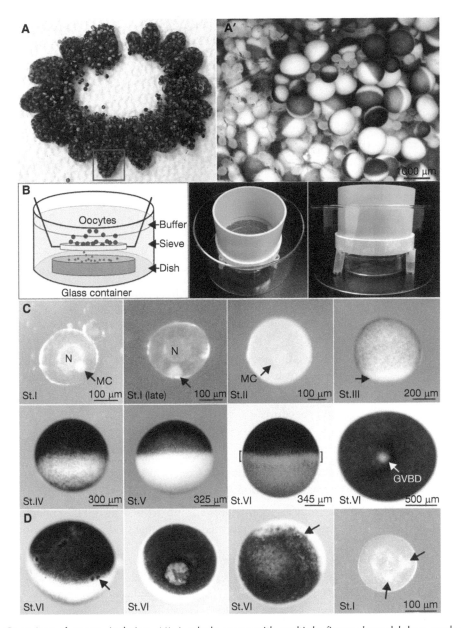

FIGURE 1. Overview of oocyte isolation. (*A*) A whole ovary with multiple finger-shaped lobes, each containing hundreds of oocytes, and (*A′*) a partially opened lobe showing oocytes in follicles at various oogenic stages. (*B*) The sieving apparatus: a schematic depiction, and top and lateral photographic views with a 1000-µm sieve ideal to isolate stage VI oocytes. A smaller culture dish is placed beneath the sieve within a larger glass dish, and medium (O-R2 or MBS) is added until the mesh is submerged. (*C*) Healthy oocytes (*left* to *right*): Stage (St.) I (50–300 µm), a white, compact Balbiani body or mitochondrial cloud (MC) (arrow) in close contact with the nucleus (N); lateral view of late stage I with MC integrated with the vegetal cortex (arrow); Stage II (300–450 µm), yolk deposition begins and MC (arrow) associated with vegetal cortex; Stage III (450–600 µm) lateral view, melanogenesis begins, vegetal pole (arrow); Stages IV through VI (600–1250 µm), uniform color, smooth "velvet" texture at the animal pole and relatively unpigmented vegetal pole, Stage VI color ranges from pale to dark brown, often displays a light equatorial band (brackets); Matured stage VI oocyte (animal pole view) showing germinal vesicle breakdown (GVBD). (*D*) Dying/unhealthy oocytes (*left* to *right*): Stages III through VI, dark (microbial contamination) or white spots (cellular degeneration or necrosis) evident (arrows) with uneven pigmentation; Stage I diffuse MC (arrows).

6. Digest the tissue pieces at room temperature (22°C–24°C) with rotation. Monitor digestion every 30 min to assess progress, as overtreatment with the enzyme can irreversibly damage the oocytes making them unsuitable for use.

This process typically takes 1.5–2.5 h. Digesting at a higher temperature (25°C–30°C) speeds the process, however temperatures above 24°C activate heat-shock processes. It is uncertain if oocytes fully recover from heat-shock even after overnight recovery times (Davis and King 1989).

Stage V–VI oocytes are released first by enzymatic treatment. If earlier stage oocytes are desired, the clumps of tissue can be further digested.

7. End digestion by pipetting the freed oocytes into 150-mm plastic culture dishes containing 100 mL of medium.

8. Wash oocytes twice with 100 mL of medium per wash.

Isolation of Oocyte Stages

9. Sort the different stages of oocytes by hand or using a sequential sieving process as follows.

   i. Place a 100-mm glass culture dish beneath a sieve within a larger glass bowl (Fig. 1B). Add medium until the mesh filter of the sieving apparatus is submerged.

   ii. Gently pipette the cells onto the submerged mesh filter.

   The glass culture dish catches the oocytes that pass through the filter. Larger oocytes (Stages V–VI) are collected on a 1000-µm filter with Stage I–IV oocytes passing through. Stages III–IV are collected on a 475-µm filter with Stage I–II and some early Stage III oocytes passing through.

   iii. Retrieve the oocytes in the culture dish by removing the buffer from the glass bowl until the culture dish can be removed without disturbing the cells.

   iv. Transfer oocytes collected on the filter to culture dishes containing medium throughout the procedure.

Oocyte Maintenance in Culture

10. Culture defolliculated oocytes overnight in medium at 18°C to allow recovery from the effects of the enzymatic digestion.

11. Culture oocytes in medium for up to 4 d at 18°C, changing medium twice daily. Remove degrading cells with each medium change to prevent deleterious effects on the healthy cells.

   The addition of antibiotics and BSA to the culture medium (as included in OCM) is useful to maintain healthy oocytes during longer-term culturing.

   Overcrowding will cause the oocytes to degrade and die. Generally, a 100-mm culture dish holds thousands of stages I–II oocytes, ~1000 oocytes at stages III–IV, and up to 500 oocytes at stages V–VI.

   Under a dissecting microscope with epi-illumination (Fig. 1C), healthy stage I oocytes will have a white, compact mitochondrial cloud (also called Balbiani body); stage II oocytes will have a disc/cone of mitochondrial cloud material attached to the vegetal cortex; the animal hemisphere of stages III through VI will have a smooth, velvety color; stage VI oocytes will often display a lighter band between the animal and vegetal pole. The color of the animal pole can vary from a light gray in stage III oocytes to almost black in later stages; healthy stage VI oocytes can range in color from pale to dark brown. Dying or unhealthy oocytes in stages III through VI will present dark or white spots and mottled or uneven pigmentation at the animal pole. Unhealthy stages I and II oocytes will have diffuse mitochondrial clouds and/or a deflated appearance (Fig. 1D).

   See Troubleshooting.

TROUBLESHOOTING

Problem (Step 2): Harvested oocyte quality is poor.
Solution: Inadequate frog maintenance (water quality, crowded tanks, unsuitable food) (Hilken et al. 1995) or parasitic infestation, such as flukes, are common reasons for poor oocyte quality. Proceed with a different frog.

Cite this protocol as *Cold Spring Harb Protoc*; doi:10.1101/pdb.prot095851

Problem (Step 11): Oocyte degradation is observed during culture.

Solution: Solutions at the incorrect pH, extended enzymatic treatment during defolliculation, crowded culture dishes, or microbial contamination (Elsner et al. 2000; O'Connell et al. 2011) can contribute to oocyte degradation in culture. Add antibiotics to the medium, such as streptomycin or gentamycin (100 µg/mL), and check the pH. Reduce the cell density in the culture dishes. Remove degrading cells with each change of medium.

RECIPES

Modified Barth's Saline (MBS) for Oocytes

1. Prepare a 0.1 M CaCl₂ solution by dissolving 1.11 g of $CaCl_2$ in deionized H_2O for a final volume of 100 mL.

2. Prepare a 10× MBS salt solution by dissolving the following reagents in deionized H_2O for a final volume of 1 L. Adjust the pH to 7.8 with 10 M NaOH, and sterilize by autoclaving. Store indefinitely at room temperature.

HEPES	11.92 g
NaCl	51.43 g
KCl	0.75 g
$NaHCO_3$	2.10 g
$MgSO_4 \cdot 7H_2O$	2.46 g

3. To prepare 1× MBS, combine 100 mL of 10× MBS salt solution and 7 mL of 0.1 M $CaCl_2$ with 893 mL of autoclaved deionized H_2O. Add antibiotics. Store for up to 1–2 mo at 16°C–18°C (the same temperature used for oocyte incubation). Check before use.

   The final reagent concentrations in 1× MBS are 0.7 mM $CaCl_2$, 5 mM HEPES, 88 mM NaCl, 1 mM KCl, 2.5 mM $NaHCO_3$, and 1 mM $MgSO_4$.

Oocyte Culture Medium (OCM)

Reagent	Amount to add	Final concentration
Leibovitz's L-15 medium with L-glutamine	0.35 L	70%
Bovine serum albumin (fraction V)	0.2 g	0.4 mg/mL
Penicillin-streptomycin (100×/100×, P4333 Sigma-Aldrich)	5 mL	100 U/mL penicillin
		0.1 mg/mL streptomycin

Bring the volume to 0.5 L with deionized H_2O, and adjust the pH to 7.8 with NaOH. Filter to sterilize. Store at 4°C–8°C for up to 1–2 mo. Check before use.

Oocyte Reagent 2 (O-R2)

1. Prepare Stock Solution A by dissolving the following reagents in deionized H_2O for a final volume of 1 L. Store indefinitely at 4°C–8°C.

HEPES	11.915 g
NaCl	48.221 g
KCl	1.864 g
$CaCl_2 \cdot 2H_2O$	1.47 g
$MgCl_2 \cdot 6H_2O$	2.03 g
NaOH	1.520 g

2. Prepare Stock Solution B by dissolving 1.420 g of Na_2HPO_4 in deionized H_2O for a final volume of 1 L. Store indefinitely at 4°C–8°C.

3. Combine 100 mL of Stock Solution A, 100 mL of Stock Solution B, and 800 mL of deionized H_2O. Add antibiotics. Store for up to 1–2 mo at 16°C–18°C (the same temperature used for oocyte incubation). Check before use.

The pH should be 7.8. The final reagent concentrations are 5 mM HEPES, 82.5 mM NaCl, 2.5 mM KCl, 1 mM $CaCl_2$, 1 mM $MgCl_2$, 3.8 mM NaOH, and 1 mM Na_2HPO_4.

REFERENCES

Crane RF, Ruderman JV. 2006. Using *Xenopus* oocyte extracts to study signal transduction. *Methods Mol Biol* **322:** 435–443.

Davis RE, King ML. 1989. The developmental expression of the heat-shock response in *Xenopus laevis*. *Development* **105:** 213–222.

Dumont JN. 1972. Oogenesis in *Xenopus laevis* (Daudin) I. Stages of oocyte development in laboratory maintained animals. *J Morph* **136:** 153–180.

Elsner HA, Honck HH, Willmann F, Kreienkamp HJ, Iglauer F. 2000. Poor quality of oocytes from *Xenopus laevis* used in laboratory experiments: Prevention by use of antiseptic surgical technique and antibiotic supplementation. *Comp Med* **50:** 206–211.

Gagnon JA, Mowry KL. 2010. Visualizing RNA localization in *Xenopus* oocytes. *J Vis Exp* **35:** 1704.

Heinrich B, Deshler JO. 2009. RNA localization to the Balbiani body in *Xenopus* oocytes is regulated by the energy state of the cell and is facilitated by kinesin II. *RNA* **15:** 524–536.

Hilken G, Dimigen J, Iglauer F. 1995. Growth of *Xenopus laevis* under different laboratory rearing conditions. *Lab Anim* **29:** 152–162.

Iwabuchi M, Ohsumi K, Yamamoto TM, Sawada W, Kishimoto T. 2000. Residual Cdc2 activity remaining at meiosis I exit is essential for meiotic M-M transition in *Xenopus* oocyte extracts. *EMBO J* **19:** 4513–4523.

Miyamoto K, Simpson D, Gurdon JB. 2015. Manipulation and in vitro maturation of *Xenopus laevis* oocytes, followed by intracytoplasmic sperm injection, to study embryonic development. *J Vis Exp* **96:** e52496.

Novak B, Tyson JJ. 1993. Numerical analysis of a comprehensive model of M-phase control in *Xenopus* oocyte extracts and intact embryos. *J Cell Sci* **106:** 1153–1168.

O'Connell D, Mruk K, Rocheleau JM, Kobertz WR. 2011. *Xenopus laevis* oocytes infected with multi-drug-resistant bacteria: Implications for electrical recordings. *J Gen Physiol* **138:** 271–277.

Opresko L, Wiley HS, Wallace RA. 1980. Proteins iodinated by the chloramine-T method appear to be degraded at an abnormally rapid rate after endocytosis. *Proc Natl Acad Sci* **77:** 1556–1560.

Schneider PN, Hulstrand AM, Houston DW. 2010. Fertilization of *Xenopus* oocytes using the host transfer method. *J Vis Exp* **45:** e1864.

Wallace RA, Misulovin Z. 1978. Long-term growth and differentiation of *Xenopus* oocytes in a defined medium. *Proc Natl Acad Sci* **75:** 5534–5538.

Wallace RA, Jared DW, Dumont JN, Sega MW. 1973. Protein incorporation by isolated amphibian oocytes. 3. Optimum incubation conditions. *J Exp Zool* **184:** 321–334.

Wylie C, Kofron M, Payne Anderson R, Hosobuchi M, Joseph E, Heasman J. 1996. Maternal β-catenin establishes a 'dorsal signal' in early *Xenopus* embryos. *Development* **122:** 2987–2996.

Cite this protocol as *Cold Spring Harb Protoc*; doi:10.1101/pdb.prot095851

Isolation and Analysis of *Xenopus* Germinal Vesicles

Garry T. Morgan[1]

School of Life Sciences, University of Nottingham, Nottingham NG7 2UH, United Kingdom

The giant nucleus or germinal vesicle (GV) of *Xenopus* oocytes provides an unusual opportunity to analyze nuclear structure and function in exquisite detail by light microscopy. Detailed here are two rapid procedures for using manually isolated GVs in combination with fluorescent reporter proteins to investigate the lampbrush chromosomes and nuclear bodies of oocytes. One procedure provides spreads of nuclear components in an unfixed and life-like, although not living, form. The other describes the isolation of intact, functional GVs directly into mineral oil offering possibilities for direct observation of nuclear dynamics.

MATERIALS

It is essential that you consult the appropriate Material Safety Data Sheets and your institution's Environmental Health and Safety Office for proper handling of equipment and hazardous materials used in this protocol.

RECIPES: Please see the end of this protocol for recipes indicated by <R>. Additional recipes can be found online at http://cshprotocols.cshlp.org/site/recipes.

Reagents

GV dispersal medium <R>
GV isolation medium <R>
Mineral oil (Sigma-Aldrich M5904)
Modified Barth's/GTP saline (MBS) <R>
Paraffin wax (solidification point 51°C–53°C)
Vaseline (petroleum jelly)
Xenopus laevis ovary

Equipment

Coverslips (glass, 18 × 18 mm, No. $1^1/_2$)
Differential interference contrast (DIC) microscope (optional; see Step 8)
Dispersal/observation chambers

Use either a microscope slide or Perspex disc (24 mm diameter, 1.5 mm thick Plexiglas) with a 6-mm diameter hole drilled in the center. To form the floor of the chamber seal a coverslip over the hole using a molten drop of a 1:1 mixture of Vaseline and paraffin wax. See Figure 1D.

Filter paper (Whatman #1)

[1]Correspondence: garry.morgan@nottingham.ac.uk

Cite this protocol as *Cold Spring Harb Protoc*; doi:10.1101/pdb.prot096958

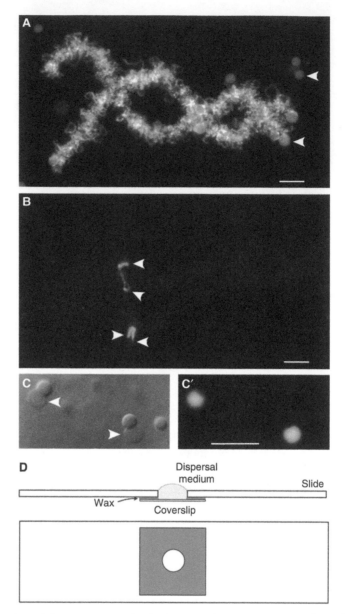

FIGURE 1. (*A–C*) Targeting of fluorescent proteins in *Xenopus* GV preparations and (*D*) diagram of a dispersal/ observation chamber. (*A*) Aqueous spread preparation showing widespread targeting of an RNA-binding protein fused to mCherry both to transcription loops of a lampbrush chromosome and to a type of nuclear body corresponding to somatic cell splicing speckles (examples indicated by arrowheads). Gray-scale fluorescent image. (*B*) Aqueous spread showing specific targeting of another RNA-binding protein fused to GFP (green) and contrasted with the low-level chromosomal fluorescence provided by a generally localized protein fused to mCherry (red). The four transcription loops corresponding to a single lampbrush chromosome locus targeted by the GFP fusion are indicated by arrowheads. Combined pseudocolored fluorescent images. (*C*) Nuclear bodies in an intact GV prepared in oil and observed by DIC microscopy. The two objects indicated by arrowheads are histone locus bodies (HLBs) while those attached or adjacent to them are splicing speckles. (*C'*) Gray-scale fluorescent image showing the specific targeting of a GFP fusion protein to these HLBs. All scale bars = 10 µm. (*D*) Views in section and plan of a chamber constructed from a standard microscope slide that has had a 6-mm diameter hole bored in its center. The main components of a prefilled chamber are labeled in the section view.

Inverted and upright fluorescence microscopes
Inverted phase contrast microscope
Microscope slides (glass, standard)
Pasteur pipettes (glass; 150 mm)

Cite this protocol as *Cold Spring Harb Protoc*; doi:10.1101/pdb.prot096958

Pull some glass Pasteur pipettes in a Bunsen flame from just above the shank to produce a narrower capillary section that can be broken at about 6 cm from the shank so as to leave a tip with a diameter of 0.8–0.9 mm. Use a 2-mL rubber teat to aspirate GVs in and out of the capillary section of these pipettes.

Petri dishes (plastic; 35 mm diameter)
Stereo microscope and fiber optic light source
Syringe needle (25G)
Tungsten needle (sharpened)
Watchmaker's forceps (Dumont #5)

METHOD

Two different procedures are described for the isolation and handling of GVs, either in aqueous solutions or in mineral oil. See the Discussion for the applicability of the two procedures.

Isolation and Observation of GV Contents in Aqueous Spreads

1. Manually dissect individual oocytes from ovary fragments in MBS using watchmaker's forceps.

   All manipulations and solutions should be at room temperature, 18°C–20°C.

   Removal of the oocyte follicle, either manually or by collagenase treatment should also be performed if the microinjection of expression constructs before GV isolation is planned.

2. Size-select the separated oocytes according to the stage of oocyte development and optimal chromatin decondensation. Pick oocytes in late stage IV to early stage V (i.e., about 0.8 to 1.1 mm in diameter). Let separated oocytes recover overnight in MBS before GV isolation to allow reextension of lampbrush loops and the identification of unhealthy oocytes.

   To express fluorescent protein fusions use standard procedures to inject 2–20 ng of synthetic capped RNA into the cytoplasm of each oocyte 24–48 h before GV isolation.

3. Fill a dispersal/observation chamber with sufficient dispersal medium to form a slightly convex meniscus (Fig. 1D).

4. Using a standard Pasteur pipette transfer several oocytes to a 35 mm Petri dish containing GV isolation medium under a stereo microscope. Use lateral illumination from the light source and a black background. With two pairs of sharp watchmaker's forceps tear an oocyte apart from the vegetal towards the animal pole. (The spherical GV sits in the animal hemisphere and can often be seen as a partial clearing in the mass of yolk.) Immediately suck the GV into the capillary section of a stretched Pasteur pipette prefilled with GV isolation medium and transfer the GV swiftly (within 10–20 sec) and with minimal medium or yolk into a dish prefilled with GV dispersal medium.

   In all GV transfers avoid the presence of air bubbles and sharp or broken edges to the pipette to prevent premature rupture of the GV.

5. Immediately after transferring the GV, pick it up with a pair of watchmaker's forceps so as to create a firm grip but not puncture the nuclear envelope. Holding the GV just above the surface of the dish, tear open the nuclear envelope over a third to a half of its circumference using either the finely sharpened point of a tungsten needle or a second pair of forceps. Hold the punctured GV perfectly still in the forceps until the gel-like GV contents spill out of the envelope.

6. Working quickly so that the GV contents do not become too liquid, use a stretched Pasteur pipette prefilled with GV dispersal medium to separate the GV contents from the envelope and transfer them in a small volume to the prefilled dispersal/observation chamber from Step 3.

   The GV contents should sink swiftly to the floor of the chamber.

7. Place a coverslip on top of the chamber. Monitor the extent of dispersal of GV contents in phase contrast with an inverted microscope.

   After 20 min to 1 h in the dispersal medium the gel-like GV contents should have completely liquefied so that lampbrush chromosomes and GV bodies lie flat on the coverslip forming the chamber floor.

8. When GV contents are fully dispersed, blot excess medium from around the top coverslip with filter paper and seal the edges with molten Vaseline. Immediately undertake detailed observation of unfixed GV structures by phase contrast or DIC microscopy. Alternatively, monitor fusion proteins by fluorescence microscopy (Fig. 1A,B).

   Spread preparations can be kept for several days at 4°C.

   To more firmly attach the GV contents and to help place most of the chromosome loops in the same focal plane, the preparation can be centrifuged (descriptions of centrifuge adapters for dispersal chambers and centrifugation conditions are described in Morgan 2008).

Isolation and Analysis of Intact GVs in Oil

9. Prepare oocytes as described in Steps 1–2. Transfer a few oocytes in a drop of MBS to a stack of several pieces of filter paper cut to about 1.0×0.5 cm. As soon as most of the liquid has drained from the oocytes, take the top piece of paper and submerge it and the attached oocytes in 4–5 mL of mineral oil in a 35 mm Petri dish.

10. Orient the oocytes so the animal pole is uppermost and make a small puncture/slash in the top with a sharp needle (e.g., 25G syringe needle). Depending on the viscosity of the cytoplasm, after a few seconds the GV may begin to emerge spontaneously from the mass of yolk; if not, gently squeeze the oocyte with forceps to encourage the emergence of the GV. With a pipetting device set to 5–8 μL gently suck the GV together with some clean oil into the pipette tip. The GV may be coated with a certain amount of yolk, and some gentle aspiration of the GV in and out of the pipette tip may displace it. Even if some yolk platelets remain attached, transfer the GV to a standard microscope slide and mount in the oil by gently lowering a coverslip in place.

    If studying nuclear bodies use a clean slide. If studying the more delicate chromosomes, use a slide with a circular dam of a 1:1 mixture of Vaseline and paraffin wax about 50 μm high to surround the GV (Patel et al. 2008).

11. Examine immediately after preparation.

    Phase contrast or DIC microscopy through yolk-free regions of the intact GV will reveal nuclear bodies at low contrast, while specific targeting of fluorescent fusion proteins to nuclear bodies or chromatin is more striking (Fig. 1C,C′).

    GVs retain normal physiological activity at room temperature for at least several hours after isolation in oil (Paine et al. 1992).

DISCUSSION

The two procedures described here provide rapid and straightforward means to assess the targeting of fluorescently labeled proteins in transcriptionally active chromatin and nuclear bodies at high levels of morphological detail (see Fig. 1). The first procedure provides aqueous spread preparations of nuclear structures in an unfixed although nonfunctional state and with a massively diluted nucleoplasmic background. These spreads provide a clear appreciation of the number and location of structures targeted by fluorescent protein fusions in the context of the entire *Xenopus* nuclear genome and its nuclear bodies. Where use of fluorescent protein fusions is not appropriate/possible, detailed molecular characterization of spread GV contents using immunofluorescence and in situ hybridization is also well established. However, additional steps, including centrifugation, are required to produce more permanent, fixed preparations on microscope slides; detailed instructions and advice for these procedures are provided in Gall and Wu (2010) and Gall and Nizami (2016). The second procedure for isolating and maintaining intact GVs in mineral oil enables the analysis of physiologically active nuclei. This procedure was first used to study molecular dynamics of nuclear bodies (Handwerger et al. 2003; Deryusheva and Gall 2004) and was then adapted (Patel et al. 2008; Austin et al. 2009) to provide a direct visual approach for examining transcription loops and chromatin in real time. Additionally, oil-based manual GV isolation can be used to obtain nuclear material for biochemical investigations (Sommerville 2010).

RECIPES

GV Dispersal Medium

Reagent	Concentration
GV isolation medium <R>	25%
Paraformaldehyde stock solution (20%) <R>	0.1%
$MgCl_2$	1.0 mM*

*Adjusted final concentration.

Adjust the final pH to 6.6–6.8 with 100 mM KH_2PO_4. Store at 4°C. Just before use, add DTT to 1 mM, and filter through 0.45-μm nitrocellulose.

GV Isolation Medium

Reagent	Concentration
KCl	83 mM
NaCl	17 mM
Na_2HPO_4	6.5 mM
KH_2PO_4	3.5 mM
$MgCl_2$	1 mM

Check that the final pH is 6.9–7.0. Store at 4°C. Just before use, add DTT to 1 mM, and filter through 0.45-μm nitrocellulose.

Modified Barth's/GTP Saline (MBS)

Reagent	Concentration
NaCl	96 mM
KCl	2 mM
$CaCl_2$	1.8 mM
HEPES	5.0 mM
Pyruvic acid	2.5 mM
Theophylline	0.5 mM

Adjust the pH to 7.5 with NaOH. Autoclave, and store at room temperature. Before use, add gentamycin to 50 μg/mL.

REFERENCES

Austin C, Novikova N, Guacci V, Bellini M. 2009. Lampbrush chromosomes enable study of cohesin dynamics. *Chromosome Res* 17: 165–184.

Deryusheva S, Gall JG. 2004. Dynamics of coilin in Cajal bodies of the *Xenopus* germinal vesicle. *Proc Natl Acad Sci* 101: 4810–4814.

Gall JG, Nizami ZF. 2016. Isolation of giant lampbrush chromosomes from living oocytes of frogs and salamanders. *J Vis Exp* 118: e54103.

Gall JG, Wu Z. 2010. Examining the contents of isolated *Xenopus* germinal vesicles. *Methods* 51: 45–51.

Handwerger KE, Murphy C, Gall JG. 2003. Steady-state dynamics of Cajal body components in the *Xenopus* germinal vesicle. *J Cell Biol* 160: 495–504.

Morgan GT. 2008. Working with oocyte nuclei: Cytological preparations of active chromatin and nuclear bodies from amphibian germinal vesicles. *Methods Mol Biol* 463: 55–66.

Paine PL, Johnson ME, Lau YT, Tluczek LJ, Miller DS. 1992. The oocyte nucleus isolated in oil retains in vivo structure and functions. *Biotechniques* 13: 238–246.

Patel S, Novikova N, Beenders B, Austin C, Bellini M. 2008. Live images of RNA polymerase II transcription units. *Chromosome Res* 16: 223–232.

Sommerville J. 2010. Using oocyte nuclei for studies on chromatin structure and gene expression. *Methods* 51: 157–164.

Microinjection of *Xenopus* Oocytes

Tristan Aguero, Karen Newman, and Mary Lou King[1]

Department of Cell Biology, University of Miami School of Medicine, Miami, Florida 33136

Microinjection of *Xenopus* oocytes has proven to be a valuable tool in a broad array of studies that require expression of DNA or RNA into functional protein. These studies are diverse and range from expression cloning to receptor–ligand interaction to nuclear programming. Oocytes offer a number of advantages for such studies, including their large size (~1.2 mm in diameter), capacity for translation, and enormous nucleus (0.3–0.4 mm). They are cost effective, easily manipulated, and can be injected in large numbers in a short time period. Oocytes have a large maternal stockpile of all the essential components for transcription and translation. Consequently, the investigator needs only to introduce by microinjection the specific DNA or RNA of interest for synthesis. Oocytes translate virtually any exogenous RNA regardless of source, and the translated proteins are folded, modified, and transported to the correct cellular locations. Here we present procedures for the efficient microinjection of oocytes and their subsequent care.

MATERIALS

It is essential that you consult the appropriate Material Safety Data Sheets and your institution's Environmental Health and Safety Office for proper handling of equipment and hazardous materials used in this protocol.

RECIPES: Please see the end of this protocol for recipes indicated by <R>. Additional recipes can be found online at http://cshprotocols.cshlp.org/site/recipes.

Reagents

Antibiotics

    Gentamycin (1000× stock solution [10 mg/mL in H_2O])

    Streptomycin (1000× stock solution [100 mg/mL in H_2O])

Ficoll 400 (Sigma-Aldrich) (optional; see Step 4)

Medium for oocyte culture

    Modified Barth's Saline (MBS) for oocytes <R>

    Oocyte culture medium (OCM) <R>

    A number of formulations exist for OCM (Wallace and Misulovin 1978; Opresko et al. 1980; Wylie et al. 1996). The formulation cited here is used for host transfer studies (Schneider et al. 2010a).

    Oocyte reagent 2 (O-R2) <R> (Wallace et al. 1973)

    The choice of medium depends upon what the oocytes will be used for experimentally.

Xenopus laevis oocytes (defolliculated)

---

[1]Correspondence: mking@med.miami.edu

Oocytes should be obtained surgically from anesthetized adult females or sourced from commercial vendors (Nasco, EcoCyte Bioscience, Xenoocyte) and defolliculated. See Protocol 1: Isolation of Xenopus Oocytes (Newman et al. 2018) for oocyte isolation and defolliculation procedures.

Equipment

Capillary tubes (Drummond Scientific 3-000-203G/X [borosilicate glass, 1.14-mm outer diameter, 0.53-mm inner diameter, 90 mm long] or Sutter Instrument BF100-50-10 [borosilicate glass with filament, 1.0-mm outer diameter, 0.5-mm inner diameter, 100 mm long])

Dissecting microscope and dual gooseneck LED illuminators (low heat)

Forceps (Dumont #4 or #5)

Hair loop

Incubator at 18°C–20°C

Mesh (nylon or polypropylene; 0.5–0.8 mm) (as needed; see Step 4)

Microinjector (e.g., Drummond Nanoject II Auto-Nanoliter Injector or Narishige IM-300)

An automated injection system (roboinject) is optional; see http://www.multichannelsystems.com/products/xenopus-oocyte-research.

Micromanipulator (e.g., Warner Instruments MM-33 or Singer Mk1)

Petri dishes (35-mm) and/or six-well microplates (35-mm/well) (plastic)

Pipette puller (e.g., Narishige PC-10 or Sutter Instrument P-97)

METHOD

Preparing for Microinjection

1. Prepare pipettes for microinjection from glass capillary tubes using a micropipette puller. Using forceps and with the aid of a dissecting microscope, gently break the pipette tip to produce a clean open tip. Measure the diameter.

   Capillary tube dimensions may vary according to the injector to be used; check the manufacturer's specifications. The length of the pipette should be 5–8 cm. A final tip diameter of <20 µm is optimal.

   Pipette quality is critical for postinjection oocyte viability. A good quality pipette has a fine, beveled tip, and will slide in and out of the oocyte leaving a small injection mark without leakage of sample or cellular material (Fig. 1B,C). A poor quality pipette has a wide and/or blunt ended tip which is difficult to insert, and will leave a large hole in the oocyte that will leak ooplasm (Fig. 1D; Maldifassi et al. 2016).

2. To perform sample loading, refer to the manufacturer's instructions for specific details for your injection apparatus.

   Loading of the sample commonly involves backfilling the pipette with mineral oil after breaking the tip as described in Step 1, and then front-filling with sample. Injection volumes vary according to the sample, the oocyte stage and the microinjector. In general, maximum injection volumes are: stage V/VI, up to 50 nL; stage IV, 20 nL; stage III, 10 nL; stage II, 4.6 nL; stage I, 1–2.3 nL.

3. Secure the injection apparatus to a micromanipulator which moves the pipette for injection. Position the pipette at a 35–55 degree angle.

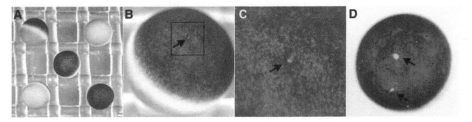

FIGURE 1. (A) Stage VI oocytes aligned in various orientations over a 0.5-mm nylon mesh for microinjection. (B) Example of nuclear injection using a beveled needle (*arrow*). (C) An enlargement of the area surrounding the injection (*boxed area in B*) shows how the injection hole constricts and eventually heals. (D) Injection holes produced by a too broad or blunt-tip pipette (*arrows*); compare with (B).

Performing Microinjection

Steps 4.i and 4.ii describe the two predominant techniques used to microinject defolliculated oocytes. Either technique can be used in Step 4.iii, depending on the preference and skill of the user.

Microinjectors generally have two options for injections: pressing a button or using a foot pedal. Use of a foot pedal frees the hands and is more efficient.

A general injection procedure can be viewed at https://www.youtube.com/watch?v=jv-BgKTvqfU (Oocyte 2009). A microinjection procedure for electrophysiological experiments is described and shown by Maldifassi et al. (2016).

4. Microinject defolliculated oocytes using the desired technique.

   i. Immobilize oocytes by positioning them into the openings of a nylon or polypropylene mesh (0.5–0.8 mm) at the bottom of a Petri dish containing culture medium (Fig. 1A). Carefully align oocytes for microinjection with the desired pole facing upward using a hair loop or blunt-end forceps. Prepare 50–100 oocytes at a time for microinjection. After each injection is complete (Step iii), gently withdraw the pipette and move to the next oocyte by moving the microinjection dish.

   ii. Alternatively, place 20–50 oocytes in the center of a Petri dish filled with culture medium. Carefully orient one oocyte at a time with the desired pole facing up using forceps, and hold it in place while performing each injection (Step iii). Gently withdraw the pipette, move the injected oocyte to another position in the dish (e.g., from center to top), and bring the next oocyte to the pipette.

   iii. Using either of the above methods, position the selected oocyte under the loaded injection pipette and move the pipette down using the micromanipulator. (The oocyte surface will dent under pressure from the pipette until penetration; further movement is unnecessary.) Inject the sample, then wait a few seconds before withdrawing the pipette to minimize sample leakage. Periodically release one injection shot into solution to ensure the pipette has not clogged.

      The use of 4% Ficoll 400 in the culture medium is optional, but may prevent leakage of sample from the injection site.

      For nuclear injections, the germinal vesicle (GV) is found just beneath the surface of the animal pole (darkly pigmented half). We recommend practicing nuclear injections into the animal pole using a dye solution (bromophenol blue 0.1% or FITC 0.1%) to ensure injections are targeting the GV. Another method for targeting the nucleus is described by Bataillé et al. (1990) and Dworetzky and Feldherr (1988). This method involves centrifugation of defolliculated oocytes at 650g for 8–10 min causing migration of the GV to a position just underneath the plasma membrane at the animal pole, which is easily visualized as a targetlike structure due to the displacement of pigment granules in the cortex. Nuclear injections can then be accomplished, preferably within 15 min after centrifugation. The maximum volume suggested for nuclear injections is 20 nL.

      See Troubleshooting.

5. When microinjection is complete, carefully separate the oocytes into individual dishes or microplate wells (up to 50 cells per 35-mm dish or well) containing the same medium used during injection, including Ficoll (if used). Return the oocytes to the 18°C incubator. If Ficoll was used, replace the medium with Ficoll-free medium after >2 h.

   Oocytes can remain in medium containing Ficoll 400 overnight.

6. Twice a day, monitor the oocytes for viability, removing damaged or dying oocytes to prevent deleterious effects on healthy oocytes. Replace medium with fresh medium each time. Include antibiotics, such as gentamycin or streptomycin, in the culture medium to prevent bacterial infection (see Protocol 1: Isolation of *Xenopus* Oocytes [Newman et al. 2018]).

   See Troubleshooting.

Cite this protocol as *Cold Spring Harb Protoc*; doi:10.1101/pdb.prot096974

TROUBLESHOOTING

Problem (Step 4): The pipette clogs during injection.

Solution: Remove the pipette from the medium and empty it, using pressure to force out the particulate material. As a last resort, use forceps to carefully break off slightly more of the tip.

Problem (Step 4): There is leakage of injected material.

Solution: Change the pipette, as the quality may be poor. Include 4% Ficoll 400 in the medium used for injections.

Problem (Step 4): Oocytes are soft and difficult to penetrate.

Solution: Only inject relatively firm oocytes. If too many are soft, use oocytes from another female. Note that using Ficoll 400 in the medium may initially make the oocytes seem soft.

Problem (Step 6): The viability of injected oocytes is poor.

Solution: Sample toxicity, old solutions with low pH, or damage to the oocytes due to a poor quality pipette are the most common reasons for poor viability. Use freshly made buffers and inject a range of sample volumes and concentrations to control for sample toxicity.

Problem (Step 6): There is low protein expression after injection.

Solution: Poor quality oocytes, oocytes insufficiently recovered from defolliculation processes, a degraded sample (particularly if working with RNA), or an insufficient amount of injected sample can contribute to low expression. Use oocytes that have been allowed to recover overnight from defolliculation and from several females to control for oocyte variability. Inject a range of sample concentrations to control for expression variability.

DISCUSSION

Examples of studies using oocyte injection are listed in Table 1. Based on PubMed searches, the major uses of *Xenopus* oocyte microinjection are heterologous transporter and channel expression followed

TABLE 1. Example studies using oocyte injection

Application	Reference
Cell cycle regulation	Daldello et al. 2015
Cellular polarization	Peter et al. 1991
Cryopreservation	Guenther et al. 2006; Yamaji et al. 2006
Dominant negative constructs	Reed et al. 2016
Electrophysiology	Miledi et al. 1982; Fujita et al. 2007; Hansen and Brauner-Osborne 2009; Papke and Smith-Maxwell 2009; Wang et al. 2014
Enzymatic activity	Sato et al. 1996; Hehl et al. 2001; Sallacz and Jantsch 2005; Ota et al. 2011
Genome editing	Miyamoto et al. 2015; Nakajima and Yaoita 2015; Ratzan et al. 2017
Germline specification	Aguero et al. 2016
Lipid metabolism	Stifani et al. 1990
mRNA knockdown	Hulstrand et al. 2010
Nuclear import/export pathways	Lane et al. 1980; Terns et al. 1993; Smillie et al. 2004; Loop and Pieler 2005
Nuclear reprogramming	Jullien 2015
Oocyte host transfer	Schneider et al. 2010a; Mei et al. 2013
Protein expression (homologous/ heterologous)	Bartnik and de Robertis 1983; Duchesne et al. 2003; Bernareggi et al. 2011; Clemencon et al. 2015; Crespin et al. 2016; Miller and Zhou 2000; Aoshima et al. 2005; Maciver et al. 2008; Reed et al. 2016; Plautz et al. 2016.
Protein–protein interaction	Schneider et al. 2010b. Nijjar and Woodland 2013
RNA–protein interaction	Mabuchi et al. 2008; Matsumoto et al. 2000
Signal transduction	Lin-Moshier and Marchant 2013; Sato 2015
Transcript/protein localization	Chang et al. 2004; Nijjar and Woodland 2013

by electrophysiological or tracer flux analysis. Oocytes can be injected with up to 10 ng of mRNA, 3 ng of cDNA or 80 ng of Morpholino antisense oligonucleotides (Hulstrand et al. 2010). In general, the efficiency rate of protein expression using mRNA is higher than using cDNA (Sigel 2001). Microinjection of dominant negative constructs, antisense Morpholinos or transcripts encoding mutant proteins are extensively used in loss-of-function and gain-of-function experiments. Genetic information can be introduced or depleted with high efficiency by microinjection into either the cytoplasmic or nuclear compartments, thus avoiding the complications of transfection. For some experiments, it may be desirable to culture oocytes. Oocytes can be cultured and remain in a healthy, stable condition for several days if not weeks (Wallace and Misulovin 1978). One cautionary note: oocytes exist within an ovarian follicle comprising different somatic cells (i.e., blood vessels and thousands of follicle cells). These somatic cells can be a source of molecular contamination, depending on the goal of the experiment. Defolliculation does not remove 100% of these somatic cells.

RECIPES

Modified Barth's Saline (MBS) for Oocytes

1. Prepare a 0.1 M $CaCl_2$ solution by dissolving 1.11 g of $CaCl_2$ in deionized H_2O for a final volume of 100 mL.
2. Prepare a 10× MBS salt solution by dissolving the following reagents in deionized H_2O for a final volume of 1 L. Adjust the pH to 7.8 with 10 M NaOH, and sterilize by autoclaving. Store indefinitely at room temperature.

HEPES	11.92 g
NaCl	51.43 g
KCl	0.75 g
NaHCO₃	2.10 g
MgSO₄ · 7H₂O	2.46 g

3. To prepare 1× MBS, combine 100 mL of 10× MBS salt solution and 7 mL of 0.1 M $CaCl_2$ with 893 mL of autoclaved deionized H_2O. Add antibiotics. Store for up to 1–2 mo at 16°C–18°C (the same temperature used for oocyte incubation). Check before use.

    The final reagent concentrations in 1× MBS are 0.7 mM $CaCl_2$, 5 mM HEPES, 88 mM NaCl, 1 mM KCl, 2.5 mM $NaHCO_3$, and 1 mM $MgSO_4$.

Oocyte Culture Medium (OCM)

Reagent	Amount to add	Final concentration
Leibovitz's L-15 medium with L-glutamine	0.35 L	70%
Bovine serum albumin (fraction V)	0.2 g	0.4 mg/mL
Penicillin-streptomycin (100×/100×, P4333 Sigma-Aldrich)	5 mL	100 U/mL penicillin
		0.1 mg/mL streptomycin

Bring the volume to 0.5 L with deionized H_2O, and adjust the pH to 7.8 with NaOH. Filter to sterilize. Store at 4°C–8°C for up to 1–2 mo. Check before use.

Cite this protocol as *Cold Spring Harb Protoc*; doi:10.1101/pdb.prot096974

Oocyte Reagent 2 (O-R2)

1. Prepare Stock Solution A by dissolving the following reagents in deionized H_2O for a final volume of 1 L. Store indefinitely at 4°C–8°C.

HEPES	11.915 g
NaCl	48.221 g
KCl	1.864 g
$CaCl_2 \cdot 2H_2O$	1.47 g
$MgCl_2 \cdot 6H_2O$	2.03 g
NaOH	1.520 g

2. Prepare Stock Solution B by dissolving 1.420 g of Na_2HPO_4 in deionized H_2O for a final volume of 1 L. Store indefinitely at 4°C–8°C.
3. Combine 100 mL of Stock Solution A, 100 mL of Stock Solution B, and 800 mL of deionzed H_2O. Add antibiotics. Store for up to 1–2 mo at 16°C–18°C (the same temperature used for oocyte incubation). Check before use.

The pH should be 7.8. The final reagent concentrations are 5 mM HEPES, 82.5 mM NaCl, 2.5 mM KCl, 1 mM $CaCl_2$, 1 mM $MgCl_2$, 3.8 mM NaOH, and 1 mM Na_2HPO_4.

REFERENCES

Aguero T, Zhou Y, Kloc M, Chang P, Houliston E, King ML. 2016. Hermes (Rbpms) is a critical component of RNP complexes that sequester germline RNAs during oogenesis. *J Dev Biol* **4:** 2.

Aoshima H, Okita Y, Hossain SJ, Fukue K, Mito M, Orihara Y, Yokoyama T, Yamada M, Kumagai A, Nagaoka Y, et al. 2005. Effect of 3-O-octanoyl-(+)-catechin on the responses of GABA(A) receptors and Na^+/glucose cotransporters expressed in *Xenopus* oocytes and on the oocyte membrane potential. *J Agric Food Chem* **53:** 1955–1959.

Bartnik E, de Robertis EM. 1983. Mitochondrial transfer RNA genes from fungi (*Aspergillus nidulans*) and plants (*Lupinus luteus*) are transcribed in *Xenopus laevis* oocyte nuclei. *J Mol Biol* **168:** 439–444.

Bataillé N, Helser T, Fried HM. 1990. Cytoplasmic transport of ribosomal subunits microinjected into the *Xenopus laevis* nucleus: A generalized, facilitated process. *J Cell Biol* **111:** 1571–1582.

Bernareggi A, Grilli M, Marchi M, Limatola C, Ruzzier F, Eusebi F. 2011. Characterization of GABA(A) receptors expressed in glial cell membranes of adult mouse neocortex using a *Xenopus* oocyte microtransplantation expression system. *J Neurosci Methods* **198:** 77–83.

Chang P, Torres J, Lewis RA, Mowry KL, Houliston E, King ML. 2004. Localization of RNAs to the mitochondrial cloud in *Xenopus* oocytes through entrapment and association with endoplasmic reticulum. *Mol Biol Cell* **15:** 4669–4681.

Clemencon B, Fine M, Schneider P, Hediger MA. 2015. Rapid method to express and purify human membrane protein using the *Xenopus* oocyte system for functional and low-resolution structural analysis. *Methods Enzymol* **556:** 241–265.

Crespin L, Legros C, List O, Tricoire-Leignel H, Mattei C. 2016. Injection of insect membrane in *Xenopus* oocyte: An original method for the pharmacological characterization of neonicotinoid insecticides. *J Pharmacol Toxicol Methods* **77:** 10–16.

Daldello EM, Le T, Poulhe R, Jessus C, Haccard O, Dupre A. 2015. Control of Cdc6 accumulation by Cdk1 and MAPK is essential for completion of oocyte meiotic divisions in *Xenopus*. *J Cell Sci* **128:** 2482–2496.

Duchesne L, Hubert JF, Verbavatz JM, Thomas D, Pietrantonio PV. 2003. Mosquito (*Aedes aegypti*) aquaporin, present in tracheolar cells, transports water, not glycerol, and forms orthogonal arrays in *Xenopus* oocyte membranes. *Eur J Biochem* **270:** 422–429.

Dworetzky SI, Feldherr CM. 1988. Translocation of RNA-coated gold particles through the nuclear pores of oocytes. *J Cell Biol* **106:** 575–584.

Fujita R, Kimura S, Kawasaki S, Watanabe S, Watanabe N, Hirano H, Matsumoto M, Sasaki K. 2007. Electrophysiological and pharmacological characterization of the K(ATP) channel involved in the K+-current responses to FSH and adenosine in the follicular cells of *Xenopus* oocyte. *J Physiol Sci* **57:** 51–61.

Guenther JF, Seki S, Kleinhans FW, Edashige K, Roberts DM, Mazur P. 2006. Extra- and intra-cellular ice formation in Stage I and II *Xenopus laevis* oocytes. *Cryobiology* **52:** 401–416.

Hansen KB, Brauner-Osborne H. 2009. *Xenopus* oocyte electrophysiology in GPCR drug discovery. *Methods Mol Biol* **552:** 343–357.

Hehl S, Stoyanov B, Oehrl W, Schonherr R, Wetzker R, Heinemann SH. 2001. Phosphoinositide 3-kinase-γ induces *Xenopus* oocyte maturation via lipid kinase activity. *Biochem J* **360:** 691–698.

Hulstrand AM, Schneider PN, Houston DW. 2010. The use of antisense oligonucleotides in *Xenopus* oocytes. *Methods* **51:** 75–81.

Jullien J. 2015. Analysis of nuclear reprogramming following nuclear transfer to *Xenopus* oocyte. *Methods Mol Biol* **1222:** 71–82.

Lane CD, Colman A, Mohun T, Morser J, Champion J, Kourides I, Craig R, Higgins S, James TC, Applebaum SW, et al. 1980. The *Xenopus* oocyte as a surrogate secretory system. The specificity of protein export. *Eur J Biochem* **111:** 225–235.

Lin-Moshier Y, Marchant JS. 2013. The *Xenopus* oocyte: A single-cell model for studying Ca^{2+} signaling. *Cold Spring Harb Protoc* doi: 10.1101/pdb.top066308.

Loop S, Pieler T. 2005. Nuclear import of mPER3 in *Xenopus* oocytes and HeLa cells requires complex formation with mPER1. *FEBS J* **272:** 3714–3724.

Mabuchi N, Masuyama K, Ohno M. 2008. Immunoprecipitation analysis to study RNA-protein interactions in *Xenopus* oocytes. *Methods Mol Biol* **488:** 257–265.

Maciver B, Smith CP, Hill WG, Zeidel ML. 2008. Functional characterization of mouse urea transporters UT-A2 and UT-A3 expressed in purified *Xenopus laevis* oocyte plasma membranes. *Am J Physiol Renal Physiol* **294:** F956–F964.

Maldifassi MC, Wongsamitkul N, Baur R, Sigel E. 2016. *Xenopus* oocytes: Optimized methods for microinjection, removal of follicular layers, and fast solution changes in electrophysiological experiments. *J Vis Exp* **118:** e55034.

Matsumoto K, Aoki K, Dohmae N, Takio K, Tsujimoto M. 2000. CIRP2, a major cytoplasmic RNA-binding protein in *Xenopus* oocytes. *Nucleic Acids Res* **28:** 4689–4697.

Mei W, Jin Z, Lai F, Schwend T, Houston DW, King ML, Yang J. 2013. Maternal Dead-End1 is required for vegetal cortical microtubule assembly during *Xenopus* axis specification. *Development* **140:** 2334–2344.

Miledi R, Parker I, Sumikawa K. 1982. Properties of acetylcholine receptors translated by cat muscle mRNA in *Xenopus* oocytes. *EMBO J* **1**: 1307–1312.

Miller AJ, Zhou JJ. 2000. *Xenopus* oocytes as an expression system for plant transporters. *Biochim Biophys Acta* **1465**: 343–358.

Miyamoto K, Suzuki KT, Suzuki M, Sakane Y, Sakuma T, Herberg S, Simeone A, Simpson D, Jullien J, Yamamoto T, et al. 2015. The expression of TALEN before fertilization provides a rapid knock-out phenotype in *Xenopus laevis* founder embryos. *PLoS One* **10**: e0142946.

Nakajima K, Yaoita Y. 2015. Highly efficient gene knockout by injection of TALEN mRNAs into oocytes and host transfer in *Xenopus laevis*. *Biol Open* **4**: 180–185.

Newman D, Aguero T, King ML. 2018. Isolation of *Xenopus* oocytes. *Cold Spring Harb Protoc* doi: 10.1101/pdb.prot095851.

Nijjar S, Woodland HR. 2013. Localisation of RNAs into the germ plasm of vitellogenic *Xenopus* oocytes. *PLoS One* **8**: e61847.

Oocyte injections. YouTube video posted by "Frank Mari" on September 1, 2009. https://www.youtube.com/watch?v=jv-BgKTvqfU.

Opresko L, Wiley HS, Wallace RA. 1980. Proteins iodinated by the chloramine-T method appear to be degraded at an abnormally rapid rate after endocytosis. *Proc Natl Acad Sci* **77**: 1556–1560.

Ota R, Kotani T, Yamashita M. 2011. Possible involvement of Nemo-like kinase 1 in *Xenopus* oocyte maturation as a kinase responsible for Pumilio1, Pumilio2, and CPEB phosphorylation. *Biochemistry* **50**: 5648–5659.

Papke RL, Smith-Maxwell C. 2009. High throughput electrophysiology with *Xenopus* oocytes. *Comb Chem High Throughput Screen* **12**: 38–50.

Peter AB, Schittny JC, Niggli V, Reuter H, Sigel E. 1991. The polarized distribution of poly(A+)-mRNA-induced functional ion channels in the *Xenopus* oocyte plasma membrane is prevented by anticytoskeletal drugs. *J Cell Biol* **114**: 455–464.

Plautz CZ, Williams HC, Grainger RM. 2016. Functional cloning using a *Xenopus* oocyte expression system. *J Vis Exp* e53518. doi: 10.3791/53518.

Ratzan W, Falco R, Salanga C, Salanga M, Horb ME. 2017. Generation of a *Xenopus laevis* F1 albino J strain by genome editing and oocyte host-transfer. *Dev Biol* **426**: 188–193.

Reed AP, Bucci G, Abd-Wahab F, Tucker SJ. 2016. Dominant-negative effect of a missense variant in the TASK-2 (KCNK5) K+ channel associated with balkan endemic nephropathy. *PLoS One* **11**: e0156456.

Sallacz NB, Jantsch MF. 2005. Chromosomal storage of the RNA-editing enzyme ADAR1 in *Xenopus* oocytes. *Mol Biol Cell* **16**: 3377–3386.

Sato K. 2015. Transmembrane signal transduction in oocyte maturation and fertilization: Focusing on *Xenopus laevis* as a model animal. *Int J Mol Sci* **16**: 114–134.

Sato K, Aoto M, Mori K, Akasofu S, Tokmakov AA, Sahara S, Fukami Y. 1996. Purification and characterization of a Src-related p57 protein-tyrosine kinase from *Xenopus* oocytes. Isolation of an inactive form of the enzyme and its activation and translocation upon fertilization. *J Biol Chem* **271**: 13250–13257.

Schneider PN, Hulstrand AM, Houston DW. 2010a. Fertilization of *Xenopus* oocytes using the host transfer method. *J Vis Exp*. doi: 10.3791/1864.

Schneider H, Dabauvalle MC, Wilken N, Scheer U. 2010b. Visualizing protein interactions involved in the formation of the 42S RNP storage particle of *Xenopus* oocytes. *Biol Cell* **102**: 469–478.

Sigel E. 2001. Microinjection into *Xenopus* oocytes. *Encyclopedia of Life Sciences*. pp 1–5. John Wiley and Sons, Ltd.

Smillie DA, Llinas AJ, Ryan JTP, Kemp GD, Sommerville J. 2004. Nuclear import and activity of histone deacetylase in *Xenopus* oocytes is regulated by phosphorylation. *J Cell Sci* **117**: 1857–1866.

Stifani S, Nimpf J, Schneider WJ. 1990. Vitellogenesis in *Xenopus laevis* and chicken: Cognate ligands and oocyte receptors. The binding site for vitellogenin is located on lipovitellin I. *J Biol Chem* **265**: 882–888.

Terns MP, Lund E, Dahlberg JE. 1993. A pre-export U1 snRNP in *Xenopus laevis* oocyte nuclei. *Nucleic Acids Res* **21**: 4569–4573.

Wallace RA, Misulovin Z. 1978. Long-term growth and differentiation of *Xenopus* oocytes in a defined medium. *Proc Natl Acad Sci* **75**: 5534–5538.

Wallace RA, Jared DW, Dumont JN, Sega MW. 1973. Protein incorporation by isolated amphibian oocytes. 3. Optimum incubation conditions. *J Exp Zool* **184**: 321–334.

Wang J, Ambrosi C, Qiu F, Jackson DG, Sosinsky G, Dahl G. 2014. The membrane protein Pannexin1 forms two open-channel conformations depending on the mode of activation. *Sci Signal* **7**: ra69.

Wylie C, Kofron M, Payne C, Anderson R, Hosobuchi M, Joseph E, Heasman J. 1996. Maternal β-catenin establishes a 'dorsal signal' in early *Xenopus* embryos. *Development* **122**: 2987–2996.

Yamaji Y, Valdez DM Jr, Seki S, Yazawa K, Urakawa C, Jin B, Kasai M, Kleinhans FW, Edashige K. 2006. Cryoprotectant permeability of aquaporin-3 expressed in *Xenopus* oocytes. *Cryobiology* **53**: 258–267.

Cite this protocol as *Cold Spring Harb Protoc*; doi:10.1101/pdb.prot096974

Oocyte Host-Transfer and Maternal mRNA Depletion Experiments in *Xenopus*

Douglas W. Houston[1]

Department of Biology, The University of Iowa, Iowa City, Iowa 52242-1324

This protocol details the oocyte host-transfer method in *Xenopus*, using transplantation by intraperitoneal injection. This approach is suitable for the overexpression of mRNAs and for the use of antisense oligonucleotides to deplete maternal mRNAs, which are not replaced until zygotic genome activation in the mid-blastula transition. *Xenopus* oocyte host-transfer can also be used for highly efficient mutagenesis in the F_0 generation by prefertilization injection of genome editing reagents.

MATERIALS

It is essential that you consult the appropriate Material Safety Data Sheets and your institution's Environmental Health and Safety Office for proper handling of equipment and hazardous material used in this protocol.

RECIPES: Please see the end of this protocol for recipes indicated by <R>. Additional recipes can be found online at http://cshprotocols.cshlp.org/site/recipes.

Reagents

Anesthetic solution (buffered MS222) <R>
Human chorionic gonadotropin (hCG; 10,000 IU/vial; Sigma-Aldrich CG10)
L-15 oocyte culture medium (OCM; 70%) <R>
Marc's modified Ringer's (MMR) solution–triple HEPES (10×) <R>
Nucleic acids for microinjection according to experimental goals

> *Details regarding oligonucleotide design, modification and use in oocytes for host-transfer are described elsewhere (Hulstrand et al. 2010; Olson et al. 2012).*

Progesterone (10 mM in 100% ethanol, stored at −20°C; Sigma-Aldrich P8783)
Tap water conditioned with AmQuel or other chloramine remover
Vital dye stocks <R>
Xenopus females for use as oocyte donors and as transfer hosts
Xenopus ovarian tissue (Nasco; optional, see Step 1)

Equipment

Containers for frogs (buckets or small tubs)
Dumont #4 or #5 forceps (Fine Science Tools)

---

[1]Correspondence: douglas-houston@uiowa.edu

Glass syringe with Luer lock adaptor (2 mL; Tomopal Inc. 140-1502)
Incubator (18°C)
Kimwipes (large, sterilized)
Microcentrifuge tubes (1.5 mL)
Pasteur pipettes (fire-polished glass, sterile)
Petri dishes (plastic 100 and 60 mm diameter)
Rocking platform
Sterile syringe needles (16 gauge, 1 inch [16G1]; BD 305197)
Teflon pestles for 1.5 mL microcentrifuge tubes (RPI 199228)

METHOD

The steps for the entire procedure are outlined in Figure 1.

Culture of Experimental Oocytes

1. Prepare ovarian tissue surgically from anesthetized females as described in Schneider et al. (2010) and Olson et al. (2012) and place in OCM at 18°C.

   Alternatively, Xenopus ovarian tissue can be obtained commercially.

2. Manually defolliculate 300–500+ oocytes using Dumont forceps. Dissect oocytes by tearing the follicle layer near the region where oocytes are attached to the ovary, using very light grip pressure on the forceps. Store oocytes in OCM at 18°C and begin experiments within 1–2 d.

   Frogs stimulated with hGC within 3 mo should not be used as oocyte donors. Healthy, high-quality oocytes are essential for successful host-transfer experiments. These should be fully grown and free of any damage to the membrane. Practice is essential for the rapid collection of many good quality oocytes and is necessary for good results with this method. For detailed descriptions of manual defolliculation as well as videos see: Smith et al. (1991), www.youtube.com/watch?v=us8rDNG69Sk (Manual defolliculation 2009), Hulstrand et al. (2010), Schneider et al. (2010), Sive et al. (2010a), Olson et al. (2012), and Protocol 1: Isolation of Xenopus Oocytes (Newman et al. 2018).

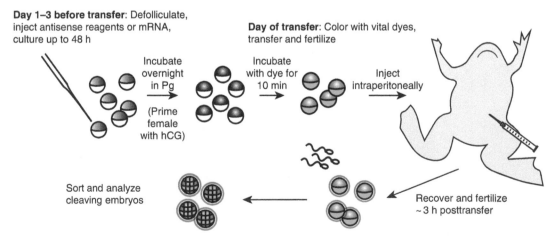

FIGURE 1. Steps for culturing and transferring oocytes by intraperitoneal injection. Up to several days before the day embryos are desired, isolate oocytes by manual defolliculation and inject with experimental antisense oligonucleotides and/or mRNAs. On the evening before host-transfer, stimulate the oocytes to mature by adding progesterone (Pg) to the medium. At the same time, inject presumptive host females with human chorionic gonadotropin (hCG) to induce ovulation. On the morning of the transfer day, sort healthy mature oocytes with visible "white spots" from germinal vesicle breakdown and color with desired vital dyes. The chosen host female is anesthetized and colored oocytes are transplanted using intraperitoneal (IP) injection. Three hours later, experimental oocytes are recovered and analyzed according to experimental goals.

Cite this protocol as *Cold Spring Harb Protoc*; doi:10.1101/pdb.prot096982

3. Microinject oocytes directly while in OCM; Ficoll is not needed. Transfer the oocytes to fresh OCM after injection and culture for up to 3 d at 18°C. Include appropriate controls, such as uninjected oocytes and mRNA rescued oocytes, in the case of oligonucleotide depletions, or control mRNAs for overexpression experiments.

> *For a detailed description of microinjection see Sive et al. (2010b). Antisense oligonucleotides are typically injected 36–48 h before fertilization, whereas mRNAs can be injected the day before host transfer and fertilization. Typical doses for antisense oligonucleotides are 2–6 ng for modified DNA-based oligonucleotides and 10–50+ ng for morpholino oligonucleotides. The optimal dose should be empirically determined in pilot studies. For overexpression/gain-of-function studies, high doses of mRNAs of 1 ng or more can be used. When rescuing a mRNA knockdown effect, minimal doses are used (20–200 pg), ideally below the phenotypic threshold, so that rescue of the depletion can be accurately assessed as opposed to an ectopic overexpression effect. For genome editing, equimolar amounts (~2 µM each) of Cas9 protein and appropriate guide RNAs are assembled for 10 min at 37°C and ~300 pg sgRNA/1.5 ng Cas9 protein are injected.*

In Vitro Oocyte Maturation and Preparation of Host Female

4. At the end of the day before transfer, add progesterone to each dish of oocytes (~75–200 in 8 mL OCM per dish). Use 1.6 µL of 10 mM progesterone stock per 8 mL OCM (2 µM final concentration). Swirl to mix and incubate overnight at 18°C. Also, place the anesthetic solution to warm overnight at 18°C.

5. Prime two-to-three prospective host females by injecting human chorionic gonadotropin (hCG, ~1000 U) into the dorsal lymph sac to induce ovulation. This is best done 10–12 h before oocyte implantation.

   > *Albino females may be used as hosts to facilitate identification of transferred oocytes. Albinos may also be used as oocyte donors if desired, and if identifying animal-vegetal poles is not critical.*

6. Check the oocytes for maturation, white spots indicating germinal vesicle breakdown.

   > *If oocytes are stuck together, gently tease them apart otherwise this will cause blockage during the transfer.*

   > *See Troubleshooting.*

7. Thaw the vital dye stocks and centrifuge briefly when thawed. Add 80 µL of dye stock to each group of oocytes in 8 mL OCM and swirl gently to mix.

   > *Only six different colors can be reliably made using standard vital dyes. For up to three groups, use the single dyes, Brown, Blue, and Red. Dyes can be combined to generate additional colors (Brown + Blue = Green [good for uninjected controls], Blue + Red = Mauve, Brown + Red = Orange [often hard to distinguish from either plain Brown or Red]).*

8. Incubate the oocytes in vital dyes with rocking for 10+ min at room temperature and then wash all oocytes together in a larger volume of OCM. However, make sure the colors are easily distinguishable before doing this.

   > *If oocyte colors are not easily distinguished, extend the staining time or wash separately and transfer into different host females.*

9. Place a host female into anesthetic solution for 5–10 min until she is nonresponsive to a toe-pinch or to being held upside-down.

   > *This can be done while oocytes are coloring. Choose a host that is laying healthy eggs and that has not yet ovulated many of her own eggs. Avoid females with stringy eggs or that are not obviously extruding eggs when squeezed.*

Intraperitoneal Injection of Donor Oocytes

10. Place the anesthetized female in dorsal recumbency on a damp Kimwipe.

11. Prepare the injection syringe by fitting a new 16G1 needle to the syringe. Set aside the plunger. Wash the inside of the syringe and needle with OCM to coat the surface. This reduces the potential for oocyte sticking.

12. Hold the empty syringe with needle next to the abdomen of the frog. With the beveled edge up, insert the needle tip at a 45° angle, making sure to penetrate skin and muscle. The skin surface can be braced with a gloved finger adjacent to the injection site if the needle does not pass smoothly.

13. Hold the inserted syringe with the nondominant hand. Collect colored oocytes using a sterile, fire-polished glass Pasteur pipette in the dominant hand and introduce into the open end of the syringe. Oocytes will drain down the side and pool at the bottom but should not flow into the needle right away.

14. Elevate the end of the syringe so it is perpendicular to the frog surface, keeping the needle in the frog. Oocytes should immediately flow by gravity into the needle and body cavity. If not, tap the side of the syringe gently. If oocytes are trapped around the outlet port flush them with additional OCM. Keep the overall volume to a minimum because excess OCM (>3–4 mL) can cause the frog to cease egg laying.

15. Return the syringe to its original insertion angle and withdraw the needle. The oocytes should not spill out from the incision, although some OCM may leak out initially. The needle hole does not need to be sutured. Rinse the host female in water and allow recovery in a shallow volume of water. Once recovered, move to a larger container. Recovery is indicated by gulping and eye-bulging.

16. Observe the host for resumption of egg-laying. This should occur within 20–30 min and the donor colored eggs should appear 2–3 h after implantation.

In Vitro Fertilization of Host-Transferred Oocytes

17. Isolate testes from a male frog through nonsurvival surgery as described in Sive et al. (2007a). Store the testes in a Petri dish of OCM at room temperature for use the same day or at 4°C for up to 1 wk.

18. Prepare a sperm suspension by homogenizing ∼1/4 of a testis in 2–3 mL 0.3× MMR.

19. Manually squeeze eggs from the female into a Petri dish or place the female in high salt MMR (1.2× MMR) solution to collect eggs. Rinse eggs collected in 1.2× MMR in 0.3× MMR. Blot off excess liquid and fertilize by adding the sperm suspension (from Step 18). Incubate fertilized eggs at the desired temperature and handle as normal embryos.

    See Troubleshooting.

20. Dejelly cleaving eggs at the 2–4 cell stage using 2% cysteine as described in Sive et al. (2007b), and sort when colors are easily distinguished.

    See Troubleshooting.

TROUBLESHOOTING

Problem (Step 6): Oocytes fail to mature after progesterone treatment.
Solution: Suspect poor oocyte quality (see Protocol 1: Isolation of *Xenopus* Oocytes [Newman et al. 2018]). Oocytes can be tested for maturation by treating a small batch with progesterone immediately after isolation before proceeding. If the ovary is deemed unsuitable, select another female and perform another surgery.

Problem (Step 19): Experimental oocytes fail to be oviposited by the host female.
Solution: This can occur if the female stops laying eggs for any reason. There is no good recourse for recovering eggs from the body cavity or oviducts; therefore, this situation is best prevented by selecting a host female that has just begun to oviposit eggs on her own (without squeezing) and that is laying high quality eggs. Rapid changes in water temperature and remaining under anesthesia for too long are also correlated with early cessation of egg-laying.

     Cite this protocol as *Cold Spring Harb Protoc*; doi:10.1101/pdb.prot096982

Problem (Step 20): Experimental oocytes fail to fertilize and develop.

Solution: This is the most common problem with this method and can generally be avoided by using only healthy, undamaged donor oocytes obtained by careful manual defolliculation. Also, avoid contamination of the OCM and be sure that MMR is correctly made and is at pH 7.6–7.8. Host-transferred oocytes also require higher amounts of sperm and/or longer incubation in the sperm suspension compared to a typical in vitro fertilization in *Xenopus.*

DISCUSSION

The early development of amphibians relies on maternal factors stored and localized in the egg during oogenesis (for review, see Houston 2013) and *Xenopus* is advantageous for studying the roles of these maternal gene products. Classic studies in other amphibians (Rugh 1935; Humphries 1956; Aplington 1957; Arnold and Shaver 1962; Lavin 1964; Smith et al. 1968) laid the foundation for *Xenopus* oocytes to be cultured, manipulated in vitro, and fertilized following transfer into the body cavity of a host female (Brun 1975). These host-transferred eggs are transported by peritoneal cilia to the ostium and oviduct, acquiring surface modifications and jelly coats needed for fertilization (Rugh 1935). This method was revived by Heasman and Wylie (Holwill et al. 1987; Heasman et al. 1991; Torpey et al. 1992) and coupled with antisense oligonucleotide injection into oocytes to elucidate basic mechanisms of axis formation, germ layer patterning and germline specification (for review, see Heasman 2006). Additionally, prefertilization injection of genome editing reagents has recently been used to generate highly efficient mutagenesis in the F_0 generation (Miyamoto et al. 2015a,b; Nakajima and Yaoita 2015; Ratzan et al. 2017; Aslan et al. 2017), providing a renewed rationale for performing the host-transfer method.

RECIPES

Anesthetic Solution (Buffered MS222)

Dissolve 2 g of MS222 (3-aminobenzoic acid ethyl ester; Sigma-Aldrich A5040) in 2 L of AmQuel-treated tap water. Add 1.4 g sodium bicarbonate; the resulting pH should be ~7.0. Store in a foil-covered container at 4°C for up to 1 mo. Warm to room temperature before use.

L-15 Oocyte Culture Medium (OCM; 70%)

Using an autoclaved glass cylinder, add 214 mL of deionized distilled water (16–18 MΩ·cm at 25°C) to 500 mL of Leibovitz's L-15 medium containing 2 mM L-glutamine (Gibco/Thermo-Fisher 11415064) in a sterile bottle or in the sterile L-15 bottle itself. Add 0.3 g of bovine serum albumin (BSA, fraction V) and 3.6 mL of 200× Penicillin-Streptomycin solution (10,000 U/mL, 200× stock; Gibco/Thermo-Fisher 15140122). Adjust the pH to ~7.6–7.8 with 5 N NaOH (3–4 drops). (The phenol red indicator dye should be dark red/light purple. The pH is best checked by measuring a small sample in a separate beaker to avoid contaminating the OCM with the pH electrode.) Make fresh weekly and store at 16°C–18°C (or at 4°C for longer term storage).

This version of OCM with 70% L-15 results in more robust oocytes compared with other OCM recipes. The BSA in this recipe can be substituted with 0.05% poly(vinyl alcohol) (Sigma-Aldrich P8136) with equivalent results.

Marc's Modified Ringer's (MMR) Solution–Triple HEPES (10×)

Reagent	Quantity	Final concentration (10×)
NaCl	58.44 g	1 M
KCl	1.34 g	18 mM
$CaCl_2 \cdot 2H_2O$	2.94 g	20 mM
$MgCl_2 \cdot 6H_2O$	2.03 g	10 mM
HEPES (molecular grade acid, not the sodium salt)	35.75 g	150 mM

Combine the above solids in ~700 mL of deionized distilled water (16–18 MΩ·cm at 25°C). Adjust pH to 7.6–7.8 with NaOH, and bring the volume to 1 L. Filter-sterilize but do not autoclave. Store at 4°C. Readjust to pH 7.6–7.8 after dilution to 1× (or to 1.2× for high-salt MMR).

Vital Dye Stocks

Add the following amounts of dye powder to 50 mL of sterile deionized water in separate 50-mL Falcon tubes:

   0.5 g/50 mL Bismarck Brown (Sigma-Aldrich B2759),
   0.125 g/50 mL Neutral Red (Sigma-Aldrich 861251), and
   0.05 g/50 mL Nile Blue A (Nile Blue sulfate; Sigma-Aldrich N0766).

Rock for 2 h to dissolve, and then centrifuge at ~3000*g*. (The final concentrations will be 1%, 0.25%, and 0.1% for Bismarck Brown, Neutral Red, and Nile Blue A, respectively.) Aliquot the supernatant in 1.5-mL microcentrifuge tubes (~1-mL aliquots) and store at −20°C.

ACKNOWLEDGMENTS

Work in my laboratory is supported by National Institutes of Health (NIH) grant GM083999 (D.W.H.) and by The University of Iowa (D.W.H.).

REFERENCES

Aplington H. 1957. The insemination of body cavity and oviducal eggs of amphibia. *Ohio J Sci* **57**: 91–99.

Arnold J, Shaver J. 1962. Interfemale transfer of eggs and ovaries in the frog. *Exp Cell Res* **27**: 150–153.

Aslan Y, Tadjuidje E, Zorn AM, Cha SW. 2017. High-efficiency non-mosaic CRISPR-mediated knock-in and indel mutation in F0 *Xenopus*. *Development* **144**: 2852–2858.

Brun RB. 1975. Oocyte maturation in vitro: Contribution of the oviduct to total maturation in *Xenopus laevis*. *Experientia* **31**: 1275–1276.

Heasman J. 2006. Maternal determinants of embryonic cell fate. *Semin Cell Dev Biol* **17**: 93–98.

Heasman J, Holwill S, Wylie CC. 1991. Fertilization of cultured *Xenopus* oocytes and use in studies of maternally inherited molecules. *Methods Mol Biol* **36**: 213–230.

Holwill S, Heasman J, Crawley C, Wylie CC. 1987. Axis and germ line deficiencies caused by u.v. irradiation of *Xenopus* oocytes cultured in vitro. *Development* **100**: 735–743.

Houston DW. 2013. Regulation of cell polarity and RNA localization in vertebrate oocytes. *Int Rev Cell Mol Biol* **306**: 127–185.

Hulstrand AM, Schneider PN, Houston DW. 2010. The use of antisense oligonucleotides in *Xenopus* oocytes. *Methods* **51**: 75–81.

Humphries A. 1956. A study of meiosis in coelomic and oviductal oocytes of *Triturus virisescens*, with particular emphasis on the origin of spontaneous polyploidy and the effects of heat shock on the first meiotic division. *J Morphol* **99**: 97–135.

Lavin L. 1964. The transfer of coelomic eggs between frogs. *J Embryol Exp Morph* **12**: 457–463.

Manual defolliculation of *Xenopus* oocytes. 2009. YouTube video posted by "uiowadblab" on March 31, 2009. www.youtube.com/watch?v=us8r DNG69Sk.

Miyamoto K, Simpson D, Gurdon JB. 2015a. Manipulation and in vitro maturation of *Xenopus laevis* oocytes, followed by intracytoplasmic sperm injection, to study embryonic development. *J Vis Exp* e52496. doi: 10.3791/52496.

Miyamoto K, Suzuki K-IT, Suzuki M, Sakane Y, Sakuma T, Herberg S, Simeone A, Simpson D, Jullien J, Yamamoto T, et al. 2015b. The expression of TALEN before fertilization provides a rapid knock-out phenotype in *Xenopus laevis* founder embryos. *PLoS One* **10**: e0142946.

Nakajima K, Yaoita Y. 2015. Highly efficient gene knockout by injection of TALEN mRNAs into oocytes and host transfer in *Xenopus laevis*. *Biol Open* **4**: 180–185.

Newman K, Aguero T, King ML. 2018. Isolation of *Xenopus* oocytes. *Cold Spring Harb Protoc* doi: 10.1101/pdb.prot095851.

Olson DJ, Hulstrand AM, Houston DW. 2012. Maternal mRNA knock-down studies: Antisense experiments using the host-transfer technique in *Xenopus laevis* and *Xenopus tropicalis*. *Methods Mol Biol* **917**: 167–182.

Ratzan W, Falco R, Salanga C, Salanga M, Horb ME. 2017. Generation of a *Xenopus laevis* F1 albino J strain by genome editing and oocyte host-transfer. *Dev Biol* **426**: 188–193.

Cite this protocol as *Cold Spring Harb Protoc*; doi:10.1101/pdb.prot096982

Rugh R. 1935. Ovulation in the frog. II. Follicular rupture to fertilization. *J Exp Zool* **71**: 163–194.

Schneider P, Hulstrand A, Houston D. 2010. Fertilization of *Xenopus* oocytes using the host transfer method. *J Vis Exp* e1864. doi: 10.3791/1864.

Sive HL, Grainger RM, Harland RM. 2007a. Isolating *Xenopus laevis* testes. *Cold Spring Harb Protoc* doi: 10.1101/pdb.prot4735.

Sive HL, Grainger RM, Harland RM. 2007b. Dejellying *Xenopus laevis* embryos. *Cold Spring Harb Protoc* doi: 10.1101/pdb.prot4731.

Sive HL, Grainger RM, Harland RM. 2010a. Isolation of *Xenopus* oocytes. *Cold Spring Harb Protoc* doi: 10.1101/pdb.prot5534.

Sive HL, Grainger RM, Harland RM. 2010b. Microinjection of *Xenopus* oocytes. *Cold Spring Harb Protoc* doi: 10.1101/pdb.prot5536.

Smith LD, Ecker RE, Subtenly S. 1968. In vitro induction of physiological maturation in Rana pipiens oocytes removed from their ovarian follicles. *Dev Biol* **17**: 627–643.

Smith LD, Xu WL, Varnold RL. 1991. Oogenesis and oocyte isolation. *Methods Cell Biol* **36**: 45–60.

Torpey N, Wylie CC, Heasman J. 1992. Function of maternal cytokeratin in *Xenopus* development. *Nature* **357**: 413–415.

Heterologous Protein Expression in the *Xenopus* Oocyte

Jonathan S. Marchant[1,2]

Department of Pharmacology, University of Minnesota, Minneapolis, Minnesota 55455

The *Xenopus* oocyte is a specialized single cell of colossal size (>1 mm diameter) that is highly amenable for microinjection and a stalwart model for heterologous expression. Oocytes are easily obtainable, robust in vitro, and faithfully express injected constructs. Their large size translational capacity provides a huge canvas for observing and recording integrated cellular responses—from studies of single molecules within single cells to medium-throughput drug-screening applications. Most eukaryotic promoters suffice for *Xenopus* expression, and the oocyte can functionally express proteins from many diverse organisms. This protocol provides a basic introduction for scientists keen to perform nuclear microinjections of cDNA constructs. These are easy methods to master, do not require elaborate equipment, and make accessible a wonderful model cell system for studying signaling, transport, cell architecture, and protein function.

MATERIALS

It is essential that you consult the appropriate Material Safety Data Sheets and your institution's Environmental Health and Safety Office for proper handling of equipment and hazardous materials used in this protocol.

RECIPES: Please see the end of this protocol for recipes indicated by <R>. Additional recipes can be found online at http://cshprotocols.cshlp.org/site/recipes.

Reagents

cDNA (50–200 µg/mL)

Mineral oil (Sigma-Aldrich M5310)

Modified Barth's solution (MBS) <R>

In my laboratory, we use MBS not only for oocyte incubation but also for oocyte preparation simply owing to paranoia over oocyte quality. Ringer's solution is an acceptable surrogate during routine oocyte preparation. Solutions may be supplemented with antibiotics other than gentamycin (for example, 1% penicillin/streptomycin).

Xenopus (adult females)

Source adult female frogs from Xenopus Express (www.xenopus.com), Xenopus 1 (www.xenopus1.com), or Nasco (www.enasco.com). Guidance for maintaining a frog colony can be found elsewhere (Koustubhan et al. 2008; Delpire et al. 2011).

Oocytes pre-prepared for injection are also available (http://ecocyte-us.com); this can be a convenient option for infrequent or novice users. Oocytes from species other than Xenopus laevis (e.g., X. tropicalis or X. borealis) can be used. They are smaller and harder to inject compared with X. laevis oocytes, but advantages have been noted (Marchant and Parker 2001; Cristofori-Armstrong et al. 2015). Xenopus resource centers provide additional strains/ lines (Pearl et al. 2012). Xenbase (www.xenbase.org) is an extensive portal collating Xenopus community resources.

---

[1]Correspondence: JMarchant@mcw.edu

[2]Present address: Department of Cell Biology, Neurobiology and Anatomy, Medical College of Wisconsin, Milwaukee, Wisconsin 53226.

Copyright © 2023 Cold Spring Harbor Laboratory Press; all rights reserved
Cite this protocol as *Cold Spring Harb Protoc*; doi:10.1101/pdb.prot096990

Equipment

Borosilicate glass capillary tubes for making pulled pipettes (type dependent on microinjection equipment)

Dissection equipment

    Surgical scissors

    Watchmaker's forceps (Dumont #5 and #55)

Glass vials (20 mL; Research Products International 121001)

Incubator (16°C–20°C)

Light box with gooseneck illuminator

Microinjection setup (see Step 7)

    Microinjection apparatus with nanoliter precision (with foot pedal, preferred)

    Micromanipulator (three-axis, preferred)

Micropipette puller (see Step 5)

Nutator (optional; see Step 3)

Oocyte holders for microinjection (see Step 8)

In my laboratory, we use customized plates (Petri dishes with secured nylon/polypropylene mesh [0.5–0.8 mm, Small Parts Inc.]).

Petri dishes

Stereomicroscopes

Separate stations for oocyte preparation and microinjection are needed.

METHOD

Bypass Steps 1 and 2 by purchasing prepared oocytes from commercial vendors.

1. Dissect ovarian lobes from donor frogs (see Sive et al. 2010) and place in a Petri dish containing MBS.

2. Manually isolate at least 200 (or more, depending on the number/combination of constructs for injection) stage V–VI oocytes under a stereomicroscope by carefully "plucking" the larger (>1 mm diameter) oocytes from surrounding tissue while removing the enveloping follicular cell layers, which can impede clean penetration of the microinjection needle. Use two pairs of fine watchmaker's forceps to grasp the tissue connecting individual oocytes to the ovarian lobe, and gently move the forceps apart to tease back the follicular cell layers, like peeling skin from a grape.

   Each oocyte will deform and pop through the resulting opening as the enveloping cell layer is ruptured and removed by forceps passing around the cell circumference. This takes practice, and many cells will be damaged or burst. Periodically restart with a fresh lobe in a new Petri dish.

   This is the preferred method for oocyte isolation in my laboratory. Alternatively, bulk preparation of oocytes can be achieved by enzymatic defolliculation as described in Protocol 1: Isolation of Xenopus Oocytes (Newman et al. 2018).

   See Troubleshooting.

3. Separate the oocytes into 20-mL glass vials containing MBS (~50 oocytes per vial). Allow the oocytes (whether isolated manually or enzymatically) to recover in MBS overnight in an incubator at 16°C–20°C.

   If the oocytes were isolated enzymatically, perform overnight recovery on a nutator.

4. After overnight incubation, remove damaged oocytes. Discard cells with mottled pigment around the animal pole. Replace the MBS.

   Healthy oocytes retain a clearly demarked asymmetry in pigment between the animal (pigmented) and vegetal hemisphere.

   See Troubleshooting.

5. Prepare multiple needles for microinjection according to the instructions of the micropipette puller used.

   A sharp needle preserves oocyte viability, especially for nuclear microinjection. Needles should have a reasonable shank (up to 900 µm taper from tip) and a tip of ~15 µm (outer diameter) after breakage under a microscope. A basic micropipette puller will meet these requirements.

6. Prepare the oocytes for microinjection. Under a stereomicroscope, align ~100 oocytes per construct within a nylon/polypropylene restraining mesh secured within a Petri dish containing MBS. Manipulate the oocytes to sit vegetal side down, with the animal (pigmented) pole upwards.

   For cDNA injections, the oocyte nucleus (or germinal vesicle) must be injected. The germinal vesicle is large (40 nL in volume) and rests within the animal hemisphere. Oocytes must therefore be oriented animal pole up for microinjection.

   For expression studies, oocytes can be injected with mRNA into the vegetal cytoplasm (see Protocol 3: Microinjection of Xenopus Oocytes [Aguero et al. 2018]) or with cDNA into the nucleus. Cytoplasmic injections are straightforward and well tolerated; injection is even possible before defolliculation (Maldifassi et al. 2016). Compared with cytoplasmic injection, germinal vesicle injection requires more time and results in poorer oocyte viability because of germinal vesicle damage, but cDNA constructs require less preparation and express through endogenous targeting pathways.

7. Load the needle on the microinjector with the solution containing cDNA (<3 ng of cDNA is injected into each oocyte) according to the instructions of the microinjector used.

   Depending on the equipment used, this step commonly involves back-filling the needle with mineral oil, breaking the tip, and then front-filling with injection solution. Different microinjectors are available, ranging from handheld injectors (for mRNA injection), to widely used and recommended plunger-based systems (e.g., Drummond Nanoject II and III), and even automated injectors (Schnizler et al. 2003; Papke and Stokes 2010). Any stably mounted microinjector that reproducibly dispenses nanoliter volumes is suitable.

8. Align an oocyte under the injection needle, and manipulate the needle to the oocyte surface. Observe the oocyte surface dimple under pressure before penetration, and then watch as the needle disappears into the cell upon penetration. (At this point, the needle will be far enough into the cell nucleus for injection.) Inject once, pause, and then gently withdraw the needle. Move to the next oocyte and repeat. Periodically withdraw the tip of the needle from the solution to check that the needle remains unclogged.

   The large volume of the oocyte nucleus (~40 nL) can tolerate a surprisingly large injection volume (<15 nL). We secure the Petri dish containing the oocytes on a moveable stage that can be ratcheted to position cells for injection. This maximizes throughput and minimizes risk of inadvertent needle damage.

   See Troubleshooting.

9. After microinjection, place 20–30 oocytes into individual 20-mL glass vials containing 10–15 mL of MBS, and return to the 16°C–20°C incubator. Remove apoptotic oocytes daily to prevent deleterious effects on healthy cells.

   See Troubleshooting.

10. Screen oocytes for expression of injected cDNA.

    The method used for examining expression will depend on experimental goals and may involve imaging, electrophysiology, radioisotope flux, or western blotting. Oocytes can be screened as early as 24 h after microinjection, although expression usually peaks over the following days.

    See Troubleshooting.

TROUBLESHOOTING

Problem (Step 2): The quality of harvested oocytes is poor.
Solution: Oocyte quality is critical for success. If oocyte quality from multiple donor frogs is consistently poor, but better from freshly sourced animals, there is likely to be a husbandry problem (e.g., inappropriate water composition/pH, nutrition, or the presence of infectious agents) (Kous-

Cite this protocol as *Cold Spring Harb Protoc*; doi:10.1101/pdb.prot096990

tubhan et al. 2008; Delpire et al. 2011). If a husbandry issue is suspected, consult veterinary staff. Even under seemingly identical housing conditions, seasonal variations in oocyte quality occur. During these periods it may be simpler to directly source commercially prepared oocytes.

Problem (Step 4): The oocytes display poor viability in vitro.

Solution: Monitor the viability of uninjected oocytes in parallel with mock-injected oocytes to discriminate between problems with culture media/conditions and poor microinjection technique. If necessary, remake buffers with careful attention to buffer osmolarity/pH. If black spots appear on the oocyte surface, or the pigment shows excessive marbling, microbial contamination is likely (O'Connell et al. 2011). In this case, supplement with fresh antibiotics. Poor viability after isolation by enzymatic digestion can result from excessive collagenase exposure. This can be prevented by the manual defolliculation of oocytes, and although this procedure takes practice, oocyte quality is better.

Problem (Step 8): The needle does not dispense fluid.

Solution: Ensure that no air bubbles or clogged material became trapped within the needle during filling. The newer Nanoject III (Drummond) streamlines needle mounting procedures. If the needle is blocked, prepare a new needle.

Problem (Step 9): The cells die after microinjection.

Solution: A poorly fashioned microinjection needle or excessive injection volume will cause significant cell death the day after injection. When injecting, check for evidence of damage to the oocyte surface (a persistent white wound). If this occurs, then the needle is too broad or blunt, and new needles should be prepared. Contaminants in the injected material can cause cell death. Check the purity of injected material; plasmid DNA preparations can be purified with an endotoxin-free plasmid preparation kit or a polymerase chain reaction (PCR) purification kit. Expressed proteins may also prove deleterious to oocyte viability.

Problem (Step 10): There is weak construct expression.

Solution: Low expression efficiency will result from a clogged microinjection needle. Any precipitates in the injected solution can be removed by centrifugation before backfilling. For cDNA injections, better visualization of the nucleus will help. This can be achieved by giving oocytes a brief pulse of centrifugation in a benchtop microcentrifuge to raise the nucleus toward the surface, or by using albino oocytes to visualize the nucleus under transillumination. Expression does vary from donor to donor (in any batch, 20%–80% of injected oocytes will express the injected construct) and from cell to cell (e.g., in observed fluorophore intensity or peak current magnitude).

RELATED INFORMATION

This protocol is an update to previously published procedures (see Lin-Moshier and Marchant 2013a,b,c).

Further information on the isolation of *Xenopus* oocytes can be found in Sive et al. (2000) and in Protocol 1: Isolation of *Xenopus* Oocytes (Newman et al. 2018). Different laboratories use iterated versions of these core methods, which can be consulted in conjunction with cytoplasmic oocyte microinjection protocols (see Protocol 3: Microinjection of *Xenopus* Oocytes [Aguero et al. 2018]) and helpful video resources that are available online (Cohen et al. 2009; Maldifassi et al. 2016).

RECIPE

Modified Barth's Solution (MBS)

Reagent	Final concentration (1×)
NaCl	88 mM
KCl	1 mM
NaHCO$_3$	2.4 mM
MgSO$_4$·7H$_2$O	0.82 mM
Ca(NO$_3$)$_2$·4H$_2$O	0.33 mM
CaCl$_2$·2H$_2$O	0.41 mM
HEPES	5 mM

Prepare a stock solution (5×, pH 7.4), which can be stored for several months at −20°C. Prepare working solutions by dilution, and autoclave prior to use. After autoclaving, add gentamycin (final concentration in 1× solution, 50 µg/mL).

ACKNOWLEDGMENTS

Work in my laboratory is supported by the National Institutes of Health (NIH) (GM088790).

REFERENCES

Aguero T, Newman K, King ML. 2018. Microinjection of *Xenopus* oocytes. *Cold Spring Harb Protoc* doi:10.1101/pdb.prot096974.

Cohen S, Au S, Panté N. 2009. Microinjection of *Xenopus laevis* oocytes. *J Vis Exp* 24: e1106.

Cristofori-Armstrong B, Soh MS, Talwar S, Brown DL, Griffin JD, Dekan Z, Stow JL, King GF, Lynch JW, Rash LD. 2015. *Xenopus borealis* as an alternative source of oocytes for biophysical and pharmacological studies of neuronal ion channels. *Sci Rep* 5: 14763.

Delpire E, Gagnon KB, Ledford JJ, Wallace JM. 2011. Housing and husbandry of *Xenopus laevis* affect the quality of oocytes for heterologous expression studies. *J Am Assoc Lab Anim Sci* 50: 46–53.

Koustubhan P, Sorocco D, Levin MS. 2008. Establishing and maintaining a *Xenopus laevis* colony for research laboratories. In *Sourcebook of Models for Biomedical Research* (ed. Conn PM), pp. 139–160. Humana Press, Totowa, NJ.

Lin-Moshier Y, Marchant JS. 2013a. Nuclear microinjection to assess how heterologously expressed proteins impact Ca^{2+} signals in *Xenopus* oocytes. *Cold Spring Harb Protoc* doi:10.1101/pdb.prot072785.

Lin-Moshier Y, Marchant JS. 2013b. A rapid western blotting protocol for the *Xenopus* oocyte. *Cold Spring Harb Protoc* doi:10.1101/pdb.prot072793.

Lin-Moshier Y, Marchant JS. 2013c. The *Xenopus* oocyte: A single-cell model for studying Ca^{2+} signaling. *Cold Spring Harb Protoc* doi:10.1101/pdb.top066308.

Maldifassi MC, Wongsamitkul N, Baur R, Sigel E. 2016. *Xenopus* oocytes: Optimized methods for microinjection, removal of follicular cell layers, and fast solution changes in electrophysiological experiments. *J Vis Exp* 118: e55034.

Marchant JS, Parker I. 2001. *Xenopus tropicalis* oocytes as an advantageous model system for the study of intracellular Ca^{2+} signalling. *Br J Pharmacol* 132: 1396–1410.

Newman K, Aguero T, King ML. 2018. Isolation of *Xenopus* oocytes. *Cold Spring Harb Protoc* doi:10.1101/pdb.prot095851.

O'Connell D, Mruk K, Rocheleau JM, Kobertz WR. 2011. *Xenopus laevis* oocytes infected with multi-drug-resistant bacteria: Implications for electrical recordings. *J Gen Physiol* 138: 271–277.

Papke RL, Stokes C. 2010. Working with OpusXpress: Methods for high volume oocyte experiments. *Methods* 51: 121–133.

Pearl EJ, Grainger RM, Guille M, Horb ME. 2012. Development of *Xenopus* resource centers: The national *Xenopus* resource and the European *Xenopus* resource center. *Genesis* 50: 155–163.

Schnizler K, Kuster M, Methfessel C, Fejtl M. 2003. The roboocyte: Automated cDNA/mRNA injection and subsequent TEVC recording on *Xenopus* oocytes in 96-well microtiter plates. *Receptors Channels* 9: 41–48.

Sive HL, Grainger RM, Harland RM. 2000. *Early development of* Xenopus laevis: *A laboratory manual.* Cold Spring Harbor Laboratory Press, Cold Spring Harbor, N.Y.

Sive HL, Grainger RM, Harland RM. 2010. Isolation of *Xenopus* oocytes. *Cold Spring Harb Protoc* doi:10.1101/pdb.prot5534.

Cite this protocol as *Cold Spring Harb Protoc*; doi:10.1101/pdb.prot096990

Patch-Clamp and Perfusion Techniques to Study Ion Channels Expressed in *Xenopus* Oocytes

Guohui Zhang and Jianmin Cui[1]

Department of Biomedical Engineering, Center for the Investigation of Membrane Excitability Disorders, Cardiac Bioelectricity and Arrhythmia Center, Washington University, St. Louis, Missouri 63130

The *Xenopus* oocyte expression system is ideal for electrophysiological characterization of voltage-dependent and ligand-dependent ion channels because of its relatively low background of endogenous channels and the large size of the cell. Here, we present a protocol to study voltage- and ligand-dependent activation of ion channels expressed in *Xenopus* oocytes using patch-clamp techniques designed to control both the membrane voltage and the intracellular solution. In this protocol, the large conductance voltage- and Ca^{2+}-activated K^+ (BK) channel is studied as an example. After injection of BK channel mRNA, oocytes are incubated for 2–7 d at 18°C. Inside-out membrane patches containing single or multiple BK channels are excised with perfusion of different solutions during recording. The protocol can be used to study structure–function relations for ion channels and neurotransmitter receptors.

MATERIALS

It is essential that you consult the appropriate Material Safety Data Sheets and your institution's Environmental Health and Safety Office for proper handling of equipment and hazardous materials used in this protocol.

RECIPES: Please see the end of this protocol for recipes indicated by <R>. Additional recipes can be found online at http://cshprotocols.cshlp.org/site/recipes.

Reagents

Basal internal solution <R>

To generate perfusion solutions containing free Ca^{2+} concentrations ($[Ca^{2+}]_i$) between 0 and 500 μM, add different amounts of 1 M $CaCl_2$ to this base solution. See the user's manual for the calcium combination ion-selective electrode (see Equipment, below) for how to calculate the amount of $CaCl_2$ to add and how to measure the final $[Ca^{2+}]_i$ of the solution.

Bath solution (Ca^{2+}-, Mg^{2+}-free) <R>

Xenopus oocytes express a large endogenous Ca^{2+}-activated Cl^- current that can interfere with BK channel currents when the intracellular solution contains free Ca^{2+}. To avoid this contamination, a Cl^--free solution usually is preferred. However, 2 mM KCl is included in this solution to maintain steady electric connection with the Ag/AgCl ground electrode.

Bleach

---

[1]Correspondence: jcui@wustl.edu

CaCl$_2$ (1 M)

Deionized (DI) water

Messenger RNA (mRNA) of the gene of interest (0.001–1.0 µg/µL)

Mineral oil

ND96 solution <R>

Oocyte dissociation solution <R>

Oocyte stripping solution <R>

Pipette solution for patch clamps <R>

Xenopus oocytes

> *Oocytes can be purchased commercially (e.g., from EcoCyte Bioscience or Nasco) or extracted from Xenopus ovaries (Sive et al. 2000; Green 2009; see also Sive et al. 2010a,b and Protocol 1: Isolation of Xenopus Oocytes [Newman et al. 2018]).*

Equipment

Acquisition interfaces (e.g., HEKA Elektronik InstruTECH ITC-18)

Acquisition software (e.g., HEKA Elektronik Patchmaster)

Calcium combination ion selective electrode (ISE) (Hanna Instruments, Inc. #HI4104)

Cell culture plates (flat-bottom, 48-well, with lid)

Dissection microscope

Forceps (Dumont #5)

Glass capillaries (3.5-in.; Drummond Scientific #3-000-203-G/X)

Headstage (e.g., Molecular Devices #CV 203BU)

Incubator (16°C–20°C)

Manipulator controller (e.g., Siskiyou Corp. MC1000e controllers)

Microcentrifuge tubes (1.5-mL)

Microforge (e.g., World Precision Instruments)

Microinjector (e.g., Drummond Scientific Co. Nanoject II Auto-Nanoliter Injector)

Micromanipulator

Micropipette puller (e.g., Sutter Instruments P-97)

Micropipettes (100-µL, disposable, calibrated; VWR International #53432-921)

Microscope (inverted; e.g., Olympus Corp. Model CKX31)

Nylon mesh (optional; see Step 7)

Patch-clamp amplifier (e.g., Molecular Devices Axopatch 200B Amplifier)

Perfusion system (e.g., ALA Scientific Instruments OctaFlow II Multi-function Multi-valve)

Petri dishes

Sticky wax (Kerr Australia Pty. Ltd. #00625)

METHOD

Oocyte Preparation

If fewer than 100 oocytes are needed each time, it might be more efficient to order them from a commercial source or to defolliculate them manually using watchmaker's forceps (Sive et al. 2000; see also Sive et al. 2010a).

1. Take 5–15 mL of oocytes from *Xenopus* ovaries through dissection as described in Sive et al. (2010b). Separate them into small clusters (~10–20 oocytes/cluster) using forceps.

   See Troubleshooting.

2. Incubate the clusters in oocyte dissociation solution for 30–90 min at room temperature, until the follicles of some oocytes are broken (as assessed by visual inspection under a microscope).

3. Wash the cells with ND96 solution at least five times to remove any residual enzyme.

Cite this protocol as *Cold Spring Harb Protoc*; doi:10.1101/pdb.prot099051

4. Using a dissection microscope, select stage V and VI oocytes (>0.8 mm in diameter) (Wasserman et al. 1984) that are visibly healthy.

   Injection can be performed on the day of harvesting or the next day, with oocytes incubated overnight in ND96 solution at 18°C.

   See Troubleshooting.

Microinjection Needle Preparation

5. Pull microinjection needles from capillary tubes in one step at 300°C using a P-97 puller according to the manufacturer's instructions.

   The needle should have a long shank (≥5 mm).

6. Using forceps and working under a microscope, break the tip of the needle to a final outer diameter ≤20 μm.

   If the needle size is too big, it can damage oocytes. If it is too small, it can be blocked easily and mRNA cannot be injected.

Microinjection of mRNA into Oocytes

Exercise caution to prevent contamination with ribonuclease (e.g., wear gloves, take care not to allow the injection needles or pipette tips to contact unclean surfaces, etc.). For a more detailed discussion of this procedure, see Sive et al. (2010c) and Protocol 3: Microinjection of Xenopus Oocytes (Aguero et al. 2018).

7. To prevent oocytes from moving during injection, insert a piece of nylon mesh (or scratch a grid using a steel needle) on the bottom of a Petri dish.

8. Load the oocytes in ND96 solution into the dish.

9. According to the microinjector manufacturer's instructions, load 0.5–2.0 μL of mRNA solution into a microinjection needle that has been back-filled with mineral oil.

10. Adjust the needle to the desired injection volume for each oocyte (e.g., 46 nL).

11. Position the needle against one oocyte in the Petri dish. Push the needle in.

    The needle should penetrate the oocyte through the animal pole, or where the vegetal and animal poles meet.

12. Inject the mRNA using the foot pedal or an injection button. Pause for 3 sec to allow the mRNA to flow into oocyte. Gently withdraw the needle.

13. Move to the next oocyte by moving the Petri dish. Repeat the injection on 1–2 dozen oocytes for each type of mRNA.

    Mineral oil will come out when the mRNA solution is used up. Monitor the needle carefully during the procedure to ensure that this does not happen.

14. Rinse the injected oocytes twice with ND96 solution.

15. Place each oocyte into a well of a 48-well culture dish loaded two-thirds full with ND96 solution.

16. Depending on the desired expression level of the ion channels, incubate for 2–7 d at 18°C.

    Monitor oocytes daily to minimize solution evaporation or condensation on the lid. Change the ND96 solution after a couple of days if prolonged incubation is required.

Preparing the Perfusion System

17. Turn on the multichannel perfusion system (Fig. 1). Open the valve of a cylinder of pressurized nitrogen.

18. Wash each solution reservoir tube three times with DI water. Add the desired perfusion solution to the tube.

19. Connect tubing to the perfusion pencil. Open the reservoir valve.

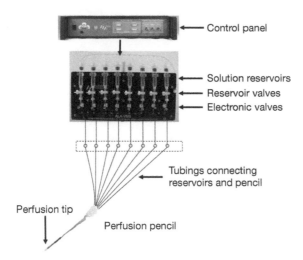

FIGURE 1. The OctaFlow II multi-function multi-valve perfusion system from ALA Scientific Instruments.

20. Use the perfusion control software to turn on the electronic valves one at a time.

    The solution should emerge from the pencil. Allow it to run for 5–10 min to ensure that all of the tubing is filled with solution, and that there are no air bubbles in the system.

21. Use a fixture (or modeling clay) to attach the perfusion pencil to the bath (Fig. 2) that is mounted on the stage of the inverted microscope.

22. Adjust the assembly to position the perfusion tip in the bath. Add DI water to the bath to confirm the perfusion tip is submerged.

23. Monitoring the perfusion tip under the microscope, turn on each valve, one at a time, to ensure the solutions come out of the tips correctly.

    A jet of fluid should be visible coming out of the perfusion tip.

    See Troubleshooting.

24. When all the perfusion solutions flow correctly, replace the DI water in the dish with the bath solution.

    The perfusion system is now ready for use.

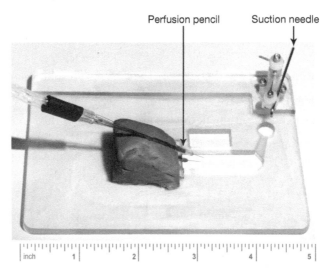

FIGURE 2. Recording bath. The perfusion pencil is mounted with modeling clay. The suction needle is connected to a vacuum and used to control the solution level in the bath.

Cite this protocol as *Cold Spring Harb Protoc*; doi:10.1101/pdb.prot099051

Patch Clamping System Preparation

25. Prepare electrodes for patch clamp recordings:

    i. Remove the silver electrode wire from the pipette holder.

    ii. Immerse three-quarters of the tip of the wire in a 1.5-mL microcentrifuge tube containing fresh bleach for at least 30 min.

    > This will deposit a layer of AgCl on the wire.

    iii. Rinse the wire with DI water. Blot dry. Return the Ag/AgCl electrode to the pipette holder.

    iv. At the same time, connect the ground (i.e., reference) electrode to the headstage. Stabilize it in the bath with a fixture or clay and tape.

26. Prepare the patch pipettes:

    i. Pull glass pipettes using the P97 puller.

    > Refer to the user's manual for the design and set up of a program to prepare the ideal shape for patch clamps.

    ii. Inspect the pipettes under a microscope to determine the shape and opening diameter of the tip.

    > The outer diameter of the opening should be 2–4 μm, and the ideal resistance of the pipette filled with solution is ∼1–1.5 MΩ.

    iii. Dip the pipette tip into melted wax to a depth of 2–5 mm.

    > Coating the pipette tip with melted wax reduces capacitive current during recordings. Sylgard 184 is another good material for coating pipette tips (Rae and Levis 1992; Levis and Rae 1998).

    iv. Fire-polish the pipette tip using a microforge.

27. Dip the tip into the pipette solution under negative pressure for 10 sec until the tip is filled with solution. Then, using a syringe and needle, back-fill the pipette to two-thirds full.

28. Place the pipette in the pipette holder with the Ag/AgCl electrode inserted inside and in contact with the pipette solution.

29. Turn on the patch clamp amplifier.

30. Start the acquisition software in the computer. Load the protocol file and adjust the configuration settings for the experiment according to the manufacturer's instructions.

31. Prepare the oocytes for patch clamping:

    i. Transfer an mRNA-injected oocyte to a Petri dish containing oocyte stripping solution for 5–10 min.

    > The stripping solution detaches the vitelline membrane from the plasma membrane, making it easier to remove.

    ii. Working under a dissection microscope, use two pairs of Dumont #5 forceps to gently strip the vitelline membrane from the oocyte.

    > Some people prefer not to use stripping solution and to directly strip the vitelline membrane using forceps.

    iii. Transfer a devitellinized oocyte to a Petri dish containing bath solution that serves as the recording chamber.

    > The oocyte is extremely fragile after devitellinizing and must be handled with extreme care.
    > See Troubleshooting.

Patch Clamping

32. Mount the recording chamber on the stage of the inverted microscope. Move the stage to find a clear edge of the oocyte under the microscope (Fig. 3).

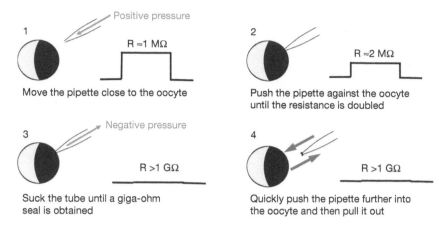

FIGURE 3. Steps to form an inside-out patch. The right side of each panel shows the current step in response to a voltage pulse on the patch clamp monitor.

33. Using a syringe (or by mouth through a suction tube connected to the pipette holder), apply a gentle positive pressure in the pipette to prevent obstructions from attaching to the pipette tip.

34. Maintain the positive pressure in the pipette by closing the suction tube with a valve switch.

35. Use the micromanipulator to move the patch pipette close to the oocyte.

36. Slowly push the pipette against the oocyte until the measured resistance of the patch clamp circuit approximately doubles.

37. Exert gentle and steady suction to switch the positive pressure in the pipette to a negative pressure until a giga-ohm seal is obtained.

    This is a critical step requiring practice to gain skills.

38. Establish the desired patch configuration, as appropriate for the experiment:

    To Excise an Inside-Out Patch

    i. Stabilize the patch by holding it at −50 mV (i.e., close to the resting potential of the oocyte) for 1 min.

    ii. Excise the patch by quickly pushing the pipette further into the oocyte and then withdrawing it to the bath solution.

       The inside-out patch is ready for patch clamp experiments.

    To Excise an Outside-Out Patch

    To study ion channels that are activated by extracellular ligands, such as neurotransmitter receptors, the outside-out patch configuration is preferred.

    iii. After forming a giga-ohm seal, apply a brief strong negative pressure to break the membrane patch.

    iv. Withdraw the pipette slowly.

       The membrane attached to the pipette follows the retreating pipette, and a giga-ohm seal should form again when the membrane breaks and an outside-out patch is excised.
       See Troubleshooting.

39. Move the pipette containing the patch to ∼100 μm away from the opening of the perfusion tip.

40. Turn on both the reservoir and electrode valves to begin the perfusion with the desired solution.

41. Run voltage protocol and record currents across the membrane patch.

    See Troubleshooting.

Cite this protocol as *Cold Spring Harb Protoc*; doi:10.1101/pdb.prot099051

42. Repeat the recordings as appropriate for the experiment using different perfusion solutions and running different voltage protocols.

   After finishing all recordings, turn off all valves of the perfusion system and switch off the microscope and amplifier. Make sure to clean the perfusion tubing, pencil and tip by pushing DI water through to avoid salt precipitation that could clog the system.

TROUBLESHOOTING

Problem (Step 1): Poor initial oocyte quality.

Solution: If the oocytes are discolored or immature (i.e., maturation stage less than level V) right after dissection, it is possible that the frog from which they were obtained is not healthy. Unhealthy frogs can result from poor animal care (water and food quality, feeding schedule, population density, infections, etc.; see Sive et al. 2000; Green 2009) or by microbial contamination. Improvements in husbandry and antibiotic supplementation might solve the problem (O'Connell et al. 2011).

Problem (Step 4): Oocyte quality decreases postdigestion.

Solution: Monitor the digestion process closely to prevent overdigestion. If necessary, decrease the enzyme concentration.

Problem (Step 23): The perfusion solution does not emerge from the perfusion pencil.

Solution: Apply vacuum to the tip of pencil to remove any air bubbles that might be blocking the system. If an air bubble is visible around the reservoir valve, it can be removed by tapping the tubing with fingers. If there is still no solution coming out, the system could be clogged; clear both the tubing and the pencil by sonication in DI water.

Problem (Step 31): Oocyte quality decreases postinjection.

Solution: Make sure the tip of the injection needle is not too large. Do not overinject with mRNA, as this can cause overexpression. Prolonged incubation postinjection without changing the solution can result in altered salt concentration caused by solute evaporation. Microbial contamination can also damage oocytes. Caution and good laboratory practices can be used to avoid these problems.

Problem (Steps 38–42): Patches are difficult to obtain or keep.

Solution: Poor oocyte quality is the most probable reason; see above. Also, if the pipette tip is not fire-polished well or if there is a leak in the pipette holder or suction tubing, it can be hard to obtain a patch successfully. The pipette solution should be fresh (i.e., no older than 18 mo) and filtered. Finally, excessively high perfusion pressure or the presence of air bubbles in the perfusion flow can damage patches.

Problem (Step 41): No current detected.

Solution: If the membrane in the pipette tip forms a sealed vesicle, it can break when the pipette tip is rapidly lifted out of the bath solution and then immediately lowered back into solution. If the concentration of the mRNA was too low (or if the sample has degraded), an insufficient number of channels might have been expressed. It is also possible that a mutation in a known, well-expressed ion channel could destroy that channel's function or shift the voltage dependence of channel activation to extreme positive voltages. The well-expressed wild-type channel can be studied as a positive control.

DISCUSSION

This protocol is based on the study of voltage- and Ca^{2+}-activated BK channels exogenously expressed in *Xenopus laevis* oocytes using patch-clamp techniques (Zhang et al. 2017). By excising inside-out patches, the cytosolic side of the channel can be exposed to the desired solutions, while the voltage across the membrane can be clamped at command potentials. BK channels sense various intracellular signal molecules besides Ca^{2+} (Yang et al. 2015), such as Mg^{2+} (Shi et al. 2002), ethanol (Dopico et al. 1998), and omega-3 fatty acids (Hoshi et al. 2013). This protocol provides an effective way to study the function of the channel by controlling both voltage and intracellular solutions and can be used readily for the study of other voltage- or ligand-dependent ion channels. Furthermore, the oocyte expression system provides an efficient method for the study of structure–function relations of the channel by combining it with mutagenesis and chemical modifications of the channel. Because of their large size, oocytes are also ideal for the study of ion channels using other electrophysiological techniques, such as two-electrode voltage-clamp and voltage-clamp fluorometry (Zaydman et al. 2014), which is an advantage when various functional aspects of an ion channel need to be examined. However, it should be noted that there might be differences in the signaling pathways for posttranslational modification between oocytes and mammalian cells (Kyle and Braun 2014), and oocytes may or may not contain specific protein or chemical factors that associate with the ion channel of interest and modify channel function (Sanguinetti et al. 1996). Therefore, the results obtained from patch clamp experiments in *Xenopus* oocytes should be compared with those obtained from other cell types.

RECIPES

Basal Internal Solution

Reagent	Quantity	Final concentration
KOH	7.854 g	140 mM
HEPES	4.766 g	20 mM
KCl	0.1491 g	2 mM
EGTA	0.3804 g	1 mM
Methanesulfonic acid	13.314 g	140 mM
H_2O	to 1 L	

Use methanesulfonic acid to adjust the pH to 7.1–7.2. Store for up to 3 years at room temperature (22°C–24°C).

Bath Solution (Ca^{2+}-, Mg^{2+}-Free)

Reagent	Quantity	Final concentration
KOH	7.854 g	140 mM
HEPES	4.766 g	20 mM
KCl	0.1491 g	2 mM
EGTA	1.902 g	5 mM
Methanesulfonic acid	13.314 g	140 mM
H_2O	to 1 L	

Use methanesulfonic acid to adjust the pH to 7.1–7.2. Store for up to 3 years at room temperature (22°C–24°C).

Cite this protocol as *Cold Spring Harb Protoc*; doi:10.1101/pdb.prot099051

ND96 Solution

Reagent	Quantity	Final concentration
NaCl	5.61 g	96 mM
KCl	0.1491 g	2 mM
$CaCl_2 \cdot 2H_2O$	0.2646 g	1.8 mM
$MgCl_2 \cdot 6H_2O$	0.2033 g	1 mM
HEPES	4.766 g	20 mM
Na pyruvate	0.276 g	2.5 mM
Penicillin–streptomycin	100,000 U	100 U/mL
H_2O	to 1 L	

Use 4 M NaOH to adjust the pH to 7.5–7.6. Store for 6 mo at 4°C.

Oocyte Dissociation Solution

Reagent	Final concentration
NaCl	88 mM
KCl	2.5 mM
Na_2HPO_4	10 mM
HEPES	5 mM

Supplement with collagenase (~1–2 mg/mL; Type 1A, Sigma-Aldrich) when ready to use. Oocyte viability may be better preserved with lower collagenase concentrations and longer incubation periods at lower temperature.

Oocyte Stripping Solution

Reagent	Quantity	Final concentration
N-methyl-D-glucamine	39.044 g	200 mM
Aspartate	26.62 g	200 mM
HEPES	4.766 g	20 mM
KCl	0.1491 g	2 mM
$MgCl_2 \cdot 6H_2O$	0.2033 g	1 mM
EGTA	3.804 g	10 mM
H_2O	to 1 L	

Use 4 M NaOH to adjust the pH to 7.4. Store for 2 years at 4°C.

Pipette Solution for Patch Clamps

Reagent	Quantity	Final concentration
KOH	7.854 g	140 mM
HEPES	4.766 g	20 mM
KCl	0.1491 g	2 mM
$MgCl_2 \cdot 6H_2O$	0.4066 g	2 mM
Methanesulfonic acid	13.314 g	140 mM
H_2O	to 1 L	

Use methanesulfonic acid to adjust pH to 7.1–7.2. Store no longer than 18 mo at room temperature (22°C–24°C).

Cite this protocol as *Cold Spring Harb Protoc*; doi:10.1101/pdb.prot099051

ACKNOWLEDGMENTS

Work in our laboratory is supported by National Institutes of Health grants HL70393, NS092570, and GM114694.

REFERENCES

Aguero T, Newman K, King ML. 2018. Microinjection of *Xenopus* oocytes. *Cold Spring Harb Protoc* doi: 10.1101/pdb.prot096974.

Dopico AM, Anantharam V, Treistman SN. 1998. Ethanol increases the activity of Ca^{++}-dependent K^+ (*mslo*) channels: Functional interaction with cytosolic Ca^{++}. *J Pharmacol Exp Ther* 284: 258–268.

Green SL. 2009. *The Laboratory Xenopus sp.* CRC Press, Boca Raton, FL.

Hoshi T, Tian Y, Xu R, Heinemann SH, Hou S. 2013. Mechanism of the modulation of BK potassium channel complexes with different auxiliary subunit compositions by the omega-3 fatty acid DHA. *Proc Natl Acad Sci* 110: 4822–4827.

Kyle BD, Braun AP. 2014. The regulation of BK channel activity by pre- and post-translational modifications. *Front Physiol* 5: 316.

Levis RA, Rae JL. 1998. Low-noise patch-clamp techniques. *Methods Enzymol* 293: 218–266.

Newman K, Aguero T, King ML. 2018. Isolation of *Xenopus* oocytes. *Cold Spring Harb Protoc* doi: 10.1101/pdb.prot095851.

O'Connell D, Mruk K, Rocheleau JM, Kobertz WR. 2011. *Xenopus laevis* oocytes infected with multi-drug-resistant bacteria: Implications for electrical recordings. *J Gen Physiol* 138: 271–277.

Rae JL, Levis RA. 1992. Glass technology for patch clamp electrodes. *Methods Enzymol* 207: 66–92.

Sanguinetti MC, Curran ME, Zou A, Shen J, Spector PS, Atkinson DL, Keating MT. 1996. Coassembly of K_VLQT1 and minK (IsK) proteins to form cardiac I_{Ks} potassium channel. *Nature* 384: 80–83.

Shi J, Krishnamoorthy G, Yang Y, Hu L, Chaturvedi N, Harilal D, Qin J, Cui J. 2002. Mechanism of magnesium activation of calcium-activated potassium channels. *Nature* 418: 876–880.

Sive HL, Grainger RM, Harland RM. 2000. *Early development of Xenopus laevis: A laboratory manual.* Cold Spring Harbor Laboratory Press, Cold Spring Harbor, NY.

Sive HL, Grainger RM, Harland RM. 2010a. Isolation of *Xenopus* oocytes. *Cold Spring Harb Protoc* doi: 10.1101/pdb.prot5534.

Sive HL, Grainger RM, Harland RM. 2010b. Defolliculation of *Xenopus* oocytes. *Cold Spring Harb Protoc* doi: 10.1101/pdb.prot5535.

Sive HL, Grainger RM, Harland RM. 2010c. Microinjection of RNA and preparation of secreted proteins from *Xenopus* oocytes. *Cold Spring Harb Protoc* doi: 10.1101/pdb.prot5538.

Wasserman WJ, Houle JG, Samuel D. 1984. The maturation response of stage IV, V, and VI *Xenopus* oocytes to progesterone stimulation in vitro. *Dev Biol* 105: 315–324.

Yang H, Zhang G, Cui J. 2015. BK channels: Multiple sensors, one activation gate. *Front Physiol* 6: 29.

Zaydman MA, Kasimova MA, McFarland K, Beller Z, Hou P, Kinser HE, Liang H, Zhang G, Shi J, Tarek M, et al. 2014. Domain–domain interactions determine the gating, permeation, pharmacology, and subunit modulation of the I_{Ks} ion channel. *Elife* 3: e03606.

Zhang G, Geng Y, Jin Y, Shi J, McFarland K, Magleby KL, Salkoff L, Cui J. 2017. Deletion of cytosolic gating ring decreases gate and voltage sensor coupling in BK channels. *J Gen Physiol* 149: 373–387.

Protocol 7

Whole-Mount In Situ Hybridization of *Xenopus* Oocytes

Diana Bauermeister and Tomas Pieler[1]

Department of Developmental Biochemistry, Göttingen Center for Molecular Biosciences (GZMB), University of Göttingen, 37077 Göttingen, Germany

Whole-mount in situ hybridization (WMISH) is a common approach that is used to visualize spatial and temporal gene expression in embryos. In this process, digoxygenin-labeled antisense RNA is hybridized to the complementary transcript of interest and RNA hybrids are immunohistochemically detected using an alkaline phosphatase-conjugated antibody against digoxigenin. During *Xenopus laevis* oogenesis, certain RNAs localize to the animal or vegetal pole laying the foundation for germ cell development and germ layer formation of the future embryo. Here we present a WMISH protocol for *Xenopus laevis* oocytes allowing for the efficient detection of localized RNAs in a large number of oocytes during different stages of oogenesis. The application of this approach might be combined with microinjection of tagged reporter RNAs and/or a gain- or loss-of-function background, allowing for the functional analysis of single protein factors involved in RNA localization.

MATERIALS

It is essential that you consult the appropriate Material Safety Data Sheets and your institution's Environmental Health and Safety Office for proper handling of equipment and hazardous materials used in this protocol.

RECIPES: Please see the end of this protocol for recipes indicated by <R>. Additional recipes can be found online at http://cshprotocols.cshlp.org/site/recipes.

Reagents

Acetic anhydride
Agarose (1%)
Alkaline phosphatase buffer (APS) <R>
  Store at room temperature.

Anti-digoxigenin-AP, Fab fragments
Blocking solution for *Xenopus* oocytes <R>
Color reaction solution <R>
Digoxygenin-11-UTP
Dithiothreitol (DTT)
DNA template, linear
  Use linearized vector DNA containing the antisense sequence of interest and a T7/T3/SP6 promoter site.

Ethanol
In situ hybridization buffer <R>
  Store at −20°C.

[1]Correspondence: tpieler@gwdg.de

MAB (10×) <R>

Prepare 1× fresh before use.

MBSH <R>

MEMFA <R>

Methanol

PBT <R>

Store at room temperature.

Proteinase K

Ribolock RNase inhibitor (Thermo Fisher Scientific)

Ribonucleotides (rATP, rGTP, rCTP, and rUTP; 10 mM each)

RNA Polymerase (T7/T3/SP6, Promega)

RNase A

RNase T1

RNA purification kit (e.g., QIAGEN RNeasy Mini Kit)

SSC <R>

Transcription Optimized 5× Buffer (Promega)

Triethanolamine (0.1 M, pH 7.5)

TURBO DNase

Water (RNase-free)

Xenopus oocytes

We recommend using albino Xenopus laevis oocytes to allow for pigment-free visualization of expression patterns.

Equipment

Dissection microscope

Glass vials (5-mL)

Incubator

Microcentrifuge tubes (1.5-mL)

Petri dishes

Rocking table

Spectrophotometer, ultraviolet (e.g., NanoDrop 2000c)

Stereomicroscope (e.g., Carl Zeiss SteREO Lumar.V12)

Water bath, shaking

METHOD

Transcription of Digoxygenin-Labeled Probes

1. Mix the 25-µL transcription reaction in a 1.5-mL tube using the following reagents:

Linear DNA template	5 µL
Transcription Optimized 5× Buffer	5 µL
rATP (10 mM)	1 µL
rCTP (10 mM)	1 µL
rGTP (10 mM)	1 µL
rUTP (10 mM)	0.64 µL
Digoxygenin-11-UTP	0.36 µL
DTT (750 mM)	1 µL
Ribolock RNase inhibitor	1 µL
T7/T3/SP6 polymerase	1 µL
RNase-free water	8 µL

2. Incubate the reaction for 3 h at 37°C.

Cite this protocol as *Cold Spring Harb Protoc*; doi:10.1101/pdb.prot097014

3. Add 1 µL of TURBO DNase and mix. Incubate for 15 min at 37°C.

4. Purify the transcribed RNA using an RNA purification kit. Elute the RNA in 20 µL of RNase-free water.

   The recovery of shorter RNA fragments can be increased by adding 675 µL of ethanol instead of the 250 µL recommended in the manufacturer's protocol.

5. Analyze the quantity and quality of the RNA by UV spectrometry and agarose gel electrophoresis.

6. Dilute the RNA probes in hybridization buffer to ~0.05–0.5 ng/mL. Store at −20°C.

   Higher dilutions will work, but the color reaction will take longer. Depending on the number of samples, we usually store 2–8 mL in 10-mL tubes.

Isolation and Preparation of Oocytes

7. Isolate, defolliculate, and stage albino *Xenopus laevis* oocytes as described in Protocol 1: Isolation of *Xenopus* Oocytes (Newman et al. 2018).

8. Select oocytes of the desired stage. Manually sort out damaged or dead oocytes under a dissection microscope. Maintain oocytes in Petri dishes at 18°C until use.

   Low-quality albino oocytes can be recognized by their whitish color or transparent spots, whereas good quality oocytes are uniformly shaped and slightly yellowish.

9. To detect endogenous RNA, proceed to Step 10. To visualize tagged reporter RNA or overexpress certain proteins, first perform RNA injections and incubation of oocytes as described (Arthur et al. 2009; Bauermeister et al. 2015; see also Protocol 3: Microinjection of *Xenopus* Oocytes [Aguero et al. 2018]).

Fixation of Oocytes

Perform the following steps in volumes of ~4 mL unless indicated differently, with gentle agitation on a rocking table. Solutions can be removed carefully using a water-jet vacuum pump, although for early-stage oocytes, we recommend to remove solutions by pipette.

10. Transfer oocytes into labeled 5-mL glass vials.

    See Troubleshooting.

11. Wash oocytes twice with MBSH.

12. Fix oocytes in MEMFA for 1 h.

13. Wash oocytes four times for 10 min in PBT.

14. To perform whole mount in situ hybridization immediately, proceed to Step 16. Alternatively, dehydrate oocytes stepwise in ethanol (30%, 50%, 75%, and 100%). Store at −20°C for later use.

    Rehydrate as described in Step 15 before proceeding with the WMISH procedure.

Whole-Mount In Situ Hybridization

15. Rehydrate oocytes stepwise in PBT and wash three times for 10 min in PBT (for dehydrated oocytes only; see Step 14).

16. Permeabilize oocytes with 10-µg/mL Proteinase K in 0.5 mL PBT for 2 min.

    Avoid extended proteinase K treatments as these tend to make the oocytes fragile.

17. Wash oocytes two times for 2 min in 4 mL of 0.1 M triethanolamine (pH 7.5).

18. Add 12.5 µL of acetic anhydride to each vial. After 5 min, add an additional 12.5 µL of acetic anhydride. Gently shake until the drop is dissolved.

19. Refix the oocytes in MEMFA for 30 min. Wash five times in PBT for 5 min.

20. Wash the oocytes once in 0.5 mL of hybridization buffer. Prehybridize the oocytes in 1 mL of hybridization buffer for 5–6 h at 65°C in a water bath with gentle agitation.

21. Add preheated (i.e., to 65°C) to diluted DIG-labeled antisense probe in hybridization buffer. Hybridize overnight at 65°C with gentle agitation.

    We usually hybridize in 0.5–2 mL of hybridization solution depending on the number of samples and available probe. The hybridization temperature of 65°C work for most of our probes, but can be adjusted depending on your probe sequence.

22. Remove the hybridization solution containing the antisense probe.

    The probe solution can be stored at −20°C and reused several times.

23. Wash oocytes once in preheated hybridization buffer at 65°C.

24. Wash three times in preheated 2× SSC for 20 min at 65°C.

25. Digest nonhybridized RNA probes by incubating in 2× SSC containing RNase A (10 µg/mL) and RNase T1 (10 units/mL) for 30 min at 37°C.

26. Wash once in 2× SSC for 10 min at room temperature.

27. Wash twice in 0.2× SSC for 30 min at 65°C.

28. Wash twice in MAB for 15 min.

29. Replace the MAB with blocking solution. Incubate the oocytes for 1 h at room temperature.

30. Replace the blocking solution with fresh blocking solution containing the alkaline phosphatase-coupled anti-DIG antibody (diluted 1:5000). Incubate the oocytes overnight at 4°C.

    Alternatively, antibody incubation can be done for 4–5 h at room temperature.

31. Remove the antibody-containing solution.

    The solution can be stored at −20°C and reused up to three times.

32. Wash the oocytes six to eight times with MAB for 15 min at room temperature.

    If the antibody incubation was done for 4–5 h at room temperature, include a washing step overnight at 4°C.

33. Wash the oocytes twice in APS for 5 min.

34. Add the color reaction solution. Incubate the oocytes with gentle agitation at 4°C in the dark until staining becomes visible.

    Depending on the gene of interest, staining can develop within 30 min up to 24 h. Incubation at room temperature can speed up the color reaction.

35. To stop the color reaction, rinse the oocytes once in distilled water. Add 100% methanol to reduce background staining.

36. Wash with 75%, 50%, and 25% methanol, 1-min each.

37. Fix oocytes in MEMFA for at least 1 h. Store in MEMFA at 4°C until imaging.

Imaging Oocytes

38. Prepare Petri dishes coated with an ∼5 mm layer of 1% agarose.

39. Transfer oocytes into PBT.

40. Manually inspect oocytes under a dissection microscope. Remove oocytes that are deformed. Transfer oocytes in PBT into the agarose plates.

    To obtain a colored background, a plastic piece of the desired color can be placed under the dish and covered with a thin water film.

41. Document staining under a stereomicroscope.

    Examples of the lacZ-tagged localizing reporter RNAs dnd1 and grip1 visualized using this protocol are shown in Figure 1.

Cite this protocol as *Cold Spring Harb Protoc*; doi:10.1101/pdb.prot097014

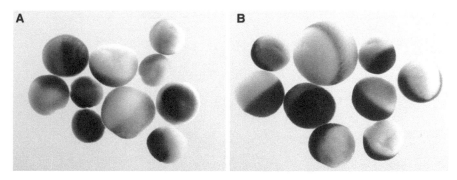

FIGURE 1. Visualization of vegetal RNA localization in reporter-RNA–injected oocytes by whole mount in situ hybridization. Oocytes were injected with *lacZ-dnd1*-LE RNA (A) or *lacZ-grip1*-LE RNA (B) and vegetal localization was assayed by whole mount in situ hybridization against the lacZ-tag.

TROUBLESHOOTING

Problem (Steps 10–37): Oocytes stick to the glass vial.

Solution: During incubation times that include horizontal agitation of the glass vial, oocytes tend to stick to the glass wall. These oocytes can be easily detached in Tween-20–containing washing solutions by gentle rinsing using a pipette. However, in solutions without detergent or with small volumes (e.g., during fixation, dehydration, permeabilization, prehybridization), agitation of the glass vial in a vertical position is advantageous, as it keeps the oocytes at the bottom of the glass vial.

DISCUSSION

A major requirement for the success of the experiment is oocyte quality. Although the quality of pigmented oocytes is most often visible by alterations in pigmentation, albino oocytes of low quality are more difficult to distinguish. Low-quality albino oocytes can be recognized by their whitish color or transparent spots, whereas good quality oocytes are uniformly shaped and slightly yellowish. An untreated oocyte batch that contains a lot of transparent, damaged or deformed oocytes might already indicate low quality. Most often, a proportion (10%–50%) of oocytes that appear healthy after defolliculation die within 1 d or after injection and should be rejected. If the proportion of dead oocytes is >50%, the experiment might be aborted. Although low quality is most often inherent to the oocytes, careful handling might optimize the oocyte survival, for example, defolliculation time should not be extended and live oocytes should be kept in flat dishes at 18°C as much as possible.

We have used in situ hybridization of localizing reporter RNAs in oocytes to investigate the roles of protein factors in RNA localization (Arthur et al. 2009; Bauermeister et al. 2015). To address the functional roles of single protein factors in the process of RNA localization more directly, one might want to achieve a knockdown of the respective endogenous protein. One major problem is the presence of maternal protein with low turnover rates in the oocyte. A clear argument for the participation of a certain protein could be given by a genetic knockout in the mother. However, a partial tetraploid genome and a long generation time hampers genetic manipulations of *Xenopus laevis*. Given that the respective protein has a similar role in the closely related diploid *Xenopus tropicalis*, a genetic knockout frog that is able to produce oocytes might be a future application that—combined with the reporter-based technique described here—could give considerable impact in the knowledge of the RNA localization process.

Alkaline Phosphatase Buffer (APS)

50 mM $MgCl_2$
100 mM NaCl
100 mM Tris, pH 9.5
0.1% Tween 20

Blocking Solution for Xenopus Oocytes

Reagent	Final concentration
MAB (10×) <R>	1×
Blocking Reagent (Roche 11096176001)	2%
Horse serum	20%

Prepare fresh before use.

Color Reaction Solution

Reagent	Final concentration
Alkaline phosphatase buffer (APS) <R>	1×
4-nitro blue tetrazolium chloride	80 μg/mL
5-bromo-4-chloro-3-indolyl-phosphate	175 μg/mL

Prepare fresh before use.

In Situ Hybridization Buffer

Reagent	Amount to add	Final concentration
Formamide	50 mL	50%
SSC (20×)	25 mL	5×
Torula RNA (200 mg/mL)	0.5 mL	1 mg/mL
Heparin (100 mg/mL)	100 μL	100 μg/mL
Denhart's (100×)	1 mL	1×
Tween 20	100 μL	0.1%
CHAPS (10%, w/v)	1 mL	0.1%
EDTA (500 mM)	2 mL	10 mM
H_2O	to 100 mL	

MAB (10×)

Reagent	Amount to add	Final concentration (10×)
Maleic acid	116 g	1 M
NaCl	87 g	1.5 M
NaOH	40 g	
H_2O	800 mL	

To make 1 L of 10× MAB, combine the above reagents, and adjust the pH to 7.5. Beware that this is a weak buffer; pH 7.5 is easily missed. A 10× stock of MAB is preferred to a 5× stock because it does not get contaminated as easily.

Cite this protocol as *Cold Spring Harb Protoc*; doi:10.1101/pdb.prot097014

MBSH

Reagent	Final concentration
NaCl	80 mM
KCl	1 mM
NaHCO$_3$	2.4 mM
HEPES–KOH (pH 7.6)	20 mM
MgSO$_4$	0.82 mM
Ca(NO$_3$)$_2$	0.33 mM
CaCl$_2$	0.41 mM

Sterilize by autoclaving. Store at room temperature.

MEMFA

MOPS
EGTA
MgSO$_4$
Formaldehyde (37%)

Prepare a 10× stock solution of 1 M MOPS (pH 7.4), 20 mM EGTA, and 10 mM MgSO$_4$. On the day of use, prepare fresh 1× MEMFA by combining 8 parts H$_2$O, 1 part 10× MEMFA stock solution, and 1 part formaldehyde (37%).

Final concentrations are 100 mM MOPS (pH 7.4), 2 mM EGTA, 1 mM MgSO$_4$, 3.7% (v/v) formaldehyde.

PBT

1× PBS
0.1% Tween 20

Prepare PBT with DEPC-treated H$_2$O.

SSC

For a 20× solution: Dissolve 175.3 g of NaCl and 88.2 g of sodium citrate in 800 mL of H$_2$O. Adjust the pH to 7.0 with a few drops of a 14 N solution of HCl. Adjust the volume to 1 L with H$_2$O. Dispense into aliquots. Sterilize by autoclaving. The final concentrations of the ingredients are 3.0 M NaCl and 0.3 M sodium citrate.

REFERENCES

Aguero T, Newman K, King ML. 2018. Microinjection of *Xenopus* oocytes. *Cold Spring Harb Protoc* doi: 10.1101/pdb.prot096974.

Arthur PK, Claußen M, Koch S, Tarbashevich K, Jahn O, Pieler T. 2009. Participation of *Xenopus* Elr-type proteins in vegetal mRNA localization during oogenesis. *J Biol Chem* **284:** 19982–19992.

Bauermeister D, Claußen M, Pieler T. 2015. A novel role for Celf1 in vegetal RNA localization during *Xenopus* oogenesis. *Dev Biol* **405:** 214–224.

Newman K, Aguero T, King ML. 2018. Isolation of *Xenopus* oocytes. *Cold Spring Harb Protoc* doi: 10.1101/pdb.prot095851.

Whole-Mount Immunofluorescence for Visualizing Endogenous Protein and Injected RNA in *Xenopus* Oocytes

Samantha P. Jeschonek and Kimberly L. Mowry[1]

Department of Molecular Biology, Cell Biology & Biochemistry, Brown University, Providence, Rhode Island 02912

Asymmetric distribution of mRNA and protein is a hallmark of cell polarity in many systems. The *Xenopus laevis* oocyte provides many technical advantages to studying such polarity. Thousands of oocytes at different stages of maturity can be harvested from a single ovary and, owing to their relatively large size, even the youngest oocytes can be manually microinjected. Microinjection of fluorescently labeled RNA combined with immunofluorescence of endogenous proteins can provide insight into the cytoplasmic interactions contributing to polarity. Here, we present an updated method to image endogenous protein and microinjected RNA in *X. laevis* oocytes.

MATERIALS

It is essential that you consult the appropriate Material Safety Data Sheets and your institution's Environmental Health and Safety Office for proper handling of equipment and hazardous materials used in this protocol.

RECIPES: Please see the end of this protocol for recipes indicated by <R>. Additional recipes can be found online at http://cshprotocols.cshlp.org/site/recipes.

Reagents

Ammonium acetate (7 M, RNase-free)

Collagenase solution (3 mg/mL collagenase I [Sigma-Aldrich C0130] and 0.1 M KPO_3^+ [pH 7.4]), freshly prepared

Cy5-UTP or Cy3-UTP (GE Healthcare Life Sciences PA55026/PA53026)

EDTA (200 mM, pH 8.0)

Ethanol (70%, RNase-free)

Ethanol (100%, RNase-free)

Glycogen (20 mg/mL; Thermo Scientific R0561)

MBSH Buffer (10×) <R>

MEMFA <R>

Methanol (anhydrous, 100%)

mMessage machine SP6/T7/T3 transcription kit (Ambion AM1340/AM1344/AM1348)

Murray's clear <R>

---

[1]Corresponding author: Kimberly_Mowry@brown.edu

Nuclease-free H$_2$O (Thermo Fisher AM9914G) or H$_2$O treated with DEPC (Sigma-Aldrich D5758)

PBT/PBT-plus for *Xenopus* oocytes <R>

Phenol/chloroform/isoamylalcohol (25:24:1; Fisher Scientific AC327115000)

Plasmid containing sequence of interest downstream from an RNA polymerase promoter (T7, SP6, or T3)

Primary antibody

Proteinase K buffer for *Xenopus* oocytes <R>

Restriction enzyme appropriate for plasmid linearization

RNase-ZAP (Ambion AM9780)

Secondary antibody, fluorescently labeled

Xenopus oocyte culture medium (XOCM) <R>

Equipment

Aluminum foil

Barrier pipette tips

Dissecting microscope

Fluorodishes (10 mm inside diameter; World Precision Instruments FD3510-100)

Heat blocks (37°C and 72°C)

Ice bucket

Illustra ProbeQuant G-50 Micro Columns (GE Healthcare 29-9034-08)

Inverted confocal microscope (such as Zeiss LSM 510 Meta Confocal Laser Scanning Microscope)

Microcentrifuges at room temperature and 4°C

Microcentrifuge tubes (1.5 mL, RNase-free)

Microcentrifuge tubes (1.5 mL, siliconized or low-retention)

Microinjection setup (such as Harvard Apparatus PLI-100 PicoInjector and Narishige MN-151 Joystick Micromanipulator)

NanoDrop spectrophotometer

Pasteur pipettes

Pulled glass pipette with a fine tip

Rocking platform

Surgical instruments for removing oocytes from *Xenopus laevis* females

Tissue culture plate (24-well)

Vortex mixer

X. laevis females for oocyte preparation

   For imaging oocytes beyond stage II, albino frogs must be used.

METHOD

This protocol covers co-imaging of injected RNA and protein. Alternatively, RNA or protein can be visualized independently of each other. If imaging only endogenous protein, proceed to Step 16.

Preparation of Fluorescently Labeled RNA

1. Linearize plasmid DNA containing the sequence of interest cloned downstream from a T7, SP6, or T3 promoter with an appropriate restriction enzyme.

2. Purify the plasmid DNA by phenol chloroform extraction followed by ethanol precipitation, and resuspend in nuclease-free H$_2$O at 500–1000 ng/µL.

3. Prepare an RNase-free working area by treating all surfaces and gloves with RNase-ZAP.

   For subsequent steps, up to Step 15, use only barrier pipette tips and RNase-free reagents.

TABLE 1. Reagents to combine in Step 4

Reagent	Volume	Final concentration	Notes
10× transcription buffer	2.0 µL	1×	
2× NTP/cap mixture	6.0 µL	3.3×	The amount of NTP/cap mixture is altered from the manufacturer's protocol to facilitate incorporation of fluorescent UTP
Cy3-UTP or Cy5-UTP	1.5 µL	0.375 mM	ChromaTide Alexa-Fluor UTPs (Thermo-Fisher) may also be used at a final concentration of 0.3 mM
Linearized DNA	(variable)	1 µg	Less than 1 µg can be used per reaction but will reduce yield
10× enzyme mix	2.0 µL	1×	Reactions using SP6 RNA polymerase require longer incubation times (4 h) compared with those using T7 or T3 RNA polymerases (2 h) for similar yields
Nuclease-free H_2O	(variable)	—	
Total	20 µL		Reaction may be scaled up for increased RNA production.

4. Modify the Ambion mMessage machine transcription kit by adding the reagents in Table 1, in the order shown, to an RNase-free microcentrifuge tube.

5. Mix reaction by gentle pipetting.

6. Incubate for 2–4 h at 37°C. Protect tubes from light by covering with foil.

7. Pulse spin in a microcentrifuge. Add 1 µL Turbo DNase (included in mMessage machine kit).

8. Incubate for 15 min at 37°C with protection from light.

9. Add 179 µL 200 mM EDTA, pH 8.0, to stop the reaction. Mix gently by pipetting.

10. Add 200 µL phenol/chloroform/isoamylalcohol (1:1 ratio) to the reaction. Vortex for 15 sec. Centrifuge for 10 min at maximum speed at room temperature.

11. Carefully remove the upper aqueous phase containing the RNA, and transfer to an RNAse-free microcentrifuge tube. Avoid transferring any of the interphase.

12. To remove unincorporated nucleotides (including free fluorescent UTP) pass the RNA through an Illustra ProbeQuant G-50 Micro Column, following manufacturer's instructions (GE Healthcare).

13. Concentrate the RNA by ethanol precipitation.

    i. Add 500 µL 100% ethanol, (2.5× volume), 20 µL 7 M ammonium acetate (0.1× volume) and 1 µL 20 mg/mL glycogen. Mix well by vortexing, and incubate for 1 h at −80°C or overnight at −20°C.

    ii. Centrifuge at maximum speed for 20 min at 4°C. Remove and discard the supernatant and add 1 mL 70% ethanol (stored at −20°C).

    iii. Centrifuge at maximum speed for 5 min at 4°C. Remove and discard the supernatant, wash again with 70% ethanol. Centrifuge at maximum speed at 4°C for 5 min.

    iv. Remove and discard the supernatant. Allow the RNA pellet to dry for 10 min at room temperature, or until most residual 70% ethanol has evaporated. Do not let the pellet dry completely.

    v. Resuspend RNA in 15 µL nuclease-free H_2O.

14. Determine RNA concentration using a NanoDrop spectrophotometer, according to the manufacturer's instructions. (The RNA should generate a curve with a single peak at 280 nm, and a 260/280 nm ratio of ∼2.0.) Calculate RNA molarity, and adjust to a final concentration of 1 µM.
    See Troubleshooting.

15. Store RNA at −80°C in 2 µL aliquots (for single use only).

    RNA will be stable for several months at −80°C.

Cite this protocol as *Cold Spring Harb Protoc*; doi:10.1101/pdb.prot097022

Preparation of Oocytes

For a detailed procedure see Protocol 1: Isolation of Xenopus *Oocytes (Newman et al. 2018).*

16. Surgically remove oocytes from *X. laevis* females.

17. Defolliculate oocytes using collagenase solution.

18. Wash oocytes 3–5 times in 1× MBSH buffer, until the wash buffer is clear.

19. Sort oocytes to obtain stages of interest and place in XOCM.

20. Allow oocytes to recover in XOCM for at least 1 h at 18°C.

     Overnight recovery is preferable.

     If imaging only endogenous protein, proceed to Step 28.

Microinjection

For a detailed microinjection protocol, see Protocol 3: Microinjection of Xenopus *Oocytes (Aguero et al. 2018).*

21. Thaw an aliquot of fluorescently labeled RNA (from Step 15) on ice. Dilute the RNA to a final concentration of 125–500 nM using nuclease-free H_2O.

22. Denature RNA for 5 min at 72°C. Immediately place on ice for 3 min.

23. Centrifuge RNA at maximum speed for 10 min at 4°C to remove any particulates. Carefully pipette RNA into a fresh microcentrifuge tube and place on ice.

24. Calibrate a beveled glass capillary tube (3.5 inch Drummond 3-000-203-G/X) with nuclease-free H_2O to deliver 2 nL per injection.

25. Place oocytes into an injection dish containing 1× MBSH.

26. Load RNA into the beveled and calibrated capillary tube and inject each oocyte with 2 nL of RNA.

27. Culture oocytes for 8–48 h at 18°C in a 24-well dish containing 500 µL XOCM containing antibiotics.

Fixation and Immunofluorescence

28. Examine cultured oocytes for survival.
     If more than 25% of oocytes have lysed, consider repeating microinjections with new oocytes.

29. Place cultured oocytes in a siliconized or low-retention microfuge tube. If visualizing only fluorescent RNA and not co-imaging protein, proceed to Step 33.

30. Remove XOCM and wash oocytes with proteinase K buffer (without enzyme added).

31. Add 500 µL proteinase K solution. Gently rock oocytes for 5 min at room temperature.

32. Immediately remove proteinase K solution.

33. To fix oocytes, add 500 µL MEMFA. Incubate for 1 h, with rocking, in the dark.

34. Remove MEMFA and add 500 µL PBT. Wash oocytes for 15 min in the dark with rocking.

35. Wash twice more in PBT, for a total of three washes.
     If only visualizing fluorescent RNA, proceed to Step 41.

36. Add 500 µL PBT-Plus blocking solution. Rock oocytes in the dark for at least 2 h.

37. Remove PBT-Plus. Incubate oocytes with primary antibody diluted in 500 µL PBT-Plus, overnight at 4°C with rocking. If antibody is limiting, oocytes may be incubated in a 250 µL dilution of primary antibody.

38. Replace primary antibody with 500 µL PBT and rock in the dark for 1.5–2 h at room temperature. Repeat wash twice for a total of three washes.

39. Add 500 µL of PBT-Plus containing an appropriate fluorescently labeled secondary antibody. Incubate overnight at 4°C with rocking.

> *Xenopus oocytes show strong autofluorescence at 488 nm, limiting the choice of secondary antibody. When possible, use a secondary antibody with emission at higher wavelengths.*

40. Wash oocytes three times in PBT for 1.5–2 h at room temperature to remove secondary antibody. After the final wash, resuspend oocytes in 1 mL PBT.

41. Dehydrate oocytes stepwise into 100% anhydrous methanol:

    i. Remove 100 µL of PBT solution and replace with 100 µL anhydrous methanol. Rock oocytes for 5–10 min at room temperature. Repeat twice.

    ii. Remove 200 µL and replace with 200 µL anhydrous methanol. Rock oocytes for 5–10 min at room temperature. Repeat twice.

    iii. Remove 300 µL and replace with 300 µL anhydrous methanol. Rock for 5–10 min at room temperature. Repeat twice.

    iv. Remove 400 µL and replace with 400 µL anhydrous methanol. Rock for 5–10 min at room temperature. Repeat twice.

    v. Remove 500 µL and replace with 500 µL anhydrous methanol. Rock for 5–10 min at room temperature. Repeat twice.

    vi. Remove 750 µL and replace with 750 µL anhydrous methanol. Rock for 5–10 min at room temperature. Repeat twice.

    vii. Remove all buffer, and perform three washes with 1 mL 100% anhydrous methanol. At the end of dehydration, the solution should be clear. If solution is cloudy or Schlieren lines are visible, continue washing with anhydrous methanol.

    > *Oocytes must be completely dehydrated in anhydrous methanol to be compatible with Murray's clear for imaging.*

42. Store oocytes in 100% anhydrous methanol at −20°C until ready for imaging. Oocytes may be stored for 2–3 wk in methanol.

Confocal Imaging

43. Use a Pasteur pipette to transfer oocytes to a fluorodish. Try to keep oocytes in the center of the dish and transfer as little methanol as possible.

44. Using a pulled glass pipette with a fine tip, remove any methanol transferred to the fluorodish.

45. With a Pasteur pipette, add Murray's clear dropwise to the fluorodish, until oocytes are covered completely.

46. Wait at least 5 min for oocytes to optically clear. Larger oocytes (stages III–VI) will take longer to clear than younger oocytes.

47. Image the oocytes using an inverted confocal microscope (Fig. 1).

    i. Locate individual oocytes using a 488 nm laser and a 10× objective. Oocytes can be easily located due to autofluorescence at 488 nm.

    ii. Once located, image oocytes at the desired magnification. It may be necessary to open the pinhole >1 airy unit to detect an RNA fluorescence signal.

    iii. To decrease cross-excitation between channels, image with each laser separately (not simultaneously).

    > *See Troubleshooting.*

Cite this protocol as *Cold Spring Harb Protoc*; doi:10.1101/pdb.prot097022

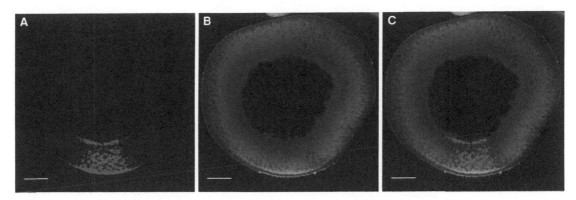

FIGURE 1. Colocalization of endogenous Staufen protein and injected *VLE* RNA. Stage II oocytes were injected with 500 nM Cy5-UTP-labeled *VLE* RNA (Mowry and Melton 1992) and cultured for 18 h to allow for vegetal localization. Immunofluorescence was performed using an antiXStau antibody (Yoon and Mowry 2004) at 1:250 dilution and an Alexa-546 secondary antibody (Thermo-Fisher) at 1:500 dilution. Oocytes were imaged on a Zeiss LSM 510 Meta Confocal Laser Scanning Microscope using a 20× objective. Shown is a confocal section with (*A*) *VLE* RNA in red, (*B*) Staufen protein in green, and (*C*) merged red and green channels, showing Staufen protein and *VLE* RNA colocalization. The oocyte is oriented with the vegetal pole at the bottom and the scale bars = 50 µm.

TROUBLESHOOTING

Problem (Step 14): The yield of RNA is low.

Solution: Increase the concentration of linear template in the in vitro transcription reaction and/or increase incubation time to 6 h at 37°C. If necessary, multiple reactions can be pooled together before ethanol precipitation.

Problem (Step 47): The RNA Fluorescence signal is dim.

Solution: If using Cy3-UTP, switch to the brighter Cy5-UTP. Increase concentration of RNA injected, or standardize injection concentration based on fluorescence intensity (e.g., 5000 RFU, which can be measured with a NanoDrop 3300 Fluorospectrometer).

Problem (Step 47): The antibody does not penetrate the interior of the oocyte.

Solution: Increase incubation with proteinase K. A time-series may be necessary to determine optimal digestion time for a particular antibody. Increase incubation period with primary antibody to 48 h.

DISCUSSION

The *X. laevis* oocyte provides an important model to study cell polarity. Oogenesis in *X. laevis* proceeds through six distinct stages, during which the cytoplasm becomes increasingly polarized (Dumont 1972). Both mRNAs and proteins can be asymmetrically distributed in the oocyte cytoplasm, in many cases acting to specify embryonic polarity (for review, see Houston 2013). Thus, analysis of protein and mRNA distributions in vertebrate oocytes is important for understanding establishment of developmental polarity. The *X. laevis* oocyte, in particular, offers many technical advantages for studying polarity. A single *X. laevis* ovary provides thousands of oocytes at different stages of oogenesis (Dumont 1972). Due to their large size, even immature oocytes can be manually microinjected (Yisraeli and Melton 1988; Chang et al. 2004). Microinjection of fluorescently labeled RNA can recapitulate the localization pattern of endogenous mRNAs and can be combined with immunofluorescence analysis of proteins of interest (Yoon and Mowry 2004). Furthermore, imaging by confocal microscopy allows a high-resolution view of RNA-protein interactions. Such analyses can provide insight into the cytoplasmic interactions contributing to polarity. The protocol provided here presents an updated method for imaging endogenous protein and microinjected RNA in immature *X. laevis* oocytes.

RECIPES

MBSH Buffer (10×)

Reagent	Final concentration
NaCl	880 mM
KCl	10 mM
NaHCO$_3$	24 mM
MgSO$_4$·7H$_2$O	8.2 mM
Ca(NO$_3$)$_2$·4H$_2$O	3.3 mM
CaCl$_2$·6H$_2$O	4.1 mM
HEPES (pH 7.6)	100 mM

Filter-sterilize and store as a 10× stock solution. Prepare fresh 1× MBSH prior to use.

PBT/PBT-Plus for Xenopus Oocytes

Reagent	Final concentration
For PBT:	
NaCl	137 mM
KCl	2.7 mM
Na$_2$HPO$_4$	10 mM
KH$_2$PO$_4$	1.8 mM
BSA (Sigma-Aldrich A2153)	0.2% (w/v)
Triton X-100	0.1% (v/v)
For PBT-Plus, supplement PBT with:	
Goat serum	2% (v/v)
BSA	2% (w/v; for a total of 2.2% BSA)

Mix and sterilize through a 0.22-µm filter. Store at 4°C.

Proteinase K Buffer for Xenopus Oocytes

EDTA (pH 8.0)	10 mM
Tris-HCl (pH 7.5)	100 mM
Proteinase K	50 µg/mL

Prepare fresh. Do not reuse.

Xenopus Oocyte Culture Medium (XOCM)

Reagent	Final concentration
Leibovitz's L-15 Medium (Thermo Fisher Scientific 11415064)	50%
HEPES (pH 7.6)	15 mM
Insulin	1 mg/mL
Nystatin (Sigma-Aldrich N1638)	50 U/mL
Penicillin/streptomycin (Life Technologies 15070063)	100 U/mL
Gentamicin (Thermo Fisher Scientific 15750060)	0.1 mg/mL

Sterilize the first three ingredients though a 0.22-µm filter. Add the antibiotics to XOCM just prior to culture. Store XOCM without antibiotics for up to 2 mo at 4°C. After antibiotics have been added, do not store or reuse.

Cite this protocol as *Cold Spring Harb Protoc*; doi:10.1101/pdb.prot097022

ACKNOWLEDGMENTS

Development of this method was supported by National Institutes of Health (NIH) grant GM071049 to K.L.M.

REFERENCES

Aguero T, Newman K, King ML. 2018. Microinjection of *Xenopus* oocytes. *Cold Spring Protoc* doi: 10.1101/pdb.prot096974.

Chang P, Torres J, Lewis RA, Mowry KL, Houliston E, King ML. 2004. Localization of RNAs to the mitochondrial cloud in *Xenopus* oocytes through entrapment and association with endoplasmic reticulum. *Mol Biol Cell* **15:** 4669–4681.

Dumont JN. 1972. Oogenesis in *Xenopus* laevis (Daudin). I. Stages of oocyte development in laboratory maintained animals. *J Morph* **136:** 153–179.

Houston DW. 2013. Regulation of cell polarity and RNA localization in vertebrate oocytes. *Int Rev Cell Mol Biol* **306:** 127–185.

Mowry KL, Melton DA. 1992. Vegetal messenger RNA localization directed by a 340-nt RNA sequence element in *Xenopus* oocytes. *Science* **255:** 991–994.

Newman K, Aguero T, King ML. 2018. Isolation of *Xenopus* oocytes. *Cold Spring Protoc* doi: 10.1101/pdb.prot095851.

Yisraeli JK, Melton DA. 1988. The material mRNA Vg1 is correctly localized following injection into *Xenopus* oocytes. *Nature* **336:** 592–595.

Yoon YJ, Mowry KL. 2004. *Xenopus* Staufen is a component of a ribonucleoprotein complex containing Vg1 RNA and kinesin. *Development* **131:** 3035–3045.

Fluorescence In Situ Hybridization of Cryosectioned *Xenopus* Oocytes

Christopher R. Neil and Kimberly L. Mowry[1]

Department of Molecular Biology, Cell Biology & Biochemistry, Brown University, Providence, Rhode Island 02912

Xenopus laevis oocytes are widely used to study mechanisms of RNA function and biogenesis. While the large size of *Xenopus* oocytes is amenable to both biochemical and imaging approaches, the relative opacity of the yolk-rich cytoplasm has limited high-resolution imaging of endogenous RNAs. Here, we present a protocol that combines multi-probe fluorescence in situ hybridization with cryosectioning to provide a highly sensitive means of imaging endogenous oocyte RNAs.

MATERIALS

It is essential that you consult the appropriate Material Safety Data Sheets and your institution's Environmental Health and Safety Office for proper handling of equipment and hazardous materials used in this protocol.

RECIPES: Please see the end of this protocol for recipes indicated by <R>. Additional recipes can be found online at http://cshprotocols.cshlp.org/site/recipes.

Reagents

Dry ice
Ethanol (95%)
Formamide (Fisher Scientific, BP228-100)
MBSH buffer (10×) <R>
MEMFA <R>
> *Prepare fresh, and protect from light.*

Nail polish
O.C.T. compound (Fisher Scientific 23-730)
PBT for *Xenopus* oocyte FISH <R>
PBT-30S (30% sucrose [w/v] in PBT)
> *Store at 4°C.*

Phosphate-buffered saline (PBS) <R>
Prolong Gold antifade mountant (Thermo Fisher Scientific P36934)
Richard-Allan Scientific Cytocool II Aerosol (Thermo Fisher Scientific 8323)
RNA FISH probe set for gene of interest
> *These are available commercially (e.g., LGC Biosearch Technologies) or can be synthesized (see Raj et al. [2008]).*

---

[1]Correspondence: Kimberly_Mowry@brown.edu

Silicone adhesive sealant (e.g., Loctite 595)
Stellaris RNA FISH hybridization buffer (LGC Biosearch Technologies SMF-HB1-10)
Stellaris RNA FISH Wash buffer A (LGC Biosearch Technologies SMF-WA1-60)
Stellaris RNA FISH Wash buffer B (LGC Biosearch Technologies SMF-WB1-20)
Xenopus oocyte culture medium (XOCM) <R>
Xenopus oocytes

Equipment

Applicators (cotton-tipped, sterile)
Cryostat (e.g., Leica CM3050 S Research Cryostat)
Embedding molds (disposable, Peel-A-Way [Truncated-T12]; Polysciences 18986)
Eppendorf centrifuge tubes (5-mL, RNase-free; Fisher Scientific 14-282-300)
Forceps
Glass insert antiroll plate (70-mm)
Microscope coverslips (24 × 60 mm)
Microscope, fluorescence (widefield or confocal)
Microscope slide box
Microscope slides (Superfrost Plus Gold, 25 × 75 × 1.0 mm; Thermo Fisher Scientific 15-188-48)
Microtome blades (high-profile, disposable, 818; Leica Biosystems 14035838926)
Rotator
Slide chamber (humidified)
Specimen disc (30-mm)

METHOD

Fixation and Freezing

1. Defolliculate and sort oocytes of the desired stage (see Protocol 1: Isolation of *Xenopus* Oocytes [Newman et al. 2018]).

2. Incubate the oocytes for at least 1 h in XOCM to allow for recovery.

   Oocytes can be cultured overnight if desired.

3. Transfer a settled volume of 300–500 μL of cultured oocytes in XOCM to a 5-mL Eppendorf tube. Remove the XOCM. Wash the oocytes with 5 mL of MBSH.

4. Remove the MBSH. Add 5 mL of MEMFA. Rotate the oocytes for 1 h at room temperature.

5. Remove the MEMFA and replace with 5 mL of PBT. Wash the oocytes for 20 min, rotating, at room temperature. Repeat the PBT washes twice (i.e., a total of three washes).

6. Following the final wash, allow the oocytes to settle to the bottom of the Eppendorf tube.

   If oocytes adhere to the side of the tube, prod gently with a pipette tip to dislodge.

7. Remove 2.5 mL of PBT and replace with 2.5 mL of PBT-30S. Allow oocytes to equilibrate for 30 min at room temperature.

   The oocytes will float on addition of the PBT-30S and sink slowly as they equilibrate.

8. Remove 5 mL of PBT/PBT-30S solution and replace with 5 mL of PBT-30S. Allow the oocytes to equilibrate for 30 min at room temperature. Replace the PBT-30S three more times (for a total of four equilibrations).

9. Transfer the oocytes to an embedding mold. Allow the oocytes to settle to the base of the mold. Using a Pasteur pipette, carefully remove the PBT-30S from the mold.

10. Gently add O.C.T. over the oocytes, filling the embedding mold approximately half-way. Allow the oocytes to equilibrate in O.C.T. for 30 min at room temperature.

    Avoid generating bubbles in the O.C.T., particularly near the sample.

11. Snap-freeze samples in a dry ice/ethanol bath by submerging the embedding mold to its midway point until the O.C.T. becomes uniformly solid. Store at −80°C.

Cryosectioning

12. Remove samples from −80°C storage. Maintain on dry ice until sectioning.

13. Place the oocytes (in embedding molds) into a precooled (−20°C) cryostat. Equilibrate for ∼30 min.

14. Remove the sample from the mold. Freeze it to a specimen disc using liquid O.C.T. and Cytocool II aerosol spray.

15. Cut 30–50-μm sections under a glass antiroll plate.

16. Transfer each section to a room-temperature microscope slide by touching the slide to the sample.

    Each slide can hold 2–4 sections.

17. To cure the sections to the slides, store overnight at 18°C–20°C, shielded from light.

RNA Fluorescence In Situ Hybridization (FISH)

18. Using a sterile cotton-tipped applicator coated in silicone adhesive sealant, create a hydrophobic barrier around the sections on the slide (Fig. 1A).

19. Coat the sections by applying 1 mL of PBS inside the well. Rehydrate the sample for 20 min at room temperature.

    Avoid applying liquid directly to the sectioned samples.

    See Troubleshooting.

20. Remove the PBS. Apply 1 mL of PBT to the sample. Incubate for 20 min at room temperature.

21. Remove the PBT and apply 1 mL of wash buffer A (prepared with 10% formamide, final concentration) to the sample. Incubate for 5 min at room temperature. Repeat once.

22. Remove the second wash of buffer A (with 10% formamide). Add 50–100 μL of hybridization buffer (prepared with 10% formamide, final concentration) containing the probe(s) of interest at a concentration of 125–500 nM.

23. Carefully place a clean coverslip over the hybridization buffer. Make sure the buffer is distributed evenly and covers the sample completely.

24. Place the slide in a humidified slide chamber. Incubate in the dark for 16–20 h at 37°C.

25. Immerse the slide in wash buffer A (with 10% formamide). Use forceps to gently remove the coverslip.

26. Remove the slide from the wash buffer. Remove excess buffer surrounding the sections.

27. Apply 1 mL of wash buffer A (with 10% formamide) to the slide. Incubate in the dark for 30 min at 37°C.

28. Remove wash buffer A (with 10% formamide) and apply 1 mL of wash buffer B. Incubate in the dark for 5 min at room temperature. Repeat the wash with buffer B once, shielded from light.

29. Remove wash buffer B. Add ∼50–100 μL of Prolong Gold antifade mountant to the sample sections. Cover with a clean coverslip.

30. Seal coverslip corners using nail polish. Cure the samples for 24 h at room temperature, shielded from light.

Cite this protocol as *Cold Spring Harb Protoc*; doi:10.1101/pdb.prot097030

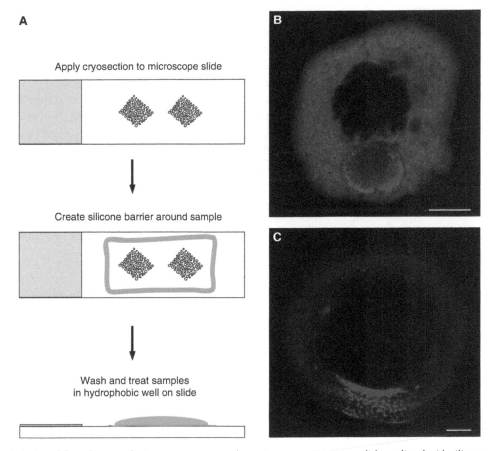

FIGURE 1. (A) Workflow diagram depicting cryosectioned oocytes on a microscope slide outlined with silicone adhesive sealant (Step 18), creating a hydrophobic well over the samples in which FISH treatments (Steps 19–28) are conducted. (B–C) Detection and localization of endogenous Vg1 (Weeks and Melton 1987), Xpat (Hudson and Woodland 1998), and GAPDH mRNA in a Stage I–II oocytes were performed using Stellaris gene specific probe sets designed and provided by LGC BioSearch Technologies. The Vg1 (Quasar® 570), Xpat (Quasar 570), and GAPDH (Quasar 670) probe sets were applied at a concentration of 250 μM and imaged using a Zeiss LSM 800 Confocal Laser Scanning Microscope with a 20× objective. (B) A stage I oocyte is shown with ubiquitously distributed GAPDH mRNA (green) and Xpat mRNA (red), which is localized to the Balbiani body. (C) A stage II oocyte is shown with Vg1 mRNA (red), which is enriched in the vegetal hemisphere. Oocytes are oriented with the vegetal pole at the bottom, and scale bars = 50 μM.

31. Seal the coverslip edges with nail polish. Allow to dry, shielded from light.

32. Samples are now suitable for imaging using a fluorescence microscope (examples are shown in Fig. 1B,C). Slides are stable up to a week at 4°C.

TROUBLESHOOTING

Problem (Steps 19–29): There is poor retention of oocytes on slides during RNA FISH treatment.

Solution: Increasing the density of oocytes on the slide can significantly improve sample retention during processing. Ensuring that oocytes are uniformly settled along the base of the embedding mold before snap-freezing generates an optimal cutting surface to produce sections that are more durable.

DISCUSSION

Xenopus laevis oocytes provide an excellent resource for the study of RNA biology in vertebrate cells. Their large size provides advantages for both biochemical and imaging-based approaches. This strength has been applied to the study of RNA-protein interactions (Hake and Richter 1994; Gall et al. 1999; Yoon and Mowry 2004), polyadenylation (McGrew et al. 1989), translational control (Colegrove-Otero et al. 2005), subcellular localization (Melton 1987; Mowry and Melton 1992), and transport of mRNA (Messitt et al. 2008; Gagnon et al. 2013). The ease with which distinct stages can be separated based on the morphology of the oocyte (Dumont 1972) allows for investigation of specific events in oogenesis in order to probe key developmental questions. However, the yolk-rich oocyte cytoplasm can present a barrier to high-resolution imaging of endogenous RNAs. Approaches to overcome these limitations include both electron microscopy and optical clearing of whole-mount preparations of oocytes and embryos (Kloc et al. 2005; Bilinski et al. 2010; Agricola and Cha 2016). Advances in fluorescence in situ hybridization (FISH), using probe sets of multiple fluorescently labeled oligonucleotides, has allowed for increased sensitivity and higher specificity (Raj et al. 2008). In addition, direct comparison of multiple RNA species in a single cell are possible through multiplexing of probe sets. This protocol combines the strengths of the multi-probe FISH approach with cryosectioning as a means of circumventing the need for clearing agents and increasing target accessibility in *Xenopus* oocytes.

RECIPES

MBSH Buffer (10×)

Reagent	Final concentration
NaCl	880 mM
KCl	10 mM
NaHCO$_3$	24 mM
MgSO$_4$·7H$_2$O	8.2 mM
Ca(NO$_3$)$_2$·4H$_2$O	3.3 mM
CaCl$_2$·6H$_2$O	4.1 mM
HEPES (pH 7.6)	100 mM

Filter-sterilize and store as a 10× stock solution. Prepare fresh 1× MBSH prior to use.

MEMFA

MOPS
EGTA
MgSO$_4$
Formaldehyde (37%)

Prepare a 10× stock solution of 1 M MOPS (pH 7.4), 20 mM EGTA, and 10 mM MgSO$_4$. On the day of use, prepare fresh 1× MEMFA by combining 8 parts H$_2$O, 1 part 10× MEMFA stock solution, and 1 part formaldehyde (37%).
Final concentrations are 100 mM MOPS (pH 7.4), 2 mM EGTA, 1 mM MgSO$_4$, 3.7% (v/v) formaldehyde.

PBT for Xenopus *Oocyte FISH*

1× phosphate-buffered saline (PBS) <R>
0.2 mg/mL bovine serum albumin (Sigma-Aldrich A2153)
0.1% (v/v) Triton X-100 (Fisher Scientific BP151)

Mix thoroughly. Sterilize through a 0.22-μm filter. Store at 4°C.

Cite this protocol as *Cold Spring Harb Protoc*; doi:10.1101/pdb.prot097030

Phosphate-Buffered Saline (PBS)

Reagent	Amount to add (for 1× solution)	Final concentration (1×)	Amount to add (for 10× stock)	Final concentration (10×)
NaCl	8 g	137 mM	80 g	1.37 M
KCl	0.2 g	2.7 mM	2 g	27 mM
Na$_2$HPO$_4$	1.44 g	10 mM	14.4 g	100 mM
KH$_2$PO$_4$	0.24 g	1.8 mM	2.4 g	18 mM

If necessary, PBS may be supplemented with the following:

CaCl$_2$·2H$_2$O	0.133 g	1 mM	1.33 g	10 mM
MgCl$_2$·6H$_2$O	0.10 g	0.5 mM	1.0 g	5 mM

PBS can be made as a 1× solution or as a 10× stock. To prepare 1 L of either 1× or 10× PBS, dissolve the reagents listed above in 800 mL of H$_2$O. Adjust the pH to 7.4 (or 7.2, if required) with HCl, and then add H$_2$O to 1 L. Dispense the solution into aliquots and sterilize them by autoclaving for 20 min at 15 psi (1.05 kg/cm^2) on liquid cycle or by filter sterilization. Store PBS at room temperature.

Xenopus *Oocyte Culture Medium (XOCM)*

Reagent	Final concentration
Leibovitz's L-15 Medium (Thermo Fisher Scientific 11415064)	50%
HEPES (pH 7.6)	15 mM
Insulin	1 mg/mL
Nystatin (Sigma-Aldrich N1638)	50 U/mL
Penicillin/streptomycin (Life Technologies 15070063)	100 U/mL
Gentamicin (Thermo Fisher Scientific 15750060)	0.1 mg/mL

Sterilize the first three ingredients though a 0.22-µm filter. Add the antibiotics to XOCM just prior to culture. Store XOCM without antibiotics for up to 2 mo at 4°C. After antibiotics have been added, do not store or reuse.

ACKNOWLEDGMENTS

Work on development of this method was supported by National Institutes of Health (NIH) grant GM071049 to K.M.

REFERENCES

Agricola Z, Cha S-W. 2016. Two-color fluorescence in situ hybridization using chromogenic substrates in Xenopus. *Biotechniques* 61: 263–268.

Bilinski SM, Jaglarz MK, Dougherty MT, Kloc M. 2010. Electron microscopy, immunostaining, cytoskeleton visualization, in situ hybridization, and three-dimensional reconstruction of Xenopus oocytes. *Methods* 51: 11–19.

Colegrove-Otero LJ, Devaux A, Standart N. 2005. The Xenopus ELAV protein ElrB represses Vg1 mRNA translation during oogenesis. *Mol Cell Biol* 25: 9028–9039.

Dumont JN. 1972. Oogenesis in Xenopus laevis (Daudin). I. Stages of oocyte development in laboratory maintained animals. *J Morphol* 136: 153–179.

Gagnon JA, Kreiling JA, Powrie EA, Wood TR, Mowry KL. 2013. Directional transport is mediated by a Dynein-dependent step in an RNA localization pathway. *PLoS Biol* 11: e1001551.

Gall JG, Bellini M, Wu Z, Murphy C. 1999. Assembly of the nuclear transcription and processing machinery: Cajal bodies (coiled bodies) and transcriptosomes. *Mol Biol Cell* 10: 4385–4402.

Hake LE, Richter JD. 1994. CPEB is a specificity factor that mediates cytoplasmic polyadenylation during Xenopus oocyte maturation. *Cell* 79: 617–627.

Hudson C, Woodland HR. 1998. Xpat, a gene expressed specifically in germ plasm and primordial germ cells of Xenopus laevis. *Mech Dev* 73: 159–168.

Kloc M, Wilk K, Vargas D, Shirato Y, Bilinski S, Etkin LD. 2005. Potential structural role of non-coding and coding RNAs in the organization of the cytoskeleton at the vegetal cortex of Xenopus oocytes. *Development* 132: 3445–3457.

McGrew LL, Dworkin-Rastl E, Dworkin MB, Richter JD. 1989. Poly(A) elongation during Xenopus oocyte maturation is required for translational recruitment and is mediated by a short sequence element. *Genes Dev* 3: 803–815.

Melton DA. 1987. Translocation of a localized maternal mRNA to the vegetal pole of *Xenopus* oocytes. *Nature* **328:** 80–82.

Messitt TJ, Gagnon JA, Kreiling JA, Pratt CA, Yoon YJ, Mowry KL. 2008. Multiple kinesin motors coordinate cytoplasmic RNA transport on a subpopulation of microtubules in *Xenopus* oocytes. *Dev Cell* **15:** 426–436.

Mowry KL, Melton DA. 1992. Vegetal messenger RNA localization directed by a 340-nt RNA sequence element in *Xenopus* oocytes. *Science* **255:** 991–994.

Newman K, Aguero T, King ML. 2018. Isolation of *Xenopus* oocytes. *Cold Spring Protoc* doi :10.1101/pdbprot095851.

Raj A, van den Bogaard P, Rifkin SA, van Oudenaarden A, Tyagi S. 2008. Imaging individual mRNA molecules using multiple singly labeled probes. *Nat Methods* **5:** 877–879.

Weeks DL, Melton DA. 1987. A maternal mRNA localized to the vegetal hemisphere in *Xenopus* eggs codes for a growth factor related to TGF-β. *Cell* **51:** 861–867.

Yoon YJ, Mowry KL. 2004. Xenopus Staufen is a component of a ribonucleoprotein complex containing Vg1 RNA and kinesin. *Development* **131:** 3035–3045.

Cite this protocol as *Cold Spring Harb Protoc*; doi:10.1101/pdb.prot097030

Analysis of Cell Fate Commitment in *Xenopus* Embryos

Sally A. Moody

Department of Anatomy and Regenerative Biology, George Washington University, Washington, D.C. 20037

The fates of individual cleavage-stage blastomeres and of groups of cells at the blastula through gastrula stages of *Xenopus* embryos have been mapped in great detail. These studies identified the major contributors of the three germ layers as well as a variety of tissues and organs and several specific cell types. One can use these fate maps to test the commitment of single cells or groups of cells to produce their normal repertoire of descendants, to identify the genes that regulate fate commitment, and to modulate the levels of gene expression in specific lineages to determine gene function in a variety of developmental processes. Here we introduce methods that include how to identify specific blastomeres, inject them with lineage tracers, and alter gene expression levels. We also discuss methods for assaying protein and mRNA expression in situ and for providing novel embryonic environments to test fate commitment. These techniques draw upon classical approaches that are quite easy to perform in the versatile *Xenopus* embryo.

FATE MAPPING AND LINEAGE TRACING

Fate mapping techniques have been used extensively in *Xenopus* to identify what kinds of tissues descend from a single blastomere or a specific region of the embryo. By precisely defining the regions of the embryo from which the germ layers, tissues and organs, and specific cell types arise, one can discover the mechanisms that control the organization of the body plan and specify cell fate. The creation of fate maps was recognized from the beginnings of experimental embryology as an essential tool (for review, see Klein and Moody [2016]). However, fate maps only describe the developmental path taken by a cell in the normal, intact embryo; they do not identify the full developmental potential of a cell or the mechanisms by which its fate is determined. Since cell fate can be influenced by a number of intrinsic and extrinsic factors, lineage tracing is an essential technique to identify the progeny of a cell as they develop under different experimental conditions. It also is an important tool for labeling donor tissue in tissue recombination experiments that are used to define inductive interactions. For example, the ability to recognize the descendants of the transplanted "organizer" in the host embryo was critical for interpreting the pioneering experiments of primary embryonic induction (Spemann and Mangold 1924).

Lineage studies, fate mapping and tests of the fate determination of single blastomeres have been very informative in species in which the cell division patterns are identical across embryos. The invariant lineages allow the exact same cell to be marked in a large number of embryos (e.g., Sulston et al. 1983; Nishida 1997). In most vertebrate embryos, however, cleavage patterns are notably irregular (e.g., Kelly 1977; Ziomek et al. 1982; Johnson and Ziomek 1983; Kimmel and Law 1985). In contrast, *Xenopus* embryos often cleave in nearly identical patterns, and a few common

Correspondence: samoody@gwu.edu

cleavage patterns have been described (Fig. 1; Hirose and Jacobson 1979). Another essential element for accurate fate mapping is the ability to identify the body axes (dorsal–ventral, left–right, anterior–posterior) at early cleavage stages so that a blastomere's position can be ascertained. Due to unique movements of pigment granules after fertilization, this is possible in *Xenopus*. The first cleavage furrow, which always defines the mid-sagittal plane that separates the right and left sides, naturally bisects the lightly pigmented gray crescent region of the animal hemisphere in about 70% of embryos. By picking these embryos at the two-cell stage, one can use the dorsal–ventral differences in pigmentation to identify specific blastomeres at later stages (Klein 1987; Masho 1990). By choosing embryos from this subset that also cleave in a "stereotypic" pattern, one can study nearly identical lineages across a number of embryos (see Protocol 1: Lineage Tracing and Fate Mapping in *Xenopus* Embryos [Moody 2018a]). This is of significant experimental advantage because it decreases variability between embryos and enables the construction of robust qualitative and quantitative fate maps.

Detailed fate maps of the different cleavage-stage blastomeres of *Xenopus* using vital dyes and intracellular lineage tracers have been published (Nakamura and Kishiyama 1971; Hirose and Jacobson 1979; Jacobson and Hirose 1981; Dale and Slack 1987; Moody 1987a,b, 1989; Masho 1988; Masho and Kubota 1988; Moody and Kline 1990). They show that when the same, stereotypic blastomere is traced in a large number of embryos, its fate is quite predictable, in contrast to most other vertebrates (e.g., Kelly 1977; Ziomek et al. 1982; Johnson and Ziomek 1983; Kimmel and Law 1985). This predictability allows one to rigorously test how and when during development a particular fate is determined.

The *Xenopus* blastomere fate maps demonstrate some important features of fate restriction during cleavage stages. First, there is no germ layer restriction at the two- through 32-cell stages; all blastomeres give rise to some cells that reside in the endoderm, mesoderm, and ectoderm. However, as expected from the gastrula fates maps (Keller 1975, 1976), vegetal blastomeres contribute more prominently to endoderm, equatorial blastomeres more prominently to mesoderm, and animal blastomeres more prominently to ectoderm. A second important feature is that no tissue or organ has been found to descend from a single blastomere, i.e., none is monogenic. However, when one places the fate map data from two-cell to 32-cell embryos in a true lineage map, i.e., a map of mitotic relationships within the clone, some apparent restrictions are revealed (Moody and Kline 1990). For example, within the ectoderm, the cement gland, lens, olfactory placode, cranial ganglia, and otocyst are derived from animal (D1, V1) but not from vegetal (D2, V2) eight-cell blastomeres. Rostral CNS

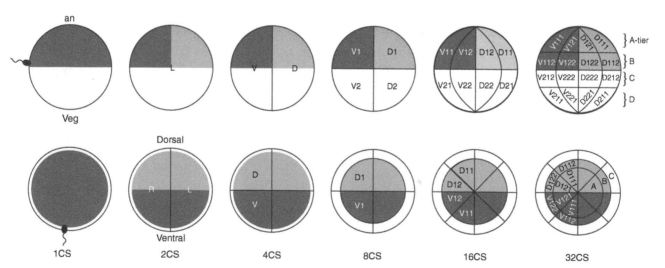

FIGURE 1. Diagram indicating the pigmentation differences and nomenclature of cells from the one-cell to the 32-cell stage. The top row represents a *left* (L) side view. The bottom row represents an animal pole view. The point of sperm entry is indicated at the one-cell stage. an, animal pole; veg, vegetal pole; CS, cell stage; D, dorsal; V, ventral; L, *left*; R, *right*. Nomenclature is that of Jacobson (Hirose and Jacobson 1979; Jacobson and Hirose 1981). Tiers on the 32-cell embryos refer to the Nakamura and Kishiyama (1971) nomenclature.

Cite this introduction as *Cold Spring Harb Protoc*; doi:10.1101/pdb.top097246

structures (retina, forebrain, and midbrain) are restricted to the D cell at the four-cell stage, and further restricted to its anterior daughter (D1) at the four-cell stage. But at the following two cell divisions (the 16-cell and 32-cell stages), all daughters and granddaughters contribute to all three structures. Thus, germ layer, tissue, and organ fates are not restricted to a single cleavage-stage blastomere, at least through the 32-cell stage. Nonetheless, there are regional differences that allow one to target manipulations to major organs or regions. For example, at the 16-cell stage one can target different ectodermal structures. The major contributor to the retina is blastomere D1.1 and the major contributor to the epidermis is blastomere V1.1. The neural crest derives mostly from blastomere D1.2 and the cranial placodes mostly from blastomere V1.2 (Fig. 1; Moody 1987a).

One also can test cell fate restriction at the cell phenotype level by labeling cells with cell type–specific markers (see Protocol 5: Whole-Mount Immunocytochemistry in *Xenopus* [Klymkowsky 2018]). However, even with these very specific approaches we have found no evidence for a monoclonal origin of several neuronal phenotypes. For example, primary motor neurons and primary sensory neurons, which can be identified by their very characteristic morphologies when filled with a cytoplasmic lineage tracer (Moody and Jacobson 1983; Jacobson and Moody 1984), each descend from a large number of 32-cell blastomeres (Moody 1989). We thought that this polyclonal origin might be related to their extensive anterior–posterior distribution along the length of the spinal cord, so we next studied several neuronal phenotypes that are spatially restricted. However, these cells also descend from multiple blastomeres. For example, four different neurotransmitter types of retinal amacrine cells each descend from several 32-cell blastomeres (Huang and Moody 1995, 1997). Even a small dopaminergic nucleus in the hypothalamus, which contains only 30–40 cells at early tadpole stages, descends from as many as seven of the 32 blastomeres (Huang and Moody 1992). Further studies using blastomere transplantation and deletion approaches (see Protocol 7: Cleavage Blastomere Deletion and Transplantation to Test Cell Fate Commitment in *Xenopus* [Moody 2018b]) showed that the ability of a blastomere to contribute to the retina depends on both intrinsic factors and environmental cues (Yan and Moody 2007). These studies illustrate the power of combining single cell lineage labeling with cell phenotype markers to explain how cell fate is specified.

TESTING FATE COMMITMENT

Fate maps, however, do not describe the full developmental potential of a blastomere or indicate whether it will express its specific cell fate regardless of the surrounding cellular environment. A cell that is committed to its fate will produce its normal repertoire of descendants even when it develops in other regions of the embryo. Alternatively, a cell that is not yet committed will express different cell fates depending upon the environment that it occupies. There are several ways to test the commitment of a blastomere to produce its mapped fate. One approach is to delete a neighboring blastomere to test if its presence is necessary for the normal fate of the cell of interest, and to determine whether the deleted lineage can be reconstituted (e.g., Huang and Moody 1993; Gallagher and Moody 2004) (see Protocol 7: Cleavage Blastomere Deletion and Transplantation to Test Cell Fate Commitment in *Xenopus* [Moody 2018b]). Another approach is to transplant a labeled blastomere or group of cells (e.g., Spemann and Mangold 1924; Yan and Moody 2007) into a novel region of the embryo to test whether it maintains its original set of descendants, or alters its progeny in accord with its new environment. Explanting a single blastomere or a group of cells (e.g., the animal cap) out of the embryo and into culture prevents the cell from communicating with its normal neighbors and allows one to experimentally manipulate extrinsic signals (see Protocol 6: Cleavage Blastomere Explant Culture in *Xenopus* [Moody 2018c] and Protocol 8: Dissecting and Culturing Animal Cap Explants [Dingwell and Smith 2018]). For each approach, one can analyze gene expression and the production of specific cell types using standard techniques of lineage tracing and in situ detection of cell type–specific mRNAs and proteins (see Protocol 4: Whole-Mount In Situ Hybridization of *Xenopus* Embryos [Saint-Jeannet 2017] and Protocol 5: Whole-Mount Immunocytochemistry in *Xenopus* [Klymkowsky 2018]).

ALTERING GENE EXPRESSION

Fate maps also do not identify the genes that regulate the development of a lineage. To do this, one needs to modify the gene expression of the precursor cell of interest. In *Xenopus*, this can be easily accomplished by microinjecting molecules into an identified blastomere of known fate to affect gene expression specifically within its lineage. One can microinject a number of different kinds of constructs, including mRNAs, plasmid DNAs, and antisense oligonucleotides (see Protocol 2: Microinjection of mRNAs and Oligonucleotides [Moody 2018d]) and Protocol 3: Microinjection of DNA Constructs into *Xenopus* Embryos for Gene Misexpression and *cis*-Regulatory Module Analysis [Yasuoka and Taira 2018]). With these reagents one can ascertain whether a gene promotes or represses target genes or specific tissue formation, or converts tissues to novel fates when ectopically expressed. One can identify functional domains by expressing mutated forms and determine the requirement of a gene in a specific tissue or process by injecting dominant-negative forms or antisense oligonucleotides that block either the mRNA splicing or translation of endogenous proteins. For all of these constructs, it is essential to mark the blastomere that is injected with a lineage label to ascertain whether the effect is confined to the descendants of the injected blastomere (i.e., is cell autonomous) or involves surrounding cells (i.e., involves cell-to-cell interactions).

CONCLUSION

Together, the methods described here provide simple yet powerful ways to demonstrate the fundamental developmental mechanisms that regulate how the different cells and regions of the embryo produce specific cell types, tissues, and organs.

REFERENCES

Dale L, Slack JMW. 1987. Fate map of the 32-cell stage of *Xenopus laevis*. *Development* **100:** 279–295.

Dingwell KS, Smith JC. 2018. Dissecting and culturing animal cap explants. *Cold Spring Harb Protoc* doi:10.1101/pdb.prot097329.

Gallagher BC, Moody SA. 2004. Regulation of primary spinal neuron lineages after deletion of a major progenitor. *Biol Cell* **96:** 539–544.

Hirose G, Jacobson M. 1979. Clonal organization of the central nervous system of the frog. I. Clones stemming from individual blastomeres of the 16-cell and earlier stages. *Dev Biol* **71:** 191–202.

Huang S, Moody SA. 1992. Does lineage determine the dopamine phenotype in the tadpole hypothalamus: A quantitative analysis. *J Neurosci* **12:** 1351–1362.

Huang S, Moody SA. 1993. The retinal fate of *Xenopus* cleavage stage progenitors is dependent upon blastomere position and competence: Studies of normal and regulated clones. *J Neurosci* **13:** 3193–3210.

Huang S, Moody SA. 1995. Asymmetrical blastomere origin and spatial domains of dopamine and Neuropeptide Y amacrine cells in *Xenopus* tadpole retina. *J Comp Neurol* **360:** 2–13.

Huang S, Moody SA. 1997. Three types of serotonin-containing amacrine cells in the tadpole retina have distinct clonal origins. *J Comp Neurol* **387:** 42–52.

Jacobson M, Hirose G. 1981. Clonal organization of the central nervous system of the frog. II. Clones stemming from individual blastomeres of the 32- and 64-cell stages. *J Neurosci* **1:** 271–284.

Jacobson M, Moody SA. 1984. Quantitative lineage analysis of the frog's nervous system. I. Lineages of Rohon-Beard neurons and primary motoneurons. *J Neurosci* **4:** 1361–1369.

Johnson MH, Ziomek CA. 1983. Cell interactions influence the fate of mouse blastomeres undergoing the transition from the 16- to the 32-cell stage. *Dev Biol* **95:** 211–218.

Keller RE. 1975. Vital dye mapping of the gastrula and neurula of *Xenopus laevis*. I. Prospective areas and morphogenetic movements of the superficial layer. *Dev Biol* **42:** 222–241.

Keller RE. 1976. Vital dye mapping of the gastrula and neurula of *Xenopus laevis*. II. Prospective areas and morphogenetic movements of the deep layer. *Dev Biol* **51:** 118–137.

Kelly SJ. 1977. Studies of the developmental potential of the 4- and 8-cell stage mouse blastomeres. *J Exp Zool* **200:** 365–376.

Kimmel CB, Law RD. 1985. Cell lineage of zebrafish blastomeres. I. Cleavage pattern and cytoplasmic bridges between cells. *Dev Biol* **108:** 78–85.

Klein SL. 1987. The first cleavage furrow demarcates the dorsal-ventral axis in *Xenopus* embryos. *Dev Biol* **120:** 299–304.

Klein SL, Moody SA. 2016. When family history matters: The importance of lineage analyses and fate maps in explaining animal development. *Curr Top Dev Biol* **117:** 93–112.

Klymkowsky MW. 2018. Whole-mount immunocytochemistry in *Xenopus*. *Cold Spring Harb Protoc* doi:10.1101/pdb.prot097295.

Masho R. 1988. Fates of animal-dorsal blastomeres of eight-cell stage *Xenopus* embryos vary according to the specific patterns of the third cleavage plane. *Dev Growth Differ* **30:** 347–359.

Masho R. 1990. Close correlation between the first cleavage plane and the body axis in early *Xenopus* embryos. *Dev Growth Diff* **32:** 57–64.

Masho R, Kubota HY. 1988. Developmental fates of blastomeres of the eight-cell stage *Xenopus* embryo. *Dev Growth Diff* **30:** 347–359.

Moody SA. 1987a. Fates of the blastomeres of the 16-cell stage *Xenopus* embryo. *Dev Biol* **119:** 560–578.

Moody SA. 1987b. Fates of the blastomeres of the 32-cell stage *Xenopus* embryo. *Dev Biol* **122:** 300–319.

Moody SA. 1989. Quantitative lineage analysis of the origin of frog primary motor and sensory neurons from cleavage stage blastomeres. *J Neurosci* **9:** 2919–2930.

Moody SA. 2018a. Lineage tracing and fate mapping in *Xenopus* embryos. *Cold Spring Harb Protoc* doi:10.1101/pdb.prot097253.

Moody SA. 2018b. Cleavage blastomere deletion and transplantation in *Xenopus*. *Cold Spring Harb Protoc* doi:10.1101/pdb.prot097311.

Moody SA. 2018c. Cleavage blastomere explant culture in *Xenopus*. *Cold Spring Harb Protoc* doi:10.1101/pdb.prot097303.

Moody SA. 2018d. Microinjection of mRNAs and oligonucleotides. *Cold Spring Harb Protoc* doi:10.1101/pdb.prot097261.

Moody SA, Jacobson M. 1983. Compartmental relationships between anuran primary spinal motoneurons and somitic muscle fibers that they first innervate. *J Neurosci* **3:** 1670–1682.

Moody SA, Kline MJ. 1990. Segregation of fate during cleavage of frog (*Xenopus laevis*) blastomeres. *Anat Embryol* **182:** 347–362.

Nakamura O, Kishiyama K. 1971. Prospective fates of blastomeres at the 32-cell stage of *Xenopus laevis* embryos. *Proc Japan Acad* **47:** 407–412.

Nishida H. 1997. Cell lineage analysis in ascidian embryos by intracellular injection of a tracer enzyme. III. Up to the tissue restricted stage. *Dev Biol* **121:** 526–541.

Saint-Jeannet JP. 2017. Whole-mount in situ hybridization of *Xenopus* embryos. *Cold Spring Harb Protoc* doi:10.1101/pdb.prot097287.

Spemann H, Mangold H. 1924. Induction of embryonic primordia by implantation of organizers from a different species. In *Foundations of experimental embryology* (ed. Willier BH, Oppenheimer JM), pp. 144–184. Hafner, New York.

Sulston JE, Schierenberg E, White JG, Thomson JN. 1983. The embryonic lineage of the nematode *Caenorhabditis elegans*. *Dev Biol* **100:** 64–119.

Yan B, Moody SA. 2007. The competence of *Xenopus* blastomeres to produce neural and retinal progeny is repressed by two endo-mesoderm promoting pathways. *Dev Biol* **305:** 103–119.

Yasuoka Y, Taira M. 2018. Microinjection of DNA constructs into *Xenopus* embryos for gene misexpression and *cis*-regulatory module analysis. *Cold Spring Harb Protoc* doi:10.1101/pdb.prot097279.

Ziomek CA, Johnson MH, Handyside AH. 1982. The developmental potential of mouse 16-cell blastomeres. *J Exp Zool* **221:** 345–355.

Lineage Tracing and Fate Mapping in *Xenopus* Embryos

Sally A. Moody

Department of Anatomy and Regenerative Biology, George Washington University, Washington, D.C. 20037

Fate mapping approaches reveal what types of cells, tissues, and organs are derived from specific embryonic cells. Classical fate maps were made by microscopic techniques using embryos comprising small numbers of transparent cells. More complex and opaque embryos require use of a vital or lipophilic dye that labels small groups of cells. Intracellular injection of a lineage tracer that labels the injected cell and all of its descendants can be used to mark a single cell in *Xenopus* embryos, whose large cells are easy to microinject and usually cleave in regular patterns. Intracellular lineage tracers must be neutral compounds that do not interact with cellular processes that might change the developmental fate of the injected cell, be small enough to diffuse quickly throughout the cytoplasm before the cell divides so that all descendants are labeled, and be large enough to not diffuse to adjacent cells via gap junctions. They should not be diluted by cell division or intracellular degradation, and should be easily detected by histochemical reactions (enzymes) or direct imaging (fluorescent compounds). Several types of lineage tracers have been used, including small, fluorescently tagged dextrans and mRNAs encoding enzymes or fluorescent proteins, described here. Many lineage tracers can be combined with cell type–specific mRNA and protein expression assays, making lineage tracing a powerful tool for testing the function of genes and cell fate commitment.

MATERIALS

It is essential that you consult the appropriate Material Safety Data Sheets and your institution's Environmental Health and Safety Office for proper handling of equipment and hazardous materials used in this protocol.

RECIPES: Please see the end of this protocol for recipes indicated by <R>. Additional recipes can be found online at http://cshprotocols.cshlp.org/site/recipes.

Reagents

Benzocaine (10% in ethanol) (optional; see Step 8)

Store 10% benzocaine stock solution in the refrigerator. Just before use, prepare a working solution of 0.5% benzocaine by adding (drop-wise, with constant agitation to prevent precipitation) 50 µL of 10% stock to 10 mL of 0.5× culture solution.

Culture medium (e.g., Steinberg's solution [100%] <R> [1× SS], Marc's modified Ringer's [MMR] [1×] <R>, or modified Barth's saline [MBS] <R> [1×])

Different laboratories prefer different media.

Dejellying reagents (see Sive et al. 2007a and Moody 2012a)
Fixative

Correspondence: samoody@gwu.edu

Paraformaldehyde (4% in PBS) <R> (for fluorescently labeled specimens)

Paraformaldehyde (4% in MEM) <R> (for specimens labeled with β-Gal)

Glycerol (e.g., Sigma G5516) or commercial aqueous mounting medium

Dilute glycerol stock solution to 70% in PBS (pH 7.4).

Phosphate-buffered saline (PBS) (0.1 M, pH 7.4) <R>

Reagents for β-Gal staining

β-Gal reaction buffer <R>

BA/BB clearing solution (optional; see Step 15)

Prepare a solution of 2 parts benzyl alcohol, 1 part benzyl benzoate. Mix thoroughly in a glass bottle (do not use plastic) and store at room temperature.

Ethanol (100%) (optional; see Step 15)

MEM salts stock buffer (10×) <R>

Reagents for natural mating or in vitro fertilization in *Xenopus* (see Sive et al. 2007b and Moody 2012a)

Because obtaining eggs by squeezing the female sometimes reduces egg quality, an alternate method for in vitro fertilization is to have the hormone-injected female naturally release eggs into High-salt MBS <R>. Collect and maintain eggs in this buffer for several hours. Wash twice with 0.1× MBS before adding sperm.

Reagents for preparing fluorescent dextran lineage tracers

Fluorescent dextran (10,000–40,000 MW; lysine-fixable)

See https://www.thermofisher.com/us/en/home/references/molecular-probes-the-handbook/fluorescent-tracers-of-cell-morphology-and-fluid-flow/fluorescent-and-biotinylated-dextrans.html.

KCl (0.2 N, pH 6.8) <R>

Reagents for preparing mRNA lineage tracers

In vitro transcription kit (e.g., Ambion mMessage mMachine kit)

Plasmid containing the open-reading frame of bacterial β-galactosidase (β-*gal*), green fluorescent protein (*gfp*) or red fluorescent protein (*rfp*) (e.g., Clontech)

Subclone the β-gal, gfp or rfp coding region into a Xenopus-appropriate expression vector (e.g., pCS2+) containing upstream and downstream sequences that allow for efficient translation in blastomeres after intracellular injection. Alternatively, Xenopus expression plasmids already containing β-gal, gfp, or rfp coding regions can be obtained from a number of labs (http://www.xenbase.org/).

Equipment

Costar Spin-X cellulose acetate filter units for aqueous solutions (Corning 8160) (if preparing fluorescent dextran)

Cryostat or Vibratome

Epifluorescence or confocal microscope

Incubator at 14°C–16°C

Incubator at 37°C with rotating platform (if labeling with β-Gal)

Microcentrifuge

Microinjection equipment (see Protocol 2: Microinjection with mRNAs and Oligonucleotides [Moody 2018])

Nutator rotator

Slides (1-inch × 3-inch glass, with depressions) or coverslips for mounting samples

Stereomicroscope with camera system and morphometric software

Transfer pipettes, plastic or glass (6-inch, autoclaved)

Vials, glass (screw-capped)

Vortex mixer (if preparing fluorescent dextran)

METHOD

Preparing the Lineage Tracer

See Discussion for lineage tracer selection.

Preparing Fluorescently Tagged Dextran

1. Prepare fluorescent dextran as follows.

   i. Make ~100 µL of a 1.0% fluorescent dextran solution in 0.2 N KCl.

   ii. Vortex the solution vigorously to dissolve.

   iii. Transfer the solution to a Costar Spin-X cellulose acetate filter.

   iv. Centrifuge at 14,000 rpm for 15–20 min at room temperature.

   v. Prepare 20-µL aliquots of the flow-through in tightly sealed tubes.
      Aliquots can be stored in the dark for up to 6 mo at −20°C.

Preparing mRNA from β-gal, gfp, and rfp-Containing Plasmids

2. Prepare mRNA from a *Xenopus* expression plasmid containing the β-*gal*, *gfp*, or *rfp* coding region as follows.

   i. Synthesize capped mRNA with a polyA tail using an in vitro transcription kit according to the manufacturer's instructions.

   ii. Dissolve the mRNA pellet in sterile, RNase-free distilled H_2O.
      mRNA stocks (usually ~1.0–3.0 µg/µL) can be stored for about 6–12 mo at −80°C.

   iii. Dilute the mRNA to a working solution of ~50–100 pg/nL for intracellular injections.

Labeling Blastomeres with Lineage Tracer

To label a specific blastomere in order to label a particular organ progenitor, it is essential to identify the body axes (animal–vegetal; left–right; dorsal–ventral) of the embryo (see Figs. 1, 2, and Discussion). The animal and vegetal hemispheres (the pigmented and unpigmented halves of the embryo, respectively; Fig. 2C) approximate the anterior–posterior axis. The first cleavage furrow at the two-cell stage denotes the mid-sagittal plane that separates the left side from the right side (Fig. 1; Klein 1987; Masho 1990). For in vitro fertilized eggs, the sperm entry point, which can be seen shortly after fertilization and marked with a vital dye (Vincent and Gerhart 1987; Moody 2012b), identifies the ventral midline. The dorsal midline will develop about 180° opposite the sperm entry point. At fertilization, the animal hemisphere pigmentation contracts towards the sperm entry point on the ventral side, causing the dorsal equatorial region to become slightly less pigmented (Figs. 1; 2A). For naturally fertilized eggs, if the first cleavage furrow bisects this lightly pigmented region equally between the two cells, then the dorsal side of the embryo can be predicted very accurately (>90%) (Klein 1987; Masho 1990).

If the experimental goal is to globally express the lineage tracer, for example to create a totally labeled embryo to use as a donor for a tissue transplantation experiment, then the cleavage furrow patterns will not matter.

Selecting Embryos

3. Set up a natural mating or an in vitro fertilization (see Sive et al. 2007b and Moody 2012a) to obtain fertilized eggs.

4. Remove the jelly coats from fertilized eggs (see Sive et al. 2007a and Moody 2012a). Store the dejellied embryos at 14°C–16°C in 1× culture medium (SS, MMR, or MBS) until they reach the desired injection stage.

   Cell cleavages occur approximately every 30 min.

   Low temperatures slow down the rate of cleavage, which provides more time to select embryos and set up microinjection.

Cite this protocol as *Cold Spring Harb Protoc*; doi:10.1101/pdb.prot097253

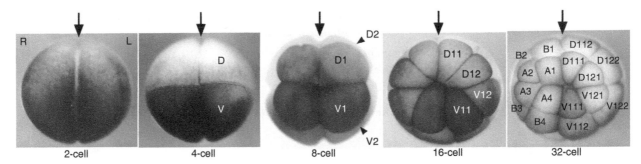

2-cell   4-cell   8-cell   16-cell   32-cell

FIGURE 1. Pigmentation patterns help to identify the body axes. At the two-cell stage, pick embryos in which the first cleavage furrow (arrow) bisects the more lightly pigmented region (*top*); this region will indicate the dorsal side (see also Fig. 2A). One can also pick four-cell embryos in which the first cleavage furrow is well defined (arrow) and the second cleavage furrow is only starting to constrict (see also Fig. 2C). The pigmentation difference between dorsal (D) and ventral (V) cells often is more accentuated at the four-cell stage. When embryos are selected in this manner, when they reach the eight- through 32-cell stages, pigmentation differences reliably indicate dorsal and ventral cells. At the eight- and 16-cell stages, we use the nomenclature of Jacobson and Hirose (*right* side of image). At the 32-cell stage, one can use the nomenclature of Jacobson and Hirose (*right* side of image), or the nomenclature of Nakamura and Kishiyama (*left* side of image). All embryos are animal views.

5. Identify the body axes (animal–vegetal; left–right; dorsal–ventral) of the embryos (Figs. 1, 2), and select embryos for injection. For specific blastomere labeling, select a large number of embryos in which the first cleavage furrow bisects the lighter, dorsal-animal area, and from this group only use embryos that divide with the regular cleavage patterns used to create the published fate maps (Fig. 1; Dale and Slack 1987; Klein 1987; Moody 1987a,b, 1989; Moody and Kline 1990). Do not use embryos in which the first cleavage furrow separates a darkly pigmented region from a light region (Fig. 2B); blastomere identification will be inaccurate in these embryos.

Microinjecting and Culturing Embryos

6. Inject the identified blastomeres with the prepared lineage tracer as described elsewhere (see Protocol 2: Microinjection with mRNAs and Oligonucleotides [Moody 2018]; Moody 2012a). For global labeling, inject both blastomeres of the two-cell embryo. Place injections just animal to the equator to facilitate diffusion of the marker towards both animal and vegetal poles

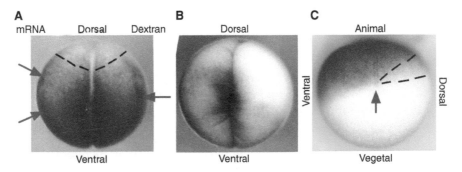

FIGURE 2. How to label the entire embryo. (*A*) If using a dextran lineage tracer, a single injection in the middle of each cell of the two-cell embryo (red arrow on *right*) is sufficient because dextrans diffuse rapidly. If using an mRNA lineage tracer, it is best to inject in two different regions (red arrows on *left*) because mRNAs do not diffuse as well. Dashes outline the lightly pigmented region that indicates the dorsal side. (*B*) Do not pick two-cell embryos in which one cell is dark and one is light; in these poorly pigmented embryos the first cleavage separates *right* and *left* sides but the lightly pigmented region does not indicate the dorsal side. (*C*) Side view of an embryo as the second cleavage furrow (red arrow) begins to form. This furrow will divide the embryo into dorsal and ventral halves. If using a dextran lineage tracer, inject at about the red arrow at the two-cell stage. If using an mRNA lineage tracer, it is best to inject once to the dorsal side of the cleavage furrow and once to its ventral side. Dashes outline the lightly pigmented region in the animal hemisphere that indicates the dorsal side.

(Figs. 2A,C). If using an mRNA tracer, inject in more than one spot per cell because mRNAs do not diffuse as well as the smaller dextran tracers (Fig. 2A).

7. Culture the embryos in 0.5× culture medium (SS, MMR, or MBS) at 14°C–22°C to allow proper gastrulation movements. If using a fluorescent tracer, culture embryos in the dark to prevent the fluorophore from releasing free radicals that can damage living cells.

Culturing at a lower temperature (14°C–16°C) can improve survival.

Fixing Embryos

8. If the embryos are swimming, anesthetize them before fixation by cooling on ice or transferring to freshly prepared 0.5% benzocaine in 0.5× culture medium. Upon loss of swimming reflex, proceed to fixation (Step 9).

9. To fix the specimens, pick up the embryos in a small volume of culture medium using a plastic or glass transfer pipette and expel them into a large volume of fixative in a glass vial with a tightly sealing screw cap. Use approximately 40× volume of fixative per volume of embryo.

    i. Fix specimens labeled with fluorescent dextrans, GFP or RFP in 4% paraformaldehyde in 0.1 M PBS (pH 7.4).

    ii. Fix specimens labeled with β-Gal in 4% paraformaldehyde in 1× MEM salt buffer (pH 7.4).

    iii. Place the vials on a rotator for 30–60 min (depending on their size) at room temperature.

10. Remove and dispose of the fixative from each vial with a transfer pipette, and replace it with the appropriate buffer (0.1 M PBS for fluorescent dextrans, GFP or RFP; 1× MEM for β-Gal).

11. Proceed to sample viewing as follows.

    i. Proceed to Steps 12–13 for immediate viewing of fluorescent specimens, or store samples in PBS for up to a few weeks at 4°C before proceeding.

    ii. Proceed to Step 14 to prepare samples labeled with β-Gal for viewing.

Viewing Specimens

Fluorescent tracers are easily combined with other fluorescent compounds, particularly UV excitable nuclear markers and fluorescently tagged antibodies. Anti-GFP and anti-β-Gal antibodies are commercially available for double-labeling.

Specimens can be sectioned for histology using standard cryostat or vibratome methods (Moody 2012a,b).

Viewing Fluorescent Specimens

12. Mount fluorescent specimens in a depression slide or sandwiched between two coverslips. Use 70% glycerol in PBS (pH 7.4) or a commercial aqueous mounting medium designed for fluorophores as the mounting medium.

    Alternatively, place samples in wells cut into an agarose bed that has been gelled onto a microscope slide.

13. View samples with epifluorescence or confocal microscopy.

Viewing β-Gal-Labeled Specimens

14. Prepare fixed β-Gal-labeled specimens for analysis as follows.

    i. Wash the specimens three times in PBS.

    ii. Incubate the specimens at 37°C on a rotator in freshly prepared β-Gal reaction buffer for 5–10 min, depending upon the strength of the enzyme activity. Check the color reaction frequently using a stereomicroscope.

Cite this protocol as *Cold Spring Harb Protoc*; doi:10.1101/pdb.prot097253

    iii. Wash the specimens three times in PBS.

    iv. Fix the specimens again, as described in Step 9, for 30 min on a rotator to stabilize the colored reaction product.

    v. Wash the specimens once in PBS.

       If specimens are to be analyzed only for cell fate, they can be viewed immediately by proceeding to Step 15.

       If specimens are to be analyzed by in situ hybridization or immunocytochemistry, refer to Protocol 5: Whole-Mount Immunocytochemistry in Xenopus *(Klymkowsky 2018) and Protocol 4: Whole-Mount In Situ Hybridization of Xenopus Embryos (Saint-Jeannet 2017).*

15. Mount and view specimens labeled with β-Gal similarly to fluorescent specimens (Step 12). Alternatively, dehydrate the specimens through a series of ethanol or methanol and clear in BA/BB clearing solution to reveal labeled cells in the internal organs. After viewing the cleared specimens, transfer the samples back to 100% ethanol to prevent loss of label, and store in the refrigerator.

    BB/BA should be used cautiously because it corrodes any surface other than glass, especially microscope parts. Spills can be cleaned with 100% ethanol.

DISCUSSION

While vital and lipophilic dyes are well-suited to staining small groups of cells and following their movements and ultimate fates in the embryo, they do not provide single-cell resolution. To accomplish this, intracellular injection of a lineage tracer that labels the injected cell and all of its descendants was developed for leech embryos (Weisblat et al. 1978). This very simple approach of microinjecting a molecule, which can be detected either by a histochemical reaction or by fluorescence microscopy, into single, identifiable blastomeres was subsequently adapted to *Xenopus* embryos (Jacobson and Hirose 1978; Hirose and Jacobson 1979; Gimlich and Braun 1985). This method has been used to create complete fate maps of the early cleavage cells of *Xenopus laevis* (Dale and Slack 1987; Hirose and Jacobson 1979; Jacobson and Hirose 1981; Klein 1987; Masho 1990; Moody 1987a,b; Moody and Kline 1990). This was feasible because *Xenopus* embryos are large, their pigment pattern denotes the three body axes, and they tend to cleave in patterns that are very similar between embryos (Fig. 1). Being able to label the same blastomere in different embryos (Figs. 1, 2) was critical for constructing the fate maps of each of the early blastomeres from two- through 32-cell. These maps showed that single blastomeres give rise to highly predictable sets of embryonic cells, tissues, and organs. They have become extremely useful in directing the microinjection of compounds for altering gene expression to the precursors of the specific tissue of interest, thus avoiding nonspecific effects of the manipulation. Intracellular lineage tracers also can be combined with immunofluorescent detection of cell type–specific proteins to study cell fate determination.

Lineage Tracer Selection

It is important to consider which lineage tracer to use for different types of experiments. Fluorescent dextrans are small (10,000–40,000 MW) and therefore diffuse rapidly upon microinjection. We found that they accurately track all of the descendants of the microinjected cell (Moore and Moody 1999). We also found that mRNA lineage tracers do not diffuse as well and do not always mark all the descendants of the injected cell. This probably occurs because they are large and likely get captured quickly onto ribosomes. Thus, mRNAs are less accurate as lineage tracers for fate mapping studies. However, if one is microinjecting mRNA to change gene expression, lineage tracer mRNAs are ideal because they are very likely to accurately mark the cells that inherited the test transcript. Another important consideration is how long it will take for the lineage tracer to be detectable. Lineage tracer mRNAs need to be translated by the cells for a few hours before the protein can be detected. In contrast, one can detect fluorescently tagged dextrans and oligonucleotides immediately. All of the commonly used tracers are relatively long-lasting. Since *Xenopus* cells do not grow, and in fact get smaller with each cell division until after the blastula stages, tracers are not diluted. Lineage tracer mRNAs are thought to be degraded by the end of gastrulation, but β-Gal, GFP, and RFP proteins are very stable and can be detected at least through tadpole stages. Interestingly, once tadpoles begin to feed, labels can become granular and unevenly distributed, suggesting they are being packaged into lysosomes.

Identification of Body Axes

If the experimental goal is to target a specific cell/tissue precursor, it is critical to accurately identify the body axes of the embryo. The dorsal side of the embryo can be predicted with high confidence by three methods detailed elsewhere (Vincent and Gerhart 1987; Moody 2012b). In our experience, selecting embryos in which the first cleavage furrow bisects the lightly pigmented side of the animal hemisphere is very effective (Klein 1987; Masho 1990) and easy (Figs. 1, 2). Embryos also can be selected at the early part of the second cleavage that forms the 4-cell embryo if the first furrow is completed and the second furrow is not yet complete (Fig. 2). To target specific blastomeres, particularly after the 8-cell stage, it is very helpful to frequently monitor whether the cleavage furrows are dividing the cytoplasm in a regular pattern. However, not every blastomere in the embryo has to perfectly match the standard cleavage patterns—just the one you are targeting.

Blastomere Nomenclature

Xenopus blastomere nomenclature requires an historical explanation. Nakamura and Kishiyama (1971) made a fate map of the 32-cell embryo using vital dye labeling. They labeled four tiers (A–D; animal to vegetal) and four rows (1–4; dorsal midline to ventral midline) (Fig. 1). This nomenclature can only be used for the 32-cell embryo. Using intracellular microinjection of an enzyme lineage tracer (horseradish peroxidase), Marcus Jacobson mapped the nervous system lineages at all of the different cleavage stages (Hirose and Jacobson 1979; Jacobson and Hirose 1981). Influenced by the nomenclature used in sea urchins and ascidians that related labeled cells to their mothers, grandmothers, and descendants (see Klein and Moody [2016]), they devised a different nomenclature that spans all stages and denotes lineal relationships (Fig. 1). All cells whose names start with "V" are on the ventral side and all cells whose names start with "D" are on the dorsal side of the embryo. The number that follows the letter denotes whether the cell is in the animal (1) or vegetal (2) hemisphere. Thus, at the 8-cell stage, D1 is the dorsal-animal cell and D2 is the dorsal-vegetal cell. At the 16-cell stage, another number is added to denote whether the daughter is located at the midline or at the plane separating dorsal and ventral. Thus, D1.1 is the midline daughter of D1, and D1.2 is its lateral daughter. At the 32-cell stage, another number is added to denote whether the daughter is located at the pole (1) or the equator (2). Thus, D1.1.1 is the animal pole daughter of D1.1, and D2.1.1 is the vegetal pole daughter of D2.1. This nomenclature is less simple to remember but represents lineal relationships that become important when tracking changes in gene and protein expression between the different cleavage stages.

RECIPES

β-Gal Reaction Buffer

20 mM $K_3Fe(CN)_6$
20 mM $K_4Fe(CN)_6 \cdot 3H_2O$
2 mM $MgCl_2$
1 mg/mL X-gal (5-bromo-4-chloro-3-indolyl-β-D-galactopyranoside)
0.1 M phosphate-buffered saline with 0.05% Tween (PBT)

Prepare immediately before use. X-gal will create a blue reaction product that nicely contrasts with red in situ hybridization chromogens. Alternatively, Magenta-gal (5-Bromo-6-chloro-3-indolyl β-D-galactopyranoside) or Red-gal (6-Chloro-3-indolyl-β-D-galactopyranoside) can be used to contrast with blue/purple chromogens.

Cite this protocol as *Cold Spring Harb Protoc*; doi:10.1101/pdb.prot097253

High-Salt MBS

Reagent	Volume to add
$CaCl_2$ (0.1 M)	7 mL
MBS salts (10×)	100 mL
NaCl (5 M)	4 mL
H_2O	888 mL

KCl (0.2 N, pH 6.8)

Prepare 100 mL of 0.2 N KCl. Adjust the pH to 6.8 with 0.05 M KOH. (The pH will overshoot and then gradually fall, so it can take about 3 h to stabilize to pH 6.8.) Filter-sterilize the solution. Store at room temperature for several months.

Marc's Modified Ringer's (MMR) (1×)

0.1 M NaCl
2 mM KCl
1 mM $MgSO_4$
2 mM $CaCl_2$
5 mM HEPES (pH 7.8)
0.1 mM EDTA

Sterilize by autoclaving, and store at room temperature.
Common alternative formulations of MMR omit EDTA and are adjusted to pH 7.4.

MBS Salts (10×)

NaCl (880 mM)
KCl (10 mM)
$MgSO_4$ (10 mM)
HEPES (50 mM, pH 7.8)

> Omit HEPES if MBS is to be used for oocyte maturation.

$NaHCO_3$ (25 mM)

Adjust pH to 7.8 with NaOH. Autoclave. Store at room temperature.

MEM Salts Stock Buffer (10×)

1 M MOPS
20 mM EGTA
10 mM $MgSO_4$

Add the salts one at a time to distilled H_2O with constant stirring. Adjust the pH to 7.4 with 10 N NaOH solution. (The solution will be cloudy until it reaches the correct pH.) Bring the solution to the final volume and filter to sterilize. Store at room temperature in a foil-wrapped bottle.

Modified Barth's Saline (MBS)

$CaCl_2$ (0.1 M)
MBS salts (10×)

For a 1× solution of MBS, mix 100 mL of 10× MBS salts with 7 mL of 0.1 M $CaCl_2$. Adjust the volume up to 1 liter with H_2O. Store at room temperature.

Cite this protocol as Cold Spring Harb Protoc; doi:10.1101/pdb.prot097253

Paraformaldehyde (4% in PBS)

1. Add 4 g of paraformaldehyde to a beaker containing 40 mL of distilled H_2O. Cover with foil and heat to 60°C in a chemical fume hood with constant stirring. Do not let the temperature rise above 65°C.
2. Add 1 N NaOH drop-wise, while stirring, until the solution just clears. Cool the solution on ice to room temperature.
3. Add 0.9 g of NaCl and stir until dissolved.
4. Add 40 mL of 0.2 M dibasic (Na_2HPO_4) phosphate buffer (PB) stock (pH 7.4) and 10 mL of monobasic (NaH_2PO_4) PB stock (pH 7.4).

    Prepared PB stocks can be stored at room temperature for months.

5. Bring the volume to 100 mL with distilled H_2O.
6. Mix thoroughly and store in the refrigerator or freezer (−20°C) in small, single-use aliquots for months.

Paraformaldehyde (4% in MEM)

1. Add 4 g of paraformaldehyde to a beaker containing 80 mL of distilled H_2O. Cover with foil and heat to 60°C in a chemical fume hood with constant stirring. Do not let the temperature rise above 65°C.
2. Add 1 N NaOH drop-wise, while stirring, until the solution just clears. Cool the solution on ice to room temperature.
3. When cool, add 10 mL of MEM salts stock buffer (10×) <R>.
4. Bring the volume to 100 mL with dH_2O.

Phosphate-Buffered Saline (PBS) (0.1 M, pH 7.4)

1. Mix 40 mL of 0.2 M dibasic (Na_2HPO_4) phosphate buffer (PB) stock with 10 mL of monobasic (NaH_2PO_4) PB stock.

    Prepared PB stocks can be stored at room temperature for months.

2. Add 0.9 g NaCl and stir until dissolved.
3. Bring the volume to 100 mL with distilled H_2O and adjust the pH to 7.4.

Steinberg's Solution (100%)

Reagent	Amount to add (for 1 L)
NaCl	3.4 g
KCl	0.05 g
$Ca(NO_3)_2 \cdot 4H_2O$	0.08 g
$MgSO_4 \cdot 7H_2O$	0.205 g
Tris	0.56 g
H_2O	to 1 L

Adjust pH to 7.4 with HCl.

ACKNOWLEDGMENTS

The development of this protocol was supported by National Institutes of Health (NIH) grant DE022065.

Cite this protocol as *Cold Spring Harb Protoc*; doi:10.1101/pdb.prot097253

REFERENCES

Dale L, Slack JMW. 1987. Fate map of the 32-cell stage of *Xenopus laevis*. *Development* **100**: 279–295.

Gimlich RL, Braun J. 1985. Improved fluorescent compounds for tracing cell lineage. *Dev Biol* **109**: 509–514.

Hirose G, Jacobson M. 1979. Clonal organization of the central nervous system of the frog. I. Clones stemming from individual blastomeres of the 16-cell and earlier stages. *Dev Biol* **71**: 191–202.

Jacobson M, Hirose G. 1978. Origin of the retina from both sides of the embryonic brain: A contribution to the problem of crossing at the optic chiasma. *Science* **202**: 637–639.

Jacobson M, Hirose G. 1981. Clonal organization of the central nervous system of the frog. II. Clones stemming from individual blastomeres of the 32- and 64-cell stages. *J Neurosci* **1**: 271–284.

Klein SL. 1987. The first cleavage furrow demarcates the dorsal–ventral axis in *Xenopus* embryos. *Dev Biol* **120**: 299–304.

Klein SL, Moody SA. 2016. When family history matters: The importance of lineage analyses and fate maps for explaining animal development. *Curr Top Dev Biol* **117**: 93–112.

Klymkowsky MW. 2018. Whole-mount immunocytochemistry in *Xenopus*. *Cold Spring Harb Protoc* doi:10.1101/pdb.prot097295.

Masho R. 1990. Close correlation between the first cleavage plane and the body axis in early *Xenopus* embryos. *Dev Growth Diff* **32**: 57–64.

Moody SA. 1987a. Fates of the blastomeres of the 16-cell stage *Xenopus* embryo. *Dev Biol* **119**: 560–578.

Moody SA. 1987b. Fates of the blastomeres of the 32-cell stage *Xenopus* embryo. *Dev Biol* **122**: 300–319.

Moody SA. 1989. Quantitative lineage analysis of the origin of frog primary motor and sensory neurons from cleavage stage blastomeres. *J Neuroscience* **9**: 2919–2930.

Moody SA. 2012a. Targeted microinjection of synthetic mRNAs to alter retina gene expression in *Xenopus* embryos. *Methods Mol Biol* **884**: 91–111.

Moody SA. 2012b. Testing retina fate commitment in *Xenopus* by blastomere deletion, transplantation and explant culture. *Methods Mol Biol* **884**: 115–127.

Moody SA. 2018. Microinjection with mRNAs and oligonucleotides. *Cold Spring Harb Protoc* doi:10.1101/pdb.prot097261.

Moody SA, Kline MJ. 1990. Segregation of fate during cleavage of frog (*Xenopus laevis*) blastomeres. *Anat Embryol* **182**: 347–362.

Moore KB, Moody SA. 1999. Animal-vegetal asymmetries influence the earliest steps in retina fate commitment in *Xenopus*. *Dev Biol* **212**: 25–41.

Nakamura O, Kishiyama K. 1971. Prospective fates of blastomeres at the 32-cell stage of *Xenopus laevis* embryos. *Proc Japan Acad* **47**: 407–412.

Saint-Jeannet JP. 2017. Whole-mount in situ hybridization of *Xenopus* embryos. *Cold Spring Harb Protoc* doi:10.1101/pdb.prot097287.

Sive HL, Grainger RM, Harland RM. 2007a. Dejellying *Xenopus laevis* embryos. *Cold Spring Harb Protoc* doi:10.1101/pdb.prot4731.

Sive HL, Grainger RM, Harland RM. 2007b. *Xenopus laevis* in vitro fertilization and natural mating methods. *Cold Spring Harb Protoc* doi:10.1101/pdb.prot4737

Vincent J-P, Gerhart JC. 1987. Subcortical rotation in *Xenopus* eggs: An early step in embryonic axis specification. *Dev Biol* **123**: 526–539.

Weisblat DA, Sawyer RT, Stent GS. 1978. Cell lineage analysis by intracellular injection of a tracer enzyme. *Science* **202**: 1295–1298.

Microinjection of mRNAs and Oligonucleotides

Sally A. Moody

Department of Anatomy and Regenerative Biology, George Washington University, Washington, D.C. 20037

Microinjecting lineage tracers into a single blastomere in the normal, intact embryo identifies the repertoire of cell types derived from it. In order to reveal the full developmental potential of that blastomere or identify the mechanisms by which its fate is determined, one needs to modify its gene expression under controlled experimental conditions. One method by which this is easily accomplished in *Xenopus* is by microinjecting synthetic mRNAs or antisense oligonucleotides into an identified blastomere to target altered gene expression specifically to its lineage. *Xenopus* blastomeres are robust and tolerate pressure-driven microinjection up to a few hundred cells, and they efficiently translate exogenously supplied mRNAs. Targeted microinjections, described here, significantly reduce off-target effects of the mRNAs or oligonucleotides. Many types of constructs can be synthesized to provide specific information about gene function. For example, microinjecting mRNA encoding the wild-type gene in its normal expression domain or in an ectopic site tests whether it promotes or represses target genes or alters the formation of tissues of interest. Mutant forms of a gene transcript can illuminate the function of different domains of the encoded protein or show the developmental consequences of a mutation found in a human disease. mRNAs encoding dominant-negative forms of a protein can elicit a functional knockdown and thereby establish the necessity for that gene in a developmental process. Microinjecting antisense morpholino oligonucleotides (MOs) that are designed to block either endogenous mRNA translation or splicing is an effective method to reduce the levels of endogenous protein.

MATERIALS

It is essential that you consult the appropriate Material Safety Data Sheets and your institution's Environmental Health and Safety Office for proper handling of equipment and hazardous materials used in this protocol.

RECIPES: Please see the end of this protocol for recipes indicated by <R>. Additional recipes can be found online at http://cshprotocols.cshlp.org/site/recipes.

Reagents

Agarose, 2% in 1× culture medium (if preparing agarose injection dishes in Step 1)

Add 2 g of electrophoresis-grade agarose to 100 mL of 1× culture medium in a screw-cap glass bottle. Autoclave to dissolve agarose. This can be stored in the refrigerator, and microwaved to liquefy the agarose when it is next needed.

Culture medium (Steinberg's solution [100%] <R> [1× SS], Marc's modified Ringer's [MMR] [1×] <R>, or modified Barth's saline [MBS] <R> [1×])

Dejellying reagents (see Sive et al. 2007a and Moody 2012a)

Correspondence: samoody@gwu.edu

Ficoll (3%–5% in 1× culture solution)

Mineral oil, heavy (sterile, RNase-/DNase-free)

mRNA or oligonucleotide solution for microinjection

Morpholino antisense oligonucleotides (MOs)

MOs should be diluted and stored in the dark according to manufacturer's instructions. We routinely use fluorescently tagged MOs to accurately lineage trace those cells affected by the protein knockdown.

mRNAs encoding lineage tracers (β-Gal, GFP, RFP) and of genes of interest

Synthesize mRNAs using an in vitro transcription kit, dissolve them in RNase-free H$_2$O, and then store as stock solutions as described in detail elsewhere (see Protocol 1: Lineage Tracing and Fate Mapping in Xenopus Embryos [Moody 2018] and Moody [2012a]). For a working solution, the concentration of the tracer mRNA should be about 50–100 pg/nL per blastomere. The concentration of the test mRNA should be based on effective doses reported in the literature. For novel genes, perform a dose-response experiment. The concentration used should consider the size of the blastomere to be injected: at each cleavage, a blastomere is reduced to about half its mother cell's volume. For an mRNA lineage tracer to accurately show which cells inherited the test mRNA, both should be mixed in the same solution and co-injected (Step 12).

Reagents for natural mating or in vitro fertilization in *Xenopus* (see Sive et al. 2007b and Moody 2012a)

Because obtaining eggs by squeezing the female sometimes reduces egg quality, an alternate method for in vitro fertilization is to have the hormone-injected female naturally release eggs into High-salt MBS <R>. Collect and maintain eggs in this buffer for several hours. Wash twice with 0.1× MBS before adding sperm.

Equipment

Capillary glass, borosilicate, without filament (OD 0.8–1.0 mm [e.g., Sutter Instrument B100–50-10; 1.0-mm OD, 0.50-mm ID, 10-cm length])

Capillary glass may need to be coated with silicon prior to pulling micropipettes to prevent samples from beading and forming air pockets during back-filling in Step 3.

Centrifuge, tabletop

Eyepiece micrometer

Forceps (e.g., Dumont #5 biologie)

Sharpen forceps to a very fine tip by gently stroking them across a piece of Alumina abrasive film (Thomas Scientific) under a stereomicroscope. We use medium film (3-μm; 6775E-46) for fine sharpening and fine film (0.3-μm; 6775E-54) for a final polish. Coarse film (12-μm; 6775E-38) is used to repair bent tips.

Glass bead tool (if preparing agarose or plasticine injection dishes in Step 1)

Melt the tip of an autoclaved 6-inch Pasteur pipette into a ~2-mm ball in the flame of a Bunsen burner (Fig. 1B).

Hair loop

Place both ends of a fine hair (~6 inch long, collected near the nape of the neck from a family member or laboratory partner; baby hair works well but tends to be too short to manipulate easily) into the narrow tip of a 6-inch Pasteur pipette to form a 2- to 3-mm loop. Secure in place with nail polish or cyanoacrylate glue (Fig. 1A). Sterilize before use by dipping in 70% ethanol and air-drying. Make several at one time to have a reserve; they break when dropped. As an alternative, purchase a superfine eyelash hair attached to a handle (Ted Pella #113) or an ultrafine single deer hair attached to a handle (Ted Pella #119).

Incubator at 14°C–20°C

Microinjector (Nanoject II [Drummond Scientific], PLI-100 Pico-Injector [Harvard Apparatus] or IM-300 [Narashige])

See Discussion.

Micromanipulator (e.g., Narashige three-axis, direct-drive [with a stand] or Singer Instruments Mk1 manipulator)

Micropipette puller

Injections of oocytes and two- to four-cell embryos can be performed using micropipettes with long tips, which are easily made with a vertical puller (e.g., Kopf Instruments 720 needle pipette puller). Injections of later stage embryos require micropipettes with short, stiff tips; these are most easily manufactured with a horizontal puller (e.g., Sutter Instrument P-97 Flaming/Brown micropipette puller).

Parafilm

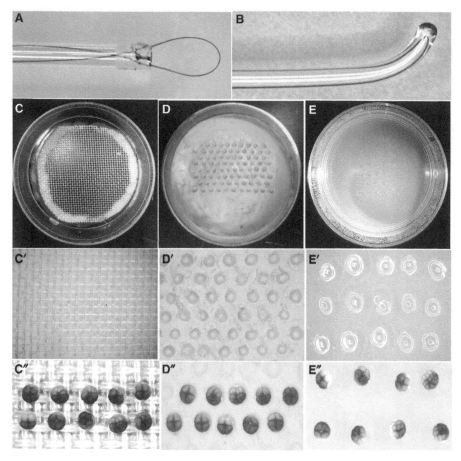

FIGURE 1. Microinjection tools. (*A*) Hair loop. (*B*) Glass bead tool. (*C*) Polypropylene mesh injection dish (*D*) Plasticine injection dish. (*E*) Agarose injection dish. (*C′–E′*) Higher magnification of each dish type showing the depressions in which the embryos will sit. (*C″–E″*) Embryos sitting in the depressions in each injection dish type.

Petri dishes (35-, 60-mm)

Plasticine modeling clay (nontoxic, nonhardening) (if preparing plasticine injection dishes in Step 1, and for preparation of micropipettes)

> *Claytoon and Nicole Crafts clays have been tested in* Xenopus *labs.*

Ruler (optional; see Step 4)

Spectra Mesh (500-µm polypropylene mesh; www.spectrumlabs.com) and cyanoacrylate glue (if preparing mesh injection dishes in Step 1)

Stage micrometer

Stereomicroscope, placed on an area that is level and free of vibrations, equipped with epi-illumination

Syringe (Hamilton 701N, or a syringe provided by the microinjection apparatus company) or gel-loading pipette (if back-filling micropipettes in Step 3)

Transfer pipettes, plastic

METHOD

Preparing Injection Tools

> *Three types of injection dishes are commonly used.*

Cite this protocol as *Cold Spring Harb Protoc*; doi:10.1101/pdb.prot097261

1. Prepare an injection dish according to one of the following methods.

For a Mesh Injection Dish

i. Glue a piece of polypropylene mesh to the bottom of a Petri dish (35- or 60-mm) with a few drops of cyanoacrylate glue (Fig. 1C,C′). Test the dish for glue toxicity before using it for an experiment by soaking it in several changes of distilled H_2O over several days, and then growing control embryos in its wells in culture medium overnight.

For a Plasticine Injection Dish

ii. Cover the bottom of a Petri dish (35- or 60-mm) with nontoxic, nonhardening modeling clay. Make several rows of shallow depressions, ∼1.5-mm wide, with a cool glass bead tool (Fig. 1D,D′). Test for clay toxicity as in i.

For an Agarose Injection Dish

iii. Cover the bottom of a Petri dish (35- or 60-mm) with liquefied 2% agarose. Allow the agarose to harden. Melt shallow depressions in the solidified agarose, ∼1.5-mm wide, using a heated glass bead tool (Fig. 1E,E′).

2. Prepare the micropipettes as follows.

i. Refer to the manufacturer's instructions to pull a glass capillary tube into a micropipette with dimensions approximating those in Figure 2A.

Glass capillary tubes must be pulled into fine tips that are strong enough to puncture the vitelline membrane of the embryo yet fine enough to not cause damage to the injected blastomere. The parameters of the pull (e.g., heat, pull time, pull strength, etc.) vary depending upon the puller.

The heat of the pull will close the tip of the micropipette.

ii. To manually break open the pipette tip, make a holder by pressing a strip of plasticine clay to the top of a 35-mm Petri dish and then press the micropipette into the clay in a horizontal position (Fig. 2B). Place the unit on the stage of a stereomicroscope and focus on the tip at high magnification. Rotate the unit to adjust the reflection of the lighting

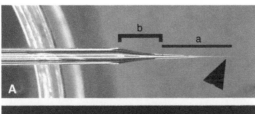
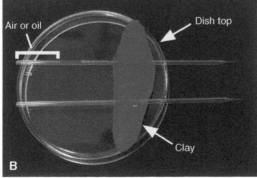

FIGURE 2. Making micropipettes. (*A*) A pulled glass micropipette should have a short primary taper (b) and a very fine secondary taper (a), each of which should be about 3–4 mm in length. The arrowhead denotes the location in the tip at which one can snip it open at an angle with finely sharpened forceps. (*B*) To manipulate micropipettes, place them on a bead of clay pressed onto the top of a 35-mm Petri dish. To back-fill a micropipette, inject the sample in the region indicated by the red bar. The air space behind the sample can be left open to air in some equipment or filled with heavy mineral oil if using a hydraulic system.

to see the tip clearly. Use fine forceps to snip off the very tip at about a 45-degree angle (Fig. 2A).

Alternatively, use a commercial beveling device (e.g., Narashige EG-3 grinder) to open the tip.

iii. To make sure the diameter of the tip is not too large, use a stage or eyepiece micrometer to measure its outer diameter; ~10–20 µm in diameter is a useful range.

Tips that are too large damage the cell and cause the cytoplasm to leak out; tips that are too small often clog with the yolky cytoplasm or viscous nucleic acid samples; see Discussion.

Filling and Calibrating Micropipettes

The method used to fill the micropipette depends upon the type of microinjection equipment. Back-filling is used if the microinjector does not draw fluid into the tip of the micropipette by back-pressure. Front-filling can be done when using an air-pressure microinjection system that has a "fill" function. This allows one to suck the solution into the micropipette through the open tip.

The volume of the droplet delivered by the micropipette determines the concentration of the nucleic acid that is injected during microinjection. Because volume depends upon the inner diameter of the open micropipette tip, each micropipette must be individually calibrated. Two methods for calibration are described in Step 4.

3. Fill the micropipette for microinjection as follows.

To Back-Fill the Pipette

i. Stretch a clean sheet of Parafilm over a 35-mm Petri dish. Place a small aliquot (~5–10 µL) of sample solution (H_2O, mRNA, or MO) onto its surface, avoiding any dust particles.

ii. Draw the aliquot into either a Hamilton syringe, a gel-loading pipette, or a syringe provided by the microinjection apparatus company. Prevent dust particles and air bubbles from being sucked into the syringe by monitoring the process under a stereomicroscope.

iii. Holding a micropipette with its tip already opened in the clay holder, insert the tip/syringe needle into its back end, with tip of the needle about 10 mm from the open end. Slowly deliver the aliquot without introducing air bubbles into the tracer/mRNA/MO (Fig. 2B).

For some microinjectors, the back-filled micropipette can now be directly attached to the equipment. For hydraulic systems, proceed to iv.

iv. If using a hydraulic system, replace the air space behind the lineage tracer with heavy mineral oil using a similar approach (Fig. 2B).

v. Attach the back-filled micropipette to the microinjector and immediately submerge its tip in a dish of culture solution (1× SS, 0.5× MMR, or 0.5× MBS) or heavy mineral oil.

This prevents the tracer/mRNA/MO from drying in the tip and clogging it.

To Front-Fill the Pipette

vi. Mount an opened micropipette onto the microinjector, and secure onto the micromanipulator.

vii. Stretch a clean sheet of Parafilm over a 35-mm Petri dish, and place a small aliquot (~1– 2 µL) of sample solution (H_2O, mRNA, or MO) onto its surface, avoiding dust particles.

Alternatively, place the aliquot (1–2 µL) under mineral oil at the bottom of a sterile 35-mm Petri dish to avoid dust particles and prevent evaporation.

viii. Using a micromanipulator, place the tip of the micropipette in the center of the aliquot and activate the "fill" mode of the microinjector according to the manufacturer's instructions. Watch the filling process carefully under the dissecting microscope to avoid formation of air bubbles.

ix. As soon as most of the droplet has been sucked into the micropipette, submerge the tip in a dish of culture solution or heavy mineral oil as above.

4. Calibrate the micropipette using one of the following methods.

Cite this protocol as *Cold Spring Harb Protoc*; doi:10.1101/pdb.prot097261

Using a Ruler

i. Place the micropipette on a strip of clay (Fig. 2B) and then place it over a ruler with 1-mm divisions (Fig. 3A). Mark several 1-mm lengths along the shank, distal to the shoulder of the primary taper, with a permanent marker.

ii. Fill the micropipette with about 5 μL of RNase-free H_2O as described in Step 3, and attach it to the microinjector.

iii. Viewing the micropipette under the stereomicroscope, expel droplets (i.e., hit the "inject" button on the microinjector) until the meniscus of the water lines up with one of the marks. Count how many deliveries it takes to move the meniscus to the next 1-mm mark. For accuracy, repeat a few times.

iv. Determine the volume of the column of water that was moved 1 mm using the following equation: $1 \text{ mm} \times \Pi r^2$, where r = ½ the inner diameter (ID) of the capillary glass and ID is provided by the capillary glass manufacturer. Divide the volume of the column of water by the number of deliveries it took to move the meniscus 1 mm; this number is the volume of each delivery.

v. Based on the volume calculated in iv, use the microinjector instructions to adjust the delivery time to deliver the desired amount of sample.

   Typically this amount is 1.0–5.0 nL.

vi. Briefly centrifuge (<1 min) the mRNA/MO working solution in a tabletop microfuge to pellet any particulate material that might clog the tip.

vii. Expel the water from the micropipette and replace it with mRNA/MO working solution as described in Step 3.

Using a Stage Micrometer

viii. Fill the micropipette with about 5 μL of RNase-free H_2O as described in Step 3, and attach it to the microinjector.

   We typically perform this calibration with the sample solution instead of H_2O to save time.

ix. Place ~10 μL of sterile, RNase-free heavy mineral oil on a stage micrometer. Viewing the micropipette under the dissecting microscope, lower the tip of the micropipette tip into the oil, just above the calibration lines of the micrometer.

x. Expel a droplet (i.e., hit the "inject" button on the microinjector).

   The aqueous sample will form a ball in the oil (Fig. 3B).

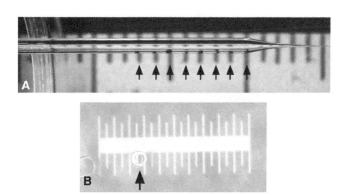

FIGURE 3. Calibrating micropipettes. (*A*) Starting at the shoulder of the primary taper (*right* arrow) mark off 1-mm segments with a black marker (arrows). (*B*) Place a drop of heavy mineral oil on a stage micrometer (white lines). Expel a droplet of water from the micropipette (arrow) and measure its diameter.

xi. Measure the ball's diameter with the stage micrometer (Fig. 3B); repeat about five times to ensure that the measurements are consistent.

xii. Calculate the volume of the ball using the following equation: $(4/3)\pi r^3$.

xiii. Based on the volume calculated in xii, use the microinjector instructions to adjust the delivery time to deliver the desired amount of sample.

   Typically this amount is 1.0–5.0 nL.

xiv. Briefly centrifuge (<1 min) the mRNA/MO working solution in a tabletop microfuge to pellet any particulate material that might clog the tip.

xv. Expel the fluid from the micropipette and replace it with mRNA/MO working solution as described in Step 3.

Selecting and Microinjecting Embryos

To deliver the mRNAs or MOs to a specific precursor blastomere, it is essential to know where the dorsal side of the embryo will be located, as detailed elsewhere (see Protocol 1: Lineage Tracing and Fate Mapping in Xenopus *Embryos [Moody 2018] and Moody [2012a]).*

For some experiments, such as animal cap assays (see Protocol 8: Dissecting and Culturing Animal Cap Explants [Dingwell and Smith 2018]), the goal is to globally express the mRNA/MO. In this case, microinject both blastomeres of a two-cell embryo (Fig. 4A) or all four blastomeres of a four-cell embryo (Fig. 4B). Injections should be placed just animal to the equator to facilitate diffusion of the mRNA or MO throughout the embryo, or targeted to the region of interest if performing explants: animal pole for animal cap explants, vegetal pole for vegetal explants, etc.

5. To obtain fertilized eggs, setup a natural mating or an in vitro fertilization (see Sive et al. 2007b and Moody 2012a).

6. Remove the jelly coats from fertilized eggs (see Sive et al. 2007a and Moody 2012a).

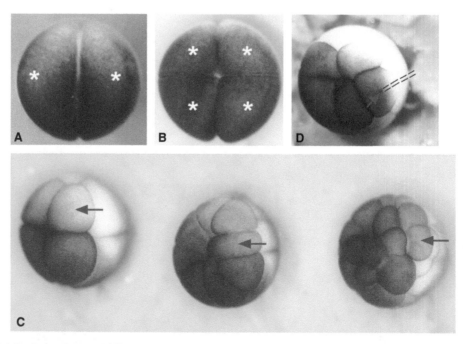

FIGURE 4. Injecting embryos at different stages. (*A*) To completely label an embryo, inject (asterisks) both blastomeres of the two-cell embryo or (*B*) all four cells of the four-cell embryo. (*C*) Position embryos in a depression in the injection dish so that the blastomere to be injected (red arrows) is facing the micropipette (*not shown, coming in from the right at a ~45° angle*). (*D*) The micropipette (dashes) should cause only a small puncture hole that darkens into a spot as the membrane heals (red arrow).

Cite this protocol as *Cold Spring Harb Protoc*; doi:10.1101/pdb.prot097261

7. Collect two-cell embryos whose body axes can be accurately identified (as described in Protocol 1: Lineage Tracing and Fate Mapping in *Xenopus* Embryos [Moody 2018] and Moody [2012a; 2012b]). Place the embryos in sterile 1× culture solution (SS, MMR, or MBS) and incubate at 14°C–20°C until reaching the desired stage for injection.

    The lower the temperature, the slower the cleavage cycle. It is often advantageous to split a clutch of embryos into two batches, one incubated at a lower temperature (e.g., 14°C) and one at a higher temperature (e.g., 18°C–20°C), to allow more time for injections.

8. Fill the injection dish with 3%–5% Ficoll solution and transfer the selected embryos to individual wells (Fig. 1C″–E″).

9. Gently angle each embryo, using a hair loop, so that the cell to be injected is en face with the tip of the micropipette (Fig. 4C).

    This helps prevent tearing of the cell membrane.

10. Advance the tip of the micropipette toward the blastomere to be injected with the micromanipulator.

    When the micropipette touches the embryo, there should be a small indentation and resistance from vitelline membrane.

11. Using the Z-axis control of the manipulator, advance the tip into the target blastomere. (A black dot should be visible as the membrane contracts around the micropipette [Fig. 4D].) Do not advance deeply into the cell.

12. Inject the tracer/mRNA/MO working solution according to the microinjector instructions.

13. Wait a few seconds before withdrawing the micropipette to prevent the sample from leaking out of the puncture wound. Move to the next embryo and repeat.

14. When all of the embryos in the dish have been injected, leave the embryos undisturbed for about 1 h to help them recover from any damage. After this period, use a plastic transfer pipette to move them to a Petri dish containing 3%–5% Ficoll solution until they reach blastula stages (~4–5 h after injection).

Culturing Embryos

15. When the embryos reach blastula stages, transfer them to clean Petri dishes filled with diluted culture solution (0.5× SS, 0.1× MMR, or 0.1× MBS). To provide adequate oxygen diffusion, do not overcrowd the dishes: Place about 15 embryos per 35-mm dish, 50 per 60-mm dish or 100 per 100-mm dish.

    The diluted solution prevents abnormal gastrulation movements.

16. Culture the embryos at room temperature (20°C–22°C) or in an incubator (14°C–18°C). If embryos are injected with a fluorescent compound, culture them in the dark to minimize phototoxicity.

    Room temperature speeds up development but cooler temperatures promote survival.

17. Change the medium to a lower dilution (0.1× SS, 0.05× MMR, or 0.05× MBS) after 24 h. Remove dead embryos and cellular debris to minimize protease damage and bacterial growth.

18. Culture embryos to the appropriate stage of development.

    After culture, embryos can be fixed according to the methods of gene expression to be used for analysis (e.g., see Protocol 5: Whole-Mount Immunocytochemistry in Xenopus [Klymkowsky 2018] and Protocol 4: Whole-Mount In Situ Hybridization of Xenopus Embryos [Saint-Jeannet 2017]).

DISCUSSION

There are several commercially available microinjectors. They range in complexity, but the most important feature to consider is the typical injection volumes that you will use. Oocytes, fertilized eggs and two- to four-cell embryos can be injected with 10-nL (or sometimes larger) volumes, which can be accurately supplied by simple equipment (e.g., the hydraulic or Drummond systems). For 16- and 32-cell stages, injection volumes should be about 1 nL per blastomere because the cells are significantly smaller. We found, for example, that injecting >10 nL of tracer alone into 16-cell blastomere precursors of the epidermis caused the cells to aberrantly contribute to the brain (Hainski and Moody 1992).

Learning to break the micropipette tip to the correct size takes practice. If the micropipette tip requires great effort to puncture the cell, then the tip is too blunt and will likely cause damage. If the puncture hole is too large, cytoplasm (and some of your tracer/mRNA/MO) will leak out in the form of a bleb; discard these embryos. This micropipette should be discarded and a new one set up. If the micropipette tip bends when it touches the embryo, the taper is too long and flexible to puncture the vitelline membrane. Break the tip to a shorter length and recalibrate it; adjust the pipette puller parameters to forge shorter tips (Fig. 2A).

Microinjection can cause cell death by either physical damage or too high a concentration of tracer/mRNA/MO. Observe the embryos ∼1 h after injection to detect blebs or cells that are not dividing. For example, you may observe one or two large cells in a field of smaller ones. Indications of cell damage at later points in development include: larger-than-normal labeled cells incorporated into the organs of the embryo; correctly sized labeled cells that are spherical, rather than differentiated; and labeled cells in the lumens of organs, particularly the gut, liver, heart, and nervous system. If you observe any of these signs of damage, discard the embryos. Try to forge better micropipettes and test smaller volumes or concentrations of tracer/mRNA/MOs.

There has been a recent discussion regarding the validity of using MOs to reduce mRNA translation and thus protein concentration because of the toxicity observed in some other animal models. However, in *Xenopus* this approach is quite specific and effective when the proper biochemical validation of the MOs is performed. This is discussed in detail elsewhere (Blum et al. 2015).

RECIPES

High-Salt MBS

Reagent	Volume to add
$CaCl_2$ (0.1 M)	7 mL
MBS salts (10×)	100 mL
NaCl (5 M)	4 mL
H_2O	888 mL

Marc's Modified Ringer's (MMR) (1×)

0.1 M NaCl
2 mM KCl
1 mM $MgSO_4$
2 mM $CaCl_2$
5 mM HEPES (pH 7.8)
0.1 mM EDTA

Sterilize by autoclaving, and store at room temperature.
Common alternative formulations of MMR omit EDTA and are adjusted to pH 7.4.

Cite this protocol as *Cold Spring Harb Protoc*; doi:10.1101/pdb.prot097261

MBS Salts (10×)

NaCl (880 mM)
KCl (10 mM)
$MgSO_4$ (10 mM)
HEPES (50 mM, pH 7.8)

> Omit HEPES if MBS is to be used for oocyte maturation.

$NaHCO_3$ (25 mM)

Adjust pH to 7.8 with NaOH. Autoclave. Store at room temperature.

Modified Barth's Saline (MBS)

$CaCl_2$ (0.1 M)
MBS salts (10×)

For a 1× solution of MBS, mix 100 mL of 10× MBS salts with 7 mL of 0.1 M $CaCl_2$. Adjust the volume up to 1 liter with H_2O. Store at room temperature.

Steinberg's Solution (100%)

Reagent	Amount to add (for 1 L)
NaCl	3.4 g
KCl	0.05 g
$Ca(NO_3)_2 \cdot 4H_2O$	0.08 g
$MgSO_4 \cdot 7H_2O$	0.205 g
Tris	0.56 g
H_2O	to 1 L

Adjust pH to 7.4 with HCl.

ACKNOWLEDGMENTS

The development of this protocol was supported by National Institutes of Health (NIH) grant DE022065. I gratefully acknowledge the excellent suggestions provided by the reviewers.

REFERENCES

Blum M, DeRobertis EM, Wallingford JB, Niehrs C. 2015. Morpholinos: Antisense and sensibility. *Dev Cell* 35: 145–149.

Dingwell KS, Smith JC. 2018. Dissecting and culturing animal cap explants. *Cold Spring Harb Protoc* doi:10.1101/pdb.prot097329.

Hainski AM, Moody SA. 1992. *Xenopus* maternal RNAs from a dorsal animal blastomere induce a secondary axis in host embryos. *Development* 116: 347–355.

Klymkowsky MW. 2018. Whole-mount immunocytochemistry in *Xenopus*. *Cold Spring Harb Protoc* doi:10.1101/pdb.prot097295.

Moody SA. 2012a. Targeted microinjection of synthetic mRNAs to alter retina gene expression in *Xenopus* embryos. *Methods Mol Biol* 884: 91–111.

Moody SA. 2012b. Testing retina fate commitment in *Xenopus* by blastomere deletion, transplantation and explant culture. *Methods Mol Biol* 884: 115–127.

Moody SA. 2018. Lineage tracing and fate mapping in *Xenopus* embryos. *Cold Spring Harb Protoc* doi:10.1101/pdb.prot097253.

Saint-Jeannet JP. 2017. Whole-mount in situ hybridization of *Xenopus* embryos. *Cold Spring Harb Protoc* doi:10.1101/pdb.prot097287.

Sive HL, Grainger RM, Harland RM. 2007a. Dejellying *Xenopus laevis* embryos. *Cold Spring Harb Protoc* doi:10.1101/pdb.prot4731.

Sive HL, Grainger RM, Harland RM. 2007b. *Xenopus laevis* in vitro fertilization and natural mating methods. *Cold Spring Harb Protoc* doi:10.1101/pdb.prot4737

Microinjection of DNA Constructs into *Xenopus* Embryos for Gene Misexpression and *cis*-Regulatory Module Analysis

Yuuri Yasuoka[1,3] and Masanori Taira[2,4,5]

[1]*Marine Genomics Unit, Okinawa Institute of Science and Technology Graduate University, Onna-son, Okinawa 904-0495, Japan;* [2]*Department of Biological Sciences, Graduate School of Science, University of Tokyo, Bunkyo-ku, Tokyo 113-0033, Japan*

Introducing exogenous DNA into an embryo can promote misexpression of a gene of interest via transcription regulated by an attached enhancer-promoter. This protocol describes plasmid DNA microinjection into *Xenopus* embryos for misexpression of genes after zygotic gene expression begins. It also describes a method for coinjecting a reporter plasmid with mRNA or antisense morpholinos to perform luciferase reporter assays, which are useful for quantitative analysis of *cis*-regulatory sequences responding to endogenous or exogenous stimuli in embryos.

MATERIALS

It is essential that you consult the appropriate Material Safety Data Sheets and your institution's Environmental Health and Safety Office for proper handling of equipment and hazardous materials used in this protocol.

RECIPES: Please see the end of this protocol for recipes indicated by <R>. Additional recipes can be found online at http://cshprotocols.cshlp.org/site/recipes.

Reagents

Chloroform:isoamyl alcohol (24:1) (Sigma-Aldrich C0549)

Ethanol

> *For Step 14, prepare ice-cold 70% ethanol/30% 0.1 M NaCl.*

Ficoll (e.g., GE 17-0300-10)

H_2O (nuclease-free)

Injection buffer for *Xenopus* <R>

Luciferase cell culture lysis reagent (Promega E1531) and luciferase assay system (Promega E1501)

mRNA solution and/or morpholino antisense oligonucleotides (optional, for coinjection with reporter plasmid DNA)

> *mRNA solution should be stored at −80°C. The concentration of the stock solution is usually 100–1000 ng/μL. Antisense morpholinos are typically stored at 2 mM.*

NaCl (5 M) (nuclease-free)

---

[3]Present address: Laboratory for Comprehensive Genomic Analysis, RIKEN Center for Integrative Medical Sciences, Tsurumi-ku, Yokohama, 230-0045, Japan

[4]Present address: Department of Biological Sciences, Faculty of Science and Engineering, Chuo University, Bunkyo-ku, Tokyo 112-8551, Japan

[5]Correspondence: m_taira@bs.s.u-tokyo.ac.jp or m-taira.183@g.chuo-u.ac.jp

Cite this protocol as *Cold Spring Harb Protoc*; doi:10.1101/pdb.prot097279

Phenol:chloroform:isoamyl alcohol (25:24:1) (Wako 311-90151)

Plasmid DNA miniprep kit (e.g., Sigma-Aldrich PLN70)

Plasmid DNA vectors (see Table 1)

pCS-based plasmid, for misexpression of genes (e.g., pCSf107mT or pCS2)

pCS-based plasmid for nuclear β-galactosidase (nβ-gal) (optional, for coinjection as a tracer)

pGL4-based plasmid, for luciferase reporter assays

Proteinase K (1 mg/mL)

Dilute proteinase K (20 mg/mL, Promega MC5005) to 1 mg/mL with nuclease-free H₂O.

Steinberg's solution (10×, pH 7.4) <R>

TE buffer for *Xenopus* <R> (nuclease-free)

Nuclease-free H₂O can be used as an alternative to TE buffer, but note that TE buffer is better for longer storage of plasmids.

Xenopus embryos (dejellied)

X. laevis or X. tropicalis embryos can be harvested and dejellied according to Sive et al. (2007).

Equipment

Capillary sequencer (as needed; see Step 1)

Centrifuge (refrigerated, tabletop) for microcentrifuge tubes (e.g., Eppendorf 5415R)

Glass capillaries (e.g., Narishige G-100)

Prepare injection needles from glass capillaries using a needle puller (e.g., Narishige PN-31 or Sutter P-1000).

Liquid nitrogen

Low-temperature incubators at 14°C–17°C (for *X. laevis*), 25°C–28°C (for *X. tropicalis*), and 37°C (e.g., EYELA LTE-1000)

Luminometer (e.g., Berthold Technologies, Centro LB960)

Microcentrifuge tubes (1.5-mL)

Microinjector (e.g., Narishige IM-300)

Micromanipulator (e.g., Narishige MN-153)

Mineral oil

Petri dishes, plastic (100-mm)

Plates, plastic (six-well)

Plates, white plastic (96-well, flat-bottom) (e.g., Greiner Bio-One 655075)

Spectrophotometer (e.g., Nanodrop)

Stereomicroscope (e.g., Leica M80) and micrometer

Syringe filter units (Millex-GS) (Millipore SLGS025NB)

After proteinase K treatment (Step 2), all solutions should be nuclease-free (free of DNase and RNase). To prepare nuclease-free solutions, filtration through two sequentially connected Millex-GS syringe filter units is empirically sufficient.

Vortex mixer

TABLE 1. Vector plasmids for DNA injection

Name	Purpose	Promoter for expression	Genes encoded
pCSf107mT[a]	Misexpression of genes	CMV	None
pCS2	Misexpression of genes	CMV	None
pGL4.23	Reporter assay	MinP	Firefly luciferase
pGL4SV40	Reporter assay	SV40	Firefly luciferase

[a]pCSf107mT can also be used for in vitro transcription of mRNA by the SP6 promoter and for transcription of antisense RNA probes by the T7 promoter (Mii and Taira 2009). Conveniently, mRNA synthesis can be performed without linearization of plasmids by restriction enzyme digestion because of the insertion of 4× SP6/T7 terminator sequences after the SV40 polyadenylation signal. The backbone of pCSf107mT is pCS107, but pCS107 in some preparations including ours have a deletion in the CMV promoter. Therefore, we replaced the deleted CMV promoter of "bad pCS107" with the intact CMV promoter from pCS2. This "repaired pCS107mT" is designated as pCSf107mT, in which "f, m, and T" stands for "full-length CMV, multicloning sites, and terminators."

METHOD

Purifying Plasmid DNA for Injection

Linearization of plasmids is not necessary for misexpression of genes. Compared to circular plasmids, linear plasmids show much lower expression (Fig. 1).

For mRNA coinjection with plasmid DNA, proteinase K treatment is necessary for making RNase-free DNA solution before mixing with mRNA. For misexpression using pCS-based constructs (i.e., without mRNA coinjection), Steps 2–20 can be skipped.

1. Prepare the plasmid DNA as follows.

    i. For misexpression of genes, prepare pCS-based plasmids, such as pCSf107mT or pCS2, using a plasmid miniprep kit.

       pCS-based plasmids contain the CMV promoter for expression.

    ii. For luciferase reporter assays, prepare pGL4-based plasmids using a plasmid miniprep kit.

       See Discussion for reporter vector selection.

    iii. If a PCR fragment was inserted into a plasmid, confirm the plasmid sequence using a capillary sequencer.

2. For an RNase-free plasmid solution, mix the following reagents in a 1.5-mL microcentrifuge tube and incubate for 30 min at 37°C.

Plasmid DNA	5 µg
TE	to 49 µL
Proteinase K (1 mg/mL)	1 µL

3. Add 50 µL of phenol:chloroform:isoamyl alcohol to the plasmid DNA sample. Mix vigorously.

4. Centrifuge at 16,000g for 5 min at room temperature.

5. Transfer the aqueous phase to a new tube.

6. Repeat Steps 3–5.

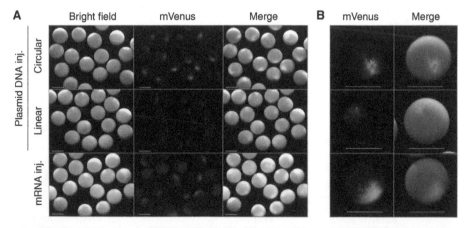

FIGURE 1. DNA injection exerts narrower expression areas as compared to mRNA injection. (*A*) Misexpression of mVenus by injection of closed-circular or linearized plasmid DNA (pCSf107mVenus_T) or synthetic mRNA. Injections were performed into a ventral-animal blastomere of albino *X. laevis* embryos, for which albino eggs were fertilized with wild-type sperm, at the four-cell stage. Injected embryos were observed at the early gastrula stage (stage 10.5) under a fluorescence stereomicroscope (Leica M205 FA). Plasmid DNA linearized with AscI restriction enzyme showed much weaker expression than did closed-circular plasmid DNA. DNA injection appeared to exert narrower expression areas as compared to mRNA injection. The amounts of DNA and mRNA injected were 200 and 100 pg, respectively. (*B*) Representative DNA- and mRNA-injected embryos. Note that mVenus expression levels in individual cells varied more among DNA injections (*upper* panel), as compared to those receiving mRNA injections (*lower* panel). *Scale bars*, 1 mm.

Cite this protocol as *Cold Spring Harb Protoc*; doi:10.1101/pdb.prot097279

7. Add 50 μL of chloroform:isoamyl alcohol to the sample. Mix vigorously.

8. Centrifuge at 1000*g* for 3 min at room temperature.

9. Transfer the aqueous phase to a new tube.

10. Repeat Steps 7–9.

11. Add 1 μL of 5 M NaCl (nuclease-free) and 150 μL of ethanol to the sample. Mix well.

12. Keep on ice for 10 min.

13. Centrifuge at 16,000*g* for 20 min at 4°C.

14. Remove the supernatant and wash the DNA pellet with 300 μL of ice-cold 70% ethanol/30% 0.1 M NaCl.

> *The DNA pellet is usually invisible.*

15. Centrifuge at 16,000*g* at 4°C for 3 min.

16. Remove the supernatant and air-dry the pellet for 10 min at room temperature.

17. Add 4 μL of TE buffer (nuclease-free). Mix well.

18. Keep the tube on ice for 30 min to dissolve the DNA pellet completely.

19. Check the concentration of the DNA with a spectrophotometer.

20. If the concentration is >1 μg/μL, adjust the concentration to 1 μg/μL with TE buffer (nuclease-free).

> *The yield of DNA after these purification steps is usually around 80% (i.e., about 4 μg of DNA). Therefore, 4 μL of TE buffer is used for resuspension of the DNA pellet.*

Preparing Injection Solution

Note that a large amount of injected DNA (usually more than 500 pg/embryo) is toxic to Xenopus embryos. Therefore, for misexpression with pCS-based constructs, the final concentration of DNA should be 50 ng/μL at most, resulting in 500 pg/embryo when 10 nL is injected. For reporter assays, the final concentration of DNA for injection is usually 5 ng/μL or below, resulting in ≤50 pg/embryo when 10 nL is injected. Starting at 5 ng/μL, the best concentration of DNA should be determined empirically for proper response to effectors. To trace the DNA-injected area, a pCS-based plasmid for nuclear β-galactosidase (nβ-gal) can be coinjected at 5–20 ng/μL.

21. Dilute the plasmid solution for injection as follows.

   i. For misexpression with pCS-based constructs, dilute the plasmid solution with injection buffer to an appropriate concentration (≤50 ng/μL).

   ii. For luciferase reporter assays, dilute the purified plasmid solution (Steps 2–20) to 10 ng/μL with injection buffer.

   > *Because TE buffer is toxic to the embryo, more than a 10-fold dilution of the plasmid DNA solution with injection buffer is important. If the concentration of purified plasmid is below 100 ng/μL, concentrate the plasmid or purify again to avoid injecting more TE buffer into the embryo.*

22. For mRNA coinjection with reporter plasmid DNA, dilute the mRNA solution with H₂O (nuclease-free) to an appropriate concentration (e.g., 20 ng/μL). Mix equal volumes of the DNA solution and mRNA solution. For injection of reporter plasmids alone as a control, mix equal volumes of the DNA solution and H₂O (nuclease-free).

   > *To perform loss of function analysis together with DNA injection, antisense morpholinos (MOs) also can be coinjected with plasmid DNA. MOs are usually injected at 0.1–0.5 mM.*

Microinjecting Embryos

For luciferase reporter assays, 20 embryos are typically injected because 15 embryos are used in the reporter assay (Step 30).

23. Load the injection solution prepared in Steps 21–22 into a glass needle.

24. Measure the volume of an injected drop and adjust the injection time to inject 5 nL of injection solution as follows.

   ### For Large-Volume Glass Needles

   i. Measure the diameter of an injected drop in mineral oil using a micrometer and stereomicroscope.
   ii. Calculate the volume of the drop.

   ### For Small-Volume Glass Needles

   iii. Mark a glass needle at 1-mm intervals.
   iv. Measure the time needed to eject a 1-mm volume in mineral oil.
   v. Calculate the time needed to eject 5 nL according to the volume of a 1-mm length of the needle.
   vi. Confirm the calculated injection time by counting the number of microinjection steps needed to discharge the solution in 1 mm of the needle length.

25. Align dejellied *Xenopus* embryos in a Petri dish containing 3%–5% Ficoll/1X Steinberg's solution.

26. Inject 5 nL (for *X. laevis*, or 1.5 nL for *X. tropicalis*) of injection solution into one blastomere of each embryo.

   i. To misexpress genes encoded in pCS-based plasmids, inject the solution into specific regions of the embryos.
   ii. To test the responsiveness of a reporter plasmid against coinjected mRNA, inject the solution into two blastomeres of two- to four-cell stage embryos at the animal pole region.

   Because the ventral blastomeres at the four-cell stage are usually bigger than the dorsal blastomeres, the animal pole region of a two-cell stage embryo is relatively ventral in character. In addition, the dorsoanimal region of the blastula includes a signaling center, called the blastula Chordin- and Noggin-expressing (BCNE) center (Kuroda et al. 2004; Sudou et al. 2012), which expresses "dorsal genes" such as the BMP antagonist genes chordin and noggin, and the transcription factor genes siamois and goosecoid. Therefore, to avoid the influence of ventral character (BMP signaling) or dorsal character, one should assess which character is more suitable for the purpose of the experiment. To examine reporter activity in pluripotent ectodermal cells, we recommend injecting the solution into the ventral blastomeres to compare activity between samples injected at the two-cell and four-cell stages. However, the activity of the SV40 late promoter for both dorsoanimal and ventroanimal injection at the four-cell stage is comparable to that of animal injection at the two-cell stage (see Fig. 2D), suggesting fewer differences between them.

   iii. To compare the responsiveness of a reporter plasmid in different embryonic tissues, inject the solution into specific sides of the embryos, such as the dorsal or ventral equatorial regions of four-cell stage embryos.

27. Transfer the injected embryos into one well of a six-well plastic plate filled with 3%–5% Ficoll/1× Steinberg's solution. Incubate for >30 min at an appropriate temperature (14°C–17°C for *X. laevis*; 25°C–28°C for *X. tropicalis*).

28. Change the medium to 0.1× Steinberg's solution. Incubate the embryos at an appropriate temperature (14°C–17°C for *X. laevis*; 25°C–28°C for *X. tropicalis*) until the embryos reach the desired stage.

   For misexpression experiments, injected embryos are subjected to subsequent experiments such as in situ hybridization, immunostaining, and Western blotting.

   See Troubleshooting.

Cite this protocol as *Cold Spring Harb Protoc*; doi:10.1101/pdb.prot097279

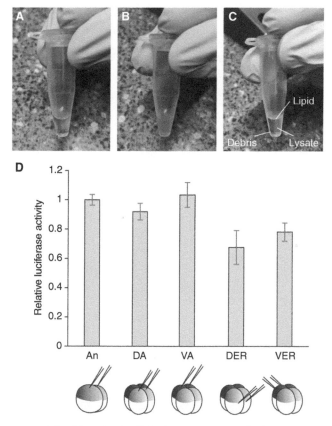

FIGURE 2. Luciferase reporter activity driven by the SV40 late promoter varies between tissues or regions of *Xenopus* gastrulae. (*A–C*) Embryonic lysate preparation. (*A*) Three *Xenopus laevis* embryos in 100 µL of 1× luciferase cell culture lysis reagent. (*B*) A uniform embryonic lysate after vortexing. (*C*) An embryonic lysate after centrifugation. The lysate separated into a lipid layer, a clear lysate, and embryonic debris, as indicated. (*D*) Activity of the SV40 late promoter in mesoderm is slightly lower than in the ectoderm. The same volume of pGL4-SV40 plasmid was injected into two-cell or four-cell stage embryos, as indicated by schematics under the graph. Luciferase activity was measured at the gastrula stage (stage 11). The luciferase activity of embryos injected with reporter DNA into animal pole regions at the two-cell stage (An) was almost the same as the activity of those injected into the animal pole regions of the two dorsal or ventral blastomeres at the four-cell stage (DA or VA). However, the luciferase activity of embryos injected into the equatorial region of the two dorsal or ventral blastomeres at the four-cell stage (DER or VER) was slightly, but significantly, lower than others (*) $P < 0.05$, indicating that SV40 promoter activity varies depending on tissues or regions of *Xenopus* gastrula embryos.

Performing Luciferase Reporter Assays

29. [Optional] To examine the responsiveness of the reporter plasmids in the animal pole regions (naive ectoderm) of later stage embryos, such as neurula (stage 13) and later, dissect animal caps at stage 8–9 and culture them until sibling embryos reach the stage of interest.

30. Pool three injected embryos or caps in a 1.5-mL microcentrifuge tube and remove as much saline as possible. Prepare five pools from each set of injected embryos (15 embryos or caps in total).

31. Prepare three pools of uninjected embryos as described in Step 30.

32. [Optional] Freeze the embryos in liquid nitrogen or on dry ice, or transfer them to a freezer at −80°C.

    Frozen embryos can be kept at −80°C indefinitely. The freezing step may help protein extraction from embryos.

33. Add 100 µL of 1× luciferase cell culture lysis reagent to each sample. Homogenize the embryos by vortexing for 10–20 sec.

34. Centrifuge the samples at 16,000g for 15 min at 4°C.

35. Transfer 30 µL of each supernatant to a new tube.

    After centrifugation, each tube will contain a pellet of embryonic debris, with a lipid layer on the surface, and a clear lysate between (see Fig. 2A–C). Carefully pipette the clear lysate and remove lipids from the pipette tip by attaching them to inside wall of the tube. The lysate can be kept at −80°C indefinitely.

36. Transfer 7.5 µL of each lysate to one well of a 96-well plate. Load the plate into a luminometer.

37. Measure the luminescence using following settings: Amount of luciferase assay reagent, 50 µL; Detection time, 20 sec.

38. Analyze the data (Fig. 2D).

    i. Average the results from the three uninjected embryo samples to calculate the background.

    ii. Subtract the background from each result.

    iii. Calculate the mean and standard deviation of the results from each set of five pools of injected embryos.

    iv. Calculate the statistical significance (*P*-value) between sample means using Student's or Welch's *t*-test after comparison of the variances with an *F*-test.

    See Troubleshooting.

TROUBLESHOOTING

Problem (Step 28): Gene expression is not (or not fully) detected.

Solution: The CMV promoter or SV40 polyadenylation signal may have mutated in the plasmid. Confirm the plasmid sequence not only for inserted sequences, but also for the vector backbone. It is also possible that injection buffer is contaminated with DNase. Confirm the DNA quality in the injection buffer by agarose gel electrophoresis after incubation for several hours at 37°C. If the DNA is degraded, nuclease-free injection buffer and DNA stock solution should be prepared again.

Problem (Step 28): Embryos injected with DNA die prematurely.

Solution: If the uninjected control embryos also die, the batch of embryos was not good. If embryos injected with H_2O or injection buffer alone do not die, the injected DNA solution is toxic, probably for one of the following reasons. (i) The DNA solution may be contaminated with phenol. Extract the plasmid DNA with chloroform:isoamyl alcohol again. (ii) The DNA concentration may be too high. Generally, the amount of DNA should be <1 ng/embryo. Confirm the DNA concentration of the injection solution and use lower concentrations of DNA for the next experiment if necessary.

Problem (Step 28): DNA injection results in mosaic expression patterns.

Solution: As shown in Figure 1, linearization of plasmids does not improve localized and mosaic expression patterns. Unfortunately, we do not have any solution to overcome this problem, but we suggest mRNA injection as an alternative in some cases. For example, mRNA injection of a blastomere in later cleavage stages (e.g., 16–32 cell stages) can result in localized and uniform expression of genes. To examine the functions of nuclear proteins after the mid-blastula transition (MBT) or at later embryonic stages, mRNA injection of hormone inducible protein (e.g., GR-fusion constructs as shown in Kolm and Sive [1995]) can be substituted for DNA injection.

Problem (Step 38): The reporter gene does not respond to coinjected mRNA.

Solution: The DNA solution may be contaminated with RNase, so that coinjected mRNAs are degraded. RNase contamination can be detected by incubating the DNA solution with total

RNA for several hours at 37°C and then checking for degradation of the RNA by agarose gel electrophoresis. If the DNA solution is contaminated with RNase, purify the plasmid DNA again.

Problem (Step 38): The reporter gene does not respond to coinjected antisense morpholino oligos (MOs).

Solution: Since MOs can bind to RNA in a sample tube before injection, separate injections of MO and other molecules (DNA and RNA) into the same blastomere can improve the results, especially for rescue experiments by mRNA injection.

Problem (Step 38): The results of the luciferase assays are not reproducible.

Solution: Since luciferase activities vary between different batches of embryos (Mochizuki et al. 2000; Yamamoto et al. 2003), absolute luciferase activity cannot be compared between biologically independent experiments. Also, activities vary according to the temperature used to raise the embryos and the sizes of the embryos used. Empirically, smaller embryos develop faster, possibly due to lower metabolic requirements. To obtain reproducible results, it is important to use embryos of the same size throughout an experiment and to raise embryos at a constant temperature. It is also important to pay attention to the position of the plastic plates within the incubator, because the temperature of each well can vary depending on the position. Usually we place embryo plates at the center of incubators. In general, larger incubators are more stable than smaller ones. If the difference between samples is subtle, but reproducible, results from independent experiments can be merged using activities relative to those of samples injected with plasmid alone (Yasuoka et al. 2009, 2014).

DISCUSSION

This protocol describes microinjection of DNA constructs into *Xenopus* embryos. Genes encoded in injected plasmid DNA are transcribed and translated into protein at later embryonic stages along with zygotic transcription after MBT (Newport and Kirschner 1982a,b). This method does not require integration of injected DNA into the genome (actually, injected DNA is rarely integrated into the genome in *Xenopus* embryos) and differs from methods for transgenesis, such as sperm nuclear transfer (Kroll and Amaya 1996; Ishibashi et al. 2008), φC31 integrase coinjection (Allen and Weeks 2005), I-SceI meganuclease coinjection (Ogino et al. 2006; Pan et al. 2006), and transposase coinjection (Hamlet et al. 2006; Sinzelle et al. 2006; Shibano et al. 2007; Yergeau et al. 2009).

Microinjection of DNA for Gene Misexpression

Microinjection of DNA constructs into *Xenopus* embryos is quite different from misexpression of genes employing mRNA injection, because injected mRNAs are immediately translated into protein. In addition, DNA is relatively more stable than mRNA in the embryo, so DNA injection typically provides longer misexpression of genes (Wallingford et al. 1998; Cleaver and Krieg 1999). While mRNA injection can affect early embryonic events, such as gastrulation and axis formation, DNA injection is useful to investigate gene roles after MBT or at later embryonic stages. Another difference between the two techniques is that expression of injected DNA constructs tends to be regionally restricted and mosaic, compared to that induced by mRNA injection, possibly due to the lower diffusion capacity of DNA constructs compared to mRNA, as well as uneven segregation of DNA constructs into sister cells (Fig. 1; Etkin and Pearman 1987; Christian and Moon 1993; Cleaver and Krieg 1999). Thus, the choice of method, DNA or mRNA injection, depends on the experimental objective.

A good example of employing DNA injection to misexpress genes is the *wnt8*-encoding DNA construct (Christian and Moon 1993; Hoppler et al. 1996; Kazanskaya et al. 2004; Mii and Taira 2009). Wnt8, zygotically expressed after DNA injection, affects anterior-posterior patterning of the embryo. Wnt8 misexpression resulting from mRNA injection affects dorso-ventral patterning and gastrulation movements in the embryo. Thus, DNA injection is useful to distinguish Wnt pathway functions of early stage embryos from those of later stage embryos.

Reporter Vectors and Assays

Another purpose of DNA injection is to examine *cis*-regulatory sequences for transcriptional activity using DNA constructs containing a reporter gene such as a luciferase or a fluorescent protein gene. Luciferase reporter assays are typically used for identification of transcription factor binding sites in *cis*-regulatory modules for gene regulation (Watabe et al. 1995; Latinkic et al. 1997; Rebbert and Dawid 1997; Mochizuki et al. 2000; Yao and Kessler 2001; Yamamoto et al. 2003; Karaulanov et al. 2004; von Bubnoff et al. 2005; Rankin et al. 2011; Sudou et al. 2012; Yasuoka et al. 2014). In addition, levels of specific cell-signaling input, such as Wnt (Kiecker and Niehrs 2001; Hikasa and Sokol 2004; Mii and Taira 2009; Luehders et al. 2015), Bmp (Hata et al. 2000; Karaulanov et al. 2004; von Bubnoff et al. 2005; Onai et al. 2010; Luehders et al. 2015), Nodal (Takebayashi-Suzuki et al. 2003; Onai et al. 2010; Luehders et al. 2015), and FGF (Haremaki et al. 2003; Takada et al. 2009) are measured by injecting specific reporters into embryos. Using a GAL4-UAS reporter, one can also examine transactivation/transrepression activity of transcription factors or portions of them by fusing them to the GAL4 DNA-binding domain (Hiratani et al. 2001; Yasuoka et al. 2009). Furthermore, evolutionarily conserved activity of *cis*-regulatory modules derived from other organisms such as amphioxus (Kozmikova et al. 2011) and coral (Yasuoka et al. 2016) can be examined using DNA reporter assays in *Xenopus* embryos.

In cell culture and other embryonic systems, dual luciferase assays have been used frequently. In such assays, another reporter construct encoding a different type of luciferase connected with a standard promoter, such as CMV or SV40, is cointroduced to normalize the data. However, in *Xenopus* embryos, single assays are much better than dual assays because promoter activity of control reporter constructs may vary between embryonic tissues and mRNA coinjection conditions. In fact, the activity of the SV40 late promoter in mesoderm is slightly weaker than that in ectoderm (Fig. 2D). The single luciferase assays described here are sufficient because an exact volume of solution can be injected into the *Xenopus* embryos.

We have used the pGL4.23 vector (Promega) for luciferase assays to examine the activity of *cis*-regulatory modules in *Xenopus* embryos (Yasuoka et al. 2014, 2016). This vector includes an artificial minimal promoter connected to the luciferase gene. Its low basal activity is useful to see the enhancer activity of the *cis*-regulatory module. However, the minimal promoter activity is too low to examine the silencer activity of the *cis*-regulatory module. Therefore, we developed the vector pGL4SV40, in which the minimal promoter of pGL4.23 is replaced with the SV40 late promoter to increase basal transcriptional activity (Yasuoka et al. 2014).

Recently, a method using a new plasmid construct was reported (Wang and Szaro 2015, 2016). This construct harbors a ϕC31 integrase-targeted attB element and two dual β-globin HS4 insulators flanking a reporter transgene in opposite orientations. Microinjection of this construct yields persistent expression of a reporter gene with sufficiently high penetrance to characterize activity of *cis*-regulatory modules without coinjection of ϕC31 integrase mRNA. Therefore, using such well-designed plasmids, our DNA injection method could be useful to examine the activity of *cis*-regulatory modules by transgenesis.

RECIPES

Injection Buffer for Xenopus

10 mM HEPES-NaOH (pH 7.6)
88 mM NaCl

Sterilize by autoclaving (for 20 min at 121°C), and store for several months to years at room temperature in the dark.

Cite this protocol as *Cold Spring Harb Protoc*; doi:10.1101/pdb.prot097279

Steinberg's Solution (10×, pH 7.4)

Reagent	Amount
NaCl	34 g
KCl	0.5 g
$Ca(NO_3)_2 \cdot 4H_2O$	0.8 g
$MgSO_4 \cdot 7H_2O$	2.06 g
Tris	5.6 g
HCl	3.3 mL
Phenol red	3 mg
H_2O	to 1 L

Sterilize by autoclaving (for 20 min at 121°C), and store for several months to years at room temperature.

TE Buffer for Xenopus

10 mM Tris-HCl (pH 8.0)
1 mM EDTA (pH 8.0)

Sterilize by autoclaving (for 20 min at 121°C), and store for several months to years at room temperature.

ACKNOWLEDGMENTS

We thank Yusuke Mii for his comments on DNA injection for misexpressing genes, Steven D. Aird for technical editing of the manuscript, and OIST Graduate University for its generous support of the Marine Genomics Unit and Animal Resources Section for our animal experiments (Approval No. 2016-169).

REFERENCES

Allen BG, Weeks DL. 2005. Transgenic *Xenopus laevis* embryos can be generated using phiC31 integrase. *Nat Methods* 2: 975–979.

Christian JL, Moon RT. 1993. Interactions between Xwnt-8 and Spemann organizer signaling pathways generate dorsoventral pattern in the embryonic mesoderm of *Xenopus*. *Genes Dev* 7: 13–28.

Cleaver O, Krieg PA. 1999. Expression from DNA injected into *Xenopus* embryos. *Methods Mol Biol* 127: 133–153.

Etkin LD, Pearman B. 1987. Distribution, expression and germ line transmission of exogenous DNA sequences following microinjection into *Xenopus laevis* eggs. *Development* 99: 15–23.

Hamlet MR, Yergeau DA, Kuliyev E, Takeda M, Taira M, Kawakami K, Mead PE. 2006. Tol2 transposon-mediated transgenesis in *Xenopus tropicalis*. *Genesis* 44: 438–445.

Haremaki T, Tanaka Y, Hongo I, Yuge M, Okamoto H. 2003. Integration of multiple signal transducing pathways on Fgf response elements of the *Xenopus* caudal homologue Xcad3. *Development* 130: 4907–4917.

Hata A, Seoane J, Lagna G, Montalvo E, Hemmati-Brivanlou A, Massague J. 2000. OAZ uses distinct DNA- and protein-binding zinc fingers in separate BMP-Smad and Olf signaling pathways. *Cell* 100: 229–240.

Hikasa H, Sokol SY. 2004. The involvement of Frodo in TCF-dependent signaling and neural tissue development. *Development* 131: 4725–4734.

Hiratani I, Mochizuki T, Tochimoto N, Taira M. 2001. Functional domains of the LIM homeodomain protein Xlim-1 involved in negative regulation, transactivation, and axis formation in *Xenopus* embryos. *Dev Biol* 229: 456–467.

Hoppler S, Brown JD, Moon RT. 1996. Expression of a dominant-negative Wnt blocks induction of MyoD in *Xenopus* embryos. *Genes Dev* 10: 2805–2817.

Ishibashi S, Kroll KL, Amaya E. 2008. A method for generating transgenic frog embryos. *Methods Mol Biol* 461: 447–466.

Karaulanov E, Knochel W, Niehrs C. 2004. Transcriptional regulation of BMP4 synexpression in transgenic *Xenopus*. *EMBO J* 23: 844–856.

Kazanskaya O, Glinka A, del Barco Barrantes I, Stannek P, Niehrs C, Wu W. 2004. R-Spondin2 is a secreted activator of Wnt/β-catenin signaling and is required for *Xenopus* myogenesis. *Dev Cell* 7: 525–534.

Kiecker C, Niehrs C. 2001. A morphogen gradient of Wnt/β-catenin signalling regulates anteroposterior neural patterning in *Xenopus*. *Development* 128: 4189–4201.

Kolm PJ, Sive HL. 1995. Efficient hormone-inducible protein function in *Xenopus laevis*. *Dev Biol* 171: 267–272.

Kozmikova I, Smolikova J, Vlcek C, Kozmik Z. 2011. Conservation and diversification of an ancestral chordate gene regulatory network for dorsoventral patterning. *PLoS One* 6: e14650.

Kroll KL, Amaya E. 1996. Transgenic *Xenopus* embryos from sperm nuclear transplantations reveal FGF signaling requirements during gastrulation. *Development* 122: 3173–3183.

Kuroda H, Wessely O, De Robertis EM. 2004. Neural induction in *Xenopus*: Requirement for ectodermal and endomesodermal signals via Chordin, Noggin, β-Catenin, and Cerberus. *PLoS Biol* 2: E92.

Latinkic BV, Umbhauer M, Neal KA, Lerchner W, Smith JC, Cunliffe V. 1997. The *Xenopus* Brachyury promoter is activated by FGF and low concentrations of activin and suppressed by high concentrations of activin and by paired-type homeodomain proteins. *Genes Dev* 11: 3265–3276.

Luehders K, Sasai N, Davaapil H, Kurosawa-Yoshida M, Hiura H, Brah T, Ohnuma S. 2015. The small leucine-rich repeat secreted protein Asporin induces eyes in *Xenopus* embryos through the IGF signalling pathway. *Development* 142: 3351–3361.

Mii Y, Taira M. 2009. Secreted Frizzled-related proteins enhance the diffusion of Wnt ligands and expand their signalling range. *Development* 136: 4083–4088.

Mochizuki T, Karavanov AA, Curtiss PE, Ault KT, Sugimoto N, Watabe T, Shiokawa K, Jamrich M, Cho KW, Dawid IB, et al. 2000. Xlim-1 and LIM domain binding protein 1 cooperate with various transcription factors in the regulation of the *goosecoid* promoter. *Dev Biol* 224: 470–485.

Newport J, Kirschner M. 1982a. A major developmental transition in early *Xenopus* embryos: I. Characterization and timing of cellular changes at the midblastula stage. *Cell* 30: 675–686.

Newport J, Kirschner M. 1982b. A major developmental transition in early *Xenopus* embryos: II. Control of the onset of transcription. *Cell* 30: 687–696.

Ogino H, McConnell WB, Grainger RM. 2006. High-throughput transgenesis in *Xenopus* using I-SceI meganuclease. *Nat Protoc* 1: 1703–1710.

Onai T, Yu JK, Blitz IL, Cho KW, Holland LZ. 2010. Opposing Nodal/Vg1 and BMP signals mediate axial patterning in embryos of the basal chordate amphioxus. *Dev Biol* 344: 377–389.

Pan FC, Chen Y, Loeber J, Henningfeld K, Pieler T. 2006. I-SceI meganuclease-mediated transgenesis in *Xenopus*. *Dev Dyn* 235: 247–252.

Rankin SA, Kormish J, Kofron M, Jegga A, Zorn AM. 2011. A gene regulatory network controlling *hhex* transcription in the anterior endoderm of the organizer. *Dev Biol* 351: 297–310.

Rebbert ML, Dawid IB. 1997. Transcriptional regulation of the *Xlim-1* gene by activin is mediated by an element in intron I. *Proc Natl Acad Sci* 94: 9717–9722.

Shibano T, Takeda M, Suetake I, Kawakami K, Asashima M, Tajima S, Taira M. 2007. Recombinant Tol2 transposase with activity in *Xenopus* embryos. *FEBS Lett* 581: 4333–4336.

Sinzelle L, Vallin J, Coen L, Chesneau A, Du Pasquier D, Pollet N, Demeneix B, Mazabraud A. 2006. Generation of trangenic *Xenopus laevis* using the Sleeping Beauty transposon system. *Transgenic Res* 15: 751–760.

Sive HL, Grainger RM, Harland RM. 2007. Dejellying *Xenopus laevis* embryos. *Cold Spring Harb Protoc* doi:10.1101/pdb.prot4731.

Sudou N, Yamamoto S, Ogino H, Taira M. 2012. Dynamic in vivo binding of transcription factors to *cis*-regulatory modules of *cer* and *gsc* in the stepwise formation of the Spemann–Mangold organizer. *Development* 139: 1651–1661.

Takada H, Kawana T, Ito Y, Kikuno RF, Mamada H, Araki T, Koga H, Asashima M, Taira M. 2009. The RNA-binding protein Mex3b has a fine-tuning system for mRNA regulation in early *Xenopus* development. *Development* 136: 2413–2422.

Takebayashi-Suzuki K, Funami J, Tokumori D, Saito A, Watabe T, Miyazono K, Kanda A, Suzuki A. 2003. Interplay between the tumor suppressor p53 and TGF β signaling shapes embryonic body axes in *Xenopus*. *Development* 130: 3929–3939.

von Bubnoff A, Peiffer DA, Blitz IL, Hayata T, Ogata S, Zeng Q, Trunnell M, Cho KW. 2005. Phylogenetic footprinting and genome scanning identify vertebrate BMP response elements and new target genes. *Dev Biol* 281: 210–226.

Wallingford JB, Carroll TJ, Vize PD. 1998. Precocious expression of the Wilms' tumor gene *xWT1* inhibits embryonic kidney development in *Xenopus laevis*. *Dev Biol* 202: 103–112.

Wang C, Szaro BG. 2015. A method for using direct injection of plasmid DNA to study *cis*-regulatory element activity in F0 *Xenopus* embryos and tadpoles. *Dev Biol* 398: 11–23.

Wang C, Szaro BG. 2016. Using *Xenopus* embryos to study transcriptional and posttranscriptional gene regulatory mechanisms of intermediate filaments. *Methods Enzymol* 568: 635–660.

Watabe T, Kim S, Candia A, Rothbacher U, Hashimoto C, Inoue K, Cho KW. 1995. Molecular mechanisms of Spemann's organizer formation: Conserved growth factor synergy between *Xenopus* and mouse. *Genes Dev* 9: 3038–3050.

Yamamoto S, Hikasa H, Ono H, Taira M. 2003. Molecular link in the sequential induction of the Spemann organizer: Direct activation of the *cerberus* gene by Xlim-1, Xotx2, Mix.1, and Siamois, immediately downstream from Nodal and Wnt signaling. *Dev Biol* 257: 190–204.

Yao J, Kessler DS. 2001. Goosecoid promotes head organizer activity by direct repression of Xwnt8 in Spemann's organizer. *Development* 128: 2975–2987.

Yasuoka Y, Kobayashi M, Kurokawa D, Akasaka K, Saiga H, Taira M. 2009. Evolutionary origins of blastoporal expression and organizer activity of the vertebrate gastrula organizer gene *lhx1* and its ancient metazoan paralog *lhx3*. *Development* 136: 2005–2014.

Yasuoka Y, Suzuki Y, Takahashi S, Someya H, Sudou N, Haramoto Y, Cho KW, Asashima M, Sugano S, Taira M. 2014. Occupancy of tissue-specific *cis*-regulatory modules by Otx2 and TLE/Groucho for embryonic head specification. *Nat Commun* 5: 4322.

Yasuoka Y, Shinzato C, Satoh N. 2016. The mesoderm-forming gene *brachyury* regulates ectoderm-endoderm demarcation in the coral *Acropora digitifera*. *Curr Biol* 26: 2885–2892.

Yergeau DA, Johnson Hamlet MR, Kuliyev E, Zhu H, Doherty JR, Archer TD, Subhawong AP, Valentine MB, Kelley CM, Mead PE. 2009. Transgenesis in *Xenopus* using the Sleeping Beauty transposon system. *Dev Dyn* 238: 1727–1743.

Cite this protocol as *Cold Spring Harb Protoc*; doi:10.1101/pdb.prot097279

Protocol 4

Whole-Mount In Situ Hybridization of *Xenopus* Embryos

Jean-Pierre Saint-Jeannet[1]

Department of Basic Science and Craniofacial Biology, College of Dentistry, New York University, New York, New York 10010

Historically, techniques to analyze the localized distribution of mRNAs during development were performed on sectioned embryos using radioactively labeled riboprobes. The processing of the tissues and the use of emulsion autoradiography were laborious and time-consuming, leading to the development of more direct approaches. The nonradioactive whole-mount in situ hybridization method was first introduced in *Drosophila* embryos, and later adapted to *Xenopus* embryos for abundant transcripts such as muscle actin. Since then, the technique has been improved and is now broadly used for the spatial detection of even less abundant transcripts in *Xenopus*. The technique has been especially powerful in the analysis of changes in gene expression in embryos manipulated by mRNA or antisense oligonucleotides microinjection, and in animal cap explants exposed to exogenous factors. The protocol described here provides an excellent signal-to-noise ratio for most labeled probes. It also is relatively high-throughput: With a little practice, approximately 50 samples can easily be processed simultaneously.

MATERIALS

It is essential that you consult the appropriate Material Safety Data Sheets and your institution's Environmental Health and Safety Office for proper handling of equipment and hazardous materials used in this protocol.

RECIPES: Please see the end of this protocol for recipes indicated by <R>. Additional recipes can be found online at http://cshprotocols.cshlp.org/site/recipes.

Reagents

Acetic anhydride

Agarose

Anti-digoxigenin (DIG)–alkaline phosphatase (AP), Fab fragments (Sigma-Aldrich, 11093274910) (optional; for DIG labeling only)

Anti-fluorescein (FITC)–alkaline phosphatase (AP), Fab fragments (Sigma-Aldrich, 11426338910) (optional; for FITC labeling only)

Benzyl alcohol (optional; see Step 63)

Benzyl benzoate (optional; see Step 63)

Blocking solution for *Xenopus* embryos <R>

BM purple (Sigma-Aldrich, 11442074001)

Dithiothreitol (100 mM)

[1]Correspondence: jsj4@nyu.edu

DNA of interest, linearized

DNase I, RNase-free (10 U/µL)

Dry color pigment (e.g., ultramarine blue or cadmium yellow; Gamblin Artists Colors) (optional; see Step 59)

EDTA (0.2 M)

Formaldehyde (37%)

Hybridization buffer for *Xenopus* embryos (HB) <R>

Hydrogen peroxide (H_2O_2), 30%

MEMFA, prepared in sterile milliQ H_2O <R>

Methanol

PBT (2 mg/mL bovine serum albumin [fraction V] and 0.1% Triton X-100, prepared in PBS, filtered, and stored at 4°C)

Phosphate-buffered saline (PBS) (pH 7.4), prepared in sterile milliQ H_2O <R>

Proteinase K (5 µg/mL, prepared in PTw; Sigma-Aldrich, 3115828001), freshly prepared before use

PTw (0.1% Tween 20, prepared in PBS, filtered, and stored at 4°C)

Recombinant RNAsin (20 U/µL)

RNA labeling mix, Digoxigenin (DIG) (Sigma-Aldrich, 11277073910) (optional; for DIG labeling only)

RNA labeling mix, Fluorescein (FITC) (Sigma-Aldrich, 11685619910) (optional; for FITC labeling only)

RNA polymerase, SP6 (20 U/µL) (optional; see Step 11)

RNA polymerase, T3 (20 U/µL) (optional; see Step 11)

RNA polymerase, T7 (20 U/µL) (optional; see Step 11)

RNeasy MinElute Cleanup Kit (QIAGEN, 74204)

SSC <R>

Staining Buffer for *Xenopus* Embryos <R>

Triethanolamine (TEA; 0.1 M in sterile milliQ H_2O, pH ~7–8), freshly prepared before use

Water, sterile, milliQ-filtered

Xenopus laevis embryos

Equipment

Agarose gel electrophoresis apparatus

Aluminum foil

Beakers

Bunsen burner

Depression slides, glass (VWR, 413057) (optional; see Step 62.iv)

Dissection microscope

Forceps, fine (Roboz; Dumont #5)

Hybridization oven (Shake 'N Bake; Boekel, 136400)

Microcentrifuge

Microcentrifuge tubes, 1.5-mL

Microwave oven

Nutator (Clay Adams, BD 421125)

Pasteur pipettes, glass

Petri dishes, plastic, 60-mm

Pipette, glass, equipped with rubber bulb for embryo transfer (Fisher Scientific, S13974)

Pipettes, sterile, disposable, 10- and 25-mL

Spectrophotometer

Suction system, vacuum-driven

Superfine eyelash with handle, No. 1 (Ted Pella, 113)

Cite this protocol as *Cold Spring Harb Protoc*; doi:10.1101/pdb.prot097287

Test tube support rack (Fisher Scientific, 1478115)
Vials, glass, screw-cap, 4-mL (Fisher Scientific, 06-408B)
Vortex mixer
Water bath, 37°C

METHOD

The protocol described here is adapted from the protocol first reported by Harland (1991).

Antibody Preadsorption

1. Fix ~100 *Xenopus* embryos (NF stage 25–35; Nieuwkoop and Faber 1967) in MEMFA for 1 h in a 4-mL glass vial on a nutator.

2. Replace the fixative with methanol. Wash for 15 min.

3. Wash the embryos in fresh methanol for 5 min at room temperature.

4. Wash in 75% methanol in H_2O for 5 min at room temperature.

5. Wash in 50% methanol in H_2O for 5 min at room temperature.

6. Wash in 25% methanol in PTw for 5 min at room temperature.

7. Wash twice in PTw for 5 min each at room temperature.

8. Wash in PTw for 15 min each at room temperature.

9. Add 4 mL of blocking solution and 40 µL of AP-conjugated Fab fragments against DIG or FITC, as appropriate (1:100 final dilution). Incubate overnight at 4°C on a nutator.

10. Decant the antibody solution. Store in a clean tube at 4°C.

Riboprobe Synthesis

11. Combine the following reagents in a microcentrifuge tube:

Linearized DNA	1 µg
RNA polymerase buffer (10×)	2 µL
RNA labeling mix (DIG or FITC) (10×)	2 µL
RNA polymerase (SP6, T7, or T3)	2 µL
Dithiothreitol	2 µL
Recombinant RNasin	2 µL
Sterile milliQ H_2O	to 20 µL

12. Mix well. Incubate for 2–3 h in a 37°C water bath.

13. Add 2 µL of RNase-free DNase I. Mix well.

14. Incubate for 20 min in a 37°C water bath.

15. Stop the reaction by adding 2 µL of 0.2 M EDTA and 80 µL of sterile H_2O.

16. Purify the RNA-labeled probe using the RNeasy MinElute Cleanup Kit according to the manufacturer's instructions.

17. Measure the concentration using a spectrophotometer. Confirm the size and integrity of the riboprobe on a denaturing gel.

18. Store labeled probes at −80°C.

Embryo Preparation

All steps are performed in 4-mL glass vials rocking on a nutator. Position the vials horizontally on the nutator, rotating end over end. For smaller volumes (1 mL or less), place the tubes vertically on a test tube support rack positioned directly on the top of the nutator. Remove solutions from the vials using a Pasteur pipette attached to a vacuum-driven suction system; add fresh solutions using 10- or 25-mL disposable pipettes. Twenty-four to 48 vials (i.e., 1–2 test tube support racks) can be processed simultaneously in this fashion.

19. Place the selected embryos in a dish containing culture medium under a dissection microscope. Using fine forceps, remove the vitelline membrane.

    At early embryonic stages (i.e., blastula/gastrula) the vitelline membrane is tightly associated with the embryo. It might be necessary to puncture the embryo at the vegetal pole with one forceps, then grasp and tear the vitelline membrane away from the embryo with the a second pair. Alternatively, for large numbers of embryos, this step can be omitted; the vitelline membrane is then removed after fixation and rehydration (i.e., after Step 27) by treating the embryos with a cocktail of proteases (10 µg/mL proteinase K, 2 mg/mL collagenase A and 20 U/mL hyaluronidase) for 10 min at room temperature (Islam and Moss 1996).

20. Using a glass pipette, transfer the selected embryos to a 4-mL glass vial filled with 4 mL of MEMFA.

    Approximately 50 embryos can be processed in one vial.

21. Rock the vial on the nutator for 1 h at room temperature.

22. Replace the MEMFA with methanol. Wash the embryos twice with methanol (15 min/wash) at room temperature.

    Embryos can be stored in methanol at −20°C for several months.

Embryo Rehydration

23. Wash the embryos in fresh methanol for 5 min at room temperature.

24. Wash in 75% methanol in H_2O for 5 min at room temperature.

25. Wash in 50% methanol in H_2O for 5 min at room temperature.

26. Wash in 25% methanol in PTw for 5 min at room temperature.

27. Wash twice in PTw for 5 min each at room temperature.

Proteinase K Treatment

28. Replace with the PTw with 1 mL of Proteinase K. Incubate for 5 min at room temperature.

29. Replace the Proteinase K with 4 mL of TEA solution. Add 11.25 µL of acetic anhydride. Wash for 5 min at room temperature.

30. Add another 11.25 µL of acetic anhydride to the vial. Wash an additional 5 min at room temperature.

31. Wash twice in PTw, 5 min each, at room temperature.

32. Postfix the embryos with 3.7% formaldehyde in PTw for 20 min at room temperature.

33. Wash twice in PTw, 5 min each, at room temperature.

Prehybridization and Hybridization

34. Remove all but 1 mL of PTw from the vial. Add 250 µL of HB. Swirl the vial gently. Allow the embryos to settle to the bottom.

35. Replace the solution with 500 µL of HB. Incubate at 60°C in the hybridization oven for 10 min, rocking gently.

    Maintain the vials in a vertical position on the test tube rack during the prehybridization and hybridization (Steps 35–37).

36. Replace with 500 µL of fresh HB. Incubate at 60°C in the hybridization oven for at least 6 h (or overnight), rocking gently.

Cite this protocol as *Cold Spring Harb Protoc*; doi:10.1101/pdb.prot097287

37. Replace the solution with 500 μL of labeled riboprobe (from Step 18) diluted in HB to a final concentration of 0.5 μg/mL. Incubate overnight (i.e., 12–18 h) at 60°C in the hybridization oven, with gentle rocking.

Posthybridization Washes

38. Remove the probe. Store in a glass vial at −20°C.

    The probe is stable in HB for several months at −20°C and can be reused up to 8–10 times.

39. Add 500 μL of HB to the embryos. Incubate 10 min at 60°C.

    Maintain the vials in a vertical position on the test tube rack during the washes in the hybridization oven (Steps 39–44).

40. Replace with 1 mL of a HB:2×SSC solution (1:1). Incubate 10 min at 60°C.

41. Replace with 1 mL of a HB:2×SSC solution (1:3). Incubate 10 min at 60°C.

42. Wash embryos twice with 4 mL of 2×SSC, 20 min each wash, at 60°C.

43. Wash embryos twice with 4 mL of 0.2×SSC, 30 min each wash, at 60°C.

44. Wash the embryos twice with 4 mL of PTw, 10 min each, at 60°C.

45. Wash the embryos with 4 mL of PTw for 10 min at room temperature on the nutator.

Immunodetection

46. Wash the embryos in PBT for 15 min at room temperature.

47. Replace the PBT with 1 mL of blocking solution. Incubate for 1 h at room temperature.

48. Replace the blocking solution with 1 mL of preadsorbed antiDIG-AP antibody (1:2000 final dilution) or preadsorbed antiFITC-AP antibody in blocking solution (1:10000 final dilution). Incubate overnight at 4°C on a nutator.

49. Remove the used antibody. Store at 4°C in a clean tube.

    The antibody is stable for several weeks at 4°C and can be reused up to three times.

50. Wash off excess antibody by incubating with six changes of PBT, 1 h each wash, at room temperature.

Chromogenic Reaction

51. Wash the embryos twice in staining buffer, 10 min each wash, at room temperature.

52. Replace with 1 mL of BM purple. Cover the test tube support rack with aluminum foil. Incubate overnight at room temperature in the dark.

    Alternatively, the chromogenic reaction can be performed after washing overnight in PBT at 4°C. This allows the reaction to be monitored closely and stopped at earlier time points, if needed.

53. If needed to enhance the staining, replace with fresh BM purple the following day.

54. Fix in MEMFA for 2 h at room temperature.

Bleaching of Pigment

55. Transfer the embryos to methanol for 15 min.

56. Bleach the embryos in a solution containing 10% H_2O_2 in methanol for 48–72 h at room temperature on a nutator.

57. Transfer to methanol for 15 min.

58. Transfer to PBS. Store at 4°C.

    Alternatively, wash fixed embryos several times with 70% ethanol in PTw, and then bleach for a few hours in a solution containing 5% formamide and 1% H_2O_2 in 0.5×SSC (Mayor et al. 1995). Although more rapid, this procedure attenuates the staining, decreasing the signal-to-noise ratio.

Embryo Imaging

Embryos are best examined and photographed in PBS, in dishes coated with agarose.

59. Using a microwave oven, dissolve 1% agarose (w/v) in PBS. Pour 5 mL immediately into a 60-mm plastic Petri dish.

    Dry color pigments such as ultramarine blue or cadmium yellow (0.5%–1.0% [w/v]), can be added to the melted agarose to provide a contrasting background during imaging of the embryos.

60. Using a Bunsen burner, melt the end of a Pasteur pipette into a rounded tip. Use the hot tip of the pipette to melt wells of different sizes and depths into the surface of the solidified agarose.

61. Pour the embryos directly from the vial into a dish filled with PBS. Using a superfine eyelash mounted on a handle, position the embryos into the wells for photography.

    See Troubleshooting.

62. Use a glass pipette to return the embryos to the vial. Store in PBS at 4°C.

63. Optionally, clear the embryos before being photographed:

    i. Wash embryos twice in methanol for 5 min each.

    ii. Replace with 500 µL of a solution of benzyl benzoate:benzyl alcohol (BB:BA [2:1]). Swirl the tube gently. Allow the embryos to settle at the bottom.

    iii. Replace with 500 µL of a fresh solution of BB:BA.

    iv. Using a glass pipette, transfer the embryos to a glass depression slide in a few drops of BB:BA solution.

TROUBLESHOOTING

Problem (Step 61): The signal is indistinct and/or there is high background.

Solution: Add the detergent CHAPS (3-[(3-Cholamidopropyl)dimethylammonio]-1-propanesulfonate hydrate) to the HB at a concentration of 0.1% for hybridization (Steps 34–37) and 0.3% for the subsequent posthybridization washes (Steps 39–43).

DISCUSSION

Whole-mount in situ hybridization is a powerful tool to analyze the spatiotemporal expression of genes during embryogenesis. Although major improvements have been made over the years, the technique remains demanding: From embryo collection to the imaging step, the process takes ~1 wk. Automated in situ hybridization platforms show promise to perform routine staining as well as high throughput screens, although these devices remain costly. Several laboratories now use the InsituPro VSi automated unit commercialized by Intavis (Tubingen, Germany) for standard in situ hybridization. Flogentec (Valence, France) has developed the Flo400-W, a unit that substantially reduces processing time while preserving signal quality. However, its applicability to *Xenopus* embryos has not yet been showed. Such standardized platforms are suitable for routine in situ hybridization to detect relatively abundant transcripts using well-characterized probes, although probes that are more "difficult" will always benefit from customized protocols.

Two-Color Whole-Mount In Situ Hybridization

By mixing the FITC- and DIG-labeled probes in the HB (Step 37), the protocol above can be used to detect two different mRNAs in the same embryo. The probes are detected sequentially using AP-conjugated antiFITC– and antiDIG–antibodies, respectively (Jaurena et al. 2015). FITC-labeled probe is typically visualized first using Magenta-phos (5-bromo-6-chloro-3-indoxyl phosphate; Biosynth) as

   Cite this protocol as *Cold Spring Harb Protoc*; doi:10.1101/pdb.prot097287

a substrate (producing a purple color) by adding 10 µL of Magenta-phos stock solution (25 mg/mL in dimethyl formamide) per mL of staining buffer. This substrate provides little-to-no background. However, the chromogenic reaction can last from several days to a few weeks, although fresh substrate can be added as needed to enhance staining. The AP is then inactivated by treatment with glycine (0.1 M, pH 2.2) for 30 min, after which the embryos are washed and incubated with the antiDIG antibody. The DIG-labeled probe is visualized using BCIP (5-bromo-4-chloro-3-indolyl-phosphate) as a substrate (producing a turquoise color) by adding 3.5 µL of BCIP stock solution (50 mg/mL in dimethyl formamide) per mL of staining buffer. Depending on the abundance of the bound label, staining generally can take overnight to a few of days and—as with the FITC label—fresh substrate can be added as needed to enhance staining. The reaction is stopped by fixing the embryos in MEMFA supplemented with 0.2% glutaraldehyde for 2 h, followed by two washes in PBS, 15 min each. Because the products of the enzymatic reaction are methanol-soluble, the embryos should be bleached in 10% H_2O_2 in PBS for 24–48 h at room temperature.

RECIPES

Blocking Solution for Xenopus Embryos

1. Mix 4.6 g maleic acid and 3.5 g NaCl in 200 mL of sterile milliQ H_2O in a 1000-mL beaker. Adjust the pH to 7.5 with 10 N NaOH.
2. Add 8 g of Blocking Reagent (Sigma-Aldrich, 11096176001). Heat the solution in a microwave using 30-sec pulses until the Blocking Reagent is fully dissolved. (The solution should appear milky.)
3. Cool on ice. Adjust the volume to 380 mL with sterile milliQ H_2O.
4. Add 20 mL of heat-inactivated lamb serum (ThermoFisher, 16070096). (Inactivate the serum by heating in a water bath at 55°C for 30 min, swirling every 5–10 min. Store at −20°C in 20-mL aliquots.)
5. Store the blocking solution at −20°C in 20-mL aliquots.

> The final concentrations of the reagents are 100 mM maleic acid, 150 mM NaCl, 5% lamb serum, 2% (w/v) blocking reagent.

Hybridization Buffer (HB) for Xenopus Embryos

Reagent	Final concentration
Formamide	50%
Saline–sodium citrate (SSC)	5×
Torula yeast RNA	1 mg/mL
Heparin	100 µg/mL
Denhart's	1×
Tween 20	0.1%
EDTA	5 mM

Prepare in sterile milliQ H_2O. Store at −20°C.

MEMFA

MOPS
EGTA
$MgSO_4$
Formaldehyde (37%)

Prepare a 10× stock solution of 1 M MOPS (pH 7.4), 20 mM EGTA, and 10 mM $MgSO_4$. On the day of use, prepare fresh 1× MEMFA by combining 8 parts H_2O, 1 part 10× MEMFA stock solution, and 1 part formaldehyde (37%).

Final concentrations are 100 mM MOPS (pH 7.4), 2 mM EGTA, 1 mM $MgSO_4$, 3.7% (v/v) formaldehyde.

Phosphate-Buffered Saline (PBS)

Reagent	Amount to add (for 1× solution)	Final concentration (1×)	Amount to add (for 10× stock)	Final concentration (10×)
NaCl	8 g	137 mM	80 g	1.37 M
KCl	0.2 g	2.7 mM	2 g	27 mM
Na_2HPO_4	1.44 g	10 mM	14.4 g	100 mM
KH_2PO_4	0.24 g	1.8 mM	2.4 g	18 mM

If necessary, PBS may be supplemented with the following:

$CaCl_2 \cdot 2H_2O$	0.133 g	1 mM	1.33 g	10 mM
$MgCl_2 \cdot 6H_2O$	0.10 g	0.5 mM	1.0 g	5 mM

PBS can be made as a 1× solution or as a 10× stock. To prepare 1 L of either 1× or 10× PBS, dissolve the reagents listed above in 800 mL of H_2O. Adjust the pH to 7.4 (or 7.2, if required) with HCl, and then add H_2O to 1 L. Dispense the solution into aliquots and sterilize them by autoclaving for 20 min at 15 psi (1.05 kg/cm^2) on liquid cycle or by filter sterilization. Store PBS at room temperature.

SSC

For a 20× solution: Dissolve 175.3 g of NaCl and 88.2 g of sodium citrate in 800 mL of H_2O. Adjust the pH to 7.0 with a few drops of a 14 N solution of HCl. Adjust the volume to 1 L with H_2O. Dispense into aliquots. Sterilize by autoclaving. The final concentrations of the ingredients are 3.0 M NaCl and 0.3 M sodium citrate.

Staining Buffer for Xenopus Embryos

Reagent	Final concentration
Tris-HCl (pH 9.5)	0.1 M
NaCl	0.1 M
$MgCl_2$	50 mM
Tween 20	0.1%
Levamisole	0.065%

Prepare fresh in sterile milliQ H_2O from filtered stock solutions.

ACKNOWLEDGMENTS

Work in the Saint-Jeannet Laboratory is supported by the National Institutes of Health (NIH), grant numbers R01DE25806 and R01DE25468.

REFERENCES

Harland RM. 1991. In situ hybridization: An improved whole-mount method for *Xenopus* embryos. *Methods Cell Biol* **36:** 685–695.

Islam N, Moss T. 1996. Enzymatic removal of the vitelline membrane and other protocol modifications for whole mount *in situ* hybridization of *Xenopus* embryos. *Trends Genet* **12:** 459.

Jaurena MB, Juraver-Geslin H, Devotta A, Saint-Jeannet J-P. 2015. Zic1 controls placode progenitor formation non-cell autonomously by regulating retinoic acid production and transport. *Nat Commun* **6:** 7476.

Mayor R, Morgan R, Sargent MG. 1995. Induction of the prospective neural crest of *Xenopus*. *Development* **121:** 767–777.

Nieuwkoop PD, Faber J (eds.). 1967. *Normal table of* Xenopus laevis *(Daudin): A systematical & chronological survey of the development from the fertilized egg till the end of metamorphosis.* North Holland Publishing Company, Amsterdam.

Cite this protocol as *Cold Spring Harb Protoc*; doi:10.1101/pdb.prot097287

Whole-Mount Immunocytochemistry in *Xenopus*

Michael W. Klymkowsky[1]

Molecular, Cellular, and Developmental Biology, University of Colorado Boulder, Boulder, Colorado 80309

To visualize the effects of experimental perturbations on normal cellular behavior, morphology, and intracellular organization, we use a simple whole-mount immunocytochemical method with *Xenopus* oocytes, explants, or embryos. This method is applicable to a wide range of systems, including human-induced pluripotent stem cell-derived organoids, and can be used with both chromogenic (horseradish peroxidase/diaminobenzidine) and fluorescent imaging.

MATERIALS

It is essential that you consult the appropriate Material Safety Data Sheets and your institution's Environmental Health and Safety Office for proper handling of equipment and hazardous materials used in this protocol.

RECIPES: Please see the end of this protocol for recipes indicated by <R>. Additional recipes can be found online at http://cshprotocols.cshlp.org/site/recipes.

Reagents

Aldehyde fixative (MEMFA) <R> (as an alternative to Dent's fixative; see Step 1)
Antibody dilution solution

Prepare a solution of 20% (v/v) blocking solution in TBS.

BABB clearing agent (1 part benzyl alcohol, 2 parts benzyl benzoate [v/v])
Benzocaine solution (10% benzocaine in 100% ethanol [w/v]) (for use with MEMFA; see Step 1)
Blocking solution

We use calf or horse serum with 10% DMSO (v/v) and 0.1% thimerosol for blocking samples. This solution can be stored at 4°C for extended periods of time (>1 yr). We use these less expensive reagents, as opposed to fetal calf serum, because of cost and the fact that they contain immunoglobulin, which serves as a useful blocking/background reduction agent.

Dent's bleach

Prepare a solution of 10% H_2O_2 by diluting 30% H_2O_2 in Dent's fixative (v/v).

Dent's fixative (1 part DMSO, 4 parts 100% methanol)
Methanol (100%, for dehydrating fixed specimens)
Methyl salicylate (wintergreen oil) (as an alternative clearing agent; see Step 11)
Primary antibody, according to experiment
Reagents for chromogenic staining

---

[1]Correspondence: klym@colorado.edu

Cite this protocol as *Cold Spring Harb Protoc*; doi:10.1101/pdb.prot097295

Diaminobenzidine tetrahydrochloride (DAB)

We routinely make up stocks of DAB (10 mg/mL in 30% H_2O_2) and store them at −20°C, at which temperature they are stable for extended periods (years). Once diluted (Step 9), unused DAB should be destroyed with bleach and discarded.

Hydrogen peroxide

Secondary antibody (HRP-conjugated)

Reagents for fluorescent staining

Fluoromount G (ThermoFisher) or equivalent

Secondary antibody (fluorochrome-conjugated)

For immunofluorescence staining and imaging, we typically use Alexa-conjugated secondary antibodies.

Tris-buffered saline (TBS) (pH 7.6) <R>

Xenopus oocytes, explants, or embryos

Eggs and early stage (prehatching) embryos are surrounded by a jelly coat. In most fertilization protocols, this jelly coat is removed after fertilization by treating with dejellying solution <R> followed by washing with 20% Marc's modified Ringer's solution (pH 7.8) <R> before fixation.

Equipment

Coverslips

Use No. 1.5 coverslips for fluorescent imaging. Generally, no coverslip is necessary for HRP/DAB-stained samples in deeper glass depression slides; however, if a coverslip is used, we use No. 1.

Glass depression slides (both shallow [∼0.8 mm] and deep [∼5 mm] for later-stage embryos [Fig. 1A])

Glass vials (2-mL screw-top; Wheaton)

FIGURE 1. (*A*) Glass depression slides for whole-mount immunocytochemistry. (*B*) A reversed Pasteur pipette for manipulation of specimens during clearing.

Cite this protocol as *Cold Spring Harb Protoc*; doi:10.1101/pdb.prot097295

Microscopes and image-capture cameras, according to experiment
Pasteur pipettes

METHOD

This protocol is based on the method developed by Dent et al. (1989) and uses chromogenic staining. For fluorescent staining and imaging, see Steps 13–15. This method is also compatible with Alcian blue staining of cartilage.

All steps should be performed in clear glass vials.

Whole-Mount Immunoperoxidase Staining

1. Fix the specimens overnight at room temperature in Dent's fixative.

   We fix oocytes, explants, and embryos overnight, although 2 h is generally sufficient. We have also used Dent's fixative successfully for high-resolution immunostaining and imaging of human-induced pluripotent cell-derived cerebral organoids. Alternatively, specimens can be fixed using a formaldehyde-based fixative, such as MEMFA, for 2 h at room temperature. In later embryonic stages, we find that aldehyde fixation can cause larvae and later-stage tadpoles to writhe, something not observed when using Dent's fixative. To avoid this, we pretreat later-stage embryos to be fixed with MEMFA with a 1:10 dilution of benzocaine solution for 10 min, which is then removed and replaced with fixative. Where aldehyde-based fixatives such as MEMFA are used, the fixed specimen should be extracted overnight with Dent's fixative.

   Samples can be stored indefinitely at 4°C in Dent's fixative.

2. Bleach pigment by incubating the specimens in Dent's bleach.

   Gentle rocking under a bright light speeds the bleaching reaction. Complete bleaching can take from 1 to 4 d; be patient! Excessive rocking movement can destroy the specimen completely, or remove its outer cortical layers.

3. Once bleached, rinse the specimens three times in TBS for 15 min each.

4. (Optional) Block the samples by incubating in blocking solution for 15 min at room temperature with rocking.

   Alternatively, proceed to Step 5 to incubate the samples directly with the primary antibody.

5. Incubate the samples with primary antibody in antibody dilution solution.

   We typically incubate in primary antibody overnight at room temperature with gentle rocking.

   It is critical to titer the antibodies to determine the range of strong staining with minimal background. Typically this will result in a lower dilution than that used for standard immunofluorescence staining of cells growing on coverslips.

6. Wash the samples five times in TBS for 1 h each with gentle rocking.

7. Incubate the samples overnight in HRP-conjugated secondary antibody in antibody dilution solution.

   Again, it is critical to check effective antibody concentration. We have had good luck with commercial antibodies (e.g., from Sigma-Aldrich) at dilutions of 1:500 to 1:1000, but sometimes dilutions as low as 1:200 may be necessary.

8. Wash the samples five times in TBS for 1 h each with gentle rocking

9. React for 15 min–2 h in DAB (0.5 mg/mL in TBS) plus 0.02% hydrogen peroxide.

10. Stop the HRP/DAB reaction by dehydration with methanol or Dent's fixative (two changes for 15–30 min each).

11. If the samples are to be cleared, wash twice with 100% methanol for 5–10 min each and then add BABB.

    Alternatively, one can test whether the less toxic methyl salicylate is adequate for clearing.

    If clearing is not complete, it is likely that the specimens were not adequately dehydrated. They can be returned to 100% methanol and then cleared again.

    Clearing agents cause specimens to become brittle; take care when manipulating them. We use a Pasteur pipette with the tip removed (after scoring with a diamond pencil) to make an aperture of adequate

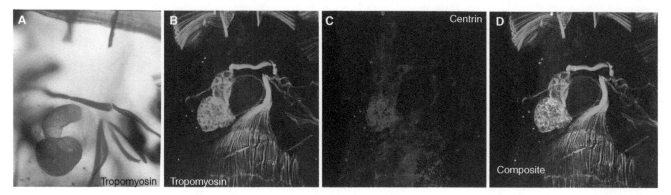

FIGURE 2. (*A*) A *Xenopus laevis* (stage ∼46–48) tadpole stained for tropomyosin and visualized using HRP-labeled secondary antibody. (*B–D*) Stage ∼42 embryos stained for tropomyosin (*B*) and centrin (*C*) and visualized using Alexa-conjugated secondary antibodies. The channel overlap image is shown in (*D*). The heart is clearly seen in all panels.

diameter for the specimen. Alternatively, the cut pipette can be reversed to generate a wide opening (Fig. 1B).

12. For imaging (Fig. 2), transfer the samples in BABB (or methyl salicylate) to deep depression slides.

    For sectioning of stained specimens, see Discussion.

Fluorescent Staining and Imaging of Embryos or Explant Surfaces

13. Follow the procedure in Steps 1–8, but omit bleaching as appropriate and substitute fluorochrome-conjugated secondary antibodies for HRP-conjugated secondary antibodies.

    For fluorescently stained samples, both bleaching and clearing are often unnecessary, particularly when imaging cells located on the embryonic surface, such as the multiciliated cells of the epidermis. In fact, the presence of pigment can reduce background autofluorescence. We have not yet tested how the use of "light sheet" and related microscopes might improve imaging in deeper embryonic layers, or how bleaching and clearing influence image quality.

    Given secondary antibodies of adequate species-specificity, it is possible to combine primary antibodies derived from different species in the first staining step, followed by combined secondary antibodies.

14. Dehydrate the samples with methanol or Dent's fixative (two changes for 15–30 min each).

15. For viewing (Fig. 2), move specimens to shallow (0.8-mm) depression slides. Place in Fluoromount G, cover with a coverslip, and allow to dry before imaging. Consult Werner and Mitchell (2013) and Brooks and Wallingford (2015) for more details of higher resolution imaging.

    Late-stage Xenopus oocytes and early embryos are ∼1–1.2 mm in diameter; explants are significantly smaller (0.2–0.5 mm in diameter).

    Particularly for late-stage oocyte and embryos, the specimens can be flattened against the coverslip, bringing large expanses of their surface into a single focal plane. This facilitates the use of automated image acquisition programs, such as that developed by Nick Galati (Shi et al. 2014). A version of this software has been used to analyze cilia patterns in Tetrahymena (Galati et al. 2016).

    For sectioning of stained specimens, see Discussion.

DISCUSSION

Sectioning

Stained specimens can be examined by sectioning to provide a more complete characterization of tissue details. In earlier studies we used Paraplast as an embedding medium (Klymkowsky and Hanken 1991; Seufert et al. 1994), but more recently we have embedded embryos in 2%–4% low melting agarose and generated 50- to 300-µm sections using a Vibratome.

 Cite this protocol as *Cold Spring Harb Protoc*; doi:10.1101/pdb.prot097295

RECIPES

Aldehyde Fixative (MEMFA)

Reagent	Final concentration
MOPS	0.1 M
EGTA	2 mM
MgSO$_4$	1 mM
Formaldehyde (dissolved solid or commercially purchased solution)	3.7%

Adjust the pH to 7.4. Store at 4°C for up to 1 mo. This formulation was developed in the Harland laboratory (Harland 1991).

Dejellying Solution

Reagent	Final concentration
Cysteine-HCl	2%
Marc's modified Ringer's solution (pH 7.8) <R>	33%

Adjust pH to 8.0. Use within an hour. Ringer's can be stored at room temperature for at least 4–6 mo if tightly sealed. See below.

Marc's Modified Ringer's Solution (pH 7.8)

Reagent	Final concentration
NaCl	110 mM
KCl	2 mM
MgCl$_2$	1 mM
CaCl$_2$	2 mM
NaHCO$_3$	2 mM
HEPES	5 mM

Dissolve in Milli-Q or reverse osmosis H$_2$O. Adjust the pH to 7.8. Store at room temperature in a tightly sealed bottle for up to 4–6 mo.

Tris-Buffered Saline (TBS) (pH 7.6)

Reagent	Final concentration
Tris base	50 mM
NaCl	150 mM

Adjust pH to 7.6 with HCl. Store at 4°C for up to 6 mo.

ACKNOWLEDGMENTS

I thank Joe Dent for his work in developing the original whole mount-staining method, Dan Chu for his work on characterizing acetylated-α-tubulin staining, and Nick Galati and Jianli Shi for the development of quantitative methods for our studies on ciliogenesis in ectodermal explants.

REFERENCES

Brooks ER, Wallingford JB. 2015. In vivo investigation of cilia structure and function using *Xenopus*. *Methods Cell Biol* **127**: 131–159.

Dent JA, Polson AG, Klymkowsky MW. 1989. A whole-mount immunocytochemical analysis of the expression of the intermediate filament protein vimentin in *Xenopus*. *Development* **105**: 61–74.

Galati DF, Abuin DS, Tauber GA, Pham AT, Pearson CG. 2016. Automated image analysis reveals the dynamic 3-dimensional organization of multi-ciliary arrays. *Biol Open* **5**: 20–31.

Harland RM. 1991. In situ hybridization: An improved whole-mount method for *Xenopus* embryos. *Methods Cell Biol* **36**: 685–695.

Klymkowsky MW, Hanken J. 1991. Whole-mount staining of *Xenopus* and other vertebrates. In Xenopus laevis: *Practical uses in cell and molecular biology* (ed. Kay BK, Peng HB), pp. 419–441.

Seufert DW, Hanken J, Klymkowsky MW. 1994. Type II collagen distribution during cranial development in *Xenopus laevis*. *Anat Embryol* **189**: 81–89.

Shi J, Zhao Y, Galati N, Winey M, Klymkowsky MW. 2014. Chibby functions in *Xenopus* ciliary assembly, embryonic development, and the regulation of gene expression. *Dev Biol* **395**: 287–298.

Werner ME, Mitchell BJ. 2013. Using *Xenopus* skin to study cilia development and function. *Methods Enzymol* **525**: 191.

Cite this protocol as *Cold Spring Harb Protoc*; doi:10.1101/pdb.prot097295

Protocol 6

Cleavage Blastomere Explant Culture in *Xenopus*

Sally A. Moody

Department of Anatomy and Regenerative Biology, George Washington University, Washington, D.C. 20037

The individual blastomeres of *Xenopus* two- to 32-cell embryos have been fate mapped. This work identified the precursors of most of the embryonic cell types, tissues and organs; however, the maps do not reveal the cell interactions or signaling pathways that are required for establishing cell fates. This protocol describes an explant culture approach for culturing blastomeres in isolation to test whether a cell's fate has been determined. Cleavage blastomeres can be cultured in a simple salt medium without added factors because they contain intracellular yolk platelets, which provide an intrinsic energy source. This method allows one to test whether an isolated blastomere can produce specific cell types or express tissue-specific genes independent of interactions with other cells or specific signaling pathways. The role of cell–cell interactions can be revealed by co-culturing different sets of blastomeres. One can identify the molecules that are required for those cell fates by applying knockdown approaches to the isolated cell. One also can determine the developmental time at which cell fates are committed by explanting blastomere lineages at different stages.

MATERIALS

It is essential that you consult the appropriate Material Safety Data Sheets and your institution's Environmental Health and Safety Office for proper handling of equipment and hazardous materials used in this protocol.

RECIPES: Please see the end of this protocol for recipes indicated by <R>. Additional recipes can be found online at http://cshprotocols.cshlp.org/site/recipes.

Reagents

Agarose, 2% in 1× culture medium

> *Add 2 g of electrophoresis-grade agarose to 100 mL of 1× culture medium in a screw-cap glass bottle. Autoclave to dissolve agarose. This can be stored in the refrigerator and microwaved to liquefy the agarose when it is next needed.*

Carrier protein (e.g., bovine serum albumin [BSA]) (optional; see Step 8)

Culture medium (Steinberg's solution [100%] <R> [1× SS], Marc's Modified Ringer's [MMR] [1×] <R>, or Modified Barth's Saline [MBS] <R> [1×])

> *L-15 Leibovitz culture medium <R> is commonly used for testing signaling pathways (Step 8).*

Dejellying reagents (see Sive et al. 2007a and Moody 2012a)

Gentamicin (50 µg/mL) (optional; see Step 8)

Ligand or inhibitor for signaling pathway of interest (optional; see Step 8)

Correspondence: samoody@gwu.edu

Lineage tracer, mRNA, and/or oligonucleotides for microinjection (see Protocol 1: Lineage Tracing and Fate Mapping in *Xenopus* Embryos [Moody 2018a] and Protocol 2: Microinjection of mRNAs and Oligonucleotides [Moody 2018b]) (optional; see Steps 6–7)

Reagents for natural mating or in vitro fertilization in *Xenopus* (see Sive et al. 2007b and Moody 2012a)

> *Because obtaining eggs by squeezing the female sometimes reduces egg quality, an alternate method for in vitro fertilization is to have the hormone-injected female naturally release eggs into high-salt MBS <R>. Collect and maintain eggs in this buffer for several hours. Wash twice with 0.1× MBS before adding sperm.*

Equipment

Culture dish (24-well)

Forceps (e.g., Dumont #5 biologie)

> *Sharpen forceps to a very fine tip by gently stroking them across a piece of Alumina abrasive film (Thomas Scientific) under a stereomicroscope. We use medium film (3-μm; 6775E-46) for sharpening and fine film (0.3-μm; 6775E-54) for a final polish. It is prudent to sharpen extra forceps in case they are accidentally damaged during the experiment. Forceps that are accidentally bent can be repaired using coarse film (12-μm; 6775E-38).*

Glass bead tool

> *Melt the tip of an autoclaved 6-inch Pasteur pipette into a ~2-mm ball in the flame of a Bunsen burner (Fig. 1B).*

Hair loop

> *Place both ends of a fine hair (~6 inch long, collected near the nape of the neck from a family member or laboratory partner; baby hair works well but tends to be too short to manipulate easily) into the narrow tip of a 6-inch Pasteur pipette to form a 2- to 3-mm loop. Secure in place with nail polish or cyanoacrylate glue (Fig. 1A). Sterilize before use by dipping in 70% ethanol and air-drying. Make several in advance in case of breakage during the experiment. As an alternative, purchase a superfine eyelash hair attached to a handle (Ted Pella #113) or an ultrafine single deer hair attached to a handle (Ted Pella #119).*

Incubator at 14°C–20°C

Microinjection equipment (see Protocol 2: Microinjection with mRNAs and Oligonucleotides [Moody 2018b]; Moody 2012a) (optional)

Pasteur pipettes, glass (6-inch, autoclaved), and rubber bulb

Petri dishes (60-mm)

Stereomicroscope equipped with a fiber optic or LED lamp

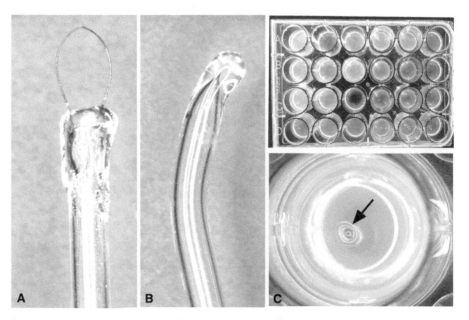

FIGURE 1. Microsurgery tools. (*A*) Hair loop. (*B*) Glass bead tool. (*C*) Explant dish. (*Top*) 24-well culture plate, with each well coated with agarose. (*Bottom*) a single well in the plate with a depression (arrow), melted with a glass bead tool, in which the explant will develop.

Cite this protocol as *Cold Spring Harb Protoc*; doi:10.1101/pdb.prot097303

METHOD

Preparing for Microsurgery

Embryos released from their vitelline membrane will stick to clay and plastic; therefore, Petri dishes to be used for microsurgery are coated with agarose.

1. Prepare microsurgery dishes as follows.

    i. Coat a 60-mm Petri dish with a thin (~1 mm) film of liquefied 2% agarose in culture medium. Allow the agarose to harden.

    ii. Briefly flame a glass bead tool and lightly touch it to the hardened agarose surface. Make ~10 shallow depressions per dish.

    iii. Make several dishes in advance of microsurgery. Wrap the dishes in plastic film and store in the refrigerator so they do not dehydrate.

2. Prepare an explant dish as follows.

    i. Pipette about 0.5 mL of liquefied 2% agarose onto the bottom of each well of a 24-well tissue culture dish.

    ii. Once the agarose has hardened, briefly flame a glass bead tool and lightly touch it to the surface to create one shallow depression in each well (Fig. 1C).

Selecting and Labeling Embryos

If the experiment requires a single cultured blastomere or a group of blastomeres of the same type, lineage labeling (Steps 6 and 7) is not necessary.

3. To obtain fertilized eggs, set up a natural mating or an in vitro fertilization (see Sive et al. 2007b and Moody 2012a).

4. Remove the jelly coats from fertilized eggs (see Sive et al. 2007a and Moody 2012a).

5. Collect two-cell embryos whose body axes (animal–vegetal; left–right; dorsal–ventral) can be accurately identified (as described in Protocol 1: Lineage Tracing and Fate Mapping in *Xenopus* Embryos [Moody 2018a]; Moody 2012a, 2012b; Grant et al. 2013). Place the embryos in sterile 1× culture medium and incubate at 14°C–20°C until they reach the desired stage.

    A lower temperature is preferred because it will slow down cleavage and provide more time to perform the dissections.

6. If a blastomere is to be recombined with other cells to test for interactions between blastomere types (Step 12), label one set of embryos with a lineage tracer as described in Protocol 1: Lineage Tracing and Fate Mapping in *Xenopus* Embryos (Moody 2018a) to be donors of one type of blastomere. Leave a second set of embryos unlabeled to be donors of the second type of blastomere.

7. If altering levels of gene expression is part of the experimental design, inject mRNAs or oligo-nucleotides (including lineage tracer) as described in Protocol 2: Microinjection of mRNAs and Oligonucleotides (Moody 2018b). Inject the blastomeres at least one cell cycle earlier than the desired explant stage. For example, microinject at the eight-cell stage if a 16-cell blastomere explant is desired.

Culturing Blastomere Explants

8. Add 1× culture medium to each well in the explant dish. If testing for a signaling pathway, add ligand or inhibitor molecule to the medium in the presence of a carrier protein (such as 1 mg/mL of BSA).

Ligand treatment that does not include a carrier protein typically shows significantly decreased ligand activity. Carrier protein also will prevent ligands from sticking to the plastic dish. For this purpose, an L-15 Leibovitz-based culture medium is commonly used.

Gentamicin can be added to 0.1% final concentration, but we generally do not need it.

9. Add diluted culture medium (0.5× SS, 0.1× MBS or 0.1× MMR) to a microsurgery dish to facilitate blastomere separation. Add an embryo to each depression.

10. Remove the vitelline membranes from the embryos, and then dissect each blastomere of interest free from the rest of the embryo by pulling away its neighbors (described in detail in Protocol 7: Cleavage Blastomere Deletion and Transplantation to Test Cell Fate Commitment in *Xenopus* [Moody 2018c]) (Fig. 2A).

11. Using a sterile glass Pasteur pipette, pick up the dissected blastomere and transfer to a well in the explant dish. Use a hair loop to gently push the blastomere into the shallow depression in the well (Fig. 2D). If culturing groups of cells, do not move the dish until the cells adhere (∼1 h).

    Blastomeres stick to plastic, so only use glass pipettes. The cells are very fragile, so avoid air bubbles and excessive pressure. Transferring the blastomere should remove most of the cellular debris, and a healthy transferred cell show not show any cytoplasmic leaking (Fig. 2D). After ∼30–60 min, a single blastomere should be a cluster of several adherent cells (Fig. 2E).

    More than one blastomere can be combined into a single explant (Fig. 2B).

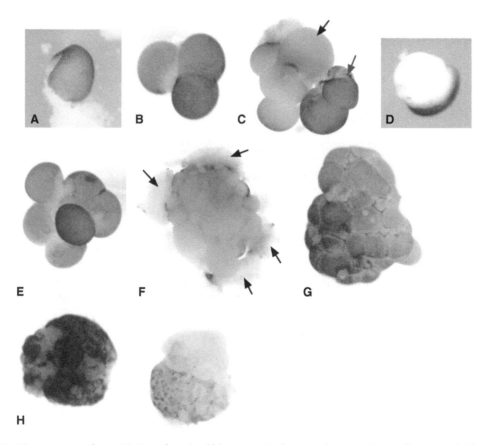

FIGURE 2. Blastomere explants. (*A*) A single animal blastomere before transfer to explant well, surrounded by cellular debris (white). (*B*) Two animal blastomeres (the right one has divided) in an explant well. (*C*) A vegetal explant (black arrow) co-cultured with an animal explant (red arrow). (*D*) Same blastomere as in *A*, with cellular debris removed upon transfer to the explant well. (*E*) An explant that has been in culture about 2 h and has divided to form a cluster of several cells. (*F*) An explant that has been in culture for several hours, surrounded by some cellular debris (arrows). (*G*) Same explant after removal of cellular debris. (*H*) Two explants after 24 h of culture.

Cite this protocol as *Cold Spring Harb Protoc*; doi:10.1101/pdb.prot097303

12. [Optional] To test for interactions between different types of blastomeres, explant a lineage-labeled blastomere (Step 7) into a depression. Then explant an unlabeled blastomere of the second type into the same depression. Push the cells together with a hair loop so they adhere (Fig. 2C).

13. After a few hours, use a sterile, glass Pasteur pipette with a rubber bulb to aspirate away any cellular debris in the dish (Fig. 2F,G).

14. Culture the explants overnight at 14°C–20°C next to a dish that contains sibling embryos; these will indicate the developmental stage of the blastomere explants.

> After 24 h, explants should form a small, sturdy ball of cells (Fig. 2H). If there is a lot of cellular debris, a hair loop can be used to move the debris to determine if there is a healthy explant buried within. For longer culture, explants survive best if transferred into fresh culture medium in a fresh agarose-coated culture well.

15. When siblings reach the desired developmental stage, harvest the explants by picking up each explant with a glass Pasteur pipette and gently expelling it into the appropriate fixative or lysis buffer.

> Although explants are small, they can be processed by standard whole mount or tissue section analysis of lineage label, gene or protein expression as described elsewhere (see Protocol 5: Whole-Mount Immunocytochemistry in Xenopus [Klymkowsky 2018], Protocol 4: Whole-Mount In Situ Hybridization of Xenopus Embryos [Saint-Jeannet 2017]; Moody 2012a, 2012b). They survive these assays as well as animal cap explants (see Protocol 8: Dissecting and Culturing Animal Cap Explants [Dingwell and Smith 2018]).

DISCUSSION

Cell fates are determined during embryogenesis by several mechanisms, including differential inheritance of mRNAs and proteins, local cell–cell interactions and gradients of a variety of signaling factors that induce specific tissue and organ fates. A classical approach that has been extremely successful in amphibian embryos is to remove a small region of the embryo and culture it either in isolation to determine whether it has already acquired a specific fate, or in the presence of other small groups of cells to identify the interactions that give rise to the germ layers and organ primordia. These approaches are presented in Protocol 8: Dissecting and Culturing Animal Cap Explants (Dingwell and Smith 2018) and elsewhere in this collection. These types of multicellular explants have provided enormous amounts of information about developmental mechanisms and the genes that regulate them.

The protocol described here extends this multicellular approach to the level of the single blastomere. The individual blastomeres of Xenopus 2- to 32-cell embryos have been fate mapped (Dale and Slack 1987a; Moody 1987a,b, 1989; Moody and Kline 1990). This work has identified the precursors of most of the embryonic cell types, tissues, and organs. However, these maps do not reveal the cell interactions or signaling pathways that are required for establishing the fates of single blastomeres. Herein an explant culture approach for testing whether a single blastomere's fate, as opposed to a group of cells, is determined by intrinsic versus extrinsic information is described. Cleavage blastomeres can be cultured in a simple salt medium for a few days without nutritional or growth factor supplements because they contain intracellular yolk platelets, which provide an intrinsic energy source. They are large enough to be manually separated, and the individual blastomeres have highly distinct and stereotypic fates (Dale and Slack 1987a; Moody 1987a,b, 1989; Moody and Kline 1990). Signaling factors can be added to the medium to determine which cell-to-cell communication pathway might influence the fate of the explanted single blastomere (e.g., Hainski and Moody 1996; Pandur et al. 2002; Gaur et al. 2016). Explants of single blastomeres from different regions of the embryo have revealed the relative roles of intrinsic versus signaling information during early developmental events. For example, Godsave and Slack (1991) showed that mesoderm specification of 32-cell blastomeres relies on both signaling and intrinsic information. Other studies showed that dorsal-animal blastomeres use maternally inherited mRNAs to autonomously express neural and dorsal mesodermal genes (Gallagher et al. 1991; Hainski and Moody 1992, 1996; Gaur et al. 2016), and that different vegetal blastomeres induce different types of mesoderm when co-cultured with animal caps (Dale and Slack 1987b). In addition, one can perform gain- and loss-of-function analyses by microinjecting mRNA or DNA constructs to identify which molecules are required for an autonomously expressed fate (e.g., Grant et al. 2013; Gaur et al. 2016).

This protocol requires practice and some manual dexterity, so be prepared for a low success rate (e.g., 10% is a great day!). In our experience, animal and equatorial blastomeres are much easier to dissect and culture than vegetal blastomeres, which are very fragile. We also find that single blastomeres from embryos older than the 32-cell stage survive better when cultured in small groups. Despite these challenges, this approach is a powerful, simple way to distinguish between intrinsic and extrinsic mechanisms that regulate how a single embryonic cell develops into specific tissues. The specificity of testing cell fate mechanisms at the level of a single cell has been very informative (e.g., Dale and Slack 1987b; Godsave and Slack 1991; Sullivan et al. 1999).

RECIPES

High-Salt MBS

Reagent	Volume to add
CaCl$_2$ (0.1 M)	7 mL
MBS salts (10×)	100 mL
NaCl (5 M)	4 mL
H$_2$O	888 mL

L-15 Leibovitz Culture Medium

67% Leibovitz L-15 medium (Invitrogen)
1 mg/mL bovine serum albumin (BSA)
7 mM Tris-HCl (pH 7.5)
50 mg/mL gentamicin

Marc's Modified Ringer's (MMR) (1×)

0.1 M NaCl
2 mM KCl
1 mM MgSO$_4$
2 mM CaCl$_2$
5 mM HEPES (pH 7.8)
0.1 mM EDTA

Sterilize by autoclaving, and store at room temperature.
Common alternative formulations of MMR omit EDTA and are adjusted to pH 7.4.

MBS Salts (10×)

NaCl (880 mM)
KCl (10 mM)
MgSO$_4$ (10 mM)
HEPES (50 mM, pH 7.8)

Omit HEPES if MBS is to be used for oocyte maturation.

NaHCO$_3$ (25 mM)

Adjust pH to 7.8 with NaOH. Autoclave. Store at room temperature.

Cite this protocol as *Cold Spring Harb Protoc*; doi:10.1101/pdb.prot097303

Modified Barth's Saline (MBS)

CaCl$_2$ (0.1 M)

MBS salts (10×)

For a 1× solution of MBS, mix 100 mL of 10× MBS salts with 7 mL of 0.1 M CaCl$_2$. Adjust the volume up to 1 liter with H$_2$O. Store at room temperature.

Steinberg's Solution (100%)

Reagent	Amount to add (for 1 L)
NaCl	3.4 g
KCl	0.05 g
Ca(NO$_3$)$_2$ · 4H$_2$O	0.08 g
MgSO$_4$ · 7H$_2$O	0.205 g
Tris	0.56 g
H$_2$O	to 1 L

Adjust pH to 7.4 with HCl.

ACKNOWLEDGMENTS

The development of this protocol was supported by National Institutes of Health (NIH) grant DE022065. S.A.M. gratefully acknowledges the excellent suggestions provided by the reviewers.

REFERENCES

Dale L, Slack JM. 1987a. Fate map of the 32-cell stage of *Xenopus laevis*. *Development* 99: 527–551.

Dale L, Slack JM. 1987b. Regional specification within the mesoderm of early embryos of *Xenopus laevis*. *Development* 100: 279–295.

Dingwell KS, Smith JC. 2018. Dissecting and culturing animal cap explants. *Cold Spring Harb Protoc* doi:10.1101/pdb.prot097329.

Gallagher BC, Hainski AM, Moody SA. 1991. Autonomous differentiation of dorsal axial structures from an animal cap cleavage stage blastomere in *Xenopus*. *Development* 112: 1103–1114.

Gaur S, Mandelbaum M, Herold M, Majumdar HD, Neilson KM, Maynard TM, Mood K, Daar IO, Moody SA. 2016. Neural transcription factors bias cleavage stage blastomeres to give rise to neural ectoderm. *Genesis* 54: 334–349.

Godsave SF, Slack JM. 1991. Single cell analysis of mesoderm formation in the *Xenopus* embryo. *Development* 111: 523–530.

Grant PA, Herold MB, Moody SA. 2013. Blastomere explants to test for cell fate commitment during embryonic development. *J Visual Exp* 71: e4458.

Hainski AM, Moody SA. 1992. *Xenopus* maternal RNAs from a dorsal animal blastomere induce a secondary axis in host embryos. *Development* 116: 347–355.

Hainski AM, Moody SA. 1996. An activin-like signal activates a dorsal-specifying RNA between the 8- and 16-cell stages of *Xenopus*. *Dev Genet* 19: 210–221.

Klymkowsky MW. 2018. Whole-mount immunocytochemistry in *Xenopus*. *Cold Spring Harb Protoc* doi:10.1101/pdb.prot097295.

Moody SA. 1987a. Fates of the blastomeres of the 16-cell stage *Xenopus* embryo. *Dev Biol* 119: 560–578.

Moody SA. 1987b. Fates of the blastomeres of the 32-cell stage *Xenopus* embryo. *Dev Biol* 122: 300–319.

Moody SA. 1989. Quantitative lineage analysis of the origin of frog primary motor and sensory neurons from cleavage stage blastomeres. *J Neurosci* 9: 2919–2930.

Moody SA. 2012a. Targeted microinjection of synthetic mRNAs to alter retina gene expression in *Xenopus* embryos. *Methods Mol Biol* 884: 91–111.

Moody SA. 2012b. Testing retina fate commitment in *Xenopus* by blastomere deletion, transplantation and explant culture. *Methods Mol Biol* 884: 115–127.

Moody SA. 2018a. Lineage tracing and fate mapping in *Xenopus* embryos. *Cold Spring Harb Protoc* doi:10.1101/pdb.prot097253.

Moody SA. 2018b. Microinjection with mRNAs and oligonucleotides. *Cold Spring Harb Protoc* doi:10.1101/pdb.prot097261.

Moody SA. 2018c. Cleavage blastomere deletion and transplantation in *Xenopus*. *Cold Spring Harb Protoc* doi:10.1101/pdb.prot097311.

Moody SA, Kline MJ. 1990. Segregation of fate during cleavage of frog (*Xenopus laevis*) blastomeres. *Anat Embryol* 182: 347–362.

Pandur PD, Sullivan SA, Moody SA. 2002. Multiple maternal influences on dorsal-ventral fate in *Xenopus* animal blastomeres. *Dev Dyn* 225: 581–587.

Saint-Jeannet JP. 2017. Whole-mount in situ hybridization of *Xenopus* embryos. *Cold Spring Harb Protoc* doi:10.1101/pdb.prot097287.

Sive HL, Grainger RM, Harland RM. 2007a. Dejellying *Xenopus laevis* embryos. *Cold Spring Harb Protoc* doi:10.1101/pdb.prot4731.

Sive HL, Grainger RM, Harland RM. 2007b. *Xenopus laevis* in vitro fertilization and natural mating methods. *Cold Spring Harb Protoc* doi:10.1101/pdb.prot4737.

Sullivan SA, Moore KB, Moody SA. 1999. Early events in blastomere fate determination. In *Cell lineage and cell fate determination* (ed. Moody SA), pp. 297–321. Academic Press, San Diego.

Cleavage Blastomere Deletion and Transplantation to Test Cell Fate Commitment in *Xenopus*

Sally A. Moody

Department of Anatomy and Regenerative Biology, George Washington University, Washington, D.C. 20037

Fate maps identify the precursors of an organ, and tracing the members of a blastomere lineage over time shows how its descendants come to populate that organ. The fates of the individual blastomeres of the two- to 32-cell *Xenopus* embryo have been fully mapped to reveal which cells are the major contributors to various cell types, tissues, and organs. However, because these fate maps were produced in the normal embryo, they do not reveal whether a precursor blastomere is competent to give rise to additional tissues or is already committed to its fate-mapped repertoire of descendants. To identify the mechanisms by which a cell's fate is committed, one needs to expose the cell to different experimental environments. If the cell's fate is determined, it will express its normal fate or gene expression profile in novel environments, whereas if it is not yet determined it will express different fates or gene expression profiles when exposed to novel external factors or neighboring cells. This protocol describes two techniques for testing cell fate commitment: single cell deletion and single cell transplantation. Deleting a blastomere allows one to test whether the deleted cell is required for the remaining cells to produce their normal, specific cell fates. Transplanting a blastomere to a novel location in a host embryo allows one to test whether the transplanted cell is committed to produce its normal fate-mapped repertoire, or whether it is still competent to respond to novel cell–cell interactions.

MATERIALS

It is essential that you consult the appropriate Material Safety Data Sheets and your institution's Environmental Health and Safety Office for proper handling of equipment and hazardous materials used in this protocol.

RECIPES: Please see the end of this protocol for recipes indicated by <R>. Additional recipes can be found online at http://cshprotocols.cshlp.org/site/recipes.

Reagents

Agarose, 2% in 1× culture medium

> *Add 2 g of electrophoresis-grade agarose to 100 mL of 1× culture medium in a screw-cap glass bottle. Autoclave to dissolve agarose. This can be stored in the refrigerator, and microwaved to liquefy the agarose when it is next needed.*

Culture medium (Steinberg's solution [100%] <R> [1× SS], Marc's modified Ringer's [MMR] [1×] <R>, or modified Barth's saline [MBS] <R> [1×])

Dejellying reagents (see Sive et al. 2007a and Moody 2012a)

Correspondence: samoody@gwu.edu

Ethanol (70%)

Gentamicin (50 mg/mL) (optional; see Step 9)

Lineage tracer (see Protocol 1: Lineage Tracing and Fate Mapping in *Xenopus* Embryos [Moody 2018a])

Reagents for natural mating or in vitro fertilization in *Xenopus* (see Sive et al. 2007b and Moody 2012a)

> *Because obtaining eggs by squeezing the female sometimes reduces egg quality, an alternate method for in vitro fertilization is to have the hormone-injected female naturally release eggs into high-salt MBS <R>. Collect and maintain eggs in this buffer for several hours. Wash twice with 0.1× MBS before adding sperm.*

Equipment

Forceps (e.g., Dumont #5 biologie)

> *Sharpen forceps to a very fine tip by gently stroking them across a piece of Alumina abrasive film (Thomas Scientific) under a stereomicroscope. We use medium film (3-μm; 6775E-46) for sharpening and fine film (0.3-μm; 6775E-54) for a final polish. Two forceps are required, but it is prudent to sharpen extra forceps in case they are accidentally damaged during the experiment. Forceps that are accidentally bent can be repaired using coarse film (12-μm; 6775E-38).*

Glass bead tool

> *Use a Bunsen burner (Fig. 1B) to melt the tip of an autoclaved 6-inch Pasteur pipette into a smooth 2-mm ball (Fig. 1C).*

Hair loop

> *Place both ends of a fine hair (~6 inch long, collected near the nape of the neck from a family member or laboratory partner; baby hair works well but tends to be too short to manipulate easily) into the narrow tip of a 6-inch Pasteur pipette to form a 2- to 3-mm loop. Secure in place with nail polish or cyanoacrylate glue (Fig. 1A). Sterilize before use by dipping in 70% ethanol and air-drying. Make several in advance in case of breakage during the experiment. As an alternative, purchase a superfine eyelash hair attached to a handle (Ted Pella #113) or an ultrafine single deer hair attached to a handle (Ted Pella #119).*

Incubator at 14°C–20°C

Microinjection equipment (see Protocol 2: Microinjection with mRNAs and Oligonucleotides [Moody 2018b]; Moody 2012a)

Pasteur pipettes, glass (6-inch, autoclaved)

Petri dishes (60-mm)

Stereomicroscope equipped with a fiber optic or LED lamp

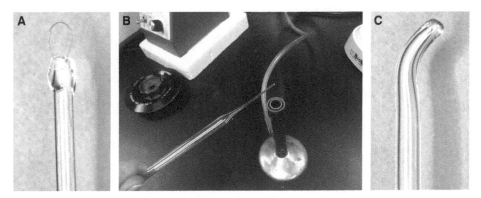

FIGURE 1. Microsurgery tools. (*A*) Hair loop. (*B*) Flaming a Pasteur pipette with a Bunsen burner to melt its tip. (*C*) Melted tip forming a glass bead tool.

METHOD

Preparing for Microsurgery

Embryos released from their vitelline membrane will stick to clay and plastic; therefore, Petri dishes to be used for microsurgery are coated with agarose.

1. Prepare microsurgery dishes as follows.

    i. Coat a 60-mm Petri dish with a thin (~1-mm) film of liquefied 2% agarose in culture medium. Allow the agarose to harden.

    ii. Briefly flame a glass bead tool and lightly touch the heated ball to the hardened agarose surface to create ~10 shallow depressions arranged in a circle.

    iii. Make several dishes in advance of microsurgery. Wrap the dishes in plastic film and store in the refrigerator so they do not dehydrate.

Selecting and Labeling Embryos

2. To obtain fertilized eggs, set up a natural mating or an in vitro fertilization (see Sive et al. 2007b and Moody 2012a).

3. Remove the jelly coats from fertilized eggs (see Sive et al. 2007a and Moody 2012a).

4. Collect two-cell embryos whose body axes (animal–vegetal; left–right; dorsal–ventral) can be accurately identified (as described in Protocol 1: Lineage Tracing and Fate Mapping in *Xenopus* Embryos [Moody 2018a]; Moody 2012a,b; Grant et al. 2013). Place the embryos in sterile 1× culture medium and incubate at 14°C–20°C until they reach the desired stage.

    A lower temperature is preferred because it will slow down cleavage and provide more time to inject embryos and perform the microsurgeries.

Performing Blastomere Deletion

5. For single-cell deletions, identify the cell you wish to delete (Fig. 2A). Inject one of its neighbors with a lineage tracer (see Protocol 1: Lineage Tracing and Fate Mapping in *Xenopus* Embryos [Moody 2018a], Protocol 2: Microinjection of mRNAs and Oligonucleotides [Moody 2018b]; Moody 2012a).

6. After microinjection, transfer the labeled embryo to a microsurgery dish filled with diluted culture solution (0.5× SS, 0.1× MMR, or 0.1× MBS). Position the embryo in a depression in the microsurgery dish so that the transparent vitelline membrane is visible above the surface of the cells of the embryo (Fig. 2B).

    Diluted medium is used to help the cells pull apart.

7. Perform deletion as follows.

    i. Grasp the vitelline membrane with forceps over the cell you intend to delete, because you are likely to damage it (Fig. 2C). With a second forceps, grasp the membrane close to the first forceps, and gently pull in opposite directions.

    When the membrane has been removed, the embryo will flatten.

    ii. If you damaged the cell to be deleted (the "victim" cell) (Fig. 2C), remove the remaining cellular debris with forceps until a clean gap is observed (Fig. 2D).

    iii. If the victim is not damaged—good job! Use a hair loop to orient the embryo with the victim with facing up. Grab the middle of the victim with one forceps and hold down the

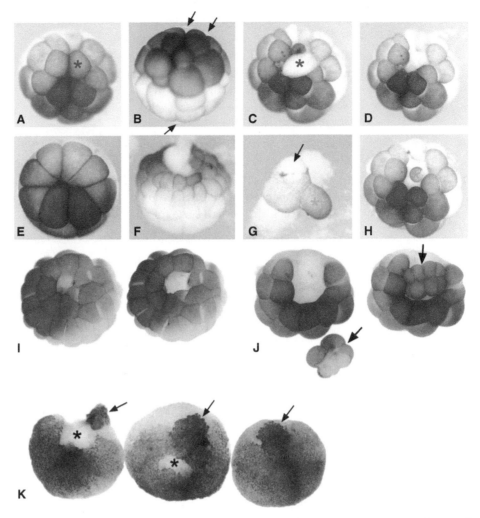

FIGURE 2. Steps in blastomere deletion and transplantation. (*A*) Identify the cell to be deleted (asterisk) and then microinject one of its neighbors. (*B*) Tilting an embryo reveals the clear vitelline membrane covering the cleavage furrows (arrows). (*C*) Embryo in A with the vitelline membrane removed is flatter. Asterisk indicates the cell to be deleted that was damaged when the vitelline membrane was removed. (*D*) Same embryo after dissecting out the cell, leaving a clean gap. (*E*) When transplanting a cell, either label the entire embryo or a precursor of the cell to be transplanted. In this case, the 16-cell precursor (green asterisk) of a 64-cell blastomere (shown in *F*) was labeled. (*F*) When the vitelline membrane was removed, a neighboring cell was damaged (white puff), but the labeled 64-cell blastomere (green asterisk) is intact. (*G*). The labeled 64-cell blastomere (green asterisk) and two neighbors were removed. One neighbor (arrow) was used as the "handle" cell. (*H*) The donor cell, removed from its handle, was placed in the gap in the embryo shown in D. Note the lack of cellular debris in the gap. (*I*) If the deletion gap has begun to heal while you are dissecting the donor cell (arrow, *left*), place forcep tines in the gap and stretch it open (*right*). (*J*) More than one blastomere can be transplanted. In this case, two 16-cell blastomeres were dissected without a handle, and they have since divided once into four cells (arrow on *left*). On right, they were floated into the gap in the host embryo with a hair loop and have divided to eight cells. (*K*) Three embryos a few hours after receiving transplants (arrows). (*Left*) the transplant (arrow) has fallen out of the gap in the host (asterisk). (*Middle*) the transplant (arrow) has healed, but there is still an unhealed gap in the host (asterisk). (*Right*) the transplant (arrow) has completed integrated into the host.

neighboring cells with the other forceps so they do not pull free with the victim. Gently pull the victim away from the embryo.

> *For best results, operate toward the end of the cell cycle (i.e., when you can start to see the beginning of the next cleavage furrow) because cytoplasmic bridges will have closed and the cells are less adhesive. In some batches of eggs, the victim will lift out of the embryo cleanly with a single tug.*
>
> *See Troubleshooting.*

iv. Use a sterile, glass Pasteur pipette with a rubber bulb to aspirate away any cellular debris in the dish.

8. Repeat the deletions on embryos located in the remaining depressions in the dish (about 10 per 60-mm dish).

9. Once all the embryos in the dish have healed (~30 min), use a sterile, glass Pasteur pipette with a rubber bulb to aspirate away any remaining debris plus about half of the culture solution around each embryo. Add fresh culture solution.

> *A fresh, sterile glass Pasteur pipette should be used for each embryo. The hair loop can be sterilized between embryos by dipping in 70% ethanol and air drying.*

> *Gentamicin (50 mg/mL) can be added to the culture solution at a 1:1000 dilution for a final concentration of 50 µg/mL.*

10. Culture the deleted embryos at 14°C–18°C next to a dish containing unoperated, sibling embryos. Transfer both the deleted embryos and the controls to fresh culture medium each day until they reach the desired stage for analysis.

> *The control (sibling) embryos will indicate whether the deletion causes developmental delays.*

> *Deleted and control embryos can be fixed and processed by standard whole mount or tissue section analysis of lineage label, gene, or protein expression as described elsewhere (see Protocol 5: Whole-Mount Immunocytochemistry in Xenopus [Klymkowsky 2018], Protocol 4: Whole-Mount In Situ Hybridization of Xenopus Embryos [Saint-Jeannet 2017]; Moody 2012a,b).*

Transplanting Blastomeres

11. To produce donor embryos for blastomere transplantation, either label an entire embryo by injecting a lineage tracer into both blastomeres of the 2-cell embryo, or inject a lineage tracer into the precursor blastomere of the cell to be transplanted (Fig. 2E) (see Protocol 1: Lineage Tracing and Fate Mapping in *Xenopus* Embryos [Moody 2018a], Protocol 2: Microinjection of mRNAs and Oligonucleotides [Moody 2018b]; Moody 2012a).

12. Transfer uninjected host embryos at the same developmental stage as the donor embryos to the depressions melted into the microsurgery dish. Transfer an equal number of labeled, donor embryos to the center of the circle of hosts.

13. Delete the cell in the host embryo that occupies the position into which you wish to transplant a cell as described in Step 7.

14. Perform transplantation as follows:

    i. Remove the vitelline membrane from one donor embryo. In this case, grasp the vitelline membrane distant from the cell to be transplanted so that you do not damage the donor cell (Fig. 2F).

    ii. Grasp a cell that neighbors the donor cell with forceps, and use it as a "handle"; avoid touching the cell to be transplanted (Fig. 2G). With the other forceps, gently pull the remaining neighboring cells away from the donor.
    > *The handle cell is likely to leak during this maneuver, which is fine.*

    iii. Once the donor cell and the remains of the handle cell are isolated (Fig. 2G), quickly check that the deletion site in the host embryo has not healed. If the deletion site is too small to accept the donor cell (Fig. 2I, left), place the tips of a closed forceps in the site and gently open them to stretch open the gap (Fig. 2I, right). Grab the handle cell with forceps and move the cells over the deletion site in the host. (A sterile hair loop also can be used to float the donor cell over to the host.) Dissect away any handle cell remnants with forceps so only the donor cell remains (Fig. 2H).
    > *More than one cell can be transplanted using this procedure (Fig. 2J).*

    iv. Use a sterile Pasteur pipette with a rubber bulb to aspirate away any cellular debris and the remains of the donor embryo from the microsurgery dish.

15. Repeat the transplantation procedure on the remaining embryos in the dish (about 10 per 60-mm dish).

16. Periodically check that the transplanted cell has not fallen out of the host and has begun to divide (Fig. 2J).

    After ~60 min, the transplant should be healed in place; after several hours, a fully integrated clone should be apparent in the host (Fig. 2K).

17. Culture hosts and unoperated siblings as described in Step 10.

    Hosts and control embryos can be fixed and processed by standard whole mount or tissue section analysis of lineage label, gene or protein expression as described elsewhere (see Protocol 5: Whole-Mount Immuno-cytochemistry in Xenopus [Klymkowsky 2018], Protocol 4: Whole-Mount In Situ Hybridization of Xenopus Embryos [Saint-Jeannet 2017]; Moody 2012a,b).

TROUBLESHOOTING

Problem (Step 7): Cells fall apart during blastomere deletion.
Solution: If cells fall apart without tugging on them, the cation concentration of the medium is too low: Add 1× culture medium to the dish during blastomere deletion.

Problem (Step 7): Cells are impossible to separate during blastomere deletion.
Solution: Lower the cation concentration of the medium. Do not use calcium-magnesium free medium because the embryos will dissociate.

DISCUSSION

Several mechanisms contribute to cell fate commitment. Intrinsic factors, such as differentially inherited cytoplasmic mRNAs or proteins, determine which cells will contribute to the germ line and the endoderm, and establish the dorsal–ventral body axis (for review, see Sullivan et al. [1999]). Extrinsic factors include short-range signals from neighboring cells, such as Notch-Delta, that specify cell fate, and long-range signals that can induce specific tissues and/or pattern an organ system, such as anti-BMP and anti-Wnt factors secreted from the organizer/node. Although fate maps identify the types of tissues and organs to which a blastomere normally gives rise, they do not reveal whether intrinsic or extrinsic factors are responsible, or the full developmental potential of the blastomere. To gather this information, the blastomere needs to develop in isolation (see Protocol 6: Cleavage Blastomere Explant Culture in *Xenopus* [Moody 2018c]) or in a novel environment as described here. Deleting and transplanting blastomeres provides an ideal approach to accomplish this.

Tests of blastomere fate choices are feasible in *Xenopus* because the body axes can be reproducibly identified, cleavage patterns are regular enough to create reproducible fate maps (Dale and Slack 1987a; Moody 1987a,b, 1989; Moody and Kline 1990), and the cells are large enough to be manually separated. For example, numerous studies have deleted single cells from embryos and transplanted cells to novel locations in host embryos to test for fate changes (Heasman et al. 1984; Dale and Slack 1987b). This has proven particularly useful for testing retinal fate commitment in the retinal lineages (for reviews, see Moore and Moody 1999; Yan and Moody 2007; Moody 2012a,b). An additional and powerful approach is to microinject the blastomere with mRNA prior to transplantation to express a mutated or dominant-negative gene construct, or with anti-sense oligonucleotides to prevent the translation of endogenous mRNAs. These gain- and loss-of-function analyses can identify the molecules that are required for the identified interaction. Keep in mind that these procedures require practice, manual dexterity and a bit of luck, so be prepared to perform several microsurgeries to obtain a sufficient number of successful cases (Fig. 2K).

RELATED INFORMATION

Explanting a single blastomere to define its full potential in isolation is described in Protocol 6: Cleavage Blastomere Explant Culture in *Xenopus* (Moody 2018c).

RECIPES

High-Salt MBS

Reagent	Volume to add
CaCl$_2$ (0.1 M)	7 mL
MBS salts (10×)	100 mL
NaCl (5 M)	4 mL
H$_2$O	888 mL

Marc's Modified Ringer's (MMR) (1×)

0.1 M NaCl
2 mM KCl
1 mM MgSO$_4$
2 mM CaCl$_2$
5 mM HEPES (pH 7.8)
0.1 mM EDTA

Sterilize by autoclaving, and store at room temperature.
Common alternative formulations of MMR omit EDTA and are adjusted to pH 7.4.

MBS Salts (10×)

NaCl (880 mM)
KCl (10 mM)
MgSO$_4$ (10 mM)
HEPES (50 mM, pH 7.8)
   Omit HEPES if MBS is to be used for oocyte maturation.

NaHCO$_3$ (25 mM)

Adjust pH to 7.8 with NaOH. Autoclave. Store at room temperature.

Modified Barth's Saline (MBS)

CaCl$_2$ (0.1 M)
MBS salts (10×)

For a 1× solution of MBS, mix 100 mL of 10× MBS salts with 7 mL of 0.1 M CaCl$_2$. Adjust the volume up to 1 liter with H$_2$O. Store at room temperature.

Cite this protocol as *Cold Spring Harb Protoc*; doi:10.1101/pdb.prot097311

Steinberg's Solution (100%)

Reagent	Amount to add (for 1 L)
NaCl	3.4 g
KCl	0.05 g
$Ca(NO_3)_2 \cdot 4H_2O$	0.08 g
$MgSO_4 \cdot 7H_2O$	0.205 g
Tris	0.56 g
H_2O	to 1 L

Adjust pH to 7.4 with HCl.

ACKNOWLEDGMENTS

The development of this protocol was supported by National Institutes of Health (NIH) grant DE022065. S.A.M. gratefully acknowledges the excellent suggestions provided by the reviewers.

REFERENCES

Dale L, Slack JM. 1987a. Fate map of the 32-cell stage of *Xenopus laevis*. *Development* 99: 527–551.

Dale L, Slack JM. 1987b. Regional specification within the mesoderm of early embryos of *Xenopus laevis*. *Development* 100: 279–295.

Grant PA, Herold MB, Moody SA. 2013. Blastomere explants to test for cell fate commitment during embryonic development. *J Visual Exp* 71: e4458.

Heasman J, Wylie CC, Hausen P, Smith JC. 1984. Fates and states of determination of single vegetal pole blastomeres of *X. laevis*. *Cell* 37: 185–194.

Klymkowsky MW. 2018. Whole-mount immunocytochemistry in *Xenopus*. *Cold Spring Harb Protoc* doi:10.1101/pdb.prot097295.

Moody SA. 1987a. Fates of the blastomeres of the 16-cell stage *Xenopus* embryo. *Dev Biol* 119: 560–578.

Moody SA. 1987b. Fates of the blastomeres of the 32-cell stage *Xenopus* embryo. *Dev Biol* 122: 300–319.

Moody SA. 1989. Quantitative lineage analysis of the origin of frog primary motor and sensory neurons from cleavage stage blastomeres. *J Neuroscience* 9: 2919–2930.

Moody SA. 2012a. Targeted microinjection of synthetic mRNAs to alter retina gene expression in *Xenopus* embryos. *Methods Mol Biol* 884: 91–111.

Moody SA. 2012b. Testing retina fate commitment in *Xenopus* by blastomere deletion, transplantation and explant culture. *Methods Mol Biol* 884: 115–127.

Moody SA. 2018a. Lineage tracing and fate mapping in *Xenopus* embryos. *Cold Spring Harb Protoc* doi:10.1101/pdb.prot097253.

Moody SA. 2018b. Microinjection of mRNAs and oligonucleotides. *Cold Spring Harb Protoc* doi:10.1101/pdb.prot097261.

Moody SA. 2018c. Cleavage blastomere explant culture in *Xenopus*. *Cold Spring Harb Protoc* doi:10.1101/pdb.prot097303.

Moody SA, Kline MJ. 1990. Segregation of fate during cleavage of frog (*Xenopus laevis*) blastomeres. *Anat Embryol* 182: 347–362.

Moore KB, Moody SA. 1999. Animal-vegetal asymmetries influence the earliest steps in retinal fate commitment in *Xenopus*. *Dev Biol* 212: 25–41.

Saint-Jeannet JP. 2017. Whole-mount in situ hybridization of *Xenopus* embryos. *Cold Spring Harb Protoc* doi:10.1101/pdb.prot097287.

Sive HL, Grainger RM, Harland RM. 2007a. Dejellying *Xenopus laevis* embryos. *Cold Spring Harb Protoc* doi:10.1101/pdb.prot4731.

Sive HL, Grainger RM, Harland RM. 2007b. *Xenopus laevis* in vitro fertilization and natural mating methods. *Cold Spring Harb Protoc* doi:10.1101/pdb.prot4737.

Sullivan SA, Moore KB, Moody SA. 1999. Early events in blastomere fate determination. In *Cell lineage and cell fate determination* (ed. Moody SA), pp. 297–321. Academic Press, San Diego.

Yan B, Moody SA. 2007. The competence of *Xenopus* blastomeres to produce neural and retinal progeny is repressed by two endo-mesoderm promoting pathways. *Dev Biol* 305: 103–119.

Dissecting and Culturing Animal Cap Explants

Kevin S. Dingwell[1,2] and James C. Smith[1,2]

[1]*The Francis Crick Institute, Developmental Biology Laboratory, London NW1 1AT, United Kingdom*

The animal cap explant is a simple but adaptable tool available to developmental biologists. The use of animal cap explants in demonstrating the presence of mesoderm-inducing activity in the *Xenopus* embryo vegetal pole is one of many elegant examples of their worth. Animal caps respond to a range of growth factors (e.g., Wnts, FGF, TGF-β), making them especially useful for studying signal transduction pathways and gene regulatory networks. Explants are also suitable for examining cell behavior and have provided key insights into the molecular mechanisms controlling vertebrate morphogenesis. In this protocol, we outline two methods to isolate animal cap explants from *Xenopus laevis*, both of which can be applied easily to *Xenopus tropicalis*. The first method is a standard manual method that can be used in any laboratory equipped with a standard dissecting microscope. For labs planning on dissecting large numbers of explants on a regular basis, a second, high throughput method is described that uses a specialized microcautery surgical instrument.

MATERIALS

It is essential that you consult the appropriate Material Safety Data Sheets and your institution's Environmental Health and Safety Office for proper handling of equipment and hazardous materials used in this protocol.

RECIPES: Please see the end of this protocol for recipes indicated by <R>. Additional recipes can be found online at http://cshprotocols.cshlp.org/site/recipes.

Reagents

Agarose
Bovine serum albumin (BSA) (Sigma-Aldrich)
Calcium magnesium-free medium (CMFM; 10×) <R>
> Use this solution at 1×.

Danilchick's for Amy (DFA) <R>
> Filter-sterilize. Store in aliquots of 50 mL at −20°C.

Gentamycin
High-pressure vacuum grease (Dow)
Marc's modified Ringer's (MMR) <R>
> Use this solution at either 0.1× or 0.7× supplemented with 50 µg/mL gentamycin.

Recombinant human E-cadherin protein (R&D Systems 8505-EC-050)

[2]Correspondence: kevin.dingwell@crick.ac.uk; jim.smith@crick.ac.uk

Reusable adhesive putty
Xenopus laevis embryos (dejellied)

   See Sive et al. (2007a) for dejellying embryos and Sive et al. (2010) for embryo microinjection.

Equipment

Dissecting microscope with a 0.75×–10× objective
Dumont #5 forceps (Fine Science Tools)
Glass coverslips (#1.5)
Glass-bottom microwell dishes with No. 1.5 coverglass (35 mm; MatTek P35G-1.5-14-C)
Microcautery instrument with footswitch and pencil tip holder (MC-2010; Protech International Inc.)
Pasteur pipettes (glass)
Petri dishes (35–50 mm)
Wire tip cautery electrode, 1 mm loop (13 µm; 13-Y1; Protech International Inc.)

METHOD

We routinely use two methods for harvesting animal cap explants. The first is a manual method using fine forceps to dissect the animal cap. The second is a high throughput method that uses a surgical cautery electrode loop to slice off the animal cap. Both methods start with dejellied Xenopus embryos. All dissections are conducted at room temperature (18°C–22°C).

1. Transfer embryos (stage 8) to a 1% agarose coated Petri dish filled with 0.7× MMR. Keep several embryos aside in 0.1× MMR in a standard Petri dish in order to stage the explants. Proceed to Step 2 for the manual method or Step 5 for the electrode method.

   The 1% agarose should be dissolved in the same culture medium that is used to culture the caps (i.e., 0.7× MMR).

Manual Forceps Method

2. Remove the vitelline envelope from 10–15 embryos using fine forceps (see Sive et al. 2007b). The vitelline envelope is a translucent membrane and is difficult to see so it is best to grasp the envelope with one set of forceps from the vegetal part of the embryo and then pull it away with the other pair. This limits any damage to the animal cap.

3. Dissect an animal cap starting at a region halfway between the animal pole and equator (Fig. 1). This will ensure that no marginal zone tissue (i.e., mesoderm) is acquired during the dissection that could inadvertently affect the differentiation and behavior of the animal cap cells. Using the forceps as "scissors," cut out a "square" animal cap ($\sim$0.5 mm^2) by a series of successive cuts to free the cap from the embryo. Alternatively, make two simultaneous cuts with both pairs of forceps on opposite sides of the cap, rotate the embryo 90°, and then make the second set of cuts in order to remove the explant.

4. Once all the caps have been excised, wash them by carefully drawing the caps into a glass Pasteur pipette and transferring them to a new 1% agarose coated dish filled with 0.7× MMR. It is important not to let the caps come in contact with the surface of the medium as they are extremely delicate and will be destroyed by the surface tension of the solution. Proceed to Step 9, 10, 11, or 12.

   When cultured in 0.7× MMR, the caps "heal" by rounding up into a ball (Fig 1E,F). Since it is the inner cells of the explant that respond to growth factors it is critical to transfer the caps quickly to the culture medium if performing growth factor experiments (Step 10). This ensures that the inner cells are sufficiently exposed to growth factor prior to the cap completely healing (Fig 1).

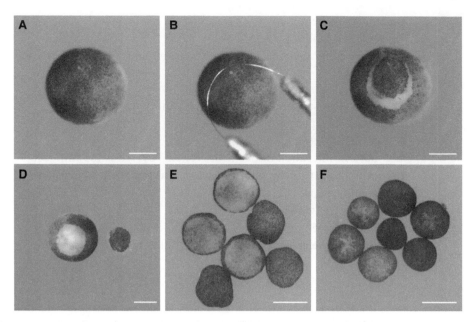

FIGURE 1. Dissecting *Xenopus laevis* animal caps. Animal cap from a stage 8.5 embryo (*A*) excised using a 13 µm wire tip cautery electrode with a 1 mm loop (*B,C*). Care must be taken to excise only the top half of animal pole (*C,D*) to ensure that mesoderm is not included with the ectodermal cap. When cultured in 0.7× MMR, caps (*E*) heal by rounding up into a ball (*F*). Scale bar: 0.5 mm.

Electrode Method

This is a high throughput method that enables one to cut hundreds of caps in a single session, and which does not require the removal of the vitelline envelope. If dissociating caps into single cells or live imaging when using this method, the vitelline envelope must be removed before cap cutting. If left on, it tends to stick to the dissociating cells, limiting the number of cells that can be collected.

5. Attach a 13-µm loop to the pencil tip that is attached to the main microcautery instrument. When not in use the pencil tip and loop can be held to the top of the instrument using a small amount of reusable adhesive putty.

    The loop is extremely delicate and great care should be taken to limit any contact with it.

6. Select power setting 2. A setting higher than 2 limits the life span of the loop.

7. Using a pair of forceps, grasp and hold an embryo submarginally (i.e., just below the equator of the embryo). This keeps the forceps out of the way of the electrode loop. With the loop immersed in the medium, press and hold the foot pedal while simultaneously and in one continuous motion drawing the loop across the top of the embryo to remove the cap (Fig 1). Release the foot pedal. If the cap sticks to the electrode, press the foot pedal until the caps fall off. Repeat for each embryo.

    Depending on the speed of cutting, embryonic tissue and/or the vitelline envelope may stick to the loop. It is important to remove this tissue on a regular basis (i.e., after 5 caps) or else the loop cutting efficacy will deteriorate. To remove the tissue, hold the loop in the medium and pulse by pressing and holding the foot pedal until all the tissue is dislodged. We usually pulse between 1 and 3 sec. In rare cases when tissue remains stubbornly associated with the loop even after several pulses, it is possible as a last resort to pulse the loop in the air. The "air" pulse should be done for less than half a second or else the tip will be ruined. Pulse again in the medium. Repeat several times if necessary.

8. It is possible to cut explants in batches of 20–30 or even higher depending on the operator's proficiency and their intended use (see note with Step 4). Wash the caps by carefully transferring the caps using a glass Pasteur pipette to a new 1% agarose coated dish filled with 0.7× MMR, and then transfer the caps to their final culture dishes (Step 9, 10, 11, or 12).

Cite this protocol as *Cold Spring Harb Protoc*; doi:10.1101/pdb.prot097329

Culturing Explants

9. For standard experiments, culture the explants on 1% agarose coated dishes at 18°C–21°C (in an incubator or on the benchtop) in 0.7× MMR supplemented with 50 µg/mL gentamycin to inhibit bacterial growth. Keep the caps well separated or they will heal together. Monitor the stage of development by comparing to the control embryos (Step 1).

10. For experiments with growth factors (Jones and Smith 2008), supplement the culture medium (0.7× MMR) with 0.1% BSA and culture at 18°C–21°C until the control embryos have reached the appropriate stage.

    The agarose layer should be made as thin as possible as it can act as a sink; too thick and the agarose will deplete the growth factor from the medium and lower its effective concentration.

11. For live imaging experiments, transfer the explant with a Pasteur pipette to a glass bottom microwell dish (these are 35 mm culture dishes that have a 1.5 coverslip on the bottom) filled with DFA. Hold the cap in place with a coverslip "chip" (see note below) secured with a small amount of high pressure vacuum grease. Special care must be taken when placing the glass chip over the explant so as not to damage the cap with excessive pressure.

    DFA (rather than MMR) is used here as its ionic concentration matches the blastocoel fluid of the embryo, and is ideal for maintaining normal cellular behavior and morphogenesis (Keller et al. 1985). Make the glass chips slightly larger than the cap by breaking up a standard coverslip.

12. To dissociate the animal caps into individual cells, transfer them to 1× CMFM in a 35 mm 1% agarose coated dish (agarose should be made with 1× CMFM). The inner unpigmented cells should dissociate from the outer pigmented epithelium within 10–20 min. Caps release calcium as they dissociate so limit the number of caps/dish (5–10 caps/35 mm dish). Pipetting the caps through a glass pipette can aid dissociation. Once the inner cells have dissociated, remove the pigmented epithelium and then swirl the medium to aggregate the cells to the center of the dish. Cells can then be either cultured as is, reaggregated in fresh 0.7× MMR, or cultured in 0.7× MMR on 3 µg/mL E-cadherin coated glass bottom microwell dishes for imaging.

    To coat the coverslip in the bottom of the microwell dish, treat overnight at 4°C with 3 µg/mL of recombinant human E-cadherin protein in 0.7× MMR, and then wash 1× in 0.7× MMR.

RECIPES

Calcium Magnesium-Free Medium (CMFM)

Reagent	Amount to add	Final concentration
5 M NaCl	17.6 mL	88 mM
1 M KCl	1 mL	1 mM
1 M NaHCO$_3$	2.4 mL	2.4 mM
1 M Tris-Cl (pH 7.6)	7.5 mL	7.5 mM
H$_2$O	971.5 mL	

Prepare CMFM from sterile stocks. Store it at room temperature.

Danilchik's for Amy (DFA) Medium

Reagent	Final concentration
NaCl	53 mM
Na$_2$CO$_3$	5 mM
Potassium gluconate	4.5 mM
Sodium gluconate	32 mM
CaCl$_2$	1 mM
MgSO$_4$	1 mM
Bovine serum albumin	1 g/L

Adjust pH to 8.3 with 1 M bicine before adding bovine serum albumin. (This recipe is adapted from Sater et al. 1993.)

Marc's Modified Ringer's (MMR) (10×, pH 7.4)

Reagent	Quantity (for 1 L)	Final concentration (10×)
NaCl	58.440 g	1 M
KCl	1.491 g	20 mM
MgSO$_4$	1.204 g	10 mM
CaCl$_2$, dihydrate	2.940 g	20 mM
HEPES	11.915 g	50 mM
H$_2$O	to 1 L	

Adjust the pH to 7.4 using 10 M NaOH. Sterilize by autoclaving. Store at room temperature indefinitely.

REFERENCES

Jones CM, Smith JC. 2008. Mesoderm induction assays. In *Molecular Embryology. Methods in molecular biology*™ (ed. Sharpe PT, Mason I), Vol. 461. Humana Press.

Keller RE, Danilchik M, Gimlich R, Shih J. 1985. The function and mechanism of convergent extension during gastrulation of *Xenopus laevis*. *J Embryol Exp Morphol* **89:** 185–209.

Sater AK, Steinhardt RA, Keller R. 1993. Induction of neuronal differentiation by planar signals in *Xenopus* embryos. *Dev Dyn* **197:** 268–280.

Sive HL, Grainger RM, Harland RM. 2007a. Dejellying *Xenopus laevis* embryos. *Cold Spring Harb Protoc* doi: 10.1101/pdb.prot4731.

Sive HL, Grainger RM, Harland RM. 2007b. Removing the vitelline membrane from *Xenopus laevis* embryos. *Cold Spring Harb Protoc* doi: 10.1101/pdb.prot4732.

Sive HL, Grainger RM, Harland RM. 2010. Microinjection of *Xenopus* embryos. *Cold Spring Harb Protoc* doi: 10.1101/pdb.ip81.

Cite this protocol as *Cold Spring Harb Protoc*; doi:10.1101/pdb.prot097329

SECTION 5: Embryology II

Xenopus Explants and Transplants

Sally A. Moody[1]

Department of Anatomy and Cell Biology, The George Washington University, School of Medicine and Health Sciences, Washington, D.C. 20037, USA

There is a long tradition of testing the developmental potential and competency of different regions of the embryo by transplanting tissue to novel locations or by growing isolated tissue in explant culture. The protocols introduced here describe several types of embryological manipulations in *Xenopus* that are particularly useful for analyzing tissue inductive signals, cell migration, and organogenesis. These techniques draw upon classical approaches that are quite easy to perform in both *Xenopus laevis* and *Xenopus tropicalis* embryos.

THE BEGINNINGS OF EXPERIMENTAL EMBRYOLOGY

Late in the nineteenth century, embryologists began to test the insights they had gained by describing the cytological and morphological changes that occur during development by experimentally manipulating embryos. Embryos were dissected into pieces that were grown in explant cultures to test whether their fate was already determined or these pieces were transplanted to unique locations in a host embryo to test the influence of the local environment. Wilhelm Roux is often credited as being the "father" of this new field of experimental embryology. He combined his observational embryological skills with a Newtonian physics point of view to found the discipline of "developmental mechanics"—the study of explaining the driving force of development. He advocated studying developmental processes by breaking them down into simple steps that could be analyzed with physical or chemical methods. He started a new scientific journal devoted to such studies: *Archiv für Entwicklungsmechanik der Organismen*, which published its first volume in 1894. From 1975 to 1996 the journal was published as *Wilhelm Roux' Archives of Developmental Biology*, and since 2000 it has been published as *Development Genes and Evolution*. Even though technology has advanced a great deal over the past century, modern developmental biology remains indebted to Roux and strives to understand developmental processes from a mechanistic perspective.

In 1888, Roux published one of his most famous experiments using amphibian embryos (reprinted in Willier and Oppenheimer 1964). To test whether embryonic cells differentiate according to an intrinsic program or are influenced by interactions with other cells, he destroyed one blastomere of a two-cell frog embryo. Amphibians were a favored system for this type of experiment because they develop external to the mother and are accessible, they are large enough to be easily manipulated with simple tools like hair loops and forceps, and they contain intracellular yolk for nutrition and thus can be maintained in a simple culture system. He published that the remaining cell developed into a half-embryo, supporting the idea that each cell contains only the information for making its part of the

[1]Correspondence: samoody@gwu.edu

embryo. However, the opposite results were found in the sea urchin embryo (Driesch 1891): When the two cells of a recently cleaved zygote were physically separated, they developed into two separate and fully formed larvae. Separation experiments using different methods were repeated by several authors using various amphibian species (for review, see Morgan 1924), including a frog in which the two cells were separating by gently tying them off with a strand of hair (McClendon 1910). These experiments confirmed that each cell could give rise to a fully formed embryo. Morgan (1924) argued that the damaged material left next to the surviving cell by Roux' technique interfered with the surviving cells' ability to reconstitute an entire embryo. Although Roux' experimental approach was flawed and his conclusion incorrect, this experiment spurred many other researchers to make similar "experimental embryology" manipulations to answer developmental questions, such as defining the importance of position, induction by cell contact and diffusible chemicals, and the role of cytoplasmic determinants.

WATERSHED DISCOVERIES

Several watershed discoveries were made by employing similar, simple dissections, explants, and transplantation. Early in the twentieth century, Spemann published a series of papers that manipulated the physical relationship between the presumptive lens ectoderm and the optic vesicle, discovering that one tissue influences the development of a nearby tissue—a process that we now call "induction" (for a review, see Saha 1991). In another instance, by transplanting the dorsal midline mesoderm to a different location in the embryo, Hilde Mangold and Hans Spemann discovered primary embryonic induction (Spemann and Mangold 1924). By recombining animal cap ectoderm with vegetal pole endoderm, Peter Nieuwkoop discovered the interaction required for mesoderm induction (Nieuwkoop 1969). The success of these experiments relied on being able to identify the transplanted tissue within the host tissue. This was particularly easy to accomplish by recombining pieces from different species of amphibians that have different patterns of pigmentation. For example, Ross Harrison used this technique extensively to show that nerve fibers arose from the nervous system (Harrison 1898a,b; Abercrombie 1961). Likewise, Twitty used interspecies neural crest grafts to study the origin of pigment cells (Twitty and Niu 1948).

IN RECENT TIMES

Since the 1980s, these classical approaches have been fruitfully applied, particularly in *Xenopus*, to identify the molecules involved in many different developmental interactions. However, the use of lineage tracing (see Sec. 4, Protocol 1: Lineage Tracing and Fate Mapping in *Xenopus* Embryos [Moody 2018a]) has made it unnecessary to use two different species for identifying host versus transplanted tissue. Most importantly, the use of microinjection techniques to alter endogenous gene expression (see Sec. 4, Protocol 2: Microinjection of mRNAs and Oligonucleotides [Moody 2018b] and Sec. 4, Protocol 3: Microinjection of DNA Constructs into *Xenopus* Embryos for Gene Misexpression and *cis*-Regulatory Module Analysis [Yasuoka and Taira 2019]) has allowed researchers to manipulate the signaling molecules, receptors, and transcription factors in host versus transplant tissues and in explant cultures. As genomic, transcriptomic, and proteomic technologies identify new molecules of unknown function, the combination of these new approaches with the classical embryological manipulations will provide great insights into the mechanics of development.

The accompanying protocols provide simple ways to isolate and recombine tissues to test gene function and identify new interactions. The described manipulations are tried and true by the *Xenopus* research community and are described in detail so that those not familiar with these approaches will be able to apply this powerful system to their own projects. These protocols are used extensively, even today, to elucidate the developmental mechanisms that regulate how the different cells and regions of

Cite this introduction as *Cold Spring Harb Protoc*; doi:10.1101/pdb.top097337

the embryo produce specific cell types, tissues, and organs and for analyzing the molecular bases of tissue inductive signals, cell migration, and organogenesis. Protocols for specific embryonic regions include the organizer, which induces the neural ectoderm and patterns the mesoderm (see Protocol 2: Einsteck Transplants [Cousin 2019a] and Protocol 1: Spemann–Mangold Grafts [Cousin 2019b]), the dorsal marginal zone precursors of the mesoderm (Protocol 3: Microsurgical Methods to Isolate and Culture the Early Gastrula Dorsal Marginal Zone [Davidson 2022a]), the mesoendoderm (see Protocol 4: Microsurgical Manipulations to Isolate Collectively Migrating Mesendoderm [Davidson 2022b] and Protocol 5: Microsurgical Methods to Make the Keller Sandwich Explant and the Dorsal Isolate [Davidson 2022c]), and the cranial neural crest, precursors of the cranial skeleton, and sensory organs (see Protocol 7: Cranial Neural Crest Transplants [Cousin 2018] and Protocol 6: Cranial Neural Crest Explants [Cousin and Alfandari 2018]). Protocols for studying organogenesis are particularly useful for analyzing interactions between stem cells and their environment (see Protocol 8: In Vitro Induction of *Xenopus* Embryonic Organs Using Animal Cap Cells [Ariizumi et al. 2017]). Finally, although *Xenopus laevis* has been the dominant species for these approaches, simple modifications to protocols make the diploid *Xenopus tropicalis* also amenable to explants and transplants (see Protocol 9: Special Considerations for Making Explants and Transplants with *Xenopus tropicalis* [Fisher and Grainger 2019]).

REFERENCES

Abercrombie M. 1961. Ross Granville Harrison, 1870–1959. *Biograph Mem Fellows R Soc* **7:** 110–126.

Ariizumi T, Michiue T, Asashima M. 2017. In vitro induction of *Xenopus* embryonic organs using animal cap cells. *Cold Spring Harb Protoc* doi:10.1101/pdb.prot097410

Cousin H. 2018. Cranial neural crest transplants. *Cold Spring Harb Protoc* doi:10.1101/pdb.prot097402

Cousin H. 2019a. Einsteck transplants. *Cold Spring Harb Protoc* doi:10.1101/pdb.prot097352

Cousin H. 2019b. Spemann–Mangold grafts. *Cold Spring Harb Protoc* doi:10.1101/pdb.prot097345

Cousin H, Alfandari D. 2018. Cranial neural crest explants. *Cold Spring Harb Protoc* doi:10.1101/pdb.prot097394

Davidson LA. 2022a. Microsurgical methods to isolate and culture the early gastrula dorsal marginal zone. *Cold Spring Harb Protoc* doi:10.1101/pdb.prot097360

Davidson LA. 2022b. Microsurgical manipulations to isolate collectively migrating mesendoderm. *Cold Spring Harb Protoc* doi:10.1101/pdb.prot097378

Davidson LA. 2022c. Microsurgical methods to make the Keller sandwich explant and the dorsal isolate. *Cold Spring Harb Protoc* doi:10.1101/pdb.prot097386

Driesch H. 1891. Entwicklungsmechanische Studien: I. Der Werthe der beiden ersten Furchungszellen in der Echinogdermenentwicklung. Experimentelle Erzeugung von Theil- und Doppelbildungen. II. Über die Beziehungen des Lichtez zur ersten Etappe der thierischen Formbildung. *Z Wiss Zool* **53:** 160–184. Translation: 1964. The potency of the first two cleavage cells in echinoderm development. Experimental production of partial and double formations, in *Foundations of experimental embryology* (eds. Willier BH, Oppenheimer JM), pp. 38–50. Hafner Press, New York.

Fisher M, Grainger RM. 2019. Special considerations for making explants and transplants with *Xenopus tropicalis*. *Cold Spring Harb Protoc* doi:10.1101/pdb.prot097428

Harrison R. 1898a. Grafting experiments on tadpoles, with special reference to the study of the growth and regeneration of the tail. *Science* **7:** 198–199.

Harrison R. 1898b. The growth and regeneration of the tail of the frog larva. Studied with the aid of Born's method of grafting. *Arch Entw Mech Org* **7:** 430–485.

McClendon JF. 1910. The development of isolated blastomeres of the frog's egg. *Amer J Anat* **10:** 425–430.

Moody SA. 2018a. Lineage tracing and fate mapping in *Xenopus* embryos. *Cold Spring Harb Protoc* doi:10.1101/pdb.prot097253

Moody SA. 2018b. Microinjection of mRNAs and oligonucleotides. *Cold Spring Harb Protoc* doi:10.1101/pdb.prot097261

Morgan TH. 1924. Two embryos from one egg. *Sci Monthly* **18:** 529–546.

Nieuwkoop PD. 1969. The formation of mesoderm in urodelean amphibians: I. induction by the endoderm. *Wilhelm Roux Arch Entwickl Mech Org* **162:** 341–373.

Saha M. 1991. Spemann seen through a lens. In *A conceptual history of modern embryology. Developmental biology (a comprehensive synthesis)* (ed. Gilbert SF), Vol. 7, pp. 91–108. Springer, Boston.

Spemann H, Mangold H. 1924. Induction of embryonic primordia by implantation of organizers from a different species. *Int J Dev Biol* **1964:** 146–184.

Twitty VC, Niu MC. 1948. Causal analysis of chromatophore migration. *J Exp Zool* **108:** 405–437.

Willier BH, Oppenheimer JM. 1964. *Foundations of experimental embryology*. Prentice Hall Inc, Englewood Cliffs, NJ.

Yasuoka Y, Taira M. 2019. Microinjection of DNA constructs into *Xenopus* embryos for gene misexpression and *cis*-regulatory module analysis. *Cold Spring Harb Protoc* doi:10.1101/pdb.prot097279

Spemann–Mangold Grafts

Hélène Cousin[1]

Department of Veterinary and Animal Sciences, University of Massachusetts, Amherst, Massachusetts 01003

In 1924, Hans Spemann and Hilde Mangold (née Pröscholdt) published their famous work describing the transplantation of dorsal blastopore lip of one newt gastrula embryo onto the ventral side of a host embryo at the same stage. They performed these grafts using two newt species with different pigmentation (*Triturus taeniatus* and *Triturus cristatus*) to follow the fate of the grafted tissue. These experiments resulted in the development of conjoined twins attached through their belly. Because of the difference in embryo pigmentation between the two *Triturus* species, they determined that the bulk of the secondary embryo arose from the host embryo while the grafted tissue per se gave increase to the notochord and a few somitic cells. This meant that the dorsal blastopore lip was able to organize an almost complete embryo out of ventral tissue. The dorsal blastopore lip is now called the Spemann–Mangold organizer. Here, we describe a simple yet efficient protocol to perform these grafts using the anuran *Xenopus laevis*.

MATERIALS

It is essential that you consult the appropriate Material Safety Data Sheets and your institution's Environmental Health and Safety Office for proper handling of equipment and hazardous materials used in this protocol.

RECIPES: Please see the end of this protocol for recipes indicated by <R>. Additional recipes can be found online at http://cshprotocols.cshlp.org/site/recipes.

Reagents

Ethanol (70%)

H_2O (reverse osmosis [RODI] or distilled [dH_2O])

Modified Barth's Saline (MBS) (1×) <R>

> *For grafting medium, use 1× MBS containing 50 µg/mL of gentamycin. For recovery medium, use 0.1× MBS containing 50 µg/mL of gentamycin.*

Xenopus laevis embryos (stage 10 or 10+, dejellied)

> *Dejellied embryos are staged according to Nieuwkoop and Faber (1967); see Step 3.*

Equipment

Bridges

> *Cut coverslips into rectangles (3-mm × 1-cm) using a diamond pen.*

Dissecting microscope equipped with a gooseneck lighting system

[1]Correspondence: hcousin@vasci.umass.edu

Eyelash knife

Select a human eyelash with the desired thickness and curvature (Fig. 1A) and thread it through a 23-gauge needle fitted on a 1-cc syringe. For safety purposes, the tip of the needle can be cutoff with scissors before the threading. Fix the eyelash with a drop of nail polish or cyanoacrylate glue.

Forceps (blunt, to manipulate bridges)
Forceps (fine, to remove vitelline envelope)
Glass bead tool

Thin out the end of a Pasteur pipette under a benzene burner and melt the end into a ball roughly the size of gastrula-stage embryo (about 2 mm).

Hair loop

Cut a human hair into 3-inch sections. Thread both ends of a section into a 23-gauge needle fitted on a 1-cc syringe. Push the loop into the needle until the desired stiffness is reached (Fig. 1A). Fix the hair with a drop of nail polish or cyanoacrylate glue.

Petri dishes (60-mm, plastic), coated with a 4-mm layer of 1% agarose
Petri dishes (60-mm, plastic), coated with plasticine

Roll 2 tsp of plasticine (nondrying, toxin free, appropriate for young children) into a ball. Flatten it out into a plastic Petri dish by hand.

Transfer pipette (disposable plastic or glass) with an opening of ≥2 mm, for transferring embryos

METHOD

The discovery of the Spemann–Mangold organizer (Spemann and Mangold 1924, 2001) earned Hans Spemann the Nobel Prize in Medicine in 1935. Spemann–Mangold grafts are still performed for research purposes (De Robertis 2006; Inui et al. 2012). They also are an incredible teaching experience for students discovering the fundamental principles of developmental biology.

The grafting procedure should be conducted between 15°C and 19°C.

1. Sterilize a plasticine-coated dish with 70% ethanol for 10 min. Rinse for 30 sec with RODI H_2O and then fill with grafting medium.

2. Using the glass bead tool, make two depressions in the plasticine as deep as one-third or one-half the diameter of an embryo (or about 1-mm).

3. Select a donor and a host gastrula between stage 10 (the forming blastopore lip appears as a dotted line with a light gray color) and 10+ (the blastopore lip is now a line of a dark gray color) and transfer the embryos to the plasticine-coated dish. Remove the vitelline envelope from each embryo using the fine forceps. Ensure that the embryos never breach the liquid surface once the vitelline envelope is removed.

    See Troubleshooting.

4. Move the donor and host embryos into the plasticine depressions and turn the embryos on their animal (pigmented) sides.

    This will ensure that the blastopore lip and dorsal marginal zone (the area between the lip and the pigmentation edge) are clearly visible (Fig. 1B).

5. On the donor embryo, insert the tip of the eyebrow knife at the commissure (corner) of the blastopore lip and thread it all the way to the animal pole. Cut the tissue by pressing the hair loop along the length of the eyelash knife. Repeat the process from the other commissure (Fig. 1B). Cut perpendicular to the first two cuts to free the pigmented edge of the explant (Fig. 1C) and peel the piece of tissue toward the vegetal pole (Fig. 1D).

    The peeled tissue is made of superficial pigmented ectoderm with underlying deep mesoderm (Fig. 1D, orange tissue) and some endoderm.

6. To free the explant from the embryo, insert the knife under the mesoderm and cut perpendicularly (Fig. 1D). Lay the explant with is superficial side down (Fig. 1E).

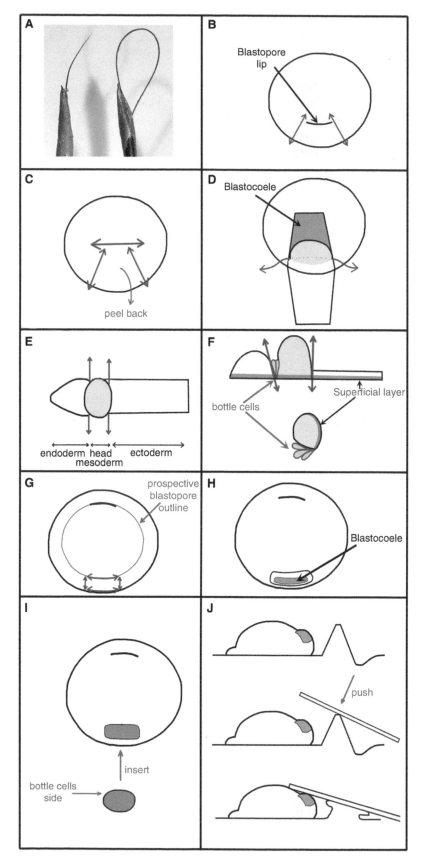

FIGURE 1. Key tools and steps for transplantation of the Spemann–Mangold organizer. (*A*) Eyelash knife (*left*) and hair loop (*right*). (*B–J*) Diagram representing the key steps of the grafting procedure.

Cite this protocol as *Cold Spring Harb Protoc*; doi:10.1101/pdb.prot097345

The freed explant now contains ectoderm (thin, long and pigmented tissue), head mesoderm (Fig. 1D–F, orange tissue) and the involuting bottle cells (Fig. 1F, purple) that created the blastopore lip. Some endoderm cells (much larger cells than the other tissues) may also be present.

7. Trim the explant so that it contains the head mesoderm and the bottle cells.

   i. To trim the ectoderm, slide the length of the eyebrow knife under the dorsal side of the head mesoderm mass (i.e., cleft of Brachet) and press firmly down until the tissue is separated.

   ii. To trim the mass of endoderm cells, insert the knife between the head mesoderm and endoderm and press firmly down until the tissue is separated (Fig. 1E,F).

8. On the host embryo, cut out a piece of tissue roughly the size of the newly trimmed explant from the ventral side of the embryo using the same cutting techniques described in Steps 5–6 (Fig. 1G).

   Ideally, the vegetal-most incision should coincide with the outline of the future blastopore that will form once the embryo reaches stage 10.5 or 11 (Fig. 1G, blue line). The blastocoel cavity should be seen on the equatorial side of the incision (Fig. 1H).

9. Graft the explant obtained in Step 7 into the cavity created in Step 8, ensuring that the bottle cells are aligned with the projected line of the blastopore and respecting the superficial/deep orientation of the explant (Fig. 1I).

   See Troubleshooting.

10. Using the hair loop, gently move the grafted embryo within the plasticine depression so that the grafted blastopore lip is located above the plasticine line (Fig. 1J, top). Using blunt forceps or a pipette tip, dig up some plasticine to form a fulcrum close to the graft (Fig. 1J, top).

11. Using blunt forceps, put a coverglass bridge on the fulcrum such that two-thirds of the bridge's length is behind the plasticine fulcrum and the other one-third covers the grafted area (Fig. 1J, middle).

12. Using blunt forceps, push gently down on the part of bridge that contacts the fulcrum. Stop pushing once the bridge touches the graft and slightly flattens it (Fig. 1J, bottom).

13. Allow the epidermal ectoderm to heal for 20 min.

    See Troubleshooting.

14. Remove the bridge with the blunt forceps and transfer the grafted embryo into an agarose-coated dish filled with recovery medium. Grow the grafted embryo between 15°C and 18°C.

    Secondary axes are visible as soon as the early neurula stage (the day after the graft).

    See Troubleshooting.

TROUBLESHOOTING

Problem (Step 3): The plasticine is sticky and wounds the embryo once the vitelline envelope is removed.

Solution: This problem will occur if the plasticine is new, but will disappear after a few uses. Until then, 1% BSA (w/v) can be added to the grafting medium to coat the surface of the plasticine. Alternatively, dishes coated with agarose can be used. When bridges need to be applied, a fulcrum made of silicon grease can be applied directly onto the bridge. It is best to fill a 5- to 10-mL syringe with the silicon grease and fit the syringe with a P200 micropipette tip for easier application.

Problem (Step 9): The embryos develop too fast and the experimenter does not have time to perform the grafts.

Solution: Keep the embryos, media and plasticine for at least a few hours at 15°C before grafting is scheduled. Perform the grafts on a cooling table set up at 15°C if available.

Problem (Step 13): The host embryo disaggregates shortly after the graft is completed.
Solution: The medium or the type of plasticine used could be the culprit. Make fresh media, add gentamicin and avoid using HCl to adjust the pH (as excess Cl^- ions may affect the healing of the explant). Make sure the plasticine used is nondrying and adequate for young children. Plasticine commonly found in art supplies stores contains chemicals that will kill the embryos.

Problem (Step 14): The secondary axis is incomplete (lacks the head).
Solution: Xenopus laevis develops faster than the urodele embryos used by Spemann and Mangold. Consequently, the window of time available to perform good quality Spemann–Mangold grafts in *Xenopus* is quite narrow (1 to 1.5 h, depending on the temperature). It is worthwhile to select embryos at stage 10 so that by the time the embryos are prepared for surgery, they are at stage 10+. Operating at the lower range of temperature (15°C) slows development and gives more time to the experimenter. Spemann–Mangold grafts can be performed at stage 10.5 or even later, but will result in incomplete twinning (a secondary embryo lacking the anterior-most structures such as the head).

RECIPE

Modified Barth's Saline (MBS) (1×)

$CaCl_2$	0.41 mM
$CaNO_3$	0.3 mM
HEPES-NaOH	15 mM
KCl	1 mM
$MgSO_4$	0.82 mM
NaCl	88 mM
$NaHCO_3$	2.4 mM

Adjust pH to 7.6. Store at room temperature for up to 1 mo.

ACKNOWLEDGMENTS

The author thanks Professor Ray Keller for his invaluable insight into *Xenopus* transplantation in general and Spemann–Mangold graft in particular. H.C. is supported by National Institutes of Health/National Institute of Dental and Craniofacial Research (NIH/NIDCR) DE 025691.

REFERENCES

De Robertis EM. 2006. Spemann's organizer and self-regulation in amphibian embryos. *Nat Rev Mol Cell Biol* 7: 296–302.
Inui M, Montagner M, Ben-Zvi D, Martello G, Soligo S, Manfrin A, Aragona M, Enzo E, Zacchigna L, Zanconato F, et al. 2012. Self-regulation of the head-inducing properties of the Spemann organizer. *Proc Natl Acad Sci* 109: 15354–15359.
Nieuwkoop PD, Faber J. 1967. *Normal table of* Xenopus laevis *(Daudin),* 2nd ed. North-Holland, Amsterdam.

Spemann H, Mangold H. 1924. Über Induktion von Embryonalanlagen durch Implantation artfremder Organisatoren. *Archiv für Mikroskoposhe Anatomie und Entwicklungsmechanik.* 599–638.
Spemann H, Mangold H. 2001. Induction of embryonic primordia by implantation of organizers from a different species. 1923. *Int J Dev Biol* 45: 13–38.

Einsteck Transplants

Hélène Cousin[1]

Department of Veterinary and Animal Sciences, University of Massachusetts, Amherst, Massachusetts 01003

Einsteck procedure refers to a method whereby the experimenter inserts material into the blastocoel cavity of an early amphibian embryo. This procedure is simpler to perform than other types of grafts, such as Spemann–Mangold, and with practice yields a sizable amount of data suitable for statistical analysis. This protocol for Einsteck transplantation in *Xenopus* describes the insertion of the gastrula-stage blastopore lip into the blastocoel cavity of a host embryo.

MATERIALS

It is essential that you consult the appropriate Material Safety Data Sheets and your institution's Environmental Health and Safety Office for proper handling of equipment and hazardous materials used in this protocol.

RECIPES: Please see the end of this protocol for recipes indicated by <R>. Additional recipes can be found online at http://cshprotocols.cshlp.org/site/recipes.

Reagents

Ethanol (70%)

H_2O (reverse osmosis [RODI] or distilled [dH_2O])

Modified Barth's saline (MBS) (1×) <R>

For grafting medium, use 1× MBS containing 50 µg/mL of gentamycin. For recovery medium, use 0.1× MBS containing 50 µg/mL of gentamycin.

Xenopus laevis embryos (stages 8, 9, 10, or 10+; see Step 3)

Embryos are staged according to Nieuwkoop and Faber (1967).

Equipment

Dissecting microscope equipped with a gooseneck lighting system

Eyelash knife

Select a human eyelash with the desired thickness and curvature (Fig. 1A) and thread it through a 23-gauge needle fitted on a 1-cc syringe. For safety purposes, the tip of the needle can be cut off with scissors before the threading. Fix the eyelash with a drop of nail polish or cyanoacrylate glue.

Forceps (fine, to remove vitelline envelope)

Glass bead tool

Thin out the end of a Pasteur pipette under a benzene burner and melt the end into a ball roughly the size of gastrula-stage embryo (about 2 mm).

---

[1]Correspondence: hcousin@vasci.umass.edu

Cite this protocol as *Cold Spring Harb Protoc*; doi:10.1101/pdb.prot097352

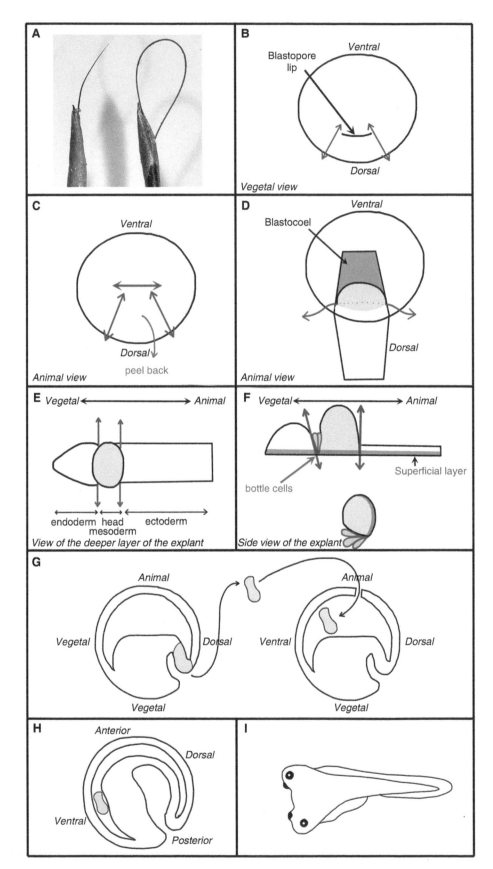

FIGURE 1. Key tools and steps for Einsteck transplantation. (*A*) Eyelash knife (*left*) and hair loop (*right*). (*B–I*) Diagram representing the key steps of the grafting procedure.

Cite this protocol as *Cold Spring Harb Protoc*; doi:10.1101/pdb.prot097352

Hair loop

Cut a human hair into 3-inch sections. Thread both ends of a section into a 23-gauge needle fitted on a 1-cc syringe. Push the loop into the needle until the desired stiffness is reached (Fig. 1A). Fix the hair with a drop of nail polish or cyanoacrylate glue.

Petri dishes (60-mm, plastic), coated with a 4-mm layer of 1% agarose

Petri dishes (60-mm, plastic), coated with plasticine

Roll 2 tsp of plasticine (nondrying, toxin free, appropriate for young children) into a ball. Flatten it out into a plastic Petri dish by hand.

Transfer pipette (disposable plastic or glass) with an opening of ≥2 mm, for transferring embryos

METHOD

The Einsteck technique has been used to transplant various kinds of material, including archenteron roof, dorsal blastopore lip (Marx 1925), neurula-stage neural plate (Mangold 1933), or a bead coated with signaling molecules (Slack and Isaacs 1994); see Discussion. The technique was heavily used in the 1930s in the rush to discover the neural inducer (Mangold 1933; Needham 1942) and again in the 1980s and 1990s by scientists investigating the specification of the antero–posterior axis (Ruiz i Altaba and Melton 1989; Cho and De Robertis 1990; Blum et al. 1992; Slack and Isaacs 1994). This protocol is based on the method by Saxén and Toivonen (1962).

The grafting procedure should be conducted between 15°C and 19°C.

1. Sterilize a plasticine-coated dish with 70% ethanol for 10 min. Rinse for 30 sec with RODI H$_2$O and then fill with grafting medium.

2. Using the glass bead tool, make two depressions in the plasticine as deep as one-third or one-half the diameter of an embryo (or about 1-mm).

3. Select a donor and a host embryo and transfer the embryos to the plasticine-coated dish. Select a donor embryo between stage 10 (the forming blastopore lip appears as a dotted line with a light gray color) and 10+ (the blastopore lip is now a line of a dark gray color). Select a host embryo at late blastula stage (stage 8 or 9) or early gastrula stage (10+).

4. Remove the vitelline envelope from each embryo using the fine forceps. Ensure that the embryos never breach the liquid surface once the vitelline envelope is removed.

   See Troubleshooting.

5. Move the donor and host embryos into the plasticine depressions. Turn a donor embryo onto its animal (pigmented) side with the blastopore lip and dorsal marginal zone (the area between the lip and the pigmentation edge) clearly visible (Fig. 1B).

6. On the donor embryo, insert the tip of the eyelash knife at the commissure (corner) of the blastopore lip and thread it all the way to the animal pole. Cut the tissue by pressing the hair loop along the length of the eyelash knife. Repeat the process from the other commissure (Fig. 1B). Cut perpendicular to the first two cuts to free the pigmented edge of the explant (Fig. 1C) and peel the piece of tissue toward the vegetal pole (Fig. 1D).

   The peeled tissue is made of superficial pigmented ectoderm with underlying deep mesoderm (Fig. 1D, orange tissue) and some endoderm.

7. To free the explant from the embryo, insert the knife under the mesoderm and cut perpendicularly (Fig. 1D). Lay the explant with its superficial side down (Fig. 1E).

   The freed explant now contains ectoderm (thin, long and pigmented tissue), head mesoderm (Fig. 1D–F, orange tissue) and the involuting bottle cells (Fig. 1F, purple) that created the blastopore lip. Some endoderm cells (much larger cells than the other tissues) may also be present.

8. Trim the explant into a tissue containing the head mesoderm and the bottle cells.

   i. To remove the neural ectoderm, insert the length of the eyelash knife under the dorsal side of the head mesoderm mass (i.e., cleft of Brachet) and press firmly onto the bottom of the dish.

ii. To cut the mass of endoderm cells attached to the head mesoderm, align the knife-edge perpendicular to the head mesoderm mass (Fig. 1E,F).

9. Position a host embryo with its animal hemisphere (pigmented area) up. Using the eyelash knife, make an incision slightly larger than the width of the explant in the animal hemisphere.

    The blastocoel cavity should be visible.

10. Insert the explant into the blastocoel cavity (Fig. 1G).

    See Troubleshooting.

11. Let the incision heal for 20 min and then transfer the grafted embryo to an agarose-coated dish filled with recovery medium.

    The incision should heal itself without resorting to glass cover bridges as described in other transplantation protocols (see Protocol 7: Cranial Neural Crest Transplants [Cousin 2018]). Because the blastocoel fluid escapes when the experimenter performs the incision, the edges of the incision should come together naturally and heal.

    By the end of gastrulation, the explant is pushed into an antero–ventral position by the involuted mesoderm (Fig. 1H). A secondary axis is visible as soon as the neurula stage (the day after the graft). At the early tailbud stage, a secondary head should be clearly visible (Fig. 1I).

    See Troubleshooting.

TROUBLESHOOTING

Problem (Step 4, Step 11): The embryo disaggregates.
Solution: The medium is the culprit. Make fresh medium, add gentamycin and avoid using HCl to adjust the pH as excess Cl- ions may affect the healing of the embryo.

Problem (Steps 4–10): The embryos develop too fast and the experimenter does not have the time to perform the grafts.
Solution: Keep the embryos, media and dishes at 15°C at least a few hours before grafting is scheduled. Perform the grafts on cooling table set up at 15°C if available.

DISCUSSION

This particular protocol has been used successfully in a classroom environment. Students that failed to induce secondary anterior structures with the Spemann–Mangold organizer transplants were usually able to do so using the Einsteck technique. The nature and extent of the induced axis will vary depending on the type of tissue or amount of growth factor used. In the case of the gastrula blastopore lip, a stage 10+ explant will be able to induce a secondary embryo with a fully formed head while an explant taken later (stage 10.5 or 11) will induce more posterior structures such as trunk and tail (Saxén and Toivonen 1962). The quality of the secondary axis induced depends on two variables: the age of the host embryos at the time of transplantation and the position of the explant after the host gastrulation is complete. If the host is at blastula stage, the grafted tissue will be able to induce a full secondary embryo attached to the ventral or lateral side of the primary embryo. If the embryo is already at the gastrula stage, the secondary axis may only consist of trunk and tail tissues (although we obtained full head-twinning when grafting CNC at stage 10+). This difference is caused by a change in the host tissue's competency to respond to the signals secreted by the Spemann–Mangold organizer. The results may also vary depending on where the grafted tissue ends up at the conclusion of gastrulation. Some explants may end up on the anterior-most part of the embryo and therefore result in a well-formed head in the secondary embryo. Others may end up on a more lateral side, resulting in a smaller secondary embryo or a secondary embryo lacking a head. If one maintains the host embryo with its ventral side down during gastrulation, gravity will help maintain the grafted tissue in the ventral part of the blastocoel cavity, resulting in more uniform results across transplanted embryos

Generally speaking, Einsteck procedures are performed to assess the capacity of a transplanted tissue to change either the fate or the patterning of surrounding tissues (phenomenon of induction). For example, Otto Mangold performed Einsteck procedures with archenteron (primitive gut lumen) roof taken from various locations along the antero–posterior axis of a neurula-stage donor (Mangold 1933). Others showed that the treatment of donor embryos with retinoid acid before transplanting the Spemann organizer changes its patterning capabilities (Sive and Cheng 1991). For such experiments, one must design the appropriate negative and positive controls. The former may consist of transplantation of an unrelated tissue such as blastula-stage ectoderm tissue (i.e., animal cap). The latter may consist of transplantation of an untreated Spemann–Mangold organizer.

The inductive capacity of a signaling molecule can also be assessed using this technique by implanting beads coated with the molecules of interest (Slack and Isaacs 1994). If molecules cannot be coated on beads, one can graft small animal cap from embryos injected with the mRNA encoding such proteins. For each of these cases, proper negative controls must be designed. For bead experiments, implanting an uncoated bead will be sufficient. If modified animal caps are used, the grafting of unmodified animal cap is appropriate. The experimenter will need to design positive controls appropriate for the inquiry. If one is assessing the ability of a compound or tissue to induce neural tissue, a positive control could consist of the graft of a Spemann–Mangold organizer or a bead coated with a well-known neural inducer.

Depending on the experimenter's inquiry, the results can be analyzed by either qualitative methods such as gross morphology (i.e., apparition of a secondary head) and histology or quantitative methods such as in situ hybridization, western blot, and quantitative polymerase chain reaction (Q-PCR).

RECIPE

Modified Barth's Saline (MBS) (1×)

$CaCl_2$	0.41 mM
$CaNO_3$	0.3 mM
HEPES-NaOH	15 mM
KCl	1 mM
$MgSO_4$	0.82 mM
NaCl	88 mM
$NaHCO_3$	2.4 mM

Adjust pH to 7.6. Store at room temperature for up to 1 mo.

ACKNOWLEDGMENTS

The author would like to thank Professor Ray Keller for his invaluable insight into *Xenopus* transplantations. H.C. is supported by National Institutes of Health/National Institute of Dental and Craniofacial Research (NIH/NIDCR) DE025691.

REFERENCES

Blum M, Gaunt SJ, Cho KW, Steinbeisser H, Blumberg B, Bittner D, De Robertis EM. 1992. Gastrulation in the mouse: the role of the homeobox gene goosecoid. *Cell* 69: 1097–1106.

Cho KW, De Robertis EM. 1990. Differential activation of *Xenopus* homeo box genes by mesoderm-inducing growth factors and retinoic acid. *Genes Dev* 4: 1910–1916.

Cousin H. 2018. Cranial neural crest transplants. *Cold Spring Harb Protoc* doi: 10.1101/pdb.prot097402.

Mangold O. 1933. Uber die Induktionsfähigkeit der verscheidenen Bezirke der Neurula von Urodelen. *Naturwissenschaften* **43:** 761–766.

Marx A. 1925. Experimentelle Untersuchungen zur Frage der Determonation der Medullarplatte. *Wilhelm Roux Arch Entwicklungsmeck Org* **105:** 20–44.

Needham J. 1942. *Biochemistry and morphogenesis.* Cambridge University Press, Cambridge.

Nieuwkoop PD, Faber J. 1967. *Normal table of* Xenopus laevis *(Daudin)*, 2nd ed. North-Holland, Amsterdam.

Ruiz i Altaba A, Melton DA. 1989. Interaction between peptide growth factors and homoeobox genes in the establishment of antero-posterior polarity in frog embryos. *Nature* **341:** 33–38.

Saxén L, Toivonen S. 1962. *Primary embryonic induction.* Academic Press, London.

Sive HL, Cheng PF. 1991. Retinoic acid perturbs the expression of Xhox.lab genes and alters mesodermal determination in *Xenopus laevis. Genes Dev* **5:** 1321–1332.

Slack JM, Isaacs HV. 1994. The Einsteck-method: position and structure of projections formed by implants of a ventral character. *Dev Biol* **161:** 313–317.

Cite this protocol as *Cold Spring Harb Protoc*; doi:10.1101/pdb.prot097352

Protocol 3

Microsurgical Methods to Isolate and Culture the Early Gastrula Dorsal Marginal Zone

Lance A. Davidson[1,2,3,4]

[1]*Department of Bioengineering,* [2]*Department of Developmental Biology, and* [3]*Department of Computational and Systems Biology, University of Pittsburgh, Pittsburgh, Pennsylvania 15213, USA*

Marginal zone explants from *Xenopus* embryos can be used to expose cell behaviors and tissue movements that normally operate in dorsal tissues. Dorsal explants comprise the diverse set of progenitor cells found in dorsal tissues including mesendoderm, head mesoderm, prechordal mesoderm, endoderm with bottle cells, axial mesoderm of the prospective notochord, paraxial mesoderm of the somites, lateral plate mesoderm, neural ectoderm, and ectoderm. Unlike an organoid, the dorsal marginal zone (DMZ) explant is "organotypic" in that microsurgery does not disrupt native tissue organization beyond manipulations needed to dissect the tissue from the embryo. An organotypic early gastrula DMZ explant preserves boundaries and close tissue associations in the native marginal zone. Depending on the stage, patterning and cell identities can be maintained in explants and tissue isolates. Local cell movements and behaviors may also be preserved; however, the large-scale biomechanical impact of their collective movements may be altered from those in the native marginal zone. For instance, involution is typically inhibited in the DMZ explant, precluding the two-layer association of mesoderm and prospective neural ectoderm normally achieved during gastrulation. DMZ explants may be mounted and imaged in a variety of ways, exposing interesting cell behaviors or collective movements such as mediolateral cell intercalation in the axial and paraxial mesoderm, apical constriction of bottle cells, and directional migration of mesendoderm. The flattened DMZ explant can also be used to study emergence of new tissue-defining boundaries such as the notochord–somite boundary, the ectoderm–mesoderm boundary, and the mesendoderm–mesoderm boundary.

MATERIALS

It is essential that you consult the appropriate Material Safety Data Sheets and your institution's Environmental Health and Safety Office for proper handling of equipment and hazardous materials used in this protocol.

RECIPES: Please see the end of this protocol for recipes indicated by <R>. Additional recipes can be found online at http://cshprotocols.cshlp.org/site/recipes.

Reagents

Danilchik's for Amy (DFA) medium (pH 8.3) <R>

DFA explant culture medium is not a random collection of odd buffers and ions but rather a solution created to match the interstitial composition of the Xenopus laevis embryo that is relatively low in chloride ions and has a slightly basic pH (Gillespie 1983). This suitable culture medium replaces conventional chloride counter ions of Na^+ and K^+ with gluconate ions. Endogenous interstitial buffering factors are replaced by the Good buffering agent bicine to minimize the amount of hydroxide counter ions needed to create pH 8.3.

[4]Correspondence: lad43@pitt.edu

Supplemental material is available at cshprotocols.cshlp.org

Cite this protocol as *Cold Spring Harb Protoc*; doi:10.1101/pdb.prot097360

Avoid adding phenol red if preparing tissues for fluorescence imaging. Phenol red will introduce nonspecific fluorescence that can interfere with red or near-infrared fluorophores.

Embryo culture medium (1/3× MBS)

1/3× MBS is the de facto embryo culture medium used in the Keller, DeSimone, and Davidson laboratories where the explant techniques described here were developed and used (Keller et al. 1985; Shih and Keller 1992a,b; Domingo and Keller 1995, 2000; Davidson et al. 2002; Marsden and DeSimone 2003; Davidson et al. 2004a, 2006; Kim and Davidson 2011). Calcium concentrations in embryo culture media vary (Sive et al. 2000) and may yield qualitative differences in abilities of tissue to remain cohesive during microsurgical manipulations, resulting in cross-tissue contamination.

Avoid adding phenol red if preparing tissues for fluorescence imaging. Phenol red will introduce nonspecific fluorescence that can interfere with red or near-infrared fluorophores.

Fibronectin (Corning 356008; 1 mg/mL) (optional; see Step 1)

Aliquot and store at −20°C.

Modified Barth's solution (MBS) (pH 7.4), diluted to a concentration of 1/3× <R>

The recipe is for a 1× solution. 3× stocks of MBS are pH stable and can be stored at 4°C. 10× stocks do not maintain their pH. When used for embryo culture, we dilute our 3× stock to 1/3×.

Xenopus laevis embryos at early gastrula stage (stage 10− to 10+ to 11) (Nieuwkoop and Faber 1967), dejellied as in Sive et al. (2007) using 2% cysteine (L-cysteine hydrochloride monohydrate; Sigma-Aldrich C7880) in 1/3× MBS that has been adjusted to a pH of 8.0 with NaOH

Equipment

Coverslip glass (24 × 40 mm)

Custom-fabricated cold plate dissecting stage (recommended for holding temperatures constant; see Method)

The cold plate is a large-footprint aluminum and acrylic base mount for the stereoscope that passes chilled or heated water produced by a chiller-circulating water bath. Lower temperatures on the dissecting stage slow developmental time as well as unintentional wound healing in dissected explants.

Diamond pencil (retractable; Ted Pella 54468)

Disposable pipette

Forceps (Dumont #5) (two; see Step 7)

Glass-bottomed imaging chamber (e.g., Cellvis or Mattek dishes, Invitrogen chambers, custom acrylic well chambers, or custom nylon washer chambers)

See Sec. 9, Protocol 5: Chambers for Culturing and Immobilizing Xenopus Embryos and Organotypic Explants for Live Imaging (Chu and Davidson 2022) and Kim and Davidson (2013).

Hair knife (Sive et al. 2000)

Hair loop (Sive et al. 2000)

Incubator (e.g., MyTemp Mini Digital Incubator; Benchmark Scientific) (recommended for holding temperatures constant; see Method)

Microscope slide

Oversize glass coverslips (45 × 50 mm, #1.5; Brain Research Laboratories 4550-1.5D)

Petri dishes (25- and 60-mm-diameter)

Silicone grease (high-vacuum silicone grease; Dow Chemical) loaded into a 10-mL syringe

Stereomicroscope

Here, we use an Olympus SZX7 equipped with a video port, Canon T3i, and a video coupler MDSLR (Martin Microscope Co.) with off-axis lighting provided by a Schott Fiber Optic Light Source with dual gooseneck.

ThermoWorks EasyLog USB Logger (TW-USB-2+) (recommended for confirming room temperatures; see Method)

Cite this protocol as *Cold Spring Harb Protoc*; doi:10.1101/pdb.prot097360

METHOD

Xenopus laevis embryos may be raised in incubators at a range of temperatures within the normal permissive range of 14°C–26°C. Similarly, microsurgical manipulations listed below can be carried out either at room temperature or on a temperature-controlled cold plate. Rates of morphogenesis and intracellular processes are very sensitive to temperature so it is critical that temperature is held constant and recorded (e.g., ThermoWorks EasyLog USB Logger).

Preparing the Embryos and the Imaging Chamber

Figure 1 shows microsurgical tools for dissection and schematics of the dorsal marginal zone explant.

1. (Optional) Coat substrates in the imaging chambers with fibronectin by incubating the substrate with 20 µg/mL of fibronectin in 1/3× MBS overnight at 4°C. Rinse three times in 1/3× MBS.

2. Fill the imaging chamber with 3–4 mm DFA with antibiotic-antimycotic.

3. Set up the operating dish. Place DFA into three Petri dishes ∼3–4 mm deep.

    One dish is for microsurgery, one dish is for rinsing fresh explants, and the last dish is for culturing or holding explants until needed.

4. (Optional) Prepare a dish of coverslip bridges and silicone grease (Fig. 1A) as follows:

    i. Score and break 3 × 8-mm coverslip bridges with the diamond pencil using a microscope slide as a straight edge. Store the coverslip bridges in a clean Petri dish.

    ii. Dispense silicone grease onto the inner surface of a clean Petri dish lid and smooth into a thin layer (1-mm-thick).

    Broken glass and fresh coverslip bridges are extremely sharp. Handle with care. If periodically re-stocked, the Petri dish with the bridge fragments and silicone grease will last indefinitely and provide a ready source during microsurgery.

5. (Optional) Use forceps to pick up a single coverslip bridge from the dish in Step 4 and dab some silicone grease along the long edge. Place the coverslip bridge grease side down in the center of the Petri dish to be used for microsurgery.

    This bridge serves as a "backstop" or aid during embryo and explant manipulation and defines a forceps-free region of the microsurgery dish.

    Forceps can create sharp edges in the surface of the Petri dish that can unintentionally wound embryos and explants as they are being manipulated.

6. Select dejellied embryos from early to mid-gastrula (stage 10− to 10+ to 11; Nieuwkoop and Faber 1967) and transfer to the operating dish with a disposable pipette (Fig. 2A).

    See also online Supplemental Movie S1, which shows a recording of a microsurgical dissection of the DMZ explant. The locations of microsurgical cuts are shown schematically in Figure 1B (arrowheads) together with the fate map of the dorsal marginal zone (Fig. 1C). Older embryos may be used but may need additional microsurgical maneuvers (see below).

    The stage of the embryos used as a source for these explants may affect gene expression patterning and cell behaviors present in the explant. For instance, Keller sandwiches (e.g., two DMZ explants) show limited patterning of the hindbrain and different cell behaviors if stage 10− embryos are used in place of stage 10+ embryos (Poznanski and Keller 1997; Poznanski et al. 1997).

7. Remove the vitelline membrane by grabbing the membrane at the same location with two forceps (Fig. 2B). Open the membrane by slowly moving the forceps apart, creating a tear in the vitelline membrane. Once the opening is large enough, release one forceps and gently shake the embryo out of the membrane with the other. Do not damage the embryo in the dorsal marginal zone. If the vitelline membrane cannot be removed without damage, try puncturing the embryo first in the animal pole. After the blastocoel collapses the vitelline membrane can be more easily removed.

    One blunt and one sharp forceps can be helpful; the blunt forceps is used to hold or pinch the vitelline membrane, and the sharp forceps is used to tear the membrane. Use forceps to remove the vitelline membrane in a different region of the Petri dish than you plan to carry out microsurgery since small scratches caused by forceps can inadvertently wound embryos or explants.

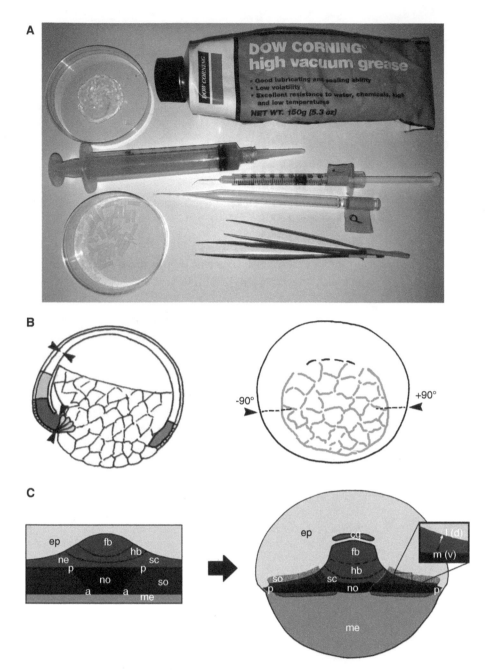

FIGURE 1. Microsurgical tools for dissection and schematics of the dorsal marginal zone explant. (*A*) Microsurgical tools including silicone grease loaded into syringe, silicone grease, hair knife mounted in a 1-mL syringe, and hair loop mounted in a Pasteur pipette, Petri dish containing precut bridges and a smear of silicone grease, and pair of forceps. (*B*) Sagittal and en face diagrams of embryo source for the 180° dorsal marginal zone (DMZ) explant. Arrows indicate incisions for the DMZ at + and −90° (after Keller et al. 2000.) (*C*) Diagram of the fate maps of the early DMZ and patterning after culture (after Davidson et al. 2004). (a) Anterior, (cg) cement gland, (ep) epidermis, (fb) forebrain, (hb) hindbrain, (me) mesendoderm, (ne) neural ectoderm, (no) notochord, (p) posterior, (sc) spinal cord, (so) somite. *Inset* on the *right* shows how the medial-to-lateral axis is distorted in the DMZ explant after 3D morphogenetic movements. Planar polarity of the anterior–posterior axis in early stage explant results in a "split axis" at later stages with posterior ends of the notochord, somites, and spinal cord located at the two lateral margins.

Microsurgery

The following steps involve microsurgery to remove the dorsal 180° of the marginal zone. This region contains the anterior-most prospective tissues of the embryo including anterior presomitic mesoderm, anterior neural ectoderm, and anterior endoderm. In addition, the explant will contain all the notochord and some ectoderm. The dorsal isolate can be easily narrowed to 90° or expanded to the full 360° of the marginal zone.

Cite this protocol as *Cold Spring Harb Protoc*; doi:10.1101/pdb.prot097360

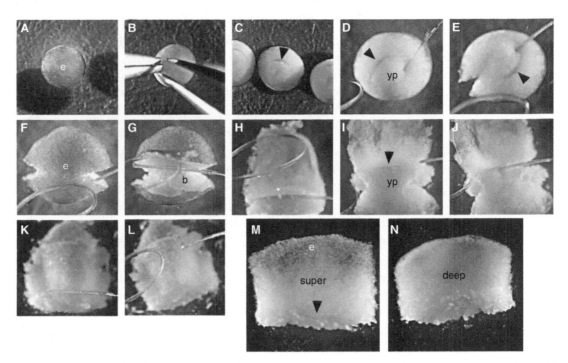

FIGURE 2. Dissection of dorsal marginal zone explant. (*A*) Select an embryo at a suitable stage. (*B*) Remove the vitelline membrane. (*C*) Select embryos at similar stages of the same experiment by the extent of contracted bottle cells and the blastopore lip (arrowhead). (*D*) Use the hair knife to make the first incision at +90° cutting into the yolk plug (yp). (*E*) Rotate the embryo to make the second incision at −90° from the center of the contracted bottle cells (arrowhead). (*F*) Turn the embryo to expose the animal cap ectoderm (e). Make an incision across the animal cap ectoderm to expose the blastocoel (b). (*G*) Open the blastocoel and use both the hair loop and knife in coordination to begin pulling the mesendoderm from the dorsal cleft. (*H*) Continue to pull the mesendoderm from the cleft. When the base of the cleft is reached use the hair knife to cut deeper toward the blastopore lip. (*I*) Turn the embryo over, exposing the ectoderm surface. Flatten the embryo and identify contracted bottle cells (arrowhead aligned with hair knife). (*J*) Make an incision either below, to retain bottle cells, or slightly above the level of the bottle cells to remove them. Separate the explant from the remainder of the embryo. (*K*) Use the hair knife to scrape deep endoderm cells from the vegetal end of the explant. Use the "back-plate" coverslip fragment as a guide to remove cells and thin the vegetal domain of the explant. (*L*) Remove animal cap ectoderm from the DMZ. (*M*) A view of the superficial cells of the DMZ explant (super). Animal is up and vegetal is down. (*N*) A view of the deep cells of the DMZ explant (deep). Animal is up and vegetal is down. The deep surface of the animal domain of the explant for (e.g., prospective neural) should be free of loose cells as these are likely contaminating cells from mesoderm or mesendoderm.

8. Roll the embryo onto its animal pole and identify the dorsal midline by the center of the contracted bottle cells (Fig. 2C). Use a stiff hair knife to make two radial incisions at +90° and −90° from the midline (Fig. 2D,E). The inner limit of the cut will just barely cross the line of prospective or extant bottle cells but will not extend to the center of the vegetal pole (Fig. 2D).

9. Roll the embryo onto its vegetal pole and open the blastocoel (Fig. 2F). Use the hair knife to connect the incisions across the animal cap (Fig. 2G). This cut will open the blastocoel of the embryo.

10. Gently pull the postinvolution mesendoderm from the preinvolution mesoderm. Insert the hair loop and hair knife into the blastocoel. Use one hair tool to gently press down on the mesendoderm and the other to lift up on the overlying ectoderm and preinvolution mesoderm (Fig. 2H). Continue to separate the two tissues, pulling the mesoderm from the mesendoderm in a slow motion. These two tissues should separate cleanly along the cleft of Brachet without leaving cells from the other compartment.

11. Roll the embryo over onto the open blastocoel and flip the dorsal marginal zone away from the embryo. Flip the preinvolution mesoderm out and directed away from the embryo (Fig. 2I). Use the hair loop to "pat" the mesoderm so that the marginal zone is flattened and no longer aligned with the remainder of the embryo.

12. Cut the dorsal marginal zone free of the embryo. Use the hair knife to make two small incisions just vegetal of the line of bottle cells. Flatten the marginal zone so that the bottle cells form a line. Use the hair knife to complete the cut by connecting the two small incisions. Discard the rest of the embryo (Fig. 2J).

   The isolated embryo fragment contains some subblastoporal endoderm, bottle cells, a thick marginal zone, and a wedge including animal cap ectoderm. Leave the animal cap ectoderm fragment in place until the final trimming cuts. Use hair tools on the ectoderm to move or position the explant so as not to damage the marginal zone. Remove the ectoderm before transferring the explant to the imaging chamber.

13. Remove layers of cells from the inner surface of the marginal zone to thin the vegetal-most domain of the explant. Position the explant with the deep cells up and the ectoderm facing the surface of the dish. Move the vegetal-most edge of the explant to the edge of the backstop. Use a hair knife remove excess deep cells from the marginal zone, slicing from animal to vegetal, level with the glass coverslip (Fig. 2K).

14. Trim the explant to the desired mediolateral width and animal to vegetal dimensions. Use the hair loop and knife in combination to trim the explant, removing animal cap ectoderm and removing tears at the margin (Fig. 2L). At this time, the line of bottle cells can also be removed. Exercise caution in removing bottle cells as it is easy to remove the anterior-most mesoderm and endoderm along with the bottle cells.

15. Transfer the explant to the imaging chamber. Use a pipette to move the explant first to a clean region of the dish, and then to a separate dish containing fresh DFA to remove adherent cells and lysate. Transfer the explant to the imaging chamber. Transfer as little contaminating lysate, cells, or yolk platelets to the imaging chamber as these can foul the fibronectin coating or obstruct imaging.

16. Mount the explant. Move the explant into position with hair tools taking care not to touch the fibronectin coated substrate with the hair tools. Superficial apical (Fig. 2M) and basal-most surfaces (Fig. 2N) can be seen by apical pigment in superficial ectoderm cells (e) and pebble-like appearance of the deep basal surface. Position the explant so the basal-most surface faces the fibronectin substrate.

17. Prepare a coverslip bridge by applying a small dab of silicone grease at either end. Too little grease will not allow the bridge to hold the explant firmly in place, and too much grease may spread and can obstruct or contact the explant.

18. Compress the explant under the glass coverslip bridge. Place the coverslip bridge over the explant and press the ends over the grease iteratively to gently compress the explant. Too much compression can break the bridge or damage the tissue; too little compression will not encourage cells to attach or spread on fibronectin.

19. Seal the imaging chamber. Lay a thin layer of grease onto the top of the chamber. Add DFA so that there is a slight bubble of media above the lip of the chamber. Add a large coverglass to the top of the chamber and compress against the lip of the chamber. Use a Kimwipe to remove excess media.

   We generally use custom acrylic imaging chambers; see Sec. 9, Protocol 5: Chambers for Culturing and Immobilizing Xenopus Embryos and Organotypic Explants for Live Imaging (Chu and Davidson 2022) for more details on the use of these and other imaging chambers. A leaky chamber may dry out or spill media onto the microscope.

   To perfect this technique is it helpful to take images or sketches of each DMZ explant when it has been completed, after the coverslip bridge is removed, and at the final time point of the experiment. Contrast successful explant preparations with those that include hidden involution or that are too curved for quantification.

DISCUSSION

This protocol describes preparation of explants detailed in many studies using marginal zone explants (Keller et al. 1985, 2000; Wallingford et al. 2000; Davidson et al. 2004a,b, 2006; Davidson and

Wallingford 2005; DeSimone et al. 2005; Goto et al. 2005; Kim and Davidson 2011; Shindo and Wallingford 2014; Shindo et al. 2019; Huebner et al. 2021). DMZ explants can be used to assess cell shapes, cell identities, morphogenetic movements, and intracellular protein dynamics within dorsal tissues of mesoderm, neural, mesendoderm, and endoderm progenitors. Skills needed to microsurgically isolate the DMZ explant are greater than those needed to dissect animal cap ectoderm but can be mastered in a few weeks. The DMZ explant complements the Keller sandwich explant by enabling the use of high-resolution light microscopy techniques to study cell and subcellular processes that polarize cell shapes and behaviors. When cultured on a fibronectin substrate (e.g., glass adsorbed with 20 µg/mL), mesodermal tissues in this explant have limited capacity for large-scale movements. Mesoderm tissues in DMZ explants can lengthen and narrow but do not reach levels seen in vivo or in Keller sandwiches. However, mesodermal cells in the DMZ explant exhibit mediolateral intercalation behaviors (MIBs) characterized by mediolaterally polarized cell shape change and mediolaterally extended cell protrusions. MIBs in DMZ explants are considered a response to polarity cues present in the explant and are thought to model in vivo cell behaviors. It should be recognized that cell shape changes in the DMZ explant may be more exaggerated than in vivo; for example, notochord and presomitic mesoderm cells elongate more and attain mediolateral to anterior–posterior aspect ratios (e.g., length to width ratios) in excess of 4. The earliest example of the DMZ explant was cultured under agarose (Keller et al. 1985). More recently, DMZ explants have been cultured on untreated glass substrates (e.g., Shih and Keller 1992a; Domingo and Keller 1995) or fibronectin-coated glass (e.g., Goto et al. 2005; Davidson et al. 2006; Kim and Davidson 2011; Shindo and Wallingford 2014; Pfister et al. 2016; Huebner et al. 2021). DMZ explants can be used to study intracellular protein dynamics and functions including: polarization of PCP factors, cytoskeleton, and cell behaviors. Cells exposed in DMZ explants include ectoderm, neural ectoderm, neural crest, axial and paraxial mesoderm, and mesendoderm including head mesoderm.

A successful explant should exhibit stage-consistent and position-specific morphogenetic movements (e.g., stages of MIBs) and differentiation of tissue boundaries (e.g., notochord–somite) and differentiate for up to 48 h with minimal dissociation or cell death. Successful isolation and culture of a DMZ explant can be assayed with antibodies to notochord (Tor70) or paraxial mesoderm (12/101) or by RNA in situ hybridization (Davidson et al. 2004a). The cell behaviors that can be quantitated by live imaging include (1) collective vegetal-directed migration of mesendoderm and animal-ward migration of ectoderm away from the explant, (2) boundary formation between notochord and presomitic mesoderm, (3) boundary formation between neural and endoderm, (4) mediolateral elongation and intercalation in both axial and paraxial mesoderm cells, and (5) alignment of axial and paraxial mesoderm cells into arc-like patterns. Additionally, cells at the edge of the explant, including both ectoderm and mesendoderm, should develop large single lamellipodia directed outward from the explant. By contrast, mesoderm cells should develop bipolar protrusions in medial and lateral directions.

Because of microsurgical isolation, some tissue interactions and induction events that occur late in gastrulation may be disrupted or missing. It is important to note that compression of the coverslip bridge forces tissues that would normally involute to form bilayer structures in the embryo to be constrained to remain in a planar configuration in the explant. However, even with this constraint, 3D movements continue. Rigid substrates such as glass or Petri plastic adsorbed with fibronectin typically hold cells in a 2D configuration and inhibit large-scale deformation that occurs with compliant substrates found in the embryo. However, cells and tissues further from that surface can undergo more complex movements. For instance, if the coverslip bridge is removed after 1 h, axial and paraxial mesoderm will coordinate movements and extend an "anterior" structure perpendicular to the substrate from the center of the DMZ. If the coverslip bridge is too loose, bottle cell expansion can drive local involution toward the animal pole and form a blastopore-like structure and a short archenteron. More typically, local movements of the presomitic mesoderm around the posterior end of the notochord will create a thicker, multilayered tissue along boundaries that would normally demarcate the limit of involution. This thickening will be present on the left and right sides of the explant and will include posterior notochord, differentiated somites, neural ectoderm, and neural crest.

Recently, the DMZ explant has been used to perform a temporal analysis of gene expression (Kakebeen et al. 2021). Bioinformatic resources from this study together with genomic sequencing of *Xenopus* species (Hellsten et al. 2010; Session et al. 2016) and single cell atlases of gene expression (Briggs et al. 2018) and protein abundance should provide many new insights into the genes that control cell behaviors and emergence of tissue boundaries in the developing gastrula. Detailed hypotheses can then be tested with both gain- and loss-of-function studies combined with live-cell imaging of DMZ explants.

RECIPES

Danilchik's for Amy (DFA) Medium (pH 8.3)

Reagent	Final concentration
NaCl	53 mM
Na$_2$CO$_3$	5 mM
Potassium gluconate	4.5 mM
Sodium gluconate	32 mM
CaCl$_2$	1 mM
MgSO$_4$	1 mM

Adjust the pH to 8.3 with granular bicine. Filter with a 0.22 μm filter. Include 0.1% bovine serum albumen (BSA; Sigma-Aldrich A7906) to prevent nonspecific adhesion to substrates or glass coverslips. Just prior to use, add 1% of antibiotic/antimycotic (Sigma-Aldrich A5955). Prepare the DFA with BSA, aliquot, and store at −20°C. Keep aliquots of antibiotic/antimycotic at −20°C and add to thawed DFA before use. Replace DFA with medium containing fresh antibiotic/antimycotic if culturing explants for >24 h. (This recipe is adapted from Sater et al. 1993.)

Modified Barth's Solution (MBS) (pH 7.4)

Reagent	Final concentration (1×)
NaCl	88 mM
KCl	1 mM
NaHCO$_3$	2.4 mM
MgSO$_4$·7H$_2$O	0.82 mM
Ca(NO$_3$)$_2$·4H$_2$O	0.33 mM
CaCl$_2$·2H$_2$O	0.41 mM
HEPES	5 mM

Adjust the pH to 7.4 with NaOH; 3× stocks of MBS are pH-stable and can be stored at 4°C. Prepare a working solution (1/3×) by dilution.

ACKNOWLEDGMENTS

I thank the members of the Davidson, Keller, and DeSimone laboratories for helpful discussions and patience when I hogged the dissection stations. I also thank the National *Xenopus* Resource (RRID: SCR-013731) and Xenbase (RRID:SCR-003280) for their invaluable support to the model organism community. L.A.D. is funded by the Eunice Kennedy Shriver Institute of Child Health and Human Development at the National Institutes of Health (NIH) (R01 HD044750 and R21 HD106629).

REFERENCES

Briggs JA, Weinreb C, Wagner DE, Megason S, Peshkin L, Kirschner MW, Klein AM. 2018. The dynamics of gene expression in vertebrate embryogenesis at single-cell resolution. *Science* 360: eaar5780.

Chu C-W, Davidson LA. 2022. Chambers for culturing and immobilizing *Xenopus* embryos and organotypic explants for live imaging. *Cold Spring Harb Protoc* doi:10.1101/pdb.prot107649

Davidson LA, Wallingford JB. 2005. Visualizing morphogenesis in the frog embryo. In *Imaging in neuroscience and development: a laboratory manual* (ed. Yuste R, Konnerth A). Cold Spring Harbor Laboratory Press, Cold Spring Harbor, New York.

Davidson LA, Hoffstrom BG, Keller R, DeSimone DW. 2002. Mesendoderm extension and mantle closure in *Xenopus laevis* gastrulation: combined roles for integrin α5β1, fibronectin, and tissue geometry. *Dev Biol* 242: 109–129. doi:10.1006/dbio.2002.0537

Davidson LA, Keller R, DeSimone D. 2004a. Patterning and tissue movements in a novel explant preparation of the marginal zone of *Xenopus laevis*. *Gene Expr Patterns* 4: 457–466. doi:10.1016/j.modgep.2004.01.001

Davidson LA, Keller R, Desimone DW. 2004b. Assembly and remodeling of the fibrillar fibronectin extracellular matrix during gastrulation and neurulation in *Xenopus laevis*. *Dev Dyn* 231: 888–895. doi:10.1002/dvdy.20217

Davidson LA, Marsden M, Keller R, Desimone DW. 2006. Integrin α5β1 and fibronectin regulate polarized cell protrusions required for *Xenopus* convergence and extension. *Curr Biol* 16: 833–844. doi:10.1016/j.cub.2006.03.038

DeSimone DW, Davidson L, Marsden M, Alfandari D. 2005. The *Xenopus* embryo as a model system for studies of cell migration. *Methods Mol Biol* 294: 235–245. doi:10.1385/1-59259-860-9:235

Domingo C, Keller R. 1995. Induction of notochord cell intercalation behavior and differentiation by progressive signals in the gastrula of *Xenopus laevis*. *Development* 121: 3311–3321. doi:10.1242/dev.121.10.3311

Domingo C, Keller R. 2000. Cells remain competent to respond to mesoderm-inducing signals present during gastrulation in *Xenopus laevis*. *Dev Biol* 225: 226–240. doi:10.1006/dbio.2000.9769

Gillespie JI. 1983. The distribution of small ions during the early development of Xenopus *laevis* and *Ambystoma mexicanum* embryos. *J Physiol (Lond)* 344: 359–377. doi:10.1113/jphysiol.1983.sp014945

Goto T, Davidson L, Asashima M, Keller R. 2005. Planar cell polarity genes regulate polarized extracellular matrix deposition during frog gastrulation. *Curr Biol* 15: 787–793. doi:10.1016/j.cub.2005.03.040

Hellsten U, Harland RM, Gilchrist MJ, Hendrix D, Jurka J, Kapitonov V, Ovcharenko I, Putnam NH, Shu S, Taher L. 2010. The genome of the Western clawed frog *Xenopus tropicalis*. *Science* 328: 633–636. doi:10.1126/science.1183670

Huebner RJ, Malmi-Kakkada AN, Sarıkaya S, Weng S, Thirumalai D, Wallingford JB. 2021. Mechanical heterogeneity along single cell-cell junctions is driven by lateral clustering of cadherins during vertebrate axis elongation. *Elife* 10: e65390. doi:10.7554/eLife.65390

Kakebeen AD, Huebner RJ, Shindo A, Kwon K, Kwon T, Wills AE, Wallingford JB. 2021. A temporally resolved transcriptome for developing "Keller" explants of the *Xenopus laevis* dorsal marginal zone. *Dev Dyn* 250: 717–731. doi:10.1002/dvdy.289

Keller RE, Danilchik M, Gimlich R, Shih J. 1985. The function and mechanism of convergent extension during gastrulation of *Xenopus laevis*. *J Embryol Exp Morphol* 89: 185–209. doi:10.1242/dev.89.Supplement.185

Keller R, Davidson L, Edlund A, Elul T, Ezin M, Shook D, Skoglund P. 2000. Mechanisms of convergence and extension by cell intercalation. *Philos Trans R Soc Lond B* 355: 897–922. doi:10.1098/rstb.2000.0626

Kim HY, Davidson LA. 2011. Punctuated actin contractions during convergent extension and their permissive regulation by the non-canonical Wnt-signaling pathway. *J Cell Sci* 124: 635–646. doi:10.1242/jcs.067579

Kim HY, Davidson LA. 2013. Assembly of chambers for stable long-term imaging of live *Xenopus* tissue. *Cold Spring Harb Protoc* doi:10.1101/pdb.prot073882

Marsden M, DeSimone DW. 2003. Integrin-ECM interactions regulate cadherin-dependent cell adhesion and are required for convergent extension in *Xenopus*. *Curr Biol* 13: 1182–1191. doi:10.1016/S0960-9822(03)00433-0

Nieuwkoop PD, Faber J. 1967. *Normal tables of* Xenopus laevis *(Daudin)*. Elsevier North-Holland Biomedical Press, Amsterdam.

Pfister K, Shook DR, Chang C, Keller R, Skoglund P. 2016. Molecular model for force production and transmission during vertebrate gastrulation. *Development* 143: 715–727. doi:10.1242/dev.128090

Poznanski A, Keller R. 1997. The role of planar and early vertical signaling in patterning the expression of *Hoxb-1* in *Xenopus*. *Dev Biol* 184: 351–366. doi:10.1006/dbio.1996.8500

Poznanski A, Minsuk S, Stathopoulos D, Keller R. 1997. Epithelial cell wedging and neural trough formation are induced planarly in *Xenopus*, without persistent vertical interactions with mesoderm. *Dev Biol* 189: 256–269. doi:10.1006/dbio.1997.8678

Session AM, Uno Y, Kwon T, Chapman JA, Toyoda A, Takahashi S, Fukui A, Hikosaka A, Suzuki A, Kondo M. 2016. Genome evolution in the allotetraploid frog *Xenopus laevis*. *Nature* 538: 336–343. doi:10.1038/nature19840

Shih J, Keller R. 1992a. Cell motility driving mediolateral intercalation in explants of *Xenopus laevis*. *Development* 116: 901–914. doi:10.1242/dev.116.4.901

Shih J, Keller R. 1992b. Patterns of cell motility in the organizer and dorsal mesoderm of *Xenopus laevis*. *Development* 116: 915–930. doi:10.1242/dev.116.4.915

Shindo A, Wallingford JB. 2014. PCP and septins compartmentalize cortical actomyosin to direct collective cell movement. *Science* 343: 649–652. doi:10.1126/science.1243126

Shindo A, Inoue Y, Kinoshita M, Wallingford JB. 2019. PCP-dependent transcellular regulation of actomyosin oscillation facilitates convergent extension of vertebrate tissue. *Dev Biol* 446: 159–167. doi:10.1016/j.ydbio.2018.12.017

Sive HL, Grainger RM, Harland RM. 2000. *Early development of* Xenopus laevis: *a laboratory manual*. Cold Spring Harbor Laboratory Press, Cold Spring Harbor, New York.

Sive HL, Grainger RM, Harland RM. 2007. Dejellying *Xenopus laevis* embryos. *Cold Spring Harb Protoc* doi:10.1101/pdb.prot4731

Wallingford JB, Rowning BA, Vogeli KM, Rothbacher U, Fraser SE, Harland RM. 2000. Dishevelled controls cell polarity during *Xenopus* gastrulation. *Nature* 405: 81–85. doi:10.1038/35011077

Microsurgical Manipulations to Isolate Collectively Migrating Mesendoderm

Lance A. Davidson[1,2,3,4]

[1]*Department of Bioengineering,* [2]*Department of Developmental Biology,* and [3]*Department of Computational and Systems Biology, University of Pittsburgh, Pittsburgh, Pennsylvania 15213, USA*

Mesendoderm mantle closure completes the gastrulation movements of the *Xenopus laevis* embryo and provides an unparalleled opportunity to study collective cell behaviors within a mesenchymal tissue. Free-edge sheet-like collective movements of these tissues contrast with movements of epithelial tissues in that mesendodermal cells are not constrained by tight junctions or adherens junctions, yet migrate in a coherent and persistent mode over several hours. Mesendoderm cells are the largest motile cells in the *Xenopus* embryo and complete a 500-μm migratory path. When mesendoderm is cultured on rigid glass substrates, these cells can exceed 100 μm in length and show a highly persistent leading lamellipodia that can exceed 20 μm from tip to base. These large collectively migrating cells provide a unique imaging opportunity to visualize polarized adhesive and cytoskeletal structures with high-numerical-aperture objectives. Mesendodermal cells in the early embryo originate from around the entirety of the marginal zone and may also be distinguished by their source along the animal–vegetal axis. Here we use the term mesendoderm but note alternative terms for these cells can include head mesoderm, endo-mesoderm, and prechordal mesoderm. This protocol summarizes microsurgical preparation of mesendoderm tissue explants and "windowed" embryos. Skills needed to dissect fragments of the mesendoderm mantle are marginally greater than those needed to isolate animal cap ectoderm and can be mastered within 2 weeks; skills needed to isolate the mesendoderm "donut" or "ring" or to prepare windowed embryos are significantly greater and may require several additional weeks of training.

MATERIALS

It is essential that you consult the appropriate Material Safety Data Sheets and your institution's Environmental Health and Safety Office for proper handling of equipment and hazardous materials used in this protocol.

RECIPES: Please see the end of this protocol for recipes indicated by <R>. Additional recipes can be found online at http://cshprotocols.cshlp.org/site/recipes.

Reagents

Danilchik's for Amy (DFA) Medium (pH 8.3) <R>

DFA explant culture medium is not a random collection of odd buffers and ions but rather a solution created to match the interstitial composition of the Xenopus laevis *embryo that is relatively low in chloride ions and has a slightly basic pH (Gillespie 1983). This suitable culture medium replaces conventional chloride counter ions of* Na^+ *and* K^+ *with gluconate ions. Endogenous interstitial buffering factors are replaced by the Good buffering agent bicine to minimize the amount of hydroxide counter ions needed to create pH 8.3.*

Avoid adding phenol red if preparing for fluorescence imaging. Phenol red will introduce nonspecific fluorescence that can interfere with red or near-infrared fluorophores.

---

[4]Correspondence: lad43@pitt.edu

Supplemental material is available at cshprotocols.cshlp.org

Cite this protocol as *Cold Spring Harb Protoc*; doi:10.1101/pdb.prot097378

Embryo culture medium (1/3× MBS)

1/3× MBS is the de facto embryo culture media used in the Keller, DeSimone, and Davidson laboratories where the mesendoderm explant techniques described here were devised (Davidson et al. 2002; DeSimone et al. 2007; Ichikawa et al. 2020). The Winklbauer laboratory, also known for working with mesendoderm fragments, uses 1/10× MBS (Winklbauer 1988) for embryo culture. Calcium concentrations in embryo culture media vary (Sive et al. 2000) and may yield qualitative differences in the abilities of tissue to remain cohesive during microsurgical manipulations.

Avoid adding phenol red if preparing for fluorescence imaging. Phenol red will introduce nonspecific fluorescence that can interfere with red or near-infrared fluorophores.

Fibronectin (1-mg/mL) (Corning 356008)

Aliquot and store at −20°C.

Modified Barth's solution (MBS) (pH 7.4) <R>

This recipe is for a 1× solution. 3× stocks of MBS are pH-stable and can be stored at 4°C. 10× stocks do not maintain their pH. When used for embryo culture, we dilute our 3× stock to 1/3×.

Xenopus laevis embryos at middle to late gastrula stage (stages 11 to 13) (Nieuwkoop and Faber 1967), dejellied as in Sive et al. (2007) using 2% cysteine (L-cysteine hydrochloride monohydrate) (Sigma-Aldrich C7880) in 1/3× MBS that has been adjusted to a pH of 8.0 with NaOH

Experiments involving mid-gastrula stages may require attention to internal staging criteria described in Nieuwkoop and Faber and whether the vitelline membrane has been removed. Removal of the vitelline can cause changes in 3D morphology, such as deepening of the groove formed by bottle cells. Such changes may make embryos appear more advanced.

Equipment

Coverslip glass (24 × 40-mm, #1.5 thickness) (Fig. 1A)

Custom-fabricated cold plate dissecting stage (recommended for holding temperatures constant; see Method)

The cold plate is a large-footprint aluminum and acrylic base mount for the stereoscope that passes chilled or heated water produced by a chiller-circulating water bath.

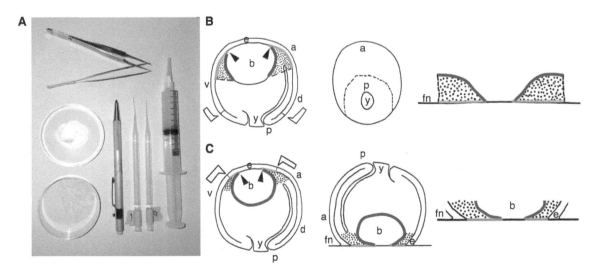

FIGURE 1. Tools and schematics for ring explant and windowed embryo. (*A*) Tools for microdissection including forceps, silicone grease dispenser, hair tools, diamond pencil, precut coverslip glass bridges, and silicone grease. (*B*) Schematic of microsurgical maneuvers to make the ring explant. Sagittal and enface views of late gastrula (Stage 12 to 12.5) with a large blastocoel and broad mantle. (*C*) Schematic of microsurgical maneuvers to make the "windowed" embryo. Sagittal view of gastrula (Stage 12.5 to 13) shows narrowing mantle. The microenvironment and tissue interactions (e.g., continuity of mesendoderm with dorsal and ventral tissue contacts) between leading mesendoderm and more vegetal mesendoderm, blastocoel, and lateral ectoderm are preserved in the windowed embryo. Mesendoderm mantle (stippled) with leading edge (arrowheads). (a) Anterior, (b) blastocoel, (d) dorsal, (e) ectoderm, (p) posterior, (v) ventral, (y) yolk plug. (*Right*) Ring explant schematic on fibronectin-coated substrate (fn).

Diamond pencil (retractable; Ted Pella 54468) (Fig. 1A)

Disposable pipette

Forceps (Dumont #5) (Fig. 1A) (two; see Step 6)

Glass-bottomed imaging chamber (e.g., Cellvis or Mattek dishes, Invitrogen chambers, custom acrylic well chambers, or custom nylon washer chambers)

See Sec. 9, Protocol 5: Chambers for Culturing and Immobilizing Xenopus *Embryos and Organotypic Explants for Live Imaging (Chu and Davidson 2022) and Kim and Davidson (2013).*

Hair knife (Sive et al. 2000) (Fig. 1A)

Hair loop (Sive et al. 2000) (Fig. 1A)

Incubator (e.g., MyTemp Mini Digital Incubator; Benchmark Scientific) (recommended for holding temperatures constant; see Method)

Microscope slide

Petri dishes (25- and 60-mm-diameter)

Pasteur pipette, plastic, cut to remove the narrow region

Silicone grease (high-vacuum silicone grease; Dow Chemical) loaded into a 10-mL syringe

Stereomicroscope

Here, we use an Olympus SZX7 equipped with a video port, Canon T3i, and a video coupler MDSLR (Martin Microscope Co.) with off-axis lighting provided by a Schott Fiber Optic Light Source with dual gooseneck.

ThermoWorks EasyLog USB Logger (TW-USB-2+) (recommended for confirming room temperatures; see Method)

METHOD

Xenopus laevis embryos may be raised in incubators at a range of temperatures within the normal permissive range of 14°C–26°C. Similarly, microsurgical manipulations listed below can be performed either at room temperature or on a temperature controlled cold plate. Rates of morphogenesis and intracellular processes are very sensitive to temperature so it is critical that temperature is held constant and recorded (e.g., ThermoWorks EasyLog USB Logger).

Preparing the Embryos and the Imaging Chamber

Figure 1 shows microsurgical tools for dissection and schematics of the ring explant and the windowed embryo.

1. Prepare a dish of coverslip bridges as follows:

   i. Score and break 3×8 mm coverslip bridges with the diamond pencil using a microscope slide as a straight edge. Store the coverslip bridges in a clean Petri dish (Fig. 1A).

   ii. Dispense silicone grease onto the inner surface of a clean Petri dish lid and smooth into a thin layer (1-mm-thick; Fig. 1A).

   Broken glass and fresh coverslip bridges are extremely sharp. Handle with care. If periodically re-stocked, the Petri dish with the bridge fragments and silicone grease will last indefinitely and provide a ready source during microsurgery.

2. Coat the imaging chamber with fibronectin. Add 20 µg/mL of fibronectin in 1/3× MBS to the chamber and incubate overnight at 4°C. Rinse three times in 1/3× MBS and fill the chamber with DFA with antibiotic–antimycotic.

3. Set up operating dish. Place sufficient DFA into three Petri dishes ∼3–4 mm deep.

   One dish is for microsurgery, one dish is for rinsing fresh explants, and the last dish is for culturing or holding explants until needed.

4. Use forceps to pick up a single coverslip bridge from the dish in Step 1 and dab some silicone grease along the long edge. Place the coverslip bridge grease side down in the center of the Petri dish to be used for microsurgery.

Cite this protocol as *Cold Spring Harb Protoc*; doi:10.1101/pdb.prot097378

This bridge serves as a "backstop" or aid during embryo and explant manipulation and defines a forceps-free region of the microsurgery dish.

Forceps can scar the plastic of the Petri dish that can tear wounds on embryos and explants as they are being manipulated.

5. Select dejellied embryos from the middle to late gastrula stage (Stages 11–13; Nieuwkoop and Faber 1967) and transfer them to the operating dish with a disposable pipette.

    The goals of the experiment will dictate the precise stages needed. For instance, investigations of collective cell migration during mantle closure would require using late gastrula stages at which time the mesendoderm around the entire margin has migrated and the mantle has formed a cup with well-defined walls surrounding the floor of the blastocoel. Alternatively, if the goals of the experiment involve characterizing leading edge migration of mesendoderm, then earlier stages may suffice, as long as the mesendoderm has begun to migrate on the overlying ectoderm. The size of the yolk plug and other external criteria may not correlate with interior progressive movements of the mesendoderm mantle. With side or glancing illumination from a gooseneck style lamp, it is possible to observe the position of the mesendoderm mantle within intact embryos (arrowheads in Fig. 2A). Under these conditions the blastocoel is dark and the edge of the mantle is detected as a lighter region. The progress of mantle closure can be more easily observed in albino embryos; see Figure 1A in Davidson et al. (2002).

6. Remove the vitelline membrane by grabbing the membrane at the same location with two forceps (Fig. 2B). The vitelline membrane is a thin flexible transparent membrane enclosing the embryo after the jelly coat has been removed. Open or tear the membrane by slowly moving the forceps apart. This movement will create a tear in the vitelline membrane. Once the opening is large enough, release one forceps and gently shake the embryo out of the membrane with the other. Do not damage the embryo in the marginal zone or animal pole. If the vitelline membrane cannot be removed without damage, try puncturing the embryo first in the ventral pole.

    One blunt and one sharp forceps can be helpful where the blunt forceps is used to hold, or pinch the vitelline membrane, and the sharp forceps used to tear the membrane. Remove the vitelline membrane in a different region of the Petri dish than where microsurgery will be performed since small scratches caused by forceps can inadvertently wound embryos or explants.

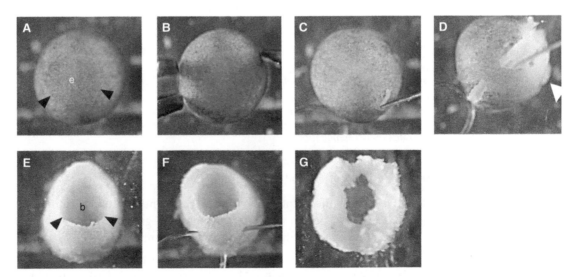

FIGURE 2. Steps of microdissection. (*A*) Select embryos. (*B*) Remove vitelline membrane. (*C*) Begin incision through ectoderm. (*D*) Complete incision around 360° of the embryo, the full extent of the ectoderm from dorsal to ventral. Note the slight anterior-ward offset of the incision (white arrowhead, yolk plug) due to the ventral-ward position of the blastocoel. Remove the ectoderm (e) by peeling from underlying mesoderm (not shown, see also Supplemental Movie S1, which shows a complete recording of microsurgical isolation of the mesendoderm mantle "ring" or "donut"). (*E*) Removal of the ectoderm reveals mesendoderm mantle (leading edge, arrowheads) and blastocoelic space (b). (*F*) Remove the mantle through a 360° incision that cuts across the mesendoderm into the blastocoelic space. (*G*) Separate the mesendoderm mantle from the remaining embryo. Full microsurgical manipulation can be followed in the Supplemental Movie S1.

Depending on the microsurgeon's speed, only remove the vitelline membrane from embryos needed within 15 min. Embryos out of their vitelline membrane for longer than 15 min will sag and may appear to be in a different stage than those remaining in the vitelline.

7. Make an incision fully encircling the pigmented region of the animal cap ectoderm. Insert the tip of a fine hair knife between the involuted mesendoderm and the overlying ectoderm and "flick" outward to make a small incision (Fig. 2C). The mesendoderm should be clearly visible as large cells closely associated with ectoderm indicated by small cells. Insert the hair knife into the hole and push the tip into the embryo to emerge at another point around the margin and "flick" outward; repeat this operation through 360° of the ectoderm (Fig. 2D). Use the hair loop or glass bridge fragment to brace the embryo during the flick so the embryo does not move. Rotate the embryo and continue the incision around the marginal zone. Adjust this circular incision to expose some of the mantle for intravital imaging (e.g., with the rest of the embryo "intact," see "Microsurgical Variations" section below) or most of the involuted mesoderm if a full "ring" explant is desired. See also Supplemental Movie S1, which shows a complete recording of microsurgical isolation of the mesendoderm mantle.

   "Flick cutting" is a microsurgical maneuver that slowly pushes the tip of a hair knife into a tissue followed by a rapid retraction movement that uses the side of the knife to cut through the tissue. Repeated "flick cuts" can be used to produce and lengthen incisions.

8. Gently remove the animal cap from the remainder of the mantle (Fig. 2E). Use the hair knife to gently lift the animal cap from the mesendoderm.

   Removing the cap will expose the leading edge of the mantle and the open cavity of the blastocoel. Some damage to mesendoderm or mesoderm can be tolerated as long as the leading edge is not wounded.

Microsurgical Variations

At this point the protocol branches based on specific goals as follows: (1) Single fragments of the leading edge can be isolated for the study of collective migration or multiple fragments can be arranged for confrontation (e.g., where two opposing leading edges collide, needed to test contact inhibition of locomotion) (Ichikawa et al. 2020) (Step 9). (2) The entire mantle can be isolated from the embryo for studies of closure in the absence of the blastocoel (Step 10). (3) The entire mantle and remainder of the embryo, with blastocoel intact, can be positioned for intravital imaging (Step 11). This configuration preserves contact between more vegetal mesendoderm and ectoderm and may retain the contents of blastocoel. (4) The mesendoderm can be isolated together with the marginal zone explant (Step 12).

In all of these cases, mesendoderm and accompanying tissues are cultured on a fibronectin-coated substrate. Protein localization and dynamics in lamellipodia, mediobasal cortex, cell–substrate and cell–cell adhesions are accessible to live-cell confocal imaging by a high numerical aperture objective (e.g., 63× or 40× oil-immersion objective) with short working distances (<150 µm). Simple cell boundaries or labeled nuclei can be imaged with 20× air or long working distance water-immersion objectives. Time lapses collected from a simple stereoscope can be used to record and quantify rates of tissue migration.

9. Isolate fragments of mesendoderm. For this preparation, select embryos when the mantle has progressed within 30°–60° of the animal pole.

   i. Use a hair knife to cut segments from the mantle from the leading edge to the vegetal margin of the blastocoel floor. Separate the mantle segment from the embryo by rolling the embryo and "flick cutting" the embryo from the mantle segment.

   ii. Immediately transfer the explant using a plastic Pasteur pipette that has been trimmed to remove the narrow region. Move the explant to a dish of fresh DFA and then transfer to the imaging chamber. Do not "expel" the explant from the pipette; instead let the explant fall into the dish or chamber under gravity (if needed rotate the pipette to dislodge a stuck explant).

   iii. Use a glass coverslip bridge with a dab of silicone grease on both ends to gently compress the mantle segment so that the face of the explant that had previously faced the blastocoel roof is brought into contact with the fibronectin substrate. Position the bridge sideways over the explant and then pivot the fragment and iteratively compress each end of the bridge over the

Cite this protocol as *Cold Spring Harb Protoc*; doi:10.1101/pdb.prot097378

dab. Too much compression can cause lysis and cell death. Too little compression will fail to bring the mesendoderm into contact with the fibronectin substrate. Adequate compression typically produces a maximum of 10%–20% strain as measured by the lengthening or widening of the compressed explant with minimal cell rupture.

The ectoderm-contacting surface can be identified throughout manipulations as the flatter surface of the mesendoderm. Leading edge migration rates of fragments are nearly 1/2 to 1/3 the rate of the intact ring; however, increased rates can be achieved by positioning fragments so they converge (e.g., "in-the-round" configuration in Fig. 10 from Davidson et al. 2002).

Proceed to Step 13.

10. Isolate the entire rim of mesendoderm mantle, a.k.a. the "donut" or "ring" explant. For this preparation, allow the mantle to progress within 25°–45° of the animal pole. With proper lighting from the side and a dark background, mantle progression may be observed as a lighter region surrounding a dark shadow (e.g., the blastocoel) under the animal cap. However, mantle progress can also be tracked by sampling embryos at regular time points and removing their animal cap ectoderm.

    i. Remove the animal ectoderm of the blastocoel roof so that the edge of the incision is even with the floor of the blastocoel. Insert a fine hair loop into the mesendoderm with the tip emerging from the blastocoel floor (Fig. 2F). Use a gentle sawing motion, removing the mesendoderm by using the hair tool to cut as it is pulled out of the tissue. This maneuver will cut the rim of the mesendoderm free from the vegetal half of the embryo.

    ii. Rotate the embryo through the full 360° to dissect the entire mesendoderm mantle from the embryo (Fig. 2G). The ring explant is exceptionally delicate and easily torn. Immediately transfer the explant to a dish of fresh DFA and then transfer to the culture or imaging chamber.

    iii. Use a glass coverslip bridge to gently compress the explant so that the face that had bound the blastocoel roof (typically smooth) is brought into contact with the fibronectin substrate. Application of too much compression will cause fluid in the center of the explant to "blow out" and break the ring of the mesendoderm mantle.

    Proceed to Step 13.

11. Isolate an intravital or "windowed" embryo preparation of mesendoderm mantle with the remainder of embryo intact. For this preparation the mantle should have progressed within 20°–30° of the animal pole. Progress can be tracked as described in Step 10.

    i. For best results remove only a small patch of ectoderm to expose 100–300 μm of the mesendoderm. The mantle and embryo are extremely delicate and may be transferred immediately to the imaging chamber without an intervening rinse in fresh DFA.

    ii. Position the embryo with the mesendoderm mantle facing the fibronectin substrate. Take a broad coverslip bridge with large dollop of silicone grease on either end and gently compress the embryo. The mantle should adhere to the fibronectin with the cell face that had been contacting the blastocoel roof brought to contact the fibronectin substrate. Too much compression will cause the blastocoelic fluid to "blow out" a portion of the mesendoderm mantle.

    Proceed to Step 13.

12. Isolate mesendoderm together with the marginal zone explant.

    i. This manipulation is better described as a variation of the marginal zone explant (see Protocol 3: Microsurgical Methods to Isolate and Culture the Early Gastrula Dorsal Marginal Zone [Davidson 2022]). In this variation, bottle cells and a few rows of subblastoporal endoderm are not removed from the marginal zone explant. Deep cells under the retained endoderm are still considered mesendoderm and adhere strongly to a fibronectin-coated substrate. Within the embryo, the smaller size and anterior fate of these cells differ from the larger, ventral-fated mesendoderm at the leading edge cells of the mantle. The mesendoderm isolated with the marginal zone expresses genes common to head mesoderm and anterior endoderm (Davidson et al. 2004).

ii. Mesendoderm cells at the vegetal edge of marginal zone explants, like preparations of the mesendoderm mantle, are larger than mesoderm cells and migrate in a collective manner similar to the leading edge mesendoderm from the mantle. Leading edge mesendoderm cells in marginal zone explants extend broad lamellipodia and migrate collectively away from the marginal zone, showing qualitatively similar morphologies, polarity, and collective migratory behaviors to leading edge mesendoderm cells isolated from the mantle preparations.
Proceed to Step 13.

Imaging Mesendoderm

13. Because of the rapid closure of 360° mantle preparations from Steps 10 and 11, perform imaging as soon as possible. Leading edge mesendoderm cells within these geometrically confined preparations can migrate at 300 µm/h and will resume migration in a highly persistent manner with large stable lamellipodia immediately after they are brought in contact with fibronectin. Once dissected, either as a tissue ring or in a windowed embryo, the mesendoderm mantle will collapse unless provided a fibronectin coated substrate. Collective migration will continue until cells at the leading edge contact cells migrating from the other side. In a classic display of contact inhibition of locomotion, leading edge cells that contact opposing cells will retract their lamellipodia within seconds, cease directed migration, and extend short-lived protrusions in random directions. Image collective migration of mesendoderm cells in tissues prepared from Steps 9 and 12 within 15–30 min after they have attached to the fibronectin substrate. Leading edge mesendoderm cells in these preparations will continue to show persistent directed migration at 100 µm/h for 5–10 h after isolation.

DISCUSSION

Mesendoderm explants may be dissected together with marginal zone tissues, as an isolated fragment, as an isolated ring, or nearly intact in a windowed embryo (Davidson et al. 2002) and can be used as a source of mesenchymal cells to study collective migration distinct from migration of epithelia. Mesendoderm cells in explants show collective migration that is thought to occur in vivo on the inner surface of the blastocoel roof (Nakatsuji and Johnson 1982; Winklbauer and Nagel 1991; Winklbauer et al. 1992; Winklbauer and Keller 1996; Ichikawa et al. 2020). High-resolution imaging of migratory cells in the mantle can reveal both adhesive and cytoskeletal dynamics because of the remarkably large size of mesendoderm cells, often longer than 100 µm during later stages of migration and mantle closure. Portions of the mesendoderm can be excised and used to study migration of cell clusters in gradients of extracellular matrix or in gradients of chemo-attractant (Nagel et al. 2004). Remarkably, mesendoderm masses can also be separated into single cells using calcium- and magnesium-free culture media to investigate single cell migration and contact inhibition (Ichikawa et al. 2020).

Mesendoderm cells are large and easily distinguished from smaller lateral plate or dorsal mesodermal derivatives. Furthermore, mesendoderm can be identified by expression of Sox17 (Hudson et al. 1997). It is possible to inadvertently include dorsal mesodermal progenitors with the mantle, but this contamination can be readily detected in explants cultured to equivalent stage 23 by immunofluorescence with tor70 (notochord) or 12/101 (somitic mesoderm) at later stages. Thus, mesendoderm explants offer the opportunity to study mass mesenchymal migration, collective movements, and single-cell movements from an easily isolated source.

From their origins around the marginal zone, mesendoderm cells migrate on the inner fibronectin-coated surface of the blastocoel as the leading edge of the mesendoderm "mantle" converges toward the animal pole (Ewald et al. 2004). As the mesendoderm mantle converges, cells at the leading edge form a perimeter shaped like a shrinking circle or ellipse. Mantle closure completes the major tissue internal movements of gastrulation by positioning mesodermal cells into the ventral-most regions of the gastrula and early neurula. After mantle closure, mesendoderm seals off the connection between the blastocoel and the animal cap ectoderm.

There is ongoing debate concerning the nature of cues driving directional migration of mesendoderm (Davidson et al. 2002; Nagel et al. 2004). It has been suggested that mechanical cues drive directional migration through "cohesotaxis" as polarized protrusions from the leading free edge of mesendoderm cells can be initiated by tension directed in the opposite direction at the rear of the cell (Weber et al. 2012; Sonavane et al. 2017). C-cadherin (CDH3) regulates both cell–cell adhesion and nonjunctional signaling to alter persistence, velocity, and contact inhibition of locomotion (Ichikawa et al. 2020). The native substrate of the mesendoderm (i.e., the extracellular matrix coat of the blastocoel animal cap ectoderm) includes fibrillar fibronectin assembled by deep ectoderm cells. It has been proposed that mesendoderm may follow directional cues embedded in this extracellular matrix using a form of haptotaxis (Winklbauer and Nagel 1991). Alternatively, mesendoderm may also follow chemotactic cues secreted by the animal cap ectoderm (Nagel et al. 2004; Smith et al. 2009; Nagel and Winklbauer 2018). Mesendoderm explants offer the unique opportunity to test these and other hypotheses and elucidate principles that integrate molecular pathways, biomechanics, and mechanobiology.

RECIPES

Danilchik's for Amy (DFA) Medium (pH 8.3)

Reagent	Final concentration
NaCl	53 mM
Na_2CO_3	5 mM
Potassium gluconate	4.5 mM
Sodium gluconate	32 mM
$CaCl_2$	1 mM
$MgSO_4$	1 mM

Adjust the pH to 8.3 with granular bicine. Filter with a 0.22 μm filter. Include 0.1% bovine serum albumen (BSA; Sigma-Aldrich A7906) to prevent nonspecific adhesion to substrates or glass coverslips. Just prior to use, add 1% of antibiotic/antimycotic (Sigma-Aldrich A5955). Prepare the DFA with BSA, aliquot, and store at −20°C. Keep aliquots of antibiotic/antimycotic at −20°C and add to thawed DFA before use. Replace DFA with medium containing fresh antibiotic/antimycotic if culturing explants for >24 h. (This recipe is adapted from Sater et al. 1993.)

Modified Barth's Solution (MBS) (pH 7.4)

Reagent	Final concentration (1×)
NaCl	88 mM
KCl	1 mM
$NaHCO_3$	2.4 mM
$MgSO_4 \cdot 7H_2O$	0.82 mM
$Ca(NO_3)_2 \cdot 4H_2O$	0.33 mM
$CaCl_2 \cdot 2H_2O$	0.41 mM
HEPES	5 mM

Adjust the pH to 7.4 with NaOH; 3× stocks of MBS are pH-stable and can be stored at 4°C. Prepare a working solution (1/3×) by dilution.

ACKNOWLEDGMENTS

I thank the members of the Davidson, Keller, and DeSimone laboratories past and present for their support and collective knowledge. The ability to carry out repeatable microsurgical manipulations in the early embryo requires patience and skills akin to a master potter or a neurosurgeon; thus, I owe a deep debt of gratitude to video recordings and descriptions of microsurgical manipulations in other amphibian, avian, and teleost embryos by Drs. Gary Schoenwolf, Ray Keller, Rudy Winklbauer, and John Trinkaus. I also thank the National Xenopus Resource (RRID:SCR-013731) and Xenbase (RRID: SCR-003280) for their invaluable support to the model organism community. L.A.D. is funded by the Eunice Kennedy Shriver Institute of Child Health and Human Development at the National Institutes of Health (NIH) (R01 HD044750 and R21 HD106629).

REFERENCES

Chu C-W, Davidson LA. 2022. Chambers for culturing and immobilizing *Xenopus* embryos and organotypic explants for live imaging. *Cold Spring Harb Protoc* doi:10.1101/pdb.prot107649

Davidson LA. 2022. Microsurgical methods to isolate and culture the early gastrula dorsal marginal zone. *Cold Spring Harb Protoc* doi:10.1101/pdb.prot097360

Davidson LA, Hoffstrom BG, Keller R, DeSimone DW. 2002. Mesendoderm extension and mantle closure in *Xenopus laevis* gastrulation: combined roles for integrin α5β1, fibronectin, and tissue geometry. *Dev Biol* 242: 109–129. doi:10.1006/dbio.2002.0537

Davidson LA, Keller R, DeSimone D. 2004. Patterning and tissue movements in a novel explant preparation of the marginal zone of *Xenopus laevis*. *Gene Expr Patterns* 4: 457–466. doi:10.1016/j.modgep.2004.01.001

DeSimone DW, Dzamba B, Davidson LA. 2007. Using *Xenopus* embryos to investigate integrin function. *Methods Enzymol* 426: 403–414. doi:10.1016/S0076-6879(07)26017-3

Ewald AJ, Peyrot SM, Tyszka JM, Fraser SE, Wallingford JB. 2004. Regional requirements for Dishevelled signaling during *Xenopus* gastrulation: separable effects on blastopore closure, mesendoderm internalization and archenteron formation. *Development* 131: 6195–6209. doi:10.1242/dev.01542

Gillespie JI. 1983. The distribution of small ions during the early development of *Xenopus laevis* and *Ambystoma mexicanum* embryos. *J Physiol (Lond)* 344: 359–377. doi:10.1113/jphysiol.1983.sp014945

Hudson C, Clements D, Friday RV, Stott D, Woodland HR. 1997. Xsox17-α and -β mediate endoderm formation in *Xenopus*. *Cell* 91: 397–405. doi:10.1016/S0092-8674(00)80423-7

Ichikawa T, Stuckenholz C, Davidson LA. 2020. Non-junctional role of Cadherin3 in cell migration and contact inhibition of locomotion via domain-dependent, opposing regulation of Rac1. *Sci Rep* 10: 17326. doi:10.1038/s41598-020-73862-y

Kim HY, Davidson LA. 2013. Assembly of chambers for stable long-term imaging of live *Xenopus* tissue. *Cold Spring Harb Protoc* doi:10.1101/pdb.prot073882

Nagel M, Winklbauer R. 2018. PDGF-A suppresses contact inhibition during directional collective cell migration. *Development* 145: dev162651. doi:10.1242/dev.162651

Nagel M, Tahinci E, Symes K, Winklbauer R. 2004. Guidance of mesoderm cell migration in the *Xenopus* gastrula requires PDGF signaling. *Development* 131: 2727–2736. doi:10.1242/dev.01141

Nakatsuji N, Johnson KE. 1982. Cell locomotion in vitro by *Xenopus laevis* gastrula mesodermal cells. *Cell Motil* 2: 149–161. doi:10.1002/cm.970020206

Nieuwkoop PD, Faber J. 1967. *Normal tables of* Xenopus laevis *(Daudin)*. Amsterdam, Elsevier North-Holland Biomedical Press

Sive HL, Grainger RM, Harland RM. 2000. *Early development of* Xenopus laevis: *a laboratory manual*. Cold Spring Harbor Laboratory Press, Cold Spring Harbor, New York.

Sive HL, Grainger RM, Harland RM. 2007. Dejellying *Xenopus laevis* embryos. *Cold Spring Harb Protoc* doi:10.1101/pdb.prot4731

Smith EM, Mitsi M, Nugent MA, Symes K. 2009. PDGF-A interactions with fibronectin reveal a critical role for heparan sulfate in directed cell migration during *Xenopus* gastrulation. *Proc Natl Acad Sci* 106: 21683–21688. doi:10.1073/pnas.0902510106

Sonavane PR, Wang C, Dzamba B, Weber GF, Periasamy A, DeSimone DW. 2017. Mechanical and signaling roles for keratin intermediate filaments in the assembly and morphogenesis of mesendoderm tissue at gastrulation. *Development* 144: 104–115. doi:10.1242/dev.155200

Sater AK, Steinhardt RA, Keller R. 1993. Induction of neuronal differentiation by planar signals in *Xenopus* embryos. *Dev Dyn* 197: 268–280. doi:10.1002/aja.1001970405

Weber GF, Bjerke MA, DeSimone DW. 2012. A mechanoresponsive cadherin-keratin complex directs polarized protrusive behavior and collective cell migration. *Dev Cell* 22: 104–115. doi:10.1016/j.devcel.2011.10.013

Winklbauer R. 1988. Differential interaction of *Xenopus* embryonic cells with fibronectin *in vitro*. *Dev Biol* 130: 175–183. doi:10.1016/0012-1606(88)90424-1

Winklbauer R, Keller RE. 1996. Fibronectin, mesoderm migration, and gastrulation in *Xenopus*. *Dev Biol* 177: 413–426. doi:10.1006/dbio.1996.0174

Winklbauer R, Nagel M. 1991. Directional mesoderm cell migration in the *Xenopus* gastrula. *Dev Biol* 148: 573–589. doi:10.1016/0012-1606(91)90275-8

Winklbauer R, Selchow A, Nagel M, Angres B. 1992. Cell interaction and its role in mesoderm cell migration during *Xenopus* gastrulation. *Dev Dyn* 195: 290–302. doi:10.1002/aja.1001950407

Microsurgical Methods to Make the Keller Sandwich Explant and the Dorsal Isolate

Lance A. Davidson[1,2,3,4]

[1]*Department of Bioengineering,* [2]*Department of Developmental Biology,* [3]*Department of Computational and Systems Biology, University of Pittsburgh, Pittsburgh, Pennsylvania 15213, USA*

This protocol summarizes preparation of the dorsal marginal zone sandwich explant (a.k.a. the "Keller sandwich") and the dorsal isolate from *Xenopus* embryos. The Keller sandwich is assembled from two early gastrula stage dorsal marginal zone (DMZ) explants. DMZ explants isolated before involution maintain planar patterning processes and block radial signals that might be exchanged between pre- and postinvolution tissues. DMZ explants isolated later in gastrulation, but subsequently opened and flattened may have both planar and radial patterning. The epithelial margins of DMZ explants in Keller sandwiches heal and basal contacts form between the deep layers of the two DMZ explants. The dorsal isolate is dissected from mid- to late-gastrula stage embryos after involution and archenteron formation. Germ-layer contacts between dorsal endoderm, mesoderm, and ectoderm generated by gastrulation movements are maintained in the dorsal isolate. These two explants can be used to study tissue, cell, and subcellular processes relevant to convergent extension, from patterning to cell behaviors, and their collective biomechanics. Skills needed to dissect the Keller sandwich are greater than those needed to dissect animal cap ectoderm and can be mastered in a few weeks; skills needed to dissect the dorsal isolate are similar to those needed to dissect animal caps and can be learned in a week.

MATERIALS

It is essential that you consult the appropriate Material Safety Data Sheets and your institution's Environmental Health and Safety Office for proper handling of equipment and hazardous materials used in this protocol.

RECIPES: Please see the end of this protocol for recipes indicated by <R>. Additional recipes can be found online at http://cshprotocols.cshlp.org/site/recipes.

Reagents

Agarose, electrophoresis grade (Bio-Rad 1613102) (optional; see Step 2)

Antibodies recognizing notochord (mAb tor70) or somitic mesoderm (mAb 12/101) (Developmental Studies Hybridoma Bank)

Danilchik's for Amy (DFA) Medium (pH 8.3) <R>

> *DFA explant culture medium is not a random collection of odd buffers and ions but rather a solution created to match the interstitial composition of the Xenopus laevis embryo that is relatively low in chloride ions and has a slightly basic pH (Gillespie 1983). This suitable culture medium replaces conventional chloride counter ions of*

[4]Correspondence: lad43@pitt.edu

Supplemental material is available at cshprotocols.cshlp.org

Na⁺ and K⁺ with gluconate ions. Endogenous interstitial buffering factors are replaced by the Good buffering agent bicine to minimize the amount of hydroxide counter ions needed to create pH 8.3.

Avoid adding phenol red if preparing for fluorescence imaging. Phenol red will introduce nonspecific fluorescence that can interfere with red or near-infrared fluorophores.

Embryo culture medium (1/3× MBS)

1/3× MBS is the de facto embryo culture medium used in the Keller, DeSimone, and Davidson labs where explant techniques described here were developed and used (e.g., Keller et al. 1985a,b; Keller and Danilchik 1988; Wilson et al. 1989; Wilson 1990; Marsden and DeSimone 2003; DeSimone et al. 2005, 2007; Davidson et al. 2006; Zhou et al. 2009, 2010, 2015; Shawky et al. 2018; Shook et al. 2018). Calcium concentrations in embryo culture media vary (Sive et al. 2000) and may yield qualitative differences in the ability of tissues to remain cohesive during microsurgical manipulations.

Avoid adding phenol red if preparing for fluorescence imaging. Phenol red will introduce nonspecific fluorescence that can interfere with red or near-infrared fluorophores.

Modified Barth's solution (MBS) (pH 7.4) <R>

This recipe is for a 1× solution. 3× stocks of MBS are pH stable and can be stored at 4°C. 10× stocks do not maintain their pH. When used for embryo culture, we dilute our 3× stock to 1/3×.

Xenopus laevis embryos at early gastrula stage (10+ to 11) (Nieuwkoop and Faber 1967) for Keller sandwiches or at mid-gastrula to post-neurula stage (12 to 21) for dorsal isolates, dejellied as in Sive et al. (2007) using 2% cysteine (L-cysteine hydrochloride monohydrate; Sigma-Aldrich C7880) in 1/3× MBS that has been adjusted to a pH of 8.0 with NaOH

Experiments involving mid-gastrula stages may require attention to the internal staging criteria described in Nieuwkoop and Faber (1967) and whether the vitelline membrane has been removed. Removal of the vitelline can cause changes in 3D morphology, such as deepening of the groove formed by bottle cells. Such changes may make embryos appear more advanced.

Equipment

CMOS camera (e.g., FLIR Blackfly BFLY-U3-23S6M-C based on compatibility with MicroManager; live video and stills recorded with a Canon T3i and video coupler MDSLR from Martin Microscope Co.)

Computer for microscope control (e.g., Windows 11 compatible with 64 GB)

Coverslip glass (24 × 40-mm, #1.5 thickness; Fig. 1A)

Custom-fabricated cold plate dissecting stage (recommended for holding temperatures constant; see Method)

The cold plate is a large-footprint aluminum and acrylic base mount for the stereoscope that passes chilled or heated water, produced by a chiller-circulating water bath. Lower temperatures on the dissecting stage slow developmental time as well as unintentional wound healing in dissected explants.

Diamond pencil (retractable; Ted Pella 54468)

Disposable pipettes

Forceps (Dumont #5; Fig. 1A) (two; see Step 5)

Glass-bottomed imaging chamber (e.g., Cellvis or Mattek dishes, Invitrogen chambers, or custom acrylic well chambers, or custom nylon washer chambers)

See Sec. 9: Protocol 5: Chambers for Culturing and Immobilizing Xenopus Embryos and Organotypic Explants for Live Imaging (Chu and Davidson 2022) and Kim and Davidson (2013).

Hair knife (Fig. 1A; Sive et al. 2000)

Hair tools can be customized for specific functions. For instance, a stiff hair tool can be used for major cutting and a fine-tipped hair tool used for removal of single cell layers.

Hair loop (Fig. 1A; Sive et al. 2000)

Incubator (e.g., MyTemp Mini Digital Incubator; Benchmark Scientific) (recommended for holding temperatures constant; see Method)

Microscope automation and acquisition control (e.g., MicroManager, Edelstein et al. 2014)

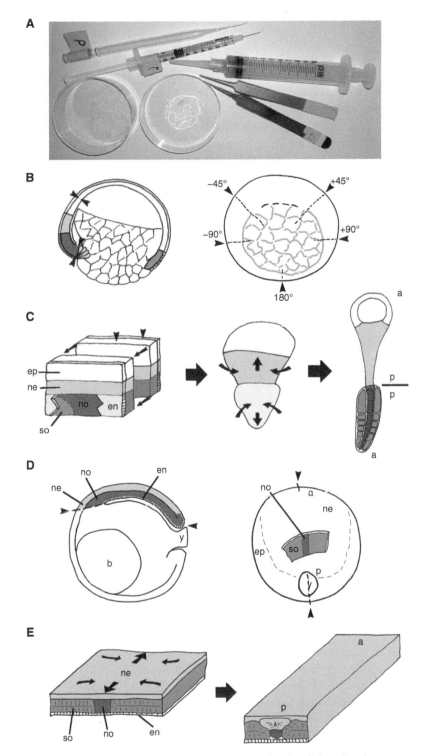

FIGURE 1. Microsurgical tools and schematics of the sandwich explant and dorsal isolate. (*A*) Microsurgical tools including silicone grease, syringe "caulking gun" of silicone grease, hair knife (1-mL syringe) and hair tool (Pasteur pipette), glass bridge kit with precut bridges and smear of silicone grease, and pair of forceps. (*B*) Sagittal and en face diagrams of embryo source for sandwich explant. Arrows indicate incisions needed to create a "Keller" sandwich (±45°), wider sandwich (±90°), and "giant" sandwich (180°; i.e., ventral midline). (*C*) Combining two marginal zone explants into a sandwich, intermediate stage of elongation, and later stage of elongation (after Keller et al. 2000). Arrows in *middle* panel indicate mediolateral convergence (small) and anterior–posterior extension (large). Narrowing should occur at the border between mesoderm and neural ectoderm (line). The orientation of the elongated mesoderm and neural ectoderm corresponds to in vivo anterior–posterior axes but are joined at their respective posterior ends. (*D*) Sagittal and en face diagrams of embryo source for dorsal isolate. Arrows indicate sites of entry and exit of hair knife. (*E*) Diagram of early and late stages of dorsal isolate elongation. Arrows indicate mediolateral convergence (small) and anterior–posterior extension (large). (b) Blastocoel, (en) endoderm, (ep) epidermis, (ne) neural ectoderm, (no) notochord, (y) yolk plug, (a) anterior, (p) posterior.

Microscope slide
Multiwell dishes (optional; see Step 14)
Off-axis lighting (e.g., Schott Fiber Optic Light Source with dual gooseneck)
Petri dishes (10- and 60-mm-diameter)
Silicone grease (high-vacuum silicone grease; Dow Chemical) loaded into a 10-mL syringe
Stereomicroscope (e.g., Olympus SZX7 equipped with video port)
Temperature logger (e.g., ThermoWorks EasyLog USB Logger, TW-USB-2+) (recommended for confirming room temperatures; see Method)

METHOD

Xenopus laevis embryos may be raised in incubators at a range of temperatures between 14°C and 26°C. Similarly, microsurgical manipulations listed below can be performed either at room temperature or on a temperature-controlled cold plate. Rates of morphogenesis and intracellular processes are very sensitive to temperature so it is critical that temperature is held constant and recorded (e.g., ThermoWorks EasyLog USB Logger). To accurately describe the developmental stage of explants, it is important, whenever possible, to coculture whole embryos intact within their vitelline in the same dish or chamber as the explants described below.

Preparing the Embryos

1. Prepare a dish of coverslip bridges as follows:

   i. Score and break 3×8-mm coverslip bridges with the diamond pencil using a microscope slide as a straight edge. Store the coverslip bridges in a clean Petri dish (Fig. 1A).

   ii. Dispense silicone grease onto the inner surface of a clean Petri dish lid and smooth into a thin layer (1-mm-thick; Fig. 1A).

      Broken glass and fresh coverslip bridges are extremely sharp. Handle with care. If periodically re-stocked, the Petri dish with the bridge fragments and silicone grease will last indefinitely and provide a ready source during microsurgery.

2. Set up the operating dish. Fill three Petri dishes ~3–4-mm-deep with DFA. One dish is for microsurgery, one dish is for rinsing fresh explants, and the last dish is for culturing or holding explants until needed.

      The operating dish can be coated with 2% agarose. Melt electrophoresis grade agarose in 1/3× MBS and pour a thin layer into a Petri dish. Use the coated Petri dish for dissection and transfer explants to a non-coated dish to assemble sandwich explants.

3. (Optional) Use forceps to pick up a single coverslip bridge from the dish in Step 1 and dab some silicone grease along the long edge. Place the coverslip bridge grease side down in the center of the microsurgery dish. This bridge serves as a "backstop" or aid during embryo and explant manipulation and defines a forceps-free region of the microsurgery dish.

      Forceps can create sharp edges in the Petri dish that can tear wounds on embryos and explants as they are being manipulated.

4. Select dejellied embryos according to the desired stage (Nieuwkoop and Faber 1967) and transfer them to the microsurgery dish with a disposable pipette.

      The stage of the embryos used as source for these explants may change gene expression patterns and cell behaviors in the explant. For instance, Keller sandwiches—for example, two DMZ explants—show limited patterning of the hindbrain and different cell behaviors if stage 10 embryos are used in place of stage 10+ embryos (Poznanski and Keller 1997; Poznanski et al. 1997).

5. Remove the vitelline membrane. Use forceps to remove the vitelline membrane by grabbing the membrane at the same location with two forceps. Open, or tear the membrane by slowly moving the forceps apart. This movement will create a tear in the vitelline membrane. Once the opening is large enough, release one forceps and gently shake the embryo out of the membrane with the other. Do not damage the embryo in the region to be observed. If the vitelline membrane cannot

Cite this protocol as *Cold Spring Harb Protoc*; doi:10.1101/pdb.prot097386

be removed without damage, puncture the embryo in the animal pole. After the blastocoel collapses the vitelline membrane can be more easily removed.

> *One blunt and one sharp forceps can be helpful where the blunt forceps is used to hold or pinch the vitelline membrane and the sharp forceps used to tear the membrane.*

> *Use forceps to remove the vitelline membrane in a different region of the Petri dish than where microsurgery will be performed because scratches in plastic caused by forceps can inadvertently wound embryos or explants.*

6. Proceed to Step 7 to prepare a Keller sandwich or Step 17 to prepare a dorsal isolate.

Keller Sandwiches

> *There are numerous variations on Keller sandwiches, including the "pita" and "giant" sandwiches. These explants and their application are best described in Poznanski and Keller (1997).*

7. Use microsurgical manipulations to separate the marginal zone from the early gastrula stage embryo (Fig. 1B) following instructions from Protocol 3: Microsurgical Methods to Isolate and Culture the Early Gastrula Dorsal Marginal Zone (Davidson 2022). Prepare the Keller sandwich explant from two narrow preparations of the marginal zone with animal-to-vegetal cuts at +45° and −45° from the midline; larger sandwiches can be made with wider preparations (e.g., +90° and −90° from the midline). The giant sandwich explant is prepared from two full marginal zone explants that are each "unwrapped" from a single incision at 180° (Fig. 1B). All of these sandwich explants will elongate in the anterior to posterior directions and narrow in the mediolateral direction (Fig. 1C). Production of a larger sandwich explant begins with the dissection of two marginal zone explants from stage-matched embryos (Fig. 2A).

   > *Typically, embryos in a clutch appear at slightly different stages. It is important that the two donor embryos show the same external criteria.*

8. With the two explants side by side in the field of view, trim the lateral and animal margins of the explant so that they match in size (Fig. 2B). The conventional Keller sandwich consists of two DMZ explants that extend ±45° from the midline. This region contains the entire prospective notochord but only a few prospective somites. Take care to keep margins of the explants straight and free of microtears.

9. Make the sandwich. Using a disposable pipette, move the explants to a region of the dish that is free of debris or into a separate dish containing DFA. Place the explants in a sandwich configuration with the animal vegetal axes aligned with the deep cells of one explant facing the deep cells of the second. Gently press them together with a hair loop (Fig. 2C).

10. Prepare a coverslip bridge by applying a small dab of silicone grease to both ends.

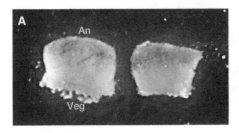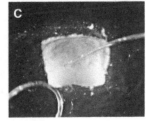

FIGURE 2. Assembly of sandwich explant from two dorsal marginal zone explants following dissection of marginal zone explants. (*A*) Two dorsal marginal zone (DMZ) explants immediately following dissection as described in Protocol 3: Microsurgical Methods to Isolate and Culture the Early Gastrula Dorsal Marginal Zone (Davidson 2022). Pigmented cells are found at the animal domain (an), whereas lighter pigmented cells are found at the vegetal domain (veg). (*B*) Trim the two explants so they match in size. Explants before "stacking" are positioned so their deep layers (deep) face each other and the superficial layers face outward (super). (*C*) Align the two explants so they have the same orientation. See Supplemental Movie S1.

11. Compress the sandwich under the coverslip bridge. Place the coverslip bridge over the sandwich and alternately press the ends over the grease dabs to compress the two explants together while keeping them aligned.

12. Let the margins of the sandwich explant heal for 30–60 min.

13. Remove the coverslip bridge, which will allow the sandwich to converge. Using forceps, grasp the coverslip bridge at one end where there is silicone grease and slowly lift the bridge from the sandwich. Move to the other end and lift the bridge from the sandwich and remove the bridge from the dish. Take care not to compress or crush the explant.

14. Culture sandwich explants in DFA where they will undergo convergent extension (e.g., narrowing along their mediolateral axis and lengthening along their anterior–posterior axis; Fig. 1C). Culture sandwich explants at room temperature or in temperature-controlled incubators used for whole-embryo culture. The rate of morphogenesis in sandwich explants will parallel movements in whole embryos. Culture explants for 10–20 h until late neurula or tailbud stages. Sandwich explants can adhere to each other so take care to maintain space between the explants or culture them separately in multiwell dishes or agarose wells.

    If agarose wells are used, prepare them as described in Step 2.

    To perfect this technique, it helpful to take images or sketch each individual sandwich explant when it has been completed, after the coverslip bridge is removed, and at the conclusion after extension. Include several whole embryos in the dish or chamber with explants to determine the "equivalent stage" of embryonic development.

    A successful sandwich will first heal around the margins of the two DMZ explants and undergo convergent extension (CE) in both the animal neural ectoderm half, and the vegetal, mesodermal half (Fig. 1C). There will be a distinctive narrowing where these two tissues are joined; this site is analogous to the limit of involution, where ectoderm and endoderm meet at the blastopore in the embryo. This narrowing marks the posterior end of both the neural and mesodermal halves; the anterior ends of these tissues lie at opposite ends of the elongated sandwich.

    See Troubleshooting.

15. Quantify sandwich explant morphogenesis using either time-lapse imaging or via end point analysis. Collect both end point images and time-lapse sequences using low-magnification stereomicroscopes equipped with CMOS cameras suitable for image acquisition (see Wallingford 2010). Typically, control and experimentally manipulated explants can be contrasted by their anterior-to-posterior length.

16. Confirm neural and mesodermal differentiation either with immunofluorescence or by RNA in situ hybridization (e.g., Keller and Danilchik 1988; Domingo and Keller 1995; Davidson et al. 2004). Confirm notochord and somitic mesodermal differentiation using antibodies recognizing notochord (mAb tor70) or somitic mesoderm (mAb 12/101).

    This step is often needed to confirm that molecular manipulations have not disrupted patterning or to confirm that microsurgery successfully targeted desired tissues.

Dorsal Isolates

17. Position the embryo next to the bridge coverslip "backstop."

18. Use hair tools to identify the dorsal blastopore lip and the archenteron (Fig. 3A; anterior to the left). Assess the degree of archenteron formation by probing the line of bottle cells around the blastopore ring. From stage 10.5, the yolk plug is encircled by a continuous line of bottle cells (Lee and Harland 2007, 2010). Archenteron formation occurs dorsally as bottle cells expand to cover postinvolution mesoderm (Ewald et al. 2004). Deduce involution by the lack of visible bottle cells in the dorsal cleft of the blastopore lip (Feroze et al. 2015). If beginning with a mid-gastrula to post-neurula stage (12 to 21) embryo, the archenteron at this point should extend the full anterior–posterior axis.

    Xenopus laevis developmental anatomy can be found in Nieuwkoop and Faber (1967).

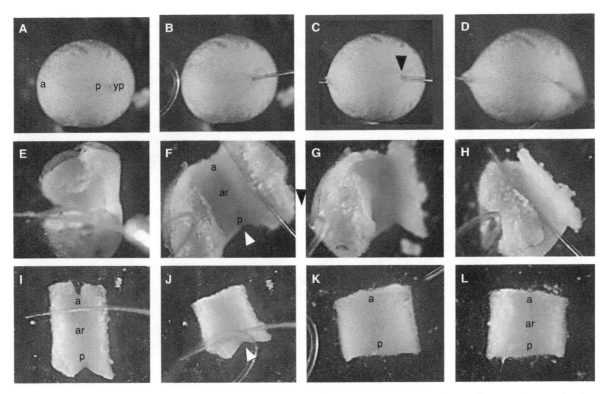

FIGURE 3. Dissection of the dorsal isolate for mid-gastrula stage embryo. (A) Remove the vitelline envelope and select an embryo of the desired stage. Identify the dorsal blastopore lip and yolk plug (yp) and confirm extension of the archenteron. (B) Insert the hair knife in the archenteron between the dorsal blastopore lip and the yolk plug (arrowhead) and extend to emerge from the anterior end of the embryo (C). (D) Press down with the hair knife pushing through the ventral surface of the embryo. Next, slide the hair loop between the hair knife and the bottom of the dish and lift the loop upward. This will simultaneously complete the cut started by the knife and open the archenteron. (E) Turn the embryo so ventral is up. If necessary, cut through the thin floor of the archenteron. (F) Use the loop and knife to open the archenteron by pushing apart the lateral mesendoderm. (G) Find the lateral margin of the archenteron where the wall of the embryo thickens with large cell endoderm. Press the knife downward along the margin to separate the dorsal isolate from the endoderm. (H) Find the contralateral margin of the archenteron and again cut to separate the dorsal isolate from the endoderm. The resulting dorsal isolate shows the anterior site where the hair knife emerged (top) and the smooth surface of the dorsal blastopore lip (bottom). (I) Cut off the anterior portion of the isolate. (J) Cut off the blastopore lip. (K) The dorsal side of the dorsal isolate. The surface contains neural ectoderm and possible epidermis. (L) The ventral side of the dorsal isolate. The surface contains the archenteron roof plate endoderm. (a) Anterior, (p) posterior, (ar) archenteron, (yp) yolk plug. See Supplemental Movie S2.

19. Insert a stiff hair knife into the archenteron (Fig. 3B) so that it emerges from the anterior end of the embryo (Fig. 3C).

20. Position the embryo with the dorsal surface up and press the embedded hair knife down to contact the Petri dish (Fig. 3D).

21. Slide the hair loop below the embryo and gently lift to complete the cut (Fig. 3D). The cut may need to be extended through the floor of the archenteron.

22. Using the hair loop, flip the embryo to expose the archenteron roof. The dorsal tissues along the archenteron roof plate will be notably thinner than lateral plate mesoderm and large cell vegetal endoderm (Fig. 3E). Trim lateral and ventral tissue from the dorsal isolate at the site where the isolate thickens (Fig. 3F–H). With both lateral sides removed the dorsal isolate will span the axis from anterior neural to the blastopore lip (Fig. 3I).

23. Trim the dorsal isolate to remove both the anterior neural and posterior blastopore lip (Fig. 3I,J). Inspect surfaces of the dorsal ectoderm (Fig. 3K) and ventral endoderm (Fig. 3L) for damage.

24. Use a disposable pipette to transfer the isolate to a fresh dish of DFA to remove debris.

25. Prepare a coverslip bridge by applying a small amount of silicone grease at either end.

26. Compress the dorsal isolate under a glass coverslip bridge. Place the coverslip bridge over the isolate and press the ends over the grease alternately to gently compress and flatten the isolate.

27. Let the margins of the isolate heal for 30–60 min.

28. Remove the coverslip bridge so that the isolate can converge and extend. Using forceps, grasp the coverslip bridge at one end where there is silicone grease and slowly lift the bridge from the sandwich. Move to the other end and lift the bridge from the isolate and remove the bridge from the dish.

29. Culture dorsal isolates in DFA. Dorsal isolates can be cultured at room temperature or in temperature-controlled incubators used for whole-embryo culture. Isolates should rest for 60 min before imaging and should undergo convergent extension (Fig. 1E) for >6 h and may be cultured for 24–48 h.

    A successful dorsal isolate will develop in a manner analogous to the intact embryo. Continuing morphogenetic processes will shape somites, fold the neural tube, and drive stiffening of the notochord. Inclusion of sufficient ectoderm can result in the dorsal isolate developing a fin.

    The above steps involve microsurgery to remove dorsal tissues from a post-gastrula embryo. This region contains all three germ layers of the embryo. Depending on the precise location of the anterior cuts, the dorsal isolate can include anterior tissues including the prospective brain and eye field. Depending on the lateral cuts, the dorsal isolate can include far lateral mesodermal tissues. The blastopore domain with five to 10 cell rows is removed from the posterior end of the isolate to remove preinvolution mesoderm and make the final explant more rectangular shaped. Additionally, several variations on the dorsal isolate include removal of the endoderm to expose the ventral surface of the presomitic mesoderm and removal of the neural ectoderm to expose the dorsal surface of the presomitic mesoderm (Wilson 1990). Because Xenopus tissues heal quickly, various "mutant" tissue configurations can also be created—for instance, the midline including notochord can be removed or dorsal isolates containing multiple notochords may be constructed from two different embryos (Zhou et al. 2009).

    See Troubleshooting.

30. Quantify dorsal isolate morphogenesis using either time-lapse imaging or via end point analysis. Both end point images and time-lapse sequences can be collected using low magnification stereomicroscopes equipped with CMOS cameras (see Wallingford 2010). Typically, control and experimentally manipulated explants can be contrasted by their anterior-to-posterior length.

31. Confirm tissue anatomy and differentiation either with immunofluorescence or by RNA in situ hybridization (Wilson 1990; Zhou et al. 2009, 2010, 2015).

    This step is often needed to confirm that molecular manipulations have not disrupted patterning or to confirm that microsurgery successfully targeted desired tissues.

    Isolates can adhere to each other so care should be taken to maintain space between them. Alternatively, isolates may be cultured separately in multiwell dishes or agarose wells.

TROUBLESHOOTING

Problem (Steps 14 and 29): If improperly made, dorsal isolates and sandwich explants may not elongate properly. In some cases, these tissues may curl or curve during elongation. Mesoderm and neural precursor tissues should remain distinct as the sandwich explant elongates (Fig. 1C). However, if not properly made, mesoderm can often migrate under and into the neural portion of the sandwich explant. In such cases, explant elongation is greatly diminished.

Cite this protocol as *Cold Spring Harb Protoc*; doi:10.1101/pdb.prot097386

Solution: Failure of the sandwich explant to elongate may reflect initial misalignment of the two DMZ explants. Maintenance of distinct neural and mesodermal regions may reflect the presence of bottle cells or invasion of mesendoderm cells that may undergo involution and invade the ectodermal half of the sandwich. The consequences of invasion can be observed by failure of the narrowing at the posterior ends of the neural and mesoderm junction or the presence of notochord or somitic mesoderm along the full length of the explant. The dorsal isolate is highly robust in convergent extension. In some cases, following neurulation the epidermis that has closed over the neural tube may tear with large wounds. This may be due to overly zealous trimming of the lateral edges of the isolate that fails to leave sufficient prospective epidermis outside the margins of the neural plate.

DISCUSSION

Sandwich explants of the marginal zone have been used for analysis of convergence and extension, or alternatively CE (Keller et al. 2000). CE is a generic term describing tissue-scale movements in which the tissue narrows along one axis and lengthens along another. *Xenopus* dorsal tissues lengthen along the animal-to-vegetal axis and narrow along the mediolateral axis from early gastrulation. The sandwich explant of dorsal tissue exposes the role of autonomously acting factors and distinguishes those influences from exogenous contributors to CE. For instance, planar propagating signals can be isolated from radial propagating signals based on the timing of microsurgical manipulations.

Xenopus organotypic explants can be used uniquely to isolate and identify processes that pattern cell identities in the early embryo and processes that are responsible for physically reshaping the blastula during gastrulation and axis elongation. Organotypic explants preserve the tissue patterning and tissue interfaces of the native microenvironment. In contrast to embryoid bodies or organoids, organotypic dorsal isolates retain germ layer interactions between mesoderm and ectoderm and between endoderm and mesoderm that are found in the embryo. The Keller sandwich explant has been used to study patterning and morphogenetic processes that shape the dorsal analgen. The sandwich explant is often used as a rapid assay of CE, in which signaling or patterning pathways have been disrupted. Sandwich explants can be used to identify and test the role of planar patterning in establishing anterior neural identities—for example, signals propagating through the plane of the tissue rather than from one layer to the next (see Doniach et al. 1992; Poznanski and Keller 1997). The sandwich explant has also been used to measure mechanical properties of and forces produced in the early gastrula by CE (Moore 1994; Moore et al. 1995). The sandwich explant in a larger form, the "giant sandwich," has been used to measure mechanics and circumferential tensions generated during blastopore closure (Shook et al. 2018). In contrast to the narrow 90° explants used in the sandwich explant, the giant sandwich explant contains a full 360° of the marginal zone. Both explants are centered about the midline and undergo CE. It is thought that CE in the sandwich explant reflects underlying cell behaviors in the native axial and paraxial mesoderm; however, this may be an oversimplification given that radial polarity within the two layers of the sandwich explant may be disrupted, as cells become radially aligned around the core of the explant (Poznanski et al. 1997). During early gastrulation movements, the tissue scale events of convergent extension can be separated from other internal movements by observation of the sandwich explant (Keller et al. 1985b; Keller 1986; Shook et al. 2018).

The dorsal isolate was first used to describe cell behaviors in the notochord and paraxial mesoderm during axis elongation and segmentation (Wilson et al. 1989; Wilson 1990). Tissue scale events that drive convergent extension at these later, postinvolution stages and during segmentation can be studied within the dorsal isolate (Wilson et al. 1989; Zhou et al. 2009; Zhou et al. 2015; Shawky et al. 2018). The sandwich explant and dorsal isolate can be used to study how cell and tissue mechanics drive morphogenesis. Force production and mechanical resistance can be quantitatively determined using standard engineering parameters such as a time-varying Young's modulus or stress (Moore et al. 1995; Zhou et al. 2009, 2010, 2015; Shawky et al. 2018). Dorsal isolates consist of dorsal endoderm, paraxial mesoderm, axial mesoderm, neural ectoderm, and ectoderm arrayed in three germ layers by earlier movements of gastrulation (Fig. 1D). Neural ectoderm is contiguous with nonneural ectoderm and is comprised of an outer epithelial cell layer (e.g., superficial layer) and a deep single inner layer of mesenchymal-like cells (e.g., deep or sensorial layer). The deep layer of the neural ectoderm contains neuronal precursors and neural crest at the boundary with nonneural ectoderm. The mesoderm consists of axial notochord flanked by presomitic mesoderm and lateral plate mesoderm. Whereas the presomitic mesoderm and lateral plate mesoderm are two clearly defined layers of mesenchymal cells, the notochord is not organized into discrete layers but forms a compact central rod of thin sector-shaped cells (e.g., "pizza slices"). The endoderm forms a single cell layered simple epithelium between the mesoderm and the archenteron.

If properly made, the dorsal isolate will undergo large-scale morphogenetic movements that parallel movement of the same tissues in vivo. CE of the dorsal isolate will continue from stage 13 to stage 16 and will be joined by neurulation movements that can result in bending along the anterior–posterior direction. Additionally, presomitic or paraxial mesoderm will progressively segment into somites (Wilson et al. 1989; Wilson 1990; Wilson and Keller 1991). Neural ectoderm in dorsal isolates can fold and fuse into a neural tube. Wide dorsal isolates may include epidermal progenitors resulting in formation of fin-like structures in equivalently aged isolates. Similarly, inclusion of anterior tissues can result in the formation of retina, otic, and other anterior sensory placodes in aged isolates.

RECIPES

Danilchik's for Amy (DFA) Medium (pH 8.3)

Reagent	Final concentration
NaCl	53 mM
Na$_2$CO$_3$	5 mM
Potassium gluconate	4.5 mM
Sodium gluconate	32 mM
CaCl$_2$	1 mM
MgSO$_4$	1 mM

Adjust the pH to 8.3 with granular bicine. Filter with a 0.22 µm filter. Include 0.1% bovine serum albumen (BSA; Sigma-Aldrich A7906) to prevent nonspecific adhesion to substrates or glass coverslips. Just prior to use, add 1% of antibiotic/antimycotic (Sigma-Aldrich A5955). Prepare the DFA with BSA, aliquot, and store at −20°C. Keep aliquots of antibiotic/antimycotic at −20°C and add to thawed DFA before use. Replace DFA with medium containing fresh antibiotic/antimycotic if culturing explants for >24 h. (This recipe is adapted from Sater et al. 1993.)

Modified Barth's Solution (MBS) (pH 7.4)

Reagent	Final concentration (1×)
NaCl	88 mM
KCl	1 mM
NaHCO$_3$	2.4 mM
MgSO$_4$·7H$_2$O	0.82 mM
Ca(NO$_3$)$_2$·4H$_2$O	0.33 mM
CaCl$_2$·2H$_2$O	0.41 mM
HEPES	5 mM

Adjust the pH to 7.4 with NaOH; 3× stocks of MBS are pH-stable and can be stored at 4°C. Prepare a working solution (1/3×) by dilution.

Cite this protocol as *Cold Spring Harb Protoc*; doi:10.1101/pdb.prot097386

ACKNOWLEDGMENTS

I thank the members of the Davidson, Keller, and DeSimone laboratories past and present for their support, helpful discussions, and instructive methods sections; a special thank you is owed to Dr. Mary Constance Lane for teaching me how to make my first sandwich explants. I also thank the National Xenopus Resource (RRID:SCR-013731) and Xenbase (RRID:SCR-003280) for their invaluable support to the model organism community. L.A.D. is funded by the Eunice Kennedy Shriver Institute of Child Health and Human Development at the National Institutes of Health (NIH) (R01 HD044750 and R21 HD106629).

REFERENCES

Chu C-W, Davidson LA. 2022. Chambers for culturing and immobilizing *Xenopus* embryos and organotypic explants for live imaging. *Cold Spring Harb Protoc* doi:10.1101/pdb.prot107649.

Davidson LA. 2022. Microsurgical methods to isolate and culture the early gastrula dorsal marginal zone. *Cold Spring Harb Protoc* doi:10.1101/pdb.prot097360.

Davidson LA, Keller R, DeSimone D. 2004. Patterning and tissue movements in a novel explant preparation of the marginal zone of *Xenopus laevis*. *Gene Expr Patterns* 4: 457–466. doi:10.1016/j.modgep.2004.01.001

Davidson LA, Marsden M, Keller R, DeSimone DW. 2006. Integrin α5β1 and fibronectin regulate polarized cell protrusions required for *Xenopus* convergence and extension. *Curr Biol* 16: 833–844. doi:10.1016/j.cub.2006.03.038

DeSimone DW, Davidson L, Marsden M, Alfandari D. 2005. The *Xenopus* embryo as a model system for studies of cell migration. *Methods Mol Biol* 294: 235–245. doi:10.1385/1-59259-860-9:235

DeSimone DW, Dzamba B, Davidson LA. 2007. Using *Xenopus* embryos to investigate integrin function. *Methods Enzymol* 426: 403–414. doi:10.1016/S0076-6879(07)26017-3

Domingo C, Keller R. 1995. Induction of notochord cell intercalation behavior and differentiation by progressive signals in the gastrula of *Xenopus laevis*. *Development* 121: 3311–3321. doi:10.1242/dev.121.10.3311

Doniach T, Phillips CR, Gerhart JC. 1992. Planar induction of anteroposterior pattern in the developing central nervous system of *Xenopus laevis*. *Science* 257: 542–545. doi:10.1126/science.1636091

Edelstein AD, Tsuchida MA, Amodaj N, Pinkard H, Vale RD, Stuurman N. 2014. Advanced methods of microscope control using µManager software. *J Biol Methods* 1: e10. doi:10.14440/jbm.2014.36

Ewald AJ, Peyrot SM, Tyszka JM, Fraser SE, Wallingford JB. 2004. Regional requirements for Dishevelled signaling during *Xenopus* gastrulation: separable effects on blastopore closure, mesendoderm internalization and archenteron formation. *Development* 131: 6195–6209. doi:10.1242/dev.01542

Feroze R, Shawky JH, von Dassow M, Davidson LA. 2015. Mechanics of blastopore closure during amphibian gastrulation. *Dev Biol* 398: 57–67. doi:10.1016/j.ydbio.2014.11.011

Gillespie JI. 1983. The distribution of small ions during the early development of *Xenopus laevis* and *Ambystoma mexicanum* embryos. *J Physiol (Lond)* 344: 359–377. doi:10.1113/jphysiol.1983.sp014945

Keller RE. 1986. The cellular basis of amphibian gastrulation. *Dev Biol* 2: 241–327. doi:10.1007/978-1-4613-2141-5_7

Keller R, Danilchik M. 1988. Regional expression, pattern and timing of convergence and extension during gastrulation of *Xenopus laevis*. *Development* 103: 193–209. doi:10.1242/dev.103.1.193

Keller R, Danilchik M, Gimlich R, Shih J. 1985a. Convergent extension by cell intercalation during gastrulation in *Xenopus laevis*. In *Molecular determinants of animal form* (ed. Edelman GM). Alan R. Liss, New York.

Keller RE, Danilchik M, Gimlich R, Shih J. 1985b. The function and mechanism of convergent extension during gastrulation of *Xenopus laevis*. *J Embryol Exp Morphol* 89: 185–209. doi:10.1242/dev.89.Supplement.185

Keller R, Davidson L, Edlund A, Elul T, Ezin M, Shook D, Skoglund P. 2000. Mechanisms of convergence and extension by cell intercalation. *Philos Trans R Soc Lond B* 355: 897–922. doi:10.1098/rstb.2000.0626

Kim HY, Davidson LA. 2013. Assembly of chambers for stable long-term imaging of live *Xenopus* tissue. *Cold Spring Harb Protoc* doi:10.1101/pdb.prot073882.

Lee JY, Harland RM. 2007. Actomyosin contractility and microtubules drive apical constriction in *Xenopus* bottle cells. *Dev Biol* 311: 40–52. doi:10.1016/j.ydbio.2007.08.010

Lee JY, Harland RM. 2010. Endocytosis is required for efficient apical constriction during *Xenopus* gastrulation. *Curr Biol* 20: 253–258. doi:10.1016/j.cub.2009.12.021

Marsden M, DeSimone DW. 2003. Integrin–ECM interactions regulate cadherin-dependent cell adhesion and are required for convergent extension in *Xenopus*. *Curr Biol* 13: 1182–1191. doi:10.1016/S0960-9822(03)00433-0

Moore SW. 1994. A fiber optic system for measuring dynamic mechanical properties of embryonic tissues. *IEEE Trans Biomed Eng* 41: 45–50. doi:10.1109/10.277270

Moore SW, Keller RE, Koehl MAR. 1995. The dorsal involuting marginal zone stiffens anisotropically during its convergent extension in the gastrula of *Xenopus laevis*. *Development* 121: 3130–3140. doi:10.1242/dev.121.10.3131

Nieuwkoop PD, Faber J. 1967. *Normal tables of* Xenopus laevis *(Daudin)*. Elsevier North-Holland Biomedical, Amsterdam.

Poznanski A, Keller R. 1997. The role of planar and early vertical signaling in patterning the expression of *Hoxb-1* in *Xenopus*. *Dev Biol* 184: 351–366. doi:10.1006/dbio.1996.8500

Poznanski A, Minsuk S, Stathopoulos D, Keller R. 1997. Epithelial cell wedging and neural trough formation are induced planarly in *Xenopus*, without persistent vertical interactions with mesoderm. *Dev Biol* 189: 256–269. doi:10.1006/dbio.1997.8678

Shawky J, Balakrishnan UL, Stuckenholz C, Davidson LA. 2018. Multiscale analysis of architecture, cell size and the cell cortex reveals cortical F-actin density and composition are major contributors to mechanical properties during convergent extension. *Development* 145: dev161281. doi:10.1242/dev.161281

Shook DR, Davidson L, Kasprowicz EM, Keller R. 2018. Large, long range tensile forces drive convergence during *Xenopus* blastopore closure and body axis elongation. *Elife* 7: e26944. doi:10.7554/eLife.26944

Sive HL, Grainger RM, Harland RM. 2000. *Early development of* Xenopus laevis: *a laboratory manual*. Cold Spring Harbor Laboratory Press, Cold Spring Harbor, New York.

Sive HL, Grainger RM, Harland RM. 2007. Dejellying *Xenopus laevis* embryos. *Cold Spring Harb Protoc* doi:10.1101/pdb.prot4731.

Wallingford JB. 2010. Low-magnification live imaging of *Xenopus* embryos for cell and developmental biology. *Cold Spring Harb Protoc* doi:10.1101/pdb.prot5425.

Wilson P. 1990. *The development of the axial mesoderm in* Xenopus laevis. *Biophysics*. University of California, Berkeley.

Wilson P, Keller R. 1991. Cell rearrangement during gastrulation of *Xenopus*: direct observation of cultured explants. *Development* 112: 289–300. doi:10.1242/dev.112.1.289

Wilson PA, Oster G, Keller R. 1989. Cell rearrangement and segmentation in *Xenopus*: direct observation of cultured explants. *Development* 105: 155–166. doi:10.1242/dev.105.1.155

Zhou J, Kim HY, Davidson LA. 2009. Actomyosin stiffens the vertebrate embryo during critical stages of elongation and neural tube closure. *Development* 136: 677–688. doi:10.1242/dev.026211

Zhou J, Kim HY, Wang JH-C, Davidson LA. 2010. Macroscopic stiffening of embryonic tissues via microtubules, Rho-GEF, and assembly of contractile bundles of actomyosin. *Development* 137: 2785–2794. doi:10.1242/dev.045997

Zhou J, Pal S, Maiti S, Davidson LA. 2015. Force production and mechanical adaptation during convergent extension. *Development* 142: 692–701. doi:10.1242/dev.116533

Cite this protocol as *Cold Spring Harb Protoc*; doi:10.1101/pdb.prot097386

Cranial Neural Crest Explants

Hélène Cousin[1] and Dominique Alfandari

Department of Veterinary and Animal Sciences, University of Massachusetts, Amherst, Massachusetts 01003

The cranial neural crest (CNC) explant assay was originally designed to assess the basic requirements for CNC migration in vitro. This protocol describes the key parameters of CNC explants in *Xenopus laevis*, with a focus on how to extirpate CNC cells and assay their migration in vitro. The protocol can be adapted according to the needs of the experimenter, some examples of which are discussed here.

MATERIALS

It is essential that you consult the appropriate Material Safety Data Sheets and your institution's Environmental Health and Safety Office for proper handling of equipment and hazardous materials used in this protocol.

RECIPES: Please see the end of this protocol for recipes indicated by <R>. Additional recipes can be found online at http://cshprotocols.cshlp.org/site/recipes.

Reagents

Danilchik medium <R> containing 50 µg/mL of gentamycin
Ethanol (70%)
Fibronectin, bovine (Sigma-Aldrich)
H_2O (reverse osmosis [RODI] or distilled [dH_2O])
Modified Barth's saline (MBS) (1×) <R>

For dissection medium, use 1× MBS containing 50 µg/mL of gentamycin.

Phosphate-buffered saline (PBS) <R>

In addition, prepare PBS containing 1% (w/v) bovine serum albumin (BSA).

Pork gelatin (optional for glass-bottom plates; see Step 2)
Xenopus laevis eggs

Equipment

Dissecting microscope equipped with a gooseneck lighting system
Embryo incubators at the desired temperature (between 15°C and 19°C for *X. laevis*)
Eyelash knife

Select a human eyelash with the desired thickness and curvature and thread it through a 23-gauge needle fitted on a 1-cc syringe. For safety purposes, the tip of the needle can be cut off with scissors before the threading. Fix the eyelash with a drop of nail polish or cyanoacrylate glue.

---

[1]Correspondence: hcousin@vasci.umass.edu

Forceps (fine, to remove vitelline envelope)

Glass bead tool

Thin out the end of a Pasteur pipette under a benzene burner and melt the end into a ball roughly the size of gastrula-stage embryo (~2 mm).

Hair loop

Cut a human hair into 3-inch sections. Thread both ends of a section into a 23-gauge needle fitted on a 1-cc syringe. Push the loop into the needle until the desired stiffness is reached. Fix the hair with a drop of nail polish or cyanoacrylate glue.

Petri dishes (60-mm, plastic), coated with a 4-mm layer of 1% agarose

Petri dishes (60-mm, plastic), coated with plasticine

Roll 2 tsp of plasticine (nondrying, toxin free, appropriate for young children) into a ball. Flatten it out into a plastic Petri dish by hand.

Plates (multiwell, plastic or glass)

See Step 2.

Transfer pipette (disposable plastic or glass) with an opening of ≥2 mm, for transferring embryos

METHOD

CNC explantation can be performed as soon as stage 14 and as late as stage 18; see Step 8. Embryos are staged according to Nieuwkoop and Faber (1967).

The following procedure should be conducted between 15°C and 19°C.

1. One or two days before explantation, fertilize the eggs in the morning (see Sive et al. 2007).

   Embryos kept at 18°C will be ready for explantation in the afternoon the day after fertilization. Alternatively, embryos can be raised at 14°C, in which case they will be ready for explantation two days after fertilization, in the morning.

2. On the day before explantation, coat a multiwell plate with fibronectin.

   - For plastic plates, dilute fibronectin to 5–10 µg/mL in PBS, distribute the solution into the appropriate wells (e.g., 50 µL per well for a 96-well plate), and let the fibronectin adsorb overnight at 4°C.

   - For glass-bottom plates, follow the above procedure but increase the concentration of fibronectin to 50 µg/mL.

   Alternatively, precoat glass-bottom wells with a solution of 0.1% pork gelatin for 1 h at 37°C, remove the gelatin solution, and then coat with 10 µg/mL of fibronectin overnight at 4°C or 1 h at 37°C. This treatment leads to explants that migrate as well as if plastic plates were used.

3. On the day of explantation, when the embryos reach stage 14, sterilize a plasticine-coated dish with 70% ethanol for 10 min. Rinse briefly with RODI H₂O and then fill with dissection medium.

4. Remove the fibronectin from the multiwell plate prepared in Step 2. Block the wells with PBS containing 1% BSA (e.g., 200 µL per well for a 96-well plate) for 1 h at room temperature.

   This blocking step may be omitted for glass-bottom plates.

5. Remove the blocking solution from the multiwell plate and replace with Danilchik medium containing 50 µg/mL of gentamicin.

6. Add Danilchik medium containing 50 µg/mL of gentamicin to an agarose-coated dish.

7. Using a glass bead tool, dig trenches in the plasticine-coated dish from Step 3 corresponding to the length of the embryos: The depth of each trench should correspond to the two-thirds of the embryos' width. The depression should be perpendicular relative to the experimenter or slightly angled toward the bottom left (3° to 5°).

8. Select an embryo between stages 14 and 15 and transfer the embryo to the plasticine-coated dish. Remove the vitelline envelope using fine forceps. Ensure that the embryo does not breach the liquid surface once the vitelline envelope is removed.

> *If one wishes to observe the migration of all the CNC, stage 14 or 15 embryos are ideal as the segments (mandibular hyoid and branchial) have not yet formed and the entirety of the tissue can be explanted without losing any of the segments in the process. However, stage 14 CNC may be difficult to distinguish from the surrounding neuroepithelium and placodal ectoderm. In that case, CNC explants can be performed at a later stage. However, by stage 17, the segments are already formed and the loss of one of the segments will likely occur during the dissection.*

> *See Troubleshooting.*

9. Move the embryo into a trench and orient it according to the experimenter's preference.

> *The goal is to orient the embryo so that the CNC directly faces the experimenter. In Figure 1A, the embryo is positioned in a ¾ anterior view with the anterior neural plate slanted slightly to the left.*

> *Being able to recognize the location of the CNC before immobilizing the embryos is critical but sometimes difficult for the novice. The experimenter should be familiar with the Xenopus developmental table. The CNC makes up the bulk of the anterior neural plate borders. They appear as two bulges on each side of the anterior neural plate and can therefore be easily identified if the gooseneck guide lights are oriented in a manner that highlights the three-dimensional conformation of the embryo (Fig. 1A, anterior neural fold bulge).*

10. To prevent the embryo from moving during the surgery, tighten the plasticine around it by pushing gently on the plasticine with the glass bead.

> *This step is optional for experimenters that prefer performing microdissection on free-moving embryos.*

11. Insert the tip of the eyelash knife under the ectoderm at the ventral and posterior end of the anterior neural fold (see online Movie 1 at cshprotocols.cshlp.org). Thread the knife under the ectoderm, move it slightly anteriorly, and cut the ectoderm by lifting the knife swiftly. Repeat the motion anteriorly along the ventral edge of the anterior neural fold until the ventral edge of the CNC is completely uncovered (see online Movie 1 at cshprotocols.cshlp.org).

> *The goal of this step is to cut the ectoderm ventrally to the anterior neural fold to peel it dorsally (i.e., toward the neural plate) in Step 12.*

12. Peel off the ectoderm dorsally, to expose the underlying neural fold, and drape it over the other side of the embryo using the hair loop (Fig. 1B, see online Movie 1 at cshprotocols.cshlp.org).

13. Identify the location of the CNC (Sadaghiani and Thiebaud 1987). Using the knife tip, cut the CNC outline. Cut superficially to avoid damaging or contaminating your explant with the underlying mesoderm and endoderm (see online Movie 1 at cshprotocols.cshlp.org).

> *In some cases, the CNC inherits some of the pigment deposited in the egg during oogenesis, which makes it easy to distinguish from the neural plate and mesoderm. However, CNC are unpigmented the majority of the time. The experimenter should be familiar with landscape of the embryos and learn to recognize the bulge formed by the neural folds. Adequate lighting (described in Step 9) is critical.*

14. Lift the explant off the embryo with the knife and hair loop and inspect it for any contaminating mesoderm and endoderm (Fig. 1C). Scrape off contaminating cells with the hair loop and knife.

> *CNC explants are made of small cells which typically give the tissue a blue hue. The large size and high yolk content of mesoderm and endoderm cells give them a white to yellow coloration.*

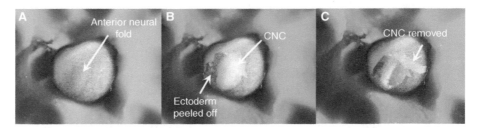

FIGURE 1. Key steps of the CNC explantation procedure.

15. Prepare a P20 micropipette fitted with a yellow tip for explant transfer. To prevent explants from sticking to the walls of the tip, aspirate 10 µL of Danilchik medium or PBS containing 1% BSA through the tip a few times to coat the plastic.

    This step should be performed only once.

16. Use the pipette to transfer the explant to the agarose-coated dish from Step 6, taking care that the explant remains submerged in the medium and does not breach the liquid surface at any time. Leave the explant in the dish until all the explants needed for the experiment have been dissected and transferred (20 min).

17. Remove any debris from the explants by moving them around the dish.

    Gentle flushing of medium out of the P20 pipette is recommended for moving the explants.

18. Transfer each explant, one at a time, into a fibronectin-coated well from Step 5. To avoid diluting the medium in the wells (which may contain a drug or other compound) and for better control over where the explant will land, do not pipette the explant into the well. Instead, insert the pipette containing the explant into the medium and let the explant fall out of the pipette tip.

19. Allow the CNC cells to adhere to the substrate for 15–30 min at 18°C.

    See Troubleshooting.

20. Record a time-lapse movie between 18°C and 20°C.

    Typically, one frame every 3 min for 8 to 10 h provides a good compromise between file size and temporal resolution (see online Movie 2 at cshprotocols.cshlp.org). Magnification can vary between 50× and 250×, depending on the cellular resolution required and on whether or not one wishes to observe all the individual cells once they migrate out of the explants.

    See Troubleshooting.

TROUBLESHOOTING

Problem (Step 8): The plasticine is sticky and wounds the embryo once the vitelline envelope is removed.

Solution: This problem will occur if the plasticine is new, but will disappear after a few uses. Until then, 1% BSA (w/v) can be added to the dissection medium to coat the surface of the plasticine. Alternatively, dishes coated with agarose can be used.

Problem (Step 19): The CNC does not attach to the fibronectin-coated wells.

Solution: It is possible that the explanted tissue is not the CNC. Ensure that the correct territory is dissected (Sadaghiani and Thiebaud 1987; Alfandari et al. 2003). Another possibility is that the well was not coated properly with fibronectin (often a problem on glass-bottom dishes). Try coating with fibronectin at a higher concentration, or coat the wells with 0.1% pork gelatin before coating with fibronectin.

Problem (Step 20): The CNC attaches to the fibronectin-coated wells, but cells do not emigrate or emigrate very poorly.

Solution: Ensure that the Danilchik medium has the correct pH and has not been adjusted with HCl. This medium contains low chlorine ions and was specifically developed to match the physiology of *Xenopus* nonepithelial tissue. Alternatively, try another type of Danilchik solution, in particular DFA (Danilchik for Amy), which was developed for culturing other type of explants (Sater et al. 1993).

Cite this protocol as *Cold Spring Harb Protoc*; doi:10.1101/pdb.prot097394

DISCUSSION

The use of the CNC explant assay (Alfandari et al. 2003) has led to many discoveries, including the findings that CNC cells use fibronectin and integrin α5β1 to migrate, and that they perform the majority of their migration by keeping cell–cell contact, a phenomenon called collective cell migration (Carmona-Fontaine et al. 2008; Friedl and Mayor 2017). Today, this assay is used to investigate the basic cellular and molecular mechanisms of CNC migration and of collective migration in general (Friedl and Mayor 2017). While an incredibly powerful system, experimenters should be aware that it may not always be useful for studying the functions of certain genes during CNC migration. For example, the metalloprotease ADAM13 is required for CNC migration in the context of the embryo but is dispensable in vitro (Cousin et al. 2012).

This protocol can be used with *Xenopus tropicalis* as well as *X. laevis*, provided the experimenter adapts the timing to that of *X. tropicalis* development. *X. tropicalis* tissues are slightly stiffer than those of *X. laevis*, and while our eyelash knife still works fine, the thickness and tapered end of human eyelashes vary widely. If cutting through the tissues proves troublesome, we suggest experimenting with various hair as the knife. For example, eyebrow hairs tend to be stiffer but have blunter ends, which may not be ideal for cutting *X. tropicalis* embryonic tissues.

Time–lapse movies of CNC explants can yield a large amount of quantifiable information on the effects of the loss or gain of function of a particular gene on CNC migration. Loss- or gain-of-function experiments can be performed by injecting embryos with antisense morpholino, or with mRNA encoding either a constitutively active or dominant negative version of the gene of interest. In addition, experimenters can treat the explants with inhibitors using concentrations similar to what has been described in tissue culture experiments. Various assays can be done to test CNC fitness to migrate. For example, experimenters can assess the trajectory, persistence and speed of the migrating cells using software such as FIJI (Carmona-Fontaine et al. 2008). CNC are also known to perform Contact Inhibition of Locomotion (CIL), a phenomenon whereby two CNC cells colliding with one another will reverse the direction of their migration. CIL is essential for CNC migration both in vivo and in vitro. To assess CIL, one can perform a collision assay, in which two explants are placed in close proximity to one another (Becker et al. 2013). CNC have been shown to be attracted by the growth factor Sdf-1, which is secreted by placodal ectoderm and is essential for their migration in vivo (Theveneau et al. 2013). To assay the ability of the CNC to respond to this chemoattractant, the experimenter can add a source of Sdf-1, such as a bead coated with Sdf-1 or an explant of placodal ectoderm (Theveneau and Mayor 2011; Cousin et al. 2012; Theveneau et al. 2013), in proximity to the explant. Attraction to Sdf-1 can be used to test the fluidity of the CNC explants. For example, explant and Sdf-1 can be deposited in two separate microchambers connected only through openings of defined sizes. By varying the size of these openings, one can test the ability of the CNC to squeeze through openings and therefore challenge the fluidity of the explant (Cousin et al. 2012). The concept of fluidity of tissues undergoing collective cell migration is important, and CNC cells are one of the few models that allow it to be tested and quantified (Friedl and Mayor 2017).

RECIPES

Danilchik Medium

Bicine	17.5 mM
Bovine serum albumin	1 mg/mL
CaCl$_2$	1 mM
MgSO$_4$	2 mM
Na$_2$CO$_3$	11.7 mM
NaCl	53 mM
Potassium gluconate	4.25 mM

Adjust pH to 8.3. Store at 4°C for no more than 2 wk.

Modified Barth's Saline (MBS) (1×)

$CaCl_2$	0.41 mM
$CaNO_3$	0.3 mM
HEPES-NaOH	15 mM
KCl	1 mM
$MgSO_4$	0.82 mM
NaCl	88 mM
$NaHCO_3$	2.4 mM

Adjust pH to 7.6. Store at room temperature for up to 1 mo.

Phosphate-Buffered Saline (PBS)

Reagent	Amount to add (for 1× solution)	Final concentration (1×)	Amount to add (for 10× stock)	Final concentration (10×)
NaCl	8 g	137 mM	80 g	1.37 M
KCl	0.2 g	2.7 mM	2 g	27 mM
Na_2HPO_4	1.44 g	10 mM	14.4 g	100 mM
KH_2PO_4	0.24 g	1.8 mM	2.4 g	18 mM

If necessary, PBS may be supplemented with the following:

$CaCl_2 \cdot 2H_2O$	0.133 g	1 mM	1.33 g	10 mM
$MgCl_2 \cdot 6H_2O$	0.10 g	0.5 mM	1.0 g	5 mM

PBS can be made as a 1× solution or as a 10× stock. To prepare 1 L of either 1× or 10× PBS, dissolve the reagents listed above in 800 mL of H_2O. Adjust the pH to 7.4 (or 7.2, if required) with HCl, and then add H_2O to 1 L. Dispense the solution into aliquots and sterilize them by autoclaving for 20 min at 15 psi (1.05 kg/cm^2) on liquid cycle or by filter sterilization. Store PBS at room temperature.

ACKNOWLEDGMENTS

H.C. and D.A. are supported by National Institutes of Health/National Institute of Dental and Craniofacial Research (NIH/NIDCR) DE025691 and DE016289, respectively.

REFERENCES

Alfandari D, Cousin H, Gaultier A, Hoffstrom BG, DeSimone DW. 2003. Integrin α5β1 supports the migration of *Xenopus* cranial neural crest on fibronectin. *Dev Biol* **260:** 449–464.

Becker SF, Mayor R, Kashef J. 2013. Cadherin-11 mediates contact inhibition of locomotion during *Xenopus* neural crest cell migration. *PLoS One* **8:** e85717.

Carmona-Fontaine C, Matthews HK, Kuriyama S, Moreno M, Dunn GA, Parsons M, Stern CD, Mayor R. 2008. Contact inhibition of locomotion in vivo controls neural crest directional migration. *Nature* **456:** 957–961.

Cousin H, Abbruzzese G, McCusker C, Alfandari D. 2012. ADAM13 function is required in the 3 dimensional context of the embryo during cranial neural crest cell migration in *Xenopus laevis. Dev Biol* **368:** 335–344.

Friedl P, Mayor R. 2017. Tuning collective cell migration by cell–cell junction regulation. *Cold Spring Harb Perspect Biol* **9:** a029199.

Nieuwkoop PD, Faber J. 1967. *Normal table of* Xenopus laevis *(Daudin),* 2nd ed. North-Holland, Amsterdam.

Sadaghiani B, Thiebaud CH. 1987. Neural crest development in the *Xenopus laevis* embryo, studied by interspecific transplantation and scanning electron microscopy. *Dev Biol* **124:** 91–110.

Sater AK, Steinhardt RA, Keller R. 1993. Induction of neuronal differentiation by planar signals in *Xenopus* embryos. *Dev Dyn* **197:** 268–280.

Sive HL, Grainger RM, Harland RM. 2007. *Xenopus laevis* in vitro fertilization and natural mating methods. *Cold Spring Harb Protoc* doi:10.1101/pdb.prot4737.

Theveneau E, Mayor R. 2011. Beads on the run: Beads as alternative tools for chemotaxis assays. *Methods Mol Biol* **769:** 449–460.

Theveneau E, Steventon B, Scarpa E, Garcia S, Trepat X, Streit A, Mayor R. 2013. Chase-and-run between adjacent cell populations promotes directional collective migration. *Nat Cell Biol* **15:** 763–772.

Cranial Neural Crest Transplants

Hélène Cousin[1]

Department of Veterinary and Animal Sciences, University of Massachusetts, Amherst, Massachusetts 01003

The transplantation of cranial neural crest (CNC) expressing green fluorescent protein (GFP) in *Xenopus laevis* has allowed researchers not only to assess CNC migration in vivo but also to address many other experimental questions. Coupled with loss- or gain-of-function experiments, this technique can be used to characterize the function of specific genes during CNC migration and differentiation. Although targeted injection can also be used to assess gene function during CNC migration, CNC transplantation allows one to answer specific questions, such as whether a gene's function is tissue autonomous, cell autonomous, or exerted in the tissues surrounding the CNC. Here we describe a protocol for performing simple CNC grafts.

MATERIALS

It is essential that you consult the appropriate Material Safety Data Sheets and your institution's Environmental Health and Safety Office for proper handling of equipment and hazardous materials used in this protocol.

RECIPES: Please see the end of this protocol for recipes indicated by <R>. Additional recipes can be found online at http://cshprotocols.cshlp.org/site/recipes.

Reagents

Ethanol (70%)
H_2O (reverse osmosis [RODI] or distilled [dH_2O])
Lineage tracer of choice (e.g., 200 pg of GFP mRNA)
Modified Barth's saline (MBS) (1×) <R>

For grafting medium, use 1× MBS containing 50 µg/mL of gentamycin. For recovery medium, use 0.1× MBS containing 50 µg/mL of gentamycin. For imaging medium, use 0.1× MBS containing 50 µg/mL of gentamycin and 30 µg/mL of benzocaine.

Xenopus laevis embryos (at the one- or two-cell stage)

Embryos are staged according to Nieuwkoop and Faber (1967).

Equipment

Bridges

Cut coverslips into rectangles (3-mm × 1-cm) using a diamond pen.

Dissecting microscope, equipped with a gooseneck lighting system
Embryo incubators at the desired temperature (for *Xenopus laevis*, between 15°C and 19°C)

[1]Correspondence: hcousin@vasci.umass.edu

Supplemental material is available at cshprotocols.cshlp.org

Eyelash knife

Select a human eyelash with the desired thickness and curvature and thread it through a 23-gauge needle fitted on a 1-cc syringe. For safety purposes, the tip of the needle can be cut off with scissors before the threading. Fix the eyelash with a drop of nail polish or cyanoacrylate glue.

Forceps (blunt, to manipulate bridges)

Forceps (fine, to remove vitelline envelope)

Glass bead tool

Thin out the end of a Pasteur pipette under a benzene burner and melt the end into a ball roughly the size of gastrula-stage embryo (~2 mm).

Glass plate (see Step 19.iv)

Hair loop

Cut a human hair into 3-inch sections. Thread both ends of a section into a 23-gauge needle fitted on a 1-cc syringe. Push the loop into the needle until the desired stiffness is reached. Fix the hair with a drop of nail polish or cyanoacrylate glue.

Petri dishes (60-mm, plastic), coated with a 4-mm layer of 1% agarose

Petri dishes (60-mm, plastic), coated with plasticine

Roll 2 tsp of plasticine (nondrying, toxin free, appropriate for young children) into a ball. Flatten it out into a plastic Petri dish by hand.

For time-lapse imaging (Step 19), choose plasticine with a low level of autofluorescence in the desired channel. White plasticine usually works best in most channels.

Transfer pipette (disposable plastic or glass) with an opening of ≥2 mm, for transferring embryos

METHOD

Tissue graft and lineage tracing, including transplantation of tissue containing neural crest, have been routinely performed in urodele such as axolotl (Horstadius 1950; Hall and Horstadius 1988). In 1987, the first interspecies cranial neural crest grafts between Xenopus laevis and Xenopus borealis were reported (Sadaghiani and Thiebaud 1987). The first grafts of cranial neural crest (CNC) expressing GFP in Xenopus laevis were reported a decade later (Carl et al. 1999; Borchers et al. 2000).

The groups who originally grafted CNC used agarose-coated dishes during the extirpation and grafting steps, which allowed the embryos to move freely (Carl et al. 1999; Borchers et al. 2000). The protocol detailed below was modified to use plasticine-coated dishes instead (Alfandari et al. 2001). This offers the option of immobilizing the embryo during the transplantation procedure, which can be an advantage for many experimenters.

CNC grafting can be performed as soon as stage 14 and as late as stage 18; see Step 4.

The grafting procedure should be conducted between 15°C and 19°C.

1. One or two days before grafting, inject embryos at the one- or two-cell stage with the lineage tracer of your choice (e.g., 200 pg of GFP mRNA).

    Embryos kept at 18°C can be grafted in the afternoon of the day after injection. Embryos raised at 14°C can be grafted 48 h after injection.

2. On the day of grafting, when the embryos reach stage 14, sterilize a plasticine-coated dish with 70% ethanol for 10 min. Rinse briefly with RODI H_2O and then fill with grafting medium.

3. Using a glass bead tool, dig two trenches corresponding to the length of the embryos. (The depth of the trench should correspond to about half of the embryo width.)

    The purpose of these trenches is to immobilize the embryos in a position that will allow the experimenter to dissect out and graft in the CNC.

4. Select donor and host embryos between stages 14 and 15 and transfer the embryos to the plasticine-coated dish. Remove the vitelline envelope from each embryo using fine forceps. Ensure that the embryos do not breach the liquid surface once the vitelline envelope is removed.

    If one wishes to graft and observe the migration of all the CNC, stage 14 or 15 embryos are ideal as the segments (mandibular hyoid and branchial) have not yet formed and the entirety of the tissue can be grafted

Cite this protocol as *Cold Spring Harb Protoc*; doi:10.1101/pdb.prot097402

without losing any of the segments in the process. However, stage 14 CNC may be difficult to distinguish from the surrounding neuroepithelium and placodal ectoderm. In that case, CNC grafts can be performed at a later stage. However, by stage 17, the segments are already formed and the loss of a segment will likely occur during the grafting procedure.

5. Move the donor and host embryos into the trenches. Orient the embryos according to the experimenter's preference.

    The goal is to orient the embryos so that the CNC directly faces the experimenter. In Figure 1A, the embryo is positioned in a ¾ anterior view with the anterior neural plate slanted slightly to the left.

    Being able to recognize the location of the CNC before immobilizing the embryos is critical but sometimes difficult for the novice. The experimenter should be familiar with the Xenopus developmental table. The CNC makes up the bulk of the anterior neural plate borders. They appear as two bulges on each side of the anterior neural plate and can therefore be easily identified if the gooseneck guide lights are oriented in a manner that highlights the three-dimensional conformation of the embryo. In Figure 1A, the incident light highlights the CNC.

6. Once the embryos are oriented, tighten the plasticine around them by pushing gently on the plasticine with the glass bead.

    This step is optional for experimenters that prefer performing microdissection on free-moving embryos.

7. On the donor embryo, insert the tip of the eyelash knife under the ectoderm at the ventral and posterior end of the anterior neural fold (see online Movie 1 at cshprotocols.cshlp.org). Thread the knife under the ectoderm, move it slightly anteriorly, and cut the ectoderm by lifting the knife swiftly. Repeat the motion anteriorly along the ventral edge of the anterior neural fold until the ventral edge of the CNC is completely uncovered (see online Movie 1 at cshprotocols.cshlp.org).

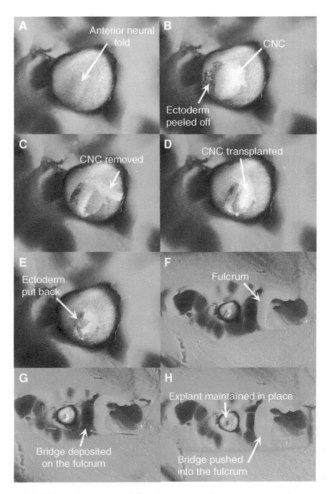

FIGURE 1. Key steps of the CNC transplantation procedure.

The goal of this step is to cut the ectoderm ventrally to the anterior neural fold in order to peel it dorsally (i.e., toward the neural plate) in Step 8. However, the experimenter may find it more convenient to cut off the ectoderm dorsally and peel it off ventrally.

Cutting a section of ectoderm larger than the size of the CNC is recommended: The experimenter can better see the outline of the CNC and neural plate once the ectoderm is peeled, and the peeled ectoderm is easier to manipulate with the hair loop.

8. Peel off the ectoderm dorsally, to expose the underlying neural fold, and drape it over the other side of the embryo using the hair loop (Fig. 1B, see online Movie 1 at cshprotocols.cshlp.org).

9. Identify the location of the CNC (Sadaghiani and Thiebaud 1987). Using the knife tip, cut the CNC outline. Cut superficially to avoid damaging or contaminating the explant with the underlying mesoderm and endoderm (see online Movie 1 at cshprotocols.cshlp.org).

    In some cases, the CNC inherits some of the pigment deposited in the egg during oogenesis, which makes it easy to distinguish from the neural plate and mesoderm. However, CNC are unpigmented the majority of the time. The experimenter should be familiar with landscape of the embryos and learn to recognize the bulge formed by the neural folds. Adequate lighting (described in Step 5) is critical.

10. Lift the explant off the embryo with the knife and hair loop and inspect it for any contaminating mesoderm and endoderm (Fig. 1C). Scrape off contaminating cells with the hair loop and knife.

    CNC explants are made of small cells which typically give the tissue a blue hue. The large size and high yolk content of mesoderm and endoderm cells give them a white to yellow coloration.

11. Place the donor explant adjacent to the host embryo. To move the explant, spear it on the eyelash knife, move it close to the host embryo and push it gently off the knife with the hair loop (see online Movie 1 at cshprotocols.cshlp.org).

    This technique helps the experimenter to keep track of the antero–posterior and dorso–ventral orientation of the CNC.

12. Remove the host CNC by repeating Steps 7–10 on the host embryo. Replace the host CNC with the donor CNC (Fig. 1D, see online Movie 1 at cshprotocols.cshlp.org). Ensure that orientation of the explant matches that of the host embryo (see online Movie 1 at cshprotocols.cshlp.org).

13. Using the hair loop, pull the ectoderm back over the explant as far as possible (Fig. 1E, see online Movie 1 at cshprotocols.cshlp.org).

14. Using blunt forceps, mount a fulcrum of plasticine next to the newly grafted embryo (Fig. 1F).

15. Using blunt forceps, install a cover glass bridge across the fulcrum (Fig. 1G).

16. Using blunt forceps, push gently down on the part of bridge that contacts the fulcrum. Stop pushing once the bridge touches the graft and slightly flattens it (Fig. 1H, see online Movie 1 at cshprotocols.cshlp.org).

17. Allow the epidermal ectoderm to heal for a minimum of 10–20 min.

18. Remove the bridge with the blunt forceps and transfer the grafted embryo into an agarose-coated dish filled with recovery medium. Grow the grafted embryo between 15°C and 18°C.

19. Perform a time-lapse movie of CNC migration.

    i. Transfer the embryo into a new plasticine-coated dish containing imaging medium. Ensure that the plasticine chosen has a low level of autofluorescence in the desired channel.

        The benzocaine in the imaging medium prevents somitic contraction once the embryos reach tailbud stage. It can be replaced with other anesthetics, such as tricain.

    ii. Using the glass bead tool, dig a trench in the plasticine as deep as the embryo is wide and as long as a stage 16 embryo. While making the depression, ensure that one end of the trench is slightly deeper than the other end.

    iii. Move the embryo into the trench with its anterior side in the deepest part of the trench.

        Because gravity will compensate for the movement caused by the ectodermal cilia, the incline ensures that the embryo will move minimally during the movie.

Cite this protocol as *Cold Spring Harb Protoc*; doi:10.1101/pdb.prot097402

iv. To prevent optical distortions caused by evaporation, fill the dish with imaging medium until the liquid bulges over. Slide a glass cover over the dish (i.e., for a 60-mm dish, use an 8- × 10-cm glass plate used to cast SDS-PAGE gels). Ensure that no air is trapped below the plate.

v. Record a time-lapse movie between 18°C and 20°C, at the rate of one image every 3 min for 8–12 h (see online Movie 2 at cshprotocols.cshlp.org).

DISCUSSION

Respecting the polarity of the CNC in relation to the host yields the best results in terms of quality and timing of the migration. While CNC grafted with any of their polarity reversed will migrate, they may be delayed and the migration of individual segments may be perturbed (H Cousin, unpubl.).

The CNC grafting experiment is invaluable to assessing whether a gene's function is cell autonomous, CNC autonomous, or exerted in the tissues surrounding the CNC (ectoderm, placode, or mesoderm). Grafting modified crest (i.e., in which the gene function has been perturbed) into a wild-type host and grafting wild-type CNC into a modified host allows the experimenter to assess whether the function of the studied gene is necessary in the CNC, the surrounding tissues, or both (Alfandari et al. 2001; Cousin et al. 2011, 2012). To assess whether the function of a particular gene is cell autonomous or tissue autonomous, one can co-graft unmodified and modified CNC that express different fluorescent markers (Cousin et al. 2012).

RECIPE

Modified Barth's Saline (MBS) (1×)

CaCl$_2$	0.41 mM
CaNO$_3$	0.3 mM
HEPES-NaOH	15 mM
KCl	1 mM
MgSO$_4$	0.82 mM
NaCl	88 mM
NaHCO$_3$	2.4 mM

Adjust pH to 7.6. Store at room temperature for up to 1 mo.

ACKNOWLEDGMENTS

H.C. is supported by National Institutes of Health/National Institute of Dental and Craniofacial Research (NIH/NIDCR) DE025691.

REFERENCES

Alfandari D, Cousin H, Gaultier A, Smith K, White JM, Darribere T, DeSimone DW. 2001. *Xenopus* ADAM 13 is a metalloprotease required for cranial neural crest-cell migration. *Curr Biol* **11**: 918–930.

Borchers A, Epperlein HH, Wedlich D. 2000. An assay system to study migratory behavior of cranial neural crest cells in *Xenopus*. *Dev Genes Evol* **210**: 217–222.

Carl TF, Dufton C, Hanken J, Klymkowsky MW. 1999. Inhibition of neural crest migration in *Xenopus* using antisense slug RNA. *Dev Biol* **213**: 101–115.

Cousin H, Abbruzzese G, Kerdavid E, Gaultier A, Alfandari D. 2011. Translocation of the cytoplasmic domain of ADAM13 to the nucleus is essential for Calpain8-a expression and cranial neural crest cell migration. *Dev Cell* **20**: 256–263.

Cousin H, Abbruzzese G, McCusker C, Alfandari D. 2012. ADAM13 function is required in the 3 dimensional context of the embryo during cranial neural crest cell migration in *Xenopus laevis*. *Dev Biol* **368**: 335–344.

Hall BK, Horstadius S. 1988. *The neural crest*. Oxford University Press.

Horstadius S. 1950. *The neural crest: Its properties and derivatives in the light of experimental research*. Oxford University Press.

Nieuwkoop PD, Faber J. 1967. *Normal table of* Xenopus laevis *(Daudin)*, 2nd ed. North-Holland, Amsterdam.

Sadaghiani B, Thiebaud CH. 1987. Neural crest development in the *Xenopus laevis* embryo, studied by interspecific transplantation and scanning electron microscopy. *Dev Biol* **124**: 91–110.

Protocol 8

In Vitro Induction of *Xenopus* Embryonic Organs Using Animal Cap Cells

Takashi Ariizumi,[1] Tatsuo Michiue,[2] and Makoto Asashima[3,4,5]

[1]*Department of Agri-Production Sciences, Tamagawa University, Machida, Tokyo 194-8610, Japan;* [2]*Department of Life Sciences (Biology), Graduate School of Arts and Sciences, The University of Tokyo, Meguro, Tokyo 153-8902, Japan;* [3]*Research Institute for Science and Technology, Tokyo University of Science, Shinjuku, Tokyo 162-8601, Japan;* [4]*Biotechnology Research Institute for Drug Discovery, National Institute of Advanced Industrial Science and Technology (AIST), Tsukuba, Ibaraki 305-8568, Japan*

The animal cap—the presumptive ectoderm of the blastula embryo—can differentiate into a variety of tissues belonging to the three germ layers following exposure to specific inducers. The "animal cap assay" was devised based on the pluripotency of presumptive ectodermal cells and enabled many important discoveries in the field of embryonic induction and cell differentiation. Using this system, investigators can test multiple factors in solution simultaneously to determine their inducing activities qualitatively, quantitatively, and synergistically. Furthermore, after dissociation and induction, reaggregated animal cap cells can be induced to form higher-order organs. This protocol details preoperative preparations, followed by the basic animal cap assay. Advanced protocols for the induction of kidney, pancreas, and heart are also described.

MATERIALS

It is essential that you consult the appropriate Material Safety Data Sheets and your institution's Environmental Health and Safety Office for proper handling of equipment and hazardous materials used in this protocol.

RECIPES: Please see the end of this protocol for recipes indicated by <R>. Additional recipes can be found online at http://cshprotocols.cshlp.org/site/recipes.

Reagents

Activin solution (0.5–100 ng/mL human recombinant activin A [e.g., Merck Millipore, 114700] in BSA-CM)

Store stock solutions (>100 ng/mL) at −80°C. Avoid repeated freeze–thaw cycles.

BSA-CM (0.1% [w/v] bovine serum albumin [BSA] in CM)

Adding BSA to CM reduces adsorption of inducing substances or embryonic tissues to the surfaces of plastic culture plates.

BSA-CMFCM

Ca^{2+}/Mg^{2+}-free BSA-CM is used to dissociate animal cap cells.

Dejelling solution

[5]Correspondence: asashi3786@gmail.com

Dissolve 4.5% (w/v) cysteine-HCl or 1% (w/v) sodium thioglycollate in CM. Adjust the pH to 7.8–8.0 with NaOH. Prepare fresh before use.

Embryonic *Xenopus* culture media (CM) <R>

There are a number of media suitable for culturing embryonic Xenopus tissues.

Retinoic acid stock solution (10 mM all-trans retinoic acid [e.g., Sigma-Aldrich, R2625] in dimethyl sulfoxide) (RA)

Store 1-mL aliquots at −80°C. Avoid repeated freeze–thaw cycles.

Xenopus laevis embryos

Detailed descriptions of the methods used to obtain embryos are presented in Ariizumi et al. (2000, 2009).

Equipment

Beakers, 50- and 100-mL, sterilized

Illuminator

Fiber-optic LED lights are preferred.

Incubator, 20°C–22°C

Microscope, binocular, equipped with 10× oculars and 1× to 4× objectives

Operation dishes

To prevent embryonic tissues from adhering to glass or plastic surfaces, line Petri dishes with autoclaved 3% (w/v) agar.

Pipette bulb, silicon

Sterilize in 70% ethanol before use.

Tissue culture plates, low-adhesion, 24-well (e.g., Sumitomo Bakelite, MS-80240Z)

Tissue culture plates, low-adhesion, round-bottomed, 96-well (e.g., Sumitomo Bakelite, MS-3096UZ)

Transfer pipettes

To prepare, flame a Pasteur pipette at its center and draw it out at a 30°–40° angle. Cut pipettes such that the opening is 0.5–2-mm diameter. Flame the tip briefly to smooth the cut edges. Heat-sterilize for 2 h at 180°C.

Tungsten needles

Insert a 0.2-mm tungsten wire (2-cm length) into a transfer pipette (2-mm i.d.). Flame to seal the wire in place. Bend the wire at a right angle; cut 3–5-mm from its end. Sharpen the wire electrolytically using 5 M NaOH and a 9V-dry cell (submerge the negative pole on a carbon point in the NaOH solution; attach the positive pole to the tungsten wire). Heat-sterilize for 2 h at 180°C.

Watchmaker's forceps (e.g., Fontax no. 5)

Heat-sterilize for 2 h at 180°C.

METHOD

Membrane Removal

To access the animal cap, the jelly coat (see online Movie 1 at cshprotocols.cshlp.org) and vitelline membrane (see online Movie 2 at cshprotocols.cshlp.org) surrounding the embryo must be removed.

1. Collect embryos in a sterilized 100-mL beaker. Add 50 mL of dejelling solution.

2. Swirl for 5–10 min. Decant the dejelling solution when the jelly-free embryos begin to pack together.

3. Rinse 10 times by swirling gently with the CM of choice.

4. Select blastula embryos at the desired stage of development (see Nieuwkoop and Faber 1994). Use a 2-mm diameter transfer pipette and a sterilized pipette bulb to place the embryos into an agar-lined Petri dish containing CM.

5. Holding the embryo upside-down, use fine forceps to grasp the vitelline membrane. Gently tear the membrane from the embryo.

Animal Cap Dissection

The procedure is shown in online Movie 3 at cshprotocols.cshlp.org.

6. Place the blastula embryo with the animal pole facing upward in an operation dish filled with the CM of choice.

7. Using a tungsten needle, trim both sides of the embryo.

8. Insert the needle into the blastocoel from one side. Divide the vegetal hemisphere (i.e., the endodermal region) by pushing down the needle.

   This produces a sheet of blastocoel roof with cell masses at each end.

9. Reverse the sheet. Trim away vegetal yolky cells and marginal zone cells.

10. Trim the sheet (i.e., the animal cap) into a 0.5×0.5-mm square.

Animal Cap Culture

The procedure is shown schematically in online Movie 4 at cshprotocols.cshlp.org. Use a transfer pipette (0.5- to 1-mm diameter) to handle the animal caps.

11. Transfer five to 10 animal caps into activin solution of the desired concentration. Place them such that the inner blastocoel side faces upward. Incubate for a defined period (e.g., 3 h).

    Depending on the concentration used, activin can induce a variety of tissues and organs (Table 1).

12. Transfer the animal caps to a dish filled with BSA-CM. Wash them by pipetting gently.

13. Incubate the caps in fresh BSA-CM for 3–4 d at 20°C–22°C.

    The explants will show obvious histodifferentiation patterns at the end of the culture.

Kidney Induction

Simultaneous treatment with RA and activin can induce the generation of the pronephros (i.e., the embryonic kidney) (see online Movie 5 at cshprotocols.cshlp.org). The original protocol (Moriya et al. 1993) used Steinberg's Solution as the CM.

14. Transfer 10 animal caps from late blastulae (i.e., stage 9) to a well of a 24-well tissue culture plate containing "Test Solution K" (i.e., 10 μL of RA stock solution and 990 μL of 10-ng/mL activin solution). Incubate for 3 h at 20°C.

15. Wash the caps in BSA-CM. Culture in fresh BSA-CM for 3 d at 20°C.

    After 3 d, pronephric tubules should be observable inside the thin epidermal vesicle.

Pancreas Induction

Sequential treatment with activin and RA can induce formation of the pancreas (see online Movie 5 at cshprotocols. cshlp.org).

16. Transfer 10 animal caps from late blastulae (i.e., stage 9) to a well of a 24-well tissue culture plate containing "Test Solution P1" (100-ng/mL activin solution). Incubate for 1 h at 20°C.

17. After washing briefly with BSA-CM, incubate in fresh BSA-CM for 5 h at 20°C.

18. Transfer the animal caps to "Test Solution P2" (i.e., 10 μL of RA stock solution in 990 μL of BSA-CM). Incubate for 1 h at 20°C.

19. Wash the animal caps with BSA-CM. Culture in fresh BSA-CM for 3 d at 20°C.

    Pancreatic differentiation can be characterized by histological examination and/or the expression of molecular markers (e.g., pdx1, insulin).

Cite this protocol as *Cold Spring Harb Protoc*; doi:10.1101/pdb.prot097410

TABLE 1. Examples of tissues and organs induced from *Xenopus* animal caps by activin and other factors

Tissues and organs	Methods	Notes	Markers	References
Ectoderm				
Epidermis	Animal cap assay (Steps 11–13)	Induced by 0.5–50 ng/mL activin	*xk81a1*	Asashima et al. 1990; Ariizumi et al. 1991
Brain (forebrain, hindbrain); Spinal cord	Combination of treated and untreated animal caps	Induced from untreated animal caps by animal caps treated with >50 ng/mL activin	*otx2, sox2, foxg1*	Ariizumi and Asashima 1994
Eye (lens, neural retina, retinal pigment epithelium); Ear vesicle	As above	Induced secondarily from brain	*pax6, six3, rax*	Ariizumi and Asashima 1994; Sedohara et al. 2003
Mesoderm				
Notochord	Animal cap assay (Steps 11–13)	Induced by high concentrations (>50 ng/mL) of activin	*chrd, not*	Asashima et al. 1990; Ariizumi et al. 1991
Skeletal muscle	As above	Induced by intermediate concentrations (5–50 ng/mL) of activin	*myod, myf5, mrf4*	Asashima et al. 1990; Ariizumi et al. 1991
Coelomic epithelium	As above	Induced by low concentrations (0.5–5 ng/mL) of activin	—	Asashima et al. 1990; Ariizumi et al. 1991
Blood cells; Blood vessel	As above	Induced by low concentrations (0.5–1 ng/mL) of activin	*runx2, spib, mpo*	Asashima et al. 1990; Ariizumi et al. 1991; Miyanaga et al. 1999; Nagamine et al. 2007
Pronephros (glomus, tubule, duct)	Multiple treatments with activin and retinoic acid (Steps 14–15)	Induced by simultaneous treatment with 10 ng/mL activin and 0.1 mM retinoic acid	*wnt4, pax2, sall1, wt1*	Moriya et al. 1993; Chan et al. 1999; Osafune et al. 2002
Heart	Dissociation and reaggregation of animal cap cells (Steps 21–24)	Individual cells are treated with 100 ng/mL activin during reaggregation	*nkx2.5, gata4, tbx5, tnni3*	Ariizumi et al. 2003
Craniofacial cartilage	Combination of treated and untreated animal caps	Induced from untreated animal caps by animal caps treated with 100 ng/mL activin	*dll4, col2a1, alx1*	Furue et al. 2002
Endoderm				
Pharynx; Intestine	Combination of treated and untreated animal caps	Animal caps treated with >50 ng/mL activin form pharynx and intestine in recombinants	*fabp2, darmin*	Ariizumi and Asashima 1994
Pancreas	Multiple treatments with activin and retinoic acid (Steps 16–19)	Induced by sequential treatment with 100 ng/mL activin and 0.1 mM retinoic acid	*pdx1, mist1, ins*	Moriya et al. 2000
Liver	Animal cap assay (Steps 11–13)	Yolk-rich endoderm induced by 100 ng/mL activin forms liver when transplanted into larvae	*hhex, ambp, transferrin, albumin*	Ariizumi et al. 2003

Dissociation and Reaggregation of Animal Caps for Heart Induction

Cell-to-cell interactions can be studied by examining the activity of animal caps dissociated into individual cells. Here, we describe a dissociation/reaggregation protocol to induce heart formation (see online Movie 6 at cshprotocols. cshlp.org). The original method (Ariizumi et al. 2003) used Holtfreter's Solution as the CM.

20. Dissect five to 10 animal caps from mid-blastulae (i.e., stage 8) in an operation dish filled with CM.

21. Transfer the caps to a 35-mm diameter Petri dish containing BSA-CMFCM.

    This is crucial to eliminate Ca^{2+} and Mg^{2+} cations that might be transferred from the operation dish.

22. Transfer five to 10 animal caps into a single well of a 96-well tissue culture plate containing 100 µL of BSA-CMFCM. Incubate for 20 min.

23. Replace the BSA-CMFCM with 100 µL of 100-ng/mL activin solution. Disperse the cells by pipetting gently. Incubate for 5 h.

24. Transfer the newly formed spherical "reaggregates" into a single well filled with 200 µL of BSA-CM.

    The reaggregated cells will begin to beat rhythmically within 3 d at 20°C.

DISCUSSION

When treated with a mesoderm-inducer such as activin, animal caps are competent for in vitro organogenesis from stage 7 (early blastula) up to stage 9 (late blastula). Thus, accurate staging of embryos is vital to the success of in vitro organogenesis. The late blastula (stage 9) is used as the standard for the animal cap assay, although for the heart induction described in this protocol, mid-blastula embryos (stage 8) are most suitable. The size of the animal caps is also critical. The number of cells in the explant will affect the efficacy of tissue differentiation and organ formation. A large cap might be contaminated with cells from the marginal zone, which can differentiate autonomously into mesodermal tissues. The most reliable animal cap size is 0.5 × 0.5 mm. Finally, the duration of the exposure of the animal caps to an inducer can influence the differentiation pattern. For example, a brief exposure (10 min) to 10-ng/mL activin causes the differentiation of ventral mesoderm, whereas a longer exposure (>3 h) to the same dose induces dorsal mesoderm differentiation (Ariizumi et al. 1991).

RECIPE

Embryonic Xenopus Culture Media (CM)

Components (mM)	Holtfreter's solution	Modified Barth's solution[a]	Marc's modified Ringers[a]	Normal amphibian medium[a]	Steinberg's solution
NaCl	60.00[b]	88.00	100.00	110.00	60.00[b]
KCl	0.67	1.00	2.00	2.00	0.67
CaCl$_2$[c]	0.90	0.41	2.00	—	—
Ca(NO$_3$)$_2$[c]	—	0.33	—	1.00	0.34
MgCl$_2$[c]	—	—	1.00	—	—
MgSO$_4$[c]	—	0.82	—	1.00	0.83
EDTA	—	—	—	0.10	—
Sodium phosphate, pH 7.50	—	—	—	2.00	—
NaHCO$_3$	0.20[d]	2.40	—	1.00	—
HEPES	4.60[d]	10.00	5.00	—	10.00
pH	7.60	7.40	7.40	7.40	7.40

[a]Use at half-strength (i.e., 0.5×) for culture media.

[b]To prevent animal caps from curling, increase the NaCl concentration from 60 mM to 90 mM. However, note that animal cap cells will differentiate into neural tissue autonomously when exposed to high NaCl concentrations.

[c]Omit these components when preparing Ca^{2+}/Mg^{2+}-free culture media (CMFCM).

[d]For Holtfreter's solution, use *either* NaHCO$_3$ *or* HEPES.

Sterilize the CM of choice by autoclaving for 20 min at 121°C. Add antibiotics such as kanamycin sulfate (0.1 g/L) to avoid bacterial contamination.

ACKNOWLEDGMENT

We thank Dr. Shuji Takahashi for technical support and many insightful discussions over the years.

REFERENCES

Ariizumi T, Asashima M. 1994. *In vitro* control of the embryonic form of *Xenopus laevis* by activin A: Time and dose-dependent inducing properties of activin A-treated ectoderm. *Dev Growth Differ* **36:** 499–507.

Ariizumi T, Sawamura K, Uchiyama H, Asashima M. 1991. Dose and time-dependent mesoderm induction and outgrowth formation by activin A in *Xenopus laevis. Int J Dev Biol* **35:** 407–414.

Ariizumi T, Takano K, Asashima M, Malacinski GM. 2000. Bioassays of inductive interactions in amphibian development. *Methods Mol Biol* **135:** 89–112.

Ariizumi T, Kinoshita M, Yokota C, Takano K, Fukuda K, Moriyama N, Malacinski GM, Asashima M. 2003. Amphibian in vitro heart induction: A simple and reliable model for the study of vertebrate cardiac development. *Int J Dev Biol* **47:** 405–410.

Ariizumi T, Takahashi S, Chan TC, Ito Y, Michiue T, Asashima M. 2009. Isolation and differentiation of *Xenopus* animal cap cells. *Curr Protoc Stem Cell Biol* **Chapter 1:** Unit: 1D.5.

Asashima M, Nakano H, Shimada K, Kinoshita K, Ishii K, Shibai H, Ueno N. 1990. Mesodermal induction in early amphibian embryos by activin A (erythroid-differentiation factor). *Roux Arch Dev Biol* **198:** 330–335.

Chan TC, Ariizumi T, Asashima M. 1999. A model system for organ engineering: Transplantation of in vitro induced embryonic kidney. *Naturwissenschaften* **86:** 224–227.

Furue M, Myoishi Y, Fukui Y, Ariizumi T, Okamoto T, Asashima M. 2002. Activin A induces craniofacial cartilage from undifferentiated *Xenopus* ectoderm *in vitro. Proc Natl Acad Sci* **99:** 15474–15479.

Miyanaga Y, Shiurba R, Asashima M. 1999. Blood cell induction in *Xenopus* animal cap explants: Effects of fibroblast growth factor, bone morphogenetic proteins, and activin. *Dev Genes Evol* **209:** 69–76.

Moriya N, Uchiyama H, Asashima M. 1993. Induction of pronephric tubules by activin and retinoic acid in presumptive ectoderm of *Xenopus laevis. Dev Growth Differ* **35:** 123–128.

Moriya N, Komazaki S, Takahashi S, Yokota C, Asashima M. 2000. In vitro pancreas formation from *Xenopus* ectoderm treated with activin and retinoic acid. *Dev Growth Differ* **42:** 593–602.

Nagamine K, Furue M, Fukui A, Matsuda A, Hori T, Asashima M. 2007. Blood cell and vessel formation following transplantation of activin-treated explants in *Xenopus. Biol Pharm Bull* **30:** 1856–1859.

Nieuwkoop PD, Faber J (eds.). 1994. *Normal table of* Xenopus laevis *(Daudin): A systematical & chronological survey of the development from the fertilized egg till the end of metamorphosis.* Garland Publishing, New York.

Osafune K, Nishinakamura R, Komazaki S, Asashima M. 2002. In vitro induction of the pronephric duct in *Xenopus* explants. *Dev Growth Differ* **44:** 161–167.

Sedohara A, Komazaki S, Asashima M. 2003. In vitro induction and transplantation of eye during early *Xenopus* development. *Dev Growth Differ* **45:** 463–471.

Special Considerations for Making Explants and Transplants with *Xenopus tropicalis*

Marilyn Fisher and Robert M. Grainger[1]

Biology Department, University of Virginia, Charlottesville, Virginia 22904

Although *Xenopus laevis* is an important model organism for embryological experimentation, the smaller, more genetically tractable, and faster developing *Xenopus tropicalis* provides advantages for using genetic approaches to understand developmental mechanisms. Explant cultures and transplants of *X. tropicalis* embryonic tissues present unique opportunities to examine embryonic tissue determination in a simplified setting. Here we demonstrate preparation of explants and transplants of preplacodal head ectoderm in order to illustrate these approaches; however, these methods apply broadly to tissues throughout the embryo. We focus on technical adjustments to accommodate the differences in size, tissue character, and rate of development between *X. laevis* and *X. tropicalis*. With only modest modifications, *X. tropicalis* embryos are quite amenable to the same kinds of experimental manipulations as *X. laevis*.

MATERIALS

It is essential that you consult the appropriate Material Safety Data Sheets and your institution's Environmental Health and Safety Office for proper handling of equipment and hazardous materials used in this protocol.

RECIPES: Please see the end of this protocol for recipes indicated by <R>. Additional recipes can be found online at http://cshprotocols.cshlp.org/site/recipes.

Reagents

Bovine serum albumin (BSA), fraction V (10 mg/mL stock solution in H_2O)

Dilute to 20–100 µg/mL in 1× MBS for coating dishes and pipettes.

Cysteine-HCl (2% [w/v] in H_2O)

Adjust the pH to 8.5 using 10 N NaOH. Alternatively, cysteine free base can be dissolved in 0.1× MBS and adjusted to pH 8.0.

Gentamicin (100 mg/mL in H_2O)

Dilute the stock solution to a working concentration of 50 µg/mL.

Modified Barth's saline (MBS) <R>

Alternatively, Steinberg's solution (2×) <R> can be used in place of MBS during dissection; see Step 7.

X. tropicalis embryos

A procedure for obtaining embryos via natural matings or in vitro fertilization can be found in Hirsch et al. (2002). X. tropicalis eggs are relatively sticky (compared to X. laevis embryos), both before and after dejellying, and must be handled using coated pipettes and dishes until the hatching stage.

All embryo staging is performed according to Nieuwkoop and Faber (1956).

---

[1]Correspondence: rmg9p@virginia.edu

Supplemental material is available at cshprotocols.cshlp.org

Equipment

Beaker, glass (150-mL)

Cover glasses (optional)

If ectoderm explants are to be maintained as flat sheets (Step 9), they must be immobilized, e.g., under small pieces of glass. To prepare small cover glasses, use a diamond pencil to score and then break small rectangles (3 × 5 mm) from microscope cover glasses. Clean the pieces by washing in alcohol and then allowing to air dry. Store glass pieces in a Petri dish.

To prepare glass bridges to hold transplants in place while healing (Step 15), briefly hold small glass pieces prepared as described above over the flame of a micro Bunsen burner with forceps to fire polish the edges. This will simultaneously cause the glass to curve somewhat, which helps it conform to the shape of the embryo. Alternatively, clay fingers can be used in place of glass bridges; see Step 15.

Dissecting microscope with magnification to 50×

Forceps, blunt-tipped (for immobilizing embryos in clay)

Forceps, fine-tipped (two pairs, for removing membranes from embryos)

Dumont #5 Biologie-grade forceps are sharper and optimal for delicate manipulations.

Hair tools (for manipulating embryos and tissue pieces without damage)

Prepare eyebrow hairs and/or hair loops fixed into glass capillary tubes with nail polish; see Sive et al. (2007).

Low-temperature incubator(s) set at 22°C and/or 25°C

Microcapillary tubes for transferring dissected tissues (e.g., 25-µL Drummond Microcaps [Drummond 1-000-0250])

Drummond Microcaps come with convenient bulb dispensers included. Alternatively, a pipette pump (e.g., Bel-Art SP Scienceware F37898-0000) can be used for controlling flow when glass Pasteur pipettes are used for transferring embryos or tissue pieces. Commercial Pasteur pipettes typically have too wide an opening to be useful for transferring explants from operating dish to culture dish, but they can be narrowed by drawing the ends out after heating delicately over a Bunsen burner flame. Break off the pulled ends of Pasteur pipettes such that the diameter of the openings are only slightly larger than the tissue pieces to be transferred, and fire-polish if necessary.

Operating dishes (clay-lined)

To prepare operating dishes, line the bottoms of 35-mm Petri dishes with clay to a depth of about 1/3. Use clay that is nontoxic, not water-soluble, and is firm enough to hold X. tropicalis embryos and tissues in place (e.g., Claytoon). Prior to use, sterilize the clay dishes with 70% EtOH and allow to air dry.

Create wells in the clay for securing the embryos during operations using a small glass bulb. To prepare the bulb, draw out the end of a Pasteur pipette over a flame and then hold the drawn-out end over the flame to melt the tip into a ball at the end of the pipette. Control the starting size of the tip and how much of it melts over the flame to form a ball that is roughly the size of an embryo. Once formed, press the glass bulb into the clay to create a well deep enough to almost submerge an embryo. Dishes can be reused after rinsing and sterilization with 70% EtOH.

Petri dishes

Any dishes used to hold embryos after dejellying must be coated to prevent tissues from sticking to plastic. To coat with 1× MBS containing ≥20 µg/mL BSA, add a sufficient volume of solution to completely cover the bottom of the dish, allow to stand for several minutes and then pour off. To coat with a thin layer of agarose (1% [w/v] in 0.1× MBS), pour sufficient melted agarose solution into the dish to cover the bottom and immediately pour off the excess. Allow agarose-coated dishes to dry before using.

Silicone grease (chemical-resistant, heat stable, inert) (Dow-Corning high-vacuum grease)

Transfer pipettes (plastic)

Prior to use, any plastic transfer pipettes that will be used to collect embryos should be coated with 1× MBS containing ≥20 µg/mL BSA. Fill each pipette and expel the solution several times to coat.

Tungsten needles

We use tungsten wire with 0.008 inch diameter (#214071 from Leico Industries). Needles are electrolytically sharpened using a small plastic beaker of 1 N NaOH in which the needle and needle holder complete a circuit using a 6 V power supply (e.g., see Moore and Kennedy [2008]). We use a modified microscope lamp power supply (for 6 V lamps); the wires that would normally attach to the bulb are instead attached to alligator clips. One wire is clipped to the metal needle holder into which a piece of tungsten wire is inserted, and the other wire is clipped to the edge of the beaker such that its tip is submerged. The power supply is turned on and the tip of the tungsten wire is submerged in the beaker until the current erodes the tungsten to form a fine tip. Once a good point is obtained on the end of the needle, it helps to use forceps to bend the needle just behind the tip, such that the tip is more parallel than perpendicular with the surface of the embryo when dissection is begun. This allows one to make shallow cuts into the embryo more easily, and prevents damage to the needle tip.

METHOD

Preparing Tissue Explants

Here we describe how to prepare explant cultures from preplacodal head ectoderm, specifically the presumptive lens ectoderm (PLE), of embryos from neurula stages 14–21. Supplemental Movie S1 demonstrates PLE dissection from a neurula-stage embryo for explantation. This is not intended to be an exact illustration of a PLE dissection, but rather to illustrate the principal points needed for explant preparation, which can be used as a guide for isolation of tissues from many regions of the embryo. Note that tissues from other germ layers in the early embryo will be far more fragile than the ectoderm, and may require use of tools like an eyebrow knife or glass needle once the ectoderm is removed to allow access to these tissues. Similarly, explants of tissues from other germ layers may be more fragile when cultured directly or placed under glass, and may require higher salt, e.g., 1× MBS, to maintain tissue integrity during early postoperation phase of culture. Each dissection requires honing the methods used for isolating and maintaining tissues, though the variations described here provide a very good starting point.

Preparing the Embryos

1. Dejelly *X. tropicalis* embryos by gently swirling them in a small dish or beaker of 2% cysteine (pH 8.5) for ~2.5 min at room temperature.

   Dejellying in a Petri dish allows easy monitoring of the process with a dissecting microscope, so the time needed for complete dejellying can be determined in the context of the specific conditions.

2. When jelly coats have come off, transfer the embryos to a small glass beaker. Rinse the embryos ~10 times with 30–50 mL of 0.1× MBS per rinse.

   Thorough removal of cysteine is essential to allow embryos to develop normally.

3. Select embryos at a slightly younger stage than desired and transfer them to a clay-lined operating dish containing 1× MBS with 20 μg/mL BSA.

   The embryos will stick to the clay just enough such that it will be easier to remove the vitelline membranes in Step 4.

   Do not prepare too many embryos in advance, as they age fairly rapidly at these stages. The exact rate at which the embryos develop will depend on ambient temperature and their starting stage. A novice should begin with 2 or 3 embryos at a time.

4. Remove the vitelline membranes from the embryos using fine forceps.

   This procedure is similar in X. tropicalis and X. laevis embryos, but it is more difficult to see a gap between the membrane and the embryo surface in X. tropicalis, especially before the neurula stages. With either Xenopus species, it is easier to accurately stage embryos with the vitelline envelope removed, since this allows the embryos to reveal subtle morphological features which are otherwise obscured by the constraint of the envelope, and which then generally match the morphologies seen in the Nieuwkoop and Faber (1956) staging series.

Dissecting the PLE

5. Transfer the demembranated embryos to a fresh clay-lined dish containing 1× MBS with 20 μg/mL BSA for operation. Orient the embryos in wells for PLE dissection: Place the embryos on their sides with the edge between anterior neural tissue and ectoderm (the approximate site of the PLE) facing upward (see Supplemental Movie S1).

   Precise illustrations of the PLE site are shown in Jin et al. (2012).

   Orientation will vary depending on the piece of tissue being isolated and the age of the donor embryos.

Cite this protocol as *Cold Spring Harb Protoc*; doi:10.1101/pdb.prot097428

6. Immobilize the embryos by using blunt forceps to press the clay gently but firmly all around the exposed parts of the embryos to hold them in place (see Supplemental Movie S1). Be sure to also press in gently below the surface level from all sides, such that the clay is touching all sides of the embryos.

   The embryos should not be loose in the wells or they will move when manipulated.

7. Dissect the PLE using a tungsten needle and hair loop (see Supplemental Movie S1).

   i. Using the needle as if it were a knife, stick the tip down into the tissue (just deep enough to cut through the layer or layers you want to remove) and advance the tip under the tissue in the direction of your outline.

   ii. Lift upward, with the tip of the needle leading, and pull the needle up out of the tissue.

   iii. Repeat Steps 7.i and 7.ii as many times as necessary until you have completed cutting around the desired piece. (Do not try to cut too much at a time.)

   iv. Using a hair loop, gently peel the ectoderm away from underlying mesendoderm and/or neural crest.

      X. tropicalis embryos have relatively tough ectoderm compared with X. laevis. At these stages with X. laevis embryos, one can use glass needles for equivalent dissections, or even eyebrow hair tools for younger stage embryos. The tougher X. tropicalis ectoderm is more compatible with the firmness of the tungsten needle, and you will need to exert some force with your needle to cut through the tissue. Lifting cut tissue from its host site can be accomplished with a hair tool, which by being less sharp is also less likely to damage the tissue.

      In the case of the PLE, the degree of difficulty of separating the ectoderm from the underlying tissues will increase with the increasing age of embryo and the degree of interaction between the PLE and ultimately the optic vesicle which comes to underlie it. By stage 21, coincidently the age of determination of the PLE (Jin et al. 2012), it can be extremely difficult to separate the ectoderm from the underlying optic vesicle. 2× Steinberg's solution can be used in place of MBS during dissection to ease tissue separation.

      See Troubleshooting.

8. Using a BSA-coated glass capillary, carefully transfer the explants from the operating dish to a fresh Petri dish containing 0.5× MBS with 0.2% BSA and 50 µg/mL gentamicin.

   PLE explants (or any ectoderm) will naturally round up during culture, and X. tropicalis ectoderm is especially prone to this. Depending on your purpose, you may choose to culture the explants as vesicles floating free in the medium or you may hold them as flat sheets under cover glasses. To prepare flat ectoderm explants, immediately transfer the explants to a fresh Petri dish as described and proceed to Step 9.

9. (Optional) To prepare flat ectoderm explants, quickly restrain the individual explants under small cover glasses as follows.

   i. Dab a small amount of silicone grease into a clean, empty Petri dish.

   ii. Holding the middle of a small cover glass with forceps, carefully apply a very small amount of grease to each end of one side of the cover glass by dipping each end into the grease.

      Do not use so much grease that it will spread enough to contact the tissue upon compression in Step 9.iv.

   iii. Carefully submerge and move the cover glass into position above an explanted tissue piece.

   iv. Carefully lower the cover glass down on top of the explant so that the grease "legs" make contact with the dish, but not the tissue. Very gently press straight down on the cover glass to compress the grease until the glass is lowered just enough to hold the tissue immobile. Do not press too hard or the tissue will disintegrate.

      Alternatively, if you do not press hard enough, the tissue will round up and make its way out from under the glass due to the action of the surface cilia.

10. Culture the explants at 22°C–25°C.

    While X. laevis explants (or transplants) are quite tolerant of temperatures between 16°C and 25°C, X. tropicalis are more temperature-sensitive, preferring to be kept within the range of 22°C–25°C.

Preparing Transplants

For some studies it might be desirable to transplant tissues to another embryo, either hetero- or homotopically. In many cases, it is advisable to use host and donor marking to distinguish transplanted tissue from host tissue (e.g., see Jin et al. [2012]). Here the PLE is dissected from a neurula-stage embryo and then transplanted to a neurula-stage host. This procedure is demonstrated in Supplemental Movie S1.

In general, surgical manipulations like the following are performed in 1× MBS, which promotes tissue healing, but embryos develop more normally in 0.5× or 0.1× MBS when the culture period is extended longer than overnight (see Step 17). It is important to note that holding embryos in the relatively high salt 1× MBS can cause exogastrulation if embryos are being manipulated during gastrula stages.

11. Prepare a clay-lined operating dish as described, using blunt forceps to draw a dividing line across the surface of the clay such that there is a row of wells along either side of the line: one set for donor embryos and one set for hosts.

    Wells for the host embryos should be deep enough so that the appropriate site on the embryo surface barely protrudes above the level of the clay surface.

12. Select, prepare and immobilize host and donor embryos in the operating dish containing 1× MBS with 20 µg/mL BSA as described in Steps 3–6. Prepare and bury only as many host and donor embryos at one time as can be processed before the embryos age too far.

    For the beginner, this may be only one or two sets of embryos.

13. Prepare the host embryo by removing a piece of ectoderm (or other tissue from the region of the body chosen for the transplant site) as described in Step 7. Discard the dissected tissue.

    The piece removed should be close in size to the piece you plan to transplant, but it is often valuable to make the host site slightly smaller than the donor tissue since the wound site will often expand slightly from the pressure of holding the embryo in place with clay. If the host site begins to heal, however, it is usually straightforward to open the site slightly by pulling at the corners with a hair tool.

14. Harvest a piece of tissue to be transplanted from the adjacent donor embryo as described in Step 7. Quickly transfer the tissue to the prepared host site using a tungsten needle or hair tool to gently push the tissue piece through the medium into position. Take care not to let the tissue come into contact with the liquid/air interface or it will immediately disintegrate.

15. Place a glass bridge over the transplant so that it is initially above the surface of the transplant site, and then gently press its ends into the clay to hold the transplanted tissue in place.

    Alternatively, as shown in Supplemental Movie S1, using fine forceps one can fashion small clay fingers adjacent to the embryo that can be folded over and pressed onto the tissue to hold it in place. The clay fingers take practice to master but are simpler to use in that there is no advance preparation of glass bridges required, and the clay is more easily custom fitted to each individual case. Also, because of the smaller size of X. tropicalis embryos compared with X. laevis, making and working with very small glass bridges can be challenging. An additional benefit of clay fingers is that they will generally not stick to and potentially tear off the transplanted tissue if BSA is included in the operating medium.

16. Allow the transplant to heal for ~15 min and then lift the clay fingers or glass bridge off the host embryo. Loosen the clay around the host embryos so that subsequent development is not hindered by compression.

17. After 30–60 min of healing in the operating dish, carefully pipette the host embryos into a Petri dish containing 0.5× MBS with 20 µg/mL BSA. Allow the embryos to develop to the appropriate stage for the experiment.

    If it is desirable to keep track of individual host/donor pairs, embryos can be raised in separate wells of 24-well plates.

    The concentration of 0.5× MBS is useful because it promotes healing to some degree but does not cause abnormal development as can occur with long culture times in 1× MBS. Also, we find that this intermediate concentration provides a good transitional level of salt for preventing shock caused by going from an initial dissection in 1× MBS directly to 0.1× MBS. If the culture period will be longer than overnight, hosts should be transferred to 0.1× MBS the following day.

TROUBLESHOOTING

Problem (Step 7): It is difficult to separate the ectoderm from the underlying optic vesicle.

Solution: In such sticky situations, there are a few tricks one can try: (1) Cut around the outline of the tissue to be removed, but wait a couple of minutes before trying to lift off the ectoderm. (2) Chill the embryos briefly (place the dish on ice for a couple of minutes) before making the cuts. Too much chilling, however, will make the ectoderm tear too easily. (3) Switch from 1× MBS to 2× Steinberg's saline (since short-term exposure to higher salt media can be helpful for tissue separation). (4) Combine use of Steinberg's saline with option 1 and/or 2. Tissues can also be separated by very mild trypsin treatment (Sigma-Aldrich T8253 or equivalent; free of contaminating proteases), with the exact concentration depending on how strongly tissues are adhering to one another. For the PLE, as an example, from stage 21 onward, 0.001%–0.01% trypsin is sufficient for as long as required for tissue separation, followed by several rinses in soybean trypsin inhibitor (Sigma-Aldrich T9003; typically 0.02%). Other tissues may require trypsin treatment for clean isolation, e.g., to remove the neural plate from underlying mesoderm.

DISCUSSION

The techniques described here, although focused on the PLE, apply to any preplacodal head ectoderm as well as to early neural plate/neural tube regions, and with slight modifications, to any embryonic stage. We have used these techniques to prepare PLE explants and transplants for studying aspects of PLE determination in wild-type and transgenic *X. laevis* (e.g., see Grainger et al. [1988]) and in wild-type, mutant and transgenic *X. tropicalis* (e.g., see Jin et al. [2012]). The key point is that PLE from both species responds similarly. Although *X. tropicalis* embryos are smaller, require sturdier dissecting tools, and have a narrower tolerance for temperature manipulation, they are well suited to the same types of experimental manipulations as *X. laevis* and bring to the table advantages for genetic manipulations.

RECIPES

$CaCl_2$ (0.1 M)

11.1 g $CaCl_2$

Dissolve 11.1 g $CaCl_2$ in 1 L H_2O. Autoclave and store in aliquots at −20°C or 4°C.

MBS Salts (10×)

NaCl (880 mM)
KCl (10 mM)
$MgSO_4$ (10 mM)
HEPES (50 mM, pH 7.8)

Omit HEPES if MBS is to be used for oocyte maturation.

$NaHCO_3$ (25 mM)

Adjust pH to 7.8 with NaOH. Autoclave. Store at room temperature.

Modified Barth's Saline (MBS)

$CaCl_2$ (0.1 M)
MBS salts (10×)

For a 1× solution of MBS, mix 100 mL of 10× MBS salts with 7 mL of 0.1 M $CaCl_2$. Adjust the volume up to 1 liter with H_2O. Store at room temperature.

Steinberg's Solution (2×)

Reagent	Quantity needed for 50 mL of stock solution	Final concentration of stock solution	Volume of stock solution needed for 2× Steinberg's solution
NaCl	8.5 g	17%	2 mL
KCl	0.25 g	0.5%	1 mL
$Ca(NO_3)_2 \cdot 4H_2O$	0.4 g	0.8%	1 mL
$MgSO_4 \cdot 7H_2O$	1.03 g	2.05%	1 mL
Tris	2.8 g	100×	1 mL

Prepare 50 mL each of NaCl, KCl, $Ca(NO_3)_2$, $MgSO_4$, and Tris stock solutions in dH_2O, and filter to sterilize. Store stock solutions at room temperature. Before each use, freshly prepare 2× Steinberg's solution (50 mL) by combining the volumes of stock solutions listed, adjust the pH to 7.5 with sterile 1 N HCl, and then bring to a final volume to 50 mL with sterile dH_2O.

ACKNOWLEDGMENTS

Support from the National Institutes of Health (NIH) grants R01 EY006675, R01 EY017400, R01 EY018000, R01 EY022954, and R01 RR013221, as well as from the Sharon Stewart Aniridia Trust and Vision for Tomorrow, is gratefully acknowledged.

REFERENCES

Grainger RM, Henry JJ, Henderson RA. 1988. Reinvestigation of the role of the optic vesicle in embryonic lens induction. *Development* 102: 517–526.

Hirsch N, Zimmerman LB, Gray J, Chae J, Curran KL, Fisher M, Ogino H, Grainger RM. 2002. *Xenopus tropicalis* transgenic lines and their use in the study of embryonic induction. *Dev Dyn* 225: 522–535. doi:10.1002/dvdy.10188

Jin H, Fisher M, Grainger RM. 2012. Defining progressive stages in the commitment process leading to embryonic lens formation. *Genesis* 50: 728–740. doi:10.1002/dvg.22038

Moore SW, Kennedy TE. 2008. Dissection and culture of embryonic spinal commissural neurons. *Curr Protoc Neurosci* Chapter 3: Unit 3 20. doi:10.1002/0471142301.ns0320s45

Nieuwkoop PD, Faber J. 1956. *Normal table of* Xenopus laevis. North Holland Publishing, Amsterdam.

Sive HL, Grainger RM, Harland RM. 2007. Embryo dissection and micromanipulation tools. *Cold Spring Harb Protoc* doi:10.1101/pdb.top7

Cite this protocol as *Cold Spring Harb Protoc*; doi:10.1101/pdb.prot097428

SECTION 6: Cell Biology

The Use of *Xenopus* for Cell Biology Applications

Anna Philpott[1,2,3]

[1]*Department of Oncology, University of Cambridge, Cambridge, CB2 0XZ, United Kingdom;* [2]*Wellcome-MRC Cambridge Stem Cell Institute, Jeffrey Cheah Biomedical Centre, Cambridge Biomedical Campus, Cambridge, CB2 0AW, United Kingdom*

Problems of cell biology and the molecular controls underpinning them have been studied in the remarkably versatile *Xenopus* systems for many years. This versatility is showcased in several accompanying protocols, which are introduced here. One protocol demonstrates how the *Xenopus* embryonic ectoderm can be used to study the effects of mechanical cell deformation; another illustrates how the developing eye can be used as a platform for determining cell-cycle length. Two protocols show how extracts from *Xenopus* embryos can be exploited to characterize the behavior of specific intracellular proteins—specifically, to determine protein phosphorylation status and the ability to bind to chromatin. Finally, because specific antibodies to *Xenopus* proteins are pivotal reagents for cell biology and biochemistry applications, four protocols describing how to generate, purify, and assay the specificity of antibodies raised against *Xenopus* proteins are included in hopes of stimulating the expansion of these critical resources across the *Xenopus* community.

BACKGROUND

Xenopus has provided a remarkably versatile array of biological systems for the study of molecular controls, through biochemistry and cell biology all the way to investigation of the intricate interactions of cells and tissues in the developing and metamorphosing frog. Indeed, it is its exceptional versatility, and the ability to use it to study processes across such a wide range of biological scales, that has resulted in studies in *Xenopus* providing the underpinning of a surprisingly large amount of what we understand about other biological systems including humans. Moreover, we continue to develop new uses for the frog egg and embryo to study interesting biology, and different facets of this system can be readily adapted to shed further light on the biochemistry of developmental control. To understand how gene function feeds through to the behavior of tissues and organs, we must look to integrate our organism-level studies with an understanding of behavior of individual cells and the molecular controls that underpin this. Indeed, with the ever-expanding use of *Xenopus* in fields such as biomechanics and detailed molecular mechanisms, it is important to disseminate these new methodologies for others to use and adapt to address new questions.

The accompanying protocols present cell-based approaches that can be used to shed light on aspects of biology as diverse as cell-cycle dynamics and cell-shape response through to protein modifications and behavior. Moreover, methods are described for the selection of antigens, as well as the generation and purification of antibodies to *Xenopus* proteins that are the essential tool in a huge range of applications in modern biology.

---

[3]Correspondence: ap113@cam.ac.uk

Cite this introduction as *Cold Spring Harb Protoc*; doi:10.1101/pdb.top105528

XENOPUS EMBRYOS AND TISSUES AS MODELS FOR CELL DYNAMICS

Effects of mechanical force on cell behavior are emerging as an important influence on processes including division, cell fate, and differentiation (Wyatt et al. 2016), although direct analysis of stretching and compressive forces on cells in living vertebrates can be challenging. The resilience of the blastula embryo animal pole ectoderm, the animal cap, and its ability to be cultured ex vivo in simple media combine with ready accessibility to make this tissue an excellent system for such studies (Stooke-Vaughan et al. 2017). In Protocol 2: Applying Tensile and Compressive Force to *Xenopus* Animal Cap Tissue, Goddard et al. (2019) take a classic cell biological approach to show how the effects of tensile force on cell behavior can be assayed in *Xenopus* animal cap ectodermal cells using confocal microscopy. The ready ability to apply very controlled forces to animal cap cells, and immunostaining of these tissues under stress, allows measurements to be taken using advanced microscopy under different conditions. Illustrating nicely the versatility of this approach, mathematical modeling has been combined with the ability to quantitate cell stresses and changes in cell shape to elegantly show that shape and stress have distinct effects on the cueing and orientation of cell division (Nestor-Bergmann et al. 2019).

Cell size and shape can influence cell division and these links can be probed experimentally, but even fundamental questions such as defining the length of a cell cycle at different stages of embryogenesis can be hard to undertake in many organisms. However, because of the ready accessibility of its embryos, *Xenopus* has long been the model of choice to study how patterns of cell division change during embryonic development (Saka and Smith 2001; Vernon and Philpott 2003; Thuret et al. 2015). Although precise measurement of the length of phases of the cell cycle as well as the overall rate of cell division in the changing environment of the developing embryo is complex, it is probably most readily achieved using the special features we find in *Xenopus*. In Protocol 1: In Vivo Assessment of Neural Precursor Cell Cycle Kinetics in the Amphibian Retina, Locker and Perron (2019) show how cell staining and labeling for markers of different cell-cycle phases can again be combined with powerful mathematical modeling approaches to determine how cell-cycle time changes during retinal development. This methodology takes advantage of the accessibility of the *Xenopus* eye and the compartmentalization of proliferative and differentiation zones that is a key feature of *Xenopus* retina (Perron and Harris 2000), although the method should be readily adaptable to other tissues.

ANALYZING PROTEIN PHOSPHORYLATION AND CHROMATIN BINDING IN *XENOPUS* EMBRYOS

Xenopus egg and embryo extracts provide complex mixtures of proteins that can recapitulate a wide range of cell biological and macromolecular behaviors. (This is also explored in Sec. 7, Introduction: The Use of Cell-Free *Xenopus* Extracts to Investigate Cytoplasmic Events [Gibeaux and Heald 2018].) As well as providing accessible systems to study coordination of macromolecular events such as mitotic spindle generation and nuclear fission, extracts from eggs and embryos can also be used to provide complex protein mixtures to probe protein behavior and modification in vivo. In Protocol 3: Analysis of Phosphorylation Status of Ectopically Expressed Proteins in Early *Xenopus* Embryos, Hardwick and Philpott (2019a) give a very simple method to assay the phosphorylation status of ectopically expressed proteins in *Xenopus* embryos, an approach agnostic to the kinases that are bringing about post-translational modifications. Embryonic cytoplasm provides a rich environment of such enzymes, particularly cyclin-dependent kinases, that may act together to bring about extensive phosphorylation of individual proteins, as is seen in the case of multisite phosphorylation of bHLH transcription factors in early embryos (Ali et al. 2011, 2014). The first steps in this protocol also give a more general, simple, and robust method for extracting proteins expressed in embryos for subsequent analysis by western blotting (see also the related antibody testing protocol introduced below).

Transcription factors and other chromatin-bound proteins can also be isolated from the DNA of developing embryos using Protocol 4: Analysis of Chromatin Binding of Ectopically Expressed Pro-

Cite this introduction as *Cold Spring Harb Protoc*; doi:10.1101/pdb.top105528

teins in Early *Xenopus* Embryos (Hardwick and Philpott 2019b), contributed by Laura Hardwick and Anna Philpott. Isolating proteins from developmentally staged embryos in this way would allow changes in DNA binding of individual factors across developmental time to be analyzed. In some cases, chromatin binding may be accompanied by post-translational modification. Combining the use of the chromatin isolation protocol with the protocol designed to analyze protein phosphorylation status was used to show that DNA-bound proneural transcription factors are phosphorylated on multiple sites after expression in *Xenopus* (Hardwick et al. 2019; Hardwick and Philpott 2019c). This type of analysis has formed the bedrock of further studies to uncover the molecular regulation by phosphorylation of this class of proteins (Ali et al. 2011, 2014; Azzarelli et al. 2017). These studies give a good example of how different facets of the use of *Xenopus* as a model system can be used together to reveal how biochemical regulation affects developmental function of control proteins.

GENERATING ANTIBODIES TO *XENOPUS* PROTEINS: CRITICAL ANALYTICAL TOOLS

Cell biology often requires the visualization of subcellular structures usually by specific detection of their protein components. Indeed, the use of antibodies to detect specific proteins underpins most studies of cell morphology and function. Specific antibodies are also vital tools in a wide range of studies of biochemistry, molecular biology, and embryology. However, the *Xenopus* community has lagged considerably behind many other model organism user groups in the generation of antibodies that recognize *Xenopus* proteins. The previously lower priority given to developing specific antibodies to frog proteins is partly due to the fact that classical uses of *Xenopus* systems such as gross morphological, embryological, and fate studies rely less on a mechanistic understanding of individual proteins than systems in which a more reductionist approach has been more prominent.

Of course, in recent decades, the functions of different genes have been extensively studied in *Xenopus* embryos, but when gene expression is assayed, this is typically undertaken by using a mechanism that detects mRNA rather than by using antibodies to detect proteins. However, the detection of mRNA transcripts will only ever give part of the story. Protein half-lives can vary enormously, so direct extrapolation from the RNA level to the protein level is often not possible (Peshkin et al. 2015). Furthermore, post-translational modification can also affect protein half-life and activity (Hindley et al. 2012). However, the need for reliable antibody resources to a large range of *Xenopus* proteins has not gone unnoticed; for more than 10 years, a number of *Xenopus* Principal Investigator forums have highlighted improving antibody resources for use by the *Xenopus* community as a very high priority. Increasing the number of antibodies validated for activity against *Xenopus* proteins can be achieved either by generating antibodies to *Xenopus* epitopes and/or by screening antibodies raised to proteins from other species for cross-reactivity with *Xenopus*. Both approaches will be important avenues to exploit if we are to move forward toward the ultimate goal of having ready availability of validated antibodies to all *Xenopus* proteins under investigation by our community.

Although there are many widely available protocols to generate antigens for inoculation that are not specific to *Xenopus* proteins, Protocol 5: Raising Antibodies for Use in *Xenopus* (Piccinni and Guille 2020a), by Maya Piccinni and Matthew Guille, describes a simple method for the generation of *Xenopus* proteins in HEK cells for immunization and gives tips on how to select an appropriate antigenic region. The method they describe has an advantage over the use of bacterially expressed proteins for host immunization because some post-translational modifications only found in eukaryotic cells may be present, and protein folding and stability is more likely to mimic that found in *Xenopus* cells. Further details are also given for antigen purification and immunization of hosts in this protocol.

Once immunization has been undertaken, it is important to separate the antibodies from extracted serum and to further purify against the selected antigen to enhance the specificity and reduce the background in subsequent antibody uses. Protocol 8: Purifying Antibodies Raised against *Xenopus* Peptides (Piccinni and Guille 2020b), also by Maya Piccinni and Matthew Guille, describes one way to isolate IgG immunoglobulin from whole serum using a caprylic acid extraction protocol. This is followed by a method for affinity purification and desalting of specific antibodies using a peptide antigen column. After antigen-specific antibody purification, a robust method is required to determine the specificity and potency of the generated antibody against the antigen in question. In Protocol 7: Confirming Antibody Specificity in *Xenopus*, Martin et al. (2020) describe a strategy that can be used to verify that the new antibody recognizes its target. This strategy involves knocking down or enhancing the expression of the target in embryos and then performing a western blotting analysis with the new antibody to evaluate its specificity for the target. In Protocol 6: Assessing the Immune Response When Raising Antibodies for Use in *Xenopus*, Piccinni and Guille (2020c) describe a simple method to determine whether a strong and specific antibody response has been generated to a specific protein expressed in *Xenopus* embryos, as assayed by western blotting. Taken together, these four protocols give a roadmap for generating new antibody resources for *Xenopus* proteins, but there are many additional and well-established ways to generate antigens for immunization, purify antibodies, and test their affinity and specificity that are equally applicable across all species.

CONCLUSION

Overall, the protocols introduced here give a snapshot of the many versatile ways that *Xenopus* can be used to study a large range of cell biological questions, complementing other protocols in the collection that describe the use of *Xenopus* oocytes and eggs, embryos, and frogs. *Xenopus* is truly an organism of remarkable versatility, and we continue to make great strides in using the many facets of the *Xenopus* system to study problems in innovative ways, as well as continually developing new reagents like antibodies to make those studies possible.

ACKNOWLEDGMENTS

I thank Ron Laskey, Marc Kirschner, and John Gurdon for teaching me almost everything I know about frogs and the many past and present members of the Philpott laboratory for stimulating discussions. I am supported by Cancer Research UK Programme Grant A25636 and Wellcome Trust Investigator Award 212253/Z/18/Z.

REFERENCES

Ali F, Hindley C, McDowell G, Deibler R, Jones A, Kirschner M, Guillemot F, Philpott A. 2011. Cell cycle–regulated multi-site phosphorylation of Neurogenin 2 coordinates cell cycling with differentiation during neurogenesis. *Development* **138:** 4267–4277. doi:10.1242/dev.067900

Ali FR, Cheng K, Kirwan P, Metcalfe S, Livesey FJ, Barker RA, Philpott A. 2014. The phosphorylation status of Ascl1 is a key determinant of neuronal differentiation and maturation in vivo and in vitro. *Development* **141:** 2216–2224. doi:10.1242/dev.106377

Azzarelli R, Hurley C, Sznurkowska MK, Rulands S, Hardwick L, Gamper I, Ali F, McCracken L, Hindley C, McDuff F, et al. 2017. Multi-site Neurogenin3 phosphorylation controls pancreatic endocrine differentiation. *Dev Cell* **41:** 274–286 e275. doi:10.1016/j.devcel.2017.04.004

Gibeaux R, Heald R. 2018. The use of cell-free *Xenopus* extracts to investigate cytoplasmic events. *Cold Spring Harb Protoc* doi:10.1101/pdb.top097048

Goddard GK, Tarannum N, Sarah Woolner S. 2019. Applying tensile and compressive force to *Xenopus* animal cap tissue. *Cold Spring Harb Protoc* doi:10.1101/pdb.prot105551

Hardwick LJA, Philpott A. 2019a. Analysis of phosphorylation status of ectopically expressed proteins in early *Xenopus* embryos. *Cold Spring Harb Protoc* doi:10.1101/pdb.prot105569

Hardwick LJA, Philpott A. 2019b. Analysis of chromatin binding of ectopically expressed proteins in early *Xenopus* embryos. *Cold Spring Harb Protoc* doi:10.1101/pdb.prot105577

Hardwick LJA, Philpott A. 2019c. N-terminal phosphorylation of xHes1 controls inhibition of primary neurogenesis in *Xenopus*. *Biochem Biophys Res Commun* **509:** 557–563. doi:10.1016/j.bbrc.2018.12.135

Hardwick LJA, Davies JD, Philpott A. 2019. Multi-site phosphorylation controls the neurogenic and myogenic activity of E47. *Biochem Biophys Res Commun* **511:** 111–116. doi:10.1016/j.bbrc.2019.02.045

Hindley C, Ali F, McDowell G, Cheng K, Jones A, Guillemot F, Philpott A. 2012. Post-translational modification of Ngn2 differentially affects transcription of distinct targets to regulate the balance between progenitor maintenance and differentiation. *Development* **139:** 1718–1723. doi:10.1242/dev.077552

Locker M, Perron M. 2019. In vivo assessment of neural precursor cell cycle kinetics in the amphibian retina. *Cold Spring Harb Protoc* doi:10.1101/pdb.prot105536

Martin SA, Page SJ, Piccinni MZ, Guille MJ. 2020. Confirming antibody specificity in *Xenopus*. *Cold Spring Harb Protoc* doi:10.1101/pdb.prot105601

Nestor-Bergmann A, Stooke-Vaughan GA, Goddard GK, Starborg T, Jensen OE, Woolner S. 2019. Decoupling the roles of cell shape and mechanical

Cite this introduction as *Cold Spring Harb Protoc*; doi:10.1101/pdb.top105528

stress in orienting and cueing epithelial mitosis. *Cell Rep* **26:** 2088–2100 e2084. doi:10.1016/j.celrep.2019.01.102

Perron M, Harris WA. 2000. Retinal stem cells in vertebrates. *Bioessays* **22:** 685–688. doi:10.1002/1521-1878(200008)22:8<685::AID-BIES1>3.0 .CO;2-C

Peshkin L, Wuhr M, Pearl E, Haas W, Freeman RM Jr, Gerhart JC, Klein AM, Horb M, Gygi SP, Kirschner MW. 2015. On the relationship of protein and mRNA dynamics in vertebrate embryonic development. *Dev Cell* **35:** 383–394. doi:10.1016/j.devcel.2015.10.010

Piccinni MZ, Guille MJ. 2020a. Raising antibodies for use in *Xenopus*. *Cold Spring Harb Protoc* doi:10.1101/pdb.prot105585

Piccinni MZ, Guille MJ. 2020b. Purifying antibodies raised against *Xenopus* peptides. *Cold Spring Harb Protoc* doi:10.1101/pdb.prot105619

Piccinni MZ, Guille MJ. 2020c. Assessing the immune response when raising antibodies for use in *Xenopus*. *Cold Spring Harb Protoc* doi:10.1101/pdb .prot105593

Saka Y, Smith JC. 2001. Spatial and temporal patterns of cell division during early *Xenopus* embryogenesis. *Dev Biol* **229:** 307–318. doi:10.1006/dbio .2000.0101

Stooke-Vaughan GA, Davidson LA, Woolner S. 2017. *Xenopus* as a model for studies in mechanical stress and cell division. *Genesis* **55:** 10.1002/ dvg.23004.

Thuret R, Auger H, Papalopulu N. 2015. Analysis of neural progenitors from embryogenesis to juvenile adult in *Xenopus laevis* reveals biphasic neurogenesis and continuous lengthening of the cell cycle. *Biol Open* **4:** 1772–1781. doi:10.1242/bio.013391

Vernon AE, Philpott A. 2003. The developmental expression of cell cycle regulators in *Xenopus laevis*. *Gene Expr Patterns* **3:** 179–192. doi:10 .1016/S1567-133X(03)00006-1

Wyatt T, Baum B, Charras G. 2016. A question of time: tissue adaptation to mechanical forces. *Curr Opin Cell Biol* **38:** 68–73. doi:10.1016/j.ceb .2016.02.012

In Vivo Assessment of Neural Precursor Cell Cycle Kinetics in the Amphibian Retina

Morgane Locker[1,2] and Muriel Perron[1,2]

[1]*Paris-Saclay Institute of Neuroscience, CERTO-Retina France, CNRS, Univ Paris-Sud, University Paris-Saclay, 91405 Orsay, France*

Cell cycle progression is intimately linked to cell fate commitment during development. In addition, adult stem cells show specific proliferative behaviors compared to progenitors. Exploring cell cycle dynamics and regulation is therefore of utmost importance, but constitutes a great challenge in vivo. Here we provide a protocol for evaluating in vivo the length of all cell cycle phases of neural stem and progenitor cells in the post-embryonic *Xenopus* retina. These cells are localized in the ciliary marginal zone (CMZ), a peripheral region of the retina that sustains continuous neurogenesis throughout the animal's life. The CMZ bears two tremendous advantages for cell cycle kinetics analyses. First, this region, where proliferative cells are sequestered, can be easily delineated. Second, the spatial organization of the CMZ mirrors the temporal sequence of retinal development, allowing for topological distinction between retinal stem cells (residing in the most peripheral margin), and amplifying progenitors (located more centrally). We describe herein how to determine CMZ cell cycle parameters using a combination of (i) a cumulative labeling assay, (ii) the percentage of labeled mitosis calculation, and (iii) the mitotic index measurement. Taken together, these techniques allow us to estimate total cell cycle length (T_C) as well as the duration of all cell cycle phases ($T_{S/G2/M/G1}$). Although the method presented here was adapted to the particular system of the CMZ, it should be applicable to other tissues and developmental stages as well.

MATERIALS

It is essential that you consult the appropriate Material Safety Data Sheets and your institution's Environmental Health and Safety Office for proper handling of equipment and hazardous materials used in this protocol.

RECIPES: Please see the end of this protocol for recipes indicated by <R>. Additional recipes can be found online at http://cshprotocols.cshlp.org/site/recipes.

Reagents

1-butanol (100%)
5-ethynyl-2′-deoxyuridine (EdU) (Thermo Fisher Scientific)
Alexa-coupled secondary antibody (1 mg/mL) (e.g., Thermo Fischer Scientific)
Antifade mounting medium (e.g., FluorSave reagent [Thermo Fischer Scientific])
Anti-phospho-Histone H3 (PH3) antibody (1 mg/mL) (e.g., Millipore 06-570)

[2]Correspondence: morgane.locker@u-psud.fr; muriel.perron@u-psud.fr
Supplemental material is available for this article at cshprotocols.cshlp.org.

Cite this protocol as *Cold Spring Harb Protoc*; doi:10.1101/pdb.prot105536

Blocking solution (PBT with 10% goat serum)

Click-iT EdU Alexa Fluor imaging kit (Thermo Fisher Scientific)

Follow the manufacturer's instructions for preparing stock solutions.

Ethanol (100%)

Prepare 95%, 70%, and 50% solutions in 1× PBS.

Glycerin albumen (100% stock solution) (e.g., VWR)

Store stock solution at −20°C. Immediately before use, prepare a fresh 3% solution in dH$_2$O.

HCl (2 N) (optional, if using bromo-2′-deoxyuridine [BrdU] instead of EdU)

Hoechst 33342 fluorescent stain (5 μM)

Modified Barth's saline (MBS) <R>

Paraffin wax (e.g., 00403 HISTOWAX [Histolab])

Paraformaldehyde (PFA) (4%)

PBT (1× PBS, 0.2% Triton X-100)

Phosphate-buffered saline (PBS) (1×, pH 7.2–7.6)

Tricaine methanesulfonate solution (MS222) (Sigma-Aldrich) (0.4 g/L buffered to pH 7.0–7.5 with sodium bicarbonate)

Unmasking citrate solution (10 mM sodium citrate, 0.05% Tween 20 in dH$_2$O)

Xylene (100%)

Equipment

Coplin jars (glass)

Embedding molds (plastic, disposable)

Fluorescence microscope

Heating plate at 50°C

Humidified chamber

Image processing software (e.g., FIJI or Adobe Photoshop)

Incubators at 18°C–20°C, 37°C, and 60°C

Microscope slides and coverslips

Microtome (e.g., Microm HM 340E)

Parafilm or GelBond (Lonza)

Petri dishes

Pipettes (plastic)

Razor blades

Roller shaker

Vials (glass)

METHOD

For the principles of the methods described here for determining the total cell cycle length (T_C) and the duration of the cell cycle phases ($T_{S/G2/M/G1}$), see Box 1. Note that if determination of all cell cycle parameters is required, it is preferable to deduce them from a unique combined experiment on sibling tadpoles.

Labeling Tadpoles with EdU

1. Raise tadpoles in 0.1× MBS in Petri dishes in an incubator at 18°C–20°C until stage 41–42.

2. Transfer the tadpoles to 0.1× MBS containing 1 mM EdU.

If analysis is to be performed at embryonic stages (i.e., before or until stage 39/40), microinjection into the yolk is preferable to bathing, since it produces more homogeneous labeling. After microinjection, bathing in EdU solution is sufficient.

3. Incubate the tadpoles for the desired duration of EdU exposure.

Typical time periods are indicated in Figure 1A. For long exposure times, the medium should be renewed every other day so that EdU remains constantly available.

BOX 1. PRINCIPLES OF THE METHODS

Principle of T_C and T_S Determination through Cumulative EdU Labeling

Total cell cycle length (T_C) and S-phase length (T_S) can be determined using cumulative labeling with thymidine analogs such as bromo-2'-deoxyuridine (BrdU) or 5-ethynyl-2'-deoxyuridine (EdU). Theoretically, this method presupposes that all cells studied are proliferating (or that nonproliferative cells can be distinguished and excluded) and uniformly distributed in the cell cycle, and that the population grows at a steady state (Nowakowski et al. 1989). The principle relies on the sequential administration of the chosen thymidine analog for a time that exceeds the presumed cell cycle length. The longer the pulse is, the more cells will go through S-phase and be labeled, until all proliferative cells are labeled (Fig. 1A,B). When determined for each pulse duration and plotted as a function of time, the proportion of labeled cells (also called the labeling index [LI]) increases linearly until a plateau is reached (Fig. 1C). The LI at the plateau allows estimation of the proportion of proliferative cells in the considered population (or growth fraction [GF]). As depicted in Figure 1A, the time needed to reach the plateau ($T_{plateau}$) is equivalent to $T_C - T_S$. A second equation derives from the fact that in a population of asynchronous cycling cells, the fraction of cells in a given phase is directly proportional to the length of that phase relative to total cell cycle length (Nowakowski et al. 1989). Thus, T_S/T_C can be deduced from the proportion of S-phase cells among proliferative cells at T_0, which corresponds to the LI at T_0 (LI_0; extrapolated from the intercept of the best-fit line with the y-axis) corrected by the growth fraction. It is therefore equal to LI_0/GF. T_C and T_S can then be calculated from these two equations:

$$T_C - T_S = T_{plateau},$$

$$T_S/T_C = LI_0/GF.$$

Principle of T_{G2} Determination Using the PLM Method

T_{G2} determination relies on a paradigm called the percentage of labeled mitosis (PLM) (Quastler and Sherman 1959). Here again, the idea is to expose the tissue of interest to a thymidine analog, such as EdU, with increasing exposure times. One should then focus on the appearance of EdU-labeled cells among mitotic cells (immunostained with the late G2/M-phase marker phospho-Histone H3 or PH3 [Hendzel et al. 1997]). These represent the cells that were in S-phase and then went through G2 to finally reach M-phase during the EdU pulse (Fig. 1D,E). PLM increases sigmoidally as a function of time until all cells are double-labeled. Importantly, the duration of the EdU pulse required to reach the plateau is here much lower than in the cumulative labeling experiment. As depicted in Figure 1D, it grossly corresponds to $T_{G2} + T_M$. After plotting the results (Fig. 1F), the time required for half-maximal appearance of EdU labeling in the mitotic population is taken as an estimation of the average T_{G2} (Arai et al. 2011).

Principle of T_M Determination Using the Mitotic Index

M phase length determination relies again on the fact that the proportion of cells in a given phase is proportional to the ratio between the corresponding phase length and total cell cycle duration. Thus, with the mitotic index (MI) representing the percentage of mitotic cells in a given population, we can deduce the following equation:

$$T_M/T_C = MI/GF.$$

Evaluation of the mitotic index represents an easy way to qualitatively compare T_C in different experimental conditions under the hypothesis that M-phase duration is constant. If T_C and GF have been determined through EdU cumulative labeling, then T_M can be calculated.

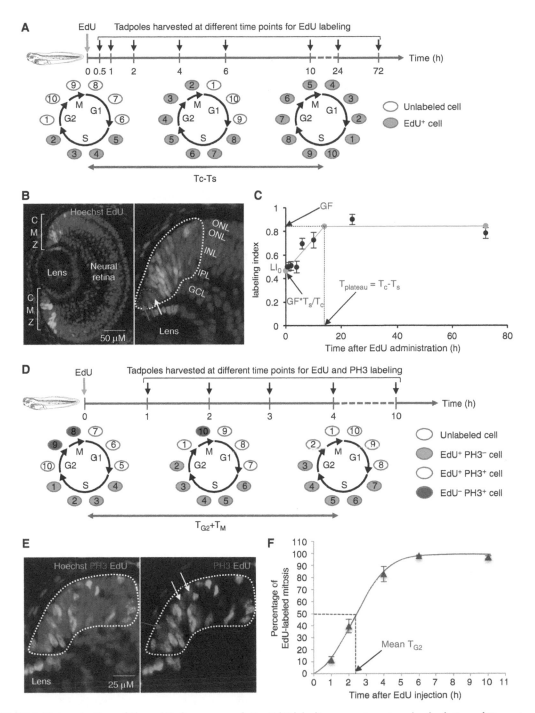

FIGURE 1. Determination of T_C and T_S from a cumulative EdU labeling experiment and calculation of T_{G2} using the percentage of labeled mitosis (PLM) paradigm. (*A*) Outline of the experiment and schematic representation of cumulative labeling during cell cycle progression. Tadpoles are exposed to EdU for different durations and then fixed, sectioned and subjected to EdU labeling. Ten cells (numbered 1 to 10) are represented on the schematic. Cells become labeled (green) when they reach S-phase. Thus, the longer the EdU pulse is, the higher is the number of EdU-positive cells. As shown on the schematic, the time needed for all cells to be labeled corresponds to $T_C - T_S$ (follow, for instance, cell number 1). For simplification, the doubling of cells after mitosis is not represented. (*B*) Typical retinal section stained for EdU following a 4-h pulse at stage 41. Nuclei are counterstained with Hoechst. The panel on the *right* is a higher magnification of the dorsal CMZ. Red and white arrows point to EdU$^-$ and EdU$^+$ cells, respectively. Quantification can be performed in the stem cell compartment (delineated in orange and corresponding to the 3–5 most peripheral cells), in progenitors only, or in the whole CMZ (delineated in white using the plexiform layers as frontiers with the differentiated neural retina). (*Legend continues on following page.*)

Fixing, Dehydrating, and Sectioning Tadpoles

Here we describe microtome sectioning of tadpoles; alternatively, cryostat sectioning can be performed.

4. Anesthetize the tadpoles in 0.4 g/L MS222. Transfer them to glass vials using plastic pipettes.

5. Fix the tadpoles in 4% PFA with gentle agitation on a roller shaker for 2 h at room temperature or overnight at 4°C.

6. Wash the tadpoles three times for 5 min per wash in 1× PBS.

7. Dehydrate the tadpoles as follows.

   i. Incubate three times for 30 min in 70% ethanol.

   ii. Incubate once for 30 min in 95% ethanol.

   iii. Incubate three times for 1 h in 100% ethanol.
   Tadpoles can be stored in 100% ethanol at −20°C.

   iv. Incubate overnight in 100% 1-butanol at room temperature.

8. Transfer the tadpoles to embedding molds. Remove the excess butanol and replace it with melted paraffin wax.

9. Incubate the tadpoles for ∼6 h at 60°C. Change the paraffin wax 3–4 times during this incubation.

10. Align the tadpoles in each mold and let the paraffin solidify at room temperature.

11. Remove a paraffin block from the mold and cut around the tadpole with a razor blade. Place the embedded tadpole on a microtome (ventral side down, anterior in your direction). Trim at 50 μM to remove excess paraffin, and then section the tadpole at 10–12 μM.

12. Arrange successive ribbons containing eye sections on annotated slides covered with 3% glycerin albumen. Place the slides on a heating plate at 50°C. Wait for 20–30 sec and then remove the excess liquid. Allow the slides to dry for 10 sec on the heating plate.

13. Transfer the slides to a rack and incubate overnight at 37°C to dry.

   After this step, slides can be stored at room temperature in a dry place.

FIGURE 1. (*Continued*) (CMZ) ciliary marginal zone, (GCL) ganglion cell layer, (INL/ONL) inner/outer nuclear layer, (IPL/OPL) inner/outer plexiform layers. (*C*) When labeling index (LI) is plotted as a function of time, a linear increase is observed until a plateau is reached. This allows determination of the proportion of cycling cells, or growth fraction (GF). The intercept of the linear part of the curve with the y-axis (LI_0) corresponds to $GF \times T_S/T_C$. The time taken for LI to reach the plateau ($T_{plateau}$) corresponds to $T_C − T_S$. The graph shown here is related to an EdU cumulative labeling experiment analyzed in the stem cell compartment and was adapted from Cabochette et al. (2015). (*D*) Outline of the experiment and schematic representation of PH3 and EdU labeling along with cell cycle progression. Tadpoles are exposed to EdU for different durations and then fixed, sectioned and subjected to double EdU/PH3 labeling. Ten cells (numbered 1 to 10) are represented on the schematic. During the EdU pulse, only cells that go through S-phase are EdU-labeled (green). This cohort then progresses through the cell cycle. If the duration of the EdU pulse is less than the G2-phase length, there will be no EdU-labeled cells among PH3-positive cells (red). When the duration increases, EdU-labeled cells appear among mitotic cells (yellow), as they had enough time to go through S and reach late G2/M phase. The minimal time required to obtain 100% double-labeled cells corresponds to the sum of G2- and M-phase lengths (follow, for instance, cell number 1). For simplification, the doubling of cells after mitosis is not represented. (*E*) Typical retinal sections (zoom on the dorsal CMZ; delineated in white) stained for both PH3 and EdU following a 2-h pulse at stage 41. Nuclei are counterstained with Hoechst. Red and yellow arrows point to $PH3^+$–EdU^- and $PH3^+$–EdU^+ cells, respectively. Note that the shortness of the M-phase and thus the weak probability of finding mitotic cells in the stem cell compartment complicates the calculation of their G2-phase duration with this method. (*F*) Percentage of EdU-positive cells among mitotic cells plotted as a function of time. The time required for half-maximal appearance of EdU labeling in the mitotic population is taken as an estimation of the mean T_{G2}. The graph shown here is related to an experiment analyzed in the whole CMZ and was adapted from Cabochette et al. (2015).

Cite this protocol as *Cold Spring Harb Protoc*; doi:10.1101/pdb.prot105536

Rehydrating Sections, Unmasking Epitopes, and Saturating Unspecific Fixation Sites

Citrate unmasking (Step 17) is necessary for PH3 detection on paraffin sections, but can be skipped if using cryostat sections or if performing EdU labeling only.

14. Transfer the slides to glass Coplin jars. Bathe twice for 10 min in 100% xylene to remove the paraffin wax.

15. Rehydrate the sections for 5 min per solution in 100%, 95%, 70%, and 50% ethanol.

16. Wash the slides three times for 5 min per wash in 1× PBS.

17. Microwave the slides in the unmasking citrate solution: Use maximal power (around 900 W) until boiling, and then use the defrost program (around 300 W) for 10 min. Cool the slides for 20 min at room temperature.

18. Wash the slides twice in dH$_2$O followed by three times in 1× PBS for 5 min per wash.

    If using BrdU instead of EdU, add a denaturation step (45 min treatment in 2 N HCl followed by three washes in 1× PBS for 5 min per wash) before proceeding to permeabilization (Step 19).

19. Permeabilize the tissues for 20 min in PBT.

20. Transfer the slides to a horizontal surface in a humidified chamber. (Do not allow them to dry.) Add 800 µL per slide of blocking solution. Incubate for 20 min at room temperature.

Staining EdU-Labeled Cells

EdU staining relies on a copper-catalyzed covalent reaction between the fluorescent dye azide and the EdU alkyne (the "Click-iT" reaction). In contrast to BrdU detection, no DNA denaturation is required, since the small size of the dye azide allows it to easily gain access to DNA. Of note, EdU staining can be followed by immunofluorescence against a marker of interest in order to specifically label a given cell population. However, depending on the antibody used (e.g., anti-GFP[green fluorescent protein]), immunolabeling should be performed before the Click-iT reaction.

21. Discard the blocking solution from the slides and replace it with 200 µL per slide of the Click-iT reaction cocktail prepared according to the manufacturer's instructions.

22. Cover the slides with Parafilm or GelBond. Incubate for 30 min in a humidified chamber at room temperature in the dark.

23. Transfer the slides to glass Coplin jars and wash three times for 5 min per wash in PBT. (Allow the Parafilm or GelBond to detach by itself and then remove it.)

    If performing both EdU and PH3 labeling, proceed to Step 24. If performing only EdU staining, proceed directly to Step 31.

Labeling PH3 and Counterstaining Nuclei

24. Repeat blocking as described in Step 20.

25. Discard the blocking solution from the slides and replace it with 200 µL per slide of primary antibody solution (anti-PH3 diluted 1:500 in blocking solution).

    If using BrdU as a thymidine analog, perform a double immunostaining against both PH3 and BrdU.

26. Cover the slides with Parafilm or GelBond. Incubate in a humidified chamber overnight at 4°C or for 2 h at room temperature.

27. Transfer the slides to glass Coplin jars and wash three times for 5 min per wash in 1× PBS. (Allow the Parafilm or GelBond to detach by itself and then remove it.)

28. Transfer the slides to a horizontal surface and add 400 µL per slide of ALEXA-coupled secondary antibody (diluted 1:1000 in blocking solution).

29. Incubate the slides for 2 h in a humidified chamber at room temperature in the dark.

30. Transfer the slides to glass Coplin jars and wash three times for 5 min per wash in 1× PBS.

31. Incubate the slides for 10 min in Hoechst solution to counterstain the nuclei.

32. Wash the slides twice for 5 min per wash in 1× PBS.

33. Transfer the slides to a horizontal surface and apply a few drops of FluorSave reagent (or an equivalent antifade mounting medium) to each slide. Place coverslips carefully on top of the mounting medium.

34. Allow the slides to dry overnight at 4°C.

35. Proceed to imaging using a fluorescence microscope.

Counting Manually in the Ciliary Marginal Zone

Automatized counting using appropriate software such as FIJI is difficult in the ciliary marginal zone (CMZ). Because of the high nucleo-cytoplasmic ratio, nuclei appear packed together, rendering their automatic identification challenging. Manual counting must thus be performed, which probably represents the most arduous part of the described experiment. For each experimental condition, we recommend analyzing 6–10 sections (transverse sections that pass through the lens) per retina on 3–5 different retinas. This is usually enough for cell cycle kinetics experiments in the whole CMZ. If restricting the analysis to the stem cell compartment, anticipate needing a larger number of tadpoles to ensure a sufficient number of counted cells and reliable results.

36. Open the images using FIJI or Adobe Photoshop.

37. On each section, delineate the dorsal and ventral CMZ using the outer and inner plexiform layers as frontiers to separate CMZ cells from neuronal layers (Fig. 1B).

38. For all considered retinas, count on each CMZ section the number of PH3-positive cells, the number of EdU-labeled cells and the number of PH3-positive cells colabeled with EdU. Count the number of Hoechst-positive nuclei to estimate the total number of cells considered.

39. Report the results in an Excel spreadsheet. For each retina at each desired time point, calculate the percentage of (i) PH3-positive cells among total counted cells, (ii) EdU-labeled cells among total counted cells, and (iii) EdU-positive cells among total counted mitotic cells. Then, calculate mean percentages per retina and standard deviations for all these parameters.

40. If comparing different experimental conditions (e.g., gain or loss of function of your favorite gene vs. control), proceed to statistical analysis for each time point using the nonparametric Mann–Whitney test.

Calculating Cell Cycle Parameters

An explanation of the calculations is provided in Box 1.

Calculating T_C and T_S

41. Plot the average percentage of EdU-positive cells as a function of time (Fig. 1C).

42. Determine the growth fraction (GF) which corresponds to the percentage of labeled cells at the time at which the plateau is reached ($T_{plateau}$).

43. Determine the labeling index (LI_0) at T_0 by extrapolating the y intercept of the best-fit line.

44. Determine T_C and T_S using the following equations:

$$T_{plateau} = T_C - T_S,$$

$$LI_0 = GF \times (T_S/T_C).$$

Of note, more accurate calculations can be obtained using the spreadsheet developed by R. Nowakowski (Nowakowski et al. 1989), which allows drawing the best-fit line and calculates GF, T_C, and T_S using a nonlinear regression method. With his kind authorization, the spreadsheet is included with this protocol as a supplemental file (see Supplemental File S1). Report your results for each retina at each time point as

Cite this protocol as *Cold Spring Harb Protoc*; doi:10.1101/pdb.prot105536

recommended and follow the included instructions. If publishing data gathered with this spreadsheet, Nowakowski et al. (1989) should be cited.

Calculating T_{G2}

45. Plot the average percentage of EdU-positive cells among mitotic cells as a function of time (Fig. 1F).

46. Determine the time at which 50% of cells are labeled.

   This corresponds to an estimation of T_{G2}.

Calculating T_M and T_{G1}

47. Using the average percentage of PH3$^+$ cells among total cells (MI) and the data gathered from EdU-cumulative labeling, calculate T_M using the following equation:

$$T_M = (MI/GF) \times T_C.$$

48. Deduce T_{G1} using the following equation:

$$T_{G1} = T_C - (T_S + T_{G2} + T_M).$$

DISCUSSION

During development, neural stem/progenitor cells undergo extensive proliferation to give rise to a nervous system with a defined size and precise cellular diversity. It is now well established that cell cycle progression is tightly linked to cell fate decisions (Ohnuma and Harris 2003; Molina and Pituello 2016). In the developing retina, cell cycle speed was shown to determine whether early or late neuronal cell types are produced (Decembrini et al. 2006, 2009). This led to the idea of a cellular clock dependent on cell cycle progression that would measure the length of the last cell cycle rather than time (Pitto and Cremisi 2010). In addition, cell cycle lengthening is known to causally contribute to the switch from proliferative to neurogenic divisions in the developing brain (Calegari et al. 2005; Lange et al. 2009; Salomoni and Calegari 2010). It has been proposed that neurogenic factor action may depend on whether G1 length provides enough time to do it (Calegari and Huttner 2003). However, G1 is not the only phase subjected to length variation. G2-phase duration was found to be important for the proper control of neuronal production in the developing spinal cord (Peco et al. 2012; Molina and Pituello 2016). Finally, S-phase length was also highlighted as a key feature distinguishing self-expanding stem/progenitor cells (long S) from progenitors committed to a neurogenic lineage (short S) (Arai et al. 2011; Cabochette et al. 2015; Turrero García et al. 2016). The mechanisms underlying such modulations of cell cycle kinetics remain quite elusive. Indeed, apart from cell cycle regulators such as cyclin and cyclin-dependent kinases, only a few factors and signaling pathways have been shown to regulate the cell cycle progression of neural precursors (Martynoga et al. 2005; Locker et al. 2006; Komada et al. 2008; Uribe and Gross 2010; El Yakoubi et al. 2012; Peco et al. 2012; Cabochette et al. 2015). The ability to measure the length of the different phases of the cell cycle in vivo following gene perturbation is thus of utmost importance to further understanding this process. The CMZ represents a powerful system for such studies since its location and spatial organization allows one to (i) easily delineate it from the nonproliferative neural retina, and (ii) topologically distinguish stem versus progenitor cells (Perron et al. 1998) and thus specifically measure their respective cell cycle kinetics.

   Although the percentage of labeled mitosis (PLM) and EdU cumulative labeling represent efficient tools for in vivo examination of cell cycle kinetics, it is fundamental to keep in mind their intrinsic limitations. First, and as mentioned in Box 1, EdU cumulative labeling and the PLM method presuppose that the studied proliferative population is in steady state. This criterion is barely fulfilled in developing tissues where total cell cycle duration and the lengths of the different phases vary dynamically throughout embryogenesis. It is, however, more or less met in the *Xenopus* CMZ, at least in late tadpole stages when retinal growth considerably slows down. A nonsteady state status is an important bias when calculating T_C, T_S, or T_{G2} because cells at the end of the labeling period might have a different cell cycle than those at its beginning. If significant variations in cell cycle speed are suspected during the labeling time period (for instance, based on the nonlinear aspect of the LI curve), an alternative method named "dual pulse labeling" can be considered. This technique is based on sequential administration of two different thymidine analogs such as EdU and IdU, CldU or BrdU (for the principle of this method, see Martynoga et al. 2005; for a protocol in *Xenopus*, see Auger et al. 2012). It allows estimating T_C and T_S using only one short time interval between injection of the two analogs (which diminishes bias from the nonsteady state of the considered population). A fundamental requirement for undertaking this experiment is to have antibodies able to discriminate between the two thymidine analogs in use.

A second and important drawback of PLM and EdU cumulative labeling (also true for dual pulse labeling) is that the calculated cell cycle parameters represent mean values which do not take into account the heterogeneity of the considered cell population. Live imaging of individual cells on neural tube slices from early chick embryos revealed striking heterogeneity, with T_C values ranging from ~9 to 28 h (Wilcock et al. 2007). The average value was found to be similar to that calculated from fixed tissue. This perfectly illustrates how analysis of populations may hide a high inter-individual variability of cycling behaviors. Although this has not been addressed yet, it is very probable that CMZ progenitors show such cell cycle heterogeneity. The same might be true for retinal stem cells as well. In addition, rare dormant cells have recently been described at the extreme tip of the zebrafish CMZ (Tang et al. 2017), suggesting that the *Xenopus* retinal stem cell cohort might be composed of both "activated" proliferating cells and "resting" quiescent ones, as shown in neurogenic zones of the adult mammalian or zebrafish brain (Than-Trong and Bally-Cuif 2015; Chaker et al. 2016). Finally, our personal observations suggest that the dorsal and ventral parts of the CMZ do not proliferate at the same rate, indicating an additional regional heterogeneity in cell cycle progression. Live imaging in the zebrafish CMZ has recently been established (Wan et al. 2016). In association with genetic tools allowing for visualization of cell cycle phases (such as the Fucci technology, Sakaue-Sawano et al. 2008), this may help in the close future to assess cell cycle kinetics at the single cell level and to estimate variability among stem/progenitor cells.

RECIPES

CaCl₂ (0.1 M)

11.1 g CaCl$_2$

Dissolve 11.1 g CaCl$_2$ in 1 L H$_2$O. Autoclave and store in aliquots at −20°C or 4°C.

MBS Salts (10×)

NaCl (880 mM)
KCl (10 mM)
MgSO$_4$ (10 mM)
HEPES (50 mM, pH 7.8)

   Omit HEPES if MBS is to be used for oocyte maturation.

NaHCO$_3$ (25 mM)

Adjust pH to 7.8 with NaOH. Autoclave. Store at room temperature.

Cite this protocol as *Cold Spring Harb Protoc*; doi:10.1101/pdb.prot105536

Modified Barth's Saline (MBS)

CaCl$_2$ (0.1 M)
MBS salts (10×)

For a 1× solution of MBS, mix 100 mL of 10× MBS salts with 7 mL of 0.1 M CaCl$_2$. Adjust the volume up to 1 liter with H$_2$O. Store at room temperature.

ACKNOWLEDGMENTS

The methods described here have been developed and optimized by several past members of the laboratory. We thank in particular Warif El Yakoubi, Caroline Borday, and Pauline Cabochette. We are very grateful to R. Nowakowski for providing the Excel spreadsheet and for his authorization to make it accessible to whomever requests to use it. M.P.'s laboratory is supported by grants from the FRM, Association Retina France and Fondation Valentin Haüy.

REFERENCES

Arai Y, Pulvers JN, Haffner C, Schilling B, Nüsslein I, Calegari F, Huttner WB. 2011. Neural stem and progenitor cells shorten S-phase on commitment to neuron production. *Nat Commun* 2: 154. doi:10.1038/ncomms1155

Auger H, Thuret R, El Yakoubi W, Papalopulu N. 2012. A bromodeoxyuridine (BrdU) based protocol for characterizing proliferating progenitors in *Xenopus* embryos. In Xenopus *protocols: post-genomic approaches* (ed. Hoppler S, Vize PD), pp. 461–475. Humana Press, Totowa, NJ.

Cabochette P, Vega-Lopez G, Bitard J, Parain K, Chemouny R, Masson C, Borday C, Hedderich M, Henningfeld KA, Locker M, et al. 2015. Yap controls retinal stem cell DNA replication timing and genomic stability. *Elife* 4: e08488. doi:10.7554/eLife.08488

Calegari F, Huttner WB. 2003. An inhibition of cyclin-dependent kinases that lengthens, but does not arrest, neuroepithelial cell cycle induces premature neurogenesis. *J Cell Sci* 116: 4947–4955. doi:10.1242/jcs.00825

Calegari F, Haubensak W, Haffner C, Huttner WB. 2005. Selective lengthening of the cell cycle in the neurogenic subpopulation of neural progenitor cells during mouse brain development. *J Neurosci* 25: 6533–6538. doi:10.1523/JNEUROSCI.0778-05.2005

Chaker Z, Codega P, Doetsch F. 2016. A mosaic world: puzzles revealed by adult neural stem cell heterogeneity. *Wiley Interdiscip Rev Dev Biol* 5: 640–658. doi:10.1002/wdev.248

Decembrini S, Andreazzoli M, Vignali R, Barsacchi G, Cremisi F. 2006. Timing the generation of distinct retinal cells by homeobox proteins. *PLoS Biol* 4: e272. doi:10.1371/journal.pbio.0040272

Decembrini S, Bressan D, Vignali R, Pitto L, Mariotti S, Rainaldi G, Wang X, Evangelista M, Barsacchi G, Cremisi F. 2009. MicroRNAs couple cell fate and developmental timing in retina. *Proc Natl Acad Sci* 106: 21179–21184. doi:10.1073/pnas.0909167106

El Yakoubi W, Borday C, Hamdache J, Parain K, Tran HT, Vleminckx K, Perron M, Locker M. 2012. Hes4 controls proliferative properties of neural stem cells during retinal ontogenesis. *Stem Cells* 30: 2784–2795. doi:10.1002/stem.1231

Hendzel MJ, Wei Y, Mancini MA, Van Hooser A, Ranalli T, Brinkley BR, Bazett-Jones DP, Allis CD. 1997. Mitosis-specific phosphorylation of histone H3 initiates primarily within pericentromeric heterochromatin during G2 and spreads in an ordered fashion coincident with mitotic chromosome condensation. *Chromosoma* 106: 348–360. doi:10.1007/s004120050256

Komada M, Saitsu H, Kinboshi M, Miura T, Shiota K, Ishibashi M. 2008. Hedgehog signaling is involved in development of the neocortex. *Development* 135: 2717–2727. doi:10.1242/dev.015891

Lange C, Huttner WB, Calegari F. 2009. Cdk4/cyclinD1 overexpression in neural stem cells shortens G1, delays neurogenesis, and promotes the generation and expansion of basal progenitors. *Cell Stem Cell* 5: 320–331. doi:10.1016/j.stem.2009.05.026

Locker M, Agathocleous M, Amato MA, Parain K, Harris WA, Perron M. 2006. Hedgehog signaling and the retina: insights into the mechanisms controlling the proliferative properties of neural precursors. *Genes Dev* 20: 3036–3048. doi:10.1101/gad.391106

Martynoga B, Morrison H, Price DJ, Mason JO. 2005. Foxg1 is required for specification of ventral telencephalon and region-specific regulation of dorsal telencephalic precursor proliferation and apoptosis. *Dev Biol* 283: 113–127. doi:10.1016/j.ydbio.2005.04.005

Molina A, Pituello F. 2016. Playing with the cell cycle to build the spinal cord. *Dev Biol* 432: 14–23. doi:10.1016/j.ydbio.2016.12.022

Nowakowski RS, Lewin SB, Miller MW. 1989. Bromodeoxyuridine immunohistochemical determination of the lengths of the cell cycle and the DNA-synthetic phase for an anatomically defined population. *J Neurocytol* 18: 311–318. doi:10.1007/BF01190834

Ohnuma S, Harris W. 2003. Neurogenesis and the cell cycle. *Neuron* 40: 199–208. doi:10.1016/S0896-6273(03)00632-9

Peco E, Escude T, Agius E, Sabado V, Medevielle F, Ducommun B, Pituello F. 2012. The CDC25B phosphatase shortens the G2 phase of neural progenitors and promotes efficient neuron production. *Development* 139: 1095–1104. doi:10.1242/dev.068569

Perron M, Kanekar S, Vetter ML, Harris WA. 1998. The genetic sequence of retinal development in the ciliary margin of the *Xenopus* eye. *Dev Biol* 199: 185–200. doi:10.1006/dbio.1998.8939

Pitto L, Cremisi F. 2010. Timing neurogenesis by cell cycle? *Cell Cycle* 9: 434–435. doi:10.4161/cc.9.3.10762

Quastler H, Sherman FG. 1959. Cell population kinetics in the intestinal epithelium of the mouse. *Exp Cell Res* 17: 420–438. doi:10.1016/0014-4827(59)90063-1

Sakaue-Sawano A, Kurokawa H, Morimura T, Hanyu A, Hama H, Osawa H, Kashiwagi S, Fukami K, Miyata T, Miyoshi H, et al. 2008. Visualizing spatiotemporal dynamics of multicellular cell-cycle progression. *Cell* 132: 487–498. doi:10.1016/j.cell.2007.12.033

Salomoni P, Calegari F. 2010. Cell cycle control of mammalian neural stem cells: putting a speed limit on G1. *Trends Cell Biol* 20: 233–243. doi:10.1016/j.tcb.2010.01.006

Tang X, Gao J, Jia X, Zhao W, Zhang Y, Pan W, He J. 2017. Bipotent progenitors as embryonic origin of retinal stem cells. *J Cell Biol* 216: 1833–1847. doi:10.1083/jcb.201611057

Than-Trong E, Bally-Cuif L. 2015. Radial glia and neural progenitors in the adult zebrafish central nervous system. *Glia* 63: 1406–1428. doi:10.1002/glia.22856

Turrero García M, Chang Y, Arai Y, Huttner WB. 2016. S-phase duration is the main target of cell cycle regulation in neural progenitors of devel-

oping ferret neocortex. *J Comp Neurol* **524:** 456–470. doi:10.1002/cne
.23801

Uribe RA, Gross JM. 2010. Id2a influences neuron and glia formation in the
zebrafish retina by modulating retinoblast cell cycle kinetics. *Development* **137:** 3763–3774. doi:10.1242/dev.050484

Wan Y, Almeida AD, Rulands S, Chalour N, Muresan L, Wu Y, Simons BD,
He J, Harris W. 2016. The ciliary marginal zone of the zebrafish retina:

clonal and time-lapse analysis of a continuously growing tissue. *Development* **143:** 1099–1107. doi:10.1242/dev.133314

Wilcock AC, Swedlow JR, Storey KG. 2007. Mitotic spindle orientation
distinguishes stem cell and terminal modes of neuron production in
the early spinal cord. *Development* **134:** 1943–1954. doi:10.1242/dev
.002519

Applying Tensile and Compressive Force to *Xenopus* Animal Cap Tissue

Georgina K. Goddard, Nawseen Tarannum, and Sarah Woolner[1]

Wellcome Trust Centre for Cell-Matrix Research, Faculty of Biology, Medicine and Health, Manchester Academic Health Science Centre, University of Manchester, Oxford Road, Manchester M13 9PT, United Kingdom

Over many years, the *Xenopus laevis* embryo has provided a powerful model system to investigate how mechanical forces regulate cellular function. Here, we describe a system to apply reproducible tensile and compressive force to *X. laevis* animal cap tissue explants and to simultaneously assess cellular behavior using live confocal imaging.

MATERIALS

It is essential that you consult the appropriate Material Safety Data Sheets and your institution's Environmental Health and Safety Office for proper handling of equipment and hazardous materials used in this protocol.

RECIPES: Please see the end of this protocol for recipes indicated by <R>. Additional recipes can be found online at http://cshprotocols.cshlp.org/site/recipes.

Reagents

Bovine serum albumin (BSA; Sigma-Aldrich A7906)

Capped RNA encoding appropriate probes tagged with fluorescent protein, such as GFP-α-tubulin or Cherry-histone2B (see Woolner et al. 2009)

Danilchik's for Amy (DFA) medium for explant culture <R>

Ethanol (70%)

Fibronectin (from bovine plasma, cell culture tested [Sigma-Aldrich F1141]; 10 µg/mL, freshly prepared in 1× PBS)

High-vacuum silicone grease (Dow Corning; Sigma-Aldrich Z273554)

Marc's Modified Ringers (MMR) solution for embryos <R>

MilliQ water

Phosphate buffered saline (PBS) (Sigma-Aldrich D1408; 10×, diluted to 1× with MilliQ water)

Silicone Sylgard 184 kit (Scientific Laboratory Supplies 63416.5S)

Xenopus laevis embryos (1- or 2-cell stage; Woolner et al. 2009)

Equipment

Air duster/compressed air

Bidirectional stretch apparatus (Deben UK Ltd.)

[1]Correspondence: sarah.woolner@manchester.ac.uk

The bidirectional stretch apparatus, cell tester software, and membrane chamber mold were all custom-made by Deben UK Ltd. based on a previously published stretch system (Jungbauer et al. 2008). We hold no proprietary claims over the Deben UK design, so interested users should contact the company directly to place orders for similar equipment.

Cell tester software (Deben UK Ltd.)
Cooling incubator
Diamond pencil (Solmedia DIM100)
Disposable plastic cup (or similar, with enough depth for vigorous stirring)
Dissecting stereoscope
Dissection needle(s) and/or hair tools (optional; see Step 14)
Dumont #5 Forceps standard Inox (Interfocus Ltd. 11251-20)
Dumont #55 Forceps Biologie Dumoster (Interfocus Ltd. 11295-51)
Glass coverslips
Graduated Pasteur pipette (3 mL)
Laboratory oven (set at 65°C)
Laboratory scales
Membrane chamber mold (Deben UK Ltd.)
Metal spatula, small
Petri dishes (60 mm and 100 mm)
Picospritzer III microinjector
Pipette, disposable plastic
Sharp needle
Upright confocal microscope (Leica SP8 or similar) with a large stage and 20× water immersion objective
UV box or tissue culture hood with UV light
Vacuum desiccator (Duran 2478261)
Vacuum pump (Welch WOB-L Piston Pump 2522)

METHOD

Making Elastomeric Membrane Chambers

Elastomeric membrane chambers are required for these experiments, which are made as described here.

1. Weigh the Sylgard liquid and curing agent from the silicone Sylgard 184 kit at a 10:1 ratio into a disposable plastic cup.

   For our experiments, 40 mL Sylgard liquid: 4 mL curing agent will make eight membrane chambers.

2. Stir the mixture continuously, by hand, for 5 min.

3. After stirring, pour the mixture into a 100 mm Petri dish. Place the dish in a vacuum desiccator and apply vacuum pumping for 1 h to remove all air bubbles.

4. Clean the membrane chamber mold using 70% ethanol and an air duster/compressed air to remove any traces of previous membrane chambers. Ensure all the ethanol has evaporated before adding the Sylgard mixture.

5. Use a disposable 3 mL plastic pipette to transfer the Sylgard mixture into the mold, starting around the edge and working inwards (Fig. 1A). Try not to overfill the mold as this makes it harder to remove the membrane chambers when set. Remove any bubbles using a sharp needle.

6. Place the mold in the laboratory oven preheated to 65°C and leave to cure for 2.5 h. Once cured, leave the mold to cool overnight at room temperature before removing the membrane chambers carefully using a small metal spatula. Apply 70% ethanol to the membrane chambers in the mold to assist with their removal if necessary.

Cite this protocol as *Cold Spring Harb Protoc*; doi:10.1101/pdb.prot105551

A Preparation of membrane chambers

B Elastomeric membrane chamber coated with fibronectin

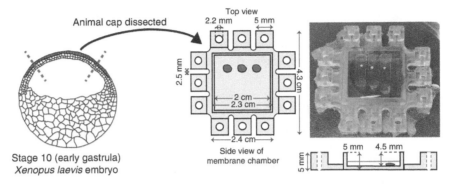

C Uniaxial stretch applied to membrane chamber

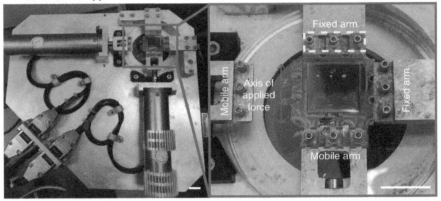

D Live imaging of animal cap tissue under stretch or compression

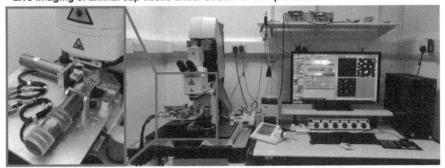

FIGURE 1. Experimental setup for application of tensile force to animal cap explants. (*A*) Membrane chambers are prepared using the membrane chamber mold (the process of loading the mold with liquid PDMS is shown). (*B*) Animal cap tissue is dissected from Stage 10 (early gastrula stage) *Xenopus laevis* embryos and adhered to a fibronectin-coated elastomeric membrane chamber by gently pushing down using glass coverslips. Representative photograph of animal caps prepared in this way from albino *Xenopus laevis* embryos is shown (superficial cell layer is shaded orange). (*C*) Before imaging, the coverslips are removed and the membrane chamber is mounted on the stretch device and secured in place. The top arm of the device holding the membrane chamber is fixed, while the lower arm is mobile. Using a computer (not shown), the stretch device is programmed to apply a specific displacement by movement of the mobile arm. This displacement can be applied to the membrane chamber either uniaxially (axis of applied force is indicated by double arrows) or biaxially. (*D*) Live imaging of animal cap tissue is performed on a Leica SP8 upright confocal microscope which has a large stage and a dipping water immersion objective. Scale bars are 2 cm.

7. Wash the membrane chambers with 1× PBS then place into a new, dry Petri dish and store ready for use.

Animal Cap Dissections for Stretch Experiments

8. Microinject *X. laevis* embryos at the 1- or 2-cell stage with capped RNA encoding appropriate probes tagged with fluorescent protein, as described previously (Woolner et al. 2009).

    We routinely use GFP-α-tubulin (4.2 nL/cell at a needle concentration of 0.5 mg/mL) and Cherry-histone2B (4.2 nL/cell at a needle concentration of 0.1 mg/mL) and inject approximately 30 embryos for each sample.

9. Once injected, incubate embryos in a 60 mm Petri dish containing 0.1× MMR and allow to develop until stage 10 (~20 h at 16°C).

Preparations Prior to Animal Cap Dissections

10. Prepare small fragments of glass coverslips (~10 mm by 3 mm) using a diamond pencil. Place the fragments in a 60 mm Petri dish and soak in a solution containing 0.1× MMR and 1% BSA overnight at 4°C.

11. Place the required number of elastomeric membrane chambers (in 100 mm Petri dishes with the lids off) under the UV light in a tissue culture hood for 30 min.

    UV irradiation is used to activate the PDMS surface to prepare the elastomeric membrane chambers for fibronectin coating.

12. Prepare 10 µg/mL fibronectin in 1× PBS. Add 1 mL to each membrane chamber and incubate overnight at 4°C. The following day, before use in Step 18, remove the fibronectin solution and rinse three times with 1× PBS and once with DFA in quick succession to prevent drying out. Fill the membrane chamber with DFA. The chamber is now ready for the mounting of animal caps.

Animal Cap Dissections

For animal cap dissections, we describe a modified version of a previously published video protocol (Joshi and Davidson 2010).

13. Transfer stage 10 microinjected embryos into a 100-mm Petri dish containing ~30 mL DFA.

14. Under a standard dissecting stereoscope, use two sets of forceps to carefully remove the vitelline membrane, starting on the underside of the embryo and taking care not to damage the animal cap.

    We find that the combination of one #5 forceps to hold the embryo in place and one #55 forceps to remove the vitelline membrane works particularly well. Dissection needle(s) and hair tools can also be used.

15. Dissect the animal cap by making gentle pinch incisions with the #55 forceps all around the edge of the animal cap.

16. Once the animal cap is separate from the rest of the embryo, gently trim any thick layers of cells from the edges and try to make the animal cap a more uniform rectangular shape using either the fine forceps or hair tools. It is important to ensure that the animal cap remains submerged in the DFA throughout the process.

17. If applying tensile force, proceed to Step 18. If applying compressive force, see the protocol modifications beginning at Step 30.

18. Using a disposable plastic pipette, transfer the animal cap onto the elastomeric membrane chamber containing DFA (Fig. 1B).

19. Carefully position the animal cap, with the superficial cell layer pointing upwards, onto the elastomeric membrane chamber (Fig. 1B, cells shaded orange).

    For our system setup, the position of the immersion objective makes it necessary to place the animal caps closer to the fixed arm of the stretch apparatus (in the top third of the chamber). Keeping the animal caps in the same position on the membrane chamber also ensures the most consistent and reproducible results (e.g., compressive force will vary down the stretched membrane chamber).

Cite this protocol as *Cold Spring Harb Protoc*; doi:10.1101/pdb.prot105551

20. Select a small piece of glass coverslip (previously cut to size and soaked in 0.1× MMR + 1% BSA overnight) and apply a small amount of vacuum grease to each of the narrow ends. Using forceps, gently apply the coverslip over the animal cap and press lightly to help the animal cap to adhere to the underlying fibronectin-coated elastomeric membrane chamber (Fig. 1B).

21. Incubate for 2 h at 18°C prior to imaging or applying force.

Applying Tensile Force to an Animal Cap

For our stretch experiments we attach a custom-designed apparatus built by Deben UK (Fig. 1C,D; see the Equipment list) to the stage of an upright confocal microscope (we use a Leica SP8 confocal) and image using a dipping water immersion objective.

22. Attach the stretch apparatus to the stage by passing a bolt through a designated hole on the apparatus which then slots into a hole on the microscope stage. Secure the bolt in place using a compatible washer and nut ensuring that the stretch device is held tightly to the stage.

    The stretch device has four arms, two rigid and two movable, onto which the elastomeric membrane chamber is secured by carefully inserting the pegs of the stretch apparatus arm into the associated holes in the teeth protruding from the top and the bottom of the membrane chamber (Fig. 1C). The movable arms of the apparatus are controlled using custom software supplied by the manufacturer (Deben UK, Ltd.) to apply a specific displacement to one or both of the arms in order to apply a uniaxial or biaxial strain, respectively. In our experiments we apply an 8.6 mm displacement of one arm, which results in a 35% uniaxial stretch of the elastomeric membrane chamber. For "unstretched" experiments, we apply a small, 0.5 mm, displacement to prevent sagging of the elastomeric membrane chamber. Obviously, protocols will vary depending on the stretch apparatus to be used but we provide a few useful pointers below.

23. Ensure that the stretch apparatus is securely attached to the microscope stage and that the axis of strain remains constant, so that data from multiple collections can be easily compared.

24. Initialize the equipment using an empty elastomeric membrane chamber before mounting the experimental membrane chamber.

25. Prior to mounting the experimental elastomeric membrane chamber, carefully remove coverslip fragments from the top of the animal caps using blunt forceps. Also use the forceps to remove any excess vacuum grease.

26. Before imaging begins, top up the level of DFA in the elastomeric membrane chamber, to ensure that the animal caps remain fully covered.

27. Apply stretch appropriate for your experiment. We apply the stretch in two increments; first setting the X arm to a 4.3 mm displacement, followed immediately by a second 4.3 mm displacement to give a total displacement of 8.6 mm. For our experiments, we perform uniaxial stretch, keeping the Y arm set at a 0 mm displacement.

28. Once a stretch has been applied to the membrane chamber, refocus on the animal cap. This is necessary because applying strain moves the position and focal plane of the tissue. It is possible (although challenging) to find the same field of cells poststretch and we are increasingly doing this by using regions with distinct morphological landmarks to aid identification of a specific region poststretch. This enables the same cells to be tracked before and after stretch.

29. Setup confocal image acquisition as follows: scan using a 512×512 format with the sequential acquisition function and bidirectional scanning; set the top and bottom slices with a step size of 5 μm; capture a stack every 20 sec (Fig. 1D).

Applying Compressive Force to an Animal Cap

The procedure for applying a compressive force to an animal cap is very similar to that of applying a tensile force with some key differences.

30. Prepare the animal cap explant as detailed in Steps 13–16 but adhere the cap to a stretched elastomeric membrane chamber attached to the stretch apparatus (we use an 8.6 mm displace-

ment of one arm). Position the animal cap as detailed in Step 19. Place a fragment of coverslip over the animal cap to encourage adherence to the fibronectin-coated membrane chamber (as in Step 20). Incubate for 2 h at room temperature.

See Troubleshooting.

31. Setup the stretch apparatus on the confocal as described above (Steps 22–26). To apply compressive force, the applied (8.6 mm) displacement should be reduced (e.g., to a 0.5 mm displacement).

32. Refocus the microscope onto the animal cap and image as described in Step 29.

In some cases the animal cap can be loosened by the release of the stretch, which may cause some issues with drifting while imaging. It is always best to select the healthiest looking animal cap to image, which may reduce any issues.

TROUBLESHOOTING

Problem (Step 30): Sometimes animal caps do not adhere well to the fibronectin-coated membrane chamber and peel away before or during imaging.

Solution: This can be due to the health of the embryo from which the explant was made and ensuring that the healthiest embryos are chosen (e.g., no large cells, etc.) can help. Another common cause for animal caps peeling away is the fibronectin itself. We find that this occurs more commonly once the fibronectin solution has been opened and stored for more than 3 mo at 4°C.

DISCUSSION

The *X. laevis* embryo is a powerful model system for studying the effects of mechanical forces on cellular function (Keller et al. 2003; Stooke-Vaughan et al. 2017). The system that we described here involves the application of reproducible tensile and compressive force to *X. laevis* animal cap tissue explants and the simultaneous assessment of cellular behavior (e.g., cell division, cell signaling, gene expression, or force transmission) using live confocal imaging. The animal cap explant provides an excellent system for these studies because it is a resilient tissue that maintains its in vivo architecture for the duration of the experiment and is highly amenable to live imaging. In this way, detailed analyses of changes in cell morphology and cell behavior in response to mechanical force can be made within a tissue context (Nestor-Bergmann et al. 2017). Moreover, this biological data can be allied with mathematical approaches to further interrogate how a tissue responds to mechanical stress (Nestor-Bergmann et al. 2018a,b).

RECIPES

Danilchik's for Amy (DFA) Medium

Reagent	Final concentration
NaCl	53 mM
Na$_2$CO$_3$	5 mM
Potassium gluconate	4.5 mM
Sodium gluconate	32 mM
CaCl$_2$	1 mM
MgSO$_4$	1 mM
Bovine serum albumin	1 g/L

Adjust pH to 8.3 with 1 M bicine before adding bovine serum albumin. (This recipe is adapted from Sater et al. 1993.)

Marc's Modified Ringer's Solution (MMR) for Embryos

	To make 5 L of 10× (diluted in MilliQ water)
1 M NaCl	292.2 g
20 mM KCl	7.45 g
10 mM MgCl$_2$.6H$_2$O	10.165 g
20 mM CaCl$_2$.2H$_2$O	14.7 g
1 mM EDTA disodium salt	1.86 g
50 mM HEPES	59.7 g

Dissolve ingredients in 4 L MilliQ water on a magnetic stirrer, heated gently. Adjust pH to 8.0 with 10 M NaOH to dissolve the EDTA and then adjust to pH 7.8 using HCl. Adjust the volume to 5 L with MilliQ water.

REFERENCES

Joshi SD, Davidson LA. 2010. Live-cell imaging and quantitative analysis of embryonic epithelial cells in *Xenopus laevis*. *J Vis Exp* doi:10.3791/1949

Jungbauer S, Gao H, Spatz JP, Kemkemer R. 2008. Two characteristic regimes in frequency-dependent dynamic reorientation of fibroblasts on cyclically stretched substrates. *Biophys J* 95: 3470–3478. doi:10.1529/biophysj.107.128611

Keller R, Davidson LA, Shook DR. 2003. How we are shaped: the biomechanics of gastrulation. *Differentiation* 71: 171–205. doi:10.1046/j.1432-0436.2003.710301.x

Nestor-Bergmann A, Stooke-Vaughan GA, Goddard GK, Starborg T, Jensen OE, Woolner S. 2017. Decoupling the roles of cell shape and mechanical stress in orienting and cueing epithelial mitosis. *Cell Rep* 26: 2088–2100. e4. doi:10.1016/j.celrep.2019.01.102

Nestor-Bergmann A, Goddard G, Woolner S, Jensen OE. 2018a. Relating cell shape and mechanical stress in a spatially disordered epithelium using a vertex-based model. *Math Med Biol* 35: 1–27. doi:10.1093/imammb/dqx008

Nestor-Bergmann A, Johns E, Woolner S, Jensen OE. 2018b. Mechanical characterization of disordered and anisotropic cellular monolayers. *Phys Rev E* 97: 052409. doi:10.1103/PhysRevE.97.052409

Sater AK, Steinhard RA, Keller R. 1993. Induction of neuronal differentiation by planar signals in *Xenopus* embryos. *Dev Dyn* 197: 268–280.

Stooke-Vaughan GA, Davidson LA, Woolner S. 2017. *Xenopus* as a model for studies in mechanical stress and cell division. *Genesis* 55. doi:10.1002/dvg.23004

Woolner S, Miller AL, Bement WM. 2009. Imaging the cytoskeleton in live *Xenopus laevis* embryos. *Methods Mol Biol* 586: 23–39. doi:10.1007/978-1 60761-376-3_2

Analysis of Phosphorylation Status of Ectopically Expressed Proteins in Early *Xenopus* Embryos

Laura J.A. Hardwick[1,3] and Anna Philpott[1,2,4]

[1]*Hutchison/MRC Research Centre, Department of Oncology, University of Cambridge, Cambridge CB2 0XZ, United Kingdom;* [2]*Wellcome–MRC Cambridge Stem Cell Institute, University of Cambridge, Cambridge CB2 0AF, United Kingdom;* [3]*Peterhouse, University of Cambridge, Cambridge CB2 1RD, United Kingdom*

Xenopus embryos provide a rapid and accessible in vivo model, expressing a plethora of endogenous kinase and phosphatase enzymes that control protein phosphorylation and, in turn, affect physiological function. Traditionally, the detection of protein phosphorylation has been achieved by radioisotope phosphate labeling of proteins, sometimes with in vitro assays using recombinant proteins or with site-specific phospho-antibodies. However, the target phospho-sites and kinases responsible are often unknown, and the use of radioactive isotopes is not always desirable. Therefore, as a first step in determining the functional significance of potential phosphorylation, it is useful to show that a protein can be phosphorylated in vivo in *Xenopus* eggs and/or embryos. This protocol describes a nonradioactive method to visualize protein phosphorylation by exposing the protein to the egg/embryo kinase environment and then observing a difference in protein migration (as assessed by sodium dodecyl sulfate-polyacrylamide gel electrophoresis [SDS-PAGE] and western blot analysis) with and without treatment with the lambda phosphatase enzyme. Subsequent investigation of the ability of site-specific phospho-mutant proteins to recapitulate the effect of phosphatase treatment can be used to explore the identity of the phosphorylated sites. Moreover, the detection of multiple bands of the protein of interest even after phosphatase treatment points to the presence of other types of posttranslational modifications.

MATERIALS

It is essential that you consult the appropriate Material Safety Data Sheets and your institution's Environmental Health and Safety Office for proper handling of equipment and hazardous materials used in this protocol.

RECIPES: Please see the end of this protocol for recipes indicated by <R>. Additional recipes can be found online at http://cshprotocols.cshlp.org/site/recipes.

Reagents

Embryo lysis buffer (*Xenopus*) <R>

Fertilization and embryo culture reagents (see Sec. 4, Protocol 2: Microinjection of mRNAs and Oligonucleotides [Moody 2018])

Lambda protein phosphatase kit containing enzyme, 10× PMP Buffer, and 10 mM MnCl$_2$ (e.g., from NEB)

The protein phosphatase enzyme is very sensitive to temperature so must be stored at −80°C and only defrosted on ice for the minimum time possible.

---

[4]Correspondence: ap113@cam.ac.uk

Loading buffer with SDS and reducing agent for SDS-PAGE (see Step 14)

mRNA encoding protein(s) of interest (see Sec. 4, Protocol 2: Microinjection of mRNAs and Oligonucleotides [Moody 2018])

Protein quantification assay (e.g., BCA assay) materials

SDS-PAGE and western blotting materials (see Lin-Moshier and Marchant 2013)

> *The gel percentage and composition required for optimal separation of phospho-forms by SDS-PAGE will depend on the protein of interest.*

Western blotting antibodies including antibody for the protein of interest and loading control antibody (e.g., anti-α-tubulin)

Xenopus embryos (one- or two-cell stage)

Equipment

Benchtop centrifuge (prechilled to 4°C)

Dry ice (optional; see Step 4)

Glass pipettes (wide- and narrow-bore) with rubber suction

Heat block (70°C)

Ice bucket with wet ice

Microcentrifuge tubes (1.5-mL)

Microinjection equipment for *Xenopus* embryos (see Sec. 4, Protocol 2: Microinjection of mRNAs and Oligonucleotides [Moody 2018])

Water bath (30°C)

METHOD

> *The protocol can be conducted in continuous or multiple discrete steps as described below.*

Preparation of Embryos

1. Microinject mRNA encoding each protein of interest into a separate dish of one- or two-cell stage *Xenopus* embryos, keeping one dish as uninjected control embryos (see Sec. 4, Protocol 2: Microinjection of mRNAs and Oligonucleotides [Moody 2018]).

   > *The amount of mRNA to be injected will require individual optimization for the protein of interest.*

2. Culture embryos to the relevant stage for analysis of the protein of interest (see Sec. 4, Protocol 2: Microinjection of mRNAs and Oligonucleotides [Moody 2018]).

3. Carefully transfer 10–15 embryos to a clean 1.5-mL microcentrifuge tube using a wide-bore glass pipette.

   > *The number of embryos collected can be altered if larger volumes of protein extract are required.*

4. Remove as much of the culture medium as possible using a narrow-bore glass pipette. Proceed directly to Step 5, or snap-freeze the embryos in dry ice and then store them at −80°C.

   > *Additional buffer may be removed with a 10-µL pipette but take care not to lyse embryos prematurely.*

Protein Extraction and Quantification from Whole Embryos (1 h)

> *Conduct this part of the protocol on ice.*

5. If the embryos were frozen, thaw them on ice while preparing fresh embryo lysis buffer and prechilling the centrifuge to 4°C.

6. Homogenize the embryos by repeated pipetting in embryo lysis buffer, using 4–10 μL per embryo.

   Use 4 μL per embryo for unstable proteins or for proteins expressed at a low level to increase the total protein concentration in the extract.

7. Centrifuge the homogenized embryos for 6 min at 1200g at 4°C. Collect the middle fraction, which is above the pellet and below the lipid layer.

8. Centrifuge the collected fraction for 6 min at 1200g at 4°C and collect the supernatant. Use this protein extract immediately or store at −20°C.

9. Use a 1-μL aliquot of each extract to determine the protein concentration in the extract.

   For example, use a BCA assay, following the manufacturer's protocol.

10. (Optional) Conduct SDS-PAGE and western blot on protein extracts to optimize loading amounts required for adequate visualization and to determine the gel conditions for optimal protein separation.

    The total protein amounts required for detection will vary depending on the protein of interest and may range from 20 to 80 μg. See Troubleshooting.

Lambda Protein Phosphatase Treatment (1.5 h)

This part of the procedure can be performed in isolation using prefrozen protein extract and before SDS-PAGE.

11. Select the amount of total protein from each extract to be used for SDS-PAGE and western blot (e.g., by optimization in Step 10).

12. Set up protein phosphatase reactions and control reactions without enzyme (i.e., with an equivalent amount of water) on ice, according to the manufacturer's instructions. Prepare duplicate reactions for each protein of interest.

    Leave the phosphatase enzyme at −80°C until immediately before use, and snap-freeze on dry ice as soon as possible after use. Due to the glycerol content in the enzyme preparation, keep the concentration of enzyme to <10%.

13. Incubate all reactions (with and without phosphatase) for 1 h in a water bath at 30°C.

14. Add the appropriate amount of loading buffer with SDS and reducing agent for SDS-PAGE and heat for 10 min on a heat block at 70°C.

    These samples can be run directly on SDS-PAGE or frozen at −80°C. If samples are stored, repeat heat treatment for 10 min at 70°C before gel loading.

Analysis by SDS-PAGE and Western Blotting

15. Separate proteins by SDS-PAGE and analyze by western blotting.

    The pair of samples for each protein of interest (with and without phosphatase treatment) should be run side-by-side to enable direct comparison of migration. An equivalent amount of total protein from uninjected embryos can be used to show antibody specificity for the overexpressed protein. A loading control such as α-tubulin can be used to show equal loading.

16. Compare protein migration in the sample lanes to determine the effect of phosphatase treatment.

    Wild-type proteins may be phosphorylated on one or more sites, resulting in the appearance of a broad-band or multiple discrete bands on western blot. The removal of these phosphate groups by phosphatase enzyme results in the appearance of a narrower, faster-migrating band. Example images to illustrate these results are shown in Hardwick and Philpott (2015) and Hardwick et al. (2016). Discrete bands may still be present if the protein undergoes alternative posttranslational modification such as ubiquitination or sumoylation.

    See Troubleshooting.

Cite this protocol as *Cold Spring Harb Protoc*; doi:10.1101/pdb.prot105569

TROUBLESHOOTING

Problem (Step 10): The concentration of protein in the whole-embryo extract is too low to visualize on a western blot.

Solution: If the protein concentration needs to be increased to enable sufficient total protein in the loading volume, embryos can be lysed in a reduced volume in Step 6.

Problem (Step 16): There is no apparent difference between the protein samples with and without phosphatase treatment.

Solution: Different SDS-PAGE conditions may be required for optimal separation of protein phospho-forms and demonstration of phospho-shift with phosphatase treatment. A higher percentage gel such as an 18% gel can aid separation. Alternatively, Phos-tag gels may provide superior resolution of discrete phospho-forms (see Horinouchi et al. 2016). If Phos-tag gels are used, an equivalent amount of $MnCl_2$ must be added to the samples before phosphatase treatment to bind the EDTA in the embryo lysis buffer. Otherwise, the EDTA will chelate the Mn^{2+} in the gel and prevent detection of the phospho-forms.

DISCUSSION

This protocol describes a rapid, nonradioactive method to visualize potential in vivo phosphorylation of an overexpressed protein of interest in *Xenopus* embryos. We compare the migration of proteins from embryo extracts after SDS-PAGE and western blotting with and without lambda protein phosphatase treatment, avoiding the need for phospho-site specific antibodies. Lambda protein phosphatase is a Mn^{2+}-dependent protein phosphatase with activity toward phosphorylated serine, threonine, and tyrosine residues. This approach complements in vitro kinase assays with recombinant target protein and kinase enzymes, and it has the added benefit of using the endogenous kinase and phosphatase enzymes in the embryo to allow determination of phosphorylation in vivo at the developmentally relevant stage. This method can also be used to show the importance of specific phosphorylation sites by comparing migration of wild-type and mutant versions of the protein of interest, where individual or a combination of potential phosphorylation sites have been mutated to a nonphosphorylatable amino acid such as alanine (e.g., Hardwick and Philpott 2015; Hardwick et al. 2016).

RECIPE

Embryo Lysis Buffer (Xenopus)

Reagent	Final concentration (1×)
Tris (pH 8.0)	50 mM
NaCl	100 mM
EDTA (pH 8.0)	5 mM
Triton X-100	0.10%
β-glycerophosphate[a]	50 mM
Protease Inhibitor Cocktail (PIC; Roche)	1×

[a]Additional phosphatase inhibitors or competitors may be required depending on the stability of the phosphorylation status of the protein of interest.
Prepare fresh and store on ice.

ACKNOWLEDGMENTS

We thank Daniel Marcos Corchado for advice on Phos-tag SDS-PAGE. L.J.A.H. was funded by a Medical Research Council Studentship and Peterhouse Research Fellowship. Research in A.P.'s laboratory is supported by Medical Research Council grants MR/K018329/1 and MR/L021129/1, as well as through core support from the Wellcome Trust and MRC to the Wellcome Trust–Medical Research Council Cambridge Stem Cell Institute.

REFERENCES

Hardwick LJ, Philpott A. 2015. Multi-site phosphorylation regulates NeuroD4 activity during primary neurogenesis: a conserved mechanism amongst proneural proteins. *Neural Dev* 10: 15. doi:10.1186/s13064-015-0044-8

Hardwick LJ, Davies JD, Philpott A. 2016. MyoD phosphorylation on multiple C terminal sites regulates myogenic conversion activity. *Biochem Biophys Res Commun* 481: 97–103. doi:10.1016/j.bbrc.2016.11.009

Horinouchi T, Terada K, Higashi T, Miwa S. 2016. Using Phos-tag in western blotting analysis to evaluate protein phosphorylation. *Methods Mol Biol* 1397: 267–277. doi:10.1007/978-1-4939-3353-2_18

Lin-Moshier Y, Marchant JS. 2013. A rapid western blotting protocol for the *Xenopus* oocyte. *Cold Spring Harb Protoc* doi: 10.1101/pdb.prot072793

Moody SA. 2018. Microinjection of mRNAs and oligonucleotides. *Cold Spring Harb Protoc* doi: 10.1101/pdb.prot097261

Analysis of Chromatin Binding of Ectopically Expressed Proteins in Early *Xenopus* Embryos

Laura J.A. Hardwick[1,3] and Anna Philpott[1,2,4]

[1]*Hutchison/MRC Research Centre, Department of Oncology, University of Cambridge, Cambridge CB2 0XZ, United Kingdom;* [2]*Wellcome–MRC Cambridge Stem Cell Institute, University of Cambridge, Cambridge CB2 0AF, United Kingdom;* [3]*Peterhouse, University of Cambridge, Cambridge CB2 1RD, United Kingdom*

Xenopus embryos have long been used to show phenotypic effects following overexpression of proteins of interest such as transcription factors. Posttranslational modification of these proteins can dramatically alter the extent of the observed phenotype by inhibiting or enhancing protein activity. To determine the mechanisms controlling transcription factor activity, it is useful to compare relative levels of chromatin-bound protein, as this can reveal altered chromatin association in addition to changes in overall protein accumulation seen in the cytoplasm. Assaying protein binding to the bulk DNA described here compliments alternative assays such as electrophoretic mobility shift assay (EMSA) and chromatin immunoprecipitation (ChIP) that measure site-specific DNA binding. This protocol describes a method to prepare and analyze chromatin and cytoplasmic extracts from embryos overexpressing proteins of interest, and it uses a robust fractionation procedure that results in clear separation of cytoplasmic tubulin from histone-H3 enriched chromatin. This assay for relative chromatin-bound protein is most suitable for comparing modified forms of a single protein (e.g., to investigate the effects of point mutations on chromatin association). Optimization is required for the specific protein of interest but guide ranges are provided.

MATERIALS

It is essential that you consult the appropriate Material Safety Data Sheets and your institution's Environmental Health and Safety Office for proper handling of equipment and hazardous materials used in this protocol.

RECIPES: Please see the end of this protocol for recipes indicated by <R>. Additional recipes can be found online at http://cshprotocols.cshlp.org/site/recipes.

Reagents

Cross-linking solution (1% formaldehyde in 0.1× MBS)
> *Prepare fresh and keep at room temperature.*

Distilled water
DNase (e.g., RNase-free DNase from QIAGEN)
Extraction buffers

> Extraction buffer E1 <R>
> Extraction buffer E2 <R>
> Extraction buffer E3 <R>

[4]Correspondence: ap113@cam.ac.uk

Fertilization and embryo culture reagents (see Sec. 4, Protocol 2: Microinjection of mRNAs and Oligonucleotides [Moody 2018])

Loading buffer with SDS and reducing agent for SDS-PAGE (see Step 29)

Modified Barth's saline (MBS) <R>

mRNA encoding protein(s) of interest (see Sec. 4, Protocol 2: Microinjection of mRNAs and Oligonucleotides [Moody 2018])

*As this protocol is a **relative** comparison of two or more overexpressed proteins, it is advisable that proteins have a common tag that can be detected with a single antibody where a protein-specific antibody is not available.*

Quenching solution (125 mM glycine in 0.25× MBS)

Prepare fresh and keep at room temperature.

Washing solution (0.25× MBS)

Prepare fresh and keep at room temperature.

Western blotting antibodies including anti-α-tubulin, anti-histone-H3, and antibody for protein of interest

SDS-PAGE and western blotting materials (see Lin-Moshier and Marchant 2013)

Xenopus embryos (one- or two-cell stage)

Equipment

Benchtop centrifuge (prechilled to 4°C)

Conical tubes (15- and 50-mL)

Dry ice (optional; see Step 12)

Glass pipettes (wide- and narrow-bore) with rubber suction

Glass vials with screw tops (4-mL)

Heat block (70°C)

Ice bucket with wet ice

Microcentrifuge tubes (1.5-mL)

Microinjection equipment for *Xenopus* embryos (see Sec. 4, Protocol 2: Microinjection of mRNAs and Oligonucleotides [Moody 2018])

Paper tissue

Rocking platform

METHOD

This protocol provides a visual means to assess relative chromatin association of a protein of interest, in addition to its relative protein accumulation in the cytoplasm, thus complementing alternative electrophoretic mobility shift assay (EMSA) and chromatin immunoprecipitation (ChIP) assay of DNA binding affinity. This assay is most suitable for comparing modified versions of a single protein. For example, it can be used to investigate the effects of point mutations on chromatin association, where these may have differential effects on protein stability and/or chromatin association (e.g., Hardwick and Philpott 2015; Hardwick et al. 2016). Two or more proteins can be compared, limited only by the number of lanes available on sodium dodecyl sulfate-polyacrylamide gel electrophoresis (SDS-PAGE). All proteins of interest should be detected with a single antibody for equivalent comparison on western blot.

Preparation of Embryos

1. Microinject mRNA encoding each protein of interest into a separate dish of one- or two-cell stage *Xenopus* embryos, keeping one dish as uninjected control embryos (see Sec. 4, Protocol 2: Microinjection of mRNAs and Oligonucleotides [Moody 2018]).

   Transcription factors are often highly unstable proteins and the amounts of mRNA to be injected require individual optimization. The highest tolerated dose should be used in the first instance with equivalent amounts for each experimental category. Injection can be performed in a one-cell stage embryo, but for improved embryo survival, inject a half dose into each cell of a two-cell embryo.

Cite this protocol as *Cold Spring Harb Protoc*; doi:10.1101/pdb.prot105577

Inject sufficient numbers of embryos to enable cross-linking of 40–50 embryos per experimental condition (see Steps 4–11), although it may be possible to detect chromatin-bound protein with fewer embryos.

2. Culture embryos to the relevant stage for the protein of interest (see Sec. 4, Protocol 2: Micro-injection of mRNAs and Oligonucleotides [Moody 2018]).

    For example, if investigating a protein involved in primary neurogenesis, culture embryos to early neural plate stages. There is usually adequate overexpressed protein for detection in embryos from stage 10.5 onward.

Cross-Linking of Embryos (2.5 h)

This part of the procedure must be conducted when the embryos reach the appropriate stage. Once cross-linked, the embryos can be fractionated immediately, or snap-frozen and stored at −80°C before subsequent fractionation.

3. Prepare cross-linking, quenching, and wash solutions fresh and store at room temperature.

4. Using a wide-bore glass pipette, transfer embryos from each category to respective glass vials.

5. Draw off residual culture medium with a narrow-bore glass pipette, and wash embryos three times by filling the vial with room temperature distilled water for 5 min per wash at room temperature on a platform rocker.

    Embryos are very delicate and can easily lyse prematurely.

6. Draw off the water with a narrow bore glass pipette and completely fill the vial with cross-linking solution, securing the vial caps.

7. Incubate the vials horizontally on a platform rocker for 30 min, ensuring all embryos are individually spaced along the vial and not clumped together.

8. Remove the cross-linking solution and fill the vial with quenching solution.

9. Incubate the vials horizontally on a platform rocker for 30 min at room temperature.

10. Remove the quenching solution and wash embryos twice with washing solution, incubating the vials horizontally on a platform rocker for 15 min per wash.

11. Using a wide-bore glass pipette, carefully transfer an equal number of embryos (approximately 45 embryos) from each experimental condition to a clean 1.5-mL microcentrifuge tube.

    It is essential that all categories contain the same number of embryos to avoid loading errors later.

12. Taking each vial in turn, remove as much of the wash buffer as possible and either proceed directly to chromatin extraction (Step 13) or snap-freeze embryos in dry ice and store at −80°C.

    It is essential to draw off all wash buffer, as residual buffer will lead to the dilution of the cytoplasmic fraction in Step 15. Embryos are very delicate so care must be taken not to lyse embryos prematurely. A 10-µL pipette can be used to draw liquid from around embryos.

Fractionation and Chromatin Extraction (1.5 h)

This part of the procedure can be done in isolation using frozen cross-linked embryos and before SDS-PAGE with storage of protein extracts at −20°C. All steps should be conducted on ice with centrifugation at 4°C.

13. Prechill a bench centrifuge to 4°C and prepare all solutions fresh on ice.

    For each vial of cross-linked embryos, ∼2.5 mL of buffer E1, 2.5 mL of buffer E2 and 150 µL of buffer E3 will be required.

14. If embryos have been prefrozen from cross-linking, thaw on ice. Homogenize each aliquot of embryos in 50 µL of buffer E1 by pipetting within the microcentrifuge tube.

15. Centrifuge each tube at 1200g for 3 min at 4°C and collect the middle fraction (supernatant below the lipid layer and above the pigmented pellet; Fig. 1A) as the cytoplasmic fraction. Maintain this fraction on ice for immediate use or store it by freezing at −20°C.

16. Use a rolled-up tissue to remove the white lipid residue above the remaining pellet.

17. Resuspend the remaining pigmented pellet in 1 mL of buffer E1.

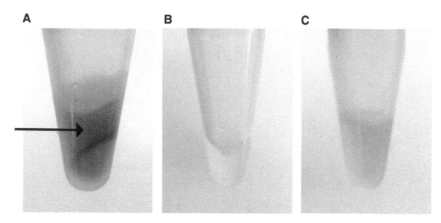

FIGURE 1. Images of the three extracts collected during fractionation. (*A*) Cytoplasmic extract from Step 15 (middle fraction, black arrow). (*B*) Nucleoplasmic extract from Step 22. (*C*) Final chromatin extract from Step 27.

18. Centrifuge at 1200*g* for 2 min at 4°C and discard the supernatant.

19. Resuspend the pellet in 1 mL of buffer E1, and incubate on ice for 10 min.

20. Centrifuge at 1200*g* for 2 min at 4°C, and discard the supernatant.

21. Resuspend the pellet in 50 µL of buffer E2.

22. Centrifuge at 1200*g* for 2 min at 4°C, and collect the supernatant as the nucleoplasmic fraction if required (Fig. 1B). Maintain this fraction on ice for immediate use or store it by freezing at −20°C.

23. Resuspend the pellet in 1 mL of buffer E2.

24. Centrifuge at 1200*g* for 2 min at 4°C, and discard the supernatant.

25. Resuspend the pellet in 1 mL of buffer E2, and incubate on ice for 10 min.

26. Centrifuge at 1200*g* for 2 min at 4°C, and discard the supernatant.

27. Resuspend the pellet in 100 µL of buffer E3, and save this as the chromatin fraction (Fig. 1C). Maintain this fraction on ice for immediate use or store it by freezing at −20°C.

> *The pellet is very thick and sticky, so resuspension is difficult and care must be taken not to discard material stuck to the pipette tip.*

Analysis of Fractions by SDS-PAGE and Western Blotting

28. Treat 20 µL of each chromatin fraction with DNase for 20 min at room temperature or according to the recommendations of the manufacturer.

> *The chromatin fractions require DNase treatment before SDS-reducing conditions to avoid clumping and solidification. If the protein of interest is abundant on DNA, the chromatin fraction can be diluted in sterile water before DNase treatment.*

29. Prepare samples of equivalent volume from each cytoplasmic and DNase-treated chromatin fraction so that, after adding loading buffer with SDS and reducing agent for SDS-PAGE, the total volume of each sample (including loading buffer) will be 20 µL. Heat the samples on a heating block for 10 min at 70°C. Once cooled, check the chromatin fraction for clumping, which will prevent loading on a gel.

> *See Troubleshooting.*

> *The amounts to load must be optimized for the individual protein of interest but will be in the range of 5–10 µL of extract from 45 embryos, either diluted directly in sterile water for the cytoplasmic fractions or from the diluted DNase-treated chromatin sample.*

30. Separate proteins by SDS-PAGE, and then use western blotting for protein detection. Confirm successful fractionation by probing for α-tubulin (wholly cytoplasmic) and histone H3 (predominantly in the chromatin fraction but some present in cytoplasm).

Cite this protocol as *Cold Spring Harb Protoc*; doi:10.1101/pdb.prot105577

See Troubleshooting.

31. (Optional) Analyze the densities of the autoradiography bands (after western blotting) to quantify the protein of interest relative to the tubulin (cytoplasmic) or histone H3 (chromatin) loading controls.

Examples to illustrate the results of this protocol are provided in Hardwick and Philpott (2015) and Hardwick et al. (2016).

TROUBLESHOOTING

Problem (Step 29): The chromatin fraction remains clumped.
Solution: Repeat the DNase treatment with a reduced amount of starting material.

Problem (Step 30): Chromatin proteins are difficult to detect.
Solution: Increase the number of embryos lysed. Do not resuspend the pellet in Step 27 in a volume of <100 µL, because the sample will be too viscous to run smoothly during SDS-PAGE.

Problem (Step 30): The bands of α-tubulin, histone H3, and/or the protein of interest are difficult to distinguish.
Solution: A gradient gel (e.g., 4%–20%) can provide better protein resolution across the large range of molecular sizes (i.e., 15 kDa for histone H3 and 50 kDa for α-tubulin, plus protein of interest).

RECIPES

Extraction Buffer E1

Reagent	Final concentration (1×)
HEPES-KOH (pH 7.5)	50 mM
NaCl	140 mM
EDTA (pH 8.0)	1 mM
Glycerol	10%
NP-40	0.5%
Triton X-100	0.25%
Dithiothreitol (DTT)	1 mM
Protease Inhibitor Cocktail (PIC; Roche)	1×
Phenylmethanesulfonyl fluoride (PMSF)	0.2 mM

Prepare fresh in sterile H_2O and store on ice.

Extraction Buffer E2

Reagent	Final concentration (1×)
Tris (pH 8.0)	10 mM
NaCl	200 mM
EDTA (pH 8.0)	1 mM
EGTA (pH 8.0)	0.5 mM
Protease Inhibitor Cocktail (PIC; Roche)	1×
Phenylmethanesulfonyl fluoride (PMSF)	0.2 mM

Prepare fresh in sterile H_2O and store on ice.

Extraction Buffer E3

Reagent	Final concentration (1×)
Tris (pH 6.8)	500 mM
NaCl	500 mM
Protease Inhibitor Cocktail (PIC; Roche)	1×

Prepare fresh in sterile H_2O and store on ice.

Modified Barth's Saline (MBS) (1×)

$CaCl_2$	0.41 mM
$CaNO_3$	0.3 mM
HEPES-NaOH	15 mM
KCl	1 mM
$MgSO_4$	0.82 mM
NaCl	88 mM
$NaHCO_3$	2.4 mM

Adjust pH to 7.6. Store at room temperature for up to 1 mo.

ACKNOWLEDGMENTS

We have derived this protocol using kind advice from Matt Guille on cross-linking and from a ChIP protocol developed by Jerome Jullien. L.J.A.H. was funded by a Medical Research Council Studentship and Peterhouse Research Fellowship. Research in A.P.'s laboratory is supported by Medical Research Council grants MR/K018329/1 and MR/L021129/1, as well as through core support from the Wellcome Trust and MRC to the Wellcome Trust–Medical Research Council Cambridge Stem Cell Institute.

REFERENCES

Hardwick LJ, Philpott A. 2015. Multi-site phosphorylation regulates NeuroD4 activity during primary neurogenesis: a conserved mechanism amongst proneural proteins. *Neural Dev* **10:** 15. doi:10.1186/s13064-015-0044-8

Hardwick LJ, Davies JD, Philpott A. 2016. MyoD phosphorylation on multiple C terminal sites regulates myogenic conversion activity. *Biochem Biophys Res Commun* **481:** 97–103. doi:10.1016/j.bbrc.2016.11.009

Lin-Moshier Y, Marchant JS. 2013. A rapid western blotting protocol for the *Xenopus* oocyte. *Cold Spring Harb Protoc* doi:10.1101/pdb.prot072793

Moody SA. 2018. Microinjection of mRNAs and oligonucleotides. *Cold Spring Harb Protoc* doi:10.1101/pdb.prot097261

Cite this protocol as *Cold Spring Harb Protoc*; doi:10.1101/pdb.prot105577

Protocol 5

Raising Antibodies for Use in *Xenopus*

Maya Z. Piccinni and Matthew J. Guille[1]

European Xenopus Resource Centre, Institute of Biological and Biomedical Sciences, University of Portsmouth, Portsmouth, Hampshire PO1 2DY, United Kingdom

For work in *Xenopus*, frog-specific antibodies must usually be raised, although a few antibodies against mammalian proteins cross-react. To produce an immunogen for antibody production, human embryonic kidney (HEK) expression systems can be used as described here. For most laboratories, the actual method of raising the antibody is determined by local ethical regulations controlling the adjuvant and injection protocols used. Because these steps are often outsourced, they are not included in this protocol.

MATERIALS

It is essential that you consult the appropriate Material Safety Data Sheets and your institution's Environmental Health and Safety Office for proper handling of equipment and hazardous materials used in this protocol.

RECIPES: Please see the end of this protocol for recipes indicated by <R>. Additional recipes can be found online at http://cshprotocols.cshlp.org/site/recipes.

The reagents and equipment needed depend entirely upon the choice of antigen and host. The materials listed here are specifically for HEK cell expression.

Reagents

Branched polyethylenimine (PEI; Sigma-Aldrich 408727)
ddH_2O
Dulbecco's modified eagle's medium (DMEM; Sigma-Aldrich D6546)
Ethanol (20%)
Fetal bovine serum (FBS)
Human embryonic kidney (HEK) 293T cells
Host animal and immunization reagents (see Step 5 and Steps 39–41)
JetPRIME reagent (Polyplus 114-01)
L-glutamine (100 mM)
Liquid nitrogen
M-PER™ Mammalian Protein Extraction Reagent (ThermoFisher Scientific 7850)
Nonessential amino acid solution (NEAAs; 100×; Sigma-Aldrich M7145)
Purification solution A <R>

[1]Correspondence: matthew.guille@port.ac.uk

Purification solution B <R>

Reagents for cloning, sequencing, transfection, western blotting, and cell culture (see Steps 6–16)

A comprehensive list of suitable plasmid expression vectors is provided by OPPF (http://www.oppf.rc-harwell.ac.uk/OPPF/).

Equipment

AKTA Start chromatography system (GE Healthcare Life Sciences)

BLAST program (https://blast.ncbi.nlm.nih.gov/Blast.cgi)

Clustal Omega program (http://www.clustal.org/omega/)

Cryovials

Equipment for cloning, sequencing, transfection, western blotting, and cell culture (see Steps 6–16)

Filters (0.22 μm)

Freezer (−80°C)

HisTrap FF column (1 mL; GE Healthcare Life Sciences, UK) or other appropriate column for the protein of interest

Incubator (37°C; 5% CO_2; 95% air)

Protein structure prediction program (e.g., Phyre [http://www.sbg.bio.ic.ac.uk/phyre2/html/page.cgi?id=index] or psipred [http://bioinf.cs.ucl.ac.uk/psipred/])

T175 flasks

Table top centrifuge

METHOD

The immunogenicity of the antigen depends on its own properties as much as on the choice of the host animal and the immunization protocol. To obtain the best result with antibody production, multiple approaches can be taken to produce the antigen and to purify the resulting serum, and the host can be varied (Bird et al. 2014; Delahaut 2017). All of these factors should be considered in order to obtain high-titer, high-affinity, and specific antisera.

Choosing the Best Target Sequence

When the full-length target protein cannot be generated in sufficient quantity or purity needed to immunize the chosen host, a peptide sequence can be used. To choose the most appropriate peptide sequence, three aspects have to be considered: its immunogenicity, the availability of the epitope in the protein's native form, and its uniqueness within the genome. Please note, some peptides might need to be coupled to a carrier protein produce a sufficient immune response (Trier et al. 2012).

1. BLAST (https://blast.ncbi.nlm.nih.gov/Blast.cgi) the target protein sequence against the *Xenopus* proteome (six-way translated genome); reject regions that have high identities found in other proteins to avoid cross-reaction.

2. Determine the target region's secondary structure using prediction programs (e.g., Phyre [http://www.sbg.bio.ic.ac.uk/phyre2/html/page.cgi?id=index] or psipred [http://bioinf.cs.ucl.ac.uk/psipred/]). Ensure that the protein's unique sequences are on the exposed protein surface.

   These programs predict the probable spatial arrangement of the target sequence but are dependent on published knowledge about the sequence.

3. Run the final antigen sequence through BLAST once more to ensure there are no cross-reacting proteins.

4. Run a Clustal Omega (http://www.clustal.org/omega/) alignment for protein family members to ensure that BLAST did not miss any of these.

Cite this protocol as *Cold Spring Harb Protoc*; doi:10.1101/pdb.prot105585

Choosing the Host Animal

5. Select an animal to be used as the host for antibody production (commonly used animals are listed below). Consider the final yield, cost, ethics, and ease of access to determine which one to use.

- Mouse (>6 wk old; 40 µg of <18 kDa antigen or 15 µg of >18 kDa; 50–70 µL preimmune bleed; 300 µL final bleed; animal culled)

- Rat (>6 wk old; 50 µg of <18 kDa antigen or 30 µg of >18 kDa; 1 mL preimmune bleed; 7–10 mL final bleed; animal culled)

- Guinea pig (>3 mo old; 50 µg of <18 kDa antigen or 30 µg of >18 kDa; 1 mL preimmune bleed; 7–10 mL final bleed; animal culled)

- Chicken and rabbit (>3 mo old; 200 µg of <18 kDa antigen or 100 µg of >18 kDa; 2 mL preimmune bleed; 50–70 mL final bleed; animal culled)

  Polyclonal antibodies transmitted to chicken eggs are five to six times more concentrated than those in the blood but are difficult to extract and not widely used despite their advantages (Michael et al. 2010).

- Goat and sheep (6–7 mo old; 400 µg of <18 kDa antigen or 200 µg of >18 kDa; 2 mL preimmune bleed; 1 L final bleed; animal generally culled)

- Llama and horse (>18 mo–2 yr old; 400 µg of <18 kDa antigen or 200 µg of >18 kDa; 2 mL preimmune bleed; 1 L final bleed; animal can be reused)

  Delahaut (2017) provides a concise and useful guide for immunizations.

Expressing the Immunogen

6. Clone the identified target sequence into a suitable vector, containing the appropriate promoter and C- or N-terminal fusion tags for enhanced expression and affinity purification.

7. Verify the sequence by Sanger sequencing.

8. Transfect HEK293T on a small scale using the JetPRIME reagent according to the manufacturer's instructions.

9. Confirm transfection efficiency by western blot and proceed with larger-scale expression.

10. Seed HEK293T cells at 7.5×10^5 cells/mL in 50 mL of complete DMEM and incubate in a T175 flask overnight at 37°C in a 5% CO_2, 95% air atmosphere.

11. When the cells reach 80% confluence, mix 87.5 µg of plasmid DNA with 2.6 mL of DMEM supplemented with 1× NEAAs and 1 mM L-glutamine at room temperature.

12. Separately, mix 154 µL of 1 mg/mL PEI with 2.6 mL of DMEM containing 1 mM L-glutamine at room temperature.

13. Mix the DNA-containing DMEM and the PEI-containing DMEM and incubate for 10 min at room temperature.

14. Remove the medium from the T175 flask and replace with 45 mL of fresh DMEM containing 2% FBS, 1× NEAAs, and 1 mM L-glutamine.

15. Incubate 10 min at room temperature.

16. Add the DNA/PEI solution to the cells and incubate at 37°C in a 5% CO_2, 95% air atmosphere for 3 d.

    Proceed to Steps 17–20 for harvesting secreted proteins or Steps 21–23 for harvesting intracellular proteins.

Harvesting Expressed Proteins

Secreted Proteins

17. Collect the cell supernatant containing the protein.

18. Centrifuge at 6000g for 15 min at 4°C.

19. Filter the supernatant through a 0.22-μm pore filter.

20. Store filtrate at 4°C.

Intracellular Proteins

21. To collect the cells from the medium, centrifuge at 6000g for 15 min at 4°C, and discard the supernatant.

22. Freeze the pelleted cells and the T175 flask containing the attached cells at −80°C.

23. Use M-PER™ Mammalian Protein Extraction Reagent or equivalent to lyse the cells and isolate the intracellular proteins according to the manufacturer's instructions.

Affinity-Purifying the Immunogen

Protein purification of tagged proteins via affinity chromatography can be performed in a wide variety of manners. Here we describe the use of HisTrap columns on an AKTA Start chromatography system. A similar protocol can be run manually, using gravity filtration, starting at Step 27.

All solutions that pass through the AKTA chromatography system must be filtered with a 0.22-μm filter.

AKTA Start Setup

24. Filter purification solutions A&B with a 0.22-μm filter.

25. Set the AKTA Start on a "manual run" and flush with H_2O at a rate of 5 mL/min for 2 min.

26. Flush all lines with filtered H_2O.

27. Connect the column (here, a 1-mL HisTrap FF column), making sure no air bubbles are introduced.

28. Flush the column with five column volumes of purification solution A at <3 mL/min or at the rate specified by the column manufacturer.

29. Flush only the "elution line" with five column volumes of purification solution B at <3 mL/min or at the rate specified by the column manufacturer.

30. Prime the system by flushing with solution A, avoiding the "elution line."

Sample Loading and Purification

31. Load the filtered sample, allowing it to flow through the column to bind.

32. Wash the column with at least five volumes of purification solution A at <3 mL/min or at the rate specified by the column manufacturer.

33. Set a gradient of 0%–100% solution B over 10 min.

34. Collect 1-mL fractions.

> *A peak on the UV absorbance shows elution of the protein.*

35. Keep the flowthrough.

Cleaning and Storing the AKTA Start

36. Wash the column with purification solution A for 5 min and switch to solution B for 3 min.

37. Flush 20% ethanol through the AKTA Start and the column for 10 min.

38. Store the AKTA Start filled with 20% ethanol.

Cite this protocol as *Cold Spring Harb Protoc*; doi:10.1101/pdb.prot105585

TABLE 1. Antibodies previously developed against *Xenopus* proteins

Target	Host	Type	Laboratory of origin	References	Availability[a]
Cell cycle components, focusing on protein kinases that regulate mitosis (cyclins, Cdc proteins, Atr, Bub, etc.)	Rabbit	Polyclonal	Maller laboratory	Gautier et al. 1988; Gautier et al. 1990; Maller 2012; Qian et al. 2001	EXRC
Cell cycle components (cyclin A, B1, B2, etc.)	Rabbit	Polyclonal	Hunt laboratory	Kobayashi et al. 1991; Minshull et al. 1990; Strausfeld et al. 1994	n/a
Key markers for the kidney, heart, nervous system, and muscle	Mice	Monoclonal	Jones laboratory	Balakrishna et al. 2011; Jones 2003; Lyons et al. 2009; Miller et al. 2011; Naylor et al. 2009; Simrick et al. 2005	EXRC
Cdc6, Orc2, Mcm3, etc.	Rabbit	Polyclonal	Various laboratories	Coleman et al. 1996	n/a
Uhrf1 and Dnmt1	Rabbit	Polyclonal	Lindsay laboratory	Taylor et al. 2013	n/a
Svv1, Svv2, Eiap, and Xiap	Rabbit	Unknown	Yamashita laboratory	Tsuchiya et al. 2005	n/a
Velo1	Goat	Polyclonal	Woodland laboratory	Nijjar and Woodland 2013	EXRC
Ilf3 and Nap1	Sheep	Polyclonal	Guille laboratory	Steer et al. 2003; Cazanove et al. 2008	EXRC

[a]EXRC, European *Xenopus* Resource Centre.

Immunizing the Host

For most laboratories, the actual method for raising the antibody is determined by local ethical regulations controlling the adjuvant used and injection protocols. Often now these steps are outsourced, particularly as the preferred animal in which to raise polyclonal antibodies has changed from rabbits to sheep. For this reason, this protocol does not detail these steps.

39. (Optional) Consider coupling proteins <8–12 kDa to a carrier protein to enhance immunogenicity (Trier et al. 2012).

    For octomeric MAPs and long peptides, this is unnecessary.

40. Determine the need and type of adjuvant, bearing in mind that this is usually regulated by national and local rules (Delahaut 2017).

41. Arrange to immunize the host animal.

    Generally, immunizations are carried out by companies or bioresource units with the appropriate licences in place.

Assessing the Immune Response

42. Proceed to Protocol 6: Assessing the Immune Response When Raising Antibodies for Use in *Xenopus* (Piccinni and Guille 2020).

DISCUSSION

To produce an immunogen, three expression systems are mainly used: in vitro cell-free extracts, *E. coli*, and HEK expression. In vitro expression depends solely on the manufacturer's instructions. These protocols can produce a wide range of soluble products but are costly and usually low yield, precluding production of immunogens for large animals (Delahaut 2017). Expression in *E. coli* produces large amounts of immunogen that have been successfully used for making several polyclonal *Xenopus*-specific antibodies (e.g., Steer et al. 2003; Cazanove et al. 2008). *E. coli*-produced proteins are poorly suited as antigens for monoclonal antibody production because contaminating *E. coli* proteins are likely to generate a stronger immune response than the desired *Xenopus* protein, resulting in the library of antibody-producing clones being dominated by those recognizing the *E. coli* contaminants (Joosten et al. 2003). More recently, expression in mammalian cells has been used to produce immunogens (e.g., Nettleship et al. 2015), and although more demanding and expensive than using *E. coli*, this method can produce sufficient protein for immunizing any host. Further, mammalian contaminants are not as immunogenic as those found in bacteria. As such, expression in HEK cells is described above.

Many antibodies, such as those in Table 1 and at http://www.xenbase.org/reagents/antibody.do, have already been developed specifically against *Xenopus* proteins and will be critical for understanding protein function and control in the *Xenopus* model. Some have been deposited and are available to the community through the European *Xenopus* Resource Centre (EXRC).

RECIPES

Purification Solution A

50 mM Tris-HCl, pH 7.5
500 mM NaCl
20 mM Imidazole

Store for up to 1 yr at room temperature.

Purification Solution B

50 mM Tris-HCl, pH 7.5
500 mM NaCl
500 mM Imidazole

Store for up to 1 yr at room temperature.

REFERENCES

Balakrishna S, Saravia J, Thevenot P, Ahlert T, Lominiki S, Dellinger B, Cormier SA. 2011. Environmentally persistent free radicals induce airway hyperresponsiveness in neonatal rat lungs. *Part Fibre Toxicol* **8:** 1693–1704. doi:10.1186/1743-8977-8-11

Bird LE, Rada H, Flanagan J, Diprose JM, Gilbert RJ, Owens RJ. 2014. Application of In-Fusion™ cloning for the parallel construction of *E. coli* expression vectors. *Methods Mol Biol* **1116:** 209–234. doi:10.1007/978-1-62703-764-8_15

Cazanove O, Batut J, Scarlett G, Mumford K, Elgar S, Thresh S, Neant I, Moreau M, Guille M. 2008. Methylation of Xilf3 by Xprmt1b alters its DNA, but not RNA, binding activity. *Biochemistry* **47:** 8350–8357. doi:10.1021/bi7008486

Coleman TR, Carpenter PB, Dunphy WG. 1996. The *Xenopus* Cdc6 protein is essential for the initiation of a single round of DNA replication in cell-free extracts. *Cell* **87:** 53–63. doi:10.1016/S0092-8674(00)81322-7

Delahaut P. 2017. Immunisation—choice of host, adjuvants and boosting schedules with emphasis on polyclonal antibody production. *Methods* **116:** 4–11. doi:10.1016/j.ymeth.2017.01.002

Gautier J, Minshull J, Lohka M, Glotzer M, Hunt T, Maller JL. 1990. Cyclin is a component of maturation-promoting factor from *Xenopus*. *Cell* **60:** 487–494. doi:10.1016/0092-8674(90)90599-A

Gautier J, Norbury C, Lohka MI, Nurse P, Maller JL. 1988. Purified maturation-promoting factor contains the product of a *Xenopus* homolog of the fission yeast cell cycle control gene *cdc2*⁺. *Cell* **54:** 433–439. doi:10.1016/0092-8674(88)90206-1

Jones EA. 2003. Molecular control of pronephric development: an overview. *Kidney* 93–118. doi:10.1016/B978-012722441-1/50010-5.

Joosten V, Lokman C, van den Hondel CA, Punt PJ. 2003. The production of antibody fragments and antibody fusion proteins by yeasts and filamentous fungi. *Microb Cell Fact* **2:** 1. doi:10.1186/1475-2859-2-1

Kobayashi H, Minshull J, Ford C, Golsteyn R, Poon R, Hunt T. 1991. On the synthesis and destruction of A- and B-type cyclins during oogenesis and meiotic maturation in *Xenopus laevis*. *J. Cell Biol* **114:** 755–765. doi:10.1083/jcb.114.4.755

Lyons JP, Miller RK, Zhou X, Weidinger G, Deroo T, Denayer T, Park JI, Ji H, Hong JY, Li A, et al. 2009. Requirement of Wnt/β-catenin signaling in pronephric kidney development. *Mech Dev* **126:** 142–159. doi:10.1016/j.mod.2008.11.007

Maller JL. 2012. Pioneering the *Xenopus* oocyte and egg extract system. *J Biol Chem* **287:** 21640–21653. doi:10.1074/jbc.X112.371161

Michael A, Meenatchisundaram S, Parameswari G, Subbraj T, Selvakumaran R, Ramalingam S. 2010. Chicken egg yolk antibodies (IgY) as an alternative to mammalian antibodies. *Indian J Sci Technol* **3:** 468–474.

Miller RK, Canny SG, Jang CW, Cho K, Ji H, Wagner DS, Jones EA, Habas R, McCrea PD. 2011. Pronephric tubulogenesis requires Daam1-mediated planar cell polarity signaling. *J Am Soc Nephrol* **22:** 1654–1664. doi:10.1681/ASN.2010101086

Minshull J, Golsteyn R, Hill CS, Hunt T. 1990. The A- and B-type cyclin associated *cdc2* kinases in *Xenopus* turn on and off at different times in the cell cycle. *EMBO J* **9:** 2865–2875. doi:10.1002/j.1460-2075.1990.tb07476.x

Naylor RW, Collins RJ, Philpott A, Jones EA. 2009. Normal levels of p27^Xic1 are necessary for somite segmentation and determining pronephric organ size. *Organogenesis* **5:** 201–210. doi:10.4161/org.5.4.9973

Nettleship JE, Watson PJ, Rahman-Huq N, Fairall L, Posner MG, Upadhyay A, Reddivari Y, Chamberlain JM, Kolstoe SE, Bagby S, et al. 2015. Transient expression in HEK 293 cells: an alternative to *E. coli* for the production of secreted and intracellular mammalian proteins. *Methods Mol Biol* **1258:** 209–222. doi:10.1007/978-1-4939-2205-5_11

Nijjar S, Woodland HR. 2013. Protein interactions in *Xenopus* germ plasm RNP particles. *PLoS One* **8:** e80077. doi:10.1371/journal.pone.0080077

Piccinni MZ, Guille MJ. 2020. Assessing the immune response when raising antibodies for use in *Xenopus*. *Cold Spring Harb Protoc* doi:10.1101/pdb.prot105593

Qian Y-W, Erikson E, Taieb FE, Maller JL. 2001. The polo-like kinase Plx1 is required for activation of the phosphatase Cdc25C and cyclin B-Cdc2 in *Xenopus* oocytes. *Mol Biol Cell* **12:** 1791–1799. doi:10.1091/mbc.12.6.1791

Cite this protocol as *Cold Spring Harb Protoc*; doi:10.1101/pdb.prot105585

Simrick S, Massé K, Jones EA. 2005. Developmental expression of *Pod1* in *Xenopus laevis*. *Int J Dev Biol* **49**: 59–63. doi:10.1387/ijdb.051982ss

Steer WM, Abu-Daya A, Brickwood SJ, Mumford KL, Jordanaires N, Mitchell J, Robinson C, Thorne AW, Guille MJ. 2003. *Xenopus* nucleosome assembly protein becomes tissue-restricted during development and can alter the expression of specific genes. *Mech Dev* **120**: 1045–1057. doi:10.1016/S0925-4773(03)00176-X

Strausfeld UP, Howell M, Rempel R, Maller JL, Hunt T, Blow JJ. 1994. Cip1 blocks the initiation of DNA replication in *Xenopus* extracts by inhibition of cyclin-dependent kinases. *Curr Biol* **4**: 876–883. doi:10.1016/S0960-9822(00)00196-2

Taylor EM, Bonsu NM, Price RJ, Lindsay HD. 2013. Depletion of Uhrf1 inhibits chromosomal DNA replication in *Xenopus* egg extracts. *Nucleic Acids Res* **41**: 7725–7737. doi:10.1093/nar/gkt549

Trier NH, Hansen PR, Houen G. 2012. Production and characterization of peptide antibodies. *Methods* **56**: 136–144. doi:10.1016/j.ymeth.2011.12.001

Tsuchiya Y, Murai S, Yamashita S. 2005. Apoptosis-inhibiting activities of BIR family proteins in *Xenopus* egg extracts. *FEBS J* **272**: 2237–2250. doi:10.1111/j.1742-4658.2005.04648.x

Assessing the Immune Response When Raising Antibodies for Use in *Xenopus*

Maya Z. Piccinni and Matthew J. Guille[1]

European Xenopus Resource Centre, Institute of Biological and Biomedical Sciences, University of Portsmouth, Portsmouth, Hampshire PO1 2DY, United Kingdom

Frog-specific antibodies usually must be raised for work in *Xenopus*. Selecting a host animal whose immune system will respond to a target antigen with an antibody response is essential to obtaining high-quality antibodies. To determine whether an immunized animal has produced antibodies against an antigen, western blotting using *Xenopus* embryo or egg extract as the protein source can be performed as described here. When a protein of the expected size is detected by western blotting in the immune sera but not the preimmune sera, the antibody has been successfully raised.

MATERIALS

It is essential that you consult the appropriate Material Safety Data Sheets and your institution's Environmental Health and Safety Office for proper handling of equipment and hazardous materials used in this protocol.

RECIPES: Please see the end of this protocol for recipes indicated by <R>. Additional recipes can be found online at http://cshprotocols.cshlp.org/site/recipes.

Reagents

1,1,1-trichlorotrifluorethane (optional; see Step 4)
Antibodies (select one of the following; see Steps 13–14)

Primary antibody (HRP-conjugated)

Primary antibody (unconjugated) and species-appropriate HRP-conjugated secondary antibody

See Protocol 5: Raising Antibodies for Use in Xenopus *(Piccinni and Guille 2020a). To reduce background on the western blot, the antibodies can be purified as described in Protocol 8: Purifying Antibodies Raised against* Xenopus *Peptides (Piccinni and Guille 2020b).*

Blocking buffer (select one of the following based on empirical testing)

1% Blocking Reagent (Roche 11096176001) in maleic acid buffer (100 mM maleic acid, 150 mM NaCl)

5% BSA (Fisher Scientific BPE9701) in TBSTw

5% (w/v) nonfat dry milk (Marvel) in TBSTw

Some blocking buffers work better than others for certain antibodies.

dH$_2$O

[1]Correspondence: matthew.guille@port.ac.uk

Detection buffers

ECLI buffer for chemiluminescent detection of HRP-conjugated antibodies <R>

ECLII buffer for chemiluminescent detection of HRP-conjugated antibodies <R>

Alternatively, use Clarity Western ECL Substrate A and Substrate B (Bio-Rad, 170-5061) instead of the ECL buffers.

Embryo buffer for testing antibodies (EB) <R>

Gel-loading solution for protein samples (select one of the following; see Step 6)

SDS protein sample buffer (2×) <R>

2× β-mercaptoethanol sample buffer <R>

Phosphate buffered saline (PBS; Sigma-Aldrich P4417-50TAB [1 tablet/200 mL])

Protease inhibitor cocktail tablets, cOmplete EDTA-free (Roche)

SDS-PAGE gel and running buffer (see Step 7)

Tris-buffered saline with detergent (TBSTw) for western blotting <R>

Western blotting transfer buffer <R>

Xenopus embryos (sets of 10 in microcentrifuge tubes)

Refer to advice at Xenbase.org for collecting the embryos, the correct stage of development to use for the test can be predicted using available stage-specific transcriptomes (Owens et al. 2016). Caution is needed, however, because differences between protein and mRNA levels exist in embryos (Peshkin et al. 2015; Sun et al. 2014). A positive control protein, often the immunogen, is thus recommended on the blot, and using extract from multiple developmental stages can also mitigate uncertain expression patterns.

Equipment

Heating block at 95°C

ImageQuant LAS4000 or equivalent

Microcentrifuge (at room temperature or at 4°C; see Step 4)

Microcentrifuge tubes

Micropipette with 200-µL tip (see Step 3)

Nitrocellulose membrane

Rocker

SDS-PAGE equipment (see Step 7)

Vortex

Western blotting equipment (see Steps 8–10)

Whatman 3 MM paper or equivalent

METHOD

Detailed western blotting protocols are available (e.g., Mahmood and Yang 2012). The advantages of including a positive control antigen and purifying antibodies are clearly seen in Figure 1. It is difficult to identify the additional band in the immune serum before purification but nonetheless the response is clear (compare the antigen lanes in the preimmune and immune samples).

Embryo Extract Preparation

1. Remove all buffer from each microcentrifuge tube containing a set of 10 embryos.

   As a general rule of thumb, use two embryos per lane of the gel.

2. Add 150 µL of EB supplemented with cOmplete EDTA-free protease inhibitor cocktail tablets to each set of 10 embryos.

3. Manually disrupt the embryos by pipetting them up and down with 200-µL tip until the solution is homogeneous.

   At this point, the homogenate can be stored at −80°C if necessary.

4. Extract the proteins with or without 1,1,1-trichlorotrifluorethane as follows. Keep in mind that yolk solubility may be increased if 1,1,1-trichlorotrifluorethane extraction is omitted.

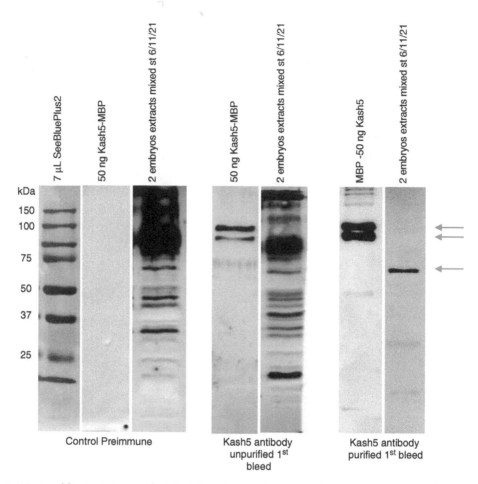

FIGURE 1. Western blot to test a purified Kash5 antibody. The three gels were run and treated under the same conditions, the only variable being the primary antibody used. Proteins from embryos at stages 6, 11, and 21 were isolated and run using extracts from two embryos per lane. The positive control, an MBP-Kash5 protein expressed in *E.coli* (calculated MW approximately 90 kDa) runs as two major bands at 90–95 kDa (*upper* arrows) while the native Kash5 from embryo extract runs as a single band at 59 kDa (*lower* arrow). The latter is only detected as a unique, clear band once the immune serum has been purified using the unpurified and purified forms of the antibody. MBP-Kash5 is not seen when the preimmune serum was used. The high background levels observed before purification of the antibody were minimized and Kash5 (*bottom* arrow at 59 kDa) was detected as a single protein of the correct size once the antibody was purified.

Extraction with 1,1,1-Trichlorotrifluorethane

i. Add an equal volume of 1,1,1-trichlorotrifluorethane, and vortex very thoroughly.

ii. Centrifuge at 16,000*g* in a microcentrifuge for 5 min at room temperature.

Extraction without 1,1,1-Trichlorotrifluorethane

iii. Centrifuge the lysate at 16000*g* for 30 min at 4°C.

iv. Aspirate the lipid layer from the top of the clear lysate.

5. Pipette the clear top layer (~200 µL), which contains yolk-depleted protein extract, and place it in a fresh tube.

6. Add one of the gel-loading solutions to a final concentration of 1×. Incubate for 3 min at 95°C.

7. Immediately separate the proteins by SDS-PAGE using any standard procedure.

Alternatively, store the protein extracts at −80°C until use.

   Cite this protocol as *Cold Spring Harb Protoc*; doi:10.1101/pdb.prot105593

Membrane Transfer

8. Insert the following items into the western blot holder in order:

   i. a fiber pad,

   ii. two Whatman 3MM paper pads,

   iii. the gel,

   iv. the nitrocellulose membrane,

   v. two Whatman 3MM paper pads covering the membrane, and

   vi. another fiber pad.

9. Place the "sandwich" into the tank, ensuring that the gel is on the negative side, and fill the tank with western transfer buffer.

10. Transfer proteins at 30 mA overnight or at 300 mA for 2 h.

Membrane Blocking

11. Wash the nitrocellulose membrane with dH_2O for 5 min at room temperature.

12. Incubate the membrane in 25 mL of the blocking buffer of choice for 1 h at room temperature or overnight at 4°C.

Antibody Incubations

13. Replace the blocking buffer with fresh blocking buffer containing the primary antibody at the appropriate dilution. Incubate with gentle agitation for 1–2 h at room temperature or overnight at 4°C.

    Serial dilutions of the new antibody can be used to determine the appropriate concentration.

14. Wash the membrane as follows depending on if the primary antibody is unconjugated or conjugated to a label. Conduct all membrane washing steps at room temperature.

    For Unconjugated Primary Antibodies

    i. Wash the membrane three times for 5 min each with 10–15 mL of TBSTw.

    ii. Incubate the membrane with fresh blocking buffer containing the species-appropriate HRP-conjugated secondary antibody according to the manufacturer's recommended dilution with gentle agitation for 1 h.

       For example, a 1:2000 dilution is used for the Anti-Rabbit IgG-Peroxidase conjugated secondary antibody (Sigma-Aldrich A0545).

    iii. Wash the membrane twice for 5 min each with 15 mL of blocking solution.

    iv. Wash once for 5 min with TBSTw.

    v. Rinse once with PBS for 5 min.

    For HRP-Conjugated Primary Antibodies

    vi. Wash the membrane twice for 5 min each with 15 mL of blocking solution.

    vii. Wash the membrane once for 5 min with TBSTw.

    viii. Rinse once with PBS for 5 min.

Protein Detection

15. Incubate the membrane at room temperature for 1 min with 10 mL of ECLI buffer and 10 mL of ECLII buffer.

    Alternatively, incubate the membrane at room temperature for 5 min with 7 mL Clarity Western ECL Substrate Solution A and 7 mL of Clarity Western ECL Substrate Solution B.

16. Visualize using an ImageQuant LAS4000 or equivalent following the manufacturer's instructions, varying the exposure from 1 sec to 10 min as necessary.

    See Troubleshooting.

TROUBLESHOOTING

Problem (Step 16): The proteins were not fully denatured prior to running the gel, causing bands to be seen at twice or three times the size of the monomer on the membrane.

Solutions: The ability of the proteins to denature under particular conditions varies. Both the loading buffer and denaturing temperature can be optimized. The 2× SDS loading dye or 2× β-mercaptoethanol sample buffer are commonly used. Potential temperature treatments include for 3 min at 95°C, 20 min at 55°C, and the combination of 20 min at 55°C followed by for 3 min 95°C.

Problem (Step 16): The antibody concentration was not optimized, causing the western blot to have high background.

Solutions: To obtain a clear and low-background western blot, optimize the antibody concentration as follows.

1. Adjust the gel's comb so that there are only two lanes, a single well for the size marker and a joint, long well containing the sample.

2. Run the gel and transfer as normal.

3. After transfer, cut the membrane containing the sample proteins into strips, leaving the first strip to contain the size marker and some of the sample.

4. Incubate the first strip with the preimmune serum.

5. Incubate the rest of the strips under the same conditions but with varying antibody concentrations.

Problem (Step 16): The antibody stability was poor, resulting in western blot results that were unclear or not reproducible.

Solutions: Avoid freeze-and-thaw cycles. Store the antibody lyophilized at −20°C for long-term storage (years) and at 4°C for short-term storage (weeks/months). Purified antibodies tend to be more stable than the raw serum; therefore, purification of serum is suggested if results with raw serum vary.

DISCUSSION

To determine whether an immunized animal has produced antibodies against an antigen, western blotting (Mahmood and Yang 2012) or an ELISA (Hornbeck 1992) are performed. Further information on the effectiveness of the antibody is then obtained by determining fixation conditions on cultured cells expressing the antigen. The final test of the antibody is by immunohistochemistry on embryos or tissue sections (see Lee et al. 2008). Detected protein size is not provided by the ELISA, so western blotting, which has been made more quantitative with fluorescent secondary antibodies and fluorescence readers, is described here.

Cite this protocol as *Cold Spring Harb Protoc*; doi:10.1101/pdb.prot105593

RECIPES

2× β-Mercaptoethanol Sample Buffer

100 mM Tris-HCl, pH 6.8
2% SDS
4% β-mercaptoethanol
20% glycerol
0.01% bromophenol blue

Store for up to 1 yr at 20°C.

ECLII Buffer for Chemiluminescent Detection of HRP-Conjugated Antibodies

0.021% Hydrogen peroxide
10 mM Tris, pH 8.5

Store for up to 1 yr at 4°C.

ECLI Buffer for Chemiluminescent Detection of HRP-Conjugated Antibodies

2.5 mM luminol
0.9 mM p-coumaric acid
10 mM Tris, pH 8.5

Store for up to 1 yr at 4°C.

Embryo Buffer for Testing Antibodies (EB)

20 mM HEPES (pH 7.9)
2 mM $MgCl_2$
10 mM β-glycerophosphate
2 mM levamisole

Store for up to 3 mo at 20°C.

SDS Protein Sample Buffer (2×)

0.125 M Tris-HCl, pH 6.8
4% SDS
0.15 M DTT
20% glycerol
0.01% bromophenol blue

Store for up to 1 yr at 20°C.

Tris-Buffered Saline with Detergent (TBSTw) for Western Blotting

20 mM Tris
150 mM NaCl
0.1% Tween 20

Store at room temperature for up to 2 mo.

Western Blotting Transfer Buffer

200 mM Tris base
150 mM glycine
0.1% SDS
20% methanol

Make fresh.

REFERENCES

Hornbeck P. 1992. Enzyme-linked immunosorbent assays. *Curr Protoc Immunol* 1: 2.1.1–2.1.22. doi:10.1002/0471142735.im0201s01

Lee C, Kieserman E, Gray RS, Park TJ, Wallingford J. 2008. Whole-mount fluorescence immunocytochemistry on *Xenopus* embryos. *Cold Spring Harb Protoc* doi:10.1101/pdb.prot4957

Mahmood T, Yang P-C. 2012. Western blot: technique, theory, and trouble shooting. *N Am J Med Sci* 4: 429–434. doi:10.4103/1947-2714 .94940

Owens NDL, Blitz IL, Lane MA, Patrushev I, Overton JD, Gilchrist MJ, Cho KWY, Khokha MK. 2016. Measuring absolute RNA copy numbers at high temporal resolution reveals transcriptome kinetics in development. *Cell Rep* 14: 632–647. doi:10.1016/j.celrep.2015.12.050

Peshkin L, Wühr M, Pearl E, Haas W, Freeman RM Jr, Gerhart JC, Klein AM, Horb M, Gygi SP, Kirschner MW. 2015. On the relationship of protein and mRNA dynamics in vertebrate embryonic development. *Dev Cell* 35: 383–394. doi:10.1016/j.devcel.2015.10.010

Piccinni MZ, Guille MJ. 2020a. Raising antibodies for use in *Xenopus*. *Cold Spring Harb Protoc* doi:10.1101/pdb.prot105585

Piccinni MZ, Guille MJ. 2020b. Purifying antibodies raised against *Xenopus* peptides. *Cold Spring Harb Protoc* doi:10.1101/pdb.prot105619

Sun L, Bertke MM, Champion MM, Zhu G, Huber PW, Dovichi NJ. 2014. Quantitative proteomics of *Xenopus laevis* embryos: expression kinetics of nearly 4000 proteins during early development. *Sci Rep* 4: 4365. doi:10.1038/srep04365

Cite this protocol as *Cold Spring Harb Protoc*; doi:10.1101/pdb.prot105593

Confirming Antibody Specificity in *Xenopus*

Sian A. Martin,[1] Suzannah J. Page,[1] Maya Z. Piccinni,[2] and Matthew J. Guille[1,2,3]

[1]Molecular Embryology Laboratory, University of Portsmouth, Portsmouth, Hampshire PO1 2DY, United Kingdom; [2]European Xenopus Resource Centre, Institute of Biological and Biomedical Sciences, University of Portsmouth, Portsmouth, Hampshire PO1 2DY, United Kingdom

Verifying that a new antibody recognizes its target can be difficult. In this protocol, expression of a target protein in *Xenopus* embryos is either knocked down using CRISPR–Cas9 technology (for zygotic proteins) or enhanced by microinjection of a synthetic mRNA (for maternal proteins). Western blotting analysis is then performed. If the antibody recognizes the target protein, the western blot will show a relatively weak band for CRISPR-injected embryos and a relatively strong band for RNA-injected embryos. This represents a straightforward, powerful strategy for confirming antibody specificity in *Xenopus*.

MATERIALS

It is essential that you consult the appropriate Material Safety Data Sheets and your institution's Environmental Health and Safety Office for proper handling of equipment and hazardous materials used in this protocol.

RECIPES: Please see the end of this protocol for recipes indicated by <R>. Additional recipes can be found online at http://cshprotocols.cshlp.org/site/recipes.

Reagents

Agarose gel electrophoresis reagents (i.e., 1%–2% agarose gels, loading dye, running buffer, and ethidium bromide)

Antibody to test

See Protocol 5: Raising Antibodies for Use in Xenopus (Piccinni and Guille 2020).

Cysteine (2% [w/v], pH 7.8; Sigma-Aldrich C7352)

Embryo extraction buffer <R>

Ficoll (3% [w/v], prepared in 1× MBS; Sigma-Aldrich GE17-0300-50)

Freon (1,1,2-trichloro-1,2,2-trifluoroethane; Fisher Scientific T1781)

Modified Barth's saline (MBS) (1×, pH 7.8) <R>

Nuclease-free dH_2O (Fisher Scientific 15303711)

Reagents for CRISPR analysis only (see Steps 1–9 and Steps 25–33)

Cas9 protein (EnGen Cas9 NLS, *Streptococcus pyogenes* [NEB M0646])

DNeasy Blood & Tissue Kit (QIAGEN) or other appropriate kit to prepare genomic DNA from embryos

---

[3]Correspondence: matthew.guille@port.ac.uk

gDNA oligo (100 μM, dissolved in H$_2$O)

MEGAshortscript T7 Transcription Kit (Thermo Fisher Scientific AM1354)

MgCl$_2$ (500 μM)

NEBuffer2 (NEB B7002S)

PCR master mix (2×; see Table 1)

PCR primers designed to amplify 600–800 bp containing the CRISPR target site close to the amplicon's center (for Step 26)

Q5 Hot Start High-Fidelity 2× Master Mix (NEB M0491S)

T7 Endonuclease I (NEB M0302)

Universal 3′ gRNA template (5′ AAAAGCACCGACTCGGTGCCACTTTTTCAAGTTGATAAC GGACTAGCCTTATTTTAACTTGCTATTTCTAGCTCTAAAAC 3′) (100 μM)

Reagents for overexpression analysis only (see Steps 10–17)

Plasmid encoding the target ORF surrounded by the globin UTRs driven by a T7, SP6, or T3 promoter (e.g., pCS2 [Rupp et al. 1994])

Such plasmids are available from the European Xenopus Resource Centre (EXRC), which also holds ORFeomes (representing more than 14000 genes).

Restriction enzyme appropriate to cut plasmid encoding the target ORF (see Step 11)

SuperScript kit (Ambion) for long transcripts

Choose a kit based on which promoter is in the plasmid encoding the protein.

SDS-PAGE reagents (i.e., sample buffer, gel, and running buffer; see Step 35)

Western blotting reagents (see Step 36)

Xenopus eggs, fertilized (see Guille 1999)

If targeting a specific protein with CRISPR, consider your choice of species (see Step 1). If Xenopus tropicalis is available, its rarity of gene duplications makes it preferable to Xenopus laevis for the ~60% of genes that have homeologs in the latter.

Equipment

Access to a sequencer (see Step 33)

Agarose gel electrophoresis apparatus

Benchtop centrifuge

Filter tips for pipettes

Flame-polished glass Pasteur pipettes, with a bore slightly larger than embryos

Glass needles for use in a microinjector

Incubators (18°C for *X. laevis* embryos; 25°C for *X. tropicalis* embryos)

Injection dish (mesh or agarose; prepared in house as described by Guille [1999])

Microinjector

Parafilm

PCR clean-up columns (for overexpression analysis only; see Steps 10–17)

Programs for designing sgRNAs (e.g., CRISPRscan; see Step 1) and TIDE analysis (see Step 33) (for CRISPR analysis only)

SDS-PAGE gel electrophoresis apparatus

SigmaSpin Sequencing Reaction Clean-Up columns (for CRISPR analysis only; see Steps 1–9)

Spectrophotometer capable of measuring microliter volumes

Sterile microcentrifuge tubes

Sterile PCR tubes

Thermocycler

Vortex mixer

Western blot transfer and imaging systems

Cite this protocol as *Cold Spring Harb Protoc*; doi:10.1101/pdb.prot105601

METHOD

Preparation of Single Guide RNA (sgRNA) or Synthetic RNA

To synthesize sgRNA for knocking down a target protein in Xenopus *embryos by CRISPR, begin with Steps 1–9. To prepare synthetic RNA for overexpressing a target protein in* Xenopus *embryos, begin with Steps 10–17. The former strategy is best for zygotic proteins, whereas the latter is best for maternal proteins (see Discussion).*

Design and In Vitro Transcription of sgRNA (for CRISPR Only)

This is best suited to zygotic proteins.

1. Design sgRNAs.

   This can be done using a variety of programs. We have had great success using CRISPRscan (http://www .crisprscan.org) to target the first or second translated exon of the gene encoding the protein against which the antibody was raised. If working in X. laevis, *then both homeologs will almost certainly need to be targeted; it is, however, worth checking that both of the genes are transcribed at significant levels by using the expression tab on Xenbase. We normally design three sgRNAs for each gene because there is variability in their effectiveness.*

2. Prepare the PCR mix by combining reagents as outlined in Table 1. Remove 3 µL before reaction and place on ice.

3. Anneal and extend the sgRNA template (ssDNA) oligos in a thermocycler using the program outlined in Table 1.

4. Run pre- and postreaction (3-µL) samples on a 2% agarose gel using standard procedures.

   Ensure that the postreaction band is much stronger with slightly lower mobility compared to the prereaction band.

5. Perform the in vitro transcription reaction according to the MEGAshortscript T7 Transcription Kit instructions, using 2 µL of the PCR from Step 3. Ensure that the reaction is set up at room temperature and not on ice.

   Because only 2 µL of the PCR is used, the remaining DNA can be stored at −20°C.

6. Remove small molecules from the reaction by spinning through a SigmaSpin Sequencing Reaction Clean-Up column following the manufacturer's instructions.

7. Run 3 µL of sgRNA plus loading buffer on a 2% agarose gel for 15 min.

   A single band should be visible.

8. Quantify the sgRNA by measuring the A_{260} of 1 µL of sgRNA using a suitable spectrophotometer.

9. Store the sgRNA at −80°C in 300-ng aliquots.

   Proceed to Step 18.

TABLE 1. Components and steps to synthesize dsDNA from ssDNA by an annealing-extension reaction

PCR components				
	PCR master mix (2×)	50 µL		
	gDNA oligo (100 µM)	2 µL		
	Universal gRNA template (100 µM)	2 µL		
	$MgCl_2$ (500 µM)	2 µL		
	Nuclease-free dH_2O	44 µL		
	Reaction total	100 µL		
Reaction steps	1 cycle	Initial denaturing	94°C	5 min
	20 cycles	Denaturing	94°C	20 sec
		Annealing	58°C	15 sec
		Extension	68°C	15 sec
	1 cycle	Final extension	68°C	5 min
	1 cycle	Hold	4°C	

Preparation of Synthetic RNA (for Overexpression Analysis Only)

This is best suited to maternal proteins.

10. Prepare plasmid encoding the target ORF surrounded by the globin UTRs driven by a T7, SP6, or T3 promoter (e.g., pCS2 [Rupp et al. 1994]).

11. Linearize 10 μg of plasmid with a unique restriction enzyme that cuts downstream from the synthetic poly(A) site or polyadenylation signal (depending on the plasmid).

12. Purify the DNA using a PCR clean-up column and resuspend it at 1 mg/mL in nuclease-free water.

13. Use this DNA as a template to synthesize RNA with the appropriate Ambion SuperScript kit following the manufacturer's instructions. Ensure that the reaction is set up at room temperature.

14. Remove small molecules from the reaction by spinning through a SigmaSpin sequencing reaction cleanup column following the manufacturer's instructions.

15. Run 1 μL of RNA on a 1% agarose gel.

    A bright band should be visible. If multiple bands are seen, these most often represent conformers of the RNA and can be resolved to a single band in a denaturing gel.

16. Quantify the synthetic RNA by measuring A_{260} of 1 μL of the sample using a suitable spectrophotometer.

17. Store the RNA in single-use 2-μL aliquots at −80°C.

    Proceed to Step 18.

Injection into Fertilized *Xenopus* Eggs

18. Calibrate an injection needle to inject 4 nL of liquid. Keep the needle tip hydrated in a dish of nuclease-free dH_2O when it is not in use.

19. When fertilized *Xenopus* eggs have turned their animal poles uppermost, add a solution of 2% (w/v) cysteine (pH 7.8) prepared in water. Agitate the embryos gently for 5 min at room temperature until they have been dejellied, and then wash them five times at room temperature in 1× MBS.

    Each wash should be of sufficient length to allow the embryos to settle after swirling to mix them with the MBS.

20. Prepare the injection solution.

    #### For CRISPR–Cas9

    i. Prepare the injection mixture by combining 16 μM Cas9 protein, 200 ng sgRNA, and nuclease-free H_2O to 4 μL.

        A second injection mix without sgRNA may be prepared and injected as a control.

    ii. Mix gently, and centrifuge at full speed in a benchtop centrifuge for 1 min at room temperature.

    #### For synthetic RNA

    iii. Prepare 1 μg of synthetic RNA in 4 μL of nuclease-free H_2O.

    iv. Mix gently, and centrifuge at full speed in a benchtop centrifuge for 1 min at room temperature.

21. Place a 1- to 3-μL droplet of injection solution onto a square of Parafilm, and fill the needle from the droplet.

22. Fill an injection dish with 3% (w/v) Ficoll in 1× MBS. Transfer 50–100 embryos to the injection dish using a precut glass pipette.

23. Inject the solution into the animal pole of each embryo.

24. Once injected, transfer the embryos into an incubator at the appropriate temperature and wait 3–4 h before transferring them to 0.1× MBS and removing damaged embryos.

Continue with Step 25 if the CRISPR analysis was performed, or proceed to western blotting (Steps 34–36) if synthetic RNA was injected for overexpression of the protein of interest.

T7 Endonuclease Assay and TIDE Analysis (for CRISPR Only)

The T7 endonuclease assay is used to confirm specific cutting by Cas9, validating the sgRNA used. TIDE analysis gives an approximate efficiency reading for the cutting of a specific site using a specific sgRNA.

T7 Endonuclease Assay

25. Prepare genomic DNA from embryos at stage 12 or later. Use an appropriate kit and follow the manufacturer's instructions. Store the genomic DNA at $-20°C$.

We use QIAGEN's DNeasy Blood & Tissue Kit.

26. Perform PCR by combining 100 ng of genomic DNA, 10 μM of forward primer, 10 μM of reverse primer, 25 μL of Q5 Hot Start High-Fidelity 2× Master Mix (which includes enzyme), and nuclease-free dH_2O to 50 μL. Include a reaction with DNA from an uninjected embryo as a control. Amplify using a basic amplification program with primer-specific annealing temperatures.

27. Run 3 μL of the amplicon mixed with 1 μL of loading solution on a 1% agarose gel.

28. Use spectrophotometry to measure the amplicon concentration.

29. Combine 2 μL of NEBuffer 2, ∼200 ng of amplicon, and nuclease-free dH_2O for a final volume of 19 μL.

30. Denature the amplicon for 5 min at 95°C. Reanneal by cooling to 25°C at a rate of 0.1°C sec^{-1} in a thermocycler, and then hold at 4°C.

31. Add 1 μL of T7 endonuclease I and incubate for 15 min at 37°C.

32. Visualize the products of the digestion on a 1.5% agarose gel.

T7 endonuclease will only cut at indels. Successful cutting will show two bands on a gel (or, if the cut site is in the center of the amplicon, there will appear to be one band at one-half the size of the amplicon).

TIDE Analysis

33. Sequence the amplicon obtained from Step 26 and analyze using TIDE, which uses algorithms to distinguish the different sequences that occur after the CRIPSR cut site to predict where and how often the main indels have occurred (https://tide.deskgen.com/).

Once cut site digestion has been achieved at a good efficiency rate, proceed to western blot experiments (Steps 34–36) to test antibodies.

Western Blot Analysis

To choose which developmental stages to target for the western analysis, use the mRNA expression information on Xenbase as a guide. Although it is clear that translational control is used in early development, using stages at which mRNA levels are highest is the best starting point. This method is described in detail in Guille and Robinson (1999).

34. Prepare a crude protein extract.

   i. For each experimental condition, gather 25 embryos.

   ii. Add 250 μL of embryo extraction buffer, and pipette up and down with a 1-mL pipette until the homogenate is uniform.

   iii. Add an equal amount of Freon, and vortex for 30 sec.

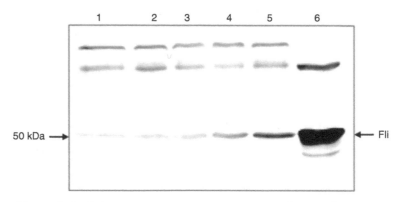

FIGURE 1. Western blot analysis of Fli1 protein overexpression in *Xenopus laevis* embryos. One-cell-stage *Xenopus laevis* embryos were injected with either H_2O (as a control; lane *1*) or increasing amounts of *fli1* mRNA (lanes *2–5*) and then allowed to develop until stage 17. After the embryos were homogenized, the proteins were Freon-extracted, separated by SDS-PAGE, and transferred to a nitrocellulose membrane. The western blot was probed with a 1/500 dilution of Fli1 anti-serum. After removal of the primary antibody, the blot was incubated with a 1/5000 dilution of donkey anti-sheep secondary antibody conjugated to horseradish peroxidase. The blot was developed using ECL Prime Western Blotting Detection Reagent (GE Healthcare). The band corresponding to Fli1 protein increases in intensity with increasing *fli1* mRNA injection amounts. The last lane (lane 6) is the protein from a *fli1* in vitro translation (Promega), showing that the Fli1 protein produced by in vitro translation is the same size as Fli1 protein in *Xenopus laevis* embryos.

    iv. Centrifuge the sample at 21,000*g* for 3 min at 4°C.

    v. Collect the upper (protein) phase.

35. Separate proteins by SDS-PAGE, loading 0.5–1 embryo equivalent of crude protein extract per well in the SDS-PAGE sample buffer.

36. Follow standard western blotting protocols to test the specificity of the antibody of interest.

> *Noninjected embryo samples should produce a band at the size expected for your protein of interest. If the antibody is detecting the target protein, CRISPR-injected embryos will show a weaker band, and the RNA-injected embryos will show a stronger band. For an example of the latter, see Figure 1.*

DISCUSSION

Once an antibody has been raised against a target protein (see Protocol 5: Raising Antibodies for Use in *Xenopus* [Piccinni and Guille 2020]), various strategies can be used to verify that the new antibody recognizes its target. Sometimes using immunoprecipitation, tryptic digest, and mass spectroscopy of the target protein is the only convincing approach (Persson et al. 2017). In a well-established model like *Xenopus*, however, inducing loss or gain of function of the target protein in embryos followed by western blotting is a straightforward, powerful alternative strategy. Target protein can be knocked down using either CRISPR–Cas9 technology (Nakayama et al. 2014), as described here, or an anti-sense morpholino oligonucleotide (AMO; Heasman et al. 2000). Both have good penetrance when the target is not maternal. CRISPR–Cas9 is cheaper than AMO, and it is straightforward to confirm that insertion or deletion of sequence at the target site has occurred by using sequencing or T7 endonuclease. However, for AMOs targeting translation, there is no such simple positive control. For these reasons, knockdown by CRISPR–Cas9 is described here. For maternal proteins, microinjection of a synthetic mRNA encoding the target to increase its expression is preferable (Guille 1999; see Fig. 1).

Cite this protocol as *Cold Spring Harb Protoc*; doi:10.1101/pdb.prot105601

RECIPES

Embryo Extraction Buffer

10 mM HEPES (pH 8.5)
2 mM MgCl$_2$
1 mM EDTA
10 mM β-glycerophosphate
1 mM dithiothreitol (DTT)

Store for up to 1 wk at 4°C.

Modified Barth's Saline (MBS) (1×, pH 7.8)

Reagent	Quantity (for 1 L)	Final concentration (1×)
NaCl	5.143 g	88 mM
KCl	0.075 g	1 mM
MgSO$_4$	0.120 g	1 mM
HEPES	1.192 g	5 mM
NaHCO$_3$	0.210 g	2.5 mM
CaCl$_2$, dihydrate	0.103 g	0.7 mM
H$_2$O	to 1 L	

Adjust the pH to 7.8 with 10 M NaOH and sterilize by autoclaving. Store at room temperature indefinitely.

REFERENCES

Guille M. 1999. Microinjection into *Xenopus* oocytes and embryos. *Methods Mol Biol* **127**: 111–123. doi:10.1385/1-59259-678-9:111

Guille M, Robinson C. 1999. Immunohistochemistry of *Xenopus* embryos. *Methods Mol Biol* **127**: 89–97. doi:10.1385/1-59259-678-9:89

Heasman J, Kofron M, Wylie C. 2000. β-catenin signaling activity dissected in the early *Xenopus* embryo: a novel antisense approach. *Dev Biol* **222**: 124–134. doi:10.1006/dbio.2000.9720

Nakayama T, Blitz I, Fish M, Odeleye A, Manohar S, Cho K, Grainger R. 2014. Cas9-based genome editing in *Xenopus tropicalis*. *Methods Enzymol* **546**: 355–375. doi:10.1016/B978-0-12-801185-0.00017-9

Persson H, Preger C, Marcon E, Lengqvist J, Gräslund S. 2017. Antibody validation by immunoprecipitation followed by mass spectrometry analysis. *Methods Mol Biol* **1575**: 175–187. doi:10.1007/978-1-4939-6857-2_10

Piccinni MZ, Guille MJ. 2020. Raising antibodies for use in *Xenopus*. *Cold Spring Harb Protoc* doi:10.1101/pdb.prot105585.

Rupp RA, Snider L, Weintraub H. 1994. *Xenopus* embryos regulate the nuclear localization of XMyoD. *Genes Dev* **8**: 1311–1323. doi:10.1101/gad.8.11.1311

Purifying Antibodies Raised against *Xenopus* Peptides

Maya Z. Piccinni and Matthew J. Guille[1]

European Xenopus Resource Centre, Institute of Biological and Biomedical Sciences, University of Portsmouth, Portsmouth, Hampshire PO1 2DY, United Kingdom

Antibody production for work in *Xenopus* involves the immunization of a host with an antigen, usually a *Xenopus* protein or peptide alien to the host. The antibody-containing serum, normally returned to the investigator by the company/bioresource unit where it was raised, is comprised of all proteins not used in blood clotting (coagulation) and all the electrolytes, antibodies, antigens, hormones, and any exogenous substances, such as drugs and microorganisms, that were in the blood. It is often necessary to separate the target antibody from the rest of the serum components to minimize nonspecific protein–antibody interactions in downstream applications (e.g., when performing western blotting). Most antibody production companies provide a column containing the peptide coupled to glass beads. A purification procedure for using this type of column (i.e., one that is based on controlled-pore glass beads) is described here.

MATERIALS

It is essential that you consult the appropriate Material Safety Data Sheets and your institution's Environmental Health and Safety Office for proper handling of equipment and hazardous materials used in this protocol.

Reagents

Antisera (see Protocol 5: Raising Antibodies for Use in *Xenopus* [Piccinni and Guille 2020a])
Caprylic acid (100%)
dH$_2$O
Ethanol (100%)
Guanidinium chloride (6 M)
HCl (10 mM)
NaCl (0.5 M in 1× PBS)
PBS-azide (0.02% sodium azide in 1× PBS)
Phosphate buffered saline (PBS; Sigma-Aldrich P4417-50TAB [1 tablet/200 mL])
Potassium thiocyanate (3.5 M)
Sodium acetate (60 mM, pH 4.0)
Sodium bicarbonate (10 mM)

Equipment

Beakers and magnetic stirrers
Columns containing peptides coupled to glass beads

---

[1]Correspondence: matthew.guille@port.ac.uk

The column used for manufacturing the antigen is supplied to the user in the case of multiple antigenic peptides (MAPs) or can be made at the user's request for normal peptides. See Discussion.

Dialysis tubing closure clips (one for each end)
G-25 column (1.5 × 75 cm; stored in PBS-azide)
Liquid nitrogen
Lyophilizer
Tabletop centrifuge
UV spectrometer
Visking tubing (18/32 inch diameter; preboiled and stored in 70% ethanol)

METHOD

Antibody purification is usually carried out in three series of steps. First, the total IgG in the serum is isolated by caprylic acid extraction. Second, specific IgGs are isolated by affinity purification. Last, the affinity-purified antibodies are desalted on a G25 column.

Caprylic Acid Extraction

1. Mix 20 mL of 60 mM sodium acetate (pH 4.0) with 10 mL of antiserum in a beaker at room temperature.

2. Add 750 µL of 100% caprylic acid to the mixture (for a final concentration of 2.5%) while stirring.

3. Leave the mixture stirring until it turns white (20–30 min), and then centrifuge at 5100g for 20 min at room temperature.

4. Transfer the supernatant into a fresh tube, and assay the total IgG present by UV spectroscopy.

   [One A_{275} unit is 0.714 mg/mL.]

Dialysis

5. Rinse 20–25 cm of preboiled 18/32 inch diameter Visking tubing with distilled H_2O for every 10 mL of starting material.

6. Clip one end of the tubing, and pipette the cleared supernatant through the other end, before securing the top with a second clip.

7. Place the tubing containing the IgG into 1 L of 1× PBS, and stir gently overnight at 4°C.

8. On the following day, measure the IgG concentration of the solution inside of the tubing using UV spectroscopy (A_{275}).

   Use the sample immediately for affinity purification or store it at −20°C.

Affinity Column Purification

The affinity column is prepared and run at room temperature by gravity filtration.

Preparing the Column

Before using the column for the first time, prepare it as described in Steps 10–15. This intensive cleaning procedure should also be used if the column clogs or stops binding IgG.

9. Equilibrate the column with 5 mL of 1× PBS.

10. Wash the column with five bead volumes of 6 M guanidinium chloride.

11. Wash the column with five bead volumes of 10 mM HCl.

12. Wash the column with five bead volumes of 100% ethanol.

13. Wash the column with five bead volumes of 1× PBS.

14. Store the column at 4°C in PBS-azide, capped at both ends to avoid drying and inhibit bacterial growth.

Purifying the Antibody

15. Before use, equilibrate the column by washing it with three column volumes of 1× PBS.

16. Pass the IgG (sample from Step 8) through the column three times.

17. Rinse the column with 10 bead volumes of 1× PBS.

18. Rinse the column with 10 bead volumes of 0.5 M NaCl in 1× PBS.

19. Finally, rinse the column with 10 bead volumes of 1× PBS.

Eluting the Antibody

20. Add 3.5 M potassium thiocyanate directly onto the beads 1 mL at a time, for a total of 6 mL.

Immediately desalt the eluted antibody by passing it through a G-25 column (see Step 25).

Cleaning the Affinity Column

21. Following the antibody elution, wash the affinity column with 10 bead volumes of 10 mM HCl.

22. Wash the column with 10 bead volumes of 1× PBS.

23. Store the column in PBS-azide at 4°C.

G-25 Column Purification

Desalting the Antibodies

The column separates the IgG from the salt, allowing the IgG to elute first.

24. To equilibrate the column, pass at least two column volumes of 10 mM sodium bicarbonate through the G-25 column.

25. Load the eluted antibody from Step 21 onto the equilibrated column at a flow rate of 2 mL/min.

26. Wash the G-25 column with 10 mM sodium bicarbonate until a peak in UV absorbance is visible, which corresponds to the antibody being eluted.

27. Allow the G-25 column to run with 10 mM sodium bicarbonate until a second UV absorbance peak is recorded, corresponding to the salt being washed off.

Figure 1 shows a representative elution graph.

28. Wash the column by passing an extra five bead volumes of 10 mM sodium bicarbonate through it.

29. Run two bead volumes of PBS-azide although the column, and store it in this solution at room temperature.

Quantifying and Storing the Collected Fractions

30. Assay the fractions collected by UV spectroscopy (275 nm) to identify those containing the IgG and to measure the IgG concentration.

One A_{275} unit is 0.714 mg/mL. The expected yield is 1–6 mg of antibody for every 10 mL of starting material. See Troubleshooting.

31. Pool fractions that have a considerable amount of antibody, and aliquot into 100-mg lots.

32. Snap-freeze the aliquots in liquid nitrogen, and lyophilize overnight (Ó'Fágáin 2004).

Cite this protocol as *Cold Spring Harb Protoc*; doi:10.1101/pdb.prot105619

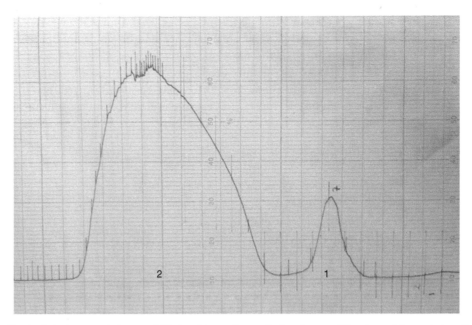

FIGURE 1. Typical elution graph from a G25 desalting column. Following equilibration of the column, the caprylic acid-treated serum was loaded onto the G25 column, in this case at a 2 mL/min rate. The column separated the IgG, which was eluted first (1) from the salt (2). The eluted antibody was quantified by UV spectroscopy (at 275 nm).

33. Store the lyophilized antibodies at −20°C until further use.

> *The lyophilized antibodies will last for many years. They have many applications; for example, they can be used in the associated western blotting protocol that describes how to test the immune response of a host animal during antibody production (see Protocol 6: Assessing the Immune Response When Raising Antibodies for Use in Xenopus [Piccinni and Guille 2020b]).*

TROUBLESHOOTING

Problem (Step 31): The yield of antibody is >6 mg.

Solutions: A yield of >6 mg may be cause for concern, as there could be contamination from other proteins binding to the column. If this occurs, then proceed as follows.

1. Store either the flowthrough or the bound fraction containing your IgG on ice.

2. Wash the column as described in Steps 9–13.

3. Rerun the fraction containing your IgG on the newly washed column.

Problem (Step 31): The yield of antibody is <1 mg.

Solutions: Should the IgG fail to bind to the affinity column, then a yield <1 mg would be expected. If this occurs, then wash the column as described in Steps 9–13.

DISCUSSION

The exact approach taken to purify antibodies varies only slightly depending on the type of immunogen, and hence column, used. Because most antibody production companies provide columns containing peptides coupled to glass beads, an antibody purification procedure for using this type of column (i.e., one that is based on controlled-pore glass beads) is described here. The convenience of having very robust columns based around controlled-pore glass beads is an advantage of using peptide antigens; however, the disadvantages of this approach are the risk of losing specificity due to the antigen's small size and the possibility that the peptide adopts conformations that do not resemble the conformation of the native protein (Trier et al. 2012; Lee et al. 2016). For larger proteins, the immunogen can be immobilized on a wide range of column matrices specifically designed for the purpose (e.g., beads carrying small organic compounds or inorganic metals [Hermanson 2013]); however, doing so does rely on sufficient immunogen having been produced initially.

REFERENCES

Hermanson GT. 2013. *Bioconjugate techniques*, 3rd ed. Elsevier, Amsterdam.

Lee B-S, Huang J-S, Jayathilaka LP, Lee J, Gupta S. 2016. Antibody production with synthetic peptides. *Methods Mol Biol* **1474:** 25–47.

Ó'Fágáin C. 2004. Lyophilization of proteins. In *Protein purification protocols. Methods in molecular biology* (ed. Cutler P.), Vol. 244. Humana Press, New York.

Piccinni MZ, Guille MJ. 2020a. Raising antibodies for use in *Xenopus. Cold Spring Harb Protoc* doi:10.1101/pdb.prot105585.

Piccinni MZ, Guille MJ. 2020b. Assessing the immune response when raising antibodies for use in *Xenopus. Cold Spring Harb Protoc* doi:10.1101/pdb.prot105593.

Trier NH, Hansen PR, Houen G. 2012. Production and characterization of peptide antibodies. *Methods* **56:** 136–144.

Cite this protocol as *Cold Spring Harb Protoc;* doi:10.1101/pdb.prot105619

SECTION 7: *Xenopus* Extract Systems

The Use of Cell-Free *Xenopus* Extracts to Investigate Cytoplasmic Events

Romain Gibeaux[1,2] and Rebecca Heald[1,3]

[1]*Department of Molecular and Cell Biology, University of California, Berkeley, Berkeley, California 94720-3200*

Experiments using cytoplasmic extracts prepared from *Xenopus* eggs have made important contributions to our understanding of the cell cycle, the cytoskeleton, and cytoplasmic membrane systems. Here we introduce the extract system and describe methods for visualizing and manipulating diverse cytoplasmic processes, and for assaying the functions, dynamics, and stability of individual factors. These in vitro approaches uniquely enable investigation of events at specific cell cycle states, including the assembly of actin- and microtubule-based structures, and the formation of the endoplasmic reticulum. Maternal stockpiles in extracts recapitulate diverse processes in the near absence of gene expression, and this biochemical system combined with microscopy empowers a wide range of mechanistic investigations.

BACKGROUND

Research utilizing *Xenopus laevis* oocytes, eggs, and cleavage-stage embryos has provided a number of important insights into the fundamental mechanisms of the meiotic and early mitotic cell cycles. The large size of the eggs, the ease of obtaining them in abundant quantities, their synchronous development, and the ability to microinject and analyze them biochemically led to the discovery of an intrinsic cytoplasmic activity with the same periodicity as the cell cycle (Hara et al. 1980; Gerhart et al. 1984). This activity, termed maturation-promoting factor (MPF), was identified as a highly conserved cell cycle regulatory complex containing cyclin-dependent kinase 1 (cdk1) and cyclin B (Maller et al. 1989). Activation of MPF induces both meiosis and mitosis, whereas its loss induces anaphase and subsequent entry into interphase. The fluctuation of MPF in oogenesis and early *Xenopus* embryos is shown in Figure 1. Hormone treatment of immature oocytes induces MPF activity and drives progression through meiosis I and arrest in metaphase of meiosis II. These eggs maintain high MPF due to cytostatic factor (CSF) activity, eventually identified as Emi2 (Rauh et al. 2005; Schmidt 2006). Upon fertilization, the fusion of the sperm and the egg induces a transient increase in calcium levels that inactivates both CSF and MPF. The first somatic cell cycle lasts 75 min, and subsequent cycles last 25 min. Unfertilized eggs can also be induced to exit CSF metaphase arrest by electrical shock or by treatment with a calcium ionophore.

Characterization of MPF in eggs and embryos was transformative for the cell cycle field. However, microinjection limited the variety of experimental perturbations, and the abundance of refractory yolk

[2]Present address: Universite de Rennes, National Center for Scientific Research, Institute of Genetics and Development of Rennes, UMR 6290, F-35000 Rennes, France

[3]Correspondence: bheald@berkeley.edu

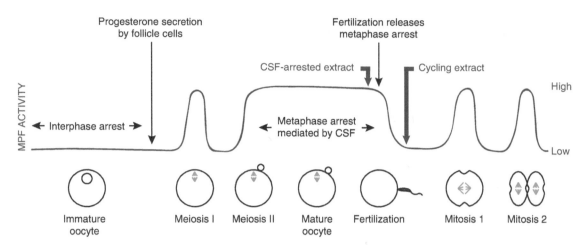

FIGURE 1. Schematic of the cell cycle of the *Xenopus* oocyte and early embryo. Immature oocytes stimulated with progesterone progress through meiosis I and arrest as unfertilized eggs in metaphase of meiosis II with high MPF and CSF activity. Fertilization induces oscillations in MPF activity that peak during mitosis. Points at which extracts are prepared are indicated. (Adapted from Murray [1991].)

platelets made examination of intracellular events in living frog embryos impossible. The development of methods for preparing cell-free extracts from *Xenopus* eggs by Lohka and Masui was a breakthrough that enabled investigation of the cell cycle and downstream events in vitro (Lohka and Masui 1983). By including the calcium chelator EGTA (ethylene glycol-bis(β-aminoethyl ether)-N,N,N′,N′-tetraacetic acid) in the buffer, egg extracts that stably maintained CSF arrest could be prepared (Lohka and Maller 1985). As a postdoc in Marc Kirschner's lab, Andrew Murray painstakingly optimized extract preparation, and succeeded in preparing synchronized extracts capable of multiple cell cycles in vitro, as well as CSF extracts that could be induced to enter interphase. He used these systems to show that cyclin B synthesis and degradation control MPF activity (Murray and Kirschner 1989; Murray et al. 1989; Murray 1991). Egg extracts were subsequently used to determine that cyclin B is degraded by the ubiquitin pathway (Glotzer et al. 1991), and to study downstream events including DNA replication (Blow and Laskey 1986), microtubule dynamics and spindle assembly (Belmont et al. 1990; Verde et al. 1990; Sawin and Mitchison 1991), and vesicle fusion (Tuomikoski et al. 1989). The protocols introduced here present the latest approaches that illustrate how the extract system can be used to examine a variety of cytoplasmic events (Fig. 2). Extract-based investigation of nuclear events is described elsewhere in this collection.

PREPARING EXTRACTS AND PERFORMING CELL CYCLE AND BIOCHEMICAL ASSAYS

The key to preparing robust egg extracts is to prevent dilution of the cytoplasm by packing dejellied eggs using a low-speed centrifugation step, which is followed by a high-speed spin to crush the eggs and fractionate them so that the cytoplasm can be collected. Protocol 1: Preparation of Cellular Extracts from *Xenopus* Eggs and Embryos (Good and Heald 2018) describes the basic method to prepare CSF-arrested *X. laevis* extracts and is adapted for eggs from the smaller *Xenopus* species, *X. tropicalis*, as well as for preparation of metaphase-arrested embryo extracts. These approaches have facilitated the identification of factors that modulate spindle and nuclear size across species and during the reductive divisions of embryonic development (Levy and Heald 2010; Loughlin et al. 2011; Wilbur and Heald 2013). In Protocol 2: Robustly Cycling *Xenopus laevis* Cell-Free Extracts in Teflon Chambers (Chang and Ferrell 2018), spontaneously cycling extracts are prepared from eggs treated with a calcium ionophore to degrade CSF and induce progression through interphase and into the next cell cycle. Interestingly, the materials to which extracts are exposed strongly affects their activity. The robust Teflon tube system could be used to elucidate how MPF activation propagates through the egg

Cite this introduction as *Cold Spring Harb Protoc*; doi:10.1101/pdb.top097048

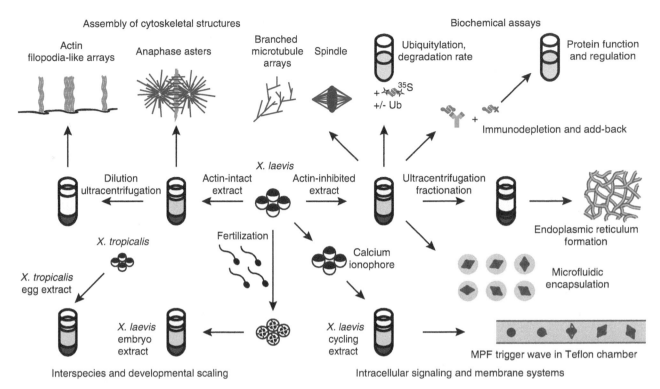

FIGURE 2. Schematic of the cytoplasmic processes reconstituted using the *Xenopus* in vitro systems described in the associated protocols.

extract in the form of a trigger wave (Chang and Ferrell 2013). Reconfinement of egg extract has emerged as a useful approach to evaluate the effects of compartment volume, shape, and boundary conditions on cellular processes under highly controlled conditions. In Protocol 8: Microfluidic Encapsulation of Demembranated Sperm Nuclei in *Xenopus* Egg Extracts (Oakey and Gatlin 2018), the authors describe how confinement technologies can be applied to generate synthetic cell-like systems, which have uniquely enabled evaluation of the impact of cytoplasmic volume on spindle size (Good et al. 2013; Hazel et al. 2013).

Importantly, the functional roles of the proteins involved in any process reconstituted in egg extracts can be determined by depleting a protein of interest using antibodies coupled to beads. Wild-type and mutant versions can be added back to verify function and test domain activities. Protocol 3: Protein Immunodepletion and Complementation in *Xenopus laevis* Egg Extracts (Jenness et al. 2018) describes how this complementation can be achieved by adding either recombinant proteins or mRNAs that are translated in the extract. This approach has been instrumental in defining the functions of cell cycle and other proteins and their regulatory mechanisms. For example, the amino-terminal 90 amino acids of cyclin B were shown to be required for its ubiquitylation and degradation, and sufficient to confer M-phase-specific degradation when fused to another protein (Glotzer et al. 1991). In Protocol 10: Assessing Ubiquitylation of Individual Proteins Using *Xenopus* Extract Systems (McDowell and Philpott 2018a) and Protocol 9: Calculating the Degradation Rate of Individual Proteins Using *Xenopus* Extract Systems (McDowell and Philpott 2018b), methods are presented that capitalize on the robust ability of the egg extract to induce protein degradation in order to quantitatively measure the ubiquitylation and half-life of exogenously added proteins. In addition to ubiquitylation, other posttranslational modifications, such as phosphorylation, have been characterized using *Xenopus* extract systems. Recent proteomic and phosphoproteomic analyses of *Xenopus* eggs and embryos provide an important resource for cell biologists using extract systems (Wühr et al. 2014, 2015; Peuchen et al. 2017; Presler et al. 2017).

ASSEMBLING CELLULAR CYTOSKELETAL AND MEMBRANE STRUCTURES IN VITRO

The ability to control a switch-like transition between metaphase and interphase, and arrest in either state, has facilitated experiments assessing the effects of cell cycle state on the morphology and dynamics of subcellular structures. Protocol 4: Assembly of Spindles and Asters in *Xenopus* Egg Extracts (Field and Mitchison 2018) provides methods for examining the microtubule cytoskeleton in extracts and illustrates its similarity to structures formed in vivo, as well as the use of fluorescent cytoskeletal proteins. Spindle assembly and the role of motor proteins and other factors have been investigated extensively in egg extracts (e.g., Walczak et al. 1998), and many experiments now focus on elucidating the mechanisms underlying the subprocesses of microtubule organization and dynamics. Protocol 5: Dissecting Protein Complexes in Branching Microtubule Nucleation Using Meiotic *Xenopus* Egg Extracts (Song and Petry 2018) provides methods to assay a key step in spindle assembly using immunodepletion and total internal reflection fluorescence (TIRF) microscopy. This protocol describes how to examine the role of a particular protein complex and investigate its assembly by mass spectrometry analysis of immunoprecipitates. The actin cytoskeleton can also be examined in egg extracts. Prior to 2011, most egg extract preparations utilized cytochalasin to inhibit actin polymerization and prevent cytoplasmic gelation and contraction that complicated analysis by microscopy and biochemistry. Field et al. (2011) showed that bulk contraction driven by actomyosin was cell cycle regulated and specific to metaphase extract. Cytochalasin-free actin-intact egg extract has subsequently been used to investigate the physics of symmetry breaking by actin (Abu Shah and Keren 2014). Strikingly, when combined with lipid bilayers, actin-intact extract could reconstitute cytokinesis signaling from microtubules to the plasma membrane (Nguyen et al. 2014). Methods for preparation of actin-intact egg extracts are described elsewhere (Field et al. 2014, 2017). Protocol 6: Filopodia-Like Structure Formation from *Xenopus* Egg Extracts (Fox and Gallop 2018) describes the assembly of one type of actin architecture, and recapitulates an actin-membrane interaction in vitro. Finally, Protocol 7: Endoplasmic Reticulum Network Formation with *Xenopus* Egg Extracts (Wang et al. 2018) illustrates how endoplasmic reticulum (ER) morphology is affected by cell cycle state, and describes the fractionation and labeling of membranes that can be combined with cytosol to reconstitute the organelle. Egg extract-based assays were instrumental in identifying cellular components that that define ER morphology (Dreier and Rapoport 2000; Voeltz et al. 2006; Hu et al. 2009; Wang et al. 2013).

Over the past several decades, *Xenopus* extracts have contributed to our knowledge in many areas of cell biology, and no other cytoplasmic in vitro system has simultaneously enabled both physiological reconstitution and biochemical approaches. To summarize the key features: (1) Extracts faithfully recapitulate complex cellular events in vitro; (2) The synchronized cell cycle state is preserved and can be controlled; (3) Extracts can easily be manipulated biochemically and observed by fluorescence microscopy; (4) Sufficient volumes can be obtained to perform a variety of assays.

ACKNOWLEDGMENTS

We thank the members of the Heald Lab for continuous support. R.H. was supported by National Institutes of Health (NIH) R35 GM118183 and the Flora Lamson Hewlett Chair. R.G. was supported by a Human Frontier Science Program (HFSP) long-term fellowship, LT 0004252014-L, and by the "Fondation pour la Recherche Médicale," ARF20170938684.

REFERENCES

Abu Shah E, Keren K. 2014. Symmetry breaking in reconstituted actin cortices. *Elife* **3**: e01433.

Belmont LD, Hyman AA, Sawin KE, Mitchison T. 1990. Real-time visualization of cell cycle-dependent changes in microtubule dynamics in cytoplasmic extracts. *Cell* **62**: 579–589.

Blow JJ, Laskey RA. 1986. Initiation of DNA replication in nuclei and purified DNA by a cell-free extract of *Xenopus* eggs. *Cell* **47**: 577–587.

Chang JB, Ferrell JE. 2013. Mitotic trigger waves and the spatial coordination of the *Xenopus* cell cycle. *Nature* **500**: 603–607.

Chang JB, Ferrell JE. 2018. Robustly cycling *Xenopus laevis* cell-free extracts in teflon chambers. *Cold Spring Harb Protoc* doi: 10.1101/pdb.prot097212.

Dreier L, Rapoport TA. 2000. In vitro formation of the endoplasmic reticulum occurs independently of microtubules by a controlled fusion reaction. *J Cell Biol* **148**: 883–898.

Field CM, Mitchison TJ. 2018. Assembly of spindles and asters in *Xenopus* egg extracts. *Cold Spring Harb Protoc* doi: 10.1101/pdb.prot099796.

Field CM, Wühr M, Anderson GA, Kueh HY, Strickland D, Mitchison TJ. 2011. Actin behavior in bulk cytoplasm is cell cycle regulated in early vertebrate embryos. *J Cell Sci* 124: 2086–2095.

Field CM, Nguyen PA, Ishihara K, Groen AC, Mitchison TJ. 2014. *Xenopus* egg cytoplasm with intact actin. *Methods Enzymol* 540: 399–415.

Field CM, Pelletier JF, Mitchison TJ. 2017. *Xenopus* extract approaches to studying microtubule organization and signaling in cytokinesis. *Methods Cell Biol* 137: 395–435.

Fox HM, Gallop JL. 2018. Filopodia-like structure formation from *Xenopus* egg extracts. *Cold Spring Harb Protoc* doi: 10.1101/pdb.prot100545.

Gerhart J, Wu M, Kirschner M. 1984. Cell cycle dynamics of an M-phase-specific cytoplasmic factor in *Xenopus laevis* oocytes and eggs. *J Cell Biol* 98: 1247–1255.

Glotzer M, Murray AW, Kirschner MW. 1991. Cyclin is degraded by the ubiquitin pathway. *Nature* 349: 132–138.

Good MC, Heald R. 2018. Preparation of cellular extracts from *Xenopus* eggs and embryos. *Cold Spring Harb Protoc* doi: 10.1101/pdb.prot097055.

Good MC, Vahey MD, Skandarajah A, Fletcher DA, Heald R. 2013. Cytoplasmic volume modulates spindle size during embryogenesis. *Science* 342: 856–860.

Hara K, Tydeman P, Kirschnert M. 1980. A cytoplasmic clock with the same period as the division cycle in *Xenopus* eggs. *Cell Biol* 77: 462–466.

Hazel J, Krutkramelis K, Mooney P, Tomschik M, Gerow K, Oakey J, Gatlin JC. 2013. Changes in cytoplasmic volume are sufficient to drive spindle scaling. *Science* 342: 853–856.

Hu J, Shibata Y, Zhu P-P, Voss C, Rismanchi N, Prinz WA, Rapoport TA, Blackstone C. 2009. A class of dynamin-like GTPases involved in the generation of the tubular ER network. *Cell* 138: 549–561.

Jenness C, Wynne DJ, Funabiki H. 2018. Protein immunodepletion and complementation in *Xenopus laevis* egg extracts. *Cold Spring Harb Protoc* doi: 10.1101/pdb.prot097113.

Levy DL, Heald R. 2010. Nuclear size is regulated by importin α and Ntf2 in *Xenopus*. *Cell* 143: 288–298.

Lohka MJ, Maller JL. 1985. Induction of nuclear envelope breakdown, chromosome condensation, and spindle formation in cell-free extracts. *J Cell Biol* 101: 518–523.

Lohka MJ, Masui Y. 1983. Formation in vitro of sperm pronuclei and mitotic chromosomes induced by amphibian ooplasmic components. *Science* 220: 719–721.

Loughlin R, Wilbur JD, McNally FJ, Nédélec FJ, Heald R. 2011. Katanin contributes to interspecies spindle length scaling in *Xenopus*. *Cell* 147: 1397–1407.

Maller JL, Gautier J, Langan TA, Lohka MJ, Shenoy S, Shalloway D, Nurse P. 1989. Maturation-promoting factor and the regulation of the cell cycle. *J Cell Sci Suppl* 12: 53–63.

McDowell GS, Philpott A. 2018a. Assessing ubiquitylation of individual proteins using *Xenopus* extract systems. *Cold Spring Harb Protoc* doi: 10.1101/pdb.prot104513.

McDowell GS, Philpott A. 2018b. Calculating the degradation rate of individual proteins using *Xenopus* extract systems. *Cold Spring Harb Protoc* doi: 10.1101/pdb.prot103481.

Murray AW. 1991. Cell cycle extracts. *Methods Cell Biol* 36: 581–605.

Murray AW, Kirschner M. 1989. Cyclin synthesis drives the early embryonic cell cycle. *Nature* 339: 275–280.

Murray AW, Solomon MJ, Kirschner MW. 1989. The role of cyclin synthesis and degradation in the control of maturation promoting factor activity. *Nature* 339: 280–286.

Nguyen PA, Groen AC, Loose M, Ishihara K, Wühr M, Field CM, Mitchison T. 2014. Spatial organization of cytokinesis signaling reconstituted in a cell-free system. *Science* 346: 244–247.

Oakey J, Gatlin JC. 2018. Microfluidic encapsulation of demembranated sperm nuclei in *Xenopus* egg extracts. *Cold Spring Harb Protoc* doi: 10.1101/pdb.prot102913.

Peuchen EH, Cox OF, Sun L, Hebert AS, Coon JJ, Champion MM, Dovichi NJ, Huber PW. 2017. Phosphorylation dynamics dominate the regulated proteome during early *Xenopus* development. *Sci Rep* 7: 15647.

Presler M, Van Itallie E, Klein AM, Kunz R, Coughlin ML, Peshkin L, Gygi SP, Wühr M, Kirschner MW. 2017. Proteomics of phosphorylation and protein dynamics during fertilization and meiotic exit in the *Xenopus* egg. *Proc Natl Acad Sci* 114: E10838–E10847.

Rauh NR, Schmidt A, Bormann J, Nigg EA, Mayer TU. 2005. Calcium triggers exit from meiosis II by targeting the APC/C inhibitor XErp1 for degradation. *Nature* 437: 1048–1052.

Sawin KE, Mitchison T. 1991. Mitotic spindle assembly by two different pathways in vitro. *J Cell Biol* 112: 925–940.

Schmidt A. 2006. Cytostatic factor: an activity that puts the cell cycle on hold. *J Cell Sci* 119: 1213–1218.

Song J, Petry S. 2018. Dissecting protein complexes in branching microtubule nucleation using meiotic *Xenopus* egg extracts. *Cold Spring Harb Protoc* doi: 10.1101/pdb.prot100958.

Tuomikoski T, Felix MA, Dorée M, Gruenberg J. 1989. Inhibition of endocytic vesicle fusion in vitro by the cell-cycle control protein kinase cdc2. *Nature* 342: 942–945.

Verde F, Labbé JC, Dorée M, Karsenti E. 1990. Regulation of microtubule dynamics by cdc2 protein kinase in cell-free extracts of *Xenopus* eggs. *Nature* 343: 233–238.

Voeltz GK, Prinz WA, Shibata Y, Rist JM, Rapoport TA. 2006. A class of membrane proteins shaping the tubular endoplasmic reticulum. *Cell* 124: 573–586.

Walczak CE, Vernos I, Mitchison T, Karsenti E, Heald R. 1998. A model for the proposed roles of different microtubule-based motor proteins in establishing spindle bipolarity. *Curr Biol* 8: 903–913.

Wang S, Romano FB, Field CM, Mitchison TJ, Rapoport TA. 2013. Multiple mechanisms determine ER network morphology during the cell cycle in *Xenopus* egg extracts. *J Cell Biol* 203: 801–814.

Wang S, Romano FB, Rapoport TA. 2018. Endoplasmic reticulum network formation with *Xenopus* egg extracts. *Cold Spring Harb Protoc* doi: 10.1101/pdb.prot097204.

Wilbur JD, Heald R. 2013. Mitotic spindle scaling during *Xenopus* development by kif2a and importin α. *Elife* 2: e00290.

Wühr M, Freeman RM, Presler M, Horb ME, Peshkin L, Gygi S, Kirschner MW. 2014. Deep proteomics of the *Xenopus laevis* egg using an mRNA-derived reference database. *Curr Biol* 24: 1467–1475.

Wühr M, Güttler T, Peshkin L, McAlister GC, Sonnett M, Ishihara K, Groen AC, Presler M, Erickson BK, Mitchison TJ, et al. 2015. The nuclear proteome of a vertebrate. *Curr Biol* 25: 2663–2671.

Preparation of Cellular Extracts from *Xenopus* Eggs and Embryos

Matthew C. Good[1,3] and Rebecca Heald[2,3]

[1]*Department of Cellular and Developmental Biology and Department of Bioengineering, University of Pennsylvania, Philadelphia, Pennsylvania 19104;* [2]*Department of Molecular and Cell Biology, University of California Berkeley, Berkeley, California 94720*

Cell-free cytoplasmic extracts prepared from *Xenopus* eggs have been used extensively to recapitulate and characterize intracellular events in vitro. Egg extracts can be induced to transit the cell cycle and reconstitute assembly of dynamic structures including the interphase nucleus and the mitotic spindle. In this protocol, methods are described for preparing crude cytoplasmic extracts from *Xenopus* eggs and embryos that are arrested in metaphase of the cell cycle. The basic protocol uses unfertilized *Xenopus laevis* eggs, which are crushed by centrifugation in the presence of EGTA to preserve the natural cytostatic factor (CSF) activity that maintains high levels of Cdk1/cyclin B kinase and metaphase arrest. In the second method, the basic procedure is adapted for *Xenopus tropicalis* eggs with minor modifications to accommodate differences in frog size, timing of egg laying, and temperature and salt sensitivity. The third variation takes advantage of the synchronous divisions of fertilized *X. laevis* eggs to generate extracts from embryos, which are arrested in metaphase by the addition of nondegradable cyclin B and an inhibitor of the anaphase-promoting complex (APC) that together stabilize Cdk1/cyclin B kinase activity. Because they are obtained in much smaller amounts and their cell cycles are less perfectly synchronized, extracts prepared from embryos are less robust than egg extracts. *X. laevis* egg extracts have been used to study a wide range of cellular processes. In contrast, *X. tropicalis* egg extracts and *X. laevis* embryo extracts have been used primarily to characterize molecular mechanisms regulating spindle and nuclear size.

MATERIALS

It is essential that you consult the appropriate Material Safety Data Sheets and your institution's Environmental Health and Safety Office for proper handling of equipment and hazardous materials used in this protocol.

RECIPES: Please see the end of this protocol for recipes indicated by <R>. Additional recipes can be found online at http://cshprotocols.cshlp.org/site/recipes.

Reagents

Agarose (1.5% in 0.1× MMR)

Microwave before using to coat the bottom of a Petri dish. Dishes should be prepared the day before embryo extract preps, covered with 0.1× MMR and stored at RT. Discard the 0.1× MMR before use.

CSF-XB <R>

This buffer is used for X. laevis *eggs and embryos.*

[3]Correspondence: bheald@berkeley.edu, mattgood@upenn.edu

CSF-XB+ (CSF-XB supplemented with 10 µg/mL LPC)

Dilute LPC protease inhibitor cocktail 1:1000 into CSF-XB. Prepare fresh and use immediately. This buffer is used for X. laevis eggs and embryos.

CSF-XB solution #2 for *X. tropicalis* eggs <R>

This buffer is used for X. tropicalis eggs and contains additional EGTA and magnesium chloride.

CSF-XB+ solution #2 (CSF-XB solution #2 supplemented with 10 µg/mL LPC)

Dilute LPC protease inhibitor cocktail 1:1000 into CSF-XB solution #2. Prepare fresh and use immediately. This buffer is used for X. tropicalis eggs and contains additional EGTA and magnesium chloride.

Cytochalasin D or B (10 mg/mL, Sigma-Aldrich C8273 or Sigma-Aldrich C6762, respectively)

Dissolve in DMSO and store in 50 µL aliquots at −20°C.

Dejelly solution #1 for *Xenopus laevis* eggs <R>

Dejelly solution #2 for *Xenopus tropicalis* eggs and *Xenopus laevis* embryos <R>

Unlike the X. laevis dejelly solution, the X. tropicalis buffer does not contain salts.

Dimethyl sulfoxide (DMSO) (anhydrous, high purity such as Sigma-Aldrich)

Energy mix (50×) <R>

Note that some labs do not add energy mix to egg extracts.

Human chorionic gonadotropin (HCG; Sigma-Aldrich CG10) or chorulon (A to Z Vet Supply), at a concentration of 1000 IU/mL

Dissolve in sterile MilliQ H₂O and store at 4°C.

LPC protease inhibitor cocktail (10 mg/mL each of leupeptin, pepstatin, chymostatin; EMD Millipore)

Dissolve in DMSO and store in 50 µL aliquots at −20°C.

Marc's modified Ringer's (MMR; 20×) <R>

Dilute to desired concentration with MilliQ H₂O before use.

Mature female *Xenopus laevis* or *Xenopus tropicalis* frogs

Frogs are obtained from NASCO. Females are used to obtain eggs as described below.

Mature male *Xenopus laevis* frogs

Frogs are obtained from NASCO. Males are killed to obtain fresh testes for fertilization to obtain embryos (see Sec. 8, Protocol 1: Isolation and Demembranation of Xenopus Sperm Nuclei [Hazel and Gatlin 2018]).

Pregnant mare serum gonadotropin (PMSG; 200 IU/mL; Prospec HOR-272)

Dissolve in sterile MilliQ H₂O and store at 4°C.

Purified proteins: Aliquots of 0.5 mg/mL (∼ 13 µM) Δ90-CyclinB1 in XB (Glotzer et al. 1991); 15 mg/mL (∼750 µM) human UbcH10-C114S in XB (Rape et al. 2006)

XB (extract buffer) <R>

Equipment

Beakers, plastic (0.5 and 4 L) and glass (500 mL)

Centrifuge (floor size, high speed, refrigerated) set to 16°C

Centrifuge tubes (round-bottom 13 mL polypropylene 16.8 × 95 mm; Sarstedt 55.518)

Clinical centrifuge (refrigerated, swinging-bucket) set to either 16°C or 25°C.

Dissection stereomicroscope

Forceps

Ice bucket with ice

Microcentrifuge tubes (1.5 and 2 mL)

Petri dishes (glass, 6 cm diameter)

Plastic containers for frogs (4 and 6 L; Corning) with tight-fitting lids and holes punched for air exchange

Plastic pestle for 1.5 mL microfuge tube (USA Scientific)

Room (16°C) or large nonair tight incubator set to 16°C for *Xenopus laevis*

Rotor (swinging bucket Sorvall HB-6) with rubber (Kimble-Chase) and microcentrifuge tube adapters (Sorvall)

Syringes (disposable; 1 mL)

Syringe needles (18-gauge 1.5 inch length, and 30-gauge, 0.5 inch length)

Transfer pipettes (plastic, draw up to 3.4 mL per squeeze; Fisher)

Ultracentrifuge tubes (SW-55 ultraclear thin-wall 5 mL; 13 × 51 mm, Beckman 344057)

METHOD

Three different extract preparation procedures are outlined, which have been optimized for X. laevis *eggs,* X. tropicalis *eggs, or* X. laevis *embryos.*

Xenopus laevis Egg Extract Procedure

1. Prime *Xenopus laevis* females at least 3 d before extract preparation by injecting 0.5 mL (100 IU) of PMSG subcutaneously into the region between the top of a hind leg and the cloaca (dorsal lymph sac) using a 30-gauge needle and 1 mL syringe.

   Typically, 3–4 females are used for each extract preparation. Priming the frogs increases egg yield and egg quality. Primed frogs should be used within 2 wk. Depending on hormone source, the injected amount may need to be titrated. If the frogs lay eggs following priming, subsequently reduce the amount injected by 50%.

2. Induce ovulation by injecting primed frogs 16–18 h before extract preparation time with 0.5 mL (500 IU) of HCG. Store each frog individually in 2 L of 1× MMR in a 4 L plastic container overnight at 16°C. On average, each female should lay more than a thousand eggs (Fig. 1A).

   Housing frogs individually prevents mixing of egg clutches of variable quality.

3. Remove frogs from containers and analyze egg quality. Good eggs have clearly delineated animal (dark) and vegetal (light) poles. Remove lysed (white and puffy), mottled or abnormal looking eggs from containers using a transfer pipette. Store eggs at 16°C until all buffers and supplies are prepared.

   Clutches that contain a significant number of eggs (>2%) that are activated or lysed (white and puffy appearance) or connected in long strings should not be used.

4. For eggs from three to four females, prepare 1 L Dejelly solution #1, 1 L XB, and 400 mL CSF-XB. Ten minutes before the extract prep, prepare 100 mL CSF-XB+, dispense 1 mL CSF-XB+ to 4 ultracentrifuge tubes, and add 10 µL of cytochalasin D (final concentration of 100 µg/mL) to each tube. For supplies: trim tips from four transfer pipettes to increase opening diameter to ~2 mm, and leave four uncut. Gather 4 L plastic and 500 mL glass beakers and all required reagents and supplies. Make sure that clinical and high-speed centrifuges are set to 16°C.

   One frog typically produces enough eggs to fill one tube.

   It is crucial to be prepared for all subsequent steps before starting.

   Dejelly and egg-washing steps can be performed at room temperature but extract quality may be higher if steps are performed in a 16°C room.

5. In the next steps (6–8), gently pour buffers down the side of the beaker, and swirl eggs while avoiding excess turbulence. Between steps, pour off as much buffer as possible into a 4 L beaker without exposing the eggs to air, which will lyse them. Manipulate eggs carefully and once eggs are dejellied, carry out subsequent washing and centrifugation steps in rapid succession. Throughout the procedure, remove abnormal or lysed eggs with a transfer pipette.

Cite this protocol as *Cold Spring Harb Protoc*; doi:10.1101/pdb.prot097055

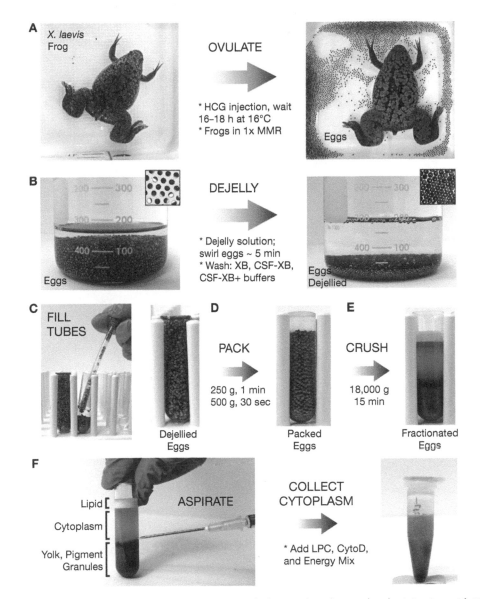

FIGURE 1. *Xenopus laevis* egg extract preparation. (*A*) Female frogs induced to ovulate by injection with HCG. Laid eggs are collected 16–18 h after the injection. (*B*) Eggs are dejellied using a cysteine solution. The eggs pack tightly together once their jelly coats have been removed. *Inset:* egg spacing before and after dejelly step. (*C*) Dejellied eggs are carefully transferred to an ultracentrifuge tube using a plastic transfer pipette whose tip has been cut to create a wider bore. (*D*) A series of low-speed spins tightly packs eggs without lysing them, allowing excess buffer to be removed. (*E*) A high-speed spin crushes and fractionates the eggs. (*F*) The cytoplasm layer is collected by puncturing the tube wall using a needle and syringe. Once aspirated, the cytoplasm is transferred to a new tube and supplemented with additional reagents.

6. Clean the eggs: Pour off 1× MMR from the plastic container containing laid eggs. Gently pour in 250–500 mL fresh 1× MMR and repeat washes until all dirt and debris are removed.

7. Dejelly eggs: Combine best clutches of eggs into a single 500 mL glass beaker and pour off remaining MMR. Add ~250 mL of Dejelly solution #1. Swirl the eggs gently, every 20 sec, and every 2 min pour off and replenish the solution. Dejellying is complete once eggs pack together tightly, without gaps (Fig. 1B). The close packing is easiest to discern visually when beaker is tipped at a 45° angle. Total time to dejelly will vary, but should be around 5 min. Once completed, pour off the dejelly solution and immediately add ~300 mL of XB.

   Alternatively, each egg clutch can be transferred to an individual beaker and dejellied and washed separately using smaller buffer volumes.

8. Wash eggs: Pour off the first addition of XB and wash again with 300 mL. Pour off the XB and wash two times with ~200 mL of CSF-XB. Pour off the CSF-XB and add 100 mL of CSF-XB+. At this point the eggs should appear very clean with their dark animal poles facing up. Sort one final time to remove any lysed or abnormal eggs.

9. Transfer eggs very carefully to prepared ultracentrifuge tubes containing CSF-XB+ and cytochalasin. Use a trimmed transfer pipette, and avoid exposing the eggs to air by first drawing up 0.5 mL of CSF-XB+ buffer and then gently drawing up the eggs, and expelling them below the liquid surface. Typically 2–3 fillings of the transfer pipette are required to fill one tube with eggs (Fig. 1C).

    Do not overfill tubes, which exposes eggs to air.

10. Packing spin: Using forceps, place each tube filled with eggs inside a 13 mL Sarstedt tube. Transfer tubes to clinical centrifuge at 16°C. Balance rotor and spin at 250g for 1 min, and then at 500g for an additional 30 sec to pack the eggs (Fig. 1D). After spinning, remove all buffer from the top of the eggs using a transfer pipette.

    This step prevents buffer from diluting the extract, so it is worth sacrificing a few eggs to ensure that all buffer is removed.

11. Crushing spin: Transfer tubes filled with packed eggs to a Sorvall HB-6 rotor with rubber adapters in a high-speed centrifuge. Crush and fractionate the eggs by spinning at 18,000g for 15 min at 16°C (Fig. 1E).

12. Collect cytoplasm: Immediately upon completion of centrifugation, remove thin-wall tubes containing fractionated eggs from Sarstedt tubes and place them on ice. Using an 18-gauge needle attached to a 1 mL syringe, puncture the thin-wall tube at the bottom of the cytoplasmic layer (Fig. 1F). Aspirate the cytoplasm without taking up any of the surrounding yolk or lipid layers. Remove the needle and expel the cytoplasm into a prechilled tube on ice.

    To avoid piercing your finger and to keep the tube cool while withdrawing the extract, hold the tube near the top up against the inside of the ice bucket, keeping the bottom of the tube on ice.

    It is good to withdraw some of the fluffy whitish layer underneath the cytoplasmic layer as well, which can be achieved by rotating the needle so that the opening is facing downward.

    Expected yield is ~1 mL of crude cytoplasm from each tube.

13. Supplement the extract with a final concentration of: 10 µg/mL LPC (1:1000 dilution of stock), 10 µg/mL cytochalasin D (1:1000 dilution of stock). Add Energy mix stock at 1:50. Mix by inverting the microfuge tube or pipetting gently with a cut off tip. The freshly prepared extract can be used for 6–8 h.

    The final molar concentrations of reagents supplemented to the extract from this step are: 16.67 µM LPC, 20 µM Cytochalasin D, 3.8 mM creatine phosphate, 0.5 mM ATP, 0.5 mM MgCl$_2$ and 0.05 mM EGTA. Not all labs supplement the extract with energy mix.

Xenopus tropicalis Egg Extract Procedure

Major differences from X. laevis egg extract preparation accommodate the smaller frog size, its higher physiological temperature, and sensitivity of the eggs to salt.

14. Dilute HCG stock to 100 IU/mL in sterile MilliQ water and prime *X. tropicalis* females 20–21 h before boosting by injecting 250 µL (25 IU) HCG subcutaneously into the dorsal lymph sac using a 30-gauge needle and 1 mL syringe. Store frogs in 6 L plastic container filled with 4 L of deionized water.

    Typically, six to eight females are used for one extract preparation with a yield of 0.1–0.2 mL cytoplasm per egg clutch.

    Larger tanks are used for X. tropicalis to facilitate squeezing and because they tend to jump.

    If working with both X. laevis and X. tropicalis, always use dedicated containers for each species.

Cite this protocol as *Cold Spring Harb Protoc*; doi:10.1101/pdb.prot097055

15. Induce ovulation by boosting primed frogs ~4 h before extract preparation by injecting 250 µL (250 IU) HCG. To control for egg quality, make sure to separate frogs into individual containers filled with deionized water.

    Note that the time from boosting injection to ovulation is much shorter than for X. laevis.

16. The first eggs should be laid 2.5–3 h following the second injection. Once eggs are observed, every 15 min accelerate egg laying by gently clasping each frog and allowing her to kick her legs under the water surface. The frogs will wiggle, which further accelerates egg laying. After four to five "squeezes," analyze egg quality. Good eggs have clearly delineated animal (dark) and vegetal (light) poles.

    Stringy eggs are okay for X. tropicalis. Clutches of eggs that are activated or lysed (white and puffy appearance) should not be used.

17. Prepare buffers and set up equipment during the clasping period. For eggs from six to eight females, prepare 0.5 L Dejelly Solution #2 and 1 L CSF-XB solution #2. Ten min before the start of experiments, prepare 50 mL CSF-XB+ solution #2, and dispense 1 mL CSF-XB+ solution #2 supplemented with 10 µL of 10 mg/mL cytochalasin D into a single ultracentrifuge tube. For supplies: trim tips from four transfer pipettes to increase opening diameter to ~2 mm, and leave four uncut, gather 4 L and 500 mL plastic beakers, and all required reagents and supplies. Make sure that clinical centrifuge is set to 25°C and high-speed centrifuge is set to 16°C.

18. In the next steps (19–21), gently pour buffers down the side of the beaker, and swirl eggs gently to avoid turbulence. Between steps, pour off as much buffer as possible into a 4 L beaker without exposing the eggs to air, which will lyse them. Manipulate eggs carefully and once eggs are dejellied, carry out subsequent washing and centrifugation steps in rapid succession. Throughout the procedure, remove abnormal or lysed eggs with a transfer pipette.

19. Remove the frogs from the containers. Detach the eggs stuck to the container by gently sweeping them with your gloved finger or by generating water flow using a transfer pipette to put them back into suspension in the water. Before they settle and stick again, combine the best clutches to a single 500 mL plastic beaker.

    Note that eggs stick to some glove brands.

20. Dejelly eggs: pour off remaining water. Add ~250 mL of Dejelly solution #2. Swirl the eggs gently, every 20 sec, and after 2 min pour off and replenish dejelly solution. Dejellying is complete once eggs pack together tightly, without gaps, when beaker is tipped to a 45° angle. Total time to dejelly will vary but should take less than 10 min. Once completed, pour off the dejelly solution and immediately add 250 mL CSF-XB solution #2.

21. Wash eggs: Pour off the first addition of CSF-XB solution #2 and wash several more times with ~250 mL CSF-XB solution #2. Pour off as much buffer as possible and add 50 mL of CSF-XB+ solution #2. At this point the eggs should be clean and their dark animal poles should be facing up. Sort one final time to remove any bad eggs.

22. Transfer eggs very carefully to prepared ultracentrifuge tube containing CSF-XB+ solution #2 and cytochalasin. Use trimmed transfer pipette, and avoid exposing the eggs to air by first drawing up 0.5 mL of CSF-XB+ buffer and then gently drawing up the eggs, and expelling them below the liquid surface.

    A single tube can accommodate eggs from ~ six females, and to avoid dilution it is better not to divide eggs into two tubes.

23. Packing spin: Using forceps, place ultracentrifuge tube filled with packed eggs inside a Sarstedt tube. Transfer to a refrigerated clinical centrifuge set to 25°C. Balance rotor and spin at 500*g* for 1 min, to pack the eggs. After spinning, remove all buffer from the top of the eggs using a transfer pipette.

    This step prevents buffer from diluting the cytoplasm, so it is worth sacrificing a few eggs to ensure all buffer is removed.

24. Crushing spin: Transfer tube filled with packed eggs and balance tube to high-speed centrifuge containing a Sorvall HB-6 rotor with rubber adapters. Crush and fractionate the eggs by centrifuging at 18,000g for 15 min at 16°C.

25. Collect cytoplasm: Upon completion of centrifugation remove ultracentrifuge tubes from Sarstedt tubes and place them at room temperature or 16°C. Using an 18-gauge needle attached to a 1 mL syringe, puncture the thin-wall tube at the bottom of the cytoplasmic layer and carefully aspirate the cytoplasmic layer without taking up any of the surrounding yolk or lipid layers. Expected yield is ~1 mL of crude cytoplasm from one tube filled with eggs from six females. Remove the needle and expel the cytoplasm to 1.5 mL microfuge tube.

26. Supplement the extract with a final concentration of 10 µg/mL LPC (1:1000 dilution of stock), 10 µg/mL cytochalasin D (1:1000 dilution of stock). Add Energy mix stock at 1:50. Mix by inverting the microfuge tube or pipetting gently with a cut off tip. The freshly prepared extract should be kept at 16°C and can be used for 5–6 h.

> Note that unlike X. laevis egg extracts, X. tropicalis egg extracts should not be placed on ice, as this will compromise activity.

Xenopus laevis Embryo Extract Procedure

Extracts arrested in mitosis can be prepared from X. laevis embryos through blastula Stage 8, after which cell cycle synchrony is lost. The amount of extract that can be obtained from embryos is much lower than from eggs. Therefore, this procedure has not yet been applied X. tropicalis, due to its smaller egg size and lower yield.

27. Prime *X. laevis* females 3–7 d before performing in vitro fertilization (IVF) by injecting 0.5 mL (100 IU) of PMSG subcutaneously into the dorsal lymph sac using a 30-gauge needle and 1 mL syringe.

28. Induce ovulation 14–15 h before collecting embryos by injecting primed frogs with 0.5 mL (500 IU) HCG. Store each frog individually in 2 L of 1× MMR in a 4 L plastic container overnight at 16°C.

> Usually three out of four females will lay high-quality eggs, allowing the preparation of three separate single-frog embryo extracts.

29. Analyze egg quality 14 h after HCG booster injection. Avoid using females that have laid lysed or stringy eggs, or that have laid no eggs at all.

30. Collect freshly laid eggs: For selected frogs, promote egg-laying by gently squeezing females in a manner that mimics amplexus (Fig. 2A), and collect freshly laid eggs from each frog in individual 6 cm glass dishes that have been coated with ~5 mL of 1.5% agarose in 0.1× MMR. Repeat every 15 min (up to four total squeezes) or until each dish is full. Place females in deionized water in between rounds of egg collection.

31. Prepare sperm slurry: For each IVF reaction, use 1/8 of a testis recently isolated from a *Xenopus* male (see Sec. 8, Protocol 1: Isolation and Demembranation of *Xenopus* Sperm Nuclei [Hazel and Gatlin 2018]). Use scissors to cut the testis into appropriately sized pieces, and place each in its own 1.5 mL microfuge tube containing 1 mL of MilliQ H_2O. Crush each piece using a plastic pestle for 1–2 min.

> Testes can be used for up to 1–2 wk following dissection, but fertilization efficiency may drop over time.

32. Fertilize eggs: For each IVF reaction in a separate dish, pipette 1 mL of a slurry that contains approximately one eighth of a testis onto the eggs and incubate for 5–10 min at 16°C. Flood each dish with 0.1× MMR and incubate an additional 10–20 min or until all of the zygotes have rotated so that the animal cap (dark) is facing up (Fig. 2B), indicating successful fertilization.

33. Dejelly eggs: This should be performed just after rotation, ~15 min postfertilization, and before the first cleavage, which occurs at ~90 min. Pour off 0.1× MMR and add 20 mL of freshly prepared Dejelly solution #2 to each dish. Swirl gently for 2–3 min, until embryos pack tightly

Cite this protocol as *Cold Spring Harb Protoc*; doi:10.1101/pdb.prot097055

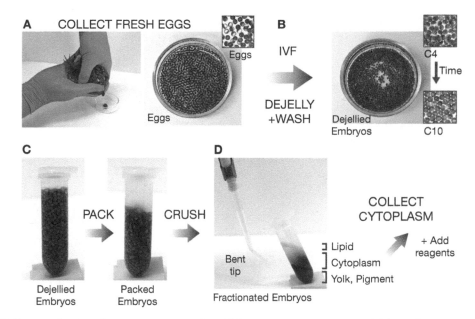

FIGURE 2. *Xenopus laevis* embryo extract preparation. (*A*) In contrast to egg extracts which can be prepared from laid eggs, embryo extracts require freshly squeezed eggs. (*B*) Eggs are fertilized by sperm in vitro and subsequently dejellied before the first embryonic division. Embryos are synchronized and collected at specific developmental stages. Inset shows embryos that have undergone four cell divisions (C4) or 10 divisions (C10). (*C*) Dejellied embryos are packed tightly together by centrifugation to remove excess buffer. These packed embryos are then crushed and fractionated by high-speed centrifugation. (*D*) The cytoplasm layer is collected by threading a bent pipette tip through the lipid layer, and supplemented with reagents to generate the final crude embryo cytoplasm.

together when dish is tilted. Pour off Dejelly solution and wash 3–5 times with 0.1× MMR. Add 15 mL 0.1× MMR and continue incubation. Let embryo development progress at 16°C until the desired developmental state.

> At room temperature the first cytokinesis should take place ∼90 min post fertilization.

> At 16°C, embryos reach mid-Stage 4 at ∼4.5 h, or mid-Stage 8 at ∼13.5 h.

34. Sort embryos: Using a dissecting stereomicroscope, transfer embryos that are cleaving nicely to a fresh dish coated in 1.5% agarose in 0.1× MMR. Continue to monitor developmental progression by recording the timing of the second and third divisions. Remove embryos that are lagging or beginning to lyse.

35. Prepare remaining buffers and carry out final sort 30 min before appropriate time point for embryo collection. Prepare 0.2 L of XB solution, and 100 mL of CSF-XB+ supplemented with 100 µL of 10 mg/mL cytochalasin D. Perform final sort to ensure that the remaining ∼1000 embryos in each dish are homogeneous.

36. Wash embryos: Pour off 0.1× MMR and add in succession to each dish: 2×25 mL of XB, then 2×15 mL of CSF-XB+ with cytochalasin. Gently swirl and pour off buffers between each addition. Tilt dish so that surface tension helps to carefully pull the solution over the lip of the dish. This ensures that embryos will not be poured off during buffer exchanges.

37. Pack embryos: After pouring off remaining buffer, and without exposing embryos to air, carefully transfer the embryos from a single dish into a 2 mL microfuge tube using a cut transfer pipette. A full dish of embryos, unpacked, will fill the tube to a level of ∼1.7 mL (Fig. 2C). Centrifuge at 200*g* for 1 min and then at 500*g* for 30 sec in a benchtop microcentrifuge at room temperature. Remove remaining buffer with a pipette, first using a P1000 and then a P200 tip.

38. Crush embryos: Transfer 2 mL tubes containing packed eggs to a Sorvall HB-6 rotor containing microfuge tube adapters, in a superspeed centrifuge set to 16°C. Centrifuge at 18,000g for 12 min.

    Alternatively, eggs can be crushed by spinning for 12 min at 18,000g using a benchtop refrigerated microcentrifuge set to 16°C.

39. Collect embryo cytoplasm: After crushing the embryos into stratified layers, place the 2 mL microfuge tube on ice. Using a bent P200 pipette tip (Fig. 2D), insert the tip from above, through the lipid layer and carefully withdraw the cytoplasm, making sure to avoid surrounding layers. Dispense the cytoplasm into a prechilled 1.5 mL microfuge tube. Expect a yield of ∼ 400–500 µL of crude cytoplasm for each full dish of embryos. The layers correspond to, top: lipid (white-yellow), middle: cytoplasm (gold), bottom: yolk granules and pigment (black/brown) (Fig. 2D).

40. Supplement the embryo extract with a final concentration of 10 µg/mL LPC (1:1000 dilution of stock) and 10 µg/mL cytochalasin D (1:1000 dilution of stock). Add Energy mix at 1:50. Mix by inverting the microfuge tube. To ensure that the extracts are synchronized and arrested in mitosis, add a nondegradable form of cyclin B along with a dominant-negative form of the ubiquitinating E2, UbcH10 (UbcH10-C114s), to final concentrations of ∼0.35 µM Δ90-CyclinB1 and ∼18 µM UbcH10-C114S. Mix by inverting tube or pipetting gently with a cut off tip. Embryo extracts stored on ice are functional for up to 6 h.

    To further enhance metaphase arrest and improve spindle forming activity, the embryo can be supplemented with 5%–10% CSF-arrested cytoplasmic egg extract from Xenopus laevis (Good 2016).

DISCUSSION

These procedures are based on the method first described by Lohka and Masui (Lohka and Maller 1985; Lohka and Masui 1983), optimized initially by Andrew Murray (Murray 1991) and further refined over the years (Desai et al. 1998; Maresca and Heald 2006; Good 2016). Central to the success of all extract preparations is the quality of the eggs. Crude or further fractionated *Xenopus laevis* egg extract have been used widely for a variety of protocols. A major strength of in vitro systems such as these comes from the ability to manipulate them biochemically by immunodepletion or by adding recombinant proteins or fluorescent probes.

Xenopus tropicalis frogs are considerably smaller than their *X. laevis* relative, and lay proportionally smaller eggs. Therefore, they are less suitable for generating the large volumes of egg extract that form the basis of many assays. However, the observation that the size of nuclei and spindles formed in *X. tropicalis* egg extract are smaller than those formed in *X. laevis* and recapitulate in vivo differences has uniquely enabled investigation of mechanisms of intracellular scaling between species (Brown et al. 2007).

Extracts prepared from *X. laevis* embryos must be arrested in metaphase using exogenously added regulators of the cell cycle machinery. Cyclin B Δ 90 is a nondegradable version of cyclin that activates Cdk1 (Glotzer et al. 1991). Mitotic arrest is further enhanced by addition of UbcH10-C114S, which functions as a dominant negative inhibitor of the Anaphase Promoting Complex (APC) (Townsley et al. 1997; Rape et al. 2006). These extracts have been used to compare spindle morphologies between meiosis and the early mitotic divisions (Wuhr et al. 2008; Wilbur and Heald 2013), and formed the basis of assays to examine spindle scaling during development (Wuhr et al. 2008; Good et al. 2013; Wilbur and Heald 2013). Embryo extracts have also been prepared that are arrested in interphase that enable investigation of nuclear scaling (Levy and Heald 2010). Like egg extracts, embryo extracts largely recapitulate in vivo size differences, but can be prepared only in very small amounts. Their activity is therefore reduced compared to egg extracts, likely due in part to increased dilution of the cytoplasm with buffer.

Cite this protocol as *Cold Spring Harb Protoc*; doi:10.1101/pdb.prot097055

RECIPES

CSF-XB

Reagent	Quantity
XB (extract buffer) <R>	400 mL
EGTA stock (K salt; 0.5 M; pH 7.7; filter-sterilized and stored in aliquots at −20°C)	4 mL
MgCl$_2$ solution (2 M; e.g., Sigma-Aldrich 68475)	200 µL

Make fresh on the day of extract preparation.

CSF-XB Solution #2 for X. tropicalis *Eggs*

Reagent	Volume
Extract Buffer (XB) salts (20×) <R>	50 mL
Sucrose (2 M; filter-sterilized and stored in aliquots at −20°C)	25 mL
K-HEPES (1 M; K salt; pH 7.7; filter-sterilized and stored in aliquots at −20°C)	10 mL
MgCl$_2$ (2 M; e.g., Sigma-Aldrich 68475)	1 mL
EGTA (0.5 M; K salt; pH 7.7; filter-sterilized and stored in aliquots at −20°C)	20 mL

Make fresh on the day of extract preparation. Combine and bring to 1 L with MilliQ H$_2$O. Adjust pH to 7.7 with KOH.

Dejelly Solution #1 for X. laevis *Eggs*

Reagent	Quantity	Concentration
L-cysteine (e.g., Sigma-Aldrich C7352)	20 g	2%
Extract buffer (XB) salts (20×) <R>	50 mL	1×

Combine with 900-mL MilliQ H$_2$O, adjust pH to 7.8 with 10-M NaOH, and bring to 1 L with MilliQ H$_2$O. Make fresh on day of preparation, and store at 16°C until use.

Dejelly Solution #2 for X. tropicalis *Eggs and* X. laevis *Embryos*

Combine 15 g of L-cysteine (e.g., Sigma-Aldrich C-7352) with 450-mL MilliQ H$_2$O. Adjust the pH to 7.8 with NaOH, and bring to 0.5 L (to make a 3% L-cysteine solution). Make fresh on day of prep and store at room temperature until use.

Energy Mix (50×)

Reagent	Quantity
Creatine phosphate disodium salt tetrahydrate (MW 327.14)	0.622 g
Adenosine 5′-triphosphate (ATP) disodium salt hydrate (MW 511.14)	0.138 g
EGTA stock (K salt; 0.5 M; pH 7.7; filter-sterilized and stored in aliquots at −20°C)	0.05 mL
MgCl$_2$ stock (2 M; e.g., Sigma-Aldrich 68475)	0.125 mL

Combine with MilliQ H$_2$O to a volume of 10 mL. Make 0.1-mL aliquots, and store at −20°C.

Extract Buffer (XB) Salts (20×)

Reagent	Quantity	Concentration
KCl	149.12 g	2 M
CaCl$_2$	0.294 g	2 mM
MgCl$_2$ (2 M; e.g., Sigma-Aldrich 68475)	10 mL	20 mM

Combine with MilliQ H$_2$O to a volume of 1 L. Autoclave and store at 4°C.

Marc's Modified Ringer's (MMR; 20×)

Reagent	Quantity	Final concentration
HEPES, free acid	95.32 g	100 mM
EDTA	2.98 g	2 mM
NaCl	467.52 g	2 M
KCl	11.93 g	40 mM
CaCl$_2$	23.52 g	40 mM
MgCl$_2$ (e.g., Sigma-Aldrich 68475)	40 mL of 2 M stock	20 mM

Combine with MilliQ H$_2$O, adjust pH to 7.8–7.9 using ~8.5 g of NaOH pellets, and bring to 4 L. Autoclave and store at room temperature.

XB (Extract Buffer)

Reagent	Volume
Extract buffer (XB) salts (20×) <R>	50 mL
Sucrose (2 M; filter-sterilized and stored in aliquots at −20°C)	25 mL
HEPES (K salt; 1 M, pH 7.7; filter-sterilized and stored in aliquots at −20°C)	10 mL

Combine and bring to 1 L with MilliQ H$_2$O. Adjust pH to 7.7 with KOH. Make fresh on day of preparation, and store at 16°C until use.

ACKNOWLEDGMENTS

We thank Romain Gibeaux and Lily Einstein for helpful comments on the manuscript. Work in our laboratories is supported by National Institutes of Health (NIH) R35 GM118183 (R.H.) and a Burroughs Wellcome Fund Career Award (M.C.G.).

REFERENCES

Brown KS, Blower MD, Maresca TJ, Grammer TC, Harland RM, Heald R. 2007. Xenopus tropicalis egg extracts provide insight into scaling of the mitotic spindle. J Cell Biol 176: 765–770.

Desai A, Murray A, Mitchison TJ, Walczak CE. 1998. The use of Xenopus egg extracts to study mitotic spindle assembly and function in vitro. Methods Cell Biol 61: 385–412.

Glotzer M, Murray AW, Kirschner MW. 1991. Cyclin is degraded by the ubiquitin pathway. Nature 349: 132–138.

Good MC. 2016. Encapsulation of Xenopus egg and embryo extract spindle assembly reactions in synthetic cell-like compartments with tunable size. Methods Mol Biol 1413: 87–108.

Good MC, Vahey MD, Skandarajah A, Fletcher DA, Heald R. 2013. Cytoplasmic volume modulates spindle size during embryogenesis. Science 342: 856–860.

Hazel JW, Gatlin JC. 2018. Isolation and demembranation of Xenopus sperm nuclei. Cold Spring Harb Protoc doi: 10.1101/pdb.prot099044.

Levy DL, Heald R. 2010. Nuclear size is regulated by importin α and Ntf2 in Xenopus. Cell 143: 288–298.

Lohka MJ, Maller JL. 1985. Induction of nuclear envelope breakdown, chromosome condensation, and spindle formation in cell-free extracts. J Cell Biol 101: 518–523.

Lohka MJ, Masui Y. 1983. Formation in vitro of sperm pronuclei and mitotic chromosomes induced by amphibian ooplasmic components. Science 220: 719–721.

Maresca TJ, Heald R. 2006. Methods for studying spindle assembly and chromosome condensation in Xenopus egg extracts. Methods Mol Biol 322: 459–474.

Murray AW. 1991. Cell cycle extracts. Methods Cell Biol 36: 581–605.

Rape M, Reddy SK, Kirschner MW. 2006. The processivity of multiubiquitination by the APC determines the order of substrate degradation. Cell 124: 89–103.

Townsley FM, Aristarkhov A, Beck S, Hershko A, Ruderman JV. 1997. Dominant-negative cyclin-selective ubiquitin carrier protein E2-C/ UbcH10 blocks cells in metaphase. Proc Natl Acad Sci 94: 2362–2367.

Wilbur JD, Heald R. 2013. Mitotic spindle scaling during Xenopus development by kif2a and importin α. Elife 2: e00290.

Wuhr M, Chen Y, Dumont S, Groen AC, Needleman DJ, Salic A, Mitchison TJ. 2008. Evidence for an upper limit to mitotic spindle length. Curr Biol 18: 1256–1261.

Robustly Cycling *Xenopus laevis* Cell-Free Extracts in Teflon Chambers

Jeremy B. Chang[1,3,4] and James E. Ferrell, Jr.[1,2,4]

[1]Department of Chemical and Systems Biology, Stanford University School of Medicine, Stanford, California 94305-5174; [2]Department of Biochemistry, Stanford University School of Medicine, Stanford, California 94305-5307

A central advantage of studying *Xenopus laevis* is the manipulability of its cell-free extracts, which perform the cell cycle in vitro. However, these extracts are known to be experimentally temperamental and will often complete at most one or two cycles. Over the course of developing systems for imaging cell cycle events in extracts in real time, we unexpectedly found that when standard *Xenopus* extracts are placed in Teflon tubes, they cycle extremely robustly; in one series of experiments, over 90% ($n = 13$) of extracts cycled an average of seven and as many as 14 times. Extracts incubated in other materials, such as glass and polydimethylsiloxane, do not cycle as robustly. Here we present protocols for preparing *Xenopus* extracts and imaging them in Teflon tubes. This method extends the usefulness of this powerful model organism.

MATERIALS

It is essential that you consult the appropriate Material Safety Data Sheets and your institution's Environmental Health and Safety Office for proper handling of equipment and hazardous materials used in this protocol.

RECIPES: Please see the end of this protocol for recipes indicated by <R>. Additional recipes can be found online at http://cshprotocols.cshlp.org/site/recipes.

Reagents

ATP regeneration/energy mix <R> (optional; see Step 19)

Calcium ionophore (A23187) solution <R>
Make fresh on the day of extract preparation.

Cytochalasin B (10 mg/mL, MP Biomedicals ICN19511910)
Dissolve in DMSO and store in 50 µL aliquots at −20°C.

Dejelly solution #1 for *X. laevis* eggs <R>
Make fresh on the day of extract preparation.

Egg laying buffer (20×) <R>

---

[3]Present address: Department of Cellular and Molecular Pharmacology, University of California, San Francisco, California 94158-2140
[4]Correspondence: james.ferrell@stanford.edu; jeremy.chang@ucsf.edu

Cite this protocol as *Cold Spring Harb Protoc*; doi:10.1101/pdb.prot097212

Ethanol (70% v/v in water)

Extract buffer (XB) <R>

Extract buffer+ (XB+; XB with 10 µg/mL each LPC)

Dilute LPC protease inhibitor cocktail 1:1000 in XB. Prepare fresh and use immediately.

Extract buffer++ (XB++; XB+ with 100 µg/mL cytochalasin B)

Dilute 10 mg/mL cytochalasin B 1:100 in XB+. Prepare fresh and use immediately.

Human chorionic gonadotropin (HCG; Sigma-Aldrich CG10) or chorulon (A to Z Vet Supply), at a concentration of 1000 IU/mL

Dissolve in sterile MilliQ H_2O and store at 4°C.

LPC protease inhibitor cocktail <R>

Marc's modified Ringer's solution (MMR; 20×) <R>

Mature female *Xenopus laevis* frogs

Frogs are obtained from NASCO. Females are used to obtain eggs as described below.

Mineral oil (Macron #MK6357500)

MMR solution (0.2×)

Make fresh on the day of extract preparation.

Nyosil M25 lubricant

Pregnant mare serum gonadotropin (PMSG; 200 IU/mL; Prospec HOR-272)

Dissolve in sterile MilliQ H_2O and store at 4°C.

Purified proteins: GST-GFP-NLS (~1 µM final concentration) and/or fluorescently labeled tubulin (~0.1 µM final concentration)

GST-GFP-NLS can be purified using a bacterial expression vector encoding green fluorescent protein (GFP) tagged with both glutathione S-transferase and a nuclear localization sequence (Chatterjee et al. 1997). Bovine or porcine tubulin can be labeled on lysines with an NHS ester dye (Hyman et al. 1991) or purchased from Cytoskeleton Inc.

Equipment

Beakers, plastic (0.5 and 4 L) and glass (500 mL)

Beckman swinging-bucket rotor JS 13.1 and suitable centrifuge

Disposable transfer pipettes (Fisher 13-711-7M)

Dissection microscope

Eppendorf 5417R tabletop centrifuge with fixed-angle rotor FA-45-24-11

Fluorodish (World Precision Instruments FD5040-100)

Forceps

Inverted fluorescence microscope (e.g., a Leica DMI6000 B) with 5× objective

Needles (16- and 30-gauge)

Parafilm

Pasteur pipette or micropipette

Petri dish

Plastic containers for frogs (4 L; Corning) with tight-fitting lids and holes punched for air exchange

PTFE Teflon tubing (100 µm diameter; Cole-Parmer WU-06417-72)

Razor blade

Round-bottom Falcon tubes for JS 13.1 (14 mL; Falcon 352059)

Syringes (disposable; 1–5 mL)

Transfer pipettes (plastic, draw up to 3.4 mL per squeeze; Fisher)

Ultracentrifuge tubes (Beckman 326819)

Cite this protocol as *Cold Spring Harb Protoc*; doi:10.1101/pdb.prot097212

METHOD

The cycling extract procedure that follows is adapted from Murray's protocol (Murray 1991), references Good and Heald's description (see Protocol 1: Preparation of Cellular Extracts from Xenopus Eggs and Embryos [Good and Heald 2018]), and expands on the method described in Chang and Ferrell (2013). Treating the dejellied eggs with calcium ionophore releases them from cytostatic factor-mediated metaphase arrest. The prepared cytoplasm continues to transit the cell cycle, driven by cycles of synthesis and degradation of cyclin B (Murray and Kirschner 1989).

See online Movie 1 at cshprotocols.cshlp.org, which demonstrates the preparation of extract after collecting eggs. The steps in the movie do not correspond directly to those written below.

Preparing Extract

1. Induce ovulation of *Xenopus laevis* females and dejelly eggs as described in Steps 1–7 of Good and Heald's Egg Extract Procedure (see Protocol 1: Preparation of Cellular Extracts from *Xenopus* Eggs and Embryos [Good and Heald 2018]) with the following modifications:

    i. After inducing ovulation with HCG, place frogs into 1× Egg laying buffer rather than 1× MMR solution.

       Extracts have been prepared successfully using either buffer. The authors more frequently use 1× Egg laying buffer. Use as many frogs (typically 2–3) as needed to obtain enough high quality dejellied eggs to fill one ultracentrifuge tube.

    ii. Replace Step 4 with the following: prepare 1 L 0.2× MMR solution, 200 mL Dejelly solution, 100 mL Calcium ionophore solution, 1 L XB, 25 mL XB+, and 1 mL XB++. Gather a few transfer pipettes. Set centrifuge and JS 13.1 rotor to 4°C.

    iii. Wash eggs as written in Step 6, or alternatively, rely on the dejellying step and subsequent washes to remove debris.

    iv. In Step 7, to dejelly, add ∼50–75 mL of Dejelly solution and agitate for ∼1 min, then decant Dejelly solution and repeat several times. Once jelly coats are removed, do not add XB.

2. Wash: After decanting the remaining Dejelly solution, immediately wash eggs with 1 L 0.2× MMR solution over the course of ∼5 washes. Decant as much of the MMR solution as possible between washes.

    Throughout the washing and activation steps, egg quality can be checked using a dissection microscope.

3. Activate eggs: Add Calcium ionophore solution to eggs. Incubate for precisely 2 min with occasional swirling. A contraction of the animal pole pigment is often visible toward the end of the 2 min. During this period, use a transfer pipette to remove any irregular eggs.

    Without agitation, the eggs may stick to each other and the bottom of the glass beaker, but this will not noticeably impact extract quality.

4. Wash: Decant the ionophore solution and wash eggs immediately 4 times with a total of 1 L 1× XB. Continue using a transfer pipette to remove irregular eggs.

    Most often, the vast majority of eggs rotate with their animal poles up. Nevertheless, extracts made from batches that do not show completely uniform rotation often still execute the cell cycle.

5. Wash: Wash eggs twice in a total of 24 mL of 1× XB+.

6. Add 1 mL of XB++ to an ultracentrifuge tube.

7. Trim tips from transfer pipettes to increase opening diameter to >2 mm. Transfer ∼4 mL eggs into the ultracentrifuge tube using a trimmed transfer pipette. To minimize dilution of the cytochalasin B, transfer as little buffer into the ultracentrifuge tube as possible. Remove excess buffer with a Pasteur or P-1000 pipette without disturbing the eggs.

    Aside from buffer composition, the remaining steps of this protocol are similar to Steps 9–13 of Good and Heald's Xenopus laevis Egg Extract Procedure (see Protocol 1: Preparation of Cellular Extracts from Xenopus Eggs and Embryos [Good and Heald 2018]).

8. Layer 1 mL of Nyosil M25 lubricant on top of the eggs.

9. Transfer ultracentrifuge tube with forceps into a 14 mL round-bottom Falcon tube.

10. Pack eggs: 15 min after beginning calcium activation (Step 3), centrifuge tube in a Beckman JS 13.1 rotor at 4°C at 1000 RPM (157g) for 60 sec, then 600g for 30 sec.

11. Aspirate remaining buffer/oil using a Pasteur or P-1000 pipette. Tilt the tube to remove as much oil/buffer as possible.

12. Place tube on ice for 15 min.

13. Crush eggs by centrifugation at 10,000 RPM (~16,000g) in JS 13.1 rotor for 10 min at 4°C.

14. Wipe outside of the tube clean with 70% ethanol and allow to dry.

15. Pierce ultracentrifuge tube with a 16-gauge needle attached to a 5 mL syringe. Entry point should be toward the bottom of the straw-colored cytosolic layer (the cytosolic layer is between the bottom darker yolk layer and the top lipid layer. Immediately underneath the lipid layer there are typically droplets of remaining Nyosil oil, which mark the top of the desired cytosolic layer). Slowly withdraw crude extract over the course of ~5 min. ~4 mL of packed eggs typically yields ~1 mL crude extract.

16. Expel extract into a 1.5 mL Eppendorf tube and supplement with 1:1000 each of leupeptin, pepstatin, chymostatin, and cytochalasin B (10 μg/mL final each). Mix several times by inversion.

17. Centrifuge crude extract at 16,000g (13,000 RPM) in a tabletop centrifuge for 6 min at 4°C.

18. Remove the top residual lipid layer by aspiration with a 30-gauge needle connected to a vacuum.

19. Trim a P-1000 pipette tip with scissors, and transfer the extract to a fresh Eppendorf tube, leaving behind the darkly pigmented portion at the bottom.

    Depending on the purpose of the experiment, extracts may be supplemented with ATP regeneration/energy mix. 30 μL of the energy mix is added to 970 μL of extract. We found that this energy mix sometimes prevents extracts from cycling, so it is not used unless attempting to do a very long (>8 cycles) experiment.

    Extracts are often supplemented with demembranated Xenopus sperm chromatin, which may be prepared using a protocol from the laboratory of Johannes Walter (Lebofsky et al. 2009). See also Sec. 8, Protocol 1: Isolation and Demembranation of Xenopus Sperm Nuclei (Hazel and Gatlin 2018).

Imaging Extracts in Teflon Tubing

20. Cut PTFE Teflon tubing into sections several cm long.

21. Add desired purified fluorophores (e.g., GST-GFP-NLS [~1 μM final concentration] and rhodamine-labeled tubulin [~0.1 μM final concentration] for monitoring cell cycle state) to extract. ~100 μL total volume is appropriate for the subsequent steps of loading into PTFE Teflon tubes, which can involve trial and error.

22. Pipette a ~10 μL droplet of the extract onto a piece of Parafilm.

23. Load extract into PTFE Teflon tube by dipping one end of the tube into the droplet of extract and using a P-20 pipette tip connected to a vacuum to aspirate the extract into the tube.

24. Immerse tubes filled with extract in a bath of mineral oil in a Petri dish. Cut tubes to desired length, usually ~1 cm.

    Sealing tubes is not necessary because the mineral oil keeps the extract from leaking out.

25. Using clean forceps, transfer cut tubes to an imaging dish such as a Fluorodish filled with mineral oil.

26. Image on an inverted fluorescence microscope.

    See Figure 1 for an example of monitoring the cell cycle in these tubes. Images were taken once every minute.

 Cite this protocol as *Cold Spring Harb Protoc*; doi:10.1101/pdb.prot097212

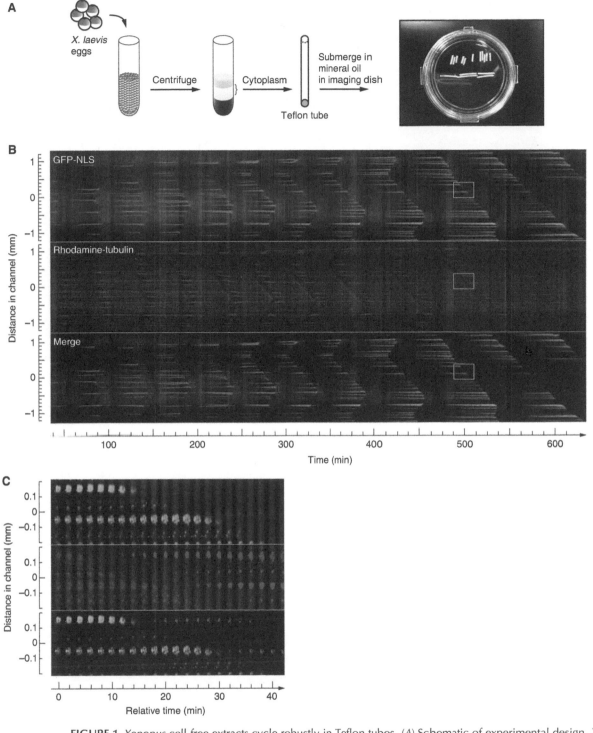

FIGURE 1. *Xenopus* cell-free extracts cycle robustly in Teflon tubes. (*A*) Schematic of experimental design. *Xenopus* eggs are crushed by centrifugation, and the cytoplasmic fraction is removed and loaded into Teflon tubes. Many tubes are shown in the dish. (*B*) Kymograph of fluorescence microscopy images of an extract executing multiple cell cycles. Green corresponds to GST-GFP-NLS, which labels the nuclei; red corresponds to rhodamine-tubulin. The kymograph shows ∼35 nuclei cycling in a metachronous and coordinated fashion across the tube. Various events may be observed in real-time: nuclear envelope breakdown (disappearing green signal), accumulation of microtubules at sites of nuclear envelope breakdown (increasing red signal), anaphase (disappearing red signal), and reformation of the nuclear envelope (reappearing green signal). (*C*) An enlarged version of the region indicated in *B*.

TABLE 1. Comparison of number of cycles observed for extracts incubated in different materials

Channel material	Additives	Number of cycles	Notes
PTFE Teflon	—	7.2 ± 3 ($n = 13$)	Eight of 13 extracts were still cycling at end of experiment.
PTFE Teflon	ATP regeneration mix	>14 ($n = 1$)	Extract was still cycling at end of experiment.
Glass	—	1 ± 0 ($n = 6$)	
Glass + PLL-PEG	—	1 ($n = 1$)	
PDMS	—	1.5 ± 0.7 ($n = 11$)	
PDMS + PLL-PEG	—	1 ($n = 1$)	

DISCUSSION

Cell-free *Xenopus laevis* extracts are described in Murray and Kirschner (1989) and Smythe and Newport (1992).

Using this protocol, we found that the extracts cycled robustly in Teflon but not in other materials. We prepared extract and loaded it into channels made from PTFE Teflon, glass, polydimethylsiloxane (PDMS), and glass and PDMS coated with poly-L-lysine-polyethyleneglycol. In aggregate, over many different preparations, extracts in glass on average completed 1 ± 0 ($n = 6$) round of nuclear envelope breakdown, and extracts in PDMS completed an average of 1.5 ± 0.7 ($n = 11$) rounds, whereas extracts completed an average of 7.2 ± 3 ($n = 13$) rounds in Teflon (Table 1). Taken together, these data suggest that Teflon has an undetermined characteristic—perhaps its hydrophobic and nonbiofouling nature—that enables the extract to cycle more robustly than in other materials.

Finally, we note that Teflon tubes are suboptimal for high resolution imaging. This is due to their shape, the poor optical transparency of PTFE, and a mismatch between the index of refraction of water (~1.33), mineral oil (~1.47), and the glass coverslip (~1.52). To address this deficiency, we coated a glass coverslip with Teflon AF, a type of Teflon with an index of refraction of ~1.32, nearly matching that of water. We were then able to take confocal images of extracts (data not shown). Further refinement of this technique may allow for higher resolution microscopy studies of extracts.

RECIPES

ATP Regeneration/Energy Mix

Reagents	Quantity
Creatine phosphate (Sigma-Aldrich P7936)	383 mg
ATP (Sigma-Aldrich A6419)	110 mg
EGTA (potassium salt; 0.5 M; pH 7.7 with KOH; filter-sterilized and stored at 4°C)	40 μL
$MgCl_2$ (2 M)	100 μL

Combine with MilliQ H_2O to a volume of 10 mL. Split into 1-mL aliquots, and store at −20°C.

Calcium Ionophore Solution

Add 5 μL of calcium ionophore A23187 (Sigma-Aldrich C7522; 10 mg/mL stock in DMSO; stored at 4°C protected from light) to 100 mL of 0.2× Marc's modified Ringer's (MMR; from 20× stock <R>). Prepare fresh and use immediately.

Cite this protocol as *Cold Spring Harb Protoc*; doi:10.1101/pdb.prot097212

Dejelly Solution #1 for X. laevis *Eggs*

Reagent	Quantity	Concentration
L-cysteine (e.g., Sigma-Aldrich C7352)	20 g	2%
Extract buffer (XB) salts (20×) <R>	50 mL	1×

Combine with 900-mL MilliQ H_2O, adjust pH to 7.8 with 10-M NaOH, and bring to 1 L with MilliQ H_2O. Make fresh on day of preparation, and store at 16°C until use.

Egg Laying Buffer (20×)

Reagent	Concentration
NaCl	2 M
KCl	40 mM
$MgSO_4$	20 mM
$CaCl_2$	50 mM
HEPES (potassium salt; 1 M; pH 7.7 with KOH; filter-sterilized and stored at −20°C)	10 mM
EDTA	2 mM

Adjust pH of 20× buffer to 7.4 with 10 M NaOH. Store at room temperature, and dilute to 1× with frog tank water before use.

Extract Buffer (XB) Salts (20×)

Reagent	Quantity	Concentration
KCl	149.12 g	2 M
$CaCl_2$	0.294 g	2 mM
$MgCl_2$ (2 M; e.g., Sigma-Aldrich 68475)	10 mL	20 mM

Combine with MilliQ H_2O to a volume of 1 L. Autoclave and store at 4°C.

LPC Protease Inhibitors

Leupeptin (Sigma-Aldrich L2023)
Pepstatin (Sigma-Aldrich P5318)
Chymostatin (MP Biomedicals ICN15284525)

Prepare each inhibitor as a 10 mg/mL stock in DMSO, and store aliquots at −20°C. Alternatively, combine inhibitors in a single stock, and store at −20°C.

Marc's Modified Ringer's (MMR; 20×)

Reagent	Quantity	Final concentration
HEPES, free acid	95.32 g	100 mM
EDTA	2.98 g	2 mM
NaCl	467.52 g	2 M
KCl	11.93 g	40 mM
$CaCl_2$	23.52 g	40 mM
$MgCl_2$ (e.g., Sigma-Aldrich 68475)	40 mL of 2 M stock	20 mM

Combine with MilliQ H_2O, adjust pH to 7.8–7.9 using ~8.5 g of NaOH pellets, and bring to 4 L. Autoclave and store at room temperature.

XB (Extract Buffer)

Reagent	Volume
Extract buffer (XB) salts (20×) <R>	50 mL
Sucrose (2 M; filter-sterilized and stored in aliquots at −20°C)	25 mL
HEPES (K salt; 1 M, pH 7.7; filter-sterilized and stored in aliquots at −20°C)	10 mL

Combine and bring to 1 L with MilliQ H$_2$O. Adjust pH to 7.7 with KOH. Make fresh on day of preparation, and store at 16°C until use.

ACKNOWLEDGMENTS

J.B.C. thanks Carsten Langrock and Tom Carver for guidance on coating glass with Teflon AF, and Joe Pomerening, Sun Young Kim, Andy Poon, Xianrui Cheng, Tek Hyung Lee, and members of the Ferrell and Straight laboratories for advice. This work was supported by grants from the National Institutes of Health (R01 GM110564 and P50 GM107615).

REFERENCES

Chang JB, Ferrell JE Jr. 2013. Mitotic trigger waves and the spatial coordination of the *Xenopus* cell cycle. *Nature* 500: 603–607.

Chatterjee S, Javier M, Stochaj U. 1997. In vivo analysis of nuclear protein traffic in mammalian cells. *Exp Cell Res* 236: 346–350.

Good MC, Heald R. 2018. Preparation of cellular extracts from *Xenopus* eggs and embryos. *Cold Spring Harb Protoc* doi: 10.1101/pdb.prot097055.

Hazel JW, Gatlin JC. 2018. Isolation and demembranation of *Xenopus* sperm nuclei. *Cold Spring Harb Protoc* doi: 10.1101/pdb.prot099044.

Hyman A, Drechsel D, Kellogg D, Salser S, Sawin K, Steffen P, Wordeman L, Mitchison T. 1991. Preparation of modified tubulins. *Methods Enzymol* 196: 478–485.

Lebofsky R, Takahashi T, Walter JC. 2009. DNA replication in nucleus-free *Xenopus* egg extracts. *Methods Mol Biol* 521: 229–252.

Murray AW. 1991. Cell cycle extracts. *Method Cell Biol* 36: 581–605.

Murray AW, Kirschner MW. 1989. Cyclin synthesis drives the early embryonic cell cycle. *Nature* 339: 275–280.

Smythe C, Newport JW. 1992. Coupling of mitosis to the completion of S phase in *Xenopus* occurs via modulation of the tyrosine kinase that phosphorylates p34cdc2. *Cell* 68: 787–797.

Cite this protocol as *Cold Spring Harb Protoc*; doi:10.1101/pdb.prot097212

Protein Immunodepletion and Complementation in *Xenopus laevis* Egg Extracts

Christopher Jenness,[1] David J. Wynne,[1,2,3] and Hironori Funabiki[1,4]

[1]*Laboratory of Chromosome and Cell Biology, The Rockefeller University, New York, New York 10065;*
[2]*Department of Biology, The College of New Jersey, Ewing, New Jersey 08628*

The *Xenopus* egg extract system has been widely used to study cell cycle events, including DNA replication, nuclear envelope formation, spindle assembly, chromosome condensation and kinetochore formation. The functional roles of the proteins involved in these processes can be determined by immunodepleting a protein of interest from the extract. As immunodepletion may result in co-depletion of other proteins, the protein of interest can be added back to the extract to verify its function. Additionally, proteins harboring point mutations or domain deletions may be added to assess their functions. Here we outline the immunodepletion procedure and two separate methods for restoring a protein of interest: addition of either a recombinant protein or an mRNA that supports translation in egg extracts. The tradeoffs between these two methods are discussed.

MATERIALS

It is essential that you consult the appropriate Material Safety Data Sheets and your institution's Environmental Health and Safety Office for proper handling of equipment and hazardous materials used in this protocol.

RECIPES: Please see the end of this protocol for recipes indicated by <R>. Additional recipes can be found online at http://cshprotocols.cshlp.org/site/recipes.

Reagents

Antibody against protein of interest
Antibody wash/coupling buffer <R>
CSF-arrested *Xenopus laevis* egg extract (see Protocol 1: Preparation of Cellular Extracts from *Xenopus* Eggs and Embryos [Good and Heald 2018]; Murray 1991)
IgG (purified)
Protein A Dynabeads (Life Technologies 10001D) (30 mg/mL)
Reagents for mRNA expression method (Steps 16–30)

4× SDS sample buffer <R>
CaCl$_2$ (10 mM)
Chloroform

---

[3]Present address: Department of Biology, University of Portland, Portland, Oregon 97203
[4]Correspondence: funabih@rockefeller.edu

Copyright © 2023 Cold Spring Harbor Laboratory Press; all rights reserved
Cite this protocol as *Cold Spring Harb Protoc*; doi:10.1101/pdb.prot097113

EDTA (500 mM [pH 8.0])

Ethanol (100%)

Isopropanol

mMESSAGE mMACHINE T7 or SP6 transcription kit (Ambion AM1344 or AM1340) (including nuclease-free H$_2$O, 2× NTP/CAP solution, 10× Reaction buffer, T7 or SP6 RNA polymerase, and TURBO DNase)

Phenol:chloroform:isoamyl alcohol (25:24:1 [pH 6.7])

Plasmid DNA template containing SP6 or T7 promoter and gene of interest

Include the Kozak sequence at the start codon. We prefer to include the 3'-UTR region from its gene, or from another gene known to translate efficiently (e.g., Xkid). The pCS2-based vector is a good choice for recloning, as it contains the SP6 promoter and SV40 poly(A) terminator. Commercially available Xenopus cDNAs are often cloned in vectors containing SP6 and T7 promoters, and may be used directly for in vitro transcription.

Restriction enzyme and buffer

Select an enzyme that cuts the plasmid DNA template downstream of the insert, and not within the template or promoter sequences.

Sodium acetate (3 M [pH 5.2])

Reagents for recombinant protein add-back method (Steps 13–15)

Phosphate or HEPES buffer compatible with egg extract (e.g., Sperm dilution buffer <R>)

Purified protein of interest

rProtein A Sepharose Fast Flow resin (GE 17127901)

Sperm dilution buffer <R>

Equipment

Centrifugal filter unit for protein concentration (e.g., Amicon UFC803024) (for recombinant protein add-back method)

Heat block at 95°C (for mRNA expression method)

Magnetic particle concentrator for 1.5-mL tubes (e.g., ThermoFisher A13346)

Microcentrifuge (for mRNA expression method)

Microcentrifuge tubes (1.5-mL)

Rotisserie rotator for 1.5-mL tubes

Spectrophotometer (for mRNA expression method)

Vortex mixer (for mRNA expression method)

Western blot system

METHOD

Cytostatic factor (CSF) extracts prepared from Xenopus laevis and arrested in metaphase II are used to study a variety of cell cycle processes. Here we outline immunodepletion and two procedures for protein add-back using recombinant protein or mRNA transcribed separately in vitro.

Antibody Bead Coupling

1. Add 100 µL of resuspended Protein A Dynabead slurry to a 1.5-mL microcentrifuge tube.

2. Collect the beads on a magnet for 30 sec and remove the supernatant.

3. Wash the beads by adding 500 µL of antibody wash/coupling buffer, resuspending by pipetting, and then collecting on a magnet. Repeat 2 times.

4. Resuspend the beads in 100 µL of antibody wash/coupling buffer containing 1–25 µg of antibody against the protein of interest.

Antibody amounts sufficient for depletion vary substantially between targets. Begin with 5 μg and evaluate depletion by western blot. For abundant proteins or low-affinity antibodies, it may be beneficial to perform two successive rounds of depletion with 50 μL of beads. For very abundant proteins (greater than 1 μM), high-capacity rProtein A sepharose resin can be used for depletion (Zierhut et al. 2014).

Antibodies nonspecifically attached to beads may leach into extracts during depletion, potentially causing a problem in some experiments. If this is the case, antibodies can be covalently cross-linked to beads (Kelly et al. 2007). Alternatively, excess Protein A Dynabeads can be added during immunodepletion to capture leached antibodies.

To control for the depletion protocol, mock depletion can be performed by coupling IgG purified from preimmune serum to Dynabeads.

5. Rotate the sample on a rotisserie rotator for 1 h at room temperature.

6. Collect the beads on a magnet for 30 sec, resuspend the beads in 500 μL of antibody wash/coupling buffer by pipetting, and then transfer the sample to a new 1.5-mL microcentrifuge tube.

7. Wash the beads three times with 500 μL of antibody wash/coupling buffer, followed by two times with sperm dilution buffer.

8. Transfer the Dynabeads to a new 1.5-mL microcentrifuge tube.

The beads can be stored for up to 1 wk at 4°C, until needed.

Performing Immunodepletion

9. Collect the antibody-coupled beads on a magnet and remove all supernatant.

10. Add 50 μL of *Xenopus* egg extract to the tube and resuspend the beads by flicking the tube firmly 3–10 times, trying to avoid generating froth.

11. Incubate the tube on ice for 1.5 h, flicking the tube after 45 min to resuspend the settled beads.

12. Collect the beads on a magnet for 5 min at 4°C and then transfer the extract to a new tube. Repeat two more times so that all beads are removed.

The depleted extract can be stored on ice until needed, for up to 4 h.

Restoring the Protein of Interest Using Recombinant Protein Add-Back

13. Dialyze the purified protein of interest into a buffer compatible with *Xenopus* egg extract.

Phosphate or HEPES buffers such as sperm dilution buffer are advised. Tris buffers are not advised (Maresca and Heald 2006).

14. Concentrate the protein until it will be at least 20× the endogenous concentration in egg extract to avoid diluting the extract.

Endogenous concentrations of most proteins in Xenopus egg extract have been reported (Wuhr et al. 2014).

15. Add the purified protein to the depleted extract and mix by flicking.

Analyze the level of added protein by western blot.

Restoring the Protein of Interest Using mRNA Expression

mRNA Synthesis

16. Linearize 25 μg of plasmid DNA template in a 200-μL restriction digest using 15 U of the appropriate enzyme.

The digest can be verified by agarose gel electrophoresis.

17. Add 10 μL of 500 mM EDTA and 20 μL of 3 M sodium acetate to the linearized template.

18. Add 200 μL of phenol:chloroform:isoamyl alcohol. Vortex the mixture for 5 sec, and then centrifuge in a microcentrifuge at 16,000 RCF for 5 min at room temperature.

19. Transfer the supernatant containing DNA to a new 1.5-mL microcentrifuge tube containing 200 µL of phenol:chloroform:isoamyl alcohol. Vortex the mixture for 5 sec, and then centrifuge in a microcentrifuge at 16,000 RCF for 5 min at room temperature.

20. Transfer the aqueous (top) layer containing DNA to a new 1.5-mL microcentrifuge tube containing 200 µL of chloroform. Vortex the mixture for 5 sec, and then centrifuge in a microcentrifuge at 16,000 RCF for 5 min at room temperature.

21. Transfer the aqueous (top) layer containing DNA to a new 1.5-mL microcentrifuge tube containing 500 µL of 100% ethanol. Vortex the mixture for 5 sec, and then incubate for 20 min at −20°C.

    DNA can be stored for at least several months at −20°C.

22. Pellet the DNA by centrifugation in a microcentrifuge at 16,000 RCF for 20 min at 4°C.

23. Completely remove the supernatant. Dry the pellet for 5 min at room temperature.

24. Resuspend the pellet in 15 µL of RNase-free H_2O.

25. Determine the concentration of DNA by measuring the optical density at 260 nm.

    DNA can be stored for 12 h at 4°C or at −20°C for long-term storage.

26. Transcribe and purify mRNA from the linearized plasmid as per the protocol provided with the mMessage Machine kit, including the following.

    i. Isolate RNA by phenol:chloroform extraction and isopropanol precipitation.

    ii. Resuspend the purified RNA in RNase-free H_2O at a final concentration of 10 mg/mL.

    iii. Prepare 2-µL aliquots and store long-term at −80°C.

mRNA Expression Add-Back

27. To 40 µL of CSF-arrested *Xenopus* egg extract, add purified mRNA from Step 26 to a final concentration of 100 ng/µL.

28. Add 1.2 µL of 10 mM $CaCl_2$ to release the extract into interphase.

    Depending on the 3′-UTR of the source mRNA, translation efficiency may vary by cell-cycle stages.

29. Allow translation from the added mRNA by incubating the extract for 90 min at 20°C.

30. To determine the amount of protein expressed, add 1 µL of extract to 19 µL of 1X SDS sample buffer. Vortex the solution for 5 sec, and then heat on a heat block for 5 min at 95°C. Perform western blotting on the sample using standard protocols.

    Expression levels around 100 nM can be achieved with mRNA add-back.

DISCUSSION

Here we discuss the tradeoffs between addition of recombinant protein and mRNA-dependent protein synthesis in *Xenopus* egg extract. Adding purified protein allows greater control of protein concentration, which is useful when multiple different proteins will be compared (for example, different point mutations). mRNA expression is less reproducible between extracts since translation efficiency depends on extract quality. mRNA addition is advantageous for proteins that are difficult to purify, or for protein complexes. For example, the chromosomal passenger complex comprising Aurora B, INCENP, Survivin and Dasra can be expressed with mRNA, avoiding the need to purify the entire complex (Kelly et al. 2007; Tseng et al. 2010). The original protocol for mRNA addition included a step to degrade endogenous RNAs using RNase A, followed by RNase A inhibition by RNAsin (Murray 1991). We found that this treatment sometimes affects the translational capacity of egg extracts, and is usually not necessary since endogenous mRNA is typically insufficient to complement depletion phenotypes.

Cite this protocol as *Cold Spring Harb Protoc*; doi:10.1101/pdb.prot097113

RECIPES

Antibody Wash/Coupling Buffer

Reagent	Amount to add
1 M HEPES (pH 8.0)	500 µL
5 M NaCl	1.5 mL

To prepare 50 mL of buffer, combine the above reagents and bring to 50 mL with MilliQ H_2O. Store at 4°C.

4× SDS Sample Buffer

Reagent	Amount to add
0.5 M Tris-Cl (pH 6.8)	15 mL
Sodium dodecyl sulfate (SDS)	1.2 g
80% glycerol	9.4 mL
Bromophenol blue	3 mg

To prepare 30 mL of buffer, combine the above reagents and bring to 30 mL with MilliQ H_2O. Store at room temperature. To use, dilute to 1× in H_2O and add 1:10 volume (of 4× buffer used) of 2-mercaptoethanol.

Sperm Dilution Buffer

Reagent	Amount to add (for 5× solution)	Starting concentration	Final concentration (5×)
Sucrose	5.625 mL	2 M	750 mM
K-HEPES (pH 7.7)	0.75 mL	1 M	50 mM
$MgCl_2$	75 µL	1 M	5 mM
KCl	2.5 mL	3 M	500 mM

To prepare 15 mL of 5× sperm dilution buffer, combine the reagents listed above and then add H_2O to 15 mL. Filter through a 0.22-µm filter. Dilute to 1× in H_2O for use. Store at −20°C.

ACKNOWLEDGMENTS

H.F. is supported by a grant from the National Institutes of Health (NIH) (R01GM075249).

REFERENCES

Good MC, Heald R. 2018. Preparation of cellular extracts from *Xenopus* eggs and embryos. *Cold Spring Harb Protoc* doi: 10.1101/pdb.prot097055.

Kelly AE, Sampath SC, Maniar TA, Woo EM, Chait BT, Funabiki H. 2007. Chromosomal enrichment and activation of the aurora B pathway are coupled to spatially regulate spindle assembly. *Dev Cell* 12: 31–43.

Maresca TJ, Heald R. 2006. Methods for studying spindle assembly and chromosome condensation in *Xenopus* egg extracts. *Methods Mol Biol* 322: 459–474.

Murray AW. 1991. Cell cycle extracts. *Methods Cell Biol* 36: 581–605.

Tseng BS, Tan L, Kapoor TM, Funabiki H. 2010. Dual detection of chromosomes and microtubules by the chromosomal passenger complex drives spindle assembly. *Dev Cell* 18: 903–912.

Wuhr M, Freeman RM Jr, Presler M, Horb ME, Peshkin L, Gygi SP, Kirschner MW. 2014. Deep proteomics of the *Xenopus laevis* egg using an mRNA-derived reference database. *Curr Biol* 24: 1467–1475.

Zierhut C, Jenness C, Kimura H, Funabiki H. 2014. Nucleosomal regulation of chromatin composition and nuclear assembly revealed by histone depletion. *Nat Struct Mol Biol* 21: 617–625.

Assembly of Spindles and Asters in *Xenopus* Egg Extracts

Christine M. Field[1,2,3] and Timothy J. Mitchison[1,2]

[1]*Department of Systems Biology, Harvard Medical School, Boston, Massachusetts 02115;* [2]*Marine Biological Laboratory, Woods Hole, Massachusetts 02543*

Here, we provide methods for assembly of mitotic spindles and interphase asters in *Xenopus laevis* egg extract, and compare them to spindles and asters in the egg and zygote. Classic "cycled" spindles are made by adding sperm nuclei to metaphase-arrested cytostatic factor (CSF) extract and inducing entry into interphase extract to promote nucleus formation and DNA replication. Interphase nuclei are then converted to cycled spindles arrested in metaphase by addition of CSF extract. Kinetochores assemble in this reaction and these spindles can segregate chromosomes. CSF spindles are made by addition of sperm nuclei to CSF extract. They are less physiological and lack functional kinetochores but suffice for some applications. Large interphase asters are prepared by addition of artificial centrosomes or sperm nuclei to actin-intact egg extract. These asters grow rapidly to hundreds of microns in radius by branching microtubule nucleation at the periphery, so the aster as a whole is a network of short, dynamic microtubules. They resemble the sperm aster after fertilization, and the asters that grow out of the poles of the mitotic spindle at anaphase. When interphase asters grow into each other they interact and assemble aster interaction zones at their shared boundary. These zones consist of a line (in extract) or disc (in zygotes) of antiparallel microtubule bundles coated with cytokinesis midzone proteins. Interaction zones block interpenetration of microtubules from the two asters, and signal to the cortex to induce cleavage furrows. Their reconstitution in extract allows dissection of the biophysics of spatially regulated cytokinesis signaling.

MATERIALS

It is essential that you consult the appropriate Material Safety Data Sheets and your institution's Environmental Health and Safety Office for proper handling of equipment and hazardous materials used in this protocol.

RECIPES: Please see the end of this protocol for recipes indicated by <R>. Additional recipes can be found online at http://cshprotocols.cshlp.org/site/recipes.

Reagents

Actin-intact CSF extract for aster assembly (Field et al. 2017)

Calcium solution (100×; 40 mM $CaCl_2$ in dH_2O or 1× sperm dilution buffer)

CSF extract containing cytochalasin D for spindle assembly (see Protocol 1: Preparation of Cellular Extracts from *Xenopus* Eggs and Embryos [Good and Heald 2018]; Desai et al. 1999; Maresca and Heald 2006).

Ethanol 70%

Extract fix <R>

[3]Correspondence: Christine_Field@hms.harvard.edu

Fluorescently labeled tubulin (~10 mg/mL; ~100 μM)

Bovine brain tubulin is labeled on lysines with an ALEXA dye NHS ester as described (Hyman et al. 1991). High quality labeled tubulin can also be purchased from Cytoskeleton Inc. Labeling tubulin from Xenopus laevis eggs (Widlund et al. 2012) can create a probe with less background fluorescence since it incorporates more efficiently into microtubules in the extract. Other useful probes for imaging tubulin are Tau-based fluorescent protein fusions (Mooney et al. 2017) which can also provide a better signal to noise ratio.

Fluorescent protein probes

We use several labeling methods: pure green fluorescent protein (GFP) fusion proteins, pure proteins labeled with reactive dyes, and directly labeled antibodies to visualize endogenous proteins see (Field et al. 2017).

Formaldehyde (37% solution)

HEPES free acid (1 M stock; pH 7.4 with KOH)

Microtubule organizing centers (MTOCs)

Two types are used: (1) Aurora kinase A antibody on magnetic beads (AurkA beads); which nucleate large symmetric asters (Tsai and Zheng 2005; Ishihara et al. 2014a; Nguyen et al. 2014; Ishihara et al. 2016), and (2) sperm nuclei (see Sec. 8, Protocol 1: Isolation and Demembranation of Xenopus Sperm Nuclei [Hazel and Gatlin 2018]). Sperm nuclei contain a centrosome as well as DNA. Chromatin can nucleate microtubules (Heald et al. 1996), and it also sends an asymmetric signal via Aurora kinase B (AurkB) activity. As a result, sperm plus centrosomes nucleate asters that are polarized in their structure and active early in the cell cycle (Field et al. 2015).

Milli-Q deionized water

Poly-L-lysine-g-polyethylene glycol (PLL-g-PEG) stock solution (10 mg/mL in 10 mM HEPES, pH 7.4; PLL(20)-g[3.5]-PEG(2) from SuSoS, Dübendorf, Switzerland)

Store in small aliquots (~30–50 μL) at −80°C; dilute to 0.1 mg/mL in 10 mM Hepes pH 7.4 the day of use. Do not refreeze aliquots.

Sperm dilution buffer <R>

Can be prepared as 5× stock and stored at −20°C.

Sperm nuclei (1–5 × 10⁷/mL stock; see Sec. 8, Protocol 1: Isolation and Demembranation of *Xenopus* Sperm Nuclei [Hazel and Gatlin 2018])

VALAP <R>

Equipment

Bunsen burner

Cotton tipped applicators

Coverslips (22 × 22 and 18 × 18 mm, # 1.5; Plain glass or passivated)

Fluorescence microscope

Forceps (Dumoxel N2A from VWR # 101413-582 or something similar)

These reverse action, rounded tip forceps are very good for working with coverslips.

Glass beakers, 500 and 250 mL

Ice bucket

Metal support slides

We machined stainless steel or aluminum slides with 18 mm circular holes in the center. Slides are 2 mm thick. A single-hole slide is 24 × 75 mm and a 4-hole slide is 50 × 75 mm and 2 mm thick. These are assembled into imaging chambers (below) that can be viewed using both upright and inverted microscopes. See (Field et al. 2017).

Microfuge tubes (1.5 mL) Microscope slides Nitrogen or house air Parafilm

Teflon rack for coverslips (Life Technologies C14784)

Temperature block with metal inserts (VWR 12621-090 or something similar)

Water bath (20°C; We generally make our own with an ice bucket, water, and ice)

Wide bore pipette tips

METHOD

Cycled Spindle Assembly Procedure

1. Add 40 µL of CSF extract to a 1.5 µL microfuge tube on ice.

2. Add fluorescently labeled tubulin to 200 nM and sperm nuclei to a final concentration of 100–300 sperm nuclei/µL of extract.

3. Add 0.4 µL of 100× Calcium Solution (final concentration of 0.4 mM). Immediately flick the tube ~8 times carefully.

   It is important to mix in the calcium rapidly and completely. This releases the CSF arrest and the extract starts going into interphase.

4. Incubate for 90 min at 20°C, flicking tube occasionally.

5. Monitor progression of the reaction using a Squash Fixation: Pipette a 1 µL aliquot of the reaction on a microscope slide, overlay with 3 µL of Extract Fix and cover gently with an 18 × 18 coverslip. View by fluorescence microscopy.

   At 90 min, nuclei (visualized by Hoechst) should be large and round indicating successful interphase nuclear envelope assembly, which is essential for DNA replication. Microtubules (visualized by fluorescent tubulin) are typically disorganized at this time point. See Figure 1A.

6. Add 60 µL of CSF extract to the tube from Step 4. Flick the tube ~8 times carefully.

   This forces the extract back into CSF-mediated metaphase arrest.

7. Incubate for 30–60 min at 20°C, flicking tube occasionally.

8. Monitor reaction progression using a Squash Fix (see Step 5).

   At 30 min after CSF add-back (120 min total) nuclear envelopes should be broken down with chromatin starting to condense into individual aggregates. At 60 min after CSF add-back the extract should be dominated by robust bipolar spindles with chromatin at the metaphase plate. In a typical extract, 40%–60% of the total structures are bipolar spindles with the remainder either monopolar or large multipolar aggregates. See Figure 1A.

9. Add additional fluorescent probes as desired.

   For further information about probes see Field et al. 2017.

   Fluorescent probes can be added at any time in the reaction, since most spindle components turn over rapidly. We typically assemble spindles containing no probes, and add probes to smaller aliquots of the reaction then incubate for a few minutes to allow incorporation. In this way, multiple team members imaging different proteins can share the reaction. We typically initiate new assembly reactions every hour for ~6 h.

10. For live imaging proceed to Imaging Spindles and Asters Procedure (Step 23) below.

CSF Spindle Assembly Procedure

11. Add 40 µL of CSF extract to a 1.5 mL microfuge tube.

12. Add fluorescently labeled tubulin to 200 nM and sperm nuclei to a final concentration of 100–300 sperm nuclei/µL of extract. Gently mix by flicking.

13. Incubate at 20°C. Continue to flick tube occasionally during incubation.

14. At 15, 30, 45, and 60 min take 1 µL samples to test the progress of the reaction via Squash Fixation (Step 5).

    At 10 min, a small microtubule aster emanates from the sperm centrosome. By 20–30 min the chromatin has migrated away from the centrosome, and microtubules are polarized toward the chromatin (these structures are termed half spindles). They achieve bipolarity either by fusion (Sawin and Mitchison 1991) or spontaneous bipolarization (Mitchison et al. 2004). In a typical extract, 40%–60% of the total structures are bipolar spindles by 60 min. This number can vary from <10% in very poor extracts to >90% in the best extracts. See Figure 1B.

15. For live imaging proceed to Imaging Spindles and Asters Procedure (Step 23) below.

Cite this protocol as *Cold Spring Harb Protoc*; doi:10.1101/pdb.prot099796

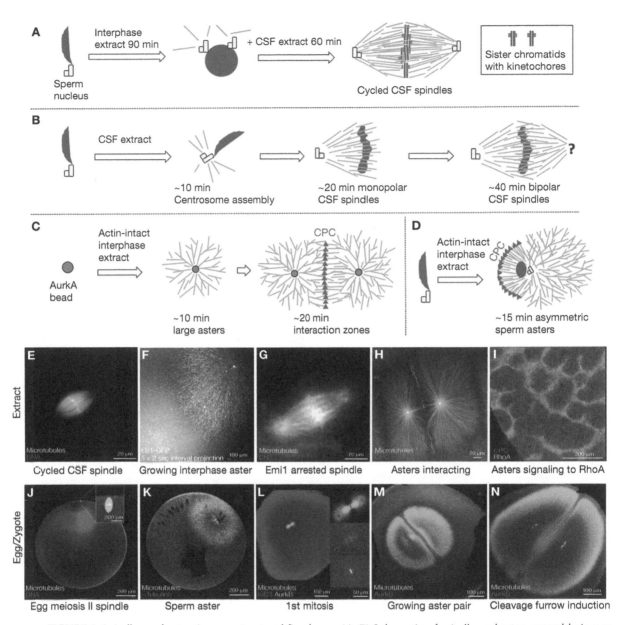

FIGURE 1. Spindles and asters in egg extract and fixed eggs. (*A–D*) Schematic of spindle and asters assembly in egg extract. Microtubules green, DNA blue, centrosomes yellow, AurkA beads purple, CPC red. (*E–I*) Spindles and interphase asters in egg extract by live widefield (*E–H*) or TIRF(I). (*J–N*) Spindles and interphase asters in fixed, cleared eggs by confocal immunofluorescence. (*A*) Cycled CSF spindles with replicated sister chromatids and functional kinetochores. (*B*) CSF spindles which lack functional kinetochores. (*C*) Interphase bead asters grow rapidly by branching nucleation, and form CPC-positive interaction zones when they meet. (*D*) Interphase sperm asters are similar to (*C*), but they grow asymmetrically due to CPC recruitment on the chromatin-proximal boundary (Field et al. 2015). (*E*) Cycled CSF spindle. Note lack of asters at poles indicating meiosis-II morphology. (*F*) Growing interphase bead aster. This image shows a projection of 5 sequential EB1-GFP images for plus end tracking (Ishihara et al. 2014b). (*G*) Cycled spindle arrested with Emi1 to model zygote mitotic spindles. Note asters at poles (Wühr et al. 2008). (*H*) Interphase bead asters interacting and recruiting CPC between them. (*I*) Signaling from aster interaction zones between bead asters to a RhoA.GTP reporter on a supported lipid bilayer (Nguyen et al. 2014). (*J*) Unfertilized egg with meiosis-II spindle. (*K*) Sperm aster during pronuclear migration. (*L*) 1st mitosis, early anaphase. Note asters at spindle poles. (*M*) Expanding sister asters after 1st anaphase. Note CPC-positive interaction zone. (*N*) Sister asters just before furrow induction. The furrow forms where the interaction zone touches the cortex.

Interphase Aster and Aster Interaction Zone Assembly Procedure

16. Add 25–100 µL of actin-intact CSF extract to a 1.5 mL microfuge tube.

17. Add fluorescently labeled tubulin to 200 nM and/or protein probes and mix well on ice.

    To observe aster interaction zone formation the probes should include one for imaging the chromosome passenger complex (CPC). For this we use GFP-DasraA, Alexa labeled anti-AurkB or anti-INCENP (Field et al. 2015). The antibodies tend to promote CPC activation, and should be added at the minimal concentration needed for visualization.

18. Add MTOCs (sperm nuclei or AurkA beads) and mix on ice.

    The final concentration must be experimentally determined so the asters are spaced as desired.

19. Add CaCl2 to a final concentration of 0.4 mM from 100× stock. Immediately flick the tube approximately eight times, or gently pipette the entire solution up and down three to five times using a wide bore tip to ensure mixing.

    It is important to mix well after calcium addition. This converts the reaction to interphase.

20. Incubate for 2–10 min at 20°C, depending on the timing of the process you wish to observe.

21. Incubate on ice for at least 2 min and flick the tube several times to break up any contracted gel within the extract.

    At 20°C, actin polymerizes and the extract forms a gel. The gel in CSF extract is contractile. The gel in interphase extract is not. After calcium addition, it takes several min for the extract to exit mitosis. During this time, the extract remains contractile. Incubation at 4°C and flicking helps break down contracted material. (Field et al. 2011).

22. For live imaging proceed to Imaging Spindles and Asters Procedure (Step 23) below.

    Upon warming reactions to room temperature (i.e., on a slide), asters start to grow. Interaction zones form when asters grow large enough to interact. Interaction zones recruit microtubule signaling complexes CPC and Centralspindlin. See Figure 1C,D.

Imaging Spindles and Asters Procedure

For more information about passivation please see The importance of coverslip passivation section of the Discussion. Perform all passivation steps (23–27) at room temperature. Treat each coverslip individually.

23. Hold coverslips (18 and 22 mm) with forceps and individually dip into a small beaker containing 70% ethanol. Swirl coverslips for 2–3 sec in ethanol, remove, flame with a Bunsen burner, and cool in air for 5 sec.

24. Place cooled coverslips onto droplets of PLL-g-PEG on a piece of Parafilm. Coverslips 22 × 22 mm in size require 110 µL of PEG solution, 18 × 18 mm coverslips require 90 µL. Allow coverslips to incubate for 15–30 min at room temperature.

    Do not incubate coverslips for much longer than 30 min, as they can dry out.

25. Wash coverslips 3× in Milli-Q water. For the first two washes place each coverslip, PEG side down, in a droplet of Milli-Q water and incubate for 5 min at room temperature. For the last wash, use forceps to dip each coverslip into a beaker of Milli-Q water and swirl.

26. Place coverslips in slots in a Teflon rack so they do not contact one another. Keep track of which surface is PEG treated.

27. Dry coverslips with a jet of nitrogen gas, to prevent residue from water droplets. Use coverslips the same day, or store for up to a week at room temperature in the dark (Field et al. 2017).

28. Assemble imaging chamber. Using a cotton-tipped applicator, dab four small drops of molten VALAP around the opening on a metal slide. The drops should be about 2–3 mm in diameter and spaced around and close to the opening. Let the drops harden. Place a 22 × 22 mm square passivated coverslip on top of the metal slide, covering the VALAP drops around the opening. Place the passivated (PEG-treated) side upward—facing away from the slide. Place the metal slide on a 68°C–70°C temp block. The drops of VALAP will melt and spread under the coverslip.

    Cite this protocol as *Cold Spring Harb Protoc*; doi:10.1101/pdb.prot099796

When this has occurred (~1 min), remove the metal slide and cool to room temperature. The hardened VALAP will seal the coverslip to the slide.

29. Pipette 4–12 µL of extract onto the center of the 22 mm^2 coverslip. We generally use 7.5 µL.

    A volume of 7.5 µL extract gives a squash nominally 23 µm in thickness. Lower volumes may improve imaging of MT by lowering background at the cost of more thickness variation across the coverslip. Higher volumes up to 25 µL increase three-dimensionality and may be more physiological. For imaging actin, thinner squashes are better, especially for CSF extract without cytochalasin addition, which is contractile.

30. Immediately and gently place an 18 × 18 mm coverslip on top with passivated side towards the extract.

    Try to avoid air bubbles. A few air bubbles will not hurt, but they may move and destroy nearby asters or spindles. They also slow depletion of oxygen in a zone ~300 µm wide around each bubble.

31. After the droplet spreads out fully (~5 sec) seal edges with VALAP so the sample does not dry out.

32. Observe by fluorescence microscopy. Figure 1E–I shows examples of microtubule assemblies in *Xenopus* egg extract. Figure 1J–N shows the corresponding assemblies in fixed *Xenopus* eggs or zygotes. The egg and zygote fixation protocol is described in Field et al. 2015.

    See Troubleshooting.

    Because cycled and CSF spindles are at steady state in structure and dynamics, time is not critical. However, it takes a few minutes for the mitochondria in extract to deplete oxygen, longer if bubbles are present. Photobleaching of chemical chromofluors will be fast until oxygen depletion has occurred. Note that ATP supply is not affected by deoxygenation (Niethammer et al. 2008). Interphase asters grow rapidly after warming to room temperature, so it is important to initiate imaging as soon as possible. To image early steps in aster growth we prepare unsealed squashes with the slide resting on a metal block in an ice bucket, to keep the squash cold until imaging. Condensed water vapor on the coverslip can be removed with a stream of dry air or nitrogen.

    The ideal temperature for preparing the squashes and imaging is 16°C–18°C. Adequate results can be obtained at temperatures up to ~25°C, although spindles are more likely to enter anaphase spontaneously at higher temperatures.

TROUBLESHOOTING

Problem (Step 32): Occasionally an extract preparation will be nonfunctional for either the interphase or mitotic reaction. Typical problems include fragmentation of chromatin, which may be due to induction of apoptosis, or an extract that gets stuck in interphase or mitosis.

Solution: If this occurs in the first reaction it is wise to terminate the experiment. After more than 4 h in the ice bucket even initially good extract may stop working for further assembly reactions. Probably the most important variables for a functional extract are the health of the frogs, care taken over "gardening" the eggs, i.e., removing dead or aberrant eggs before extract preparation, and the initial egg packing spin which removes excess buffer before crushing the eggs. Excess buffer dilutes the extract.

Problem (Step 32): The extract does not release from CSF arrest and enter interphase. This is a common problem for novices and can be detected via microscopy. Microtubules appear as spindle-like structures rather than large asters.

Solution: The ER rapidly uptakes calcium, decreasing the cytosolic concentration, so immediate and thorough mixing after calcium addition is crucial to release CSF arrest. Larger volumes may cause problems with mixing and gas exchange. Mix reactions in 1.5 mL tubes and limit reaction volumes to 100 µL at most.

DISCUSSION

General Considerations for Working with Extracts

Xenopus extracts are a powerful tool but their use can be frustrating because of high variability. The extracts are sensitive to physical perturbations and must be treated gently to ensure best results. Avoid vortexing or pipetting up and down vigorously. Use wide bore pipette tips (these can be purchased or created by cutting)—an ~2 mm diameter opening is optimal. Use a fresh tip for each aliquot, since extracts are viscous.

It is very important to avoid diluting extracts more that 10% of original volume. Keep in mind that some buffers and reagents can inhibit spindle assembly. When adding small molecules, we have found that the EC50 for drug action is often 10- to 1000-fold higher in extract than in tissue culture cells, although mechanism and specificity are maintained. We believe there are two reasons: (1) tissue culture cells are surrounded by a large reservoir of drug in the medium, and can accumulate drugs to a higher internal concentration; (2) hydrophobic drugs partition into lipid reservoirs and/or acidic compartments present in extract. For these reasons, EC_{50} for drug action in extract must be determined empirically, and we aim to use them at ~5× the EC_{50} to maximize specificity. If drugs are added from a dimethyl sulfoxide (DMSO) stock, it is important to keep the total percent DMSO at <0.5%. Drugs stocks are often prepared as concentrated stocks in DMSO, e.g., 1000×. These should be diluted in small volumes of egg extract to give 10×–50× final stocks. Dilution in aqueous buffer can cause hydrophobic drugs to precipitate, in which case the final concentration is unreliable.

The Importance of Coverslip Passivation

For live imaging experiments, we typically squash small volumes of extract between two coverslips to provide a 4–25 μm thick preparation. Cleaning and preparation of the glass surface to reduce nonspecific adsorption of molecules (passivation) is very important. Untreated glass tends to bind proteins nonspecifically and affect results. For example, nonspecific recruitment of motor proteins such as dynein can artificially influence microtubule organization, or sequestration of proteins to a coverslip can deplete the cytoplasmic pool. After cleaning, coverslips are typically passivated by treatment with PLL-g-PEG. Many PEG passivation schemes have been described in the literature, reviewed in (Field et al. 2017). Above we have included our current favorite, which is simple and fast. Cycled spindles are extremely robust, and not much affected by squashing under uncoated glass surfaces. Pre-2010, work from the Mitchison laboratory on cycled spindles generally used squashes between uncoated slides and coverslips. Passivation is much more important for interphase asters, which can be torn up by dynein bound to uncoated glass surfaces.

RECIPES

Extract Fix

Glycerol	60% (v/v)
Marc's modified Ringer's (MMR; 20×) <R>	1×
Hoechst (for live imaging of DNA, use Sigma-Aldrich BisBenzimide H 33342 trihydrochloride, B2261)	1 μg/mL
Formaldehyde	10%

Prepare fresh.

Cite this protocol as *Cold Spring Harb Protoc*; doi:10.1101/pdb.prot099796

Marc's Modified Ringer's (MMR; 20×)

Reagent	Quantity	Final concentration
HEPES, free acid	95.32 g	100 mM
EDTA	2.98 g	2 mM
NaCl	467.52 g	2 M
KCl	11.93 g	40 mM
CaCl$_2$	23.52 g	40 mM
MgCl$_2$ (e.g., Sigma-Aldrich 68475)	40 mL of 2 M stock	20 mM

Combine with MilliQ H$_2$O, adjust pH to 7.8–7.9 using ~8.5 g of NaOH pellets, and bring to 4 L. Autoclave and store at room temperature.

Sperm Dilution Buffer (SDB)

10 mM HEPES-KOH pH 7.8
1 mM MgCl$_2$
100 mM KCl
150 mM sucrose

Store at −20°C in 1-mL aliquots.

Valap

Add equal weight of lanolin, Vaseline (or other petroleum jelly), and paraffin together in a 1-liter beaker. Heat mixture at a low setting on a hot plate. Stir occasionally until thoroughly blended. Aliquot into several small screw-capped jars (~50-mL capacity). Store at room temperature.

ACKNOWLEDGMENTS

Work in our laboratories is supported by National Institutes of Health (NIH) grant GM39565 (T.J.M.), Marine Biological Laboratory (MBL) fellowships from the Evans Foundation, MBL Associates, and the Colwin Fund (T.J.M. and C.M.F.). Authors thank the Nikon Imaging Center at Harvard Medical School for microscopy support and the National *Xenopus* Resource at MBL for *Xenopus* animals and care.

REFERENCES

Desai A, Murray A, Mitchison TJ, Walczak CE. 1999. The use of *Xenopus* egg extracts to study mitotic spindle assembly and function in vitro. *Methods Cell Biol* 61: 385–412.

Field CM, Wühr M, Anderson GA, Kueh HY, Strickland D, Mitchison TJ. 2011. Actin behavior in bulk cytoplasm is cell cycle regulated in eary vertegrate embryos. *J Cell Sci* 124: 2086–2095.

Field CM, Groen AC, Nguyen PA, Mitchison TJ. 2015. Spindle-to-cortex communication in cleaving, polyspermic *Xenopus* eggs. *Mol Biol Cell* 26: 3628–3640.

Field CM, Pelletier JF, Mitchison TJ. 2017. *Xenopus* extract approaches to studying microtubule organization and signaling in cytokinesis. *Methods Cell Biol* 137: 395–435.

Good MC, Heald R. 2018. Preparation of cellular extracts from *Xenopus* eggs and embryos. *Cold Spring Harb Protoc* doi: 10.1101/pdb.prot097055.

Hazel JW, Gatlin JC. 2018. Isolation and demembranation of *Xenopus* sperm nuclei. *Cold Spring Harb Protoc* doi: 10.1101/pdb.prot099044.

Heald R, Tournebize R, Blank T, Sandaltzopoulos R, Becker P, Hyman A, Karsenti E. 1996. Self-organization of microtubules into bipolar spindles around artificial chromosomes in *Xenopus* egg extracts. *Nature* 382: 420–425.

Hyman A, Drechsel D, Kellogg D, Salser S, Sawin K, Steffen P, Wordeman L, Mitchison T. 1991. Preparation of modified tubulins. *Methods Enzymol* 196: 478–485.

Ishihara K, Nguyen PA, Groen AC, Field CM, Mitchison TJ. 2014a. Microtubule nucleation remote from centrosomes may explain how asters span large cells. *Proc Natl Acad Sci* 111: 17715–17722.

Ishihara K, Nguyen PA, Groen AC, Field CM, Mitchison TJ. 2014b. Microtubule nucleation remote from centrosomes may explain how asters span large cells. *Proc Natl Acad Sci* 111: 17715–17722.

Ishihara K, Korolev KS, Mitchison TJ. 2016. Physical basis of large microtubule aster growth. *Elife* 5: e19145.

Maresca TJ, Heald R. 2006. Methods for studying spindle assembly and chromosome condensation in *Xenopus* egg extracts. *Methods Mol Biol* **322**: 459–474.

Mitchison TJ, Maddox P, Groen A, Cameron L, Perlman Z, Ohi R, Desai A, Salmon ED, Kapoor TM. 2004. Bipolarization and poleward flux correlate during *Xenopus* extract spindle assembly. *Mol Biol Cell* **15**: 5603–5615.

Mooney P, Sulerud T, Pelletier JF, Dilsaver MR, Miroslav T, Geisler C, Gatlin JC. 2017. Tau-based fluorescent protein fusions to visualize microtubules. *Cytoskeleton* **74**: 221–232.

Nguyen PA, Groen AC, Loose M, Ishihara K, Wühr M, Field CM, Mitchison TJ. 2014. Spatial organization of cytokinesis signaling reconstituted in a cell-free system. *Science* **346**: 244–247.

Niethammer P, Kueh HY, Mitchison TJ. 2008. Spatial patterning of metabolism by mitochondria, oxygen, and energy sinks in a model cytoplasm. *Curr Biol* **18**: 586–591.

Sawin KE, Mitchison TJ. 1991. Mitotic spindle assembly by two different pathways in vitro. *J Cell Biol* **112**: 925–940.

Tsai M-Y, Zheng Y. 2005. Aurora A kinase-coated beads function as microtubule-organizing centers and enhance RanGTP-induced spindle assembly. *Curr Biol* **15**: 2156–2163.

Widlund PO, Podolski M, Reber S, Alper J, Storch M, Hyman AA, Howard J, Drechsel DN. 2012. One-step purification of assembly-competent tubulin from diverse eukaryotic sources. *Mol Biol Cell* **23**: 4393–4401.

Wühr M, Chen Y, Dumont S, Groen AC, Needleman DJ, Salic A, Mitchison TJ. 2008. Evidence for an upper limit to mitotic spindle length. *Curr Biol* **18**: 1256–1261.

Cite this protocol as *Cold Spring Harb Protoc*; doi:10.1101/pdb.prot099796

Dissecting Protein Complexes in Branching Microtubule Nucleation Using Meiotic *Xenopus* Egg Extracts

Jae-Geun Song and Sabine Petry[1]

Department of Molecular Biology, Princeton University, Princeton, New Jersey 08544

The mitotic spindle is the microtubule-based apparatus that reliably segregates chromosomes during cell division. Recently, it was discovered that microtubules originate within the mitotic spindle by nucleating off of existing spindle microtubules. This mechanism, termed branching microtubule nucleation, allows the efficient amplification of microtubules while preserving their original polarity as required in the spindle. Three molecular players are known to be involved in this process, namely, the protein TPX2, the protein complex augmin, and the gamma-tubulin ring complex; however, little is known about the assembly of the protein complexes. Here, we use the eight-subunit augmin complex as an example of how to dissect the function and assembly of a protein complex using meiotic *Xenopus* egg extracts. Specifically, immunodepletion combined with total internal reflection fluorescence (TIRF) microscopy is used to identify the role of the protein complex. In parallel, immunoprecipitation (IP) and tandem mass spectrometry (MS/MS) are used to infer how it is assembled. This approach can be applied to investigate the assembly of other multisubunit protein complexes that function in branching microtubule nucleation and mitotic spindle assembly.

MATERIALS

It is essential that you consult the appropriate Material Safety Data Sheets and your institution's Environmental Health and Safety Office for proper handling of equipment and hazardous materials used in this protocol.

RECIPES: Please see the end of this protocol for recipes indicated by <R>. Additional recipes can be found online at http://cshprotocols.cshlp.org/site/recipes.

Reagents

Antibodies targeting augmin subunits

> *Generate antibodies targeting augmin subunits, or order from a commercial vendor. Use immunoglobulin G (IgG) from rabbit serum (Sigma-Aldrich) as a control antibody.*

CSF-arrested *Xenopus laevis* egg extract

> *Prepare CSF-arrested* Xenopus laevis *egg extract as described in Protocol 1: Preparation of Cellular Extracts from* Xenopus *Eggs and Embryos (Good and Heald 2018) with the following modification: The packing spin, during which the interstitial buffer is removed without lysing the eggs, should be performed in a table centrifuge at 150g for 60 sec, followed by 598g for 25 sec.*

CSF-XB <R>

---

[1]Correspondence: spetry@princeton.edu

Cy5-labeled porcine tubulin (20 μM in CSF-XB) (Peloquin et al. 2005)

Dynabeads Protein A (Thermo Scientific)

Glycine (0.1 M [pH 2.5]) (optional; see Step 24)

mCherry-EB1 (2 μM in CSF-XB, purified as described by Petry et al. [2011])

RanQ69L (200 μM in CSF-XB, purified as described by Weis et al. [1996] and Petry et al. [2013])

Sodium orthovanadate (Vanadate) (10 mM in CSF-XB)

TBS-T (Tris-buffered saline [pH 7.4] with 0.1% Tween 20)

Equipment

Forceps

Glass coverslips (22 mm × 22 mm)

Immersion oil

Magnetic separation rack

Microcentrifuge tubes (1.5-mL)

Microscope slides

Nail polish

Tape (double-sided)

TIRF (total internal reflection fluorescence) microscope, equipped with a 100 × 1.49 NA objective, an electron-multiplying charge-coupled device (EM-CCD) or complementary metal-oxide semiconductor (CMOS) camera, and laser power of at least 20 mW out-of-fiber

METHOD

Coupling Antibodies to Magnetic Protein A Beads

1. Transfer 150 μL of Dynabeads Protein A suspension to a microcentrifuge tube.

2. Retrieve the beads using a magnetic separation rack. Exchange the buffer with 150 μL of TBS-T and resuspend the beads.

3. Repeat Step 2 twice.

4. Prepare a suspension of 35 μg of antibody targeting augmin subunits in TBS-T. Add the suspension to the beads and gently mix using a rotator overnight at 4°C or for 2 h at room temperature.

   This antibody amount is above the maximum binding capacity of the beads.

   After mixing, keep beads at 4°C until ready to proceed with Step 19.

Preparing the Flow Cell

5. Place two strips of double-sided tape onto a microscope slide to make a flow channel with a 2- to 3-mm gap, which will accommodate ~5 μL of extract (Fig. 1A).

6. Place a glass coverslip over the flow channel.

7. Seal the flow cell by gently pressing onto the glass coverslip with the blunt end of a forceps.

Performing the Branching Microtubule Nucleation Assay

8. Aliquot 8.0 μL of CSF-arrested egg extract into a 1.5-mL tube on ice, pipetting as gently as possible.

9. Gently add 0.5 μL of mCherry EB1, 0.5 μL of Vanadate, 0.5 μL of RanQ69L, and 0.5 μL of Cy5-tubulin to the extract to prepare a 10.0-μL reaction mixture.

   Reagents should be individually and freshly added to the extract. Do not make a premixture of reagents.

10. Gently pipette the extract up and down once or twice on ice to mix the reagents.

Cite this protocol as *Cold Spring Harb Protoc*; doi:10.1101/pdb.prot100958

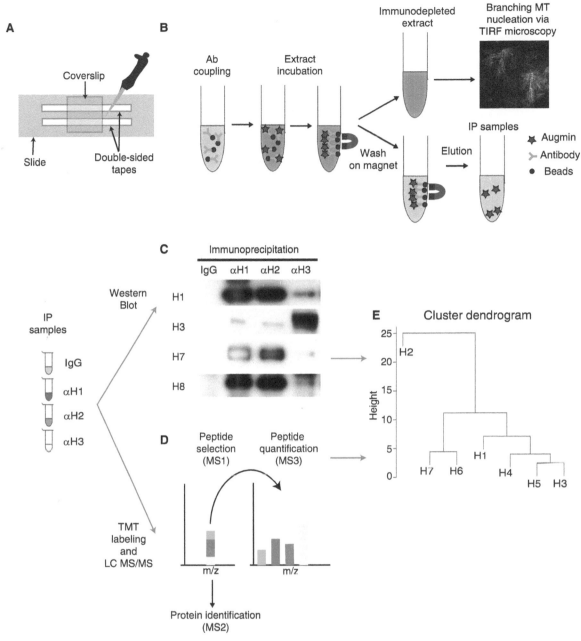

FIGURE 1. Schematic diagram of the protocol. (*A*) Preparation of the flow cell. (*B*) Immunoprecipitation (IP) and immunodepletion (ID) using Protein A Dynabeads and *Xenopus* egg extracts, followed by the branching microtubule nucleation assay via TIRF microscopy. (*C*) IP samples obtained using IgG, anti-HAUS1 (αH1), -anti-HAUS2 (αH2), and anti-HAUS3 (αH3) antibodies are analyzed by western blot to quantify the intensities of augmin subunits (H1-8). (*D*) Quantitative LC MS/MS using IP samples. Sequential mass spectrometry analyses allow identifying augmin subunits and relative amounts in each IP sample. (*E*) Hierarchical clustering dendrogram is generated by hclust package of R using normalized intensities of augmin subunits in each sample. Seven augmin subunits are depicted as H1-7 (H8 is not identified), showing a putative assembly model for augmin.

11. Add the extract mixture to the flow cell.

12. (Optional) Time the start of the reaction.

   The reaction begins as soon as the extract mixture is introduced to the flow channel and warmed to room temperature.

13. (Optional) Seal the flow cell with nail polish to prevent drying.

14. Place the coverslip face-down on the 100× 1.49N objective of a TIRF microscope with immersion oil. Image the sample.

15. Optimize the TIRF angle, laser power, and exposure time to visualize mCherry-EB1 and Cy5-microtubules at the best contrast and without photobleaching.

16. Begin imaging the branching microtubule nucleation (Petry et al. 2013).

    Microtubule nucleation activity can vary based on the surrounding temperature. A constant temperature of 18°C is ideal for imaging this reaction.

17. (Optional) Collect images every 2 sec for 15–30 min.

18. If the extract quality is sufficient to activate the branching reaction while maintaining low background nucleation, proceed to immunodepletion.

Performing Immunodepletion and Immunoprecipitation

A schematic diagram is provided in Figure 1B.

19. Using the magnetic separation rack, wash the antibody-coupled beads from Step 4 three times with 150 µL of TBS-T followed by twice with 150 µL of CSF-XB.

20. Remove the CSF-XB from the beads and place the tubes on ice.

21. Add 65 µL of CSF-arrested egg extract to the antibody-conjugated beads. Mix the extract with the beads by gentle pipetting.

    Wide-bore tips are preferred for mixing the extract.

22. Incubate the extract with the beads on ice for a total of 45 min, gently mixing every 15 min.

    Do not use a rotator to mix the extract, as mechanical stress to the extract should be minimized.

23. Using the magnetic separation rack, retrieve the immunodepleted extract from the beads and transfer it to a fresh tube on ice until ready to proceed with Step 25.

    Because of the viscosity of the extract, it takes 10 min to completely separate the extract from the beads on the magnetic separation rack.

24. (Optional) For further analysis of the immunoprecipitation samples, wash the beads three times with 150 µL of TBS-T and then elute with 100 µL of 0.1 M glycine (pH 2.5).

    Samples can be further analyzed by SDS-PAGE and western blot or mass spectrometry (see Discussion).

Performing Branching Microtubule Nucleation Using Immunodepleted Extracts

25. Prepare 8.0-µL aliquots of immunodepleted extract samples in tubes on ice.

26. Perform the branching reaction and initiate imaging using TIRF microscopy as described in Steps 9–17.

27. Compare the activities of the branching reactions among immunodepleted extract samples.

DISCUSSION

This method provides an initial approach to analyzing the endogenous assembly of augmin (or other protein complexes), which is essential for branching microtubule nucleation, in *Xenopus laevis* egg extracts. Additional approaches include the analysis of immunoprecipitated samples by western blot, which can be used to calculate the relative intensities of subunits in a protein complex (Fig. 1C). For better quantification, liquid chromatography (LC) MS/MS analysis can be performed. The best technique for this purpose is multiplexed tandem mass tag labeling, in which each IP sample is individually labeled with the TMTsixplex isobaric label reagent set (Thermo Scientific). Orbitrap

LC MS/MS is performed to identify peptides and quantify proteins (Fig. 1D; Eng et al. 1994; Ting et al. 2011; Wuhr et al. 2012; McAlister et al. 2014). After normalizing the protein signals to the total protein intensity across all samples, hierarchical clustering analysis can be performed using the normalized intensities of augmin subunits by each IP sample to understand the assembly of the protein complex (Fig. 1E; here hclust package of R is used to generate the hierarchical clustering dendrogram). Last, the LC MS/MS results can be compared with the quantified Western blot data to obtain a more accurate model for the architecture of the protein complex studied.

With a single experimental setup, this protocol includes immunodepletion for functional investigation and immunoprecipitation for assembly analysis of the protein complex. Furthermore, these results guide the design of subunit constructs for in vitro reconstitution and structural studies. Although this protocol is described for augmin, it is applicable for any multisubunit protein complex in *Xenopus* extract that is involved in mitotic spindle assembly.

RECIPES

CSF-XB

Reagent	Amount to add
1 M HEPES (pH 7.7, adjusted with KOH)	5 mL
XB salts (20×) <R>	25 mL
0.5 M EGTA (pH 8.0, adjusted with KOH)	5 mL
Sucrose	50 g

Combine the listed reagents in a beaker. Add β-mercaptoethanol to a final concentration of 6 mM and bring the volume to 500 mL with deionized H_2O (18.2 MΩ-cm). Stir with a magnetic bar to dissolve sucrose and mix reagents. Adjust to pH 7.8 if necessary. Store at 4°C.

XB Salts (20×)

2 mM $CaCl_2$
2 M KCl
20 mM $MgCl_2$

Filter-sterilize. Store at 4°C

ACKNOWLEDGMENTS

We thank Matthew R. King for critical reading of the manuscript and helpful comments. This work was supported by the National Institutes of Health New Innovator Award (DP2), the Pew Scholars Program in the Biomedical Sciences, and the David and Lucile Packard Foundation (all to S.P.).

REFERENCES

Eng JK, McCormack AL, Yates JR. 1994. An approach to correlate tandem mass-spectral data of peptides with amino-acid-sequences in a protein database. *J Am Soc Mass Spectrom* 5: 976–989.

Good MC, Heald R. 2018. Preparation of cellular extracts from *Xenopus* eggs and embryos. *Cold Spring Harb Protoc* doi: 10.1101/pdb.prot097055.

McAlister GC, Nusinow DP, Jedrychowski MP, Wuhr M, Huttlin EL, Erickson BK, Rad R, Haas W, Gygi SP. 2014. MultiNotch MS3 enables accurate, sensitive, and multiplexed detection of differential expression across cancer cell line proteomes. *Anal Chem* 86: 7150–7158.

Peloquin J, Komarova Y, Borisy G. 2005. Conjugation of fluorophores to tubulin. *Nat Methods* 2: 299–303.

Petry S, Pugieux C, Nedelec FJ, Vale RD. 2011. Augmin promotes meiotic spindle formation and bipolarity in *Xenopus* egg extracts. *Proc Natl Acad Sci* 108: 14473–14478.

Petry S, Groen AC, Ishihara K, Mitchison TJ, Vale RD. 2013. Branching microtubule nucleation in *Xenopus* egg extracts mediated by augmin and TPX2. *Cell* 152: 768–777.

Ting L, Rad R, Gygi SP, Haas W. 2011. MS3 eliminates ratio distortion in isobaric multiplexed quantitative proteomics. *Nat Methods* 8: 937–940.

Weis K, Dingwall C, Lamond AI. 1996. Characterization of the nuclear protein import mechanism using Ran mutants with altered nucleotide binding specificities. *EMBO J* 15: 7120–7128.

Wuhr M, Haas W, McAlister GC, Peshkin L, Rad R, Kirschner MW, Gygi SP. 2012. Accurate multiplexed proteomics at the MS2 level using the complement reporter ion cluster. *Anal Chem* 84: 9214–9221.

Protocol 6

Filopodia-Like Structure Formation from *Xenopus* Egg Extracts

Helen M. Fox and Jennifer L. Gallop[1]

Wellcome Trust/Cancer Research UK Gurdon Institute and Department of Biochemistry, Cambridge CB2 1QN, United Kingdom

The actin cytoskeleton comprises many different architectures of filaments, including branched networks, parallel bundles and antiparallel fibers. A current challenge is to elucidate how the diverse array of actin regulators, which controls the growth, assembly and turnover of actin filaments, is used to orchestrate cytoskeletal organization and in turn cell shape and movement. Long observed to assemble at cell membranes, actin in *Xenopus* egg extracts recapitulates membrane-triggered assembly at specific lipid and membrane environments. The use of *Xenopus* egg extracts has contributed greatly to identifying how constitutively autoinhibited regulatory pathways are activated, which converge on activation of the Arp2/3 complex. Here we describe a protocol for making parallel actin bundles using *Xenopus* egg extracts from supernatants prepared by high-speed centrifugation. These filopodia-like actin bundles emanate from clusters of actin regulators that self-assemble at phosphatidylinositol (4,5)-bisphosphate-containing supported lipid bilayers. Forming a plasma membrane-mimicking bilayer on glass allows easy, optimizable, high signal-to-noise microscopy at high spatial and temporal resolution. The use of *Xenopus* egg extracts yields large quantities of active material that can be flexibly tailored to address specific questions, for example, by dilution, addition of fluorescent proteins, antibodies or protein fragments, immunodepletion, addition of small molecule inhibitors, or biochemical fractionation.

MATERIALS

It is essential that you consult the appropriate Material Safety Data Sheets and your institution's Environmental Health and Safety Office for proper handling of equipment and hazardous materials used in this protocol.

RECIPES: Please see the end of this protocol for recipes indicated by <R>. Additional recipes can be found online at http://cshprotocols.cshlp.org/site/recipes.

Reagents

Actin, fluorophore-labeled (Thermo Fisher)
Actin from rabbit muscle (available from Cytoskeleton [AD99] or see Pardee and Spudich [1982])
Bradford assay reagents
Dithiothreitol (DTT; 200 mM)
Energy mix (10×) <R>
Extract buffer (XB) (1×) <R>
G-buffer <R>

[1]Correspondence: j.gallop@gurdon.cam.ac.uk

Lissamine rhodamine phosphatidylethanolamine 18:1 (Avanti Polar Lipids, 810150C) (optional; see Step 8)

Methanol

Phosphatidylcholine, chloroform stock (Avanti Polar Lipids, 840053C)

Phosphatidylinositol (4,5)-bisphosphate, chloroform/methanol/water stock (Avanti Polar Lipids, 840046X)

Phosphatidylserine, chloroform stock (Avanti Polar Lipids, 840032C)

Sucrose (2 M)

Equipment

Bath sonicator

BluTack or modeling clay

Coverslips (to suit your microscope; e.g., no. 1.5, 22 mm × 50 mm)

Filopodia-like structures grow less well on very well cleaned (e.g., plasma-treated or detergent-boiled) coverslips. A light wash with dishwashing detergent and extensive rinsing is sufficient. Different batches can produce variable results so it is advisable to test several for their effectiveness.

Disposable inoculation loops

Filter paper

Float to hold glass test tube

Glass test tubes (e.g., 75 mm × 12 mm)

Glass vials (2 mL) with PTFE lined caps (Supelco, 507601 and 27091-U) (optional; Step 4)

Ice

Ice bucket

Inverted fluorescence microscope (e.g., spinning disk confocal or total internal reflection fluorescence)

Liquid nitrogen

Needles (19 G)

Nitrogen or argon compressed gas cylinder with regulator to allow a gentle flow

Measuring cylinder (100 mL)

Parafilm

Petri dishes

Polypropylene tubes (2 mL)

Silicone gaskets (Sigma-Aldrich, no. GBL103250)

Spin concentrator (molecular weight cutoff: 10,000)

Syringes (disposable, 5 mL)

Ultracentrifuge tubes

Ultracentrifuge with swinging bucket and fixed angle rotors (e.g., Beckman Coulter Optima XPN-100 with SW40 Ti and Ti70 rotors)

Vacuum desiccator

Vortex mixer (optional; Step 8)

Water bath (60°C)

METHOD

High-Speed Supernatant (HSS) Extracts Preparation Procedure

1. Prepare crude cytostatic factor (CSF) egg extracts from 5–10 superovulated *X. laevis* frogs as described in Protocol 1: Preparation of Cellular Extracts from *Xenopus* Eggs and Embryos (Good and Heald 2018) with modifications as described in Walrant et al. (2015).

Do not add cytochalasin D to preserve the actin. HSS retains its activity well with storage at −20°C; therefore, large preparations sufficient for 2–3 mo of experiments can typically be made.

2. Aspirate the crude cytoplasmic layer from the crude CSF with a needle and syringe and transfer to a measuring cylinder. Dilute 1:10 with ice-cold 1× XB. Centrifuge at 260,000g (50,000 rpm in the Ti 70 rotor) for 1 h.

3. Collect the supernatant from the ultracentrifuge tube. Concentrate the HSS extracts to ~25 mg/ mL with a spin concentrator, using a pipette to mix every 10 min. Recover the concentrated extracts and measure the protein concentration using a Bradford Assay. Add sucrose to 200 mM, transfer to a 2 mL polypropylene tube and snap freeze. Store at −80°C.

Liposome Preparation Procedure

4. Allow the lipids listed in Table 1 to equilibrate to room temperature for 2–3 min then mix them together sequentially in a glass test tube in the order in which they are listed in Table 1.

   Pipette as quickly and as accurately as possible to avoid the chloroform dripping from the tip. If the solution goes cloudy, add 5 µL methanol to re-solubilize the lipids. Lipids can be stored in glass vials with PTFE lined caps at −80°C (long term) or −20°C (short term). To guard against oxidation, flood the vials with nitrogen or argon before returning to the freezer.

5. Dry the lipids under a gentle flow of nitrogen or argon. To obtain an even lipid film, continuously rotate the tube or move the hose in a circular motion until the solvent has evaporated.

6. Dry the lipid film by placing the test tube in a vacuum desiccator for at least 1 h.

7. Resuspend the lipid film by adding 100 µL 1× XB buffer prewarmed to 60°C to create a 2 mM final lipid concentration. Leave for 10 min to hydrate in the 60°C water bath.

   Warming the buffer increases the temperature above the phase transition of lipids within the mixture.

8. Flick the tube or briefly vortex to dislodge the hydrated lipid lamellae and cover the top of the tube with Parafilm. Place the tube in a float and sonicate the sample for 15 min in a bath sonicator at room temperature. Place the tube containing the now formed liposomes in an ice bucket for use the same day. Alternatively, store the liposomes for up to 1 wk at 4°C.

   To make the supported bilayer visible under the microscope, a small amount of rhodamine-phosphatidyl-ethanolamine from the chloroform stock can be added to the liposome suspension by letting a small amount move into the tip of a P2-pipette by capillary action, then touching it into the lipid film suspension before sonication. If liposomes are used the following day or later, a brief resonication helps resuspend and separate the liposomes.

Supported Lipid Bilayer Procedure

Lipid bilayers should be freshly made for use the same day. Make a sufficient number of bilayers for the planned experiments and store them in the humidified chamber made in Step 9 until use.

9. Prepare a humidified chamber with a small rack. Cut inoculation loop handles to size and secure them to the bottom of a Petri dish with BluTack adhesive. Line the bottom of a Petri dish with wet filter paper. (Fig. 1A).

TABLE 1. Composition of 10% PI(4,5)P_2 liposomes

	Stock concentration (mg/mL)	Average molecular weight (g/mol)	% by moles	Volume (µL) for 100 µL of 2 mM lipids
Phosphatidylcholine	10	771	60	9
Phosphatidylserine	10	825	30	5
Phosphatidylinositol 4,5-bisphosphate	1	1096	10	22
Optional: rhodamine-phosphatidylethanolamine	1	1302	Trace	~0.1

Cite this protocol as *Cold Spring Harb Protoc*; doi:10.1101/pdb.prot100545

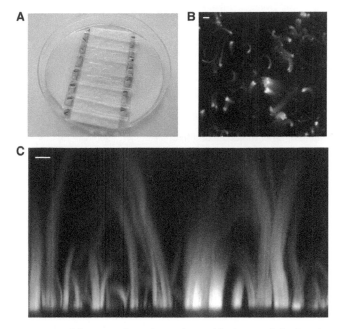

FIGURE 1. The FLS assay. (*A*) A small homemade rack in a humidified Petri dish chamber is used to support loose coverslips when making the supported bilayers. (*B*) *x–y* (en face) view and (*C*) *x–z* (side-on) view of the actin bundles in the FLS assay. Scale bars = 3 μm.

10. Rest a glass coverslip on the rack and place silicon gaskets onto the coverslip to create ∼50 μL wells. Four wells typically fit onto one large rectangular coverslip. Mark a black dot in one corner of the coverslip to help locate each well (Fig. 1A).

11. Pipette ∼30 μL liposomes into each well. Wait a few moments for liposomes to fuse to form a lipid bilayer and then remove the excess fluid without allowing the well to dry or an air–water interface to form across the bilayer. Pipette 50 μL 1× XB on top to dilute the liposomes.

    Leaving the liposomes in the well for longer (up to 40 min) increases the sizes of phase separated regions of membrane.

12. Wash the wells with 1× XB a further five times by pipetting up and down to remove excess liposomes.

Filopodia-Like Structure (FLS) Formation Procedure

13. The day before the experiment, thaw frozen aliquots of fluorescently labeled and unlabeled actin between your fingers, and add G-buffer to make stock concentrations of 0.3 mg/mL. Leave on ice overnight for the F-actin to depolymerize.

    These stocks keep for ∼1 wk at 4°C.

14. Calculate the total volume of assay mix required (Table 2). Thaw aliquots of DTT and energy mix at room temperature and place in an ice bucket. Thaw an HSS aliquot between your fingers, then place in the ice bucket. Add energy mix and DTT to the HSS.

15. Add the other reagents in the order shown in Table 2. Leave for ∼10 min at room temperature to give time for any F-actin to depolymerize.

16. Using fingertips, carefully transfer the coverslip from Step 12 to the microscope taking care not to crack the glass.

    We have designed a magnetic stage holder to secure loose coverslips to a microscope stage (described in Walrant et al. 2015). A stage insert with a simple round hole and a piece of tape is also feasible. We find that better FLSs form on loose coverslips compared with chambered coverglasses. Wet filter paper or paper

TABLE 2. Mix for one 50-µL assay

	Stock concentration	Final concentration	Volume
HSS	25 mg/mL	4.2 mg/mL	8.4 µL
Energy mix	10×	1×	5 µL
DTT	200 mM	2 mM	0.5 µL
Actin (unlabeled)	50 µM	5 µM[a]	5 µL
Actin (labeled)	7.5 µM	150 nM	1
XB (10×)	10×	1×	0.6 µL
Protein or other molecule of interest	500 nM[a]	10 nM[a]	1 µL[a]
XB (1×)	1×	1×	28.5 µL

[a]These values need optimization depending on the molecule or mechanism of interest being assayed.

> towels can be placed around the stage and an appropriately sized plastic box used to create a humid environment, for example, a pipette tip box, or the laser safety box of the microscope.

17. Carefully remove excess XB then add the assay mix to the bilayer and image according to experimental goals.

    See Troubleshooting.

TROUBLESHOOTING

Problem (Step 17): Very little or no growth, or very messy FLSs, are seen.
Solution: Consider the following.

- Make up fresh buffers and energy mix.
- Try a different batch of extracts.
- Use fresh unlabeled actin.
- Try using fresh lipids, particularly PI(4,5)P$_2$.

Problem (Step 17): Rings of actin are seen.
Solution: These are liposomes that have not fused to form the bilayer. Wash the bilayer more rigorously.

Problem (Step 17): Bright spots or threads of actin that are not associated with the membrane are seen.
Solution: Centrifuge the fluorescent actin to remove particulates (e.g., 55,000 rpm, Beckman TLA 100 rotor, 15 min).

DISCUSSION

The FLS assay provides an ideal format for microscopy that allows molecular analysis of the actin regulatory machinery at the membrane interface (Fig. 1B,C; Lee et al. 2010). The open chamber arrangement allows for easy access for the addition of drugs and for pulse-chase experiments. F-actin is depleted from the extracts during the high speed centrifugation step and supplementation with up to ~15 µM unlabeled actin promotes FLS growth. The assay differs from filopodia in that the supported bilayer contains only the lipid portion of a biological membrane and the actin bundle is not encapsulated by the membrane.

Cite this protocol as *Cold Spring Harb Protoc*; doi:10.1101/pdb.prot100545

RECIPES

Energy Mix

20 mM ATP
150 mM creatine phosphate
20 mM MgCl$_2$

Store in 0.1-mL aliquots at −20°C.

Extract Buffer (XB) (1×)

Reagent	Final concentration (1×)
KCl	100 mM
CaCl$_2$	100 nM
MgCl$_2$	1 mM
Sucrose	50 mM
K-HEPES (pH 7.4)	10 mM

Prepare as a 10× stock solution and store at 4°C. Dilute to 1× with ddH$_2$O before each experiment.

G-Buffer

Reagent	Final concentration
Tris-HCl (pH 8.0)	5 mM
CaCl$_2$	0.2 mM
Na-ATP	0.2 mM

Store at −20°C in 2-mL aliquots.

ACKNOWLEDGMENTS

J.L.G. and H.M.F. are supported by the Wellcome Trust (WT095829AIA and 099740/Z/12/Z) and the European Research Council (281971). The Gurdon Institute is funded by the Wellcome Trust (203144) and Cancer Research UK (C6946/A24843). Thanks to Jonathan Gadsby for providing the FLS assay images in Figure 1.

REFERENCES

Good MC, Heald R. 2018. Preparation of cellular extracts from *Xenopus* eggs and embryos. *Cold Spring Harb Protoc* doi:10.1101/pdb.prot097055.

Lee K, Gallop JL, Rambani K, Kirschner MW. 2010. Self-assembly of filopodia-like structures on supported lipid bilayers. *Science* **329:** 1341–1345.

Pardee JD, Spudich JA. 1982. Purification of muscle actin. *Methods Cell Biol* **24:** 271–289.

Walrant A, Saxton DS, Correia GP, Gallop JL. 2015. Triggering actin polymerization in *Xenopus* egg extracts from phosphoinositide-containing lipid bilayers. *Methods Cell Biol* **128:** 125–147.

Endoplasmic Reticulum Network Formation with *Xenopus* Egg Extracts

Songyu Wang,[1,2,3] Fabian B. Romano,[1,2,3] and Tom A. Rapoport[1]

[1]*Howard Hughes Medical Institute and Department of Cell Biology, Harvard Medical School, Boston, Massachusetts 02115*

The endoplasmic reticulum (ER) consists of morphologically distinct domains, including a polygonal network of tubules that is connected by three-way junctions. This network is found in all eukaryotic cells. Extracts from *Xenopus laevis* eggs contain stockpiles of components that allow the assembly of an ER network in vitro. Here we provide protocols for assembly of ER networks in extracts that are arrested at different stages of the cell cycle. Unfertilized *Xenopus laevis* eggs contain a cytostatic factor (CSF) that keeps them in the metaphase stage of the cell cycle. Disruption of the eggs by low-speed centrifugation releases calcium and the eggs cycle into interphase. This state can then be maintained by the addition of cycloheximide, which prevents the synthesis of cyclin B. CSF extracts can be also prepared in the presence of a calcium chelator, thus keeping the extract in metaphase. In this protocol, we outline procedures for the assembly of an ER network using either interphase- or metaphase-arrested *Xenopus* egg extracts. The network assembled is strikingly similar to the network observed in tissue culture cells. The extract allows easy biochemical manipulation, permitting the effects of purified proteins or small molecules, or the depletion of cytosolic components to be tested.

MATERIALS

It is essential that you consult the appropriate Material Safety Data Sheets and your institution's Environmental Health and Safety Office for proper handling of equipment and hazardous material used in this protocol.

RECIPES: Please see the end of this protocol for recipes indicated by <R>. Additional recipes can be found online at http://cshprotocols.cshlp.org/site/recipes.

Reagents

$CaCl_2$ (20 mM in ddH$_2$O; Sigma-Aldrich)

Store at room temperature.

CSF extract

Prepared according to Protocol 1: Preparation of Cellular Extracts from Xenopus *Eggs and Embryos [Good and Heald 2018]) except that cytochalasin should not be added during preparation.*

Cycloheximide (10 mg/mL in ddH$_2$O; Sigma-Aldrich)

Store in 100 µL aliquots at −20°C.

---

[2]These authors contributed equally to this work.

[3]Correspondence: songyuwang.sw@gmail.com; Fabian_Romano-Chernac@hms.harvard.edu

Cite this protocol as *Cold Spring Harb Protoc*; doi:10.1101/pdb.prot097204

Cytochalasin B (10 mg/mL; Sigma-Aldrich)

Dissolve in DMSO and store in 50 μL aliquots at −80°C.

Dithiothreitol (DTT) (1 M in ddH$_2$O; Sigma-Aldrich)

Store in 1 mL aliquots at −20°C.

Energy regenerating system (20×) <R>

Fluorescent dye DiIC$_{18}$ (1,1'-dioctadecyl-3,3,3',3'-tetramethylindocarbocyanine perchlorate) or DiOC$_{18}$ (3,3'-dioctadecyloxacarbocyanine perchlorate) (10 mg/mL; Molecular Probes)

Dissolve in DMSO with brief heating. Store in 20 μL aliquots at −20°C.

Frog (HCG-injected to induce egg laying)

For injection of frogs with hormones refer to Protocol 1: Preparation of Cellular Extracts from Xenopus Eggs and Embryos (Good and Heald 2018).

Interphase buffer (IB) <R>

Interphase buffer with 200 mM KCl (IB200) <R>

Interphase-dejellying solution (2% L-cysteine in water)

Prepare fresh. Adjust pH to 7.8 with KOH.

Isopropanol

Leupeptin, Pepstatin and Chymostatin protease inhibitors (LPC, 10 mg/mL; Sigma-Aldrich)

Dissolve in DMSO and store in 20 μL aliquots at −80°C.

Marc's Modified Ringers (MMR) (1×) <R>

Dilute to desired concentration (1× or 0.25×) with ddH$_2$O.

Methoxy-Polyethylene Glycol-Silane, MW 5000 (100 mg/mL; JenKem Technology USA)

Prepare fresh in 100% ethanol. Briefly heat solution to 80°C until dissolved.

Octadecyl rhodamine (10 mg/mL; Molecular Probes)

Dissolve in DMSO and store in 20 μL aliquots at −20°C.

Valap (equal masses of Vaseline, paraffin, and lanolin; Sigma-Aldrich) <R>

Equipment

Aluminum foil

Aspirator

Clinical centrifuge with adaptors for 15 and 50 mL tubes

Confocal fluorescence microscope equipped with a 60× objective

Container for frog (3 L minimum)

Coverslip racks (Electron Microscopy Sciences, catalogue #70366-16 for 18 × 18 mm coverslips or #72241-01 for 22 × 22 mm coverslips)

Coverslips (No. 1.5) (square, 18 × 18 mm or 22 × 22 mm)

Desiccator

Dishes (Pyrex glass, flat-bottom, 150 × 75 mm; VWR, catalogue #89090-662)

Heat block

Ice bucket

Lens paper

Liquid nitrogen in Dewar flask

Microcentrifuge tubes (0.5 and 1.5 mL)

Microscope slides (metal, custom-made with round holes for imaging purposes)

Needle (18-G)

Oven (65°C–70°C)

Parafilm

Pipette tips (200 μL, large-orifice; USA Scientific)

Plasma etcher (Technics 500-II, Technics West)

SW40 centrifuge tubes (10 mL, Beckman, catalogue #344060)
SW40 rotor (Beckman)
SW55 centrifuge tubes (5 mL, Beckman, catalogue #344057)
SW55 rotor (Beckman)
Syringes (plastic, 3 mL)
Tabletop ultracentrifuge (Beckman Optima TL-100)
TLS55 centrifuge tubes (1 and 2.2 mL, Beckman, catalogue #343778 and #347356)
TLS55 swinging-bucket rotor (Beckman)
Transfer pipettes (disposable, wide bore) (Fisher)
Tubes (14 mL round bottom)
Ultracentrifuge (Beckman Optima L-100XP)
Ultrasonic bath (Branson)

METHOD

Preparation of Crude Interphase Extract and Fractionation into Cytosol and Membranes

The following method to generate crude interphase-arrested extracts is similar to Protocol 1: Preparation of Cellular Extracts from Xenopus Eggs and Embryos (Good and Heald 2018); however, it omits a calcium chelator and includes cycloheximide. The interphase extract can be further fractionated into cytosol and light membranes and stored as frozen aliquots.

Preparation of Crude Interphase Extract

Perform Steps 2–12 with all reagents and samples at 18°C.

1. Prepare the following buffers and prechill them to 18°C: 400 mL/frog interphase-dejellying solution, 2 L/frog 1× MMR, 200 mL/frog 0.25× MMR, 400 mL/frog IB.

2. Place an HCG-injected frog in a container with 1.5 L of 1× MMR prechilled at 18°C. Make sure the frog is completely submerged in 1× MMR. Use more 1× MMR if necessary.

3. Transfer laid eggs to a Pyrex flat-bottom dish containing 1× MMR.

   The volume of dejellied eggs obtained per frog is ~8–10 mL. We typically prepare extracts from the eggs of just one frog. Only high quality eggs are used.

4. Squeeze the frog very gently to collect more fresh eggs into the same dish.

5. Gently rinse eggs with 1× MMR to remove any dirt or debris.

6. Pour off the 1× MMR and add 50–100 mL interphase-dejellying solution. Gently swirl eggs, and remove any bad eggs with a disposable wide bore transfer pipette while dejellying. Change dejellying solution several times. Continue dejellying until all eggs are packed. This will take ~15 min at 18°C.

   Dejellying takes longer at 18°C compared to room temperature.

7. Rinse packed eggs with 50 mL 0.25× MMR several times to completely remove the dejellying solution.

8. Wash eggs with 50–100 mL IB several times to remove all 0.25× MMR solution.

9. Add 2–3 mL IB into 5 mL SW55 tubes, and transfer eggs to the tubes with a disposable wide bore transfer pipette. Remove as much buffer from the top of the eggs as possible.

   The 2–3 mL of IB in the SW55 tubes ensures that eggs are not exposed to air.

10. Place the SW55 tubes into round bottom 14 mL tubes as adapters and pack eggs by centrifugation at 300g for 1 min followed by 20 sec at 800g in a clinical centrifuge.

11. Use a wide-bore transfer pipette to remove all buffer above the packed eggs.

12. Estimate volume of eggs and add LPC to 10 μg/mL, cytochalasin B to 10 μg/mL, and cycloheximide to 50 μg/mL onto the top of the packed eggs. Mixing is not required.

Cite this protocol as *Cold Spring Harb Protoc*; doi:10.1101/pdb.prot097204

13. Crush eggs by centrifugation at 12,000*g* (10,000 rpm in SW55) in an ultracentrifuge for 15 min at 18°C.

14. Using an 18-G needle attached to a 3 mL syringe, pierce the side of the tube just above the bottom dark layer and slowly collect the middle layer, which contains the crude interphase extract. Transfer to a 1.5 mL microcentrifuge tube and place on ice.

Preparation of Interphase Cytosol and Light Membranes

15. To the crude interphase extract, add DTT to 1 mM and LPC to a final concentration of 20 µg/mL. Gently invert the tube several times.

16. Transfer the crude extract to a 2.2 mL TLS55 tube and spin in a TLS-55 swinging-bucket rotor at 55,000 rpm for 90 min at 4°C.

    The best separation is obtained with a TLS-55 rotor.

    The clarified lysate consists of the following five layers: a top yellow lipid layer, cytosol, a pale yellow light membrane layer, a dark brown heavy membrane layer that is just below the light membrane layer, and a clear ribosome and glycogen layer (Fig. 1).

17. Loosen the lipid layer with a 200 µL pipette tip and then gently remove it using a 200 µL large-orifice pipette tip connected to an aspirator.

    It is difficult to remove the lipid layer by pipetting. The best way is to use an aspirator.

18. Transfer the cytosol with a 200 µL large-orifice pipette tip into a 2.2 mL TLS55 tube, avoiding the membrane layer below.

    A 1 mL TLS55 polycarbonate tube is recommended for a volume of about 1 mL.

19. Spin the cytosol in a TLS55 rotor at 55,000 rpm for 30 min at 4°C to remove residual membranes. Aliquot the cytosol, and snap freeze the aliquots in liquid nitrogen.

Preparation of Unlabeled Light Membranes

Steps 20–23 are for preparation of unlabeled light membranes. Please refer to Steps 24–29 for preparation of prelabeled light membranes.

20. Carefully pipette the cloudy pale yellow light membrane layer (from Step 16) with a 200 µL large-orifice pipette tip into a 2.2 mL TLS55 tube. Add 20 volumes of IB200 containing 1 mM DTT and 10 µg/mL LPC. Seal the tube with Parafilm and invert several times to ensure complete membrane dispersion. Incubate on ice for 5 min.

    Avoid pipetting the heavy membrane layer that lies below light membranes.

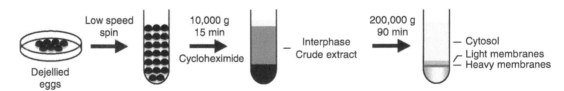

FIGURE 1. Schematic diagram showing different layers obtained during preparation of crude and fractionated interphase egg extract. Dejellied *Xenopus* eggs are packed in a buffer lacking a calcium chelator (e.g., EGTA) via a short and low-speed spin. A further centrifugation at 10,000*g* for 15 min crushes the eggs, releasing Ca^{2+}, which induces cycling of the extract from metaphase into interphase. Crushed eggs are separated into three layers with the top yellow layer being lipids, the middle yellowish brown layer being the crude extract, and the bottom dark layer being pigments and yolk particles. The addition of cycloheximide before centrifugation prevents the synthesis of the Cdc2/cyclin B complex, thereby maintaining the crude extract in interphase. Fractionation of the crude interphase extract by high-speed centrifugation separates the extract into five layers. The top layer contains the residual lipid. The second and third layers are the interphase cytosol and light membranes, respectively. The heavy membrane layer, which lies underneath the light membrane layer, consists of mitochondria and should be avoided during the collection of light membranes. Ribosomes and glycogen reside in the bottom clear layer.

21. Centrifuge the diluted light membranes in a TLS55 rotor at 25,000 rpm for 20 min at 4°C using a tabletop ultracentrifuge.

22. Use an aspirator to remove most of the supernatant. Keep ~1/20 of the original cytosol volume, which contains the light membranes. Harvest the light membranes with a 200 μL large-orifice pipette tip into a 1.5 mL microcentrifuge tube. Leave behind the dark-colored pellet (heavy membranes) at the bottom of the tube.

23. Gently pipette light membranes up and down to make them homogenous. Dispense concentrated light membranes in 5 μL aliquots (an ~20× stock), and snap freeze them in liquid nitrogen.

    While preparing the egg extract, it is important to estimate extract volumes. A good guideline is that the volume of crude extract is around one third of the volume of dejellied eggs. The volume of cytosol is roughly one tenth of the volume of dejellied eggs (Newmeyer and Wilson 1991). The volume of light membranes is one twentieth of the cytosol volume (Dreier and Rapoport 2000; Voeltz et al. 2006; Wang et al. 2013).

Prelabeling of Interphase Light Membranes

Light membranes can be prelabeled with a lipophilic dye.

24. Carefully pipette the cloudy pale yellow light membrane layer (from Step 16) with a 200 μL large-orifice pipette tip into a 10 mL SW40 centrifuge tube.

25. Add dye (DiIC$_{18}$ or DiOC$_{18}$) to 0.2 mg/mL. Gently mix well and incubate on ice for 20 min in the dark.

26. Dilute reactions with 50 volumes of IB200 containing 1 mM DTT and 10 μg/mL LPC. Incubate on ice for 5 min in the dark.

27. Centrifuge at 40,000g for 20 min at 4°C in an SW40 rotor in an ultracentrifuge.

28. Use an aspirator to remove most of the supernatant. Keep ~1/20 of the original cytosol volume, which contains the light membranes.

29. Gently pipette up and down to resuspend the membranes. Snap freeze in liquid nitrogen in 5 μL aliquots.

Passivation of Coverslips

30. Arrange all coverslips on a coverslip rack.

31. Sonicate coverslips in isopropanol for 10 min in an ultrasonic bath.

32. Thoroughly rinse coverslips with water, and sonicate them in water for 10 min.

33. Dry coverslips completely in an oven at 65°C.

34. Use a plasma etcher to clean the glass surface at 200 W for 4 min, and quickly proceed to functionalization of the surface.

35. Turn on a heat block to 80°C, and cover its surface with a clean piece of aluminum foil.

36. Apply ~6 μL of PEG-Silane solution onto the surface of a clean 18 × 18 mm coverslip. Quickly cover with a second coverslip to make a sandwich. Put the sandwiches on the heat block for several minutes, and then transfer them to a Pyrex flat-bottom dish.

    For 22 × 22 mm coverslips, apply 9 μL of PEG-Silane solution.

    PEG-Silane solution precipitates at room temperature. Briefly heat at 80°C to bring it back into solution.

    Ethanol evaporates quickly so be quick when making the sandwich.

37. Incubate the coverslips in the oven for 2–4 h at 70°C.

38. Heat coverslips on the heat block for a few minutes at 80°C, separate them and make a mark on the non-PEGylated side.

    Apply samples to the nonmarked, PEGylated side.

Cite this protocol as *Cold Spring Harb Protoc*; doi:10.1101/pdb.prot097204

39. Thoroughly rinse coverslips with water, and then sonicate them in water for 5 min.

40. Dry coverslips completely. Wrap them in lens paper and store them in a desiccator at −20°C.

Assembly of Interphase or Mitotic (Metaphase-Arrested) ER Networks Using CSF Extract

41. Prepare a prestained CSF extract (10× stained extract) by combining 30 µL CSF extract with 0.3 µL of a 10 mg/mL stock solution of either DiIC$_{18}$ or DiOC$_{18}$.

    Use only freshly prepared CSF extract. CSF extract that has been frozen or kept on ice for too long (more than 6 h) will not perform properly.

    The 1/100 dilution of dye stock is important to limit the total concentration of DMSO in the extract.

42. Incubate for 45 min at 18°C while gently mixing the reaction every 15 min.

43. Place the reaction on ice until use.

44. Set up individual endoplasmic reticulum (ER) formation reactions on ice by combining 20 µL of fresh CSF extract with 2 µL of 10× stained extract.

    Any additional reagents, such as recombinant proteins or small molecules can be added to the reaction at this point. Ensure that the final DMSO concentration is below 0.2%, and avoid diluting the extract to below 10% of the total volume.

45. To induce ER network formation in interphase of the cell cycle, supplement the reaction with 20 mM CaCl$_2$ to a final concentration of 0.5 mM and incubate the sample for 10 min at 18°C to release the metaphase arrest induced by CSF. For ER network formation at the mitotic (metaphase arrested) stage of the cell cycle, omit the addition of CaCl$_2$ and proceed to the following step.

46. Incubate the reaction for 15 min at 18°C to allow formation of the ER network.

    Network formation is fast and is usually complete within a few minutes.

47. For imaging the ER network, sandwich the reaction between two PEG-passivated coverslips. Apply 8 µL of reaction volume to 22 × 22 mm coverslips or 5 µL to 18 × 18 mm coverslips.

48. Place the coverslip sandwich on top of a metal slide. Affix and seal the coverslip sandwich using Valap around the perimeter of the sandwich.

49. Image reactions by fluorescence microscopy (Fig. 2).

    Place the ready-to-image slide on the microscope stage for at least 5 min before contacting the sample with the objective lens. This allows temperature equilibration and prevents flow within the sample from convection, which severely affects image quality.

ER Network Formation Using Membranes with or without Cytosol

50. Thaw one aliquot of cytosol (from Step 19) and one aliquot of light membranes [from Step 23 (unlabeled) or 29 (prelabeled)] and put the tubes on ice.

51. Set up a 10 µL reaction with cytosol and membranes by mixing at a 20:1 volume ratio. Add energy regenerating system to 1×. Incubate the reaction for 20–30 min at room temperature. For network formed from light membranes only, resuspend 0.5 µL of membranes in 10 µL of IB200 containing 1 mM DTT. Add energy regenerating system to 1×. Incubate the reaction for 15 min at room temperature.

    Pipette the reaction up and down gently several times to ensure complete dispersion of membranes in the reaction.

    Purified proteins or inhibitors can be added at this point but ensure their volume is less than 10% to avoid excessive dilution of the reaction. It is recommended that purified proteins are in buffer containing HEPES and KCl. Avoid buffers containing Tris and NaCl.

52. Add ∼0.5–1 µL 100 µg/mL octadecyl rhodamine to a 10 µL reaction and mix gently. Apply the reaction onto a passivated No. 1.5 coverslip sandwich and mount the coverslip onto a metal

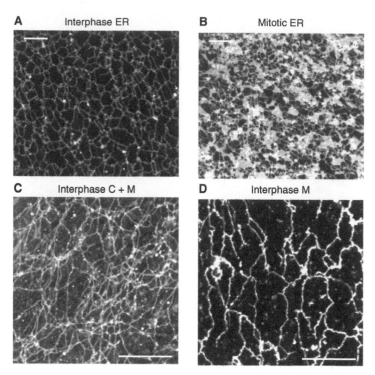

FIGURE 2. Confocal fluorescent images of ER networks assembled from *Xenopus* egg extracts. (*A*) A crude CSF extract was driven into interphase by Ca^{2+} and stained with DiIC$_{18}$. The sample was visualized by confocal fluorescence microscopy. Scale bar, 10 µm. (*B*) As in *A*, but in the absence of Ca^{2+}, the CSF extract remained in the meiotic state. Scale bar, 20 µm. (*C*) Fractionated interphase cytosol, membranes, and 1× energy regenerating system were incubated for 30 min at room temperature. The reaction was stained with octadecyl rhodamine and visualized by confocal microscopy. Scale bar, 20 µm. (*D*) Interphase light membranes prelabeled with DiOC$_{18}$ were incubated with IB200 in the presence of 1× energy regenerating system for 15 min at room temperature. The reaction was then visualized by confocal microscopy. Scale bar, 20 µm. Images are adapted from Wang et al. (2013).

microscopic slide. Seal the slide with Valap. Image the ER network with a fluorescence microscope equipped with a 60× objective (Fig. 2).

> It is important to minimize the volume of octadecyl rhodamine added to prevent dilution of the network reaction.

> If labeled membranes are used in this step, addition of octadecyl rhodamine is not required.

DISCUSSION

The establishment of an in vitro system for ER network formation using *Xenopus laevis* eggs was the starting point for identifying components involved in shaping the organelle (Dreier and Rapoport 2000; Voeltz et al. 2006; Hu et al. 2009; Wang et al. 2013). Reactions in crude egg extracts can also be used to observe the concomitant growth of astral microtubules and the formation of the nuclear envelope if demembranated sperm chromatin is added, which provides the microtubule-organizing center (Wang et al. 2013). Fractionated cytosol and light membranes are also capable of assembling a nuclear envelope in the presence of demembranated sperm chromatin (Lohka and Masui 1983; Lohka and Masui 1984; Newmeyer and Wilson 1991; Powers et al. 2001). The DiI or DiO dye used in this protocol is very bright and photostable. However, it has a long carbon chain so incorporation into membranes requires a long incubation period (Hetzer et al. 2000; Wang et al. 2013). We use PEGy-lated coverslips for all imaging because the treatment allows the glass surface to be less hydrophobic. This reduces the amount of membranes that attach to the coverslips. In summary, *Xenopus* egg extracts are a proven, robust system to study ER network formation in vitro in different cell cycle stages and permits biochemical manipulation.

Cite this protocol as *Cold Spring Harb Protoc*; doi:10.1101/pdb.prot097204

RECIPES

Energy Regenerating System (20×)

Reagent	Concentration
Creatine phosphokinase	2.5 mg/mL
Phosphocreatine	250 mM
ATP (pH 7.5)	50 mM
GTP (pH 7.5)	10 mM

Store in 20-μL aliquots at −20°C.

Interphase Buffer (IB)

Reagent	Concentration
HEPES (pH 7.7)	50 mM
KCl	50 mM
$MgCl_2$	2.5 mM
Sucrose	250 mM

Adjust pH to 7.7 with KOH, filter-sterilize, and store at 4°C.

Interphase Buffer with 200 mM KCl (IB200)

Reagent	Concentration
HEPES (pH 7.5)	50 mM
KCl	200 mM
$MgCl_2$	2.5 mM
Sucrose	250 mM

Adjust pH to 7.5 with KOH, filter-sterilize, and store at 4°C.

Marc's Modified Ringer's (MMR) (1×)

0.1 M NaCl
2 mM KCl
1 mM $MgSO_4$
2 mM $CaCl_2$
5 mM HEPES (pH 7.8)
0.1 mM EDTA

Sterilize by autoclaving, and store at room temperature.
Common alternative formulations of MMR omit EDTA and are adjusted to pH 7.4.

Valap

Add equal weight of lanolin, Vaseline (or other petroleum jelly), and paraffin together in a 1-liter beaker. Heat mixture at a low setting on a hot plate. Stir occasionally until thoroughly blended. Aliquot into several small screw-capped jars (∼50-mL capacity). Store at room temperature.

Cite this protocol as *Cold Spring Harb Protoc*; doi:10.1101/pdb.prot097204

REFERENCES

Dreier L, Rapoport TA. 2000. In vitro formation of the endoplasmic reticulum occurs independently of microtubules by a controlled fusion reaction. *J Cell Biol* **148:** 883–898.

Good MC, Heald R. 2018. Preparation of cellular extracts from *Xenopus* eggs and embryos. *Cold Spring Harb Protoc* doi: 10.1101/pdb.prot097055.

Hetzer M, Bilbao-Cortés D, Walther TC, Gruss OJ, Mattaj IW. 2000. GTP hydrolysis by Ran is required for nuclear envelope assembly. *Mol Cell* **5:** 1013–1024.

Hu J, Shibata Y, Zhu PP, Voss C, Rismanchi N, Prinz WA, Rapoport TA, Blackstone C. 2009. A class of dynamin-like GTPases involved in the generation of the tubular ER network. *Cell* **138:** 549–561.

Lohka MJ, Masui Y. 1983. Formation in vitro of sperm pronuclei and mitotic chromosomes induced by amphibian ooplasmic components. *Science* **220:** 719–721.

Lohka MJ, Masui Y. 1984. Roles of cytosol and cytoplasmic particles in nuclear envelope assembly and sperm pronuclear formation in cell-free preparations from amphibian eggs. *J Cell Biol* **98:** 1222–1230.

Newmeyer DD, Wilson KL. 1991. Egg extracts for nuclear import and nuclear assembly reactions. *Methods Cell Biol* **36:** 607–634.

Powers M, Evans EK, Yang J, Kornbluth S. 2001. Preparation and use of interphase *Xenopus* egg extracts. *Curr Protoc Cell Biol* **Chapter 11:** Unit 11.10.

Voeltz GK, Prinz WA, Shibata Y, Rist JM, Rapoport TA. 2006. A class of membrane proteins shaping the tubular endoplasmic reticulum. *Cell* **124:** 573–586.

Wang S, Romano FB, Field CM, Mitchison TJ, Rapoport TA. 2013. Multiple mechanisms determine ER network morphology during the cell cycle in *Xenopus* egg extracts. *J Cell Biol* **203:** 801–814.

Microfluidic Encapsulation of Demembranated Sperm Nuclei in *Xenopus* Egg Extracts

John Oakey[1,3] and Jesse C. Gatlin[2,3]

[1]*Department of Chemical Engineering, University of Wyoming, Laramie, Wyoming 82071;* [2]*Department of Molecular Biology, University of Wyoming, Laramie, Wyoming 82071*

The cell-free nature of *Xenopus* egg extract makes it a uniquely tractable experimental model system. The extract, effectively unconfined cytoplasm, allows the direct and relatively straight-forward addition of purified proteins and other reagents, a characteristic that renders the system amenable to many biochemical and cell biological manipulations. Accessibility to the system also facilitates the direct physical manipulation and probing of biological structures, in turn enabling mechanical properties of intracellular assemblies and organelles, such as the mitotic spindle and nucleus, to be measured. Recently, multiphase microfluidics have been combined with *Xenopus* egg extracts to encapsulate discrete cytoplasmic volumes. Described here is a protocol detailing the use of multiphase microfluidic devices to encapsulate sperm nuclei within extract droplets of defined size and shape. This protocol can also be applied more generally to encapsulation of microbeads and other particles.

MATERIALS

It is essential that you consult the appropriate Material Safety Data Sheets and your institution's Environmental Health and Safety Office for proper handling of equipment and hazardous materials used in this protocol.

RECIPES: Please see the end of this protocol for recipes indicated by <R>. Additional recipes can be found online at http://cshprotocols.cshlp.org/site/recipes.

Reagents

CSF-XB (cytostatic factor-XB) <R>

> *For preparation of CSF-XB, see Protocol 1: Preparation of Cellular Extracts from* Xenopus *Eggs and Embryos (Good and Heald 2018).*

Ethanol (optional; see Step 8)
Mineral oil (light) (optional; see Step 16)
Oil (Novec HFE-7500 Engineered Fluid, 3M; CAS Number 297730-93-9)
Polydimethylsiloxane (PDMS) elastomer kit (Sylgard 184, Dow Corning)
Sperm nuclei (1–5×10^7/mL stock)

> *For preparation of sperm nuclei, see Sec. 8, Protocol 1: Isolation and Demembranation of* Xenopus *Sperm Nuclei (Hazel and Gatlin 2018).*

Surfactant (oil-compatible; 008 FluoroSurfactant, RAN biotechnologies; or PicoSurf, Dolomite).

> *PEG-containing surfactants (Holtze et al. 2008) are trusted due to their passivity toward extracts. Ionic surfactants, such as Krytox, interact with extract immediately and irreversibly and should be avoided.*

---

[3]Correspondence: joakey@uwyo.edu; jgatlin@uwyo.edu

Cite this protocol as *Cold Spring Harb Protoc*; doi:10.1101/pdb.prot102913

Xenopus egg extract

For preparation of Xenopus egg extract, see Protocol 1: Preparation of Cellular Extracts from Xenopus Eggs and Embryos (Good and Heald 2018)

Equipment

Coverslips (#1.5, 25×40 mm or larger)

Forceps (medium)

Microfluidic master

Microscope (inverted with multimode epifluorescence)

Microscope (with phase-contrast optics)

Nail polish (acrylic)

Oven (Gravity Convection; 70°C)

Oxygen plasma cleaner (Harrick Plasma PDC-001 or equivalent)

Petri dish

Sonicator (Fisher Scientific FS20D or equivalent) (optional; see Step 8)

Syringe pump (dual control, Harvard Apparatus Pump 33 DDS or equivalent)

Syringe tips (blunt, 30 G)

Syringe tips (blunt, 20 G, sharpened with a diamond-burr conical Dremel Tool bit)

Syringes (Tuberculin syringe with Luer slip tip and low dead volume plunger)

Tape (Scotch; 3 M) (optional; see Step 8)

Tape (vinyl laboratory)

Tubing (0.01 inch ID Tygon tubing)

Tweezers

Vacuum pump and chamber (United Scientific Supplies Vacuum Desiccator)

Weighing boat (plastic)

Wipes (Kimwipe delicate task wipers)

X-Acto knife or scalpel

METHOD

Microfluidic Device Design and Fabrication

Extract-in-oil droplets are generally produced within microfluidic T-junctions, a channel intersection where immiscible fluids are merged. T-junction devices possess three dimensions that affect the size of produced droplets: oil channel width (w_o), aqueous channel width (w_{aq}), and channel depth (d). Droplet size within a device of fixed geometry can also be regulated by varying the relative volumetric flow rate of the oil and aqueous phases (Steijn et al. 2010). Given a device with fixed dimensions of $w_o = 30$ µm, $w_{aq} = 20$ µm, and $d = 20$ µm, spherical droplets or their equivalent volumes can be produced with diameters ranging from 20 µm to 60 µm. A single device may be produced with both a droplet formation junction and an incubation reservoir to accommodate large numbers of droplets that remain separate to minimize convective motion and coalescence over long time periods. To minimize the likelihood of reservoir collapse, structural supports in the form of walls and columns can be designed into device reservoirs (Fig. 1). Devices are frequently designed using standard drafting packages that can be subsequently converted to photolithography masks or CNC mill or 3D printer instructions. An example AutoCAD file for a microfluidic device is available at: http://oakeylab.com/s/Xenopus_T_Junction__Reservoir.dwg.

Integrated microfluidic devices for extract handling, encapsulation, and droplet incubation are constructed from a molded polydimethylsiloxane (PDMS) microchannel replica (Whitesides and Stroock 2001) bonded to a coverslip. Reusable microchannel masters may be fabricated from an array of materials using a variety of approaches including, for instance, photolithography, microscale CNC milling, or 3D printing. These masters may be repeatedly replicated in PDMS to create single-use, disposable microfluidic devices according to the following protocol.

1. Thoroughly mix the two part PDMS elastomer kit in a 1:10 (hardener:base) ratio in a clean plastic weighing boat.

Cite this protocol as *Cold Spring Harb Protoc*; doi:10.1101/pdb.prot102913

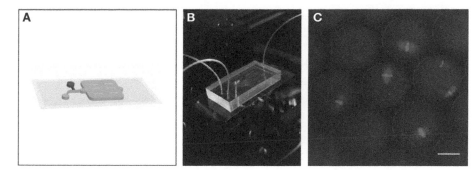

FIGURE 1. Microfluidic encapsulation of *X. laevis* nuclei to monitor spindle assembly in extract droplets. (*A*) Diagram of a standard "T-junction" type microfluidic device showing the extract phase inlet (red cylinder and stem) and oil phase inlet (blue cylinder and stem). Rounded gray rectangles in the reservoir represent supporting walls used to ensure constant reservoir depth. These features can also be seen in (*B*) a photograph of a device ready to be imaged on an inverted microscope stage. The two tubes coming in from the left side of the image are inlet tubes whereas the tube exiting to the upper right is an exit tube (an optional feature). An additional inlet is shown plugged with a piece of knotted tubing. (*C*) Shows a fluorescent image taken of spindles assembled in extract droplets within a device reservoir. Demembranated sperm nuclei (see Sec. 8, Protocol 1: Isolation and Demembranation of *Xenopus* Sperm Nuclei [Hazel and Gatlin 2018]) were allowed to assemble into interphase nuclei before the extract was induced to go back into m-phase by adding an equal volume of CSF-arrested extract. The nuclei were then encapsulated in droplets before nuclear envelope breakdown to allow spindle assembly to occur in discrete volumes. Extract was spiked with DAPI to visualize DNA (blue) and Atto-565-labeled tubulin (red) to visualize spindle microtubules. Scale bar = 50 µm.

2. Place the microfluidic master in a containment reservoir, such as a Petri dish, that allows the master to be completely covered by PDMS. Fasten the master securely to the bottom of the reservoir with vinyl laboratory tape so that it cannot shift or float away from the surface.

3. Pour the mixed PDMS over the master until completely submerged, with 2–4 mm of excess PDMS covering the master.

4. Place the PDMS-covered master into a vacuum chamber, seal, and evacuate until bubbles emerge from the PDMS. Allow the PDMS to degas until bubbles no longer appear. This process may take 1–2 h.

5. Place the degassed, but still fluid PDMS-covered master for 4 h into a 70°C oven. Care must be taken to level the oven and ensure that the reservoir sits flat in the oven, otherwise final devices will be of uneven thickness.

6. Remove the now-cured PDMS from the oven. Using a sharp scalpel or X-Acto knife carefully cut through the PDMS, completely circumnavigating the channel network. With tweezers, remove the PDMS replica and set the master aside for subsequent replication. Excess PDMS need not be removed from the master before reuse, provided the replica was removed cleanly.

7. Place the PDMS replica upon a clean, firm surface with network side facing up. Use the scalpel to cut and trim the device to fit upon a coverslip.

8. With a sharpened 20 G blunt syringe tip, punch holes through the PDMS at microchannel inlets and at the outlet. After cutting and punching, clean any PDMS debris from the surface.

   PDMS may be cleaned of dust and debris by dabbing the surface with Scotch tape. If PDMS fragments are found flowing through the network, their introduction is likely from the punched holes. These may be cleaned by sonicating the replica in ethanol for 15 min. PDMS swells considerably in ethanol, so devices must be baked for an hour between 80°C and 100°C to volatilize all residual organic liquid before bonding to a glass coverslip.

9. Place the coverslip and clean PDMS device, network facing up, into the oxygen plasma cleaner.

10. Expose the PDMS device and glass coverslip to an oxygen plasma under moderate power for about 1 min.

The optimal oxygen plasma exposure time varies greatly based upon oxygen partial pressure, RF power, and ambient atmospheric conditions (temperature and humidity, for instance). Therefore, the exposure time must be determined empirically, by trial and error, for each plasma cleaner and laboratory.

11. Remove the PDMS and coverslip from the plasma cleaner, taking care not to touch the bonding surfaces. Bring the PDMS replica into contact with the coverslip and gently form a seal by applying pressure with tweezers.

12. Place the hybrid device for 10 min into a 70°C oven to complete bonding.

Device Loading

To mitigate problems associated with premature microtubule nucleation and polymerization, perform all steps on ice (or in a 4°C cold room) unless otherwise stated.

13. Submerge devices in CSF-XB for 30–60 min before use to saturate the PDMS with buffer.

    This step is crucial to prevent water from extract droplets permeating into the device matrix, which otherwise occurs over experimental timescales (Randall and Doyle 2005).

14. Prepare needle assemblies by threading tubing over a blunt-ended needle tip (30 G) taking care not to puncture the tubing wall. Make sure that the tubing is sufficiently long to run from the syringe pump to the microfluidic device and then cut the opposite end off at a 45° angle to facilitate its insertion into one of the device's inlet ports.

15. Make inlet plugs by cutting small lengths of tubing (~4 cm) at 45° angles. Tie each with an overhand knot (alternatively a monkey fist knot can be used, but the length of tubing must be adjusted accordingly).

16. Load *Xenopus* egg extract containing demembranated sperm nuclei treated according to experimental goals into a Tuberculin syringe. For small volumes of extract (<150 µL), syringes can be preloaded with 200–300 µL light mineral oil to increase plunger travel. Mix Novec oil with 2% w/v compatible surfactant and load into a separate 1 mL syringe. Cap each syringe with a 30 G blunt syringe tip/tubing assembly and place the loaded syringes on ice.

    The proportion of drops populated with a nucleus depends upon nuclei concentration and drop size; for a given concentration, the number of populated drops will decrease with drop size. This can be compensated for by increasing nuclei concentration, but only to a certain extent before nuclei clumping begins to occur.

17. With a pair of medium forceps, insert plugs into all but one of the inlet ports. Prefill the device with the oil phase by first setting the syringe pump flow rate to 3 µL/min and pump oil until it reaches the end of the tubing. With the oil phase flowing, insert the tubing into the open inlet. Prop the device against a solid support with the outlet higher than the inlet and allow to fill.

    This ensures that the downstream reservoir (or path) is expunged of air and filled completely with oil. To conserve oil, stop the pump once oil begins to flow out of the device outlet.

18. Mount the syringe containing extract on the second pump and set the flow rate to 3 µL/min. Once the extract reaches the end of the tubing, reduce the flow rate to 1 µL/min.

19. Ensure that the oil-syringe pump is set to 3 µL/min and that the oil is flowing. With the device on a flat surface, secure it by straddling a plugged inlet port with forceps tips and applying a localized downward pressure. Pull the plug and immediately replace it with the end of the extract tubing.

20. Monitor droplet formation under a phase contrast microscope. Once the reservoir is filled with droplets, proceed by first removing the tubing at the extract phase inlet and then by removing the tubing at the oil phase inlet. Turn off both pumps to conserve oil and extract.

    See Troubleshooting.

21. Use a delicate task Kimwipe to remove oil and extract from the surface of the device and seal all inlets with fast-drying acrylic nail polish. Place the filled device on ice until ready to image on an epifluorescence microscope.

Cite this protocol as *Cold Spring Harb Protoc*; doi:10.1101/pdb.prot102913

TROUBLESHOOTING

Problem (Step 20): The formation and stability of extract-in-oil droplets can be compromised by PDMS surfaces of intermediate water contact angles (~90°) or by the adsorption of extract components to the microchannel surface.

Solution: PDMS is natively hydrophobic, but exposure to oxygen plasma in Step 10 renders it mildly and temporarily hydrophilic. To facilitate the recovery of the surface to a hydrophobic state, bake bonded devices in an oven for 2 h at 100°C. If the surface remains water-wet, the channels can be filled with Aquapel and flushed with nitrogen. Microchannel surfaces, fabricated in this manner, will be hydrophobic but can be altered by direct contact with extract; this can be avoided by prefilling devices with oil and not flooding oil channels with extract.

DISCUSSION

The open nature of *Xenopus* egg extract provides tremendous flexibility for the design of experiments requiring biochemical and cell biological manipulations and for the direct physical manipulation and probing of biological structures (Itabashi et al. 2009; Gatlin et al. 2010; Shimamoto et al. 2011). The utility of combining multiphase microfluidics with *Xenopus* egg extracts to encapsulate discrete cytoplasmic volumes has been established (Jimenez et al. 2011; Pinot et al. 2012; Good et al. 2013; Hazel et al. 2013). The protocol presented here, originally developed to address the scaling relationship between steady-state spindle size and cytoplasmic volume using *X. laevis* egg extracts (Hazel et al. 2013), can be more generally applied to encapsulate any microscale subcellular structure [e.g., acentriolar microtubule-organizing centers or synthetic microbeads (Tsai and Zheng 2005)]. In principle, extract can also be substituted with any aqueous phase mixture facilitating bottom-up, in vitro study of cytoskeletal organization in discrete volumes containing defined molecular components (e.g., Vleugel et al. 2016). Although one of the inherent advantages of cell-free extract as a model system is its open nature, there are clearly tangible experimental benefits to putting the genie back in the bottle, or recellularizing cell-free extracts, particularly when one has exquisite control of the bottle's size and shape.

RECIPES

CSF-XB

Reagent	Quantity
XB (extract buffer) <R>	400 mL
EGTA stock (K salt; 0.5 M; pH 7.7; filter-sterilized and stored in aliquots at −20°C)	4 mL
MgCl$_2$ solution (2 M; e.g., Sigma-Aldrich 68475)	200 µL

Make fresh on the day of extract preparation.

Extract Buffer (XB) Salts (20×)

Reagent	Quantity	Concentration
KCl	149.12 g	2 M
CaCl$_2$	0.294 g	2 mM
MgCl$_2$ (2 M; e.g., Sigma-Aldrich 68475)	10 mL	20 mM

Combine with MilliQ H$_2$O to a volume of 1 L. Autoclave and store at 4°C.

XB (Extract Buffer)

Reagent	Volume
Extract buffer (XB) salts (20×) <R>	50 mL
Sucrose (2 M; filter-sterilized and stored in aliquots at −20°C)	25 mL
HEPES (K salt; 1 M, pH 7.7; filter-sterilized and stored in aliquots at −20°C)	10 mL

Combine and bring to 1 L with MilliQ H₂O. Adjust pH to 7.7 with KOH. Make fresh on day of preparation, and store at 16°C until use.

ACKNOWLEDGMENTS

The authors thank Ben Noren for acquiring the included photographs and generating graphics for the included figure. We would also like to express our appreciation to the agencies and programs that fund our work, namely, the National Institutes of Health (National Institute of General Medical Sciences (NIGMS) grants R01GM102428, R01GM113028), the Wyoming INBRE program (NIGMS grant 2P20GM103432), and the Pew Biomedical Scholars program.

REFERENCES

Gatlin JC, Matov A, Danuser G, Mitchison TJ, Salmon ED. 2010. Directly probing the mechanical properties of the spindle and its matrix. *J Cell Biol* **188**: 481–489.

Good MC, Heald R. 2018. Preparation of cellular extracts from *Xenopus* eggs and embryos. *Cold Spring Harb Protoc* doi: 10.1101/pdb.prot097055.

Good MC, Vahey MD, Skandarajah A, Fletcher DA, Heald R. 2013. Cytoplasmic volume modulates spindle size during embryogenesis. *Science* **342**: 856–860.

Hazel JW, Gatlin JC. 2018. Isolation and demembranation of *Xenopus* sperm nuclei. *Cold Spring Harb Protoc* doi: 10.1101/pdb.prot099044.

Hazel J, Krutkramelis K, Mooney P, Tomschik M, Gerow K, Oakey J, Gatlin JC. 2013. Changes in cytoplasmic volume are sufficient to drive spindle scaling. *Science* **342**: 853–856.

Holtze C, Rowat AC, Agresti JJ, Hutchison JB, Angile FE, Schmitz CHJ, Koster S, Duan H, Humphry KJ, Scanga RA. 2008. Biocompatible surfactants for water-in-fluorocarbon emulsions. *Lab Chip* **8**: 1632–1639.

Itabashi T, Takagi J, Shimamoto Y, Onoe H, Kuwana K, Shimoyama I, Gaetz J, Kapoor TM, Ishiwata S. 2009. Probing the mechanical architecture of the vertebrate meiotic spindle. *Nat Methods* **6**: 167–172.

Jimenez AM, Roche M, Pinot M, Panizza P, Courbin L, Gueroui Z. 2011. Towards high throughput production of artificial egg oocytes using microfluidics. *Lab Chip* **11**: 429–434.

Pinot M, Steiner V, Dehapiot B, Yoo BK, Chesnel F, Blanchoin L, Kervrann C, Gueroui Z. 2012. Confinement induces actin flow in a meiotic cytoplasm. *Proc Natl Acad Sci* **109**: 11705–11710.

Randall G, Doyle P. 2005. Permeation-driven flow in poly (dimethylsiloxane) microfluidic devices. *Proc Natl Acad Sci* **102**: 10813–10818.

Shimamoto Y, Maeda YT, Ishiwata S, Libchaber AJ, Kapoor TM. 2011. Insights into the micromechanical properties of the metaphase spindle. *Cell* **145**: 1062–1074.

Steijn V, Kleijn CR, Kreutzer MT. 2010. Predictive model for the size of bubbles and droplets created in microfluidic T-junctions. *Lab Chip* **10**: 2513–2518.

Tsai MY, Zheng Y. 2005. Aurora A kinase-coated beads function as microtubule-organizing centers and enhance RanGTP-induced spindle assembly. *Curr Biol* **15**: 2156–2163.

Vleugel M, Roth S, Groenendijk CF, Dogterom M. 2016. Reconstitution of basic mitotic spindles in spherical emulsion droplets. *J Vis Exp.* (114), e54278, doi: 10.3791/54278.

Whitesides GM, Stroock AD. 2001. Flexible methods for microfluidics. *Phys Today* **54**: 42–48.

Calculating the Degradation Rate of Individual Proteins Using *Xenopus* Extract Systems

Gary S. McDowell[1,2,5] and Anna Philpott[3,4,5]

[1]*Future of Research, Abington, Massachusetts 02351;* [2]*Manylabs, San Francisco, California 94103;* [3]*Department of Oncology, MRC/Hutchison Research Centre, Cambridge CB21XZ, United Kingdom;* [4]*Wellcome Trust-Medical Research Centre Cambridge Stem Cell Institute, University of Cambridge, Cambridge CB21QR, United Kingdom*

The *Xenopus* extract system has been used extensively as a simple, quick, and robust method for assessing the stability of proteins against proteasomal degradation. In this protocol, methods are provided for assessing the half-life of in vitro translated radiolabeled proteins using *Xenopus* egg or embryo extracts.

MATERIALS

It is essential that you consult the appropriate Material Safety Data Sheets and your institution's Environmental Health and Safety Office for proper handling of equipment and hazardous materials used in this protocol.

RECIPES: Please see the end of this protocol for recipes indicated by <R>. Additional recipes can be found online at http://cshprotocols.cshlp.org/site/recipes.

Reagents

2× Laemmli sample buffer with β-mercaptoethanol <R>

Cycloheximide (20 mg/mL in water stored at −20°C)

Energy mix (50×) <R>

Plasmid DNA template containing an SP6 or T7 promoter and a gene of interest, dissolved in water at 250 ng/μL

> *We use pCS2 vectors, with the gene of interest downstream from the SP6 promoter. We include a Kozak sequence before the gene of interest to ensure efficient transcription, and use BamHI and EcoRI restriction sites for ligation of the gene of interest into the vector.*

SDS-PAGE reagents

^{35}S-methionine (10–11 mCi/mL)

> *Ensure ^{35}S-methionine is fresh, or arrange experiments as close to the receipt of fresh ^{35}S-methionine as possible. ^{35}S has a half-life of ∼3 mo; use beyond 3 mo is not advised for this assay.*

SP6 or T7 TnT Quick Coupled Transcription/Translation System Master mix (Promega)

Ubiquitin from bovine erythrocytes (12.5 mg/mL in water stored at −20°C, Sigma-Aldrich)

Xenopus extract (see Protocol 1: Preparation of Cellular Extracts from *Xenopus* Eggs and Embryos [Good and Heald 2018] and Step 4)

---

[5]Correspondence: ap113@cam.ac.uk; garymcdow@gmail.com

Cite this protocol as *Cold Spring Harb Protoc*; doi:10.1101/pdb.prot103481

Equipment

Gel dryer
Heat blocks set to 21°C, 30°C, and 95°C
Ice bucket
Microcentrifuge
Microcentrifuge tubes (0.5 mL)
Microcentrifuge tubes (1.5 mL)
Phosphor imaging system (e.g., Typhoon, Storm, or PhosphorImager) and ImageQuant software
(GE) or, alternatively, autoradiography film and equipment and ImageJ software
SDS-PAGE apparatus

METHOD

The purpose of this protocol is to calculate the rate of degradation of a protein of interest using cytoplasmic extract. Radiolabeled proteins are first produced to be used in the assay as they provide a simple method for measuring protein levels at various timepoints; nonradiolabeled proteins could be substituted and Western blotting used for detection but our experience is that use of radiolabeled proteins allows for more robust measurements of protein amount and so gives a more precise and reproducible measurement of protein half-life. Egg extract is prepared and supplemented with ATP to provide energy for the process, cycloheximide to prevent translation of new protein, and ubiquitin to ensure that the concentration of ubiquitin is not rate-limiting. Therefore, degradation of a protein of interest can be measured as a direct function of the protein's potential to be ubiquitylated and degraded by the ubiquitin-proteasome system.

In Vitro Transcription/Translation of Radiolabeled Protein

1. Remove TnT Quick Master Mix from −80°C storage and rapidly thaw between fingers and place on ice.

2. Place the required volume of Master Mix in a 0.5 mL microcentrifuge tube. For each 10 μL of Master Mix, add 1 μL plasmid solution and 1 μL ^{35}S-methionine. Gently mix by pipetting up and down.

   Each degradation assay requires 8 μL of in vitro translated (IVT) protein mix.

   Air bubbles reduce the efficiency of the reaction; therefore be extremely careful not to mix the reaction too vigorously. Centrifugation at a low speed for 30 sec after mixing can be performed, if necessary.

3. Incubate for 2 h at 30°C. Store at −20°C until needed.

   Volumes and reaction times described here differ from the instructions given with the TnT Quick Coupled Transcription/Translation System; we have found these conditions to be preferable but they can be adjusted as desired.

   To check for production of radiolabeled protein, add 1 μL of IVT protein mix to 19 μL of 2× Laemmli Buffer and separate using SDS-PAGE. Dry the gel and use autoradiography film or a phosphorimaging system to determine whether a single radioactive product of the appropriate molecular weight has been produced.

Degradation Assay

4. Prepare *Xenopus* extract as required (see Protocol 1: Preparation of Cellular Extracts from *Xenopus* Eggs and Embryos [Good and Heald, 2018]; Fig. 1A[a–c]).

   Each degradation assay requires 36 μL of extract.

   Variations of normal extracts can be used, such as interphase extract (see Vosper et al. 2007), mitotic extract (see Vosper et al. 2009) and neurula embryo extract (see McDowell et al. 2010). In each case, energy mix or ATP should be added before use (as described in Protocol 1: Preparation of Cellular Extracts from Xenopus Eggs and Embryos [Good and Heald 2018]).

5. Supplement extract with energy mix to 1×, cycloheximide to a final concentration of 20 μg/mL, and ubiquitin to a final concentration 1.25 mg/mL.

Cite this protocol as *Cold Spring Harb Protoc*; doi:10.1101/pdb.prot103481

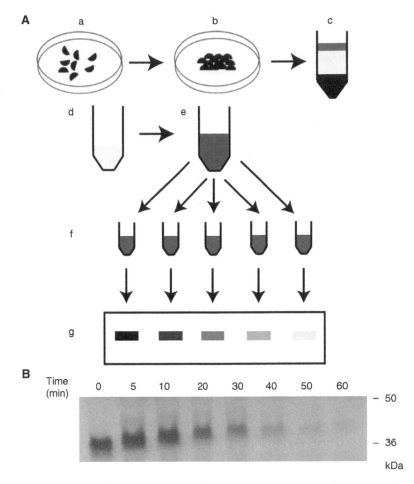

FIGURE 1. Degradation assay using *Xenopus* extract. (*A*) For interphase extracts, (a) *Xenopus laevis* eggs are harvested and activated with calcium ionophore, resulting in (b) the animal hemispheres (pigmented) presenting dorsally and contracting. (c) Eggs are packed into centrifuge tubes and centrifuged to separate out layers of lipids (shown in green), crude extract (yellow), and pigment (black). (d) The crude extract is removed, supplemented with energy mix and ubiquitin and (e) IVT radiolabeled protein is added. The sample is incubated at 21°C and at particular timepoints (f) aliquots of the degradation reaction are removed, quenched with Laemmli buffer and (g) separated by SDS-PAGE. (*B*) A representative example of degradation assay results (the protein analyzed was *Xenopus laevis* Neurogenin3; the bands shift to higher molecular weight over time because of protein phosphorylation).

6. For each assay, combine 36 μL of extract with 8 μL of the IVT radiolabeled protein. Mix by pipetting up and down, and flick the base of the tube with an index finger until the red IVT appears well mixed with the extract. Incubate the reaction at 21°C (Fig. 1A[d,e]).

   This volume of reaction mixture is sufficient to give eight timepoints, which we analyze with handcast 10-lane 15% polyacrylamide gels. If you use gels with more lanes, the number of timepoints and, therefore, volumes could be adjusted accordingly.

7. Immediately after mixing (i.e., time = 0), and at each required timepoint thereafter, remove 5 μL of the reaction mixture, and quench by adding to a 1.5 mL microcentrifuge tube containing 15 μL 2× Laemmli Sample Buffer with β-mercaptoethanol. Immediately incubate for 5 min at 95°C.

   After quenching, samples can be stored at −20°C (Fig. 1A[f]).

Analysis

8. Separate 15 μL of sample by SDS PAGE and dry the gel in a gel dryer for at least 1 h (Fig. 1A[g]). Place the dried gel against an erased phosphor screen (erase by exposure to bright light for at least 30 min) overnight, or against a piece of autoradiography film for at least 24 h.

   Keep the rest of the sample in case there is a problem with the gel and the SDS-PAGE needs to be repeated.

9. View the phosphor screen using a phosphorimaging system, or develop the autoradiography film (the image should look like the example of *Xenopus* Neurogenin3 in Fig. 1B). Use ImageQuant or ImageJ software to quantify the level of radiation at each timepoint.

> *The exposure of the phosphor screen or autoradiography film may need to be tailored; it may be necessary to reduce or increase the exposure time to get appropriate exposure levels. We have found that with our proteins of interest we get adequate exposure placing a gel against the phosphor screen overnight. The image in Figure 1B was obtained by placing the gel against autoradiography film for 2 wk. Increasing the amount of ^{35}S-methionine should be considered if relying on autoradiography alone and how fresh the isotope is should also be taken into consideration.*

10. Calculate the concentration of your protein of interest, [protein], at each timepoint using the imaging software. Use the level of radiation as a proxy for [protein].

> *When calculating the level of protein at each timepoint using the imaging software, ensure to quantify areas of identical size at each timepoint, by drawing a box around the largest band, and then copying this box multiple times and dragging a box over each corresponding band at each timepoint. Also make sure to quantify the level in a box at a uniform part of the gel elsewhere; this will be used as the background level and should be subtracted from each of the values at each timepoint to give normalized [protein] at each timepoint.*

11. Determine the protein half-life using first order kinetic analysis. Plot ln[protein] against time which should give a straight-line of gradient K. Divide ln(2) by K to give the half-life of the protein.

> *Assuming the rate of protein degradation in vitro follows first order kinetics, and using the integrated first order rate law and taking advantage of the fact that the half-life is independent of the starting concentration, we can express the half-life of a protein, t(1/2), as:*

$$t(1/2) = \ln(2)/K.$$

> *If the results do not give a good fit to a straight line, repetition of the experiment is necessary. We usually begin with a time course of 1 h, as shown in Figure 1B; however if the protein is very stable, we have used time courses of up to 4 h. Likewise if the protein degrades very quickly, the time course may need to be shorter. Where possible, we aim to run the experiment for two half-lives of the protein.*

DISCUSSION

This degradation assay is based on the method of Murray, Solomon, and Kirschner (Murray et al. 1989) with improvements by Vosper and colleagues (Vosper et al. 2007, 2009). We have used this method with a variety of extract types to analyze the stability of proteins, principally Neurogenin2 (Vosper et al. 2009; McDowell et al. 2010, 2014a,b; Ali et al. 2011; Hindley et al. 2012). Others have also used this system in a similar manner to study the stability of proteins (as reviewed in McDowell and Philpott 2016b).

The entire experiment can be performed in a single day. IVT radiolabeled protein can be prepared in the morning, followed by extract preparation and a subsequent degradation assay in the afternoon. The experienced experimentalist can prepare enough IVT radiolabeled proteins and three egg extracts from different frogs to perform the experiment in triplicate in 1 d.

As a control protein, we use wild-type *Xenopus laevis* Neurogenin2. Depending on the experimentalist, the half-life for Neurogenin2 can vary from 20 to 30 min (Vosper et al. 2007; McDowell et al. 2014b), likely because of differences in extract preparation. As a control, we recommend use of a well-characterized protein, and *Xenopus laevis* Neurogenin2 is an appropriate standard for optimizing the protocol and for comparison with other proteins.

We have directly compared results obtained from different frog egg and embryo extracts with experiments carried out in mammalian tissue culture systems, and have found the results to be identical (McDowell et al. 2010). Therefore, experiments using *Xenopus* extracts are quick and easy, cost a fraction of expensive tissue culture-based systems, and can be representative of protein stability in other model systems. This simple system enabled us to perform a comprehensive investigation of noncanonical ubiquitylation (for reviews, see Kravtsova-Ivantsiv and Ciechanover 2012; McDowell and Philpott 2013, 2016a).

It should be straightforward to adapt this *Xenopus* extract protocol for use with nonradiolabeled in vitro translated proteins followed by Western blot detection. However, this will considerably lengthen the time taken to obtain results and quantification of Western blot signals is rarely linear and not as easy as quantification of radioactive signals.

RECIPES

2× Laemmli Sample Buffer with β-Mercaptoethanol

100 mM Tris-Cl (pH 6.8)
2% SDS
20% glycerol
4% β-mercaptoethanol (added fresh)

Energy Mix (50×)

Reagent	Quantity
Creatine phosphate disodium salt tetrahydrate (MW 327.14)	0.622 g
Adenosine 5′-triphosphate (ATP) disodium salt hydrate (MW 511.14)	0.138 g
EGTA stock (K salt; 0.5 M; pH 7.7; filter-sterilized and stored in aliquots at −20°C)	0.05 mL
MgCl$_2$ stock (2 M; e.g., Sigma-Aldrich 68475)	0.125 mL

Combine with MilliQ H$_2$O to a volume of 10 mL. Make 0.1-mL aliquots, and store at −20°C.

ACKNOWLEDGMENTS

We thank Jonathan Vosper for work on optimizing the degradation assay. G.S.M. was funded by a Medical Research Council (MRC) Studentship. Research in A.P.'s laboratory is supported by MRC grants MR/K018329/1, MR/L021129/1, and by core support from the Wellcome Trust and MRC to the Wellcome Trust—MRC Cambridge Stem Cell Institute.

REFERENCES

Ali F, Hindley C, McDowell G, Deibler R, Jones A, Kirschner M, Guillemot F, Philpott A. 2011. Cell cycle-regulated multi-site phosphorylation of Neurogenin 2 coordinates cell cycling with differentiation during neurogenesis. *Development* **138:** 4267–4277.

Good MC, Heald R. 2018. Preparation of cellular extracts from *Xenopus* eggs and embryos. *Cold Spring Harb Protoc* doi:10.1101/pdb.prot097055.

Hindley C, Ali F, McDowell G, Cheng K, Jones A, Guillemot F, Philpott A. 2012. Post-translational modification of Ngn2 differentially affects transcription of distinct targets to regulate the balance between progenitor maintenance and differentiation. *Development* **139:** 1718–1723.

Kravtsova-Ivantsiv Y, Ciechanover A. 2012. Non-canonical ubiquitin-based signals for proteasomal degradation. *J Cell Sci* **125:** 539–548.

McDowell GS, Philpott A. 2013. Non-canonical ubiquitylation: Mechanisms and consequences. *Int J Biochem Cell Biol* **45:** 1833–1842.

McDowell GS, Philpott A. 2016a. New insights into the role of ubiquitylation of proteins. *Int Rev Cell Mol Biol* **325:** 33–88.

McDowell GS, Philpott A. 2016b. Ubiquitin-mediated proteolysis in *Xenopus* extract. *Int J Dev Biol* **60:** 263–270.

McDowell GS, Kucerova R, Philpott A. 2010. Non-canonical ubiquitylation of the proneural protein Ngn2 occurs in both *Xenopus* embryos and mammalian cells. *Biochem Biophys Res Commun* **400:** 655–660.

McDowell GS, Hardwick LJ, Philpott A. 2014a. Complex domain interactions regulate stability and activity of closely related proneural transcription factors. *Biochem Biophys Res Commun* **450:** 1283–1290.

McDowell GS, Hindley CJ, Lippens G, Landrieu I, Philpott A. 2014b. Phosphorylation in intrinsically disordered regions regulates the activity of Neurogenin2. *BMC Biochem* **15:** 24.

Murray AW, Solomon MJ, Kirschner MW. 1989. The role of cyclin synthesis and degradation in the control of maturation promoting factor activity. *Nature* **339:** 280–286.

Vosper JMD, Fiore-Heriche CS, Horan I, Wilson K, Wise H, Philpott A. 2007. Regulation of neurogenin stability by ubiquitin-mediated proteolysis. *Biochem J* **407:** 277–284.

Vosper JMD, McDowell GS, Hindley CJ, Fiore-Heriche CS, Kucerova R, Horan I, Philpott A. 2009. Ubiquitylation on canonical and non-canonical sites targets the transcription factor neurogenin for ubiquitin-mediated proteolysis. *J Biol Chem* **284:** 15458–15468.

Assessing Ubiquitylation of Individual Proteins Using *Xenopus* Extract Systems

Gary S. McDowell[1,2,5] and Anna Philpott[3,4,5]

[1]*Future of Research, Abington, Massachusetts 02351;* [2]*Manylabs, San Francisco, California 94103;* [3]*Department of Oncology, MRC/Hutchison Research Centre, Cambridge CB21XZ, United Kingdom;* [4]*Wellcome Trust-Medical Research Centre Cambridge Stem Cell Institute, University of Cambridge, Cambridge CB21QR, United Kingdom*

Xenopus extract systems have been used to study ubiquitylation of proteins, and to uncover some of the fundamental processes of the ubiquitylation pathway itself. They provide a simple, quick, and robust method for studying ubiquitylation. In this protocol, methods are provided for studying protein ubiquitylation using *Xenopus* egg or embryo extracts and in vitro radiolabeled proteins. These methods also enable examination of whether proteins undergo noncanonical ubiquitylation, through modification of the protein by covalent linkage to ubiquitin through residues other than lysine, such as cysteine, serine, and threonine.

MATERIALS

It is essential that you consult the appropriate Material Safety Data Sheets and your institution's Environmental Health and Safety Office for proper handling of equipment and hazardous materials used in this protocol.

RECIPES: Please see the end of this protocol for recipes indicated by <R>. Additional recipes can be found online at http://cshprotocols.cshlp.org/site/recipes.

Reagents

2× Laemmli sample buffer with β-mercaptoethanol <R> (optional; see Step 3)

Cycloheximide (20 mg/mL in water; stored at −20°C)

His-tagged ubiquitin (25 mg/mL in water; stored at −20°C; Sigma-Aldrich)

LPC protease inhibitor cocktail (10 mg/mL leupeptin, pepstatin, and chymostatin)

Dissolve in DMSO and store in 50 µL aliquots at −20°C.

MG132 proteasome inhibitor (200 µM *N*-carbobenzoxyl-Leu-Leu-Leucinal, Calbiochem)

Dissolve in DMSO and store in 10 µL aliquots at −80°C.

Nickel-charged affinity bead slurry

Nonreducing buffer solution <R>

Prepare on the day of the experiment.

---

[5]Correspondence: ap113@cam.ac.uk; garymcdow@gmail.com

Plasmid DNA template containing SP6 or T7 promoter and gene of interest, dissolved in water at 250 ng/μL

> *We use pCS2 vectors, with the gene of interest downstream from the SP6 promoter. We include a Kozak sequence before the gene of interest to ensure efficient transcription, and use BamHI and EcoRI restriction sites for ligation of the gene of interest into the vector.*

Reducing buffer solution <R>

> *Prepare on the day of the experiment.*

^{35}S-methionine, 10–11 mCi/mL

> *Ensure ^{35}S-methionine is fresh, or arrange experiments as close to the receipt of fresh ^{35}S-methionine as possible. ^{35}S has a half-life of ~3 mo; use beyond 1 mo is not advised for this assay.*

SP6 or T7 TnT Quick Coupled Transcription/Translation System Master Mix (Promega)
Ubiquitin from bovine erythrocytes (25 mg/mL in water stored at −20°C; Sigma-Aldrich)
Vosper buffer <R>

> *Prepare on the day of the experiment.*

Equipment

Autoradiography film and equipment or, alternatively, a phosphor imaging system (e.g., Typhoon, Storm, or PhosphorImager)
Gel dryer
Heat blocks set to 21°C, 30°C, and 95°C
Ice bucket
Microcentrifuge
Microcentrifuge tubes (0.5 mL)
Microcentrifuge tubes (1.5 mL, with lids cut off)
Parafilm
Rocking platform
Rotary wheel

METHOD

The purpose of this protocol is to determine whether a protein of interest is ubiquitylated on canonical sites (i.e., through amide linkages to lysine residues and the amino-terminal amine group) and/or on noncanonical sites (i.e., to cysteines through thioester linkages, or to serines and threonines through hydroxyl linkages) using Xenopus cytoplasmic egg extract. Radiolabeled proteins are first produced to be used in the assay as they provide a simple method for viewing ubiquitylated proteins. The method can, however, be adapted to use nonradiolabeled protein and Western blotting for specific protein detection. Egg extract is prepared and supplemented with the following: ATP to provide energy for the process, cycloheximide to prevent translation of new protein, MG132 to prevent ubiquitylated proteins from being degraded by the ubiquitin-proteasome system, and ubiquitin.

Various conditions are compared in a standard experiment. His-ubiquitin is used to pull down ubiquitylated proteins using nickel bead resin, but it is very important to also have a control using nontagged ubiquitin: this ensures that there is a comparison of nonspecific binding of the protein to the resin. These experiments are further subdivided into reducing and nonreducing conditions, to evaluate whether the protein is ubiquitylated and, further, whether the protein is liberated from ubiquitin in high pH, reducing conditions, indicating noncanonical ubiquitylation on cysteine, serine, and/or threonine residues. Therefore, for each protein, there should be four different conditions:

- *Control with nontagged ubiquitin, treated in reducing conditions.*
- *Control with nontagged ubiquitin, treated in nonreducing conditions.*
- *Sample with his-tagged ubiquitin, treated in reducing conditions.*
- *Sample with his-tagged ubiquitin, treated in nonreducing conditions.*

> *See Figure 1 for an example of the output expected for each protein.*

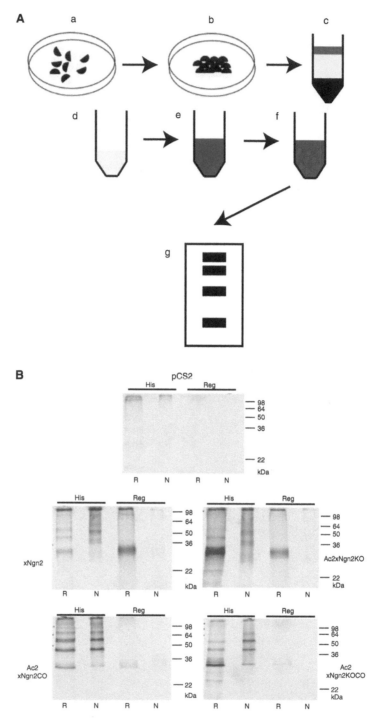

FIGURE 1. Ubiquitylation assays using *Xenopus* extract. (*A*) For interphase extracts, (a) *Xenopus laevis* eggs are harvested and treated with calcium ionophore to mimic fertilization, resulting in (b) the animal hemispheres (pigmented) presenting dorsally and contracting. (c) Eggs are packed into Eppendorf tubes and centrifuged at high speed to separate out layers of lipids (shown in green), cytoplasm (yellow), and pigment (black). (d) The cytoplasmic layer is removed, supplemented with ATP, MG132 and ubiquitin (either untagged or his-tagged) and (e) IVT, radiolabeled protein is added. The sample is incubated for 90 min at 20°C and (f) nickel charged beads are added and incubated for a further 75 min on a rotating wheel. Samples are then eluted with sodium dodecyl sulfate loading buffer (SDS-LB) and are g) separated by sodium dodecyl sulfate polyacrylamide gel electrophoresis (SDS-PAGE). (*B*) Example results of a ubiquitylation assay (the proteins analyzed were wild-type and mutant forms of *Xenopus laevis* Neurogenin2 (xNgn2). KO denotes all lysines mutated to arginines; CO denotes all cysteines mutated to alanines; Ac2 denotes mutation of the first five residues to up-regulate cotranslational amino-terminal acetylation (to block the amino-terminal amine group from ubiquitylation). (R) reducing conditions, (N) nonreducing conditions, (His) his-tagged ubiquitin, (Reg) regular, untagged ubiquitin.

Cite this protocol as *Cold Spring Harb Protoc*; doi:10.1101/pdb.prot104513

However, it should be noted that noncanonical ubiquitin on most substrates is thought to be highly unusual so in most cases the samples in nonreducing conditions may be omitted; ubiquitylated proteins detected under reducing conditions will represent linkages via canonical lysines or on the protein amino terminus.

In Vitro Transcription/Translation of Radiolabeled Protein

1. Remove TnT Quick Master Mix from −80°C storage and rapidly thaw between fingers and place on ice.

2. Place the required volume of Master Mix in a 0.5 mL microcentrifuge tube. For each 5 µL of Master Mix, add 0.5 µL plasmid solution and 0.5 µL ^{35}S-methionine. Gently mix by pipetting up and down.

    Each ubiquitylation assay requires 5 µL of in vitro translated (IVT) protein mix.

    Air bubbles reduce the efficiency of the reaction; therefore be extremely careful not to mix the reaction too vigorously. Centrifugation at a low speed for 30 sec after mixing can be performed, if necessary.

3. Incubate for 2 h at 30°C. Store at −20°C until needed.

    Volumes and reaction times described here differ from the instructions given with the TnT Quick Coupled Transcription/Translation System; we have found the conditions described here to be preferable but they can be adjusted as desired.

    To check for production of radiolabeled protein, add 1 µL of IVT protein mix to 19 µL of 2× Laemmli Sample Buffer and separate using SDS-PAGE. Dry the gel and use autoradiography film or a phosphorimaging system to determine whether a single radioactive product of the appropriate molecular weight has been produced.

Ubiquitylation

4. Prepare *Xenopus* extract as required (see Protocol 1: Preparation of Cellular Extracts from *Xenopus* Eggs and Embryos, Fig. 1A[a–c]; Good and Heald 2018).

    A total of 80 µL of extract is required for each protein being assayed under four conditions, each requiring 20 µL. Two samples are exposed to his-tagged ubiquitin, one of which is subjected to SDS-PAGE in regular nonreducing buffer, and one in reducing buffer; and two samples are exposed to untagged ubiquitin, and are also subjected to SDS-PAGE in regular or reducing buffer.

    Variations of normal extracts can be used, such as interphase extract (see Vosper et al. 2007), mitotic extract (see Vosper et al. 2009) and neurula embryo extract (see McDowell et al. 2010). In each case, energy mix or ATP should be added before use (as described in Protocol 1: Preparation of Cellular Extracts from Xenopus Eggs and Embryos (Good and Heald 2018).

5. Supplement extract with cycloheximide to a final concentration of 20 µg/mL, and with MG132 to a final concentration of 20 µM.

6. For each sample, divide 80 µL of extract into two aliquots. To one, add 3 µL of his-tagged ubiquitin per 20 µL of extract; to the other, add 3 µL of regular, untagged ubiquitin per 20 µL of extract.

    The untagged ubiquitin acts as a control, to check for nonspecific binding of radiolabeled protein to the nickel beads.

7. Prepare four 0.5 mL microcentrifuge tubes for each protein; in two tubes, combine 5 µL of radiolabeled IVT protein with 20 µL of control extract; and in the other two tubes combine 5 µL of radiolabeled IVT protein with 20 µL of his-ubiquitin extract. For each pair of tubes, mark one "reducing" and the other "nonreducing." Gently flick the tubes with an index finger until the red IVT protein mix appears mixed with the extract (Fig. 1A[d,e]).

    The "reducing" and "nonreducing" tubes for each ubiquitin condition are identical, and will be treated identically until the point of SDS-PAGE nonreducing or reducing buffer addition.

8. Incubate the reactions for 90 min at 20°C.

Binding

9. Supplement enough Vosper buffer for the number of wash steps desired (see Steps 12–15) with LPC protease inhibitor cocktail to a final concentration of 10 µg/mL.

10. Add 250 μL of Vosper buffer + LPC to each of the reactions.

11. To each reaction, add 15 μL of Nickel bead slurry using a pipette tip with the end cut off, to create a larger opening. Place these tubes inside 1.5 mL tubes with the lids cut off and secure them in place by wrapping in Parafilm. Place the tubes on a rotary wheel for 75 min at 12 revolutions per minute at room temperature (Fig. 1A[f]).

    Before dispensing, resuspend the nickel beads by gentle inversion.

Washing

12. Spin samples at 0.8 relative centrifugal force (rcf) for 30 sec in a microcentrifuge.
    Do not exceed 0.8 rcf.

13. Remove the supernatant from each tube taking care not to touch the beads. Put the tip of the pipette just under the surface of the liquid and gently lower the tip to aspirate off the supernatant. Discard the supernatant.

    The supernatant will be radioactive and should be discarded in an appropriate manner, according to local procedures.

14. Add 400 μL of Vosper buffer + LPC to each tube; mix by gently inverting the tubes. Repeat Steps 1–2.

15. Add 400 μL of Vosper buffer without LPC. Mix by gentle inversion then centrifuge at 0.8 rcf. Remove the supernatant and repeat this step at least twice.

    A greater number of washes will produce less background from nonspecific binding. Err on the side of performing more rather than fewer washes.

Elution

16. After the final wash, remove as much of the supernatant as possible, leaving ∼ 15 μL of slurry.

17. To the samples marked "nonreducing" add 15 μL of nonreducing buffer. To the samples marked "reducing" add 15 μL of reducing buffer. Gently mix, and rock on a rocking platform at room temperature for 20 min.

18. Incubate the "nonreducing" samples on ice for 5 min. Incubate the "reducing" samples for 5 min at 95°C.
    At this point samples may be stored overnight at −20°C.

19. Separate 10 μL of each sample by SDS PAGE and dry the gel in a gel dryer for at least 1 h (Fig. 1A [g]). Expose the dried gel to autoradiography film for at least 24 h.
    Keep the rest of the sample in case there is a problem with the gel and the SDS-PAGE needs to be repeated.

20. Develop the autoradiography film to see whether distinct bands are visible (see example in Fig. 1B).

    The exposure time may need to be tailored; it may be necessary to expose the film to the gel for several weeks. If phosphorimaging equipment is available, expose a phosphor screen overnight to see whether the experiment has worked (see Protocol 9: Calculating the Degradation Rate of Individual Proteins Using Xenopus Extract Systems [McDowell and Philpott 2018]).

DISCUSSION

This ubiquitylation assay is based on the method of Salic et al. (2000) with improvements by Vosper et al. (2009). We have used this method with a variety of extract types to assess the ability of proteins, principally Neurogenin2, to be ubiquitylated on canonical and noncanonical ubiquitylation sites (Vosper et al. 2009; McDowell et al. 2010, 2014). Others have used the *Xenopus* system in a similar

 Cite this protocol as *Cold Spring Harb Protoc*; doi:10.1101/pdb.prot104513

manner to study the ubiquitylation of proteins (for review, see McDowell and Philpott 2016b). This simple system has enabled us to perform a comprehensive investigation of noncanonical ubiquitylation (for reviews, see Kravtsova-Ivantsiv and Ciechanover 2012; McDowell and Philpott 2013, 2016a).

As controls, we used the pCS2 vector alone (see Fig. 1B) and untagged ubiquitin for each sample. Untagged ubiquitin is an important control that is often omitted from ubiquitylation assays, because it is often presumed that the presence of a band at the level of a wild-type (i.e., nonubiquitylated) protein is simply background binding of unmodified protein. Our assays show that this is not necessarily the case; we find that comparing reducing and nonreducing conditions shows the release of protein that was modified by ubiquitin, but through labile, easily reduced bonds that are broken under normal SDS-PAGE conditions. The absence of protein in the nontagged ubiquitin treatment further highlights that this is not simply an artifact.

It should be straightforward to adapt this *Xenopus* extract protocol for use with nonradiolabeled, in vitro translated proteins followed by western blot detection. However, this will considerably lengthen the time taken to obtain results and background from antibody cross-reactivity may be higher.

The protocol can distinguish canonical ubiquitylation (through stable amide bonds between the ubiquitin carboxyl group and lysines or the amino-terminal amine groups) from noncanonical ubiquitylation (through weaker thioester bonds to cysteine residues or hydroxyl linkages to serines, threonines, and possibly tyrosines). However as the reducing SDS PAGE buffer contains both a high concentration of β-mercaptoethanol (to reduce thioester linkages) and is of high pH (to hydrolyze hydroxyl linkages), modifications on cysteine cannot be distinguished from those on serine and threonine.

RECIPES

2× Laemmli Sample Buffer with β-Mercaptoethanol

100 mM Tris-Cl (pH 6.8)
2% SDS
20% glycerol
4% β-mercaptoethanol (added fresh)

Nonreducing Buffer Solution

Reagent	Quantity
2× Laemmli sample buffer with β-mercaptoethanol <R>	1 mL
Water	100 µL
Total volume	1.1 mL

Prepare fresh on day of use. No sterilization is necessary. Store at room temperature.

Reducing Buffer Solution

Reagent	Quantity
2× Laemmli sample buffer with β-mercaptoethanol <R>, adjusted to pH 10 with NaOH	1 mL
β-mercaptoethanol	100 µL
Imidazole	7.49 mg
Total volume	1.1 mL

Prepare fresh on day of use. No sterilization is necessary. Store at room temperature.

Vosper Buffer

Reagent	Quantity
Urea	48.05 g
Tris (1 M, pH 7.4)	10 mL
NaCl	3.51 g
Imidazole	0.136 g
IGEPAL (CA-630; Sigma-Aldrich) or Nonidet P40	1 mL

Make up to 100 mL with water. No sterilization is necessary. Store at room temperature and use on day of preparation.

ACKNOWLEDGMENTS

We thank Jonathan Vosper for work on optimizing the ubiquitylation assay. G.S.M. was funded by a Medical Research Council (MRC) Studentship. Research in A.P.'s laboratory is supported by MRC grants MR/K018329/1, MR/L021129/1, and MR/L021129/1, and by core support from the Wellcome Trust and MRC to the Wellcome Trust—MRC Cambridge Stem Cell Institute.

REFERENCES

Good MC, Heald R. 2018. Preparation of cellular extracts from *Xenopus* eggs and embryos. *Cold Spring Harb Protoc* doi: 10.1101/pdb.prot097055.

Kravtsova-Ivantsiv Y, Ciechanover A. 2012. Non-canonical ubiquitin-based signals for proteasomal degradation. *J Cell Sci* **125**: 539–548.

McDowell GS, Philpott A. 2013. Non-canonical ubiquitylation: Mechanisms and consequences. *Int J Biochem Cell Biol* **45**: 1833–1842.

McDowell GS, Philpott A. 2016a. New insights into the role of ubiquitylation of proteins. *Int Rev Cell Mol Biol* **325**: 33–88.

McDowell GS, Philpott A. 2016b. Ubiquitin-mediated proteolysis in *Xenopus* extract. *Int J Dev Biol* **60**: 263–270.

McDowell GS, Philpott A. 2018. Calculating the degradation rate of individual proteins using *Xenopus* extract systems. *Cold Spring Harb Protoc* doi: f10.1101/pdb.prot103481.

McDowell GS, Kucerova R, Philpott A. 2010. Non-canonical ubiquitylation of the proneural protein Ngn2 occurs in both *Xenopus* embryos and mammalian cells. *Biochem Biophys Res Commun* **400**: 655–660.

McDowell GS, Hardwick LJ, Philpott A. 2014. Complex domain interactions regulate stability and activity of closely related proneural transcription factors. *Biochem Biophys Res Commun* **450**: 1283–1290.

Salic A, Lee E, Mayer L, Kirschner MW. 2000. Control of β-catenin stability. *Mol Cell* **5**: 523–532.

Vosper JMD, Fiore-Heriche CS, Horan I, Wilson K, Wise H, Philpott A. 2007. Regulation of neurogenin stability by ubiquitin-mediated proteolysis. *Biochem J* **407**: 277–284.

Vosper JMD, McDowell GS, Hindley CJ, Fiore-Heriche CS, Kucerova R, Horan I, Philpott A. 2009. Ubiquitylation on canonical and non-canonical sites targets the transcription factor neurogenin for ubiquitin-mediated proteolysis. *J Biol Chem* **284**: 15458–15468.

Reconstituting Nuclear and Chromosome Dynamics Using *Xenopus* Extracts

Susannah Rankin[1,2,3]

[1]*Program in Cell Cycle and Cancer Biology, Oklahoma Medical Research Foundation, Oklahoma City, Oklahoma 73104;* [2]*Department of Cell Biology, University of Oklahoma Health Sciences Center, Oklahoma City, Oklahoma 73104*

Extracts prepared from the eggs of frogs, particularly *Xenopus* species, have provided critical material for seminal studies of nuclear and chromosome dynamics over several decades. Their usefulness for these types of analyses lies in several important characteristics: stockpiled nuclear components, absence of endogenous DNA, and intact and functioning signaling networks. These factors have allowed detailed molecular analyses of many aspects of chromosome biology, including DNA replication, checkpoint signaling, epigenetic control, and chromosome condensation, cohesion, and segregation. In this introduction, the preparation and application of *Xenopus* egg extracts for the study of chromosomes and chromatin are described in detail.

BACKGROUND

Early nuclear transfer experiments suggested that functional changes were exerted on somatic nuclei when they were incubated in the cytoplasm of a *Xenopus* egg, and that these changes represented reversal of cellular differentiation, restoring the nuclei to pluripotency (Gurdon et al. 1958; Gurdon 1962a,b). These studies eventually led to the notion that embryonic extracts might also be exploited to understand other aspects of nuclear function. In the early 1980s, Lohka and Masui showed that demembranated sperm nuclei added to extracts prepared from frog eggs, in this case *Rana pipiens*, spontaneously underwent the nuclear events that had previously been reported to occur in embryos following fertilization of the egg (Lohka and Masui 1983). These included decondensation of the sperm chromatin, nuclear envelope assembly, and DNA replication. The fact that these events occurred spontaneously in the extracts supported the notion that this system could be exploited to understand their molecular underpinnings. Murray and Kirschner (Murray et al. 1989; Murray and Kirschner 1989) showed that extracts could be prepared from *Xenopus* eggs that spontaneously cycled between the different phases of the cell cycle, and that this cycling was accompanied by and dependent on fluctuations in the level of a protein "cyclin," an essential component of the previously described mitosis promoting factor, or MPF (Gerhart et al. 1984; Minshull et al. 1989). This system has since been exploited for many groundbreaking investigations of the fundamental mechanisms of nuclear and chromosome dynamics. Drugs, proteins, nucleotides, and DNA templates can be added to the extracts with ease. In addition, proteins can be depleted from extracts using appropriate antibodies, allowing analysis of mutant recombinant proteins in rescue experiments. This ability is particularly

---

[3]Correspondence: susannah-rankin@omrf.org

Cite this introduction as *Cold Spring Harb Protoc*; doi:10.1101/pdb.top097105

useful if the proteins under investigation are essential for viability, which can make their study in somatic cells technically challenging.

MANIPULATION OF CELL CYCLE STATE IN VITRO

The ability to manipulate the cell cycle state makes *Xenopus* egg extracts a powerful tool for the in vitro study of chromatin and chromosome dynamics. Mature *Xenopus* eggs are arrested in metaphase of the second meiotic division (MII) with high levels of meiotic kinase activity. This arrest, mediated by cytostatic factor (CSF), is relatively stable. Extract prepared directly from eggs ("CSF extract") can be exploited to study the impact of M phase activities on chromosome and chromatin behavior and events (Fig. 1). Additionally, because CSF arrest can be released by the addition of calcium to the extract, CSF extract can also be used to study the impact of mitotic exit on chromosome dynamics. Extracts induced to enter interphase by the addition of calcium can be maintained in interphase by preventing translation of endogenous cyclin RNA.

CSF activity is dominant and can be used to drive interphase extracts into M phase. Freshly prepared or thawed frozen CSF extract added in equal volume to interphase extract causes the entire mixture to enter M phase, where it will again arrest. This kind of approach is particularly useful if events to be analyzed in M phase, such as sister chromatid cohesion, are dependent on prior DNA replication.

NUCLEI, CHROMOSOMES, AND CHROMATIN IN VITRO

A variety of DNA substrates can be added to egg extract to assess different aspects of chromatin function and regulation, chromosome dynamics, and nuclear assembly (Fig. 1).

Sperm nuclei can be isolated from *Xenopus* testes and demembranated with a mild detergent extraction, as described in Protocol 1: Isolation and Demembranation of *Xenopus* Sperm Nuclei (Hazel and Gatlin 2018). This method of sperm preparation enables reconstitution of chromosome reactions in vitro and allows the simultaneous addition of many nuclei to the extract. Nuclear assembly occurs spontaneously when demembranated sperm nuclei are added to interphase extract and can be analyzed as described in Protocol 6: Nucleus Assembly and Import in *Xenopus laevis* Egg Extract (Chen and Levy 2018). This approach was critical to the demonstration that nuclear envelope assembly is promoted by chromatin-dependent remodeling of the endoplasmic reticulum (Anderson and Hetzer 2007), and has enabled seminal findings about the biochemistry of nuclear import (Görlich et al. 1994). Egg extract was also used to provide insight into the mechanisms of nuclear disassembly in M phase (Mühlhäusser and Kutay 2007).

Early work from several labs established *Xenopus* egg extract as a uniquely powerful system in which to study DNA replication and its regulation (Blow and Laskey 1986; Blow and Watson 1987; Blow 1993; Walter and Newport 1997, 2000). Complete DNA replication occurs spontaneously in <2 h when sperm nuclei are added to crude interphase egg extract through the recruitment of proteins and nucleotides from the extract. The roles of particular replication proteins can be dissected by immunodepletion, and recombinant and mutant derivatives can be used to rescue function. This system has been used to elucidate how replication licensing is entrained to cell cycle progression (McGarry and Kirschner 1998; Wohlschlegel et al. 2000), and to define the mechanism by which DNA replication promotes degradation of certain replication factors (Havens and Walter 2009), as well as many other aspects of DNA replication and repair.

The in vitro study of DNA replication was further refined by Walter, Newport, and colleagues (Walter et al. 1998), who established a method for nucleus-free DNA replication. Whereas the critical early replication-licensing step occurs in clarified high-speed extract, replication initiation requires nuclear assembly. This requirement for nuclear assembly can be bypassed by sequential incubation

Cite this introduction as *Cold Spring Harb Protoc*; doi:10.1101/pdb.top097105

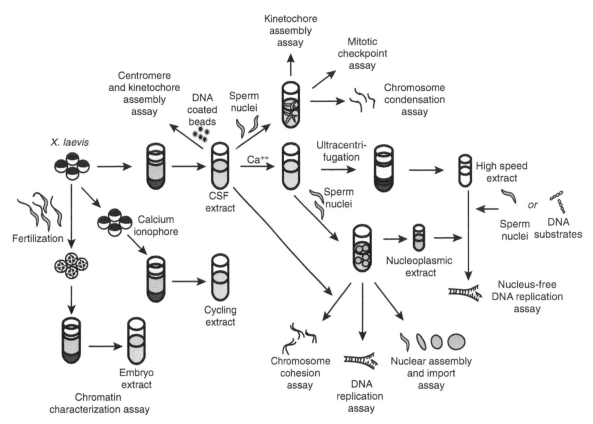

FIGURE 1. The use of *Xenopus* egg extracts to study chromatin and chromosome dynamics. CSF extracts, which contain high levels of MPF (mitosis promoting factor) activity are prepared directly from *Xenopus* eggs, which are arrested in M phase of meiosis II. CSF extracts can be used to study events that require M phase activity, such as centromere and kinetochore assembly, mitotic checkpoint function, and chromosome condensation (shown at *top*). The addition of Ca^{++} releases CSF arrest, resulting in cyclin degradation, and entry into interphase (*middle*). Interphase extract can be used to study events such as DNA replication and nuclear assembly and import (shown at *bottom*). Interphase extract, which contains abundant membrane stores, can also be further fractionated by centrifugation to study licensed DNA replication in a nucleus-free system. This requires the preparation of two different extracts, one enriched in nuclear activities (nucleoplasmic extract or NPE), and the other a membrane-free supernatant, as shown at *right*. Some studies, such as chromosome cohesion assays, require cycling of the extract through interphase and back into M phase. This is achieved by adding CSF extract to the reaction following interphase incubation. CSF promotes M phase entry and arrest in a dominant manner, resulting in condensation of replicated chromosomes. Some assays require multiple cell cycles in vitro, which can be achieved by preparation of cycling extract (center). Finally, developmental changes in chromosome and chromatin regulation can be studied by preparation of developmentally staged embryonic extracts or nuclei (*bottom left*). These are prepared from embryos collected at the desired developmental stage after fertilization. The manipulations shown in this schematic can also be applied to other questions of vertebrate chromosome and chromatin regulation as new questions arise.

of DNA substrates in clarified interphase extract to promote replication licensing, followed by incubation in concentrated nucleoplasmic extract (NPE). NPE is prepared in advance from nuclei assembled in crude interphase extract and stored frozen in aliquots. This powerful system is described in detail in Protocol 7: Extracts for Analysis of DNA Replication in a Nucleus-Free System (Sparks and Walter 2018). Importantly, this system allows the complete replication of plasmids and other small templates that are not efficiently replicated in interphase extract, and has been exploited to define intermediates in replication initiation and related events (Walter 2000; Takahashi et al. 2004).

During nuclear assembly in interphase extract, histones are recruited from the extract and nucleosomes are assembled on the DNA. The complement of histone variants and modifications deployed in different cell types is developmentally regulated. Extracts prepared from eggs and embryos at different developmental stages can be used to study these processes in detail. This approach is described in Protocol 3: Chromatin Characterization in *Xenopus laevis* Cell-Free Egg Extracts and Embryos (Wang et al. 2018). Using this method, it has been possible to correlate certain histone modification signatures with differentiation status (Shechter et al. 2009).

Egg extracts can also be used to study specialized forms of chromatin, such as centromeres, and how they are formed. Methods to tackle these questions are described in Protocol 4: Centromere and Kinetochore Assembly in *Xenopus laevis* Egg Extract (Flores Servin and Straight 2018), which includes descriptions of how to study kinetochore and centromere assembly on intact sperm chromosomes, as well as methods to prepare beads coated with model DNA substrates. Using immunodepletion and rescue, in combination with mutant protein analysis in these assays, the dependencies and functional interactions of various centromere- and kinetochore-associated factors can be assessed. The readouts can be both biochemical and microscopy based. This system has been used to show that sequences in the histone variant CENP-A that promote histone assembly are distinct from those that ensure centromere function (Guse et al. 2011).

The morphology of in vitro assembled whole condensed chromosomes can be assessed using egg extracts, as described in Protocol 2: Chromosome Cohesion and Condensation in *Xenopus* Egg Extracts (da Silva and Rankin 2018). The chromosomes of sperm nuclei added directly to CSF extracts are chromatinized and condensed to form single chromatids, all through the activity of extract-associated proteins. This approach allowed the earliest biochemical identification of chromosome condensation factors (Hirano and Mitchison 1994). The ability to cycle extracts through interphase, in which DNA replication occurs, and back into M phase also allows exploration of mechanisms that control cohesion between sister chromatids, and has been very useful for studying the relationship between DNA replication and cohesion establishment (Losada et al. 1998; Sumara et al. 2000; Gillespie and Hirano 2004; Takahashi et al. 2004; Rankin et al. 2005). Importantly, because many cohesion and condensation factors are essential, the use a depletion and rescue approach in egg extracts bypasses the complication this presents when working in somatic cells. Rapid and synchronous genome duplication in egg extracts make these studies practical.

Xenopus egg extracts can also be used to study signaling events that depend on chromatin or nuclear activity. For example, the mitotic checkpoint, or "spindle checkpoint" is a mechanism that prevents progression out of M phase in the presence of unattached kinetochores. Unattached kinetochores generate a soluble signal, which ultimately prevents cyclin degradation, resulting in sustained mitotic arrest. The mitotic checkpoint can be artificially generated in egg extract by adding a high concentration of sperm nuclei together with a microtubule depolymerizing drug, as described in Protocol 5: Analysis of Mitotic Checkpoint Function in *Xenopus* Egg Extracts (Mao 2018). In this system, mitotic arrest is monitored by assaying for mitotic kinase activity, or by assessing nuclear morphology. This method can be used to assess kinetochore function, as well as to dissect the mitotic checkpoint mechanism itself (Minshull et al. 1994). Other chromatin-dependent signaling events, such as the Ran GTPase gradient (Kalab et al. 1999; Zhang and Clarke 2000; Wiese et al. 2001), and DNA damage signaling (Lupardus et al. 2007; Costanzo et al. 2004) have also been successfully characterized using the egg extract system. This flexible system can be further adapted and applied to new questions as they arise.

SUMMARY

Since the discovery that the events of cell division and their regulation could be recapitulated in vitro using extract prepared from *Xenopus* eggs, a wide variety of chromatin and chromosome-based events and mechanisms have been elucidated using this powerful system. The protocols discussed herein provide an introduction to the use of extracts to study questions of chromatin, chromosomes, and signaling events that control their activity.

Cite this introduction as *Cold Spring Harb Protoc*; doi:10.1101/pdb.top097105

ACKNOWLEDGMENTS

I am grateful to all members of the Rankin laboratory for their continued support and enthusiasm, Romain Gibeaux for components of the illustration, and to all investigators, past and present, who helped develop the mighty *Xenopus* extract system. This work was supported by National Institutes of Health (NIH) R01 GM101250, and a grant from the Oklahoma Center for Adult Stem Cell Research, to S.R.

REFERENCES

Anderson DJ, Hetzer MW. 2007. Nuclear envelope formation by chromatin-mediated reorganization of the endoplasmic reticulum. *Nat Cell Biol* **9:** 1160–1166.

Blow J. 1993. Preventing re-replication of DNA in a single cell cycle: Evidence for a replication licensing factor. *J Cell Biol* **122:** 993–1002.

Blow JJ, Laskey RA. 1986. Initiation of DNA replication in nuclei and purified DNA by a cell-free extract of *Xenopus* eggs. *Cell* **47:** 577–587.

Blow JJJ, Watson JVJ. 1987. Nuclei act as independent and integrated units of replication in a *Xenopus* cell-free DNA replication system. *EMBO J* **6:** 1997–2002.

Chen P, Levy DL. 2018. Nucleus assembly and import in *Xenopus laevis* egg extract. *Cold Spring Harb Protoc* doi:10.1101/pdb.prot097196.

Costanzo V, Robertson K, Gautier J. 2004. *Xenopus* cell-free extracts to study the DNA damage response. *Methods Mol Biol* **280:** 213–227.

da Silva EL, Rankin S. 2018. Chromosome cohesion and condensation in *Xenopus* egg extracts. *Cold Spring Harb Protoc* doi:10.1101/pdb.prot097121.

Flores Servin JC, Straight AF. 2018. Centromere and kinetochore assembly in *Xenopus laevis* egg extract. *Cold Spring Harb Protoc* doi:10.1101/pdb.prot102509.

Gerhart J, Wu M, Kirschner M. 1984. Cell cycle dynamics of an M-phase-specific cytoplasmic factor in *Xenopus laevis* oocytes and eggs. *J Cell Biol* **98:** 1247–1255.

Gillespie PJ, Hirano T. 2004. Scc2 couples replication licensing to sister chromatid cohesion in *Xenopus* egg extracts. *Curr Biol* **14:** 1598–1603.

Görlich D, Prehn S, Laskey RA, Hartmann E. 1994. Isolation of a protein that is essential for the first step of nuclear protein import. *Cell* **79:** 767–778.

Gurdon JB. 1962a. Adult frogs derived from the nuclei of single somatic cells. *Dev Biol* **4:** 256–273.

Gurdon JB. 1962b. The developmental capacity of nuclei taken from intestinal epithelium cells of feeding tadpoles. *J Embryol Exp Morphol* **10:** 622–640.

Gurdon JB, Elsdale TR, Fischberg M. 1958. Sexually mature individuals of *Xenopus laevis* from the transplantation of single somatic nuclei. *Nature* **182:** 64–65.

Guse A, Carroll CW, Moree B, Fuller CJ, Straight AF. 2011. In vitro centromere and kinetochore assembly on defined chromatin templates. *Nature* **477:** 354–358.

Havens CG, Walter JC. 2009. Docking of a specialized PIP Box onto chromatin-bound PCNA creates a degron for the ubiquitin ligase CRL4Cdt2. *Mol Cell* **35:** 93–104.

Hazel JW, Gatlin JC. 2018. Isolation and demembranation of *Xenopus* sperm nuclei. *Cold Spring Harb Protoc* doi:10.1101/pdb.prot099044.

Hirano T, Mitchison TJ. 1994. A heterodimeric coiled-coil protein required for mitotic chromosome condensation in vitro. *Cell* **79:** 449–458.

Kalab P, Pu RT, Dasso M. 1999. The ran GTPase regulates mitotic spindle assembly. *Curr Biol* **9:** 481–484.

Lohka M, Masui Y. 1983. Formation in vitro of sperm pronuclei and mitotic chromosomes induced by amphibian ooplasmic components. *Science* **220:** 719–721.

Losada A, Hirano M, Hirano T. 1998. Identification of *Xenopus* SMC protein complexes required for sister chromatid cohesion. *Genes Dev* **12:** 1986–1997.

Lupardus PJ, Van C, Cimprich KA. 2007. Analyzing the ATR-mediated checkpoint using *Xenopus* egg extracts. *Methods* **41:** 222–231.

Mao Y. 2018. Analysis of mitotic checkpoint function in *Xenopus* egg extracts. *Cold Spring Harb Protoc* doi:10.1101/pdb.prot099853.

McGarry TJ, Kirschner MW. 1998. Geminin, an inhibitor of DNA replication, is degraded during mitosis. *Cell* **93:** 1043–1053.

Minshull J, Blow JJ, Hunt T. 1989. Translation of cyclin mRNA is necessary for extracts of activated *Xenopus* eggs to enter mitosis. *Cell* **56:** 947–956.

Minshull J, Sun H, Tonks NK, Murray AW. 1994. A MAP kinase-dependent spindle assembly checkpoint in *Xenopus* egg extracts. *Cell* **79:** 475–486.

Mühlhäusser P, Kutay U. 2007. An in vitro nuclear disassembly system reveals a role for the RanGTPase system and microtubule-dependent steps in nuclear envelope breakdown. *J Cell Biol* **178:** 595–610.

Murray A, Kirschner M. 1989. Cyclin synthesis drives the early embryonic cell cycle. *Nature* **339:** 275–280.

Murray A, Solomon M, Kirschner M. 1989. The role of cyclin synthesis and degradation in the control of maturation promoting factor activity. *Nature* **339:** 280–286.

Rankin S, Ayad NG, Kirschner MW. 2005. Sororin, a substrate of the anaphase-promoting complex, is required for sister chromatid cohesion in vertebrates. *Mol Cell* **18:** 185–200.

Shechter D, Nicklay JJ, Chitta RK, Shabanowitz J, Hunt DF, Allis CD. 2009. Analysis of histones in *Xenopus laevis*. I. A distinct index of enriched variants and modifications exists in each cell type and is remodeled during developmental transitions. *J Biol Chem* **284:** 1064–1074.

Sparks JL, Walter, J. 2018. Extracts for analysis of DNA replication in a nucleus-free system. *Cold Spring Harb Protoc* doi:10.1101/pdb.prot097154.

Sumara I, Vorlaufer E, Gieffers C, Peters BH, Peters JM. 2000. Characterization of vertebrate cohesin complexes and their regulation in prophase. *J Cell Biol* **151:** 749–762.

Takahashi TS, Yiu P, Chou MF, Gygi S, Walter JC. 2004. Recruitment of *Xenopus* Scc2 and cohesin to chromatin requires the pre-replication complex. *Nat Cell Biol* **6:** 991–996.

Walter JC. 2000. Evidence for sequential action of cdc7 and cdk2 protein kinases during initiation of DNA replication in *Xenopus* egg extracts. *J Biol Chem* **275:** 39773–39778.

Walter J, Newport JW. 1997. Regulation of replicon size in *Xenopus* egg extracts. *Science* **275:** 993–995.

Walter J, Newport J. 2000. Initiation of eukaryotic DNA replication: Origin unwinding and sequential chromatin association of Cdc45, RPA, and DNA polymerase α. *Mol Cell* **5:** 617–627.

Walter J, Sun L, Newport J. 1998. Regulated chromosomal DNA replication in the absence of a nucleus. *Mol Cell* **1:** 519–529.

Wang W-L, Onikubo T, Shechter D. 2018. Chromatin characterization in *Xenopus laevis* cell-free egg extracts and embryos. *Cold Spring Harb Protoc* doi:10.1101/pdb.prot099879.

Wiese C, Wilde A, Moore MS, Adam SA, Merdes A, Zheng Y. 2001. Role of importin-β in coupling Ran to downstream targets in microtubule assembly. *Science* **291:** 653–656.

Wohlschlegel JA, Dwyer BT, Dhar SK, Cvetic C, Walter JC, Dutta A. 2000. Inhibition of eukaryotic DNA replication by geminin binding to Cdt1. *Science* **290:** 2309–2312.

Zhang C, Clarke PR. 2000. Chromatin-independent nuclear envelope assembly induced by Ran GTPase in *Xenopus* egg extracts. *Science* **288:** 1429–1432.

Isolation and Demembranation of *Xenopus* Sperm Nuclei

James W. Hazel[1,2] and Jesse C. Gatlin[1,3]

[1]*Department of Molecular Biology, University of Wyoming, Laramie, Wyoming 82071*

The inherent experimental advantages of intact amphibian eggs have been exploited for several decades to advance our understanding of fundamental developmental processes and the cell cycle. Characterization of these processes at the molecular level has been greatly advanced by the use of cell-free extracts, which permit the development of biochemically tractable approaches. Demembranated *Xenopus laevis* sperm nuclei have been used with cell-free extracts to recapitulate cell cycle progression and to control the cell cycle state of the egg extract. This system has become an invaluable and widely used tool for studies of cell cycle regulation and many downstream events. Here, we describe a protocol, derived in part from other published protocols and modified over time, for the preparation of *Xenopus* sperm nuclei that can be used in a variety of in vitro assays.

MATERIALS

It is essential that you consult the appropriate Material Safety Data Sheets and your institution's Environmental Health and Safety Office for proper handling of equipment and hazardous materials used in this protocol.

RECIPES: Please see the end of this protocol for recipes indicated by <R>. Additional recipes can be found online at http://cshprotocols.cshlp.org/site/recipes.

Reagents

Benzocaine (0.5%)

> *Dissolve in ddH$_2$O; adjust pH to 7.8–8.0 with NaHCO$_3$.*

DMSO (Sigma-Aldrich 472301)
Frog husbandry water <R>
Hormones

> Human chorionic gonadotropin (HCG) (1000 IU/mL) (Sigma-Aldrich CG10)
> Pregnant mare serum gonadotropin (PMSG) (100 IU/mL) (Prospec HOR-272)
> > *Dissolve hormones in sterile ddH$_2$O and store at 4°C.*

LPC protease inhibitor cocktail (10 mg/mL leupeptin, pepstatin, chymostatin; EMD Millipore)

> *Dissolve in DMSO and store in 50 µL aliquots at −20°C.*

Lysolecithin (10 mg/mL; Sigma-Aldrich L4129)

> *Prepare freshly in ddH$_2$O.*

---

[2]Present address: Center for Biomedical Ethics and Society, Institute for Medicine and Public Health, Vanderbilt University Medical Center, Nashville, Tennessee 37203

[3]Correspondence: jgatlin@uwyo.edu

Copyright © 2023 Cold Spring Harbor Laboratory Press; all rights reserved
Cite this protocol as *Cold Spring Harb Protoc*; doi:10.1101/pdb.prot099044

Marc's modified Ringer's (MMR) for *Xenopus* (1×) <R>
Nuclear preparation buffer (NPB) <R>
NPB/BSA <R>
S/S Buffer (20×) <R>
X. laevis male frogs

> *This procedure can be adapted to any* Xenopus *species by adjusting for differences in frog and testis size. Alterations in the protocol for the smaller frog,* X. tropicalis, *are indicated.*

Equipment

Buckets for frogs (4 L, plastic, with tight-fitting lids that have holes punched for air exchange; Rubbermaid)
Centrifuge (clinical, capable of spinning 15-mL conical tubes at 1500*g*)
Conical tubes (plastic; 15 mL)
Filter paper
Fine forceps (two)
Hemocytometer
Ice bucket
Incubator (16°C)
Liquid nitrogen
Microcentrifuge (tabletop)
Microcentrifuge tubes (0.5 mL)
Microcentrifuge tubes (1.5 mL)
Microscope (for tissue culture)
Pellet pestles (disposable, plastic; Fisher 12-141-364)
Petri dish (glass, 100 mm × 15 mm)
Serological pipettes (plastic, 10 mL) and pipette aid
Surgical instruments (pithing needle, scalpel, blunt-ended forceps, and fine scissors)
Syringes (disposable; 1 mL)
Syringe needles (27–30 gauge)

METHOD

1. Inject four male *X. laevis* frogs subcutaneously in the dorsal lymph sac with 250 µL PMSG (25 IU) using a 1 mL syringe and 27–30 gauge needle ∼72 h before sperm nuclei preparation. Maintain frogs in 4 L frog buckets containing 2 L frog husbandry water in a dark incubator at 16°C.

   > *For* X. tropicalis, *inject six males similarly with 100 µL PMSG (10 IU) 48 h before sperm nuclei preparation and keep at room temperature.*

2. Inject the same frogs with 125 µL HCG (125 IU) ∼24 h before testes isolation using the procedure described in Step 1.

   > *For* X. tropicalis, *inject 50 IU HCG ∼24 h before testes isolation.*

3. On the day of sperm nuclei preparation, ensure all buffers are cold and that the centrifuges are at 4°C. Prepare lysolecithin solution and NPB solution.

   > *With the exception of Step 11, all steps are performed with ice-cold buffers kept on ice.*

4. Prepare 2 L of 0.05% benzocaine solution (pH 7.8–8.0) in a 4 L frog bucket. Anesthetize the frogs by immersing them in the benzocaine solution for ∼20–25 min (5–10 min after all movement has ceased). Remove the sedated frogs from the solution and pith or decapitate to ensure death.

5. To remove the testes, use a scalpel to make a vertical incision through the skin and musculature along the midline of the abdomen. Make two orthogonal incisions, one at each end of the initial cut, to create an "I" like shape and resect the flaps to expose the abdominal cavity (Fig. 1A). The testes are located behind the yellow fat pads on either side of the midline and present as white egg-shaped bodies ~1–2 cm in length (Fig. 1A inset). Use a pair of fine scissors to remove the testes from each frog, taking care to avoid contamination with fat tissue and place in a Petri dish containing cold 1× MMR on ice.

6. Remove any remaining vessels and fatty tissue using a pair of fine forceps and dry filter paper, being careful not to pierce or rupture the testes (compare Fig. 1B,C).

7. Rinse testes three times in cold 1× MMR in the Petri dish. Pour the buffer into the dish (~20 mL per rinse) and swirl. Use a pipette aid and a 10 mL serological pipette to remove as much buffer as possible between washes.

8. Wash two times in cold NPB (~20 mL per rinse), making sure to remove as much buffer as possible between washes.

9. Using forceps, transfer two testes per 1.5 mL microcentrifuge tube and add 0.5 mL cold NPB. Keeping the tubes on ice, macerate using fine scissors and crush using a blunt, plastic pellet pestle. Homogenize the tissue vigorously.

10. Centrifuge the tubes containing the mashed testes in a tabletop microcentrifuge for 10–20 sec to pellet the testicular tissue (take care to ensure the tubes remain cold; actively cool the centrifuge if possible). Transfer all supernatants (which contain the sperm) to a single 15 mL conical tube on ice. Resuspend the pelleted testicular tissue in another 0.5 mL of cold NPB and triturate vigorously with a cut off pipette tip (to avoid excessive shearing forces). Repeat the 10–20 sec spin and again transfer the supernatants to the same 15 mL conical tube.

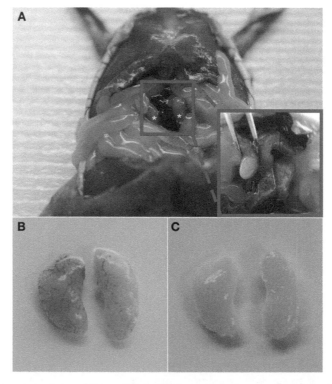

FIGURE 1. Frog testes at various stages of the protocol. (A) photograph showing the anatomical location of the testes after partial resection (asterisks mark the testes and the inset shows a zoomed image with forceps tips providing scale). Photographs in B & C are of resected testes before and after removal of vessels and fat, respectively.

          Cite this protocol as *Cold Spring Harb Protoc*; doi:10.1101/pdb.prot099044

11. Centrifuge the 15 mL conical tube in a clinical centrifuge at 1500*g* for 10 min at 4°C to pellet the sperm.

12. To demembranate the sperm, resuspend the sperm pellet in 1 mL room-temperature NPB. Add 50 µL freshly prepared 10 mg/mL lysolecithin. Mix thoroughly by pipetting up and down and incubate for exactly 5 min at room temperature.

    It is important to incubate for exactly 5 min because these conditions yield a high percentage (typically >95%) of demembranated sperm nuclei (as assessed by Hoechst staining) without compromising nuclei integrity.

13. Add 4 mL cold NPB/BSA 3% to quench the lysolecithin and then spin at 1500*g* for 10 min at 4°C.

14. Aspirate the supernatant and resuspend the demembranated nuclei pellet in 2 mL cold NPB/BSA 0.3% and spin at 1500*g* for 10 min at 4°C.

15. Aspirate the supernatant and resuspend the pellet in 500 µL cold NPB/BSA 0.3% + 30% glycerol. Count sperm density using a hemocytometer.

    For X. tropicalis, resuspend pellet in 300 µL.

    For X. laevis, this protocol typically yields about 50,000,000 sperm nuclei per testis.

16. Dilute nuclei to the desired density in cold NPB/BSA 0.3% + 30% glycerol and flash freeze aliquots in liquid nitrogen in 0.5 mL or smaller microcentrifuge tubes. Store at −80°C until use.

    For spindle assembly reactions, use a 100× concentration equivalent to 10^5 sperm/µL. Aliquots of 3–5 µL are typically stored (a volume generally sufficient for a day of experiments), but storage in larger volumes can make aliquoting more convenient. When storing in larger volumes, care should be taken to avoid excessive cycles of thawing and refreezing.

DISCUSSION

Early efforts to elucidate the cell cycle capitalized on the inherent experimental advantages of intact amphibian eggs, leading to vertical advances in our collective understanding of this fundamental biological process (Murray and Hunt 1993). However, further characterization of the cell cycle at the molecular level required different, more biochemically tractable approaches. In 1983, Manfred Lohka and Yoshio Masui introduced cell-free extracts derived from *R. pipiens* eggs as a new experimental system to study cell cycle progression and used demembranated sperm nuclei isolated from *X. laevis* to assess changes in the cell cycle state of the extract (Lohka and Masui 1983). They showed that the system was capable of recapitulating several hallmarks of cell cycle progression, including chromatin decondensation, nuclear envelope assembly, and DNA replication (see also Lohka and Masui 1984; Blow and Laskey 1986). Subsequent work showed that the cell cycle state of the egg extracts could be controlled (Lohka and Maller 1985), leading to the discovery that exogenous nuclei (or DNA/ chromatin) could induce nuclear assembly during interphase, and serve as local sites of microtubule nucleation and spindle formation in metaphase (Forbes et al. 1983; Karsenti et al. 1984; Heald et al. 1996, 1997).

The protocol presented here is based largely on methods described later by Andrew Murray (Murray 1991). Although other sources of chromatin can be used, demembranated frog sperm nuclei typically serve as a source of chromatin for nucleus and spindle assembly reactions in *Xenopus* egg extracts (see Sec. 7, Protocol 4: Assembly of Spindles and Asters in *Xenopus* Egg Extracts [Field and Mitchison 2018] and Protocol 6: Nucleus Assembly and Import in *Xenopus laevis* Egg Extract [Chen and Levy 2018]). For assembly of interphase nuclei and "cycled" spindles, demembranated sperm nuclei are added to extract, which is then induced to progress into interphase, resulting in DNA synthesis and nuclear assembly (Sawin and Mitchison 1991; Desai et al. 1999). Subsequent addition of an equal volume of metaphase-arrested extract forces the extract back into mitosis, which causes chromosome condensation, disassembly of the nuclear envelope, and assembly of spindles containing chromosomes with duplicated kinetochores.

Frog Husbandry Water

Instant Ocean Sea Salt (Instant Ocean)	0.88 g/L
Cichlid Lake Salt (SeaChem)	0.33 g/L
Alkaline Buffer (SeaChem)	0.84 g/L
Acidic Buffer (SeaChem)	0.33 g/L
Equilibrium Buffer (SeaChem)	0.033 g/L

Prepare fresh. Dissolve components in double-distilled, reverse-osmosis, or carbon-filtered water. Sterilization is not required. Adjust the amounts of salts and buffers to ensure frog husbandry water quality parameters fall within acceptable ranges. For example, in some *X. laevis* aquatic husbandry systems, water temperature is kept at $\sim$18°C, pH at 7.0–7.5, conductivity at 1.2–1.8 mS, and hardness at $\sim$200 ppm ($CaCO_3$).

Marc's Modified Ringer's (MMR) for Xenopus *(1×)*

HEPES (free acid)	5 mM
EDTA	0.1 mM
NaCl	100 mM
KCl	2 mM
$MgCl_2$	1 mM
$CaCl_2$	2 mM

Prepare as a 25× stock and adjust pH to 7.8 with NaOH. Sterilize by autoclaving and store at 4°C.

Nuclear Preparation Buffer (NPB) (1×)

Sucrose (2 M)	25 mL
HEPES (1 M, pH 7.4)	3 mL
EDTA (0.5 M, pH 8.0)	400 µL
S/S Buffer (20×) <R>	10 mL

Add ddH_2O to 200 mL. Make fresh, and add DTT to 1 mM and LPC protease inhibitor cocktail (10 mg/mL leupeptin, pepstatin, and chymostatin stock in DMSO; EMD Millipore) at 1/1000 before use.

Nuclear Preparation Buffer (NPB) Solutions Containing Bovine

Serum Albumin (BSA)

1. To prepare a solution of 10% BSA in NPB, dissolve 2 g of BSA in 15 mL of 1× NPB <R>. Adjust the pH to 7.6 with KOH, and bring the volume to 20 mL with 1× NPB.

2. To prepare a solution of 3% BSA in NPB, combine 10.5 mL of 1× NPB with 4.5 mL of 10% BSA in NPB for a total of 15 mL.

3. To prepare a solution of 0.3% BSA in NPB, combine 9 mL of 1× NPB with 1 mL of 3% BSA in NPB for a total of 10 mL.

4. To prepare a solution of 0.3% BSA in NPB plus 30% glycerol, combine 3.5 mL of 0.3% BSA in NPB with 1.9 g of glycerol, and bring the total volume to 5 mL with 0.3% BSA in NPB.

Prepare all NPB/BSA solutions fresh, and use the same day. Store on ice until use.

Cite this protocol as *Cold Spring Harb Protoc*; doi:10.1101/pdb.prot099044

S/S Buffer (20×)

Spermidine	127.5 mg
Spermine	69.5 mg

Add ddH$_2$O to 50 mL, filter-sterilize, and store in 10-mL aliquots at −20°C.

ACKNOWLEDGMENTS

The authors thank Kelly Miller for advice and comments on the manuscript and Jitender Bisht and Miroslav Tomschik for acquiring the included photographs. We would also like to express our gratitude to the programs and agencies that fund our work, namely, the National Institutes of Health (National Institute of General Medical Sciences (NIGMS) grants R01GM102428, R01GM113028), the Wyoming IDeA Networks for Biomedical Research Excellence (INBRE) program (NIGMS grant 2P20GM103432), and the Pew Biomedical Scholars program.

REFERENCES

Blow JJ, Laskey RA. 1986. Initiation of DNA replication in nuclei and purified DNA by a cell-free extract of *Xenopus* eggs. *Cell* **47:** 577–587.

Chen P, Levy DL. 2018. Nucleus assembly and import in *Xenopus laevis* egg extract. *Cold Spring Harb Protoc* doi: 10.1101/pdb.prot097196.

Desai A, Murray AW, Mitchison T, Walczak CE. 1999. The use of *Xenopus* egg extracts to study mitotic spindle assembly and function. *Methods Cell Biol* **61:** 385–412.

Field CM, Mitchison TJ. 2018. Assembly of spindles and asters in *Xenopus* egg extracts. *Cold Spring Harb Protoc* doi: 10.1101/pdb.prot099796.

Forbes DJ, Kirschner MW, Newport JW. 1983. Spontaneous formation of nucleus-like structures around bacteriophage DNA microinjected into *Xenopus* eggs. *Cell* **34:** 13–23.

Heald R, Tournebize R, Blank T, Sandaltzopoulos R, Becker P, Hyman A, Karsenti E. 1996. Self-organization of microtubules into bipolar spindles around artificial chromosomes in *Xenopus* egg extracts. *Nature* **382:** 420–425.

Heald R, Tournebize R, Habermann A, Karsenti E, Hyman A. 1997. Spindle assembly in Xenopus egg extracts: Respective roles of centrosomes and microtubule self-organization. *J Cell Biol* **138:** 615–628.

Karsenti E, Newport J, Hubble R, Kirschner M. 1984. Interconversion of metaphase and interphase microtubule arrays, as studied by the injection of centrosomes and nuclei into *Xenopus* eggs. *J Cell Biol* **98:** 1730–1745.

Lohka MJ, Maller JL. 1985. Induction of nuclear envelope breakdown, chromosome condensation, and spindle formation in cell-free extracts. *J Cell Biol* **101:** 518–523.

Lohka MJ, Masui Y. 1983. Formation in vitro of sperm pronuclei and mitotic chromosomes induced by amphibian ooplasmic components. *Science* **220:** 719–721.

Lohka MJ, Masui Y. 1984. Roles of cytosol and cytoplasmic particles in nuclear envelope assembly and sperm pronuclear formation in cell-free preparations from amphibian eggs. *J Cell Biol* **98:** 1222–1230.

Murray AW. 1991. Cell cycle extracts. *Methods Cell Biol* **36:** 581–605.

Murray AW, Hunt T. 1993. *The cell cycle: An introduction.* W.H. Freeman, New York. **xii,** 251 pp.

Sawin KE, Mitchison TJ. 1991. Mitotic spindle assembly by two different pathways in vitro. *J Cell Biol* **112:** 925–940.

Protocol 2

Chromosome Cohesion and Condensation in *Xenopus* Egg Extracts

Eulália M.L. da Silva[1] and Susannah Rankin[1,2,3]

[1]*Program in Cell Cycle and Cancer Biology, Oklahoma Medical Research Foundation, Oklahoma City, Oklahoma 73104;* [2]*Department of Cell Biology, University of Oklahoma Health Sciences Center, Oklahoma City, Oklahoma 73104*

Chromosome structure in both interphase and M-phase cells is strongly influenced by the action of the cohesin and condensin protein complexes. The cohesin complex tethers the identical copies of each chromosome, called sister chromatids, together following DNA replication and promotes normal interphase chromosome structure and gene expression. In contrast, condensin is active largely in M phase and promotes the compaction of individual chromosomes. The *Xenopus* egg extract system is uniquely suited to analyze the functions of both complexes. Egg extracts, in which the cell cycle state can be manipulated, contain stockpiles of nuclear proteins (including condensin and cohesin) sufficient for the assembly of thousands of nuclei per microliter. Extract prepared from unfertilized eggs is arrested by the presence of cytostatic factor (CSF) in a state with high levels of M-phase kinase activity, but can be stimulated to enter interphase, in which DNA replication occurs spontaneously. For cohesion assays, demembranated sperm nuclei are incubated in interphase extract, where they undergo rapid and synchronous DNA replication and cohesion establishment through the recruitment of proteins and other factors (e.g., nucleotides) from the extract. Sister chromatid cohesion is assessed by then driving the extract into M phase by the addition of fresh CSF-arrested extract. In contrast, because chromosome condensation occurs spontaneously in M-phase extracts, sperm nuclei are added directly to CSF extracts to assay condensation.

MATERIALS

It is essential that you consult the appropriate Material Safety Data Sheets and your institution's Environmental Health and Safety Office for proper handling of equipment and hazardous materials used in this protocol.

RECIPES: Please see the end of this protocol for recipes indicated by <R>. Additional recipes can be found online at http://cshprotocols.cshlp.org/site/recipes.

Reagents

Prepare all solutions using ultrapure (i.e., ~18.2 MΩ·cm at 25°C) water.

4′,6-diamidino-2-phenylindole dihydrochloride (DAPI; 1 µg/mL in H_2O) (optional; see Step 17)
Antibodies to *Xenopus* chromosomal proteins of interest

Antibodies to condensin, topoisomerase II, and CENP-A are ideal.

Antibody dilution buffer <R>
Biotin-14-dATP (Invitrogen 19524016) (optional; see Step 3)

---

[3]Correspondence: susannah-rankin@omrf.org

CaCl$_2$ (30 mM)

> Prepare fresh from 1 M stock solution.

Chromosome dilution buffer <R>

Chromosome fix solution <R>

Cushion buffer (40% glycerol in 1× MMR)

Energy mix (EM) (35×) <R>

Fluorescein-12-dUTP (Invitrogen C7604) (optional; see Step 3)

Marc's modified Ringer's (MMR) for *Xenopus* (1×) <R>

Mounting medium (with anti-fade reagent) <R>

Nail polish

Quick fix solution <R>

Sperm nuclei (>10^8 nuclei/mL)

> Prepare according to Protocol 1: Isolation and Demembranation of Xenopus Sperm Nuclei (Hazel and Gatlin 2018) or Tutter and Walter (2006). Store in 5- to 10-μL aliquots at −80°C.

Tris-buffered saline (TBS; 150 mM NaCl in 20 mM Tris-HCl, pH 7.4)

Xenopus laevis egg extracts, cytostatic factor (CSF)-arrested

> Xenopus laevis egg extracts, cytostatic factor (CSF)-arrested. We use frozen CSF-arrested extract prepared as in Gillespie (Gillespie et al. 2012). Alternatively, freshly prepared CSF can be used. See Sec. 7, Protocol 1: Preparation of Cellular Extracts from Xenopus Eggs and Embryos (Good and Heald 2018) for preparation of fresh CSF-arrested extract.

Equipment

Centrifuge, equipped with a swinging bucket rotor (e.g., Beckman Avanti J25 with either the JS13.1 or JS 7.5 rotor)

> Use rubber adapters to accommodate Kimax tubes.

Cover glasses (12 mm round, #1 thickness)

Cover glasses (18-mm square, #1 thickness)

Epifluorescence microscope (preferably equipped with a 40× phase contrast air objective lens)

Forceps (fine-tipped)

Glass slides

Moist chamber

> We use 15-cm disposable cell culture dishes lined with Parafilm, covered and placed inside of a glass dish lined with wet paper towels and covered with clear plastic wrap. This is further covered with aluminum foil when using fluorescent reagents.

Spin-down tubes

> These are constructed by cutting 0.5-in dia. × ~0.5-inch long cylinders from a single acrylic rod (e.g., McMaster-Carr 8528K32; polish the ends of each cylinder with fine sandpaper before use) and securing them to the inside bases of thick-walled Kimax 15-mL glass tubes (Kimble 45500-15) using ~1 mL of a silicone adhesive (e.g., Sylgard 184; Dow Corning). Each setup also requires an additional loose cylinder to serve as a chock on which the cover glass will rest. When the tube is inverted, the chock slides out, bringing the cover glass with it.

Water bath (18°C–21°C)

METHOD

> The following protocols describe basic assays for cohesin (Steps 1–21) and condensin (Steps 22–30) function using Xenopus egg extracts (Fig. 1).

Assay for Sister Chromatid Cohesion

> Cohesion is analyzed by measuring the distance between sister chromatids. This is most easily done when the chromosome cores are stained with antibodies against condensin subunits or topoisomerase II.

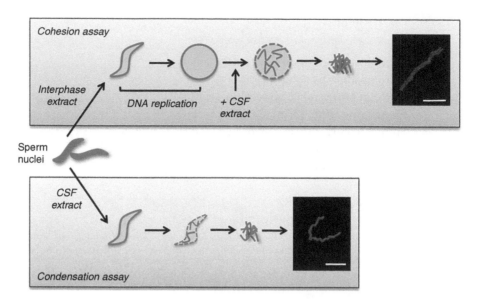

FIGURE 1. Schematic illustration of protocols for cohesion and condensation assays. For cohesion assays (*top*), in which the tethering together of sister chromatids is analyzed, nuclei are added to interphase extracts to allow DNA replication to occur. After DNA replication, the extracts are driven into M phase by the addition of CSF-arrested extract. The resulting condensed chromosomes are collected and analyzed for sister chromatid cohesion by measuring the distance between sister chromatids. In the chromosome condensation assay (*bottom*), demembranated sperm are added directly to M-phase–arrested CSF extract, where the haploid genome spontaneously condenses into individual chromatids. In both cases, chromosome cores can be identified by immunolocalization of the CapG subunit of Condensin I (*red*) and the DNA is counterstained with DAPI (*blue*). Scale bars = 5 μm.

1. Hand-thaw an aliquot of egg extract. Place on ice. Add 1/35th volume of EM. Mix by flicking the tube gently.

   Volumes of 40–100 μL per sample work best. Smaller volumes can be affected by evaporative loss, whereas larger volumes can suffer from inadequate mixing.

2. Add nuclei to the desired final concentration. Mix by gently pipetting the mixture up and down two to three times using a standard 200-μL pipette tip. Avoid excess shear force by pipetting slowly, with the pipette tip well clear of the tube bottom.

   This is the nuclear assembly mixture. To ensure complete and timely replication, keep the concentration of nuclei below 3000/μL.

3. Optionally, supplement extracts with 1 μM fluorescein-dUTP or biotinylated dATP to identify replicated chromatids during subsequent imaging.

4. Stimulate CSF release by adding 1/50th volume of 30 mM CaCl₂. Mix by flicking. Immediately place in a water bath at 18°C –21°C.

5. After 45 min, check the nuclear morphology to confirm that the extract is in interphase:

   i. Place 3 μL of Quick Fix on a glass slide.

   ii. Add 0.5 μL of the nuclear assembly mixture to the middle of the drop of fix.

   iii. Cover with an 18-mm square coverslip.

   iv. Observe by epifluorescence microscopy.

      By 45 min, the nuclei should appear round, and a well-developed nuclear envelope will be evident by phase contrast microscopy. If the extract failed to release from CSF then there will be no nuclear envelope, and the mass of DNA will have a bumpy and uneven edge, indicative of chromosome condensation.

6. Allow replication to proceed for 120 min.

   At this stage, nuclei can become quite large, and sometimes the DNA will have a reticular staining pattern.

Cite this protocol as *Cold Spring Harb Protoc*; doi:10.1101/pdb.prot097121

7. Freshly thaw an aliquot of CSF extract immediately before use, supplement it with 1/35th volume energy mix, and prewarm to 19°C. Add an equal volume of this CSF extract to the sample.

8. Incubate the sample an additional 90–120 min. Monitor the nuclear morphology (as described in Step 5) for entry into M phase.

   CSF activity is dominant and will drive the interphase extract into M phase, causing condensation of the chromosomes. This will be evident by loss of the nuclear envelope, chromosome condensation, and collapse of chromosomes into discrete clumps.

9. During the above incubation, prepare the spin-down tubes:

   i. Place one loose acrylic chock in each spin-down tube.

   ii. Add 4 mL of cushion buffer to each tube.

   iii. Add one clean 12-mm round coverslip to each tube.

      Use a Pasteur pipette to push down the coverslip so that it is resting on the loose chock, under the cushion buffer.

10. Add one volume of the chromosome assembly mix (from Step 8) to four volumes of chromosome dilution buffer. Incubate for 15–20 min at 18°C–21°C.

11. Fix by the addition of 20 volumes of chromosome fix solution. Incubate for 5 min.

12. Layer the fixed mixtures onto the cushions in the spin-down tubes. Centrifuge at 6000 rpm (6600g) for 20 min.

    The samples can be split to accommodate different downstream analyses. For example, if immunostaining with different antibody combinations is desired, then use a separate spin-down tube for each, dividing samples accordingly. The equivalent of 5 µL of chromosome assembly mix per cover glass provides ample material for morphological analysis.

13. Aspirate and discard the top third of the liquid from the tube.

    This removes diluted extract and minimizes background in subsequent immunolocalization experiments.

14. Decant and discard the cushion. Invert the tubes to ~45° and tap on a clean paper towel until the loose chock begins to slide toward the opening of the tube. Collect the chock in a gloved hand, avoiding the coverslip.

15. Use forceps to grasp the coverslip. Place face-up on clean Parafilm. Cover with ~200 µL of antibody dilution buffer. Repeat for the remaining coverslips.

16. Wash the coverslips three times, 5 min each, with antibody dilution buffer.

17. Stain with primary and secondary antibodies against the protein of interest using standard immunofluorescence procedures. Counterstain with DAPI (if so desired).

18. Wash the coverslips in TBS.

19. Using fine-tipped forceps, invert each coverslip onto an ~2-µL drop of mounting medium on a slide. Allow the coverslip to settle for 10 min.

20. Aspirate any liquid from around the coverslip. Seal with nail polish.

21. Store at −20°C. Image within 2 wk.

    See Troubleshooting.

Assay for Chromosome Condensation

To assay condensation, sperm nuclei are added directly to M-phase extract. Following incubation, during which chromatin assembly and chromosome condensation occur, the chromosomes are collected on coverslips for analysis.

22. Hand-thaw an aliquot of CSF extract. Place on ice. Add 1/35th volume EM. Mix by flicking the tube gently.

    Do not warm above 23°C.

23. Add sperm nuclei to a final concentration of <3000 nuclei/µL. Mix by gently pipetting up and down using a standard 200-µL pipette tip. Avoid excess shear force by pipetting slowly, with the pipette tip well clear of the tube bottom.

24. Incubate in a water bath for 90–120 min at 19°C–21°C.

25. Monitor nuclei for entry into M phase (as described in Step 5).

> As chromosomes condense and individualize in M-phase extract, the mass of DNA will have a bumpy and uneven edge and there will be no nuclear envelope.

26. While the samples are incubating, prepare spin-down tubes (as described in Step 9).

27. Add one volume of the chromosome assembly mix to four volumes of room-temperature chromosome dilution buffer. Incubate for 15–20 min.

28. Fix by the addition of 20 volumes of chromosome fix solution. Incubate for 5 min.

29. Layer the samples onto the cushions in the spin-down tubes. Centrifuge at 6600g for 20 min.

30. Process the coverslips for immunofluorescence labeling of chromosomes (as described in Steps 13–21).

> The general level of condensation can be determined by measuring the length and width of chromosomes, as well as the loading of condensin proteins onto chromatin. The impact of DNA replication and cohesion establishment on subsequent chromosome condensation can be assessed by analyzing chromosomes that have been "cycled" through interphase extract (as in the cohesion assay protocol).
>
> See Troubleshooting.

TROUBLESHOOTING

Problem (Steps 21 and 30): Incomplete DNA replication can occur in poor extracts, and can give an appearance similar to aberrant chromosome cohesion and result in tangled chromatids.
Solution: Use only excellent CSF extracts prepared from high-quality eggs.

DISCUSSION

The *Xenopus* egg extract system provides a powerful approach for the in vitro analysis of cohesion and condensation mechanisms. Importantly, the synchrony and rapid cell cycles of the *Xenopus* embryonic extracts allow us to probe the function of essential conserved proteins in chromosome dynamics, something that is inherently difficult in somatic cells. The assays described here can be used to test the effects of recombinant or mutant proteins on chromosome cohesion and condensation, and to perform depletion-and-rescue experiments.

For depletion experiments, freshly thawed CSF extract is incubated with antibody-coated beads before the CSF is released with CaCl$_2$ and the above procedures are followed (see Sec. 7, Protocol 3: Protein Immunodepletion and Complementation in *Xenopus laevis* Egg Extracts [Jenness et al. 2018] for details). In addition to protein depletion and addition, the extracts can also be treated with pharmacological agents to manipulate DNA replication or enzyme activity, allowing analysis of their impacts on chromosome condensation and cohesion.

The protocols described here represent a synthesis from contributions made by numerous investigators over many years to the study of DNA replication and chromosome dynamics in *Xenopus* egg extracts (Hirano et al. 1997; Losada et al. 1998; Walter et al. 1998; Funabiki and Murray 2000; Takahashi et al. 2004). Another method similar to that described here should prove useful to others particularly interested in cohesion assays (Shintomi and Hirano 2017). For our laboratory, the protocol for preparation of frozen CSF-arrested extract as described by Gillespie et al. (2012) provided a critical technical breakthrough because it allows for routine control of cell cycle transitions without the requirement for freshly prepared extracts, which can be maddeningly inconsistent.

Cite this protocol as *Cold Spring Harb Protoc*; doi:10.1101/pdb.prot097121

RECIPES

Antibody Dilution Buffer (Abdil)

Reagent	Amount to add	Final concentration
NaCl (5 M)	15 mL	150 mM
Tris-HCl (1 M; pH 7.4)	10 mL	20 mM
Triton X-100 (20%)	2.5 mL	0.10%
Bovine serum albumin	10 g	2%
NaN$_3$ (20%)	2.5 mL	0.10%

To prepare 500 mL of Abdil, combine the reagents listed above and then add H$_2$O to 500 mL. Filter the solution through a 0.22-μm filter. Store at 4°C.

Chromosome Dilution Buffer

Reagent	Amount to add	Final concentration (1×)
K-HEPES (1 M, pH 7.6)	1 mL	10 mM
KCl (2 M)	10 mL	200 mM
MgCl$_2$ (1 M)	0.1 mL	0.5 mM
K-EGTA (0.5 M)	0.2 mL	0.5 mM
Sucrose	8.57 g	250 mM

Mix and dissolve the reagents listed above in H$_2$O to a final volume of 100 mL. Store in 10-mL aliquots at −20°C.

Chromosome Fix Solution

Reagent	Amount to add	Final concentration (1×)
Marc's modified Ringers (MMR) (25×) <R>	2 mL	1×
Triton X-100 (20%)	1.25 mL	0.5%
Glycerol (100%)	10 mL	20%
Formaldehyde (∼37% ACS grade, VWR Intl.)	3.65 mL	2.7%

Prepare fresh by mixing the above ingredients to a final volume of 50 mL with H$_2$O. Alternatively, mix and freeze aliquots of the first three ingredients, and add the formaldehyde just before use.

Creatine Phosphate (Phosphocreatine) (1 M)

Prepare a stock solution in 10 mM KHPO$_4$ buffer, pH 7.0 from the disodium salt, tetrahydrate. Store the working stock at −20°C and the remainder at −80°C.

Creatine Phosphokinase (5 mg/mL)

Prepare a stock solution of 5 mg/mL creatine phosphokinase (type 1, from rabbit muscle; Sigma-Aldrich C3755) in 10 mM HEPES (pH 7.5), 50 mM NaCl, and 50% glycerol. Store in 200-μL aliquots at −20°C.

Energy Mix (EM) (35×)

Reagent	Amount to add	Final concentration (35×)
ATP (200 mM)	10 μL	~67 mM
Creatine phosphate (phosphocreatine) (1 M) <R>	20 μL	~667 mM
Creatine phosphokinase (5 mg/mL) <R>	0.8 μL	0.13 mg/mL

Prepare fresh 35× EM by mixing the stock solutions as indicated. Keep on ice. Discard unused portion at the end of the experiment.

Marc's Modified Ringers (MMR) (25×)

Reagent	Amount to add	Final concentration (25×)	Final concentration (1×)
HEPES free acid	119.2 g	125 mM	5 mM
NaCl	584.5 g	2.5 M	100 mM
KCl	14.91 g	50 mM	2 mM
$MgCl_2$	20.33 g	25 mM	1 mM
$CaCl_2$	29.4 g	50 mM	2 mM
EDTA (0.5 M, pH 8.0)	20 mL	2.5 mM	0.1 mM

To prepare a 25× stock, dissolve the above ingredients in 3.8 L of H_2O, and adjust the pH to pH 7.8 with 10 M NaOH. Bring the final volume to 4 L, and store at room temperature.

Mounting Medium (with Anti-Fade Reagent)

Reagent	Amount to add	Final concentration
Tris-HCl (1 M, pH 8.8)	0.8 mL	20 mM
p-phenylenediamine	0.20 g	0.5%
Glycerol (100%)	36 mL	90%

Mix the above ingredients together in a 50-mL conical tube. Adjust volume to 40 mL with H_2O. Dissolve the solids by slowly bubbling nitrogen gas through a Pasteur pipette inserted deep into the mixture. Filter through a 0.8-μm filter at room temperature. Store at −20°C in 1- or 3-mL syringes devoid of air bubbles. Discard when dark brown. Handle this reagent with gloves; it causes stains when spilled.

Quick Fix Solution

Reagent	Amount to add	Final concentration
Marc's modified Ringers (MMR) (25×) <R>	80 μL	1×
DAPI (1 mg/mL)	2 μL	1 μg/mL
Glycerol (100%)	1.2 mL	60%
Formaldehyde (37%)	0.59 mL	11%
H_2O	128 μL	

Prepare to a final volume of 2 mL. Store in 0.5-mL aliquots for up to a year at −20°C.

REFERENCES

Funabiki H, Murray AW. 2000. The *Xenopus* chromokinesin Xkid is essential for metaphase chromosome alignment and must be degraded to allow anaphase chromosome movement. *Cell* 102: 411–424.

Gillespie PJ, Gambus A, Blow JJ. 2012. Preparation and use of *Xenopus* egg extracts to study DNA replication and chromatin associated proteins. *Methods* 57: 203–213.

Good MC, Heald R. 2018. Preparation of cellular extracts from *Xenopus* eggs and embryos. *Cold Spring Harb Protoc* doi: 10.1101/pdb.prot097055.

Hazel JW, Gatlin JC. 2018. Isolation and demembranation of *Xenopus* sperm nuclei. *Cold Spring Harb Protoc* doi: 10.1101/pdb.prot099044.

Hirano T, Kobayashi R, Hirano M. 1997. Condensins, chromosome condensation protein complexes containing XCAP-C, XCAP-E and a *Xenopus* homolog of the *Drosophila* Barren protein. *Cell* 89: 511–521.

Jenness C, Wynne DJ, Funabiki H. 2018. Protein immunodepletion and complementation in *Xenopus laevis* egg extracts. *Cold Spring Harb Protoc* doi: 10.1101/pdb.prot097113.

Losada A, Hirano M, Hirano T. 1998. Identification of *Xenopus* SMC protein complexes required for sister chromatid cohesion. *Genes Dev* 12: 1986–1997.

Shintomi K, Hirano T. 2017. A sister chromatid cohesion assay using *Xenopus* egg extracts. *Methods Mol Biol* 1515: 3–21.

Takahashi TS, Yiu P, Chou MF, Gygi S, Walter JC. 2004. Recruitment of *Xenopus* Scc2 and cohesin to chromatin requires the pre-replication complex. *Nat Cell Biol* 6: 991–996.

Tutter AV, Walter JC. 2006. Chromosomal DNA replication in a soluble cell-free system derived from *Xenopus* eggs. *Methods Mol Biol* 322: 121–137.

Walter J, Sun L, Newport J. 1998. Regulated chromosomal DNA replication in the absence of a nucleus. *Mol Cell* 1: 519–529.

Chromatin Characterization in *Xenopus laevis* Cell-Free Egg Extracts and Embryos

Wei-Lin Wang,[1,3] Takashi Onikubo,[1,2,3] and David Shechter[1,4]

[1]*Department of Biochemistry, Albert Einstein College of Medicine, Bronx, New York 10461*

Xenopus laevis development is marked by accelerated cell division solely supported by the proteins maternally deposited in the egg. Oocytes mature to eggs with concomitant transcriptional silencing. The unique maternal chromatin state contributing to this silencing and subsequent zygotic activation is likely established by histone posttranslational modifications and histone variants. Therefore, tools for understanding the nature and function of maternal and embryonic histones are essential to deciphering mechanisms of regulation of development, chromatin assembly, and transcription. Here we describe protocols for isolating pronuclear sperm chromatin from *Xenopus* egg extracts and hydroxyapatite-based histone purification from this chromatin. The histones purified through this method can be directly assembled into chromatin through in vitro assembly reactions, providing a unique opportunity to biochemically dissect the effect of histone variants, histone modifications, and other factors in chromatin replication and assembly. We also describe how to isolate chromatin from staged embryos and analyze the proteins to reveal dynamic developmental histone modifications. Finally, we present protocols to measure chromatin assembly in extracts, including supercoiling and micrococcal nuclease assays. Using these approaches, analysis of maternal and zygotic histone posttranslational modifications concomitant with cell-cycle and developmental transitions can be tested.

MATERIALS

It is essential that you consult the appropriate Material Safety Data Sheets and your institution's Environmental Health and Safety Office for proper handling of equipment and hazardous materials used in this protocol.

RECIPES: Please see the end of this protocol for recipes indicated by <R>. Additional recipes can be found online at http://cshprotocols.cshlp.org/site/recipes.

Reagents

Agarose/TAE gels (standard; Lonza SeaKem Agarose LE is preferable for MNase assays as the gels have a clearer appearance)
Chromatin assembly stop buffer <R>
DNA loading buffer containing RNase A <R>
Egg lysis buffer—chromatin isolation buffer (ELB-CIB) <R>
ELB-CIB with 0.5 M sucrose final concentration
ELB-CIB with 250 m M KCl final concentration

[2]Current address: Laboratory of Biochemistry and Molecular Biology, The Rockefeller University, New York, NY 10065
[3]These authors contributed equally to this work.
[4]Correspondence: david.shechter@einstein.yu.edu

Embryo lysis buffer (EmLB) <R>

Ethanol (100% and 70%)

Ethidium bromide

GlycoBlue (15 mg/mL)

HDB 2000 Buffer <R>

High salt buffer <R>

High salt phosphate buffer <R>

Hydroxyapatite resin (DNA Grade Bio-Gel HTP, Bio-Rad)

Laemmli SDS-PAGE sample buffer (1.5×)

Low salt phosphate buffer <R>

Medium salt buffer <R>

Micrococcal nuclease (MNase; Sigma-Aldrich)

MNase reaction buffer <R>

Use 90 μL of the buffer containing 1 unit of MNase (Sigma-Aldrich) for each reaction. Prepare the mixture fresh.

MNase stop buffer <R>

NaOAc (3 M; pH 5.2)

Phenol/chloroform/isoamyl alcohol

Proteinase K (20 mg/mL)

RNase A (10 mg/mL)

SDS-PAGE gel (15% for checking histone fractions) and appropriate buffer

Spermine and spermidine (optional; see Step 3)

Topoisomerase-I relaxed pG5ML or similar plasmid (625 ng/μL)

Xenopus laevis egg (low-speed interphase supernatant/LSS or high-speed interphase supernatant/HSS) or oocyte extracts (Banaszynski et al. 2010), or fertilized embryos (see Sec. 7, Protocol 1: Preparation of Cellular Extracts from *Xenopus* Eggs and Embryos [Good and Heald 2018])

Xenopus laevis demembranated sperm chromatin (see Protocol 1: Isolation and Demembranation of *Xenopus* Sperm Nuclei [Hazel and Gatlin 2018])

Equipment

Agarose gel electrophoresis apparatus

Beaker, 1L

Centrifuge with Fiberlite F15-8×50c, F14-14, or SS34 rotors (Thermo Scientific)

Cold room

Digital imaging system

Dounce homogenizer (5 mL or larger) with type B pestle

Econo-Pac chromatography column (Bio-Rad)

Falcon tubes (15 mL)

HB-6 swinging bucket rotor or equivalent

Liquid nitrogen and dewar

Microcentrifuge, fixed-angle and swinging-bucket microcentrifuge rotors with temperature control and cooling

Microcentrifuge tube heating/cooling block

Microcentrifuge tubes (1.5 mL, flip-top)

6–8 kDa MWCO dialysis tubing (10 mm)

Pellet Pestle Cordless Motor

Pellet Pestles (Fisher)

Peristaltic pump and tubing

Protein concentrator with 3000 Da molecular weight cut off

SDS-PAGE gel electrophoresis apparatus

Sonicator with microtip

Water bath

METHOD

Five procedures are outlined here: three to isolate chromatin and histones from extract or embryos for immunoblotting (Wang et al. 2014), mass spectrometry of modifications and variants (Nicklay et al. 2009; Shechter et al. 2009; Wang et al. 2014), or for use in chromatin assembly (Onikubo et al. 2015); and two procedures for measuring histone deposition and chromatin assembly in cell-free extract assays (Wang and Shechter 2016). Low-protein binding tubes and tips substantially reduce variability in these assays. The hydroxyapatite purification of histones was based on (Schnitzler 2001). Refer to (Banaszynski et al. 2010) for Xenopus egg extract and sperm preparation.

Isolation of Sperm Pronuclear Chromatin

1. Split 10 mL of fresh interphase-arrested LSS with energy mix (Banaszynski et al. 2010) into five 15 mL falcon tubes and add demembranated sperm chromatin (see Protocol 1: Isolation and Demembranation of *Xenopus* Sperm Nuclei [Hazel and Gatlin 2018]) to a final concentration of 4000/µL.

2. Incubate in a water bath for 60 to 90 min at 22°C to form sperm pronuclei.

3. Lyse pronuclei by mixing each 2 mL of assembly reaction with ELB-CIB for a 10 mL total reaction volume. Incubate for 10 min on ice. Used chilled buffers and ice from this step on.

   Adding 1 mM spermine and spermidine to ELB-CIB buffer results in a tighter pellet that allows more facile collection of chromatin but is more difficult to sheer in the subsequent steps for hydroxyapatite purification of histones. These polyamines should be included when an isolation of pronuclear chromatin is performed for immunoblotting, acid extraction, or mass spectrometry.

4. Carefully underlay the suspension with 1 mL ELB-CIB containing 0.5 M sucrose and centrifuge at 4000 RPM in an HB-6 swinging-bucket rotor for 5 min at 4°C.

5. Carefully remove the entire supernatant with a 1 mL pipet without disrupting the chromatin pellet at the bottom of the tube. Wash the pellet once with 1 mL ELB-CIB containing 250 mM KCl and spin at 13,000 RPM for 2 min at 4°C. Retain the pellet which contains the pronuclear chromatin and histones (Fig. 1A).

   At this point, histones can be collected from the chromatin pellet using acid extraction for immunoblotting or mass spectrometry (Shechter et al. 2007; Wang et al. 2014). Alternatively, Steps (6–18) can be used to isolate intact, native histone octamers and subcomplexes as substrates for chaperone chromatin assembly activity assays (Fig. 1B), remodeling assays, or posttranslational modifying-enzyme assays (Wilczek et al. 2011).

Hydroxyapatite Purification of Pronuclear Histone Complexes

6. Resuspend and wash the chromatin pellet in 10 mL of medium salt buffer. Centrifuge at 10,000 RPM for 10 min at 4°C in a fixed-angle Fiberlite F15-8×50c or equivalent rotor. Retain the pellet.

7. Resuspend the pellet with 1 mL of high salt buffer and remove the solution to a Dounce homogenizer with type B pestle.

8. Rigorously Dounce 50 times on ice to sheer the chromatin.

9. Transfer the solution into a 1.5 mL microcentrifuge tube and centrifuge at 10,000 RPM for 20 min at 4°C in a microcentrifuge. Retain the supernatant that contains the solubilized chromatin. Move immediately to Step 10 or immediately freeze at −80°C.

   It is critical that the chromatin is not over-sheared as it leads to high DNA contamination in the final product. Running the sample on a 0.8% agarose gel after phenol-chloroform purification can be used to measure the efficiency of chromatin shearing, which should appear as a smear at the top of the gel; if lower molecular weight DNA is observed, reduce homogenization accordingly.

10. Dialyze solubilized chromatin in 6–8 kDa MWCO dialysis tubing against 1 L of Low salt phosphate buffer for 4 h. Perform all procedures in a cold room with chilled buffers from this step on.

11. Hydrate 5 g of hydroxyapatite resin with 30 mL of low salt phosphate buffer and remove the fines by pouring off the top suspension after the bulk of the resin settles; repeat this step 3–4 times.

12. Connect the Econo-Pac chromatography column *outlet* to a peristaltic pump.

Cite this protocol as *Cold Spring Harb Protoc*; doi:10.1101/pdb.prot099879

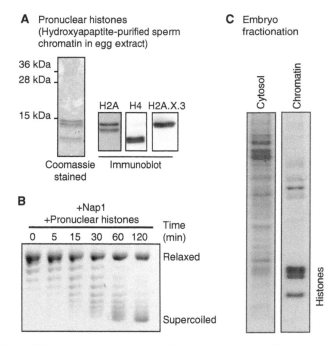

FIGURE 1. Isolation and use of *Xenopus* egg and embryo histones. (*A*) SDS-PAGE of hydroxyapatite purified histones from pronuclei assembled in *Xenopus* cell-free egg extract. A Coomassie-stained gel is shown on the *left*, while immunoblots for histones H2A, H4, and H2A.X.3 (also known as H2A.X-F) are shown on the *right*. (*B*) Recombinant *Xenopus* Nap1 chaperone was added to hydroxyapatite purified histones in the presence of a relaxed plasmid for the indicated times. The DNA was deproteinized and run on an agarose gel to measure supercoiling. (*C*) Fractionated stage 42 embryo cytoplasmic and nuclear protein was run on an SDS-PAGE, transferred to a membrane, and stained by Direct Blue 71 for subsequent immunoblotting. Enriched embryo histones are indicated on the *right*. Immunoblotting of chromatin-bound histones is possible from early (pre MBT) embryos, depending upon the antibody (Wang et al. 2014).

> *It is essential that the column is connected to a pump as it will not flow by gravity. From this point on, it is crucial not to let the resin dry out.*

13. Pour the hydrated hydroxyapatite resin into the chromatography column and wash with 10 mL of Low salt phosphate buffer.

14. Adjust the peristaltic pump to 0.5–1 mL/min flow rate. Stop the pump as all the wash buffer enters the resin.

15. Load the solubilized chromatin onto the resin by directly applying the sheared chromatin sample over the resin. Start the pump to load onto resin and stop the pump as all the solution enters the resin.

16. Wash the resin with 50 mL of low salt phosphate buffer using the pump.

> *This step is crucial for removing noncore histone-bound proteins, including linker histones. Adjust the volume of wash if linker histone contamination is observed in the final product. Alternatively, linker histones can be isolated from these washes. Separation of H2A-H2B and H3-H4 can also be achieved by eluting the histones using the phosphate buffer with a gradient of NaCl from 0.6 to 2.5 M. Since histones elute in a narrow range of NaCl (0.9 M for H2A-H2B and 1.1 M for H3-H4), the use of a prepacked hydroxyapatite column connected to a FPLC system is highly recommended for this purpose, as this system allows more facile monitoring of eluted proteins.*

17. Elute core histones with four 8 mL volumes of high salt phosphate buffer.

> *Run each fraction on a SDS-PAGE gel to confirm the presence of histones, which primarily elute in the first two fractions.*

18. Collect the fractions containing histones. Exchange the buffer to HDB2000 and concentrate using a protein concentrator per the manufacturer's instructions. Histones can be stored at −80°C for subsequent chromatin assembly activity assays (Fig. 1B).

Isolation of Chromatin from Staged Embryos

This procedure can be used to isolate chromatin and histones from staged embryos for immunoblotting of histone modifications and transcription factors (Wang et al. 2014).

19. Fertilize and culture embryos (see Sec. 7, Protocol 1: Preparation of Cellular Extracts from *Xenopus* Eggs and Embryos [Good and Heald 2018]).

20. Collect embryos according to Niewkoop and Faber stages (Faber and Niuwkoop 1994; Karpinka et al. 2015) into a microcentrifuge tube and remove excess buffer.

21. Flash-freeze aliquots of five embryos in a microcentrifuge tube in liquid nitrogen; at this point the samples can be saved at −80°C or used immediately in the next steps.

22. Add 100 μL of cold EmLB to each tube of five-staged and flash-frozen embryos and homogenize using a motorized Pellet Pestle within the tube.

23. Incubate the lysate on ice for 10 min.

24. Centrifuge the lysate in a microcentrifuge swinging-bucket rotor for 3 min at 1000*g* at 4°C to separate the nuclei.

    Eggs, oocytes and embryos before stage 10 contain abundant egg yolk that should be moved out of the way with the pipet tip, and the tip wiped off with Kimwipes to avoid transferring the yolk, before collecting the supernatant after centrifugation.

25. Collect 80 μL supernatant as the cytoplasmic fraction (four embryo equivalent), dilute to 200 μL in 1.5× Laemmli sample buffer, boil, and store at −80°C.

26. Wash the pellet in 200 μL of EmLB twice, spinning for 3 min at 1000*g* at 4°C after each wash and suspend the pellet in 200 μL of 1.5× Laemmli buffer as the nuclear fraction.

    Alternatively, the nuclear fraction from embryos after stage 8–9 can be saved for chromatin immunoprecipitation experiments using standard techniques.

27. Sonicate the nuclear fractions three times with 30 sec on, 30 sec off at 30% amplitude to solubilize chromatin on ice in the cold room. Boil the nuclear fractions for 5 min at 95°C and store at −80°C.

    Dilute the cytoplasmic supernatant and chromatin fractions five- to 10-fold prior to loading on SDS-PAGE for immunoblots (Fig. 1C).

Plasmid Chromatin Assembly Reaction and Supercoiling Analysis in Extract

These next two procedures can be used to study how chromatin assembly pathways function in cell-free extract (Wang and Shechter 2016). There are batch-to-batch variations of egg extract chromatin assembly efficiency. Test each batch of egg extract after preparation using a time course from 30 min to 3 h of DNA incubation.

28. Add Topoisomerase-I relaxed pG5ML vector to a final concentration of 25 ng/μL in egg or oocyte extract containing 1× energy mix (Banaszynski et al. 2010). Tap the capped tube to mix and incubate at 23°C.

    The typical reaction volume is 25–100 μL depending on the type and number of assay points desired, as in Step 29.

29. At desired time points (e.g., 0, 15, 30, 60, 90, and 120 min), aliquot 5 μL of reaction into 200 μL of Chromatin assembly stop buffer and incubate for 30 min at 37°C. Alternatively, or in parallel, continue to Step 34 for micrococcal nuclease analysis (skip Steps 30–33).

30. Add 5 μL of 20 mg/mL proteinase K (final 50 μg/mL) and incubate for 1 h at 56°C.

31. Purify plasmid DNA through standard procedures of Phenol/Chloroform extraction and Ethanol/NaOAc precipitation in the presence of 1 μL GlycoBlue.

32. Dissolve the DNA pellet in 12 μL of DNA loading buffer containing RNase A.

33. Analyze the DNA supercoiling on a 0.8% agarose-TAE gel at 25 V overnight in the cold room. Continue with Step 39 (Fig. 2A).

    See Troubleshooting.

  Cite this protocol as *Cold Spring Harb Protoc*; doi:10.1101/pdb.prot099879

A

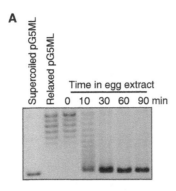

Supercoiled pG5ML
Relaxed pG5ML

Time in egg extract

0 10 30 60 90 min

B Egg extract
chromatinized plasmid

1 Unit MNase

0 5 10 15 20 min

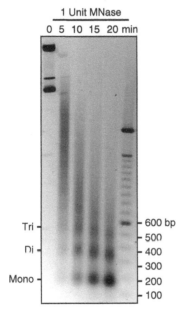

Tri —

Di —

Mono —

— 600 bp
— 500
— 400
— 300
— 200
— 100

FIGURE 2. Chromatin assembly in egg extract. (*A*) A relaxed plasmid assembled in interphase extract for the indicated times was deproteinized and run on an agarose gel to measure supercoiling. (*B*) A plasmid assembled in interphase egg extract was digested with 1 Unit of MNase for the times indicated. Extracted DNA was run on an agarose gel and visualized with ethidium bromide. Size standards are shown on the *right*. Observed nucleosomes (indicated on *left*) had a repeat length <200 bps.

Micrococcal Nuclease Analysis

34. (Continued from Step 29). Transfer 11 µL of the chromatin assembly reaction into a fresh microcentrifuge tube with 90 µL MNase reaction buffer containing 1 unit MNase and incubate for 5 to 6 time points at room temperature for an empirically determined time.

    See Troubleshooting regarding determination of MNase time course.

    The MNase reaction is very sensitive. It is important to start the timer right after the addition of the reaction product into MNase reaction buffer and simply invert the tube twice to mix.

35. Add 110 µL MNase stop buffer to stop the reaction.

36. Add 8 µL RNase A (10 mg/mL) to each reaction and incubate overnight at 37°C.

37. Follow Steps 30–32 above.

38. Load the final product on a 2% agarose gel and run at 100 V for 60 min at room temperature.

    See Troubleshooting.

39. Stain the gel with ethidium bromide and image on a digital imaging system for quantification (Fig. 2B).

    See Troubleshooting.

TROUBLESHOOTING

Problem (Step 33 or 38): Speckles or shadows on 2% agarose gel.

Solution: Since the quantity of DNA for mono- or di-nucleosomes is <25 ng, background staining can lower the resolution of the nucleosome ladder. It is important to clean the electrophoresis apparatus thoroughly before casting the gel. Make sure to filter the buffer and dissolve the agarose completely, swirling the flask until it dissolved completely and avoid bubbles when pouring the agarose.

Problem (Steps 34 and 39): No nucleosome ladder observed after MNase digestion.

Solution: Perform a time course analysis of the supercoiling assay and use the optimal time point for assembly that produces multiple nucleosomal repeats in the MNase assay (e.g., 0 to 30 min in MNase). If the MNase digestion is too fast or too slow, one can fine tune the reaction by adjusting the concentration of $CaCl_2$ or MNase.

DISCUSSION

The methods presented here are appropriate for both in vitro and in vivo analysis of histones and chromatin from *Xenopus laevis* oocyte and egg extracts as well as from staged embryos. The extracts can be further manipulated by immunodepletion, drugs, or other means prior to probe specific roles of targeted proteins on chromatin assembly pathways. Embryos can be microinjected with morpholinos or drugs to similarly target developmentally regulated chromatin assembly pathways and these techniques used to probe their developmental function on chromatin. Finally, these methods can be combined with chromatin immunoprecipitation (ChIP) qPCR or ChIP-sequencing and qRT-PCR or RNA-sequencing to test chromatin-based mechanisms of the regulation of gene expression and genome integrity.

RECIPES

Chromatin Assembly Stop Buffer

Reagent	Concentration
Tris (pH 8.0)	10 mM
EDTA	20 mM
SDS	0.5%
RNase A	0.1 mg/mL

Store for up to 2 yr at 25°C (without RNase A).

DNA Loading Buffer Containing RNase A

Reagent	Concentration
Tris (pH 7.5)	10 mM
EDTA	1 mM
Glycerol	5%
RNase A	10 μg/mL

Add a very small amount of bromophenol blue if desired. (This can make the loading easier, but make sure that the color does not interfere with DNA signal.) Store for up to 2 yr at 25°C (without RNase A).

Cite this protocol as *Cold Spring Harb Protoc*; doi:10.1101/pdb.prot099879

Egg Lysis Buffer–Chromatin Isolation Buffer (ELB–CIB)

Reagent	Concentration
HEPES (pH 7.8)	10 mM
KCl	50 mM
MgCl$_2$	2.5 mM
Sucrose	0.25 M
DTT	1 mM
Triton X-100	0.1%
HALT Phosphatase Inhibitor Cocktail (Pierce)	1×
Protease Inhibitor Cocktail (Roche)	1×
Sodium butyrate	10 mM

Prepare fresh for each use.

Embryo Lysis Buffer (EmLB)

Reagent	Concentration
Tris (pH 7.5)	10 mM
NaCl	200 mM
MgCl$_2$	5 mM
NP-40	0.5%
Na butyrate	5 mM
HALT Phosphatase Inhibitor Cocktail (Pierce)	1×
Protease Inhibitor Cocktail (Roche)	1×
Cycloheximide	100 µg/mL

Prepare fresh for each use.

HDB 2000 Buffer

Reagent	Concentration
Tris (pH 8.0)	20 mM
NaCl	2 M
EDTA	1 mM
Glycerol	5%
β-mercaptoethanol	10 mM

Prepare fresh for each use.

High Salt Buffer

Reagent	Concentration
HEPES (pH 7.5)	20 mM
NaCl	400 mM
EDTA	1 mM
Glycerol	5%
PMSF	1 mM

Prepare fresh for each use.

Cite this protocol as *Cold Spring Harb Protoc*; doi:10.1101/pdb.prot099879

High Salt Phosphate Buffer

Reagent	Concentration
Sodium phosphate (pH 6.8)	20 mM
NaCl	2.5 M
β-mercaptoethanol	5 μM
PMSF	1 mM

Prepare fresh for each use.

Low Salt Phosphate Buffer

Reagent	Concentration
Sodium phosphate (pH 6.8)	20 mM
NaCl	600 mM
β-mercaptoethanol	5 μM
PMSF	1 mM

Prepare fresh for each use.

Medium Salt Buffer

Reagent	Concentration
HEPES (pH 7.5)	20 mM
NaCl	400 mM
EDTA	1 mM
Glycerol	5%
PMSF	1 mM

Prepare fresh for each use.

MNase Reaction Buffer

Reagent	Concentration
HEPES (pH 8.0)	10 mM
KCl	50 mM
$MgCl_2$	5 mM
$CaCl_2$	3 mM
DTT	1 mM
NP-40	0.1%
Glycerol	8%

Prepare fresh for each use.

MNase Stop Buffer

Reagent	Concentration
EDTA	40 mM
EGTA	2 mM
SDS	1%

Store for up to 2 yr at 25°C.

Cite this protocol as *Cold Spring Harb Protoc*; doi:10.1101/pdb.prot099879

ACKNOWLEDGMENTS

This work was supported by The American Cancer Society-Robbie Sue Mudd Kidney Cancer Research Scholar Grant (124891-RSG-13-396-01-DMC) and National Institutes of Health (NIH) grant R01GM108646 (both to D.S.).

REFERENCES

Banaszynski LA, Allis CD, Shechter D. 2010. Analysis of histones and chromatin in *Xenopus laevis* egg and oocyte extracts. *Methods* **51**: 3–10.

Faber J, Niuwkoop PD. 1994. Normal Table of *Xenopus Laevis* (Daudin). *J Anat* **103**: 578.

Good MC, Heald R. 2018. Preparation of cellular extracts from *Xenopus* eggs and embryos. *Cold Spring Harb Protoc* doi: 10.1101/pdb.prot097055.

Hazel JW, Gatlin JC. 2018. Isolation and demembranation of *Xenopus* sperm nuclei. *Cold Spring Harb Protoc* doi: 10.1101/pdb.prot099044.

Karpinka JB, Fortriede JD, Burns KA, James-Zorn C, Ponferrada VG, Lee J, Karimi K, Zorn AM, Vize PD. 2015. *Xenbase*, the *Xenopus* model organism database; new virtualized system, data types and genomes. *Nucleic Acids Res* **43**: D756–D763.

Nicklay JJ, Shechter D, Chitta RK, Garcia BA, Shabanowitz J, Allis CD, Hunt DF. 2009. Analysis of histones in *Xenopus laevis*. II. Mass spectrometry reveals an index of cell type-specific modifications on H3 and H4. *J Biol Chem* **284**: 1075–1085.

Onikubo T, Nicklay JJ, Xing L, Warren C, Anson B, Wang WL, Burgos ES, Ruff SE, Shabanowitz J, Cheng RH, et al. 2015. Developmentally regulated post-translational modification of nucleoplasmin controls histone sequestration and deposition. *Cell Rep.* doi: 10.1016/j.celrep.2015.02.038.

Schnitzler GR. 2001. Isolation of histones and nucleosome cores from mammalian cells. *Curr Protoc Mol Biol* **21**: 21.25.

Shechter D, Dormann HL, Allis CD, Hake SB. 2007. Extraction, purification and analysis of histones. *Nat Protoc* **2**: 1445–1457.

Shechter D, Chitta RK, Xiao A, Shabanowitz J, Hunt DF, Allis CD. 2009. A distinct H2A.X isoform is enriched in *Xenopus laevis* eggs and early embryos and is phosphorylated in the absence of a checkpoint. *Proc Natl Acad Sci* **106**: 749–754.

Wang WL, Shechter D. 2016. Chromatin assembly and transcriptional cross-talk in *Xenopus laevis* oocyte and egg extracts. *Int J Dev Biol* **60**: 315–320.

Wang WL, Anderson LC, Nicklay JJ, Chen H, Gamble MJ, Shabanowitz J, Hunt DF, Shechter D. 2014. Phosphorylation and arginine methylation mark histone H2A prior to deposition during *Xenopus laevis* development. *Epigenetics Chromatin* **7**: 22.

Wilczek C, Chitta R, Woo E, Shabanowitz J, Chait BT, Hunt DF, Shechter D. 2011. Protein arginine methyltransferase Prmt5-Mep50 methylates histones H2A and H4 and the histone chaperone nucleoplasmin in *Xenopus laevis* eggs. *J Biol Chem* **286**: 42221–42231.

Protocol 4

Centromere and Kinetochore Assembly in *Xenopus laevis* Egg Extract

Julio C. Flores Servin[1] and Aaron F. Straight[1,2]

[1]*Department of Biochemistry, Stanford University Medical School, Stanford, California 94305*

During cell division, chromosomes must be equally segregated to daughter cells. Centromeres, the primary interaction site between chromosomes and microtubules, mediate faithful chromosome segregation during mitosis. Functional studies of centromere proteins in cells have proven difficult, as mutation or deletion of most centromeric proteins often results in cell lethality. In this protocol, sperm chromatin or reconstituted chromatin arrays, together with *Xenopus laevis* egg extracts, are used to overcome these limitations and study centromere and kinetochore assembly in vitro. *X. laevis* egg extract is a powerful tool, as it can be readily cycled in vitro by addition of calcium and easily modified biochemically. Coupled with the addition of customizable reconstituted chromatin arrays or sperm chromatin, *X. laevis* egg extract provides distinct advantages over cell-based approaches in which similar experiments would not be feasible. Following incubation in egg extract, reconstituted centromeric chromatin arrays and sperm chromatin specifically assemble core centromere and kinetochore components that can be analyzed via immunofluorescence.

MATERIALS

It is essential that you consult the appropriate Material Safety Data Sheets and your institution's Environmental Health and Safety Office for proper handling of equipment and hazardous materials used in this protocol.

RECIPES: Please see the end of this protocol for recipes indicated by <R>. Additional recipes can be found online at http://cshprotocols.cshlp.org/site/recipes.

Reagents

Antibodies (to protein of interest)

Antibodies, secondary (fluorophore-conjugated)

Antibody dilution buffer (Abdil) <R>

CaCl$_2$ (30 mM in H$_2$O) (for cycling extracts)

CSF-arrested *X. laevis* egg extract (freshly prepared) (see Sec. 7, Protocol 1: Preparation of Cellular Extracts from *Xenopus* Eggs and Embryos [Good and Heald 2018]; Desai et al. 1999)

Cycloheximide solution (40×; 4 mg/mL in H$_2$O) (as needed; see Steps 11 and 36)

DNA dye (bisBenzimide/Hoechst 33258 [Sigma-Aldrich B2883]) (as needed; see Step 33)

Formaldehyde solution (37% [w/v]) (J.T. Baker 2106)

Mounting medium for immunofluorescent staining <R>

[2]Correspondence: astraigh@stanford.edu

Nail polish

Phosphate-buffered saline (PBS) <R> (pH 7.2)

Propidium iodide (1 mg/mL) (VWR IC19545880) (as needed; see Step 21)

Reagents for the chromatin array method

Chromatin bead arrays (freshly prepared) (Guse et al. 2012)

Chromatin arrays are typically assembled at a final DNA concentration of 0.15 or 0.2 µg/µL. In brief, biotinylated pUC18 plasmid containing 19 repeats of the 601 nucleosome positioning sequence and recombinant histones (2.2 times the DNA amount) are combined in a high-salt buffer (2 M NaCl, 10 mM Tris-HCl [pH 7.5], 0.25 mM EDTA) and dialyzed in a dialysis button into a low-salt buffer (2.5 mM NaCl, 10 mM Tris-HCl [pH 7.5], 0.25 mM EDTA) over a 36-h period. Chromatin arrays are then bound to streptavidin-coated magnetic beads at the beginning of each experiment.

Cytostastic factor extract buffer (CSF-XB) <R>

Prepare CSF-XBT by adding 0.05% (v/v) Triton X-100 to CSF-XB <R>. Chill to 18°C–20°C before use.

Reagents for sperm chromatin method

Cushion buffer

Prepare 1× BRB80 <R> containing 40% glycerol (w/v).

Dilution buffer

Prepare 1× BRB80 <R> containing 0.05% Triton X-100 (v/v) and 30% glycerol (w/v).

Fixation buffer

Prepare 1× BRB-80 <R> containing 0.5% Triton-X 100 (v/v) and 30% glycerol. Chill to 18°C –20°C before adding formaldehyde in Step 26.v.

Sperm dilution buffer <R>

Sperm nuclei (see Protocol 1: Isolation and Demembranation of *Xenopus* Sperm Nuclei [Hazel and Gatlin 2018] and Desai et al. [1999])

Equipment

Centrifuge and JS4.2 rotor (with tube adaptors) (for sperm chromatin method)

Coverslips (round glass, No. 1.5, 12-mm), coated with poly-L-Lysine

Forceps (jewelers microforceps, straight and smooth)

Glass tubes (Corex) and components for constructing spin-down adaptors (TAP Plastics) (for sperm chromatin method; see Step 26)

Chock (grooved)

Chock (groove-less)

Glass half-sphere

Metal hook or extended paperclip

Humidified chamber for coverslips

For example, prepare a Petri dish coated with Parafilm and covered with aluminum foil to protect it from light, with wet paper towels inside to create a humid environment.

Magnetic particle concentrator (Dynal MPC; Invitrogen) (for chromatin array method)

Microcentrifuge tubes (1.5- and 2.0-mL, flip-top)

Microscope slides

Water bath at 18°C–20°C

METHOD

Two different methods, one using chromatin arrays and one using sperm chromatin, are outlined here (see Fig. 1). Immunofluorescent staining is used for analysis. Chromatin beads make it possible to manipulate the chromatin substrate directly, while sperm chromatin provides the physiological substrate for centromere assembly. In addition, taking advantage of CSF-arrested extracts versus cycled extracts enables the study of metaphase centromere function

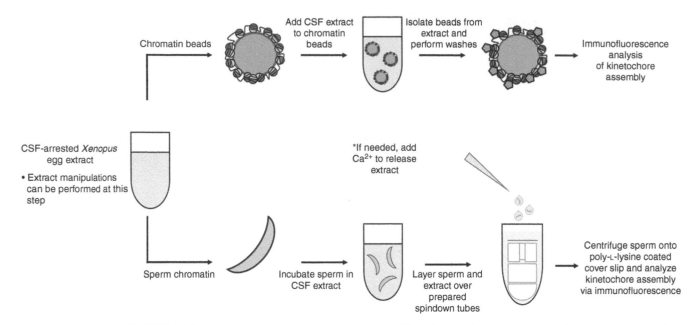

FIGURE 1. Schematic showing general experimental workflow for assaying kinetochore and centromere assembly on chromatin bead arrays and sperm chromatin.

(using the former) or how centromere function is altered by cell cycle transitions (using the latter). Select the appropriate procedure and materials according to the desired experiment (see Discussion).

Centromere and Kinetochore Assembly Using Chromatin Bead Arrays

Using Chromatin Bead Arrays with CSF-Arrested Extracts

1. For each reaction in a separate tube, use a magnetic particle concentrator to wash 0.5 µL of freshly prepared chromatin beads twice in 200–300 µL of CSF-XBT.

2. Aspirate the supernatant from the beads and remove the tubes from the concentrator. Mix the chromatin beads with 20 µL of freshly prepared CSF-arrested extract per tube on ice by pipetting the extract-bead mixture two or three times up and down. Transfer the tubes to a water bath at 18°C–20°C.

3. Incubate the tubes for 75 min at 18°C–20°C to allow centromere and kinetochore assembly. Flick the tubes every 15 min to keep the beads in solution.

4. Wash the beads by adding 200 µL of cold CSF-XBT to the reaction mix. Transfer the tubes to the magnetic particle concentrator at room temperature to collect the chromatin beads.

5. Remove the supernatant from the chromatin beads and quickly wash the beads twice more in 200–300 µL of cold CSF-XBT, removing the supernatant from the beads using the particle concentrator between washes.

6. To fix the chromatin beads, resuspend the beads in 100 µL of CSF-XBT plus 2% formaldehyde (v/v) by pipetting up and down. Incubate for 5 min at room temperature.

7. Wash the fixed beads three times with 200–300 µL of Abdil.

8. Remove the supernatant and resuspend the beads in 40 µL of Abdil.

   To split a reaction on two coverslips, bring the total volume to 80 µL. Further dilution is not recommended as it results in fewer chromatin beads attaching to coverslips.

   Fixed samples may be stored in Abdil for up to 2 d at 4°C.

 Cite this protocol as *Cold Spring Harb Protoc*; doi:10.1101/pdb.prot102509

Using Chromatin Bead Arrays with Cycled Extracts

9. Follow Steps 1–2 to prepare samples.

10. Incubate the tubes for 5–10 min at 18°C–20°C, and then add $CaCl_2$ to a final concentration of 0.75 mM.

    The extracts may be kept in interphase (see Step 11) by addition of cycloheximide (together with $CaCl_2$) to a final concentration of 0.1 μg/mL.

11. Incubate the extracts for 75 min at 18°C–20°C to drive the CSF-arrested extracts into interphase. Flick the tubes every 15 min to keep the chromatin beads in solution.

12. To analyze centromere and kinetochore assembly during interphase, wash and fix the samples as described in Steps 4–8 and then proceed to Step 15. To drive the extracts into metaphase, proceed to Step 13.

13. To drive the extracts into metaphase, add one volume (20 μL) per tube of fresh CSF-arrested extract and then incubate the tubes for 90 min at 18°C–20°C. Flick the tubes every 15 min to keep the chromatin beads in solution.

14. Wash and fix the samples as described in Steps 4–8, and then proceed to Step 15.

Immunofluorescence Staining

15. Place dry poly-L-lysine-coated coverslips in a humidified chamber.

16. Pipette 40 μL of chromatin beads in Abdil (from Steps 8, 12, or 14) onto a dry coverslip. Allow the beads to attach to the coverslip for 30 min at room temperature. Leave the sample-side facing up through Step 22.

17. Remove the Abdil by aspiration and incubate each coverslip with antibodies to the proteins of interest diluted in Abdil. Incubate for 1 h at room temperature or overnight at 4°C.

    The concentration of each antibody must be determined individually.

18. Wash each coverslip three times using 30–50 μL of Abdil.

19. Stain the samples using the appropriate fluorophore-conjugated secondary antibodies for 30 min at room temperature.

20. Wash each coverslip three times using 30–50 μL of Abdil.

21. [Optional] If DNA staining is required, incubate each sample with 30–40 μL of propidium iodide (1 μL/mL in Abdil) for 10 min at room temperature. Wash each coverslip three times using 30–50 μL of Abdil.

22. Wash each coverslip twice using 30–50 μL of 1× PBS.

23. For each coverslip, place 5–7 μL of mounting medium on a labeled microscope slide. Using forceps and a pipette tip, place a coverslip, sample-side down, onto the mounting medium. If needed, carefully aspirate off the excess mounting medium. Seal the slides with nail polish.

    Slides may be stored at −20°C until analysis.

Centromere and Kinetochore Assembly Using Sperm Chromatin

Using Sperm Chromatin with CSF-Arrested Extracts

24. For each reaction, add demembrenated sperm to freshly prepared CSF-arrested extract to a final concentration of $\sim 3 \times 10^5$ sperm per 20 μL of extract. If the sperm is too concentrated, dilute it using sperm dilution buffer.

    It is recommended to add 1–2 μL of diluted sperm per 20 μL of CSF-arrested extract.

25. Allow centromere and kinetochore assembly by incubating the tubes in a water bath for 60 min at 18°C–20°C, flicking them every 15 min.

26. During the incubation, prepare spin-down tubes, processing tubes and fixative for the next steps.

    i. In a glass Corex tube, place a clear glass half-sphere, followed by groove-less chock, followed by a grooved chock (grooved-side down) (Fig. 2).

    ii. Add 5 mL of cushion buffer to each tube. Remove air bubbles by gently tapping down each tube. Alternatively, use an extended paperclip coupled to a 200 μL pipette tip on its end by hooking it onto the grooved chock and moving it up and down until air bubbles rise (Fig. 3).

    iii. Place a 12-mm poly-L-lysine-coated coverslip on top of the grooved chock using forceps and allow it to settle onto the top of the chock by gravity.

    iv. Prepare an extract dilution tube for each reaction by adding 1 mL of dilution buffer to a 2-mL microcentrifuge tube. Keep the tubes at 4°C.

    v. Add formaldehyde to a final concentration of 4% to chilled fixation buffer to prepare fixative.

27. At the end of the incubation, pipette each full reaction volume into a prepared dilution tube and invert several times. Incubate on ice for 5 min.

28. Add 1 mL of fixative to each dilution tube reaction. Invert several times to mix and keep on ice.

29. Using a transfer pipette, gently layer the fixed diluted reactions over the prepared spin-down tubes.

30. Centrifuge the spin-down tubes in a JS4.2 rotor (using the correct tube adapters) at 3500 rpm for 20 min at 4°C.

31. Aspirate the cushion buffer. Using a paperclip coupled to a 200-μL pipette tip, gently remove the top chock. Carefully slide the coverslip off with forceps and place it into the staining chamber.

32. Wash the coverslips twice with Abdil and then block in Abdil for 30 min at room temperature.

33. Wash and stain the samples as described in Steps 17–23.

> *If DNA staining is required, samples can be stained for 5 min using Hoechst (10 μg/mL in Abdil).*

> *Slides may be stored at −20°C until analysis.*

Using Sperm Chromatin with Cycled Extracts

34. Follow Step 24 to prepare samples.

35. Incubate the samples for 5–10 min in a water bath at 18°C–20°C, and then add $CaCl_2$ to a final concentration of 0.75 mM.

> *The extracts may be kept in interphase (see Step 36) by addition of cycloheximide (together with $CaCl_2$) to a final concentration of 0.1 μg/mL.*

36. Incubate the extracts for 75 min at 18°C–20°C to drive the CSF-arrested extracts into interphase. Flick the tubes every 15 min.

37. To analyze centromere and kinetochore assembly during interphase, process the samples as described in Steps 26–33. To drive the extracts into metaphase, proceed to Step 38.

38. To drive the extracts into metaphase, add one volume (20 μL) per tube of fresh CSF-arrested extract and then incubate the tubes for 90 min at 18°C–20°C. Flick the tubes every 15 min.

39. Process the samples as described in Steps 26–33.

Cite this protocol as *Cold Spring Harb Protoc*; doi:10.1101/pdb.prot102509

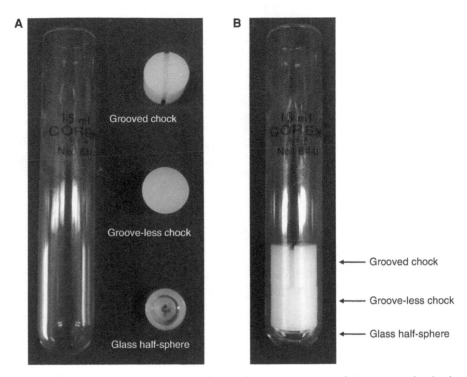

FIGURE 2. Pictures illustrating (*A*) individual spin-down tube components and (*B*) proper chock placement and orientation in spin-down tubes.

DISCUSSION

Xenopus egg extracts are a powerful in vitro system for investigating the molecular mechanisms of centromere and kinetochore assembly. The addition of demembranated *Xenopus* sperm or reconstituted chromatin beads will support recruitment of kinetochore and centromere components (Fig. 4), as well as spindle assembly (Guse et al. 2011). Combined with biochemical manipulations (see Sec. 7, Protocol 3: Protein Immunodepletion and Complementation in *Xenopus laevis* Egg Extracts [Jenness

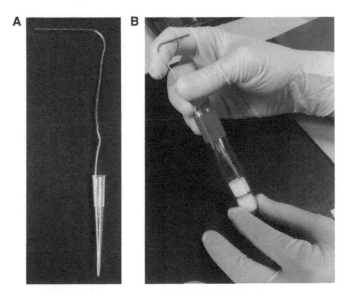

FIGURE 3. Pictures illustrating (*A*) an extended paper clip coupled to a p200 tip and (*B*) a method for using the paper clip tool to remove the top chock and air bubbles from the spin-down tube.

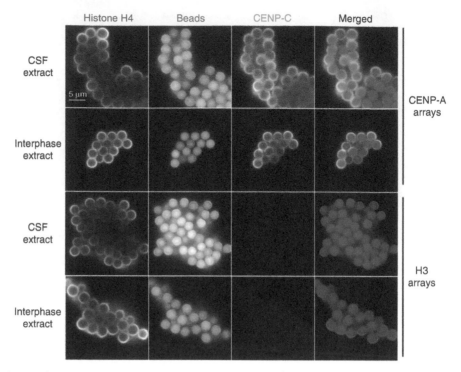

FIGURE 4. Image showcasing CENP-C recruitment to CENP-A and H3 arrays when exposed to CSF-arrested or interphase-released *Xenopus* egg extract.

et al. 2018]) and cycling of extracts, it is possible to dissect the underlying molecular mechanisms governing kinetochore and centromere assembly in ways that would otherwise be difficult using living systems or biochemical reconstitution.

Sperm chromatin contains centromeres that can assemble functional kinetochores; however, it is difficult to manipulate the underlying sperm chromatin. The use of reconstituted chromatin on beads allows one to overcome these limitations and directly manipulate the underlying chromatin to understand its role in centromere and kinetochore assembly. Coupled with in vitro cycling of *Xenopus* egg extracts, the use of chromatin beads has allowed us to define molecular mechanisms of kinetochore assembly that would be difficult or impossible using sperm chromatin (Guse et al. 2011; Westhorpe et al. 2015).

RECIPES

Antibody Dilution Buffer (Abdil)

Reagent	Amount to add	Final concentration
NaCl (5 M)	15 mL	150 mM
Tris-HCl (1 M; pH 7.4)	10 mL	20 mM
Triton X-100 (20%)	2.5 mL	0.10%
Bovine serum albumin	10 g	2%
NaN$_3$ (20%)	2.5 mL	0.10%

To prepare 500 mL of Abdil, combine the reagents listed above and then add H$_2$O to 500 mL. Filter the solution through a 0.22-μm filter. Store at 4°C.

Cite this protocol as *Cold Spring Harb Protoc*; doi:10.1101/pdb.prot102509

BRB80

Reagent	Amount to add (for 5× solution)	Final concentration (for 5× solution)
K-PIPES (0.5 M, pH 6.8)	800 mL	400 mM
MgCl$_2$ (2 M)	2.5 mL	5 mM
K-EGTA (0.5 M)	10 mL	5 mM

To prepare 1 L of 5× BRB80, combine the reagents listed above and then add H$_2$O to 1 L. Filter solution through a 0.22-μm filter. Store at 4°C.

Cytostastic Factor Extract Buffer (CSF-XB)

Reagent	Amount to add	Final concentration
XB salts (20×) <R>	25 mL	1×
Sucrose (2 M)	12.5 mL	50 mM
K-HEPES (1 M, pH 7.7)	5 mL	10 mM
K-EGTA (0.5 M)	5 mL	5 mM
MgCl$_2$ (1 M)	0.5 mL	1 mM

To prepare 500 mL of CSF-XB, combine the reagents listed above and then add H$_2$O to 500 mL. Filter through a 0.22-μm filter. Store at 4°C.

Mounting Medium for Immunofluorescent Staining

Reagent	Amount to add	Final concentration
p-Phenylenediamine	0.1 g	0.5% (w/v)
Tris-HCl (1 M, pH 8.8)	0.4 mL	20 mM
Glycerol	18 mL	90%

To prepare 20 mL of mounting medium, combine Tris-HCl (pH 8.8) and glycerol together with 1.6 mL of H$_2$O. Dissolve p-phenylenediamine in the Tris-HCl and glycerol solution by bubbling nitrogen through the containing vial for 3 to 4 h. Store at −20°C. Discard mounting medium when it turns dark brown.

Phosphate-Buffered Saline (PBS)

Reagent	Amount to add (for 1× solution)	Final concentration (1×)	Amount to add (for 10× stock)	Final concentration (10×)
NaCl	8 g	137 mM	80 g	1.37 M
KCl	0.2 g	2.7 mM	2 g	27 mM
Na$_2$HPO$_4$	1.44 g	10 mM	14.4 g	100 mM
KH$_2$PO$_4$	0.24 g	1.8 mM	2.4 g	18 mM

If necessary, PBS may be supplemented with the following:

Reagent	Amount to add (for 1× solution)	Final concentration (1×)	Amount to add (for 10× stock)	Final concentration (10×)
CaCl$_2$·2H$_2$O	0.133 g	1 mM	1.33 g	10 mM
MgCl$_2$·6H$_2$O	0.10 g	0.5 mM	1.0 g	5 mM

PBS can be made as a 1× solution or as a 10× stock. To prepare 1 L of either 1× or 10× PBS, dissolve the reagents listed above in 800 mL of H$_2$O. Adjust the pH to 7.4 (or 7.2, if required) with HCl, and then add H$_2$O to 1 L. Dispense the solution into aliquots and sterilize them by autoclaving for 20 min at 15 psi (1.05 kg/cm^2) on liquid cycle or by filter sterilization. Store PBS at room temperature.

Sperm Dilution Buffer

Reagent	Amount to add (for 5× solution)	Starting concentration	Final concentration (5×)
Sucrose	5.625 mL	2 M	750 mM
K-HEPES (pH 7.7)	0.75 mL	1 M	50 mM
$MgCl_2$	75 μL	1 M	5 mM
KCl	2.5 mL	3 M	500 mM

To prepare 15 mL of 5× sperm dilution buffer, combine the reagents listed above and then add H_2O to 15 mL. Filter through a 0.22-μm filter. Dilute to 1× in H_2O for use. Store at −20°C.

XB Salts (20×)

2 mM $CaCl_2$

2 M KCl

20 mM $MgCl_2$

Filter-sterilize. Store at 4°C

REFERENCES

Desai A, Murray A, Mitchison TJ, Walczak CE. 1999. The use of *Xenopus* egg extracts to study mitotic spindle assembly and function in vitro. *Methods Cell Biol* **61**: 385–412.x.

Good MC, Heald R. 2018. Preparation of cellular extracts from *Xenopus* eggs and embryos. *Cold Spring Harb Protoc* doi: 10.1101/pdb.prot097055.

Guse A, Carroll CW, Moree B, Fuller CJ, Straight AF. 2011. In vitro centromere and kinetochore assembly on defined chromatin templates. *Nature* **477**: 354–358.

Guse A, Fuller CJ, Straight AF. 2012. A cell-free system for functional centromere and kinetochore assembly. *Nat Protoc* **7**: 1847–1869.

Hazel JW, Gatlin JC. 2018. Isolation and demembranation of *Xenopus* sperm nuclei. *Cold Spring Harb Protoc* doi: 10.1101/pdb.prot099044.

Jenness C, Wynne DJ, Funabiki H. 2018. Protein immunodepletion and complementation in *Xenopus laevis* egg extracts. *Cold Spring Harb Protoc* doi: 10.1101/pdb.prot097113.

Westhorpe FG, Fuller CJ, Straight AF. 2015. A cell-free CENP-A assembly system defines the chromatin requirements for centromere maintenance. *J Cell Biol* **209**: 789–801.

Analysis of Mitotic Checkpoint Function in *Xenopus* Egg Extracts

Yinghui Mao[1]

Department of Pathology and Cell Biology, Columbia University Medical Center, New York, New York 10032

Accurate sister chromatid segregation is pivotal in the faithful transmission of genetic information during each cell division. To ensure accurate segregation, eukaryotic organisms have evolved a "mitotic (or spindle assembly) checkpoint" to prevent premature advance to anaphase before successful attachment of every chromosome to the microtubules of the mitotic spindle. An unattached kinetochore generates a diffusible signal that inhibits ubiquitination of substrates such as cyclin B and securins. This protocol presents an in vitro assay for studying the mitotic checkpoint using *Xenopus laevis* egg extracts. Meiotic spindles assembled around nuclei added to egg extracts are synchronized at metaphase by an endogenous activity known as cytostatic factor (CSF). Normally, the mitotic checkpoint results in continued metaphase arrest following inactivation of CSF activity (by addition of calcium) and disassembly of spindle microtubules (with a microtubule inhibitor such as nocodazole). Simple DAPI staining for chromatin structure or biochemical analysis of Cdc2/cyclin B (cyclin-dependent kinase) histone H1 kinase activity can be used to evaluate the stages of the cell cycle and the status of the mitotic checkpoint. This cell-free system derived from *Xenopus* eggs has been successfully used to unravel the mechanisms of mitosis, and it provides a distinct advantage over cell-based studies in which perturbing kinetochore functions often results in lethality.

MATERIALS

It is essential that you consult the appropriate Material Safety Data Sheets and your institution's Environmental Health and Safety Office for proper handling of equipment and hazardous material used in this protocol.

RECIPES: Please see the end of this protocol for recipes indicated by <R>. Additional recipes can be found online at http://cshprotocols.cshlp.org/site/recipes.

Reagents

12% SDS-polyacrylamide gels

$CaCl_2$ (10 mM)

CSF-arrested egg extracts prepared from unfertilized *Xenopus laevis* eggs (see Sec. 7, Protocol 1: Preparation of Cellular Extracts from *Xenopus* Eggs and Embryos [Good and Heald 2018]; Murray 1991).

> *Extracts should contain LPC (leupeptin, pepstatin, and chymostatin) (10 µg/mL), cytochalasin B (10 µg/mL), and 1× energy-mix.*

---

[1]Correspondence: ym2183@cumc.columbia.edu

Cite this protocol as *Cold Spring Harb Protoc*; doi:10.1101/pdb.prot099853

Demembranated sperm nuclei from *Xenopus laevis* (180,000–200,000 sperm/μL)
   (see Protocol 1: Isolation and Demembranation of *Xenopus* Sperm Nuclei [Hazel and Gatlin 2018]
   and Murray [1991])
Fixative with DAPI <R>
Histone H1 kinase assay cocktail <R>
Nocodazole (5 mg/mL) (Sigma-Aldrich M1404)
SDS gel-loading buffer (2×) <R>

Equipment

Autoradiography equipment (X-ray film, cassette and developer, or phosphorimager)
Coverslips (18 × 18 mm)
Epifluorescence microscope (equipped with 40× objective)
Gel drying equipment
Liquid nitrogen
Microcentrifuge tubes (0.5- and 1.5-mL, flip-top)
Microscope slides
Western blot system

METHOD

Generating a Mitotic Checkpoint in CSF-Arrested *Xenopus* Egg Extract

1. Add 20 μL of CSF-arrested egg extract to a 1.5-mL microcentrifuge tube on ice.

2. Add demembranated sperm nuclei to a final concentration of 9000 sperm/μL.

   As the checkpoint inhibitors are rapidly released from the unattached kinetochore, the strength of the mitotic checkpoint is dependent on the number of kinetochores. A large number of unattached kinetochore is required in Xenopus *egg extracts to maintain the checkpoint signaling; thus, the number of sperm nuclei needed for checkpoint assays is much higher than for spindle assembly reactions.*

3. Supplement the egg extracts with nocodazole to a final concentration of 10 μg/mL. Mix thoroughly by tapping on the side of the tube. Incubate for at least 30 min at room temperature (23°C), mixing periodically.

Releasing CSF-Arrested *Xenopus* Egg Extracts and Monitoring Cell Cycle Progression

4. Add $CaCl_2$ to the extracts to a final concentration of 0.4 mM. Mix thoroughly by tapping on the side of the tube. Incubate at room temperature.

5. To observe the nuclear morphology at 0, 30, and 60 min, withdraw 1 μL of extract at each time point, transfer it onto a slide, and mix with an equal volume of fixative. Gently lower a coverslip on top.

6. Visualize sperm chromatin morphology by fluorescence microscopy.

   Condensed chromatin is the hallmark of an active mitotic checkpoint (Fig. 1, bottom panel, right). In contrast, if mitotic arrest cannot be maintained, the chromatin decondenses and forms round interphase nuclei (Fig. 1, bottom panel, left).

7. Transfer 1 μL of extract from each reaction into a 0.5-mL microcentrifuge tube and snap-freeze the tube in liquid nitrogen. Store the samples in a freezer at −80°C for subsequent assay of histone H1 kinase.

   Samples should be assayed within a couple of months.

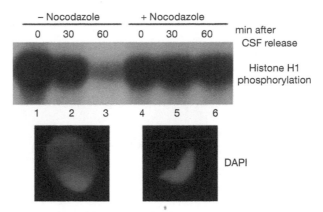

FIGURE 1. Activation and maintenance of the mitotic checkpoint in *Xenopus* egg extracts. After addition of sperm nuclei and nocodazole for 30 min to initiate the mitotic checkpoint in CSF-arrested egg extracts, CSF is inactivated upon addition of calcium. Aliquots were assayed at the indicated time for Cdc2 kinase activity (assayed on histone H1) (*top* panel) and chromatin decondensation by visualization with DAPI (*bottom* panel).

Assaying Histone H1 Kinase Activity

8. Thaw on ice each 1-μL sample of frozen extract to be assayed. To each sample, add 9 μL of histone H1 kinase assay cocktail. Incubate for 10 min at room temperature.

9. Stop the reactions by adding 10 μL of 2× SDS gel-loading buffer to each tube.

10. Electrophorese the samples on 12% SDS-polyacrylamide gels.

11. Dry the gels and expose to film or a phosphorimager plate to visualize phosphorylated histone H1 (Cdc2 kinase activity).

> There is a continued high level of Cdc2 kinase activity following calcium-mediated inactivation of CSF (30 and 60 min) with an activated mitotic checkpoint (Fig. 1, top panel, lanes 4–6). Without an activated mitotic checkpoint, extracts return to interphase and rapidly lose Cdc20 kinase activity after inactivation of CSF (Fig. 1, top panel, lanes 1–3).

DISCUSSION

These procedures can be used to provide an informative assessment of mitotic checkpoint status. In combination with immunodepletion and reconstitution, as well as antibody addition, *Xenopus* egg extracts provide a means to evaluate whether any protein/protein complex plays a direct role in mitotic checkpoint activation and/or maintenance. Upon addition of high numbers of sperm nuclei and nocodazole to disrupt spindle microtubule assembly in CSF-arrested egg extracts, the presence of an activated mitotic checkpoint can be assessed by continuous chromosome condensation and Cdc2/cyclin B (cyclin-dependent kinase) histone H1 kinase activity following addition of calcium to release the CSF arrest (Abrieu et al. 2000, 2001; Mao et al. 2003). Furthermore, the strength of the mitotic checkpoint activation can be analyzed semiquantitatively either by using various concentrations of sperm nuclei (e.g., 5000, 7500, or 10,000 nuclei/μL of egg extract) to produce different numbers of unattached kinetochores or by using various concentrations of nocodazole (e.g., 1, 3, or 10 ng/μL of egg extract) to disrupt kinetochore microtubule attachment to different degrees. For example, the mitotic checkpoint can be triggered and maintained in the presence of nocodazole at 1 and 10 ng/μL at a sperm nuclei concentration of 10,000 nuclei/ μL of egg extract (Chen 2004). However, immunodepletion of a specific kinetochore component results in an activated mitotic checkpoint with 10 ng/μL but not 1 ng/μL of nocodazole, suggesting that the mitotic checkpoint function is partially compromised without this kinetochore component.

RECIPES

Fixative with DAPI

Reagent	Amount to add
37% formaldehyde	300 µL
80% glycerol	600 µL
MMR solution (10×) <R>	100 µL
4′,6′-diamidino-2-phenylindole (DAPI)	1 µg/mL

Freeze aliquots at −20°C. Store for up to 6 mo at −20°C.

Histone H1 Kinase Assay Cocktail

Reagent	Amount to add
Histone H1	5 mg
β-glycerophosphate	111 mg
1 M $MgCl_2$	76 µL
0.5 M EGTA	204 µL
50 mM ATP	6.4 µL

Bring volume to 6.4 mL with H_2O. Keep frozen in 500-µL aliquots. Combine with 50 µCi [γ-^{32}P]ATP just prior to use.

MMR Solution (10×)

1 M NaCl
20 mM KCl
10 mM $MgCl_2$
20 mM $CaCl_2$
5 mM HEPES (adjusted to pH 7.8 with NaOH)
1 mM EDTA

Store for up to 6 mo at room temperature.

SDS Gel-Loading Buffer (2×)

100 mM Tris-Cl (pH 6.8)
4% (w/v) SDS (sodium dodecyl sulfate; electrophoresis grade)
0.2% (w/v) bromophenol blue
20% (v/v) glycerol
200 mM DTT (dithiothreitol)

Store the SDS gel-loading buffer without DTT at room temperature. Add DTT from a 1 M stock just before the buffer is used. 200 mM β-mercaptoethanol can be used instead of DTT.

Cite this protocol as *Cold Spring Harb Protoc*; doi:10.1101/pdb.prot099853

REFERENCES

Abrieu A, Kahana JA, Wood KW, Cleveland DW. 2000. CENP-E as an essential component of the mitotic checkpoint in vitro. *Cell* **102:** 817–826.

Abrieu A, Magnaghi-Jaulin L, Kahana JA, Peter M, Castro A, Vigneron S, Lorca T, Cleveland DW, Labbe JC. 2001. Mps1 is a kinetochore-associated kinase essential for the vertebrate mitotic checkpoint. *Cell* **106:** 83–93.

Chen RH. 2004. Phosphorylation and activation of Bub1 on unattached chromosomes facilitate the spindle checkpoint. *EMBO J* **23:** 3113–3121.

Good MC, Heald R. 2018. Preparation of cellular extracts from *Xenopus* eggs and embryos. *Cold Spring Harb Protoc* doi: 10.1101/pdb.prot097055.

Hazel JW, Gatlin JC. 2018. Isolation and demembranation of *Xenopus* sperm nuclei. *Cold Spring Harb Protoc* doi: 10.1101/pdb.prot099044.

Mao Y, Abrieu A, Cleveland DW. 2003. Activating and silencing the mitotic checkpoint through CENP-E-dependent activation/inactivation of BubR1. *Cell* **114:** 87–98.

Murray AW. 1991. Cell cycle extracts. *Methods Cell Biol* **36:** 581–605.

Protocol 6

Nucleus Assembly and Import in *Xenopus laevis* Egg Extract

Pan Chen and Daniel L. Levy[1]

Department of Molecular Biology, University of Wyoming, Laramie, Wyoming 82071

Xenopus egg extract represents a powerful cell-free biochemical tool for studying organelle assembly and function. Large quantities of cytoplasm can be isolated, and biochemical manipulation of extract composition and cell cycle state is relatively straightforward. In this protocol, we describe the reconstitution of nuclear assembly by adding a chromatin source to interphasic *X. laevis* egg extract. Intact nuclei assemble within 30–45 min of initiating the reaction, followed by nuclear growth. We also describe methods for imaging and quantifying nuclear import kinetics. Recombinant proteins or small molecules of interest can be added to the extract before or after nuclear assembly, and immunodepletion allows for removal of specific proteins from the extract. This approach will continue to inform mechanisms of nuclear assembly, nuclear pore complex assembly and function, nucleocytoplasmic transport, DNA replication, nuclear envelope breakdown, and nuclear size and shape regulation.

MATERIALS

It is essential that you consult the appropriate Material Safety Data Sheets and your institution's Environmental Health and Safety Office for proper handling of equipment and hazardous materials used in this protocol.

RECIPES: Please see the end of this protocol for recipes indicated by <R>. Additional recipes can be found online at http://cshprotocols.cshlp.org/site/recipes.

Reagents

Calcium stock solution <R>

Cycloheximide stock solution <R>

Demembranated *Xenopus* sperm (see Protocol 1: Isolation and Demembranation of *Xenopus* Sperm Nuclei [Hazel and Gatlin 2018])

Energy mix for *Xenopus* egg extracts <R>

Ethanol (70% or 95%)

FITC-dextran stock solution (150 kD; 20 mg/mL in ddH$_2$O; Sigma-Aldrich 46946)

Store at −20°C in 100-μL aliquots.

GFP-NLS stock solution

Purify recombinant GST-GFP-NLS and dialyze into 1× XB (Levy and Heald 2010). Store at −80°C in 10-μL aliquots. Our stock concentrations range from 2 to 28 mg/mL.

---

[1]Correspondence: dlevy1@uwyo.edu

Cite this protocol as *Cold Spring Harb Protoc*; doi:10.1101/pdb.prot097196

Hoechst (10 mg/mL in ddH$_2$O)

Store at −20°C in 40 µL aliquots.

Nuclear fix solution <R>

Oxygen scavenging mix (Aitken et al. 2008)

Sperm dilution buffer (SDB) <R>

Valap sealant <R>

XB buffer (10×) <R>

X. laevis metaphase-arrested egg extract (see Sec. 7, Protocol 1: Preparation of Cellular Extracts from *Xenopus* Eggs and Embryos [Good and Heald 2018])

Equipment

Cotton tipped applicators

Epifluorescence microscope, equipped with DAPI and GFP/FITC filters and 20× or 40× objective

Eppendorf tubes (0.65 and 1.5 mL)

Glass slides and 22 mm × 22 mm coverslips

Clean slides and coverslips immediately before use with Kimwipes saturated with 70% or 95% ethanol and let dry.

Hot plate

Kimwipes

Pap pen (Scientific Device Laboratory, 9804-2)

Software for image quantification (e.g., ImageJ, Metamorph, cellSens)

Water bath set between 16°C–20°C

Wide bore pipette tips (~1.5 mm diameter opening for 200 µL pipette tips, VWR 89079-456; ~2 mm diameter opening for 1 mL pipette tips, VWR 89049-162) or normal pipette tips with the tips cutoff

METHOD

1. Prepare *X. laevis* metaphase-arrested egg extract (see Sec. 7, Protocol 1: Preparation of Cellular Extracts from *Xenopus* Eggs and Embryos [Good and Heald 2018]) (Hannak and Heald 2006; Maresca and Heald 2006) and demembranated *Xenopus* sperm (see Protocol 1: Isolation and Demembranation of *Xenopus* Sperm Nuclei [Hazel and Gatlin 2018]) (Murray 1991).

Assemble Nuclei

2. For every 100 µL of freshly prepared egg extract, add 2 µL energy mix, 1.5 µL cycloheximide stock solution, and 6 µL calcium stock solution in a 1.5 mL Eppendorf tube on ice. To mix, gently tap the tube or pipette using wide bore pipette tips. The reaction volume can be scaled between 25–500 µL. For smaller reactions, use of 0.65 mL Eppendorf tubes is preferred.

   Calcium induces progression of the metaphase-arrested extract into interphase, and cycloheximide keeps the extract arrested in interphase. Excessive pipetting or mixing can inactivate the extract. While we always add energy mix, some researchers do not. We suggest that the requirement for energy mix be determined empirically.

3. Add demembranated *Xenopus* sperm nuclei at 1000/µL (e.g., 1 µL of a 100,000 sperm nuclei/µL stock for a 100 µL reaction). If necessary, sperm can first be diluted in SDB or extract.

4. Incubate the reaction in a water bath at 16°C–20°C. Invert or tap gently every 15 min to ensure that nuclei remain suspended in the extract. Initial nuclear assembly usually occurs within 30–45 min.

   For subsequent steps, it is best to use wide bore pipette tips to avoid damaging the nuclei.

Image Nuclei

5. Pipette 2 µL of the nuclear assembly reaction onto a glass slide, directly pipette 2 µL of nuclear fix solution into the extract droplet, and overlay with a 22 mm × 22 mm coverslip.

6. Image nuclei using the DAPI filter on an epifluorescence microscope. A high proportion of round chromatin masses is indicative of successful nuclear assembly, while S-shaped or elongated slug-shaped chromatin indicates that longer incubation is required or that the extract quality is poor (Fig. 1A).

    This step assesses if the nuclear assembly reaction is proceeding.

    See Troubleshooting.

7. To test if nuclei are intact, pipette 2 µL of the nuclear assembly reaction onto a glass slide, add 0.5 µL of FITC-dextran (150 kD) stock solution, and overlay with a 22 mm × 22 mm coverslip. Seal the coverslip using a Pap pen or Valap to minimize extract flow. Image nuclei using the GFP/FITC filter. Intact nuclei should exclude the FITC signal due to the diffusion size exclusion limit of the nuclear pore complex (Fig. 1B,F).

    Melt Valap using a hot plate and apply with cotton tipped applicators.

    This step is a more stringent test for nuclear assembly than Step 6; 150 kD FITC-Dextran will only be excluded from the nucleus if a fully intact nuclear envelope has formed.

    FITC-dextran can photobleach rapidly depending on the light source. In our experience, use of a light-emitting diode (LED) light source and reduction of the light intensity result in negligible photobleaching.

8. To test if nuclei are functional, pipette 2 µL of the nuclear assembly reaction onto a glass slide, add 0.5 µL of GFP-NLS stock solution, and overlay with a 22 mm × 22 mm coverslip. Image nuclei using the GFP/FITC filter. Functional nuclei should show intranuclear GFP-NLS signal (Fig. 1C,F).

    Nuclei generally begin to show intranuclear GFP-NLS signal around 20–30 min after initiating the reaction. Nuclei continue to grow over time, with large round nuclei observed by 75–90 min (Fig. 1A,C). Nuclei can also be fixed, spun onto coverslips, and processed for immunofluorescence (Edens and Levy 2016).

Image Nuclear Import

9. Add GFP-NLS to the nuclear assembly reaction; we generally use a final concentration of 0.04–0.14 µg/µL. Gently tap or invert the tube to mix.

    The concentration of GFP-NLS added to the reaction can vary, but the added volume should not exceed 10% of the reaction volume. Other fluorescently labeled import cargos of interest can also be used.

10. Immediately pipette 5 µL of the reaction onto a clean glass slide and overlay with a 22 mm × 22 mm coverslip. Seal the coverslip using a Pap pen or Valap.

    Melt Valap using a hot plate and apply with cotton tipped applicators.

11. Image using the GFP/FITC filter at 30 sec intervals using the same exposure time for ~60 min. To minimize photodamage, oxygen scavenging mix can be added to the reaction immediately before imaging (see online Movie 1 at cshprotocols.cshlp.org, Fig. 1C). To correct for photobleaching during quantification, acquire a similar time-lapse using extract without sperm added.

Quantify Nuclear Import

12. Using an appropriate software program (e.g., Metamorph), set the image distance calibration based on the objective used. Threshold nuclei, and quantify average nuclear GFP-NLS intensity (i.e., total integrated intensity divided by area) at each time point.

13. For the time-lapse without nuclei, select a region in the middle and quantify average GFP-NLS intensity over time to obtain the rate of photobleaching, which is the slope (m) (Fig. 1E).

Cite this protocol as *Cold Spring Harb Protoc*; doi:10.1101/pdb.prot097196

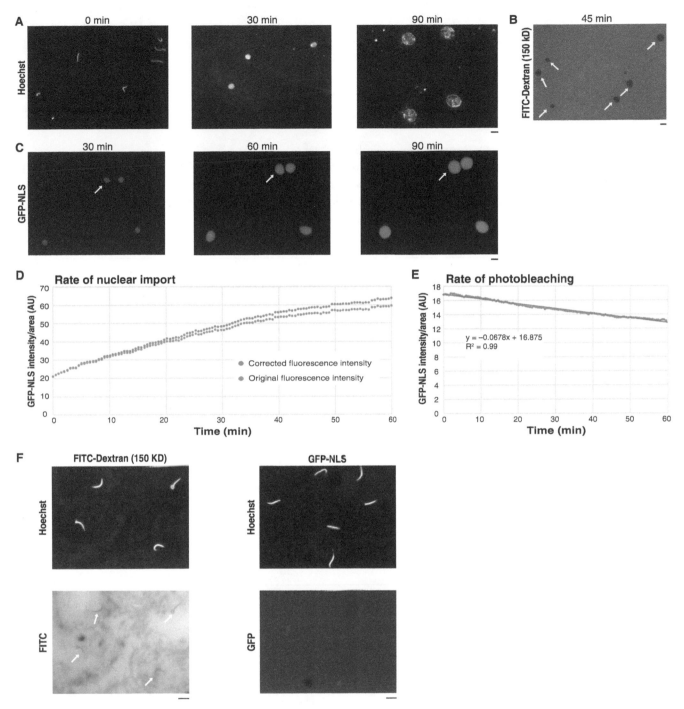

FIGURE 1. Nuclear assembly and import quantification. (*A*) Nuclei were assembled in *X. laevis* egg extract, fixed at different time points with nuclear fix solution, and visualized with Hoechst. (*B*) Nuclei were incubated with FITC-Dextran (150 kDa), and intact nuclei excluding FITC signal are indicated with arrows. (*C*) GFP-NLS (0.04 µg/µL) was added to nuclei 30 min after the initiation of nuclear assembly. A slide was prepared and GFP-NLS images were acquired every 30 sec with the same exposure time. Also see online Movie 1 at cshprotocols.cshlp.org. (*D*) For the nucleus indicated with an arrow in (*C*), nuclear GFP-NLS fluorescence intensity per unit area was measured at each time point using Metamorph (orange points), starting 30 min after the initiation of nuclear assembly. Using the trend line shown in (*E*), nuclear GFP-NLS fluorescence intensity per unit area was corrected for photobleaching (blue points), as described in the protocol. (*E*) Imaging and quantification were performed as in (*C*) and (*D*) except that the extract contained no nuclei. These data reflect the extent of photobleaching. The data (blue points) were fit by linear regression (orange line). (*F*) Sperm nuclei in *X. laevis* egg extract were visualized 10 min after initiating the reaction. While sperm nuclei initially exclude FITC-dextran (arrows), they fail to import GFP-NLS. After complete nuclear assembly, only intact nuclei will exclude FITC-dextran and show intranuclear GFP-NLS signal, while damaged nuclei will fail to both exclude FITC-dextran and accumulate imported GFP-NLS. Scale bars, 20 µm.

14. To correct for photobleaching, add back the appropriate nuclear GFP-NLS signal intensity as calculated by: $I(t)_{corr} = I(t)_{meas} + |m|*t$, where $I(t)_{corr}$ is the corrected intensity at time t and $I(t)_{meas}$ is the original measured intensity at time t. Plot I_{corr} as a function of time (Fig. 1D).

TROUBLESHOOTING

Problem (Step 6): Poor nuclear assembly.

Solutions: This is generally indicative of poor quality egg extract and the best course of action is to repeat the experiment with a new extract. Alternatively, the sperm stock solution may be too concentrated or sperm demembranation may have been incomplete, requiring dilution or new preparation of the sperm stock, respectively.

DISCUSSION

This method describes the in vitro assembly of nuclei in *Xenopus* egg extract and quantification of nuclear import kinetics. While GFP-NLS informs bulk nuclear import kinetics, specific import cargos of interest can be fluorescently tagged and their import similarly measured using this approach. It is worth noting that this protocol can also be performed with *X. tropicalis* egg extract, as well as extracts prepared from staged *Xenopus* embryos that allow one to study the structure and function of endogenous embryonic nuclei (Sec. 7, Protocol 1: Preparation of Cellular Extracts from *Xenopus* Eggs and Embryos [Good and Heald 2018]) (Levy and Heald 2010; Edens and Levy 2014). *Xenopus* extracts have provided mechanistic insights into nuclear assembly (Anderson and Hetzer 2007), nuclear pore complex assembly and function (D'Angelo et al. 2006), nucleocytoplasmic transport (Gorlich et al. 1994), DNA replication (Blow and Laskey 1986), nuclear envelope breakdown (Muhlhausser and Kutay 2007), and nuclear size and shape regulation (Jevtic et al. 2015; Vukovic et al. 2016). Mechanistic aspects of these various processes can be easily assessed due to the open biochemical nature of the extract, which can be supplemented with recombinant proteins, small molecules, or neutralizing antibodies and/or immunodepleted of specific proteins (Hannak and Heald 2006; Maresca and Heald 2006). For these experiments, proteins and antibodies should be dialyzed into XB buffer to be compatible with the extract.

RECIPES

Calcium Stock Solution

10 mM CaCl$_2$
0.1 M KCl
1 mM MgCl$_2$

Store at −20°C in 200-μL aliquots.

Cycloheximide Stock Solution

Prepare 10 mL of 1× XB buffer <R> and warm slightly in microwave. Add 100 mg cycloheximide and vortex to dissolve. Syringe filter to sterilize. Store at −20°C in 100-μL aliquots.

Energy Mix for Xenopus Egg Extracts

190 mM creatine phosphate disodium
25 mM ATP disodium salt
25 mM MgCl$_2$

Store at −20°C in 100-μL aliquots.

Nuclear Fix Solution

Combine 125 µL 2 M sucrose, 12.5 µL 1 M HEPES-KOH pH 7.8, 250 µL 37% formaldehyde, 112 µL ddH$_2$O, and 0.5 µL Hoechst stock solution (10 mg/mL). Store at room temperature for up to several months.

Sperm Dilution Buffer (SDB)

10 mM HEPES-KOH pH 7.8
1 mM MgCl$_2$
100 mM KCl
150 mM sucrose

Store at −20°C in 1-mL aliquots.

Valap Sealant

Add equal weight of petroleum jelly (e.g., Vaseline), lanolin, and paraffin together in a 1-L beaker. (The preparation of a few hundred milliliters of this mixture should be a lifetime supply.) Heat the mixture at a low setting on a hot plate. Stir occasionally until thoroughly blended. Take extreme care not to overheat this mixture. Warm it just enough to make the components liquefy. The final product should be golden yellow, the color of corn or canola oil. If it is dark brown, it was heated too much. In that case, start again. Aliquot the final mixture into several small screw-capped jars (∼50-mL capacity). Store at room temperature. Keep a small amount in a 25-mL beaker, and warm at a low setting on a hot plate before use.

XB Buffer (10×)

1 M KCl
1 mM CaCl$_2$
10 mM MgCl$_2$
500 mM sucrose
100 mM HEPES-KOH pH 7.8

Store at 4°C and dilute to 1× with ddH$_2$O before use.

ACKNOWLEDGMENTS

Research in the Levy laboratory is supported by the National Institutes of Health (NIH) (R01GM113028) and American Cancer Society (RSG-15-035-01-DDC). P.C. is supported by a Wyoming INBRE graduate assistantship through the NIH (P20GM103432).

All *Xenopus* procedures and studies illustrated in Figure 1 and Movie 1 (see online Movie 1 at cshprotocols.cshlp.org) were conducted in compliance with the U.S. Department of Health and Human Services Guide for the Care and Use of Laboratory Animals. Protocols were approved by the University of Wyoming Institutional Animal Care and Use Committee (Assurance # A-3216-01).

REFERENCES

Aitken CE, Marshall RA, Puglisi JD. 2008. An oxygen scavenging system for improvement of dye stability in single-molecule fluorescence experiments. *Biophys J* 94: 1826–1835.

Anderson DJ, Hetzer MW. 2007. Nuclear envelope formation by chromatin-mediated reorganization of the endoplasmic reticulum. *Nat Cell Biol* 9: 1160–1166.

Blow JJ, Laskey RA. 1986. Initiation of DNA replication in nuclei and purified DNA by a cell-free extract of *Xenopus* eggs. *Cell* 47: 577–587.

D'Angelo MA, Anderson DJ, Richard E, Hetzer MW. 2006. Nuclear pores form de novo from both sides of the nuclear envelope. *Science* 312: 440–443.

Edens LJ, Levy DL. 2014. cPKC regulates interphase nuclear size during *Xenopus* development. *J Cell Biol* **206**: 473–483.

Edens LJ, Levy DL. 2016. A cell-free assay using *Xenopus laevis* embryo extracts to study mechanisms of nuclear size regulation. *J Vis Exp* **114**. doi: 10.3791/54173.

Good MC, Heald R. 2018. Preparation of cellular extracts from *Xenopus* eggs and embryos. *Cold Spring Harb Protoc* doi: 10.1101/pdb.prot097055.

Gorlich D, Prehn S, Laskey RA, Hartmann E. 1994. Isolation of a protein that is essential for the first step of nuclear protein import. *Cell* **79**: 767–778.

Hannak E, Heald R. 2006. Investigating mitotic spindle assembly and function in vitro using *Xenopus laevis* egg extracts. *Nat Protoc* **1**: 2305–2314.

Hazel JW, Gatlin JC. 2018. Isolation and demembranation of *Xenopus* sperm nuclei. *Cold Spring Harb Protoc* doi: 10.1101/pdb.prot099044.

Jevtic P, Edens LJ, Li X, Nguyen T, Chen P, Levy DL. 2015. Concentration-dependent effects of nuclear lamins on nuclear size in *Xenopus* and mammalian cells. *J Biol Chem* **290**: 27557–27571.

Levy DL, Heald R. 2010. Nuclear size is regulated by importin α and Ntf2 in *Xenopus*. *Cell* **143**: 288–298.

Maresca TJ, Heald R. 2006. Methods for studying spindle assembly and chromosome condensation in *Xenopus* egg extracts. *Methods Mol Biol* **322**: 459–474.

Muhlhausser P, Kutay U. 2007. An in vitro nuclear disassembly system reveals a role for the RanGTPase system and microtubule-dependent steps in nuclear envelope breakdown. *J Cell Biol* **178**: 595–610.

Murray AW. 1991. Cell cycle extracts. *Methods Cell Biol* **36**: 581–605.

Vukovic LD, Jevtic P, Zhang Z, Stohr BA, Levy DL. 2016. Nuclear size is sensitive to NTF2 protein levels in a manner dependent on Ran binding. *J Cell Sci* **129**: 1115–1127.

Extracts for Analysis of DNA Replication in a Nucleus-Free System

Justin Sparks[1] and Johannes C. Walter[1,2,3]

[1]Department of Biological Chemistry and Molecular Pharmacology, Harvard Medical School, Boston, Massachusetts 02115; [2]Howard Hughes Medical Institute, Department of Biological Chemistry and Molecular Pharmacology, Harvard Medical School, Boston, Massachusetts 02115

Frog egg extracts represent a powerful approach with which to dissect molecular mechanisms of vertebrate DNA replication and repair. In the classical approach, sperm chromatin is added to a crude egg lysate to form replication-competent nuclei. We subsequently described a procedure that bypasses the requirement for nuclear assembly in DNA replication. In this method, DNA is first added to a high-speed supernatant (HSS) of egg lysate, which mimics the G_1 phase of the cell cycle in that it supports replication licensing. Subsequent addition of a concentrated nucleoplasmic extract (NPE) leads to replication initiation followed by a single complete round of DNA replication. The advantage of the nucleus-free system is that it supports efficient replication of model DNA templates such as plasmids and lambda DNA that can be modified with specific features such as LacI arrays, DNA protein cross-links, or DNA interstrand cross-links. Here, we describe our current protocol for preparation of HSS and NPE. Methods for their use in DNA replication and repair are described elsewhere.

MATERIALS

It is essential that you consult the appropriate Material Safety Data Sheets and your institution's Environmental Health and Safety Office for proper handling of equipment and hazardous materials used in this protocol.

RECIPES: Please see the end of this protocol for recipes indicated by <R>. Additional recipes can be found online at http://cshprotocols.cshlp.org/site/recipes.

Reagents

Adenosine 5′-triphosphate disodium salt hydrate (0.2 M; Sigma-Aldrich A5394)

Dissolve in sterile water and adjust to pH ~7 with 10 M NaOH using pH indicator paper. Store aliquots at −20°C.

Aprotinin/leupeptin stock solution (10 mg/mL each; Roche 11583794001, 11529048001)

Dissolve in filter-sterilized water, and store aliquots at −80°C.

Benzocaine (10% (w/v) stock solution in ethanol)

Stable for 6 mo at room temperature.

Creatine phosphokinase solution <R>

---

[3]Correspondence: johannes_walter@hms.harvard.edu

Cite this protocol as *Cold Spring Harb Protoc*; doi:10.1101/pdb.prot097154

Cycloheximide (10 mg/mL; Calbiochem 239763)

Make fresh by dissolving in filter-sterilized water.

Cytochalasin B (5 mg/mL; Sigma-Aldrich C6762)

Dissolve in DMSO, and store aliquots at −20°C.

Dejelly solution (1 L for HSS; 2 L for NPE) <R>
Dithiothreitol (DTT; 1 M; Bio-Rad 161-0611)

Dissolve in filter-sterilized water, and store aliquots at −20°C.

ELB-sucrose (1 L for HSS; 2 L for NPE) <R>
Ethanol (70%)
Hoechst stock solution <R>
Hormones for inducing ovulation (see Lebofsky et al. 2009)
Ice
Liquid nitrogen
Marc's modified Ringer's (MMR) for *Xenopus* (20×) <R>
MMR (0.5×; 2 L for HSS; 4 L for NPE)

Dilute from 20× MMR with dH$_2$O; make fresh on the day of extract preparation.

Mature female *Xenopus laevis* frogs (six for HSS; 20 for NPE)
Mature male *Xenopus laevis* frogs (six to eight)
NaCl (100 mM; 15 L for HSS; 50 L for NPE)
Nocodazole (5 mg/mL; Sigma-Aldrich M14104)

Dissolve in DMSO. Store aliquots at −20°C.

Phosphocreatine disodium salt (1 M, Sigma-Aldrich P7936)

Dissolve in 10 mM sodium phosphate pH 7. Store aliquots at −20°C.

Sucrose buffers for preparation of demembranated sperm chromatin <R>

It is important to make the sucrose solutions well before beginning the protocol as the sucrose can take an hour or more to dissolve.

Triton X-100 (20% stock solution in sterile water)

Equipment

Aspiration setup
Benchtop centrifuge with swinging-bucket rotor, such as IEC Centra CL2, or equivalent, with adapters to hold 1.5-mL microcentrifuge tubes and Falcon 2059 15-mL tubes
Centrifuge (Sorvall Lynx 4000) and swinging-bucket rotor TH13-6X50 or equivalent, with adapters to hold Falcon 2059 15-mL tubes
Conical tubes with screw caps (15 and 50 mL)
Coverslips
Dissection equipment (i.e., forceps, scissors, pithing needle)
Gel-loading tips
Glass beakers (100-mL; 6)
Glass beakers (2-L)
Glass test tubes (13 × 100 mm; Kimble Glass Inc. 73500 13100)
Hemocytometer
Ice buckets, 10 L
Microcentrifuge tubes (1.5 and 2 mL)
Microscope, epifluorescent (e.g., an Eclipse E600POL [Nikon])
Microscope slides
Needles (18-gauge)
Paper towels

Pasteur pipettes, glass

Petri dishes

Plastic transfer pipettes

Polypropylene tubes (15-mL; Falcon 2059 tubes or similar)

Razor blades (2–3)

Rotating wheel, end-over-end

Spray bottle for ethanol

Ultracentrifuge, such as Optima Max-E (Beckman), with swinging-bucket rotor, such as the TLS-55 (Beckman), or equivalent

> It is important that the ultracentrifuge can be chilled and the configuration can reach and withstand speeds of up to 260,000g.

Ultracentrifuge tubes (2.5-mL; thick-walled; Beckman 343778)

> Use for NPE if yield is <1.5 mL.

Ultracentrifuge tubes (2.5-mL; thin-walled and clear; Beckman 347356)

> Use for sperm chromatin and HSS. Use also for NPE if yield is greater than 1.5 mL.

Vortex

Waste beaker (4-L)

METHOD

Preparation of Demembranated Sperm Chromatin

All steps are performed at room temperature unless otherwise stated.

1. In a 2 L beaker, dilute 6 mL of 10% benzocaine in 800 mL of water (0.075%).

2. Euthanize a male frog by placing it in the benzocaine solution for ∼6 min or until it no longer rights itself when turned over and lacks a sucking reflex (frog does not suck on gloved finger placed in its mouth).

3. Remove frog from benzocaine and ensure euthanasia by pithing the frog's spinal cord at the junction with its skull.

4. Recover the testes by first making an incision across the stomach and continuing up both sides of the frog's abdomen. The testes are attached to the intestine wall. *Xenopus* testes are white or cream in color, bean shaped, and ∼5–8 mm in length and 3–4 mm in width. Using scissors, cut the testes away from the connective tissue and place them on a dry paper towel.

   Once the testes are removed, place the next male frog in the benzocaine solution.

5. While waiting for the next frog to be euthanized, blot the blood from the testes with a dry paper towel and remove excess connective tissue and blood vessels on the surface of the testes. It is not essential to remove all excess tissue at this point.

6. Place clean testes in a Petri dish with 2–3 mL of buffer #1 (Sucrose buffers for preparation of demembranated sperm chromatin).

7. Start dissection of next frog.

8. Once testes have been recovered from all six frogs, cleaned, and collected in the petri dish containing buffer #1, remove the testes from the petri dish and reclean them by removing any leftover connective tissue or blood with forceps and scissors. Place the clean testes in a new Petri dish with 1 mL of buffer #1. At this step, it is essential to remove all visible connective tissue and blood cells.

   Adding more buffer will make the chopping in the next step less efficient.

9. Mince the testes into tiny pieces by extensive chopping with a razor blade for 10 min. Change blades every 3–4 min.

10. Transfer the minced testes into a 15 mL screw cap conical tube with a p1000 tip. If there are pieces of testes too big to be removed with a p1000 tip, the testes are not minced enough.

11. Rinse the petri dish with small volumes of buffer #1 and combine the buffer with the minced testes to a final volume of ~6 mL.

    Use the same p1000 tip to avoid losing material.

12. Vortex the tube vigorously for 1 min and pellet the large pieces of tissue by centrifugation for 10 sec at 200g at room temperature in a benchtop centrifuge with a swinging-bucket rotor, such as the IEC Centra CL2 or equivalent, with adapters to hold 15 mL screw cap conical tubes.

13. Transfer the supernatant (cloudy) to a 15 mL Falcon 2059 tube at room temperature.

14. Add 2 mL of buffer #1 to the pellet from Step 12 and vortex for 1 min.

15. Centrifuge the tube for 15 sec at 200g at room temperature in the same centrifuge used in Step 12. Combine the supernatant with the supernatant from Step 13. Repeat Steps 13–14 two to three times until the supernatant is no longer cloudy.

16. Centrifuge the combined supernatants for 50 sec at 380g at room temperature in the same centrifuge used in Step 12. Transfer the supernatant to a new 15 mL Falcon 2059 tube using a p1000 tip. Be careful to avoid transfer of the pelleted tissue in this step. Remaining supernatant will be recovered in Step 17.

17. Add 12 mL of buffer #1 to the pellet from Step 16. Vortex for 1 min and centrifuge for 50 sec at 380g at room temperature in the same centrifuge used in Step 12. Keep the supernatant and transfer it to a new 15 mL Falcon 2059 tube. Be careful to avoid transfer of the pelleted tissue in this step.

18. Centrifuge both 2059 tubes to pellet the sperm for 15 min at 2600g at 4°C in a Sorvall Lynx 4000 centrifuge and swinging-bucket rotor TH13-6X50 (or equivalent), with adapters to hold Falcon 2059 15-mL tubes. After centrifugation, there should be a visible gray pellet in the tubes. This pellet contains the sperm.

    From now on, the sperm should be kept on ice.

19. During the above centrifugation, prepare sucrose step gradients in four 2.5 mL thin-walled ultracentrifuge tubes. Pour gradients on ice with a cut p1000 tip. In each tube, underlay 1.7 mL of buffer #2 (2.3 M sucrose) with 0.25 mL of buffer #3 (2.5 M sucrose). First add 1.7 mL of buffer #2 to the tubes and then underlay with buffer #3 by positioning the pipette tip at the bottom of the tube and slowly expelling buffer #3 under buffer #2.

20. Following centrifugation (Step 18), discard the supernatant and thoroughly resuspend the sperm pellet in a total of 1.2 mL (600 µL in each tube) of buffer #4 (2 M sucrose) using a P1000 tip, while keeping the solution on ice.

    It is essential to fully resuspend the sperm without leaving any clumps. This can take up to 15 min.

21. Overlay each of the four sucrose step gradients with the sperm preparation (0.2–0.4 mL/gradient). Thoroughly mix the interface between the sperm and the buffer #2 with a sealed p200 tip (smash the tip with forceps) by alternately pushing the tip into the sperm and sucrose gradient and stirring the tip in the sperm layer until the sperm layer increases to about 2.5× its original depth.

22. Centrifuge the sucrose gradients in an Optima Max-E ultracentrifuge (Beckman) with a swinging-bucket rotor, such as the TLS-55 (Beckman), or equivalent for 45 min at 93,000g at 2°C.

    Red blood cells should band on the top of the 2.3 M sucrose layer. The sperm will form a pellet on the bottom of the tube. If there is still a cloudy layer above the red blood cells, remix the upper layers with a sealed p200 tip and centrifuge for another 20 min.

Cite this protocol as *Cold Spring Harb Protoc*; doi:10.1101/pdb.prot097154

23. Aspirate away the top half of the gradient containing the red blood cells. Recover the rest of the gradient with a p1000 tip and transfer it to a 15 mL 2059 Falcon tube. The bottom half of the gradient may contain sperm.

24. Resuspend each sperm pellet with 300 μL of buffer #1 by pipetting up and down with a p1000 tip (avoid the upper wall of the tube to avoid contamination by red blood cells). Transfer the sperm to the 15 mL 2059 Falcon tube from Step 23.

25. Dilute the sperm to a final volume of 12 mL with buffer #1 and mix by inversion. Pellet the sperm by centrifugation for 15 min at 3000g at 4°C in a Sorvall Lynx 4000 centrifuge and swinging-bucket rotor TH13-6X50, with adapters to hold Falcon 2059 15-mL tubes.

26. Remove the supernatant and resuspend the sperm pellet in 0.5 mL of buffer #1 plus 10 μg/mL aprotinin/leupeptin and 1 mM DTT. Transfer sperm to a 2 mL microfuge tube.

27. Adjust the volume to 1.66 mL with buffer #1 plus aprotinin/leupeptin (10 μg/mL) and DTT (1 mM). Mix thoroughly.

28. Add 34 μL of 20% Triton x-100 for a final concentration of 0.4% TX-100. Gently mix by inversion.

29. Incubate on an end-over-end rotating wheel for 45 min at 4°C.

30. Prepare four 1.5 mL microfuge tubes each containing 0.5 mL buffer #5 plus 3.0% BSA, aprotinin/leupeptin (10 μg/mL), and DTT (1 mM).

31. After the 45 min incubation, overlay each sucrose cushion from Step 30 with ¼ of the sperm preparation.

32. Centrifuge the tubes for 10 min at 750g at room temperature in a benchtop centrifuge with swinging-bucket rotor, such as IEC Centra CL2 or equivalent, with adapters to hold 1.5 mL microcentrifuge tubes. Place tubes on ice between spins.

33. Remove and discard the supernatant by aspiration and resuspend each pellet in 0.2 mL of buffer #6 plus 3.0% BSA, aprotinin/leupeptin (10 μg/mL), and DTT (1 mM). It is essential to avoid the sides of the tubes containing residual Triton X-100.

    Make sure to resuspend the sperm pellet very well in the tube.

34. Transfer the sperm chromatin to four new 1.5 mL microcentrifuge tubes and dilute to 0.7 mL with buffer #6 plus 3.0% BSA, aprotinin/leupeptin (10 μg/mL), and DTT (1 mM). Mix and centrifuge for 5 min at 750g at room temperature in the same centrifuge used in Step 32.

35. Repeat Steps 33–34.

36. Remove the supernatant. Resuspend each sperm chromatin pellet in 0.2 mL of buffer #6 plus 3.0% BSA, aprotinin/leupeptin (10 μg/mL), and DTT (1 mM). Combine and adjust the volume to around 1–1.5 mL with the same buffer. Store on ice.

37. To determine the sperm concentration, mix 3 μL of the sperm prep with 30 μL Hoechst solution and 67 μL of water to a final volume of 100 μL. Count the number of sperm using a hemocytometer with the UV/DAPI channel on an epifluorescent microscope and calculate the concentration of the undiluted preparation.

    The sperm chromatin sediment quickly. Therefore, it is essential to mix the sperm by inverting the tube just prior to taking a sample for counting as well as prior to distributing the sperm into the aliquots.

38. Dilute the sperm to a final concentration of 220,000/μL using buffer #6 plus 3.0% BSA, aprotinin/leupeptin (10 μg/mL), and DTT (1 mM), and snap freeze in liquid nitrogen in 90 μL aliquots in 1.5 mL microfuge tubes. Store at −80°C. Alternatively, do not dilute the sperm and aliquot sperm in an appropriate volume to yield aliquots of 19.8 million sperm chromatin/aliquot. Snap freeze in liquid nitrogen in 1.5 mL microfuge tubes. Store at −80°C.

    The important factor in aliquotting is not the concentration but the number of sperm chromatin per aliquot that will be used for nuclear assembly.

Preparation of LSS (Low-Speed Supernatant) Extract

For preparation of a high-speed supernatant (HSS), we harvest unfertilized eggs from six female frogs. The unfertilized eggs are arrested in metaphase II of meiosis. cytostatic factor (CSF) blocks cell cycle progression in the unfertilized eggs (Lohka et al. 1988). The eggs are first crushed at low centrifugal force which allows for release of intracellular Ca^{2+} to promote MPF destruction, facilitating entry into interphase (Murray et al. 1989), and making low-speed supernatant (LSS). The LSS is recentrifuged at high centrifugal force to make HSS.

39. Induce ovulation in six female *Xenopus laevis* frogs according to standard protocols (Lebofsky et al. 2009). Collect eggs from each frog separately into 2.5 L of 100 mM NaCl.

40. Prepare 1 L Dejelly solution, 2 L 0.5× MMR, and 1 L of ELB-sucrose. Prepare MMR and ELB-sucrose in graduated cylinders for easy pouring. Prepare 1 mL of 10 mg/mL cycloheximide. Ensure the swinging-bucket rotor is at room temperature for egg crushing.

41. Pour the majority of the 100 mM NaCl out of the egg laying bucket. Gently collect eggs from each frog into a separate 100 mL beaker, pouring the eggs down the wall of the container. Remove "bad" eggs, including white apoptotic, activated, variegated, and stringy eggs using a transfer pipette. If a batch contains more than ∼10% "bad" eggs, discard it.

42. Combine the good batches of eggs in a 2 L glass beaker and pour off as much liquid as possible into a 4 L waste beaker.

    It is important to move through the HSS and NPE protocols as quickly as possible to achieve active extracts. From dejellying eggs to flash freezing the final extract, the NPE protocol should not take longer than 6 h. It is helpful to start a timer at the point of addition of the Dejelly solution. Approximate timing is indicated at each step in ().

43. (Time: 0 min) To dejelly the eggs, add one-third of the Dejelly solution and mix thoroughly with the eggs using a plastic transfer pipette or glass Pasteur pipette. Start a timer that counts up at the moment of Dejelly solution addition. Continue stirring gently for ∼3 min. Allow the eggs to settle, pour off the Dejelly solution, add another third of the Dejelly solution, stir for 2 min, and allow the eggs to settle. After once again decanting the Dejelly solution, add the last of the Dejelly solution and stir gently until the 7 min time point. Decant as much Dejelly solution as possible.

    It is essential to thoroughly dejelly the eggs, which is seen by dramatic compaction of the eggs. The solution should become completely clear.

44. (Time: 7 min) Rinse the dejellied eggs three times with 0.6 L of 0.5× MMR each time, giving a quick stir and then allowing the eggs to settle. Decant excess MMR before the next addition.

    Pour the buffer down the side of the beaker to prevent harming the eggs, which are fragile once dejellied.

45. After decanting the last 0.5× MMR rinse, wash the eggs three times with 300 mL ELB-sucrose each time. The eggs are more stable in ELB-sucrose, which allows time to remove any bad or apoptotic eggs. The bad white eggs collect in the center of the beaker after stirring, allowing convenient removal with a plastic transfer pipette.

46. (Time: 20 min) Decant as much ELB-sucrose as possible and transfer the eggs into 15-mL Falcon 2059 tubes. Allow the eggs to settle, and aspirate excess buffer. Pack the eggs by centrifugation for 1 min at 200*g* at room temperature in a swinging-bucket rotor such as IEC Centra CL2, and aspirate excess buffer. Make a second addition of eggs to each tube, adding as many as possible, and repeat the packing centrifugation and excess buffer removal.

    A normal yield at this stage is 12 tubes of eggs.

47. After buffer removal, onto the surface of the packed eggs, add 0.5 µL of aprotinin/leupeptin stock solution (10 mg/mL) and 0.5 µL of cytochalasin B stock solution (5 mg/mL) per mL packed eggs. No mixing is necessary as the centrifugation in the next step is sufficient.

48. (Time 40 min) Crush the eggs by centrifugation for 20 min at 20,000*g* in a Sorvall Lynx 4000 centrifuge and swinging-bucket rotor TH13-6X50 in a centrifuge that is precooled to 4°C. Although the centrifuge itself is precooled, it is essential to keep the rotors at room temperature

Cite this protocol as *Cold Spring Harb Protoc*; doi:10.1101/pdb.prot097154

until the centrifugation starts to ensure the eggs are at room temperature when crushed. This is essential to drive extracts into interphase. After centrifugation, keep the extract on ice.

49. (Time 60 min) After centrifugation, transfer the tubes with crushed eggs to an ice bucket. Clean the outside of the tube with 70% ethanol and then dry the tube with a paper towel. Insert an 18 g needle about 3 mm above the mitochondrial layer (Fig. 1B, red arrow). Remove the needle and allow the LSS to drip into a 50-mL conical tube. When the yellow lipids begin to come out, stop collecting and discard the tube. It is important to include the brown layer just below the yellow lipids during collection.

50. Repeat Step 49 for each tube of crushed eggs.

Preparation of HSS (High-Speed Supernatant)

51. (Time 90 min) To the LSS, add 50 µg/mL cycloheximide, 1 mM DTT, 10 µg/mL aprotinin/leupeptin, and 5 µg/mL cytochalasin B (final concentrations).

52. Mix by inverting gently 10 times.

53. Transfer LSS to 2.5 mL thin-walled ultracentrifuge tubes. The rotor should be kept at 4°C until use. Centrifuge the extract for 90 min at 260,000g at 2°C in a swinging-bucket rotor (55,000 rpm

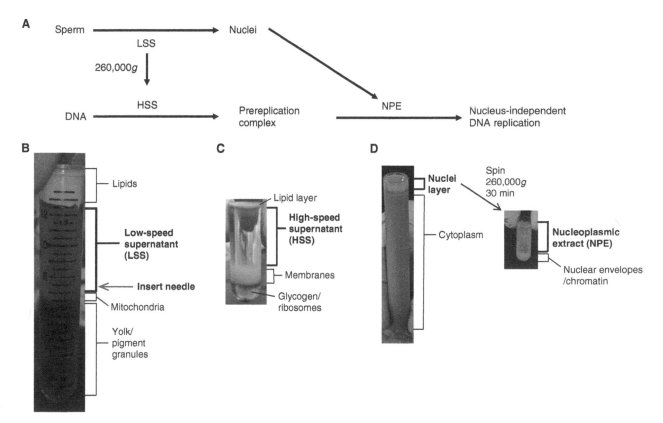

FIGURE 1. Preparation of DNA replication competent extracts from *Xenopus* eggs. (*A*) Schematic representation of nucleus-free DNA replication competent extracts, and the nucleus-free DNA replication assay. Addition of sperm to LSS extract allows for nuclei formation, which are harvested to make NPE. LSS is centrifuged at high speed to prepare HSS. For nucleus-free DNA replication, sperm chromatin or plasmid DNA is incubated in HSS to promote the assembly of prereplication complexes (pre-RCs). Addition of NPE after pre-RC assembly converts the pre-RCs to active replication forks, and replication commences. (*B*) The eggs are crushed through centrifugation to make low-speed supernatant (LSS) (Step 48). Red arrow indicates the insertion of the needle to extract the LSS. (*C*) Result of ultracentrifugation of the LSS extract to harvest high-speed supernatant (HSS) (Step 56). (*D*) Nuclei are collected after incubation of LSS with sperm chromatin. Centrifugation of the nuclei assembly reaction causes the nuclei to float and nucleoplasmic extract (NPE) can be derived from these nuclei through ultracentrifugation (Step 71).

using Optima Max-E Beckman ultracentrifuge or equivalent and swinging-bucket rotor, such as the TLS-55 Beckman or equivalent).

54. After the centrifugation, remove the top layer of lipids (Fig. 1C) by gently touching the surface with the bulb of a plastic Pasteur pipette, twisting it, and lifting it out. Aspirate the remaining lipids with an ultrathin gel-loading tip.

55. Using a p200 tip cut off a third of the way from the bottom, recover the HSS (Fig. 1C), taking care not to contaminate the preparation with the underlying membrane fraction, and transfer the extract to a new ultracentrifuge tube.

56. (Time 3.25 h) Centrifuge the extract again for 30 min at 260,000g at 2°C.

57. Remove any lipids on the surface by aspiration as before. Harvest the clear HSS layer, being careful to avoid the pellet.

58. Mix the HSS thoroughly with a pipette. Snap-freeze 33 µL aliquots in liquid nitrogen. Store aliquots at −80°C. HSS is stable for several years at −80°C.

Preparation of Nucleoplasmic Extract (NPE)

59. Begin by making LSS according to the above protocol, but use eggs collected from 20 female frogs and increase the volume of buffers used to dejelly the eggs to 2 L Dejelly solution, 4 L 0.5× MMR, and 2 L ELB-sucrose.

    To achieve maximally active extract, this protocol should be finished in less than 6 h from the time the eggs are dejellied.

60. To the harvested LSS from Step 50 (usually 35–80 mL), add 50 µg/mL cycloheximide, 1 mM DTT, 10 µg/mL aprotinin/leupeptin, 5 µg/mL cytochalasin B, and 3.3 µg/mL nocodazole (final concentrations). Mix by gentle inversion 15 times.

    If nocodazole is omitted, microtubules will bind to the nuclei (Karsenti et al. 1984) and prevent their recovery.

61. (Time: 90 min) Transfer the extract to 15 mL Falcon 2059 tubes and centrifuge for 15 min at 20,000g at 4°C in a Sorvall Lynx 4000 centrifuge and swinging-bucket rotor TH13-6X50, with adapters to hold Falcon 2059 15-mL tubes.

62. After the centrifugation, remove the pale-yellow lipid layer by gently touching the surface with the bulb of a plastic transfer pipette, twisting it, and lifting it out. Aspirate remaining lipids with a thin gel-loading tip. Remove as little extract as possible, as the dark brown layer just below the lipid layer will be helpful when harvesting the nuclei later in the protocol.

63. In one continuous, steady movement, decant the extract into a 50 mL conical tube. Leave behind about 1 mL of extract per tube to avoid transferring the black particles at the bottom of the tube, which inhibit nuclear assembly.

64. Warm the extract to room temperature by holding it for 5–10 min; this is essential for proper nuclear assembly.

65. (Time: 2 h) Add 2 mM ATP, 20 mM phosphocreatine, and 5 µg/mL creatine phosphokinase (final concentrations). Mix by gently inverting the tubes 12 times.

66. Set up the nuclear assembly reaction by pouring 4.5 mL of extract into a new 50 mL conical tube. Remove 1 mL of the 4.5 mL with a p1000 and mix it with a 90 µL aliquot of 220,000/µL demembranated sperm chromatin by pipetting up and down 15 times. Return the extract-sperm mixture to the 50 mL conical tube. Pour another 4.5 mL of extract into the same 50 mL conical tube and repeat the procedure of mixing 1 mL extract with an aliquot of sperm chromatin. Repeat this procedure until the 50 mL conical tube is filled with extract-sperm mixture or until extract is used up. If there are more than 50 mL of extract, continue the procedure with a second 50 mL tube. If a second tube is used, equalize the volume of nuclear assembly reaction in both tubes.

Cite this protocol as *Cold Spring Harb Protoc*; doi:10.1101/pdb.prot097154

67. (Time: 2.5 h) Mix the nuclear assembly reaction by inverting it gently 10 times and lay the tube on its side on the benchtop.

68. Incubate the nuclear assembly reaction for roughly 80 min from the time of the addition of the first sperm chromatin. Invert the tube a few times every 10 min, gently but thoroughly.

    Nuclear assembly should not exceed 90 min as the nuclei will eventually undergo apoptosis.

69. Check the progress of nuclear assembly at 60 min by pipetting 1 μL of the assembly reaction and 1 μL of Hoechst staining solution onto a microscope slide. Mix by gentle stirring with a pipette tip and cover the mixture with a microscope coverslip. Examine the nuclei with a fluorescence microscope using the DAPI channel. Nuclei should be ~25–30 μm in diameter. If nuclei have not reached this diameter, keep incubating until nuclei reach 25–30 μm (or more), but not longer than 90 min.

    See Troubleshooting.

70. At the 70–80 min time point, pour the assembly reaction into 13 × 100 mm glass test tubes placed inside a 15 mL Falcon 2059 tube with the void volume filled with water to prevent the glass tube from cracking during centrifugation. From now on, keep the extract on ice.

71. (Time: 4 h) Centrifuge the extract for 2.5 min at 20,000g at 4°C in a Sorvall Lynx 4000 centrifuge and swinging-bucket rotor TH13-6X50 or equivalent, with adapters to hold Falcon 2059 15-mL tubes.

    Nuclei will float to the top of the tube during centrifugation creating a clear/white viscous layer about 4–6 mm thick (Fig. 1D). This layer is separated from the golden cytoplasmic layer by a thick brown layer that is more viscous than the nuclei layer. The brown layer helps prevent contamination of the nuclei with the underlying LSS.

    See Troubleshooting.

72. Harvest nuclei using a p200 pipette with the tip cut a third of the way from the bottom. Hold the pipette at a 45° angle to more easily collect the viscous nuclei while avoiding the LSS. It is helpful to recover the nuclei by collecting along the wall of the tube while slowly rotating the tube, being careful not to disrupt the interface separating the nuclei from the cytoplasmic layer underneath. Transfer the nuclei into a thick-walled ultracentrifuge tube (or a thin-walled tube if you have more than 1.5 mL of nuclei). After no more nuclei can be removed around the edge of the tube, collect from the center of the tube while still slowly rotating. It is important to be patient and collect as much of the nuclei as possible.

73. Centrifuge the nuclei for 30 min at 260,000g at 2°C in an ultracentrifuge, such as an Optima Max-E ultracentrifuge (Beckman) or equivalent using a swinging-bucket rotor, such as the TLS-55 (Beckman) or equivalent. The tubes should be kept on ice after the centrifugation.

74. Aspirate and discard the white lipid layer (Fig. 1D) with a fine gel-loading tip, taking care not to lose much extract.

75. Harvest the clear nucleoplasmic extract with a p200 pipettor, being careful not to contaminate the extract with the insoluble pellet, which consists of nuclear envelopes and chromatin (Fig. 1D).

76. After mixing the harvested nucleoplasmic extract by pipetting up and down, snap-freeze the NPE in 20 μL aliquots with liquid nitrogen. Expected yield can vary from 100 to 2 mL. The aliquots are stored at −80°C and are stable for several years.

    See Troubleshooting.

TROUBLESHOOTING

Problem (Step 69): Nuclei grow but then disappear completely.
Solution: The extract underwent apoptosis due to contamination with and lysis of mitochondria.

Problem (Step 69): No nuclei grew as visualized by epifluorescence microscopy.

Solution: One possible reason is poor quality sperm. We have found that it is critical to completely remove the Triton X-100 from the sperm chromatin. Residual Triton prevents the formation of nuclei during the assembly reaction.

Problem (Step 71): Only small nuclear layers form.

Solution: There are several possible factors that can be considered in the next NPE preparation.

- The sperm are of poor quality or the concentration is too low or too high.
- Too many mitochondria may have been collected from the respun LSS (Step 63), which can impede growth.
- The extract must be allowed to come to room temperature (Step 64) before the nuclear assembly reaction can begin. Starting the assembly too quickly will slow the growth of the nuclei.

Problem (Step 71): Nuclei grow but no nuclear layer forms to allow harvesting.

Solution: It is possible that nocodozole was left out. Microtubule growth prevents the nuclei from floating and forming a layer.

Problem (Step 76): There are many steps that can affect the volume and the quality of an NPE preparation. Below we outline several pitfalls of the NPE preparation, and how to avoid them.

Solutions: Consider the following.

- It is important to move through the procedures as quickly as possible. The NPE preparation should not exceed 6 h to maintain the activity of the extract.
- It is also important that the centrifuge rotor is at room temperature for the egg-crushing step.
- The egg quantity and quality are major factors in determining the quantity and quality of an NPE preparation. It is important to remove as many of the apoptotic eggs as possible prior to packing of the eggs.
- Avoid mitochondrial contamination during the NPE preparation, which starts with the harvesting of the LSS (Step 49). High levels of mitochondrial contamination can lead to nuclear membrane fragmentation during nuclear assembly by the release of cytochrome C (Kluck et al. 1997). This will prevent the formation of the NPE layer.

DISCUSSION

The extracts made using these protocols allow for in vitro nucleus-free replication of fully chromatinized DNA templates including small plasmids, which is outlined in Figure 1A (Walter et al. 1998; Arias and Walter 2004). The high-speed supernatant (HSS) allows DNA licensing by loading of pre-replicative complexes (pre-RCs) randomly on the DNA template. Demembranated sperm chromatin is used to assemble nuclei that are harvested to make concentrated nucleoplasmic extract (NPE) that drives the initiation of DNA replication by supplying high concentrations of Cdk2-Cyclin E (Prokhorova et al. 2003). *Xenopus* egg extracts contain the entirety of the soluble egg proteome including all of the DNA replication and repair machinery. This extract system has been successfully used to study many aspects of DNA metabolism including DNA replication, mismatch repair (Kawasoe et al. 2016), replication-coupled DNA interstrand cross-link repair (Räschle et al. 2008), replication-coupled DNA-protein cross-link repair (Duxin et al. 2014), DNA replication termination (Dewar et al. 2015), as well as many other processes (Lebofsky et al. 2009; Knipscheer et al. 2012). How

Cite this protocol as *Cold Spring Harb Protoc;* doi:10.1101/pdb.prot097154

Xenopus extracts and the NPE system have contributed to our detailed understanding of DNA replication and repair is reviewed elsewhere (Hoogenboom et al. 2017).

RELATED INFORMATION

In our experience, almost any preparation of HSS is sufficient to support DNA licensing. However, NPE preparations can vary significantly in volume and in their ability to initiate DNA replication. Highly active NPE will complete replication of 2.5 ng/µL (final concentration) of supercoiled plasmid DNA in less than 15 min. This is a benchmark against which new NPE preparations should be compared.

RECIPES

Buffer X (10×)

Reagent	Quantity (20 mL)	Final concentration
HEPES-KOH (1 M, pH 7.4–7.6)	2 mL	100 mM
KCl (1 M)	1.6 mL	80 mM
NaCl (5 M)	0.6 mL	150 mM
MgCl$_2$ (1 M)	1 mL	50 mM
EDTA (0.5 M)	0.4 mL	10 mM
dH$_2$O	14.4 mL	

Make fresh on the day of extract preparation.

Creatine Phosphokinase Solution

Dissolve creatine phosphokinase (5 mg/mL; Sigma-Aldrich C3755) in 50 mM NaCl, 50% glycerol, and 10 mM HEPES-KOH (pH 7.5). Store for 2–6 mo at −20°C. Do not freeze at −80°C.

Dejelly Solution

Reagent	Quantity	Final concentration
L-cysteine-HCl (Fisher ICN10144601)	44 g	2.2%

Use dH$_2$O to bring volume to 2 L, and adjust pH to 7.7 with 10 M NaOH. Make fresh on the day of extract preparation.

Egg Lysis Buffer (ELB-Sucrose)

Reagent	Quantity (1 L)	Final concentration
Sucrose	85.6 g	0.25 M
ELB salts (10×) <R>	100 mL	1×
DTT (Bio-Rad 161-0611)	154 mg	0.154 mg/mL
Cycloheximide (Calbiochem 239763)	50 mg	0.05 mg/mL

Use ultrapure sucrose (≥99%; Invitrogen 15503-022). Adjust volume to 1 L with dH$_2$O. Make fresh on the day of extract preparation.

Egg Lysis Buffer Salts (ELB Salts; 10×)

Reagent	Quantity (500 mL)	Final concentration
MgCl$_2$ (1 M)	12.5 mL	25 mM
KCl	18.64 g	500 mM
HEPES	11.91 g	100 mM

Add water to a volume of 500 mL. Adjust pH to 7.7 with KOH. Filter-sterilize, and store at 4°C.

Hoechst Stock Solution

Reagent	Quantity (~400 µL)	Final concentration
Sucrose (1 M)	80 µL	200 mM
HEPES-KOH (1 M, pH 7.6)	4 µL	10 mM
Formaldehyde (37%)	80 µL	7.4%
Hoechst (1.0 mg/mL)	3.2 µL	8 µg/mL
H$_2$O	232 µL	

Dilute 10 mg/mL stock of Hoechst in water prior to use above. Store final solution at room temperature protected from light.

Marc's Modified Ringer's (MMR) for Xenopus *(20×)*

Reagent	Quantity (1 L)	Final concentration
NaCl	116.8 g	2 M
KCl	3 g	40 mM
MgSO$_4$·7H$_2$O	2.4 g	10 mM
CaCl$_2$·2H$_2$O	7.34 g	50 mM
EDTA	0.58 g	2 mM
HEPES	23.8 g	100 mM

Add water to 1 L. Adjust pH of solution to 7.8 with KOH. Filter-sterilize, and store at 4°C.

Sucrose Buffers for Preparation of Demembranated Sperm Chromatin

Reagent	Buffer #1 (0.2 M sucrose)	Buffer #2 (2.3 M sucrose)	Buffer #3 (2.5 M sucrose)	Buffer #4 (2.0 M sucrose)	Buffer #5 (0.5 M sucrose)
Buffer X (10×) <R>	10 mL	1 mL	1 mL	1 mL	1 mL
Sucrose	6.84 g	7.87 g	8.56 g	6.84 g	1.71
BSA					300 mg
H$_2$O	to 100 mL	to 10 mL	to 10 mL	to 10 mL	to 10 mL

For Buffer #6 (0.2 M sucrose), add 300 mg of BSA to 10 mL of Buffer #1. Use BSA from Sigma-Aldrich (A7906). Do not use other types of BSA, as they can ruin the sperm preparation. For all buffers, use ultrapure sucrose (≥99%; Invitrogen 15503-022). For Buffers #5 and #6 and for 5 mL of Buffer #1, include 10 µg/mL aprotinin (Roche 11583794001), 10 µg/mL leupeptin (Roche 11529048001), and 1 mM DTT (Bio-Rad 161-0611). Make all buffers fresh on the day of extract preparation.

Cite this protocol as *Cold Spring Harb Protoc*; doi:10.1101/pdb.prot097154

REFERENCES

Arias EE, Walter JC. 2004. Initiation of DNA replication in *Xenopus* egg extracts. *Front Biosci* **1:** 3029–3045.

Dewar JM, Budzowska M, Walter JC. 2015. The mechanism of DNA replication termination in vertebrates. *Nature* **525:** 345–350.

Duxin JP, Dewar JM, Yardimci H, Walter JC. 2014. Repair of a DNA-protein crosslink by replication-coupled proteolysis. *Cell* **159:** 346–357.

Hoogenboom WS, Klein Douwel D, Knipscheer P. 2017. *Xenopus* egg extract: A powerful tool to study genome maintenance mechanisms. *Dev Biol* **428:** 300–309.

Karsenti E, Newport J, Hubble R, Kirschnre M. 1984. Interconversion of metaphase and interphase microtubule arrays, as studied by the injection of centrosomes and nuclei into *Xenopus* eggs. *J Cell Biol* **98:** 1730–1745.

Kawasoe Y, Tsurimoto T, Nakagawa T, Masukata H, Takahashi TS. 2016. MutSα maintains the mismatch repair capability by inhibiting PCNA unloading. *Elife* **5:** e15155.

Kluck RM, Bossy-Wetzel E, Green DR, Newmeyer DD. 1997. The release of cytochrome c from mitochondria: A primary site for Bcl-2 regulation of apoptosis. *Science* **275:** 1132–1136.

Knipscheer P, Räschle M, Schärer OD, Walter JC. 2012. Replication-coupled DNA interstrand cross-link repair in *Xenopus* egg extracts. *Methods Mol Biol* **920:** 221–243.

Lebofsky R, Takahashi T, Walter JC. 2009. DNA replication in nucleus-free *Xenopus* egg extracts. *Methods Mol Biol* **521:** 229–252.

Lohka MJ, Hayes MK, Maller JL. 1988. Purification of maturation-promoting factor, an intracellular regulator of early mitotic events. *Proc Natl Acad Sci* **85:** 3009–3013.

Murray AW, Solomon MJ, Kirschner MW. 1989. The role of cyclin synthesis and degradation in the control of maturation promoting factor activity. *Nature* **339:** 280–286.

Prokhorova TA, Mowrer K, Gilbert CH, Walter JC. 2003. DNA replication of mitotic chromatin in *Xenopus* egg extracts. *Proc Natl Acad Sci* **100:** 13241–13246.

Räschle M, Knipscheer P, Enoiu M, Angelov T, Sun J, Griffith JD, Ellenberger TE, Schärer OD, Walter JC. 2008. Mechanism of replication-coupled DNA interstrand crosslink repair. *Cell* **134:** 969–980.

Walter J, Sun L, Newport J. 1998. Regulated chromosomal DNA replication in the absence of a nucleus. *Mol Cell* **1:** 519–529.

Imaging Methods in *Xenopus* Cells, Embryos, and Tadpoles

Lance A. Davidson[1,3] and Laura Anne Lowery[2,3]

[1]*University of Pittsburgh, Department of Bioengineering, Pittsburgh, Pennsylvania 15260, USA;* [2]*Boston University School of Medicine, Boston Medical Center, Boston, Massachusetts 02118, USA*

Xenopus is an excellent vertebrate model system ideally suited for a wide range of imaging methods designed to investigate cell and developmental biology processes. The individual cells of *Xenopus* are much larger than those in many other vertebrate model systems, such that both cell behavior and subcellular processes can more easily be observed and resolved. Gene function in *Xenopus* can be manipulated and visualized using a variety of approaches, and the embryonic fate map is stereotypical, such that microinjections can target specific tissues and cell types during development. Tissues, organotypic explants, and individual cells can also be mounted in stable chambers and cultured easily in simple salt solutions without cumbersome environmental controls. Furthermore, *Xenopus* embryonic tissues can be microsurgically isolated and shaped to expose cell behaviors and protein dynamics in any regions of the embryo to high-resolution live-cell imaging. The combination of these attributes makes *Xenopus* a powerful system for understanding cell and developmental processes as well as disease mechanisms, through quantitative analysis of protein dynamics, cell movements, tissue morphogenesis, and regeneration. Here, we introduce various methods, of both fixed and living tissues, for visualizing *Xenopus* cells, embryos, and tadpoles. Specifically, we highlight protocol updates for whole-mount in situ hybridization and immunofluorescence, as well as robust live imaging approaches including methods for optimizing the time-lapse imaging of whole embryos and explants.

BACKGROUND

For the last century, advances in cell and developmental biology have been propelled by leaps in imaging and microscopy technologies. Our knowledge of vertebrate development, generally in amphibians, and more specifically in *Xenopus*, has benefited greatly from these leaps. For instance, methods for lineage tracking, unique to amphibians (Vogt 1929; Keller 1975, 1976; Lane and Sheets 2002), have been central to our foundational understanding of fate, commitment, and specification. The emergence of electron microscopy provides unparalleled descriptions of cytoarchitecture that form the basis of our understanding of gastrulation (Keller 1980; Winklbauer et al. 1992) and neurulation (Schroeder 1970). Key arrangements of germ layers, organs, and cells in the embryo and larva were exposed in an atlas laid out after use of plastic embedding and sectioning (Hausen and Riebesell 1991). Fluorescence imaging of fluorophore-conjugated dextrans (e.g., Keller and Tibbetts 1989) followed by the introduction of confocal microscopy and fluorescent proteins (e.g., Wallingford et al. 2000) afforded advances in live-cell imaging in a range of preparations from single cells to organotypic explants to whole embryos. New imaging modalities, from magnetic resonance imaging (MRI; Jacobs and Fraser 1994) to X-ray phase contrast microscopy (Moosmann et al. 2013) to optical coherence tomography (OCT; Boppart et al. 1996) are routinely shown in

---

[3]Correspondence: lad43@pitt.edu; lalowery@bu.edu

Cite this introduction as *Cold Spring Harb Protoc*; doi:10.1101/pdb.top105627

developing *Xenopus*. Advanced imaging tools leveraged with fate maps and stage descriptions of Nieuwkoop and Faber (1967) have provided a rich framework for posing and testing mechanistic hypotheses on the genetic regulation and molecular pathways that were subsequently uncovered and that continue to be the subject of lively investigation.

Efforts to decrypt the cell and molecular mechanisms responsible for patterning, morphogenesis, and organogenesis have been made possible by advances in sample preparation. Whole-mount RNA in situ hybridization has exposed gene expression patterns, whereas the creation of synthetic fluorophores and discovery and refinement of fluorescent proteins has made possible localization and protein dynamics in vivo. Whole-mount RNA in situ hybridization methods (Harland 1991) were rigorously adopted by the field and continue to advance, revealing patterns of gene expression, making possible the construction of gene regulatory networks. Fluorophores combined with antibody technology and fluorescent proteins have revealed the effector networks that integrate these gene regulatory networks with cell–cell communication and microenvironmental cues to guide cells to their targets and effect the construction of physiologically functional organs.

The protocols in this collection reflect newer techniques and detailed methods of sample preparation that have not been well-covered in past collections. In each of the sections below, we will also recommend previously published methods for imaging and sample preparation as these remain excellent resources for improving technique and troubleshooting methods.

IMAGING OF FIXED EMBRYOS AND TADPOLES

Whole-mount, half-mount, and sectioned preparations of fixed *Xenopus* embryos provide access for high-resolution studies of gene expression patterns, tissue as well as subcellular protein localization, cell shape, and tissue microanatomy. Whole-mount RNA in situ hybridization adapted northern blot techniques for use in a permeabilized *Xenopus* embryo (Harland 1991) and was further adapted to deposit tyramide fluorescent substrates, allowing single-cell gene expression studies (Davidson and Keller 1999; Vize et al. 2009). In the first protocol, new techniques are introduced to improve methods to visualize gene expression patterns simultaneously with protein localization. Recent years have seen development of novel technologies to enhance fluorescent RNA in situ methods using RNAscope (ACD RNAscope) and Hybridization Chain Reaction (HCR3.0, Molecular Instruments). In Protocol 1: Whole-Mount RNA In Situ Hybridization and Immunofluorescence of *Xenopus* Embryos and Tadpoles (Willsey 2021), Willsey et al. (2020) includes an update on the Harland protocol as well as a protocol for HCR and whole-mount immunofluorescence. The adaptations of these protocols to late stages of *Xenopus* are particularly helpful for characterizing the molecular anatomy of the larval brain and shows great potential for other organ systems where efforts to dissect tissue–tissue interactions are aided by preserving 3D context.

High-resolution imaging of the cell and subcellular structures are enhanced when the tissues are fixed and sectioned to preserve protein–protein interactions and protein localization. Such preparations can eliminate the need for antibodies to penetrate deep tissues and reduce background fluorescence that accompanies thick-tissue imaging. In Protocol 2: Cryosectioning and Immunostaining of *Xenopus* Embryonic Tissues (Ossipova and Sokol 2021), Ossipova and Sokol provide an update on the methods of fixation, embedding, sectioning, and immunostaining that best preserve the delicate epitopes of *Xenopus* early embryos. Immunofluorescence staining of thin frozen sections remains a central tool in visualizing the location and status of endogenous proteins in *Xenopus* embryos.

IMAGING OF LIVING TISSUES

One of the greatest benefits of using *Xenopus* is that it is particularly ideal for live imaging of dynamic processes at multiple scales, ranging from whole embryos to individual cells to even subcellular

 Cite this introduction as *Cold Spring Harb Protoc*; doi:10.1101/pdb.top105627

processes (Joshi et al. 2012). Embryonic development is external, the embryos and tissues can be cultured at room temperature without incubation nor CO_2, and the cells are quite large compared to other vertebrates. Thus, *Xenopus* is a powerful model system for in vivo time-lapse studies of cell and developmental biology. Here, we summarize the various types of protocols that have been used for imaging living tissues in *Xenopus*.

Previously published protocols have described the strengths of using *Xenopus* for low-magnification imaging of whole embryos, which, in combination with manipulation of gene functions, have provided numerous insights into the control of morphogenetic processes such as gastrulation and neural tube closure (Wallingford 2010; Danilchik 2011). Moreover, previous protocols have detailed methods for high-magnification time-lapse imaging using confocal microscopy (see Kieserman et al. 2010; Kim and Davidson 2013). Protocol 5: Chambers for Culturing and Immobilizing *Xenopus* Embryos and Organotypic Explants for Live Imaging (Chu and Davidson 2021) describes the preparation of special chambers for immobilizing *Xenopus* embryos and embryonic explants for live-cell and tissue imaging.

There are many different cell and tissue types that can be imaged in living intact *Xenopus* embryos and tadpoles. For example, Sec. 14, Protocol 4: Imaging the Dynamic Branching and Synaptic Differentiation of *Xenopus* Optic Axons In Vivo (Santos et al. 2020) demonstrates how individual fluorophore-labeled neurons can be imaged in living tadpoles to visualize optic axonal arbors and formation of neuronal circuits. Not only can confocal microscopy be used to image individual cells during development, but fluorescent fusion proteins and molecules can also be imaged to follow their dynamics. For example, fluorescent probes for Rho and Rac GTPase activity have been used while imaging large blastomeres of the early *Xenopus* embryo to uncover new insights into epithelial cell biology (Stephenson and Miller 2017) and collective migration (Ichikawa et al. 2020). Furthermore, Sec. 14, Protocol 3: Whole-Brain Calcium Imaging in Larval *Xenopus* (Offner et al. 2020) describes how calcium imaging can be performed in larvae to provide information regarding the spatiotemporal activity of the entire brain.

Although live imaging of *Xenopus* embryos can be illuminating, one caveat is that the yolk opacity of *Xenopus* cells can sometimes interfere with traditional types of deep imaging in intact embryos. However, almost any type of *Xenopus* cell and/or tissue can also easily be cultured ex vivo. For example, high-resolution live imaging of mucociliary organoids from *Xenopus* have been used to investigate mechanisms of mucociliary signaling and morphogenesis (Walentek 2018; Kang and Kim 2020; Kim et al. 2020). Protocol 3: Live Imaging of Cytoskeletal Dynamics in Embryonic *Xenopus laevis* Growth Cones and Neural Crest Cells (Erdogan et al. 2020) describes how live imaging of fluorophore-tagged cytoskeletal proteins allows for monitoring the dynamic nature of cytoskeleton components in cultured embryonic neuronal growth cones and neural crest cells.

Finally, although the majority of live imaging studies of *Xenopus* have used confocal microscopy, OCT is another powerful method for live imaging tissue structures in vivo. OCT is a rapid, label-free imaging modality (Boppart et al. 1996) that can acquire 2D and 3D data in real time to assess cardiac and facial structures, as described in Protocol 4: *Xenopus* Tadpole Craniocardiac Imaging Using Optical Coherence Tomography (Deniz et al. 2021).

Over the past several decades, *Xenopus* cells, embryos, and tadpoles have been used in imaging studies to dissect key processes in cell and developmental biology. As new imaging techniques become available, the many benefits of the *Xenopus* model system (external and rapid development, fate mapping and gene manipulation, easy culture conditions) will ensure that *Xenopus* will remain an ideal system for continued investigation of the molecular and cell biological mechanisms governing cell and developmental processes.

ACKNOWLEDGMENTS

We thank the members of the Lowery and Davidson labs for helpful discussions. We also thank the National *Xenopus* Resource (RRID:SCR-013731) and Xenbase (RRID:SCR-003280) for their invaluable support to the model organism community. L.A.L. is funded by National Institutes of Health (NIH) grant R01 MH109651 and the American Cancer Society's Ellison Foundation Research Scholar Grant (RSG-16–144-01-CSM). L.A.D. is funded by the *Eunice Kennedy Shriver* Institute of Child Health and Human Development at the NIH (R01 HD044750).

REFERENCES

Boppart SA, Bouma BE, Brezinski ME, Tearney GJ, Fujimoto JG. 1996. Imaging developing neural morphology using optical coherence tomography. *J Neurosci Methods* **70:** 65–72. doi:10.1016/S0165-0270 (96)00104-5

Chu C-W, Davidson LA. 2021. Chambers for culturing and immobilizing *Xenopus* embryos and organotypic explants for live imaging. *Cold Spring Harb Protoc* doi:10.1101/pdb.prot107649

Danilchik MV. 2011. Manipulating and imaging the early *Xenopus laevis* embryo. *Methods Mol Biol* **770:** 21–54. doi:10.1007/978-1-61779-210-6_2

Davidson LA, Keller RE. 1999. Neural tube closure in *Xenopus laevis* involves medial migration, directed protrusive activity, cell intercalation and convergent extension. *Development* **126:** 4547–4556. doi:10.1242/dev.126.20.4547

Deniz E, Mis EK, Lane M, Khokha MK. 2021. *Xenopus* tadpole craniocardiac imaging using optical coherence tomography. *Cold Spring Harb Protoc* doi:10.1101/pdb.prot105676

Erdogan B, Bearce EA, Lowery LA. 2020. Live imaging of cytoskeletal dynamics in embryonic *Xenopus laevis* growth cones and neural crest cells. *Cold Spring Harb Protoc* doi:10.1101/pdb.prot104463

Harland RM. 1991. In situ hybridization: an improved whole-mount method for *Xenopus* embryos. *Methods Cell Biol* **36:** 685–695. doi:10.1016/S0091-679X(08)60307-6

Hausen P, Riebesell M. 1991. *The early development of* Xenopus laevis: *an atlas of the histology*. Springer Verlag, New York.

Ichikawa T, Stuckenholz C, Davidson LA. 2020. Non-junctional role of Cadherin3 in cell migration and contact inhibition of locomotion via domain-dependent, opposing regulation of Rac 1. *Sci Rep* **10:** 17326.

Jacobs RE, Fraser SE. 1994. Magnetic resonance microscopy of embryonic cell lineages and movements. *Science* **263:** 681–684. doi:10.1126/science.7508143

Joshi SD, Kim HY, Davidson LA. 2012. Microscopy tools for quantifying developmental dynamics in *Xenopus* embryos. *Methods Mol Biol* **917:** 477–493. doi:10.1007/978-1-61779-992-1_27

Kang HJ, Kim HY. 2020. Mucociliary epithelial organoids from *Xenopus* embryonic cells: generation, culture and high-resolution live imaging. *J Vis Exp* **161:** e61604. doi:10.3791/61604

Keller RE. 1975. Vital dye mapping of the gastrula and neurula of *Xenopus laevis*. I. Prospective areas and morphogenetic movements of the superficial layer. *Dev Biol* **42:** 222–241. doi:10.1016/0012-1606(75)90331-0

Keller RE. 1976. Vital dye mapping of the gastrula and neurula of *Xenopus laevis*. II. Prospective areas and morphogenetic movements of the deep layer. *Dev Biol* **51:** 118–137. doi:10.1016/0012-1606(76)90127-5

Keller R. 1980. The cellular basis of epiboly: an SEM study of deep-cell rearrangement during gastrulation in *Xenopus laevis*. *Development* **60:** 201–234. doi:10.1242/dev.60.1.201

Keller R, Tibbetts P. 1989. Mediolateral cell intercalation in the dorsal, axial mesoderm of *Xenopus laevis*. *Dev Biol* **131:** 539–549. doi:10.1016/S0012-1606(89)80024-7

Kieserman EK, Lee C, Gray RS, Park TJ, Wallingford JB. 2010. High-magnification in vivo imaging of *Xenopus* embryos for cell and developmental biology. *Cold Spring Harb Protoc* doi:10.1101/pdb.prot5427

Kim HY, Davidson LA. 2013. Investigating morphogenesis in *Xenopus* embryos: imaging strategies, processing, and analysis. *Cold Spring Harb Protoc* **2013:** 298–304. doi:10.1101/pdb.top073890

Kim HY, Jackson TR, Stuckenholz C, Davidson LA. 2020. Tissue mechanics drives regeneration of a mucociliated epidermis on the surface of *Xenopus* embryonic aggregates. *Nat Comm* **11:** 1–10.

Lane MC, Sheets MD. 2002. Rethinking axial patterning in amphibians. *Dev Dyn* **225:** 434–447. doi:10.1002/dvdy.10182

Moosmann J, Ershov A, Altapova V, Baumbach T, Prasad MS, LaBonne C, Xiao X, Kashef J, Hofmann R. 2013. X-ray phase-contrast in vivo microtomography probes new aspects of *Xenopus* gastrulation. *Nature* **497:** 374–377. doi:10.1038/nature12116

Nieuwkoop PD, Faber J. 1967. *Normal tables of* Xenopus laevis *(Daudin): a systematic and chronological survey of the development from the fertilized egg till the end of morphogenesis*. North-Holland Biomedical Press, Elsevier, Amsterdam.

Offner T, Daume D, Weiss L, Hassenklover T, Manzini I. 2020. Whole-brain calcium imaging in larval *Xenopus*. *Cold Spring Harb Protoc* doi:10.1101/pdb.prot106815

Ossipova O, Sokol SY. 2021. Cryosectioning and immunostaining of *Xenopus* embryonic tissues. *Cold Spring Harb Protoc* doi:10.1101/pdb.prot107151

Santos RA, del Rio R Jr, Cohen-Cory S. 2020. Imaging the dynamic branching and synaptic differentiation of *Xenopus* optic axons in vivo. *Cold Spring Harb Protoc* doi:10.1101/pdb.prot106823

Schroeder TE. 1970. Neurulation in Xenopus laevis. An analysis and model based upon light and electron microscopy. *J Embryol Exp Morphol* **23:** 427–462.

Stephenson RE, Miller AL. 2017. Tools for live imaging of active Rho GTPases in *Xenopus*. *Genesis* **55:** 10.1002/dvg.22998. doi:10.1002/dvg.22998

Vize PD, McCoy KE, Zhou X. 2009. Multichannel wholemount fluorescent and fluorescent/chromogenic in situ hybridization in *Xenopus* embryos. *Nat Protoc* **4:** 975–983. doi:10.1038/nprot.2009.69

Vogt W. 1929. Gestaltungsanalyse am Amphibienkeim mit ortlicher Vitalfarbung II. Teil: Gastrulation und Mesodermbildung bei Urodelen und Anuren. *Wilhelm Roux'Arch Entwicklungsmech Org* **120:** 384–706. doi:10.1007/BF02109667

Walentek P. 2018. Manipulating and analyzing cell type composition of the *Xenopus* mucociliary epidermis. *Methods Mol Biol* **1865:** 251–263. doi:10.1007/978-1-4939-8784-9_18

Wallingford JB. 2010. Low-magnification live imaging of *Xenopus* embryos for cell and developmental biology. *Cold Spring Harb Protoc* doi:10.1101/pdb.prot5425

Wallingford JB, Rowning BA, Vogeli KM, Rothbacher U, Fraser SE, Harland RM. 2000. Dishevelled controls cell polarity during *Xenopus* gastrulation. *Nature* **405:** 81–85. doi:10.1038/35011077

Willsey HR. 2021. Whole-mount RNA in situ hybridization and immunofluorescence of *Xenopus* embryos and tadpoles. *Cold Spring Harb Protoc* doi:10.1101/pdb.prot105635

Willsey HR, Xu Y, Everitt A, Dea J, Exner CR, Willsey AJ, Harland RM. 2020. The neurodevelopmental disorder risk gene *DYRK1A* is required for ciliogenesis and control of brain size in *Xenopus* embryos. *Development* **147:** dev189290. doi:10.1242/dev.189290

Winklbauer R, Selchow A, Nagel M, Angres B. 1992. Cell interaction and its role in mesoderm cell migration during *Xenopus* gastrulation. *Dev Dyn* **195:** 290–302. doi:10.1002/aja.1001950407

Cite this introduction as *Cold Spring Harb Protoc*; doi:10.1101/pdb.top105627

Whole-Mount RNA In Situ Hybridization and Immunofluorescence of *Xenopus* Embryos and Tadpoles

Helen Rankin Willsey[1]

Department of Psychiatry and Behavioral Sciences, Weill Institute for Neurosciences, University of California, San Francisco, San Francisco, California 94143, USA

A major advantage of experimentation in *Xenopus* is the ability to query the localization of endogenous proteins and RNAs in situ in the entire animal during all of development. Here I describe three variations of staining to visualize mRNAs and proteins in developing *Xenopus* embryos and tadpoles. The first section outlines a traditional colorimetric staining for mRNAs that is suitable for all stages of development, and the second extends this protocol for fluorescence-based detection for higher spatial and quantitative resolution. The final section details detection of proteins by immunofluorescence, optimized for tadpole stages but widely applicable to others. Finally, optimization strategies are provided.

MATERIALS

It is essential that you consult the appropriate Material Safety Data Sheets and your institution's Environmental Health and Safety Office for proper handling of equipment and hazardous materials used in this protocol.

RECIPES: Please see the end of this protocol for recipes indicated by <R>. Additional recipes can be found online at http://cshprotocols.cshlp.org/site/recipes.

Reagents

Reagents required for the colorimetric RNA in situ hybridization procedure only (Steps 1–34)
  Acetic anhydride (Sigma-Aldrich 320102)
  Alkaline phosphatase buffer with tetramisole hydrochloride <R>
  Anti-digoxygenin-AP antibody (Sigma-Aldrich 11093274910)
  Bleaching solution for in situ hybridization <R>
  BMB blocking solution <R>
  BM purple (Sigma-Aldrich 11442074001)
  Bouin's fixative for in situ hybridization <R>
  Buffered ethanol <R>
  Digoxygenin-11-UTP-labeled RNA probes, in-vitro-transcribed (see Sive et al. 2007a)
    Both an experimental probe and a control probe (either sense-transcribed, or a probe with a known, very specific pattern) should be used.

  Hybridization buffer for colorimetric in situ hybridization <R>
  Maleic acid buffer (MAB, 10×) <R>

---

[1]Correspondence: helen.willsey@ucsf.edu

MEMFA (1×) <R>

Methanol (Fisher A4544)

Molten agarose (1%) (optional; see Step 33)

Paraformaldehyde (4% in PTw)

Dilute 20% paraformaldehyde to 4% in PTw on staining day.

Paraformaldehyde (20%) <R>

PBS with Tween 20 (PTw)

Add 0.1% Tween 20 to 1× PBS. Store at room temperature.

Phosphate-buffered saline (PBS) (10×; pH 7.4) <R>

Proteinase K (10 mg/mL; Fisher EO0491)

Proteinase K (10 µg/mL in PTw)

RNase A (20 µg/mL; Fisher EN0531)

RNase T1 (10 µg/mL; Fisher EN0542)

Sodium chloride-sodium citrate buffer (SSC; 20×) <R>

Triethanolamine (0.1 M, pH 7-8; Sigma-Aldrich T1502)

Xenopus tadpoles or embryos of any stage

Reagents required for the fluorescent RNA in situ hybridization by hybridization chain reaction (HCR) procedure only (Steps 35–46)

Acetic anhydride (Sigma-Aldrich 320102)

Amplification buffer <R>

HCR probes (custom-designed from Molecular Technologies)

HCR hairpins (standard-designed from Molecular Technologies)

Hybridization buffer for fluorescent in situ hybridization (30%) <R>

MEMFA (1×) <R>

Methanol (Fisher A4544)

Molten agarose (1%) (optional; see Step 46)

Paraformaldehyde (4% in PTw)

Dilute 20% paraformaldehyde to 4% in PTw on staining day.

Paraformaldehyde (20%) <R>

PBS (10×; pH 7.4) <R>

PBS with Tween 20 (PTw)

Add 0.1% Tween 20 to 1× PBS. Store at room temperature.

Probe wash buffer (30%) <R>

Proteinase K (10 mg/mL; Fisher EO0491)

Proteinase K (10 µg/mL in PTw)

SSCT (5×) <R>

Triethanolamine (0.1 M, pH 7–8) (Sigma-Aldrich T1502)

Xenopus tadpoles or embryos of any stage

Reagents required for the immunofluorescence procedure only (Steps 47–58)

Bleaching solution for immunofluorescence <R>

CAS-Block (10% in PBT)

CAS-Block (Invitrogen 00-8120)

Gentamicin (optional; see Step 57)

Molten agarose (1%) (optional; see Step 58)

Paraformaldehyde (4% in PBS)

Dilute 20% paraformaldehyde to 4% in PBS on staining day.

Paraformaldehyde (20%)

PBS (10×; pH 7.4) <R>

TABLE 1. Common antibodies compatible with this immunostaining procedure in stage 46 tadpoles along with relevant information

Antigen	Labels	Host	Company	Product number	Dilution
β-tubulin	Neurons, microtubules	Mouse	DSHB	E7	1:100
PCNA	Cells in S phase	Mouse	Life Technologies	133900	1:50
pHH3	Cells in M phase	Rabbit	Millipore	06-570	1:250
α-tubulin	Spindles, microtubules	Mouse	DSHB	12G10	1:100
Ac-α-tubulin	Cilia	Mouse	Sigma-Aldrich	T6793	1:700
Cleaved Caspase 3	Cell death	Rabbit	BD Pharmingen	559565	1:250

PBS with Triton X-100 (PBT)

Add 0.1% Triton X-100 to 1× PBS. Store at room temperature.

Primary antibodies

Potential primary antibodies are listed in Table 1.

Secondary antibodies compatible with the primary antibodies (fluorescently conjugated) *Xenopus* tadpoles

Equipment

Aluminum foil
Equipment for basket format (high-throughput sample processing) only
   Baskets (Fig. 1A)

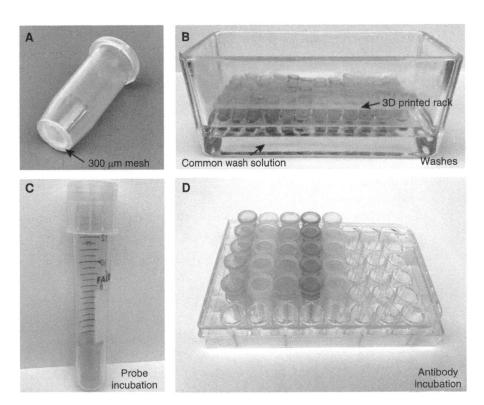

FIGURE 1. Basket format for higher-throughput staining. (*A*) Sample basket made from a 1.7-mL microcentrifuge tube (with cap and bottom cut off) with 300-µm mesh melted to the bottom. (*B*) Wash setup, with color-coded baskets arranged in a 3D-printed rack within a glass staining dish. Samples share a common wash buffer. (*C*) Probe incubation setup, in baskets within 15-mL round-bottom culture tubes with 500 µL of probe solution. (*D*) Antibody incubation setup within a 48-well culture plate with 500 µL of antibody solution in each well. These incubation setups allow for each sample to experience a different probe or antibody, if desired, and minimize the total volume required.

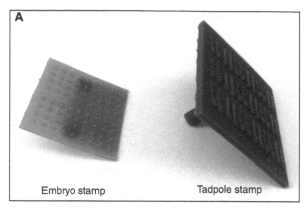
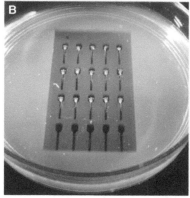

FIGURE 2. Imaging in agarose wells. (*A*) 3D-printed stamps for positioning embryos (*left*) or tadpoles (*right*). Agarose wells can be made by pressing these into molten agarose in a culture dish during cooling. (*B*) Tadpoles arranged into agarose wells. (*C*) Imaging animals using an upright stereoscope in agarose wells.

Construct the baskets from 1.7-mL tubes and 300-μm nylon mesh (Spectra Mesh 146487); for dissected tissue, use finer mesh. For basket-making instructions, see Sive et al. (2007b).

Basket racks (Fig. 1B)

Prepare the basket racks using 3D printer files available at willseyfroggers.org/resources. Basket racks can also be made by cutting the bottoms off of 1.7-mL tube racks.

Culture plates (48-well; Fisher 720086) (Fig. 1D)

Culture tubes (15-mL; Fisher 1496215E) (Fig. 1C)

Glass dish for washes (Fig. 1B)

Select a small dish (fits 24 baskets, holds 50 mL of wash buffer; Wheaton Inc. 900170), a medium dish (fits 30 baskets, holds 100 mL of wash buffer; Wheaton Inc. 900203), or a large dish (fits 60 baskets, holds 150 mL of wash buffer; Grainger 49WF37).

Lateral shaker at room temperature

Equipment for tube format (lower-throughput sample processing) only

　　Individual vials or tubes

　　Nutator at room temperature

Forceps

Glass slides and coverslips

Imaging stamps (Fig. 2A)

Prepare imaging stamps using 3D printer files available at willseyfroggers.org/resources.

Light box

Bright lights from a stereoscope can be substituted.

Syringes (1-mL)

These are used for dispensing vacuum grease onto glass slides for imaging.

Vacuum grease

Water bath (37°C) with lateral agitation for in situ RNAs

Water bath (60°C) with lateral agitation for in situ RNAs

Zeiss Axio Zoom.V16 microscope (or any appropriate microscope)

METHOD

Three separate procedures are described here: colorimetric whole-mount RNA in situ hybridization (Steps 1–34), fluorescence whole-mount RNA in situ hybridization by HCR (Steps 35–46), and whole-mount immunofluorescence (Steps 47–58). See the Discussion section to determine the best procedure for your experiment.

These procedures have been optimized for higher-throughput staining (24 to 60 samples processed in parallel), with each sample in a basket within a large rack in a glass staining dish sharing a common buffer solution (Fig. 1; see Sive et al. 2007b). This protocol is also effective, although lower-throughput, in individual vials or tubes with manual washes.

Colorimetric Whole-Mount RNA In Situ Hybridization

Perform all steps on a lateral shaker with light agitation (∼40 rpm).

Fixation and Dehydration (2.5 h)

1. Fix animals for 2 h in 1× MEMFA solution at room temperature.

    This step and the following one can be done in basket format or in individual vials or tubes if planning on long-term storage before staining. If using the basket format, limit the number of animals in a basket to 20 and make sure they are well-covered by solutions. Consult one's IACUC protocol for whether anesthetization is required before fixation.

2. Wash several times in methanol at room temperature to dehydrate. Freeze at least overnight at −20°C.

    Samples can be stored long term at −20°C.

Rehydration and Permeabilization (55 min)

3. Rehydrate stepwise into PTw at room temperature as follows:

    i. 5 min in 100% methanol,

    ii. 5 min in 75% methanol and 25% H_2O,

    iii. 5 min in 50% methanol and 50% H_2O, and

    iv. 5 min in 25% methanol and 75% PTw.

4. Wash four times for 5 min each time in PTw at room temperature.

5. Permeabilize in 10 µg/mL proteinase K in PTw for 5 min at room temperature.

    This step should be carefully monitored and not prolonged. For staining of superficial structures like epidermal cilia, omit this step. For staining of deeper structures, this step can be extended with careful testing or combined with dissection for further permeabilization.

Blocking and Hybridization (2.5 h for Xenopus tropicalis, 7.5 h for Xenopus laevis)

6. Wash twice for 5 min each time in 0.1 M triethanolamine (pH 7–8) at room temperature.

7. Wash twice for 5 min each time in 0.1 M triethanolamine with acetic anhydride (125 µL of acetic anhydride per 50 mL of 0.1 M triethanolamine) at room temperature.

8. Wash twice for 5 min each time in PTw at room temperature.

9. Refix for 20 min in 4% paraformaldehyde in PTw at room temperature.

10. Wash five times for 5 min each time in PTw at room temperature.

    For fluorescence-based detection, proceed to Step 35.

11. Preheat probes at 1 µg/mL in hybridization buffer for colorimetric in situ hybridization for several hours at 60°C to help with penetration. In addition to your experimental probe, use a control probe (either sense-transcribed, or a probe with a known, very specific pattern).

12. Prehybridize samples in hybridization solution for 1 h (*X. tropicalis*) or 6 h (*X. laevis*) at 60°C with shaking.

    Depending on the probe, this can be shortened to 1 h for X. laevis samples. Dissected tissues may also require less time than whole embryos.

13. Transfer samples into preheated probe solution overnight at 60°C with shaking.

Save the prehybridization solution to reuse the next day in Step 15. For the basket format, remove baskets from the rack and place into 15-mL round-bottom culture tubes with 500 µL of probe solution each (Fig. 1C). This allows each sample to have a different probe, if desired.

Probe Detection (Time Varies)

This is essentially an antibody staining against digoxygenin-11-UTP present in the RNA probe followed by enzymatic colorimetric detection. The procedure can be modified depending on the probe label and desired detection modality. Time to antibody incubation is 4.5 h; antibody incubation can be done overnight at 4°C or 4 h at room temperature; MAB washes can be done overnight at 4°C or for 5 h at room temperature; AP buffer washes take 15 min; and developing the stain in BM Purple varies from 1 h to days depending on the probe and sample.

14. Remove the probe solution and save at −20°C for reuse.

15. Wash samples in hybridization buffer, reused from Step 12, for 5 min at 60°C.

16. Wash twice for 3 min each time in 2× SSC at 60°C.

17. Wash three times for 20 min each time in 2× SSC at 60°C.

18. Incubate for 30 min at 37°C in 2× SSC containing 20 µg/mL RNase A and 10 µg/mL RNase T_1.

19. Wash once for 10 min in 2× SSC at room temperature.

20. Wash twice for 30 min each time in 0.2× SSC at 60°C.

21. Wash twice for 10 min each time in 1× MAB at room temperature.

22. Incubate in 2% BMB blocking solution for at least 1 h at room temperature.

23. Incubate in antibody solution (dilute anti-digoxigenin-AP antibody 1:3000 in 2% BMB blocking solution) overnight at 4°C or for 4 h at room temperature.

    For basket format in a dish, this is 16.6 µL of antibody per 50 mL of blocking solution. Alternatively, baskets can be transferred into a 48-well plate to use less total antibody, in which case each well has 500 µL of diluted antibody.

24. Wash five times for 1 h each time in 1× MAB at room temperature (or wash overnight at 4°C with multiple quick washes before and after overnight incubation).

25. Wash twice for 5 min each time in alkaline phosphatase buffer with tetramisole hydrochloride at room temperature.

26. Incubate in BM Purple reagent in wells of a 48-well plate at room temperature (Fig. 1D), protected from light with aluminum foil, and monitor until chromogenic reaction produces a stain of the desired intensity.

    Incubation time in BM Purple varies widely depending on the probe and can only be determined empirically or by comparison to published literature for a given probe.

    Depending on the stage and tissue interrogated, endogenous pigment may make the visualization of BM Purple precipitate difficult. Bleaching (see Steps 28–34) may make the signal easier to see. Consider this as the chromogenic reaction proceeds. If pigment precludes sensitive monitoring, consider using albino embryos at the start.

27. Stop the chromogenic reaction with a wash in 1× MAB at room temperature.

Postfixation and Bleaching (6 h)

28. Fix for at least 2 h in Bouin's fixative at room temperature.

29. Wash at room temperature in buffered ethanol solution 10 times for 10 min each time or until the embryos are no longer yellow.

30. Rehydrate stepwise into 1× SSC at room temperature as follows:

    i. 5 min in 75% buffered ethanol and 25% 1× SSC,

    ii. 5 min in 50% buffered ethanol and 50% 1× SSC,

    iii. 5 min in 25% buffered ethanol and 75% 1× SSC, and

 Cite this protocol as *Cold Spring Harb Protoc*; doi:10.1101/pdb.prot105635

iv. twice for 5 min each time in 100% 1× SSC.

31. Bleach in bleaching solution for in situ hybridization for 1–2 h at room temperature under a light box or until embryos are white.

32. Wash three times for 5 min each time in 1× SSC at room temperature.

    At this point, the samples are ready for imaging. The samples can be stored for years at −20°C in methanol or for months at 4°C in 1× SSC.

33. Mount the samples for imaging as follows:

    • For macroscale imaging, mount samples in agarose wells made using 3D printed stamps pressed in 1% molten agarose during cooling (Fig. 2).

        3D printer files for stamps are available at willseyfroggers.org/resources.

    • Alternatively, mount on glass slides in 1× SSC within a vacuum grease well and affix a coverslip.

34. Image using brightfield microscopy.

    See Troubleshooting.

Fluorescence Whole-Mount RNA In Situ Hybridization by Hybridization Chain Reaction (HCR)

This method is identical to the previous one until prehybridization. Perform all steps on a lateral shaker with light agitation (~40 rpm).

Fixation, Dehydration, Rehydration, Permeabilization, and Blocking

35. Carry out Steps 1–10, and then proceed to Step 36.

Hybridization (35 min until Overnight Incubation; 1.5 h until Amplification)

36. Prehybridize samples in 30% probe hybridization buffer for fluorescent in situ hybridization for 30 min at 37°C.

37. Prepare probe solution by adding 2 pmol of each probe as provided by Molecular Technologies (1 µL of 2 µM stock per probe mixture) to 500 µL of 30% probe hybridization buffer that has been prewarmed to 37°C. In addition to your experimental probe, ideally use a control probe (either sense-transcribed, or a probe with a known, very specific pattern).

    Probe volume can be reduced to the minimum required to cover samples.

38. Replace the 30% probe hybridization buffer with probe solution and incubate overnight (12–16 h) at 37°C with shaking.

    For the basket format, remove baskets from the rack and place into 15-mL round bottom culture tubes with 300–500 µL of probe solution each (Fig. 1C).

39. Wash four times for 15 min each time in 30% probe wash buffer at 37°C with shaking.

    Save probe solutions. They can be stored at −20°C and reused multiple times.

    Heat wash solutions to 37°C before use.

40. Wash samples three times for 5 min each time in 5× SSCT at room temperature with shaking.

Amplification (35 min until Overnight Incubation; 1.5 h until Mounting)

41. Incubate samples in amplification buffer for 30 min at room temperature.

42. Prepare 30 pmol hairpin solutions (10 µL of each desired 3 µM hairpin) in amplification buffer as follows:
    i. heat hairpins for 90 sec at 95°C,

    ii. cool for 30 min in a dark drawer to room temperature, and

    iii. add all hairpin solutions to amplification buffer (for a total volume of 500 μL) at room temperature.

43. Transfer samples into the hairpin solution and incubate overnight (12–16 h) in the dark at room temperature.

> *Hairpin solutions can be stored at −20°C and reused multiple times. For the basket format, remove baskets from the rack and place into 15-mL round bottom culture tubes with 300–500 μL of hairpin solution each (Fig. 1C).*

44. Wash in 5× SSCT at room temperature as follows:

    i. twice for 5 min each time,

    ii. twice for 30 min each time, and

    iii. once for 5 min.

45. Wash three times for 5 min each time in 1× SSC at room temperature.

> *At this point, the samples are ready for imaging. The samples can be stored for weeks in the dark at 4°C in 1× SSC.*

46. Mount and image as follows:

- For macroscale imaging, place in 1× SSC in agarose wells made using 3D printed stamps pressed in 1% molten agarose during cooling (Fig. 2). Image on an upright stereomicroscope with fluorescence.

   > *3D printer files for stamps are available at willseyfroggers.org/resources.*

- For higher-magnification imaging, mount in 1× SSC in a vacuum grease well on a glass slide, affix coverslip, and image.
   > See Troubleshooting.

Whole-Mount Immunofluorescence

> *Perform all incubations (excluding antibody incubations) on a lateral shaker (basket format) or on a nutator (tube format).*

Fixation (1 h)

47. Fix animals in 4% paraformaldehyde in PBS for 40 min at room temperature.

> *Consult one's IACUC protocol for whether anesthetization is required before fixation.*

48. Wash in PBS three times for 5 min each time at room temperature.

Bleaching and Permeabilization (2 h, 5 min)

49. Bleach samples in bleaching solution for immunofluorescence for 1 h at room temperature under a light box.

> *This step is incompatible with phalloidin staining and will quench any fluorescent proteins (e.g., GFP); therefore, it should be omitted in those cases.*

> *Bubbles are created in this step. If using tubes, transfer samples to a glass dish or open the tube tops to allow for gas release.*

> *This step will remove pigmentation and provide some permeabilization. It should not be prolonged as it can begin to disintegrate the sample if performed for too long.*

50. Permeabilize in PBT by washing three times for 20 min each time at room temperature.

> *For stages younger than 44, additional permeabilization may be required, such as dehydration.*

Blocking and Incubation with Primary Antibody (1 h until Overnight Incubation)

51. Block in 10% CAS-Block in PBT for at least 1 h at room temperature.

Cite this protocol as *Cold Spring Harb Protoc*; doi:10.1101/pdb.prot105635

52. Incubate in primary antibody diluted in 100% CAS-Block overnight at 4°C.

If using baskets, move the baskets into 48-well plates with 300 μL of antibody per well (Fig. 1D). If using tubes, use a minimum volume to cover animals completely.

A reasonable starting concentration for a new antibody is 1:100, but the concentration should be optimized empirically. For unconcentrated sera (e.g., from DSHB), start with a 1:5 dilution.

Washes and Incubation with Secondary Antibody (3 h, 10 min)

53. Wash in PBT three times for 10 min each time at room temperature.

54. Block in 10% CAS-Block in PBT for 30 min at room temperature.

55. Incubate in secondary antibody diluted in 100% CAS-Block for 2 h in the dark at room temperature.

If using baskets, move the baskets into 48-well plates with 300 μL of antibody per well (Fig. 1D). If using tubes, use a minimum volume to cover embryos completely.

If using fluorescence-conjugated secondary antibodies, cover tubes or baskets with aluminum foil to protect samples from the light for the remainder of the staining.

A typical commercial antibody dilution for this step is 1:250. Additional fluorescent dyes can be added during this step (e.g., DAPI).

Washes and Mounting (1.5 h until Mounting)

56. Wash three times for 10 min each time in PBT at room temperature.

57. Wash three times for 20 min each time in PBS at room temperature.

At this point, the samples are ready for imaging. The samples can be stored for a few weeks at 4°C in 1× PBS. If in solution, rather than mounted, add gentamicin (50 μg/mL) to the 1× PBS to extend storage time. An additional, final fixation for 40 min at room temperature in 4% paraformaldehyde in PBS (post-fixation) can also extend storage time if necessary.

58. Mount and image as follows:

- For macroscale imaging, place in 1× PBS in agarose wells made using 3D printed stamps pressed in molten agarose during cooling (Fig. 2). Image on an upright stereomicroscope.

- For higher-magnification imaging, mount in 1× PBS in a vacuum grease well on a glass slide, affix coverslip, and image.

   See Troubleshooting.

TROUBLESHOOTING

Problem (Step 34, 46, or 58): Superficial staining is observed, but there is an absence of deeper tissue staining.

Solution: Increase permeabilization by a longer or more concentrated proteinase K treatment (for RNA hybridization), a longer or more concentrated detergent treatment (for RNA or protein staining), or by physically dissecting the tissue to expose the target region. Adding a dehydration step to the immunofluorescence protocol may increase permeabilization.

Problem (Step 34, 46, or 58): Tissue disintegrates during the procedure.

Solution: Increase fixation time and/or decrease proteinase K or detergent washes.

Problem (Step 34, 46, or 58): Excessive background staining is observed.

Solution: Increase stringency steps (longer 0.2× SSC washes and increased temperature for RNA hybridization; increased blocking time and permeabilization for immunostaining).

Problem (Step 58): Weak antibody staining is observed.
Solution: Empirically test alternative fixatives (e.g., try glutaraldehyde), antibody concentration, bleaching time, and/or detergent concentration.

Problem (Step 58): No signal is seen.
Solution: Refer to other excellent protocols for additional steps (dehydration, etc.), alternative fixatives, and additional positive control antibodies (e.g., Lee et al. 2008; Brooks and Wallingford 2015).

DISCUSSION

Three separate procedures are described in this protocol. The first involves whole-mount RNA in situ hybridization with colorimetric detection by BM Purple staining (Steps 1–34; Fig. 3A). This is a cost-effective strategy for assaying mRNA expression in embryos and tadpoles of all stages using in-vitro-transcribed digoxygenin-11-UTP-labeled RNA probes (see Sive et al. 2007a). The second, whole-mount RNA in situ hybridization with fluorescent detection by HCR (Steps 35–46; Fig. 3B,C; Choi et al. 2018), is more expensive for assaying mRNA expression because it requires commercial RNA probes designed for *Xenopus* sequences (https://www.moleculartechnologies.org), but these probes can be reused. Fluorescence-based detection provides a great increase in spatial and quantitative resolution over colorimetric detection as well as the ability to label up to five RNAs in different wavelengths. Because of the nature of detection, there is also not the subjectivity of when to terminate the development of signal, which can be an advantage over colorimetric detection in some respects.

The third procedure involves whole-mount immunostaining with fluorescent detection (Steps 47–58; Fig. 3D). Although it is optimized for tadpole stages, it also works well for many epitopes in earlier stages, including before gastrulation. See Table 1 for a list of primary antibodies compatible with this procedure, particularly for stage 46 tadpoles. Because this is one of the simpler procedures available, it is a good one to try first.

The procedures in this protocol should be modified according to the developmental stage and tissue type of interest. For example, dehydration is often helpful in earlier, more yolky stages, whereas

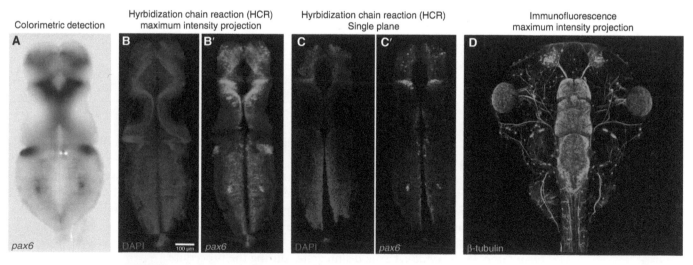

| Colorimetric detection | Hyrbidization chain reaction (HCR) maximum intensity projection | Hyrbidization chain reaction (HCR) Single plane | Immunofluorescence maximum intensity projection |

FIGURE 3. Sample RNA in situ hybridization and immunostaining micrographs. (*A*) Colorimetric staining for *pax6* RNA (purple) in a stage 46 *Xenopus tropicalis* dissected brain imaged by widefield microscopy. (*B,C*) Fluorescence staining for *pax6* RNA by HCR (green; *B'*, *C'*) costained with DAPI to label nuclei (blue; *B, C*) in a stage 46 *X. tropicalis* dissected brain imaged by confocal microscopy. (*B–B'*) Maximum intensity projection of confocal sections. (*C–C'*) Single imaging plane. Note the increased resolution potential with the fluorescence-based method. (*D*) Immunostaining for β-tubulin in the stage 46 *X. tropicalis* head region imaged by confocal microscopy.

Cite this protocol as *Cold Spring Harb Protoc*; doi:10.1101/pdb.prot105635

it can interfere with staining in later tadpole stages. Some tissues and stages require physical permeabilization (e.g., removal of skin in later tadpole stages to better permeabilize the brain), whereas superficial tissues may need less permeabilization (e.g., omit the proteinase K step for epidermal cilia staining). Further, some antibodies produce better results with a particular fixative, permeabilization condition, etc., and require empirical testing to optimize (see Troubleshooting).

On a final note, the colorimetric RNA in situ procedure was derived from a widely used contribution from Joanna Yeh and Mustafa Khokha according to Sive et al. (2000) and originally described in Harland (1991). The fluorescent RNA in situ procedure was derived from Choi et al. (2018), and the imaging stamps were derived from Truchado-Garcia et al. (2018).

ACKNOWLEDGMENTS

I thank Cameron Exner for careful editing; Richard Harland, Edivinia Pangilinan, Mustafa Khokha, Maura Lane, Emily Mis, Karen Liu, Peter Walentek, Yuxiao Xu, and Cameron Exner for expert instruction and modifications of these procedures; Yuxiao Xu for the *pax6* colorimetric image; and Albert Kim, Marta Truchado-Garcia, and Richard Harland for help with 3D-printing racks and stamps.

RECIPES

Alkaline Phosphatase Buffer with Tetramisole Hydrochloride

Reagent	Volume (for 1 L)	Final concentration
Tris (1 M, pH 9.5)	100 mL	100 mM
$MgCl_2$ (1 M)	50 mL	50 mM
NaCl (4 M)	25 mL	100 mM
Tween 20	1 mL	0.1%
Tetramisole hydrochloride (1 M; Sigma-Aldrich L9756)	2 mL	2 mM

Adjust to 1 L with H_2O and store at −20°C in 50-mL aliquots.

Amplification Buffer

Reagent	Volume (for 40 mL)	Final concentration
Sodium chloride-sodium citrate buffer (SSC; 20×) <R>	10 mL	5×
Tween 20 (10%)	400 μL	0.1%
Dextran sulfate (50% w/v in H_2O)	8 mL	10%

Adjust to 40 mL with H_2O. Store at room temperature.

Bleaching Solution for Immunofluorescence

Reagent	Volume (for 100 mL)	Final concentration
PBS (10×; pH 7.4) <R>	10 mL	1×
Formamide	5 mL	5%
Peroxide (30%)	4 mL	1.2%
H_2O	81 mL	

Add the formamide and peroxide to the larger volume of PBS, in that order only; formamide and peroxide can become explosive if mixed directly together. This solution must be made fresh. Do not store.

Bleaching Solution for In Situ Hybridization

Reagent	Volume (for 100 mL)	Final concentration
Sodium chloride-sodium citrate buffer (SSC; 20×) <R>	2.5 mL	0.5×
Formamide	5 mL	5%
Peroxide (30%)	4 mL	1.2%
H_2O	88.5 mL	

Add the formamide and peroxide to the larger volume of SSC, in that order only; formamide and peroxide can become explosive if mixed directly together. This solution must be made fresh. Do not store.

BMB Blocking Solution

Reagent	Quantity (for 500 mL)	Final concentration
BMB Blocking Agent (Sigma-Aldrich 11096176001)	50 g	10%
Maleic acid buffer (MAB; 10×) <R>	50 mL	1×

Adjust to 500 mL with H_2O. Mix with heat until dissolved. Store this 10% solution at −20°C in 50-mL aliquots. Dilute to 2% with 1× MAB on staining day.

Bouin's Fixative for In Situ Hybridization

Reagent	Volume (for 100 mL)	Final concentration
Picric acid (saturated)	70 mL	70%
Formaldehyde (37%)	25 mL	9.25%
Acetic acid (glacial)	5 mL	5%

Store in a glass container in a corrosive cabinet at room temperature.

Buffered Ethanol

Reagent	Volume (for 4 L)	Final concentration
Ethanol (95%)	3.9 L	92.625%
Tris (1 M, pH 8)	100 mL	25 mM
EDTA (0.5 M)	10 mL	1.25 mM

Adjust to 4 L with H_2O. Store in a flammables cabinet at room temperature.

Cite this protocol as *Cold Spring Harb Protoc*; doi:10.1101/pdb.prot105635

Denhardt's Solution (100×)

Reagent	Quantity (for 50 mL)	Final concentration (100×)
Bovine serum albumin (Fraction V)	1 g	2% (w/v)
Ficoll 400	1 g	2% (w/v)
Polyvinylpyrrolidone (PVP)	1 g	2% (w/v)
H_2O	to 50 mL	

Dissolve the components in the H_2O. Filter to sterilize and remove particulate matter. Divide into aliquots, and store at −20°C.

Hybridization Buffer for Colorimetric In Situ Hybridization

Reagent	Quantity (for 3 L)	Final concentration
Formamide	1.5 L	50%
Sodium chloride-sodium citrate buffer (SSC; 20×) <R>	750 mL	5×
Torula RNA Type IX (Sigma-Aldrich R3629)	3 g	1 mg/mL
Heparin (Sigma-Aldrich H3393)	0.3 g	100 µg/mL
Denhardt's solution (100×) <R>	30 mL	1×
Tween 20	3 mL	0.1%
CHAPS (Sigma-Aldrich C3023)	3 g	0.1%
EDTA	11.167 g	10 mM

Adjust to 3 L with H_2O, check that pH is ~7.5, and store at −20°C.

Hybridization Buffer for Fluorescent In Situ Hybridization (30%)

Reagent	Quantity (for 40 mL)	Final concentration
Formamide	12 mL	30%
Sodium chloride-sodium citrate buffer (SSC; 20×) <R>	10 mL	5×
Citric acid (1 M, pH 6.0)	360 µL	9 mM
Tween 20 (10%)	400 µL	0.1%
Heparin (10 mg/mL; Sigma-Aldrich H3393)	200 µL	50 µg/mL
Denhardt's solution (100×) <R>	400 µL	1×
Dextran sulfate (50%, w/v, in H_2O)	8 mL	10%

Adjust to 40 mL with H_2O. Store at −20°C.

Maleic Acid Buffer (MAB, 10×)

Reagent	Quantity (for 4 L)	Final concentration
Maleic acid (Sigma-Aldrich M0375)	464.4 g	1 M
NaCl	350.6 g	1.5 mM

Adjust the pH to 7.2 with ~280 g of NaOH, and then bring the final volume to 4 L with H_2O. Store at room temperature.

Adjusting the pH here is an exothermic acid–base reaction and often takes a long time. Mix in a 4°C room, adding 70 g of NaOH at a time.

MEMFA (1×)

Reagent	Volume (for 100 mL)	Final concentration
MEM salts (10× stock) <R>	10 mL	1×
Formaldehyde (37%)	10 mL	3.7%
H$_2$O	80 mL	

Store at room temperature.

MEM Salts (10× Stock)

Reagent	Quantity (for 1 L)	Final concentration
MOPS	209.3 g	1 M
EGTA	7.6 g	20 mM
MgSO$_4$	1.2 g	10 mM

Adjust the pH to 7.4 with NaOH, and bring the final volume to 1 L with H$_2$O. Autoclave for 20 min. (10× MEM salts turn yellow after autoclaving.) Store at room temperature.

Paraformaldehyde (20%)

Reagent	Quantity (for 500 mL)	Final concentration
Paraformaldehyde	100 g	20%
H$_2$O	500 mL	

Prepare 20% paraformaldehyde by boiling distilled water and then adding 20 g of solid paraformaldehyde per 100 mL of H$_2$O with stirring. Add NaOH pellets until the paraformaldehyde is dissolved. Aliquot in 50-mL tubes and store at −20°C.

To make a 4% paraformaldehyde fixative solution, thaw 20% aliquots (may need to heat in a water bath for solute to go back into solution) and dilute in the appropriate buffer on staining day.

PBS (10×; pH 7.4)

320 g NaCl
8 g KCl
57.6 g Na$_2$HPO$_4$
9.6 g KH$_2$PO$_4$

Dissolve in 3 L of distilled H$_2$O. Adjust the pH to 7.4. Bring to 4 L with distilled H$_2$O. Autoclave. Store at room temperature indefinitely.

Probe Wash Buffer (30%)

Reagent	Volume (for 40 mL)	Final concentration
Formamide	12 mL	30%
Sodium chloride-sodium citrate buffer (SSC; 20×) <R>	10 mL	5×
Citric acid (1 M, pH 6.0)	360 µL	9 mM
Tween 20 (10%)	400 µL	0.1%
Heparin (10 mg/mL)	200 µL	50 µg/mL

Adjust to 40 mL with H$_2$O. Store at −20°C.

Cite this protocol as *Cold Spring Harb Protoc*; doi:10.1101/pdb.prot105635

Sodium Chloride-Sodium Citrate Buffer (SSC; 20×)

Reagent	Quantity (for 4 L)	Final concentration
NaCl	701.1 g	3 M
Na$_3$ citrate•2H$_2$O	352.8 g	0.3 M

Adjust the pH to 7.0 with HCl or 10 N NaOH, and then bring the final volume to 4 L with H$_2$O. Store at room temperature.

SSCT (5×)

Reagent	Volume (for 40 mL)	Final concentration
Sodium chloride-sodium citrate buffer (SSC; 20×) <R>	10 mL	5×
Tween 20 (10%)	400 µL	0.1%

Adjust to 40 mL with H$_2$O. Store at room temperature.

REFERENCES

Brooks ER, Wallingford JB. 2015. In vivo investigation of cilia structure and function using *Xenopus*. *Methods Cell Biol* **127:** 131–159. doi:10.1016/bs.mcb.2015.01.018

Choi HMT, Schwarzkopf M, Fornace ME, Acharya A, Artavanis G, Stegmaier J, Cunha A, Pierce NA. 2018. Third-generation hybridization chain reaction: multiplexed, quantitative, sensitive, versatile, robust. *Development* **145:** dev165753. doi:10.1242/dev.165753

Harland RM. 1991. Appendix G: in situ hybridization: an improved whole-mount method for *Xenopus* embryos. *Methods Cell Biol* **36:** 685–695. doi:10.1016/s0091-679x(08)60307-6

Lee C, Kieserman E, Gray RS, Park TJ, Wallingford J. 2008. Whole-mount fluorescence immunocytochemistry on *Xenopus* embryos. *Cold Spring Harb Protoc* doi:10.1101/pdb.prot4957

Sive HL, Grainger RM, Harland RM. 2000. *Early development of* Xenopus laevis: *a laboratory manual*. Cold Spring Harbor Laboratory Press, Cold Spring Harbor, New York.

Sive HL, Grainger RM, Harland RM. 2007a. Synthesis and purification of digoxigenin-labeled RNA probes for in situ hybridization. *Cold Spring Harb Protoc* doi:10.1101/pdb.prot4778

Sive HL, Grainger RM, Harland RM. 2007b. Baskets for in situ hybridization and immunohistochemistry. *Cold Spring Harb Protoc* doi:10.1101/pdb.prot4777

Truchado-Garcia M, Harland RM, Abrams MJ. 3D-printable tools for developmental biology: improving embryo injection and screening techniques through 3D-printing technology. biorxiv doi:10.1101/376657

Cryosectioning and Immunostaining of *Xenopus* Embryonic Tissues

Olga Ossipova and Sergei Y. Sokol[1]

Department of Cell, Developmental and Regenerative Biology, Icahn School of Medicine at Mount Sinai, New York, New York 10029, USA

The *Xenopus* embryo is a classical vertebrate model for molecular, cellular, and developmental biology. Despite many advantages of this organism, such as large egg size and external development, imaging of early embryonic stages is challenging because of nontransparent cytoplasm. Staining and imaging of thin tissue sections is one way to overcome this limitation. Here we describe a step-by-step protocol that combines cryosectioning of gelatin-embedded embryos with immunostaining and imaging. The purpose of this protocol is to examine various cellular and tissue markers after the manipulation of protein function. This protocol can be performed within a 2-d period and allows detection of many antigens by immunofluorescence.

MATERIALS

It is essential that you consult the appropriate Material Safety Data Sheets and your institution's Environmental Health and Safety Office for proper handling of equipment and hazardous materials used in this protocol.

RECIPES: Please see the end of this protocol for recipes indicated by <R>. Additional recipes can be found online at http://cshprotocols.cshlp.org/site/recipes.

Reagents

Acetone
Agarose, 1% in water
Antibodies for the protein of interest
Blocking buffer for embryos <R>
DAPI (5 mg/mL in dimethylformamide [DMF]) (optional; see Step 22)
> *Store for up to 6 mo at −20°C.*

Dent's fixative <R>
Embedding solution <R>
Dry ice
Formaldehyde, 3.7% (optional; see Step 2)
Marc's modified Ringer's solution (MMR; 1×; pH 7.5) <R>
> *Dilute to 0.1× in water.*

[1]Correspondence: sergei.sokol@mssm.edu

Nail polish

Phosphate buffered saline (PBS) buffer, pH 7.4

Secondary donkey anti-rabbit IgG antibodies conjugated to Cy3 (Jackson ImmunoResearch 711-165-152; 1:200 in blocking buffer)

Secondary goat anti-mouse IgG antibodies conjugated to Alexa Fluor488 (Thermo Fisher A11029; 1:200 in blocking buffer)

Trichloroacetic acid (TCA; Sigma-Aldrich T-6399), 2% in water (optional; see Step 2)

Vectashield antifade mounting media for fluorescence (Vector Laboratories; liquid or hard-set)

Xenopus laevis embryos at the desired stage of development (Sive et al. 1998)

Equipment

Aluminum foil

Coverslips, 24 × 60 mm (StatLab SL 102460)

Cryostat (Leica CM3050 S)

Cryostat embedding molds, 10 × 10 × 5-mm (Tissue-Tek Cryomold, Sakura 4565)

Cryostat sample holders

Cryostat sectioning blades, Polycut LP (StatLab CUT8100)

Dissecting stereomicroscope

Dry ice bucket

Dumont Biologie tip #55 fine forceps (Fine Science Tools 11255-20)

Freezer (−80°C)

Gel-loading pipette tips

Glass vials (5-mL)

Histology glass slide holder and glass reservoir

Humidity chamber

> *Construct this chamber from any reusable plastic tray with lid that provides a constant humid environment for immunohistochemical staining procedures. The parallel plastic rails for horizontal incubation of slides on the bottom of chamber can be made from single use plastic pipettes and attached with tape.*

Hydrophobic barrier pen (PAP pen)

Nutator (Adams) or horizontal shaker

Paint brushes

Petri dishes

Plastic slide holders

Razor blades

Superfrost Plus Slides (Thermo Fisher) or Millennia 2.0 Adhesion Slides (StatLab SL 318L)

Tissue-Plus O.C.T. (optimal cutting temperature) compound 4586 (Scigen)

Toothpicks

Transfer pipettes

Zeiss AxioImager with the Apotome attachment or equivalent fluorescence or confocal microscope

METHOD

> *This protocol is based on Fagotto and Gumbiner (1994) and Chalmers et al. (2003) with modifications introduced by our laboratory (Dollar et al. 2005; Ossipova et al. 2007). It is suitable for immunodetection of many embryonic antigens and favorably compares with paraffin sections (Fischer et al. 2008) or cryosections after embedding into 30% sucrose-optimal cutting temperature (OCT) compound as recommended for frog oocytes (see Sec. 3, Protocol 9: Fluorescence In Situ Hybridization of Cryosectioned* Xenopus *Oocytes [Neil and Mowry 2018]), zebrafish, and chick embryos (Westerfield 2007; Khudyakov and Bronner-Fraser 2009). One advantage of cold water fish gelatin (CWFG) over the OCT compound is that tissue morphology is preserved better and that embryos can be easily oriented before sectioning. However, in situ hybridization works better with OCT-based sections.*

Fixation

1. Remove the vitelline membrane from *Xenopus* embryos at the desired stage, using forceps in a 1% agarose dish filled with 0.1× MMR.

   Detailed protocols for in vitro fertilization, Xenopus embryo culture and microinjections, and vitelline membrane removal have been described elsewhere (Sive et al. 1998).

   To prepare agarose dishes, heat 1% agarose in water according to the manufacturer's instructions. Pour the agarose into a Petri dish and let it solidify before using.

2. Using a transfer pipette, place embryos into 5-mL glass vials containing Dent's fixative that has been chilled on ice. Wash embryos twice with Dent's fixative to remove water completely. Fix overnight at −20°C with slow shaking.

   The fixed embryos can be stored at −20°C.

   Optimal immunostaining may require alternative fixatives, such as 3.7% formaldehyde (Kim et al. 2012; Chu et al. 2016) or 2% trichloroacetic acid (Nandadasa et al. 2009; Ossipova et al. 2014; Ossipova et al. 2021).

3. Rinse embryos twice in 1× PBS for 5–10 min at room temperature with slow shaking.

Embedding and Cryosectioning

4. Remove the PBS. Add the embedding solution to the vials, making sure that the embryos are completely submerged. Equilibrate the vials for 15–20 min at room temperature, then incubate for 24 h at 4°C with slow shaking.

   The embedded samples can be stored for up to 2 wk at 4°C.

5. Place fresh embedding solution into the embedding mold. Transfer five to seven embryos into the center of the mold. Fill the mold completely with embedding solution. To keep track of multiple groups, label the samples.

6. Orient the embryos under a stereoscope with a gel-loading pipette tip.

   Sectioning will start from the bottom side of the gelatin block.

7. Place the mold with embryos in a dry ice bucket. Freeze the gelatin block containing embryos on dry ice for 10–20 min. Once frozen, the samples must be sectioned within the same day.

8. Release the frozen block from the mold by making an incision with a razor blade and removing the plastic surrounding the sample.

9. Place a sample holder into the cryostat. The cryostat object temperature (OT) and chamber temperature (CT) should be set at −19°C and −25°C, respectively.

10. Place some liquid OCT on the sample holder to attach the frozen block containing embryos to the holder. Attach the gelatin block in a way that allows sectioning to start from the bottom of the block that has a smooth surface. Let the block solidify completely for 5–10 min in the cryostat.

11. Equilibrate sample for at least 30 min in the cryostat (optional).

12. Cut 10- to 12-µm sections at −19°C OT, −25°C CT as follows:

    i. Present the corner of the block to the blade, and as the sections are cut, stab them with a toothpick immediately as the first part of the section comes off the blade's edge (Fig. 1).

    ii. First, trim the block at 20- to 30-µm thickness. Once the embryos become visible, set the sectioning thickness to 10–12 µm. Adjust the sample holder to produce properly oriented uniform sections.

       See Troubleshooting.

13. Finish cutting the sections and immediately transfer each desired section to a Fisherbrand Superfrost Plus or StatLab Millennia 2.0 slide using a wooden toothpick. Keep slides close to the

   Cite this protocol as *Cold Spring Harb Protoc*; doi:10.1101/pdb.prot107151

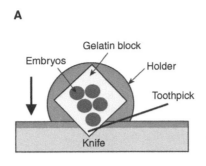

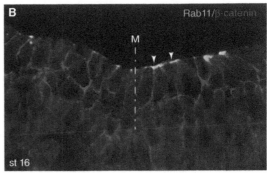

FIGURE 1. (A) Cryosectioning of gelatin-embedded *Xenopus* embryos. Section is picked with a toothpick from the sectioning blade to be transferred to a microscope slide. An arrow indicates the direction of movement of the specimen attached to the holder. (B) Double staining of the neural plate with mouse monoclonal antibodies specific for Rab11 (1:100, Invitrogen) and rabbit polyclonal antibody against β-catenin (1:200, Sigma-Aldrich). Goat secondary antibodies (1:100, Invitrogen) were against mouse or rabbit IgG conjugated to Alexa Fluor 488 or Alexa Fluor 555, respectively. Rab11 is apically localized (arrowheads) in neuroepithelial cells. Midline (M) is indicated (dashed line).

cryostat at room temperature, ready to receive the sections. (Waiting too long to transfer the section may result in section melting and rolling.) Label slides with a pencil to avoid label removal during subsequent acetone treatment.

See Troubleshooting.

14. Store slides at −80°C after sectioning. Because the immunostaining signal goes down after prolonged storage (>2 mo), use freshly prepared sectioned embryos for best results.

Indirect Immunofluorescence

15. Dry slides at room temperature for 1 h in a fume hood after transferring them from the −80°C freezer.

16. Place the slides in a histology glass slide holder and remove the gelatin by placing the slide holder in a glass reservoir filled with acetone for 5–7 min at room temperature.

17. Dry slides in a fume hood for 10 min to prevent the sections from floating off the slides during subsequent steps. To reduce the amount of primary antibody needed, use a PAP pen to draw a ring around sections and let it dry completely.

See Troubleshooting.

18. Place slides horizontally in a humidity chamber with a wet Kimwipe tissue at the bottom. Make sure there is enough liquid in the humidity chamber for overnight incubation. Rehydrate slides in PBS for 5 min at room temperature.

19. Remove the PBS. Add 300–400 μL of the blocking buffer on top of each slide using a 1-mL pipette tip. Make sure the slides are not touching each other or the Kimwipe. Block the slides for 60 min at room temperature in the humidity chamber.

20. Replace the blocking solution with 300–350 μL of the primary antibody diluted in fresh blocking solution. Incubate slides in the humidity chamber overnight at 4°C (or for 3–4 h at room temperature). Include a control sample with no primary antibody.

The optimal dilution of the primary antibody is determined in prior experiments or recommended by the manufacturer. Some commercial antibodies that are useful for Xenopus embryo staining, their dilutions, and fixation conditions are listed in Table 1.

For double immunostaining, incubate the samples simultaneously with two primary antibodies raised in different animals (e.g., mouse monoclonal anti-GFP and rabbit polyclonal anti-Myc). Use Alexa Fluor488–conjugated goat (or donkey) anti-mouse IgG and Cy3-conjugated donkey anti-rabbit IgG secondary antibodies. Blocking and washing steps remain the same.

TABLE 1. Selected antibodies that are suitable for staining of cryosectioned *Xenopus* tissues

Antigen	Dilution	Source	Species of origin	Special fixation conditions (if any)
β-catenin	1:200	Sigma-Aldrich	Rabbit	
Centrin-2	1:200	Homemade	Rabbit	
GFP	1:200	B-2, Santa Cruz	Mouse	
Protein kinase Cζ	1:200	C-20, Santa Cruz	Rabbit	
Rab11	1:100	Zymed	Rabbit	TCA fixative
		BD Biosciences	Mouse	
Sox3	1:200	5H6, Alfandari Lab/DSHB	Mouse	
Acetylated tubulin	1:200	6-11B-1, Sigma-Aldrich	Mouse	Dent's fixative
α-tubulin	1:100	B-5-1-2, Sigma-Aldrich	Mouse	Dent's fixative
γ-tubulin	1:150	GTU-88, Sigma-Aldrich	Mouse	Dent's fixative
Vangl2	1:200	Homemade	Rabbit	TCA fixative
ZO-1	1:200	Invitrogen	Mouse	
		Zymed	Rabbit	

21. Wash slides three to four times with PBS (20 min each, 1 h total time) at room temperature. Dry slides in a vertical position on a paper towel. If the slides were incubated with primary antibody for 3–4 h, reduce washing time to 30 min total.

22. Wrap the humidity chamber in aluminum foil and carry out this and subsequent steps in this dark chamber. Place the slides in the humidity chamber and cover each slide with 350 µL of properly diluted fluorophore-conjugated secondary antibody in the blocking solution. Incubate for 2 h at room temperature. If desired, add DAPI (0.5 µg/mL final concentration) to the secondary antibody mix.

23. Wash slides three times (20 min each) with PBS at room temperature.

24. Mount the samples with 2–3 drops of the Vectashield mounting medium. Care must be taken when placing the coverslip, as the frozen sections are very fragile. Avoid making air bubbles. If using the liquid set mounting medium, seal the slides with nail polish.

25. The mounted sections are now ready for imaging using regular fluorescence or confocal microscope. The immunostained samples can be stored for several weeks in the dark at 4°C.

See Troubleshooting.

TROUBLESHOOTING

Problem (Steps 12 and 13): Sections are not easily coming off the knife or are rolling.
Solution: Cool down the cryostat (it may not be cold enough) or replace the knife. Rolling of sections may indicate that the block is too large. In that case trimming the block would help. Use the correct gelatin powder to prepare the gelatin solution. Clean the blade surface and the stage from unused sections using a paintbrush, otherwise freshly cut sections will stick to the blade.

Problem (Step 17): The tissue is detaching from the slides during immunostaining.
Solution: Replace the slides with a different brand of slides containing a positively charged surface or purchase a new batch.

Problem (Step 25): The slides have high nonspecific background from antibody staining.
Solution: To reduce background, we recommended adding 0.1% Triton X-100 or 1%–5% DMSO to the washing and/or blocking solutions. When using a commercial antibody, always check man-

Cite this protocol as *Cold Spring Harb Protoc*; doi:10.1101/pdb.prot107151

ufacturer-suggested fixation and blocking conditions. Some commercial antibodies that are useful for *Xenopus* embryo staining, their dilutions, and fixation conditions are listed in Table 1.

RECIPES

Blocking Buffer for Embryos

1× phosphate-buffered saline (PBS) buffer, pH 7.4
1.2% bovine serum albumin (BSA), fraction V
6% heat-inactivated donkey or goat serum (Sigma-Aldrich)

Heat-inactivate the serum for 30 min at 60°C. For optimal blocking, use serum from the species in which the secondary antibodies have been raised. Combine the ingredients and filter-sterilize the solution using a 0.2-μm Nalgene filter. For long-term storage, aliquot the solution and store for up to 6 mo at −20°C.

Dent's Fixative

80% methanol
20% dimethylsulfoxide (DMSO)

Embedding Solution

Prepare a stock solution of 45% (w/v) cold water fish gelatin (CWFG; Sigma-Aldrich G7041) in water. Incubate overnight at 37°C. Store in 50-mL Falcon tubes for up to several months at 4°C.

Prepare the embedding solution by making a solution of 15% CWFG with 15% sucrose in water. Dissolve at 37°C until completely homogeneous with no air bubbles. For short-term storage, keep at 4°C for without sodium azide; this solution will be stable for at least 3–4 mo. For longer storage, add 10% sodium azide to a 0.02% final concentration.

Marc's Modified Ringer's Solution (MMR; 1×; pH 7.5)

100 mM NaCl
2 mM KCl
1 mM MgCl$_2$
2 mM CaCl$_2$
5 mM HEPES (pH 7.5)

Sterilize by autoclaving and store indefinitely at room temperature.

ACKNOWLEDGMENTS

We thank Nancy Papalopulu and Chris Wylie for sharing their laboratory immunostaining protocols. Work in the Sokol laboratory has been supported by grants from the National Institutes of Health (GM122492, HD092990, DE027665, and NS100759).

REFERENCES

Chalmers AD, Strauss B, Papalopulu N. 2003. Oriented cell divisions asymmetrically segregate aPKC and generate cell fate diversity in the early *Xenopus* embryo. *Development* 130: 2657–2668. doi:10.1242/dev.00490

Chu CW, Ossipova O, Ioannou A, Sokol SY. 2016. Prickle3 synergizes with Wtip to regulate basal body organization and cilia growth. *Sci Rep* 6: 24104. doi:10.1038/srep24104

Dent JA, Polson AG, Klymkowsky MW. 1989. A whole-mount immunocytochemical analysis of the expression of the intermediate filament protein vimentin in *Xenopus*. *Development* 105: 61–74. doi:10.1242/dev.105.1.61

Dollar GL, Weber U, Mlodzik M, Sokol SY. 2005. Regulation of lethal giant larvae by dishevelled. *Nature* 437: 1376–1380. doi:10.1038/nature04116

Fagotto F, Gumbiner BM. 1994. Beta-catenin localization during *Xenopus* embryogenesis: accumulation at tissue and somite boundaries. *Development* 120: 3667–3679. doi:10.1242/dev.120.12.3667

Fischer AH, Jacobson KA, Rose J, Zeller R. 2008. Paraffin embedding tissue samples for sectioning. *CSH Protoc* 2008: pdb prot4989. doi:10.1101/pdb.prot4989

Khudyakov J, Bronner-Fraser M. 2009. Comprehensive spatiotemporal analysis of early chick neural crest network genes. *Dev Dyn* 238: 716–723. doi:10.1002/dvdy.21881

Kim K, Lake BB, Haremaki T, Weinstein DC, Sokol SY. 2012. Rab11 regulates planar polarity and migratory behavior of multiciliated cells in *Xenopus* embryonic epidermis. *Dev Dyn* 241: 1385–1395. doi:10.1002/dvdy.23826

Nandadasa S, Tao Q, Menon NR, Heasman J, Wylie C. 2009. N- and E-cadherins in *Xenopus* are specifically required in the neural and non-neural ectoderm, respectively, for F-actin assembly and morphogenetic movements. *Development* 136: 1327–1338. doi:10.1242/dev.031203

Neil CR, Mowry K. 2018. Fluorescence in situ hybridization of cryosectioned *Xenopus* oocytes. *Cold Spring Harb Protoc* doi:10.1101/pdb.prot097030

Newport J, Kirschner M. 1982. A major developmental transition in early *Xenopus* embryos: I. Characterization and timing of cellular changes at the midblastula stage. *Cell* 30: 675–686. doi:10.1016/0092-8674(82)90272-0

Ossipova O, Tabler J, Green JB, Sokol SY. 2007. PAR1 specifies ciliated cells in vertebrate ectoderm downstream of aPKC. *Development* 134: 4297–4306. doi:10.1242/dev.009282

Ossipova O, Kim K, Lake BB, Itoh K, Ioannou A, Sokol SY. 2014. Role of Rab11 in planar cell polarity and apical constriction during vertebrate neural tube closure. *Nat Commun* 5: 3734. doi:10.1038/ncomms4734

Ossipova O, Mancini P, Sokol S. 2021. Direct imaging of core PCP proteins in *Xenopus* embryos. *Methods Mol Biol* (in press).

Sive H, Grainger RM, Harland RM. 1998. *The early development of* Xenopus laevis: *a laboratory manual*. Cold Spring Harbor Laboratory Press, Cold Spring Harbor, NY.

Westerfield M. 2007. *The zebrafish book. A guide for the laboratory use of zebrafish* (Danio rerio). University of Oregon Press, Eugene.

Cite this protocol as *Cold Spring Harb Protoc*; doi:10.1101/pdb.prot107151

Live Imaging of Cytoskeletal Dynamics in Embryonic *Xenopus laevis* Growth Cones and Neural Crest Cells

Burcu Erdogan,[1,4] Elizabeth A. Bearce,[2,4] and Laura Anne Lowery[3,5]

[1]Department of Stem Cell and Regenerative Biology, Harvard University, Cambridge, Massachusetts 02138, USA;
[2]Institute of Molecular Biology, Department of Biology, University of Oregon, Eugene, Oregon 97403, USA;
[3]Boston University School of Medicine, Boston Medical Center, Boston, Massachusetts 02118, USA

The cytoskeleton is a dynamic, fundamental network that not only provides mechanical strength to maintain a cell's shape but also controls critical events like cell division, polarity, and movement. Thus, how the cytoskeleton is organized and dynamically regulated is critical to our understanding of countless processes. Live imaging of fluorophore-tagged cytoskeletal proteins allows us to monitor the dynamic nature of cytoskeleton components in embryonic cells. Here, we describe a protocol to monitor and analyze cytoskeletal dynamics in primary embryonic neuronal growth cones and neural crest cells obtained from *Xenopus laevis* embryos.

MATERIALS

It is essential that you consult the appropriate Material Safety Data Sheets and your institution's Environmental Health and Safety Office for proper handling of equipment and hazardous materials used in this protocol.

RECIPES: Please see the end of this protocol for recipes indicated by <R>. Additional recipes can be found online at http://cshprotocols.cshlp.org/site/recipes.

Reagents

Ficoll (5% in 0.1× MMR)

Marc's modified Ringer's (MMR) (10×, pH 7.4) <R>

mRNA encoding fluorophore-tagged protein of interest (see Step 12 and Table 1)

Synthesize capped mRNAs in vitro using the mMessage mMachine kit (Invitrogen). Determine the concentration of mRNAs using a Nanodrop. Verify the presence and integrity of the mRNA on an RNA gel.

PBS (1×; sterile)

Dilute 10× PBS (pH 7.4, RNase-free; Invitrogen AM9625) in deionized H_2O to make 1× PBS.

Reagents for neural crest cell culture only (see Steps 5–10 and Steps 24–28)

  Fibronectin (20 µg/mL in 1× PBS; Sigma-Aldrich F1141)

  Gelatin (Sigma-Aldrich G1890)

  Plating culture medium for neural crest cells <R>

Reagents for neural tube culture only (see Steps 1–4 and Steps 16–23)

  Crystalline collagenase (2 mg/mL in Steinberg's medium)

  Laminin (10 µg/mL in 1× PBS; Sigma-Aldrich L2020)

---

[4]These authors contributed equally to this work.
[5]Correspondence: lalowery@bu.edu

Copyright © 2023 Cold Spring Harbor Laboratory Press
Cite this protocol as *Cold Spring Harb Protoc*; doi:10.1101/pdb.prot104463

TABLE 1. Useful cytoskeletal markers to visualize microtubules (MTs), actin filaments, or focal adhesions

Marker	Labels	Purpose	Concentration range per embryo	Imaging technique	Analysis software
EB1-3 (Stepanova et al. 2003)	MT plus-end	MT growth dynamics	100 pg–300 pg mRNA	Spinning disk confocal microscopy (SDCM)	plusTipTracker (Applegate et al. 2011; Stout et al. 2014)
MACF-43 (Honnappa et al. 2009)	MT plus-end	MT growth dynamics	100 pg–300 pg mRNA	SDCM	plusTipTracker (Stout et al. 2014; Applegate et al. 2011)
Fluorophore-tagged tubulin	MTs	Uniform labeling to visualize MTs	900 pg mRNA	SDCM	Fiji (ImageJ) (Schindelin et al. 2012)
		Speckle labeling to assess MT flow rates	300 pg mRNA	Total internal reflection microscopy (TIRF)	QFSM software (Danuser and Waterman-Storer 2006; Mendoza et al. 2012)
Actin monomer (G-actin)	Actin	Speckle labeling to assess actin dynamics	300 pg mRNA	TIRF	QFSM software (Danuser and Waterman-Storer 2006; Mendoza et al. 2012)
LifeAct-GFP (Riedl et al. 2008), Utrophin (Burkel et al. 2007), F-Tractin (Schell et al. 2001)	F-actin	Actin dynamics	LifeAct/utrophin may affect dynamics dependent on the concentration. Optimization may be required. 60–300 pg can be used to begin.	SDCM, TIRF	Fiji (ImageJ) (Schindelin et al. 2012)
Paxillin (Robles and Gomez 2006), PAK2 (Santiago-Medina et al. 2013), focal adhesion kinase (Myers and Gomez 2011), zyxin, vinculin, talin	Focal adhesion	Focal adhesion dynamics	50–75 pg DNA, 250–500 pg mRNA	TIRF	FAAS (Berginski and Gomez 2013)

Plating culture medium for neural tube explants <R>

This is an enriched medium that is commonly used for older Xenopus *retinal ganglion cultures, where neurons have reduced energy reserves. Young spinal cord cultures will survive and grow for >24 h in pure Ringer's solution <R> without additional growth factors; this is one of the benefits of using this system.*

Poly-L-lysine (PLL; 100 µg/mL; Sigma-Aldrich P4832)

PLL is reconstituted in sterile H_2O and diluted to working concentration (100 µg/mL) with 1× PBS.

Steinberg's medium <R>

Xenopus laevis embryos

These must be staged appropriately for neural crest (stage 18) or neural tube (stage 20–23) isolation (see Step 15; Nieuwkoop and Faber 1994; Wlizla et al. 2018). Injections are often made at the two- to four-cell stage (see Step 11).

Equipment

Equipment for neural crest cell dissections only (see Steps 24–28):

    Eyelash knife (or insect pins; see Step 26)

    Plasticine clay-coated Petri dish <R>

Equipment for neural tube dissections only (see Steps 16–23):

    Agarose-coated Petri dish <R>

Alternatively, use a Sylgard 184-coated Petri dish <R>. Agarose- and Sylgard-coated plates can be used multiple times.

    Electrolytically sharpened tungsten needles

Fine forceps (Dumont #5 or equivalent)

Cite this protocol as *Cold Spring Harb Protoc*; doi:10.1101/pdb.prot104463

Fluorescence dissecting microscope

Glass injection needles

See Lowery et al. (2012) for instructions on how to make injection needles.

Image analysis software (see Step 30 and Table 1)

Imaging setup (see Step 29)

Use a spinning disk confocal microscope (SDCM), total internal reflection fluorescence (TIRF) microscope, or widefield epifluorescence scope equipped with an appropriate objective. Imaging cytoskeletal dynamics will generally require a relatively high-magnification objective (60× or 100×). Our SDCM rig uses a 63× Plan Apo 1.4 NA objective, and our TIRF setup uses a 60×, 1.49 NA objective. For detection, a growing number of modern CCD and scMOS cameras will be suitable; ensure that the chip size, sensitivity, and camera speed are appropriate. Our SDCM uses an Orca Flash 4 CCD; our TIRF imaging is typically acquired with an Andor ELYRA scMOS.

Incubator(s) set at the appropriate temperature (see Steps 1, 3, 15, 23, and 28)

MatTek glass bottom (No. 1.5) culture dish (35-mm)

Petri dish (60-mm × 15-mm) and plastic mesh with 1-mm grid to fit in Petri dish (see Step 11)

Rotator at room temperature

METHOD

Culture Dish Coating

Coat imaging chambers with the appropriate coating reagents depending on the type of culture. For neural tube cultures, begin with Step 1; for neural crest cell cultures, begin with Step 5.

Neural Tube Culture Plates

1. Add 500 μL of 100 μg/mL poly-L-lysine (PLL) into the center of a 35-mm MatTek glass bottom (No. 1.5) culture dish and incubate for 1 h at 37°C.

2. Rinse off PLL by washing the dish with 1× sterile PBS three times.

3. Add 500 μL of 10 μg/mL laminin into the center of the PLL-coated culture dish and incubate for 1 h at 37°C.

4. Rinse off laminin by washing the dish with sterile 1× PBS three times and use immediately for neural tube culture (see Steps 16–23).

Neural Crest Cell Culture Plate

5. Heat 2 mg/mL gelatin dissolved in distilled water in the microwave until boiling; allow it to boil 5–10 sec until dissolved, then allow to cool for 5 min.

6. Add 500 μL of cooled gelatin into the center of a 35-mm MatTek glass bottom (No. 1.5) culture dish.

7. Rotate gently for 25 min at room temperature.

8. Discard the gelatin and rinse the dish with 1× sterile PBS several times.

9. Add 500 μL of 20 μg/mL fibronectin and incubate dish overnight at 4°C.

10. On the next day remove the fibronectin. Rinse the plate with sterile 1× PBS several times and use to culture cells (see Steps 24–28).

Injection of RNA Encoding Fluorescently Tagged Cytoskeleton Proteins

More information regarding embryo injections, including injection needle preparation, can be found in Lowery et al. (2012).

11. Place a plastic mesh with a 1-mm grid on the bottom of a Petri dish (60-mm × 15-mm) and fill the dish with 5% Ficoll in 0.1× MMR. Position *Xenopus* embryos on the mesh grid to hold them in place during injections.

    Injections are often made at the two- to four-cell stage.

    Each animal blastomere receives at least one injection; however, multiple injections can be performed depending on the amount of mRNA needing to be delivered. Tissue-targeted injections can be performed as well based on fate maps provided on Xenbase.

12. Inject mRNAs using glass needles according to the following guidelines.
    - To quantify changes in microtubule dynamics, inject in vitro transcribed mRNA of fluorophore-tagged EB1 or MACF-43 (a truncated protein that contains a minimal EB1-binding domain [Honnappa et al. 2009]), at 300 pg/embryo or 75–100 pg/embryo, respectively (see Table 1).

        These mRNAs can be injected alongside overexpression or knockdown strategies for your protein of interest.

        Primary mRNA stocks are not diluted. The mRNA working concentration is generally set to 500 ng/µL. However, depending on the size of the construct, the working concentration can be higher in order to reduce the injection volume. Total injection volume is generally 3 nL per embryo. Both high concentration and high volume of mRNA can be toxic to the embryo. The concentration of mRNA of interest can be determined empirically depending on the construct and the purpose of the experiment. For example, for protein localization analysis the amount of mRNA should be titrated and the minimum amount of mRNA should be used to avoid overexpression artifacts, which can be elicited as protein aggregations.

    - To monitor changes in actin filament dynamics, use fluorophore-tagged actin binding domains such as LifeAct, Utrophin, F-Tractin, or actin monomer. Begin with LifeAct or Utrophin mRNA concentrations at 100–300 pg mRNA/embryo.

        It has been reported that these markers may impact actin dynamics and architecture in a dose-dependent manner (Belin et al. 2014; Flores et al. 2019; Legerstee et al. 2019). Therefore, take care to use the probe that will least impact the dynamic behavior of interest.

    - To assess cell–extracellular matrix adhesion dynamics, label focal adhesions with 50–75 pg/embryo of fluorophore-tagged Paxillin, Zyxin, or PAK2 DNA or mRNA.

        Many other labels are also possible (Kerstein et al. 2015; Stutchbury et al. 2017). Focal adhesion proteins have notoriously high cytoplasmic fluorescence, and TIRF microscopy greatly facilitates crisp adhesion segmentation in later analyses. Advanced imaging techniques such as fluorescence recovery after photobleaching (FRAP) can also be used to monitor the adhesion dynamics involved in cell migration and adhesion (Worth and Parsons 2010).

13. Upon completion of injections, transfer embryos into a dish containing 0.1× MMR.

14. Prior to incubation, sort out dead or unhealthy-looking embryos.

    Dead or unhealthy embryos can adversely affect the quality of healthy embryos. They be detected by the change in the pigmentation or embryo shape.

15. Incubate the embryos until they reach the desired developmental stage. To accelerate the rate of development, keep injected embryos that are stored in the 0.1× MMR-containing dish at warmer temperatures (20°C–23°C). Conversely, to slow development, keep embryos in cooler temperatures such as 16°C.

    The rate of embryo development is dependent on temperature. A chart for temperature-dependent development time can be found in Xenbase (https://www.xenbase.org/anatomy/alldev.do; https://www.xenbase.org/anatomy/static/xenopustimetemp.jsp).

    Axon outgrowth analysis is performed on neurons isolated from neural tubes harvested from embryos at stage 22. Embryos kept at 20°C–23°C reach stage 22 ~24 h after fertilization. Once this stage has been achieved, proceed to Steps 16–23.

    A similar strategy can be applied to embryos to be used for neural crest cell isolation. Embryos kept at 20°C–23°C reach stage 18 ~20 h after fertilization. Once this stage has been achieved, proceed to Steps 24–28.

Cite this protocol as *Cold Spring Harb Protoc*; doi:10.1101/pdb.prot104463

Preparation of Neural Tube and Neural Crest Cell Cultures

For neural tube cultures, follow Steps 16–23; for neural crest cell cultures, follow Steps 24–28.

Neural Tube Culture

Isolation and preparation of neural tube cultures have been previously described in Lowery et al. (2012).

16. Before doing the dissections, prepare culture dishes as described in Steps 1–4. Fill the culture dishes with the neural tube explant culture medium.

17. Screen embryos with a fluorescence dissecting scope and identify which express fluorescence in the neural tube. Transfer these at stage 20–23 to an agarose-coated plastic dish containing Steinberg's medium.

    Embryos can display a mosaic of fluorescence owing to variable expression of the mRNA injected.

    It is possible, but more challenging, to dissect embryos beyond stage 23. The tissues adhere more tightly to one another, which necessitates longer collagenase treatment (in Step 19.i).

18. Isolate the entire dorsal portion of the embryo, including the neural tube, as follows.

    i. View the embryo with a dissecting scope and use fine forceps to remove the vitelline membrane.

    ii. While holding the embryo in place with one pair of forceps, use a second pair to create an incision on the side of the embryo, exposing the hollow interior.

    iii. Use both forceps to pinch along the tissue between the dorsal and ventral halves of the embryo, thereby cutting the embryo in half.

    Visual guidance for how to perform this step can be found in Lowery et al. (2012).

19. Isolate the neural tube as follows.

    i. Transfer the dorsal explant to a small tube or dish containing 2 mg/mL collagenase in Steinberg's medium and place on a rotator at room temperature for 15–20 min to loosen the tissues.

    ii. Pipette the explant into a plastic agarose-coated dish containing fresh Steinberg's medium.

    iii. Gently remove the neural tube from the dorsal epidermis and the ventral notochord using a pair of forceps. Insert the tip of the forceps in between the epidermis and the underlying tissue; then pull the epidermis back slowly to expose the neural tube. Hold the tissue with forceps and slide the tip of another pair of forceps between the notochord and neural tube.

    iv. Remove the somites on both sides using forceps.

20. Move the neural tube to a clean 35-mm dish containing plating culture medium for neural tube explants.

21. After collecting several neural tubes, transect each of them into approximately 20 thin slice explants using electrolytically sharpened tungsten wires.

22. Transfer the explants to the prepared culture dishes.

    The number of explants plated depends on the size of the culture dish used. To avoid overcrowding in the culture dish and to allow axons to grow freely, we generally plate 15–20 explants in each 35-mm culture dish and spread the explants evenly in rows (up to 10 rows) and columns (two columns). After plating, do not move the dishes, as this will disturb the attaching cells.

23. Allow explants to adhere for 12–18 h, ideally at ~20°C–22°C.

    We typically leave the dish of cells on the bench at room temperature overnight. If conditions are appropriate, X. laevis neural explants send out neurites in a robust manner by 24 h after plating on the laminin/poly-L-lysine substrate. Growth cones are highly motile on this substrate, extending outward from the explant in all directions, and the axons can achieve lengths of up to 1 mm (typical lengths are ≥100 µm). See Troubleshooting.

Axons will extend radially, allowing various live imaging of growth cone behavior and cytoskeletal dynamics. Proceed to Step 29 to observe and image neurites and growth cones at room temperature 12–24 h after plating. RNA or plasmid products typically display variable expression between cells, so a range of fluorescence levels may be seen among growth cones. Expression of fluorescent protein RNAs can persist >48 h after plating, depending on the construct.

Cranial Neural Crest Cell Culture

A very helpful and thorough guide to neural crest isolation is given by Milet and Monsoro-Burq (2014), and we advise that this be heavily relied upon for tissue identification and dissection technique. We offer minor modifications here.

24. Prior to performing the dissections, prepare the culture dishes as described in Steps 5–10. Fill the culture dishes with plating culture medium for neural crest cells.

25. Sort embryos to identify those with fluorescence in the cranial neural crest region at stage 18. Strip vitelline membranes of stage 18 embryos, and embed them gently in a plasticine clay-coated Petri dish in plating culture medium for neural crest cells with the anterior dorsal regions exposed.

    Neural crest cells emerge from the tissue just along the anterior neural fold, which is raised slightly from the surrounding tissue.

26. Remove the skin above the neural crest using an eyelash knife. Apply gentle pressure along the edge of the neural fold to allow the neural crest (two to three cell layers) to separate. Lift the explant with a lateral/ventral flicking motion.

27. Rinse explants several times by gently pipetting into a fresh culture medium for neural crest cells, and then transfer the explants into rows along a fibronectin-coated imaging chamber.

    Explants should be at least 500 microns apart to ensure clear fields of view, but the exact number or density of explants in the chamber is not critical and will be dependent on size of the glass-bottom culture dish.

28. Allow the explants to adhere to the coverslip for at least 30 min in a cool (16°C–19°C) and vibration-free place before imaging (see Step 29).

    Tissue will begin collective cell migration within an hour of plating, and will subsequently delaminate to begin single-cell movement after 8 h or more, allowing for various measurements of cytoskeletal dynamics in both individual and collectively migrating cells. See Troubleshooting.

Imaging and Analysis of Cytoskeletal Dynamics

29. Once the cells are adhered, position the culture plate on the microscope stage. Perform live imaging of intracellular cytoskeletal dynamics in either neuronal growth cones or neural crest cells using spinning disk confocal microscopy, TIRF microscopy (for dynamics at or near the cell membrane), or even widefield fluorescence microscopy (for microtubule plus-end dynamics). While imaging, take the following points into consideration.

    - To analyze growth and pause events of microtubule plus-ends, which show very rapid dynamics, perform time-lapse imaging.

        Images should be captured at least every 2 sec for a duration of 1–2 min. Given the rapid dynamics of MT plus-ends, spinning-disk confocal microscopy allows fast and automated acquisition of large region of interests (ROIs) such as whole cells. It is incredibly important that laser power and exposure time of acquisitions are optimized to minimize light toxicity. Using short exposure times and low laser power with reasonably higher gain will help reduce phototoxicity and minimize photobleaching.

    - To visualize actin retrograde flow, use low-density labeling of fluorescent actin monomers followed by quantitative fluorescent microscopy or kymograph analysis (Watanabe and Mitchison 2002; Danuser and Waterman-Storer 2006; Yamashiro et al. 2014).

        Perform an initial test with 1- to 2-sec imaging intervals as a starting point, although flow rates will vary greatly based on structure stability. Focal adhesion dynamics of neural crest cells can be captured every 1–2 min for manual tracking of general size and turnover, but more sensitive tracking measures will often require much greater sampling rates, again within the order of 5–10 sec, for extended imaging periods (20–30 min or longer). General surveys of cell morphology, perhaps to assess cell polarization or filopodial density/number following a genetic perturbation, can perhaps be reduced to every few minutes or collected at only select time points over a number of hours.

- If comparing dynamics of two labeled proteins, consider the time that passes while one is acquired and the system switches to capture the other.

  This is necessary because two fluorophore-labeled structures in separate imaging channels will not be captured simultaneously in typical single-camera/filter systems. The time that passes can be quite minimal with some setups (i.e., triggered acquisition using a small ROI) and dramatically delayed with others (i.e., averaged line scans on a large ROI). Ensure optimization of the microscope for rapid multichannel acquisition if this is of utmost importance.

- If cytoskeletal drugs are to be applied while imaging, administer treatment using a perfusion chamber.

  Many pharmacological agents are not easily soluble in aqueous media, and precipitation can occur if a concentrated solution is administered into a corner of the dish. Construct a perfusion chamber in the bottom of a typical inverted imaging chamber using vacuum grease and an additional coverslip.

  See Troubleshooting.

30. Process and analyze the collected images.

- To quantify parameters of microtubule plus-end dynamics, use plusTipTracker, a Matlab-based open-source software package that automatically detects, tracks, and analyzes time-lapse movies of fluorophore-tagged cytoskeletal proteins.

  For a detailed explanation of how to manage files to perform microtubule dynamics analysis refer to Stout et al. (2014). It should be noted that a currently supported version of this pipeline has been revamped to analyze multiple types of dynamic particles and is now available under the name u-Track (Applegate et al. 2011).

- Assess actin network flow using the quantitative fluorescent speckle microscopy (QFSM) technique and software (Mendoza et al. 2012).

- Use appropriate pipelines for focal adhesion dynamics (FAAS) (Berginski and Gomez 2013), automated filopodia tracking (FiloQuant) (Jacquemet et al. 2017), and single-particle tracking (TrackMate) (Tinevez et al. 2017).

  Customized analysis methods can be built and streamlined in Fiji, Imaris, or Matlab, as preferred.

TROUBLESHOOTING

Problem (Steps 23 and 28): No neurite extension or cell adhesion is observed.
Solution: Neurites might fail to grow if the cultured explants did not adhere well to the culture plate. To ensure cell adherence, use freshly made culture medium and PLL and laminin-coated dishes for neural tube explants, and fibronectin-coated dishes for cranial neural crest. Try not to disturb the culture dish after explants are plated. Finally, it is possible that embryos used to establish the explant cultures were unhealthy, but because the embryos developed to the neural tube stage, this is unlikely to be the case. However, it is best to set aside some whole intact embryos from the same batch of eggs fertilized and used in the neural tube and neural crest cell cultures in 0.1× MMR along with dissections to ensure that continued development occurs.

Problem (Step 29): Cultured explants do not express the fluorophore-tagged protein.
Solution: One reason for low or absent expression is that the needle tip could be clogged depending on the injected material. For example, morpholinos are highly prone to clogging needles. To ensure a successful injection, occasionally check if the solution is coming out from the needle. Alternatively, it is possible that not enough material was injected. Optimal concentration will vary based

on transcript quality and molecular weight and should be titrated accordingly. Generally, the lowest possible concentration of a fluorophore-labeled protein is the most favorable one; low-level expression can achieve a much more favorable signal-to-noise ratio and minimize the artifacts caused by exogenous protein expression. Heavily overexpressed plus-end tracking proteins can be especially hard to deal with, as the excess fluorescent protein will coat the entire microtubule lattice instead of selectively labeling distal microtubule ends, rendering automated plus-end segmentation and tracking much more difficult.

Problem (Step 29): Growth cones or cells die during imaging.
Solution: Light toxicity is the major cause of the cell death. To avoid cell death, image samples at high gain rather than high laser power and keep exposure time short.

DISCUSSION

Xenopus laevis has been a powerful embryonic tool, not only for addressing questions regarding vertebrate development, but also for deciphering key cellular events. One advantage of *Xenopus laevis* as a model is that *Xenopus* embryos can tolerate extensive manipulation, and exogenous materials can be easily introduced via microinjections, such as fluorophore-tagged cytoskeletal-associated proteins. Compared to other systems, cell culturing methods are facile and do not require expensive culture equipment. Cultures can be generated in large quantities and maintained at the bench at room temperature.

The protocol described here takes advantage of this versatile model organism to examine the dynamics of the cytoskeletal components. Measures of cytoskeletal dynamics can be taken from any tissue of interest. To gather data from heterogeneous neuronal cultures, we use neural tube explants dissected from stage 20–23 embryos. Retinal cultures from embryos at stage 24 or beyond can also be used if a more homogenous neuronal population is desired. Cranial neural crest, a multipotent motile cell with well-established migration routes and fate determination cues, can be a valuable model for identifying effectors of collective cell migration, craniofacial development, and cancer metastasis.

The scope of the high-resolution live imaging of cytoskeletal dynamics in axonal growth cones and in cells can be expanded to study different aspects of the cytoskeletal behavior. By mimicking various physiological conditions, one can study how microtubule or actin dynamicity would change within the growth cone or cell in response to guidance signals introduced in culture medium. Multiple cues can be introduced to the cells to test the spatiotemporal changes in cytoskeletal dynamics, which could mimic the pathfinding behavior of the growth cone in its native environment.

RECIPES

Agarose-Coated Petri Dish

Dissolve 1% agarose in 0.1× Marc's modified Ringer (MMR) <R> solution and microwave to melt. Pour into a Petri dish until the dish is half-full. Let the agarose harden at room temperature. Store for up to 2 wk at 4°C.

Culture Media Base

Reagent	Amount	Final concentration
L-15 medium (Sigma-Aldrich L1518)	98 mL	49%
Ringer's solution (1×) <R>	100 mL	50%

Sterilize by vacuum filtration. Add 2 mL of Antibiotic-Antimycotic (100×; Gibco 15240062). Store at 4°C.

Cite this protocol as *Cold Spring Harb Protoc*; doi:10.1101/pdb.prot104463

Marc's Modified Ringer's (MMR) (10×, pH 7.4)

Reagent	Quantity (for 1 L)	Final concentration (10×)
NaCl	58.440 g	1 M
KCl	1.491 g	20 mM
$MgSO_4$	1.204 g	10 mM
$CaCl_2$, dihydrate	2.940 g	20 mM
HEPES	11.915 g	50 mM
H_2O	to 1 L	

Adjust the pH to 7.4 using 10 M NaOH. Sterilize by autoclaving. Store at room temperature indefinitely.

Plasticine Clay-Coated Petri Dishes

To prepare dissection dishes, gather several 35-mm or 60-mm tissue-culture dishes (e.g., Falcon 353001). Push a small amount of plasticine clay (e.g., Plastilina Modeling Clay) into the middle of each dish, and then spread it evenly to fill the bottom 3–5 mm. Rinse the surface of the clay with ample water and ethanol to remove excess oily residue. UV-sterilize prior to use.

After the dissections, sterilize the dishes for reuse. Rinse them thoroughly with deionized H_2O to remove any embryo particulate, use a gloved hand to level out the depressions made for embryos, rinse with ethanol, and sterilize under UV light. Cap and store with dissection kit. Discard and make fresh dissection dishes if clay appears dirty or has trapped any debris.

Plating Culture Medium for Neural Crest Cells

Reagent	Amount	Final concentration
NaCl	1.55 g	53 mM
Na_2CO_3	0.265 g	5 mM
Potassium gluconate	0.527 g	4.5 mM
Sodium gluconate	3.82 g	35 mM
$MgSO_4$	0.060 g	1 mM
$CaCl_2$	0.055 g	1 mM

Add reagents to 400 mL of deionized (DI) H_2O. Adjust the pH to 8.0 with bicine, and bring to final volume (500 mL) with DI H_2O. Sterilize by filtration. Store frozen at −20°C in small aliquots. Add 50 µg/mL gentamicin sulfate and 1 µg/mL bovine serum albumin to the culture medium prior to use.

Plating Culture Medium for Neural Tube Explants

Reagent	Amount to add	Final concentration
Fetal bovine serum (Gibco 10438018)	20 µL	1%
Penicillin-Streptomycin (Sigma-Aldrich P4333)	20 µL	1%
BDNF (100 µg/mL; Sigma-Aldrich B3795; optional[a])	0.5 µL	25 ng/mL
NT3 (25 µg/mL; Sigma-Aldrich SRP3128; optional[a])	2 µL	25 ng/mL

Add culture media base <R> to a final volume of 2 mL (per 35-mm plate). Prepare plating culture medium fresh and maintain at 4°C.

[a]Axon outgrowth is greatly enhanced by addition of BDNF and NT3, but these may be omitted for guidance assays if cues will be added later.

Ringer's Solution (10×)

Reagent	Amount	Final concentration
NaCl	33.6 g	1.15 M
KCl	0.93 g	25 mM
CaCl$_2$	1.11 g	20 mM
EDTA (0.5 M pH 8.0)	5 mL	5 mM

Adjust the pH to 7.4 and raise the volume to 500 mL with distilled water. Sterilize by autoclaving. Store indefinitely at room temperature.

Steinberg's Medium

Reagent	Amount	Final concentration
NaCl	3.39 g	58 mM
KCl	0.05 g	0.67 mM
Ca(NO$_3$)$_2$	0.07 g	0.44 mM
MgSO$_4$	0.16 g	1.3 mM
Tris (1 M, pH 7.8)	2.3 mL	4.6 mM

Add reagents to 250 mL of deionized (DI) H$_2$O and adjust the pH to 7.8. Fill to final volume (500 mL) with DI H$_2$O. Sterilize by autoclaving. Store indefinitely at room temperature.

Sylgard 184-Coated Petri Dish

Mix 10 parts polymer with 1 part curing agent. Pour the mix into a Petri dish until the dish is half-full. Allow the mix to cure and harden overnight at room temperature. Store indefinitely at room temperature.

ACKNOWLEDGMENTS

We thank members of the Lowery laboratory for helpful discussions. We also thank the National *Xenopus* Resource (RRID:SCR-013731) and Xenbase (RRID:SCR-003280) for their invaluable support to the model organism community. L.A.L. is funded by National Institutes of Health (NIH) grants R01 MH109651 and R03 DE025824, March of Dimes (#1-FY16-220), Charles H. Hood Foundation 2018-2019 Bridge Funding Award, and the American Cancer Society's Ellison Foundation Research Scholar Grant (RSG-16–144-01-CSM).

REFERENCES

Applegate KT, Besson S, Matov A, Bagonis MH, Jaqaman K, Danuser G. 2011. plusTipTracker: quantitative image analysis software for the measurement of microtubule dynamics. *J Struct Biol* **176:** 168–184. doi:10.1016/j.jsb.2011.07.009

Belin BJ, Goins LM, Mullins RD. 2014. Comparative analysis of tools for live cell imaging of actin network architecture. *Bioarchitecture* **4:** 189–202. doi:10.1080/19490992.2014.1047714

Berginski ME, Gomez SM. 2013. The Focal Adhesion Analysis Server: a web tool for analyzing focal adhesion dynamics. *F1000Res* **2:** 68. doi:10.12688/f1000research.2-68.v1

Burkel BM, Von Dassow G, Bement WM. 2007. Versatile fluorescent probes for actin filaments based on the actin-binding domain of utrophin. *Cell Motil Cytoskeleton* **64:** 822–832. doi:10.1002/cm.20226

Danuser G, Waterman-Storer CM. 2006. Quantitative fluorescent speckle microscopy of cytoskeleton dynamics. *Annu Rev Biophys Biomol Struct* **35:** 361–387. doi:10.1146/annurev.biophys.35.040405.102114

Flores LR, Keeling MC, Zhang X, Sliogeryte K, Gavara N. 2019. Author Correction: Lifeact-TagGFP2 alters F-actin organization, cellular morphology and biophysical behaviour. *Sci Rep* **9:** 9507. doi:10.1038/s41598-019-45276-y

Honnappa S, Gouveia SM, Weisbrich A, Damberger FF, Bhavesh NS, Jawhari H, Grigoriev I, van Rijssel FJA, Buey RM, Lawera A, et al. 2009. An EB1-binding motif acts as a microtubule tip localization signal. *Cell* **138:** 366–376. doi:10.1016/j.cell.2009.04.065

Jacquemet G, Paatero I, Carisey AF, Padzik A, Orange JS, Hamidi H, Ivaska J. 2017. FiloQuant reveals increased filopodia density during breast cancer progression. *J Cell Biol* **216:** 3387–3403. doi:10.1083/jcb.201704045

Kerstein PC, Nichol RH, Gomez TM. 2015. Mechanochemical regulation of growth cone motility. *Front Cell Neurosci* **9:** 244. doi:10.3389/fncel.2015.00244

Legerstee K, Geverts B, Slotman JA, Houtsmuller AB. 2019. Dynamics and distribution of paxillin, vinculin, zyxin and VASP depend on focal adhesion location and orientation. *Sci Rep* **9:** 10460. doi:10.1038/s41598-019-46905-2

Lowery LA, Faris AE, Stout A, Van Vactor D. 2012. Neural explant cultures from *Xenopus laevis*. *J Vis Exp* e4232. doi:10.3791/4232

Mendoza MC, Besson S, Danuser G. 2012. Quantitative fluorescent speckle microscopy (QFSM) to measure actin dynamics. *Curr Protoc Cytom* **Chapter 2:** Unit2.18. doi:10.1002/0471142956.cy0218s62

Milet C, Monsoro-Burq AH. 2014. Dissection of *Xenopus laevis* neural crest for in vitro explant culture or in vivo transplantation. *J Vis Exp* 51118. doi:10.3791/51118

Myers JP, Gomez TM. 2011. Focal adhesion kinase promotes integrin adhesion dynamics necessary for chemotropic turning of nerve growth cones. *J Neurosci* **31:** 13585–13595. doi:10.1523/JNEUROSCI.2381-11.2011

Nieuwkoop PD, Faber J. 1994. *Normal table of* Xenopus laevis *(Daudin)*. Garland Publishing Inc, New York.

Riedl J, Crevenna AH, Kessenbrock K, Yu JH, Neukirchen D, Bista M, Bradke F, Jenne D, Holak TA, Werb Z, et al. 2008. Lifeact: a versatile marker to visualize F-actin. *Nat Methods* **5:** 605–607. doi:10.1038/nmeth.1220

Robles E, Gomez TM. 2006. Focal adhesion kinase signaling at sites of integrin-mediated adhesion controls axon pathfinding. *Nat Neurosci* **9:** 1274–1283. doi:10.1038/nn1762

Santiago-Medina M, Gregus KA, Gomez TM. 2013. PAK-PIX interactions regulate adhesion dynamics and membrane protrusion to control neurite outgrowth. *J Cell Sci* **126:** 1122–1133. doi:10.1242/jcs.112607

Schell MJ, Erneux C, Irvine RF. 2001. Inositol 1,4,5-trisphosphate 3-kinase A associates with F-actin and dendritic spines via its N terminus. *J Biol Chem* **276:** 37537–37546. doi:10.1074/jbc.M104101200

Schindelin J, Arganda-Carreras I, Frise E, Kaynig V, Longair M, Pietzsch T, Preibisch S, Rueden C, Saalfeld S, Schmid B, et al. 2012. Fiji: an open-source platform for biological-image analysis. *Nat Methods* **9:** 676–682. doi:10.1038/nmeth.2019

Stepanova T, Slemmer J, Hoogenraad CC, Lansbergen G, Dortland B, De Zeeuw CI, Grosveld F, van Cappellen G, Akhmanova A, Galjart N. 2003. Visualization of microtubule growth in cultured neurons via the use of EB3-GFP (end-binding protein 3-green fluorescent protein). *J Neurosci* **23:** 2655–2664. doi:10.1523/JNEUROSCI.23-07-02655.2003

Stout A, D'amico S, Enzenbacher T, Ebbert P, Lowery LA. 2014. Using plusTipTracker software to measure microtubule dynamics in *Xenopus laevis* growth cones. *J Vis Exp* e52138. doi:10.3791/52138

Stutchbury B, Atherton P, Tsang R, Wang DY, Ballestrem C. 2017. Distinct focal adhesion protein modules control different aspects of mechanotransduction. *J Cell Sci* **130:** 1612–1624. doi:10.1242/jcs.195362

Tinevez JY, Perry N, Schindelin J, Hoopes GM, Reynolds GD, Laplantine E, Bednarek SY, Shorte SL, Eliceiri KW. 2017. TrackMate: an open and extensible platform for single-particle tracking. *Methods* **115:** 80–90. doi:10.1016/j.ymeth.2016.09.016

Watanabe N, Mitchison TJ. 2002. Single-molecule speckle analysis of actin filament turnover in *Lamellipodia*. *Science* **295:** 1083–1086. doi:10.1126/science.1067470

Wlizla M, McNamara S, Me H. 2018. Generation and care of *Xenopus laevis* and *Xenopus tropicalis* embryos. *Methods Mol Biol* **1865:** 19–32. doi:10.1007/978-1-4939-8784-9_2

Worth DC, Parsons M. 2010. Advances in imaging cell-matrix adhesions. *J Cell Sci* **123:** 3629–3638. doi:10.1242/jcs.064485

Yamashiro S, Mizuno H, Smith MB, Ryan GL, Kiuchi T, Vavylonis D, Watanabe N. 2014. New single-molecule speckle microscopy reveals modification of the retrograde actin flow by focal adhesions at nanometer scales. *Mol Biol Cell* **25:** 965–1185. doi:10.1091/mbc.e13-03-0162

Xenopus Tadpole Craniocardiac Imaging Using Optical Coherence Tomography

Engin Deniz,[1,3] Emily K. Mis,[1] Maura Lane,[1] and Mustafa K. Khokha[1,2]

[1]Pediatric Genomics Discovery Program, Department of Pediatrics, [2]Department of Genetics, Yale University, New Haven, Connecticut 06520, USA

Optical coherence tomography (OCT) imaging can be used to visualize craniocardiac structures in the *Xenopus* model system. OCT is analogous to ultrasound, utilizing light instead of sound to create a gray-scale image from the echo time delay of infrared light reflected from the specimen. OCT is a high-speed, cross-sectional, label-free imaging modality, which can outline dynamic in vivo morphology at resolutions approaching histological detail. OCT imaging can acquire 2D and 3D data in real time to assess cardiac and facial structures. Additionally, during cardiac imaging, Doppler imaging can be used to assess the blood flow pattern in relation to the intracardiac structures. Importantly, OCT can reproducibly and efficiently provide comprehensive, nondestructive in vivo cardiac and facial phenotyping. Tadpoles do not require preprocessing and thus can be further raised or analyzed after brief immobilization during imaging. The rapid development of the *Xenopus* model combined with a rapid OCT imaging protocol allows the identification of specific gene/teratogen phenotype relationships in a short period of time. Loss- or gain-of-function experiments can be evaluated in 4–5 d, and OCT imaging only requires ∼5 min per tadpole. Thus, we find this pairing an efficient workflow for screening numerous candidate genes derived from human genomic studies to in-depth mechanistic studies.

MATERIALS

It is essential that you consult the appropriate Material Safety Data Sheets and your institution's Environmental Health and Safety Office for proper handling of equipment and hazardous materials used in this protocol.

RECIPES: Please see the end of this protocol for recipes indicated by <R>. Additional recipes can be found online at http://cshprotocols.cshlp.org/site/recipes.

Reagents

Anesthetic agents (optional; see Step 1)

Benzocaine base USP (0.05%)

Prepare 5% stock in 100% ethanol in fume hood. Store at −20°C. Dilute 5% stock in distilled water to 0.05%. Adjust pH to 6.8–7.0 with sodium bicarbonate. Store at room temperature for up to 7 d.

Ethyl 4-aminobenzoate hydrochloride (benzocaine HCl USP; ≥250 mg/L)

Prepare fresh. Adjust pH to 6.8–7.0 with sodium bicarbonate. Use at room temperature.

[3]Correspondence: engin.deniz@yale.edu

Cite this protocol as *Cold Spring Harb Protoc*; doi:10.1101/pdb.prot105676

Tricaine (2 g/L)

Prepare a 5× stock (10 g/L) in distilled water in amber glass bottle in fume hood. Store for up to 1 mo or until the solution becomes discolored at −4°C. Protect from light. Prepare the 2 g/L solution fresh in distilled water from 5× stock. Adjust pH to 7.0–7.4 with sodium bicarbonate. Use at room temperature.

Low-melt agarose (Bio-Rad 1613111; 1%) in 1/9× or 1/3× MR
Modified Ringer's solution (MR; 1×) <R>

Dilute 1× MR with distilled water. Throughout the protocol use 1/9× MR for X. tropicalis and 1/3× MR for X. laevis.

Xenopus tropicalis or *laevis* stage 46–47 tadpoles (Nieuwkoop and Faber 1994)

For X. tropicalis, stage 46–47 is day 3 postfertilization at 26°C. For X. laevis, stage 46–47 is day 4 postfertilization when embryos are incubated for the first 24 h at 22°C and thereafter at 25°C.

Raise X. tropicalis as in Protocol: Obtaining Xenopus tropicalis Embryos by Natural Mating (Lane and Khokha 2021a) or as in Protocol: Obtaining Xenopus tropicalis Embryos by In Vitro Fertilization Xenopus tropicalis (Lane and Khokha 2021b). Raise X. laevis as in Protocol: Obtaining Xenopus laevis Embryos (Shaidani et al. 2021).

Equipment

Fine forceps
Ganymede Series SD-OCT System (Thorlabs)

System requirements include the following: 36 kHz A-scan rate, 101 dB sensitivity, 3 μm axial resolution in air with 1.9 mm imaging depth (900 nm center wavelength), 5-axis imaging stage established with Tip, Tilt, and Rotation Stage (Thorlabs) with 4° yaw/tilt and Thorlabs-OCT-XYR1-XY Linear Translation and Rotation Stage with Removable Solid Top Plate, and ThorImageOCT Software Package (2D mode for cross-sectional imaging and 3D mode for volumetric imaging).

Polystyrene Petri dish (35-mm × 10-mm)
Stereomicroscope
Stirring hot plate with temperature probe (e.g., VWR 97042-754)
Wax-based nonhardening clay (Van Aken Claytoon) (optional; see Step 1)

METHOD

Carry out all steps at room temperature. Choose either 3D craniofacial imaging (Steps 1–6) or 2D cardiac imaging (Steps 7–15) before proceeding to image processing (Step 16).

3D Craniofacial Imaging

1. Prepare low-melt agarose and immobilize tadpoles as follows.

   i. Warm 2.5 mL of 1% low-melt agarose to 60°C and maintain it in a liquid state using a stirring hot plate equipped with a temperature sensor.

   ii. Transfer 2.5 mL of liquid agarose to a 35-mm × 10-mm polystyrene Petri dish to cool it down to ∼30°C.

   iii. Transfer a stage 46–47 tadpole into the dish containing agarose, taking care to minimize the amount of 1/9× or 1/3× MR that is transferred.

   iv. While the agar solidifies (∼1 min), use a stereomicroscope and fine forceps to adjust the tadpole to ventral or dorsal exposure (based on the experiment), aligning the eyes as symmetrically as possible (Fig. 1A).

      As an alternative, coat the bottom of the Petri dish with clay and form a well (0.5 cm diameter) at the center. Immobilize the tadpole using a paralytic agent. Follow your institution's approved methods and drugs for anesthesia. Options include benzocaine base (0.05%; USP), ethyl 4-aminobenzoate hydrochloride (benzocaine HCl, USP; ≥250 mg/L), or tricaine (2 g/L). Incubate the tadpole with the agent for 2 min at room temperature and transfer to the well. Using the stereomicroscope and fine

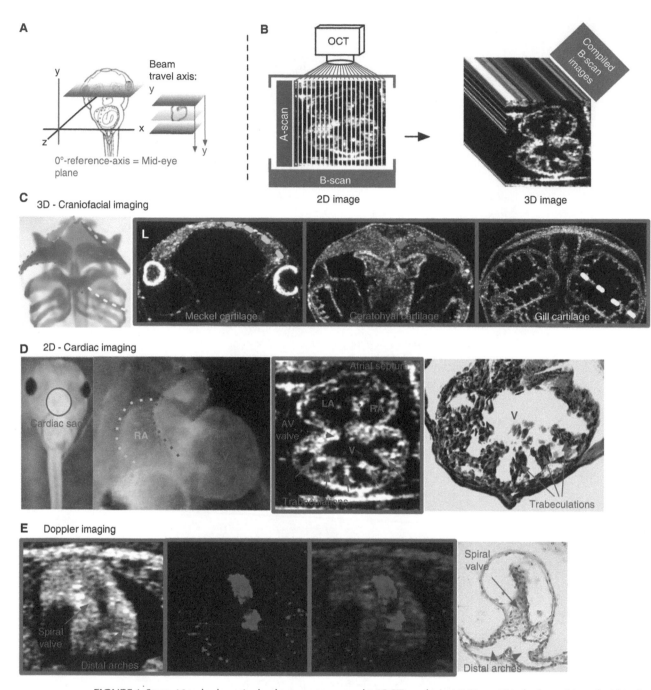

FIGURE 1. Stage 46 tadpole optical coherence tomography (OCT) analysis. (*A*) Stage 46 tadpole positioned within the clay or agar with its ventral side toward the OCT imaging plane. The imaging plane should be adjusted to capture a reference plane where the scan will intersect the tadpole's eyes symmetrically. Once the axis is aligned, optical sections can be obtained from any direction for heart or facial structures. (*B*) An A-scan (axial depth scan) contains the information of the reflected light with regard to the spatial dimension and location of the sample. When the A-scans are laterally combined, a cross-sectional image forms (B-scan). B-scans can also be obtained in series and combined to form a 3D image of the sample. (*C*) Alcian blue staining of the facial cartilage and OCT optical cross sections derived from the 3D data set shows the Meckel, Ceratohyal, and Gill cartilages of a stage 46 tadpole. (*D*) A stage 46 tadpole cardiac sac imaged under stereomicroscopy and OCT. For comparison of trabecular detail obtained with OCT, histology of the ventricle is also shown here. (*E*) OCT image of the heart focused on the outflow tract with Doppler imaging applied. Blood flow was recorded within the outflow tract. For comparison, histology of the outflow tract is also shown here. (2D) Two-dimensional, (3D) three-dimensional, (RA) right atrium, (LA) left atrium, (V) ventricle, (AV-valve) atrioventricular valve.

Cite this protocol as *Cold Spring Harb Protoc*; doi:10.1101/pdb.prot105676

TABLE 1. Standard parameters for stage 46 tadpole imaging

Standard scan parameters (stage 46 tadpole)	3D craniofacial imaging			2D cardiac imaging/Doppler imaging	
	x	y	z	x	y
Size	170–200 pixel	375–450 pixel	765 pixel	200–250 pixel	700–920 pixel
Field of view (FOV)	0.7–0.8 mm	1.5–1.8 mm	1 mm	0.8–1 mm	0.8–1.2 mm
Pixel size	4 µm	4 µm	1.31µm	4 µm	1.31 µm

forceps, gently manipulate the clay to support the tadpole from both sides to keep it straight. Adjust the head to align the eyes as symmetrically as possible.

2. Transfer the dish to the optical coherence tomography (OCT) imaging stage and position the tadpole so it is visible within the sample monitor.

   The live video image of the tadpole allows the user to adjust the focus and determine the scan position.

   See Troubleshooting.

3. Once the focus is set under the live video, use the 3D mode to draw a rectangular scan region directly onto the video image displayed by the red borders. Depending on the experiment, select either the whole cranium or the specific region of interest. Adjust the size and the field of view within the Scan Pattern controls.

   See Troubleshooting.

4. Set the parameters (see Table 1), and then view and scan the sample.

   See Troubleshooting.

5. Once the scan is complete, render a 3D data set within the software and display as a Volume View, Sectional View (Fig. 1B), or Surface View.

6. Once imaging is complete, extract tadpoles from the agar (or clay) with either multiple 1/9× or 1/3× MR washes or by gentle extraction with fine forceps.

2D Cardiac Imaging

7. Mechanically immobilize stage 46-47 tadpoles in 2.5 mL of 1% low melt agarose as described in Step 1.

8. Transfer the dish to the OCT imaging stage and position the tadpole to be visible within the sample monitor.

   The live video image of the tadpole will allow the user to adjust the focus and determine the scan position.

9. For consistency in determining cardiac structures, adjust the imaging plane to a reference axis. Here, we use the mid-eye axis as the reference axis. In 2D mode, using the live video image window embedded in the software, draw a line between the mid-eye points.

   The OCT probe beam progressively scans along this line at an angle perpendicular to the mid-eye axis, in an anterior-to-posterior (head-to-tail) direction. The probe then collects light reflected from the beam to form a reflectivity profile (A-scan) at each individual scan position. The A-scans collected along this line will be used to build a 2D cross-sectional image called the B-scan. Appropriate parameters are shown in Table 1.

10. To form the reference imaging plane, first adjust the tadpole along the linear axis and then along the other axes using the 5-axis stage (range: 4° yaw and tilt) to position the tadpole so the eyes align symmetrically along the horizontal plane and form a horizontally symmetric B-scan image.

    Once the mid-eye axis is symmetric, no further adjustments are required for the scanning parameters.

11. Using the linear stage adjustment knob, scan the cardiac sac from anterior to posterior (head-to-tail direction). As the beam scans through the cardiac sac, obtain optical sections, which can be

infinite, to analyze the intracardiac structures (e.g., outflow tract or atrioventricular valve), as well as the cardiac function.

> *See Troubleshooting.*

12. Once the region of interest has been captured, if desired, switch the 2D imaging mode to Doppler mode to measure the velocity of blood flow.

    > *By keeping the scan position in the same 2D imaging mode, an overlay of flow velocities will be apparent with flow colored to indicate positive (red) or negative (blue) flow velocity relative to the scanning direction.*

13. Two sets of data can be obtained from Doppler imaging—intensity.data and phase.data—containing intensity in decibel (dB) and phase values, respectively.

14. Compute the maximum absolute flow velocity (v_{max}) using the equation
    $$v_{max} = f \cdot \lambda/4,$$
    where f is the A-scan rate and λ is the central wavelength.

15. Once imaging is complete, extract tadpoles from the agar with either multiple 1/9× or 1/3× MR washes or by gentle extraction with fine forceps.

Image Processing

16. Analyze the images using the embedded tools within the OCT software as follows. Consult Figure 2A,B for the application to measure ventricular diameter and myocardial thickness from various angles.

    i. Using the marker tool (Fig. 2A1), measure the size of structures in the B-scan image. To do this, click on the add line marker button, mark any point on the B-scan as the start point with a second click, and define the end point with a third click.

       > *A new window (Fig. 2A4) populates each data point and shows their coordinates, length, and angle to the x-axis. The data from this window can be exported as a.csv file.*

    ii. Repeat measurements at various frames to capture the end-systolic and end-diastolic lengths.

       > *Alternatively, export the OCT files as TIFF files, and use the image-processing platform ImageJ to compute area and volume measurements. (Alternative file formats include PNG, JPG, BMP, PDF, FITS, VTK, VFF, TXT, and CSV.)*

TROUBLESHOOTING

Problem (Steps 2 and 3): 3D imaging is blurry.

Solution: In 3D mode, three-dimensional OCT scans acquire stacks of B-scans; therefore, sample movement will be detrimental and may lead to blurry images. Imaging within a well may cause tadpole movements even when using paralytic agents. An air table and/or filtered air circulation can minimize these movements within the well. The OCT imaging system is highly sensitive to vibrations in addition to tadpole movement.

Problem (Steps 4 and 11): Some tadpoles in a field of agarose produce high-quality OCT images, whereas others do not.

Solution: In both 3D and 2D imaging, tadpoles may be embedded at different depths within the agar. Those embedded too close to the surface will cause irregularities in the surface of the agar, leading to excessive light reflection and ultimately to a distorted image. Conversely, tadpoles positioned too deeply will be out of the imaging range because of limitations in the imaging depth of OCT systems (the listed system is limited to 1.9 mm imaging depth). Although 2.5 mL is ideal for

Cite this protocol as *Cold Spring Harb Protoc*; doi:10.1101/pdb.prot105676

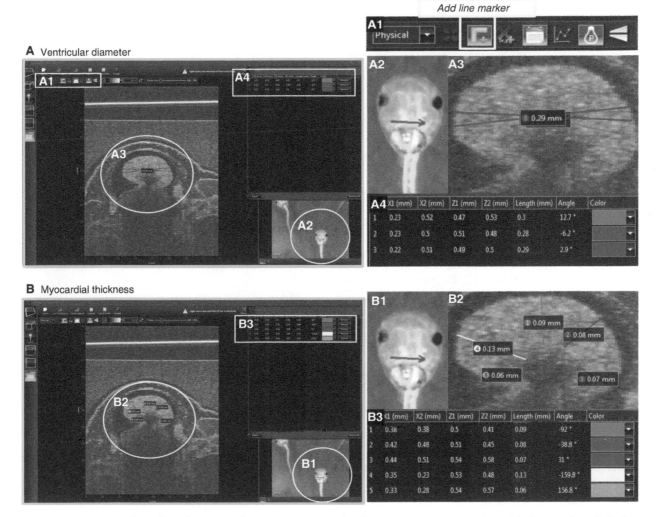

FIGURE 2. ThorLabs software interface. (*A1–4*) Ventricular diameter measurements. (*A1*) Add line marker clicked to draw lines for length measurements. (*A2*) Sample monitor shows the stage 46 tadpole positioned within the agar with the ventral side towards the OCT imaging plane. (*A3*) Each line colored and (*A4*) populated to a new window with coordinates, angle to the *x*-axis and length for ventricular diameter measurement. (*B1–3*) Myocardial thickness from various angles can also be measured with same steps.

embedding, starting with 2 mL initially can allow one to embed the tadpoles, assess image quality, and then add more agarose to the dish to achieve the optimal depth as well as a smooth imaging surface.

Problem (Step 11): Cardiac function is depressed even in control embryos.

Solution: Prolonged exposure (>5 min) to paralytics will depress tadpole cardiac function and affect cardiovascular assessment under OCT imaging. For cardiac imaging, tadpoles should not be paralyzed before embedding in agarose. Depressed cardiac function is not an issue for craniofacial imaging, thus paralytic agents can be used to improve the quality of 3D data sets.

DISCUSSION

Current advances in optical imaging enable comprehensive, high-resolution imaging in small animal models, which facilitates studying gene–phenotype relationships. OCT imaging (Boppart et al. 1996) can outline dynamic in vivo morphology to visualize craniocardiac structures in the *Xenopus* model system in a nondestructive fashion. (Deniz et al. 2017; Landim-Vieira et al. 2019; Date et al. 2020; Dur et al. 2020; Marquez et al. 2020). When coupled with OCT imaging, the *Xenopus* model can be used for cardiac phenotype evaluation of candidate genes identified in large-scale sequencing projects for craniofacial defects and congenital heart disease. Here we establish an efficient standard operating protocol for tadpole craniocardiac phenotyping.

RECIPE

Modified Ringer's Solution (MR; 1×)

0.1 M NaCl
1.8 mM KCl
2.0 mM $CaCl_2$
1.0 mM $MgCl_2$
5.0 mM HEPES (pH 7.6)

Store for up to 1 mo at room temperature.

REFERENCES

Boppart SA, Brezinski ME, Bouma BE, Tearney GJ, Fujimoto JG. 1996. Investigation of developing embryonic morphology using optical coherence tomography. *Dev Biol* 177: 54–63. doi:10.1006/dbio.1996.0144

Date P, Ackermann P, Furey C, Fink IB, Jonas S, Khokha MK, Kahle KT, Deniz E. 2020. Author Correction: Visualizing flow in an intact CSF network using optical coherence tomography: implications for human congenital hydrocephalus. *Sci Rep* 10: 2791. doi:10.1038/s41598-020-59301-y

Deniz E, Jonas S, Hooper M JNG, Choma MA, Khokha MK. 2017. Analysis of craniocardiac malformations in *Xenopus* using optical coherence tomography. *Sci Rep* 7: 42506. doi:10.1038/srep42506

Dur AH, Tang T, Viviano S, Sekuri A, Willsey HR, Tagare HD, Kahle KT, Deniz E. 2020. In *Xenopus* ependymal cilia drive embryonic CSF circulation and brain development independently of cardiac pulsatile forces. *Fluids Barriers CNS* 17: 72. doi:10.1186/s12987-020-00234-z

Landim-Vieira M, Johnston JR, Ji W, Mis EK, Tijerino J, Spencer-Manzon M, Jeffries L, Hall EK, Panisello-Manterola D, Khokha MK, et al. 2019.

Familial dilated cardiomyopathy associated with a novel combination of compound heterozygous TNNC1 variants. *Front Physiol* 10: 1612. doi:10.3389/fphys.2019.01612

Lane M, Khokha MK. 2021a. Obtaining *Xenopus tropicalis* embryos by natural mating. *Cold Spring Harb Protoc* doi:10.1101/pdb.prot106609.

Lane M, Khokha MK. 2021b. Obtaining *Xenopus tropicalis* embryos by in vitro fertilization. *Cold Spring Harb Protoc* doi:10.1101/pdb.prot106351.

Marquez J, Mann N, Arana K, Deniz E, Ji W, Konstantino M, Mis EK, Deshpande C, Jeffries L, McGlynn J, et al. 2020. DLG5 variants are associated with multiple congenital anomalies including ciliopathy phenotypes. *J Med Genet* doi:10.1136/jmedgenet-2019-106805

Nieuwkoop PD, Faber J. 1994. *Normal table of* Xenopus laevis *(Daudin): a systematical and chronological survey of the development from the fertilized egg till the end of metamorphosis.* Garland, New York.

Shaidani N-I, McNamara S, Wlizla M, Horb ME. 2021. Obtaining *Xenopus laevis* embryos. *Cold Spring Harb Protoc* doi:10.1101/pdb.prot106211.

Cite this protocol as *Cold Spring Harb Protoc*; doi:10.1101/pdb.prot105676

Protocol 5

Chambers for Culturing and Immobilizing *Xenopus* Embryos and Organotypic Explants for Live Imaging

Chih-Wen Chu[1] and Lance A. Davidson[1,2,3,4]

[1]*Department of Bioengineering, Swanson School of Engineering,* [2]*Department of Developmental Biology, School of Medicine,* [3]*Department of Computational and Systems Biology, School of Medicine, University of Pittsburgh, Pittsburgh, Pennsylvania 15213, USA*

Live imaging of *Xenopus* embryos and organotypic explants can be challenging because of their large size and slippery nature. This protocol covers the preparation of special chambers for immobilizing *Xenopus* embryos and embryonic explants for live-cell and tissue imaging. The opaque nature of *Xenopus* embryonic tissues enables simple bright-field imaging techniques for tracking surface movements across large regions. Such surface imaging of embryos or organotypic explants can directly reveal cell behaviors, obviating the need for complex postprocessing commonly required to extract this data from 3D confocal or light-sheet observations of more transparent embryos. Furthermore, *Xenopus* embryos may be filled with light-absorbing pigment granules and light-scattering yolk platelets, but these limitations are offset by the utilitarian nature of *Xenopus* organotypic explants that expose and stabilize large embryonic cells in a nearly native context for high-resolution live-cell imaging. Additionally, whole embryos can be stabilized for long-term bright-field and confocal microscopy. Simple explants can be prepared using a single cell type, and organotypic explants can be prepared in which multiple tissue types are dissected while retaining native tissue–tissue interactions. These preparations enable both in-toto imaging of tissue dynamics and super-resolution imaging of protein dynamics within individual cells. We present detailed protocols for these methods together with references to applications.

MATERIALS

It is essential that you consult the appropriate Material Safety Data Sheets and your institution's Environmental Health and Safety Office for proper handling of equipment and hazardous materials used in this protocol.

RECIPES: Please see the end of this protocol for recipes indicated by <R>. Additional recipes can be found online at http://cshprotocols.cshlp.org/site/recipes.

Reagents

Agarose, both standard molecular biology grade and ultra-low melting temperature
(Type IX-A; Sigma-Aldrich A2576)
Danilchik's for Amy (DFA) medium (pH 8.3) <R>

Use DFA with 1 mg/mL BSA and 1% antibiotic/antimycotic (Sigma-Aldrich A5955) as explant culture medium.

Fibronectin (20 µg/mL) in PBS (optional; see Step 8)
Fingernail polish

Sally Hansen Hard as Nails is more resistant to fixatives and organic solvents used during postprocessing.

---

[4]Correspondence: lad43@pitt.edu

Fluorescently labeled dextran (tetramethylrhodamine, 70-kDa, lysine fixable; Invitrogen Life Sciences D1818) (optional; see Step 14)

Follow Lane and Sheets (2005) to prepare RNase-free fluorescent dextrans.

Modeling clay (Van Aken Plastalina; black)

Modified Barth's solution (MBS) (pH 7.4) <R>

Dilute to 1/3× MBS to use as embryo culture medium.

mRNA encoding fluorescently tagged reporter (optional; see Step 14)

Silicone grease (high vacuum silicone grease, Dow Chemical)

Xenopus laevis (Nieuwkoop and Faber 1967) or *Xenopus tropicalis* (Khokha et al. 2002) embryos or explants as appropriate

Dejelly embryos as previously described (Sive et al. 2000) before use.

Equipment

Alcohol burner

Cold plate (optional; see Step 6)

3D printed "stamp" (Truchado-Garcia et al. 2018) (optional; see Step 5)

Diamond pencil (retractable; Ted Pella 54468)

Fluid sealed chamber (one of the following)

Commercial imaging chamber (e.g., Invitrogen Attofluor A7816)

Custom-milled acrylic chambers (cut from 1/4″ thick alcohol-resistant acrylic sheet; Figs. 1 and 2)

Glass-bottom Petri dish (e.g., CellVis 35-mm glass-bottom dish [Fisher Scientific NC0409658])

Forceps, both fine and coarse (Fine Science Tools)

Glass coverslips (#1.5) of various sizes (e.g., 18-mm-round [Fisher Scientific 50-948-975])

Glass slides

Gooseneck LED lamp

We prefer no-name brands from Amazon for <$100/pair.

Hair tools, custom-made to various sizes (see Kim and Davidson 2013a)

Heating block

Image acquisition software (e.g., Micromanager)

Incubator

Inverted confocal microscope, compound microscope, or stereo dissecting microscope equipped with CCD or CMOS camera (computer-controlled XYZ stage optional)

NiTex mesh (inner diameter ~1 mm; Amazon Small Parts CMN-1000) (optional; see Step 4)

Nylon monofilament suture thread #7-0 (50 µm diameter) (optional; see Step 7)

 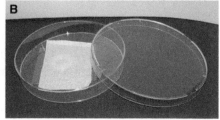

FIGURE 1. Different types of chambers for culturing embryos and organotypic explants. (*A*) Acrylic chambers can be milled from 1/4″ thick sheets. Acrylic chambers can be mounted on 45 × 50 mm coverslips using silicone grease and sealed along the top with smaller coverslips. After experiments the acrylic pieces can be recovered, cleaned in ethanol, and reused. Alternatively, a disposable chamber can be assembled by mounting a nylon washer onto a coverslip using nail polish. Once dried, the nylon washer can be filled with media and two to three explants and sealed with a circular coverslip. (*B*) Prior to imaging, the lower surface of the coverslip should be protected from dust and contamination in a Petri dish lined with a lint-free towel.

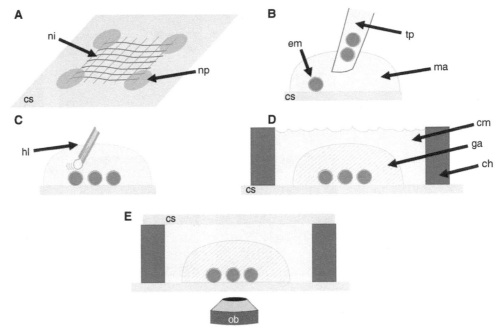

FIGURE 2. Mounting embryos in nylon mesh or molten agarose. (*A*) A small piece of nylon mesh (NiTex, ni) is glued at the corners to the coverslip (cs) with nail polish (np). NiTex mesh can be adapted to several different types of chambers. Culture media can be added after drying overnight. (*B*) Care is taken to avoid transferring culture media (yellow) by transfer pipette (tp) with embryos (em) moved into molten agarose (ma). (*C*) Embryos are manipulated with a hair loop tool (hl) to position the surface-of-interest within the working distance of the coverslip. (*D*) Additional culture media (cm) is added once the agarose has solidified (gelled agarose, ga) to fill the chamber (ch). (*E*) The chamber, gelled agarose, and media are sealed with minimal air pockets by a top coverslip (cs) to enable long-term stable imaging with an appropriate objective (ob) on an inverted microscope.

Nylon washers (e.g., Grainger Industrial Supply 4DAW9) (optional; see Step 12)
Oversize (45 × 50-mm, #1.5) glass coverslips (Brain Research Laboratories 4550-1.5D)
Petri dishes, various sizes
Pipette tips (1-mL)
Scalpel
Syringe (5-mL)
Transfer pipettes

METHOD

Preparation of Glass Coverslip Fragments

Throughout the protocol, fragments of glass coverslips are used. These can be made ahead of time to a range of sizes, whether long and narrow (e.g., Fig. 3D,E) or shorter to immobilize a single explant. Making glass coverslip fragments will produce sharp shards of glass. Take care to clear broken fragments from the work surface.

1. Prepare glass coverslip fragments.

   i. Place a large glass slide on the stage. Cut the coverslip fragments on this slide.

   ii. Select a cover glass and place it on the large glass slide.

   iii. Hold a smaller glass slide as a guide on top of the cover glass as a guide along the cut you wish to make.

   iv. Run the diamond pencil over the guide to score and break the coverslip to make the desired fragments—for example, long and narrow cover-glass fragments (Fig. 3D,E).

   v. Use forceps to transfer fragments to a Petri dish (60-mm) for storage.

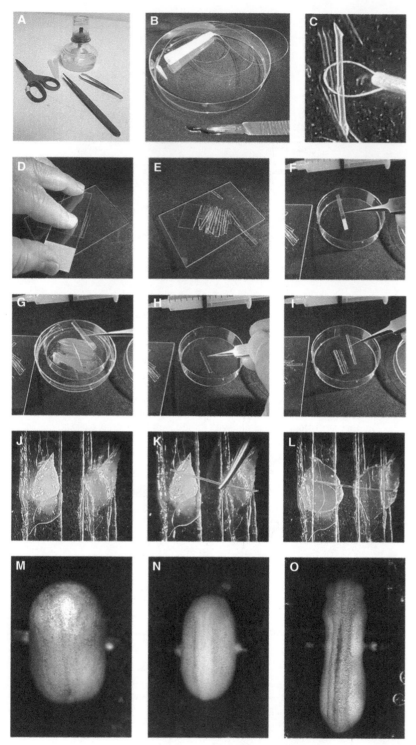

FIGURE 3. Embryos may be suspension-mounted to allow the largest range of morphogenetic movements. Mounting posts can be prepared from polystyrene 1-mL pipette tips pulled after heating. (*A*) Tools needed to pull out tips. (*B*) A pulled pipette tip can be cut to desired lengths with a scalpel and stored in a Petri dish. (*C*) The size of the mounting post can be chosen to minimize damage to the embryo and immobilize it within a narrow channel (e.g., "canyon") made of cut coverslip fragments (*D*–*F*). (*G*–*I*) The channel is assembled by gluing long strips of glass with silicone grease into stacks. (*J*–*L*) A dab of silicone grease on the top of the stack is used to secure the mounting post after addition of a final coverslip fragment. (*M*–*O*) Postgastrulation embryos are sufficiently robust to tolerate a mounting post and can develop until tadpole stages.

Cite this protocol as *Cold Spring Harb Protoc*; doi:10.1101/pdb.prot107649

Preparation of Chambers for Whole Embryos

Seven methods for preparation of whole embryos and chambers (Figs. 1 and 2) for imaging are presented here. Choose the option that best fits your experimental needs.

2. Prepare embryos for imaging loose in a dish.

    Time-lapse sequences of loose embryos can be used in a preliminary study to coarsely identify where and when a major developmental movement is altered. Sequences of loose embryos are best collected using a stereomicroscope. Embryos placed in a dish should be provided sufficient media to fill the space between the embryos and an oversize glass coverslip over the top of the dish. The coverslip provides a high-quality optical surface for the stereoscope and, furthermore, prevents evaporation of media or condensation on the underside of the lid.

    i. Select previously dejellied embryos with intact vitelline membranes of the appropriate stage for your experimental purposes (Nieuwkoop and Faber 1967). Ensure that the embryos to be imaged are synchronously staged. Use a transfer pipette to move embryos to a 25-mm Petri dish containing 1/3× MBS to a density of no more than five embryos per mL of culture media.

    ii. (Optional) Form a "fenced area" with coverslip fragments to prevent the embryos from drifting out of the camera field of view.

    iii. Fully fill the dish with 1/3× MBS.

    iv. Cover the dish with a 45 × 50-mm glass coverslip, taking care not to trap bubbles under the surface. Media may overflow, so it is helpful to place the entire dish within a larger Petri dish.

    v. Position the lamp and the embryos to capture the desired field of view with the best lighting.

    vi. Begin imaging.

3. Prepare embryos for imaging when pressed under a coverslip fragment.

    Time-lapse sequences of compressed whole embryos can reveal coarse tissue movements, cell-scale behaviors such as mitosis, intercalation, or rearrangement, and local protein dynamics such as actomyosin remodeling. Such sequences can be collected with the stereoscope, bright-field, or confocal microscopes. This step describes imaging with a stereoscope but may be altered by using an acrylic chamber with the embryo facing the bottom of the chamber.

    i. Select embryos and transfer them to a Petri dish containing 1/3× MBS as in Step 2.i.

    ii. Select a coverslip fragment that is ~2 mm × 10 mm and place dabs of silicone grease on either end.

    iii. Position the embryo and place the coverslip fragment over the top. Gently press the coverslip into place, alternating between the two ends until the embryo is firmly in place. As in Step 2, fill the dish with 1/3× MBS and cover with an oversize cover glass.

    iv. Position the dish and lighting appropriately and begin imaging with a stereoscope.

4. Prepare embryos for imaging in nylon mesh (NiTex).

    Multiple embryos from single clutches and different experimental treatments can be visualized simultaneously to provide developmental rates or identify lesions in specific events during morphogenesis. Time-lapse sequences of embryos trapped in NiTex mesh are best collected using an acrylic chamber mounted on a computer-controlled XYZ stage of a compound bright-field microscope. Multiple samples can be imaged simultaneously to provide comparative rates of morphogenesis or to contrast effects on morphogenetic movements such as blastopore closure (von Dassow et al. 2014). Embryos at some stages may rotate from the optimal alignment because of internal mechanical processes.

    i. Prepare an acrylic chamber by gluing it with silicone grease to an oversize coverslip base (Fig. 1). Dispense the grease using a syringe loaded with silicone grease to make a small watertight chamber that will hold media, NiTex mesh, and embryos.

    ii. Cut a small piece of NiTex mesh and use fingernail polish to glue the mesh to the bottom of the coverslip base (Fig. 2A). Make sure the fingernail polish does not transfer to the space to be occupied by embryos. Let the polish cure overnight. Chambers can be assembled and stored in a Petri dish indefinitely. When ready for use, fill with 1/3× MBS.

iii. As described in Step 2i, select embryos and transfer them to the chamber with the mesh. Take care to transfer only a small number of embryos into the mesh. Each embryo should settle into one mesh "cell."

iv. Gently use a hair tool to reposition embryos taking care not to puncture their vitelline membrane.

v. Fully fill the culture chamber with 1/3× MBS, dispense silicone grease along the top surface of the acrylic chamber, and place a second coverslip to seal the chamber. With practice, a small amount of overflow can be cleaned up to limit the formation of air bubbles within the chamber.

vi. Position the chamber on the stage of an inverted compound microscope with a suitably low magnification objective (e.g., 3.2×), position the LED lighting to highlight the area of interest, and begin imaging.

5. Prepare embryos for imaging in agarose wells.

Time-lapse sequences of embryos affixed in a specific location without the application of compressive forces can be used for quantitative measurement of developmental rates, cell behaviors, or protein dynamics. Images of embryos trapped in agarose wells (Truchado-Garcia et al. 2018), embryos fully embedded in agarose (see Step 6), embryos mounted on posts (Davidson and Keller 1999) (see Step 7), and embryos fixed to rigid substrates (Jackson et al. 2017) (see Step 8) can be collected with stereoscope, bright-field, or confocal microscopes. At low magnification (e.g., <10×), sequences of multiple samples can be collected simultaneously, whereas higher magnification samples may require multidimensional acquisition programs with postacquisition projection (e.g., maximum intensity) and stitching. The opaqueness of the Xenopus embryo has the advantage that cell behaviors at the surface can be directly visualized without requiring complex postprocessing to extract this data from 3D confocal or light-sheet observations of transparent embryos. In some cases when a dark background is preferred for better contrast to the whole embryo, oil-based modeling clay can be used instead of agarose to make wells to hold embryos in position.

Embryos at late stages after the body axis has elongated can be positioned precisely in agarose wells to stabilize anatomical features such as the forming mouth, fin, or tail bud for long-term imaging.

i. Prepare a mold shaped to the form of the embryo being studied using a 3D printed "stamp" (Truchado-Garcia et al. 2018).

ii. Prepare 1 mL molten agarose (standard grade) 2% w/v in 1/3× MBS, and pour into an acrylic chamber prepared as described in Step 4.i.

iii. Place the mold over the molten agarose, and wait for the agarose to harden.

iv. Add 1/3× MBS to the chamber, and remove the mold once it is loosened.

v. As in Step 2.i, select embryos, and transfer them to the chamber.

vi. Gently use a hair tool to reposition embryos into the agarose wells, taking care not to puncture their vitelline membranes.

vii. Follow Steps 4.v and 4.vi.

6. Prepare embryos embedded in agarose for imaging.

Embryos at early stages in which they are nearly spherical can be stabilized in agarose for higher-resolution imaging with short-working-distance objectives. Such imaging techniques can be used to correlate cell behaviors and protein dynamics with the production of mutant phenotypes.

i. Prepare an acrylic chamber as in Step 4.i.

ii. Dissolve ultra-low melting temperature agarose 2% w/v in ~1 mL 1/3× MBS at 65°C. Allow the solution to cool down to ~30°C–37°C, and then maintain the temperature by aliquoting and keeping the solution in a heating block. The temperature of the molten agarose should be ~25°C when mixed with the embryos, so the temperature of the heating block should be adjusted according to the actual room temperature.

iii. Transfer 250 µL molten agarose to the acrylic chamber and 250 µL in a Petri dish.

iv. Transfer selected embryos using a transfer pipette with minimal culture media to the small drop of molten agarose in the Petri dish, and gently mix them by slowly pipetting up and down (Fig. 2B). To ensure minimal media transfer, let embryos descend from the pipette tip to the agarose under the force of gravity.

v. Transfer embryos with a minimal amount of molten agarose from the Petri dish to the acrylic chamber. Use hair tools to gently position embryos in the acrylic chamber so that the field of interest is facing downward and slightly touching the bottom of the chamber (Fig. 2C). The embryos should be fully immersed in the agarose and positioned immediately adjacent to the glass coverslip—for example, within the working distance of the intended microscope objective. Make the final adjustments when the agarose starts to solidify, so that the embryos remain positioned afterward.

vi. Carefully place the chamber for 30 min or until the agarose has fully solidified in a temperature-controlled incubator set for 14°C. Alternatively, a cold plate can be used.

vii. As in the previous sections, completely fill the chamber with 1/3× MBS (Fig. 2D) and seal with a coverslip. Position on the stage of an inverted microscope (Fig. 2E), adjust lighting, and begin imaging.

7. Prepare embryos mounted on posts and suspended for imaging.

Post- or suspension-mounting embryos can be used to capture details of cell behaviors and tissue dynamics at stages where the embryo is rapidly elongating, such as postneurula, tail-bud, and early tadpole stages. At these stages, confinement in agarose typically leads to body flexure and repositioning, confounding efforts to track cell movements and dynamics.

i. Prepare an acrylic chamber as in Step 4.i.

ii. Prepare plastic posts by pulling disposable 1-mL pipette tips out over an alcohol flame (Fig. 3A–C). Take care to only melt the plastic pipette tip and not catch it on fire. With practice a single molten tip can be pulled to >1 m in length. The resulting plastic fiber will be thin (<30 μm in diameter) and rigid and can be cut with scissors and stored in a Petri dish. Alternatively, use 50-μm-diameter nylon suture thread.

iii. Fill the acrylic chamber with 1/3× MBS.

iv. Use 2 × 10-mm fragments of cut coverslip (Fig. 3D–F) and silicone grease to build a "canyon" in which the embryo will be placed. The height of the canyon should reach around half the thickness of the embryo. Because coverslip fragments are 170-μm-thick, several may need to be stacked using forceps (Fig. 3G–I). Place a thin layer of silicone grease along the top of the stack.

v. Select and transfer several embryos into the acrylic chamber prefilled with 1/3× MBS.

vi. Choose a plastic post cut to 2–3 mm in length. Hold the post with one forcep, and push it through an embryo along the mediolateral axis. Take care to leave the area of interest undamaged. Place the ends of the post on top of the canyon walls with the area of interest pressed gently to the bottom of the chamber. Then, press the post into the silicone grease (Fig. 3J,K). Repeat until the canyon is loaded with a sufficient number of samples.

vii. Place 2 × 10-mm fragments of cut coverslip on the top of the canyon walls to stabilize the post and press the embryo closer to the bottom of the chamber (Fig. 3L).

viii. As in the previous sections, fill the remainder of the chamber with 1/3× MBS, seal with a coverslip, position on the stage of an inverted microscope, adjust lighting, and begin imaging. The location pierced by the post will heal. Subsequent development may change the length of the embryo, but tissues near the position of the post will remain relatively fixed in place (Fig. 3M–O).

8. Prepare embryos fixed or adhered to a substrate at a fixed point for imaging.

As an alternative to Step 7, embryos can be affixed to a rigid substrate through cell substrate adhesion. Time-lapse sequences of embryos mounted in this manner are very stable and can be imaged over hours and days. Microscopy compatible with this mounting method is generally limited to low-resolution objectives but may be combined with motorized stage controls to collect extended depth of field (Forster et al. 2004) or large fields stitched from multi-image tiles (Hörl et al. 2019).

i. Prepare an acrylic chamber as in Step 4i but using a rectangular coverslip that matches the size of the chamber.

ii. Adsorb extracellular matrix such as fibronectin to the coverslip in the chamber. Incubate the chamber overnight at 4°C with 20 µg/mL of fibronectin in PBS. Rinse three times in 1/3× MBS and fill the chamber with 1/3× MBS.

iii. Select embryos and remove their vitelline membrane with forceps. Pinch the vitelline membrane with one forcep while using another to pinch the vitelline close by, and pull away to tear off the membrane.

iv. Intentionally wound embryos slightly with hair tools in a location 180° from the region of interest. Wounds do not need to be large or deep.

v. Immediately transfer the wounded embryo to the acrylic culture chamber.

vi. Position the embryo so the wound faces the fibronectin-coated coverslip and gently compress the embryo under a 2×10-mm coverslip fragment with silicone grease dollops at their ends. Compress the embryo by iteratively compressing either end of the coverslip fragment.

vii. Allow the embryo to rest for at least 30 min at room temperature and then gently remove the coverslip fragment.

viii. Fully fill the chamber with 1/3× MBS, seal with a large coverslip leaving little or no air bubble in the chamber, and flip the chamber so that the area of interest faces the microscope objective. Position the chamber on the stage of an inverted microscope, adjust lighting, and begin imaging. Imaging samples prepared in this manner requires the use of a long-working-distance objective or a stereomicroscope; however, the working distance may be reduced by using acrylic chambers machined from thinner sheets of acrylic or by mounting embryos with coverslip fragments that are themselves coated with fibronectin.

Preparation of Chambers for Organotypic Explants

9. Prepare explants for imaging loose in a dish.

   Time-lapse sequences of free-floating explants are typically used to assess large-scale morphogenetic movements such as convergent extension, vegetal rotation, or neurulation. In contrast to explants cultured on rigid adherent substrates, free-floating samples are not constrained mechanically and are typically able to show large motions reminiscent of those occurring in vivo.

   i. Follow Step 2 for a simple approach to collecting time-lapse sequences of free-floating explants using a stereomicroscope or low-magnification compound microscope. Use DFA explant culture media.

   ii. If desired, prepare a "fence" as described in Step 2.ii, keeping in mind that mesenchymal or wounded surfaces of explants can fuse to one another if they come into contact. Rates of tissue movement and strain rates can be collected from postacquisition analysis.

10. Prepare explants for short-term imaging under a coverslip or on an adhesive substrate.

    Time-lapse sequences of explants confined under a coverslip or attached to an adhesive substrate can be used to investigate cell behaviors and spatial and temporal protein dynamics. It is important to note that large-scale tissue rearrangements are limited in these preparations. Flat preparations such as these provide stability needed for high-resolution imaging and can stabilize either apical or basal cell surfaces so they can be imaged in a limited number of z-plane sections by confocal microscopy. Simple chambers such as those described below are sufficient for immediate imaging for proof of principle or to determine the levels of expressed proteins. The duration of time lapses collected should be minimized to 5–15 min because of evaporation and sample drift; longer-term time-lapse sequences or long-term culture should instead utilized closed-chamber methods (see Step 11 or 12) as these prevent evaporation and limit sample drift.

    i. Follow Step 3 for a simple approach to collecting time-lapse sequences from explants compressed under a coverslip or Step 8 for preparation of an adhesive substrate with the modifications described below.

Cite this protocol as *Cold Spring Harb Protoc*; doi:10.1101/pdb.prot107649

ii. Prepare explants and transfer to a drop of DFA explant culture media onto a large coverslip.

iii. Mount a smaller coverslip over the top of the explant on three or four dollops of silicone grease. Add DFA explant culture media to the edge of the small coverslip to fill the gap between the coverslips. Additional silicone grease can be added to the edge to seal the coverslips.

iv. Image explants prepared in this way immediately. Media can evaporate or wick out from between the two coverslips but that is typically enough time to confirm localization of fluorescent proteins or to track protein dynamics.

11. Prepare explants embedded in agarose for imaging.

Agarose is nonadherent and provides a nearly frictionless surface to the embedded explant. Explants embedded in agarose are physically constrained to a specific shape but are otherwise free to generate force. If the elastic modulus of the agarose is known, stresses generated by the explant that deform the agarose can be calculated (Zhou et al. 2015).

i. Follow Step 6 to prepare agarose and place explants within the gel for optimal viewing. Explants are more difficult to manipulate in the molten agarose and more subject to damage than whole embryos within their vitelline.

ii. Select agarose and culture media to match the cells and tissues being observed. For instance, DFA should be used for viewing mesenchymal cells, whereas 1/3× MBS should be used for observing the apical surface of ectoderm or epidermal progenitor cells. Two percent ultralow-melting temperature agarose should be prepared with either DFA without BSA or 1/3× MBS.

12. Prepare explants for long-term imaging under a coverslip or on an adhesive substrate.

Long-term, high-resolution imaging of cell behaviors and protein dynamics requires a highly stable fluid-sealed chamber and mount for sample tissues. Preparations of explants adhered to substrates within sealed culture chambers are required when samples must be kept in culture for >30 min or imaged over hours and days. Given the rapid development of Xenopus embryos, long-duration imaging can capture the progressive commitment, movement, and differentiation of cells from early morphogenesis through formation of organ primordia (e.g., Jackson et al. 2017).

i. Follow Step 3 for a simple approach to collecting time-lapse sequences from explants compressed under a coverslip or Step 8 above for preparation of an adhesive substrate with the modifications described below.

ii. Utilize glass-bottom dishes, custom acrylic, or commercial culture chambers for long-term (>60 min) observation of cell and tissue behaviors. Sealed chambers resist evaporation and limit drift that can accompany simple chambers made of glass. Alternatively, smaller chambers composed of nylon washers glued to coverslips with silicone grease can also be stable but hold a limited number of samples. The advantage of nylon washer chambers is the small volume, typically <100 µL. Such small volumes can preserve precious reagents such as function blocking antibodies.

Lighting and Imaging of Chambered Embryos and Explants

Three different lighting and imaging strategies are listed for embryos and explants mounted by the above techniques.

13. Image by grazing incidence or off-axis illumination microscopy.

Because of their opacity, immobilized embryos and explants require unique lighting techniques. The simplest and least-phototoxic method of imaging is to use grazing incidence, otherwise known as off-axis illumination with fiber-optic or LED light sources to collect time-lapse sequences on a stereomicroscope or inverted compound microscope. Because Xenopus embryos are not transparent, standard transmitted light illumination methods, such as phase or DIC microscopy, are of limited use. Instead, grazing incidence illumination can provide exceptional contrast over the face of an explant or surface of an embryo. Adjustable fiber-optic guides, or gooseneck lamps, have been used historically in the classical work on Xenopus convergent extension and mediolateral cell intercalation of John Shih and Ray Keller (Shih and Keller 1992a,b). Such illumination systems continue to be useful when imaging large-scale tissue movements that take place on the exterior of the embryo, such as those accompanying gastrulation (von Dassow et al. 2014; Feroze et al. 2015) or heart morphogenesis (Jackson et al. 2017). Sensitive CMOS cameras combined with modern low-power LEDs further reduce the potential for photodamage due to heat or excessive illumination that are nearly inescapable with high-intensity epifluorescence or laser-based confocal imaging modalities.

i. Prepare samples as described above and place the chambers on the stage of the compound microscope or on the stand of the stereomicroscope.

ii. Position two goosenecks below the sample, under the stage, so that they illuminate from two opposing directions. Adjust their position further so the sample is evenly illuminated on both sides. Conventional fiber-optic lamps can often generate heat at the ends of the fiber optic, so exercise caution. Keep the fiber optic 3–6 cm away from the chamber and turn the lamp power down as low as possible for the imaging.

14. Image by epifluorescence microscopy.

Epifluorescence microscopy can provide valuable information for lineage tracing and analysis of the behavior of cells at the surface of the sample. As pigment granules and yolk platelets greatly interfere with signals emitted from deeper regions, epifluorescence microscopy is only suitable for analysis of the cellular structures at the surface of the sample.

i. Prepare sample by injecting fluorescent dextran or mRNA encoding fluorescently tagged proteins into single blastomeres of embryos at the 2- to 64-cell stage as described (Sive et al. 2000).

ii. Mount whole embryo or explants in chambers that allow placement and imaging with an inverted epifluorescence microscope.

iii. Use appropriate filter blocks for the fluorescent probe.

iv. Increase the camera gain and exposure while reducing the excitation lamp intensity until a suitable image is acquired.

15. Image by confocal microscopy.

Compared to epifluorescence microscopy, confocal microscopy allows detailed analysis of signals within thick and opaque tissues. Because of the efficient capture of excitation light and independence of the signal from ballistic photons, confocal microscopy can record motions of cells from one to two cell layers (25–50 μm) deep in a scattering tissue. For development and preparation of fluorescent probes for live imaging, readers are directed to additional references (e.g., Woolner et al. 2009; Kieserman et al. 2010; Kim and Davidson 2013b).

i. Prepare samples as described in Step 14 and place them on the stage of an inverted confocal microscope.

ii. Use appropriate filter blocks for the fluorescent probe and adjust the laser power and exposure time until a suitable image is acquired. Keeping the laser power low is always preferred to avoid signal bleaching or damage of the sample.

iii. If the sample is large or heterogeneous, taking an image of the whole sample by tiling and stitching is very helpful to identify fields of interest. For instance, the plug-in SlideExplorer in Micromanager will automatically collect and present a stitched image and integrate with the identification of fields of interest.

iv. Acquire z-series of optical sections as needed. If both z-series and time-lapse sequences are needed, efforts should be made to reduce the acquisition time per time frame, so that the time stamp of the first and last z section is close enough to allow proper 3D reconstruction of the data later on.

DISCUSSION

In brief, this protocol seeks to provide different methods to culture and immobilize *Xenopus* embryos and tissue isolates for live imaging of embryonic processes. Several imaging modalities may be useful depending on the questions being asked. Furthermore, the type of mounting chamber used will further depend on the goals of the study—for instance, will short-term imaging be sufficient

Cite this protocol as *Cold Spring Harb Protoc*; doi:10.1101/pdb.prot107649

(e.g., <1 h) or will long-term time-lapse sequences be needed? A variety of chambers can be used depending on the duration and resolution desired. The simplest chamber consists of two coverslips sandwiched around a sample. Such chambers are prone to evaporation and drift of the sample from the image plane. To collect longer-duration time series we use a range of acrylic or nylon washer chambers (Fig. 1). Sealed chambers limit evaporation and increase the volume of culture media needed to support development. Successful live-cell and tissue imaging requires care, skill, and innovation, and these and other protocols (Davidson and Wallingford 2005; Kim and Davidson 2013c) are intended as a starting point to achieve those aims.

In contrast to whole embryos, tissue explants offer the advantage of bringing specific cell surfaces into the optimal working range of high-numerical-aperture objectives. Furthermore, mounted explants can stabilize cells with respect to the laboratory frame of reference by minimizing drift; in contrast, intact whole embryos within vitelline membranes often shift during morphogenesis, precluding long-term observation of single cells or tissue regions. Tissue explants are typically smaller and thinner than whole embryos and require a distinct set of approaches from those used to image whole embryos. A key point for the experimenter is the recognition that different explant preparations may constrain large-scale morphogenetic movements. For instance, explants adhered to an adhesive substrate show severely restricted convergent extension movements even while local cell behaviors and rearrangements are unaffected. Because they are unconstrained by the mechanical drag of the surrounding tissue, "free-floating" explants may show large-scale tissue shape changes that exceed those seen endogenously.

RECIPES

Danilchik's for Amy (DFA) Medium (pH 8.3)

Reagent	Final concentration
NaCl	53 mM
Na_2CO_3	5 mM
Potassium gluconate	4.5 mM
Sodium gluconate	32 mM
$CaCl_2$	1 mM
$MgSO_4$	1 mM

Adjust the pH to 8.3 with granular bicine. Filter with a 0.22 µm filter. Include 0.1% bovine serum albumen (BSA; Sigma-Aldrich A7906) to prevent nonspecific adhesion to substrates or glass coverslips. Just prior to use, add 1% of antibiotic/antimycotic (Sigma-Aldrich A5955). Prepare the DFA with BSA, aliquot, and store at −20°C. Keep aliquots of antibiotic/antimycotic at −20°C and add to thawed DFA before use. Replace DFA with medium containing fresh antibiotic/antimycotic if culturing explants for >24 h. (This recipe is adapted from Sater et al. 1993.)

Modified Barth's Solution (MBS) (pH 7.4)

Reagent	Final concentration (1×)
NaCl	88 mM
KCl	1 mM
NaHCO$_3$	2.4 mM
MgSO$_4$·7H$_2$O	0.82 mM
Ca(NO$_3$)$_2$·4H$_2$O	0.33 mM
CaCl$_2$·2H$_2$O	0.41 mM
HEPES	5 mM

Adjust the pH to 7.4 with NaOH; 3× stocks of MBS are pH-stable and can be stored at 4°C. Prepare a working solution (1/3×) by dilution.

ACKNOWLEDGMENTS

We thank Connie Lane, Ray Keller, John Shih, Paul Wilson, and many others whose lessons, both in person and in text, have informed our development of these tools for live-cell imaging. We thank members of the Davidson laboratory, past and present, especially Hye Young Kim, for refining these methods and improving their application. Our work is supported by a grant to L.A.D. from the *Eunice Kennedy Shriver* National Institute of Child Health and Human Development at the National Institutes of Health (R01 HD044750).

REFERENCES

Davidson LA, Keller RE. 1999. Neural tube closure in *Xenopus laevis* involves medial migration, directed protrusive activity, cell intercalation and convergent extension. *Development* 126: 4547–4556. doi:10.1242/dev.126.20.4547

Davidson LA, Wallingford JB. 2005. Visualizing cell biology and tissue movements during morphogenesis in the frog embryo. In *Imaging in neuroscience and development: a laboratory manual* (ed. Yuste R, Konnerth A), pp. 125–136. Cold Spring Harbor Laboratory Press, Cold Spring Harbor, NY.

Feroze R, Shawky JH, von Dassow M, Davidson LA. 2015. Mechanics of blastopore closure during amphibian gastrulation. *Dev Biol* 398: 57–67. doi:10.1016/j.ydbio.2014.11.011

Forster B, Van De Ville D, Berent J, Sage D, Unser M. 2004. Extended depth-of-focus for multi-channel microscopy images: a complex wavelet approach. In *Proceedings of the 2004 2nd IEEE International Symposium on Biomedical Imaging*, Vol. 1, pp. 660–663: *Nano to Macro (IEEE Cat No. 04EX821)*. IEEE, New York.

Hörl D, Rusak FR, Preusser F, Tillberg P, Randel N, Chhetri RK, Cardona A, Keller PJ, Harz H, Leonhardt H, et al. 2019. BigStitcher: reconstructing high-resolution image datasets of cleared and expanded samples. *Nat Methods* 16: 870–874. doi:10.1038/s41592-019-0501-0

Jackson TR, Kim HY, Balakrishnan UL, Stuckenholz C, Davidson LA. 2017. Spatiotemporally controlled mechanical cues drive progenitor mesenchymal-to-epithelial transition enabling proper heart formation and function. *Curr Biol* 27: 1326–1335. doi:10.1016/j.cub.2017.03.065

Khokha MK, Chung C, Bustamante EL, Gaw LW, Trott KA, Yeh J, Lim N, Lin JC, Taverner N, Amaya E, et al. 2002. Techniques and probes for the study of *Xenopus tropicalis* development. *Dev Dyn* 225: 499–510. doi:10.1002/dvdy.10184

Kieserman EK, Lee C, Gray RS, Park TJ, Wallingford JB. 2010. High-magnification in vivo imaging of *Xenopus* embryos for cell and developmental biology. *Cold Spring Harb Protoc* doi:10.1101/pdb.prot5427

Kim HY, Davidson LA. 2013a. Microsurgical approaches to isolate tissues from *Xenopus* embryos for imaging morphogenesis. *Cold Spring Harb Protoc* doi:10.1101/pdb.prot073874

Kim HY, Davidson LA. 2013b. Preparation and use of reporter constructs for imaging morphogenesis in *Xenopus* embryos. *Cold Spring Harb Protoc* doi:10.1101/pdb.prot073866

Kim HY, Davidson LA. 2013c. Assembly of chambers for stable long-term imaging of live *Xenopus* tissue. *Cold Spring Harb Protoc* doi:10.1101/pdb.prot073882

Lane MC, Sheets MD. 2005. Fate mapping hematopoietic lineages in the *Xenopus* embryo. *Methods Mol Med* 105: 137–148. doi:10.1385/1-59259-826-9:137

Nieuwkoop PD, Faber J. 1967. *Normal tables of* Xenopus laevis *(Daudin)*. North-Holland Biomedical Press/Elsevier, Amsterdam.

Sater AK, Steinhardt RA, Keller R. 1993. Induction of neuronal differentiation by planar signals in *Xenopus* embryos. *Dev Dyn* 197: 268–280. doi:10.1002/aja.1001970405

Shih J, Keller R. 1992a. Patterns of cell motility in the organizer and dorsal mesoderm of *Xenopus laevis*. *Development* 116: 915–930. doi:10.1242/dev.116.4.915

Shih J, Keller R. 1992b. Cell motility driving mediolateral intercalation in explants of *Xenopus laevis*. *Development* 116: 901–914. doi:10.1242/dev.116.4.901

Sive HL, Grainger RM, Harland RM. 2000. *Early development of* Xenopus laevis*: a laboratory manual*. Cold Spring Harbor Laboratory Press, Cold Spring Harbor, NY.

Truchado-Garcia M, Harland RM, Abrams MJ. 2018. 3D-printable tools for developmental biology: improving embryo injection and screening techniques through 3D-printing technology. bioRxiv doi:10.1101/376657

von Dassow M, Miller CJ, Davidson LA. 2014. Biomechanics and the thermotolerance of development. *PLoS One* 9: e95670. doi:10.1371/journal.pone.0095670

Woolner S, Miller AL, Bement WM. 2009. Imaging the cytoskeleton in live *Xenopus laevis* embryos. *Methods Mol Biol* 586: 23–39. doi:10.1007/978-1-60761-376-3_2

Zhou J, Pal S, Maiti S, Davidson LA. 2015. Force production and mechanical adaptation during convergent extension. *Development* 142: 692–701. doi:10.1242/dev.116533

Cite this protocol as *Cold Spring Harb Protoc*; doi:10.1101/pdb.prot107649

Genetics and Gene Editing Methods in *Xenopus laevis* and *Xenopus tropicalis*

Matthew J. Guille[1] and Robert M. Grainger[2,3]

[1]*European Xenopus Resource Centre, School of Biological Sciences, University of Portsmouth, Portsmouth PO1 2UP, United Kingdom;* [2]*Department of Biology, University of Virginia, Charlottesville, Virginia 22903, USA*

Our understanding of biological systems has for many years been heavily influenced by experimental approaches that exploit genetic methods. These include gain-of-function experiments that overexpress transgenes or ectopically express injected RNA and loss-of-function experiments that knock out genes or knock down RNAs. Here, we review how these methods have been applied in *Xenopus* frogs and introduce a variety of protocols for genetic manipulation of *Xenopus laevis* and *Xenopus tropicalis*.

INTRODUCTION

Xenopus laevis has been an organism in which some molecular genetic methods have been spectacularly successful—for example, producing ion channel proteins from synthetic or purified RNA injected into oocytes (for reviews, see Weber 1999; Zeng et al. 2020) or identifying a molecular component of the Spemann–Mangold organizer by using a gain-of-function (GOF) screen (Smith and Harland 1992). Loss-of-function (LOF) experiments were always more difficult than GOF experiments in *Xenopus*. Initially LOF was limited to introducing dominant interfering forms of proteins (e.g., Amaya et al. 1991; Schulte-Merker and Smith 1995); they improved with the advent of antisense morpholino oligonucleotides (AMOs) that block translation of a specific mRNA (for review, see Heasman 2002) or alter its splicing (e.g., Malartre et al. 2006; Timberlake et al. 2021). The use of AMOs requires careful control (Eisen and Smith 2008; Gentsch et al. 2018; Paraiso et al. 2019), including rescue experiments that can be very demanding. Nevertheless, key developmental and signaling processes have been revealed using these approaches in *Xenopus* (e.g., Collart et al. 2017; Singh et al. 2020). All these methods are limited to early development, however, and this had generally focused studies using *Xenopus* on early processes rather than organogenesis and events useful for modeling human genetic diseases.

MANIPULATING THE *XENOPUS* GENOME

The ability to manipulate the *Xenopus* genome itself first really became feasible with the advent of the restriction-enzyme-mediated integration (REMI) transgenesis method in 1996 (Kroll and Amaya 1996) and hundreds of transgenic lines have now been made (for review, see Horb et al. 2019).

---

[3]Correspondence: rmg9p@virginia.edu

Cite this introduction as *Cold Spring Harb Protoc*; doi:10.1101/pdb.top107045

Many of these allow particular cell types to be visualized or isolated (Kakebeen et al. 2020; Kakebeen et al. 2021) via the specific expression of fluorescent proteins. Lines that allow inducible and cell-type-specific expression at any developmental stage, including metamorphosis, have also been generated for GOF experiments (Rankin et al. 2011; Sterner et al. 2019). LOF methods that can be applied at later embryonic stages and during organogenesis are a much more recent arrival in the *Xenopus* system. One of the first improvements was the adoption of a second *Xenopus* species for LOF experiments, *Xenopus tropicalis*. This was important for the 66% of *X. tropicalis* genes in which *X. laevis* has homeologs (Session et al. 2016). A successful pilot, forward genetic screen using *N*-ethyl-*N*-nitrosourea (ENU) mutagenesis of spermatogonia was performed in *X. tropicalis* (for review, see Abu-Daya et al. 2012). The advent of high-quality genome sequences and gene editing techniques, however, completely rewrote the methodology for LOF studies in *Xenopus*.

It was apparent from work using the earliest double-stranded-DNA-cutting method, zinc-finger nucleases, that these would work well in *Xenopus* (Young et al. 2011), and like TALENS they led to very successful LOF studies (e.g., Young et al. 2017; Lei et al. 2012). In 2013, the arrival of CRISPR–Cas9 technology made knocking out genes in *Xenopus* extremely rapid and cost-efficient (Blitz et al. 2013; Nakayama et al. 2013), and with improved single-guide RNA (sgRNA) design and indel analysis programs it has become possible to analyze LOF phenotypes in F_0 animals, in which >90% of target genes can often sustain indels that cause frameshift mutations (Naert et al. 2020). Our thorough understanding of the *Xenopus* fate map (Dale and Slack 1987; Moody 1987) has enabled researchers to successfully perform targeted LOF experiments without having to resort to complex genetic approaches; these require only blastomere-specific microinjection (DeLay et al. 2018). Even animals with lethal mutations can be grown to adulthood and used to generate heterozygotes (Blitz et al. 2016). Most recently, methods that use CRISPR–Cas to target nucleotide-modifying enzymes to specific sites in the genome have also been used to make single base pair changes in the *Xenopus* genome (Park et al. 2017). These have the potential to combine CRISPR–Cas cutting with homology directed repair (HDR)-dependent DNA insertion as methods to engineer precise changes to the frog genome (Aslan et al. 2017; Nakayama et al. 2020).

These technical advances have unlocked the potential of *Xenopus* as a model for human genetic diseases (for reviews, see Sater and Moody 2017; Blum and Ott 2018), allowing it to sit comfortably between zebrafish and mice as an excellent compromise between similarity to humans and cost effectiveness. This is enabling investigation of the precise developmental mechanisms underpinning common diseases (e.g., Willsey et al. 2021), including cancers (for review, see Naert et al. 2017), and confirmation of links between specific genetic variations and rare disease phenotypes (Barbosa et al. 2020; Macken et al. 2021).

PROTOCOLS

The protocols referred to below include tools for genetic manipulation of *Xenopus* and supporting techniques that make these methods more efficient. They include transgenic approaches that can be used for labeling cells, organelles or specific proteins and for GOF studies. CRISPR–Cas methods for obtaining gene knockouts in F_0 animals and modeling human genetic disorders are described (see Protocol 1: CRISPR–Cas9 Mutagenesis in *Xenopus tropicalis* for Phenotypic Analyses in the F_0 Generation and Beyond [Blitz and Nakayama 2022]), as is the combination of CRISPR–Cas and DNA microinjection to allow investigators to make precise genomic changes—for example, to recreate human disease gene variants or perform epitope labeling of an endogenous protein.

Transgenesis can be used for a variety of purposes. For most of these, I-SceI-mediated transgenesis is ideal (see Protocol 6: I-SceI-Mediated Transgenesis in *Xenopus* [Noble et al. 2022a]). It is simple for almost all *Xenopus* researchers to perform because it involves microinjection of a nucleic-acid–protein mixture into a fertilized egg. A more technically demanding transgenesis technique involves a sperm nuclear transplantation protocol, historically known as "REMI" (restriction enzyme-mediated integration) (Kroll and Amaya 1996 [https://pubmed.ncbi.nlm.nih.gov/8898230/], Offield et al. 2000

Cite this introduction as *Cold Spring Harb Protoc*; doi:10.1101/pdb.top107045

[https://pubmed.ncbi.nlm.nih.gov/10751168/]). An updated protocol can be successfully performed without the restriction enzyme treatment that damages DNA in the earlier procedures (see Protocol 5: Production of Transgenic F_0 Animals and Permanent Lines by Sperm Nuclear Transplantation in *Xenopus tropicalis* [Nakayama et al. 2023c]); in this case a DNA-decondensed sperm nucleus mixture is injected into unfertilized eggs. The large needle required to keep the sperm nucleus intact during implantation causes significant damage to the egg. The survival rates are thus relatively low compared with macromolecule microinjection, but this approach is still used when large numbers of inserts are needed for transgenesis—for example, to produce high levels of fluorescence.

Knocking out genes in specific cells of the *Xenopus* embryo using tissue-targeted CRISPR–Cas9-mediated genome editing is a powerful way to test gene function and can be used in *X. laevis* even when the gene of interest has been duplicated (see Protocol 3: Tissue-Targeted CRISPR–Cas9-Mediated Genome Editing of Multiple Homeologs in F_0-Generation *Xenopus laevis* Embryos [Corkins et al. 2022]). This and related approaches have proven effective methods for modeling human genetic disorders (see Protocol 4: Modeling Human Genetic Disorders with CRISPR Technologies in *Xenopus* [Willsey et al. 2022]). Most of the CRISPR–Cas-based studies used so far in *Xenopus* have involved the generation of indels, but to model the majority of monogenic human diseases, precise changes, often heterozygous ones, are required. Approaches that use HDR to insert either long single-stranded (ss) DNA or single-stranded oligonucleotides into a site cut with CRISPR–Cas are becoming available to achieve this. Although oocyte-based gene editing can produce nonmosaic heterozygous embryos, it is a technique that is difficult to establish. Long ssDNA injection in eggs is much easier for most laboratories but the production of nonmosaic animals from these eggs requires breeding and holding large numbers of lines (see Protocol 2: Generating Nonmosaic Mutants in *Xenopus* Using CRISPR–Cas in Oocytes [Cha 2022] and Protocol 8: Homology-Directed Repair by CRISPR–Cas9 Mutagenesis in *Xenopus* Using Long Single-Stranded Donor DNA Templates via Simple Microinjection of Embryos [Nakayama et al. 2023b]).

Gene editing technology allows us to build a deeper understanding of the effects of mutations that affect early developmental decisions as well as those responsible for congenital diseases. The development of single-cell technology has provided a way to track the initial consequences in greater detail. A significant challenge is collecting the required number of cells from multiple mutant and wild-type embryos at early stages when morphological differences between the embryos cannot yet be detected. To accomplish this, embryos must be frozen and genotyped and then wild-type and mutant samples from multiple embryos pooled for expression studies. Although cells are destroyed during this procedure, nuclei can be isolated from separately pooled wild-type and mutant lysates and then subjected to single-nucleus RNA sequencing to determine differences in gene expression in mutant and wild-type embryos at the single-cell level (see Protocol 10: Preparation of Intact Nuclei for Single-Nucleus Omics Using Frozen Cell Suspensions from Mutant Embryos of *Xenopus tropicalis* [Nakayama et al. 2023d]).

To keep and transport genetically altered animals cost effectively, use space efficiently, and improve animal welfare, it is best to store as many lines as possible by cryopreserving sperm and then using in vitro fertilization (see Protocol 7: Cryopreservation of *Xenopus* Sperm and In Vitro Fertilization Using Frozen Sperm Samples [Noble et al. 2022b]). Gynogenetic production of embryos using a cold-shock procedure (see Protocol 9: Gynogenetic Production of Embryos in *Xenopus tropicalis* Using a Cold Shock Procedure: Rapid Screening Method for Gene Editing Phenotypes [Nakayama et al. 2023a]) produces homozygous mutants at high frequency from adult female *X. tropicalis*, often saving one round of breeding for phenotype analysis. Alongside the support offered by dedicated resource centers, these methods will make *Xenopus* an ever more effective model system for genetic studies in the future.

FUTURE CONSIDERATIONS

Constantly improving algorithms, both for design of sgRNAs and for predicting their effectiveness at indel production, are making it easier to perform efficient CRISPR–Cas experiments, and these will almost certainly continue to make LOF studies in *Xenopus* more and more successful. Less certain is which technology will dominate in the different approaches to making precise changes to the genome. Although base or prime editing have the potential to achieve such changes without the need to design, make, and coinject DNA constructs, they do have limitations in terms of targets and the size of changes that can be made. Currently preparation of ssDNA for HDR-mediated approaches can be challenging, but the methods available are becoming more straightforward, and this method is much more flexible than techniques that do not make double-stranded breaks in DNA. Insertion of large DNA fragments precisely into the genome using coinjection of lssDNA and CRISPR–Cas also has the potential to replace conventional transgenesis. Because one of the main challenges of making any transgenic line remains identifying the necessary DNA-binding elements to control the transgene, using those of an endogenous gene with the required expression pattern is attractive. An alternative strategy for transgenesis in *Xenopus* that does not require identification of DNA regulatory elements would be the use of BAC clones (Fish et al. 2012). Given that the methods for altering the *Xenopus* genome have changed completely over the past 8 years, it is difficult to predict what techniques will appear, but whatever happens *Xenopus* genetics are now well-established in the model's extensive experimental toolkit.

REFERENCES

Abu-Daya A, Khokha MK, Zimmerman LB. 2012. The hitchhiker's guide to *Xenopus* genetics. *Genesis* 50: 164–175. doi:10.1002/dvg.22007

Amaya E, Musci TJ, Kirschner MW. 1991. Expression of a dominant negative mutant of the FGF receptor disrupts mesoderm formation in *Xenopus* embryos. *Cell* 66: 257–270. doi:10.1016/0092-8674(91)90616-7

Aslan Y, Tadjuidje E, Zorn AM, Cha SW. 2017. High-efficiency non-mosaic CRISPR-mediated knock-in and indel mutation in F0 *Xenopus*. *Development* 144: 2852–2858.

Barbosa S, Greville-Heygate S, Bonnet M, Godwin A, Fagotto-Kaufmann C, Kajava AV, Laouteouet D, Mawby R, Wai HA, Dingemans AJM, et al. 2020. Opposite modulation of RAC1 by mutations in TRIO is associated with distinct, domain-specific neurodevelopmental disorders. *Am J Hum Genet* 106: 338–355. doi:10.1016/j.ajhg.2020.01.018

Blitz IL, Nakayam T. 2022. CRISPR–Cas9 mutagenesis in *Xenopus tropicalis* for phenotypic analyses in the F0 generation and beyond. *Cold Spring Harb Protoc* doi:10.1101/pdb.prot106971

Blitz IL, Biesinger J, Xie X, Cho KW. 2013. Biallelic genome modification in F(0) *Xenopus tropicalis* embryos using the CRISPR/Cas system. *Genesis* 51: 827–834. doi:10.1002/dvg.22719

Blitz IL, Fish MB, Cho KW. 2016. Leapfrogging: primordial germ cell transplantation permits recovery of CRISPR/Cas9-induced mutations in essential genes. *Development* 143: 2868–2875. doi:10.1242/dev.138057

Blum M, Ott T. 2018. *Xenopus*: an undervalued model organism to study and model human genetic disease. *Cells Tissues Organs* 205: 303–313. doi:10.1159/000490898

Cha S-W. 2022. Generating nonmosaic mutants in *Xenopus* using CRISPR–Cas in oocytes. *Cold Spring Harb Protoc* doi:10.1101/pdb.prot106989

Collart C, Smith JC, Zegerman P. 2017. Chk1 inhibition of the replication factor Drf1 guarantees cell-cycle elongation at the *Xenopus laevis* mid-blastula transition. *Dev Cell* 42: 82–96. doi:10.1016/j.devcel.2017.06.010

Corkins ME, DeLay BD, Miller RK. 2022. Tissue-targeted CRISPR–Cas9-mediated genome editing of multiple homeologs in F0-generation *Xenopus laevis* embryos. *Cold Spring Harb Protoc* doi:10.1101/pdb.prot107037

Dale L, Slack JM. 1987. Fate map for the 32-cell stage of *Xenopus laevis*. *Development* 99: 527–551. doi:10.1242/dev.99.4.527

DeLay BD, Corkins ME, Hanania HL, Salanga M, Deng JM, Sudou N, Taira M, Horb ME, Miller RK. 2018. Tissue-specific gene inactivation in *Xenopus laevis*: knockout of lhx1 in the kidney with CRISPR/Cas9. *Genetics* 208: 673–686. doi:10.1534/genetics.117.300468

Eisen JS, Smith JC. 2008. Controlling morpholino experiments: don't stop making antisense. *Development* 135: 1735–1743. doi:10.1242/dev.001115

Fish MB, Nakayama T, Grainger RM. 2012. Simple, fast, tissue-specific bacterial artificial chromosome transgenesis in *Xenopus*. *Genesis* 50: 307–315. doi:10.1002/dvg.20819

Gentsch GE, Spruce T, Monteiro RS, Owens NDL, Martin SR, Smith JC. 2018. Innate immune response and off-target mis-splicing are common morpholino-induced side effects in *Xenopus*. *Dev Cell* 44: 597–610.e10. doi:10.1016/j.devcel.2018.01.022

Heasman J. 2002. Morpholino oligos: making sense of antisense? *Dev Biol* 243: 209–214. doi:10.1006/dbio.2001.0565

Horb M, Wlizla M, Abu-Daya A, McNamara S, Gajdasik D, Igawa T, Suzuki A, Ogino H, Noble A; Centre de Ressource Biologique Xenope team in France, et al. 2019. *Xenopus* resources: transgenic, inbred and mutant animals, training opportunities, and Web-based support. *Front Physiol* 10: 387. doi:10.3389/fphys.2019.00387

Kakebeen AD, Chitsazan AD, Williams MC, Saunders LM, Wills AE. 2020. Chromatin accessibility dynamics and single cell RNA-Seq reveal new regulators of regeneration in neural progenitors. *Elife* 9: e52648. doi:10.7554/eLife.52648

Kakebeen AD, Chitsazan AD, Wills AE. 2021. Tissue disaggregation and isolation of specific cell types from transgenic *Xenopus* appendages for transcriptional analysis by FACS. *Dev Dyn* 250: 1381–1392. doi:10.1002/dvdy.268

Kroll KL, Amaya E. 1996. Transgenic *Xenopus* embryos from sperm nuclear transplantations reveal FGF signaling requirements during gastrulation. *Development* 122: 3173–3183. doi:10.1242/dev.122.10.3173

Lei Y, Guo X, Liu Y, Cao Y, Deng Y, Chen X, Cheng CH, Dawid IB, Chen Y, Zhao H. 2012. Efficient targeted gene disruption in *Xenopus* embryos using engineered transcription activator-like effector nucleases (TALENs). *Proc Natl Acad Sci* 109: 17484–17489. doi:10.1073/pnas.1215421109

Macken WL, Godwin A, Wheway G, Stals K, Nazlamova L, Ellard S, Alfares A, Aloraini T, AlSubaie L, Alfadhel M, et al. 2021. Biallelic variants in COPB1 cause a novel, severe intellectual disability syndrome with cataracts and variable microcephaly. *Genome Med* 13: 34. doi:10.1186/s13073-021-00850-w

Malartre M, Short S, Sharpe C. 2006. *Xenopus* embryos lacking specific isoforms of the corepressor SMRT develop abnormal heads. *Dev Biol* 292: 333–343. doi:10.1016/j.ydbio.2006.01.007

Moody SA. 1987. Fates of the blastomeres of the 32-cell-stage *Xenopus* embryo. *Dev Biol* 122: 300–319. doi:10.1016/0012-1606(87)90296-X

Naert T, Van Nieuwenhuysen T, Vleminckx K. 2017. TALENs and CRISPR/Cas9 fuel genetically engineered clinically relevant *Xenopus tropicalis* tumor models. *Genesis* 55. doi:10.1002/dvg.23005

Naert T, Tulkens D, Edwards NA, Carron M, Shaidani N-I, Wlizla M, Boel A, Demuynck S, Horb ME, Coucke P, et al. 2020. Maximizing CRISPR/

Cas9 phenotype penetrance applying predictive modeling of editing outcomes in *Xenopus* and zebrafish embryos. *Sci Rep* **10:** 14662. doi:10.1038/s41598-020-71412-0

Nakayama T, Fish MB, Fisher M, Oomen-Hajagos J, Thomsen GH, Grainger RM. 2013. Simple and efficient CRISPR/Cas9-mediated targeted mutagenesis in *Xenopus tropicalis*. *Genesis* **51:** 835–843. doi:10.1002/dvg.22720

Nakayama T, Grainger RM, Cha S-W. 2020. Simple embryo injection of long single-stranded donor templates with the CRISPR/Cas9 system leads to homology-directed repair in *Xenopus tropicalis* and *Xenopus laevis*. *Genesis* **58:** e23366. doi:10.1002/dvg.23366

Nakayama T, Cox A, Howell M, Grainger RM. 2023a. Gynogenetic production of embryos in *Xenopus tropicalis* using a cold shock procedure: rapid screening method for gene editing phenotypes. *Cold Spring Harb Protoc* doi:10.1101/pdb.prot107648

Nakayama T, Grainger RM, Cha S-W. 2023b. Homology-directed repair by CRISPR–Cas9 mutagenesis in *Xenopus* using long single-stranded donor DNA templates via simple microinjection of embryos. *Cold Spring Harb Protoc* doi:10.1101/pdb.prot107599

Nakayama T, Gray J, Grainger RM. 2023c. Production of transgenic F_0 animals and permanent lines by sperm nuclear transplantation in *Xenopus tropicalis*. *Cold Spring Harb Protoc* doi:10.1101/pdb.prot107003

Nakayama T, Roubroeks JAY, Veenstra GJC, Grainger RM. 2023d. Preparation of intact nuclei for single-nucleus omics using frozen cell suspensions from mutant embryos of *Xenopus tropicalis*. *Cold Spring Harb Protoc* doi:10.1101/pdb.prot107825

Noble A, Abu-Daya A, Guille MJ. 2022a. I-SceI-mediated transgenesis in *Xenopus*. *Cold Spring Harb Protoc* doi:10.1101/pdb.prot107011

Noble A, Abu-Daya A, Guille MJ. 2022b. Cryopreservation of *Xenopus* sperm and in vitro fertilization using frozen sperm samples. *Cold Spring Harb Protoc* doi:10.1101/pdb.prot107029

Offield MF, Hirsch N, Grainger RM. 2000. The development of *Xenopus tropicalis* transgenic lines and their use in studying lens developmental timing in living embryos. *Development* **127:** 1789–1797. doi:10.1242/dev.127.9.1789

Paraiso KD, Blitz IL, Zhou JJ, Cho KWY. 2019. Morpholinos do not elicit an innate immune response during early *Xenopus* embryogenesis. *Dev Cell* **49:** 643–650.e3. doi:10.1016/j.devcel.2019.04.019

Park DS, Yoon M, Kweon J, Jang AH, Kim Y, Choi SC. 2017. Targeted base editing via RNA-guided cytidine deaminases in *Xenopus laevis* embryos. *Mol Cell* **40:** 823–827. doi:10.14348/molcells.2017.0262

Rankin SA, Zorn AM, Buchholz DR. 2011. New doxycycline-inducible transgenic lines in *Xenopus*. *Dev Dyn* **240:** 1467–1474. doi:10.1002/dvdy.22642

Sater AK, Moody SA. 2017. Using *Xenopus* to understand human disease and developmental disorders. *Genesis* **55**. doi:10.1002/dvg.22997

Schulte-Merker S, Smith JC. 1995. Mesoderm formation in response to Brachyury requires FGF signalling. *Curr Biol* **5:** 62–67. doi:10.1016/S0960-9822(95)00017-0

Session AM, Uno Y, Kwon T, Chapman JA, Toyoda A, Takahashi S, Fukui A, Hikosaka A, Suzuki A, Kondo M, et al. 2016. Genome evolution in the allotetraploid frog *Xenopus laevis*. *Nature* **538:** 336–343. doi:10.1038/nature19840

Singh MD, Jensen M, Lasser M, Huber E, Yusuff T, Pizzo L, Lifschutz B, Desai I, Kubina A, Yennawar S, et al. 2020. *NCBP2* modulates neurodevelopmental defects of the 3q29 deletion in *Drosophila* and *Xenopus laevis* models. *PLoS Genet* **16:** e1008590. doi:10.1371/journal.pgen.1008590

Smith WC, Harland RM. 1992. Expression cloning of noggin, a new dorsalizing factor localized to the Spemann organizer in *Xenopus* embryos. *Cell* **70:** 829–840. doi:10.1016/0092-8674(92)90316-5

Sterner ZR, Rankin SA, Wlizla M, Choi JA, Luedeke DM, Zorn AM, Buchholz DR. 2019. Novel vectors for functional interrogation of *Xenopus* ORFeome coding sequences. *Genesis* **57:** e23329. doi:10.1002/dvg.23329

Timberlake AT, Griffin C, Heike CL, Hing AV, Cunningham ML, Chitayat D, Davis MR, Doust SJ, Drake AF, Duenas-Roque MM, et al. 2021. Haploinsufficiency of *SF3B2* causes craniofacial microsomia. *Nat Commun* **12:** 4680. doi:10.1038/s41467-021-24852-9

Weber W. 1999. Ion currents of *Xenopus laevis* oocytes: state of the art. *Biochim Biophys Acta* **1421:** 213–233. doi:10.1016/S0005-2736(99)00135-2

Willsey HR, Exner CRT, Xu Y, Everitt A, Sun N, Wang B, Dea J, Schmunk G, Zaltsman Y, Teerikorpi N, et al. 2021. Parallel *in vivo* analysis of large-effect autism genes implicates cortical neurogenesis and estrogen in risk and resilience. *Neuron* **109:** 788–804.e8. doi:10.1016/j.neuron.2021.01.002

Willsey HR, Guille MJ, Grainger RM. 2022. Modeling human genetic disorders with CRISPR technologies in *Xenopus*. *Cold Spring Harb Protoc* doi:10.1101/pdb.prot106997

Young JJ, Cherone JM, Doyon Y, Ankoudinova I, Faraji FM, Lee AH, Ngo C, Guschin DY, Paschon DF, Miller JC, et al. 2011. Efficient targeted gene disruption in the soma and germ line of the frog *Xenopus tropicalis* using engineered zinc-finger nucleases. *Proc Natl Acad Sci* **108:** 7052–7057. doi:10.1073/pnas.1102030108

Young JJ, Kjolby RAS, Wu G, Wong D, Hsu SW, Harland RM. 2017. Noggin is required for first pharyngeal arch differentiation in the frog *Xenopus tropicalis*. *Dev Biol* **426:** 245–254. doi:10.1016/j.ydbio.2016.06.034

Zeng SL, Sudlow LC, Berezin MY. 2020. Using *Xenopus* oocytes in neurological disease drug discovery. *Expert Opin Drug Discov* **15:** 39–52. doi:10.1080/17460441.2020.1682993

CRISPR–Cas9 Mutagenesis in *Xenopus tropicalis* for Phenotypic Analyses in the F_0 Generation and Beyond

Ira L. Blitz[1,3] and Takuya Nakayama[2,3]

[1]*Department of Developmental and Cell Biology, University of California, Irvine, California 92697, USA;*
[2]*Department of Biology, University of Virginia, Charlottesville, Virginia 22904, USA*

CRISPR–Cas9 mutagenesis is being widely used to create targeted loss-of-function mutations in the diploid frog *Xenopus tropicalis*. Here we describe a simple mutagenesis protocol using microinjection of Cas9 protein or mRNA, together with synthetic guide RNAs (sgRNAs) targeting specific DNA sequences, into the early embryo. Cas9-catalyzed double-strand breaks undergo error-prone repair, resulting in production of short insertions and/or deletions. Thus, careful selection of target sites can lead to mutations that impair normal function of the protein product. CRISPR–Cas9 can be used to create either mosaic loss-of-function *Xenopus* embryos that display F_0 generation phenotypes or mutant lines for downstream analysis. In addition to describing how to mutagenize genes using CRISPR–Cas9, we also discuss a simple method to determine the mutagenesis efficiency, some potential problems that can arise, and possible solutions to overcome them. The protocol described here should be applicable to other amphibians and, in principle, many other organisms.

MATERIALS

It is essential that you consult the appropriate Material Safety Data Sheets and your institution's Environmental Health and Safety Office for proper handling of equipment and hazardous materials used in this protocol.

Reagents

Use only molecular biology grade reagents.

Agarose
Cas9 protein with nuclear localization signal (e.g., PNA Bio CP01) or a linearized plasmid template for
    human codon–optimized Cas9 mRNA (see Nakayama et al. 2014)
Chloroform
DSP lysis buffer (proteinase K freshly added) <R>
Ethanol (70% in DEPC-treated H_2O)
Ethidium bromide (or equivalent reagent for DNA staining)
Gel electrophoresis buffer
Gene-specific oligonucleotide pair for polymerase chain reaction (PCR) amplification
Gene-specific nested oligonucleotide for Sanger sequencing (optional; see Step 17)
Isopropanol

---

[3]Correspondence: ilblitz@uci.edu; tn8t@virginia.edu

Cite this protocol as *Cold Spring Harb Protoc*; doi:10.1101/pdb.prot106971

MEGAscript or MEGAshortscript T7 Transcription Kit (Thermofisher AM1334 or AM1354, respectively)

mMESSAGE mMACHINE SP6 or T7 Transcription Kit (Thermofisher AM1340 or AM1344, respectively) (optional; see Step 8)

Modified Barth's saline <R> or Marc's modified Ringers (MMR) <R>, 1× (without EDTA)

Working concentrations are 0.1× modified Barth's saline or 1/9 × MMR; see Step 12.

Nuclease-free (or DEPC-treated) H_2O

PCR cleanup kit (e.g., NucleoSpin Gel and PCR Clean-Up kit (Macherey-Nagel 740609.250), QIAquick PCR Purification Kit (28104), or Zymo DNA Clean & Concentrator-5 (D4013)

Phenol

Platinum SuperFi PCR Master Mix (ThermoFisher 12358-010) or another high-fidelity polymerase

We have also had success with other thermostable polymerases, including Takara EmeraldAmp GT PCR Master Mix (RR310B) and New England Biolabs Phusion High-Fidelity PCR Master Mix with HF Buffer (M0531S).

Target site-specific oligonucleotide for generation of sgRNA template:

5′-GCAGCTAATACGACTCACTATAG(N)$_{16\text{-}19}$GTTTTAGAGCTAGAAATA-3′ (target sequence in bold; see Step 2 for sequence details)

Universal oligonucleotide for generation of sgRNA template:

5′-AAAAGCACCGACTCGGTGCCACTTTTTCAAGTTGATAA
CGGACTAGCCTTATTTTAACTTGCTATTTCTAGCTCTAAAAC-3′

Xenopus tropicalis embryos, obtained by in vitro fertilization and dejellied using cysteine-free base as previously described (Ogino et al. 2006)

Equipment

Dissecting microscope with fiber optic light source

Gel electrophoresis equipment

Heat block

Microcentrifuge

Microinjector and accessories (injection needles)

PCR machine

PCR tubes (0.2-mL; DNase- and RNase-free)

Spectrophotometer

METHOD

Identifying Target Site(s) in the Gene of Interest

1. Identify regions of the gene/protein to target and design a minimum of two synthetic guide RNAs (sgRNAs) recognizing sites to test for mutagenesis efficiency using the following guidelines.

    i. To assist in finding target sequences that are unique in the genome, we prefer the simple user interface of CRISPRdirect (https://crispr.dbcls.jp/) (Naito et al. 2015). Other useful websites supporting *Xenopus* can be found on Xenbase's genome editing support page (http://www.xenbase.org/other/static-xenbase/CRISPR.jsp).

    ii. Candidate target sites for *Streptococcus pyogenes* Cas9 are typically 20 nt in length and have the sequence 5′-N$_{20}$-NGG-3′ (where N is any base). To find sites within genes using CRISPRdirect one can enter either NCBI accession numbers, genomic coordinates, or directly enter the DNA sequence. Instructions for use of CRISPRdirect can be found at https://crispr.dbcls.jp/doc/. Using the gene instead of the cDNA sequence avoids identifying erroneous target sites that span exon–exon splice junctions, which are found in mRNA, but not represented in the intron-containing gene sequence. These sequences can be obtained at Xenbase (www.xenbase.org/).

iii. To minimize the chance of off-target mutagenesis, it is a good practice to choose target sites with unique sequence in the genome and avoid those that are highly similar to sequences elsewhere. CRISPRdirect (and many other CRISPR webtools) provides information on the similarity of target sequences to distant sites. Choose targets that do not have large skews in base content. Although some reports suggest a slight preference for higher GC content in the 3′ half of the PAM proximal "seed" sequence (Gagnon et al. 2014; Ren et al. 2014), others suggest that between 40% and 60% of GC content across the entire target might be most advantageous (Doench et al. 2014; Wang et al. 2014; Liu et al. 2016).

iv. sgRNA synthesis using bacteriophage T7 RNA polymerase requires at least one G residue at the beginning of the target site, but targets lacking a 5′ G can be accommodated via one of two approaches. One method is to replace at least the first base of the target site with a G: one or two nucleotides can be replaced with Gs for successful mutagenesis (Fu et al. 2014; Ren et al. 2014; Moreno-Mateos et al. 2015). Extending the sequence of sgRNAs at the 5′ end of the 20 nt target, by adding Gs upstream without deleting 2 nt, may reduce activity (Ren et al. 2014; Moreno-Mateos et al. 2015), although this has not been verified in *Xenopus*. A second strategy is to use an endogeneous guanine residue as a 5′ G that is internal to the identified target to start the sgRNA transcripts. This results in a slightly shorter sgRNA, and the target takes the form $G(N)_{16-18}$ (Fu et al. 2014). Such 5′ truncated sgRNAs may have lower off-target mutagenesis activity (Fu et al. 2014), although they appear to have similar binding specificity to off-target sites as full-length sgRNAs (Josephs et al. 2015). It has also been reported in cell culture that there is a risk of higher off-target mutagenesis when using 17 nt sgRNAs, but not 20 nt sgRNAs designed to the same site (Zhang et al. 2016).

Additional essential considerations for choosing target sites are in the Discussion section.

2. Synthesize target-specific oligonucleotides. The gene-specific oligo sequences have the sequence 5′-*GCAGC*T̲A̲A̲T̲A̲C̲G̲A̲C̲T̲C̲A̲C̲T̲A̲T̲A̲**G**(N)$_{16-19}$GTTTTAGAGCTAGAAATA-3′, in which the underline highlights the T7 promoter, and the 5′ italicized sequence is present for enhancement of transcription (Baklanov et al. 1996). The target sequence is in bold. The bold G in the promoter is also part of the target site. It is transcribed into the sgRNA and is required for efficient T7 promoter function.

It is essential to recognize that the NGG portion of the target site, the protospacer adjacent motif (PAM), while being critical for Cas9 recognition and cleavage, is not included in the sgRNA itself.

Synthesis of dsDNA Template Used for In Vitro Transcription of sgRNA

3. Anneal the gene-specific oligonucleotide to the universal oligo, 5′-AAAAGCACCGACTCGGT GCCACTTTTTCAAGTTGATAACGGACTAGCCTTATTTTAACTTGCT̲A̲T̲T̲T̲C̲T̲A̲G̲C̲T̲C̲T̲A̲ A̲A̲A̲C̲-3′, which is complementary to the 3′ 18 bases (underlined) of the gene-specific oligo.

i. Perform the annealing and a fill-in reaction using a thermostable polymerase and the following reaction conditions:
   2 μL of 100 μM gene-specific oligonucleotide
   2 μL of 100 μM universal oligonucleotide
   50 μL of 2× Platinum Superfi PCR Master Mix
   Water, nuclease-free, to a 100-μL final volume

ii. Use the following cycling parameters:
   98°C for 5 min.
   20 cycles of
      98°C for 20 sec
      58°C for 20 sec

72°C for 15 sec

72°C for 5 min.

To assess the completeness of the synthesis, remove 2 µL of the mix before performing the reaction to compare to 2 µL after the fill in (see below). Other thermostable polymerases can be used but may require adjustment of the extension temperature according to manufacturer's recommendations.

4. Purify the double-stranded (~120 bp) template using either a commercial PCR cleanup kit or standard phenol/CHCl$_3$ and CHCl$_3$ extractions followed by sodium acetate/isopropanol precipitation and 70% ethanol/30% DEPC-treated H$_2$O wash.

The DNA should be in a final volume of ~20–30 µL of RNAse-free column elution buffer or DEPC-treated H$_2$O. It is recommended that the concentration be quantified spectrophotometrically and the success of the synthesis is also verified by 2% agarose gel electrophoresis, alongside lanes containing the starting oligos but without synthesis (Fig. 1A), as described above.

Synthesis of sgRNA

5. Synthesize sgRNAs by in vitro transcription using either a T7 MEGAscript or MEGAshortscript kit, following kit specifications with modifications. If using the MEGAscript kit, use ~50–500 ng of template per 20 µL reaction and incubate for 6 h to overnight at 37°C. If using the MEGAshortscript kit, use ~500 ng of template per 20 µL reaction and incubate for 1 h at 37°C.

6. Digest the template DNA following kit specifications by a 15-min incubation with DNase I. Purify sgRNA using phenol-chloroform extraction followed by NH$_4$OAc precipitation as described in the kit manuals. Wash pelleted RNAs twice in 1 mL of 70% ethanol (prepared using DEPC-H$_2$O) to ensure removal of trace ammonium ions. Air dry pellets very briefly.

7. Resuspend the sgRNA pellet in 20–40 µL of DEPC-treated H$_2$O and quantitate spectrophotometrically. Evaluate RNA integrity by electrophoretic separation on a 2% agarose gel (Fig. 1B). Briefly, dilute 200–500 ng of sgRNA in 5 µL of DEPC-treated H$_2$O and add 5 µL Gel Loading Buffer II (supplied with MEGAscript and MEGAshortscript kits). Denature the RNA by heating it for 2 min to 50°C–60°C. Quickly cool on ice for 2–5 min before electrophoresis.

8. If using Cas9 mRNA instead of purified protein, prepare 5′-capped mRNA from a linearized plasmid template using a mMESSAGE MACHINE Transcription Kit according to the manufacturer's instructions. After mRNA synthesis, remove the template DNA with DNase I and purify the mRNA using the phenol-chloroform extraction method followed by NH$_4$OAc precipitation as described in the kit manuals. Wash pelleted RNAs at least twice in 70% ethanol (prepared using DEPC-H$_2$O) to ensure removal of trace ammonium ions, and briefly air dry. Resuspend the pellet in 20–40 µL of DEPC-treated H$_2$O. Quantitate the RNA spectrophotometrically and evaluate its integrity by 1% agarose gel electrophoresis.

Obtaining Embryos and Microinjection

9. Obtain dejellied *Xenopus tropicalis* embryos as previously described in Ogino et al. (2006) (see also Sec. 2, Protocol 5: Obtaining *Xenopus tropicalis* Embryos by In Vitro Fertilization [Lane and Khokha 2021]).

X. tropicalis eggs have slightly different requirements for efficient fertilization and dejellying compared to X. laevis.

10. Prepare Cas9–sgRNA injection cocktails.

i. If using Cas9 mRNA, mix 250 ng–2 µg Cas9 mRNA together with ~250 ng–1 µg of sgRNA, with the addition of DEPC-treated H$_2$O to a final volume of 4 µL.

ii. If using Cas9 protein, preform Cas9–sgRNA complexes in vitro to enhance efficiency. To precomplex Cas9 protein and sgRNA, mix 250 ng–1 µg sgRNA with up to 2 µg Cas9 protein, adding DEPC-treated H$_2$O to a final volume of 4 µL, and incubate for 5 min at 37°C.

A range of concentrations of Cas9 and sgRNA is given but typically 250 pg of sgRNA and 1 ng of Cas9 mRNA/protein injected per embryo is sufficient. If a target site appears refractory to mutagenesis, then

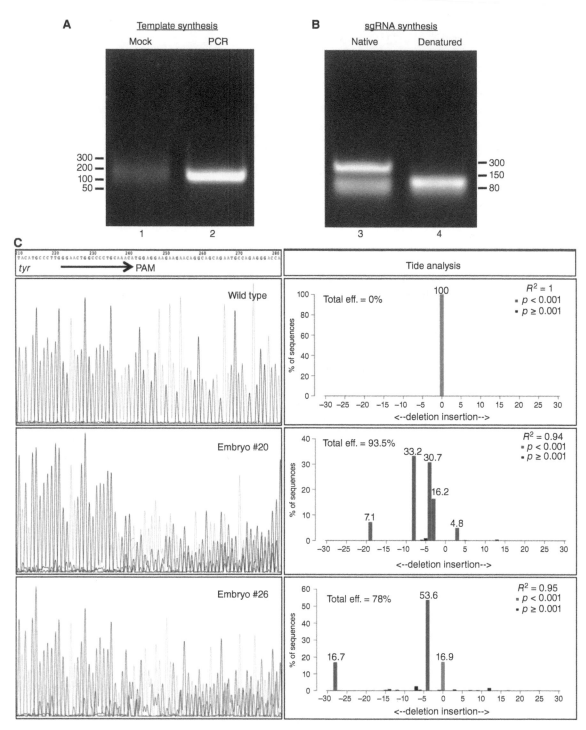

FIGURE 1. Agarose gel electrophoretic analysis of DNA template and sgRNA syntheses. (A) Lane *1* contains gene-specific and universal oligonucleotides in a mock fill-in reaction, which was not subject to the cycling conditions outlined in Step 3 of the Method. Running a mock reaction alongside an aliquot of the fill-in reaction assists in differentiating bands that represent double-stranded product from single-stranded oligonucleotides. Lane *2* shows a stronger fluorescent band migrating at the expected size (~123 bp) for the successfully synthesized template. (B) Lanes *3* and *4* show the product of a Megascript sgRNA synthesis without ("native") or with denaturation of the sgRNA, respectively, before running on a 2% agarose gel. (C) DSP analysis of amplicons from individual *tyrosinase* sgRNA (Blitz et al. 2013; Nakayama et al. 2013) injected F_0 embryos. *Left* panels contain a wild-type Sanger sequencing trace to compare to two injected embryos below. *Right* panels show TIDE outputs from the .abi files, which show that embryo #20 had ~94% mutagenesis efficiency, whereas embryo #26 had ~78%. The histograms show the prediction of percentages of each species of insertion/deletion in the pool of amplicon DNA molecules, with red bars indicating statistically significant calls (*P* value < 0.001).

higher concentrations in the ranges provided is recommended. Note that high concentrations of Cas9 protein make for a more viscous solution that can clog injection needles.

11. Inject embryos as described in Ogino et al. (2006) with a volume of 1–4 nL of injection cocktail. We recommend microinjection into a single site at the animal pole during the one-cell stage.

12. Culture *X. tropicalis* embryos in either 0.1× modified Barth's saline or 1/9× MMR at temperatures between 22°C and 25°C (Ogino et al. 2006) until they reach the desired stage.

Evaluating Mutagenesis Efficiency

To evaluate the mutagenesis efficiency in F_0 (injected) embryos, PCR is used to amplify a genomic fragment using primers flanking the CRISPR target site. Sanger sequencing of the mosaic population of amplicons is then performed and the trace (peaks) is analyzed, which we refer to as DSP for direct sequencing of PCR amplicons (Nakayama et al. 2013, 2014). If Cas9 cleavage is successful, the sequence trace will contain a mixed population of peaks starting at the target site, which is then deconvoluted to determine the extent of mutagenesis. For deconvolution, we have used either the web-based tool Tracking of Indels by DEcomposition (TIDE) (Brinkman et al. 2014) or Inference of CRISPR Edits (ICE) (Hsiau et al. 2018). These deconvolution methods provide an estimate of both the mutagenesis efficiency and percentages for different alleles. A wild-type sequence trace is needed by these webtools for comparison. Alternatively, clone the PCR products and sequence multiple clones, ideally more than 10 clones per embryo, to estimate the efficiency (e.g., Nakayama et al. 2013; Blitz et al. 2013; Guo et al. 2014).

13. Perform PCR amplification directly from embryo lysates with limited further cleanup as follows. Homogenize single embryos in a 0.2 mL PCR tube by trituration using a P200 Pipetman in a small volume of DSP lysis buffer, which contains proteinase K, and incubate for 2 h to overnight at 56°C. Heat-inactivate proteinase K by incubation for 10 min at 90°C–95°C followed by either quick or gradual cooling to 4°C. Centrifuge samples for 5 min at 4°C at maximum speed in a microcentrifuge to pellet debris and either use the supernatant immediately for PCR or store at −20°C.

    When lysing embryos, we typically use 100 µL lysis buffer per embryo for gastrula to tailbud stages, and 30 µL for tadpole tail clips.

14. Using PCR primers that flank the target site (see ThermoFisher Tm calculator for oligonucleotides used in SuperFi reactions: https://www.thermofisher.com/us/en/home/brands/thermo-scientific/molecular-biology/molecular-biology-learning-center/molecular-biology-resource-library/thermo-scientific-web-tools/tm-calculator.html?CID=fl-we120377), amplify a genomic fragment that is 200–800 bp in length.

    TIDE recommends amplicons >700 bp, http://shinyapps.datacurators.nl/tide/, whereas ICE recommends amplicons of 400–800 bp, https://www.synthego.com/help/pcr-ice-analysis, although shorter amplicons can be used. Both recommend that the target site should be located asymmetrically. Alternatively, 200–400 bp in length (or longer) are suitable for cloning followed by sequencing of individual amplicons.

    Prepare 25 µL PCR reactions as follows, per reaction:

Water, nuclease-free	9 µL
2× Platinum SuperFi PCR Master Mix	12.5 µL
5′ primer (10 µM)	1.25 µL
3′ primer (10 µM)	1.25 µL
Embryo lysate (undiluted or 1:10 diluted)	1 µL

15. Use a touchdown PCR protocol as follows:

    1 cycle of 98°C, 30 sec
    13 cycles of:
        98°C, 10 sec
        65°C, 10 sec, with a decrease in temperature of −0.5°C/cycle
        72°C, 20 sec
    30 cycles of:
        98°C, 10 sec
        58°C, 10 sec
        72°C, 20 sec
    1 cycle of 72°C, 5 min

    4°C hold

Extension steps can be modified based on amplicon length using ~15–30 sec per kb for SuperFi polymerase.

16. Purify the amplicons using a PCR cleanup kit and elute in the minimum volume recommended by the vendor. Quantify the concentration by spectrophotometry.

    See Troubleshooting.

17. Sequence the amplicons using one of the PCR oligonucleotide primers or a nested primer and similarly sequence an amplicon from an uninjected embryo to obtain a wild-type sequence trace (Fig. 1C, left panels).

    See Troubleshooting.

18. Following the instructions available on the TIDE (http://shinyapps.datacurators.nl/tide/) or ICE (https://ice.synthego.com/#/) websites, upload mutant and wild-type sequencing files and enter the nucleotide sequence of the target site. Use only those sgRNAs that yield high-efficiency mutagenesis for subsequent experiments (Fig. 1C, right panels).

    When using CRISPR–Cas9 for phenotypic analyses in F_0 generation embryos, the highest mutation rates are preferable (perhaps >75% efficiency, but this has not been tested carefully to determine efficiencies needed to obtain phenotypes). When using CRISPR–Cas9 to create mutant lines, one must consider whether the mutations might compromise viability of the F_0 generation embryos. If so, then one can reduce the concentrations of Cas9 and sgRNA injected to reduce the efficiency of mutagenesis to the 50%–60% (or lower) range. Alternatively, one can use leapfrogging (see Discussion), the transplantation of primordial germ cells from efficiently mutagenized F_0 embryos (preferably >90%), to overcome the deleterious effects of high mutation rates on somatic cells (Blitz et al. 2016; Blitz 2018). Finally, one can inject the Cas9–sgRNA cocktail into vegetal blastomeres at the 16- to 32-cell stage, which may permit production of viable animals with germline transmissible mutations.

    See Troubleshooting.

TROUBLESHOOTING

Problem (Steps 16 and 17): The genomic PCR is not working.

Solution: The first time a new region is amplified, the PCR products should be examined by gel electrophoresis to determine whether a single band is recovered or if there is contamination by nonspecific amplification. In cases where it is difficult to obtain single products, an internal "nested" primer can be designed for the subsequent sequencing step, which usually yields traces uncontaminated by low level signal from unwanted genomic fragments. Finally, negative controls (without embryo lysate) are valuable to assure that reagents are not contaminated with DNA that amplifies in these reactions.

Problem (Step 18): An sgRNA is inefficient in mutagenesis.

Solution: Try increasing the amount of Cas9 to as high as 1.5–2 ng protein or mRNA per embryo, and sgRNA to as high as 800 pg to 1 ng per embryo. If the sgRNA still fails even at these doses, abandon using that sgRNA and target a different site. It is important to note that high concentrations of Cas9 protein are more viscous and easily clog injection needles and therefore it is best to deliver higher doses in larger injection volumes to avoid this issue.

DISCUSSION

Here we describe a methodology for performing CRISPR–Cas9 mutagenesis in *Xenopus tropicalis*. These methods can be used to create mutant lines, or for phenotypic analysis in mosaic F_0 generation embryos, and are likely to be generally applicable to other frog species. We direct the reader to Protocol 3: Tissue-Targeted CRISPR–Cas9-Mediated Genome Editing of Multiple Homeologs in F_0-Generation *Xenopus laevis* Embryos (Corkins et al. 2021) for additional detail on CRISPR–Cas9

Cite this protocol as *Cold Spring Harb Protoc*; doi:10.1101/pdb.prot106971

applications in *X. laevis*. It is important to note that mutation of genes for maternally expressed gene products are not expected to yield phenotypes in F_0 embryos until this maternal pool is exhausted and zygotic transcripts are dominant. As many genes have maternal protein/RNA contributions that persist until the neurula and tailbud stages (Peshkin et al. 2015; Owens et al. 2016), checking for maternal expression using the many resources at Xenbase (www.xenbase.org/) can be instructive. Mutant lines may be needed in order to assess maternal phenotypes via genetic crosses (see below).

Strategies for Choice of Target Site Location within Genes

Different parameters should be considered to produce knockout alleles, depending on whether the goal is to make mutant lines or perform F_0 loss-of-function experiments. To create mutant lines, a common strategy is to introduce frameshift mutations interrupting the coding sequence near the $5'$ end of the coding sequence (CDS). F_1 progeny from outcrosses between adult frogs carrying potential mutant alleles and wild types are then screened to find founders that transmit the mutation(s) of interest through the germline. Intercrosses between F_1 heterozygotes produce F_2 animals bearing either homozygous mutations or compound heterozygous mutations (the two mutant alleles are not identical). An alternative approach is to intercross two F_0 adults that both transmit mutant alleles at sufficient frequencies, to yield phenotypes in the F_1s of these crosses. These approaches using either heterozygous mutant lines or F_0 intercrosses can be efficient for obtaining loss-of-function phenotypes (e.g., Nakayama et al. 2015). When designing the sgRNA, it is important to understand the structure of the gene products because it is possible that introduced frameshifts or stop codons near the $5'$ end of the CDS will be designed without regard to internal transcriptional or translational initiation, or alternative splice isoforms and therefore may result in expression of functional proteins even after mutation.

The F_0 intercross approach requires that most alleles from both parental animals are out-of-frame or stop codons. Because F_0 injected embryos are mosaic, containing a mixture of different allelic variants, one must contend with the fact that, on average approximately one-third of alleles are expected to be in-frame (e.g., see Varshney et al. 2015). Therefore, a subset of mutant alleles carried by F_0 mosaic animals might produce proteins with near-wild-type activity. One strategy to overcome this is to select target sites that lie within the encoded protein's folded domains. These are more likely to result in loss of function even when containing in-frame short indels, by disrupting domain function and/or protein folding (Nakayama et al. 2015; Shi et al. 2015; Blitz et al. 2016).

There are other strategies that can achieve loss of function. One approach is targeting of the promoter of genes as this can lead to abrogation of mRNA transcription (Nakayama et al. 2013). Successful disruption of gene expression depends on knowledge of the location of transcriptional start sites and promoter elements. Another approach is simultaneous targeting at multiple (e.g., four to five) sites within a gene, which has been estimated to produce frameshifts in nearly all mutated gene copies (Servetnick et al. 2017; Wu et al. 2018). This assumes that all four sites are cleaved at high efficiency, and, to our knowledge, this approach has not yet been tested in *Xenopus*. Analysis of mutagenesis efficiency/genotyping is made difficult in these cases as deletions, inversions, and other recombination can occur between cleavage sites (our unpublished observations). Finally, websites including Microhomology-Predictor and Indelphi that provide predictions for mutational outcomes can be consulted (Bae et al. 2014; Shen et al. 2018). Indelphi has been specifically tested in *Xenopus* and makes predictions for F_0 outcomes that correlate with ICE deconvoluted Sanger sequencing data (Naert et al. 2020). More work is needed to determine whether use of these tools optimizes recovery of predicted mutations in the germline when creating mutant *Xenopus* lines.

Many mutations lead to embryonic (or later) lethal phenotypes, making it more challenging to obtain viable adults that transmit mutant alleles through the germline at moderate to high efficiencies. An approach to circumvent this issue is "leapfrogging," which combines CRISPR–Cas9 mutagenesis with transplantation of the F_0 mutant primordial germ cells into wild-type recipient embryos (Blitz et al. 2016; Blitz 2018). Leapfrogging allows for generation of somatically wild-type adults carrying fully mutant germlines, and phenotypes are obtained in the F_1 generation following intercrosses between two leapfrogged animals. Because of the mechanistic details of meiosis, this approach is also essential for generating maternal loss of function phenotypes as simple intercrosses between heterozygotes will not reveal maternal effects. Target site choice within folded domains is also recommended for leapfrogging to obtain phenotypes directly in F_0 embryos.

Finally, the issue of off-target mutagenesis is always worth considering. One approach that has been taken is to compare F_0 phenotypes to those produced by knockdowns using morpholino antisense oligonucleotides (e.g., McQueen and Pownall 2017; DeLay et al. 2018). An alternative approach is to compare multiple different sgRNAs that target the same gene (e.g., Nakayama et al. 2013). If these all produce very similar phenotypes, then off-target mutagenesis is an unlikely explanation for the phenotypes produced. Intercrossing two animals that contain mutations at different sites within the same gene (and therefore appear as heterozygotes in sequence traces at each site) is also easier to genotype than when embryos carry two different mutant sequences at the same site (e.g., Nakayama et al. 2015).

Alternative Methods to Assess CRISPR–Cas9 Mutagenesis Efficiency

We describe above a method for assessing CRISPR–Cas9 mutagenesis efficiency in F_0 embryos. A variety of other methods exist including detection of differences in mobility between wild-type and indel-containing fragments (e.g., fragment analysis: Bhattacharya et al. 2015; Ramlee et al. 2015), identifying mutations after creating heteroduplex DNA (e.g., Surveyor, T7E1, or high melting resolution assays: Wittwer et al. 2003; Kim et al. 2009; Guschin et al. 2010), high-throughput sequencing (Bell et al. 2014; Güell et al. 2014), restriction fragment length polymorphism (RFLP) analysis (Urnov et al. 2005), and in vitro resistance to digestion by Cas9–sgRNA (Kim et al. 2014). Some of these approaches require expensive instrumentation that is not readily available in all laboratories. We find DSP analysis using TIDE/ICE to be a relatively straightforward method for assessing both the efficiency of mutagenesis and the ratios of different alleles. In addition, it is noteworthy that this general approach is as predictive of mutagenesis efficiency as high throughput sequencing of the amplicons (Sentmanat et al. 2018).

RECIPES

DSP Lysis Buffer

50 mM Tris (pH 8.8)
1 mM EDTA
0.5% Tween 20
200 µg/mL Proteinase K (e.g., Sigma-Aldrich/Roche 3115887001)

Store at room temperature. Add proteinase K immediately before use.

Marc's Modified Ringers (MMR) (1×, without EDTA)

5 mM HEPES (free acid)
100 mM NaCl
2 mM KCl
1 mM $MgCl_2$
2 mM $CaCl_2$

Prepare as a 10× stock and adjust pH to 7.4–7.8 with NaOH. Store at room temperature.

Cite this protocol as *Cold Spring Harb Protoc*; doi:10.1101/pdb.prot106971

Modified Barth's Saline (1×)

5 mM HEPES (free acid)
88 mM NaCl
1 mM $CaCl_2$
1 mM $MgSO_4$
2.5 mM $NaHCO_3$

Prepare as a 10× stock and adjust pH to 7.4–7.8 with NaOH. Store at room temperature.

ACKNOWLEDGMENTS

The authors thank Drs. Ken W.Y. Cho and Robert M. Grainger for their ongoing support and Dr. Rachel Miller for sharing her manuscript before submission. This work was funded by National Institutes of Health (NIH) grant, 1R21HD080684, awarded to I.L.B., and by NIH grant EY022954 and research awards from the Sharon Stewart Aniridia Trust and Vision for Tomorrow to Robert M. Grainger. The authors also acknowledge Xenbase (RRID:SCR_003280) for providing essential online resources for *Xenopus* research.

REFERENCES

Bae S, Kweon J, Kim HS, Kim J-S. 2014. Microhomology-based choice of Cas9 nuclease target sites. *Nat Methods* 11: 705–706. doi:10.1038/nmeth.3015

Baklanov MM, Golikova LN, Malygin EG. 1996. Effect on DNA transcription of nucleotide sequences upstream to T7 promoter. *Nucleic Acids Res* 24: 3659–3660. doi:10.1093/nar/24.18.3659

Bell CC, Magor GW, Gillinder KR, Perkins AC. 2014. A high-throughput screening strategy for detecting CRISPR–Cas9 induced mutations using next-generation sequencing. *BMC Genomics* 15: 1002. doi:10.1186/1471-2164-15-1002

Bhattacharya D, Marfo CA, Li D, Lane M, Khokha MK. 2015. CRISPR/Cas9: an inexpensive, efficient loss of function tool to screen human disease genes in *Xenopus*. *Dev Biol* 408: 196–204. doi:10.1016/j.ydbio.2015.11.003

Blitz IL. 2018. Primordial germ cell transplantation for CRISPR/Cas9-based leapfrogging in *Xenopus*. *J Vis Exp* 56035. doi:10.3791/56035

Blitz IL, Biesinger J, Xie X, Cho KW. 2013. Biallelic genome modification in F₀ *Xenopus tropicalis* embryos using the CRISPR/Cas system. *Genesis* 51: 827–834. doi:10.1002/dvg.22719

Blitz IL, Fish MB, Cho KW. 2016. Leapfrogging: primordial germ cell transplantation permits recovery of CRISPR/Cas9-induced mutations in essential genes. *Development* 143: 2868–2875. doi:10.1242/dev.138057

Brinkman EK, Chen T, Amendola M, van Steensel, B. 2014. Easy quantitative assessment of genome editing by sequence trace decomposition. *Nucleic Acids Res* 42: e168. doi:10.1093/nar/gku936

Corkins ME, DeLay BD, Miller RK. 2021. Tissue-targeted CRISPR–Cas9-mediated genome editing of multiple homeologs in F₀-generation *Xenopus laevis* embryos. *Cold Spring Harb Protoc* doi:10.1101/pdb.prot107037

DeLay BD, Corkins ME, Hanania HL, Salanga M, Deng JM, Sudou N, Taira M, Horb ME, Miller RK. 2018. Tissue-specific gene inactivation in *Xenopus laevis*: knockout of *lhx1* in the kidney with CRISPR/Cas9. *Genetics* 208: 673–686. doi:10.1534/genetics.117.300468

Doench JG, Hartenian E, Graham DB, Tothova Z, Hegde M, Smith I, Sullender M, Ebert BL, Xavier RJ, Root DE. 2014. Rational design of highly active sgRNAs for CRISPR–Cas9-mediated gene inactivation. *Nat Biotechnol* 32: 1262–1267. doi:10.1038/nbt.3026

Fu Y, Sander JD, Reyon D, Cascio VM, Joung JK. 2014. Improving CRISPR–Cas nuclease specificity using truncated guide RNAs. *Nat Biotechnol* 32: 279–284. doi:10.1038/nbt.2808

Gagnon JA, Valen E, Thyme SB, Huang P, Akhmetova L, Pauli A, Montague TG, Zimmerman S, Richter C, Schier AF. 2014. Efficient mutagenesis by Cas9 protein–mediated oligonucleotide insertion and large-scale assessment of single-guide RNAs. *PLoS One* 9: e98186. doi:10.1371/journal.pone.0098186

Güell M, Yang L, Church GM. 2014. Genome editing assessment using CRISPR Genome Analyzer (CRISPR-GA). *Bioinformatics* 30: 2968–2970. doi:10.1093/bioinformatics/btu427

Guo X, Zhang T, Hu Z, Zhang Y, Shi Z, Wang Q, Cui Y, Wang F, Zhao H, Chen Y. 2014. Efficient RNA/Cas9-mediated genome editing in *Xenopus tropicalis*. *Development* 141: 707–714. doi:10.1242/dev.099853

Guschin DY, Waite AJ, Katibah GE, Miller JC, Holmes MC, Rebar EJ. 2010. A rapid and general assay for monitoring endogenous gene modification. *Methods Mol Biol* 649: 247–256. doi:10.1007/978-1-60761-753-2_15

Hsiau, T, Maures T, Waite K, Yang J, Kelso R, Holden K, Stoner R. 2018. Inference of CRISPR edits from sanger trace data. bioRxiv doi:10.1101/251082

Josephs EA, Kocak DD, Fitzgibbon CJ, McMenemy J, Gersbach CA, Marszalek PE. 2015. Structure and specificity of the RNA-guided endonuclease Cas9 during DNA interrogation, target binding and cleavage. *Nucleic Acids Res* 43: 8924–8941. doi:10.1093/nar/gkv892

Kim HJ, Lee HJ, Kim H, Cho SW, Kim JS. 2009. Targeted genome editing in human cells with zinc finger nucleases constructed via modular assembly. *Genome Res* 19: 1279–1288. doi:10.1101/gr.089417.108

Kim JM, Kim D, Kim S, Kim JS. 2014. Genotyping with CRISPR–Cas-derived RNA-guided endonucleases. *Nat Commun* 5: 3157. doi:10.1038/ncomms4157

Lane M, Khokha MK. 2021. Obtaining *Xenopus tropicalis* embryos by in vitro fertilization. *Cold Spring Harb Protoc* doi:10.1101/pdb.prot106351

Liu X, Homma A, Sayadi J, Yang S, Ohashi J, Takumi T. 2016. Sequence features associated with the cleavage efficiency of CRISPR/Cas9 system. *Sci Rep* 6: 19675. doi:10.1038/srep19675

McQueen C, Pownall ME. 2017. An analysis of MyoD-dependent transcription using CRISPR/Cas9 gene targeting in *Xenopus tropicalis* embryos. *Mech Dev* 146: 1–9. doi:10.1016/j.mod.2017.05.002

Moreno-Mateos MA, Vejnar CE, Beaudoin JD, Fernandez JP, Mis EK, Khokha MK, Giraldez AJ. 2015. CRISPRscan: designing highly efficient sgRNAs for CRISPR–Cas9 targeting in vivo. *Nat Methods* 12: 982–988. doi:10.1038/nmeth.3543

Naert T, Tulkens D, Edwards NA, Carron M, Shaidani NI, Wlizla M, Boel A, Demuynck S, Horb ME, Coucke P, Willaert A, Zorn AM, Vleminckx K. 2020. Maximizing CRISPR/Cas9 phenotype penetrance applying predictive modeling of editing outcomes in *Xenopus* and zebrafish embryos. *Sci Rep* **10:** 14662. doi:10.1038/s41598-020-71412-0

Naito Y, Hino K, Bono H, Ui-tei K. 2015. CRISPRdirect: software for designing CRISPR/Cas guide RNA with reduced off-target sites. *Bioinformatics* **31:** 1120–1123. doi:10.1093/bioinformatics/btu743

Nakayama T, Fish MB, Fisher M, Oomen-Hajagos J, Thomsen GH, Grainger RM. 2013. Simple and efficient CRISPR/Cas9-mediated targeted mutagenesis in *Xenopus tropicalis*. *Genesis* **51:** 835–843. doi:10.1002/dvg.22720

Nakayama T, Blitz IL, Fish MB, Odeleye AO, Manohar S, Cho KW, Grainger RM. 2014. Cas9-based genome editing in *Xenopus tropicalis*. *Methods Enzymol* **546:** 355–375. doi:10.1016/B978-0-12-801185-0.00017-9

Nakayama T, Fisher M, Nakajima K, Odeleye AO, Zimmerman KB, Fish MB, Yaoita Y, Chojnowski JL, Lauderdale JD, Netland PA, Grainger RM. 2015. *Xenopus pax6* mutants affect eye development and other organ systems, and have phenotypic similarities to human aniridia patients. *Dev Biol* **408:** 328–344. doi:10.1016/j.ydbio.2015.02.012

Ogino H, McConnell WB, Grainger RM. 2006. High-throughput transgenesis in *Xenopus* using *I-SceI* meganuclease. *Nat Protoc* **1:** 1703–1710. doi:10.1038/nprot.2006.208

Owens NDL, Blitz IL, Lane MA, Patrushev I, Overton JD, Gilchrist MJ, Cho KWY, Khokha MK. 2016. Measuring absolute RNA copy numbers at high temporal resolution reveals transcriptome kinetics in development. *Cell Rep* **14:** 632–647. doi:10.1016/j.celrep.2015.12.050

Peshkin L, Wühr M, Pearl E, Haas W, Freeman RM Jr, Gerhart JC, Klein AM, Horb M, Gygi SP, Kirschner MW. 2015. On the relationship of protein and mRNA dynamics in vertebrate embryonic development. *Dev Cell* **35:** 383–394. doi:10.1016/j.devcel.2015.10.010

Ramlee MK, Yan T, Cheung AM, Chuah CT, Li S. 2015. High-throughput genotyping of CRISPR/Cas9-mediated mutants using fluorescent PCR-capillary gel electrophoresis. *Sci Rep* **5:** 15587. doi:10.1038/srep15587

Ren X, Yang Z, Xu J, Sun J, Mao D, Hu Y, Yang SJ, Qiao HH, Wang X, Hu Q, Deng P, Liu LP, Ji JY, Li JB, Ni JQ. 2014. Enhanced specificity and efficiency of the CRISPR/Cas9 system with optimized sgRNA parameters in *Drosophila*. *Cell Rep* **9:** 1151–1162. doi:10.1016/j.celrep.2014.09.044

Sentmanat MF, Peters ST, Florian CP, Connelly JP, Pruett-Miller SM. 2018. A survey of validation strategies for CRISPR–Cas9 editing. *Sci Rep* **8:** 888. doi:10.1038/s41598-018-19441-8

Servetnick MD, Steinworth B, Babonis LS, Simmons D, Salinas-Saavedra M, Martindale MQ. 2017. Cas9-mediated excision of *Nematostella brachyury* disrupts endoderm development, pharynx formation and oral-aboral patterning. *Development* **144:** 2951–2960.

Shen MW, Arbab M, Hsu JY, Worstell D, Culbertson SJ, Krabbe O, Cassa CA, Liu DR, Gifford DK, Sherwood RI. 2018. Predictable and precise template-free CRISPR editing of pathogenic variants. *Nature* **563:** 646–651. doi:10.1038/s41586-018-0686-x

Shi J, Wang E, Milazzo JP, Wang Z, Kinney JB, Vakoc CR. 2015. Discovery of cancer drug targets by CRISPR–Cas9 screening of protein domains. *Nat Biotechnol* **33:** 661–667. doi:10.1038/nbt.3235

Urnov FD, Miller JC, Lee YL, Beausejour CM, Rock JM, Augustus S, Jamieson AC, Porteus MH, Gregory PD, Holmes MC. 2005. Highly efficient endogenous human gene correction using designed zinc-finger nucleases. *Nature* **435:** 646–651. doi:10.1038/nature03556

Varshney GK, Pei W, LaFave MC, Idol J, Xu L, Gallardo V, Carrington B, Bishop K, Jones M, Li M, et al. 2015. High-throughput gene targeting and phenotyping in zebrafish using CRISPR/Cas9. *Genome Res* **23:** 1030–1042. doi:10.1101/gr.186379.114

Wang T, Wei JJ, Sabatini DM, Lander ES. 2014. Genetic screens in human cells using the CRISPR–Cas9 system. *Science* **343:** 80–84. doi:10.1126/science.1246981

Wittwer CT, Reed GH, Gundry CN, Vandersteen JG, Pryor RJ. 2003. High-resolution genotyping by amplicon melting analysis using LCGreen. *Clin Chem* **49:** 853–860. doi:10.1373/49.6.853

Wu RS, Lam II, Clay H, Duong DN, Deo RC, Coughlin SR. 2018. A rapid method for directed gene knockout for screening in G_0 zebrafish. *Dev Cell* **46:** 112–125. doi:10.1016/j.devcel.2018.06.003

Zhang JP, Li XL, Neises A, Chen W, Hu LP, Ji GZ, Yu JY, Xu J, Yuan WP, Cheng T, Zhang XB. 2016. Different effects of sgRNA length on CRISPR-mediated gene knockout efficiency. *Sci Rep* **6:** 28566. doi:10.1038/srep28566

Generating Nonmosaic Mutants in *Xenopus* Using CRISPR–Cas in Oocytes

Sang-Wook Cha[1]

School of Natural Sciences, University of Central Missouri, Warrensburg, Missouri 64093, USA

In CRISPR–Cas9 genome editing, double-strand DNA breaks (DSBs) primarily undergo repair through nonhomologous end joining (NHEJ), which produces insertion or deletion of random nucleotides within the targeted region (indels). As a result, frameshift mutation-mediated loss-of-function mutants are frequently produced. An alternative repair mechanism, homology-directed repair (HDR), can be used to fix DSBs at relatively low frequency. By injecting a DNA-homology repair construct with the CRISPR–Cas components, specific nucleotide sequences can be introduced within the target region by HDR. We have taken advantage of the fact that *Xenopus* oocytes have much higher levels of HDR than eggs to increase the effectiveness of creating precise mutations. We introduced the oocyte host transfer technique, well established for knockdown of maternal mRNA for loss-of-function experiments, to CRISPR–Cas9-mediated genome editing. The host-transfer technique is based on the ability of *Xenopus* oocytes to be isolated, injected with CRISPR–Cas components, and cultured in vitro for up to 5 d before fertilization. During these 5 d, CRISPR–Cas components degrade, preventing further alterations to the paternal or maternal genomes after fertilization and resulting in heterozygous, nonmosaic embryos. Treatment of oocytes with a DNA ligase IV inhibitor, which blocks the NHEJ repair pathway, before fertilization further improves the efficiency of HDR. This method allows straightforward generation of either nonmosaic F_0 heterozygous indel mutant *Xenopus* or *Xenopus* with efficient, targeted insertion of small DNA fragments (73–104 nt). The germline transmission of mutations in these animals allows homozygous mutants to be obtained one generation (F_1) sooner than previously reported.

MATERIALS

It is essential that you consult the appropriate Material Safety Data Sheets and your institution's Environmental Health and Safety Office for proper handling of equipment and hazardous materials used in this protocol.

RECIPES: Please see the end of this protocol for recipes indicated by <R>. Additional recipes can be found online at http://cshprotocols.cshlp.org/site/recipes.

Reagents

Agarose
BSA (0.1% in 0.1× MMR)
Collagenase VII, cell culture grade (Sigma-Aldrich C2799-15KU) (optional; see Step 13)
Dejellying solution (2% cysteine [Sigma-Aldrich] in 0.1× MMR)
  Prepare fresh; adjust the pH to 7.6–7.8 with NaOH.

---

[1]Correspondence: cha@ucmo.edu

DNA ligase IV inhibitor (Selleck Chemical SCR-7)
Ethanol (100%)
Ethyl 3-aminobenzoate methanesulfonate salt (tricaine, MS-222; 2%)
Forward and reverse polymerase chain reaction (PCR) primers to make single-guide RNA (sgRNA)
    templates (see Step 2)
Gel electrophoresis buffer
Genomic PCR primers (see Step 33)
High-salt solution <R>
Human chorionic gonadotropin (HCG; Chorulon)
Lysis buffer for tadpoles <R>
Marc's modified Ringer's (MMR; 10×; pH 7.6) <R>
MegaShortscript Kit (Ambion)
Nested PCR primer (optional; see Step 36)
Nuclease-free water
Oocyte culture medium (OCM) <R>
Phusion High-Fidelity PCR Kit (NEB)
Progesterone (e.g., Sigma-Aldrich; 1 mM in 100% ethanol)
    Store at −20°C.

QIAquick PCR purification Kit (QIAGEN)
Recombinant SpCas9 protein (PNA Bio)
Repair oligo (Ultramer, Integrated DNA Technologies)
TOPO TA Cloning Kit (Life Technologies)
Vital dye stock solutions <R>
Xenopus laevis or *tropicalis* female frogs
Xenopus laevis or *tropicalis* testis (obtained as in Sec. 2, Protocol 2: Obtaining *Xenopus laevis* Embryos
    [Shaidani et al. 2021] or Sec. 2, Protocol 5: Obtaining *Xenopus tropicalis* Embryos by In Vitro
    Fertilization [Lane and Khokha 2021])

Equipment

Catheters (BD Insyte Autoguard 381457 [for *X. laevis*] or 330003 [for *X. tropicalis*])
    (optional; see Step 22)
Gel electrophoresis apparatus
Glass bead sterilizer
Low-temperature incubators
Microinjector with stereomicroscope
Microscope slides
Pasteur pipettes
PCR machine
Petri dishes (90- and 100-mm)
Plastic spoon (perforated)
Plastic tissue grinder
Platform rocker
Surgical instruments (scissors, surgical blades, needle holders, and forceps)
Sutures (4-0 silk black braided c-17, with 12-mm 3/8 circle needle; Surgical Specialties 785B)

METHOD

This procedure is based on Tadjuidje and Cha (2018).

Preparation of CRISPR–Cas9 Component

1. Identify potential CRISPR–Cas9 target sites(N_{18-20}) by finding sites adjacent to protospacer-adjacent motif (PAM) sequences within the genomic sequence of the gene of interest. Identify sequences using CHOPCHOP v3 (https://chopchop.cbu.uib.no) for both *X. laevis* and *X. tropicalis*.

   Off-target screening is automatically performed in the algorithm (Labun et al. 2019; Montague et al. 2014).

2. To make sgRNA templates, design two long PCR primers as described below.

   i. Design a unique forward primer for each target site that contains the T7 promoter sequence followed by an initiator guanine for better transcriptional activity, followed by the targeting sequence(N_{18-20}) without the PAM sequence, and then a portion of the sgRNA backbone: 5′-GAAATTAATACGACTCACTATAGGN_{18-20}*GTTTAAGAGCTATGCTGGAAACAGCAT AGCA*-3′.

   ii. Use a universal reverse primer, with the sequence 5′-AAAAGCACCGACTCGGTGC CACTTTTTCAAGTTGATAACGGACTAGCCTTATTTAAACT*TGCTATGCTGTTTC*-3′, for all sgRNA templates.

   A region of partial complementary to the forward primer is indicated in italics. We use the sgRNA$^{(F+E)}$ backbone to improve the specificity and stability (Chen et al. 2013).

   When mixed, these two primers base pair to one another. Each primer is then extended to generate a double-stranded DNA template for sgRNA transcription.

3. Prepare a PCR reaction to generate the sgRNA templates using a high-fidelity DNA polymerase (e.g., NEB Phusion HF).

   i. Prepare a total reaction mixture of 100 µL according to the manufacturer's instructions.

   As an example, a typical reaction might contain 10 µL of 10× buffer, 1.2 µL of 25 mM dNTP mix, 2 µL of 50 mM MgSO$_4$, 2 µL each primer at 100 pmol/µL, 1 µL of DNA polymerase, and nuclease-free H$_2$O to 100 µL.

   ii. Carry out PCR with the following conditions: 5 min at 94°C, 20 cycles of 20 sec at 94°C, 20 sec at 58°C, and 15 sec at 68°C, followed by a final extension of 5 min at 68°C.

   iii. Analyze 1 µL of the reaction by agarose gel electrophoresis to confirm the synthesis of the desired product.

   iv. Purify sgRNA templates with a column-based PCR cleanup kit.

4. Synthesize sgRNA using the MegaShortscript Kit (Ambion) according to the manufacturer's instructions.

   A minimum of 300 pg of the templates is required.

5. Purify the sgRNA with the RNeasy Kit (QIAGEN).

6. If the aim of genome editing is to introduce or change DNA sequence, design the repair oligo, a ~200 base, single-stranded oligonucleotide, for the desired purpose.

   For example, if adding the sequence for a carboxy-terminus epitope tag to a gene, the oligonucleotide would typically include 30–50 base homology at either end spanning the targeted locus. An epitope tag sequence would be added just before the endogenous stop codon. To prevent repaired sequences from being cleaved by Cas9 after successful recombination, silent mutations are introduced into the sgRNA targeting sequence. Also, the efficiency of the HDR mediated knock-in can be increased by including phosphorothioate modifications in the final two bases on both ends and by having asymmetric homology arms in the donor oligo (Renaud et al. 2016; Richardson et al. 2016) including in Xenopus (Nakayama et al. 2020). To confirm the incorporation of the desired sequence, introduce a new restriction enzyme site in the oligonucleotide to facilitate the screening process. When introducing an epitope tag, the epitope sequence is usually large enough to create a second band on 2% agarose gel of a short amplicon across the target region.

Oocyte Preparation for Microinjection

Perform oocyte collection following the method described by Mir and Heasman (2008) and Olson et al. (2012) with the modifications described below.

7. Perform frog surgery under aseptic conditions. Sterilize all surgical instruments except single-use surgical blades for 5 min at 250°C in a glass bead sterilizer.

8. Submerge females that will act as ovary donors in 0.2% Tricaine (MS-222) for 5–8 min at room temperature. Closely monitor frogs until they become unresponsive to a gentle touch on the lower jaw (also known as the swallowing reflex).

9. Place the frog on its back. Use a surgical blade to make a small incision in the skin on one side of the lower abdomen. Use sterile curved scissors to extend the incision to ∼1.0 cm in length. Lift the abdominal wall away from underlying organs and make a similar incision through the muscle layer. Pull part of the ovary out of the abdominal cavity. Cut out a portion of the ovary with scissors. Place the ovary fragment in OCM and evaluate under a dissecting microscope.

10. Suture the incision, beginning with the muscle layer (three stitches) and followed by the skin (three stiches).

11. Rinse the frog briefly in running tap water to remove any residual MS-222 from her skin and return her to paper towel soaked in fresh water to ensure she neither dries out nor drowns during recovery. Monitor her until she is fully awake and alert then return her to a tank alone to prevent the stitches from being inadvertently caught on other frogs' claws. Return to a normal tank after 2–3 d.

12. Cut the ovary into 1- to 2-cm^3-sized pieces using surgical scissors. Place the pieces in OCM with approximately eight to 12 pieces per 90-mm dish. To keep oocytes healthy, change the medium every day.

13. Leave the pieces of ovary in OCM overnight at 16°C to soften the surrounding follicle tissues and make the defolliculation easier. Using a pair of forceps, manually defolliculate both *X. laevis* and *X. tropicalis* oocytes in OCM under aseptic conditions. Do this by holding the transparent stalk linking the oocyte to the ovary with forceps and using the other pair to gently pull away the oocyte, without touching the oocyte itself. Alternatively, for *X. tropicalis*, gently shake small pieces of ovary in 0.1 KU/mL collagenase VII in OCM for ∼2 h at room temperature to release the individual, full-grown oocytes from the ovarian tissue. Sort and thoroughly rinse them in fresh OCM. Store sorted oocytes in fresh OCM at room temperature while you test for maturation.

14. Before beginning a full experiment, test the maturation of each batch of defolliculated oocytes. Do this by treating a dozen defolliculated oocytes with 2 μM progesterone in OCM for 5 to 8 h (for *X. laevis*) or 30 min to 1 h (for *X. tropicalis*) at room temperature. Calculate the maturation rate as the % of oocytes undergoing germinal vesicle breakdown (GVBD). Judge GVBD by the appearance of a white spot at the animal pole of the oocyte. If the rate of GVBD is <80%, then discard the current batch and use ovaries from other donor females.

    We routinely use mixed oocytes from at least two different donor females, which can increase the overall success rate.

sgRNA Injection and Oocyte Host Transfer

15. Using standard techniques, inject oocytes with sgRNA as follows:

    - For indel mutations, inject 300 pg of sgRNA and 600 pg of Cas9 protein for *X. laevis* in up to a 10-nL injection volume or 200 pg of sgRNA and 300 pg of Cas9 protein for *X. tropicalis* in up to a 4-nL injection volume.

    - For knock-in, HDR experiments, include the repair oligo (200 pg for *X. laevis*, 30 pg for *X. tropicalis*) in the injection solution, staying within the maximum injectable volume.

Cite this protocol as *Cold Spring Harb Protoc*; doi:10.1101/pdb.prot106989

16. For indel mutations keep the injected oocytes in OCM for 72 h at 18°C (for *X. laevis*) or 23°C (for *X. tropicalis*). Change the culture medium daily and remove any damaged oocytes from the culture. For knock-in, HDR experiments, include 5 µM SCR-7, the DNA ligase IV inhibitor, in the OCM, including when the culture medium is changed daily.

17. Prepare host *X. laevis* females and inject with 1000 units of HCG into the dorsal lymph sac. Keep injected frogs for 12–14 h at 18°C before the host transfer surgery. Prepare three female frogs for each host needed to ensure appropriately timed egg laying.

    X. laevis *hosts are used for* X. laevis *oocytes as well as* X. tropicalis *oocytes.*

18. Following the 72-h incubation in Step 16, culture oocytes in 10 mL of 2 µM progesterone in OCM for 12 h at 18°C (for *X. laevis*) or 3 h at 23°C (for *X. tropicalis*). Omit SCR-7 during this incubation.

19. On the day of surgery, stain oocytes to distinguish among experimental groups. Add vital dye stock solutions to the dish containing the oocytes (in 10 mL OCM) in the amounts listed below. Stain for 10 min at room temperature with gentle rocking.

    - Red: 100 µL of Neutral Red
    - Brown: 100 µL of Bismarck Brown
    - Blue: 100 µL of Nile Blue A
    - Mauve: 75 µL of Neutral Red + 100 µL of Nile Blue A
    - Green: 100 µL of Bismarck Brown alone for 5 min + 100 µL of Nile Blue A for 10 min

20. While staining the oocytes, submerge a host female that has just begun laying good-quality eggs in 0.2% Tricaine at room temperature. Monitor the frog until it just becomes unresponsive. Once the frog becomes unresponsive, remove it from the tricaine and lay it on a water-soaked paper towel to prevent overdosing.

21. Transfer the stained oocytes to a 100-mm dish containing fresh OCM to prevent overstaining. If using more than two colors, place the eggs in separate piles in the dish until the last moment to avoid unnecessary contact between two different colored eggs.

22. Make a small incision in the frog similar to that made when removing ovaries but make the body wall incision just large enough to accommodate the tip of a glass pipette. To prevent the loss of experimental oocytes, hold one side of the body wall up until the first stitch is made after oocyte implantation.

    Alternatively, lay the anesthetized frog on its back and insert the needle of a catheter into the lower abdomen at an angle almost parallel to the surgery bench. Face the catheter tip toward the frog's midline to avoid puncturing an internal organ. Pipette the oocytes through the barrel of catheter with minimum amount of OCM. Once the eggs are transferred, tilt the frog to help the transferred oocytes spread into the abdominal cavity. Slowly remove the catheter and place the frog in 2 L of high-salt solution. Sutures are not required.

23. Load the treated oocytes into an OCM-coated Pasteur pipette. Insert the pipette tip into the opening of the body wall. Spread the oocytes in between the ovary and the body wall.

24. Suture the body wall and skin. Allow the frog to recover from anesthesia as before. When the frog is fully awake and alert (within 30 min), transfer her into 2 L of high-salt solution.

Egg Collection and Fertilization

25. To ensure the eggs' competency, allow the host frog to lay eggs into the high-salt solution. Two hours after colored eggs first appear, gently squeeze the frog in high-salt solution at intervals of 30 min or more for a maximum of two more hours to collect additional colored eggs. If colored eggs are not released after 5 h postsurgery, squeeze the frog anyway.

26. Use a pipette to place only the colored eggs into a new dish. Remove excess high-salt solution by decanting and using a pipette.

27. Place a large piece of species-appropriate testis into a 1.5-mL tube containing 200 μL of 0.1× MMR, 0.1% BSA at room temperature. Crush it with a disposable plastic tissue grinder and immediately spread the sperm around the colored eggs with a plastic transfer pipette. Wait 5 min, and then flood the dish with 0.1× MMR. Keep the embryos in jelly until the blastula stage to improve the survival rate of host transfer embryos unless you are performing early analysis. Before dejellying, keep the embryos in 0.1× MMR at 18°C for *X. laevis* and 23°C for *X. tropicalis*.

28. When the embryos reach the desired stage, remove the 0.1× MMR and add dejellying solution to the dish. Gently swirl the embryos for 2–3 min at room temperature until the jelly coats are removed. Rinse thoroughly with 0.1× MMR and keep embryos in 0.1× MMR at desired temperature, which is generally 18°C for *X. laevis* and 23°C for *X. tropicalis*.

Genotyping

Genomic DNA Extraction and PCR from Tadpoles

29. Transfer each tadpole (NF stage 42–45) to a 0.0025% Tricaine solution at room temperature and wait for 30 sec. Once it has stopped moving, take it out with a perforated plastic spoon and place it on top of a microscope slide. Cut off one-fourth of the tail with a surgical blade and place the fragment in lysis buffer (150 μL per tail for *X. laevis*; 100 μL per tail for *X. tropicalis*).

30. Return the amputated tadpole to a small volume of 0.1× MMR. From this point on, assign an identification code for each tadpole. Make sure to clean the blade and forceps between the tadpoles to minimize cross-contamination.

31. Incubate the tail in lysis buffer for 6 h (up to overnight) at 56°C.

32. To denature the proteinase K in the lysis buffer, incubate the lysate for 10 min at 95°C. Cool to room temperature. Dilute the lysate 1:5 dilution in nuclease-free water to prepare the template for the PCR reaction and store at −20°C.

33. Use the Phusion High-fidelity Kit to carry out PCR according to the manufacturer's instructions. We use the Primer3 website (http://bioinfo.ut.ee/primer3) to design primers for genomic PCR and sequencing. Use 1–4 μL of diluted lysate from Step 32 as a template in each sample.

34. Perform PCR as follows: 5 min at 96°C followed by 40 cycles of 20 sec at 96°C, 20 sec at 64°C, 20 sec at 72°C, followed by a final extension of 5 min at 72°C.

35. Analyze a fraction of the PCR reaction on a 2% agarose gel to confirm successful amplification. If further analysis is necessary, purify the PCR product with a QIAquick purification kit.

Sequencing

36. Submit the purified PCR product for direct Sanger sequencing using one of the PCR primers or one nested primer. Mixed peaks will occur in the sequencing results after the predicted sgRNA targeting site or for the insertion.

37. If your mutation was an indel, once the result has come back, use TIDE or CRISP-ID on the web to reveal the frequency and the nature of mutations (Brinkman et al. 2014; Dehairs et al. 2016). TIDE analysis will reveal the number of inserted or deleted bases and the proportion of mutations found in the PCR products. CRISP-ID analysis will differentiate between unmixed sequences and mixed peaks for up to three sequences.

38. If your mutation was a knock-in using HDR, use the TIDER algorithm to quantitate the frequency of the templated mutation and the spectrum of nontemplated indels (Brinkman et al. 2018)

39. To get the precise sequence information within the embryo, clone the PCR products with a TA Cloning Kit according to the manufacturer's instructions. Sequence at least 20 randomly selected clones for both type and frequency.

DISCUSSION

CRISPR–Cas9 genome editing is based on sequence recognition between a guiding RNA and the target DNA, combined with the catalytic ability of the Cas9 enzyme to make double-strand DNA breaks within the target sequences (Cong et al. 2013; Mali et al. 2013; Qi et al. 2013; Wang et al. 2013; Nakayama et al. 2014). These double-strand breaks are mainly repaired by nonhomologous end joining (NHEJ), which produces insertion or deletion of random nucleotides within the targeted region (indels). As a result, frameshift mutation–mediated loss-of-function mutants are easily produced (Blitz et al. 2013; Nakayama et al. 2013; Xue et al. 2014; Kotani et al. 2015; Wang et al. 2015). An alternative repair mechanism, homology-directed repair (HDR) can be used to fix the double-stranded break at relatively low frequency. By providing a DNA-homology repair construct together with the CRISPR–Cas components, specific nucleotide sequences can be introduced within the target region by HDR (Gratz et al. 2014; Miyaoka et al. 2014; Yu et al. 2014; Aslan et al. 2017).

To increase the frequency of HDR and hence the effectiveness of creating precise mutations in *Xenopus*, we have taken advantage of the fact that oocytes have much higher levels of HDR than eggs (Hagmann et al. 1996). The oocyte host transfer technique is well established for knockdown of maternal mRNA for loss of function experiments (see Sec. 3, Protocol 4: Oocyte Host-Transfer and Maternal mRNA Depletion Experiments in *Xenopus* [Houston 2018]; Mir and Heasman 2008). We have introduced this technique to CRISPR–Cas9-mediated genome editing. The method takes advantage of the fact that *Xenopus* oocytes can be isolated, injected with CRISPR–Cas components, and cultured in vitro for up to 5 d before they are fertilized using the host-transfer technique. During these 5 d, we have shown that the CRISPR–Cas components degrade, preventing further alterations to the paternal or maternal genomes after fertilization and resulting in heterozygous, nonmosaic embryos. The efficiency of HDR is further improved by treating the oocytes with a DNA ligase IV inhibitor (which inhibits the NHEJ repair pathway) before fertilization (Chu et al. 2015; Maruyama et al. 2015; Ratzan et al. 2017). The method described is a straightforward approach to generating either nonmosaic F_0 heterozygous indel mutant *Xenopus* or those with efficient, targeted insertion of small DNA fragments (73–104 nt) (Aslan et al. 2017). These have germline transmission, which offers the chance to obtain homozygous mutants one generation (F_1) earlier than previously reported (Nakayama et al. 2013; Yu et al. 2014; Ratzan et al. 2017).

RECIPES

High-Salt Solution (10× Stock)

280 g NaCl
7.2 g KCl
14.1 g CaCl$_2$
9.7 g MgCl$_2$
17.12 g HEPES

Bring to 4 L with sterile water. Adjust the pH to 7.6 with NaOH. Filter-sterilize and store at room temperature.

Lysis Buffer for Tadpoles

50 mM Tris (pH 8.8)
1 mM EDTA
0.5% Tween 20
200 µg/mL proteinase K (PCR grade; RPROTK-RO ROCHE; Sigma-Aldrich)

Store at room temperature. Add proteinase K immediately prior to use.

Marc's Modified Ringer's (MMR; 10×; pH 7.6)

58.44 g NaCl
1.5 g KCl
2.94 g CaCl$_2$
2.03 g MgCl$_2$
35.7 g HEPES

Prepare 1 L of 10× stock in sterile water. Adjust pH to 7.6 with 5 M NaOH. Filter sterilize and store at room temperature.

Oocyte Culture Medium (OCM)

60% L-15 (Sigma-Aldrich L4386)
0.1 mg/mL BSA
50 units/mL penicillin/streptomycin (Gibco 15140122)

Adjust the pH to 7.6–7.8 and filter-sterilize using a 0.22-µm filter. Store for up to 1 wk in a low-temperature incubator (12°C–23°C). If the color of the solution changes to dark red or yellow, discard it.

Vital Dye Stocks

Add the following amounts of dye powder to 50 mL of sterile deionized water in separate 50-mL Falcon tubes:

0.5 g/50 mL Bismarck Brown (Sigma-Aldrich B2759),
0.125 g/50 mL Neutral Red (Sigma-Aldrich 861251), and
0.05 g/50 mL Nile Blue A (Nile Blue sulfate; Sigma-Aldrich N0766).

Rock for 2 h to dissolve, and then centrifuge at ∼3000*g*. (The final concentrations will be 1%, 0.25%, and 0.1% for Bismarck Brown, Neutral Red, and Nile Blue A, respectively.) Aliquot the supernatant in 1.5-mL microcentrifuge tubes (∼1-mL aliquots) and store at −20°C.

REFERENCES

Aslan Y, Tadjuidje E, Zorn AM, Cha S-W. 2017. High-efficiency non-mosaic CRISPR-mediated knock-in and indel mutation in F$_0$ *Xenopus*. *Development* **144**: 2852–2858. doi:10.1242/dev.152967

Blitz IL, Biesinger J, Xie X, Cho KWY. 2013. Biallelic genome modification in F$_0$ *Xenopus tropicalis* embryos using the CRISPR/Cas system. *Genesis* **51**: 827–834. doi:10.1002/dvg.22719

Brinkman EK, Chen T, Amendola M, van Steensel B. 2014. Easy quantitative assessment of genome editing by sequence trace decomposition. *Nucleic Acids Res* **42**: e168. doi:10.1093/nar/gku936

Brinkman EK, Kousholt AN, Harmsen T, Leemans C, Chen T, Jonkers J, van Steensel B. 2018. Easy quantification of template-directed CRISPR/Cas9 editing. *Nucleic Acids Res* **46**: e58. doi:10.1093/nar/gky164

Chen B, Gilbert LA, Cimini BA, Schnitzbauer J, Zhang W, Li G-W, Park J, Blackburn EH, Weissman JS, Qi LS, et al. 2013. Dynamic imaging of genomic loci in living human cells by an optimized CRISPR/Cas system. *Cell* **155**: 1479–1491. doi:10.1016/j.cell.2013.12.001

Chu VT, Weber T, Wefers B, Wurst W, Sander S, Rajewsky K, Kühn R. 2015. Increasing the efficiency of homology-directed repair for CRISPR–Cas9-induced precise gene editing in mammalian cells. *Nat Biotechnol* **33**: 543–548. doi:10.1038/nbt.3198

Cong L, Ran FA, Cox D, Lin S, Barretto R, Habib N, Hsu PD, Wu X, Jiang W, Marraffini LA, et al. 2013. Multiplex genome engineering using CRISPR/Cas systems. *Science* **339**: 819–823. doi:10.1126/science.1231143

Dehairs J, Talebi A, Cherifi Y, Swinnen JV. 2016. CRISP-ID: decoding CRISPR mediated indels by Sanger sequencing. *Sci Rep* **6**: 28973. doi:10.1038/srep28973

Gratz SJ, Ukken FP, Rubinstein CD, Thiede G, Donohue LK, Cummings AM, O'Connor-Giles KM. 2014. Highly specific and efficient CRISPR/Cas9-catalyzed homology-directed repair in *Drosophila*. *Genetics* **196**: 961–971. doi:10.1534/genetics.113.160713

Hagmann M, Adlkofer K, Pfeiffer P, Bruggmann R, Georgiev O, Rungger D, Schaffner W. 1996. Dramatic changes in the ratio of homologous recombination to nonhomologous DNA end-joining in oocytes and early embryos of *Xenopus laevis*. *Biol Chem Hoppe-Seyler* **377**: 239–250.

Houston DW. 2018. Oocyte host-transfer and maternal mRNA depletion experiments in *Xenopus*. *Cold Spring Harb Protoc* doi:10.1101/pdb.prot096982

Kotani H, Taimatsu K, Ohga R, Ota S, Kawahara A. 2015. Efficient multiple genome modifications induced by the crRNAs, tracrRNA and Cas9 protein complex in Zebrafish. *PLoS One* **10**: e0128319. doi:10.1371/journal.pone.0128319

Labun K, Montague TG, Krause M, Torres Cleuren YN, Tjeldnes H, Valen E. 2019. CHOPCHOP v3: expanding the CRISPR web toolbox beyond genome editing. *Nucleic Acids Res* **47**: W171–W174. doi:10.1093/nar/gkz365

Lane M, Khokha MK. 2021. Obtaining *Xenopus tropicalis* embryos by in vitro fertilization. *Cold Spring Harb Protoc* doi:10.1101/pdb.prot106351

Cite this protocol as *Cold Spring Harb Protoc*; doi:10.1101/pdb.prot106989

Mali P, Yang L, Esvelt KM, Aach J, Guell M, DiCarlo JE, Norville JE, Church GM. 2013. RNA-guided human genome engineering via Cas9. *Science* **339:** 823–826. doi:10.1126/science.1232033

Maruyama T, Dougan SK, Truttmann MC, Bilate AM, Ingram JR, Ploegh HL. 2015. Increasing the efficiency of precise genome editing with CRISPR–Cas9 by inhibition of nonhomologous end joining. *Nat Biotechnol* **33:** 538–542. doi:10.1038/nbt.3190

Mir A, Heasman J. 2008. How the mother can help: studying maternal Wnt signaling by anti-sense-mediated depletion of maternal mRNAs and the host transfer technique. *Methods Mol Biol* **469:** 417–429. doi:10.1007/978-1-60327-469-2_26

Miyaoka Y, Chan AH, Judge LM, Yoo J, Huang M, Nguyen TD, Lizarraga PP, So P-L, Conklin BR. 2014. Isolation of single-base genome-edited human iPS cells without antibiotic selection. *Nat Methods* **11:** 291–293. doi:10.1038/nmeth.2840

Montague TG, Cruz JM, Gagnon JA, Church GM, Valen E. 2014. CHOP-CHOP: a CRISPR/Cas9 and TALEN web tool for genome editing. *Nucleic Acids Res* **42:** W401–W407. doi:10.1093/nar/gku410

Nakayama T, Fish MB, Fisher M, Oomen-Hajagos J, Thomsen GH, Grainger RM. 2013. Simple and efficient CRISPR/Cas9-mediated targeted mutagenesis in *Xenopus tropicalis*. *Genesis* **51:** 835–843. doi:10.1002/dvg.22720

Nakayama T, Blitz IL, Fish MB, Odeleye AO, Manohar S, Cho KWY, Grainger RM. 2014. Cas9-based genome editing in *Xenopus tropicalis*. *Methods Enzymol* **546:** 355–375. doi:10.1016/B978-0-12-801185-0.00017-9

Nakayama T, Grainger RM, Cha S-W. 2020. Simple embryo injection of long single-stranded donor templates with the CRISPR/Cas9 system leads to homology-directed repair in *Xenopus tropicalis* and *Xenopus laevis*. *Genesis* **58:** e23366. doi:10.1002/dvg.23366

Olson DJ, Hulstrand AM, Houston DW. 2012. Maternal mRNA knockdown studies: antisense experiments using the host-transfer technique in *Xenopus laevis* and *Xenopus tropicalis*. *Methods Mol Biol* **917:** 167–182. doi:10.1007/978-1-61779-992-1_10

Qi LS, Larson MH, Gilbert LA, Doudna JA, Weissman JS, Arkin AP, Lim WA. 2013. Repurposing CRISPR as an RNA-guided platform for sequence-specific control of gene expression. *Cell* **152:** 1173–1183. doi:10.1016/j.cell.2013.02.022

Ratzan W, Falco R, Salanga C, Salanga M, Horb ME. 2017. Generation of a *Xenopus laevis* F_1 albino J strain by genome editing and oocyte host-transfer. *Dev Biol* **426:** 188–193. doi:10.1016/j.ydbio.2016.03.006

Renaud J-B, Boix C, Charpentier M, De Cian A, Cochennec J, Duvernois-Berthet E, Perrouault L, Tesson L, Edouard J, Thinard R, et al. 2016. Improved genome editing efficiency and flexibility using modified oligonucleotides with TALEN and CRISPR–Cas9 nucleases. *Cell Rep* **14:** 2263–2272. doi:10.1016/j.celrep.2016.02.018

Richardson CD, Ray GJ, DeWitt MA, Curie GL, Corn JE. 2016. Enhancing homology-directed genome editing by catalytically active and inactive CRISPR–Cas9 using asymmetric donor DNA. *Nat Biotechnol* **34:** 339–344. doi:10.1038/nbt.3481

Shaidani N-I, McNamara S, Wlizla M, Horb ME. 2021. Obtaining *Xenopus laevis* embryos. *Cold Spring Harb Protoc* doi:10.1101/pdb.prot106211

Tadjuidje E, Cha SW. 2018. How to generate non-mosaic CRISPR/Cas mediated knock-in and mutations in F_0 *Xenopus* through the host-transfer technique. *Methods Mol Biol* **1865:** 105–117. doi:10.1007/978-1-4939-8784-9_8

Wang H, Yang H, Shivalila CS, Dawlaty MM, Cheng AW, Zhang F, Jaenisch R. 2013. One-step generation of mice carrying mutations in multiple genes by CRISPR/Cas-mediated genome engineering. *Cell* **153:** 910–918. doi:10.1016/j.cell.2013.04.025

Wang F, Shi Z, Cui Y, Guo X, Shi Y-B, Chen Y. 2015. Targeted gene disruption in *Xenopus laevis* using CRISPR/Cas9. *Cell Biosci* **5:** 15. doi:10.1186/s13578-015-0006-1

Xue W, Chen S, Yin H, Tammela T, Papagiannakopoulos T, Joshi NS, Cai W, Yang G, Bronson R, Crowley DG, et al. 2014. CRISPR-mediated direct mutation of cancer genes in the mouse liver. *Nature* **514:** 380–384. doi:10.1038/nature13589

Yu Z, Chen H, Liu J, Zhang H, Yan Y, Zhu N, Guo Y, Yang B, Chang Y, Dai F, et al. 2014. Various applications of TALEN- and CRISPR/Cas9-mediated homologous recombination to modify the *Drosophila* genome. *Biol Open* **3:** 271–280. doi:10.1242/bio.20147682

Tissue-Targeted CRISPR–Cas9-Mediated Genome Editing of Multiple Homeologs in F_0-Generation *Xenopus laevis* Embryos

Mark E. Corkins,[1,5] Bridget D. DeLay,[1] and Rachel K. Miller[1,2,3,4,5]

[1]*Department of Pediatrics, Pediatric Research Center, University of Texas Health Science Center McGovern Medical School, Houston, Texas 77030, USA;* [2]*Program in Genetics and Epigenetics,* [3]*Program in Biochemistry and Cell Biology, The University of Texas MD Anderson Cancer Center University of Texas Health Science Center Graduate School of Biomedical Sciences, Houston, Texas 77030, USA;* [4]*Department of Genetics, University of Texas MD Anderson Cancer Center, Houston, Texas 77030, USA*

Xenopus laevis frogs are a powerful developmental model that enables studies combining classical embryology and molecular manipulation. Because of the large embryo size, ease of microinjection, and ability to target tissues through established fate maps, *X. laevis* has become the predominant amphibian research model. Given that their allotetraploid genome has complicated the generation of gene knockouts, strategies need to be established for efficient mutagenesis of multiple homeologs to evaluate gene function. Here we describe a protocol to use CRISPR–Cas9-mediated genome editing to target either single alleles or multiple alloalleles in F_0 *X. laevis* embryos. A single-guide RNA (sgRNA) is designed to target a specific DNA sequence encoding a critical protein domain. To mutagenize a gene with two alloalleles, the sgRNA is designed against a sequence that is common to both homeologs. This sgRNA, along with the Cas9 protein, is microinjected into the zygote to disrupt the genomic sequences in the whole embryo or into a specific blastomere for tissue-targeted effects. Error-prone repair of CRISPR–Cas9-generated DNA double-strand breaks leads to insertions and deletions creating mosaic gene lesions within the embryos. The genomic DNA isolated from each mosaic F_0 embryo is sequenced, and software is applied to assess the nature of the mutations generated and degree of mosaicism. This protocol enables the knockout of genes within the whole embryo or in specific tissues in F_0 *X. laevis* embryos to facilitate the evaluation of resulting phenotypes.

MATERIALS

It is essential that you consult the appropriate Material Safety Data Sheets and your institution's Environmental Health and Safety Office for proper handling of equipment and hazardous materials used in this protocol.

RECIPES: Please see the end of this protocol for recipes indicated by <R>. Additional recipes can be found online at http://cshprotocols.cshlp.org/site/recipes.

Reagents

Use only molecular biology grade reagents.

Acid-phenol:chloroform, pH 4.5 (e.g., ThermoFisher AM9720)
Agarose

---

[5]Correspondence: Mark.E.Corkins@uth.tmc.edu; Rachel.K.Miller@uth.tmc.edu

Cite this protocol as *Cold Spring Harb Protoc*; doi:10.1101/pdb.prot107037

Anesthetic solution (0.1% benzocaine or 0.1% tricaine in 0.1× MMR or 0.1× MBS)

Anesthetic stock (5% benzocaine or 5% tricaine, dissolved in ethanol)

Cas9 protein with nuclear localization signal (e.g., PNA Bio CP01)

Chloroform

Dextran labeled with a fluorophore or mRNA tracer

> Tracers (Step 11) are normally the mRNA for a fluorophore-labeled protein (Mem::GFP mRNA; 20 ng/µL final concentration) or a fluorophore-labeled dextran (Rhodamine Dextran Fixable 10,0000 MW [Fisher Science D1817]; 2.5 mg/mL final concentration diluted from 25 mg/mL in dH₂O stock).

dH$_2$O

dNTPs (2.5 mM each)

Ethanol (100%; 70%)

Ethidium bromide

> Use ethidium bromide for staining nucleotides.

Modified Barth's saline (1×) (MBS) <R> or Marc's modified Ringers (MMR) (1×, without EDTA) <R>

MEGAshortscript T7 Transcription Kit (Thermo Fisher AM1354)

Nuclease-free (or DEPC-treated) H$_2$O

Oligonucleotides

Forward and reverse primer to amplify edited DNA for sequencing (10 mM; see Step 17)

Target-specific forward primer for sgRNA template synthesis (10 mM; see Step 4)

Universal reverse primer for sgRNA template synthesis (10 mM; see Step 5)

Phenol red (25 µM) (optional; see Step 16)

RNA loading dye (Ambion AM8546G)

Sodium hydroxide (50 mM)

TAE or TBE buffer

Taq (5000 units/mL) and 10× buffer (e.g., NEB M0273)

Tris (1 M, pH 8)

Xenopus laevis embryos obtained using standard protocols (Sive et al. 2000)

Equipment

Agarose gel electrophoresis equipment

CRISPR analysis software

ICE (https://ice.synthego.com/)

TIDE (https://tide.nki.nl/)

Dissection tools (optional; see Step 16)

Genome tools including:

Clustal Omega (https://www.ebi.ac.uk/Tools/msa/clustalo/)

Francis Crick Institute genome browser (http://genomes.crick.ac.uk/)

Lasergene

Pfam (https://pfam.xfam.org/)

UniProt (https://www.uniprot.org/)

Xenbase (http://www.xenbase.org)

Heat block

Microcentrifuge

Microcentrifuge tubes (1.5-mL; DNase and RNase-free)

Microinjector and accessories (injection needles)

PCR tubes (0.2-mL; DNase and RNase-free)

Single-guide RNA (sgRNA) design tools including:

CHOPCHOP (https://chopchop.cbu.uib.no/)
CRISPRscan (https://www.crisprscan.org/)
InDelphi (https://indelphi.giffordlab.mit.edu/)
Spectrophotometer
Stereomicroscope with overhead lighting
Stereomicroscope with fluorescent capabilities
Thermocycler
Vortex

METHOD

sgRNA Design

1. Determine whether your gene of interest has one or two homeologs in *X. laevis* using Xenbase (http://www.xenbase.org) or the genome browser at the Francis Crick Institute (http://genomes.crick.ac.uk/). If the gene has both long and short alloalleles, align them to identify conserved regions with software such as Lasergene or Clustal Omega (https://www.ebi.ac.uk/Tools/msa/clustalo/) (Fig. 1, example alignment). Translate the protein sequence and identify functional domains using protein domain prediction software, such as Pfam (https://pfam.xfam.org/) or UniProt (https://www.uniprot.org/) (Fig. 1, example translation).

2. Use one homeolog to identify potential target sequences that are *unique* to the gene(s) of interest and *do not* span multiple exons using a sgRNA design tool, such as CRISPRscan (https://www.crisprscan.org/) or CHOPCHOP (https://chopchop.cbu.uib.no/) (Bhattacharya et al. 2015; Moreno-Mateos et al. 2015; Labun et al. 2019). Both CRISPRscan and CHOPCHOP will automatically annotate intron/exon segments and provide guides accordingly. Alternately, individual exons, conserved domains, or regions that are identical between both homeologs can be quarried independently. Use machine-learning software, such as InDelphi (https://indelphi.giffordlab.mit.edu/) to narrow potential sgRNAs based upon their likelihood of inducing deleterious mutations (Shen et al. 2018; Naert et al. 2020).

   See Troubleshooting.

3. Within regions of the mRNA predicted to encode a functional protein domain, design a sgRNA to target identical sequences in both the long and short alloalleles (DeLay et al. 2018). If there are multiple protein domains with conserved sgRNA targets, design one sgRNA for each domain (Fig. 1; example sgRNA targets). These regions should include 16–19 nt and must be followed by an NGG protospacer adjacent motif (PAM) sequence at the 3′ end.

   Avoid selecting sgRNA targets with repetitive sequence, and BLAST target sequences to avoid unintended off-target mutations.

   See Troubleshooting.

sgRNA Template Synthesis

4. To generate the DNA template for sgRNA synthesis using a polymerase chain reaction (PCR) strategy, generate the following target-specific oligonucleotide (Fig. 2A; example sgRNA primers). Structure of target-specific forward primer:
   5′ T7 scaffold-T7 promoter-GG sequence (T7 processivity)-target sequence (exclude PAM)-partial overlap with Cas9 scaffold 3′
   Target-specific forward primer:

   5′ CTAG-CTAATACGACTCACTATA-GG-(N)$_{16-19}$ (exclude PAM)-GTTTTAGAGCTAGAAA TAGCAAG 3′

   DO NOT include the PAM sequence!

 Cite this protocol as *Cold Spring Harb Protoc*; doi:10.1101/pdb.prot107037

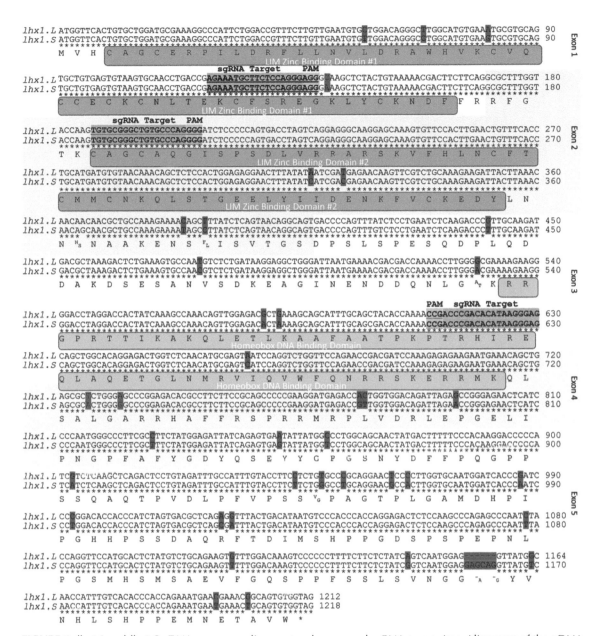

FIGURE 1. *lhx1.L* and *lhx1.S* cDNA sequence alignment and conserved sgRNA target sites. Alignment of the cDNA sequences for the short and long *lhx1* homeologs of *Xenopus laevis*. Nucleotides that are not conserved between alloalleles are highlighted red. Odd exons are white; even exons are labeled in yellow. Translation of the long and short *lhx1* homeologs (*lhx1.L* and *lhx1.S*) to show the encoded protein domains. LIM domains are marked in blue; the homeodomain is labeled green. Conserved single-guide RNAs (sgRNAs) are marked in orange, whereas the associated protospacer adjacent motif (PAM) sequences are marked in turquoise. The exon 1 and 2 sgRNAs target the regions encoding the first and second LIM domains, respectively, whereas the exon 3 sgRNA targets a region encoding the homeodomain.

5. Synthesize the following universal oligonucleotide as the reverse primer to generate the DNA template for sgRNA generation (Fig. 2A; sgRNA primers).

Structure of universal reverse primer:

5′ reverse complement of Cas9 Scaffold <u>Overlap with Forward Primer</u> 3′

Universal reverse primer (NXR-Cas9-Universal-R):

5′ AAAAGCACCGACTCGGTGCCACTTTTTCAAGTTGATAACGGACTAGCCTTATTTTAA <u>CTTGCTATTTCTAGCTCTAAAAC</u> 3′

A

Target-specific forward primer:

Structure: 5' T7 scaffold T7 promoter GG sequence (T7 processivity) Target sequence (exclude PAM) Partial overlap with Cas9 scaffold 3'

Sequence: 5' CTAG CTAATACGACTCACTATA GG (N)₁₆₋₁₉ (exclude PAM) GTTTTAGAGCTAGAAATAGCAAG 3'

Example forward primers for lhx1 sgRNA:

> *lhx1.L and .S* (Exon 1, LIM domain target), AGAAATGCTTCTCCAGGG(AGG):
> 5' CTAG CTAATACGACTCACTATA GG AGAAATGCTTCTCCAGGG (exclude AGG) GTTTTAGAGCTAGAAATAGCAAG 3'
>
> *lhx1.L and .S* (Exon 2, LIM domain target), TGTGCGGGCTGTGCCCAG(GGG):
> 5' CTAG CTAATACGACTCACTATA GG TGTGCGGGCTGTGCCCAG (exclude GGG) GTTTTAGAGCTAGAAATAGCAAG 3'
>
> *lhx1.L and .S* (Exon 3, Homeodomain target), reverse complement of (CCG)ACCCGACACATAAGGGAG:
> 5' CTAG CTAATACGACTCACTATA GG CTCCCTTATGTGTCGGGT (exclude CGG) GTTTTAGAGCTAGAAATAGCAAG 3'

Universal reverse primer (NXR-Cas9-Universal-R):

Structure: 5' Reverse complement of Cas9 scaffold <u>Overlap with forward primer</u> 3'

Sequence: 5' AAAAGCACCGACTCGGTGCCACTTTTTCAAGTTGATAACGGACTAGCCTTATTTTAA<u>CTTGCTATTTCTAGCTCTAAAAC</u> 3'

B

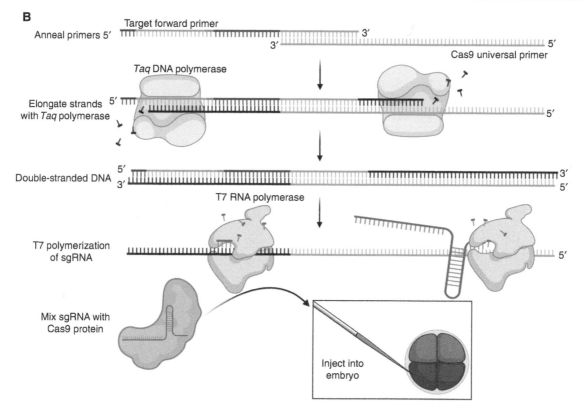

FIGURE 2. Single-guide RNA (sgRNA) design and generation. (*A*) Primer design for generation of sgRNA template. The structure of the target-specific forward primer includes a T7 scaffold (purple), the T7 promoter (orange), a GG sequence for T7 processivity (blue), the sgRNA target sequence (red), and a portion of the Cas9 scaffold. The sequence of the target-specific forward primer is given, in which N represents nucleotides of the sgRNA target site. Examples of sgRNA template primers that were designed to target exon 1, exon 2, and exon 3 of the *lhx1* consensus sequence are provided in the boxed region. The universal reverse primer consists of the Cas9 scaffold (green), including a portion that overlaps with the target-specific forward primer (underlined). (*B*) Schematic of sgRNA synthesis. The target-specific forward primer (nucleotides to scale) and universal reverse primer (a number of nucleotides have been truncated for illustration purposes) are annealed and polymerase chain reaction (PCR)-amplified to generate the DNA template for in vitro transcription of the sgRNA. The sgRNA and Cas9 protein are mixed and injected into early-stage *Xenopus laevis* embryos.

 Cite this protocol as *Cold Spring Harb Protoc*; doi:10.1101/pdb.prot107037

6. Amplify the DNA template for sgRNA synthesis (Fig. 2B; sgRNA template synthesis).

    i. Prepare the following PCR mix.

Stock concentration	Volume	Final concentration
1× dH₂O	36.5 µL	
10× *Taq* buffer	5 µL	1×
2.5 mM (each) dNTPs	4 µL	0.2 mM
10 mM target-specific forward	2 µL	0.4 mM
10 mM NXR-Cas9-Universal-R	2 µL	0.4 mM
Taq (5000 units/mL)	0.5 µL	2.5 units
Final volume	50 µL	

    ii. Carry out the following thermocycler program.

Step	Temperature	Min:sec
1	98°C	0:30
2	98°C	0:10
3	62°C	0:20
4	72°C	0:20
5	Go to step #2 10×	
6	98°C	0:10
7	72°C	0:30
8	Go to step #6 25×	
9	72°C	5:00
10	4°C	∞

    iii. Run 5 µL of the PCR product on 2% agarose gel.

    There should be a strong band at ~125 bp.

    No purification of the PCR product is needed for future steps.

sgRNA Synthesis

7. Use the PCR-amplified DNA template to transcribe the sgRNA using the T7 MegaShortScript kit (Fig. 2B; sgRNA transcription).

    i. Prepare the following transcription mix.

Stock concentration	Volume	Final concentration
1× sgTemplate (PCR)	8 µL	
ATP (75 mM)	2 µL	7.5 mM
GTP (75 mM)	2 µL	7.5 mM
CTP (75 mM)	2 µL	7.5 mM
UTP (75 mM)	2 µL	7.5 mM
10× buffer	2 µL	1×
T7 enzyme mix	2 µL	
Final volume	20 µL	

    ii. Incubate for 5 h at 37°C.

    iii. Add 1 µL of DNase and incubate for 15 min at 37°C.

    iv. Add 15 µL of ammonium acetate and 114 µL of dH₂O (150 µL total). Mix by pipetting.

8. Extract and purify the transcribed sgRNA.

    i. Extract the sgRNA by mixing with 150 μL of acid-phenol:chloroform (pH 4.5).

    ii. Centrifuge the tube at 14,000 rpm for 5 min at 4°C and move the upper layer containing the sgRNA to a new microcentrifuge tube.

    iii. Repeat Steps 8.i and 8.ii.

    iv. Mix the extracted sgRNA with 150 μL of chloroform.

    v. Centrifuge the tube at 14,000 rpm for 5 min at 4°C and move the upper layer containing the sgRNA to a new microcentrifuge tube.

    vi. Purify the sgRNA by mixing with 500 μL of 100% ethanol and precipitating at −80°C overnight.

    vii. Pellet the sgRNA by centrifuging at 14,000 rpm for 30 min at 4°C.

    viii. Carefully remove the ethanol with a pipette, leaving the sgRNA pellet (which should be visible at the bottom of the tube).

    ix. Rinse the sgRNA with 1 mL of 70% ethanol. Tap the tube to dislodge the pellet.

    x. Centrifuge the tube at 14,000 rpm for 10 min at 4°C to repellet the sgRNA.

    xi. Repeat Steps 8.viii–8.x.

    xii. Carefully remove the ethanol with a pipette without disturbing the sgRNA pellet.

    xiii. Invert the tube and air-dry the pellet (the pellet should become clear when it is completely dry).

    xiv. Resuspend the sgRNA pellet in 40–80 μL of nuclease-free H_2O and obtain the absorbance.
    The A_{260}/A_{280} value should be 1.95 or higher.

    xv. Add RNA loading dye to 2 μL of the resuspended sgRNA and heat for 10 min at 65°C.

    xvi. Run the sgRNA on a 2% agarose gel. There should be a strong band around ∼100 bp.

    xvii. Further dilute the sgRNA in nuclease-free H_2O to 1000 ng/μL for long-term storage at −80°C.

    xviii. Dilute the working stock of sgRNA to 500 ng/μL in nuclease-free H_2O and store 5-μL aliquots at −80°C.

Cas9 Protein Dilution

9. Reconstitute lyophilized Cas9 protein to 1 μg/μL with nuclease-free water. Aliquot 1-μL volumes into microcentrifuge tubes, and store at −80°C.

Embryo Microinjection

10. Obtain embryos using established protocols (Sive et al. 2000).

11. Prepare the Cas9 protein/sgRNA injection mixes based upon empirically determined concentrations (DeLay et al. 2018) as follows:

Stock concentration	Volume	Final concentration
1 μg/μL Cas9 protein	1 μL	0.1 μg/μL (1 ng/10 nL)
500 ng/μL sgRNA	1 μL	50 ng/μL (500 pg/10 nL)
dH₂O/tracer	8 μL	
Final volume	10 μL	

12. Leave the injection mix at room temperature for ∼5 min before putting on ice.

Cite this protocol as *Cold Spring Harb Protoc*; doi:10.1101/pdb.prot107037

13. Use established fate maps (Moody 1987a,b; Moody and Kline 1990) to determine which blastomere(s) should be injected to target the tissue of interest for your particular experiment (DeLay et al. 2018, 2019). For genotypic analysis of the efficacy of your sgRNA in the whole embryo, target the animal pole (near the faint colored nucleus) of the early single-cell embryo.

14. Microinject the blastomere of interest with 10 nL of the injection mix according to standard protocols (DeLay et al. 2016), which have been adapted from the original *Xenopus tropicalis* protocols (see Protocol 1: CRISPR–Cas9 mutagenesis in *Xenopus tropicalis* for phenotypic analyses in the F_0 generation and beyond [Blitz et al. 2013]; Nakayama et al. 2013; Blitz and Nakayama 2021).

    The injected embryos are defined as the F_0 generation.

15. Culture F_0 embryos in 0.1× MMR (or 0.1× MBS) to the stage of interest and assess the phenotype (Wlizla et al. 2018).

    See Troubleshooting.

DNA Extraction and Amplification from F_0 Embryos

Because CRISPR–Cas9 genome editing may occur in embryos at different times and in different lineages as the early blastomeres divide, different insertions and deletions are expected as the genome is repaired in different downstream cell lines. Therefore, the F_0 embryos are expected to be mosaic, meaning that the numerous cell lineages within these embryos will be compiled into a single DNA sequencing reaction.

16. Extract DNA from embryos and/or tissues as follows.

    i. Anesthetize embryos at the stage in which the phenotype of interest is to be evaluated in anesthetic solution (0.1× MMR or 0.1× MBS containing 0.1% benzocaine or tricaine) for ~5 min. Anesthetized embryos should not respond to prodding.

    ii. (Optional) Isolate the tissue of interest if targeting has been performed. Combine the targeted tissue from multiple embryos into a single tube to obtain an equivalent amount of tissue as would be derived from a single embryo.

    Forceps, dissection needles, hair loops, and microdissection knives can be used to isolate tissues of interest. However, techniques will vary depending on the tissue to be dissected.

    iii. Place anesthetized individual CRISPant F_0 embryos (as well as unedited wild-type embryos) into microcentrifuge tubes and remove as much MMR (or MBS) as possible from tubes by pipetting.

    iv. Add 100 μL of 50 mM NaOH to each tube.

    v. Heat for 10 min at 95°C.

    vi. Vortex the tube to homogenize the embryo.

    vii. Add 20 μL of 1 M Tris (pH 8) to each sample. Mix by vortexing to neutralize the NaOH.

    If desired, phenol red can be added to a final concentration of 25 μM to confirm that the pH of the solution has been neutralized/yellow at a pH < 6.8 and red at a pH of > 7.4.

    viii. Centrifuge the tubes at 14,000 rpm for 5 min at 4°C.

    ix. Carefully pipette out the supernatant and transfer to a clean microcentrifuge tube.

    The supernatant contains embryo DNA that is ready to use in PCR.

17. Amplify the DNA region surrounding the CRISPR editing site:

    i. Design and order genotyping primers to PCR amplify 700–900 bp of genomic DNA centered upon the sgRNA target sequence. If you are targeting a gene with two homeologs, design a set of primers in which at least one primer is unique to each homeolog (Fig. 3; example sgRNA target amplification).

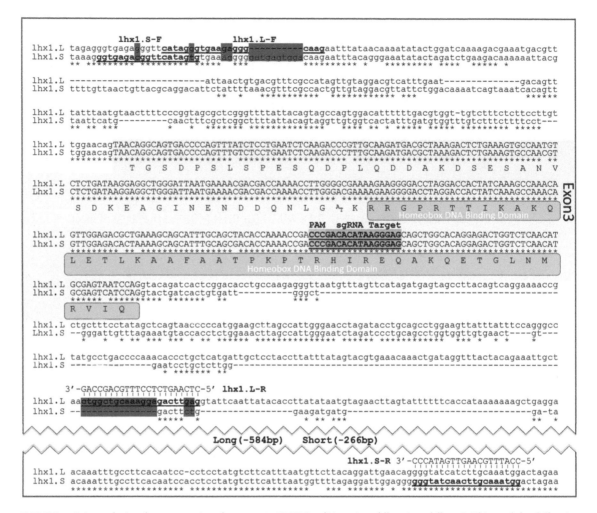

FIGURE 3. Primer design for sequencing the exon 3 CRISPR editing site of *lhx1.L* and *lhx1.S*. This and the following figure focus on the exon 3 sgRNA depicted in Figure 1, because it had a high editing efficiency (see Fig. 4). To confirm CRISPR editing the single-guide RNA (sgRNA) target (highlighted orange with the protospacer adjacent motif [PAM] in turquoise) within the homeodomain (labeled in blue) in exon 3 (labeled in yellow) was amplified using polymerase chain reaction (PCR) primer pairs that are unique to *lhx1.L* and *lhx1.S* in two adjacent introns. The forward primers and the complement of the reverse primers are underlined. The reverse primers sequences are above the genomic targets. Nucleotides that are not conserved between alloalleles are highlighted red. Notice that the *lhx1.L* reverse primer sequence is conserved in *lhx1.S*. However, when paired with the *lhx1.L* forward primer, the product is unique.

ii. Set up the following PCR mix to amplify the DNA surrounding target sequence for both wild-type and CRISPant embryos.

Stock concentration	Volume	Final concentration
dH₂O	14.875 μL	
10× *Taq* buffer	2.5 μL	1×
2.5 mM (each) dNTP	2 μL	0.2 mM
Embryo DNA	5 μL	
10 mM forward primer	1 μL	0.4 mM
10 mM reverse primer	1 μL	0.4 mM
Taq (5000 units/mL)	0.125 μL	0.625 units
Final volume:	25 μL	

One microliter of DNA from a stage 40 embryo seems to work better than 5 μL of DNA from a stage 10–12 embryo.

Cite this protocol as *Cold Spring Harb Protoc*; doi:10.1101/pdb.prot107037

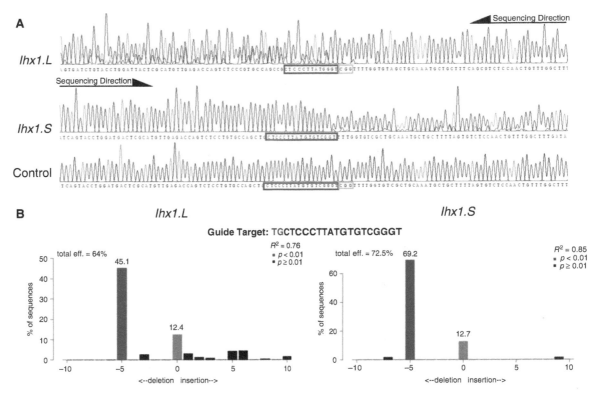

FIGURE 4. Analysis of the *lhx1.S* CRISPR editing site using TIDE analysis. Sanger sequencing of exon 3 from *lhx1.L* and *lhx1.S* around the single-guide RNA (sgRNA) target site was used to perform editing analysis TIDE software (https://tide.nki.nl/) (Brinkman et al. 2014). (*A*) Chromatogram of sequence surrounding sgRNA targeting exon 3 of *lhx1.L* and *lhx1.S* homeologs. (*B*) Indel distribution as determined by TIDE analysis for the *lhx1.L* and *lhx1.S* homeologs. Note that TIDE analysis requires a 20-bp sgRNA target input, so 2-bp 5′ of the sgRNA target have been added to the sgRNA sequence to perform this analysis. Outputs indicate that the sgRNA targeting exon 3 of *lhx1* is efficient in generating a 5-bp frameshift deletion in both of the homeologs.

iii. Carry out the following thermocycler program.

Step	Temperature	Min:sec
1	95°C	2:00
2	95°C	0:30
3	58°C	0:30
4	Go to step #3 35×	
5	72°C	1:00
6	72°C	5:00
7	4°C	∞

Annealing temperatures in step #3 may vary between primer sets.

iv. Run the PCR products on a 1%–2% agarose gel.

There should be a clear band for each sample.

See Troubleshooting.

Assessment of DNA Editing Efficiency in F_0 Embryos

18. Sanger-sequence the amplified PCR products using one of the PCR amplification primers to verify CRISPR editing. Sequence a wild-type embryo first to verify that your PCR product is

amplifying your gene of interest. If the wild-type sequence looks good, sequence the CRISPR-edited embryo DNA.

19. Obtain the chromatogram sequence files (.ab1) for both the wild-type and CRISPR–Cas9-edited F_0 embryos. Upload the sequences to CRISPR analysis software, such as TIDE (Tracking of Indels by Decomposition; https://tide.nki.nl/) (Brinkman et al. 2014) or ICE (Inference of CRISPR Edits; https://ice.synthego.com/) (Fig. 4; example chromatogram analysis results) (Hsiau et al. 2019).

See Troubleshooting.

TROUBLESHOOTING

Problem (Step 2): An sgRNA that targets both homeologs cannot be identified.
Solution: Design sgRNAs that target the individual homeologs and coinject them with the Cas9 protein.

Problem (Step 3): The designed sgRNA has homology with off-target sequences elsewhere in the genome.
Solution: To avoid off-target effects, design guides that have longer regions of homology with the target. sgRNAs that target 19 bp will be found less frequently in the genome than those that target 16 bp.

Problem (Step 15): The expected phenotype is not observed.
Solution: Be sure to sequence the embryonic DNA to assess whether a significant insertion or deletion has been generated. Scoring systems used to predict sgRNA efficiency do not always accurately predict the efficiency of cutting (DeLay et al. 2018). If the sgRNA is cutting efficiently, it may be that the insertions or deletions are not being generated early enough to see the phenotype in the F_0 generation, potentially because of the presence of maternally deposited mRNA and protein. For example, early axis phenotypes observed with β-catenin and lhx1 loss-of-function phenotypes are not observed in F_0 CRISPants (Bhattacharya et al. 2015; DeLay et al. 2018). Generation of F_1 animals may be required to study early developmental processes (Aslan et al. 2017). If the generated mutants are lethal, leapfrogging, which is a strategy to transfer a knockout's germline to a wild-type somatic embryo, should be considered (Blitz et al. 2016).

Problem (Step 17): A clean band is not observed when the target site is PCR-amplified from the extracted genomic DNA.
Solution: Be sure to test your primers first on wild-type DNA (a large deletion has the possibility of disrupting the primer binding sites). If the amplification does not work using DNA extracted from wild-type embryos, make sure that you are using a good quality polymerase (we suggest Taq Polymerase from New England Biolabs or Platinum Taq from Fisher). If you are still having trouble, extract and amplify DNA from later-stage embryos (PCR of target sites is more successful from embryos at later stages, e.g., stage 40.) If the PCR still does not work, used nested PCR primers. Amplify the region of interest with outer primers first and then use 1–2 μL of the product as a template for a pair of primers that lie inside of the outer pair (DeLay et al. 2018). It is normal not to see a band in the first PCR reaction. If you get multiple bands, rather than a single band, gel-purify the band that is the correct size of your PCR product to sequence. Alternatively, western blotting of whole embryo extracts can be used to measure the amount of full-length protein made at a particular stage (DeLay et al. 2018).

Problem (Step 19): Efficient cutting is not achieved (Fig. 5).

Solution: Design a different guide. Not all guides work well, and a good score from CRISPR design software commonly does not result in efficient cutting. In our experience, we did not see much toxicity with high levels of injected sgRNA, whereas unloaded Cas9 appears to cause exogastrulation. Based on our experience (DeLay et al. 2018), we suggest starting with roughly a 2:1 molar concentration of guide to Cas9 to saturate the Cas9. Therefore, we observe that addition of more guide does not increase cutting efficiently. Alternatively, techniques such as host transfer have been found to increase efficiency (Aslan et al. 2017). Evaluation of DNA editing efficiency can be

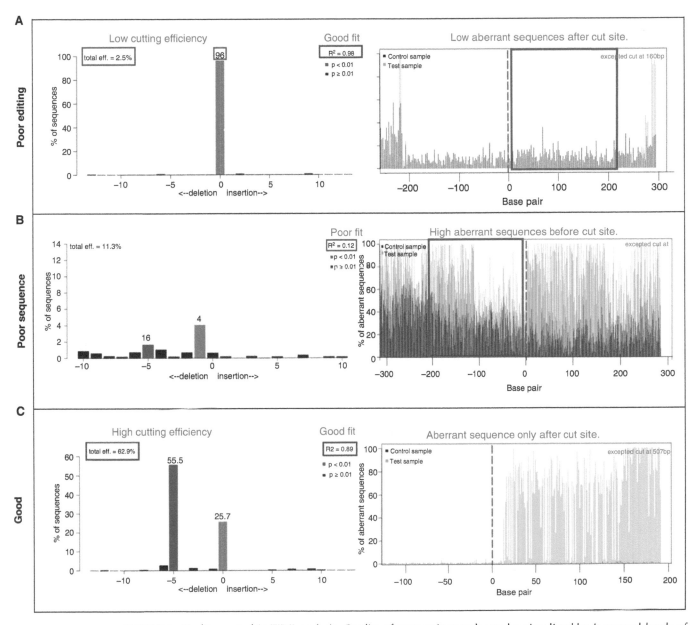

FIGURE 5. Quality control in TIDE analysis. Quality of sequencing reads can be visualized by increased levels of discordant/aberrant sequences, R^2 values, and percentage efficiency. (*A*) TIDE analysis of sgRNA targeting exon 1. Output from TIDE analysis showing features of poor genomic editing such as low efficiency and low aberrant sequence after cut site. (*B*) TIDE analysis from a poor sequencing read of the long homeolog of exon 3. Output indicates a low R^2 and high aberrant sequences before cut site. (*C*) Good output from editing of exon 3 of the short homeolog of *lhx1*.

performed at a wide range of stages, but the level of gene editing may be varied by stage or tissue tested.

Problem (Step 19): Efficient cutting of one homeolog is not observed.

Solution: Design an additional guide targeted to the homeolog that is poorly cut. Then inject multiple sgRNAs. Increasing the amount of Cas9 may not help as toxicity may become a problem (Delay et al. 2018).

Problem (Step 19): Upon tissue targeting, a low efficiency of genome editing is detected because only small amounts of tissue are edited or the targeted tissue is difficult to isolate.

Solution: Use complementary techniques to check the editing efficiency. If a good antibody exists, western blot and/or immunostaining analyses will provide a mechanism to check for loss of protein. Immunostaining has the advantage that the tissue of interest can be easily assessed for editing efficiency (DeLay et al. 2018). This is beneficial if targeted injection was performed because much of the embryo may not receive the guide and sgRNA. Analysis of the mRNA by techniques such as in situ hybridization or quantitative real-time (qRT)-PCR may not lead to predictable results as the mRNA may or may not be degraded by nonsense-mediated decay (NMD). If the mRNA is targeted for NMD then the mRNA and protein should be lost (DeLay et al. 2018). However, if the mRNA is stable, a truncated or heavily altered protein may be expressed, and the mRNA will not appear altered using these strategies.

DISCUSSION

Amphibians, including frogs and salamanders, have served as valuable models for developmental and regenerative biology. *Xenopus laevis* became the predominant amphibian model in the 1950s, because of their widespread use in pregnancy testing (Wallingford et al. 2010). Their use as a classical model facilitated the establishment of strategies to explant and transplant cells and tissues to understand developmental phenomena, such as induction, specification, and determination. *X. laevis* have numerous experimental advantages, including amenable embryo culture throughout the year. Also, they lay numerous eggs that can be fertilized in vitro, and the large size of their embryos that develop externally can be easily manipulated and visualized.

In addition to elegant classical approaches to study development in *X. laevis*, modern strategies have been developed to perform molecular manipulations. Microinjection of mRNA and morpholinos has become commonly used strategies to overexpress and knockdown genes, respectively. Additionally, established fate maps enable manipulation of gene expression in specific tissues of interest (Moody 1987a,b; Moody and Kline 1990).

Although *X. laevis* has numerous experimental advantages, their 1-yr generation time and complex allotetraploid genome have made traditional genetic approaches in this species difficult. Therefore, the diploid *Xenopus tropicalis* species with a 6-mo generation time was established as a model to facilitate chemical mutagenesis and crossing of genetic lines (Offield et al. 2000; Khokha et al. 2002). However, the recent application of efficient DNA editing approaches, including transcription activator–like effector nuclease (TALEN)-mediated (Lei et al. 2012) and CRISPR–Cas9 genome editing (Blitz et al. 2013; Nakayama et al. 2013) has made genetics in *Xenopus* feasible.

Simultaneously with the development of genome editing strategies, the use of morpholinos as a tool to generate knockdowns was under scrutiny (Schulte-Merker and Stainier 2014; Gentsch et al. 2019), creating a need for new experimental strategies to assess gene function. Although morpholino use in *Xenopus* has been validated in more recent studies (Paraiso et al. 2019), the timing of the controversy surrounding morpholinos and the application of CRISPR–Cas9 technology to *X. tropicalis* helped push forward this technology. At this point, these strategies are complementary, as recent studies indicate that morpholino knockdown phenotypes are similar to phenotypes resulting from CRISPR gene knockouts (DeLay et al. 2019).

Genome editing strategies developed in *X. tropicalis* (Blitz et al. 2013; Nakayama et al. 2013; Bhattacharya et al. 2015) have opened the door to human disease modeling (Tandon et al. 2017; Deniz et al. 2018; Naert and Vleminckx 2018). This approach for gene knockout has also provided the foundation to generate knock-ins in *Xenopus* (Aslan et al. 2017; Nakayama et al. 2020). Additionally, a strategy for transplanting the germline of lethal knockouts into the soma of a wild-type embryo, called "leapfrogging," has enabled the analysis of mutations that would be difficult to study in other organisms (Blitz et al. 2016; Blitz 2018).

The genome editing strategies developed in *X. tropicalis* have been adapted for use *X. laevis*, given the large embryo size and slower cleavage for targeted microinjection. *X. laevis* has a more complex allotetraploid genome, meaning that they have multiple homeologs of numerous genes. However, as the genome fusion responsible for these gene duplications occurred fairly recently, 17 million years ago (Session et al. 2016), the coding sequences of the gene homeologs are frequently quite similar (Fig. 1). This is an advantage in performing CRISPR–Cas9 genome editing, given that a single sgRNA, if efficient, can target both homeologs (Fig. 2) in F_0 embryos (DeLay et al. 2018, 2019). This strategy may not be feasible in other organisms with paralogs, such as fish models that underwent a whole-genome duplication 300 million years ago (Inoue et al. 2015), because the homologs have diverged more significantly and have less sequence similarity.

Previously, it was thought that targeting CRISPR mutations to early exons within a gene would result in a high rate of knockout mutations because they would produce an early frameshift mutation. However, more recently, it has become evident that because insertions and deletions of multiples of 3 nt result in frame mutations, targeting an important domain may be a better strategy to cause loss-of-function mutations. In some cases, loss of a single amino acid results in disruption of protein function, and an efficient sgRNA in a protein domain coded from a later exon may result in more deleterious effects than mutations in earlier exons (DeLay et al. 2018). However, hitting an important domain early in the coding region is the best approach. Additionally, deletions that span intron–exon junctions may interrupt splicing. Targeted key domains in the protein should still be affected if splicing is affected. Furthermore, through machine learning, it has become clear that certain sgRNAs have higher efficiencies of causing deleterious mutations, and this information has been compiled into online databases (Shen et al. 2018) to integrate into sgRNA design (Naert et al. 2020).

Because the genome of *X. laevis* is allotetraploid, making lines from knockouts may not be desirable. Therefore, to assess whether phenotypes of injected F_0 embryos are due to insertions or deletions in the targeted gene, analysis of the genomic sequence at the target sites should be performed. To assess whether the target sites of multiple alloalleles are efficiently mutated upon injection of sgRNA and Cas9 protein into the single-cell *X. laevis* embryo (analyzing embryos that are injected at later stages will dilute the assessed efficiency of the sgRNA), they should be grown up to the stage of interest to isolate the genomic DNA. The DNA surrounding the sgRNA target should be amplified using primer sets that are unique to each homeolog (Fig. 3; DeLay et al. 2018, 2019).

Upon PCR amplification of the regions surrounding the sgRNA target sites, Sanger sequencing along with TIDE analysis can be performed to assess the types of mutations that are generated in the F_0 embryos. Multiple types of insertions and deletions are generated through error-prone nonhomologous end joining repair. Therefore, distinct mutations are present in different cells as they divide. This results in mosaic embryos with discrete cell lineages having a unique genetic makeup. Thus, sequencing the genomic DNA from a single embryo will result in a chromatogram that has multiple sequences from all lineages of the mosaic embryo. Therefore, sequencing through an sgRNA target site will result in decomposition of the read around the sgRNA target site. This decomposition is visible with each peak having multiple nucleotides present at each location.

Comparison of the peak heights of the CRISPant chromatograms can be analyzed using online software called TIDE (Brinkman et al. 2014). TIDE analysis can be used to assess the percentage of each type of insertion and deletion generated by CRISPR–Cas9 in each alloallele within a mosaic F_0 *X. laevis* embryo (Fig. 4; DeLay et al. 2018, 2019). More recently, ICE analysis with Synthego software

(Hsiau et al. 2019) has been adapted for similar analysis of Sanger sequencing in *Xenopus* (Sempou et al. 2018). Although the results from TIDE and ICE analysis are similar, they are not exactly the same and are not directly comparable. This is in part because TIDE allows for the fine-tuning of parameters such as the maximum indel size and limiting the sequence window to analyze, allowing for a more accurate result by avoiding sections of sequencing with poor reads. It is also easier to download the INDEL% and *P*-values for each indel using TIDE. Although TIDE is a more powerful software tool, ICE provides an aligned chromatogram around the guide RNA that may be useful.

Because many genes are essential for early developmental processes, such as gastrulation and neurulation, analyzing later developmental phenotypes in null mutants may not be possible. Therefore, use of established *Xenopus* lineage maps can be used to target molecular manipulations to tissues of interest (Moody 1987a,b; Moody and Kline 1990). Building upon the observation that CRISPR can be targeted to subsets of blastomeres in other types of embryos, such as *Parhyale hawaiensis* (Martin et al. 2016), gene editing was achieved in specific *X. laevis* tissues (DeLay et al. 2018, 2019). This was made possible through utilization of their established, robust fate maps and targeted microinjection, despite the more rapid cell division relative to *Parhyale* embryos.

Overall, the adaptation of CRISPR–Cas9 genome editing strategies established in *X. tropicalis* for use in *X. laevis* will provide a valuable tool for the community. Strategies for editing and sequencing multiple alloalleles have been developed in *X. laevis*. Additionally, approaches for evaluating mutations in F_0 embryos and targeting mutations to tissues of interest established in *X. laevis* have become valuable in *X. tropicalis* and other organisms.

RECIPES

Marc's Modified Ringers (MMR) (1×, without EDTA)

5 mM HEPES (free acid)
100 mM NaCl
2 mM KCl
1 mM $MgCl_2$
2 mM $CaCl_2$

Prepare as a 10× stock and adjust pH to 7.4–7.8 with NaOH. Store at room temperature.

Modified Barth's Saline (1×)

5 mM HEPES (free acid)
88 mM NaCl
1 mM $CaCl_2$
1 mM $MgSO_4$
2.5 mM $NaHCO_3$

Prepare as a 10× stock and adjust pH to 7.4–7.8 with NaOH. Store at room temperature.

ACKNOWLEDGMENTS

We thank the instructors and teaching assistants of the 2015 and 2016 Cold Spring Harbor Laboratory *Xenopus* Course as well as the National *Xenopus* Resource director, Dr. Marko Horb, and the instructors of the 2017 Genome Editing Workshop for providing guidance into genome editing techniques. We are grateful to the members of the laboratories of M. Kloc, P.D. McCrea, R.K. Miller, and J. Park for their helpful suggestions and advice. Figure 2 was made with the help of BioRender.com. This work was supported by National Institutes of Health (NIH) grants (K01DK092320, R03DK118771, and R01DK115655 to R.K.M.), and Startup funding from the Department of Pediatrics, Pediatric Research Center at UTHealth McGovern Medical School (to R.K.M.).

REFERENCES

Aslan Y, Tadjuidje E, Zorn AM, Cha SW. 2017. High-efficiency non-mosaic CRISPR-mediated knock-in and indel mutation in F0 *Xenopus*. *Development* **144:** 2852–2858.

Bhattacharya D, Marfo CA, Li D, Lane M, Khokha MK. 2015. CRISPR/Cas9: an inexpensive, efficient loss of function tool to screen human disease genes in *Xenopus*. *Dev Biol* **408:** 196–204. doi:10.1016/j.ydbio.2015.11.003

Blitz IL. 2018. Primordial germ cell transplantation for CRISPR/Cas9-based leapfrogging in *Xenopus*. *J Vis Exp* **132:** 56035. doi:10.3791/53799

Blitz IL, Nakayama T. 2021. CRISPR–Cas9 mutagenesis in *Xenopus tropicalis* for phenotypic analyses in the F0 generation and beyond. *Cold Spring Harb Protoc* doi:10.1101/pdb.prot106971

Blitz IL, Biesinger J, Xie X, Cho KW. 2013. Biallelic genome modification in F0 *Xenopus tropicalis* embryos using the CRISPR/Cas system. *Genesis* **51:** 827–834. doi:10.1002/dvg.22719

Blitz IL, Fish MB, Cho KW. 2016. Leapfrogging: primordial germ cell transplantation permits recovery of CRISPR/Cas9-induced mutations in essential genes. *Development* **143:** 2868–2875. doi:10.1242/dev.138057

Brinkman EK, Chen T, Amendola M, van Steensel B. 2014. Easy quantitative assessment of genome editing by sequence trace decomposition. *Nucl Acids Res* **42:** e168. doi:10.1093/nar/gku936

DeLay BD, Krneta-Stankic V, Miller RK. 2016. Technique to target microinjection to the developing *Xenopus* kidney. *J Vis Exp* **111:** 53799. doi:10.3791/53799

DeLay BD, Corkins ME, Hanania HL, Salanga M, Deng JM, Sudou N, Taira M, Horb ME, Miller RK. 2018. Tissue-specific gene inactivation in *Xenopus laevis*: knockout of *lhx1* in the kidney with CRISPR/Cas9. *Genetics* **208:** 673–686. doi:10.1534/genetics.117.300468

DeLay BD, Baldwin TA, Miller RK. 2019. Dynamin binding protein is required for *Xenopus laevis* kidney development. *Front Physiol* **10:** 143. doi:10.3389/fphys.2019.00143

Deniz E, Mis EK, Lane M, Khokha MK. 2018. CRISPR/Cas9 F0 screening of congenital heart disease genes in *Xenopus tropicalis*. *Methods Mol Biol* **1865:** 163–174. doi:10.1007/978-1-4939-8784-9_12

Gentsch GE, Spruce T, Owens NDL, Monteiro RS, Smith JC. 2019. The innate immune response of frog embryos to antisense morpholino oligomers depends on developmental stage, GC content and dose. *Dev Cell* **49:** 506–507. doi:10.1016/j.devcel.2019.05.004

Hsiau T, Conant D, Rossi N, Maures T, Waite K, Yang J, Joshi S, Kelso R, Holden K, Enzmann BL, et al. 2019. Inference of CRISPR edits from Sanger trace data. bioRxiv doi:10.1101/251082

Inoue J, Sato Y, Sinclair R, Tsukamoto K, Nishida M. 2015. Rapid genome reshaping by multiple-gene loss after whole-genome duplication in teleost fish suggested by mathematical modeling. *Proc Natl Acad Sci* **112:** 14918–14923. doi:10.1073/pnas.1507669112

Khokha MK, Chung C, Bustamante EL, Gaw LW, Trott KA, Yeh J, Lim N, Lin JC, Taverner N, Amaya E, et al. 2002. Techniques and probes for the study of *Xenopus tropicalis* development. *Dev Dyn* **225:** 499–510. doi:10.1002/dvdy.10184

Labun K, Montague TG, Krause M, Torres Cleuren YN, Tjeldnes H, Valen E. 2019. CHOPCHOP v3: expanding the CRISPR web toolbox beyond genome editing. *Nucl Acids Res* **47:** W171–W174. doi:10.1093/nar/gkz365

Lei Y, Guo X, Liu Y, Cao Y, Deng Y, Chen X, Cheng CH, Dawid IB, Chen Y, Zhao H. 2012. Efficient targeted gene disruption in *Xenopus* embryos using engineered transcription activator-like effector nucleases (TALENs). *Proc Natl Acad Sci* **109:** 17484–17489. doi:10.1073/pnas.1215421109

Martin A, Serano JM, Jarvis E, Bruce HS, Wang J, Ray S, Barker CA, O'Connell LC, Patel NH. 2016. CRISPR/Cas9 mutagenesis reveals versatile roles of hox genes in crustacean limb specification and evolution. *Curr Biol* **26:** 14–26. doi:10.1016/j.cub.2015.11.021

Moody SA. 1987a. Fates of the blastomeres of the 16-cell stage *Xenopus* embryo. *Dev Biol* **119:** 560–578. doi:10.1016/0012-1606(87)90059-5

Moody SA. 1987b. Fates of the blastomeres of the 32-cell-stage *Xenopus* embryo. *Dev Biol* **122:** 300–319. doi:10.1016/0012-1606(87)90296-X

Moody SA, Kline MJ. 1990. Segregation of fate during cleavage of frog (*Xenopus laevis*) blastomeres. *Anat Embryol* **182:** 347–362. doi:10.1007/BF02433495

Moreno-Mateos MA, Vejnar CE, Beaudoin JD, Fernandez JP, Mis EK, Khokha MK, Giraldez AJ. 2015. CRISPRscan: designing highly efficient sgRNAs for CRISPR–Cas9 targeting in vivo. *Nat Methods* **12:** 982–988. doi:10.1038/nmeth.3543

Naert T, Vleminckx K. 2018. CRISPR/Cas9 disease models in zebrafish and *Xenopus*: the genetic renaissance of fish and frogs. *Drug Discov Today Technol* **28:** 41–52. doi:10.1016/j.ddtec.2018.07.001

Naert T, Tulkens D, Edwards NA, Carron M, Shaidani NI, Wlizla M, Boel A, Demuynck S, Horb ME, Coucke P, et al. 2020. Maximizing CRISPR/Cas9 phenotype penetrance applying predictive modeling of editing outcomes in *Xenopus* and zebrafish embryos. *Sci Rep* **10:** 14662. doi:10.1038/s41598-020-71412-0

Nakayama T, Fish MB, Fisher M, Oomen-Hajagos J, Thomsen GH, Grainger RM. 2013. Simple and efficient CRISPR/Cas9-mediated targeted mutagenesis in *Xenopus tropicalis*. *Genesis* **51:** 835–843. doi:10.1002/dvg.22720

Nakayama T, Grainger RM, Cha SW. 2020. Simple embryo injection of long single-stranded donor templates with the CRISPR/Cas9 system leads to homology-directed repair in *Xenopus tropicalis* and *Xenopus laevis*. *Genesis* **58:** e23366. doi:10.1002/dvg.23366

Offield MF, Hirsch N, Grainger RM. 2000. The development of *Xenopus tropicalis* transgenic lines and their use in studying lens developmental timing in living embryos. *Development* **127:** 1789–1797. doi:10.1242/dev.127.9.1789

Paraiso KD, Blitz IL, Zhou JJ, Cho KWY. 2019. Morpholinos do not elicit an innate immune response during early *Xenopus* embryogenesis. *Dev Cell* **49:** 643–650 e643. doi:10.1016/j.devcel.2019.04.019

Schulte-Merker S, Stainier DY. 2014. Out with the old, in with the new: reassessing morpholino knockdowns in light of genome editing technology. *Development* **141:** 3103–3104. doi:10.1242/dev.112003

Sempou E, Lakhani OA, Amalraj S, Khokha MK. 2018. Candidate heterotaxy gene *FGFR4* is essential for patterning of the left–right organizer in *Xenopus*. *Front Physiol* **9:** 1705. doi:10.3389/fphys.2018.01705

Session AM, Uno Y, Kwon T, Chapman JA, Toyoda A, Takahashi S, Fukui A, Hikosaka A, Suzuki A, Kondo M, et al. 2016. Genome evolution in the allotetraploid frog *Xenopus laevis*. *Nature* **538:** 336–343. doi:10.1038/nature19840

Shen MW, Arbab M, Hsu JY, Worstell D, Culbertson SJ, Krabbe O, Cassa CA, Liu DR, Gifford DK, Sherwood RI. 2018. Predictable and precise template-free CRISPR editing of pathogenic variants. *Nature* **563:** 646–651. doi:10.1038/s41586-018-0686-x

Sive HL, Grainger RM, Harland RM. 2000. *Early development of* Xenopus *laevis: a laboratory manual*. Cold Spring Harbor Laboratory Press, Cold Spring Harbor, NY.

Tandon P, Conlon F, Furlow JD, Horb ME. 2017. Expanding the genetic toolkit in *Xenopus*: approaches and opportunities for human disease modeling. *Dev Biol* **426:** 325–335. doi:10.1016/j.ydbio.2016.04.009

Wallingford JB, Liu KJ, Zheng Y. 2010. *Xenopus*. *Curr Biol* **20:** R263–R264. doi:10.1016/j.cub.2010.01.012

Wlizla M, McNamara S, Horb ME. 2018. Generation and care of *Xenopus laevis* and *Xenopus tropicalis* embryos. *Methods Mol Biol* **1865:** 19–32. doi:10.1007/978-1-4939-8784-9_2

Modeling Human Genetic Disorders with CRISPR Technologies in *Xenopus*

Helen Rankin Willsey,[1,4] Matthew J. Guille,[2] and Robert M. Grainger[3,4]

[1]*Department of Psychiatry and Behavioral Sciences, Weill Institute for Neurosciences, University of California, San Francisco, San Francisco, California 94143, USA;* [2]*European Xenopus Resource Centre, School of Biological Sciences, University of Portsmouth, Portsmouth PO1 2UP, United Kingdom;* [3]*Department of Biology, University of Virginia, Charlottesville, Virginia 22904, USA*

Combining the power of *Xenopus* developmental biology with CRISPR-based technologies promises great discoveries in understanding and treating human genetic disorders. Here we provide a practical pipeline for how to go from known disease gene(s) or risk gene(s) of interest to methods for gaining functional insight into the contribution of these genes to disorder etiology in humans.

MATERIALS

Equipment

Databases for disease models (e.g., modelmatcher.net)
Databases for model organisms
    Caenorhabditis elegans (wormbase.org)
    Drosophila (flybase.org)
    Mice (informatics.jax.org)
    Rats (rgd.mcw.edu)
    Zebrafish (zfin.org)
Human genomic databases
    ClinVar (ncbi.nlm.nih.gov/clinvar)
    DECIPHER (decipher.sanger.ac.uk)
    gnoMAD (gnomad.broadinstitute.org)
    GWAS Catalog (ebi.ac.uk/gwas)
Xenopus resources
    Centre de Ressources Biologiques *Xénopes*, France (xenopus.univ-rennes1.fr)
    European *Xenopus* Resource Center, UK (xenopusresource.org)
    National *Xenopus* Resource, USA (mbl.edu/xenopus)
    The National BioResource Project, Japan (xenopus.nbrp.jp)
Xenbase (xenbase.org)
Xenopus transcriptome data sets
    Briggs et al. (2018)

[4]Correspondence: helen.willsey@ucsf.edu; rmg9p@virginia.edu

Copyright © 2023 Cold Spring Harbor Laboratory Press; all rights reserved
Cite this protocol as *Cold Spring Harb Protoc*; doi:10.1101/pdb.prot106997

Owens et al. (2016)
Session et al. (2016)

METHOD

1. For disorders with well-characterized causal genes, determine whether animal models already exist and how *Xenopus* modeling could enhance understanding of underlying disease mechanisms.

   i. Consult model organism databases to find information about existing animal models and determine to what degree they replicate the human condition or mechanism.

      Some examples include flybase.org, wormbase.org, zfin.org, informatics.jax.org, and rgd.mcw.edu, in addition to the Xenopus-specific database, xenbase.org. A new database of disease models is now available at modelmatcher.net.

   ii. Determine whether a *Xenopus* model has the potential to add to mechanistic understanding of the disorder. (For example, if current models of a genetic disorder are only described in fish, consider the advantages of studying the gene in *Xenopus tropicalis*. Because *X. tropicalis* has a diploid genome, such a model would avoid potential issues caused by the genome duplication present in fish). Additionally, consider whether study in *Xenopus* is advantageous given greater similarities in aspects of organogenesis like lung and limb development in humans versus in other species like fish. Finally, consider that if knockout mice are embryonic lethal, targeted injections in *Xenopus* (see below) can bypass embryonic lethality and allow for the characterization of more severe alleles that cannot be described in germline mutant mouse lines.

2. For disorders with candidate causal genes, prioritize risk genes to study and analyze known alleles.

   i. Determine if the putative risk gene is not often varied in unaffected populations (i.e., that it is a "loss-of-function intolerant" gene; Lek et al. 2016). Consult large databases of genome sequencing data like gnoMAD (Karczewski et al. 2020) to determine whether the gene of interest is ever varied in unaffected control subjects.

      The amount of variation in unaffected populations that is acceptable for a given gene largely depends on the genetic architecture and prevalence of the disorder. For example, for a rare disorder (e.g., prevalence = 2% of the population) with many genes estimated to contribute risk mainly through a loss-of-function, haploinsufficiency mechanism, allele frequencies for a risk variant should likely be <0.1%. Therefore, variants at a 1% population frequency are unlikely to carry risk for that disorder. Alternatively, for a monogenic disorder with the same prevalence, an allele frequency of 1% could contribute to risk. These considerations are more complex when one considers recessive inheritance patterns, gain-of-function mutations, and X chromosome variants. Although an allele frequency cutoff of <0.01% is generally encouraging for risk alleles, engaging a statistical geneticist can greatly aid interpretation of putative alleles.

   ii. Consult databases of human patient conditions by genetic variant such as DECIPHER (Firth et al. 2009) and ClinVar (Landrum et al. 2020).

      These can help identify patient conditions associated with disease alleles, which can be helpful in designing experiments in Xenopus, as they can provide clues about which organ(s) and process(es) are disrupted.

   iii. Analyze known alleles.

      "Likely-gene-disrupting" alleles (i.e., frameshift, nonsense, splice-site disrupting, and insertion/deletions) are the most straightforward to study because they likely result in loss of function. Subsequent studies might take advantage of technology for homology-directed repair to more precisely model a particular human mutation (Aslan et al. 2017; Nakayama et al. 2020).

   iv. Consider whether common variants for a disorder are known (GWAS catalog; ebi.ac.uk/gwas) to see if knowledge of these variants can help refine hypotheses and assay design.

Keep in mind that common variants often have very small effect sizes, so modeling individually may not produce robust phenotypes, and that the gene(s) affected by the common variant may not be the nearest neighboring gene(s) to the common variant.

   v. Align the human and frog proteins to see if the regions containing variants are conserved in frog.

3. Incorporate gene expression data to prioritize cell types and/or tissues of interest.

F_0 CRISPR-based approaches target only the zygotic genome, and therefore will not disrupt maternally provided gene products. For this reason, F_0 CRISPR disruptions are often less severe than morpholino or germline mutant approaches. To anticipate this discrepancy, one can consult publicly available databases of gene expression data during Xenopus development.

   i. Consult time-course whole-animal bulk RNA sequencing data from *X. laevis* (Session et al. 2016) and *Xenopus tropicalis* (Owens et al. 2016).

The presence of transcripts before the zygotic transition (~stage 9), indicates a maternal RNA contribution. This does not preclude use of F_0 CRISPR analysis, but it does offer context if phenotypes observed are milder than expected. Assaying mutant animals later in development when the maternal contribution has been turned over may reveal more severe phenotypes.

   ii. Consult single-cell RNA sequencing data sets of embryonic development in *X. tropicalis* (Briggs et al. 2018) to help focus hypotheses about embryonic gene function.

If Xenbase contains RNA in situ hybridization data for the gene, this will also help the development of hypotheses.

   iii. If RNA in situ hybridization data are not available, conduct whole-mount RNA in situ hybridization using in vitro transcribed antisense RNA probes (see Sec. 9, Protocol 1: Whole-Mount RNA In Situ Hybridization and Immunofluorescence of *Xenopus* Embryos and Tadpoles [Willsey 2021]).

This technique is an excellent, cost-effective approach for pinpointing potential developmental epochs and cell types in which the gene(s) of interest is expressed. In situ data is especially useful when studying disorders with a large contributing set of risk genes and for which little is known about the affected cell types. Although each individual gene may be expressed in a wide range of tissues, when considered as a group, the overlapping expression epochs may be more limited and help pinpoint a window of risk.

4. Design the genetic perturbation to most closely model the genetic architecture of the disorder.

F_0-based approaches generate an array of mosaic mutations, depending on the efficiency of the sgRNA. In contrast, outbred germline mutants will not have mosaicism, but require a much larger time and effort commitment and may be complicated by genetic compensation (El-Brolosy et al. 2019). The decision between F_0 and germline mutagenesis is largely one of time commitment and goal. Phenotypes can be quickly screened using F_0 analysis and then followed up more closely in germline mutants if F_0 results are promising. A potential compromise between F_0 analysis and a stable line is crossing two F_0 founders and analyzing F_1 offspring.

   i. If the disorder is based on rare, de novo loss-of-function variants, inject CRISPR reagents to create mutations in F_0 animals and assess the phenotype in the F_0 generation (Nakayama et al. 2013; Bhattacharya et al. 2015). Design single-guide RNAs (sgRNAs) to target an early exon with the goal of generating a null, frameshift mutation, or target a conserved, functional domain like an enzyme active site, if one is known. Algorithms exist to design sgRNAs with minimal off-targets for *Xenopus* genomes (Moreno-Mateos et al. 2015). Empirically determine the mutational efficiency of the sgRNA by polymerase chain reaction (PCR), Sanger sequencing, and tracking of indels (TIDE) (Brinkman et al. 2014; DeLay et al. 2018). If desired, use homology-directed repair techniques to generate a specific mutation of interest (Aslan et al. 2017; Nakayama et al. 2020).

   ii. If the disorder is based on a deletion, design two sgRNAs flanking the region of interest and inject CRISPR reagents (Nakayama et al. 2013). If the variant is inherited, propagate the lesion to the next generation and confirm along the way by PCR.

In modeling disorders of larger genomic regions, it is important to consider whether synteny has been conserved, as well as conservation of noncoding regions.

Cite this protocol as *Cold Spring Harb Protoc*; doi:10.1101/pdb.prot106997

iii. If the disorder is based on genetic haploinsufficiency, consider morpholino-based experiments, in which knockdown severity can be closely titrated. Pair the morpholino approach with a validated antibody to quantify precise levels of protein abundance by morpholino dose. Alternatively, generate a heterozygous null line to appropriately model 50% reduction in gene dosage.

iv. For disorders of increased gene dosage, use a gain-of-function approach by using CRISPR technology to insert additional sequence at a locus of interest (Nakayama et al. 2020) or overexpress the gene by injecting mRNA or plasmid encoding the gene (Vize et al. 1991).

v. Consider whether targeting the perturbation to a particular lineage may be advantageous, especially in a situation in which gene mutations are embryonic lethal but need to be studied in later stage animals (DeLay et al. 2018) or in the next generation (Blitz et al. 2016). Targeting the germline by injecting the four most vegetal cells of a 32-cell embryo can also be used to generate mutations in the next generation. Targeted injections can also be used to generate unilateral F_0 mutants, in which half of the embryo is mutated and the other half serves as an internal control (Willsey et al. 2018b; Lasser et al. 2019). This is especially useful for identifying more subtle phenotypes or those that could be influenced by environment or genetic background. When performing unilateral mutagenesis, ensure that the cells are firmly at the two-cell stage to avoid the reagents crossing into both cells; alternatively, inject two cells of a four-cell stage embryo.

5. Design phenotyping assays.

i. Combine knowledge of gene expression patterns in *Xenopus* and humans with patient phenotypes to design phenotyping assays.

The vast data on gene expression patterns in Xenopus provide a rich resource for understanding gene networks that may be affected in a model of a human disease. Such a model can be used to perform expression studies on known or hypothesized targets of an affected gene under study. As a result, novel epistatic relationships among genes in key pathways may be revealed in the Xenopus model.

ii. Consider the potential for drug screening to identify molecules that could rescue disease phenotypes. In this case, a simple and higher-throughput assay is desired. Consider the use of fluorescent transgenic lines when possible, as this can speed phenotyping.

Many lines are available from Xenopus resource centers (National Xenopus Resource, United States, mbl.edu/xenopus; European Xenopus Resource Center, United Kingdom, xenopusresource.org; Centre de Ressources Biologiques Xénopes [Biological Resource Center Xenopus], France, xenopus.univ-rennes1.fr; The National BioResource Project, Japan, xenopus.nbrp.jp).

6. Test for rescue of phenotype(s) with reference cDNAs as well as patient variants.

Large plasmid repositories exist for both Xenopus and human coding sequences (e.g., ORFeomes) (Morin et al. 2006; Yang et al. 2011; Grant et al. 2015) and many are available from Xenopus resource centers (National Xenopus Resource, United States, mbl.edu/xenopus; European Xenopus Resource Center, United Kingdom, xenopusresource.org; Centre de Ressources Biologiques Xénopes [Biological Resource Center Xenopus], France, xenopus.univ-rennes1.fr; The National BioResource Project, Japan, xenopus.nbrp.jp).

i. Test whether the human cDNA of the gene of interest can rescue the phenotype. First, determine whether the perturbation reagent (sgRNA, morpholino, etc.) also targets the human cDNA sequence. If it does not, the human cDNA can be used as a potential rescue reagent. For early embryonic phenotypes, synthesize human mRNA from a cDNA clone and coinject with the perturbation reagent (Vize et al. 1991). Alternatively, for later phenotypes, inject the cDNA plasmid itself, which may result in a longer duration of expression than the mRNA and may provide later rescue. Plasmid injections, however, result in highly mosaic organisms, creating cells with very high expression and cells without any expression. In both cases, titration of the rescue construct is often required to identify an optimal dose. Such rescues do not always work and failure of a cDNA to rescue a phenotype does not necessarily indicate that the phenotype is not based on mutation of the gene of interest because the gene could be very dosage-sensitive, have a very restricted expression pattern, or rely on isoforms not represented in the cDNA. However, successful rescue is very strong evidence that the phenotype is specific and the role of that gene in the process is recapitulated by the human version. In cases where cDNA rescue is not successful, another option is a BAC clone rescue, which may give more control over the level and tissue-specificity of an expressed gene product (Fish et al. 2012, 2014; Nakayama et al. 2017).

ii. If rescue with the reference cDNA or BAC clone is successful, generate patient-derived variants and test whether they fail to rescue.

7. Ensure rigor of findings by designing appropriate controls and outcomes.

i. Target a control gene not implicated in the disorder to control for the effects of early mutagenesis and/or injection, especially in the case of F_0 analysis.

ii. Use two different sgRNAs targeting the same locus to confirm that the phenotype of interest is specific to the intended gene.

iii. Use an alternative method of perturbation to phenocopy.

This can be done using morpholino oligonucleotides, a pharmacological compound, or overexpression.

iv. Raise a stable, germline mutant line and confirm the phenotype.

8. Consider possibilities of variable penetrance and expressivity.

i. Consider the effect of genetic background.

Genetic background of different Xenopus *lines or divergence between* Xenopus *and human genes may influence the penetrance and/or expressivity of a given gene perturbation. Although genetic circuitry may be largely conserved because of the high degree of synteny between human and* Xenopus *genomes (Hellsten et al. 2010), differences in genome structure may lead to* Xenopus *models having a somewhat different phenotype from humans. Although this can be viewed as a potentially negative aspect of using a model system, these differences may not preclude the other advantages of having an amphibian model. Any variations can lead to novel insights about genetic circuitry during development and their impact on phenotype as well (S Manohar, T Nakayama, and RM Grainger, unpubl.).*

ii. Consider the influence of genetic sex.

Many genetic disorders show evidence of sex bias, and therefore a consideration of genetic sex in phenotype penetrance can be important. X. laevis *boasts a well-understood genetic mechanism of sex determination, and therefore genetic sex can be considered even in stages before overt gonadal sexual dimorphism.*

In instances in which phenotypes are only half-penetrant, consider the potential influence of sex-differential biology in the phenotypes. Genetic sex can be determined by PCR in X. laevis *(Mawaribuchi et al. 2017), whereas in* X. tropicalis *this must be determined phenotypically after gonad development. Phenotypic sex can be reversed by hormone treatment and could be used to test these interactions (Villalpando and Merchant-Larios 1990).*

9. Consider validation in human-derived models or additional patient cohorts.

i. Confirm human conservation by translating findings in *Xenopus* to human in vitro models like induced pluripotent stem cells and 3D organoids.

This is most easily done if pharmacological tools can be directly transferred from Xenopus *in vivo work to human in vitro work. When this is not possible, again effort should be similarly made to ensure the genetic architecture is reflected in the experimental strategy.*

ii. Work closely with clinician scientists to use their expertise along with the genetic and detailed human phenotype data to which they have access to validate data obtained in the frog, particularly as new patients are recruited.

DISCUSSION

Although these strategies are largely straightforward for monogenic loss-of-function genetic disorders, when considering a disorder that has a large set of contributing risk genes and for which little is known about the biology (e.g., psychiatric disorders), there are a few strategies that can be helpful in planning experiments (Willsey et al. 2018a, 2021). These genes may all have pleiotropic functions and studying any one in isolation makes it difficult to know whether the phenotypes observed are relevant to the disorder or rather a by-product of pleiotropy. Therefore, a first helpful step is to identify tissues and time points of gene coexpression to limit the hypotheses about disorder risk gene function. Next, a productive strategy is to identify phenotypic convergence. The cost-effective and higher-throughput nature of F_0 CRISPR-based approaches make this strategy attractive for studying disorders with an underlying rare, gene-disrupting genetic architecture (e.g., autism spectrum disorders, Tourette disorder, congenital heart disease).

ACKNOWLEDGMENTS

We thank Jeremy Willsey, Matthew State, Mustafa Khokha, Annie Godwin, Emily Mis, Takuya Nakayama, and Karen Liu for formative conversations over the years. Our work has been supported by the National Institute of Mental Health grant 1U01MH115747 and The Overlook International Foundation for H.R.W., by the Wellcome Trust (212942/Z/18/Z) and Biotechnology and Biological Sciences Research Council (BBSRC) (BB/R014841/1) to M.G., and the National Institutes of Health (NIH) grant EY022954 and research awards from the Sharon Stewart Aniridia Trust and Vision for Tomorrow to R.M.G.

REFERENCES

Aslan Y, Tadjuidge E, Zorn AM, Cha S-W. 2017. High-efficiency non-mosaic CRISPR-mediated knock-in and indel mutation in F_0 Xenopus. Development 144: 2852–2858.

Bhattacharya D, Marfo CA, Li D, Lane M, Khokha MK. 2015. CRISPR/Cas9: an inexpensive, efficient loss of function tool to screen human disease genes in Xenopus. Dev Biol 408: 196–204. doi:10.1016/j.ydbio.2015.11.003

Blitz IL, Fish MB, Cho KWY. 2016. Leapfrogging: primordial germ cell transplantation permits recovery of CRISPR/Cas9-induced mutations in essential genes. Development 143: 2868–2875. doi:10.1242/dev.138057

Briggs JA, Weinreb C, Wagner DE, Megason S, Peshkin L, Kirschner MW, Klein AM. 2018. The dynamics of gene expression in vertebrate embryogenesis at single-cell resolution. Science 360: eaar5780. doi:10.1126/science.aar5780

Brinkman EK, Chen T, Amendola M, van Steensel B. 2014. Easy quantitative assessment of genome editing by sequence trace decomposition. Nucleic Acids Res 42: e168. doi:10.1093/nar/gku936

DeLay BD, Corkins ME, Hanania HL, Salanga M, Deng JM, Sudou N, Taira M, Horb ME, Miller RK. 2018. Tissue-specific gene inactivation in: Xenopus laevis knockout of lhx1 in the kidney with CRISPR/Cas9. Genetics 208: 673–686. doi:10.1534/genetics.117.300468

El-Brolosy MA, Kontarakis Z, Rossi A, Kuenne C, Günther S, Fukuda N, Kikhi K, Boezio GLM, Takacs CM, Lai S-L, et al. 2019. Genetic compensation triggered by mutant mRNA degradation. Nature 568: 193–197. doi:10.1038/s41586-019-1064-z

Firth HV, Richards SM, Bevan AP, Clayton S, Corpas M, Rajan D, Van Vooren S, Moreau Y, Pettett RM, Carter NP. 2009. DECIPHER: database of chromosomal imbalance and phenotype in humans using ensembl resources. Am J Hum Genet 84: 524–533. doi:10.1016/j.ajhg.2009.03.010

Fish MB, Nakayama T, Grainger RM. 2012. Simple, fast, tissue-specific bacterial artificial chromosome transgenesis in Xenopus. Genesis 50: 307–315. doi:10.1002/dvg.20819

Fish MB, Nakayama T, Fisher M, Hirsch N, Cox A, Reeder R, Carruthers S, Hall A, Stemple DL, Grainger RM. 2014. Xenopus mutant reveals necessity of rax for specifying the eye field which otherwise forms tissue with telencephalic and diencephalic character. Dev Biol 395: 317–330. doi:10.1016/j.ydbio.2014.09.004

Grant IM, Balcha D, Hao T, Shen Y, Trivedi P, Patrushev I, Fortriede JD, Karpinka JB, Liu L, Zorn AM, et al. 2015. The Xenopus ORFeome: a resource that enables functional genomics. Dev Biol 408: 345–357. doi:10.1016/j.ydbio.2015.09.004

Hellsten U, Harland RM, Gilchrist MJ, Hendrix D, Jurka J, Kapitonov V, Ovcharenko I, Putnam NH, Shu S, Taher L, et al. 2010. The genome of the Western clawed frog Xenopus tropicalis. Science 328: 633–636. doi:10.1126/science.1183670

Karczewski KJ, Francioli LC, Tiao G, Cummings BB, Alföldi J, Wang Q, Collins RL, Laricchia KM, Ganna A, Birnbaum DP, et al. 2020. The mutational constraint spectrum quantified from variation in 141,456 humans. Nature 581: 434–443. doi:10.1038/s41586-020-2308-7

Landrum MJ, Chitipiralla S, Brown GR, Chen C, Gu B, Hart J, Hoffman D, Jang W, Kaur K, Liu C, et al. 2020. ClinVar: improvements to accessing data. Nucleic Acids Res 48: D835–D844. doi:10.1093/nar/gkz972

Lasser M, Pratt B, Monahan C, Kim SW, Lowery LA. 2019. The many faces of Xenopus: Xenopus laevis as a model system to study Wolf–Hirschhorn syndrome. Front Physiol 10: 817. doi:10.3389/fphys.2019.00817

Lek M, Karczewski KJ, Minikel EV, Samocha KE, Banks E, Fennell T, O'Donnell-Luria AH, Ware JS, Hill AJ, Cummings BB, et al. 2016. Analysis of protein-coding genetic variation in 60,706 humans. Nature 536: 285–291. doi:10.1038/nature19057

Mawaribuchi S, Takahashi S, Wada M, Uno Y, Matsuda Y, Kondo M, Fukui A, Takamatsu N, Taira M, Ito M. 2017. Sex chromosome differentiation and the W- and Z-specific loci in Xenopus laevis. Dev Biol 426: 393–400. doi:10.1016/j.ydbio.2016.06.015

Moreno-Mateos MA, Vejnar CE, Beaudoin J-D, Fernandez JP, Mis EK, Khokha MK, Giraldez AJ. 2015. CRISPRscan: designing highly efficient sgRNAs for CRISPR-Cas9 targeting in vivo. Nat Methods 12: 982–988. doi:10.1038/nmeth.3543

Morin RD, Chang E, Petrescu A, Liao N, Griffith M, Chow W, Kirkpatrick R, Butterfield YS, Young AC, Stott J, et al. 2006. Sequencing and analysis of 10,967 full-length cDNA clones from Xenopus laevis and Xenopus tropicalis reveals post-tetraploidization transcriptome remodeling. Genome Res 16: 796–803. doi:10.1101/gr.4871006

Nakayama T, Fish MB, Fisher M, Oomen-Hajagos J, Thomsen GH, Grainger RM. 2013. Simple and efficient CRISPR/Cas9-mediated targeted mutagenesis in Xenopus tropicalis. Genesis 51: 835–843. doi:10.1002/dvg.22720

Nakayama T, Nakajima K, Cox A, Fisher M, Howell M, Fish MB, Yaoita Y, Grainger RM. 2017. no privacy, a Xenopus tropicalis mutant, is a model of human Hermansky–Pudlak syndrome and allows visualization of internal organogenesis during tadpole development. *Dev Biol* **426:** 472–486. doi:10.1016/j.ydbio.2016.08.020

Nakayama T, Grainger RM, Cha S-W. 2020. Simple embryo injection of long single-stranded donor templates with the CRISPR/Cas9 system leads to homology-directed repair in *Xenopus tropicalis* and *Xenopus laevis*. *Genesis* **58:** e23366. doi:10.1002/dvg.23366

Owens NDL, Blitz IL, Lane MA, Patrushev I, Overton JD, Gilchrist MJ, Cho KWY, Khokha MK. 2016. Measuring absolute RNA copy numbers at high temporal resolution reveals transcriptome kinetics in development. *Cell Rep* **14:** 632–647. doi:10.1016/j.celrep.2015.12.050

Session AM, Uno Y, Kwon T, Chapman JA, Toyoda A, Takahashi S, Fukui A, Hikosaka A, Suzuki A, Kondo M, et al. 2016. Genome evolution in the allotetraploid frog *Xenopus laevis*. *Nature* **538:** 336–343. doi:10.1038/nature19840

Villalpando I, Merchant-Larios H. 1990. Determination of the sensitive stages for gonadal sex-reversal in *Xenopus laevis* tadpoles. *Int J Dev Biol* **34:** 281–285.

Vize PD, Melton DA, Hemmati-Brivanlou A, Harland RM. 1991. Assays for gene function in developing *Xenopus* embryos. *Methods Cell Biol* **36:** 367–387. doi:10.1016/S0091-679X(08)60288-5

Willsey HR. 2021. Whole-mount RNA in situ hybridization and immunofluorescence of *Xenopus* embryos and tadpoles. *Cold Spring Harb Protoc* doi: 10.1101/pdb.prot105635

Willsey AJ, Morris MT, Wang S, Willsey HR, Sun N, Teerikorpi N, Baum TB, Cagney G, Bender KJ, Desai TA, et al. 2018a. The psychiatric cell map initiative: a convergent systems biological approach to illuminating key molecular pathways in neuropsychiatric disorders. *Cell* **174:** 505–520. doi:10.1016/j.cell.2018.06.016

Willsey HR, Walentek P, Exner CRT, Xu Y, Lane AB, Harland RM, Heald R, Santama N. 2018b. Katanin-like protein Katnal2 is required for ciliogenesis and brain development in *Xenopus* embryos. *Dev Biol* **442:** 276–287. doi:10.1016/j.ydbio.2018.08.002

Willsey HR, Exner CRT, Xu Y, Everitt A, Sun N, Wang B, Dea J, Schmunk G, Zaltsman Y, Teerikorpi N, et al. 2021. Parallel in vivo analysis of large-effect autism genes implicates cortical neurogenesis and estrogen in risk and resilience. *Neuron* **109:** 788–804.e8. doi:10.1016/j.neuron.2021.01.002

Yang X, Boehm JS, Yang X, Salehi-Ashtiani K, Hao T, Shen Y, Lubonja R, Thomas SR, Alkan O, Bhimdi T, et al. 2011. A public genome-scale lentiviral expression library of human ORFs. *Nat Methods* **8:** 659–661. doi:10.1038/nmeth.1638

Production of Transgenic F_0 Animals and Permanent Lines by Sperm Nuclear Transplantation in *Xenopus tropicalis*

Takuya Nakayama, Jessica Gray,[1] and Robert M. Grainger[2]

Department of Biology, University of Virginia, Charlottesville, Virginia 22904, USA

Early efforts in the 1980s showed that DNA microinjected into *Xenopus* embryos could be integrated into the genome and transmitted through the germline at low efficiency. Subsequent studies revealed that transgenic lines, typically with multiple-copy inserts (e.g., to develop bright fluorescent protein-reporter lines), could be created via sperm nuclear injection protocols such as the one entitled restriction enzyme-mediated insertion, or REMI. Here we describe a refined sperm nuclear injection procedure, with a number of alterations, including elimination of a potential DNA-damaging restriction enzyme treatment, aimed at making F_0 transgenic animals and transgenic lines in *Xenopus tropicalis*. This protocol also uses an oocyte extract rather than the egg extract used in older protocols. These changes simplify and improve the efficiency of the procedure.

MATERIALS

It is essential that you consult the appropriate Material Safety Data Sheets and your institution's Environmental Health and Safety Office for proper handling of equipment and hazardous materials used in this protocol.

RECIPES: Please see the end of this protocol for recipes indicated by <R>. Additional recipes can be found online at http://cshprotocols.cshlp.org/site/recipes.

Reagents

All reagents used should be molecular biology grade. Use sterile Nanopure H_2O for solutions unless other solvent is indicated.

Bovine serum albumin (BSA; e.g., Sigma-Aldrich A3294; 10% w/v)

Adjust pH to 7.6 with KOH, aliquot, and store at −20°C.

Modified Barth's Saline for *Xenopus* (MBS) <R>

Adjust pH to 7.5 rather than 7.8 mentioned in the recipe.

Sperm dilution buffer for *Xenopus* (SDB) <R>
Tricaine-S (MS-222; e.g., Syndel Syncaine; 1:500 w/v)

Dissolve in H_2O and adjust pH to 7.0–7.5 by gradually adding sodium bicarbonate powder (~1–2 g/L).

Reagents needed for egg preparation and sperm nuclear injection only (see Steps 36–51)
    Agarose (e.g., Fisher BP160)
    Chorulon (hCG, Intervet Inc)
    Injection medium <R>

---

[1]Current address: Department of Systems Biology, Harvard Medical School, Boston, Massachusetts 02115, USA
[2]Correspondence: rmg9p@virginia.edu

I-SceI (NEB R0694) (optional; see Step 43)

L-Cysteine (nonhydrochloride form; e.g., Sigma-Aldrich C5360; 2% w/v in 0.1× MBS)

Adjust pH to 7.9 with NaOH.

Mineral oil (e.g., Sigma-Aldrich M8410)

Purified transgene DNA fragment that has been restriction enzyme-digested and gel-purified in advance; alternatively, an I-SceI-treated plasmid that has a transgene flanked by two I-SceI sites (see Ogino et al. 2006)

Sigmacote (Sigma-Aldrich SL2)

Xenopus tropicalis females (three)

See Sec. 2, Protocol 4: Obtaining Xenopus Tropicalis *Eggs (Lane et al. 2022a) for information on identifying ideal females.*

Reagents needed for oocyte extract preparation only (see Steps 1–16)

1× XB <R>

Chymostatin (e.g., Roche 11004638001; 10 mg/mL in ethanol)

Aliquot and store at −20°C.

Collagenase solution for isolation of oocytes <R>

Leupeptin (e.g., Sigma-Aldrich 11017128001; 10 mg/mL in DMSO)

Store at −20°C.

Liquid nitrogen

Modified CSF-XB (MCSF-XB) <R>

Modified *Xenopus* Oocyte Culture Medium (MXOCM) <R>

Oocyte storage solution <R>

Pepstatin A (e.g., Sigma-Aldrich 516485; 10 mg/mL in DMSO)

Aliquot and store at −20°C.

Versilube F50

Xenopus laevis females (two to three)

Use 2- to 5-yr-old females that are known to produce good eggs (e.g., pigmentation is uniform and fertilization rate is high).

Reagents needed for sperm nuclei preparation only (see Steps 17–35)

1× NPB

Dilute 2× NPB <R> in sterile Nanopure H$_2$O. Freshly prepare 30 mL. Place 1 mL at room temperature for Step 28. Store the remainder on ice.

1× NPB with 0.3% BSA <R>

25% Percoll in 1× NPB <R>

50% Percoll in 1× NPB <R>

Digitonin (e.g., Calbiochem 300410; 10 mg/mL in DMSO)

Aliquot and store at −20°C.

Hoechst 33342 (e.g., Invitrogen H1399; 10 mg/mL)

Store at −20°C. Dilute 1:100 for use.

Protease inhibitor cocktail <R>

Sperm storage buffer for *Xenopus* (SSB) <R>

Xenopus tropicalis males (two)

See Sec. 2, Protocol 5: Obtaining Xenopus tropicalis *Embryos by In Vitro Fertilization (Lane and Khokha 2022) for details on identifying ideal males.*

Equipment

Aluminum foil

Beakers, 200- and 300-mL

Cite this protocol as *Cold Spring Harb Protoc*; doi:10.1101/pdb.prot107003

Culture incubators (22°C and 25°C)

Dissection tools

Fluorescence microscope (with DAPI filter)

Gastight syringe (Hamilton 1702)

Glass capillary tube (WPI TW100-4)

Glass culture dish (~4½" × 1¾"; e.g., Flinn Scientific AB1264)

Graduated conical bottomed centrifuge tubes, 15-mL (e.g., Fisher 339650)

Hemacytometer (glass or disposable; e.g., VWR 82030-470)

High-speed, refrigerated centrifuge (e.g., Sorvall RC-5B)

Laboratory oven with a range of 120°C–220°C

Microcentrifuge (optional; see Step 34)

Microcentrifuge tubes, 1.5-mL, 0.5-mL

Micromanipulator (e.g., WPI M3301-M3-R)

Micropestle (to fit a 1.5-mL microcentrifuge tube, sterile)

Micropipette puller (e.g., Sutter Instrument Company)

Needle holder

Nitex mesh, 50-μm (e.g., Sefar America 03-50/31)

Open-top centrifuge tube, 3.5-mL (e.g., Beckman 349622)

Parafilm

Pasteur pipettes

Petri dishes (60-mm/100-mm)

Razor blade

Round-bottom 30-mL/15-mL centrifuge tubes (e.g., Corex 8445/8441)

Sieve, 1000-μm (optional; see Step 5)

Stage micrometer

Stereomicroscope with fiber optic light source

Swinging bucket rotors (e.g., Sorvall HB-6 rotor for high speed and Eppendorf S-4-104 for low speed)

Syringe pump (Harvard)

Tabletop low-speed centrifuge (e.g., HN-S, IEC)

Tabletop ultracentrifuge (e.g., Beckman Optima TLX with TLA-100.3 rotor)

Transfer pipettes

Tygon tubing (R-3603, inner diameter of 0.8 mm, outer diameter of 2.4 mm)

METHOD

Carry out all procedures at room temperature unless otherwise indicated.

The success of this method largely depends on egg quality and the colony must be maintained in a very healthy state in general (see Sec. 1, Protocol 3: Raising and Maintaining Xenopus tropicalis *from Tadpole to Adult [Lane et al. 2022b] and Sec. 2, Protocol 8: Best Practices for* Xenopus tropicalis *Husbandry [Nakayama and Grainger 2022] for further information as needed). If experiments continue to fail with frogs in one's facility, consider obtaining new frogs.*

Preparation of Oocyte Extract

1. Euthanize two (or three) female *Xenopus laevis* with tricaine-S (MS222) according to an approved animal protocol and remove the entire ovaries.

   If oocytes are mostly large and abundant, only two females are needed. If one or both females have many immature oocytes but not many full-grown oocytes, use an extra female. It is very important to have enough starting material for this preparation.

2. Cut ovaries into small pieces using scissors and digest for ~1.5–3.5 h in collagenase solution for isolation of oocytes in beakers at 28°C in a water bath with agitation.

It takes less time to complete the digestion if the oocytes are split into a few 200- to 300-mL beakers with ~100–150 mL of collagenase solution in each beaker.

Begin checking the oocytes every 15 min after about an hour. They will be ready for the next step once clumps of oocytes are no longer visible.

3. After digestion, wash the oocytes several times with oocyte storage solution.

    Swirl the oocytes in a beaker so that the small immature oocytes float to the top and can be easily poured off. In this way, most of the small oocytes will be eliminated. We aim to collect the largest oocytes (~1000–1300 µm in size, equivalent to stage V-VI oocytes, see Dumont 1972) in the following steps.

4. Transfer oocytes to modified *Xenopus* oocyte culture medium (MXOCM) with a cut-off transfer pipette. Remove any dead, dying, or tiny oocytes. We use large glass culture dishes to collect up to a 100–125-mL volume of oocytes in 200–250-mL medium per dish (i.e., the volume of medium should be minimum twice volume of oocytes).

    One can use beakers but the depth of beakers, compared to shallow culture dishes, may be a disadvantage in sorting the oocytes out of beakers in the next step. This is a good step at which to take a break in the protocol because the following steps take a full day. The oocytes can be stored here for about a week at 16°C–18°C as needed with daily medium changes and removal of dead/unhealthy oocytes; however, we recommend use within 1–2 d. One recommendation is to leave oocytes overnight and to sort the next morning.

5. Sort and collect only the largest intact oocytes and transfer to fresh MXOCM in new culture dishes with a cut-off transfer pipette.

    This is a laborious step and we recommend working with several people to shorten the time. To make the procedure easier, one can use a 1000-µm sieve, which will retain stage VI oocytes (and larger stage V oocytes) while allowing smaller oocytes to pass through; see Sec. 3, Protocol 1: Isolation of Xenopus Oocytes (Newman et al. 2018) for more details. One should aim to collect a minimum of 40-mL- up to 60-mL-equivalent volume of the largest oocytes split into several dishes to keep them healthy.

6. Collect all oocytes in a beaker and rinse them four times with 1× XB (~100 mL each rinse) followed by two rinses with modified CSF-XB (MCSF-XB) (~50 mL each rinse).

7. Gently load oocytes into 30-mL centrifuge tubes using a cut-off transfer pipette. Remove as much MCSF-XB as possible and try to balance tubes visually, with approximately equal volumes in each centrifuge tube.

    There should be at least enough oocytes to half fill two 30-mL centrifuge tubes. It is preferable to start with enough to fill each tube three-quarters full, depending on the quality of the oocytes.

8. Add ~1 mL of Versilube F50 on top of the oocytes in each centrifuge tube and centrifuge tubes in a tabletop low-speed centrifuge at room temperature for ~60 sec at 1000 rpm, and then for an additional 30 sec at 2000 rpm.

    This will pack the oocytes, but they should not be lysed or broken.

9. Remove the excess MCSF-XB and Versilube F50. Try to balance two tubes precisely by adjusting the amounts removed.

10. Centrifuge the tubes using a swinging bucket rotor at ~16,300g (when using HB-6 rotor, 10,000 rpm) for 10 min at 4°C.

    During this centrifugation the oocytes are crushed and should be separated into three layers: lipid (top), cytoplasm (center), and yolk and pigment (bottom). Any excess Versilube F50 from previous steps will appear as bubbles in the cytoplasmic layer. During this centrifugation, start cooling the ultracentrifuge to be used for Step 13.

11. Remove the cytoplasmic layer and transfer it to a 15-mL graduated conical tube on ice to measure the volume. Add a 1:1000 dilution of the 10 mg/mL protease inhibitors (leupeptin, pepstatin A, chymostatin) to the tube. Close the lid and invert the tube several times to mix the solution well.

    The cytoplasmic layer should be gray to brown/gold color, darker or lighter depending on the pigment of the oocytes. We use a drawn-out disposable glass pipette (made by pulling the Pasteur pipette over a flame to make the diameter thinner and cutting off the tip) to remove the cytoplasmic layer. A 3- to 5-mL syringe with a long needle also works well. Directly pass the long (thin) pipette or needle through the lipid layer to the cytoplasmic layer from the top. The recovered amount of cytoplasm will vary depending on the amount of starting material, but at least 11–12 mL should be recovered.

Cite this protocol as *Cold Spring Harb Protoc*; doi:10.1101/pdb.prot107003

12. Transfer the cytoplasm to a 15-mL centrifuge tube (divide it into two tubes as needed) and recentrifuge as in Step 10 (i.e., at ~16,300g for 10 min at 4°C). Collect the cytoplasmic layer as described in Step 11 and transfer to ultracentrifuge tubes.

> The number of ultracentrifuge tubes varies depending on yield but there should be enough to half fill at least two tubes. Fill each at least halfway, ensuring a minimum of 2 mL, but not more than 3.5 mL, in each tube (for Beckman 349622) because of the fixed angle rotor. Balance ultracentrifuge tubes carefully using a scale.

13. Centrifuge tubes in the ultracentrifuge at ~264,400g (when using TLA100.3 rotor, ~70,000 rpm) for 1.5 h at 4°C. It is critical not to use braking on this centrifugation.

> After this centrifugation there will be four layers, top to bottom: lipid, cytosol, membranes/mitochondria, and glycogen/ribosomes. The cytoplasmic layer should make up 80%–90% of the volume recovered in Step 12.

14. Remove the cytosolic layer as in Step 11. Place the extract in new ultracentrifuge tubes settled on ice. Rebalance using a scale and recentrifuge at ~264,400g for 20 min at 4°C. It is once again critical not to use braking on this centrifugation.

> There should still be enough extract to fill at least half of each tube.

15. Remove cytosolic layer as in Step 11 and transfer it into 1.5-mL microcentrifuge tubes on ice.

> There should be enough cytosol collected to fill at least two microcentrifuge tubes.

16. Aliquot 7–13 μL of extract into 0.5-mL microcentrifuge tubes and snap-freeze in liquid nitrogen. Store tubes at −80°C until use. The extract is stable and can be stored for more than a year.

> We use 2 μL per reaction in Step 45. The number of aliquots used will depend on how many reactions are expected in a given day. For example, 7 μL should be sufficient for three reactions.

Preparation of Sperm Nuclei

17. Euthanize two males of *Xenopus tropicalis* with tricaine-S (MS222) following an approved animal protocol.

18. Coat two 15 mL conical bottomed centrifuge tubes with 10% BSA and settle on ice until use in Steps 19 and 25.

> This can be done while waiting during Step 17. If the 15-mL tube does not fit into a swinging bucket rotor for the available centrifuge (see Step 24), simply cut the top part of the tubes to make them shorter (~12 mL) or use appropriate alternative conical tubes.

19. Attach Nitex mesh to one of the above BSA-coated tubes for filtration.

> As shown in Figure 1A, cut the middle part of a new 15-mL tube (or any other appropriate plastic tube) to make a ring (5- to 6-mm-high), then cut a piece out of the ring to make a small (~4-mm) gap. Cut a square (~3-cm × ~3-cm) of Nitex mesh (50-μm) and place it across opening of the centrifuge tube. Secure the Nitex with the slit-ring made above onto the top part of the tube, and place on ice.

20. Remove testes from killed male frogs as described in Sec. 2, Protocol 5: Obtaining *Xenopus tropicalis* Embryos by In Vitro Fertilization (Lane and Khokha 2022). Store testes temporarily in a Petri dish (60-mm) filled with 1× MBS (pH 7.5). Clean testes one by one under a stereomicroscope (i.e., remove as much of the fat bodies and blood as possible). Carefully poke holes in blood vessels and push blood out without damaging the testis and releasing sperm. Rinse twice with 1× MBS (pH 7.5) in a new dish. Rinse twice with ice-cold 1× NPB in a new dish.

21. Place cleaned testes in a 1.5-mL microcentrifuge tube with 100 μL of ice-cold 1× NPB. Homogenize testes with a micropestle (make sure no chunks of testes remain) and add 1 mL of ice-cold 1× NPB to the tube, while rinsing the micropestle to collect all attached sperm.

22. Using a cut-off 1-mL tip, mix and resuspend the homogenized tissue. Apply the suspension to the mesh prepared in Step 19 to filter through into the conical centrifuge tube on ice. Rinse the emptied microcentrifuge tube with another 1 mL of ice-cold 1× NPB and filter though the mesh into the same tube. Wash off the sperm remaining on the mesh into the tube by applying 1 mL of ice-cold 1× NPB three times. The total volume in the tube will be ~5.1 mL.

23. While wearing gloves, remove the mesh and squeeze it to recover remaining liquid on the mesh into the tube. Add ice-cold 1× NPB to a final volume of 6 mL.

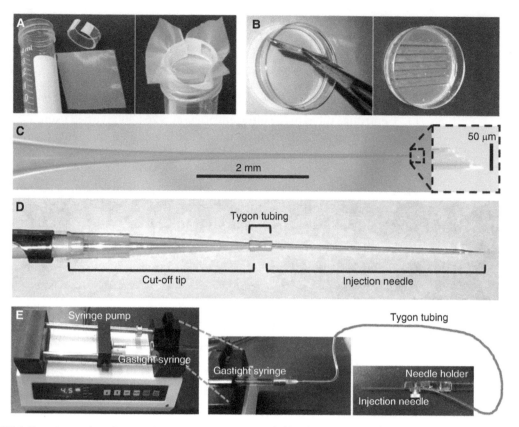

FIGURE 1. Experimental equipment. (*A*) How to prepare a mesh filter for sperm nuclei preparation. (*B*) How to prepare an injection dish. (*C*) An example of a standard needle for injection. (*D*) How to load sperm nuclei into the needle. (*E*) Setup of syringe pump and needle for injection.

24. Prepare a balance tube and centrifuge the sample in a swinging bucket rotor at ~1500*g* to 1800*g* (~3000 rpm depending on available rotors) for 15 min at 4°C.

    The resultant pellet will be white with a red (blood cells) central spot in color.

25. Prepare a Percoll gradient in the second BSA-coated tube from Step 18. Place 3 mL of 50% Percoll in 1× NPB into the tube. Carefully layer 3 mL of 25% Percoll in 1× NPB on top of the 50% Percoll in 1× NPB.

    This should be done during the centrifugation in Step 24.

26. Remove the supernatant from the centrifugation in Step 24 by carefully pipetting as much as possible without disrupting the pellet. Resuspend the pellet with 1 mL of ice-cold 1× NPB using a cut-off 1-mL tip. Carefully layer the resultant suspension on top of the Percoll gradient prepared in Step 25. Rinse the emptied centrifugation tube with an additional ice-cold 1 mL of 1× NPB and add to top layer of the gradient. As a result, the second tube will contain ~8 mL. Centrifuge the second tube as in Step 24.

27. Remove the Percoll gradient layers carefully without disrupting the pellet. Resuspend the pellet with 1 mL of ice-cold 1× NPB using a cut-off 1-mL tip. Bring final volume to 6 mL with ice-cold 1× NPB, and centrifuge as in Step 24.

    During the centrifugation, prepare the protease inhibitor cocktail and 1× NPB with 0.3% BSA and store on ice until use.

28. Carefully remove all supernatant and resuspend the pellet with 1 mL of room temperature 1× NPB using a cut-off 1-mL tip. Add 5 µL of 10 mg/mL digitonin and mix the suspension thoroughly with the cut-off 1-mL tip. Incubate for 5 min at room temperature.

Cite this protocol as *Cold Spring Harb Protoc*; doi:10.1101/pdb.prot107003

29. Add 8 mL of the protease inhibitor cocktail prepared in Step 27, which will stop the reaction above. Cover the top of the tube with Parafilm and mix the suspension by gently repeating inversions.

30. Centrifuge as in Step 24.

   After this centrifugation the red center of the pellet should be gone.

31. Quickly remove the supernatant by pipetting. Resuspend the pellet in 5 mL of $1\times$ NPB $+ 0.3\%$ BSA prepared in Step 27, and centrifuge as in Step 24.

   During the centrifugation, prepare the sperm storage buffer for Xenopus (SSB) and place on ice.

32. Quickly remove as much supernatant as possible by pipetting. Resuspend the pellet gently in 75 μL of ice-cold SSB and transfer the resuspended nuclei to a 1.5-mL microcentrifuge tube on ice using a cut-off 200-μL tip. Rinse the emptied centrifuge tube with another 75 μL of ice-cold SSB and transfer to the microcentrifuge tube on ice. Mix the suspension thoroughly with the cut-off 200-μL tip.

   At this point the protocol can be continued, or the suspension can be stored overnight at 4°C and the protocol resumed the next day.

33. Add 1 μL of the sperm suspension to 98 μL of SDB in a 1.5-mL (or smaller) microcentrifuge tube on ice. Add 1 μL of a 1:100 dilution of 10 mg/mL Hoechst 33342. Mix well with a cut-off 200-μL tip.

34. Load 8–10 μL per chamber of the suspension onto a hemacytometer (make sure to fill both chambers) and count the sperm under a fluorescence microscope. Count both chambers and average the number. Calculate the concentration of the undiluted sperm suspension.

   If the number of sperm in each chamber are not approximately equal, the suspension may not have been completely homogeneous. Gently remix, reload, and recount. We aim here to obtain an average calculated concentration of undiluted sperm nuclei suspension of 100 nuclei per nanoliter (nuc/nL). A range between 80 and 120 nuc/nL is acceptable. If the concentration is substantially higher than this, dilute the original sperm suspension further with SSB and repeat the counting. If the concentration is substantially lower than expected (e.g., 50 nuc/nL), centrifuge the original suspension at 3000 rpm for 5 min at 4°C in a microcentrifuge, remove the appropriate amount of supernatant (i.e., excess amount of SSB), and recount.

   After experience with this protocol through Step 34, one may come to be able to judge if the pellet seems large or small at Step 32. The appropriate amount of SSB to resuspend the pellet in Step 32 can then be estimated. However, it is better to have the nuclei more concentrated than less, because reconcentration always results in some loss of sperm.

35. In Step 44, use sperm nuclei from Step 32 that has been stored for up to 2 d at 4°C. Alternatively, for long-term storage, at this point or after 2 d, store sperm nuclei at −80°C. Aliquot an appropriate amount (see below) of sperm nuclei into 0.5-mL microcentrifuge tubes and slow-freeze (i.e., put tubes directly in a −80°C freezer).

   We use 4 μL (100 nuc/nL) per reaction in Step 44. The appropriate aliquot amount will depend on how many reactions will be performed. The sperm nuclei stored at −80°C are usable at least for a year.

Preparation for Injection

The following steps in this section can be performed in any order.

36. One to three days before performing injections, prime *X. tropicalis* females with hCG (10 units per frog) as described in Sec. 2, Protocol 4: Obtaining *Xenopus tropicalis* Eggs (Lane et al. 2022a). The number of females used will depend on how many rounds of injections are done. The quality of eggs is key for success and therefore it is beneficial to have the option of extra available females. We usually have at least three frogs ready per experiment.

37. Prepare four kinds of dishes, for injection, embryo collecting, egg collecting, and embryo culturing.

   i. To prepare injection and embryo collecting dishes, pour melted 1% agarose dissolved in $0.1\times$ MBS (pH 7.5) in 60-mm Petri dishes to make an ~4-mm thick agarose layer. Once solidified, half of the dishes are used directly for embryo collecting dishes. The rest are used to make injection dishes as follows: Use the nonblade side of a single-edge razor blade heated by a Bunsen burner to melt the agarose and make as many grooves as possible (see Fig. 1B).

The thickness of agarose should be 4–5 mm for injection and embryo collecting dishes. Avoid pouring too thick of an agarose layer because dishes will then hold an insufficient amount of injection medium. One should be able to make at least six rows of grooves in which eggs will be placed for injection. One should be able to place 40–50 eggs per groove—namely, approximately 300 eggs per injection dish. Do not make grooves too deep because it makes harder to recover embryos after injection. Prepare an appropriate number of dishes for the number of eggs that will be injected. Unused injection and embryo collecting dishes can be stored for several weeks at 4°C (unless they are dried up or grow mold or bacteria).

ii. Prepare egg collecting dishes (60-mm) and embryo culturing dishes (100-mm) by adding melted agarose (0.8%–1% agarose in 0.1× MBS, pH 7.5) to cover the whole bottom of each dish and immediately remove as much as possible to make a thin layer of agarose coating the bottom of the dish (also refer to Protocol 9: Gynogenetic Production of Embryos in *Xenopus tropicalis* Using a Cold Shock Procedure: Rapid Screening Method for Gene Editing Phenotypes (Nakayama et al. 2022) for agarose-coated dishes as needed).

38. Siliconize the inside of glass capillary tubes with Sigmacote, rinse with Nanopure H$_2$O, and bake in a laboratory oven for 2 h or longer at 120°C–220°C.

We treat and bake approximately 50 to 100 tubes at once. Pour Sigmacote in an appropriate container (e.g., small glass Petri dish) to a height of 2–3 mm. Dip one end of a bundle of glass capillary tubes into the Sigmacote solution, which will automatically come into the tube by capillary action. Carefully invert the bundled tubes to allow the Sigmacote solution in each tube to pass through inside to the other end. After inverting the tube several times, place one end of the bundle of tubes against a Kimwipe to remove the Sigmacote. Move the bundle into a 15-mL or 50-mL (depending on the bundle size) conical tube full of Nanopure H$_2$O to wash inside the capillary tubes. Close the conical tube lid and invert the tube repeatedly until confirming that no air bubbles are trapped inside of any capillary tubes in the conical tube, meaning that water has passed through the inside of each capillary tube. Remove the bundle of tubes and remove extra water inside by placing the ends against the Kimwipe. After drying overnight, bake the bundle in a 150-mL glass beaker covered with an aluminum foil lid. After baking, store as is. In our experience this protocol still works without siliconization, but treatment makes it easier to fill the injection mixture (see Step 47).

39. Pull needles with a micropipette puller. The conditions will be different depending on the machine used, and one may need to find the optimal condition by trial and error. The goal is to prepare long and gradually tapering needles (Fig. 1C).

40. Once needles are pulled, clip needles with a pair of forceps. Under a stereomicroscope, clip the tip of a needle on a stage micrometer to make a beveled tip of 40–50 µm diameter (Fig. 1C).

Clipping of needles is essentially done by trial and error. Start by clipping to make a very fine diameter, and then keep trimming the end of tip to get to the desired diameter, rather than trying to obtain the correct size initially.

Performing Transgenesis

41. Three to four hours before performing transgenesis, boost female frogs (that have been primed in Step 36) with 100 (up to 200 as needed) units of hCG, expecting that ovulation will start 3–4 h later. Thaw the injection medium to room temperature if it is frozen.

While waiting, perform Steps 37 and 40 as needed, also prepare 0.1× MBS (pH 7.5, for culturing embryos) and 2% (w/v) L-cysteine (nonhydrochloride form) in 0.1× MBS (pH 7.9, adjusted with NaOH, for dejellying).

If possible, prepare additional frogs boosted at a different time (i.e., at a 1- to 2-h interval) to perform multiple injections using better quality eggs throughout a day. We can usually squeeze females up to 6–7 h after boosting for multiple injections per day but egg quality will typically drop during this extended period of collection.

42. Once successful ovulation is confirmed, squeeze a small number of eggs from each frog, check for unmottled pigment, absence/presence of a germinal vesicle spot, lysing, and overall consistency in a given batch of eggs. Only use females that are laying healthy eggs.

 Cite this protocol as *Cold Spring Harb Protoc*; doi:10.1101/pdb.prot107003

43. Prepare a transgene for injection. There are two options for a transgene source: an I-SceI-treated plasmid that has a transgene flanked by two I-SceI sites (see Ogino et al. 2006) or a purified transgene DNA fragment.

    i. For an I-SceI plasmid, digest 250–1000 ng in a 10-μL reaction mixture (i.e., 25–100 ng DNA/μL final concentration) using 1 μL of I-SceI with 1 μL of 10× reaction buffer at 37°C, starting ~40 min before Step 44.

> *This is a new option we have added to the previous protocol (Hirsch et al. 2002) since we developed I-SceI transgenesis (Ogino et al. 2006). A similar approach has been reported independently for X. laevis (L'hostis-Guidet et al. 2009). Because, to the best of our knowledge, there is no endogenous I-SceI site in the genome of X. tropicalis, adding this procedure is not considered restriction enzyme–mediated insertion (REMI). If a given laboratory already uses I-SceI transgenesis or has I-SceI constructs from other laboratories, they can simply use existing materials as described here—that is, use an aliquot of the mixture containing an appropriate amount of plasmid without further purification.*

    ii. For a DNA fragment, use a transgene that has been restriction enzyme–digested and gel-purified in advance. Use at a concentration of 20–200 ng/μL in DNase-/RNase-free H$_2$O (a higher concentration is preferred).

44. Either use unfrozen sperm nuclei directly from Steps 32–35 or thaw an aliquot of frozen sperm nuclei (see Step 35). Place the sperm nuclei on ice. Mix the following in a 1.5-mL microcentrifuge tube (using a cut-off tip) and incubate for 5 min at room temperature:

    4 μL sperm nuclei (100 nuc/nL)
    x μL of DNA (75–300 ng)
    y μL SDB
Total volume is 9 μL (or more as needed)

> *If the sperm nuclei concentration is not ~100 nuc/nL, adjust the volume accordingly. The desired total volume is 9 μL (or more as needed, see below). Start by calculating necessary volumes of sperm nuclei and DNA (x μL) and make up the difference with SDB (y μL).*

> *The appropriate amount of I-SceI-treated plasmid or DNA fragment will vary from one transgene to another, and one may need to test several doses in different injections. Larger constructs usually require larger amounts and can be increased accordingly unless toxicity is seen.*

> *This reaction can be >9 μL if needed but should have a minimum of 2 μL SDB. SDB can be reduced in Step 45 if a volume >9 μL is needed here.*

> *During this incubation thaw an aliquot of oocyte extract from Step 16 and place on ice.*

45. Add the following to the tube from Step 44, mix well using a cut-off tip, and incubate for 15 min at room temperature.

    9 μL (or more, see above) mixture from Step 44
    2 μL oocyte extract
    9 μL SDB (or less, see above)
    Total 20 μL

46. When convenient during Steps 44–45, squeeze eggs into an egg collecting dish from Step 37.ii without any drippings from the frogs. Collect 600 to 1000 eggs per injection (per person injecting). Once eggs are obtained, immediately add 2% cysteine solution and dejelly eggs. Eggs do not necessarily need to be completely dejellied as is typically done for microinjection purposes. We find that it is, in fact, preferable that eggs are not completely, but almost completely, dejellied (i.e., a condition where one can still see a narrow space between packed eggs under a stereomicroscope). We also have success with completely dejellied eggs (i.e., no space between eggs). Rinse eggs with 0.1× MBS several times, and load eggs using a Pasteur pipette (smooth the end with a flame to avoid damaging eggs) coated inside with 0.1× MBS + 0.1% BSA into grooves of injection dishes that have been filled in advance with the injection medium.

> *Here, an earlier version of protocol (Hirsch et al. 2002) recommended intentionally performing partial dejellying, which requires much experience. If it is not performed properly, too much leftover jelly becomes compressed in the presence of Ficoll, which squeezes cytoplasm of eggs out of the injected portion, leading to death. For generalizing the protocol, therefore, we recommend the above method.*

Both dejellying and rinsing eggs must be gently performed and using the thin-layer agarose-coated dish is essential to avoid activation of eggs caused by physical damage. Use BSA-coated pipettes for transferring eggs. As for obtaining eggs, also refer to Protocol 9: Gynogenetic Production of Embryos in Xenopus tropicalis *Using a Cold Shock Procedure: Rapid Screening Method for Gene Editing Phenotypes (Nakayama et al. 2022) for more information as needed (e.g., for obtaining eggs from females who do not lay a sufficient number of eggs at once).*

47. After the 15 min incubation in Step 45, add 50 µL of SDB and mix well with a cut-off tip. The total volume in the tube should now be 70 µL. As seen in Figure 1D, using a cut-off tip with a piece of Tygon tubing, backfill the injection needle with sperm nuclei suspension. Avoid air being mixed with the injection solution. Therefore, do not remove the filled needle directly, which will introduce air due to back pressure caused by pulling out the tubing, but first remove the entire tip with the needle attached from the pipettor, and then remove the needle carefully just before Step 48.

    An optimal situation for injection would be to arrange timing of procedures so that, when the above steps are done, eggs from Step 46 are simultaneously becoming ready for injection.

48. Set a syringe pump with a gastight syringe connected with Tygon tubing that is filled with mineral oil in advance and set the flow rate to 4.5 µL/h. Avoid any air bubbles trapped in the line from the gastight syringe to the open end of Tygon tubing. At the end of the tubing, carefully attach the injection needle filled with sperm nuclei suspension without letting air inside, and then attach to a needle holder (Fig. 1E). Set the needle holder (rod) into a micromanipulator and place the tip of the injection needle in the injection medium of the injection dish filled with eggs to confirm that flow through the tip of the needle is occurring. After confirming that the flow has started, begin injection. Between injections, occasionally make sure that the flow continues. If flow is not seen, check that air is not trapped in the line or whether the needle is clogged.

    i. When attaching the needle, start the pump first so that mineral oil comes out from the open end of the tubing to avoid air inside.

    ii. The flow rate for the first minutes could be inaccurate because it could be affected from back pressure in the system upon attaching the needle to the tubing, but injections can be started nonetheless.

    iii. If air trapping or debris clogging leading to no flow occurs near the tip of the needle, one may be able to induce flow again by speeding up the flow rate temporarily—that is, set the pump speed to maximum and observe if air or debris could be released. Once released, set the speed back to the original level and wait until flow becomes stable before restarting injection. Alternatively, the tip of the needle can be gently clipped (without taking it out), which would result in a bigger than optimal diameter. As a result, injections would have to be performed more quickly and resultant embryos could be more damaged. If any clogging due to debris or air occurs in the middle of the line far from the tip of the needle, essentially one needs to start over.

    iv. Inject at a rate a little less than a second per egg, just below the surface of eggs (but at any position around the egg). Be careful not to poke too deeply and thus pass through the egg completely by moving the needle too much, which can easily happen. A quick movement of the needle is needed for insertion through the egg membrane but a slower movement is optimal when pulling the needle out of the egg.

49. Finish injections within 30–40 min after starting. After injection, incubate the injection dishes at 22°C.

50. After ~70 min from when first row of eggs was injected, start checking whether any cleavage has begun. Wait until the (two- to) four-cell stage (or eight-cell stage but not beyond), and then start collecting only normal cleaving embryos. Move them to the embryo collecting dishes from Step 37.i filled with injection medium using a Pasteur pipette (smooth the end with a flame to avoid damaging embryos) coated inside with 0.1× MBS + 0.1% BSA. Place up to 20 to 25 embryos per dish and incubate at 22°C until the blastula stage.

Cite this protocol as *Cold Spring Harb Protoc*; doi:10.1101/pdb.prot107003

If embryos are sticking to each other, use a fine glass rod (made by pulling the end of a Pasteur pipette in a flame) and place it between embryos and gently push down to separate them.

Embryos with more cleavage-type furrows than normal (hereafter called "multicleavers") may be observed because of injection of multiple sperm nuclei. Multicleaver embryos can survive for a while, and at later stages (i.e., 32 cell stage or later) become indistinguishable from normal embryos (because there are many blastomeres). It is therefore important to sort normal embryos during two- to eight-cell stages. The best stage is the four-cell stage because at the two-cell stage one still cannot tell for sure if an embryo will become a normal four-cell or a multicleaver embryo. At the eight-cell stage cleavages may sometimes happen in an atypical pattern, even for normal embryos, which makes it harder to distinguish between normal and abnormal cleavers.

We usually see a range of surviving embryos, from <10% (very poor) up to 40% (very good) normal cleavers at this step. Among successful cleavers, the same range of percentages of embryos typically develop to late stage 30s to early stage 40s normally, within which the same range of percentage of embryos could be transgenic. However, there is not necessarily a correlation of success rates between each point of evaluation—namely, even fewer numbers of initial cleavers could end up with greater numbers of transgenic tadpoles at the end, or vice versa. In short, from 1000 injected embryos, one may be able to generate up to approximately 50 to 60 healthy transgenic embryos, but in our hands there is no reliable way to consistently predict the outcome in a given experiment.

See Troubleshooting.

51. Once embryos reach the blastula stage, use a BSA-coated pipette (see Step 50) to move embryos to embryo culturing dishes from Step 37.ii filled with 0.1× MBS, pH 7.5. Place up to 50 embryos per dish, and keep culturing at 22°C. On the next day, sort and move normal embryos to a noncoated dish (100-mm) filled with 0.1× MBS (pH 7.5). Culture at 25°C, followed by culturing with daily sorting/medium-changing (and/or feeding as needed) to the desired stage (or raising them to mature frogs as desired). See Sec. 1, Protocol 3: Raising and Maintaining *Xenopus tropicalis* from Tadpole to Adult [Lane et al. 2022b] and Sec. 2, Protocol 8: Best Practices for *Xenopus tropicalis* Husbandry [Nakayama and Grainger 2022] for further information.

When changing to the culture medium from the injection medium, do not rinse embryos but move embryos with some amount of the injection medium together so that gradual exchange of medium surrounding embryos occurs.

See Troubleshooting.

TROUBLESHOOTING

Problem (Step 50): Not many normal cleavers and no multicleavers are observed.
Solution: Several possible solutions exist.

1. Try to inject a little more slowly.
2. Reduce the SDB volume at Step 47.
3. Increase the number of sperm nuclei at Step 44.

If you still do not see an increase in normal cleavers, then failure is at least not due to an incorrect estimation of numbers of sperm nuclei.

Problem (Step 50): More multicleavers than normal cleavers are observed.
Solution: Several possible solutions exist.

1. Inject faster to generate more normal cleavers.
2. If the same situation persists even with faster injection, decrease the amount of sperm at Step 44 or increase the SDB at Step 47.
3. If multiple people do side-by-side injections using the same batch of eggs and sperm nuclei, one can often distinguish if the injection skill or if sperm nuclei number affects the result.

Problem (Step 51): Not many normal embryos are obtained.
Solution: Several possible solutions exist.

1. Obtain new females to produce better eggs.

2. Try to use more or less oocyte extract in Step 45 and/or remake oocyte extract.

3. Reprepare sperm nuclei.

DISCUSSION

The protocol described here is based on the original "REMI" procedure developed by Kroll and Amaya (1996) for *X. laevis* and subsequently for *X. tropicalis* (Offield et al. 2000). In the present protocol we add the refinement of using oocyte extract for transgenesis in *X. tropicalis* (Hirsch et al. 2002). It is thought that nucleoplasmin in the oocyte/egg extract is responsible for decondensation of the sperm nuclei (Ohsumi and Katagiri 1991; Philpott et al. 1991), but the nucleoplasmin from fully grown oocytes (before maturation) is less phosphorylated and is less potent in decondensation of sperm nuclei than nucleoplasmin from eggs (Ohsumi et al. 1995). Therefore, we do not necessarily think more potency for decondensation of sperm nuclei would yield better efficiency of transgenesis. Using oocyte extract rather than egg extract, however, certainly contributes to recovery of healthier embryos and promotes long-term viability of transgenic embryos as described in Hirsch et al. (2002), and because of this, we think more transgenic animals will be obtained in the end. The successful use of oocyte extract for transgenesis of *X. laevis* has also been published (e.g., Ochi et al. 2012; Ochi et al. 2017). In addition, several steps in this protocol are further modified to simplify and generalize the procedure for more convenience (in the procedure) and consistency (in the results). The method described here, in addition to the original "REMI" procedures (Kroll and Amaya 1996; Offield et al. 2000) or a simplified version (e.g., Sparrow et al. 2000), all of which are transgenesis methods mediated by sperm nuclear transplantation, remain the most efficient current strategies for obtaining nonmosaic transgenic animals in the F_0 generation in *Xenopus*. In addition, because they typically generate multiple copy insertions, they are very useful for generating bright fluorescent protein-reporter transgenic lines. It is also now possible to generate F_0 transient BAC transgenic animals (Fish et al. 2012), and using the sperm-mediated protocol described here, it has become possible to obtain a germline transmitted BAC-transgenic animal (M. Fish, T. Nakayama, R. Grainger, unpubl.).

RECIPES

1× XB

1× XB salts, dilute from Extract buffer (XB) salts (20×) <R>
50 mM sucrose (e.g., Sigma-Aldrich SX1075; dilute from a 1.5 M stock solution that has been filter-sterilized and stored at −20°C)
10 mM HEPES (e.g., Sigma-Aldrich H3375; dilute from a 1 M, pH 7.7 stock solution that has been filter-sterilized and stored at −20°C)

Freshly prepare 1 L and adjust pH to 7.7 with KOH.

1× NPB with 0.3% BSA

2.5 mL 2× NPB <R>
2.35 mL sterile Nanopure H_2O
150 µL 10% bovine serum albumin (BSA; e.g., Sigma-Aldrich A3294; 10% w/v with pH adjusted to 7.6 with KOH, aliquoted, and stored at −20°C)

Freshly prepare 5 mL and store on ice.

2× NPB

Amount of reagent	Final concentration
10 mL sucrose (e.g., Sigma-Aldrich SX1075) from a 1.5 M solution that has been filter-sterilized and stored at −20°C	500 mM
900 μL HEPES (e.g., Sigma-Aldrich H3375) from a 1 M, pH 7.7 solution that has been filter-sterilized and stored at −20°C	30 mM
120 μL EDTA (e.g., J.T. Baker 8991-01) from a 0.5 M, pH 8.0 solution	2 mM
300 μL spermidine trihydrochloride (e.g., Sigma-Aldrich S2501) from a 100 mM solution that has been aliquoted and stored at −20°C	1 mM
120 μL spermine tetrahydrochloride (e.g., Sigma-Aldrich S1141) from a 100 mM solution that has been aliquoted and stored at −20°C	0.4 mM
60 μL DL-dithiothreitol (DTT; e.g., Sigma-Aldrich D9779) from a 1 M solution that has been aliquoted and stored at −20°C	2 mM
18.5 mL sterile Nanopure H_2O	

Freshly prepare 30 mL. Adjust pH to 7.7 with KOH. Store on ice.

25% Percoll in 1× NPB

2 mL 2× NPB <R>
1 mL Percoll (e.g., Sigma-Aldrich P4937; stored at 4°C)
1 mL sterile Nanopure H_2O

Freshly prepare 4 mL and store on ice.

50% Percoll in 1× NPB

2 mL 2× NPB <R>
2 mL Percoll (e.g., Sigma-Aldrich P4937; stored at 4°C)

Freshly prepare 4 mL and store on ice.

Collagenase Solution for Isolation of Oocytes

Add 3 mg collagenase Λ (e.g., Roche 11088793001) per mL of oocyte storage solution <R>.

Freshly prepare 1 L for two female frogs.

Extract Buffer (XB) Salts (20×)

Reagent	Quantity	Concentration
KCl	149.12 g	2 M
$CaCl_2$	0.294 g	2 mM
$MgCl_2$ (2 M; e.g., Sigma-Aldrich 68475)	10 mL	20 mM

Combine with MilliQ H_2O to a volume of 1 L. Autoclave and store at 4°C.

Injection Medium

6% Ficoll PM400 (e.g., Sigma-Aldrich GE17-0300-10)
0.1× MBS (pH 7.5) (diluted from Modified Barth's Saline for *Xenopus* (MBS) <R>)

Filter-sterilize and freeze at −20°C for long-term storage or store for up to a week at 4°C. For use, warm up to 22°C or room temperature. The pH may become acidic after use for a few days or after long-term storage. We add a small amount of phenol red (e.g., 4–5 µL per 100 mL) so that we can monitor pH roughly by visually checking color change. Readjust pH as needed before use.

Modified Barth's Saline for Xenopus (MBS)

Make a 10× MBS stock:
> 800 mM NaCl
> 10 mM KCl
> 10 mM $MgSO_4 \cdot 7H_2O$
> 50 mM HEPES free acid
> 25 mM $NaHCO_3$

Prepare a 10× stock that consists of the reagents listed above, adjusting the pH to 7.8 with NaOH and sterilizing by autoclaving. Prepare a separate 0.1 M stock of $CaCl_2$, and sterilize by autoclaving. Store the stocks indefinitely at room temperature.

To make 1× MBS, dilute the 10× MBS stock with type 1 ultrapure water (ASTM International 2018), and add $CaCl_2$ stock to a final concentration of 0.7 mM.

Modified CSF-XB (MCSF-XB)

1× XB salts, dilute from Extract buffer (XB) salts (20×) <R>
1 mM $MgCl_2$
50 mM sucrose (e.g., Sigma-Aldrich SX1075; dilute from a 1.5 M solution that has been filter-sterilized and stored at −20°C)
10 mM HEPES (e.g., Sigma-Aldrich H3375; dilute from a 1 M, pH 7.7 solution that has been filter-sterilized and stored at −20°C)
5 mM ethylene glycol tetraacetic acid (EGTA; e.g., Sigma-Aldrich 324626; dilute from a 0.1 M solution)

> *Freshly prepare 250 mL (or more as needed). Adjust pH to 7.7 using KOH.*

Just before use, add 1:1000 of each of:
Chymostatin (e.g., Roche 11004638001; 10 mg/mL in ethanol that has been aliquoted and stored at −20°C)
Leupeptin (e.g., Sigma-Aldrich 11017128001; 10 mg/mL in DMSO that has been stored at −20°C)
Pepstatin A (e.g., Sigma-Aldrich 516485; 10 mg/mL in DMSO that has been aliquoted and stored at −20°C)

Modified Xenopus Oocyte Culture Medium (MXOCM)

50% w/v L-15 (Leibovitz; e.g., Sigma-Aldrich L4386)
15 mM HEPES (e.g., Sigma-Aldrich H3375; dilute from a 1 M, pH 7.7 stock solution that has been filter-sterilized and stored at −20°C)
50 µg/mL Gentamicin (e.g., Corning 61-098-RF, dilute from a 100 mg/mL stock solution that has been stored at 4°C)

Freshly prepare and adjust the pH to 7.7. Prepare 500 mL or more as needed for two female frogs.

Cite this protocol as *Cold Spring Harb Protoc*; doi:10.1101/pdb.prot107003

Oocyte Storage Solution

87 mM NaCl

2.5 mM KCl

1 mM $MgCl_2$

1 mM Na_2HPO_4

5 mM HEPES (e.g., Sigma-Aldrich H3375; dilute from a 1 M, pH 7.7 solution that has been filter-sterilized and stored at −20°C)

1% (v/v) penicillin–streptomycin (Pen/Strep; e.g., Gibco 15140148; dilute 1:100 from a stock solution)

Freshly prepare and adjust the pH to 7.8. Prepare 1 L or more as needed for two female frogs.

Protease Inhibitor Cocktail

4 mL 2× NPB <R>

2.4 mL 10% bovine serum albumin (BSA; e.g., Sigma-Aldrich A3294; 10% w/v with pH adjusted to 7.6 with KOH, aliquoted, and stored at −20°C)

8 µL leupeptin (e.g., Sigma-Aldrich 11017128001; 10 mg/mL in DMSO that has been stored at −20°C)

8 µL 0.3 M phenylmethylsulfonyl fluoride (PMSF; e.g., Sigma-Aldrich 78830; that has been aliquoted and stored at −20°C)

1.6 mL sterile Nanopure H_2O

Freshly prepare 8 mL and store on ice. Thaw leupeptin and PMSF stocks in a 37°C water bath just before use (vortex in case crystals are formed). Use gloves when handling protease inhibitors.

Sperm Dilution Buffer for Xenopus (SDB)

75 mM KCl

250 mM sucrose (e.g., Sigma-Aldrich SX1075; dilute from a 1.5 M solution that has been filter-sterilized and stored at −20°C)

2 mM HEPES (e.g., Sigma-Aldrich H3375; dilute from a 1 M, pH 7.7 solution that has been filter-sterilized and stored at −20°C)

0.5 mM spermidine trihydrochloride (e.g., Sigma-Aldrich S2501; dilute from a 100 mM solution that has been aliquoted and stored at −20°C)

0.2 mM spermine tetrahydrochloride (e.g., Sigma-Aldrich S1141; dilute from a 100 mM solution that has been aliquoted and stored at −20°C)

Adjust pH to 7.4–7.6 with NaOH. Prepare a large volume (e.g., 50–100 mL), filter sterilize, aliquot (e.g., 0.5–1 mL), and store for a year at −20°C.

Sperm Storage Buffer for Xenopus (SSB)

500 µL 2× NPB <R>

300 µL glycerol (e.g., Fisher G33500)

170 µL sterile Nanopure H_2O

30 µL 10% bovine serum albumin (BSA; e.g., Sigma-Aldrich A3294; 10% w/v with pH adjusted to 7.6 with KOH, aliquoted, and stored at −20°C)

Freshly prepare 1 mL and store on ice until use. Alternatively, for a larger volume, aliquot and store for a year at −20°C.

ACKNOWLEDGMENTS

The protocol described here is a consequence of multiple modifications for improvement of the procedure and efficiency through many years in our laboratory, for which we acknowledge people who were involved: Martin F. Offield, Nicolas Hirsch, Mandy Rollins, William B. McConnell, Margaret B. Fish, and Sumanth Manohar. We also thank Hajime Ogino and Haruki Ochi for their unpublished information, and Keita Ohsumi for discussions about nucleoplasmin. This work was funded by National Institutes of Health (NIH) grants RR13221, EY017400, EY018000, and EY022954 and research awards from the Sharon Stewart Aniridia Trust and Vision for Tomorrow to R.M.G.

REFERENCES

Dumont JN. 1972. Oogenesis in *Xenopus laevis* (Daudin). I. Stages of oocyte development in laboratory maintained animals. *J Morphol* **136**: 153–179. doi:10.1002/jmor.1051360203

Fish MB, Nakayama T, Grainger RM. 2012. Simple, fast, tissue-specific bacterial artificial chromosome transgenesis in *Xenopus*. *Genesis* **50**: 307–315. doi:10.1002/dvg.20819

Hirsch N, Zimmerman LB, Gray J, Chae J, Curran KL, Fisher M, Ogino H, Grainger RM. 2002. *Xenopus tropicalis* transgenic lines and their use in the study of embryonic induction. *Dev Dyn* **225**: 522–535. doi:10.1002/dvdy.10188

Kroll KL, Amaya E. 1996. Transgenic *Xenopus* embryos from sperm nuclear transplantations reveal FGF signaling requirements during gastrulation. *Development* **122**: 3173–3183. doi:10.1242/dev.122.10.3173

Lane M, Khokha MK. 2022. Obtaining *Xenopus tropicalis* embryos by in vitro fertilization. *Cold Spring Harb Protoc* doi:10.1101/pdb.prot106351

Lane M, Mis EK, Khokha MK. 2022a. Obtaining *Xenopus tropicalis* eggs. *Cold Spring Harb Protoc* doi:10.1101/pdb.prot106344

Lane M, Slocum M, Khokha MK. 2022b. Raising and maintaining *Xenopus tropicalis* from tadpole to adult. *Cold Spring Harb Protoc* doi:10.1101/pdb.prot106369

L'hostis-Guidet A, Recher G, Guillet B, Al-Mohammad A, Coumailleau P, Tiaho F, Boujard D, Madigou T. 2009. Generation of stable *Xenopus laevis* transgenic lines expressing a transgene controlled by weak promoters. *Transgenic Res* **18**: 815–827. doi:10.1007/s11248-009-9273-0

Nakayama T, Grainger RM. 2022. Best practices for *Xenopus tropicalis* husbandry. *Cold Spring Harb Protoc* doi:10.1101/pdb.top106252

Nakayama T, Cox A, Howell M, Grainger RM. 2022. Gynogenetic production of embryos in *Xenopus tropicalis* using a cold shock procedure: rapid screening method for gene editing phenotypes. *Cold Spring Harb Protoc* doi:10.1101/pdb.prot107648

Newman K, Aguero T, King ML. 2018. Isolation of *Xenopus* oocytes. *Cold Spring Harb Protoc* doi:10.1101/pdb.prot095851

Ochi H, Tamai T, Nagano H, Kawaguchi A, Sudou N, Ogino H. 2012. Evolution of a tissue-specific silencer underlies divergence in the expression of *pax2* and *pax8* paralogues. *Nat Commun* **3**: 848. doi:10.1038/ncomms1851

Ochi H, Kawaguchi A, Tanouchi M, Suzuki N, Kumada T, Iwata Y, Ogino H. 2017. Co-accumulation of cis-regulatory and coding mutations during the pseudogenization of the *Xenopus laevis* homoeologs six6.L and six6.S. *Dev Biol* **427**: 84–92. doi:10.1016/j.ydbio.2017.05.004

Offield MF, Hirsch N, Grainger RM. 2000. The development of *Xenopus tropicalis* transgenic lines and their use in studying lens developmental timing in living embryos. *Development* **127**: 1789–1797. doi:10.1242/dev.127.9.1789

Ogino H, McConnell WB, Grainger RM. 2006. High-throughput transgenesis in *Xenopus* using I-SceI meganuclease. *Nat Protoc* **1**: 1703–1710. doi:10.1038/nprot.2006.208

Ohsumi K, Katagiri C. 1991. Characterization of the ooplasmic factor inducing decondensation of and protamine removal from toad sperm nuclei: involvement of nucleoplasmin. *Dev Biol* **148**: 295–305. doi:10.1016/0012-1606(91)90338-4

Ohsumi K, Shimada A, Okumura E, Kishimoto T, Katagiri C. 1995. Dependence of removal of sperm-specific proteins from *Xenopus* sperm nuclei on the phosphorylation state of nucleoplasmin. *Dev Growth Differ* **37**: 329–336. https://doi.org/10.1046/j.1440-169X.1995.t01-1-00011.x

Philpott A, Leno GH, Laskey RA. 1991. Sperm decondensation in *Xenopus* egg cytoplasm is mediated by nucleoplasmin. *Cell* **65**: 569–578. doi:10.1016/0092-8674(91)90089-H

Sparrow DB, Latinkic B, Mohun TJ. 2000. A simplified method of generating transgenic *Xenopus*. *Nucl Acids Res* **28**: e12. doi:10.1093/nar/28.4.e12

I-SceI-Mediated Transgenesis in *Xenopus*

Anna Noble, Anita Abu-Daya, and Matthew J. Guille[1]

European Xenopus *Resource, School of Biological Sciences, University of Portsmouth, Portsmouth PO1 2UP, United Kingdom*

Transgenic frogs can be very efficiently generated using I-SceI meganuclease, a nuclease with an 18-bp recognition site. The desired transgene must be flanked by I-SceI sites, in either a plasmid or a polymerase chain reaction (PCR) product. After a short in vitro digestion with the meganuclease, the complete reaction is injected into fertilized eggs, where the enzyme mediates genomic integration by an unknown mechanism. Posttransgenesis development is typically normal, and up to 70% of the embryos integrate the transgene.

MATERIALS

It is essential that you consult the appropriate Material Safety Data Sheets and your institution's Environmental Health and Safety Office for proper handling of equipment and hazardous materials used in this protocol.

RECIPES: Please see the end of this protocol for recipes indicated by <R>. Additional recipes can be found online at http://cshprotocols.cshlp.org/site/recipes.

Reagents

DNeasy Blood & Tissue Kit (QIAGEN 69504)

Ficoll PM400 (Sigma-Aldrich F4375; 3% in 0.1× MBS or 0.1× MMR)
Prepare fresh or the day before and store at 4°C or 18°C.

High-purity transgenic construct with one or two I-SceI sites (midi-prep quality)
See details about possible constructs in the Discussion.

Human chorionic gonadotropin (HCG; Intervet)

I-SceI meganuclease enzyme (NEB R0694)
Store I-SceI meganuclease in 2-μL aliquots at −80°C.

L-cysteine (Sigma-Aldrich 168149; 2%)
Prepare fresh; adjust pH to 7.8–8 using NaOH.

Modified Barth's saline (MBS) (1×, pH 7.8) <R> or Marc's modified Ringer's (MMR) (1×)
Dilute buffer of choice to 0.1×. Supplement 0.1× solution with 1 mL/L penicillin–streptomycin (Sigma-Aldrich P4458) solution. MMR can be used instead of MBS at the same dilution. For Xenopus tropicalis *we use a more dilute concentration of MMR than we use for* Xenopus laevis *(0.05× MMR vs. 0.1× MMR); other labs use 0.01× MMR. We use penicillin–streptomycin for both* X. laevis *and* X. tropicalis, *but you can choose to use 100 U/mL*

[1]Correspondence: matthew.guille@port.ac.uk

gentamicin for X. tropicalis embryos as it is more stable at 25°C; make sure the pH is 7.6 after adding the antibiotic. Prepare MBS fresh or the day before and store at 4°C or 18°C. Prepare MMR up to a week in advance and store at 4°C.

Molecular-biology-grade water (Sigma-Aldrich W4502)

QIAGEN Plasmid Midi Kit (12143 or equivalent)

X. laevis or *X. tropicalis* eggs (obtain as in Sec. 2, Protocol 2: Obtaining *Xenopus laevis* Embryos [Shaidani et al. 2021] or Sec. 2, Protocol 4: Obtaining *Xenopus tropicalis* Eggs [Lane et al. 2022])

X. laevis or *X. tropicalis* testis (obtain as in Sec. 2, Protocol 4: Obtaining *Xenopus laevis* Embryos [Shaidani et al. 2021] or Sec. 2, Protocol 5: Obtaining *Xenopus tropicalis* Embryos by In Vitro Fertilization [Lane and Khokha 2022])

Equipment

Agarose-coated Petri dishes (optional; used to house *X. tropicalis* from steps between dejellying and hatching)

Prepare by making 1% agarose in 0.05× MMR. Pour as little as needed to coat the base of the Petri dish. Store at 4°C for up to several weeks.

Borosilicate glass capillaries (e.g., Harvard Apparatus GC100-10 No. 30-0016; 1.0 mm O.D. × 0.58 mm I.D.)

Disposable micropipettes (1-µL; Camlab DMP 001)

These are used for calibrating injection droplet size.

Dissection microscope with cold light source (e.g., Nikon SMZ800 with SCHOTT KL1600 LED)

This microscope is used for microinjections and to check on embryo health.

Falcon tube (50-mL)

Flaming/Brown micropipette puller (e.g., Sutter Instruments Co. P-87)

Fluorescence dissection microscope (e.g., Zeiss)

Heating block (37°C)

Injection dishes <R>

Microcentrifuge (e.g., Eppendorf 5415D)

Microcentrifuge tubes (1.5-mL)

Microinjector (e.g., Digitimer PLI-100) with micromanipulator

Pasteur pipette, cut and flame-polished to an appropriate diameter to move embryos

Petri dishes (60-mm and 90-mm)

Temperature-controlled incubator (e.g., LMS cooled incubator, range 14°C–25°C)

Tweezers (Drummond no. 5)

METHOD

This protocol is based on Ogino et al. (2006). Use all solutions at 18°C.

Prepare Embryos and the I-SceI Reaction

1. In the afternoon of the day before the experiment inject two *X. laevis* or two *X. tropicalis* females with HCG according to Sec. 2, Protocol 2: Obtaining *Xenopus laevis* Embryos (Shaidani et al. 2021) or Sec. 2, Protocol 4: Obtaining *Xenopus tropicalis* Eggs (Lane et al. 2022).

   Egg laying will start in the morning of the next day.

2. Euthanize the male frog(s) painlessly according to local regulations and remove testes as in Sec. 2, Protocol 2: Obtaining *Xenopus laevis* Embryos (Shaidani et al. 2021) or Sec. 2, Protocol 5: Obtaining *Xenopus tropicalis* Embryos by In vitro Fertilization (Lane and Khokha 2022). Carefully

remove any blood vessels and excess fat tissue using tweezers. Roll the testes on paper towels to remove all traces of blood.

> *X. laevis testes can be stored in 1× MBS for up to 2 wk at 4°C. If using X. tropicalis, dissect testes when females start laying eggs or use frozen sperm (Pearl et al 2017).*

3. Pull injection needles from glass capillary tubes using a micropipette puller according to Guille (1999).

> *This may be done before the day of injection.*

4. Calibrate the injection needle for a rate of ~5 nL/sec according to Guille (1999). Use fine tweezers to open the end of the needle.

> *The injection time can be changed to alter the size of the droplet, but do not shorten it to <500 msec per injection as this means the needle is too big and will cause excessive damage to embryos.*

5. Prepare the I-SceI reaction as follows.

    i. Mix 600 ng of plasmid with 2 µL CutSmart Buffer and make up the volume to 18 µL with molecular-biology-grade water in a microcentrifuge tube.

    ii. Keep the reaction on ice without adding enzyme at this point.

    > *For constructs <10 kb in size use 300 ng plasmid. Make sure that the plasmid DNA is of high purity (prepared using QIAGEN Plasmid Midi Kit or equivalent).*

6. Macerate one-quarter of the testis with tweezers in 1 mL of 1× MBS by pulling the testis quarter apart again and again until there is very little solid material left—simply a pale white suspension of testicular material.

7. Obtain eggs by gentle massage of the female's abdomen over a 90-mm Petri dish and inspect their quality. If the eggs look healthy, fertilize them with the sperm from Step 6. Transfer sperm onto the eggs using a pipette, and then gently shake the dish to ensure good mixing and to get the eggs into a monolayer. Use pipette tips to gently tease eggs into a monolayer if necessary; this increases the fertilization rate, particularly when using frozen sperm. Incubate eggs with sperm for 5–10 min at room temperature.

> *All solutions used are at 18°C. For X. laevis, solutions can be used at 14°C, which will slow down development and allow more injections to be done before first cell division.*

8. Add 2 µL I-SceI enzyme to the reaction mix from Step 5 and mix very gently by pipetting. Centrifuge the tube for 1 min at full speed in a microcentrifuge at room temperature to pellet any particles that could block the needle. Incubate for 25 min at 37°C.

> *While the reaction is incubating, proceed to Step 9.*

> *To minimize mosaicism, it is important that embryos be injected as soon as they are ready, well before the first division. Correct timing of the I-SceI reaction is essential so that it finishes when the eggs are ready to inject. If X. tropicalis are being used, the reaction will need to be started before obtaining eggs.*

9. Flood the Petri dish containing embryos with 0.1× MBS. You should observe animal pole pigment contraction and second polar body extrusion within 10 min of fertilization. This is followed by animal pole rotation ~5 min later. The percentage of rotated embryos gives an idea of the rate of fertilization. When the embryos turn so the animal pole is uppermost, dejelly them as follows.
    i. Replace the 0.1× MBS with the cysteine solution and swirl gently.

    ii. After embryos become loose, transfer them to a 50-mL Falcon tube and keep mixing gently. After ~5 min, the jelly coat should be removed, and the embryos will become tightly packed in the conical part of the tube. Wash embryos five times with 0.1× MBS.

    > *Animal pole rotation after fertilization is not as reliable in X. tropicalis as it is in X. laevis, therefore we start dejellying X. tropicalis embryos 20 min after the flooding.*

10. Using a Pasteur pipette, sort healthy embryos into injection dishes containing 3% Ficoll in 0.1× MBS. Each dish may contain up to 150 embryos and should contain enough 3% Ficoll in 0.1× MBS to ensure the embryos do not dry out but not enough that the embryos can move.

If desired, solutions can be precooled to 14°C–16°C to slow down the development of X. laevis. Healthy embryos have uniform pigmentation on the animal pole and no mottling. They are round, support their own weight, and are not flattened or flaccid.

A good way to identify an appropriate level of liquid is to look at the surface of the buffer, which should show bumps where the embryos are, with surface tension holding the liquid in place.

Microinjections

Detailed information on microinjections can be found in Guille (1999).

11. At this point the I-SceI reaction should be complete. Load the needle with the reaction mixture.

12. As soon as possible, start injections between the animal pole and the sperm entry site, which appears as a white dot, sometimes with a dark outline. In essence, you are aiming for the forming nucleus, which lies quite close to the surface.

    Every 20 injections check the needle for blockage or breakage by injecting into the air and making sure the droplet is approximately the correct size.

13. Aim to inject about 50 embryos per construct. If possible, inject 50 embryos with 5 nL and 50 embryos with 2.5 nL.

    If using X. tropicalis inject 2 nL per embryo.

    If some embryos appear soft and resistant to penetration by the needle they should not be injected; they are probably unfertilized and trying to inject them risks breaking the needle.

14. Using a cut Pasteur pipette, transfer 50–100 two-cell embryos into a 90-mm Petri dish containing fresh 3% Ficoll in 0.1× MBS.

    Agarose-coated dishes can be used for X. tropicalis; however, in our hands keeping embryos in 3% Ficoll until the 16-cell stage is sufficient to stop them from sticking to the Petri dish and each other.

15. Around stage 6, use a cut Pasteur pipette to transfer healthy-looking 32-cell embryos (Nieuwkoop and Faber 1994) into Petri dishes containing fresh 0.1× MBS and culture them at 18°C (25°C for *X. tropicalis*). Unhealthy embryos can be identified by abnormal cleavage patterns and mottled pigmentation. Place 50–100 embryos in each 90-mm dish.

    Incubation temperature can be lowered in the first few hours to 16°C for X. laevis and 22°C for X. tropicalis to slow down development and decrease mosaicism.

16. Change the 0.1× MBS daily until tadpoles are stage 43 (Nieuwkoop and Faber 1994), are free swimming, can be observed rhythmically moving their mouths, and are ready to move into tanks.

17. Depending on the transgene, screen for fluorescent protein expression at the appropriate stage using the fluorescence microscope or genotype tail-tip DNA isolated using the DNeasy tissue extraction kit (QIAGEN) (Geach et al. 2012).

DISCUSSION

Transgenic animals are a crucial tool for manipulating gene expression and elucidating biological processes. There are more than 150 transgenic *Xenopus* lines available from the *Xenopus* stock centers; these are described in Horb et al. (2019).

I-SceI-mediated transgenesis is efficient and technically undemanding; it relies on simple microinjection, a technique that is very common in *Xenopus* laboratories. The 18-bp I-SceI target site is predicted to occur once in a 7×10^{10}-bp sequence. It does not appear to be found in either the *X. tropicalis* (version 10.0) or *X. laevis* (version 1.0 outbred, version 9.2 J strain) genomes. It is likely that transgenes integrate randomly in the genome, rather than at a unique site.

The challenge often comes in identifying transgenic carriers. This is especially true when the transgene does not lead to the expression of a fluorescent protein. We find it very helpful to include an unambiguous fluorescent marker for transgenesis and use a modular Gateway system

Cite this protocol as *Cold Spring Harb Protoc*; doi:10.1101/pdb.prot107011

specifically suited to *Xenopus* called pTransgenesis (Love et al. 2011), which is available from the European *Xenopus* Resource Centre (EXRC). The end plasmid contains I-SceI sites, as well as a fluorescent marker such as *cryga*-GFP, which results in lens fluorescence. A variety of promoter sequences and effector genes are also included with this system. To enable tissue-specific and inducible control of open-reading frame (ORF) expression, two destination vectors for transgenesis—pDXTP and pDXTR—have been generated (Sterner et al. 2019). Both are available from EXRC alongside ORFeome clones.

RECIPES

Injection Dishes

MATERIALS

Reagents

Ethanol, 95%
Methylene chloride
  Chloroform can be used as an alternative.

Equipment

Cork, silicone rubber
  The cork should have a diameter close to that of the Petri dish.
Nitex mesh, 800 µm
Petri dishes, plastic

METHOD

Injection dishes for microinjection of Xenopus *oocytes are modified Petri dishes to which a nylon mesh has been fixed to hold oocytes or embryos in place. The mesh also helps to keep track of which specimens have been injected because it holds the specimens in regular rows. Ensure that the grid fits snugly to the edge of the dish and lies flat on the surface, so that the specimens do not become trapped underneath.*

1. Cut the 800-µm Nitex mesh into circles so that it fits snugly into the Petri dishes.

2. Fix the mesh in place by melting the bottom of the plastic Petri dish with five drops of methylene chloride or chloroform.
   Use the silicone rubber cork to hold the grid flat while the plastic sets.

3. After complete evaporation of the solvent, rinse the dishes thoroughly with 95% ethanol and then with water before use.

Marc's Modified Ringer's (MMR) (1×)

0.1 M NaCl
2 mM KCl
1 mM MgSO$_4$
2 mM CaCl$_2$
5 mM HEPES (pH 7.8)
0.1 mM EDTA

Sterilize by autoclaving, and store at room temperature.
Common alternative formulations of MMR omit EDTA and are adjusted to pH 7.4.

Modified Barth's Saline (MBS) (1×, pH 7.8)

Reagent	Quantity (for 1 L)	Final concentration (1×)
NaCl	5.143 g	88 m$_M$
KCl	0.075 g	1 m$_M$
$MgSO_4$	0.120 g	1 m$_M$
HEPES	1.192 g	5 m$_M$
$NaHCO_3$	0.210 g	2.5 m$_M$
$CaCl_2$, dihydrate	0.103 g	0.7 m$_M$
H_2O	to 1 L	

Adjust the pH to 7.8 with 10 M NaOH and sterilize by autoclaving. Store at room temperature indefinitely.

REFERENCES

Geach TJ, Stemple DL, Zimmerman LB. 2012. Genetic analysis of *Xenopus tropicalis*. *Methods Mol Biol* **917:** 69–110. doi:10.1007/978-1-61779-992-1_5

Guille M. 1999. Microinjection into *Xenopus* oocytes and embryos. *Methods Mol Biol* **127:** 111–123. doi:10.1385/1-59259-678-9:111

Horb M, Wlizla M, Abu-Daya A, McNamara S, Gajdasik D, Igawa, T, Suzuki A, Ogino H, Noble A, Centre de Ressource Biologique Xenope team in France, et al. 2019. *Xenopus* resources: transgenic, inbred and mutant animals, training opportunities, and web-based support. *Front Physiol* **10:** 387. doi:10.3389/fphys.2019.00387

Lane MA, Khokha MK. 2022. Obtaining *Xenopus tropicalis* embryos by in vitro fertilization. *Cold Spring Harb Protoc* doi: 10.1101/pdb.prot106351

Lane MA, Mis EK, Khokha MK. 2022. Obtaining *Xenopus tropicalis* eggs. *Cold Spring Harb Protoc* doi: 10.1101/pdb.prot106344

Love NR, Thuret R, Chen Y, Ishibashi S, Sabherwal N, Paredes R, Alves-Silva J, Dorey K, Noble AM, Guille MJ, et al. 2011. pTransgenesis: a cross-species, modular transgenesis resource. *Development* **138:** 5451–5458. doi:10.1242/dev.066498

Nieuwkoop PD, Faber J. 1994. *Normal table of* Xenopus laevis *(Daudin)*. Garland Publishing, New York.

Ogino H, McConnell WB, Grainger RM. 2006. Highly efficient transgenesis in *Xenopus tropicalis* using I-SceI meganuclease. *Mech Dev* **123:** 103–113. doi:10.1016/j.mod.2005.11.006

Pearl EJ, Morrow S, Noble AM, Lerebours A, Horb ME, Guille M. 2017. An optimized method for cryogenic storage of *Xenopus* sperm to maximise the effectiveness of research using genetically altered frogs. *Theriogenology* **92:** 149–155. doi:10.1016/j.theriogenology.2017.01.007

Shaidani N-I, McNamara S, Wlizla M, Horb ME. 2021. Obtaining *Xenopus laevis* embryos. *Cold Spring Harb Protoc* doi: 10.1101/pdb.prot106211

Sterner ZR, Rankin SA, Wlizla M, Choi JA, Luedeke DM, Zorn AM, Buchholz DR. 2019. Novel vectors for functional interrogation of *Xenopus* ORFeome coding sequences. *Genesis* **57:** e23329. doi:10.1002/dvg.23329

Cite this protocol as *Cold Spring Harb Protoc*; doi:10.1101/pdb.prot107011

Cryopreservation of *Xenopus* Sperm and In Vitro Fertilization Using Frozen Sperm Samples

Anna Noble, Anita Abu-Daya, and Matthew J. Guille[1]

European Xenopus Research Center (EXRC), School of Biological Sciences, University of Portsmouth, Portsmouth PO1 2UP, United Kingdom

The cryopreservation of *Xenopus* sperm allows for a significant reduction of the number of animals that must be kept, more efficient archiving of genetically altered (GA) lines, and easy exchange of lines with other laboratories, leading to improvements in animal welfare and cost efficiency. In this protocol, sperm from *Xenopus laevis* or *Xenopus tropicalis* are frozen using straightforward techniques and standard laboratory equipment. Testes are macerated in Leibovitz's L-15 medium, mixed with a simple cryoprotectant made from egg yolk and sucrose, and frozen slowly overnight in a polystyrene box at −80°C. Unlike mouse sperm, *Xenopus* sperm can be stored at −80°C rather than in liquid nitrogen, further reducing costs. The frozen sperm are then used for in vitro fertilization.

MATERIALS

It is essential that you consult the appropriate Material Safety Data Sheets and your institution's Environmental Health and Safety Office for proper handling of equipment and hazardous materials used in this protocol.

RECIPES: Please see the end of this protocol for recipes indicated by <R>. Additional recipes can be found online at http://cshprotocols.cshlp.org/site/recipes.

Reagents

Cryoprotectant solution <R>, ice-cold
L-15 supplemented with 10% FBS and 2 mM L-glutamine <R>, freshly prepared and on ice
Modified Barth's saline (MBS) (1×, pH 7.8) <R> or Marc's modified Ringer's (MMR) (1×)<R>, diluted to 0.1× and supplemented with 1 mL/L penicillin-streptomycin solution (Sigma-Aldrich P4458)
Unsupplemented L-15 (Sigma-Aldrich L5520) or MBS (1×, pH 7.8) (optional; see Step 1)
Xenopus laevis or *Xenopus tropicalis* adult males and females

Equipment

Aluminum foil
Cardboard CryoBox (e.g., 136 × 136 × 32 mm [Fisher]) (optional; see Step 6)
Dissecting microscope

---

[1]Correspondence: matthew.guille@port.ac.uk

Filter tips (200-µL, large-orifice [e.g., Fisher Scientific 11947744] or with ~2 mm of the ends cut off)
Using large-orifice tips or cutting the ends off of standard filter tips is essential to eliminate mechanical shearing damage to the sperm.

Incubator set at 18°C for *X. laevis* or at 23°C for *X. tropicalis*
Liquid nitrogen (optional; see Step 7)
Microcentrifuge tubes (1.5-mL [optional; see Step 4] and 2-mL)
Paper towels
Pellet pestles (optional; see Step 4)
Petri dishes (60-mm and 90-mm)
Pipette tips (1-mL, with the ends cut off)
Polystyrene box (external measurements: ~23 × 23 × 27 cm; thickness of the wall: 4 cm)
Thin-walled PCR tubes (0.5-mL; e.g., ThermoFisher Scientific AB-0350)
Tubes (30-mL) (optional; see Step 2)
Tweezers (forceps)
Water bath at 37°C

METHOD

This protocol was adapted from the method initially developed by Sargent and Mohun (2005), with further amendments by the EXRC and NXR (Pearl et al. 2017).

Use the same volume of reagents for all procedures regardless of species (X. laevis or X. tropicalis).

Freezing Sperm

1. Euthanize the male frog(s) painlessly according to local regulations and remove testes as in Sec. 2, Protocol 2: Obtaining *Xenopus laevis* Embryos (Shaidani et al. 2021) or Sec. 2, Protocol 5: Obtaining *Xenopus tropicalis* Embryos by In Vitro Fertilization (Lane and Khokha 2021). Use tweezers to carefully remove any blood vessels and excess fat tissue. Roll the testes on paper towels to remove all traces of blood.

   X. laevis testes can be stored in 1×MBS or in L-15 (unsupplemented) for up to 1 wk at 4°C, but X. tropicalis sperm should be frozen immediately because success of freezing diminishes rapidly over time.

2. For each pair of testes, place 1 mL of ice-cold cryoprotectant solution in a 2-mL tube and leave on ice.

   Testes from 2–6 males can be pooled per batch. If pooling, scale up Steps 2–5 as necessary and use 30-mL tubes.

3. Label 16 500-µL thin-walled PCR tubes, and place them in a rack ready for rapid filling.

   Eight samples of 125 µL per testis will be frozen (16 samples per male).

4. Transfer testes to a 6-mm Petri dish placed at a 45° angle with 1 mL of L-15 medium supplemented with FBS and L-glutamine, and carefully macerate the testes with tweezers.

   X. tropicalis testes can be transferred to a microcentrifuge tube containing 0.5 mL of L-15 supplemented with FBS and L-glutamine and macerated gently using a plastic disposable pestle made for use in Eppendorf tubes. Once the testes are homogenized, add an additional 0.5 mL of ice-cold, supplemented L15 medium. Macerating X. laevis testes with pellet pestle is not recommended. We saw very low sperm recovery compared to maceration with the tweezers.

5. Transfer the sperm suspension using a 1-mL tip with a cut end to the 2-mL tube containing 1 mL of cryoprotectant on ice. Gently mix the contents by inversion.

6. With a cut-off pipette tip or a large-orifice 200-µL filter tip, transfer 125-µL aliquots of the sperm into the prepared tubes. Immediately place the samples in a polystyrene box and cover with aluminum foil (instead of the normal lid). Place the box in a −80°C freezer.

Cite this protocol as *Cold Spring Harb Protoc*; doi:10.1101/pdb.prot107029

In total, 96 tubes (sperm from six frogs) can be put in a polystyrene box in any one round of freezing. For larger numbers of aliquots, racking the sperm samples in cardboard CryoBoxes within the polystyrene box makes it easier to transfer them to long-term storage and has the advantage of ensuring the sperm suspension is at the bottom of all the tubes.

Aim to get the samples into the −80°C freezer as fast as possible. We can get 96 aliquots of sperm into the freezer in <10 min from the maceration. Label all the tubes and organize them in racks at room temperature in advance.

7. The next day, move the samples from the polystyrene box into long-term storage at −80°C.

   Samples can be also be successfully frozen in cryotubes and stored in liquid nitrogen.

In Vitro Fertilization Using Frozen Sperm

Keep tubes with frozen sperm at −80°C or on dry ice until the eggs are ready to be fertilized.

8. Extrude approximately 500 *X. laevis* or *X. tropicalis* eggs into a dry 90-mm Petri dish as described in Sec. 2, Protocol 2: Obtaining *Xenopus laevis* Embryos (Shaidani et al. 2021) or in Sec. 2, Protocol 4: Obtaining *Xenopus tropicalis* Eggs (Lane et al. 2021). Check egg quality under a dissecting microscope as there is no point wasting frozen sperm on poor-quality eggs.

   If you are using valuable GA sperm, we recommend testing the first batch of eggs produced on the fertilization day by fertilizing with fresh sperm so that precious frozen samples are not wasted.

9. Thaw frozen sperm in a water bath for 30 sec at 37°C by moving it in a figure 8 until most of the frozen sperm has thawed.

10. Immediately add 250 µL of 0.1× MBS or 0.1× MMR to the 125-µL sperm sample, and mix gently by pipetting up and down five to10 times using large-orifice or cut-off pipette tips.

11. As quickly as possible, using the same tip, apply the sperm suspension to the eggs and mix gently. Spread the eggs into a monolayer with a pair of fresh, uncut, 200-µL pipette tips. Make sure all eggs are covered by the sperm and place the lid on the Petri dish. Incubate the dish at 18°C for *X. laevis* or at 23°C for *X. tropicalis*.

12. After 5–10 min, flood the eggs with 0.1× MBS or 0.1× MMR. Incubate the dish at 18°C for *X. laevis* or at 23°C for *X. tropicalis*.

13. Within 2 h, estimate the fertilization rate by comparing divided to undivided eggs.

    See Troubleshooting.

14. Culture the embryos as described in Sec. 2, Protocol 2: Obtaining *Xenopus laevis* Embryos (Shaidani et al. 2021) or Protocol: **Obtaining *Xenopus tropicalis* Embryos by In vitro Fertilization** (Lane and Khokha 2021), changing the medium on a daily basis.

TROUBLESHOOTING

Problem (Step 13): The fertilization rate is low.

Solution: Successful fertilization following sperm cryopreservation occurs more robustly in *X. tropicalis* than in *X. laevis* (Pearl et al. 2017). In the latter species, results are more inconsistent between individual frogs, possibly because of variations in sexual maturity and differing health of animals. However, it is possible to successfully freeze sperm from both species. *X. laevis* males can be primed with 35–50 IU of PMSG 3–5 d before harvesting testes to improve sperm maturity, but this does not appear to improve *X. tropicalis* sperm. See Sec. 2, Protocol 2: Obtaining *Xenopus laevis* Embryos (Shaidani et al. 2021) for information on priming frogs.

In addition, consider the following.

- If you are recovering a GA line, consider pooling eggs from different females into one Petri dish in Step 8 to increase the success of fertilization, which may be compromised by a lack of biological compatibility.

- Be consistent in Step 10. Too many or too few pipetting actions seems to be an area in which inconsistent success arises.

- Before adding the sperm suspension in Step 11, make sure that the eggs are not in any liquid. If during egg extrusion some liquid drops to the Petri dish, blot it off with a paper towel.

- In our hands, collecting eggs in high-salt solution did not result in fertilization using frozen sperm.

RECIPES

Cryoprotectant Solution

1. Disperse one chicken egg's yolk (~15-mL; preferably organic) in an equal volume of molecular biology grade water to make Solution A. Mix by pipetting up and down.
2. Combine the following reagents and bring to 1000 mL to make Solution B, which consists of 0.4 M sucrose, 10 mM NaHCO$_3$, 2 mM pentoxyfilline.

Sucrose	136.92 g
NaHCO$_3$	0.84 g
Pentoxyfilline (Sigma-Aldrich P1784)	0.56 g

3. Dilute Solution A to 20% v/v in Solution B by adding 10 mL of Solution A to 40 mL of Solution B. Divide the mixture into two 30-mL centrifuge tubes and centrifuge at 10,000g in a Beckman Allegra 25R Refrigerated Centrifuge with a TA-14-50 rotor or equivalent for 20 min at 10°C. Discard pellets. For immediate use, store on ice. Store the rest as 1- or 5-mL aliquots for up to 1 yr at −20°C.

L-15 Supplemented with 10% FBS and 2 mM L-glutamine

Reagent	Volume
L-15 medium (Sigma-Aldrich L5520)	9 mL
Fetal bovine serum (FBS) (Sigma-Aldrich F9665)	1 mL
L-glutamine 200 mM (Sigma-Aldrich G7513)	10 µL

Store 100-µL aliquots of 200 mM L-glutamine and 1-mL aliquots of FBS at −20°C and use them to supplement L15 media on the day of sperm cryopreservation. (L-glutamine solutions degrade relatively rapidly, although the powder form is very stable.)

Marc's Modified Ringer's (MMR) (1×)

0.1 M NaCl
2 mM KCl
1 mM MgSO$_4$
2 mM CaCl$_2$
5 mM HEPES (pH 7.8)
0.1 mM EDTA

Sterilize by autoclaving, and store at room temperature.
Common alternative formulations of MMR omit EDTA and are adjusted to pH 7.4.

Cite this protocol as *Cold Spring Harb Protoc*; doi:10.1101/pdb.prot107029

Modified Barth's Saline (MBS) (1×, pH 7.8)

Reagent	Quantity (for 1 L)	Final concentration (1×)
NaCl	5.143 g	88 mM
KCl	0.075 g	1 mM
$MgSO_4$	0.120 g	1 mM
HEPES	1.192 g	5 mM
$NaHCO_3$	0.210 g	2.5 mM
$CaCl_2$, dihydrate	0.103 g	0.7 mM
H_2O	to 1 L	

Adjust the pH to 7.8 with 10 M NaOH and sterilize by autoclaving. Store at room temperature indefinitely.

REFERENCES

Lane MA, Khokha MK. 2021. Obtaining *Xenopus tropicalis* embryos by in vitro fertilization. *Cold Spring Harb Protoc* doi: 10.1101/pdb.prot106351

Lane MA, Mis EK, Khokha MK. 2021. Obtaining *Xenopus tropicalis* eggs. *Cold Spring Harb Protoc* doi: 10.1101/pdb.prot106344

Pearl EJ, Morrow S, Noble AM, Lerebours A, Horb ME, Guille M. 2017. An optimized method for cryogenic storage of *Xenopus* sperm to maximise the effectiveness of research using genetically altered frogs. *Theriogenology* 92: 149–155. doi:10.1016/j.theriogenology.2017.01.007

Sargent MG, Mohun TJ. 2005. Cryopreservation of sperm of *Xenopus laevis* and *Xenopus tropicalis*. *Genesis* 41: 41–46. doi:10.1002/gene.20092

Shaidani N-I, McNamara S, Wlizla M, Horb ME. 2021. Obtaining *Xenopus laevis* embryos. *Cold Spring Harb Protoc* doi: 10.1101/pdb.prot106211

Protocol 8

Homology-Directed Repair by CRISPR–Cas9 Mutagenesis in *Xenopus* Using Long Single-Stranded Donor DNA Templates via Simple Microinjection of Embryos

Takuya Nakayama,[1] Robert M. Grainger,[1,3] and Sang-Wook Cha[2,3]

[1]*Department of Biology, University of Virginia, Charlottesville, Virginia 22904, USA;* [2]*School of Natural Sciences, University of Central Missouri, Warrensburg, Missouri 64093, USA*

We describe a step-by-step procedure to perform homology-directed repair (HDR)-mediated precise gene editing in *Xenopus* embryos using long single-stranded DNA (lssDNA) as a donor template for HDR in conjunction with the CRISPR–Cas9 system. A key advantage of this method is that it relies on simple microinjection of fertilized *Xenopus* eggs, resulting in high yield of healthy founder embryos. These embryos are screened for those animals carrying the precisely mutated locus to then generate homozygous and/or heterozygous mutant lines in the F_1 generation. Therefore, we can avoid the more challenging "oocyte host transfer" technique, which is particularly difficult for *Xenopus tropicalis*, that is required for an alternate HDR approach. Several key points of this protocol are (1) to use efficiently active single-guide RNAs for targeting, (2) to use properly designed lssDNAs, and (3) to use 5′-end phosphorothioate-modification to obtain higher-efficiency HDR.

MATERIALS

It is essential that you consult the appropriate Material Safety Data Sheets and your institution's Environmental Health and Safety Office for proper handling of equipment and hazardous materials used in this protocol.

RECIPES: Please see the end of this protocol for recipes indicated by <R>. Additional recipes can be found online at http://cshprotocols.cshlp.org/site/recipes.

Reagents

All reagents (including kits) used should be molecular biology grade.

Agarose gel (0.7%–1%; prepared using standard techniques) and appropriate buffer
Cas9 protein with nuclear localization signal (e.g., CP01, PNA Bio)
Column-based DNA purification kit (e.g., DNA Clean & Concentrator-5, Zymo Research)
DNA cloning plasmid (e.g., Zero Blunt TOPO PCR Cloning Kit, Invitrogen; optional; see Step 7)
DNA oligonucleotides (e.g., Eurofins, Integrated DNA Technologies, see Methods for details)
DNase/RNase-free H_2O (e.g., Fisher BioReagents BP2484100)
dsDNA fragment; custom-made (e.g., gBlocks Gene Fragments, Integrated DNA Technologies; optional; see Step 7)

---

[3]Correspondence: rmg9p@virginia.edu; cha@ucmo.edu

Copyright © 2023 Cold Spring Harbor Laboratory Press; all rights reserved
Cite this protocol as *Cold Spring Harb Protoc*; doi:10.1101/pdb.prot107599

Embryo culture medium (any medium used for *Xenopus* embryo research can be used, e.g., Modified Barth's Saline for *Xenopus* (MBS) <R> or Marc's Modified Ringer's (MMR) (10×)

We use 0.1× MBS and readjust its pH to 7.4–7.5.

Embryo lysis buffer <R>

Guide-it Long ssDNA Strandase Kit (Takara Bio)

Alternatively, lambda exonuclease (e.g., NEB M0262) can be used to selectively digest the 5'-phosphorylated strand of dsDNA followed by purification of the undigested long single-stranded DNA (lssDNA), bypassing the need for this kit (see Step 11).

Another alternative is to have an lssDNA custom-made (e.g., Megamer Single-Stranded DNA Fragments, Integrated DNA Technologies; see Step 11).

High-fidelity polymerase chain reaction (PCR) polymerase (e.g., Platinum SuperFi PCR Master Mix, ThermoFisher 12358010)

L-Cysteine (e.g., Sigma-Aldrich C5360)

This is used to de-jelly embryos; nonhydrochloride form of L-cysteine is essential for Xenopus tropicalis.

Single-guide RNA (sgRNA) (designed in Step 1)

X. tropicalis and *Xenopus laevis* embryos, obtained by in vitro fertilization and dejellied using standard protocols for I-SceI transgenesis (Ogino et al. 2006)

Equipment

Equipment used routinely in laboratories performing Xenopus embryo research that is the equivalent of what is listed here can be used, unless specifically designated.

Agarose gel electrophoresis apparatus

Fluorescence stereomicroscope

Incubator appropriate for culturing embryos

Microinjector and accessories (e.g., injection needles)

PCR machine

PCR tubes (0.2-mL, DNase/RNAse-free)

Spectrophotometer (e.g., NanoDrop, ThermoFisher)

Stereomicroscope with fiber optic light source (for microinjection)

Supplies for embryo manipulation (e.g., forceps, Pasteur pipettes) and culture (e.g., Petri dishes)

Water bath

Xenbase (http://www.xenbase.org/)

METHOD

Some required procedures can be found in protocols published by ourselves or others. In such cases, the procedures are briefly described by citing such protocols to avoid redundancy. In particular, this protocol is a detailed explanation of critical points from our published protocol (Nakayama et al. 2020).

Identifying sgRNA Target Site(s) in the Region of Interest for Targeting by HDR and Synthesizing sgRNA(s)

1. Identify a CRISPR target(s) near the region to be mutated via homology-directed repair (HDR).

    i. By using lssDNA as a donor for HDR, the available distance between the target region for HDR and the CRISPR target site becomes less constrained compared to when using a short donor oligonucleotide. The ideal distance between the target region for HDR and the CRISPR target site is within 50–100 bp, and closer is better. See Figure 1A and below for details.

    ii. For designing sgRNA target sites, including web-based tools for searching target sequences, and then synthesizing sgRNAs, refer to published methods (e.g., Nakayama et al. 2013, 2014), see Protocol 1: CRISPR–Cas9 Mutagenesis in *Xenopus tropicalis* for Phenotypic

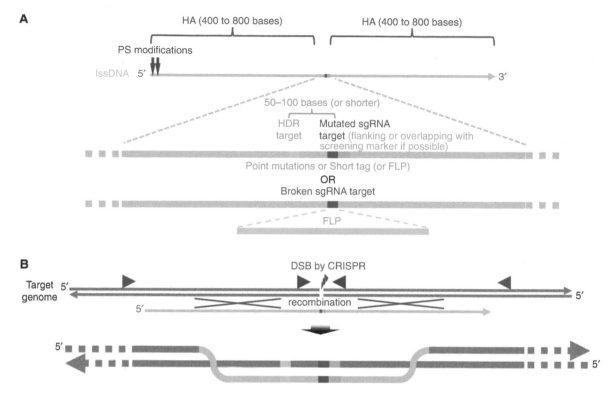

FIGURE 1. Schematic representation of experimental design. (*A*) Design of long single-stranded DNA (lssDNA) (yellow arrow) and enlarged region (yellow bars). Purple arrows indicate phosphorothioate (PS) modifications. (Green box) Homology-directed repair (HDR) target region to introduce point mutations, small tags, or fluorescent proteins (FLPs). (Red box) Mutated or broken single-guide RNA (sgRNA) target site. (Blue box) Screening marker, the position of which could overlap with the red box, or could be either the 5′ side or 3′ side (as shown here) of the red box. (*B*) Schematic representation of a recombination event. Gray arrows indicate genomic double-stranded DNA (dsDNA). Purple arrowheads indicate genomic polymerase chain reaction (PCR) primers for evaluation of sgRNA efficiency. Red arrowheads indicate genomic PCR primers for evaluation of the HDR event.

Analyses in the F_0 Generation and Beyond [Blitz and Nakayama 2022], see Protocol 2: Generating Nonmosaic Mutants in *Xenopus* Using CRISPR–Cas in Oocytes [Cha 2022], see Protocol 3: Tissue-Targeted CRISPR–Cas9-Mediated Genome Editing of Multiple Homeologs in F_0-Generation *Xenopus laevis* Embryos [Corkins et al. 2022].

iii. In short, using the tools and references above, design and synthesize a resultant target sequence that starts with G and has a total length of 17 to 20 bases (i.e., $G(N)_{16-19}$) that is followed by a PAM (NGG) sequence in the genome, in which N can be any nucleotide. If available, use already established (published) sgRNAs for a given target gene of interest as long as the targeting position is suitable for the desired HDR site. Use either the sense or antisense strand for the CRISPR target.

Evaluation of Efficiency of sgRNAs

2. Refer to Ogino et al. (2006) for obtaining embryos and microinjection of *X. tropicalis* and *X. laevis* eggs, including instructions for culture after injection. Complete injections by 40 min following fertilization for *X. tropicalis* and 45 min following fertilization for *X. laevis* to achieve the highest efficiency of CRISPR digestion.

i. Empirically determine the minimum and sufficient amounts of sgRNA and Cas9 protein (sgRNA/Cas9) to achieve the best efficiency of cleavage (see Step 5). We recommend starting

Cite this protocol as *Cold Spring Harb Protoc*; doi:10.1101/pdb.prot107599

with 200–400 pg sgRNA and 500–1000 pg Cas9 protein and testing other amounts as needed.

Up to 1 ng of sgRNA and up to 2 ng of Cas9 protein per embryo can be used for both X. tropicalis *and* X. laevis.

ii. Preincubate a mixture of sgRNA and Cas9 protein for 5 min at 37°C in a water bath and then immediately microinject into embryos.

iii. Inject sgRNA/Cas9 in a volume of up to 4 nL (for *X. tropicalis*) and 10 nL (for *X. laevis*) per embryo. Higher concentrations of Cas9 protein make a more viscous solution, which may clog the injection needle, therefore only use the minimally sufficient amount of Cas9 protein. In general, it is better to inject a bigger volume with diluted materials rather than inject a smaller volume of concentrated materials.

3. When embryos have reached the desired stage, lyse CRISPR-treated embryos (and untreated embryos as a control) with embryo lysis buffer as described in Protocol 1: CRISPR–Cas9 Mutagenesis in *Xenopus tropicalis* for Phenotypic Analyses in the F_0 Generation and Beyond (Blitz and Nakayama 2022) or Protocol 2: Generating Nonmosaic Mutants in *Xenopus* Using CRISPR–Cas in Oocytes (Cha 2022). Use 50 or 100 µL of lysis buffer for *X. tropicalis* or *X. laevis*, respectively, if using before neurula-stage embryos, or 100 and 200 µL for later-stage embryos.

4. In advance, design a primer set for genomic PCR of CRISPR-treated embryos to amplify a 300- to 700-bp region containing the target site in the middle of this region (purple arrowheads in Fig. 1B) as described (e.g., Nakayama et al. 2014, 2020), see Protocol 1: CRISPR–Cas9 Mutagenesis in *Xenopus tropicalis* for Phenotypic Analyses in the F_0 Generation and Beyond [Blitz and Nakayama 2022]. Perform genomic PCR on the lysate from Step 3 without further purification as described in Protocol 1: CRISPR–Cas9 Mutagenesis in *Xenopus tropicalis* for Phenotypic Analyses in the F_0 Generation and Beyond [Blitz and Nakayama 2022] or in Protocol 2: Generating Nonmosaic Mutants in *Xenopus* Using CRISPR–Cas in Oocytes [Cha 2022].

5. Perform assays to evaluate the cleavage efficiency by CRISPR.

i. Perform the DSP (direct sequencing of PCR amplicons) assay (Nakayama et al. 2013, 2014) followed by data analysis with a web-based tool—either tracking of indels by decomposition (TIDE) (Brinkman et al. 2014) or inference of CRISPR edits (ICE) (Conant et al. 2022)—to provide an estimate of cleavage efficiency (for details, see Protocol 1: CRISPR–Cas9 Mutagenesis in *Xenopus tropicalis* for Phenotypic Analyses in the F_0 Generation and Beyond [Blitz and Nakayama 2022]).

ii. Alternatively, clone the PCR products and sequence multiple clones (ideally more than 10 clones) to estimate the cleavage efficiency (e.g., Nakayama et al. 2013).

iii. As another alternative, use any other published method that can be used to estimate cleavage efficiency of a given sgRNA. Use only sgRNAs (and specific combinations of the amounts of sgRNA and Cas9 protein) that yield high-efficiency mutagenesis (90%–100%) for subsequent experiments.

Here, whether or not the resultant sgRNAs cause in-frame mutations resulting from indels caused by nonhomologous end joining (NHEJ) is not a concern unlike for regular CRISPR targeting, which aims to cause loss-of-function mutations.

See Troubleshooting.

Designing the lssDNA Template and Synthesis of lssDNA

6. Design the lssDNA template. Figure 1A shows a conceptual design for an lssDNA. Refer to the genome database available in Xenbase (http://www.xenbase.org/) for the genomic sequence.

i. Based on our previous work (Nakayama et al. 2020), include these two key features for an effective design of lssDNA: (1) 5′ phosphorothioate (PS) modifications between the first three nucleotides (see Step 9), which enhance HDR efficiency; and (2) 400- to 800-base homology arms (HAs) (also see Step 15 for the rationale regarding HA length) flanking the HDR target of interest. The HDR target could be, for example, a point mutation(s) or a small tag or fluorescent protein (FLP) construct to insert. The distance between the sgRNA target site and the HDR target can be up to 100 bases, but shorter is better, depending on availability of a usable sgRNA target site.

ii. Consider the following points especially when targeting a protein coding region to insert small tags or point mutations. First, the corresponding region for the sgRNA target site should have silent mutations so that the recombined region cannot be recut by Cas9. A silent mutation involving one or both Gs of the PAM (NGG) sequence is sufficient. If this cannot be accomplished, it is important to introduce silent mutations in the "seed region" so that sgRNA recognition would fail (e.g., Jiang et al. 2013) for the recombined region. Second, if possible, it is optimal to make a new restriction site (for screening purposes, see Step 15) and, simultaneously, silent mutations in the sgRNA target site (or in a region as close to the sgRNA site as possible) (e.g., see Nakayama et al. 2020). If a restriction site exists in the sgRNA target site or nearby, the same goal could be accomplished by destroying it and generating a silent mutation. If either of these options are possible, screening can be done in part by restriction digestion, thus becoming much easier. Third, when inserting a FLP, if the native protein-coding sequence flanking the FLP does not need to be maintained, it may be sufficient to design a strategy by which the FLP insertion disrupts the sgRNA target site, rather than making silent mutations in the sgRNA site or making/breaking a restriction site. For an FLP, the primary screening can be done by detecting fluorescence and does not necessarily require a molecular marker; however, a careful design should be considered, for example, such that the FLP becomes functionally active only when an in-frame insertion (as a fused protein) occurs (e.g., using an open reading frame [ORF] without a translational start ATG).

iii. Alternatively, if the sgRNA target site is in an untranslated region, simply destroy the target sequence to avoid recutting, taking care not to destroy any regulatory sequence (e.g., for splicing, transcription/translation enhancement, or polyadenylation, etc.).

7. Once the basic design of the lssDNA is determined, make a template construct for PCR amplification of what will then be a dsDNA "substrate," from which an lssDNA will be made (see Step 8 for more details).

   i. The construction of a template can be done by a conventional multiple-step PCR procedure (e.g., see strategy in Supplemental Fig. S1 of Nakayama et al. 2020), in which the exact conditions will vary depending on the target site for HDR.

   ii. Alternatively, it is possible to purchase a custom-made dsDNA fragment (e.g., gBlocks Gene Fragments, Integrated DNA Technologies). However, we have found that some sequences are challenging to synthesize by this approach and that the multistep PCR strategy was preferable.

   It is best to clone the template construct, which should have 50- to 100-bp extra 5' and 3' flanking genomic sequence (used as discussed below, see Step 9), into any convenient plasmid so that the accuracy of the template can be confirmed by sequencing before proceeding.

8. Prepare the lssDNA from a PCR-amplified dsDNA fragment (substrate) derived from the Step 7 template as described in the following steps. We use the Guide-it Long ssDNA Strandase Kit to prepare lssDNA as described in the kit manual.

   Here, an important consideration is the direction of the lssDNA, especially for a small insertion or point mutations. We have found that only a partial region of lssDNA is involved in actual recombination. Specifically, we have observed occasions in which the 3' part of designed mutations was not introduced and only the 5' half of the lssDNA was involved in recombination (Nakayama et al. 2020). This suggests that, given the direction of DNA synthesis (5' to 3') during recombination, it is better to locate the HDR target on the 5' side of the sgRNA target in the adjacent region (where mutations for avoiding recutting and for screening are made). Doing so increases the possibility of obtaining an HDR mutation because a selection marker–

positive case is automatically HDR mutation–positive (see Fig. 1B). This approach would not necessarily apply for the insertion of a large fragment (e.g., FLP) and both sense-stranded and antisense-stranded lssDNAs would work equally in that case (S.-W. Cha, unpubl.).

9. Design appropriate PCR primers in the "50- to 100-bp extra 5′ and 3′ flanking" regions mentioned above (Step 7).

   It is important to understand how single-stranded DNA is made from the PCR-amplified dsDNA substrate. Namely, using the kit mentioned above, the 5′-phosphorylated primer-derived strand of the dsDNA is specifically digested, and the resultant undigested strand (which is derived from the nonphosphorylated primer) is recovered. Therefore, depending on the direction of the desired lssDNA, based on the above criteria, PCR primers need to be designed properly.

   i. Design the 5′ primer as follows: 5′-N*N*NN…N-3′, in which N is any nucleotide (~20 bases in length) and * indicates PS bond modification, which primes the strand that will become the lssDNA.

   ii. Design the 3′ primer as follows: 5′-P-NNNN…N-3′, in which N is any nucleotide (~20 bases in length) and P indicates 5′-phosphorylation modification, which primes the strand that will be digested.

10. Perform PCR as described in the kit manual. PCR conditions will vary depending on the target length. Before purification, confirm that a single band was produced by analysis on an agarose gel (0.7%–1%, depending on the length of product). Purify the single PCR product using a column-based DNA purification kit. Elute with up to 40 µL DNase/RNase-free H_2O.

    We aim to ideally obtain 15 µg dsDNA per up to 40 µL elution for next step, to maximize the yield of lssDNA. As needed, repeat PCR to collect ~15 µg (or a minimum of ~9 µg) dsDNA product in total.

    See Troubleshooting.

11. Prepare lssDNA as described in the kit manual (saving an aliquot of dsDNA for a quality check). Purify the product using a column-based DNA purification kit. Elute with up to 20 µL DNase/RNase-free H_2O and measure its concentration. Our preparations usually result in a concentration of 60–200 ng/µL. The recovery yield varies with preparations for different dsDNAs but also even with the same dsDNA. Check ~100 ng of the resultant lssDNA together with ~100 ng dsDNA substrate side-by-side on a 0.7%–1% agarose gel. The dsDNA and lssDNA should migrate at different mobilities, and lssDNA should appear less stained. The resultant lssDNA can be used without further purification.

    Although the kit manual indicates checking the product before purification, we prefer to check after purification.

    It is possible to purchase a custom-made lssDNA (e.g., Megamer Single-Stranded DNA Fragments, Integrated DNA Technologies), thus bypassing Steps 7–11, although PS modifications may or may not be available depending on the particular sequence.

    Alternatively, to reduce costs, instead of using the above-described kit, lambda exonuclease (e.g., NEB M0262) can be used to selectively digest the 5′-phosphorylated strand of dsDNA followed by purification of the undigested lssDNA (K. Nakajima, pers. comm.). However, precise control of the digestion reaction would be required because nonphosphorylated substrate could also be degraded, although likely at a greatly reduced rate.

Evaluation of Toxicity of lssDNAs Alone and Together with sgRNA/Cas9 Complex

12. Empirically determine the maximum amount of lssDNA alone that can be injected into embryos without toxicity as done in Step 2 for sgRNA/Cas9. Begin with a range of 30–120 pg lssDNA for *X. tropicalis* and 100–200 pg for *X. laevis*. Inject 50 to 100 embryos per dose and score completely healthy embryo numbers after stage 45 (according Nieuwkoop and Faber 1967) compared to uninjected embryos. Use the highest dose at which healthy embryos can be obtained at a comparable percentage to uninjected siblings for the procedures below.

    Here, healthy embryos scored after stage 45 (around this stage embryos are about to start feeding) are expected to survive at a high rate, but we have experienced situations in which, when younger embryos are scored, even if they look healthy, they develop abnormally as they age.

13. Once the proper dose of the lssDNA is determined, combine the lssDNA with the sgRNA component (sgRNA/Cas9) at the dose determined in Step 2 to start injections. For example, make a mixture so that the maximum injection volume (e.g., 4 nL for *X. tropicalis*) contains each component at the highest dose. Then inject 4, 3, 2, and 1 nL per embryo (for *X. laevis*, use a range between 1 and 10 nL per embryo) to examine toxicity. Examine 50 to 100 embryos per dose. Here, the goal is to find a condition at which 50% of injected embryos remain healthy after stage 45.

> *If the mixture combined at the predetermined nontoxic highest doses now becomes toxic, more testing will need to be performed to find the proper dose of the combined mixture. If all tested concentrations are still toxic, simply make a less concentrated mixture and repeat these experiments. Once an appropriate combination is determined, additional adjustment may be needed to obtain higher HDR efficiency (if evaluation of HDR efficiency can be done easily, e.g., by detection of FLP, see below) as needed.*

Injection of an HDR Component and Evaluation of HDR Events

14. Once the final amount of each component is determined, prepare an appropriate mixture such that 4 nL per embryo is injected for *X. tropicalis* and 5–10 nL per embryo is injected for *X. laevis*. Use the maximum volume to avoid needle clogging (see Step 2.iii to clarify why this may be a problem). Inject as many embryos as needed to ensure production of healthy, mature frogs.

15. Evaluate the HDR event with F_0 animals or F_1 animals (or later generations) depending on the particular strategy and purpose.

    i. For FLP tagging, simply monitor fluorescence for rapid screening. For small tagging with an epitope for which an antibody is available, immunodetection (i.e., western blotting or immunohistochemistry; see, e.g., Aslan et al. 2017) can be used for rapid screening.

    ii. In these cases, and for other small tag insertions or point mutations, carry out a molecular level examination to know if the intended HDR occurs precisely. As described in Step 3, prepare embryo or tissue lysate (using a whole embryo, clipped tail tissue from a tadpole, or web-punched tissue of a froglet/frog), and perform genomic PCR as follows.

    iii. Use PCR primers (red arrowheads in Fig. 1B, "outside primers") lying outside of the theoretical recombined region if the entire lssDNA is involved in the recombination. Use a high-fidelity PCR polymerase for the PCR. PCR conditions may vary depending on the experiment and should be optimized empirically.

    > *When performing molecular screening of F_0 animals, the primers used in Step 4 (purple arrowheads, Fig. 1B) or Step 9 (i.e., "internal primers") should not be used without careful consideration. Even if combining with one of the "outside primers" (red arrowheads, Fig. 1B) and one of "internal primers," which should not cause a positive band in theory for negative cases, such a combination of primers actually could cause an artifactual PCR product originating from injected lssDNA and genomic DNA, resulting in a false-positive band (T. Nakayama and R.M. Grainger, unpubl.). A similar artifact, although using a dsDNA donor, has also been reported (e.g., Won and Dawid 2017). This is not a concern for F_1 animals or later generations and may still work even for F_0 animals when done very carefully by using several critical controls including testing embryos injected with only lssDNA as a negative control, or by nested PCR using the purified PCR product amplified first with "outside primers."*

    > *Because of the required use of "outside primers," a relatively large genomic region must be successfully amplified by PCR. For this reason, HAs that are too long should not be used even if they may increase HDR efficiency slightly. Our recommended length for HAs (see Step 6, Fig. 1A) is a reasonable size to perform both efficient HDR (Nakayama et al. 2020) as well as successful genomic PCR for screening.*

    iv. For a large insertion (e.g., FLP insertion), a larger band should be seen via agarose electrophoresis because of an insertion in addition to a smaller band for a nonrecombined locus (from DNA that is either nondigested or repaired by NHEJ). For small tags or point mutations, subject the PCR product to a restriction digestion test when a restriction marker is available (for either a newly made restriction site or destroyed restriction site).

    > *When restriction digestion for screening is not available, one possibility is to develop a PCR primer specific to HDR mutations for genomic PCR. However, careful examination with critical controls to avoid detection of false-positives is essential.*

Cite this protocol as *Cold Spring Harb Protoc*; doi:10.1101/pdb.prot107599

v. Precise confirmation of an HDR event can be done by conventional cloning/sequencing of the PCR product.

TROUBLESHOOTING

Problem (Step 5): Efficient digestion is not achieved even when injecting 1 ng sgRNA.
Solution: A different target site(s) should be designed.

Problem (Step 10): Multiple bands are detected from the PCR reaction used to make the dsDNA substrate for lssDNA.
Solution: Reduce the number of PCR cycles or, as needed, redesign new primer(s) in the "50- to 100-bp extra 5′ and 3′ flanking" regions.

DISCUSSION

We have used lssDNA as an HDR donor template, which was first successfully used for HDR in mice (Miura et al. 2015), to show proof-of-principle experiments illustrating HDR-mediated gene editing in both *X. laevis* and *X. tropicalis* with the CRISPR–Cas9 system via simple microinjection of fertilized *Xenopus* eggs (Nakayama et al. 2020). Based on knowledge obtained through this publication, we provide in this protocol a detailed explanation of critical points, which we could not necessarily describe in the original paper, but which are important to enhance the chances of success in the readers' own experiments. However, it is not necessarily possible to standardize a detailed protocol step by step because, depending on the target gene and specific aims for HDR modification, every case needs a customized protocol, and we can only explain generic examples here.

Our published model experiments (Nakayama et al. 2020) used visible markers (i.e., rescued pigmentation of albino embryos or HDR-mediated tagged GFP expression) for primary screening of HDR-positive embryos followed by further confirmation at the molecular level. If the experimental aim also uses any visible marker for primary screening (especially for FLP tagging), one can select only strongly HDR-positive F_0 embryos (i.e., selecting embryos showing the best numbers of positive cells) to raise for further screening in the next generation. For "invisible cases" such as small tags or point mutations in which only molecular screening is available, 100 or more animals will need to be raised and screened, given the percentage of molecularly detectable levels of HDR events for point mutations (see Nakayama et al. 2020), which is certainly feasible but potentially laborious.

One potential problem that is not described above and that should be considered concerns targeting a gene that would cause lethality when mutated. Because of the "functional redundancy" of many genes (due to L and S genes), *X. laevis* has an advantage that one can target only one homeologue at high efficiency for CRISPR to achieve high HDR efficiency without causing developmental defects, thus avoiding lethality. However, for a singleton gene of *X. laevis* or for the diploid species *X. tropicalis*, a lower efficiency of CRISPR-mediated digestion must be used to avoid lethal biallelic mutations, which automatically reduces the efficiency of HDR. To overcome this dilemma and/or the potential labor involved in screening "invisible cases" mentioned above, we may still need to further improve HDR efficiency to make this technique more readily feasible. Thus, this protocol is still a work in progress.

RECIPES

Embryo Lysis Buffer

50 mM Tris pH 8.8
1 mM EDTA
0.5% Tween 20
200 µg/mL Proteinase K (freshly added from a stock solution)

Autoclave and store at room temperature, before adding proteinase K, for up to a year.

Marc's Modified Ringer's (MMR) (10×)

20 mM $CaCl_2$
50 mM HEPES (pH 7.5)
20 mM KCl
10 mM $MgCl_2$
1 M NaCl

Adjust pH with NaOH to 7.5. Sterilize by autoclaving. Store at room temperature.

Modified Barth's Saline for Xenopus (MBS)

Make a 10× MBS stock:
  800 mM NaCl
  10 mM KCl
  10 mM $MgSO_4 \cdot 7H_2O$
  50 mM HEPES free acid
  25 mM $NaHCO_3$

Prepare a 10× stock that consists of the reagents listed above, adjusting the pH to 7.8 with NaOH and sterilizing by autoclaving. Prepare a separate 0.1 M stock of $CaCl_2$, and sterilize by autoclaving. Store the stocks indefinitely at room temperature.

To make 1× MBS, dilute the 10× MBS stock with type 1 ultrapure water (ASTM International 2018), and add $CaCl_2$ stock to a final concentration of 0.7 mM.

ACKNOWLEDGMENTS

The authors thank Ira Blitz and Rachel Miller for sharing their protocols before publication and Keisuke Nakajima for his unpublished experimental results. This work was funded by a National Institutes of Health (NIH) grant K01DK101618 awarded to S.-W.C., and by an NIH grant EY022954 and research awards from the Sharon Stewart Aniridia Trust and Vision for Tomorrow, and research support from the University of Virginia Department of Ophthalmology to R.M.G. The authors also acknowledge Xenbase (RRID: SCR_003280) for providing essential online resources for *Xenopus* research.

REFERENCES

Aslan Y, Tadjuidje E, Zorn AM, Cha S-W. 2017. High-efficiency non-mosaic CRISPR-mediated knock-in and indel mutation in F0 *Xenopus*. *Development* **144:** 2852–2858.

Blitz IL, Nakayama T. 2022. CRISPR–Cas9 mutagenesis in *Xenopus tropicalis* for phenotypic analyses in the F0 generation and beyond. *Cold Spring Harb Protoc* doi:10.1101/pdb.prot106971

Brinkman EK, Chen T, Amendola M, van Steensel B. 2014. Easy quantitative assessment of genome editing by sequence trace decomposition. *Nucl Acids Res* **42:** e168. doi:10.1093/nar/gku936

Cha S-W. 2022. Generating nonmosaic mutants in *Xenopus* using CRISPR–Cas in oocytes. *Cold Spring Harb Protoc* doi:10.1101/pdb.prot106989.

Conant D, Hsiau T, Rossi N, Oki J, Maures T, Waite K, Yang J, Joshi S, Kelso R, Holden K, et al. 2022. Inference of CRISPR edits from Sanger trace data. *CRISPR J* **5:** 123–130. doi:10.1089/crispr.2021.0113.

Corkins ME, DeLay BD, Miller RK. 2022. Tissue-targeted CRISPR–Cas9-mediated genome editing of multiple homeologs in F0-generation *Xenopus laevis* embryos. *Cold Spring Harb Protoc* doi:10.1101/pdb.prot107037

Jiang W, Bikard D, Cox D, Zhang F, Marraffini LA. 2013. RNA-guided editing of bacterial genomes using CRISPR–Cas systems. *Nat Biotechnol* **31**: 233–239. doi:10.1038/nbt.2508

Miura H, Gurumurthy CB, Sato T, Sato M, Ohtsuka M. 2015. CRISPR/Cas9-based generation of knockdown mice by intronic insertion of artificial microRNA using longer single-stranded DNA. *Sci Rep* **5**: 12799. doi:10.1038/srep12799

Nakayama T, Fish MB, Fisher M, Oomen-Hajagos J, Thomsen GH, Grainger RM. 2013. Simple and efficient CRISPR/Cas9-mediated targeted mutagenesis in *Xenopus tropicalis*. *Genesis* **51**: 835–843. doi:10.1002/dvg.22720

Nakayama T, Blitz IL, Fish MB, Odeleye AO, Manohar S, Cho KW, Grainger RM. 2014. Cas9-based genome editing in *Xenopus tropicalis*. *Methods Enzymol* **546**: 355–375. doi:10.1016/B978-0-12-801185-0.00017-9

Nakayama T, Grainger RM, Cha SW. 2020. Simple embryo injection of long single-stranded donor templates with the CRISPR/Cas9 system leads to homology-directed repair in *Xenopus tropicalis* and *Xenopus laevis*. *Genesis* **58**: e23366. doi:10.1002/dvg.23366

Nieuwkoop PD, Faber J. 1967. *Normal table of* Xenopus laevis *(Daudin)*. Garland, New York.

Ogino H, McConnell WB, Grainger RM. 2006. High-throughput transgenesis in *Xenopus* using I-SceI meganuclease. *Nat Protoc* **1**: 1703–1710. doi:10.1038/nprot.2006.208

Won M, Dawid IB. 2017. PCR artifact in testing for homologous recombination in genomic editing in zebrafish. *PLoS ONE* **12**: e0172802. doi:10.1371/journal.pone.0172802

Protocol 9

Gynogenetic Production of Embryos in *Xenopus tropicalis* Using a Cold Shock Procedure: Rapid Screening Method for Gene Editing Phenotypes

Takuya Nakayama,[1] Amanda Cox,[1] Mary Howell,[1] and Robert M. Grainger[1,2]

[1]*Department of Biology, University of Virginia, Charlottesville, Virginia 22904, USA*

Gynogenesis is a form of parthenogenesis in which eggs require sperm for fertilization but develop to adulthood without the contribution of paternal genome information, which happens naturally in some species. In *Xenopus*, gynogenetic diploid animals can be made experimentally. In mutagenesis strategies that only generate one allele of a recessive mutation, as might occur during gene editing, gynogenesis can be used to quickly reveal a recessive phenotype in eggs carrying a recessive mutation, thereby skipping one generation normally required to screen by conventional genetics. *Xenopus* oocytes do not complete meiosis until shortly after fertilization, and the second polar body is retained in fertilized eggs. Using ultraviolet (UV)-irradiated sperm, fertilization can be triggered without a genetic paternal contribution. Upon applying cold shock at the proper time to such embryos, ejection of the second polar body can be suppressed and both maternal sister chromatids are retained, leading to the development of gynogenetic diploid embryos. Because the genome of the resultant animals consists of recombined sister chromatids because of crossover events during meiosis, it is not completely homozygous throughout the whole genome. Nevertheless, the genome is homozygous at some loci proximal to the centromere that are unlikely to undergo recombination during meiosis and homozygous at reduced frequency if mutations are farther from the centromere, but still generally at a scorable level. Therefore, this technique is useful for rapid screening phenotypes of recessive mutations in such regions. We describe here a step-by-step protocol to achieve cold shock-mediated gynogenesis in *Xenopus tropicalis*.

MATERIALS

It is essential that you consult the appropriate Material Safety Data Sheets and your institution's Environmental Health and Safety Office for proper handling of equipment and hazardous materials used in this protocol.

RECIPES: Please see the end of this protocol for recipes indicated by <R>. Additional recipes can be found online at http://cshprotocols.cshlp.org/site/recipes.

Reagents

All reagents used should be molecular biology grade.

Agarose (e.g., BP160, Thermo Fisher Scientific) (0.8%–1% dissolved in 0.1× MBS or H_2O) (optional; see Step 16)

[2]Correspondence: rmg9p@virginia.edu

Bovine serum albumin (BSA; Fraction V, heat shock, e.g., Roche 03116956001)

Prepare a 10% stock solution, dissolved in H₂O, and store at −20°C until use.

Chorulon (hCG, Intervet Inc)

Add 5 mL of the supplied diluent to the vial (making 2 units/μL), aliquot in small volumes (e.g., 500 μL), and store at −20°C. Once thawed, store at 4°C and use ideally within 2 wk to a mo.

Cysteine (2%) dissolved in 0.1× MBS (pH 7.8–8.0)

Prepare fresh using the nonhydrochloride form of ʟ-cysteine (e.g., Sigma-AldrichC5360).

Gentamicin (e.g., Corning, 61-098-RF)

Prepare a stock solution (100 mg/mL) in sterile H₂O, store at 4°C.

MBS (0.1×) + 0.1% BSA

Prepare fresh. Adjust pH to 7.5.

MBS (1×) + 0.1% BSA

Prepare fresh. Adjust pH to 7.5.

MBS (1×) + 0.5% BSA (optional, see Step 9)

Prepare fresh. Adjust pH to 7.5.

MBS (0.1×) + 50 μg/mL gentamicin

Prepare fresh. Adjust pH to 7.5.

Modified Barth's Saline (MBS) for *Xenopus* <R>

Xenopus tropicalis female frogs

See Sec. 2, Protocol 4: Obtaining Xenopus tropicalis *Eggs (Lane et al. 2022) for information on identifying mature females.*

X. tropicalis male frogs

Choose males with dark, prominent nuptial pads.

Equipment

Conical tubes (15- and 50-mL)
Dissection tools
Lid of a micropipette tip box
Microcentrifuge tubes
Micropestle (to fit a microfuge tube, sterile)
Microwave (optional, see Step 16)
Pasteur pipettes

Smooth the end with a flame to avoid damaging embryos.

Petri dishes (60-mm, 100-mm)
Stereoscope
Transfer pipettes
UV cross-linker/light source with a device for monitoring strength of UV light
Water bath with chiller set to 1°C–3°C (optional; see Step 4)

METHOD

Refer to Figure 1 for overall flow of the experimental procedure.

Carry out all procedures at room temperature except where otherwise noted.

Priming Frogs

1. Prime females with hCG (10 units per frog) as described in Sec. 2, Protocol 5: Obtaining *Xenopus tropicalis* Embryos by In Vitro Fertilization (Lane and Kohkha 2022) 1–3 d before experimental day.

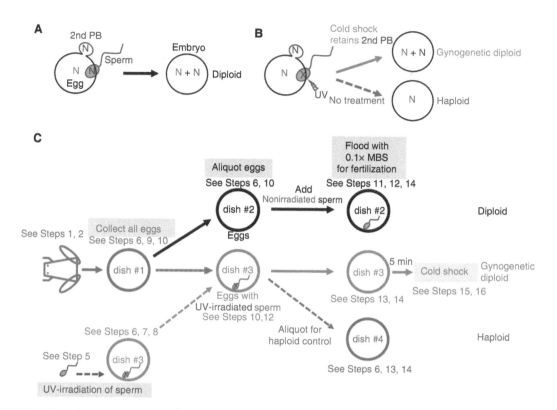

FIGURE 1. Experimental flow chart of gynogenesis. (*A*) Simplified illustration of natural fertilization. 2nd PB refers to second polar body during meiosis II. First polar body is not shown for simplicity. N means the haploid genome. (*B*) Simplified illustration of creation of gynogenetic diploid (with cold shock) and haploid (without cold shock) fertilized with using UV-irradiated sperm. (*C*) Summary of experimental procedure. See the text for corresponding steps and for each experimental dish.

The appropriate dose of hCG for priming (and boosting, see below) may vary depending on the particular sensitivity of frog populations. As needed, the dose can be increased up to 15 units or decreased to five units per frog. Four to six female frogs can reasonably be managed at once.

Experimental Day

2. Boost female frogs with 100 (up to 200 as needed) units of hCG per frog as in Step 1, expecting that ovulation will start 3–4 h later.

3. Prepare the following reagents while waiting:
   0.1× MBS (pH 7.5) + 0.1% BSA (1 L)
   1× MBS (pH 7.5) + 0.1% BSA (50 mL)
   0.1× MBS (pH 7.5) + 50 µg/mL gentamicin (1 L)

4. Prechill 2 × 50 mL of 0.1× MBS + 0.1% BSA at 1°C–3°C in conical tubes. Also prechill a micropipette tip box lid (or other vessel) by putting ice inside and setting it in an ice bucket until use. Petri dishes will be placed into this container floating on the slushy ice bath in Step 15.

   The prechilled 0.1× MBS + 0.1% BSA will be used to apply the cold shock to embryos. One hundred milliliters should be sufficient to treat embryos from four to five frogs.

   It is important to maintain the temperature between 1°C and 3°C to ensure gynogenesis occurs at high efficiency. This temperature range is critical and if not maintained during treatment, gynogenesis will be incomplete, resulting in embryo death or development of haploids. Maintaining the temperature is easiest when using a water bath with a chiller. However, if such a bath is not available, an appropriate water bath can be made using a laboratory ice bucket with ice. We have found that an adequate slushy water bath can be prepared by mixing H₂O and ice in an ice bucket placed in a 4°C refrigerator (or cold room) one hour

Cite this protocol as *Cold Spring Harb Protoc*; doi:10.1101/pdb.prot107648

before use (to stabilize the temperature). During this hour the temperature should be checked and the amount of H_2O and ice adjusted as needed.

5. Monitor females to confirm that at least one frog has started ovulation naturally. Once confirmed, prepare the sperm suspension in 1× MBS + 0.1% BSA as follows.

    i. Place 1.2 mL of 1× MBS + 0.1% BSA in a 15-mL conical tube for each male frog to be used.

    ii. Dissect the testes from one male frog as in Sec. 2, Protocol 5: Obtaining *Xenopus tropicalis* Embryos by In Vitro Fertilization (Lane and Kohkha 2022). Do not tear the testes to avoid losing sperm by leaking. Roll the testes around on a paper towel to remove excess blood and fat. Use a scalpel to cut off attached fat as needed.

       Importantly, keep the testes and sperm suspension at room temperature during the procedure. Unlike Xenopus laevis *sperm,* X. tropicalis *sperm is quickly killed by cold temperatures.*

    iii. Place testes from one male in 200 µL of 1× MBS + 0.1% BSA in a microfuge tube. Using a sterile micropestle, homogenize the testes in the tube, and transfer the sperm suspension to the 15-mL conical tube prepared above (Step 5.i) using a cut-off tip. Rinse the emptied microfuge tube and pestle with another 100 µL of 1× MBS + 0.1% BSA and collect in the same 15-mL tube.

       At the end, the resulting amount in the 15-mL conical tube will be ~1.5 mL (i.e., 1.2 mL + 200 µL + 100 µL) per one male.

    iv. After briefly hand swirling the tube, set the tube aside to allow debris to settle for ~1 min.

       Two males (i.e., two tubes) can comfortably be used for four to five females. A general rule is two females/one male.

6. Prepare four coated Petri dishes (60-mm, numbered 1 to 4) per one female frog in advance. Pour and remove 1× MBS + 0.1% BSA into and out of the dishes until the solution completely covers the dish bottoms.

    Coating dishes is essential to prevent sperm from sticking to one place in the dish, and allows sperm to move fluidly around the dish.

    Dish #1 will be used for collecting all eggs, which are subsequently divided and placed into the remaining three dishes #2–4 (see Fig. 1C). In this way, potential artifacts that might result from taking early eggs or late eggs for a given treatment that may bias the experiment due to egg quality differences between early and late egg collections can be avoided.

    Dish #2 will be used for a control fertilization with untreated sperm. Dishes #3 (for gynogenesis) and #4 (haploid control) are used for fertilization with UV-treated sperm (see Step 8 and Fig. 1C).

7. Take 400 µL of sperm suspension from the tube in which debris has been allowed to settle in Step 5 into dish #3 per female.

    Avoid taking sperm from the bottom of the tube that could contain small chunks of testis tissue. These chunks could shelter individual sperm, which could then escape irradiation.

    This procedure and the following procedures can be done easily for two females simultaneously using one sperm suspension described in Step 5.

8. Per female, irradiate dish #3 with a UV light source (be sure to remove the lid). Irradiate the sperm suspension twice at 50,000 µJ/cm^2. Swirl slightly in between each irradiation to ensure complete irradiation of all sperm cells.

9. Immediately collect eggs by gentle squeezing from one female into dish #1 as described in Sec. 2, Protocol 4: Obtaining *Xenopus tropicalis* Eggs (Lane et al. 2022). Avoid dripping from the frog skin into the dish as this could kill eggs.

    If it is difficult to squeeze dry (i.e., without dripping) or to get enough eggs from one squeeze, eggs can be squeezed into 1× MBS + 0.5% BSA a few times. Alternatively, a frog can be placed in 1× MBS + 0.5% BSA (minimum 1 L per frog) and allowed to lay eggs naturally in the medium. Eggs can then be collected periodically (a higher salt medium without BSA often used to collect X. laevis *eggs cannot be used here). After collection of enough eggs, remove all medium and follow the procedure below. Eggs may be held in this higher strength BSA solution for up to 5 h before fertilization/activation if necessary, but longer incubation before fertilization could cause maternal effects on development (Kosubek et al. 2010), thus it may be best to move to the next step within 30 min.*

Four hundred to six hundred eggs are needed at a minimum to have 200 to 300 healthy cold-shocked post-gastrula-stage embryos for proper scoring at the end. A smaller number of cold-shocked embryos can be scored if egg numbers are low, but this may reduce the statistical reliability in scoring phenotypes.

10. Aliquot collected eggs into the experimental dishes. Using a transfer pipette coated with 1× MBS + 0.1% BSA, first move a small portion of the eggs into dish #2 (for the control fertilization), then the rest of the eggs to dish #3 (that already contains UV-irradiated sperm, see Step 8). As a result, dish #1 becomes empty, and dish #4 is not yet used at this point.

    More eggs should always go into the cold shock dish (i.e., dish #3) unless a large number of animals need to be raised from controls. For example, if 1000 eggs were collected, transfer 200 eggs into dish #2 and transfer 800 eggs into dish #3, which contains UV-irradiated sperm.

11. Once the eggs are divided, using a Pipetman with a cut-off tip, aliquot 150–200 μL of nonirradiated sperm (from the tube in Step 5) into dish #2.

    For dish #2, sperm can be taken from the bottom part of the tube because the presence of tissue chunks would not matter. Take care to avoid cross-contamination between nontreated and UV-treated sperm in different dishes.

12. Swirl both dishes #2 and #3 to ensure mixing of the sperm and eggs and let stand for 2–5 min at room temperature.

13. From dish #3, use a fresh coated transfer pipette as described above to move a small portion of eggs that have been mixed with UV-treated sperm into dish #4 (e.g., approximately 100 to 200 eggs, if starting from a total of 1000 eggs as in the example above).

14. Flood dishes #2–4 with 0.1× MBS + 0.1% BSA and incubate at room temperature. Wait exactly for 5 min for dish #3 as follows.

    Start the timer at the exact moment dish #3 is flooded to ensure that the cold shock occurs during the appropriate moment of extrusion of the polar body during meiosis (see next step). The experiment will be the most successful if the cold shock is applied exactly 5 min after flooding. While waiting, ensure that your prechilled media temperature is between 1°C–3°C (see Step 4).

    Approximately 30 sec before the time is up, start removing the medium from dish #3 using a transfer pipette to get ready for the application of the cold shock for the next step. Because the medium used here has BSA in it, eggs are not attached to the dish and one cannot remove the medium by decanting.

15. Completely remove the medium from dish #3 only and replace with the cold medium that was prepared in Step 4. Place dish #3 in the prechilled micropipette tip box lid that contains slushy ice, then incubate the vessel containing the dish in the refrigerated ice water bath for 7.5 min.

    Make sure that the iced H₂O in the vessel is not mixed in the medium.

    The time course described here (i.e., 7.5 min starting at the exact time the cold medium is added) was the optimal time under our conditions for cold shocking to recover a maximum number of gynogenetic diploid embryos. However, the optimum time could be different from one laboratory to another depending on the equipment used and/or genetic background of frogs and needs to be tested empirically. Seven to 10 min is a good starting range. In general, an increase beyond the optimal cold shock time may lead to embryo death and a decrease in time may lead to an increase in haploids.

16. Completely remove the cold medium from dish #3 and replace as quickly as possible with room temperature medium (0.1× MBS + 0.1% BSA). Place all dishes (#2–4) in an incubator to maintain the proper temperature (22°C–25°C). Cleavage should start in 1–1.5 h.

    Ensure that the cold medium is left on the entire 7.5 min before exchanging with the room temperature medium.

    While waiting for cleavage, prepare 2% cysteine dissolved in 0.1× MBS (pH 7.8–8.0).

    Also, coat 100-mm Petri dishes to culture embryos. Prepare enough dishes for collecting about 50 (up to 100) embryos per dish. Dishes can be coated with 0.1× MBS + 0.1% BSA at a minimum, but coating with 0.8%–1% agarose dissolved in 0.1× MBS (or simply in H₂O) is preferable. To prepare agarose-coated dishes, pour microwaved boiled agarose solution in the dish and spread the agarose throughout the dish by swirling. Immediately remove the excess using a transfer pipette before the agarose is solidified so that only a very thin layer of agarose is left to coat the dish. For treatment of multiple dishes, keep the agarose hot by periodically microwaving or keep the melted agarose in a hot water bath.

Cite this protocol as *Cold Spring Harb Protoc*; doi:10.1101/pdb.prot107648

17. When the embryos in dishes #2–4 begin to cleave, dejelly them using 2% cysteine within 5 min, while swirling the dishes. Once dejellied, rinse embryos with room temperature 0.1× MBS + 0.1% BSA at least four times to remove the cysteine completely.

> As needed, change the cysteine solution a few times during dejellying in order to finish dejellying in 5 min. A longer incubation in cysteine could cause abnormal development of embryos. Embryos do not necessarily need to be completely dejellied, as would be the case for microinjection purposes. Under a stereoscope, a narrow space between packed embryos can be observed with partial dejellying, or no space will be observed when embryos are completely dejellied. Partially remaining jelly coats will protect embryos from physical damage, but embryos can be completely dejellied if the process is hard to control.

18. Sort the cleaving embryos into coated 100- mm Petri dishes using a transfer or Pasteur pipette coated with 0.1× MBS + 0.1% BSA in groups no larger than 100 embryos per dish. The dishes should contain 0.1× MBS + 50 µg/mL gentamicin for the first 1–2 d. Culture embryos at 25°C. Change the medium with 0.1× MBS and sort embryos daily using new dishes until embryos reach the desired stage for scoring.

> Gentamicin is a well-known ototoxic antibiotic, damaging hair cells (Lombarte et al. 1993), and therefore should not be used beyond the stage when inner ear formation begins.

> Reduce the number of embryos per dish to 50 on the second day of culture. Coated dishes are not necessary once embryos have hatched.

19. Evaluate experimental results.

   i. The embryos in dish #2 are normal diploid embryos and should thus develop normally.

   ii. The embryos in dish #4 should be haploid. In general, haploid embryos develop with a shortened body axis, microcephaly, edema, and a poorly formed gut (e.g., see Noramly et al. 2005), all of which should be easily distinguishable from normal diploid embryos in dish #2 by stage 40 (Nieuwkoop and Faber 1967).

   iii. If a homozygous transgenic frog expressing a fluorescent protein is used for the testes source, the resultant diploid embryos in dish #3 fertilized from irradiation-escaped sperm (which would be transgenic glowing animals) can be distinguished from those formed from successful gynogenesis (which would not be transgenic). If carefully examined, heterozygous transgenic animals could also be used as the source of sperm (e.g., see Noramly et al. 2005).

   > See Troubleshooting.

TROUBLESHOOTING

Problem (Step 19): Embryos in dish #2 did not develop normally.
Solution: There was a problem with either the eggs or sperm used and the experiment should be repeated.

Problem (Step 19): Normal tadpoles are seen in dish #4.
Solution: This suggests that the UV treatment of sperm failed. The UV source should be checked and the sperm suspension properly remade before repeating experiments. If the majority of embryos in dish #4 are *not* haploid, there is no point in scoring experimental dish #3.

DISCUSSION

In *X. tropicalis,* an initial attempt to perform gynogenesis used hydrostatic pressure to suppress ejection of the second polar body (Noramly et al. 2005), but in subsequent protocols used the simpler cold shock protocol described here, modified from a protocol used in *X. laevis* (Kawahara 1978). Our optimized standard gynogenesis protocol for *X. tropicalis* was previously only available through a web-based platform or as supplemental data and is of potentially renewed value for the reverse genetic screens using gene editing being performed in many laboratories.

Because the method described here is based on suppression of the second polar body, resultant gynogenetic diploid embryos are homozygous over only some of their genome because of crossover events during meiosis (e.g., Reinschmidt et al. 1985). If a given region of interest (e.g., mutation or polymorphism) is located at a locus proximal to the centromere that is unlikely to undergo recombination during meiosis, the region of interest may be homozygous in the gynogenetic diploid animals. If the locus is farther from the centromere, crossover events may reduce the amount of homozygosity, but this method nonetheless revealed a sufficient fraction of homozygotes to be a useful screening technique (Reinschmidt et al. 1985; Noramly et al. 2005). In a gene editing experiment, the overall percentage of expected mutants in the gynogenesis procedure would also depend on the level at which gene editing has occurred in F_0 animals. This procedure allows the skipping of one generation in screening a lesion of a given gene, with the proviso that the efficiency of gynogenesis depends on the location of the gene relative to the centromere. Indeed, gynogenesis has successfully been used for screening naturally occurring mutants in wild-caught females (Noramly et al. 2005) and for rapid mapping of mutations to chromosomes (Goda et al. 2006; Khokha et al. 2009; Abu-Daya et al. 2012; Nakayama et al. 2017) in *X. tropicalis*. This method was originally developed for forward genetics-type approaches but is more widely applicable for genetic screens of females for mutations in a given gene if the gene's location is at a nonrecombined region. For example, at present, because of the establishment of gene editing tools, this gynogenesis technique should be also useful for reverse genetics—that is, to screen eggs of gene-edited F_0 females to identify mutation carriers and study homozygous phenotypes, skipping one generation required for conventional genetics.

RECIPE

Modified Barth's Saline for Xenopus (MBS)

Make a 10× MBS stock:
  800 mM NaCl
  10 mM KCl
  10 mM $MgSO_4 \cdot 7H_2O$
  50 mM HEPES free acid
  25 mM $NaHCO_3$

Prepare a 10× stock that consists of the reagents listed above, adjusting the pH to 7.8 with NaOH and sterilizing by autoclaving. Prepare a separate 0.1 M stock of $CaCl_2$, and sterilize by autoclaving. Store the stocks indefinitely at room temperature.

To make 1× MBS, dilute the 10× MBS stock with type 1 ultrapure water (ASTM International 2018), and add $CaCl_2$ stock to a final concentration of 0.7 mM.

ACKNOWLEDGMENTS

This work was funded by National Institutes of Health (NIH) grants RR13221, EY017400, EY018000, and EY022954 and research awards from the Sharon Stewart Aniridia Trust and Vision for Tomorrow to R.M.G.

REFERENCES

Abu-Daya A, Khokha MK, Zimmerman LB. 2012. The hitchhiker's guide to *Xenopus* genetics. *Dev Dyn* 50: 164–175.

Goda T, Abu-Daya A, Carruthers S, Clark MD, Stemple DL, Zimmerman LB. 2006. Genetic screens for mutations affecting development of *Xenopus tropicalis*. *PLoS Genet* 2: e91. doi:10.1371/journal.pgen.0020091.

Kawahara H. 1978. Production of triploid and gynogenetic diploid *Xenopus* by cold treatment. *Dev Growth Differ* 20: 227–236. doi:10.1111/j.1440-169X.1978.00227.x

Khokha MK, Krylov V, Reilly MJ, Gall JG, Bhattacharya D, Cheung CYJ, Kaufman S, Dang KL, Macha J, Ngo C, et al. 2009. Rapid gynogenetic mapping of *Xenopus tropicalis* mutations to chromosomes. *Dev Dyn* **238:** 1398–1406. doi:10.1002/dvdy.21965

Kosubek A, Klein-Hitpass L, Rademacher K, Horsthemke B, Ryffel GU. 2010. Aging of *Xenopus tropicalis* eggs leads to deadenylation of a specific set of maternal mRNAs and loss of developmental potential. *PLoS ONE* **5:** e13532. doi:10.1371/journal.pone .0013532

Lane M, Khokha MK. 2022. Obtaining *Xenopus tropicalis* embryos by in vitro fertilization. *Cold Spring Harb Protoc* doi:10.1101/pdb.prot106351

Lane M, Mis EK, Khokha MK. 2022. Obtaining *Xenopus tropicalis* eggs. *Cold Spring Harb Protoc* doi:10.1101/pdb.prot106344

Lombarte A, Yan HY, Popper AN, Chang JS, Platt C. 1993. Damage and regeneration of hair cell ciliary bundles in a fish ear following treatment with gentamicin. *Hear Res* **64:** 166–174. doi:10.1016/0378-5955(93) 90002-I

Nakayama T, Nakajima K, Cox A, Fisher M, Howell M, Fish MB, Yaoita Y, Grainger RM. 2017. *no privacy*, a *Xenopus tropicalis* mutant, is a model of human Hermansky–Pudlak syndrome and allows visualization of internal organogenesis during tadpole development. *Dev Biol* **426:** 472–486. doi:10.1016/j.ydbio.2016.08.020

Nieuwkoop PD, Faber J. 1967. *Normal table of* Xenopus laevis *(Daudin)*. Garland, New York.

Noramly S, Zimmerman L, Cox A, Aloise R, Fisher M, Grainger RM. 2005. A gynogenetic screen to isolate naturally occurring recessive mutations in *Xenopus tropicalis*. *Mech Dev* **122:** 273–287. doi:10.1016/j.mod.2004.11 .001

Reinschmidt D, Friedman J, Hauth J, Ratner E, Cohen M, Miller M, Krotoski D, Tompkins R. 1985. Gene-centromere mapping in *Xenopus laevis*. *Jour Hered* **76:** 345–347. doi:10.1093/oxfordjournals .jhered.a110108

Preparation of Intact Nuclei for Single-Nucleus Omics Using Frozen Cell Suspensions from Mutant Embryos of *Xenopus tropicalis*

Takuya Nakayama,[1] Janou A.Y. Roubroeks,[2] Gert Jan C. Veenstra,[2,3] and Robert M. Grainger[1,3]

[1]Department of Biology, University of Virginia, Charlottesville, Virginia 22904, USA; [2]Department of Molecular Developmental Biology, Faculty of Science, Radboud Institute for Molecular Life Sciences, Radboud University, Nijmegen, 6525 GA, the Netherlands

Single-cell omics such as single-cell RNA-sequencing (RNA-seq) have been used extensively to obtain single-cell genome-wide expression data. This technique can be used to compare mutant and wild-type embryos at predifferentiation stages when individual tissues are not yet formed (therefore requiring genotyping to distinguish among embryos), for example, to determine effects of mutations on developmental trajectories or congenital disease phenotypes. It is, however, hard to use single cells for this technique, because such embryos or cells would need to be frozen until genotyping is complete to capture a given developmental stage precisely, but intact cells cannot be isolated from frozen samples. We developed a protocol in which high-quality nuclei are isolated from frozen cell suspensions, allowing for genotyping individual embryos based on a small fraction of a single embryo suspension. The remaining suspension is frozen. After genotyping is complete, nuclei are isolated from embryo suspensions with the desired genotype and encapsulated in 10× Genomics barcoded gel beads for single-nucleus RNA-seq. We provide a step-by-step protocol that can be used for single transcriptomic analysis as well as single-nucleus chromatin accessibility assays such as ATAC-seq. This technique allows for high-quality high-throughput single-nucleus analysis of gene expression in genotyped embryos. This approach may also be valuable for collection of wild-type embryonic material, for example, when collecting tissue from a particular developmental stage. In addition, freezing of tissue suspensions allows precise staging of collected embryos or tissue that may be difficult to manage when collecting and processing cells from living embryos for single-cell RNA-seq.

MATERIALS

It is essential that you consult the appropriate Material Safety Data Sheets and your institution's Environmental Health and Safety Office for proper handling of equipment and hazardous materials used in this protocol.

RECIPES: Please see the end of this protocol for recipes indicated by <R>. Additional recipes can be found online at http://cshprotocols.cshlp.org/site/recipes.

Reagents

All reagents used should be molecular biology grade (i.e., RNase-, DNase-free). Items that are equivalent to those listed here and already routinely used in one's laboratory can be used, unless specifically designated.

---

[3]Correspondence: g.veenstra@science.ru.nl; rmg9p@virginia.edu

DAPI (4′,6-diamidino-2-phenylindole) (e.g., Invitrogen D1306, or an equivalent nuclear stain)

Prepare a 5 mg/mL solution in RNase/DNase-free H_2O and store at $-20°C$.

DNase/RNase-free H_2O (e.g., Fisher BioReagents BP2484-100)
Embryo collecting buffer I <R>
Embryo collecting buffer II <R>
Embryo lysis buffer <R>
Glycerol (e.g., Sigma-Aldrich G5516-500ML)

Prepare an 80% (w/v) stock with RNase/DNase-free H_2O, filter-sterilize with a 0.2 µm filter, and store at room temperature.

Liquid nitrogen
Nuclear buffer <R>
Nuclei purification buffer <R>
Proteinase K (e.g., Sigma-Aldrich/Roche 3115887001)
Reagents for genomic PCR and genotyping

See Protocol 1: CRISPR–Cas9 Mutagenesis in Xenopus tropicalis *for Phenotypic Analyses in the F_0 Generation and Beyond (Blitz and Nakayama 2021).*

Xenopus embryo culture medium of choice (see Step 1)
Xenopus nuclear PBS <R>
Xenopus tropicalis embryos (see Step 1)

Equipment

Equipment used routinely in laboratories performing Xenopus *embryo research that is the equivalent of what is listed here can be used, unless specifically designated.*

Conical centrifuge tubes (15-mL, RNase/DNase-free; e.g., Thermo Scientific 339650)
Coverslip glass (9-mm × 9-mm, to save volume) (e.g., Bellco Glass 1916-09009A)
Disposable hemacytometer (e.g., VWR 82030-470)
Fluorescence microscope with an ultraviolet (UV) filter set for detection of DAPI staining
Ice buckets
Liquid nitrogen dewar (1–3 L)
Microcentrifuge that can accommodate a swinging bucket rotor from 400 to 3300g at 4°C
Microcentrifuge tubes (1.5- and 2-mL, RNase/DNase-free)
Micropipettes (e.g., Pipetman P2, P20, P200, and P1000, Gilson, or equivalent)

Use only RNase/DNase-free filtered tips with these pipettes.

Microscope slides
Multichannel pipettes (eight-channels, 0.5–10 µL and 10–100 µL or similar ranges available for 1, 10, and 20 µL)

Use only RNase/DNase-free filtered tips with these pipettes.

Pasteur pipettes, finely tapered

Prepare by pulling the end of the pipette in a flame, see Figure 1A.

Pasteur pipettes, smoothened

Smooth the ends with a flame to avoid damaging embryos.

PCR machine
PCR strip tubes; eight-well (0.2-mL, DNase/RNAse-free)

Test in advance whether these can be frozen in liquid nitrogen.

Rubber bulb
Stereomicroscope with fiber optic light source (for sorting embryos)
Ultracentrifuge tubes (polypropylene, Beckman Coulter 347357 or equivalent)

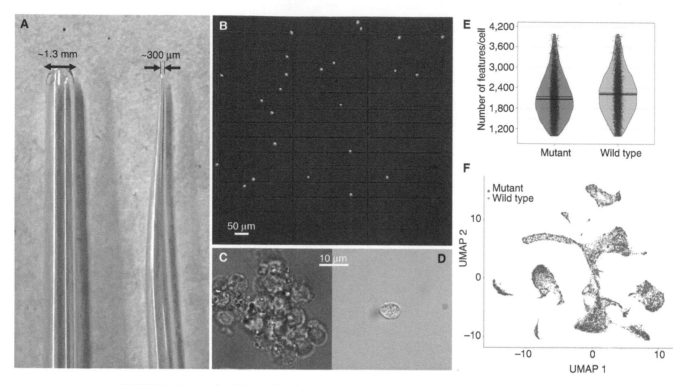

FIGURE 1. (*A*) Regular (*left*) and finely tapered (*right*) Pasteur pipettes. (*B*) Examples of DAPI-stained nuclei in hemacytometer. (*C*) Aggregated damaged nuclei. (*D*) Intact nucleus. (*E*) Example of distribution of the number of different transcripts (features) detected by single-nucleus RNA sequencing (snRNA-seq) after processing using the 10× Genomics platform. (*F*) Example of UMAP dimensionality reduction of snRNA-seq data generated from a parallel, nonintegrated analysis using the *Seurat* R package, showing high reproducibility between samples obtained from a typical *Xenopus tropicalis* mutant and wild-type sibling embryos.

Ultracentrifuge with a swinging bucket rotor (Optima MAX-TL with TLS 55 rotor, Beckman Coulter or equivalent)

Water bath

METHOD

Isolation and Lysis of Individual Embryos

The following steps describe the isolation and partial lysis of individual embryos in eight-well strip PCR tubes, followed by a sampling of aliquots for genotyping, then snap freezing of tubes.

1. Obtain embryos by in vitro fertilization or by natural mating from the *Xenopus tropicalis* mutant to be studied, and dejelly them using standard protocols (Ogino et al. 2006, see Sec. 2, Protocol 5: Obtaining *Xenopus tropicalis* Embryos by In Vitro Fertilization [Lane and Khokha 2022a] and Sec. 2, Protocol 6: Obtaining *Xenopus tropicalis* Embryos by Natural Mating [Lane and Khokha 2022b]). Any embryo culture medium appropriate for *Xenopus* embryo research can be used (e.g., see Ogino et al. 2006; see Protocol 1: CRISPR–Cas9 Mutagenesis in *Xenopus tropicalis* for Phenotypic Analyses in the F_0 Generation and Beyond [Blitz and Nakayama 2022]; Protocol 3: Tissue-Targeted CRISPR–Cas9-mediated Genome Editing of Multiple Homeologs in F_0-Generation *Xenopus laevis* Embryos [Corkins et al. 2022]; Sec. 2, Protocol 5: Obtaining *Xenopus tropicalis* Embryos by In Vitro Fertilization [Lane and Khokha 2022a]; and Sec. 2, Protocol 6: Obtaining *Xenopus tropicalis* Embryos by Natural Mating [Lane and Khokha 2022b]).

Cite this protocol as *Cold Spring Harb Protoc*; doi:10.1101/pdb.prot107825

2. Sort only fertilized eggs (i.e., embryos) and culture to the stage of interest. For a given single recessive heterozygous mutant line (+/− for a given locus), we aim to obtain 10 each of the mutant (−/−) and wild-type (+/+) embryos, which means in theory that a minimum of 40 embryos need to be collected to obtain ~25% mutant and 25% wild-type embryos. From our experience, however, we recommend collecting 72 embryos to ensure one will obtain the desired number of 10 mutant and 10 wild-type embryos. The following steps are optimized for collecting 72 embryos to accommodate tubes described later in the protocol. When collecting greater or fewer numbers of embryos, adjust the required amounts of reagents and tubes as needed.

> It is important to collect at least three batches of age-matched embryos that are ready for collection at three different times (e.g., at 1 h intervals) so that the embryos at the desired stage are not missed. This is important because the collection of embryos at a given stage may involve careful inspection and take a considerable amount of time, and because the time required for Steps 4–8 may result in embryos developing past the stage of interest.

> Keep embryos healthy until the stage of interest by careful handling and husbandry.

3. Start to prepare the following tubes and reagents 1–1.5 h before the first batch of embryos reaches the desired stage.

   i. Prepare 800 µL of embryo collecting buffer I.

   ii. Aliquot an equal volume (~100 µL) of embryo collecting buffer I per well in a new eight-well strip (stock tubes for embryo collecting buffer I). Store the stock tubes on ice.

   iii. Prepare two series of eight-well strip PCR tubes. One set, called sampling tubes, will be used for collecting and partially lysing embryos and a second set, genotyping tubes, will be used for genotyping. To save time, label only the leftmost tubes for both sampling and genotyping tubes as 1, 9, 17, 25, 33, 41, 49, 57, 65 (meaning that the first strip is for embryo *1* to 8, second one for *9* to 16, and so on).

   iv. Add proteinase K to a final concentration of 200 µg/mL to 1.5 mL embryo lysis buffer. Aliquot an equal volume (~187 µL) per well in a new eight-well strip (stock tubes for embryo lysis buffer). Using a multichannel pipette, aliquot 20 µL per well of embryo lysis buffer from the resultant stock tubes into the genotyping tubes from Step 3.iii. Store these tubes on ice.

   v. Prepare an appropriate amount (enough to use for several hours) of liquid nitrogen in the dewar in advance.

4. Collect eight embryos into a strip of sampling tubes described in Step 3.iii individually (i.e., one embryo per well/tube of the strip) at room temperature. Work with one strip of tubes at a time. Use a smooth-tipped Pasteur pipette to move embryos and try to minimize the amount of the medium carried into sampling tubes. Next, use a finely tapered Pasteur pipette (Fig. 1A) to remove almost all residual medium from the tubes. This should happen naturally by capillary action, but as needed gently suck medium using a rubber bulb.

> Unless an embryo breaks, the same pipette can be used for the entire set. If an embryo breaks, change the pipette before moving to the next tube to avoid cross-contamination of cells for genotyping. Broken embryos are not useable and additional embryos should be collected as needed.

> It is important to remove the residual amount of medium from tubes to avoid dilution of the buffer used in the next step.

5. Add 10 µL of embryo collecting buffer I from the stock tubes prepared in Step 3.ii into each tube of the strip of sampling tubes at once by using the multichannel pipette, then place the strip on ice.

6. Keeping the strip of the sampling tubes on ice, break embryos in each tube by pipetting using the multichannel pipette (set the volume at 10 µL with tips for a P20 pipette). Because it is hard to manipulate eight embryos at once, first break four embryos (i.e., attaching only four tips at a time) and for the next four embryos use four new tips. Repeat pipetting (usually 10 times is enough but use more as needed) and make sure embryos in all four tubes are broken into small pieces without

any large tissue masses. Avoid accidentally manipulating the pipette to cause cross-contamination when switching to another set of four embryos.

7. Using a multichannel pipette, remove 1 µL from each of the sampling tubes and put it into the corresponding genotyping tubes prepared in Step 3.iv. Once 1 µL has been removed from the sampling tubes, immediately snap-freeze the sampling tubes in liquid nitrogen. Store the genotyping tubes either on ice or at room temperature until all samples are done.

8. Repeat Steps 4–7 until all strips are used to collect 72 embryos in total. In the end, you should have nine frozen strips of sampling tubes in liquid nitrogen and corresponding nine strips of genotyping tubes on ice (or at room temperature). Move the strips of sampling tubes to −80°C and store until used.

> It can easily take 10 min or more to complete Steps 4–7 of a strip (i.e., eight embryos) for the first time. However, once the procedure is familiar, you can likely complete each strip in 5–6 min.

Genotyping of Each Sample to Identify Mutant and Wild-Type Embryo Suspensions

9. Place the strips of genotyping tubes (from Step 7) in a PCR machine and lyse the cells for 10 min at 56°C, followed by 10 min at 95°C for heat inactivation of the proteinase K. Continue to treat the lysates to renature the DNA, with the following short incubations: 10 sec at 80°C, 10 sec at 60°C, 10 sec at 40°C, and ending at 22°C. Upon reaching 22°C, use the lysates for genotyping or store until use at −20°C.

10. Perform genotyping to identify mutant and wild-type cell suspensions. Use 0.5–1 µL of the lysates from Step 9 to prepare 25–50 (or up to 100) µL PCR reactions. Ideally, identify 10 mutant embryos and 10 wild-type embryos and identify the specific tubes of interest based on the results of genotyping.

> The conditions for PCR will vary depending on the given mutation of interest, and the genotyping method following the genomic PCR will also vary depending on your mutants. Therefore, no universal condition can be recommended, but refer to Protocol 1: CRISPR–Cas9 Mutagenesis in Xenopus tropicalis for Phenotypic Analyses in the F_0 Generation and Beyond [Blitz and Nakayama 2021] for a general protocol for PCR of genomic DNA for genotyping.

Preparation of Nuclei from Frozen Cell Suspensions

11. Precool the ultracentrifuge and rotor to 4°C and prepare the following reagents in advance.

    i. Prepare 600 µL of embryo collecting buffer II in a 1.5-mL microcentrifuge tube and store on ice.

    ii. Prepare 4 mL of nuclei purification buffer in a 15-mL conical centrifuge tube and store on ice. Wait at least 5 min before using after placing on ice to make sure the buffer becomes sufficiently cold.

12. In a room-temperature water bath, thaw cell suspensions in the strips (sampling tubes) containing tubes of interest identified in Step 10 (do not isolate the tubes of interest but treat whole strips for quick manipulation). It takes only a few moments to thaw. Immediately place those strips on ice. Combine cell suspensions of the same genotype from the tubes of interest (aiming for 10 samples of each genotype) into a new 2-mL microcentrifuge tube for each genotype as follows.

    i. Keeping the strips on ice, set the 200-µL pipette at 110 µL. Transfer the suspension from the first tube to the second tube, then transfer all solution from the second tube to the third tube, and so on to collect suspensions from all tubes of one genotype. Remember that the suspension in each tube is ~9 µL, a much smaller volume than 110 µL, and carefully pipette to avoid having air bubbles.

    ii. Transfer the pooled samples (~90 µL at maximum in total from 10 tubes) from the last tube to a 2-mL microcentrifuge tube set on ice. Add 90 µL of fresh embryo collecting buffer II (from Step 11.i) into the first tube and repeat the above procedure to rinse and collect all

Cite this protocol as *Cold Spring Harb Protoc*; doi:10.1101/pdb.prot107825

residual suspensions left in all the tubes. Transfer the pooled material to the same 2-mL tube to add them together. Measure the total volume in the 2-mL tube (usually 160 to 180 L). Add embryo collecting buffer II to bring up the final volume in each tube to 350 µL.

At the end there should be two 2-mL tubes, one for mutant embryos and one for wild-type embryos.

13. Gradually add 1.68 mL nuclei purification buffer to each 2-mL tube on ice as follows.

    i. Add three aliquots of 560 µL nuclei purification buffer (from Step 11.ii) with a 1000-µL pipette, mixing gently by pipetting after each aliquot for a total of 1.68 mL.

    ii. After adding the third aliquot of buffer, close the lid tightly and mix solution by repeatedly inverting the tube upside down by hand until homogenous (usually 40 times or more inversions are required). One can assess if mixing is complete by determining if Schlieren lines (the visible lines separating the two different density solutions being mixed due to their differences in refractive index) are gone after these inversions. Keep the tubes cold by placing them on ice periodically while inverting tubes. Centrifuge the tubes briefly to remove the solution inside of the lids and place the tubes on the ice.

14. Place 105 µL of fresh nuclei purification buffer at the bottom of each of two polypropylene ultracentrifuge tubes set on ice. Gently layer the suspension prepared in Step 13 on top of the buffer. Place the tubes in the rotor, and centrifuge at 130,000g for 2 h at 4°C (set acceleration at 2, deceleration at 9 for the Optima MAX-TL).

15. Prepare the following reagents and tubes during ultracentrifugation.

    i. Prepare 1 mL of nuclear buffer in a 1.5-mL microcentrifuge tube and store on ice.

    ii. Prepare 1.7 mL of *Xenopus* nuclear PBS in a 2-mL microcentrifuge tube and store on ice.

    > *This protocol is written for single-nucleus RNA-seq steps appropriate for the 10× Genomics system, which requires a PBS-based buffer to suspend single nuclei. The application for different omics and systems may require different buffers and the steps using this buffer should be modified as appropriate.*

    iii. Coat two 1.5-mL microcentrifuge tubes with *Xenopus* nuclear PBS as follows: Put ~500 µL of *Xenopus* nuclear PBS into a 1.5-mL microcentrifuge tube. Vortex and briefly centrifuge the tube to collect the buffer. Transfer the buffer into the second 1.5-mL microcentrifuge tube. Similarly treat the second tube, then transfer the buffer back to the original 2-mL tube. Label the two coated tubes as mutant and wild-type. Add 10 µL of 80% (w/v) glycerol to these tubes and store them on ice until use (see Step 18).

    iv. Make 20–30 µL of two nuclear staining solutions. Prepare 10 µg/mL DAPI in *Xenopus* nuclear PBS (nuclear staining solution I) and 50 µg/mL DAPI in *Xenopus* nuclear PBS (nuclear staining solution II), and store on ice.

16. Gently remove the ultracentrifuge tubes from the buckets and place tubes on ice. Black (or dark-colored) pigment granules should be present on the bottom of the tubes (as described by Andrews and Brown 1987). These can be used to monitor the possible location of nuclei (see below). Carefully remove 1 mL of supernatant with a 1000-µL pipette and repeat for a total of 2 mL. Subsequently remove as much supernatant as possible little by little using 20- and 200-µL pipettes (for example 40–80 µL with a 200-µL pipette, followed by another 5–10 µL with a 20-µL pipette). Do not touch or disturb the pellet. In the end, roughly 60–80 µL of supernatant should be left.

17. Gradually resuspend each pellet with a total of 250 µL of nuclear buffer (prepared in Step 15.i) as follows.

    i. Add the first 80 µL with a 200-µL pipette, and then set the volume for 50–60 µL (to avoid introducing air to the suspension) to resuspend the pellet gently. Keep the tubes cold, on ice, as much as possible. Look at the dark pigment granules described above to monitor the progression of resuspending the pellet.

ii. Once the pellet is well resuspended, add another 80 µL of nuclear buffer and similarly resuspend further. If the resuspension is done well, the suspension may become darker in color because of homogenous distribution of pigment granules.

iii. Finally, add another 90 µL of nuclear buffer (total 250 µL) and gently but thoroughly resuspend with a 1000-µL pipette with the volume set to ~250 µL or less to avoid introducing air bubbles. In the end the total volume should be ~310–330 µL (measure the volume to check as needed to estimate how much supernatant could be removed in Step 19).

18. Gently layer the nuclear suspension on top of the 80% glycerol cushions prepared in Step 15.iii for each tube. Centrifuge at 3200–3400g for 10 min at 4°C.

> Use swinging bucket rotors here and in later steps, if possible. You can use fixed-angle rotors, but recovery yield will be lower.

19. Gently handle tubes after centrifugation and place them on ice. Remove the supernatant, little by little, as much as possible without removing the pellet. For example, remove ~150 µL × 2 times (~300 µL in total), followed by further removal of ~5–20 µL. Keep the tubes cold, on ice, as much as possible during the procedure.

20. Resuspend the pellet as follows:

i. Add 80 µL of nuclear buffer and resuspend the pellet gently as described in Step 17. Add another 80 µL of nuclear buffer (total 160 µL) and similarly resuspend further.

ii. Add 160 µL of *Xenopus* nuclear PBS (prepared in Step 15.ii) and leave on ice for 5 min for equilibration. After incubation, slowly mix the suspension with a 1000 µL pipette. Add another 90 µL of *Xenopus* nuclear PBS (total 250 µL) and further resuspend gently. Keep the tubes cold on ice as much as possible during the procedure. The resultant suspension will be ~420–430 µL in total.

21. Centrifuge the above suspension at 400g for 5 min at 4°C in a swinging bucket rotor.

22. Remove the supernatant (~400–410 µL), little by little as described in Step 19.

23. Add 60 µL *Xenopus* nuclear PBS and leave on ice for 5 min for equilibration. After incubation, slowly mix the suspension with a 200-µL pipette and add another 140 µL of *Xenopus* nuclear PBS to gently resuspend further. Finally add 210 µL of *Xenopus* nuclear PBS (total 410 µL) and resuspend gently with a 1000-µL pipette. Keep the tubes cold on ice as much as possible during the procedure. The resultant suspension will be ~420–430 µL in total.

24. Repeat Steps 21 and 22.

25. Add 30–40 µL of *Xenopus* nuclear PBS to the resultant pellet and gently resuspend with a 200-µL pipette. Keep the tubes cold on ice as much as possible during the procedure. The resultant suspension will be ~40–50 µL in total.

Evaluation of Yield and Quality of the Isolated Nuclei

26. To evaluate the yield of nuclei, mix 1 µL of the nuclear staining solution I from Step 15.iv, 8 µL of *Xenopus* nuclear PBS, and 1 µL of the nuclear suspension from Step 25. Load all 10 µL of the resultant mixture into the hemacytometer. Count the whole area (900 nL in total) under a fluorescence microscope with the UV filter to calculate the yield of nuclei. A typical DAPI image of successfully isolated nuclei is shown in Figure 1B.

> A regular glass hemacytometer can be used, but it takes time to prepare it properly. Here, it is important to keep nuclei fresh before downstream applications, and a shorter time for evaluation of nuclei is preferred, therefore we recommend using a disposable hemacytometer that can be used for multiple samples without any advance preparation.
> We recommend counting two samples prepared independently for more accuracy. If the concentration is too high, that is, 2000–3000 nuclei per µL or more, we recommend diluting the aliquot and recounting the diluted samples for better estimation. If the concentration is less than desired, the sample can be concentrated by recentrifugation (400g for 5 min at 4°C) to pellet nuclei and remove extra supernatant but doing so

Cite this protocol as *Cold Spring Harb Protoc*; doi:10.1101/pdb.prot107825

usually causes a loss of nuclei. For the regular application for single-nuclei RNA-seq, a concentration ~700–1000 nuclei per µL is optimal.

27. To evaluate the quality of the nuclei, mix 0.5 µL of nuclear staining solution II from Step 15.iv, 4 µL of *Xenopus* nuclear PBS, and 1–2 µL (depending on concentration calculated above, aiming to load 700–1400 nuclei in total) of the nuclear suspension from Step 25. Place the full suspension on the microscope slide and gently place the coverslip glass (9-mm × 9-mm) over the sample. Observe under the microscope using a 60× or higher objective lens. A typical successfully isolated nucleus is shown in Figure 1D, whereas failed nuclei are in Figure 1C.

If nuclei are damaged during preparation, high-magnification microscopy shows blebs emerging from the nuclei. Depending on the degree of damage, blebbing may be minor (in which case nuclei are still usable for downstream applications) or very obvious. In the latter case the nucleoplasm is released and such nuclei become sticky and make aggregates. Thus, the presence of aggregates is an indication of the presence of significantly damaged nuclei, and depending on how many nuclei are damaged, the size of the aggregate varies—namely, a higher presence of larger aggregates indicates a greater number of severely damaged nuclei. The larger aggregates can also easily be recognized under low magnification in Step 26. We recommend proceeding with downstream applications only if 80%–90% or more nuclei are of high quality and the preparation contains at most a few small aggregates.

In summary, if nuclei look like the one in Figure 1D at a reasonable yield, the preparation can be deemed successful. Examples of data from a successful nuclear preparation and application in snRNA-seq are shown in Figure 1E,F.

See Troubleshooting.

TROUBLESHOOTING

Problem (Step 27): Nuclei aggregate.
Solution: Simply start over if there are too many aggregates. Proceed with more gentle manipulation.

DISCUSSION

The nuclear isolation protocol described here is confirmed to work for embryos from stage 10 to stage 21 and has been used successfully for single-nucleus RNA-seq (Fig. 1E,F) and single-nucleus ATAC-seq (Bright et al. 2021) with 10× Genomics reagents. The reagents allow for a high-quality and relatively deep coverage of the transcriptome (number of different transcripts per nucleus, Fig. 1E) despite the fact that the nucleus only contains a fraction of total cellular RNA. Moreover, we found that the protocol can be applied robustly and reproducibly, such that samples produced in parallel produce virtually identical patterns after dimensionality reduction (Fig. 1F).

This protocol incorporates steps adapted from Mariano (1964), Andrews and Brown (1987), and Bright et al. (2021). The embryo collecting buffer I and II, nuclei purification buffer, and nuclear buffer in this protocol were modified, respectively, from E-1 buffer-0.25 M sucrose, E-1 buffer-2.2 M sucrose, and nuclear buffer described in the original publications. We have included Mg^{2+} in *Xenopus* nuclear PBS, as it helps to preserve nuclei. The salt concentration in this buffer is reduced compared to regular PBS that is used with mammalian cells.

Using the conditions we describe here, we generally obtain 30,000–120,000 nuclei in total. This suggests that we probably do not need 10 embryos as starting material, but we have not tested this protocol with a lower number of embryos. Given that the original protocols were for *Xenopus laevis*, our protocol should also be applicable for use with *X. laevis* although some small adjustments such as number of required embryos may be needed.

This protocol can simply be applied to any kind of mutation (dominant or recessive), and there are advantages of using this approach for wild-type embryos as well, or mutants in which the phenotypes are already obvious at the time of harvesting, for single-nucleus omics. For example, snRNA-seq can capture a snapshot of ongoing transcripts as opposed to a steady-state level of transcripts by single-cell RNA sequencing (scRNA-seq) allowing a more accurate assessment of the timing of transcription. Also, in assessing mutations which contain premature termination codons and may thus lead to nonsense-mediated decay (NMD) of mRNA, one may still be able to track mutant transcripts by snRNA-seq, as opposed to scRNA-seq, because NMD mainly happens in the cytoplasm (e.g., Karousis and Mühlemann 2019). In addition, if one is collecting pieces of tissue for single-cell analysis, and these need to be pooled over time to collect enough samples, this approach, which freezes samples as they are prepared, would be more likely to ensure that collected samples were all from the same stage and not aged during collection. The advantages of using snRNA-seq here for obtaining a "snapshot" of transcription and collecting sequential samples would also be beneficial for embryos treated with antisense reagents (e.g., morpholino oligo).

RECIPES

Embryo Collecting Buffer I

In advance, prepare a 7× stock solution of cOmplete, Mini, EDTA-free Protease Inhibitor Cocktail (Sigma-Aldrich 11836170001) and a 100 mM stock solution of spermine (e.g., Sigma-Aldrich S1141-1G) in DNase-, RNase-free H_2O. Store both at −20°C until use. Prepare the recipe as follows:

2× Stock Solution I <R>	400 µL
1 M DTT (e.g., Sigma-Aldrich 646563, once opened, store at −20°C)	1.6 µL
7× cOmplete, Mini, EDTA-free Protease Inhibitor Cocktail	114.4 µL
100 mM Spermine	0.8 µL
Protector RNase inhibitor (Sigma-Aldrich 3335399001) (40 U/µL)	4 µL
80% (w/v) glycerol (made from e.g., Sigma-Aldrich G5516-500ML)	279.2 µL

Prepare fresh in a 1.5-mL microcentrifuge tube (RNase/DNase-free) and store on ice. This amount of buffer is appropriate for collecting 72 embryos.

Embryo Collecting Buffer II

In advance, prepare a 7× stock solution of cOmplete, Mini, EDTA-free Protease Inhibitor Cocktail (Sigma-Aldrich 11836170001) and a 100 mM stock solution of spermine (e.g., Sigma-Aldrich S1141-1G) in DNase-, RNase-free H_2O. Store both at −20°C until use. Prepare the recipe as follows:

2× Stock Solution I <R>	300 µL
1 M DTT (e.g., Sigma-Aldrich 646563, once opened, store at −20°C)	1.2 µL
7× cOmplete, Mini, EDTA-free Protease Inhibitor Cocktail	85.8 µL
100 mM Spermine	0.6 µL
Protector RNase inhibitor (Sigma-Aldrich 3335399001) (40 U/µL)	3 µL
DNase-, RNase-free H_2O	209.4 µL

Prepare fresh in a 1.5-mL microcentrifuge tube (RNase/DNase-free) and store on ice. This amount of buffer is appropriate for collecting 20 embryos.

Cite this protocol as *Cold Spring Harb Protoc*; doi:10.1101/pdb.prot107825

Embryo Lysis Buffer

50 mM Tris pH 8.8
1 mM EDTA
0.5% Tween 20
200 µg/mL Proteinase K (freshly added from a stock solution)

Autoclave and store at room temperature, before adding proteinase K, for up to a year.

Nuclear Buffer

In advance, prepare a 7× stock solution of cOmplete, Mini, EDTA-free Protease Inhibitor Cocktail (Sigma-Aldrich 11836170001) in DNase-, RNase-free H_2O (store at −20°C until use) and prepare the recipe as follows:

2× Stock Solution for Nuclear Buffer <R>	500 µL
1 M DTT (e.g., Sigma-Aldrich 646563, once opened, store at −20°C)	2 µL
7× cOmplete, Mini, EDTA-free Protease Inhibitor Cocktail	143 µL
Protector RNase inhibitor (Sigma-Aldrich 3335399001) (40 U/µL)	5 µL
DNase-, RNase-free H_2O	350 µL

Prepare fresh in a 1.5-mL microcentrifuge tube (RNase/DNase-free) and store on ice.

Nuclei Purification Buffer

In advance, prepare a 7× stock solution of cOmplete, Mini, EDTA-free Protease Inhibitor Cocktail (Sigma-Aldrich 11836170001) and a 100 mM stock solution of spermine (e.g., Sigma-Aldrich S1141-1G) in DNase-, RNase-free H_2O. Store both at −20°C until use. Prepare the recipe as follows:

1.25× Stock Solution II <R>	3.2 mL
1 M DTT (e.g., Sigma-Aldrich 646563, once opened, store at −20°C)	8 µL
7× cOmplete, Mini, EDTA-free Protease Inhibitor Cocktail	572 µL
100 mM Spermine	4 µL
Protector RNase inhibitor (Sigma-Aldrich 3335399001) (40 U/µL)	20 µL
DNase-, RNase-free H_2O	196 µL

Prepare this solution in a 15-mL conical centrifuge tube (RNase/DNase-free). Because of the high viscosity of this solution, it is important to vortex it vigorously to mix thoroughly, followed by centrifugation at ∼3000 rpm for ∼30 sec at 4°C to remove air bubbles. Prepare fresh and store on ice.

2× Stock Solution I

220 mM KCl (diluted from 1 M solution, e.g., Sigma-Aldrich 60142)
100 mM Tris-HCl (pH 7.4) (diluted from 1 M solution, e.g., Fisher BioReagents BP1757-500)
10 mM $MgCl_2$ (diluted from 1 M solution, e.g., Sigma-Aldrich M1028-100ML)
0.2 mM EDTA (diluted from 500 mM solution, e.g., Invitrogen 15575020)
0.5 M sucrose (e.g., Sigma-Aldrich S0389-1KG)

Prepare with RNase/DNase-free H_2O and filter-sterilize the solution with a 0.2 µm filter. Store for up to 6 mo at 4°C.

1.25× Stock Solution II

137.5 mM KCl (diluted from 1 M solution, e.g., Sigma-Aldrich 60142)

62.5 mM Tris-HCl (pH 7.4) (diluted from 1 M solution, e.g., Fisher BioReagents BP1757-500)

6.25 mM MgCl$_2$ (diluted from 1 M solution, e.g., Sigma-Aldrich M1028-100ML)

0.125 mM EDTA (diluted from 500 mM solution, e.g., Invitrogen 15575020)

2.75 M sucrose (e.g., Sigma-Aldrich S0389-1KG)

Prepare with RNase/DNase-free H$_2$O. It takes time to dissolve sucrose at this concentration. As needed, heat the solution and vortex vigorously. Filter-sterilize the solution with a 0.2 μm filter. Store for up to 6 mo at 4°C. If the sucrose becomes crystalized during storage, heat the solution and dissolve the crystals completely before use.

2× Stock Solution for Nuclear Buffer

140 mM KCl (diluted from 1 M solution, e.g., Sigma-Aldrich 60142)

50 mM Tris-HCl (pH 7.4) (diluted from 1 M solution, e.g., Fisher BioReagents BP1757-500)

10 mM MgCl$_2$ (diluted from 1 M solution, e.g., Sigma-Aldrich M1028-100ML)

0.4 mM EDTA (diluted from 500 mM solution, e.g., Invitrogen 15575020)

50% (w/v) glycerol (diluted from e.g., Sigma-Aldrich G5516-500ML)

Prepare with RNase/DNase-free H$_2$O and filter-sterilize the solution with a 0.2 μm filter. Store for up to 6 mo at 4°C.

Xenopus Nuclear PBS

5% BSA (UltraPure, Fisher AM2616, store at −20°C until use)	680 μL
10× PBS (e.g., Corning 46-013-CM)	119 μL
1 M MgCl$_2$ (e.g., Sigma-Aldrich M1028-100ML)	3.4 μL
Protector RNase inhibitor (Sigma-Aldrich 3335399001) (40 U/μL)	8.5 μL
DNase-, RNase-free H$_2$O	889.1 μL

Prepare fresh in a 2-mL microcentrifuge tube (RNase/DNase-free) and store on ice.

ACKNOWLEDGMENTS

The authors thank John Campbell for his valuable information from his experience with snRNA-seq technology and David Parichy for his support. This work was funded by a National Institutes of Health (NIH) grant EY022954, the University of Virginia Department of Ophthalmology, and research awards from the Sharon Stewart Aniridia Trust and Vision for Tomorrow to R.M.G. The authors also acknowledge Xenbase (RRID:SCR_003280) for providing essential online resources for *Xenopus* research. A part of this work was performed with the help of the University of Virginia, Genome Analysis and Technology Core (RRID:SCR_018883), and we also thank facility staff members Alyson Prorock, Yongde Bao, and Katia Sol-Church for their support and guidance.

REFERENCES

Andrews MT, Brown DD. 1987. Transient activation of oocyte 5S RNA genes in *Xenopus* embryos by raising the level of the *trans*-acting factor TFIIIA. *Cell* **51**: 445–453. doi:10.1016/0092-8674(87)90640-4

Blitz IL, Nakayama T. 2022. CRISPR–Cas9 mutagenesis in *Xenopus tropicalis* for phenotypic analyses in the F_0 generation and beyond. *Cold Spring Harb Protoc* doi:10.1101/pdb.prot106971

Bright AR, van Genesen S, Li Q, Grasso A, Frölich S, van der Sande M, van Heeringen SJ, Veenstra GJC. 2021. Combinatorial transcription factor activities on open chromatin induce embryonic heterogeneity in vertebrates. *EMBO J* **40**: e104913. doi:10.15252/embj.2020104913

Corkins ME, DeLay BD, Miller RK. 2022. Tissue-targeted CRISPR–Cas9-mediated genome editing of multiple homeologs in F_0-generation *Xenopus laevis* embryos. *Cold Spring Harb Protoc* doi:10.1101/pdb.prot107037

Karousis ED, Mühlemann O. 2019. Nonsense-mediated mRNA decay begins where translation ends. *Cold Spring Harb Perspect Biol* **11**: a032862. doi: 10.1101/cshperspect.a032862

Lane M, Khokha MK. 2022a. Obtaining *Xenopus tropicalis* embryos by in vitro fertilization. *Cold Spring Harb Protoc* doi:10.1101/pdb.prot106351

Lane M, Khokha MK. 2022b. Obtaining *Xenopus tropicalis* embryos by natural mating. *Cold Spring Harb Protoc* doi:10.1101/pdb.prot106609

Mariano EE. 1964. The isolation of nuclei from *Xenopus laevis* embryonic cells. *Exp Cell Res* **34**: 201–205. doi:10.1016/0014-4827(64)90199-5

Ogino H, McConnell WB, Grainger RM. 2006. High-throughput transgenesis in *Xenopus* using *I-SceI* meganuclease. *Nat Protoc* **1**: 1703–1710. doi:10.1038/nprot.2006.208

Genomics Methods for *Xenopus* Embryos and Tissues

Michael J. Gilchrist,[1,4] Ken W.Y. Cho,[2,4] and Gert Jan C. Veenstra[3,4]

[1]*The Francis Crick Institute, London NW1 1AT, United Kingdom;* [2]*Department of Developmental and Cell Biology, University of California, Irvine, California 92697;* [3]*Radboud University, Department of Molecular Developmental Biology, 6525GA Nijmegen, The Netherlands*

High-throughput sequencing methods have created exciting opportunities to explore the regulatory landscape of the entire genome. Here we introduce methods to characterize the genomic locations of bound proteins, open chromatin, and sites of DNA–DNA contact in *Xenopus* embryos. These methods include chromatin immunoprecipitation followed by sequencing (ChIP-seq), a combination of DNase I digestion and sequencing (DNase-seq), the assay for transposase-accessible chromatin and sequencing (ATAC-seq), and the use of proximity-based DNA ligation followed by sequencing (Hi-C).

OVERVIEW

The epigenetic state of chromatin regulates gene expression—and hence cellular differentiation—by controlling the access of transcription factors to DNA. Here we introduce methods to explore the locations of bound proteins, open chromatin, and sites of DNA–DNA contact at the whole-genome level in *Xenopus* embryos and tissues. These methods of data generation are often used in conjunction with gene expression analyses, which we discuss in Sec. 12, Introduction: Transcriptomics and Proteomics Methods for *Xenopus* Embryos and Tissues (Gilchrist et al. 2019).

Chromatin immunoprecipitation (ChIP) is one of the most direct ways to identify the sites of interaction between the genome and DNA-binding proteins (typically transcription factors and histones). In its most straightforward application, ChIP relies on an initial cross-linking stage in the intact animals or dissected tissues, followed by DNA fragmentation by sonication. Complexes containing DNA fragments bound to the protein of interest are immunoprecipitated with an antibody that recognizes the protein. After dissolution of the cross-linking, the DNA fragments are recovered and subjected to direct high-throughput sequencing (hence the abbreviation ChIP-seq).

Mapping regions of relatively open chromatin is valuable for identifying regulatory elements, including promoters and enhancers. Even when bound by proteins such as transcription factors, the DNA of the regulatory element remains accessible, rendering these regions relatively sensitive to cleavage by DNaseI or Tn5 transposase. Two methods, DNase-seq (Song and Crawford 2010) and the assay for transposase-accessible chromatin (ATAC) and sequencing (ATAC-seq) (Buenrostro et al. 2013), exploit this feature to map regions of accessible DNA. In these methods, digested double-cut fragments of DNA are either directly amplified (as in ATAC-seq) or amplified after purification (as in DNase-seq) and then sequenced. These approaches provide the ability to discover important

[4]Correspondence: drmikegilchrist@gmail.com; kwcho@uci.edu; g.veenstra@science.ru.nl

new enhancers without having a priori knowledge of the location and identity of bound transcription factors.

In a rather different approach to genomic data generation, we can determine the distribution of DNA–DNA contacts within and between chromosomes to understand how chromosomal DNA is folded in the nucleus. Cross-linking is used to form bridges at the contact points, and after DNA fragmentation, the biotinylated loose ends of DNA fragments in the same complex are ligated. After removal of the cross-links and further digestion, the joined fragments are isolated and sequenced from both ends to identify the different genomic regions that were in contact. This global and relatively unbiased variant of the chromosome conformation capture methods is referred to as Hi-C. The sequence data contain information on the DNA looping structures understood to regulate transcription, and are thus rather different from, but highly complementary to, the data generated by ChIP-seq, DNase-seq, and ATAC-seq experiments.

In general, these experimental methods are not very efficient, requiring relatively large numbers of cells to provide a robust signal. The *Xenopus* system is therefore ideal for these types of applications because of the ability to collect large numbers of synchronously developing embryos after in vitro fertilization. Below we introduce ChIP-seq, DNase-seq, ATAC-seq, and Hi-C protocols that have been developed specifically for use in *Xenopus* embryos and tissues and provide examples of their applications.

The *Xenopus* community is fortunate in having two well-assembled genomes, that of the diploid *X. tropicalis* (Hellsten et al. 2010) and that of the allo-tetraploid *X. laevis* (Session et al. 2016). This is important, as the sequence fragments generated by these methods generally map to the intergenic and intragenic noncoding (introns) regions, and incomplete assemblies will cause loss of potentially valuable data. The diploid genome of *X. tropicalis* makes the high-throughput genomic data simpler to interpret when compared to data from the larger, partly duplicated genome of *X. laevis*. However, the larger *X. laevis* embryos may be preferred for ease of experimental manipulation.

PROTOCOLS

Two ChIP-seq protocols, Protocol 5: Mapping Chromatin Features of *Xenopus* Embryos (Gentsch and Smith 2019) and Protocol 1: ChIP-Sequencing in *Xenopus* Embryos (Hontelez et al. 2019), offer slightly different approaches to prepare enriched nuclei and yolk-depleted embryo lysate. In particular, ChIP with cleared lysates—as described in the latter protocol—requires less starting material. ChIP-seq has been widely used in the *Xenopus* community to understand the targets of developmentally important transcription factors—for example, Smad2/3 (Yoon et al. 2011); Foxh1 (Chiu et al. 2014); T-Box family proteins (Gentsch et al. 2013); β-catenin (Nakamura et al. 2016); Otx2, Lim1/Lhx1, and Gsc (Yasuoka et al. 2014); Vegt and Otx1 in early embryos (Paraiso et al. 2019); and Prdm12 in developing inhibitory neurons important for vertebrate locomotion (Thélie et al. 2015). ChIP-seq has also been used to identify promoters and compare them across species (van Heeringen et al. 2011) or to look at the dynamic changes in genome-wide distribution of histone modifications and RNA polymerase II occupancy (Akkers et al. 2009; Hontelez et al. 2015).

Open chromatin is generally considered a prerequisite for binding of transcription factors, and two protocols can be used to study the distribution of open chromatin regions over the *Xenopus* genome: One uses DNase-seq (Protocol 3: DNase-seq to Study Chromatin Accessibility in Early *Xenopus tropicalis* Embryos [Cho et al. 2019]), and the other uses ATAC-seq (Protocol 2: Assay for Transposase-Accessible Chromatin-Sequencing Using *Xenopus* Embryos [Bright and Veenstra 2019]). Reads from these protocols are mapped to the genome and generally produce peaks or otherwise-delineated small regions of DNA. These regions can then be analyzed for enrichment in DNA-binding motifs to predict which transcription factors may be active in particular developmental stages or tissues. If enough sequence reads are mapped, specific transcription factor footprints can be resolved at single-base resolution to determine how those factors contact the DNA (Boyle et al. 2011; Neph et al. 2012; Buenrostro et al. 2013).

Most chromosome conformation capture methods use defined viewpoints (genomic regions of interest for which interactions are determined) to build dense maps of DNA–DNA contact information (for review, see Nicoletti et al. 2018). However, it is also possible to use this approach to perform an unbiased analysis in an organism's cells to study the three-dimensional organization of chromosomes under physiological conditions. This approach is described for *Xenopus* in Protocol 4: Generating a Three-Dimensional Genome from *Xenopus* with Hi-C (Quigley and Heinz 2019). This approach has previously been used in *Xenopus* to investigate the relationship between Foxj1 binding and chromatin loops in multiciliated cells (Quigley and Kintner 2017) using tethered conformation capture (Kalhor et al. 2011).

FUTURE CONSIDERATIONS

The major change sweeping through the world of biological data generation is the switch in emphasis from the bulk analysis of whole organisms or tissues to the analysis of hundreds to many thousands of individual cells from an embryo or tissue sample. Advances have been most rapid in the field of single cell transcriptomics—with, for example, the recent launch of the Human Cell Atlas (Rozenblatt-Rosen et al. 2017), which aims to create comprehensive reference maps of cell types in all major tissues. Recently, a major single-cell transcriptomics survey was conducted in *Xenopus* specimens spanning the blastula stage to the tailbud stage; this enabled the characterization of cell types from the earliest pluripotent cells to the well-differentiated cells of early organogenesis (Briggs et al. 2018). Although not all mRNAs in each cell are currently detected, single-cell RNA sequencing (scRNA-seq) can help us define the set of cell types in a population of cells, determine the evolution of cell lineages during development, and better understand the regulatory relationships between genes. Inevitably, the very large data sets that are generated do introduce some challenges in the development and application of computational methods, but modern approaches such as deep learning may be ideal for this.

Genomics techniques will not be far behind. Already, a number of approaches to single-cell genomics analysis are being developed along the lines of the multicellular methods, although the relatively small amounts of DNA in each sample will require highly efficient experimental methods and some rethinking of the downstream analysis. New approaches include the use of the Tn5 transposase for efficient library preparation—for example, in the CUT&Tag method (Kaya-Okur et al. 2019) for efficient epigenomic profiling on low cell numbers and single cells. An ATAC-seq method has been developed for single-cell work (Chen et al. 2018) and used to resolve dynamic changes in the chromatin landscape and to uncover the *cis*-regulatory programs of *Drosophila* germ layer formation (Cusanovich et al. 2018). Last, in addition to Hi-C, Capture Hi-C (Jäger et al. 2015) has been developed to specifically enrich (for example) for promoter-containing fragments from Hi-C libraries, and it will be able to generate evidence for dynamic interactions between promoters and enhancers.

The availability of genome assemblies for two closely related *Xenopus* species (see above) provides opportunities for new insights in comparative biology. In addition, there are many potential sources of genetic variation data captured in expressed sequences from both species and from different strains of these. Analysis of such data is useful for many different purposes, such as the design of morpholinos and CRISPR guide RNAs when using outbred populations. Genetic variation is also key to analyses of quantitative trait loci (QTL) and genome evolution. Although analysis of variation in whole-genome sequencing data is outside the scope of this introduction, genomic variation in *X. laevis* and *X. tropicalis* has already been analyzed in several reports (Elurbe et al. 2017; Savova et al. 2017; Mitros et al. 2019). Further studies will likely address genetic variation and their associated developmental, gene-regulatory, and phenotypic traits, both within populations and between closely related species.

Future genome-wide studies in *Xenopus* using the techniques described above will lead to the discovery of new mechanisms and to a better understanding of the processes controlling transcription and gene activation. We look forward to seeing the *Xenopus* community employing this next generation of methods in this most tractable of model systems.

REFERENCES

Akkers RC, van Heeringen SJ, Jacobi UG, Janssen-Megens EM, Françoijs KJ, Stunnenberg HG, Veenstra GJ. 2009. A hierarchy of H3K4me3 and H3K27me3 acquisition in spatial gene regulation in *Xenopus* embryos. *Dev Cell* 17: 425–434. doi:10.1016/j.devcel.2009.08.005

Boyle AP, Song L, Lee BK, London D, Keefe D, Birney E, Iyer VR, Crawford GE, Furey TS. 2011. High-resolution genome-wide in vivo footprinting of diverse transcription factors in human cells. *Genome Res* 21: 456–464. doi:10.1101/gr.112656.110

Briggs JA, Weinreb C, Wagner DE, Megason S, Peshkin L, Kirschner MW, Klein AM. 2018. The dynamics of gene expression in vertebrate embryogenesis at single-cell resolution. *Science* 360: eaar5780. doi:10.1126/science.aar5780

Bright AR, Veenstra GJC. 2019. Assay for transposase-accessible chromatin-sequencing using *Xenopus* embryos. *Cold Spring Harb Protoc* doi:10.1101/pdb.prot098327.

Buenrostro JD, Giresi PG, Zaba LC, Chang HY, Greenleaf WJ. 2013. Transposition of native chromatin for fast and sensitive epigenomic profiling of open chromatin, DNA-binding proteins and nucleosome position. *Nat Methods* 10: 1213–1218. doi:10.1038/nmeth.2688

Chen X, Miragaia RJ, Natarajan KN, Teichmann SA. 2018. A rapid and robust method for single cell chromatin accessibility profiling. *Nat Commun* 9: 5345. doi:10.1038/s41467-018-07771-0

Chiu WT, Charney Le R, Blitz IL, Fish MB, Li Y, Biesinger J, Xie X, Cho KW. 2014. Genome-wide view of TGFβ/Foxh1 regulation of the early mesendoderm program. *Development* 141: 4537–4547. doi:10.1242/dev.107227

Cho JS, Blitz IL, Cho KWY. 2019. DNase-seq to study chromatin accessibility in early *Xenopus tropicalis* embryos. *Cold Spring Harb Protoc* doi:10.1101/pdb.prot098335

Cusanovich DA, Reddington JP, Garfield DA, Daza RM, Aghamirzaie D, Marco-Ferreres R, Pliner HA, Christiansen L, Qiu X, Steemers FJ, et al. 2018. The *cis*-regulatory dynamics of embryonic development at single-cell resolution. *Nature* 555: 538–542. doi:10.1038/nature25981

Elurbe DM, Paranjpe SS, Georgiou G, van Kruijsbergen I, Bogdanovic O, Gibeaux R, Heald R, Lister R, Huynen MA, van Heeringen SJ, et al. 2017. Regulatory remodeling in the allo-tetraploid frog *Xenopus laevis*. *Genome Biol* 18: 198. doi:10.1186/s13059-017-1335-7

Gentsch GE, Smith JC. 2019. Mapping chromatin features of *Xenopus* embryos. *Cold Spring Harb Protoc* doi:10.1101/pdb.prot100263

Gentsch GE, Owens ND, Martin SR, Piccinelli P, Faial T, Trotter MW, Gilchrist MJ, Smith JC. 2013. In vivo T-box transcription factor profiling reveals joint regulation of embryonic neuromesodermal bipotency. *Cell Rep* 4: 1185–1196. doi:10.1016/j.celrep.2013.08.012

Gilchrist MJ, Veenstra GJC, Cho KWY. 2019. Transcriptomics and proteomics methods for *Xenopus* embryos and tissues. *Cold Spring Harb Protoc* doi:10.1101/pdb.top098350

Hellsten U, Harland RM, Gilchrist MJ, Hendrix D, Jurka J, Kapitonov V, Ovcharenko I, Putnam NH, Shu S, Taher L, et al. 2010. The genome of the western clawed frog *Xenopus tropicalis*. *Science* 328: 633–636. doi:10.1126/science.1183670

Hontelez S, van Kruijsbergen I, Georgiou G, van Heeringen SJ, Bogdanovic O, Lister R, Veenstra GJC. 2015. Embryonic transcription is controlled by maternally defined chromatin state. *Nat Commun* 6: 10148. doi:10.1038/ncomms10148

Hontelez S, van Kruijsbergen I, Veenstra GJC. 2019. ChIP-sequencing in *Xenopus* embryos. *Cold Spring Harb Protoc* doi:10.1101/pdb.prot097907

Jäger R, Migliorini G, Henrion M, Kandaswamy R, Speedy HE, Heindl A, Whiffin N, Carnicer MJ, Broome L, Dryden N, et al. 2015. Capture Hi-C identifies the chromatin interactome of colorectal cancer risk loci. *Nat Commun* 6: 6178. doi:10.1038/ncomms7178

Kalhor R, Tjong H, Jayathilaka N, Alber F, Chen L. 2011. Genome architectures revealed by tethered chromosome conformation capture and population-based modeling. *Nat Biotechnol* 30: 90–98. doi:10.1038/nbt.2057

Kaya-Okur HS, Wu SJ, Codomo CA, Pledger ES, Bryson TD, Henikoff JG, Ahmad K, Henikoff S. 2019. CUT&Tag for efficient epigenomic profiling of small samples and single cells. *Nat Commun* 10: 1930. doi:10.1038/s41467-019-09982-5

Mitros T, Lyons JB, Session AM, Jenkins J, Shu S, Kwon T, Lane M, Ng C, Grammer TC, Khokha MK, et al. 2019. A chromosome-scale genome assembly and dense genetic map for *Xenopus tropicalis*. *Dev Biol* 452: 8–20. doi:10.1016/j.ydbio.2019.03.015

Nakamura Y, de Paiva Alves E, Veenstra GJ, Hoppler S. 2016. Tissue- and stage-specific Wnt target gene expression is controlled subsequent to beta-catenin recruitment to *cis*-regulatory modules. *Development* 143: 1914–1925. doi:10.1242/dev.131664

Neph S, Vierstra J, Stergachis AB, Reynolds AP, Haugen E, Vernot B, Thurman RE, John S, Sandstrom R, Johnson AK, et al. 2012. An expansive human regulatory lexicon encoded in transcription factor footprints. *Nature* 489: 83–90. doi:10.1038/nature11212

Nicoletti C, Forcato M, Bicciato S. 2018. Computational methods for analyzing genome-wide chromosome conformation capture data. *Curr Opin Biotechnol* 54: 98–105. doi:10.1016/j.copbio.2018.01.023

Paraiso KD, Blitz IL, Coley M, Cheung J, Sudou N, Taira M, Cho KWY. 2019. Endodermal maternal transcription factors establish super-enhancers during zygotic genome activation. *Cell Rep* 27: 2962–2977 e2965. doi:10.1016/j.celrep.2019.05.013

Quigley IK, Heinz S. 2019. Generating a three-dimensional genome from *Xenopus* with Hi-C. *Cold Spring Harb Protoc* doi:10.1101/pdb.prot098343

Quigley IK, Kintner C. 2017. Rfx2 stabilizes Foxj1 binding at chromatin loops to enable multiciliated cell gene expression. *PLoS Genet* 13: e1006538. doi:10.1371/journal.pgen.1006538

Rozenblatt-Rosen O, Stubbington MJT, Regev A, Teichmann SA. 2017. The human cell atlas: from vision to reality. *Nature* 550: 451–453. doi:10.1038/550451a

Savova V, Pearl EJ, Boke E, Nag A, Adzhubei I, Horb ME, Peshkin L. 2017. Transcriptomic insights into genetic diversity of protein-coding genes in *X. laevis*. *Dev Biol* 424: 181–188. doi:10.1016/j.ydbio.2017.02.019

Session AM, Uno Y, Kwon T, Chapman JA, Toyoda A, Takahashi S, Fukui A, Hikosaka A, Suzuki A, Kondo M, et al. 2016. Genome evolution in the allotetraploid frog *Xenopus laevis*. *Nature* 538: 336–343. doi:10.1038/nature19840

Song L, Crawford GE. 2010. DNase-seq: a high-resolution technique for mapping active gene regulatory elements across the genome from mammalian cells. *Cold Spring Harb Protoc* doi:10.1101/pdb.prot5384

Thélie A, Desiderio S, Hanotel J, Quigley I, Van Driessche B, Rodari A, Borromeo MD, Kricha S, Lahaye F, Croce J, et al. 2015. *Prdm12* specifies V1 interneurons through cross-repressive interactions with *Dbx1* and *Nkx6* genes in *Xenopus*. *Development* 142: 3416–3428. doi:10.1242/dev.121871

van Heeringen SJ, Akhtar W, Jacobi UG, Akkers RC, Suzuki Y, Veenstra GJ. 2011. Nucleotide composition-linked divergence of vertebrate core promoter architecture. *Genome Res* 21: 410–421. doi:10.1101/gr.111724.110

Yasuoka Y, Suzuki Y, Takahashi S, Someya H, Sudou N, Haramoto Y, Cho KW, Asashima M, Sugano S, Taira M. 2014. Occupancy of tissue-specific *cis*-regulatory modules by Otx2 and TLE/Groucho for embryonic head specification. *Nat Commun* 5: 4322. doi:10.1038/ncomms5322

Yoon SJ, Wills AE, Chuong E, Gupta R, Baker JC. 2011. HEB and E2A function as SMAD/FOXH1 cofactors. *Genes Dev* 25: 1654–1661. doi:10.1101/gad.16800511

Cite this introduction as *Cold Spring Harb Protoc*; doi:10.1101/pdb.top097915

Protocol 1

ChIP-Sequencing in *Xenopus* Embryos

Saartje Hontelez,[1,2] Ila van Kruijsbergen,[1,2] and Gert Jan C. Veenstra[1,3]

[1]*Radboud University, Department of Molecular Developmental Biology, Radboud Institute for Molecular Life Sciences, Faculty of Science, 6500 HB Nijmegen, The Netherlands*

Chromatin immunoprecipitation (ChIP) followed by deep sequencing (ChIP-seq) is a powerful technique for mapping in vivo, genome-wide DNA–protein interactions. The interplay between DNA and proteins determines the transcriptional state of the genome. Using specific antibodies for the ChIP, it is possible to generate genome-wide profiles of histone posttranslational modifications, providing insight into the epigenetic memory and developmental potential of cells. The interactions between DNA and proteins involved in epigenetic regulation and transcription are highly dynamic during embryonic development. ChIP-seq allows for a detailed analysis of these dynamic changes in DNA–protein binding during embryogenesis. ChIP-seq is performed on protein epitopes that have been cross-linked to genomic DNA. After shearing the DNA, fragments bound by the (modified) protein of interest are captured with antibodies. The genomic loci of interest are identified by sequencing. Here, we provide a step-by-step ChIP-seq protocol that efficiently captures epitopes from relatively small embryo samples.

MATERIALS

It is essential that you consult the appropriate Material Safety Data Sheets and your institution's Environmental Health and Safety Office for proper handling of equipment and hazardous materials used in this protocol.

RECIPES: Please see the end of this protocol for recipes indicated by <R>. Additional recipes can be found online at http://cshprotocols.cshlp.org/site/recipes.

Reagents

Agarose

Agencourt AMPure XP

Antibody to protein of interest

An antibody recognizing the H3K4me3 epitope (Abcam ab8580) can be used as a positive control.

Dynabeads (Protein A or Protein G)

E-gel Sizeselect 2% agarose gels

Elution buffer for ChIP-seq <R>

Formaldehyde solution (1% [w/v] in 0.1× MMR) (freshly prepared)

Immediately before use, dilute 16% formaldehyde solution (methanol-free) in 0.1× MMR to prepare a 1% solution.

---

[2]These authors contributed equally to this work.

[3]Correspondence: g.veenstra@science.ru.nl

Cite this protocol as *Cold Spring Harb Protoc*; doi:10.1101/pdb.prot097907

Glycine (125 mM in 0.25× MMR) (freshly prepared)

Immediately before use, dilute 2.5 M glycine 20-fold in 0.25× MMR.

Incubation buffer for ChIP-seq <R>
KAPA Hyper Prep Kit for Illumina sequencing (Kapa Biosystems)
Marc's Modified Ringer's (MMR) solution (10×, pH 7.4) <R>
MinElute PCR Purification Kit and QIAquick columns (QIAgen)
NaCl (5 M)
NEXTflex Illumina DNA Barcodes (Bioo Scientific 514101)
Proteinase K (10 mg/mL)

Store at −20°C.

RNase A (10 mg/mL)

Store at −20°C.

Sonication buffer <R>
TE buffer (ChIP-seq) <R>
Wash buffer 1 for ChIP-seq <R>
Wash buffer 2 for ChIP-seq <R>
Wash buffer 3 for ChIP-seq <R>
Xenopus embryos (*X. tropicalis* or *X. laevis*), dejellied (Sive et al. 2000; del Viso and Khokha 2012)

Equipment

2100 Bioanalyzer (Agilent)
Bottle-top vacuum filter (0.22-μm; Corning)

All solutions are made with demineralized H_2O in autoclaved cylinders, sterilized with 0.22-μm filters and stored at room temperature unless indicated otherwise.

Centrifuge at 4°C
Fluorometer (e.g., Qubit)
Gel electrophoresis equipment for sizing DNA fragments
Heat blocks at 37°C and 65°C
Liquid nitrogen
Magnet rack (Eppendorf)
Microcentrifuge tubes
Next generation sequencer
PCR cycler for quantitative (real-time) polymerase chain reaction (qPCR)
Rotator for microcentrifuge tubes at 4°C
Shaker for six-well plates
Sonication device (e.g., Diagenode Bioruptor Pico Ultrasonicator)
Thermomixer at 65°C
Tissue culture plates (six-well)

METHOD

Previously we described a ChIP-seq protocol for Xenopus embryos (Jallow et al. 2004; Akkers et al. 2012). This method provides an updated ChIP-seq method for capturing epitopes from relatively small embryo samples.

Chromatin Isolation

1. Fix dejellied embryos in a six-well plate (up to 300 *X. tropicalis* or *X. laevis* embryos per well) in 5 mL of 1% formaldehyde (methanol-free) in 0.1× MMR.

   Make sure that the embryos are fully submerged; they tend to float to the surface after changing solutions and are very sensitive to the surface tension of the liquid.

2. Gently shake the embryos for 30 min at room temperature.

   Optimize this step for new antibodies by varying the fixation time.

3. Transfer the embryos to 5 mL of 125 mM glycine in 0.25× MMR.

4. Gently shake the embryos for 30 min at room temperature.

5. Transfer the embryos to 5 mL of 0.25× MMR.

6. Gently shake the embryos for 15 min at room temperature.

7. Wash the embryos by repeating Steps 5 and 6 with fresh 0.25× MMR.

8. Remove all liquid.

9. Homogenize 300 embryos in 2 mL of sonication buffer by pipetting up and down. Transfer the homogenized embryos to tubes on ice for sonication.

   Adjust the volume of sonication buffer for different numbers of embryos.

Sonication

Keep the chromatin on ice after sonication.

10. Sonicate the chromatin at 4°C to obtain chromatin fragments ranging from 150 to 1000 bp.

    The majority of fragments should be around 200 bp. Fragment size and sonication efficiency is checked in Step 12.

11. Centrifuge the sample for 5 min at 13,000 rpm at 4°C. Keep the supernatant (chromatin extract).

12. Transfer a small aliquot (e.g., 10–50 μL) of chromatin extract to a new tube and check the fragment size as follows.

    i. Add 0.5 μL of proteinase K (10 mg/mL) to the aliquot. Incubate for 1 h at 65°C.

    ii. Add 2 μL of RNase A (10 mg/mL) to the aliquot. Incubate for 30 min at 37°C.

    iii. Load the DNA on a 1% agarose gel and evaluate the fragment size by electrophoresis.

       Alternatively, a Bioanalyzer (Agilent) is recommended to reduce the amount of genomic DNA required.

    iv. If sonication was insufficient, repeat Steps 10–12 to sonicate the chromatin extract for additional time. Otherwise, proceed to Step 13.

       See Troubleshooting.

13. Snap-freeze aliquots of the remaining chromatin extract in liquid nitrogen. Store at −80°C until ready to proceed with immunoprecipitation.

Immunoprecipitation

This protocol has been optimized for efficient capture of epitopes. The described method uses 15 embryo equivalents (Eeq) of chromatin and can be used as a starting point. The actual number of embryos required depends on antibody quality (see Fig. 1 for positive control of histone H3 lysine 4 trimethylation, H3K4me3), the developmental stage of the embryos, and the properties of the epitope. For scaling up to n x 15 embryos, n ChIPs with 15 embryos can be pooled in Step 40.

All steps are performed on ice or at 4°C, unless indicated otherwise, until Step 30.

Day 1

14. Set aside a 20-μL aliquot of chromatin extract as input control (20%). Store at 4°C until Step 38.

15. Mix 100 μL of chromatin extract (15 Eeq) with 100 μL of incubation buffer.

16. Add 1–5 μg of antibody to the chromatin extract solution.

    Test several amounts of antibody (e.g., 1, 2, and 5 μg) to obtain the best signal-to-noise ratio.

17. Rotate the sample overnight.

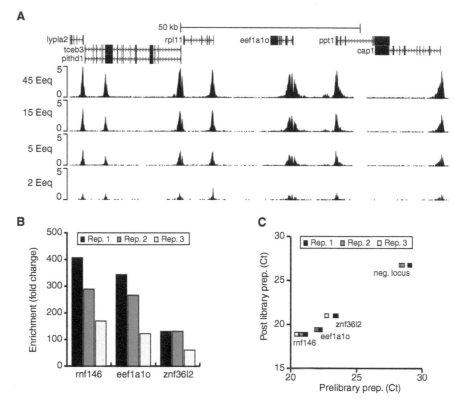

FIGURE 1. ChIP-seq quality controls. (*A*) Genome browser view showing the H3K4me3 (RPKM, normalized read counts) generated by performing ChIP-seq with 2, 5, 15, and 45 blastula embryos (Stage 9). (*B*) Before library construction and potential pooling of replicates, samples are analyzed by qPCR to determine the signal-to-noise ratio and reproducibility of the ChIP. The qPCR signal (recovery of input) of known positive loci (*rnf146*, *eef1a1o*, *znf36l2*) was plotted over the signal of a known negative locus (gene desert). (*C*) qPCR analysis is also performed after library construction to confirm that signal ratios have been retained. The qPCR signal (Ct values) of four loci (*rnf146*, *eef1a1o*, *znf36l2*, gene desert) before and after library construction were plotted against each other. In this example, replicates 1, 2, and 3 were pooled when constructing the library.

Day 2

18. Resuspend and transfer 20 µL of Dynabeads to a fresh microcentrifuge tube.

19. Place the tube on a magnet. Discard the supernatant.

20. Resuspend the beads in 1 mL of incubation buffer.

21. Repeat Step 19.

22. Resuspend the beads in 50 µL of incubation buffer.

23. Add 50 µL of the prepared Dynabeads to the chromatin extract and antibody mix from Step 17.

24. Rotate the sample for 1 h.

25. Place the tube on a magnet. Discard the supernatant.

26. Resuspend the beads in 1 mL of wash buffer 1 by inverting the tube.

27. Rotate the beads for 5 min.

28. Centrifuge the tube briefly (<2000 rpm for a few seconds) to collect liquid from the lid in the tube.

29. Wash the beads by repeating Steps 25–28 using the appropriate wash buffer as follows.

    i. Wash once using 1 mL of wash buffer 2.

    ii. Wash once using 1 mL of wash buffer 3.

Cite this protocol as *Cold Spring Harb Protoc*; doi:10.1101/pdb.prot097907

      iii. Wash once using 1 mL of wash buffer 1.

      iv. Wash once using 0.5 mL of TE buffer.

30. Transfer the tube to room temperature.

    The remainder of the procedure is performed at room temperature.

31. Centrifuge the tube briefly (<2000 rpm for a few seconds) and then place the tube on a magnet. Discard the TE buffer.

32. Resuspend the beads in 200 µL of elution buffer by inverting the tube.

33. Rotate the beads for 15 min.

34. Centrifuge the tube briefly (<2000 rpm for a few seconds).

35. Place the tube on a magnet. Transfer the supernatant to a fresh tube.

36. Repeat Step 32 to 35 and pool the supernatant with the supernatant from Step 35.

37. Add 16 µL of 5 M NaCl and 0.5 µL of proteinase K (10 mg/mL) to the pooled supernatant and mix well by vortexing.

38. Add 380 µL of elution buffer, 16 µL of 5 M NaCl and 0.5 µL of proteinase K (10 mg/mL) to the chromatin input control from Step 14. Mix by vortexing. Process the input control along with the ChIP sample from this point onward.

39. Incubate the samples for 5 h at 65°C with rotation (1000 rpm).

40. Purify the DNA using a MinElute PCR purification kit and QIAquick columns. Elute in 30 µL of nuclease-free H_2O. Measure the amount of DNA recovered (for example, using Qubit fluorometric quantification).

    Library construction (Steps 42–46) starts with 2–5 ng DNA. For scaling up to n x 15 embryos, n ChIPs with 15 embryos can be pooled in this step.

41. Analyze 5 µL of DNA by qPCR.

      i. Dilute 5 µL of the ChIP-purified DNA 25-fold for qPCR analysis.

      ii. Dilute the input equally and then prepare an input dilution series of 10%, 1%, 0.1%, and 0.001% to calculate recovery after ChIP.

      iii. Analyze by qPCR.

        To ensure a good signal-to-noise ratio, recovery of positive loci should be >10-fold over negative loci as assessed by qPCR (Fig. 1B,C).

        See Troubleshooting.

Library Construction

42. Construct the library using the KAPA Hyper Prep kit according to the manufacturer's instructions.

      i. Perform end repair, A-tailing and ligation of DNA Barcodes (600 nM for 2–5 ng of DNA, or 300 nM for <2 ng) using the KAPA Hyper Prep kit reagents and 2–5 ng of DNA from Step 40.

      ii. Perform postligation clean-up using Agencourt AMPure XP.

      iii. Amplify the library using the KAPA Hyper Prep kit reagents.

      iv. Perform postamplification cleanup using the MinElute PCR Purification Kit.

43. Select 300-bp fragments using an E-gel SizeSelect 2% agarose gel.

44. Dilute 2 µL of the isolated DNA 100- to 1000-fold and analyze by qPCR.

    The Ct values before and after sample preparation should show a good correlation (Fig. 1C).

45. Check for adapter contamination using a Bioanalyzer.

46. Store the DNA at −20°C until sequencing on a next generation sequencer.

    See Discussion.

TROUBLESHOOTING

Problem (Step 12): Fragment size or sonication efficiency is suboptimal.

Solution: If fragments are too small or too big, reduce or increase the sonication time, respectively. For specific tissues (e.g., anterior head dissections), sonication might not be effective, and other fragmentation methods can be applied, for example using MNase.

Problem (Step 41): The qPCR analysis signal-to-noise ratio is too low.

Solution: For a signal-to-noise ratio of <10, consider the following for optimization.

- *Fixation time.* If fixation is too long, molecules will be cross-linked over larger distances. The chromatin will also become refractory to sonication. This leads to a background signal in non-target regions of DNA. Over-cross-linking also reduces the availability of the epitope in the chromatin, reducing the signal. If the fixation is too short, the epitope may not be covalently linked to DNA. This may lead to reduced capture of the epitope.

- *Number of embryos per ChIP.* Increasing the number of embryos per ChIP could improve the ChIP signal. Generally, earlier developmental stages require more embryos per ChIP, but this also depends on other factors, such as how abundant the epitope is in the collected cells or embryos. Also, some antibodies are markedly more efficient in capturing epitopes than other antibodies. Some antibodies are specific (high target-signal over genome-wide noise) but do require more input chromatin for sufficient yields. See Figure 1 for the example of a highly sensitive H3K4me3 ChIP with as few as two blastula embryos: H3K4me3 is a ubiquitous (present in all cells on active promoters) and abundant (two epitopes per promoter-associated nucleosome) epitope that is stably associated with DNA and cross-links very well.

- *Amount of antibody.* The signal-to-noise ratio of an antibody depends on the amount of antibody per ChIP volume. The antibody-to-volume ratio should be optimized for each individual antibody.

- *Choice of antibody.* Ultimately the quality of a ChIP is determined by the quality of the antibody. Several antibodies should be tested if multiple antibodies are available for the protein of interest.

DISCUSSION

Common Data Analysis Steps

The ChIP library can be sequenced on different platforms in-house or using commercial sequencing service providers. The default (primary) analysis pipelines may vary accordingly. Raw sequence reads on the Illumina NextSeq or HiSeq platforms are typically obtained by real-time image processing and base calling during the run, resulting in FASTQ files. Subsequently, mapping data (indexed BAM files, containing raw and mapped reads) are generated by aligning reads to the reference genome (downloaded from Xenbase [www.xenbase.org] using BWA [Li and Durbin 2009, 2010; Li 2012]). The BAM files will be used as a starting point for follow-up analyses. Next, read-density maps (bigWig files) can be generated by counting overlapping mapped reads at nucleotide resolution. In the case of single end sequencing, the reads should be extended to the mean fragment length of DNA in the library. The bigWig files are ideal for visualization of the data in the UCSC genome browser, and should be normalized for sequencing depth, for example using RPKM (reads per kilobase per million reads sequenced). In addition, in many cases it will be useful to distinguish peaks from background (peak files, typically in BED format) using MACS2 software (Zhang et al. 2008). A random selection of peaks can be subjected to experimental verification (ChIP-qPCR) to obtain an experimentally derived False

Cite this protocol as *Cold Spring Harb Protoc*; doi:10.1101/pdb.prot097907

Discovery Rate (FDR < 0.05; if higher values are obtained, peaks are called more stringently by q-value or filtered for signal over background after peak-calling). This ensures that enriched regions can be established with a high level of confidence. Normalized read counts (RPKM) are calculated for all enriched regions. These normalized values can be used to generate an intensity profile for all peak regions across samples or developmental stages. These can be used for follow-up analyses, for example mapping to the closest gene, enrichment analysis, clustering, and regulatory motif analysis. For a further discussion, see Bogdanovic and van Heeringen (2016).

RECIPES

Elution Buffer for ChIP-Seq

0.1 M $NaHCO_3$ (pH 8.8)
1% SDS

Prepare fresh each day and store at room temperature.

Incubation Buffer for ChIP-Seq

50 mM Tris-HCl (pH 8.0)
100 mM NaCl
2 mM EDTA (pH 8.0)
1% Nonidet P40 (10%)

Combine the reagents listed above. Immediately before use, cool buffer on ice and add 1 M DTT (1 mM final concentration) and protease inhibitor cocktail (PIC) (25×) <R> (1× final concentration).

Marc's Modified Ringer's (MMR) Solution (10×, pH 7.4)

51.4 g NaCl
2.94 g $CaCl_2 \cdot 2H_2O$
2.03 g $MgCl_2 \cdot 6H_2O$
11.9 g HEPES
6.67 mL 3 M KCl

Combine the reagents listed above and bring the volume to 1 L with H_2O. Store for up to 2 years at room temperature.

Protease Inhibitor Cocktail (PIC) (25×)

To prepare a 25× stock solution, dissolve one tablet of PIC (Roche 04 693 132 001) in 2 mL of H_2O. Store long-term at −20°C. After thawing, store for up to 2 wk at 4°C.

Sonication Buffer

20 mM Tris-HCl (pH 8.0)
70 mM KCl
1 mM EDTA (pH 8.0)
10% glycerol
0.125% Nonidet P40 (10%)

Combine the reagents listed above. Immediately before use, cool buffer on ice and add 1 M DTT (5 mM final concentration) and protease inhibitor cocktail (PIC) (25×) <R> (1× final concentration).

TE Buffer (ChIP-Seq)

10 mM Tris–HCl (pH 8.0)

1 mM EDTA (pH 8.0)

Combine the reagents listed above. Immediately before use, cool on ice and add protease inhibitor cocktail (PIC) (25×) <R> to a final concentration of 1× PIC.

Wash Buffer 1 for ChIP-Seq

Incubation buffer for ChIP-seq <R>

0.1% deoxycholate

Combine the reagents listed above. Immediately before use, cool buffer on ice and add 1 M DTT (1 mM final concentration) and protease inhibitor cocktail (PIC) (25×) <R> (1× final concentration).

Wash Buffer 2 for ChIP-Seq

Incubation buffer for ChIP-seq <R>

400 mM NaCl

0.1% deoxycholate

Combine the reagents listed above. Immediately before use, cool buffer on ice and add 1 M DTT (1 mM final concentration) and protease inhibitor cocktail (PIC) (25×) <R> (1× final concentration).

Wash Buffer 3 for ChIP-Seq

Incubation buffer for ChIP-seq <R>

250 mM LiCl

0.1% deoxycholate

Combine the reagents listed above. Immediately before use, cool buffer on ice and add 1 M DTT (1 mM final concentration) and protease inhibitor cocktail (PIC) (25×) <R> (1× final concentration).

REFERENCES

Akkers RC, Jacobi UG, Veenstra GJ. 2012. Chromatin immunoprecipitation analysis of *Xenopus* embryos. *Methods Mol Biol* **917**: 279–292.

Bogdanovic O, van Heeringen SJ. 2016. ChIP-seq data processing for PcG proteins and associated histone modifications. *Methods Mol Biol* **1480**: 37–53.

del Viso F, Khokha M. 2012. Generating diploid embryos from *Xenopus tropicalis*. In Xenopus *protocols: Post-genomic approaches*, 2nd ed. (ed. Hoppler S, Vize P), Vol. **917**. Methods in molecular biology. Humana Press, Totowa, NJ.

Jallow Z, Jacobi UG, Weeks DL, Dawid IB, Veenstra GJ. 2004. Specialized and redundant roles of TBP and a vertebrate-specific TBP paralog in embryonic gene regulation in *Xenopus*. *Proc Natl Acad Sci* **101**: 13525–13530.

Li H. 2012. Exploring single-sample SNP and INDEL calling with whole-genome de novo assembly. *Bioinformatics* **28**: 1838–1844.

Li H, Durbin R. 2009. Fast and accurate short read alignment with Burrows-Wheeler transform. *Bioinformatics* **25**: 1754–1760.

Li H, Durbin R. 2010. Fast and accurate long-read alignment with Burrows-Wheeler transform. *Bioinformatics* **26**: 589–595.

Sive HL, Grainger M, Harland RM. 2000. *Early development of* Xenopus laevis: *A laboratory manual*. Cold Spring Harbor Laboratory Press, Cold Spring Harbor, NY.

Zhang Y, Liu T, Meyer CA, Eeckhoute J, Johnson DS, Bernstein BE, Nusbaum C, Myers RM, Brown M, Li W, et al. 2008. Model-based analysis of ChIP-seq (MACS). *Genome Biol* **9**: R137.

Cite this protocol as *Cold Spring Harb Protoc*; doi:10.1101/pdb.prot097907

Assay for Transposase-Accessible Chromatin-Sequencing Using *Xenopus* Embryos

Ann Rose Bright[1] and Gert Jan C. Veenstra[1,2]

[1]*Radboud University, Department of Molecular Developmental Biology, Faculty of Science, Radboud Institute for Molecular Life Sciences, Nijmegen 6500 HB, The Netherlands*

The DNA of eukaryotic genomes is packaged into chromatin by nucleosomes. This not only compacts the DNA but also plays a central role in gene regulation and establishment of cellular identity during development. Because of this packaging, the DNA is relatively inaccessible to nucleoplasmic factors; however, regulatory elements such as promoters, enhancers, and insulators are largely kept nucleosome-free. The assay for transposase-accessible chromatin (ATAC-seq) can be used to identify genomic locations of "open" chromatin, footprints of DNA-binding proteins, and positioned nucleosomes. It therefore is a powerful tool for unraveling the dynamic regulatory landscape of chromatin. The method exploits the action of hyperactive prokaryotic Tn5-transposase, which preferentially cuts DNA in accessible chromatin and tags the sites with sequencing adaptors. Here we describe an ATAC-seq protocol for use with *Xenopus tropicalis* embryos.

MATERIALS

It is essential that you consult the appropriate Material Safety Data Sheets and your institution's Environmental Health and Safety Office for proper handling of equipment and hazardous materials used in this protocol.

RECIPES: Please see the end of this protocol for recipes indicated by <R>. Additional recipes can be found online at http://cshprotocols.cshlp.org/site/recipes.

Reagents

Cysteine (3%)
   Prepare 3% cysteine solution in 0.1× MMR. Adjust to pH 7.8 using NaOH.

HiFi HotStart ReadyMix PCR Kit (KAPA KK2602)
Lysis buffer for *Xenopus* <R> (ice-cold)
Marc's Modified Ringer's (MMR) solution (1×) <R>
MinElute PCR Purification Kit (including Elution Buffer) (QIAgen 28004)
Nextera DNA Library Preparation Kit (including 2× TD Buffer and Tn5 Transposase) (Illumina FC-121-1030)
Nextera Index Kit (Illumina FC-121-1011)
qPCR primers for performing quality checks (see Discussion and Table 1)

---

[2]Correspondence: g.veenstra@science.ru.nl

TABLE 1. Primer sequences spanning hypersensitive, nonhypersensitive, and mitochondrial regions

Genomic location[a]	Forward sequence	Reverse sequence
Hypersensitive region		
Chr02:86683349-86683371	GCAGAGGTGAGTATGGGATGG	AAAAGCAGGCCAGTAAGCCA
Chr05:14131176-14131196	AGCAGGAGGGTCATGTCAAC	GTGCGTTTAGGGCTATCGAG
Chr01:141442069-141442089	GCAGAAGGGGAGGAGAACT	AACTGCAAGCCTGCTAAGGT
Chr05:118265078-118265098	GAATAGGGCAACACAAGGCT	TCTTTTCAAACACCACCCGC
Nonhypersensitive region		
Chr02:5242998-5243018	CATTCCCTACTGGGCTGGGT	CACACTGCTGGCCATCGTT
Chr05:7602266-7602286	TTTCAGTCCCGCAGATTTTC	TACAATGGCCCAATCAAAGC
Mitochondrial region		
mitochondrion:11795-11815	CCTCCACCTCATCCCTATCA	CCAGCGGAGAGACTGTTTTC
mitochondrion:8777-8797	GTCGCAGCCCTTCTACTCAC	ACTGGAGGAGTGTGGTGTCC
mitochondrion:8918-8938	CCGAGCCTACTTCACCTCAG	AGAAGAGTTGGCAAGGACGA

[a]Genomic positions refer to *X. tropicalis* genome assembly v9.0.

SYBR Green I nucleic acid gel stain (Life Technologies, S7563)

Prepare 100× SYBR Green I solution by mixing 1 µL of 10,000× SYBR Green I with 99 µL of Elution Buffer (10 mM Tris-Cl, pH 8).

Xenopus tropicalis embryos

Equipment

Bioanalyzer
Heat block at 37°C
Heratherm incubator (Thermo Scientific)
Microcentrifuge, cooling (Thermo Scientific)
Microcentrifuge tubes (1.5-mL)
PCR tubes (0.2-mL)
Petri dishes (100-mM, plastic)
Real-time quantitative polymerase chain reaction (qPCR) system
Thermal cycler (Bio-Rad T100)

METHOD

This protocol is adapted from the ATAC-seq method originally described by Buenrostro et al. (2013). For further reading about chromatin structure and development, see Perino and Veenstra (2016) and Boyle et al. (2008).

Preparing Lysates

Generally it is recommended to use fresh material for preparation of lysates. ATAC libraries can be prepared using frozen embryos (see Step 4); however, the enrichment of signal in accessible chromatin tends to be lower compared to libraries prepared with fresh material.

1. Collect *X. tropicalis* embryos in a Petri dish. Dejelly the embryos using 3% cysteine.

2. Rinse the embryos in H₂O to remove the cysteine. Raise the embryos in 0.1× MMR until the desired stage is attained.

3. Transfer embryos in a number equivalent to 25,000–75,000 cells to a 1.5-mL microcentrifuge tube for further processing.

   For example, for stage 12 embryos (~30,000 cells/embryo) we have used two embryos per reaction. More generally, the efficiency of the reaction depends on the number of cells relative to the amount of Tn5 transposase. For best results, we have found titration of the embryo lysate with a constant amount of Tn5 to be useful.

4. Remove the excess MMR solution and then add 10 µL of ice-cold lysis buffer to the embryos. Pipette the embryos up and down with a clipped P20 pipette tip until a homogenized lysate is formed.

Clipped pipette tips help to prevent shearing of nuclei and chromatin.

Embryos can be snap-frozen after removal of excess MMR solution and stored at −80°C. For subsequent ATAC-seq, the samples are thawed quickly by addition of lysis buffer followed by pipetting up and down.

5. Place the lysate on ice for 5 min to ensure proper lysis, and then proceed directly with the transposition reaction (Step 6).

Performing Transposition

6. Combine the following to prepare the transposition reaction mix. Mix gently without vortexing.

Reagent	Amount
Cell lysate from Step 5	10 µL
2× TD Buffer	25 µL
Tn5 Transposase	2.5 µL
Nuclease-free H_2O	12.5 µL
Total	50 µL

7. Incubate the reaction mix for 30 min at 37°C.

8. Immediately following transposition, purify the DNA from the reaction mix using a MinElute kit. Elute the transposed DNA in 10 µL of Elution Buffer (10 mM Tris buffer, pH 8).

Purified samples can be stored at −20°C.

Performing PCR Amplification

9. Combine the following in a PCR tube to prepare the PCR reaction mix. Mix gently by pipetting.

Reagent	Amount
Transposed DNA from Step 8	10 µL
Nuclease-free H_2O	10 µL
25 µM Index 1 (Nextera Index Kit)	2.5 µL
25 µM Index 2 (Nextera Index Kit)	2.5 µL
Kapa Hifi PCR Master mix	25 µL
Total	50 µL

10. Amplify the transposed DNA using the following cycling program.

1	72°C for 5 min
2	98°C for 30 sec
3	98°C for 10 sec
4	63°C for 30 sec
5	72°C for 1 min
6	Repeat 3–5 for 4 cycles
7	Hold at 4°C

11. Determine the number of additional cycles needed for the amplification of the library using qPCR as follows.

i. Prepare the qPCR reaction mix.

Reagent	Amount
Amplified DNA from Step 10	5 μL
Nuclease-free H_2O	4.44 μL
25 μM Index 1 (Nextera Index Kit)	0.25 μL
25 μM Index 2 (Nextera Index Kit)	0.25 μL
100× SYBR Green I	0.06 μL
PCR Master Mix (HiFi HotStart ReadyMix)	5 μL
Total	15 μL

ii. Use the following cycling program to perform qPCR.

1	98°C for 30 sec
2	98°C for 10 sec
3	63°C for 30 sec
4	72°C for 1 min
5	Repeat 2–4 for 19 cycles
6	Hold at 4°C

iii. Calculate the number of additional cycles (x) needed for the amplification of the library by plotting the resulting fluorescence versus the number of cycles and setting the threshold at one-fourth of maximum fluorescence.

> This corresponds to the number of cycles needed to reach the beginning of exponential increase of fluorescence (Buenrostro et al. 2013).

12. Amplify the remaining reaction mix from Step 10 using the following cycling program and the cycle number (x) calculated in Step 11.iii.

1	98°C for 30 sec
2	98°C for 10 sec
3	63°C for 30 sec
4	72°C for 1 min
5	Repeat 2–4 for x cycles
6	Hold at 4°C

13. Purify the amplified product using the MinElute kit. Elute in 20 μL of Elution Buffer.

> Purified samples can be stored at −20°C.

14. Perform quality checks of signal-to-noise ratio, mitochondrial contamination and sample profile (fragment size distribution) before sequencing and analysis (see Discussion and Table 1).

> See Troubleshooting.

TROUBLESHOOTING

Problem (Step 14): Cells are not adequately lysed, leading to low quality samples.
Solution: Prepare and test lysis buffer containing different concentrations of IGEPAL CA-630 (0.5%, 0.1%, 0.05%, and 0.025%).

Problem (Step 14): Large fragments (700- to 1000-bp) are present in the library due to inadequate (nonspecific) transposition, and/or low enrichment of positive regions over negative regions in qPCR quantification.

Cite this protocol as *Cold Spring Harb Protoc*; doi:10.1101/pdb.prot098327

Solution: To optimize the efficiency of specific transposition, do a titration with different concentrations of Tn5 (or use different amounts of embryo lysate with a fixed amount of Tn5). If large fragments (>1 kb, nonspecific transposition) are observed in the library, in addition to the small sizes (transposition in accessible regions), size-selecting the library for 100- to 700-bp fragments (including primer sequence) might help to enhance library quality. The Select-a-Size DNA Clean & Concentrator kit (Zymo Research D4080) can be used for purifying out the desired library size.

DISCUSSION

Performing Quality Checks

Signal-to-Noise Ratio

The signal-to-noise ratio can be determined by performing qPCR using primers spanning hypersensitive and nonhypersensitive regions. We have designed primers spanning H3K4me3-positive promoters and Ep300-positive enhancers for positive regions and for regions without any of the well-known histone modifications and genes nearby as negative (relatively inaccessible) controls (Table 1). Good quality samples have ≥10-fold enrichment for positive regions against negative regions.

Mitochondrial Contamination

The abundance of mitochondrial DNA in the accessible fraction is one of the well-known limitations of the technique: The mitochondrial DNA reduces the coverage of genomic DNA when the ATAC library is sequenced. The extent of mitochondrial contamination in the library can be quantified by qPCR, using primers spanning mitochondrial regions (Table 1). The percentage of mitochondrial reads may vary across samples; it can range from ∼10%–50% of sequenced reads.

Sample Profile

The libraries generated using ATAC-seq usually have a highly diverse fragment size distribution. Samples containing excessive large fragments (>1 kb) are relatively hard to quantify and result in reduced clustering efficiencies when sequenced (Buenrostro et al. 2013). Thus, this information is a good indicator of sample quality and can be checked using a Bioanalyzer.

ATAC-Sequencing and Analysis

Paired-End or Single-End Sequencing

The fragments generated in ATAC-seq are of different lengths. The small ones (50–150 bp, excluding primer sequences) correspond to nucleosome-free hypersensitive regions, and the longer ones (200–500 bp) correspond to open regions that are separated by one or more nucleosomes (mostly double-cut fragments from the edges of the accessible regions). Paired-end sequencing is advantageous over single-end sequencing as it gives us the information to determine both the pattern of accessibility within open regions and the positions of flanking nucleosomes based on fragment length. Paired-end sequencing of 2 × 40 bp is sufficient to map the reads to the genome.

Sequencing Depth

The depth to which the sample needs to be sequenced depends on the information to be obtained. Information on chromatin accessibility and nucleosome positioning can be attained through ∼30–40 million mapped reads for samples with decent enrichments of accessible chromatin and low mitochondrial DNA contamination (Fig. 1). For further analysis, it is advised in most cases to take only the nonduplicate (only one read per genomic position) and uniquely mapped (mapped to unique parts of the genome) reads. A good data set typically has at least 25 million nonduplicate, nonmitochondrial

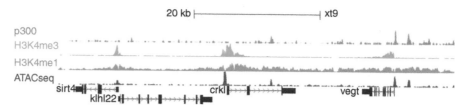

FIGURE 1. ChIP-seq and ATAC-seq profiles surrounding the *vegt* locus in stage 12 (late gastrula) *X. tropicalis* embryos.

aligned (mapped) reads per replicate. For in-depth analysis of transcription factor footprints, at least 100 million mapped reads are necessary.

Replicates

The most accessible sites will be consistent across replicates, so for a global analysis of chromatin accessibility two replicates often suffice; for high confidence analysis of weakly accessible regions and to explicitly account for biological variation, a larger number of replicates may be necessary.

Sample Analysis Pipeline

Single-end or paired-end reads can be mapped to the genome using BWA-MEM (Li and Durbin 2009) followed by filtering steps to remove duplicates using Picard's Mark Duplicates (http://broadinstitute. github.io/picard/) and unmapped reads. Shift the reads +4 bp for the +strand and −5 bp for the − strand to account for the properties of the accessible site relative to the read; this is primarily relevant when attempting to do high resolution transcription factor footprinting analysis. Peak calling can be done using MACS2 (Zhang et al. 2008). The processed samples can be used for further analyses such as differential chromatin accessibility, clustering and Motif enrichment. In the case of footprint identification, the sequence bias of Tn5 should be controlled using naked genomic DNA (Lu et al. 2017).

RECIPES

Lysis Buffer for Xenopus

10 mM Tris-HCl (pH 7.4)
10 mM NaCl
3 mM MgCl$_2$
0.1% IGEPAL CA-630

Store for up to 1 wk at 4°C.

Marc's Modified Ringer's (MMR) Solution (1×)

0.1 M NaCl
0.1 mM EDTA
1 mM MgSO$_4$
2 mM CaCl$_2$
2 mM KCl
5 mM HEPES (pH 7.8)

Autoclave. Store for up to 6 mo at room temperature.

Cite this protocol as *Cold Spring Harb Protoc*; doi:10.1101/pdb.prot098327

REFERENCES

Boyle AP, Davis S, Shulha HP, Meltzer P, Margulies EH, Weng Z, Furey TS, Crawford GE. 2008. High-resolution mapping and characterization of open chromatin across the genome. *Cell* **132**: 311–322.

Buenrostro JD, Giresi PG, Zaba LC, Chang HY, Greenleaf WJ. 2013. Transposition of native chromatin for fast and sensitive epigenomic profiling of open chromatin, DNA-binding proteins and nucleosome position. *Nat Methods* **10**: 1213–1218.

Li H, Durbin R. 2009. Fast and accurate short read alignment with Burrows-Wheeler transform. *Bioinformatics* **25**: 1754–1760.

Lu Z, Hofmeister BT, Vollmers C, Dubois RM, Schmitz RJ. 2017. Combining ATAC-seq with nuclei sorting for discovery of cis-regulatory regions in plant genomes. *Nucleic Acids Res* **45**: e41.

Perino M, Veenstra GJ. 2016. Chromatin control of developmental dynamics and plasticity. *Dev Cell* **38**: 610–620.

Zhang Y, Liu T, Meyer CA, Eeckhoute J, Johnson DS, Bernstein BE, Nusbaum C, Myers RM, Brown M, Li W, et al. 2008. Model-based analysis of ChIP-Seq (MACS). *Genome Biol* **9**: R137.

DNase-seq to Study Chromatin Accessibility in Early *Xenopus tropicalis* Embryos

Jin Sun Cho, Ira L. Blitz, and Ken W.Y. Cho[1]

Department of Developmental and Cell Biology, School of Biological Sciences, University of California, Irvine, California 92697

Transcriptional regulatory elements are typically found in relatively nucleosome-free genomic regions, often referred to as "open chromatin." Deoxyribonuclease I (DNase I) can digest nucleosome-depleted DNA (presumably bound by transcription factors), but DNA in nucleosomes or higher-order chromatin fibers is less accessible to the nuclease. The DNase-seq method uses high-throughput sequencing to permit the interrogation of DNase hypersensitive sites (DHSs) across the entire genome and does not require prior knowledge of histone modifications, transcription factor binding sites, or high quality antibodies to identify potentially active regions of chromatin. Here, discontinuous iodixanol gradients are used as a gentle preparation of the nuclei from *Xenopus* embryos. Short DNase I digestion times are followed by size selection of digested genomic DNA, yielding DHS fragments. These DNA fragments are subjected to real-time quantitative polymerase chain reaction (qPCR) and sequencing library construction. A library generation method and pipeline for analyzing DNase-seq data are also described.

MATERIALS

It is essential that you consult the appropriate Material Safety Data Sheets and your institution's Environmental Health and Safety Office for proper handling of equipment and hazardous materials used in this protocol.

RECIPES: Please see the end of this protocol for recipes indicated by <R>. Additional recipes can be found online at http://cshprotocols.cshlp.org/site/recipes.

Reagents

Agarose
Buffer A for DNase-seq <R> (4°C)
Chloroform
DNase I digestion buffer (1×) <R> (freshly prepared, equilibrated to 37°C)
DNase I stock solution (10 U/µL) <R>
Ethanol
Ethidium bromide
Gel extraction kit (e.g., NucleoSpin Gel and PCR Clean-up kit [Macherey-Nagel 740609.250])
High sensitivity DNA kit (Agilent Technologies 5067-4626)
Iodixanol solutions (20%, 25%, 30%) <R> (freshly prepared, at 4°C)
Library quantification kit (e.g., KAPA KK4824 [Roche 07960140001])

---

[1]Correspondence: kwcho@uci.edu

LightCycler 480 SYBR Green I Master Mix (Roche 04707516001)

NaCl (5 M)

NEXTflex ChIP-Seq Barcodes (Illumina-compatible barcode adaptors) (Perkin Elmer NOVA-514121)

NEXTflex ChIP-seq Library Prep kit (Perkin Elmer NOVA-5143-01)

Phenol:chloroform:isoamyl alcohol (25:24:1 [v/v])

Primers for qPCR validation (Fig. 1B)

> *eef1a1o* forward: 5′-GCTGGAATTTAAAGGGATGGA-3′
> *eef1a1o* reverse: 5′-CCGGCGTTTTATTGGAACT-3′
> *hbe1* forward: 5′-TTGCATTTGGTTCAGTGCTC-3′
> *hbe1* reverse: 5′-TGTCAGATGCTGGTTCTCCA-3′
> *otx2* forward: 5′-CAGAAAGGGCTTTGTTTTCG-3′
> *otx2* reverse: 5′-AAACTTGATTGGGGCCATTT-3′

Proteinase K (20 mg/mL)

Qubit dsDNA HS Assay kit (Invitrogen Q32851)

RNase A (10 mg/mL)

Stop buffer for DNase I <R> (freshly prepared, equilibrated to 37°C)

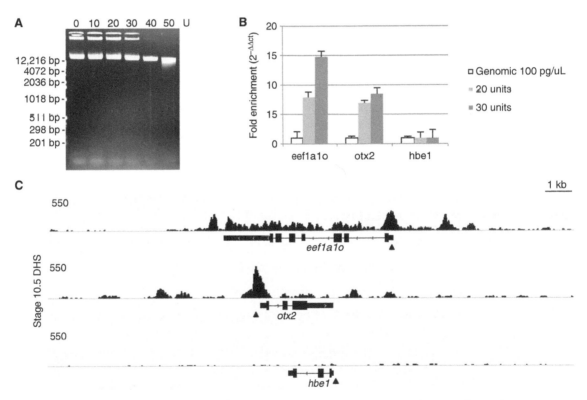

FIGURE 1. Methods to detect DHSs using *Xenopus* early gastrulae. (*A*) Gel picture of DNase-digested genomic DNA extracted from nuclei isolated from stage 10.5 *Xenopus tropicalis* embryos. The amount of high-molecular-weight DNA fragments are gradually decreased as DNase concentrations increase. (*B*) DNase hypersensitivity monitored by qPCR. DHSs are normalized by Cp value from 100 pg of genomic DNA from *Xenopus* liver using $2^{-\Delta\Delta Ct}$ method. Reference regions are selected from nonexpressing genes' promoters (e.g., *hbe1*, encoding hemoglobin subunit epsilon 1, which is not expressed until tailbud stages). (*C*) A genome-wide profile of DHSs at gastrula stage by DNase-seq on *Xenopus tropicalis* genome version 9.0. The number of sequence reads from a stage 10.5 DNase-seq library was 45.5 million. Bowtie aligned 43.7% of these reads after discarding multiply aligned reads. Samtools (Step 32) removed duplicates, and peaks were called using Homer findPeaks or by MACS2 (Zhang et al. 2008). DHSs were detected on the promoter and enhancer regions of *eef1a1o* and *otx2* but no DHSs were shown around *hbe1* (also used as a reference gene for qPCR in Fig. 1B). Regions of qPCR are marked by arrowheads.

Sucrose solution (SS) (0.3 M) <R> (at 4°C)

TE buffer (10 mM Tris-HCl, 1 mM EDTA [pH 8.0])

Tris-HCl (10 mM [pH 8.0])

Triton N-101, reduced (Sigma-Aldrich 303135)

Xenopus embryos, dejellied using standard protocols (Ogino et al. 2006)

> *This protocol is based on use of 2000 early gastrulae (stage 10.5 [Nieuwkoop and Faber 1967]), which contain ~10,000 cells per embryo.*

Equipment

Bioanalyzer (Agilent 2100)

Centrifuge (low-speed, refrigerated), with a swinging-bucket rotor

Centrifuge tubes (50-mL)

Centrifuge tubes, polyallomer (38.5-mL)

Dounce homogenizer (15-mL) with pestle B (0.025- to 0.076-mm clearance) (prechilled to 4°C)

Gel electrophoresis equipment

Microcentrifuge tubes (2-mL)

Nutator at 4°C

Nylon mesh (100-μm pore size) (Ted Pella 41-12115)

Pipette tips (P1000), made with a wide bore by clipping with scissors

Polypropylene tubes (15-mL, conical)

Qubit fluorometer and assay tubes

Real-time quantitative PCR (qPCR) system (Lightcycler 480 II [Roche])

Sequencer (Illumina)

Syringe (15-mL)

Tygon tubing

Ultracentrifuge, with swinging-bucket rotor (e.g., Beckman SW32Ti)

Water baths at 37°C and 55°C

METHOD

Isolating Nuclei from *Xenopus* Embryos

> *This nuclear isolation protocol was modified from Farzaneh and Pearson (1978) and uses iso-osmotic iodixanol gradients instead of hyperosmotic sucrose. This reduces centrifugation time and avoids the extreme depletion of water from the nucleoplasm (Graham 2002).*

> *During Steps 1–11, keep all materials and solutions on ice.*

1. Collect dejellied embryos at the desired stage and transfer to a prechilled 15-mL Dounce homogenizer.

2. Gently wash the embryos twice in 10 mL of ice-cold 0.3 M SS per wash. Remove as much solution as possible after the second wash.

3. Add 4 mL of 0.3 M SS containing 0.4% (v/v) Triton N-101.

4. Homogenize the embryos using 5–7 strokes of the pestle to release intact nuclei.

5. Filter the homogenate through nylon mesh into a 50-mL tube. Rinse the mesh with 1 mL of 0.3 M SS containing 0.4% (v/v) Triton N-101.

6. Mix the filtered homogenate with one volume of 30% iodixanol to make 15% iodixanol.

7. In a 38.5-mL polyallomer centrifuge tube, layer 12 mL each of 25% and 20% iodixanol solutions followed by 14 mL of the 15% iodixanol with homogenate.

   > *To make sharply separated discontinuous gradients (25%; 20%; 15% iodixanol), add 20% iodixanol solution to the centrifuge tube. Then, gently add the 25% iodixanol solution underneath the 20% iodixanol*

Cite this protocol as *Cold Spring Harb Protoc*; doi:10.1101/pdb.prot098335

using a syringe attached to Tygon tubing. The final layer of 15% iodixanol solution containing the homogenized embryos is gently pipetted onto the top of the 20% gradient.

8. Centrifuge the sample at 20,000*g* for 20 min at 4°C using a SW32Ti rotor with maximum braking.

9. Using a pipette, harvest ~5 mL of the solution above the visible interface between the 25% iodixanol cushion and the 20% iodixanol layer; this solution contains the nuclei. Transfer this solution to a 15-mL polypropylene tube.

10. Dilute the nuclei with two volumes of 0.3 M SS and mix by gently inverting several times. Centrifuge at 1000*g* for 5 min at 4°C using a swinging-bucket rotor. Remove the solution by pipetting.

11. Gently resuspend the pellet in 5 mL of buffer A. Centrifuge at 1000*g* for 5 min at 4°C using a swinging-bucket rotor. Remove as much of the solution as possible by gentle pipetting.

    An aliquot of the nuclei can be stained with DAPI and quantitated using a hemocytometer.

Digesting Isolated Nuclei with DNase I

This DNase I digestion procedure was modified from Neph et al. (2012).

For early gastrulae, nuclei from ~500 embryos are used in each digestion reaction. Thus, if starting from 2000 embryos, four reaction tubes are prepared in Step 12.

12. During one of the centrifugations above (Step 10 or 11), prepare tubes for DNase digestion as follows.

    i. Add 200 μL of 1× digestion buffer to each of four 2-mL microcentrifuge tubes.

    ii. Add the required volumes of DNase stock solution (10 U/μL) (e.g., 1, 2, or 4 μL for 10, 20, or 40 U/reaction, respectively).

       The DNase digestion conditions that permit the maximal release of DHS regions should be determined empirically (see Steps 24–25).

    iii. Gently flick to mix.

13. Resuspend the nuclear pellet from Step 11 in 1200 μL of 1× digestion buffer ($N \times 300$ μL, where N = number of reaction tubes) by gentle pipetting with a wide-bore pipette tip.

14. Transfer 300-μL volumes of nuclear suspension to each reaction tube prepared in Step 12 with a wide-bore pipette tip. Mix the samples by gentle pipetting.

15. Incubate the DNase digestion reactions in a water bath for exactly 3 min at 37°C.

16. Add 500 μL of stop buffer to each reaction. Mix by gently inverting the tubes and then incubate for 15 min at 37°C.

Recovering the DNase I Hypersensitive Fragments

17. Transfer the samples to 15-mL conical tubes and add 2 mL of TE to each sample.

18. Add 60 μL of RNase A (10 mg/mL) to each tube. Incubate for 1 h at 37°C.

19. Add 40 μL of proteinase K (20 mg/mL) to each tube. Incubate for 2 h to overnight at 55°C.

20. Extract the DNA using phenol:chloroform:isoamyl alcohol (25:24:1 [v/v]). Remove the organic layer, perform one chloroform extraction, and then recover the DNA by adding 1/10 volume of 5 M NaCl and two volumes of ethanol.

    To prevent shearing of genomic DNA during the organic extraction, rock on a nutator for 30 min at 4°C.

21. Resuspend the DNA pellet in 30 μL of TE.

22. Electrophoretically fractionate the DNA through a 1% agarose gel and visualize using ethidium bromide.

    Run the gel at ~10 volts/cm to permit better size resolution. The vast majority of the DNA should be more than 10 kb. DNA liberated in the 50- to 500-bp size range will not be visible to the eye (Fig. 1A).

    See Troubleshooting.

23. Isolate the gel region corresponding to 50- to 500-bp DHS fragments. Use a gel extraction kit (e.g., NucleoSpin) to recover the DNA. Elute in 20–30 µL of 10 mM Tris-HCl (pH 8.0).

[Optional] Validating DHS Fragments by qPCR

The success of DNase digestion conditions can be monitored by qPCR (Fig. 1B). The promoter regions of highly expressed genes serve as positive controls. Negative controls include genes not expressed at the desired stage. Xenopus liver DNA can be used as an external standard.

24. Dilute a small aliquot of the DNA recovered from Step 23 10-fold in 10 mM Tris-HCl (pH 8.0) for use in qPCR.

25. Perform qPCR using SYBR Green I master mix and primers for the recommended reference genes (e.g., *eef1a1o, otx2, hbe1*).

Constructing the DNase-seq Library

26. Quantitate the total amount of DNase-digested DNA from Step 23 using a Qubit fluorometer.

27. Using ~10 ng of DNA as input for library construction, build the library using a Nextflex ChIP-seq kit together with Illumina-compatible barcode adaptors under the direction of the kit manual.

    We use 11 cycles of PCR amplification.

28. Determine the size distribution of the library using a Bioanalyzer 2100 and high sensitivity DNA kit.

29. Quantitate the library concentration by qPCR (e.g., using a KAPA Library Quantification kit).

30. Sequence the library on an Illumina platform.

Analyzing Data

31. Align the sequencing reads to the *Xenopus tropicalis* v 9.0 genome assembly (ftp://ftp.xenbase.org/pub/Genomics/JGI/Xentr9.0/) using Bowtie v1.0.0 (Langmead et al. 2009) with the following command: "bowtie -S -v 2 -k 1 -m 1 –best –strata."

32. Remove the duplicate reads from a sorted BAM file using the "rmdup" command in Samtools (Li et al. 2009).

33. Create a BigWig file using deepTool2 bamCoverage (Ramírez et al. 2016) and then visualize it using the Broad Institute's Integrative Genomics Viewer genome browser (Robinson et al. 2011).

34. Call DHS peaks using Homer (Heinz et al. 2010) with the following command: "findPeaks -style dnase -gsize 1.42e09."

TROUBLESHOOTING

Problem (Step 22): DNA laddering with multiple bands in ~150-bp increments is apparent after gel electrophoresis.
Solution: The DNA is overdigested. Reduce the amount of DNase I or increase the number of nuclei.

Cite this protocol as *Cold Spring Harb Protoc*; doi:10.1101/pdb.prot098335

RECIPES

Buffer A for DNase-seq

Reagent	Final concentration
Tris-HCl (1 M, pH 8.0)	15 mM
NaCl (5 M)	15 mM
KCl (1 M)	60 mM
EDTA (0.5 M [pH 8.0])	1 mM
EGTA (50 mM [pH 8.0])	0.5 mM
Spermine (500 mM)	0.5 mM
Pefabloc SC PLUS (Roche 11873601001) (20 mg/mL)	0.1 mg/mL
Dithiothreitol (DTT) (1 M)	2 mM
Protease inhibitor tablet (cOmplete, Mini, EDTA-free; Roche 04693159001)	1 tablet for 10 mL

Combine Tris-HCl, NaCl, KCl, EDTA, EGTA, and spermine. Store the buffer at 4°C. Immediately before use, add Pefabloc SC, DTT, and protease inhibitor.

DNase I Digestion Buffer (1×)

To make 5 mL of 1× DNase I digestion buffer, add 500 µL of DNase I Digestion Buffer (10×) <R> to 4.5 mL of Buffer A for DNase-seq <R>. Prepare fresh and equilibrate to 37°C prior to use.

DNase I Digestion Buffer (10×)

Reagent	Final concentration
NaCl (5 M)	750 mM
CaCl$_2$ (1 M)	60 mM

Store for up to 1 yr at room temperature.

DNase I Stock Solution (10 U/µL)

Reagent	Final concentration
Tris-HCl (1 M, pH 7.6)	20 mM
NaCl (5 M)	50 mM
MgCl$_2$ (1 M)	2 mM
CaCl$_2$ (1 M)	2 mM
Pefabloc SC PLUS (Roche 11873601001) (20 mg/mL)	0.1 mg/mL
Dithiothreitol (1 M)	1 mM
Glycerol (100%)	50%

Combine the reagents listed above to prepare storage buffer. On ice, solubilize an entire bottle of deoxyribonuclease I (DNase I [Sigma-Aldrich D4527; 10,000 U]) with 1 mL of ice-cold storage buffer. Prepare 50- to 100-µL aliquots and store at −20°C.

Iodixanol Solutions (20%, 25%, 30%)

Reagent	Final concentration
Tris-HCl (1 M, pH 8.0)	10 mM
Sucrose (1 M)	0.3 M
MgCl$_2$ (1 M)	5 mM
KCl (1 M)	25 mM
NaF (500 mM)	10 mM
β-glycerophosphate (1 M)	5 mM
Sodium pyrophosphate (100 mM)	5 mM
Spermine (500 mM)	0.5 mM
Spermidine (500 mM)	0.5 mM
Pefabloc SC PLUS (Roche 11873601001) (20 mg/mL)	0.1 mg/mL
Dithiothreitol (1 M)	2 mM
Protease inhibitor tablet (cOmplete, Mini, EDTA-free) (Roche 04693159001)	1 tablet for 10 mL
Iodixanol (60%) (OptiPrep Density Gradient Medium; Sigma-Aldrich D1556)	20%; 25%; 30%

Prepare fresh and keep at 4°C.

Stop Buffer for DNase I

Reagent	Final concentration
Tris-HCl (1 M, pH 8.0)	50 mM
NaCl (5 M)	100 mM
SDS (20%)	0.10%
EDTA (0.5 M [pH 8.0])	100 mM
Spermine (500 mM)	1 mM
Spermidine (500 mM)	0.3 mM

Prepare fresh and equilibrate to 37°C prior to use.

Sucrose Solution (SS) (0.3 M)

Reagent	Final concentration
Tris-HCl (1 M, pH 8.0)	10 mM
Sucrose (1 M)	0.3 M
MgCl$_2$ (1 M)	5 mM
KCl (1 M)	25 mM
NaF (500 mM)	10 mM
β-Glycerophosphate (1 M)	5 mM
Sodium pyrophosphate (100 mM)	5 mM
Spermine (500 mM)	0.5 mM
Spermidine (500 mM)	0.5 mM
Pefabloc SC PLUS (Roche 11873601001) (20 mg/mL)	0.1 mg/mL
Dithiothreitol (DTT) (1 M)	2 mM
Protease inhibitor tablet (cOmplete, EDTA-free) (Roche 4693132001)	1 tablet for 50 mL

Combine Tris-HCl, sucrose, MgCl$_2$, KCl, NaF, β-glycerophosphate, sodium pyrophosphate, spermine, and spermidine. Store the buffer at 4°C. Immediately before use, add Pefabloc SC, DTT, and protease inhibitor.

REFERENCES

Farzaneh F, Pearson CK. 1978. A method for isolating uncontaminated nuclei from all stages of developing *Xenopus laevis* embryos. *J Embryol Exp Morphol* **48**: 101–108.

Graham J. 2002. Rapid purification of nuclei from animal and plant tissues and cultured cells. *Scientific World J* **2**: 1551–1554.

Heinz S, Benner C, Spann N, Bertolino E, Lin YC, Laslo P, Cheng JX, Murre C, Singh H, Glass CK. 2010. Simple combinations of lineage-determining transcription factors prime cis-regulatory elements required for macrophage and B cell identities. *Mol Cell* **38**: 576–589.

Langmead B, Trapnell C, Pop M, Salzberg SL. 2009. Ultrafast and memory-efficient alignment of short DNA sequences to the human genome. *Genome Biol* **10**: R25.

Li H, Handsaker B, Wysoker A, Fennell T, Ruan J, Homer N, Marth G, Abecasis G, Durbin R, 1000 Genome Project Data Processing Subgroup. 2009. The Sequence Alignment/Map format and SAMtools. *Bioinformatics* **25**: 2078–2079.

Neph S, Vierstra J, Stergachis AB, Reynolds AP, Haugen E, Vernot B, Thurman RE, John S, Sandstrom R, Johnson AK, et al. 2012. An expansive human regulatory lexicon encoded in transcription factor footprints. *Nature* **489**: 83–90.

Nieuwkoop PD, Faber J. 1967. *Normal table of* Xenopus laevis *(Daudin): A systematical and chronological survey of the development from the fertilized egg till the end of metamorphosis.* North-Holland Pub. Co., Amsterdam.

Ogino H, McConnell WB, Grainger RM. 2006. High-throughout transgenesis in *Xenopus* using I-SceI meganuclease. *Nat Protoc* **1**: 1703–1710.

Ramírez F, Ryan DP, Grüning B, Bhardwaj V, Kilpert F, Richter AS, Heyne S, Dündar F, Manke T. 2016. deepTools2: A next generation web server for deep-sequencing data analysis. *Nucleic Acids Res* **44**: W160–W165.

Robinson JT, Thorvaldsdóttir H, Winckler W, Guttman M, Lander ES, Getz G, Mesirov JP. 2011. Integrative genomics viewer. *Nat Biotechnol* **29**: 24–26.

Zhang Y, Liu T, Meyer CA, Eeckhoute J, Johnson DS, Bernstein BE, Nussbaum C, Myers RM, Brown M, Li W, et al. 2008. Model-based analysis of ChIP-Seq (MACS). *Genome Biol* **9**: R137.

Generating a Three-Dimensional Genome from *Xenopus* with Hi-C

Ian K. Quigley[1,3] and Sven Heinz[2,4]

[1]*Molecular Neurobiology Lab, Salk Institute for Biological Studies, La Jolla, California 92037;* [2]*Department of Medicine, University of California, San Diego, San Diego, California 92093*

Hi-C is a sequencing-based method that captures three-dimensional (3-D) genome interactions by counting the interaction frequencies of pairs of genomic loci. This protocol describes the application of in situ Hi-C to the *Xenopus* embryo. Briefly, after fixing embryos with formaldehyde, nuclei are isolated and chromatin is digested with a restriction enzyme. Restriction sites are filled in with a biotinylated nucleotide and the blunted ends are re-ligated in place, all while still contained in the nuclei (i.e., in situ). Subsequently, the re-ligated genomic DNA is isolated and fragmented by sonication. Biotinylated ligation junctions are captured with streptavidin-coated beads, and DNA fragments are amplified by ligation-mediated polymerase chain reaction (LM-PCR). The PCR product is isolated and sequenced from both ends (paired-end), and informatics methods are then applied to align the two sides of the ligation junctions to the reference genome. Because ligation occurs much more frequently intra- than interchromosomally, and with generally decreasing frequency the further away DNA loci are from each other on the linear chromosome, interaction frequency information can be used to assist in assembling genomes and to phase haplotypes, which is especially useful in the case of a tetraploid organism such as *X. laevis*. Our streamlined version of in situ Hi-C was optimized for high throughput and low cost, and enables generation of high-quality Hi-C libraries from small cell numbers (down to ~10,000 cells) in 2 d.

MATERIALS

It is essential that you consult the appropriate Material Safety Data Sheets and your institution's Environmental Health and Safety Office for proper handling of equipment and hazardous materials used in this protocol.

RECIPES: Please see the end of this protocol for recipes indicated by <R>. Additional recipes can be found online at http://cshprotocols.cshlp.org/site/recipes.

Reagents

Agarose gel (low EEO; e.g., Seakem LE) (2% in TBE)
Biotin-14-dATP (0.4 mM) (Invitrogen 19524016)
Bovine serum albumin (BSA) (10% [v/v])
BW (bind & wash) buffer (2×) <R>
BWT (bind & wash & Tween) buffer (2×) <R>
BWT100 (bind & wash & Triton X-100) buffer (1×) <R>
Chloroform
ClaI (10 U/μL) (New England BioLabs R0197S) and 10× CutSmart buffer (optional; see Step 89)
Deoxynucleotide solution set (dATP/dCTP/dGTP/dTTP; 100 mM each) (New England BioLabs N0446S)

---

[3]Present address: Recursion Pharmaceuticals, Salt Lake City, Utah 84101.
[4]Correspondence: sheinz@ucsd.edu

Cite this protocol as *Cold Spring Harb Protoc*; doi:10.1101/pdb.prot098343

DNA isolation buffer <R> (optional; see Step 82)

Dynabeads MyOne Streptavidin T1 (Invitrogen 65601)

EDTA (0.5 M, pH 8.0)

Ethanol (80%, 100%)

Formaldehyde (16%, EM grade [e.g., Pierce 28906] or 37%, methanol-stabilized [e.g., Sigma-Aldrich F8775])

GelGreen (Biotium) and/or SYBR Gold (Thermo)

Glycine (2.625 M)

GlycoBlue coprecipitant (15 µg/µL) (Invitrogen AM9515)

Klenow fragment (5 U/µL) (Enzymatics P7060L)

Klenow fragment (3′ → 5′ exo-) (5 U/µL) (Enzymatics P7010-LC-L)

Loading dye (12×) (e.g., 60% glycerol/0.05% xylene cyanol/0.05% Orange G)

Lysis buffer for *Xenopus* Hi-C <R>

MboI (25 U/µL) (New England BioLabs R0147M) or DpnII (50 U/µL) (New England BioLabs R0543T/M)

NaCl (5 M)

NEBuffer 2 (10×) (New England BioLabs B7002S)

NEXTflex DNA barcodes (Bioo Scientific 514104)

PCR master mix (NEBNext Ultra II Q5) (New England Biolabs M0544S)

PEG (20%)/1.5 M NaCl <R> (optional; see Step 84)

PEG (20%)/2.5 M NaCl <R>

Phenol:chloroform:isoamyl alcohol (25:24:1 [v/v]) (Invitrogen 15593049)

Phosphate-buffered saline (PBS)

Proteinase K (20 µg/µL) (Invitrogen AM2546)

Qubit dsDNA HS Assay Kit (Invitrogen Q32854)

RNase A (DNase-free) (10 µg/µL)

SDS (10% [v/v])

Sera-Mag SpeedBeads (carboxylate-modified magnetic particles), hydrophobic (GE Healthcare Life Sciences 65152105050250)

Solexa 1GA primer: 5′-AATGATACGGCGACCACCGA-3′ (10 µM in TE buffer)

Solexa 1GB primer: 5′-CAAGCAGAAGACGGCATACGA-3′ (10 µM in TE buffer)

T4 DNA ligase (rapid) (600 U/µL), with 2× rapid ligation buffer (Enzymatics L6030-HC-L)

T4 DNA ligase buffer (10×)

T4 DNA polymerase (3 U/µL) (Enzymatics P7080L)

T4 polynucleotide kinase (10 U/µL) (Enzymatics Y9040L)

TET buffer <R>

Triton X-100 (10% [v/v])

Trypan blue (optional; see Step 10)

TT buffer <R>

TTLoE buffer <R>

Tween 20 (10% [v/v])

Xenopus laevis (~100 dejellied embryos [stage 10.5] or animal caps [stage 18]) (Ariizumi et al. 2009; Sive et al. 2010)

Equipment

Agarose gel electrophoresis system (e.g., Labnet International ENDURO Gel XL electrophoresis system with 4-µL well-size combs)

Centrifuge (refrigerated tabletop) with swing buckets for 96-well plates (e.g., Eppendorf 5810R)

Centrifuge (refrigerated tabletop) for microcentrifuge tubes (e.g., Eppendorf 5424R)

Hybridization oven or heat block(s) at 37°C, 55°C, and 65°C

Liquid nitrogen or dry ice/methanol slurry

Microcentrifuge tubes (1.5- and 2.0-mL, polypropylene)

Microcentrifuge tubes (1.5-mL) (LoBind DNA) (Eppendorf 30108051)

Microcentrifuge tubes (2.0-mL) (Axygen MAXYmum Recovery) (Corning MCT-200-L)

Microscope and hemocytometer (optional; see Step 10)

microTUBE (AFA fiber preslit snap-cap) (Covaris 520045)

PCR 8-tube strips (0.2-mL) with 8-cap strips (USA Scientific 1402-2700)

PCR tube rack (Axygen) (Corning R-96-PCR-FSP)

Qubit fluorometer (Invitrogen Q33216) and assay tubes (Axygen PCR-05-C)

Rotator (e.g., Thermo Scientific HulaMixer) at 16°C and 37°C

Side-skirted magnet (DynaMag-96) (Invitrogen 12027)

Sonicator (e.g., Covaris Focused-Ultrasonicator M220)

TapeStation or BioAnalyzer (optional; see Steps 43, 76, and 88)

Thermal cycler with 100 μL capacity (e.g., Bio-Rad T100)

Tissue grinder, Dounce (2-mL) (Kimble 885300-0002) (optional; see Step 4)

Transilluminator (UV or non-UV, e.g., Clair Chemicals Dark Reader DR22A) and smartphone or gel documentation system

Vacuum aspirator (manifold for PCR tube strips) (BrandTech 704526)

Use with filterless 10-μL tips.

Vortex

METHOD

There are a number of experimental variations on generating sequencing libraries that capture 3D interactions via proximity ligation (Mishra and Hawkins 2017), all based on the original chromosome conformation capture protocol (3C) (Dekker et al. 2002) and its genome-wide adaptation, Hi-C (Lieberman-Aiden et al. 2009). Here, we describe the application of one of the easier-to-perform protocols for capturing 3-D genome interactions, in situ Hi-C (Rao et al. 2014). This protocol has proven to be exceptionally robust. To date, we have used it to generate several hundred high-quality libraries in different conditions and species. In the handful of instances where the protocol failed, it did not yield a library at all, e.g., in species with extreme GC content, or in apoptotic cells.

Fixing and Storing Embryos

The protocol works well with 5×10^4 to 1.5×10^6 cells, and we have had success making libraries with as little as 12,000 cells. For a first attempt, starting with at least 10^5 cells is recommended. This number of nuclei is easy to see during the different steps, which prevents accidental sample loss.

1. Transfer *Xenopus* embryos to a 2-mL microcentrifuge tube containing 1 mL 1% formaldehyde in PBS. Fill the tube to the top with 1% formaldehyde in PBS, and fix the embryos for 30 min at room temperature on a rotating platform.

   The number of caps or embryos used here should occupy a 2-mL microcentrifuge tube; top off the tube with the formaldehyde solution in this step.

2. Remove the formaldehyde and quench the fixative by washing once for 10 min with 0.125 M glycine in PBS.

3. Wash the embryos three times for 5 min per wash in ice-cold PBS. Remove the PBS and flash-freeze the pellet in a dry ice/methanol slurry or liquid nitrogen. Store at −80°C.

   Fixed cells are extraordinarily stable. We have successfully performed Hi-C on murine cells 5 yr after fixation.

Isolating Nuclei

Use filter tips for all pipetting steps to prevent adding 500 μL room-temperature lysis buffer and cross-contamination between samples and experiments.

4. Dissociate the embryonic tissue by pipetting rapidly with a 1-mL pipette tip.

   Alternatively, grind the tissue in a 2-mL Dounce tissue grinder with 10 strokes of a loose (A) pestle.

Cite this protocol as *Cold Spring Harb Protoc*; doi:10.1101/pdb.prot098343

5. After using a Dounce homogenizer, transfer the homogenate to a 2-mL microcentrifuge tube. Centrifuge at 2500*g* for 5 min at room temperature.

> *The supernatant will have a fat layer on top, which will be subsequently removed by blotting with Kimwipes and two rounds of washes.*

6. Remove as much of the top fat layer as possible by blotting with a Kimwipe, and then carefully aspirate most of the supernatant, leaving the nuclear/plasma protein pellet intact.

7. To lyse the cells, resuspend the pellet by pipetting with 1 mL of Lysis Buffer at room temperature.

8. To isolate the nuclei, centrifuge the sample at 2500*g* for 5 min at room temperature. Blot off the fat layer on top and discard the supernatant.

9. Repeat Steps 7 and 8 to wash the nuclei and remove debris.

> *Incubating fixed cells with lysis buffer containing only sodium dodecyl sulfate (SDS) lyses the cytoplasmic membrane and strips cytoplasmic proteins but leaves the nuclei intact, likely a result of formaldehyde fixation of the nuclear lamina (Nickerson et al. 1997). We found the initial nonionic detergent washes that precede SDS lysis in common Hi-C protocols to be unnecessary.*

10. (Optional) Check/count the nuclei under a microscope: After resuspending the nuclei in lysis buffer in Step 9, mix 20 μL of the nuclei suspension with an equal amount of 0.4% Trypan Blue and count the dark-blue nuclei on a hemocytometer.

> *In our hands, this procedure produces roughly 4000–5000 nuclei per embryo at stage 10.5 or per animal cap at stage 18.*

11. After the last wash is complete, remove as much of the supernatant as possible and resuspend the nuclei in 200 μL of lysis buffer.

12. Transfer the suspension to an eight-well PCR tube strip and complete lysis by incubating for 7 min at 62°C in a thermal cycler.

> *SDS and elevated temperature make restriction sites more accessible by solubilizing non-cross-linked proteins and partially reversing protein–DNA cross-links.*

13. (Optional) As an Input Control, transfer 8 μL (4%) of the well-resuspended sample to a separate 0.2-mL PCR tube.

> *When using a low number of cells, the DNA content will be difficult or impossible to detect on an ethidium bromide- or even GelGreen-stained agarose gel. For example, if 10^5 cells are used for an experiment, the 4% Input Control sample will at best contain 27 ng of DNA.*

14. Centrifuge the PCR tube strip containing the nuclei at 2500*g* for 5 min at room temperature in a tabletop centrifuge with swing buckets for 96-well plates. Discard most of the supernatant (at a minimum 190 μL [or 182 μL if the optional Input Control was set aside]), leaving the nuclei in a small volume (~10 μL) of L2 buffer.

Digesting Chromatin

15. Resuspend the nuclei in digestion buffer (25 μL 10% Triton X-100, 25 μL 10× NEBuffer 2, 195 μL H₂O) and rotate for 15 min at 37°C to dilute and sequester the remaining SDS.

16. Add 4 μL of MboI restriction enzyme (25 U/μL) and incubate for 3 h up to overnight at 37°C on a rotating platform at 8 RPM.

> *Other four-cutter restriction enzymes have been used (e.g., DpnII, HaeIII), and depending on their recognition motif produce different coverage biases (Liang et al. 2017). If employing DpnII instead of MboI, use the dedicated DpnII buffer and 2 μL of the high-concentration enzyme (50 U/μL). For previously untried enzymes, test their efficiency on a PCR product before using them in Hi-C, to ensure that they are not inhibited by the presence of the detergents used in this reaction.*

17. Inactivate the restriction digest for 20 min at 62°C in a thermal cycler, and then cool to room temperature.

> *For DpnII and MboI, omitting this heat inactivation step does not affect the Hi-C results. Both enzymes are inhibited by Dam methylation of their recognition motif. The fill-in reaction with N6-Biotin-14-dATP in the next section mimics Dam hemi-methylation and inhibits digestion of the reconstituted GATC site. This was confirmed for DpnII with a dsDNA oligonucleotide containing a central single-stranded GATC site. Fill-in*

with N6-Biotin-14-dATP (i.e., hemi-modification) blocked DpnII digestion similar to N6-Methyl-dATP hemi-methylation (data not shown).

18. (Optional) As a Digestion Control, transfer 12 μL (4.8%) of the well-resuspended sample to a 0.2-mL PCR tube.

   With 10^5 cells of starting material, the Digestion Control can contain up to 31 ng of DNA.

19. Collect the nuclei by centrifugation at 500g for 5 min at room temperature in a tabletop centrifuge with swing buckets for 96-well plates.

20. Discard 200 μL (188 μL if Digestion Control was set aside) of the supernatant, leaving the nuclei in a ~50-μL volume.

Filling in and Biotinylating Restriction Overhangs

21. To fill in the GATC overhangs generated by MboI cleavage, prepare a 2× Fill-In Master Mix of the following reagents on ice.

Reagent	Amount per reaction	Final concentration
ddH$_2$O	31.95 μL	
NEB2 (10×)	5 μL	1×
dCTP (10 mM)	0.35 μL	35 μM
dTTP (10 mM)	0.35 μL	35 μM
dGTP (10 mM)	0.35 μL	35 μM
Biotin-14-dATP (0.4 mM)	7.5 μL	30 μM
Triton X-100 (10%)	2 μL	0.2%
Klenow fragment (5 U/μL)	2.5 μL	12.5 U
Total volume	50 μL	

22. Add 50 μL of 2× Fill-In Master Mix to each sample tube from Step 20. Rotate the fill-in reactions for 40 min at room temperature (22°C–24°C).

   The lower Biotin-14-dATP concentration compared to the original protocol is offset by incubation at lower temperature, which yields similarly high-quality libraries and reduces cost. Biotin-14-dCTP may also be used but in our hands produces lower yields.
   In the meantime, prepare Ligation Mix 1 (see Step 24) on ice.

23. Place the reactions on ice.

   (Optional) Collecting the nuclei and discarding most of the supernatant (as in Step 19) might further decrease the amount of free DNA that contributes to noninformative reads. In this case, replace the discarded supernatant with an equal volume of 1× NEBuffer 2.

Performing Proximity Ligation

24. To prepare Ligation Mix 1, combine the following reagents on ice and mix. For each sample, transfer 400 μL Ligation Mix 1 into a LoBind 1.5-mL microcentrifuge tube.

Reagent	Amount per reaction	Final concentration
ddH$_2$O	322.5 μL	
T4 DNA ligase buffer (10×)	40 μL	1×
Triton X-100 (10%)	36 μL	0.72%
BSA (10%)	0.5 μL	0.01%
T4 DNA ligase (600 U/μL)	1 μL	600 U
Total volume	400 μL	

25. Transfer each sample (100 μL) to a separate LoBind tube containing Ligation Mix 1 (final volume: 500 μL). Rotate at 8 rpm for 4 h to overnight at 16°C.

26. Stop the proximity ligation reactions by adding 20 μL of 0.5 M EDTA (a 2× molar excess over 10 mM Mg^{2+}).

Reversing Cross-Links and Purifying DNA

The purification described in Steps 30–37 is required to clean the biotinylated fragments.

27. Add 1 μL of RNase A (10 μg/μL) to each reaction. Incubate for 15 min at 37°C.

28. Add 33 μL of 5 M NaCl and 55 μL of 10% SDS to each reaction. Incubate overnight at 65°C in a hybridization oven or heat block to reverse cross-links.

29. Add 10 μL of proteinase K (20 μg/μL) to each reaction. Incubate for 90 min at 55°C followed by for 30 min at 65°C.

30. Extract each reaction once with 600 μL of phenol:chloroform:isoamyl alcohol (25:24:1 [v/v]), mixing vigorously by hand or by vortexing. Centrifuge the samples at full speed in a microcentrifuge (>15,000g) for 5 min at room temperature.

31. Transfer each DNA-containing upper (aqueous) phase to a fresh 1.5-mL tube.

32. Extract residual phenol from the DNA-containing supernatants by adding 300 μL of chloroform to each sample. Mix vigorously as above and centrifuge at full speed (>15,000g) for 1 min at room temperature.

33. Transfer the DNA-containing upper phase to a 2-mL MAXYmum Recovery tube.

    The cylindrical shape of the bottom of these tubes helps collect the DNA in an easily visible pellet in the next steps.

34. Add 1.5 μL of GlycoBlue to each sample. Precipitate the DNA with 1400 μL of 100% EtOH.

35. Centrifuge at 15,000g for 20 min at 4°C to pellet the DNA. Discard the supernatant.

36. Wash each DNA pellet once with 1 mL of 80% EtOH. Centrifuge at 8000g for 5 min at 4°C.

37. Discard the supernatant and air-dry the DNA pellet.

    "Dangling" ends that are biotinylated but did not ligate to anything typically make up 1%–5% of the sequencing reads in an in situ Hi-C library. The biotin can be removed from these ends by treating the DNA with T4 DNA polymerase, which has a strong 3′–5′ exonuclease activity, to prevent their binding to streptavidin beads in the next steps. Exonuclease treatment is performed at low temperature and in the presence of dGTP and dATP to prevent complete digestion of all DNA. A detailed protocol for biotin removal from unligated ends can be found in Belaghzal et al. (2017). In our hands, T4 DNA polymerase treatment reduces dangling ends by 50% to 0.5%–2.5% of the total while reducing the total DNA amount by up to 60%. Therefore, we typically do not digest dangling ends before streptavidin capture of biotinylated DNA.

38. Dissolve each DNA pellet in 131 μL of TT buffer.

    Following DNA isolation, the use of Tween 20 or Triton X-100 in all buffers prevents sample loss due to DNA adsorption to the tube wall, even at very low (<1 pg/μL) DNA concentrations.

39. (Optional) As a Pre-Sonication Control, transfer 5 μL of each DNA solution to a separate 0.2-mL PCR tube. Make up the missing volume in the original sample with an extra 5 μL of TT buffer.

40. Measure the DNA concentration of 1 μL of the DNA solution, e.g., using a Qubit fluorometer.

    The amount of recovered DNA will be slightly lower than 6.8 pg/nucleus in the starting material.

Shearing and Immobilizing DNA

Shearing DNA

41. Transfer up to 800 ng of DNA in a total volume of 130 μL of TT buffer to a Covaris microTUBE.

    The DNA amount of 800 ng represents an upper bound; this protocol has generated high-quality data from as little as 80 ng of DNA (equivalent to ~12,000 cells).

42. Shear the DNA to a 300-bp average size in a Covaris Focused-ultrasonicator using the manufacturer-provided settings for shearing DNA to 300 bp and the required accessories for DNA shearing in 130 μL microTubes. (For Covaris M220/S220&E220, use Peak Incident Power 50/140 W, Duty Factor 20/10%, Cycles per Burst 200/200, Treatment Time 65/80 sec; see https://covaris.com/resources/protocols/dna-shearing-protocols/).

If a Covaris sonicator is unavailable, other water bath or probe sonicators (e.g., Diagenode Bioruptor, Branson Ultrasonics S-250, QSonica Q800, Active Motif EpiShear) can be used to shear the DNA to a 300-bp average size. Sonication conditions can be optimized by shearing RNA-free genomic DNA in TT buffer for increasing amounts of time and checking size on an agarose gel. DNA sonication efficiency is relatively independent of concentration. For example, a 10-µg/mL genomic DNA stock in TT buffer can be used for protocol optimization.

For probe sonicators, the minimum volume that can easily be sheared without splashing using a 1/8 inch tip in a 1.5-mL microcentrifuge tube is 200 µL. This requires inserting the tip into the tube with ∼2 mm of clearance to the tube bottom, and mounting the tube with a clamp to the sonicator stand such that the tip does not touch the tube wall. For many sonicators, optimization protocols and approximate sonication conditions are published (e.g., http://www.activemotif.com/documents/1746.pdf). Importantly, probe sonicator tips erode with continued use, leading to decreased sonication efficiency. This means that compared to a new tip, more sonication cycles will be needed to yield the same results, which needs to be considered when relying on published protocols.

As an example, to optimize conditions for a probe sonicator with a 1/8-inch-wide tip, make up 250 µL of TT buffer containing 2.5 µg of genomic DNA in a 1.5-mL LoBind microcentrifuge tube. Mount the tube to the post of the sonicator stand using a small clamp, such that the sonicator tip is immersed in the liquid with ∼2 mm of clearance from the tube bottom (placing the tip much higher than that will lead to splashing and sample loss). Immerse the tube in a wet-ice slurry in a 100-mL plastic beaker, deep enough to reach above the top of the liquid in the tube. Sonicate the sample for 5, 10, and 15 cycles of 10 sec on, 30 sec off (e.g., at 25% duty or 5–10 W output; low enough to not splash, but high enough to sonicate the DNA). Take off 15 µL (150 ng of DNA) every five cycles, and run directly on a 1× GelGreen-containing 2% agarose/TBE gel, next to a 1-kb Plus ladder. For Hi-C, use a setting that gives a 300-bp average fragment size.

43. (Optional) Determine the average fragment size on a 2% agarose/TBE gel containing GelGreen or SYBR Gold, or on a TapeStation or BioAnalyzer. Alternatively, keep 5 µL of DNA solution in a separate 0.2-mL PCR tube as a Post-Sonication Control for later testing, and make up the missing volume with an extra 5 µL of TT buffer.

Size estimation before library prep may require unacceptable amounts of material when using very low amounts of starting material. In this case, the Test PCR (Steps 60–64) after library prep will provide a first estimate of library concentration and size distribution.

If testing low amounts of DNA at this stage, we highly recommend using a TapeStation or BioAnalyzer with high-sensitivity reagents, or a DNA dye that is more sensitive than ethidium bromide, such as GelGreen (∼3 ng detection limit for a 300-bp average size DNA smear in a prestained 2% agarose/TBE gel) or SYBR Gold.

Equilibrating Streptavidin Beads

44. Per reaction, transfer 20 µL of Dynabeads MyOne Streptavidin T1 bead suspension to a 1.5-mL microcentrifuge tube (up to 300 µL total; ensure that beads are thoroughly resuspended before use). Collect beads on a magnet and discard the supernatant.

45. Wash the beads by resuspending in 1.2 mL of 1× BW buffer. Collect the beads on a magnet and discard the supernatant.

46. Resuspend the beads in 130 µL of 2× BWT buffer per sample (e.g., for 10 samples, resuspend the amount of beads contained in the 200-µL volume of the initial bead suspension in 1300 µL of 2× BWT buffer). Distribute 130-µL aliquots of the bead suspension into an eight-well PCR tube strip.

Immobilizing Biotinylated DNA on Streptavidin Beads

47. Add the sonicated DNA (130 µL) from the microTUBE to the bead suspension. Bind the DNA to the Dynabeads by rotating overhead (e.g., at 8 RPM) for 30 min at room temperature.

In the meantime, prepare the Blunting Mix (see Step 50) on ice.

In our hands, 30%–40% of the genomic DNA binds to 20 µL of T1 Dynabeads. This fraction decreases if more than 800 ng of DNA are used, indicating that the beads saturate with this amount of DNA. Conversely, lowering the DNA amount below 800 ng does not change the percentage of DNA that binds to the beads, suggesting that only a fraction of the total is labeled with biotin and/or that only a (constant) fraction of the biotinylated DNA fraction binds to the beads.

Cite this protocol as *Cold Spring Harb Protoc*; doi:10.1101/pdb.prot098343

48. Wash away the unbound DNA by collecting the beads on a magnet and discarding the supernatant. Resuspend the beads in 180 µL of 1× BWT100.

49. Rebuffer the sample and dilute the NaCl by collecting the beads on a magnet, completely aspirating the supernatant, and resuspending the beads in 180 µL of TET buffer.

Preparing the Library

Blunting

50. To prepare the Blunting Mix, combine the following reagents on ice. Collect the DNA-coated beads from Step 49 on a magnet, discard the supernatant and immediately resuspend each sample in 95 µL of Blunting Mix. Incubate for 30 min at 20°C in a thermal cycler.

Reagent	Amount per reaction	Final concentration
ddH$_2$O	76.1 µL	
T4 DNA ligase buffer (10×)	10 µL	1×
Tween 20 (10%)	0.5 µL	0.05%
dNTP mix (10 mM each)	4 µL	0.4 mM
T4 DNA polymerase (3 U/µL)	2 µL	6 U
Klenow fragment (5 U/µL)	0.4 µL	2 U
T4 polynucleotide kinase (10 U/µL)	2 µL	20 U
Total volume	95 µL	

51. Stop each reaction by adding 2.5 µL of 0.5 M EDTA.

A-Tailing

52. Collect the beads on a magnet and discard the supernatant. Wash the beads twice with 150 µL of 1× BWT100 followed by once with 180 µL of TET buffer.

53. To prepare the A-Tail Mix, combine the following reagents. Resuspend each bead sample from Step 52 in 50 µL A-Tail Mix. Incubate for 30 min at 37°C.

Reagent	Amount per reaction	Final concentration
ddH$_2$O	41.55 µL	
NEBuffer 2 (10×)	5 µL	1×
Tween 20 (10%)	0.25 µL	0.05%
dATP (100 mM)	0.2 µL	0.4 mM
Klenow (3′ → 5′ exo) (5 U/µL)	3 µL	15 U
Total volume	50 µL	

In the meantime, prepare Ligation Mix 2 (see Step 54) on ice.

Ligating Adapters

54. To prepare Ligation Mix 2, combine the following reagents on ice. Collect the beads from Step 53 on a magnet, discard the supernatant, and resuspend the beads in 48 µL of ice-cold Ligation Mix 2. Keep on ice.

Reagent	Amount per reaction	Final concentration
ddH$_2$O	21.5 µL	
Rapid ligation buffer (2×)	25 µL	1×
Tween 20 (10%)	0.5 µL	0.1%
T4 DNA ligase, rapid (600 U/µL)	1 µL	15 U
Total volume	48 µL	

55. Add 1 µL (25 pmol) of Nextflex DNA barcode adapters to each reaction on ice and mix. Incubate for 20 min at room temperature.

    When using less than 400 ng of DNA to bind to the beads, the amount of adapters can be reduced to 0.5 µL and to 0.25 µL when using <100 ng of DNA. Using too little adapter will lower the amount of amplifiable library, even if nominally in excess over bound DNA ends, probably because of ligation being less efficient when DNA is bound to beads.

56. Collect the beads on magnet and discard the supernatant.

57. Wash the beads twice in 150 µL of 1× BW buffer by completely resuspending them in buffer each time, collecting them on a magnet, and discarding the supernatant.

58. Wash the beads twice with 180 µL of TET buffer by completely resuspending them in buffer each time, collecting them on a magnet, and discarding the supernatant.

59. Resuspend the beads in 32 µL of TTLoE buffer. Store at 4°C.

    For long-term storage, the library-loaded beads can be frozen at −20°C.

(Optional) Performing Test PCR

Semiquantitative PCR using 0.5 µL of bead suspension (1/64th of the total) at 12, 16, and 20 cycles (~ 4000-, 65,000-, 1,000,000-fold amplification) gives an indication of how well sonication, bead capture and library prep worked, provides an estimate of the size distribution of the library, and can be used to estimate the minimum cycle number to run for the Library PCR (Step 66).

60. Combine the following reagents.

Reagent	Amount per 3.3 reactions	Final concentration
ddH$_2$O	4.95 µL	
NEB Ultra II Q5 Master Mix (2×)	8.25 µL	1×
Solexa 1GA primer (10 µM)	0.825 µL	0.5 µM
Solexa 1GB primer (10 µM)	0.825 µL	0.5 µM
Total volume	14.85 µL	

61. Thoroughly resuspend the library beads from Step 59. Add 1.65 µL of the bead suspension to 14.85 µL of Test PCR mix. Mix briefly by vortexing, and then distribute 5 µL of the reaction mix into three different 0.2-mL PCR tubes or tube strips.

62. Amplify the library using a thermal cycler as follows: 30 sec at 98°C; 12, 16, or 20 cycles of [10 sec at 98°C, 20 sec at 60°C, and 30 sec at 72°C]; 5 min at 72°C; cool to 4°C. Pause the cycler at the end of the 72°C extension step after 12, 16, or 20 cycles and remove a tube strip, or run the tube strips in different cycler blocks.

63. Add 0.5 µL of 12× Loading Dye (e.g., 60% glycerol/0.05% xylene cyanol/0.05% Orange G) to each sample and mix. Collect the library beads on a magnet.

64. Electrophorese the supernatant on a 2% agarose/TBE gel containing 1× GelGreen.

    The lowest cycle number at which a product can be detected as a prominent smear by GelGreen in the Test PCR can be used to determine the appropriate number of PCR cycles for the Library PCR. When using 10 µL (1/3) of the remaining library for amplification, this volume contains 20-fold more beads than the Test PCR, which is equivalent to around four fewer cycles to obtain similar levels of PCR product.

    The expected average library size is 300 bp insert size + 125 bp adapter size = 425 bp. If very low input DNA amounts are used (<100 ng), an adapter dimer band at 125 bp might appear in the Test PCR product. Adapter dimers are problematic for sequencing because they generate noninformative reads and if present at high concentrations can jeopardize the sequencing run. The final PCR cleanup employs polyethylene glycol (PEG) precipitation to size-select against adapter dimers.

    See Troubleshooting.

Cite this protocol as *Cold Spring Harb Protoc*; doi:10.1101/pdb.prot098343

Amplifying the Library

In our hands, one-third (10 µL) of the library-containing bead suspension from 100,000 nuclei of starting material in a 50-µL PCR reaction can yield up to 1 µg of library amplicon after 12 cycles.

65. Combine the following reagents.

Reagent	Amount per reaction	Final concentration
ddH$_2$O	10 µL	
NEB Ultra II Q5 Master Mix (2×)	25 µL	1×
Solexa 1GA primer (10 µM)	2.5 µL	0.5 µM
Solexa 1GB primer (10 µM)	2.5 µL	0.5 µM
DNA-bound bead suspension	10 µL	
Total volume	50 µL	

66. Amplify the library for four fewer cycles than the lowest cycle number for which a product was observed in the Test PCR; alternatively, amplify for 12 cycles: 30 sec at 98°C; 12 cycles (or cycle number for visible Test PCR result—4 cycles) of [10 sec at 98°C, 20 sec at 60°C, and 30 sec at 72°C]; 5 min at 72°C; cool to 4°C.

Performing DNA Cleanup

67. Place each PCR reaction tube in an Axygen PCR tube rack on a side-skirted 96-well magnet to collect the Dynabeads on the side of the tube. Transfer 49 µL of the supernatant to a new PCR tube strip that contains 2 µL of Sera-Mag SpeedBead suspension (unwashed, i.e., in ddH$_2$O) and 38 µL of 20% PEG/2.5 mM NaCl (final concentration, 8.5% PEG).

68. Close the tubes and mix thoroughly by vortexing for 10 sec. Centrifuge briefly and then incubate for 10 min at room temperature.

69. Collect the beads on a magnet for ~2 min and then discard supernatant with a pipette.

70. Wash the beads by adding 180 µL of 80% EtOH and sliding the rack and tubes back and forth between the magnets six times, causing the beads to "zip through" the solution from side to side.

71. Collect the beads on a magnet and aspirate the supernatant with a vacuum aspirator.

72. Wash the beads once more with 80% EtOH as in Step 70. Collect the beads and aspirate the supernatant for 30 sec, effectively air-drying the beads.

73. Allow the beads to further air-dry until cracks appear in the bead pellet.

74. Elute the DNA from the beads by resuspending them in 20 µL of TT buffer.

    Libraries can be frozen and stored at −20°C with SpeedBeads in the tubes. In this case, avoid bead carryover by placing the samples on a magnet before retrieving DNA.

75. (Optional) Collect the beads on a magnet and transfer the library-containing supernatant to a new PCR tube strip.

76. Measure the DNA concentration by Qubit using the HS DNA assay, and determine the average library size on a 2% agarose/TBE gel, TapeStation, or BioAnalyzer.

    See Troubleshooting.

77. Proceed to high-throughput paired-end sequencing of the final library, or perform optional size selection (Steps 78–81) before sequencing.

    See Discussion.

(Optional) Size-Selecting by Double-PEG Precipitation

Exact size selection of the libraries is generally unnecessary, since the 8.5% PEG precipitation (Step 67) selects against small inserts, the Covaris sonication reproducibly yields a well-defined, tight size range, and the PCR and sequencing procedure itself selects against large inserts. Additionally, size selection incurs a sample loss in the desired size range that should be avoided when using very low amounts of input material. Should size selection be desired, libraries can be size-selected to 250–450 bp either by gel extraction or by differential cleanup with different concentrations of PEG 8000.

78. For double-Speedbead size selection, first bind large (>350 bp) DNA fragments to SpeedBeads with 7% PEG: Add 5 µL of SpeedBead suspension and 29 µL of 20% PEG/2.5 M NaCl to 49 µL of PCR reaction and incubate for 10 min at room temperature.

79. Collect the beads on a magnet and transfer 80 µL of the small DNA-containing supernatant to a new tube. Discard (or keep for size determination) the SpeedBeads containing the large DNA fragments.

    The DNA bound to the SpeedBeads can be washed and eluted as described in Steps 70–74.

80. To the supernatant, add 2 µL of fresh SpeedBead suspension and 12 µL of 20% PEG/2.5 M NaCl (final concentration, 8.5% PEG). Precipitate for 5 min at room temperature.

81. Wash twice with 180 µL of 80% EtOH and elute the DNA from the beads with 20 µL of TT buffer as described in Steps 70–74.

(Optional) Analyzing Control Samples

Analyzing Control DNA

For troubleshooting purposes, a fraction of the sample can be kept where indicated, the DNA processed, and checked on a gel. To rapidly isolate the DNA from the control samples, RNA and proteins are degraded in the presence of 0.5% SDS, cross-links are reversed, and the DNA is precipitated with high levels of PEG 8000 onto SpeedBeads in the presence of a low NaCl concentration, which prevents precipitation of the SDS.

82. Add 40 µL of DNA Isolation Buffer each to the Input and Digestion Control samples. Mix, briefly centrifuge, and incubate for 15 min at 37°C in a thermal cycler.

83. Add 1 µL of proteinase K (20 µg/µL stock) to each sample. Mix and incubate for 15 min at 55°C followed by 30 min at 65°C in a thermal cycler.

84. Prepare a master mix containing 2.2 µL of SpeedBeads and 52.8 µL of 20% PEG 8000/1.5 M NaCl per sample. Add 50 µL of this master mix to each of the Input and Digestion control samples. Mix by vortexing for 10 sec and incubate at room temperature for 10 min.

85. Collect the beads on a magnet, aspirate the supernatant, and wash the beads twice with 180 µL of 80% EtOH as described in Step 70.

86. Aspirate the supernatant and air-dry the beads until cracks appear.

87. Resuspend the beads in 5 µL of TT buffer.

88. Check the size distribution of all Control samples on a TapeStation or BioAnalyzer.

    If using more than 35,000 cells total, more than 10 ng DNA are expected per Control sample (~4% of total), and their DNA size distribution can be checked on an agarose gel. Add 0.5 µL of 12× Loading Dye (e.g., 60% glycerol/0.05% xylene cyanol/0.05% Orange G), to each sample, mix, and collect the beads on magnet. Electrophorese the entire supernatant on a 1% agarose/TBE gel containing 1× GelGreen. The Input Control should generate a smear at the top of the gel above 20 kb, and the Digestion Control should be a much smaller smear from 100–600 bp in average size. Proximity ligation should increase the fragment size range of the Pre-Sonication Control to up to 10 kb, and sonication should lower the average Post-Sonication Control to 300 bp.

Digesting the Final Library

ClaI digest of the final library tests whether the fill-in, blunt-end proximity ligation and bead capture yielded tandem GATC sites ("GATCGATC") at the ligation sites within the fragments, which are cleavage sites for the restriction enzyme ClaI. Digestion will reduce the size of the average library fragment in libraries with this characteristic by almost half. For example, the average fragment size of a 425-bp-average library (300 bp ClaI-containing insert + 125 bp ClaI-less adapter) that can be cut once in every fragment will drop to (300/2 + 125) bp = 275 bp. This size drop will confirm

Cite this protocol as *Cold Spring Harb Protoc*; doi:10.1101/pdb.prot098343

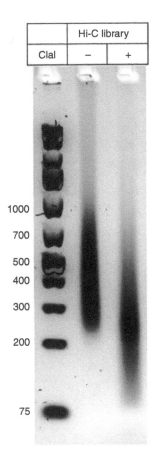

	Hi-C library	
ClaI	−	+

1000
700
500
400
300
200

75

FIGURE 1. Example of a high-performing in situ Hi-C library exhibiting the expected size drop after ClaI digest.

that these library prep steps were successful. An example of a high-quality library and the size drop resulting from ClaI digestion is shown in Figure 1.

89. Digest 50 ng of the final library (1 μL, if starting with 10^5 nuclei) in a 10-μL reaction with 1 μL of 10× CutSmart buffer and 1 μL of ClaI (10 U/μL) for 30 min at 37°C.

90. Electrophorese equal amounts of undigested and digested Hi-C library next to each other on a 2% agarose/TBE gel prestained with 1× GelGreen.

TROUBLESHOOTING

Problem (Step 64, Step 76): The library yield is low, has an unexpected size distribution, or does not display the expected patterning.
Solution: Analyzing the control samples (Steps 81–89) collected throughout the protocol should give an indication of which protocol step might have failed.

Problem (Step 64, Step 76): Biotin incorporation was insufficient.
Solution: Troubleshooting of biotin incorporation can be performed by analyzing the cleavability of a PCR product derived from the ligation of two interacting loci with different restriction enzymes. Fill-in and ligation of MboI-cleaved GATC sites gives rise to a double GATCGATC site, which can be cleaved by both MboI and ClaI (recognizes ATCGATC). Ligation products that can be cleaved by both must be the product of biotin-14-dATP incorporation. Products that can only be cleaved by MboI are likely the product of a sticky-end ligation of GATC overhangs where the fill-in failed.

Details on how to perform this type of troubleshooting experiment can be found in Belaghzal et al. (2017).

Problem (Step 76): Adapter dimers are present in the final library.

Solution: If despite size selection high concentrations of adapter dimers remain in the final library (visible as a 125 bp band on the agarose gel), and if only a faint library smear in the expected ~400 bp size range is visible, one or more of the steps preceding the library preparation may have failed, and the entire experiment should be repeated.

DISCUSSION

Hi-C is a sequencing-based method that captures 3-D genome interactions by counting the interaction frequencies of pairs of genomic loci. Other than to characterize the 3-D organization of DNA in the nucleus, Hi-C data can be used for whole-genome sequencing and assembly, which is based on two properties of chromosomes in live cells: (1) chromosomes occupy defined territories in the nucleus, characterized by more intrachromosomal interactions than interchromosomal interactions, and (2) pairwise interaction frequencies between loci on the same chromosome decrease with linear distance. Together, these features can be exploited in order to (1) phase haplotypes, i.e., determine which alleles reside on the same chromosome, and (2) estimate chromosome assignment and the distance between sequence contigs relative to each other to improve genome-scale scaffolds (Burton et al. 2013; Kaplan and Dekker 2013; Selvaraj et al. 2013). Data generated using this protocol was used to phase the two divergent genomes of *X. laevis* and to improve its genome assembly (Fig. 2).

Rules of Thumb for Sequencing Hi-C Libraries

On the high-throughput sequencer, Hi-C libraries behave similar to large ChIP-seq libraries, clustering less efficiently due to their large size, and with a slight GATC bias at the beginning of the reads, but with an otherwise uniform sequence distribution of the inserts. Loading Hi-C libraries at 2.5 pM on an Illumina NextSeq 500 and at 12 pM on a HiSeq 2500 resulted in excellent sequence yields and read qualities (high cluster densities and pass-filter percentages). For patterned flow cells on HiSeq 3000/4000 (and NovaSeq), much higher loading concentrations (>200 pM) are required to avoid Exclusion Amplification duplicates. For de novo genome assembly, long reads will be preferred,

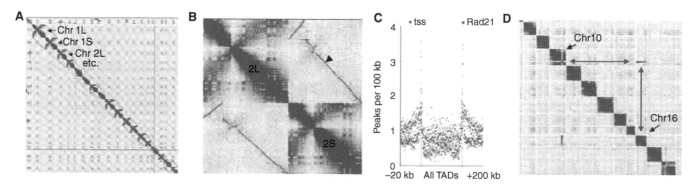

FIGURE 2. (*A*) Interaction matrix of *X. laevis* Hi-C, generated by the protocol described here. *X. laevis* has an allotetraploid genome, the consequence of an interspecies hybridization event, and the chromosomes are labeled with their number and "L" or "S," corresponding to "long" and "short" homologous chromosomes (Session et al. 2016). (*B*) Interaction matrix of *X. laevis* chromosome 2L and S. In our hands, some 0.5% of *X. laevis* Hi-C reads map one end to the L chromosome and one end to S, artifactually generating connections between the two chromosomes (*arrowhead*). (*C*) Metagene plot demonstrating enrichment of transcriptional start sites (tss) and Rad21 peaks at TAD boundaries. (Reprinted from Quigley and Kintner 2017.) (*D*) Possible misassembly in the *Tetraodon nigroviridis* genome. We applied the protocol described here to *T. nigroviridis* tissue and observed, among others, striking interactions between chromosomes 10 and 16 (*blue arrows*), suggestive of recombination or misassembly (Jaillon et al. 2004).

while for structure/chromosome conformation capture applications it is most economical to perform paired-end 42 cycle sequencing using 75 cycle high-output kits on a NextSeq, unless a NovaSeq is available. For initial quality assessment of a Hi-C library, 10–30 million reads are generally sufficient. In high-quality libraries, this sequencing depth will expose topological domains (TADs) and strong long-range interactions. For genome assembly with HiRise, 100 million read pairs per gigabase of genome is recommended (Putnam et al. 2016). To assemble the *X. laevis* genome, we used approximately 550 million read pairs (Fig. 2).

Analysis of Hi-C Libraries

To analyze 3-D genome interactions, a range of options exist for the computational processing and analysis of Hi-C data and generating Hi-C contact maps, and several programs allow for powerful visualization and direct browsing of interaction frequencies (see Yardımcı and Noble [2017] for an overview). Out of these, Juicebox (https://github.com/aidenlab/Juicebox/) (Durand et al. 2016) is a comprehensive collection of tools for Hi-C analysis, as is Galaxy HiCExplorer (https://hicexplorer .usegalaxy.eu/) (Wolff et al. 2018). Both packages enable the user to do basic quality assessment, generate Hi-C contact matrices, and call TADs and A/B compartments. Both packages also allow the user to start from files in fastq format, directly from a sequencing run.

Alternate methods to conduct these analyses can be deployed using HOMER (Lin et al. 2012), which comes equipped with extensive, user-friendly tutorials (found at http://homer.ucsd.edu/homer/ interactions/) to perform all functions. Additionally, HOMER contains a powerful set of tools for integrating features with data from other genomic experiment types, such as ChIP-seq peaks or RNA-seq, and genomic features such as TADs can be represented as peaks (e.g., bed files), allowing for easy comparison and manipulation. For example, makeMetaGeneProfile.pl, a tool included in the HOMER suite of software, can be used to show enrichment of ChIP-seq peaks at TAD boundaries (Fig. 2C).

Generating Hi-C data from understudied organisms can greatly assist in chromosome-level genome assembly (Burton et al. 2013; Putnam et al. 2016; Session et al. 2016; Dudchenko et al. 2017; Ghurye et al. 2017; Lazar et al. 2018). The basic principle underlying these approaches is that contact frequencies will increase inversely proportional to linear distance along a chromosome (along with the additional contact frequency variation that comes from DNA looping in three dimensions). These contact frequencies can be used to orient contigs and stitch together larger fragments. We have had success with HiRise (https://github.com/DovetailGenomics/HiRise_July2015_GR) (Putnam et al. 2016; Session et al. 2016), and more recent efforts have both improved on these methods computationally (https://github.com/machinegun/SALSA) (Ghurye et al. 2017) and in terms of ease-of-use (http://aidenlab.org/assembly/) (Dudchenko et al. 2017). Finally, we and others have also noticed that Hi-C data can provide evidence of chromosomal rearrangements (Chakraborty and Ay 2017; Harewood et al. 2017; Dixon et al. 2018) as well as erroneously assembled genomes (Fig. 2D).

RECIPES

40% PEG 8000

Reagent	Amount to add (for 50 mL)	Final concentration
PEG 8000 (Sigma-Aldrich 89510-250G-F)	20 g	40% (w/v)
ddH$_2$O (sterile)	Add to 50 mL	

Transfer the PEG 8000 to a 50-mL conical centrifuge tube and add ddH$_2$O to the 50-mL line. Mix vigorously to suspend the PEG 8000 in the liquid and displace air in the powder. Continue mixing (e.g., on a rotator) and adding ddH$_2$O to 50 mL until the PEG 8000 is completely dissolved. To rapidly dissolve the PEG 8000, the suspension can be briefly microwaved for no more than 15 sec with the unscrewed lid placed loosely on top of the conical tube. (Do not attempt to microwave the tube with the lid screwed on.) To prevent gradual loss of activity due to polyether peroxidation and chain shortening, store for up to 6 mo at 4°C in the dark or freeze 10-mL aliquots at −20°C for long-term storage.

BW (Bind & Wash) Buffer (2×)

Reagent	Amount to add (for 50 mL)	Final concentration
Tris-HCl (1 M, pH 7.5)	500 µL	10 mM
NaCl (5 M)	20 mL	2 M
EDTA (0.5 M, pH 8.0)	100 µL	1 mM
ddH$_2$O (sterile)	29.40 mL	

Store for up to 1 yr at room temperature.

BWT (Bind & Wash & Tween) Buffer (2×)

Reagent	Amount to add (for 50 mL)	Final concentration
Tris-HCl (1 M, pH 7.5)	500 µL	10 mM
NaCl (5 M)	20 mL	2 M
EDTA (0.5 M, pH 8.0)	100 µL	1 mM
Tween 20 (10%)	1 mL	0.2%
ddH$_2$O (sterile)	28.40 mL	

Store for up to 6 mo at room temperature.

BWT100 (Bind & Wash & Triton X-100) Buffer (1×)

Reagent	Amount to add (for 50 mL)	Final concentration
BW (bind & wash) buffer (2×) <R>	25 mL	1×
Triton X-100 (10%)	500 µL	0.1%
ddH$_2$O (sterile)	24.5 mL	

Store for up to 6 mo at room temperature.

DNA Isolation Buffer

Reagent	Amount to add (for 1 mL)	Final concentration
Lysis Buffer for *Xenopus* Hi-C <R>	909 µL	
NaCl (5 M)	66 µL	330 mM
RNase A (10 µg/µL)	25 µL	250 ng/µL

Store for up to 1 mo at 4°C. SDS precipitates at 4°C; thoroughly resuspend before using the buffer after storage at 4°C

Lysis Buffer for Xenopus Hi-C

Reagent	Amount to add (for 5 mL)	Final concentration
Tris-HCl (1 M, pH 7.5)	250 µL	50 mM
NaCl (5 M)	10 µL	10 mM
EDTA (0.5 M, pH 8.0)	10 µL	1 mM
SDS (10%)	280 µL	0.56%
ddH$_2$O (sterile)	4.4 mL	

Store for up to 1 yr at room temperature. Add 50 µL of 100× protease inhibitor cocktail (Sigma-Aldrich P8340) immediately before use.

Cite this protocol as *Cold Spring Harb Protoc*; doi:10.1101/pdb.prot098343

PEG (20%)/1.5 M NaCl

Reagent	Amount to add (for 50 mL)	Final concentration
40% PEG 8000 <R>	25 mL	20%
NaCl (5 M)	15 mL	1.5 M
ddH$_2$O (sterile)	10 mL	

Store for up to 6 mo at 4°C.

PEG (20%)/2.5 M NaCl

Reagent	Amount to add (for 50 mL)	Final concentration
40% PEG 8000 <R>	25 mL	20%
NaCl (5 M)	25 mL	2.5 M

Store for up to 6 mo at 4°C.

TE Buffer

Reagent	Quantity (for 100 mL)	Final concentration
EDTA (0.5 M, pH 8.0)	0.2 mL	1 mM
Tris-Cl (1 M, pH 8.0)	1 mL	10 mM
H$_2$O	to 100 mL	

TET Buffer

TET buffer <R>
0.05% Tween 20

TT Buffer

Reagent	Amount to add (for 50 mL)	Final concentration
Tris-HCl (1 M, pH 8.0)	500 μL	10 mM
Tween 20 (10%)	250 μL	0.05%
ddH$_2$O (sterile)	49.25 mL	

Store for up to 6 mo at 4°C.

TTLoE Buffer

Reagent	Amount to add (for 10 mL)	Final concentration
Tris-HCl (1 M, pH 8.0)	50 μL	5 mM
Tween 20 (10%)	25 μL	0.025%
EDTA (0.5 M, pH 8.0)	2 μL	0.1 mM
ddH$_2$O (sterile)	9.92 mL	

Store for up to 6 mo at 4°C.

Cite this protocol as *Cold Spring Harb Protoc*; doi:10.1101/pdb.prot098343

REFERENCES

Ariizumi T, Takahashi S, Chan T-C, Ito Y, Michiue T, Asashima M. 2009. Isolation and differentiation of *Xenopus* animal cap cells. *Curr Protoc Stem Cell Biol* **Chapter 1:** Unit 1D.5. doi:10.1002/9780470151808.sc01d05s9

Belaghzal H, Dekker J, Gibcus JH. 2017. Hi-C 2.0: an optimized Hi-C procedure for high-resolution genome-wide mapping of chromosome conformation. *Methods* **123:** 56–65. doi:10.1016/j.ymeth.2017.04.004

Burton JN, Adey A, Patwardhan RP, Qiu R, Kitzman JO, Shendure J. 2013. Chromosome-scale scaffolding of de novo genome assemblies based on chromatin interactions. *Nat Biotechnol* **31:** 1119–1125. doi:10.1038/nbt.2727

Chakraborty A, Ay F. 2017. Identification of copy number variations and translocations in cancer cells from Hi-C data. *Bioinformatics* doi:10.1093/bioinformatics/btx664.

Dekker J, Rippe K, Dekker M, Kleckner N. 2002. Capturing chromosome conformation. *Science* **295:** 1306–1311. doi:10.1126/science.1067799

Dixon JR, Xu J, Dileep V, Zhan Y, Song F, Le VT, Yardımcı GG, Chakraborty A, Bann DV, Wang Y, et al. 2018. Integrative detection and analysis of structural variation in cancer genomes. *Nat Genet* **50:** 1388–1398. doi:10.1038/s41588-018-0195-8

Dudchenko O, Batra SS, Omer AD, Nyquist SK, Hoeger M, Durand NC, Shamim MS, Machol I, Lander ES, Aiden AP, et al. 2017. De novo assembly of the *Aedes aegypti* genome using Hi-C yields chromosome-length scaffolds. *Science* **356:** 92–95. doi:10.1126/science.aal3327

Durand NC, Robinson JT, Shamim MS, Machol I, Mesirov JP, Lander ES, Aiden EL. 2016. Juicebox provides a visualization system for Hi-C contact maps with unlimited zoom. *Cell Syst* **3:** 99–101. doi:10.1016/j.cels.2015.07.012

Ghurye J, Pop M, Koren S, Bickhart D, Chin C-S. 2017. Scaffolding of long read assemblies using long range contact information. *BMC Genomics* **18:** 527. doi:10.1186/s12864-017-3879-z

Harewood L, Kishore K, Eldridge MD, Wingett S, Pearson D, Schoenfelder S, Collins VP, Fraser P. 2017. Hi-C as a tool for precise detection and characterisation of chromosomal rearrangements and copy number variation in human tumours. *Genome Biol* **18:** 125. doi:10.1186/s13059-017-1253-8

Jaillon O, Aury J-M, Brunet F, Petit J-L, Stange-Thomann N, Mauceli E, Bouneau L, Fischer C, Ozouf-Costaz C, Bernot A, et al. 2004. Genome duplication in the teleost fish *Tetraodon nigroviridis* reveals the early vertebrate proto-karyotype. *Nature* **431:** 946–957. doi:10.1038/nature03025

Kaplan N, Dekker J. 2013. High-throughput genome scaffolding from in vivo DNA interaction frequency. *Nat Biotechnol* **31:** 1143–1147. doi:10.1038/nbt.2768

Lazar NH, Nevonen KA, O'Connell B, McCann C, O'Neill RJ, Green RE, Meyer TJ, Okhovat M, Carbone L. 2018. Epigenetic maintenance of topological domains in the highly rearranged gibbon genome. *Genome Res* **28:** 983–997. doi:10.1101/gr.233874.117

Liang Z, Li G, Wang Z, Djekidel MN, Li Y, Qian M-P, Zhang MQ, Chen Y. 2017. BL-Hi-C is an efficient and sensitive approach for capturing structural and regulatory chromatin interactions. *Nat Commun* **8:** 1622. doi:10.1038/s41467-017-01754-3

Lieberman-Aiden E, van Berkum NL, Williams L, Imakaev M, Ragoczy T, Telling A, Amit I, Lajoie BR, Sabo PJ, Dorschner MO, et al. 2009. Comprehensive mapping of long-range interactions reveals folding principles of the human genome. *Science* **326:** 289–293. doi:10.1126/science.1181369

Lin YC, Benner C, Mansson R, Heinz S, Miyazaki K, Miyazaki M, Chandra V, Bossen C, Glass CK, Murre C. 2012. Global changes in the nuclear positioning of genes and intra- and interdomain genomic interactions that orchestrate B cell fate. *Nat Immunol* **13:** 1196–1204. doi:10.1038/ni.2432

Mishra A, Hawkins RD. 2017. Three-dimensional genome architecture and emerging technologies: looping in disease. *Genome Med* **9:** 87. doi:10.1186/s13073-017-0477-2

Nickerson JA, Krockmalnic G, Wan KM, Penman S. 1997. The nuclear matrix revealed by eluting chromatin from a cross-linked nucleus. *Proc Natl Acad Sci* **94:** 4446–4450. doi:10.1073/pnas.94.9.4446

Putnam NH, O'Connell BL, Stites JC, Rice BJ, Blanchette M, Calef R, Troll CJ, Fields A, Hartley PD, Sugnet CW, et al. 2016. Chromosome-scale shotgun assembly using an in vitro method for long-range linkage. *Genome Res* **26:** 342–350. doi:10.1101/gr.193474.115

Quigley IK, Kintner C. 2017. Rfx2 stabilizes Foxj1 binding at chromatin loops to enable multiciliated cell gene expression. *PLoS Genet* **13:** e1006538. doi:10.1371/journal.pgen.1006538

Rao SSP, Huntley MH, Durand NC, Stamenova EK, Bochkov ID, Robinson JT, Sanborn AL, Machol I, Omer AD, Lander ES, et al. 2014. A 3D map of the human genome at kilobase resolution reveals principles of chromatin looping. *Cell* **159:** 1665–1680. doi:10.1016/j.cell.2014.11.021

Selvaraj S, Dixon JR, Bansal V, Ren B. 2013. Whole-genome haplotype reconstruction using proximity-ligation and shotgun sequencing. *Nat Biotechnol* **31:** 1111–1118. doi:10.1038/nbt.2728

Session AM, Uno Y, Kwon T, Chapman JA, Toyoda A, Takahashi S, Fukui A, Hikosaka A, Suzuki A, Kondo M, et al. 2016. Genome evolution in the allotetraploid frog *Xenopus laevis*. *Nature* **538:** 336–343. doi:10.1038/nature19840

Sive HL, Grainger RM, Harland RM. 2010. *Early development of Xenopus laevis: a laboratory manual*. Cold Spring Harbor Laboratory Press, Cold Spring Harbor, NY.

Wolff J, Bhardwaj V, Nothjunge S, Richard G, Renschler G, Gilsbach R, Manke T, Backofen R, Ramírez F, Grüning BA. 2018. Galaxy HiCExplorer: a web server for reproducible Hi-C data analysis, quality control and visualization. *Nucleic Acids Res* **46:** W11–W16. doi:10.1093/nar/gky504

Yardımcı GG, Noble WS. 2017. Software tools for visualizing Hi-C data. *Genome Biol* **18:** 26. doi:10.1186/s13059-017-1161-y

Mapping Chromatin Features of *Xenopus* Embryos

George E. Gentsch[1,2] and James C. Smith[1,2]

[1]*Developmental Biology Laboratory, The Francis Crick Institute, London NW1 1AT, United Kingdom*

Chromatin immunoprecipitation (ChIP) combined with genomic analysis provides a global snapshot of protein–DNA interactions in the context of chromatin, yielding insights into which genome loci might be regulated by the DNA-associated protein under investigation. This protocol is an update of a previous version and describes how to perform ChIP on intact or dissected *Xenopus* embryos. The ChIP-isolated DNA fragments are suitable for both deep sequencing (ChIP-Seq) and quantitative polymerase chain reaction (ChIP-qPCR). General advice for qPCR and for making ChIP-Seq libraries is offered, and approaches for analyzing ChIP-Seq data are outlined.

MATERIALS

It is essential that you consult the appropriate Material Safety Data Sheets and your Institution's Environmental Health and Safety Office for proper handling of equipment and hazardous materials used in this protocol.

RECIPES: Please see the end of this protocol for recipes indicated by <R>. Additional recipes can be found online at http://cshprotocols.cshlp.org/site/recipes.

Reagents

Agarose gel electrophoresis reagents
Amphibian medium

 Marc's modified Ringer's (MMR) (10×, pH 7.8) <R>
 Normal amphibian medium (NAM) <R>

Antibody
Antibody control (e.g., normal rabbit IgG, Santa-Cruz sc-2027)
Chromatin extraction and wash buffer 1 (CEWB1) <R>
Dejellied *Xenopus* embryos at the desired developmental stage (Sive et al. 2007)
Dithiothreitol (DTT) (1 M)
DNA ladder (e.g., Thermo Fisher SM1331)
Formaldehyde solution (Sigma-Aldrich F8775)
Ethanol (absolute)
GlycoBlue (Thermo Fisher AM9516)
HEG solution (optional; see Step 5) <R>

[2]Correspondence: george.gentsch@crick.ac.uk; jim.smith@crick.ac.uk

Liquid nitrogen

Magnetic beads coated with Protein G (Thermo Fisher 10004D)

NaCl (5 M)

NaF (0.5 M)

Na$_3$VO$_4$ (0.1 M)

Phenol:chloroform:isoamyl alcohol (25:24:1, pH 7.9; Thermo Fisher AM9730)

Protease inhibitor tablets (e.g., Roche 05892953001)

Proteinase K (Thermo Fisher AM2548)

QIAquick PCR Purification Kit (Qiagen 28106)

RNase A (Thermo Fisher 12091039)

SDS elution buffer <R>

Sodium acetate (3 M, pH 5.2)

SYBR Gold nucleic acid stain (Thermo Fisher S11494)

TE, pH 8.0

TEN buffer <R>

Washing buffer 2 (WB2) <R>

Washing buffer 3 (WB3) <R>

Equipment

Agarose gel electrophoresis apparatus

Centrifuge for 50-mL conical tubes (e.g., Eppendorf 5810R)

Centrifuge for microcentrifuge tubes (e.g., Eppendorf 5424R)

Conical tubes (50-mL)

Glass vials with cap (8-mL; e.g., Wheaton 224884)

Magnetic rack for bead separation (Thermo Fisher AM10027)

Microcentrifuge tubes (1.5-mL, siliconized; Thermo Fisher AM12450)

Phase Lock Gel Heavy tubes (2.0-mL; VWR 733-2478)

Oven for incubations from 37°C to 65°C (e.g., Techne Hybridiser HB-1D)

Rotator (e.g., Stuart SB3)

Sonication equipment for shearing cross-linked chromatin (Steps 12–14)

> Custom-built tube for probe-mediated sonication
>> *Score and clip a 15-mL polystyrene tube at 7-mL mark.*
>
> Glass vials for focused sonication (1-mL; Covaris 520135)
>
> Holder for 1-mL glass vials (Covaris 500371)
>
> Isothermal focused sonicator (Covaris S220) or microtip-equipped sonicator (Misonix 3000 or Qsonica Q500/700)
>
> Laboratory jack (e.g., Scientific Laboratories Supplies CH0642)
>
> Plastic beaker (e.g., Azlon 600-mL plastic beaker) filled with ice water
>
> Sound enclosure and/or ear protectors
>
> Tapered microtip (1/16-inch (1.6-mm); VWR 432-0199)
>
> Thermometer clamp with a total length of 3–4 inch (Fig. 1B)

Surgical tools for dissection (optional; see Step 4)

Thermomixer (e.g., Eppendorf 5384000020)

METHOD

This is an update to Gentsch and Smith (2014).

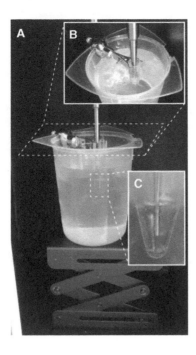

FIGURE 1. Setup of a tapered microtip-equipped sonicator.

Cross-Linking Proteins to Their Endogenous DNA Binding Sites (~30 min to 2 h)

1. Transfer dejellied embryos at the desired developmental stage to an 8-mL glass vial (capacity ~9 mL).

   The number of embryos needed to reach a minimal amount of co-immunoprecipitated DNA for ChIP-Seq or ChIP-qPCR is best determined empirically as some of the critical factors are difficult to define in practice (like epitope accessibility and cross-linking efficiency). As a starting point, collect 500,000 cells containing the chromatin feature of interest.

2. Fix whole embryos by adding 250 µL formaldehyde (1% final concentration) to the glass vial almost fully filled up with amphibian medium (e.g., 0.05× MMR for *X. tropicalis* or 0.1× NAM for *X. laevis*). Shake gently and incubate for 15–45 min (*X. tropicalis*) or 25–60 min (*X. laevis*) at room temperature without further rocking.

   Fixation time is best determined empirically by ChIP-qPCR aiming for high signal-to-noise ratios. Because of their higher yolk content, early X. tropicalis *and any stage* X. laevis *embryos tend to require longer fixation times.*

3. Remove fixative by briefly washing the embryos three times with 8 mL of ice-cold amphibian medium. If dissections (Step 4) and freezing for storage (Step 5) are not required, remove as much remaining medium as possible from the embryos before continuing with Step 6.

   At this stage, embryos are very fragile and should not come into contact with the liquid surface.

4. (Optional) Dissect embryos in cold amphibian medium to isolate the anatomical region of interest.

5. (Optional) Freeze and store the embryos as follows.

   i. Carefully transfer the embryonic tissue (e.g., groups of ~250 embryos or fewer) into prechilled 2-mL microcentrifuge tubes. Remove as much remaining medium as possible from the embryos.

   ii. Add approximately one volume of ice-cold HEG solution.

   iii. Allow the embryos to settle to the bottom of the tube and remove excess HEG.

   iv. Snap-freeze in liquid nitrogen and store at −80°C for future use.

Preparing Chromatin Extract for Sonication (~30–40 min)

The extraction of preneurula chromatin works best with no more than 60 X. tropicalis or 25 X. laevis embryos per 1-mL ice-cold CEWB1 in a single homogenization step (i.e., maximal 3000 X. tropicalis or 1250 X. laevis embryos per 50-mL Falcon tube). The number of embryos can be increased at later developmental stages. During Steps 6–11, keep all biological samples and solutions on ice.

6. Supplement CEWB1 with the appropriate amount of protease inhibitor tablet and DTT to a final concentration of 0.5 mM. If using a phospho-specific antibody for ChIP, add NaF and Na_3VO_4 to a final concentration of 2.5 and 0.5 mM, respectively.

7. Homogenize fixed embryonic tissue with 500 μL CEWB1 by pipetting up and down. Transfer homogenate to prechilled 50-mL conical tube and dilute it with the appropriate amount of CEWB1 (see recommendation above). Embryo batches of the same condition can be pooled here.

8. Keep the homogenate on ice for 10 min before centrifugation at 1000*g* for 5 min at 4°C.

9. Aspirate the supernatant and any lipids attached to the wall and lid.

10. Resuspend the pellet in the appropriate volume of CEWB1 (see recommendation above) and repeat Steps 8 and 9.

    Upon efficient extraction of cross-linked nuclei, the pellet is hardly visible or appears brown because of insolubilized pigment particles. When processing cleavage-stage embryos or extensively fixed embryos, the pellet may still contain a substantial amount of yolk.

11. Resuspend the pellet containing cross-linked nuclei in 1 mL CEWB1. If the nuclear suspension appears very viscous, dilute it further with CEWB1 to a maximum volume of 3 mL.

Shearing Cross-Linked Chromatin (~30 min to 1 h)

Here we describe the parameters used to shear chromatin with either a microtip-equipped sonicator or an isothermal focused sonicator. In principle, any sonicator achieving ultrasonic frequencies and 6–18 W can shear cross-linked chromatin.

12. Transfer the chromatin extract as follows:
    - *Probe-mediated sonication*: Transfer all the chromatin extract into a custom-built tube (see Equipment).

    - *Focused sonication*: If necessary dilute the chromatin extract with ice-cold CEWB1 to the next full mL. Transfer 1-mL chromatin extract into a designated 1-mL glass vial. Repeat focused sonication with the remaining chromatin extract if necessary.

13. Attach the tube or container to the sonicator as follows:
    - *Probe-mediated sonication*: Attach the custom-built tube with the chromatin extract to a plastic beaker filled with ice water via a short thermometer clamp (Fig. 1A,B). With a laboratory jack, move the beaker/tube vertically and horizontally so that the sonicator microtip is immersed in the sample to two-thirds of the volume depth and properly centered without contacting the tube wall (Fig. 1B,C).

    - *Focused sonication*: Fill the water tank with distilled or deionized water to level 8. Degas and cool water to 4°C–10°C, which requires about 30 min to 1 h. Clip the 1-mL glass vial into the appropriate holder, place the holder/container into the acoustic assembly and close the safety cover.

14. Sonicate the sample of cross-linked chromatin as follows:
    - *Probe-mediated sonication*: Set the program to 8–20 cycles of 30-sec sonication (i.e., 4–10 min total) followed by 1 min pauses. Close the sound enclosure or wear ear protectors during sonication. Start sonication with an intensity setting of 1.0. Depending on the volume and ionic strength of the sample, the intensity should be changed to reach 6–18 W output for chromatin shearing.

Cite this protocol as *Cold Spring Harb Protoc*; doi:10.1101/pdb.prot100263

Use a maximum of 20 cycles for extensively cross-linked chromatin (fixed for 45–60 min). If sample begins to froth, pause immediately, reposition the tube and wait until the froth disappears before resuming sonication.

- *Focused sonication*: Set duty cycle to 5%, intensity to 4, cycles per burst to 200, and processing time to 240–600 sec. This achieves ~11 W output.

  Use maximal processing time of 10 min for extensively cross-linked chromatin.

15. After sonication, transfer chromatin solution into 1.5-mL tubes, and remove debris by centrifuging at ≥15,000g for 5 min at 4°C.

16. Transfer the clear supernatant containing solubilized and sheared chromatin to fresh prechilled 1.5-mL tube(s).

17. Collect a 50-µL sample to check the efficiency of chromatin shearing (see Steps 18–21). Use the remaining chromatin for immunoprecipitation (see Steps 22–32). Keep chromatin at 4°C if used on the same or following 3 d; alternatively, snap-freeze the sample in liquid nitrogen and store it at −80°C for later use.

Checking the Efficiency of Chromatin Shearing (~1–2 d)

18. Add ~50 µL sodium dodecyl sulfate (SDS) elution buffer, 4 µL 5 M NaCl and 20 µg proteinase K to the 50 µL chromatin aliquot.

19. Incubate the chromatin solution overnight (12–16 h) in an oven set to 65°C.

20. Purify DNA fragments using the QIAquick PCR Purification Kit according to the manufacturer's instructions. Elute the DNA with 30 µL EB buffer (10 mM Tris-HCl, pH 8.5).

21. Add 8 µg RNase A and 5 µL 5× DNA loading buffer. Run the entire sample alongside a 1-kb DNA ladder on a 1.4% TAE agarose gel containing a standard nucleic acid stain at 120 V for 30 min.

    If the sample contains genomic DNA from fewer than ~100,000 nuclei, run the purified DNA on a 1.4% TAE agarose gel cast as thin as possible and without nucleic acid stain. After gel electrophoresis submerge gel in 0.5× electrophoresis buffer with 1:20,000 SYBR Gold for 10 min. An asymmetric DNA fragment distribution should be visible, mainly within the range of 100–1000 bp peaking ≤500 bp (Fig. 2).

    See Troubleshooting.

Immunoprecipitation of Sheared Chromatin (~1–2 d)

Use siliconized 1.5-mL tubes to collect chromatin at any given step. Prechill all washing buffers (CEWB1, WB2, WB3, and TEN) on ice.

22. From the sheared chromatin solution (Step 17), collect a sample that is equivalent to ~1% of the ChIP input. Transfer the "input" sample to a new 1.5-mL tube and keep it at 4°C.

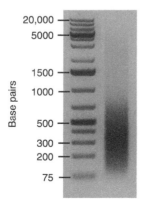

FIGURE 2. Sonication-mediated shearing of chromatin-associated DNA as measured by gel electrophoresis.

23. Add the ChIP-grade antibody to the remaining chromatin solution. If using an antibody control such as nonspecific Ig proteins or preimmune serum of the appropriate antibody isotype and host species, use half of the chromatin for the antibody and the other half for the antibody control (matching the amount of antibody in use).

    The antibody is one of the most critical factors of a ChIP experiment. Follow the ENCODE guidelines (Landt et al. 2012) to demonstrate antibody specificity. The optimal amount of antibody can be estimated by ChIP-qPCR aiming at maximal yield and signal-to-noise ratios (see Discussion). Normally, it is within the range of 1–10 µg of antibody per million cells expressing the epitope of interest.

    See Troubleshooting.

24. Incubate the chromatin/antibody solution on a vertical rotator (10 rpm) for 4–16 h (overnight) at 4°C.

25. Choose the magnetic beads that are capable of binding the antibody and antibody control and wash an appropriate volume of them once with CEWB1 for 5 min at 4°C.

    We usually use 5–20 µL protein G magnetic beads per 1 µg mouse, rabbit or goat IgG antibody. To mitigate relative loss of the beads during washes, use at least 30 µL beads in total. It is best to estimate the optimal volume of beads by ChIP-qPCR.

26. Resuspend the washed beads in 100 µL CEWB1. Add the appropriate volume of beads to the ChIP samples and incubate for another 4 h on a vertical rotator at 4°C.

27. Wash the beads twice with 1 mL of each washing buffer in the following order on a vertical rotor (10 rpm) for 5 min at 4°C: CEWB1, WB2, and WB3.

    To avoid loss of the beads, leave the beads in the magnetic rack for at least 30 sec and invert the rack a couple of times to collect any beads stuck to the underside of the lid before discarding the liquid.

28. Rinse the beads once with 1 mL TEN.

29. Resuspend the beads in 50 µL TEN, and transfer the bead suspension to a fresh tube. If several tubes of the same ChIP are in use, pool the samples at this stage.

30. Use centrifugation at 1000g for 1 min at 4°C and the magnetic rack to assemble beads at the bottom of the tube. Remove as much of the supernatant as possible while keeping the bottom of the tube adjacent to the magnetic rack.

31. Elute the immunoprecipitated chromatin by adding 50 µL SDS elution buffer to the beads. Incubate and vortex (1200 rpm; continuously or intermittently) the samples for 15 min at 65°C before centrifuging at ≥15,000g for 30 sec at room temperature. Transfer as much of the eluate as possible to a fresh tube while keeping the bottom of the tube adjacent to the magnetic rack.

32. Repeat the elution step and combine the eluates.

Recovery of Co-Immunoprecipitated DNA (~1–2 d)

33. Add 50 µL SDS elution buffer to the input sample collected at Step 22.

34. Add 4 µL 5 M NaCl to each ChIP eluate, antibody control, and input sample and incubate the chromatin solution overnight (12–16 h) in an oven set to 65°C.

35. Dilute the samples with 100 µL TE pH 8.0 and add 2-µL RNase A to achieve a final concentration of 200 µg/mL. Incubate for 1 h at 37°C.

36. Add 2 µL proteinase K to achieve a final concentration of 200 µg/mL. Incubate for 3 h in an oven set to 55°C.

37. Extract the DNA fragments using either the QIAquick PCR Purification Kit (see manufacturer's instructions) or phenol:chloroform:isoamyl alcohol purification followed by ethanol precipitation as described below.

    i. Transfer the samples to prespun 2-mL Phase Lock Gel Heavy tubes, add 200 µL phenol:chloroform:isoamyl alcohol, and mix the samples by inverting the tubes 4–6 times.

Cite this protocol as *Cold Spring Harb Protoc*; doi:10.1101/pdb.prot100263

ii. Centrifuge the samples at ≥15,000*g* for 5 min at room temperature before transferring the aqueous (top) phase to a fresh tube.

iii. Add 8 µL 5 M NaCl, 400-µL ethanol, and 15 µg GlycoBlue and mix the samples by inverting the tubes 4–6 times.

iv. Keep the samples overnight at −20°C.

v. Centrifuge the samples at ≥15,000*g* for 1 h at 4°C, and then discard the supernatant without disturbing the blue DNA pellet.

vi. Wash the pellet with 500 µL ice-cold 80% ethanol.

vii. Centrifuge the samples at >15,000*g* for 2 min at 4°C, and then discard the supernatant without disturbing the blue DNA pellet.

viii. Air-dry the DNA pellet for 10 min at room temperature.

ix. Add 11 µL Buffer EB (provided with the QIAquick PCR Purification Kit) to the dry DNA pellet, and leave samples on ice for 30 min to ensure that the DNA is completely dissolved.

   After extraction, the DNA is ready for either qPCR or for making ChIP-Seq libraries (see Discussion). ChIP-isolated and input DNA samples can be stored at −20°C.

TROUBLESHOOTING

Problem (Step 21): DNA fragments are longer than expected (i.e., average >500 bp).

Solution: Try shearing the chromatin again, beginning at Step 12. However, poor shearing efficiency often relates to inefficient nuclear extraction or prolonged cross-linking rather than inefficient sonication.

Problem (Step 23): Immunoprecipitation is poor (no or low-level enrichment) as judged by ChIP-qPCR with positive and negative controls.

Solution: Try another antibody because not all antibodies are suitable for ChIP. Consider tagging your protein of interest with an epitope for which ChIP-grade antibodies are available. Analysis of tagged proteins can be informative in the appropriate developmental context (Mazzoni et al. 2011).

DISCUSSION

This protocol is an update to Gentsch and Smith (2014) and was devised to prepare ChIP-isolated DNA from embryos with relatively high yolk content. It has since been carried out successfully in our and other laboratories with *Xenopus* embryos of various developmental stages (Gentsch et al. 2013, 2018; Miyamoto et al. 2013). It provides an alternative to other protocols (Blythe et al. 2009; Akkers et al. 2012) and has the advantage of easy upscaling to thousands of embryos, which are sometimes required to map rare chromatin features.

Analyzing ChIP Isolated DNA by qPCR

In ChIP-qPCR, DNA enrichment is quantified relative to either locus-specific standard curves or by the $2^{-\Delta\Delta Cq}$ method using real-time PCR (Livak and Schmittgen 2001; Blythe et al. 2009; Gentsch et al. 2015). DNA occupancy levels can be visualized either as a percentage of input DNA (Fig. 3A) or as a ratio of ChIP versus mock (antibody control) IP. If known, include a negative (not bound) control locus as well as a positive (bound) control locus. A successful ChIP experiment yields significant DNA enrichment at positive control loci and no or weak enrichment at negative control loci. The antibody control should not give any enrichment at positive or negative control loci. To critically evaluate ChIP-qPCR data, they should be reported according to the MIQE guidelines (Bustin et al. 2009).

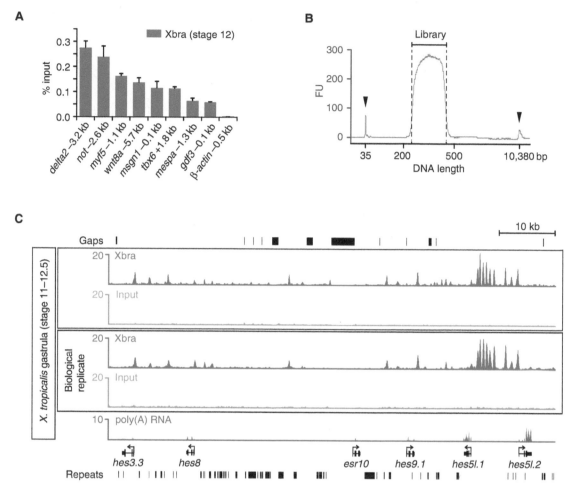

FIGURE 3. Profiling Xbra (*Xenopus* Brachyury) binding at late gastrula stage by ChIP-qPCR (*A*) and ChIP-Seq (*C*). Locus at β-actin is used as a negative ChIP-qPCR control. Error bar, SD from the mean (*n* = 2). ChIP-Seq data from Gentsch et al. (2013) and poly(A) RNA track from Akkers et al. (2009) normalized to 1 million mapped reads. (*B*) Electropherogram of a ChIP-Seq library alongside ladder markers (filled arrowheads). FU, fluorescence units.

Creating and Analyzing ChIP-Seq Data

In our hands, the minimal amount of well-fragmented DNA (Fig. 2) required to make a high-complexity ChIP-Seq library is ∼500 pg. However, we recommend aiming at a few nanograms for backup and to account for inefficient library preparation. Our protocol to make an indexed paired-end library from a few picograms to nanograms of ChIP-DNA is outlined in Gentsch and Smith (2017). Libraries should be quantified by fluorometry and quality-controlled on either a polyacrylamide gel or on a microfluidics-based platform. Successful library preparation yields 100–400 ng DNA of the expected size range (250–500 bp) without any adapter dimer (120 bp) contamination (Fig. 3B). Ten million single-end sequencing reads of at least 36 bases are sufficient for mapping chromatin features with narrow (≤500 bp) occupancies, which is commonly observed with sequence-specific transcription factors. For chromatin features spread over longer distances, such as histone modifications or RNA polymerase, consider increasing the sequencing depth.

Following sequencing of the library on a deep sequencing platform, apply tools like FASTQC (Andrews 2010) and Trim Galore/Cutadapt (Martin 2011; Krueger 2012) to verify sequence quality and library complexity and to remove any adapter contamination, respectively. A quality-approved high-complexity library produces more than 75% nonredundant reads and virtually no adapter contamination. Use the default settings of Bowtie2 (Langmead and Salzberg 2012) to map reads to

Cite this protocol as *Cold Spring Harb Protoc*; doi:10.1101/pdb.prot100263

the genome. Remove any redundant reads from the sequence alignment file with PICARD (https:// broadinstitute.github.io/picard) or HOMER (Heinz et al. 2010) to eliminate any potential PCR duplicates. This does not compromise the sensitivity of finding any enriched alignments (peaks) by various peak callers (Chen et al. 2012) like HOMER (Heinz et al. 2010) or MACS (Zhang et al. 2008). HOMER and R/Bioconductor provide many more excellent tools to further analyze the location of chromatin features. To avoid any incorrect interpretation of sequencing data, we advise ignoring sites with poor mappability and false-positive enrichments caused by repetitive and low-complexity DNA (Pickrell et al. 2011). Genome-wide profiles can be displayed on-line via the Francis Crick Institute based UCSC Genome Browser interface hosting all *Xenopus* genome assemblies (http:// genomes.crick .ac.uk) or off-line with the Integrative Genomics Viewer IGV (Fig. 3C; Robinson et al. 2011).

RECIPES

Chromatin Extraction and Wash Buffer 1 (CEWB1)

Reagent	Amount (for 500 mL)	Final concentration
Tris-HCl (1 M, pH 7.5)	25 mL	50 mM
EDTA (0.5 M, pH 8.0)	1 mL	1 mM
NaCl (5 M)	15 mL	150 mM
Igepal CA-630 (10%)	50 mL	1%
Sodium deoxycholate (10%)	12.5 mL	0.25%
SDS (10%)	5 mL	0.1%
Double-distilled water	391.5 mL	–

Store for up to 1 yr at 4°C.

HEG Solution

Reagent	Amount for 10 mL	Final concentration
HEPES–*KOH* (1 M, pH 7.5)	500 µL	50 mM
EDTA (0.5 M, pH 8.0)	20 µL	1 mM
Glycerol (100%)	2 mL	20%
Double-distilled water	7.5 mL	–

Store for up to 1 year at 4°C.

Marc's Modified Ringer's (MMR) (10×, pH 7.8)

Reagent	Concentration (10×)
NaCl	1 M
KCl	20 mM
CaCl$_2$	20 mM
MgSO$_4$	10 mM
HEPES	50 mM
EDTA	1 mM

Adjust pH to 7.8 and sterilize by autoclaving. Use the solution at 0.05×. Store for up to 1 yr at room temperature.

Normal Amphibian Medium (NAM)

Reagent	Concentration (10×)
NaCl	1.1 M
KCl	20 mM
Ca(NO$_3$)$_2$	10 mM
MgSO$_4$	10 mM
EDTA	1 mM

Use this solution at 0.1×, supplemented with 2 mM sodium phosphate (pH 7.5) from a 50× stock (0.1 M sodium phosphate buffer). Store for up to 1 year at room temperature.

SDS Elution Buffer

Reagent	Amount (for 1 mL)	Final concentration
Tris–HCl (1 M, pH 8.0)	50 μL	50 mM
EDTA (0.5 M, pH 8.0)	2 μL	1 mM
SDS (10%)	100 μL	1%
Double-distilled water	848 μL	–

Store for up to 1 week at room temperature.

TEN Buffer for ChIP

Reagent	Amount (for 50 mL)	Final concentration
Tris–HCl (1 M, pH 8.0)	500 μL	10 mM
EDTA (0.5 M, pH 8.0)	100 μL	1 mM
NaCl (5 M)	1.5 mL	150 mM
Double-distilled water	47.9 mL	–

Store for up to 1 year at 4°C.

Washing Buffer 2 (WB2)

Reagent	Amount (for 50 mL)	Final concentration
Tris-HCl (1 M, pH 8.0)	500 μL	10 mM
EDTA (0.5 mM, pH 8.0)	100 μL	1 mM
NaCl (5 M)	5 mL	500 mM
Igepal CA-630 (10%)	5 mL	1%
Sodium deoxycholate (10%)	1.25 mL	0.25%
SDS (10%)	500 μL	0.1%
Double-distilled water	37.65 mL	–

Store for up to 1 yr at 4°C.

Washing Buffer 3 (WB3)

Reagent	Amount (for 50 mL)	Final concentration
Tris-HCl (1 M, pH 8.0)	500 μL	10 mM
EDTA (0.5 mM, pH 8.0)	100 μL	1 mM
LiCl (8 M)	1.57 mL	250 mM
Igepal CA-630 (10%)	5 mL	1%
Sodium deoxycholate (10%)	5 mL	1%
Double-distilled water	37.83 mL	–

Store for up to 1 yr at 4°C.

Cite this protocol as *Cold Spring Harb Protoc*; doi:10.1101/pdb.prot100263

COMPETING INTEREST STATEMENT

The authors declare no conflict of interest or competing financial interest.

ACKNOWLEDGMENTS

G.E.G. and J.C.S. were supported by the Medical Research Council (program number U117597140) and are now supported by the Francis Crick Institute, which receives its core funding from Cancer Research UK (FC001-157), the UK Medical Research Council (FC001-157), and the Wellcome Trust (FC001-157).

REFERENCES

Akkers RC, van Heeringen SJ, Jacobi UG, Janssen-Megens EM, Françoijs KJ, Stunnenberg HG, Veenstra GJC. 2009. A hierarchy of H3K4me3 and H3K27me3 acquisition in spatial gene regulation in *Xenopus* embryos. *Dev Cell* 17: 425–434. doi:10.1016/j.devcel.2009.08.005

Akkers RC, Jacobi UG, Veenstra GJ. 2012. Chromatin immunoprecipitation analysis of *Xenopus* embryos. *Methods Mol Biol* 917: 279–292. doi: 10.1007/978-1-61779-992-1_17

Andrews S. 2010. *A quality control tool for high throughput sequence data.* http://www.bioinformatics.babraham.ac.uk/projects/fastqc.

Blythe SA, Reid CD, Kessler DS, Klein PS. 2009. Chromatin immunoprecipitation in early *Xenopus laevis* embryos. *Dev Dyn* 238: 1422–1432. doi:10.1002/dvdy.21931

Bustin SA, Benes V, Garson JA, Hellemans J, Huggett J, Kubista M, Mueller R, Nolan T, Pfaffl MW, Shipley GL, et al. 2009. The MIQE guidelines: minimum information for publication of quantitative real-time PCR experiments. *Clin Chem* 55: 611–622. doi:10.1373/clinchem.2008.112797

Chen Y, Negre N, Li Q, Mieczkowska JO, Slattery M, Liu T, Zhang Y, Kim TK, He HH, Zieba J, et al. 2012. Systematic evaluation of factors influencing ChIP-seq fidelity. *Nat Methods* 9: 609–614. doi:10.1038/nmeth.1985

Gentsch GE, Smith JC. 2014. Investigating physical chromatin associations across the *Xenopus* genome by chromatin immunoprecipitation. *Cold Spring Harb Protoc* 2014: pdb.prot080614. doi:10.1101/pdb.prot080614

Gentsch GE, Smith JC. 2017. Efficient preparation of high-complexity ChIP-seq profiles from early *Xenopus* embryos. *Methods Mol Biol* 1507: 23–42. doi:10.1007/978-1-4939-6518-2_3

Gentsch GE, Owens ND, Martin SR, Piccinelli P, Faial T, Trotter MW, Gilchrist MJ, Smith JC. 2013. In vivo T-box transcription factor profiling reveals joint regulation of embryonic neuromesodermal bipotency. *Cell Rep* 4: 1185–1196. doi:10.1016/j.celrep.2013.08.012

Gentsch GE, Patrushev I, Smith JC. 2015. Genome-wide snapshot of chromatin regulators and states in *Xenopus* embryos by ChIP-Seq. *J Vis Exp* 96: e52535. doi:10.3791/52535

Gentsch GE, Spruce T, Owens ND, Smith JC. 2018. The role of maternal pioneer factors in predefining first zygotic responses to inductive signals. *bioRxiv* 306803: doi:10.1101/306803

Heinz S, Benner C, Spann N, Bertolino E, Lin YC, Laslo P, Cheng JX, Murre C, Singh H, Glass CK. 2010. Simple combinations of lineage-determining transcription factors prime cis-regulatory elements required for macrophage and B cell identities. *Mol Cell* 38: 576–589. doi:10.1016/j.molcel.2010.05.004

Krueger F. 2012. *A wrapper tool around Cutadapt and FastQC to consistently apply quality and adapter trimming to FastQ files.* http://www.bioinformatics.babraham.ac.uk/projects/trim_galore.

Landt SG, Marinov GK, Kundaje A, Kheradpour P, Pauli F, Batzoglou S, Bernstein BE, Bickel P, Brown JB, Cayting P, et al. 2012. ChIP-seq guidelines and practices of the ENCODE and modENCODE consortia. *Genome Res* 22: 1813–1831. doi:10.1101/gr.136184.111

Langmead B, Salzberg SL. 2012. Fast gapped-read alignment with Bowtie 2. *Nat Methods* 9: 357–359. doi:10.1038/nmeth.1923

Livak KJ, Schmittgen TD. 2001. Analysis of relative gene expression data using real-time quantitative PCR and the $2^{-\Delta\Delta C}T$ method. *Methods* 25: 402–408. doi:10.1006/meth.2001.1262

Martin M. 2011. Cutadapt removes adapter sequences from high-throughput sequencing reads. *EMBnet J* 17: 10–12. doi:10.14806/ej.17.1.200

Mazzoni EO, Mahony S, Iacovino M, Morrison CA, Mountoufaris G, Closser M, Whyte WA, Young RA, Kyba M, Gifford DK, et al. 2011. Embryonic stem cell-based mapping of developmental transcriptional programs. *Nat Methods* 8: 1056–1058. doi:10.1038/nmeth.1775

Miyamoto K, Teperek M, Yusa K, Allen GE, Bradshaw CR, Gurdon JB. 2013. Nuclear Wave1 is required for reprogramming transcription in oocytes and for normal development. *Science* 341: 1002–1005. doi:10.1126/science.1240376

Pickrell JK, Gaffney DJ, Gilad Y, Pritchard JK. 2011. False positive peaks in ChIP-seq and other sequencing-based functional assays caused by unannotated high copy number regions. *Bioinformatics* 27: 2144–2146. doi:10.1093/bioinformatics/btr354

Robinson JT, Thorvaldsdóttir H, Winckler W, Guttman M, Lander ES, Getz G, Mesirov JP. 2011. Integrative genomics viewer. *Nat Biotechnol* 29: 24–26. doi:10.1038/nbt.1754

Sive HL, Grainger RM, Harland RM. 2007. Dejellying *Xenopus laevis* embryos. *CSH Protoc* 2007: pdb.prot4731. doi:10.1101/pdb.prot4731

Zhang Y, Liu T, Meyer CA, Eeckhoute J, Johnson DS, Bernstein BE, Nusbaum C, Myers RM, Brown M, Li W, et al. 2008. Model-based analysis of ChIP-Seq (MACS). *Genome Biol* 9: R137. doi:10.1186/gb-2008-9-9-r137

Transcriptomics and Proteomics Methods for *Xenopus* Embryos and Tissues

Michael J. Gilchrist,[1,2,4] Gert Jan C. Veenstra,[1,2,4] and Ken W.Y. Cho[3,4]

[1]*The Francis Crick Institute, London NW1 1AT, United Kingdom;* [2]*Department of Molecular Developmental Biology, Radboud University, 6525 GA Nijmegen, The Netherlands;* [3]*Department of Developmental and Cell Biology, School of Biological Sciences, University of California, Irvine, California 92697*

The general field of quantitative biology has advanced significantly on the back of recent improvements in both sequencing technology and proteomics methods. The development of high-throughput, short-read sequencing has revolutionized RNA-based expression studies, while improvements in proteomics methods have enabled quantitative studies to attain better resolution. Here we introduce methods to undertake global analyses of gene expression through RNA and protein quantification in *Xenopus* embryos and tissues.

OVERVIEW

In recent years, biology has become a more quantitative science, and this trend continues on the back of advances in genomics, proteomics, and analytical methods. *Xenopus*, a well-established model for vertebrate embryology and development, is benefitting from these advances. The availability of more precise data has opened the door to the dissection of intricate biological problems, including protein-interaction networks and gene regulation during cell specification and morphogenesis.

RNA sequencing (RNA-seq), the measurement of gene-specific RNA by counting RNA sequence fragments, is most frequently used to compare gene expression between biological states. Advantages over older methods such as microarrays and quantitative polymerase chain reaction (qPCR) are that it is global, relatively unbiased, and has a very large dynamic range, capable of measuring gene expression with great sensitivity over ∼5 orders of magnitude. RNA-seq requires cellular RNAs to be fragmented into short pieces, which are converted to double-stranded cDNAs. Adaptor sequences are added to the cDNAs and amplified to generate a library, which is then sequenced to generate many millions of reads. These reads are computationally mapped to genes to generate read counts per gene in each sample.

Much of the interest in transcriptomic analysis lies in its ability to compare gene expression, with considerable sensitivity, between two or more states. These states may be developmental stages or tissues; experimental treatments are often compared to controls. To generate statistically meaningful data, experiments need to be carefully designed with sufficient biological replicates. It is also advisable to carefully examine "finished" data sets, looking for consistent expression across groups of similar samples, which may help to identify and eliminate outlying, probably low-quality, samples.

[4]Correspondence: drmikegilchrist@gmail.com; g.veenstra@science.ru.nl; kwcho@uci.edu

Alternatively, gene expression can be explored by measuring global protein levels. The drawback is that these experimental methods have lower sensitivity than transcriptomic approaches. Proteomics is, however, a rapidly developing field, and we expect sensitivity to steadily improve.

In the early days of mass spectrometry, the goal was simply to detect proteins, but with increasing emphasis on understanding biological processes, protein quantification in samples is generally required. This can be achieved in tandem mass spectrometry, which fragments sample proteins in two stages for highly refined fragment identification. The first stage generates a precursor ion spectrum of the components obtained after digestion into short peptides; the second stage further fragments these peptides. The combined data from these two stages allow accurate masses to be calculated and hence carefully identify the proteins in the sample. This is, however, not a simple process and unambiguous assignment of all fragments to proteins remains challenging. Nevertheless, quantification is achievable for many proteins, and even posttranslational modifications may be identified where the mass change is sufficient. Quantitative determination of protein levels may be performed in both label-based and label-free methods.

It has recently been shown, but was long suspected, that transcript expression does not correlate well with protein expression in *Xenopus* (Smits et al. 2014; Peshkin et al. 2015). Indeed, this is to be expected in early *Xenopus* development: Many stored maternal transcripts are polyadenylated and translated shortly after fertilization, so the protein levels will lag behind the appearance of the corresponding polyadenylated transcripts. To further complicate matters, there is also a reservoir of maternal proteins, whose half-lives are not known. Combined with the decay rates of the maternal RNAs and their changing polyadenylation status, these factors add layers of complexity to the regulatory events occurring during early embryonic development. Care is clearly needed when extrapolating protein expression levels from mRNA expression levels.

PROTOCOLS

The accompanying protocols provide tools to undertake global analyses of gene expression through RNA and protein quantification in *Xenopus* embryos or tissues. These include generating transcriptomic data via RNA-seq to evaluate the differences in gene expression between biological states, measuring absolute protein levels in biological samples by mass spectrometry, and preparing tissue-specific nuclei (via isolation of nuclei tagged in specific cell types [INTACT]) to analyze the proteomes of selected tissue samples.

High-throughput sequencing can be carried out on RNA samples from embryos or adult tissues to generate counts per gene per sample. These data can then be used to investigate differences in gene expression levels between states, as described in Protocol 1: An RNA-Seq Protocol for Differential Expression Analysis (Owens et al. 2019). Similar approaches have been widely used in the *Xenopus* community: Examples include analyzing the role of Sox7 in *Xenopus* germ cell development by studying differential expression between primordial germ cells and endoderm cells (Butler et al. 2018); examining the role of maternal transcription factors such as Foxh1, VegT, and Otx1 in zygotic genome activation by injecting translation-blocking antisense morpholino oligonucleotides (Charney et al. 2017; Paraiso et al. 2019); and elucidating different signatures for the development of dorsal and ventral lips of the early *Xenopus* gastrula embryo using gene set expression analysis (GSEA) (Ding et al. 2017).

To circumvent the poor correlation between mRNA and protein levels in early stage *Xenopus* embryos, gene expression dynamics can be studied though proteome profiling, as described in Protocol 2: Mass Spectrometry-Based Absolute Quantification of Single *Xenopus* Embryo Proteomes (Lindeboom et al. 2019). Although care must be taken to work around the yolk proteins in *Xenopus*, the large eggs and cells from the early embryonic divisions contain sufficient protein (unlike those in early mammalian development) to enable quantification of thousands of proteins using mass spectrometry-based proteomics.

There have been steady developments in quantitative proteomics in *Xenopus*, especially from the Kirschner lab, which described protein expression dynamics of whole embryos during embryogenesis

(Wühr et al. 2014; Presler et al. 2017), and from the Nemes and Moody labs, which studied the proteomics of single blastomeres (Onjiko et al. 2015; Lombard-Banek et al. 2016). Together, these highlight the important contribution of the *Xenopus* system in probing early development from a protein perspective.

Pure cell populations may be isolated from embryonic tissues by tagging the nuclei of specific cell types, which are then used for the proteomic analysis of nuclear protein complexes, as described in Protocol 3: INTACT Proteomics in *Xenopus* (Wasson et al. 2019). This biotin-based assay is used to separate the nuclei of interest from those of other cell types without requiring specific antibodies, although it does require transgenic animals. Coupled with liquid chromatography tandem mass spectrometry, this approach allowed isolation of nuclei from developing *Xenopus* heart cells and has identified *Xenopus* orthologs of human genes implicated in congenital heart diseases (Amin et al. 2014).

FUTURE CONSIDERATIONS

The specific properties of the *Xenopus* system have encouraged time-resolved experiments in early development, enabling studies of the dynamics (Collart et al. 2014) and absolute quantitation (Owens et al. 2016) of gene expression. Steadily improving low-RNA capabilities of sequencing protocols have enabled the spatial resolution of gene expression between axially mapped blastomeres of the eight-cell embryo (De Domenico et al. 2015) and different regions of gastrula stage embryos (Blitz et al. 2017).

Single-cell RNA-seq developed using droplet microfluidics (Klein et al. 2015) has enabled the temporal and spatial analysis of gene expression in *Xenopus* embryos, producing a catalog of developmental cell states and a map of cell lineage differentiation spanning the blastula stage to early organogenesis (Briggs et al. 2018). We also anticipate that new technologies, such as RNA seqFISH+ (Eng et al. 2019), may be used to study the spatial distribution of transcripts in the embryo, which is not possible in current single-cell RNA-seq approaches.

The other significant development in sequencing technology is that of long-read, single-molecule sequencing from the likes of PacBio (McCarthy 2010) and Oxford Nanopore (Laver et al. 2015). The short fragments used in Illumina sequencing do not allow exon connectivity to be determined over whole transcripts, and these new long reads provide the perfect tool to properly explore transcript isoform diversity. This is very relevant to developing our understanding of tissue-specific alternative splicing. The relatively small numbers of reads generated in long-read sequencing largely preclude useful quantitation, although these technologies are steadily improving. Short-read sequencing is currently the preferred choice for quantitative biology, but long-read sequencing has the potential to revolutionize our understanding of haplotypes at the genome level and exon usage at the transcript level.

It is inevitable that these trends will come together in single-cell proteomics, and that the resolving power of such techniques will steadily improve. Already there is a study that examines the coexpression of lineage-specific transcription factors in individual cells during human erythropoiesis (Palii et al. 2019). There is no doubt that *Xenopus* will continue to make a powerful contribution to biomedicine, as the lens of biological science, while retaining the global viewpoint, focuses on ever smaller elements of the system.

REFERENCES

Amin NM, Greco TM, Kuchenbrod LM, Rigney MM, Chung MI, Wallingford JB, Cristea IM, Conlon FL. 2014. Proteomic profiling of cardiac tissue by isolation of nuclei tagged in specific cell types (INTACT). *Development* 141: 962–973. doi:10.1242/dev.098327

Blitz IL, Paraiso KD, Patrushev I, Chiu WTY, Cho KWY, Gilchrist MJ. 2017. A catalog of *Xenopus tropicalis* transcription factors and their regional expression in the early gastrula stage embryo. *Dev Biol* 426: 409–417. doi:10.1016/j.ydbio.2016.07.002

Briggs JA, Weinreb C, Wagner DE, Megason S, Peshkin L, Kirschner MW, Klein AM. 2018. The dynamics of gene expression in vertebrate embryogenesis at single-cell resolution. *Science* 360: eaar5780. doi:10.1126/science.aar5780

Butler AM, Owens DA, Wang L, King ML. 2018. A novel role for *sox7* in *Xenopus* early primordial germ cell development: mining the PGC transcriptome. *Development* **145:** dev155978. doi:10.1242/dev.155978

Charney RM, Forouzmand E, Cho JS, Cheung J, Paraiso KD, Yasuoka Y, Takahashi S, Taira M, Blitz IL, Xie X, et al. 2017. Foxh1 occupies cis-regulatory modules prior to dynamic transcription factor interactions controlling the mesendoderm gene program. *Dev Cell* **40:** 595–607 e594. doi:10.1016/j.devcel.2017.02.017

Collart C, Owens ND, Bhaw-Rosun L, Cooper B, De Domenico E, Patrushev I, Sesay AK, Smith JN, Smith JC, Gilchrist MJ. 2014. High-resolution analysis of gene activity during the *Xenopus* mid-blastula transition. *Development* **141:** 1927–1939. doi:10.1242/dev.102012

De Domenico E, Owens ND, Grant IM, Gomes-Faria R, Gilchrist MJ. 2015. Molecular asymmetry in the 8-cell stage *Xenopus tropicalis* embryo described by single blastomere transcript sequencing. *Dev Biol* **408:** 252–268. doi:10.1016/j.ydbio.2015.06.010

Ding Y, Colozza G, Zhang K, Moriyama Y, Ploper D, Sosa EA, Benitez MDJ, De Robertis EM. 2017. Genome-wide analysis of dorsal and ventral transcriptomes of the *Xenopus laevis* gastrula. *Dev Biol* **426:** 176–187. doi:10.1016/j.ydbio.2016.02.032

Eng CL, Lawson M, Zhu Q, Dries R, Koulena N, Takei Y, Yun J, Cronin C, Karp C, Yuan GC, et al. 2019. Transcriptome-scale super-resolved imaging in tissues by RNA seqFISH+. *Nature* **568:** 235–239. doi:10.1038/s41586-019-1049-y

Klein AM, Mazutis L, Akartuna I, Tallapragada N, Veres A, Li V, Peshkin L, Weitz DA, Kirschner MW. 2015. Droplet barcoding for single-cell transcriptomics applied to embryonic stem cells. *Cell* **161:** 1187–1201. doi:10.1016/j.cell.2015.04.044

Laver T, Harrison J, O'Neill PA, Moore K, Farbos A, Paszkiewicz K, Studholme DJ. 2015. Assessing the performance of the Oxford Nanopore Technologies MinION. *Biomol Detect Quantif* **3:** 1–8. doi:10.1016/j.bdq.2015.02.001

Lindeboom RGH, Smits AH, Perino M, Veenstra GJC, Vermeulen M. 2019. Mass spectrometry-based absolute quantification of single *Xenopus* embryo proteomes. *Cold Spring Harb Protoc* **2019:** pdb.prot098376. doi:10.1101/pdb.prot098376

Lombard-Banek C, Reddy S, Moody SA, Nemes P. 2016. Label-free quantification of proteins in single embryonic cells with neural fate in the cleavage-stage frog (*Xenopus laevis*) embryo using capillary electrophoresis electrospray ionization high-resolution mass spectrometry (CE-ESI-HRMS). *Mol Cell Proteomics* **15:** 2756–2768. doi:10.1074/mcp.M115.057760

McCarthy A. 2010. Third generation DNA sequencing: Pacific Biosciences' single molecule real time technology. *Chem Biol* **17:** 675–676. doi:10.1016/j.chembiol.2010.07.004

Onjiko RM, Moody SA, Nemes P. 2015. Single-cell mass spectrometry reveals small molecules that affect cell fates in the 16-cell embryo. *Proc Natl Acad Sci* **112:** 6545–6550. doi:10.1073/pnas.1423682112

Owens NDL, Blitz IL, Lane MA, Patrushev I, Overton JD, Gilchrist MJ, Cho KWY, Khokha MK. 2016. Measuring absolute RNA copy numbers at high temporal resolution reveals transcriptome kinetics in development. *Cell Rep* **14:** 632–647. doi:10.1016/j.celrep.2015.12.050

Owens NDL, De Domenico E, Gilchrist MJ. 2019. An RNA-seq protocol for differential expression analysis. *Cold Spring Harb Protoc* **2019:** pdb.prot098368. doi:10.1101/pdb.prot098368

Palii CG, Cheng Q, Gillespie MA, Shannon P, Mazurczyk M, Napolitani G, Price ND, Ranish JA, Morrissey E, Higgs DR, et al. 2019. Single-cell proteomics reveal that quantitative changes in co-expressed lineage-specific transcription factors determine cell fate. *Cell Stem Cell* **24:** 812–820 e815. doi:10.1016/j.stem.2019.02.006

Paraiso KD, Blitz IL, Coley M, Cheung J, Sudou N, Taira M, Cho KWY. 2019. Endodermal maternal transcription factors establish super-enhancers during zygotic genome activation. *Cell Rep* **27:** 2962–2977 e2965. doi:10.1016/j.celrep.2019.05.013

Peshkin L, Wühr M, Pearl E, Haas W, Freeman RM Jr, Gerhart JC, Klein AM, Horb M, Gygi SP, Kirschner MW. 2015. On the relationship of protein and mRNA dynamics in vertebrate embryonic development. *Dev Cell* **35:** 383–394. doi:10.1016/j.devcel.2015.10.010

Presler M, Van Itallie E, Klein AM, Kunz R, Coughlin ML, Peshkin L, Gygi SP, Wühr M, Kirschner MW. 2017. Proteomics of phosphorylation and protein dynamics during fertilization and meiotic exit in the *Xenopus* egg. *Proc Natl Acad Sci* **114:** E10838–E10847. doi:10.1073/pnas.1709207114

Smits AH, Lindeboom RG, Perino M, van Heeringen SJ, Veenstra GJ, Vermeulen M. 2014. Global absolute quantification reveals tight regulation of protein expression in single *Xenopus* eggs. *Nucleic Acids Res* **42:** 9880–9891. doi:10.1093/nar/gku661

Wasson L, Amin NM, Conlon FL. 2019. INTACT proteomics in *Xenopus*. *Cold Spring Harb Protoc* **2019:** pdb.prot098384. doi:10.1101/pdb.prot098384

Wühr M, Freeman RM Jr, Presler M, Horb ME, Peshkin L, Gygi S, Kirschner MW. 2014. Deep proteomics of the *Xenopus laevis* egg using an mRNA-derived reference database. *Curr Biol* **24:** 1467–1475. doi:10.1016/j.cub.2014.05.044

Cite this introduction as *Cold Spring Harb Protoc*; doi:10.1101/pdb.top098350

An RNA-Seq Protocol for Differential Expression Analysis

Nick D.L. Owens,[1,2] Elena De Domenico,[1,2] and Michael J. Gilchrist[1,3]

[1]*The Francis Crick Institute, NW1 1ST London, United Kingdom*

Here we consider RNA-Seq, used to measure global gene expression through RNA fragmentation, capture, sequencing, and subsequent computational analysis. *Xenopus*, with its large number of RNA-rich, synchronously developing, and accessible embryos, is an excellent model organism for exploiting the power of high-throughput sequencing to understand gene expression during development. Here we present a standard RNA-Seq protocol for performing two-state differential gene expression analysis (between groups of replicates of control and treated embryos) using Illumina sequencing. Samples contain multiple whole embryos, and polyadenylated mRNA is measured under relative normalization. The protocol is divided into two parts: wet-lab processes to prepare samples for sequencing and downstream computational analysis including quality control, quantification of gene expression, and differential expression.

MATERIALS

It is essential that you consult the appropriate Material Safety Data Sheets and your Institution's Environmental Health and Safety Office for proper handling of equipment and hazardous materials used in this protocol.

Reagents

Chloroform

Developmentally synchronized *Xenopus* embryos at stage(s) appropriate for the research question being addressed

This protocol is appropriate for both X. laevis *and* X. tropicalis.

Ethanol (75%–80%, freshly prepared)

Isopropanol

Liquid nitrogen

Lithium chloride (10 M)

Qubit RNA HS Assay kit (Thermo Fisher Scientific)

RNase*Zap* Decontamination Solution (Invitrogen)

TRIzol (Thermo Fisher Scientific)

TruSeq Stranded mRNA Kit (Illumina)

Water, nuclease-free

Equipment

Benchtop fluorometer (Qubit fluorometer, Thermo Fisher Scientific)

[2]These authors contributed equally to this work.

[3]Correspondence: drmikegilchrist@gmail.com

Capillary electrophoresis machine to analyze RNA, DNA, and protein (Agilent 2100 Bioanalyzer)

Glass Pasteur pipette

Magnetic stand

Microcentrifuge tubes (RNAse-free, 1.5-mL)

Microvolume spectrophotometer (NanoDrop, Thermo Fisher Scientific)

> *This item is optional (see Step 23). It can also replace the Qubit, although this is not recommended.*

PCR tubes (RNase-free)

Pestle for 1.5-mL microcentrifuge tube (optional; see Step 2)

Refrigerated microcentrifuge

Thermocycler

Vortex mixer

Xenopus genome and/or *Xenopus* transcriptome reference data

Software

Bowtie2

FastQC

Integrative Genomics Viewer (for visualization if required)

Kallisto (optional mapping software; see Step 36)

R

R packages DESeq2 and tximport

RSEM

STAR

METHOD

Wet-Lab Processes

Before starting this protocol, it is necessary to decide on the experimental design, particularly how many biological replicates in each state (experimental condition). There is a necessary tradeoff between sequencing depth and numbers of biological replicates, with additional replicates preferred over deeper sequencing (Liu et al. 2014), but a minimum of ~10 million reads per sample. We recommend 15–20 million reads per sample, with at least four biological replicates per state (see Schurch et al. 2016).

Wet lab processes include sample collection, RNA extraction, and sequencing library preparation. Care should be taken to make sure the methods and kits used match your samples' RNA content. Quality controls are available at various stages, and poorly performing samples should be excluded as necessary.

While working with RNA, it is critical to avoid RNase contamination by using sterile, RNase-free solutions and plastic ware. RNase/DNase decontamination solution must be used on all work surfaces and equipment before starting.

Sample Collection

1. Prepare embryos via in vitro fertilization as described (see Showell and Conlon 2009).

2. When the embryos have reached the desired stage(s), collect groups of 10 embryos per sample in RNAse-free tubes using a glass Pasteur pipette.

3. Remove as much liquid as possible and snap freeze the tube in liquid nitrogen. Store the sample at −80°C.

> *If samples are collected over a period of time, it may help to use an RNA stabilizing agent.*

RNA Extraction

High-quality RNA is necessary for accurate expression quantification. Avoid overamplification of low-RNA samples during library preparation as this can lead to artifacts and spurious differential expression. RNA extraction should be performed consistently, preferably in a single batch with randomized processing order. Several RNA extraction protocols are available; here we use Trizol extraction followed by precipitation with LiCl.

Cite this protocol as *Cold Spring Harb Protoc*; doi:10.1101/pdb.prot098368

4. Thaw the tubes of collected embryo samples on ice and add 1 mL TRIzol to each tube. If RNA stabilizing agent has been used during the collection process, follow the manufacturer's instructions before starting.

5. Homogenize the embryos with a pipette, a pestle, or using a vortexer. Leave homogenate at room temperature for 5 min.

6. Add 200 µL chloroform.

7. Mix by tipping the tube until the solution is homogenous.

8. Centrifuge the tube at top speed in a microcentrifuge for 30 min at 4°C.

9. Transfer the top (aqueous) layer to a new RNase-free tube; avoid touching the white layer (protein) in the middle.

10. Add an equal volume of isopropanol.

11. Mix, tipping the tube and place for at least 30 min at −20°C.

12. Centrifuge the tube at top speed in a microcentrifuge for 20 min at 4°C.

13. Pour off the liquid, taking care not to lose the pellet. Add 500 µL of 75% ethanol and mix.

14. Centrifuge the tube at top speed in a microcentrifuge for 20 min at 4°C.

15. Remove all ethanol with a pipette and allow the pellet to dry with the tube lid open until it become translucent.

16. Resuspend the pellet in 40 µL nuclease free water.

17. Add 40 µL 10 M LiCl and place overnight at −20°C (or at least 1 h).

18. Centrifuge the tube at top speed in a microcentrifuge for 30 min at 4°C.

19. Pour off the liquid, taking care not to lose the pellet. Add 200 µL of 75% ethanol and mix.

20. Centrifuge the tube at top speed in a microcentrifuge for 20 min at 4°C.

21. Air-dry the pellet as in Step 15 and resuspend, initially, in 20 µL nuclease-free water. Once the RNA concentration has been assessed the sample can be further diluted if required.

22. Measure the concentration of extracted RNA in the samples using the Qubit according to the manufacturer's instructions. Consult the library preparation guide (see below) for the acceptable range of RNA input concentrations and dilute the samples accordingly.

    If the samples cannot be brought within this range, consult the manufacturers' guidelines for a more appropriate kit.

23. Assess the quality of the extracted RNA samples using the Bioanalyzer to measure RNA Integrity Numbers (RIN) for each sample.

    Any sample with a RIN <8 should be discarded, or will run a serious risk of generating sequencing artifacts.

    In addition, the Nanodrop may be used to measure the A_{260}/A_{280} ratio, which estimates the purity of the RNA. Samples should lie in the range 1.8–2.0.

    Samples falling outside acceptable thresholds may be taken through sequencing with the risk of generating unacceptable data and should therefore be examined carefully at postquantification quality control (QC).

Sequencing Library Construction

24. Prepare the sequencing libraries.

    The number of samples to multiplex at this stage will depend on the yield per lane of the sequencer being used. The sequencing facility should be able to provide guidance based on the number of samples and the desired number of reads per sample.

    For most applications, enrichment of polyadenylated mRNA from total RNA using poly-T oligo-attached magnetic beads is suitable. The resulting data will comprise coding and noncoding polyadenylated RNAs. We recommend following the Illumina Stranded mRNA Sequencing Sample Preparation guide

(TruSeq Stranded mRNA Sample Prep Guide), which will generate libraries from samples with 100–1000 ng total RNA. Let the beads from the TruSeq kit stand at room temperature for at least 30 min before use.

Library Quality Check

25. Before sequencing, check the fragment size distribution of the library using the Bioanalyzer. Follow the manufacturer's instructions and study the graphical output for the fragment size distribution.

    There should be a well-separated broad central peak with a maximum around 250 bp. Additional significant smaller peaks around 120–130 bp may represent adapter dimers and should be removed by repeating the "Clean Up PCR" step in the TruSeq protocol.

    Library concentration can also be assessed using qRT-PCR (Hawkins and Guest 2018).

Sequencing the Library

26. Send the libraries to a sequencing facility.

    Sequencing read length will probably be determined by this facility. Sequencing depth will be determined by the number of samples multiplexed together. Helpful considerations about these choices can be found on the Illumina website: https://support.illumina.com/bulletins/2017/04/considerations-for-rna-seq-read-length-and-coverage-.html.

    Sequencing data will likely be returned to you as a set of fastq files annotated by sample name.

Downstream Computational Analysis

Computational analysis is usually performed at the command line. Reads are mapped to a reference transcriptome or genome, and then quantified to gene counts per gene or transcript. These are then used for differential expression analysis. For mapping and quantification we recommend the RSEM + STAR (Li and Dewey 2011) pipeline. This approach gives accurate quantification but is computationally intensive (Dobin et al. 2013). An alternative, less computationally intensive approach, also presented here, uses Kallisto (Bray et al. 2016) and can be run on a laptop. New methods continue to be developed, so it is advisable to read recent reviews (e.g., Conesa et al. 2016). We recommend the Linux or Mac operating systems to run the tools described here, although some tools can be run on Windows.

Postsequencing Quality Control

27. Use FastQC (Andrews and Others 2010) to detect and provide evidence of problematic sequencing data. Run this postsequencing quality assessment as follows:

    ```
    $> fastqc sample_01.fastq.gz
    ```

    Postsequencing quality assessment can identify problematic libraries: such as those with low quality base-call scores, a shift from the expected GC-content (~40% in Xenopus), or overrepresented adapter sequences.

28. Because failure of quality tests does not guarantee a low-quality sample and vice-versa, proceed with analysis and discard samples based on postquantification QC.

Read Alignment to Reference Sequence Data

Read alignment (mapping) assigns reads to their possible reference sequence locations. Mapping to a reference transcriptome is computationally efficient but restricted to defined transcript models and may perform poorly with repetitive sequences. Mapping to a reference genome can be performed with or without aid from reference gene models. Genomic alignment can be visualized with the UCSC genome browser (Kent et al. 2002) or with the integrative genomics viewer (IGV) (Thorvaldsdóttir et al. 2013).

29. Perform read alignment using the splice-aware (capable of spanning exon–exon junctions) genomic aligner, STAR (Dobin et al. 2013). If gene expression is to be quantified by RSEM, skip this section and proceed with the RSEM alignment/quantification described below. For accurate quantification it is important to retain ambiguously mapping reads from the initial mapping stage. This is particularly important for the homeologous paired genes in *X. laevis*.

Cite this protocol as *Cold Spring Harb Protoc*; doi:10.1101/pdb.prot098368

For transcriptome mapping we recommend Bowtie 2 (Langmead and Salzberg 2012). The quantification tool RSEM (Li and Dewey 2011) (described below) is able to perform transcriptomic (via Bowtie 2) or genomic (via STAR) alignments and quantification in a single command. We describe Bowtie 2 only as part of the RSEM pipeline.

30. Proceed with the two stages of alignment: one-off construction of an *index* from reference sequences, followed by mapping reads from each sample using this index.

31. Download *Xenopus* reference sequence files from Xenbase (Karimi et al. 2018). Create an index from the genome reference file:

```
$> STAR --runThreadN [N] --runMode genomeGenerate

    --genomeDir <directory> --genomeFastaFiles <xen_genome.fa>
```

In this and the following text, basic commands are shown with a placeholder for command line [options], which are expanded underneath, and <file> denotes a required input file.

32. For genomes with large numbers of small scaffolds, such as current *X. tropicalis* and *X. laevis* genome builds, STAR recommends setting "–genomeChrBinNbits=min(18, log2(GenomeLength/NumberOfReferences))".

STAR index creation requires memory proportional to the number of sequences in the genome assembly. For the X. laevis genome v9.2 of ~2.7 × 10⁹ bp over ~108,000 sequences this implies –genomeChrBinNbits 15.

33. Perform alignment as follows:

```
$> STAR [options] --outFileNamePrefix <directory_output> --genomeDir <directory_genome> \
                  --readFilesIn sample_01_r1.fastq.gz sample_01_r2.fastq.gz
```

[options]	Explanation
--runThreadN <N>	Number of threads (processors)
--readFilesCommand zcat	Compressed FASTQ files (*.gz format)
--sjdbGTFfile	Annotation for guiding genomic
<gtf_file.gtf>	alignment
--outFilterMultimapNMax 200	Report max 200 ambiguous alignments
--outSAMtype BAM Unsorted	Output alignments in BAM format
--quantMode GeneCounts	Count reads mapping uniquely to genes, see below.

Quantification of Gene Expression Data

Here we describe quantification of gene expression directly from reads and read alignments. Transcript isoform usage can also be quantified with the tools here, but the specifics are beyond the scope of this protocol.

We recommend that quantification should include statistical assignment of ambiguous reads, as performed by RSEM/ Kallisto. Alternatively, one can count total unique aligning reads for each gene, and although this will underestimate total reads mapping to certain genes (especially X. laevis homeologs) it can be robust in quantifying genes with pathogenic repeats. This is done by STAR with the option "–quantMode GeneCounts".

Recommended Quantification: RSEM + STAR

RSEM quantifies transcript/gene expression from genomic or transcriptomic alignments; the associated pipeline generates the required alignments as necessary. We recommend RSEM+STAR alignment, as it is the current gold standard for RNA-Seq quantification.

34. Build an RSEM reference as follows (required for all uses of RSEM):

```
$> rsem-prepare-reference [options] <genome/transcriptome>.fa <genome/transcriptome name>
```

[options]	Explanation
-p <N>	Number of threads
--gtf <file>	GTF file describing gene models **for genomic alignments**
--transcript-to-gene-map <file>	File with rows "gene_id<TAB>transcript_id" **for transcriptomic alignments**.
--bowtie2	Generate bowtie2 index from transcriptome (if desired)
--star	Generate STAR index from genome (if desired)

For genome builds with a large number of scaffolds such as the X. laevis v9.2 it may be necessary to set STAR options as previously described, and so the STAR genome index should be built directly with STAR, not using rsem-prepare-reference.

35. To quantify expression, run RSEM in one of three modes depending on single/paired end reads (we recommend paired end) and the requirement for alignments:

```
$> rsem-calculate-expression [options]   <reads>.fastq.gz \
                                         <genome/transcriptome name> <sample_name>

$> rsem-calculate-expression [options]   --paired-end <reads_r1>.fastq.gz
                                         <reads_r2>.fastq.gz \ <genome/transcriptome
                                         name> <sample_name>

$> rsem-calculate-expression [options]   --alignments [--paired-end] <alignments>.
                                         bam \
                                         <genome/transcriptome name> <sample_name>
```

[options]	Explanation
-p <N>	Number of threads
--paired-end	Paired end reads
--bowtie	Align with bowtie2
--star	Align with STAR
--star-gzipped-read-file	If STAR is used and reads are compressed (.gz format)
--star-output-genome-bam	Output genome bam file generated by STAR, for visualization
Advanced options:	
--calc-pme	Calculate posterior mean estimates, for assessing the impact of ambiguous reads
--calc-ci	Calculate 95% confidence intervals for the pme.
--estimate-rspd	Estimate the "read start position distribution", account for biases in read coverage over transcripts, e.g. the 3′ bias of polyA+ RNA-Seq.

Alternative Quantification: Kallisto

Kallisto is an alignment free quantification method (Bray et al. 2016), which employs pseudoalignment to identify the set of transcripts consistent with each read. This approach significantly reduces the computational burden, reducing processing times by two orders of magnitude, whilst retaining quantification accuracy.

36. Construct a kallisto index as follows:

```
$> kallisto index -i <xen_transcriptome> <xen_transcriptome>.fa
```

37. Quantify fastq files using kallisto:

```
$> kallisto quant [options] -i <xen_transcriptome> -o <sample_output_dir> \
        <sample_01_r1.fastq.gz> <sample_01_r2.fastq.gz>
```

Cite this protocol as *Cold Spring Harb Protoc*; doi:10.1101/pdb.prot098368

[options]	Explanation
-t <N>	Number of threads
--single	Input data is single end
-l	Estimated fragment length mean, **for single end reads.**
-s	Estimated fragment length standard deviation, **for single end reads.**

Normalization

38. Normalize RNA-Seq data for sequencing depth (total reads per sample) to compare a gene's expression *between* samples.

39. To compare the expression of *different* genes *within* a sample, further normalize data by transcript/gene length to units of reads (or fragments for paired reads) per kilobase per million: R(F) PKM, or as transcripts per million (TPM).

40. For differential expression analysis do not perform length normalization (Love et al. 2014). Appropriate depth normalization is provided by the package DESeq2 (Love et al. 2014), which we describe below.

Postquantification QC

41. Identify samples with quality problems by one of the following methods:

    i. Evaluate pairwise sample correlations or total genes expressed above a given threshold.

    ii. Perform principal components analysis (PCA), which can indicate the existence of outlying samples suggesting quality issues. A PCA appropriate for RNA-Seq is offered by DESeq2.

Differential Expression Analysis

Differentially expressed genes are characterized by smaller variation in gene expression within each condition than between multiple conditions.

42. Options for statistical testing appropriate to count data (counting discrete items, i.e., sequence reads) are as follows:

    i. Use a test designed for count data, such as DESeq2. We recommend DESeq2, which is also more suited to small numbers of replicates (Soneson and Delorenzi 2013; Liu et al. 2014; Zhang et al. 2014; Seyednasrollah et al. 2015).

    ii. Or transform the data to abrogate the technical effects of count data and then apply a noncount data test (Ritchie et al. 2015).

43. Install R (R Core Team 2018) and R packages DESeq2 (Love et al. 2014) and tximport (Soneson et al. 2016).

    We describe a basic DESeq2 analysis below. More information and examples of usage are provided by DESeq2. The tximport package is used to import RSEM quantifications into DESeq2; it can also import expression data from other sources.

44. Perform DESeq2 as shown in the following example. In this example, we compare two conditions A and B, with three replicates per condition and sample names Sample_N for $N = 1,2,\ldots,6$.

45. Prepare an output folder for each sample from RSEM in a path which we will refer to as dir, such that dir/Sample_N/Sample_N.genes.results is the output quantification of Sample_N.

46. Create a tab or comma delimited file dir/samples.txt that describes the identity of the samples as follows:

Sample	Condition
Sample_1	A
Sample_2	A
Sample_3	A
Sample_4	B
Sample_5	B
Sample_6	B

47. Use the following R code to load the data and run DESeq2 with default settings comparing condition A vs condition B. Save the results.

```
1   # Load packages
2   library("tximport")                              # For loading RSEM data
3   library("DESeq2")                                # Load DESeq2
4
5
6   # Load Sample tables and quantification files
7   samples <- read.table(file.path(dir, "samples.txt"),   # Load Sample table
    header=TRUE)
8   files <- file.path(dir, samples$Sample,          # Find RSEM
    paste0(samples$Sample, ".genes.results"))        results files
9   names(files) <- samples$Sample
10  txi <- tximport(files, type = "rsem", txIn = FALSE,    # Import RSEM data
    txOut = FALSE)
11  txi$length[txi$length == 0] <- 1                 # RSEM sets some transcript#
                                                     lengths to 0, set these to 1
12
13  # DESeq2
14  dds <- DESeqDataSetFromTximport(txi, colData     # Load Data into DESeq2
    = samples, design = ~ Condition)
15  dds <- DESeq(dds)                                # Run differential
                                                     expression
16  res <- results(dds, name="Condition_B_vs_A")     # Collect results
17  summary(res)                                     # Show summary of results
18
19
20  # Write results
21  resOrdered <- res[order(res$padj),]              # Order results for padj
22  write.csv(as.data.frame(resOrdered),             # Write results
    file=file.path(dir, "deseq_results.csv"))
```

The result of this analysis will be a comma-separated file describing the results of the differential expression tests for all genes ordered by the adjusted p-value. Columns in the file include: mean expression, fold change between conditions, relevant statistics on the statistical test, p-value and adjusted p-value. The adjusted p-values are derived from the p-values accounting for the number of tests performed and should be used in preference over the p-values.

RELATED INFORMATION

A two-state experiment is described here, but more complex designs are possible. For example, *Xenopus* is ideal for time-resolved experimental design to understand the developmental timing of gene expression (Collart et al. 2014; Owens et al. 2016).

Protocols are available for measuring differing types and levels of RNA. Standard polyA+ protocols can be applied with as few as five *Xenopus* embryos per sample. Low-RNA kits (SMART-Seq v4 Ultra

Cite this protocol as *Cold Spring Harb Protoc*; doi:10.1101/pdb.prot098368

Low Input RNA Kit [Takara] or Ovation SoLo RNA-Seq Library Preparation Kit [NuGEN]) and protocols are available for both single embryos and individual cells.

REFERENCES

Andrews S, et al. 2010. FastQC: A quality control tool for high throughput sequence data. https://www.bioinformatics.babraham.ac.uk/projects/fastqc/

Bray NL, Pimentel H, Melsted P, Pachter L. 2016. Near-optimal probabilistic RNA-seq quantification. *Nat Biotechnol* **34**: 525–527. doi:10.1038/nbt.3519

Collart C, Owens NDL, Bhaw-Rosun L, Cooper B, De Domenico E, Patrushev I, Sesay AK, Smith JN, Smith JC, Gilchrist MJ. 2014. High-resolution analysis of gene activity during the *Xenopus* mid-blastula transition. *Development* **141**: 1927–1939. doi:10.1242/dev.102012

Conesa A, Madrigal P, Tarazona S, Gomez-Cabrero D, Cervera A, McPherson A, Wojciech Szcześniak M, et al. 2016. A survey of best practices for RNA-seq data analysis. *Genome Biol* **17**: 13. doi:10.1186/s13059-016-0881-8

Dobin A, Davis CA, Schlesinger F, Drenkow J, Zaleski C, Jha S, Batut P, Chaisson M, Gingeras TR. 2013. STAR: ultrafast universal RNA-seq aligner. *Bioinformatics* **29**: 15–21. doi:10.1093/bioinformatics/bts635

Hawkins SFC, Guest PC. 2018. Rapid and easy protocol for quantification of next-generation sequencing libraries. *Methods Mol Biol* **1735**: 343–350. doi:10.1007/978-1-4939-7614-0_23

Karimi K, Fortriede JD, Lotay VS, Burns KA, Wang DZ, Fisher ME, Pells TJ, et al. 2018. Xenbase: a genomic, epigenomic and transcriptomic model organism database. *Nucleic Acids Res* **46**: D861–D868. doi:10.1093/nar/gkx936

Kent WJ, Sugnet CW, Furey TS, Roskin KM, Pringle TH, Zahler AM, Haussler D. 2002. The human genome browser at UCSC. *Genome Res* **12**: 996–1006. doi:10.1101/gr.229102

Langmead B, Salzberg SL. 2012. Fast gapped-read alignment with Bowtie 2. *Nat Methods* **9**: 357–359. doi:10.1038/nmeth.1923

Li B, Dewey CN. 2011. RSEM: accurate transcript quantification from RNA-seq data with or without a reference genome. *BMC Bioinformatics* **12**: 323. doi:10.1186/1471-2105-12-323

Liu Y, Zhou J, White KP. 2014. RNA-seq differential expression studies: more sequence or more replication? *Bioinformatics* **30**: 301–304. doi:10.1093/bioinformatics/btt688

Love MI, Huber W, Anders S. 2014. Moderated estimation of fold change and dispersion for RNA-seq data with DESeq2. *Genome Biol* **15**: 550. doi:10.1186/s13059-014-0550-8

Owens NDL, Blitz IL, Lane MA, Patrushev I, Overton JD, Gilchrist MJ, Cho KWY, Khokha MK. 2016. Measuring absolute RNA copy numbers at high temporal resolution reveals transcriptome kinetics in development. *Cell Rep* **14**: 632–647. doi:10.1016/j.celrep.2015.12.050

R Core Team. 2018. *R: a language and environment for statistical computing.* R foundation for statistical computing, Vienna, Austria.

Ritchie ME, Belinda Phipson DW, Hu Y, Law CW, Shi W, Smyth GK. 2015. Limma powers differential expression analyses for RNA-sequencing and microarray studies. *Nucleic Acids Res* **43**: e47. doi:10.1093/nar/gkv007

Schurch NJ, Schofield P, Gierliński M, Cole C, Sherstnev A, Singh V, Wrobel N, et al. 2016. Erratum: how many biological replicates are needed in an RNA-seq experiment and which differential expression tool should you use? *RNA* **22**: 1641. doi:10.1261/rna.058339.116

Seyednasrollah F, Laiho A, Elo LL. 2015. Comparison of software packages for detecting differential expression in RNA-seq studies. *Brief Bioinform* **16**: 59–70. doi:10.1093/bib/bbt086

Showell C, Conlon FL. 2009. Egg collection and in vitro fertilization of the western clawed frog *Xenopus tropicalis. Cold Spring Harb Protoc* **2009**: pdb.prot5293. doi:10.1101/pdb.prot5293

Soneson C, Delorenzi M. 2013. A comparison of methods for differential expression analysis of RNA-seq data. *BMC Bioinformatics* **14**: 91. doi:10.1186/1471-2105-14-91

Soneson C, Love MI, Robinson MD. 2016. Differential analyses for RNA-seq: transcript-level estimates improve gene-level inferences. *F1000Res* **4**: 1521. doi:10.12688/f1000research.7563.2

Thorvaldsdóttir H, Robinson JT, Mesirov JP. 2013. Integrative genomics viewer (IGV): high-performance genomics data visualization and exploration. *Brief Bioinform* **14**: 178–192. doi:10.1093/bib/bbs017

TruSeq Stranded mRNA Sample Prep Guide (accessed 18th December 2018) https://support.illumina.com/downloads/truseq_stranded_mrna_sample_preparation_guide_15031047.html.

Zhang ZH, Jhaveri DJ, Marshall VM, Bauer DC, Edson J, Narayanan RK, Robinson GJ, et al. 2014. A comparative study of techniques for differential expression analysis on RNA-seq data. *PLoS One* **9**: e103207. doi:10.1371/journal.pone.0103207

Mass Spectrometry–Based Absolute Quantification of Single *Xenopus* Embryo Proteomes

Rik G.H. Lindeboom,[1,4] Arne H. Smits,[2,4] Matteo Perino,[3] Gert Jan C. Veenstra,[3] and Michiel Vermeulen[1,5]

[1]*Department of Molecular Biology, Faculty of Science, Radboud Institute for Molecular Life Sciences, Radboud University, Nijmegen 6500 HB, The Netherlands;* [2]*Genome Biology Unit, European Molecular Biology Laboratory, 69117 Heidelberg, Germany;* [3]*Department of Molecular Developmental Biology, Faculty of Science, Radboud Institute for Molecular Life Sciences, Radboud University, Nijmegen 6500 HB, The Netherlands*

Early *Xenopus* development is characterized by a poor correlation between global mRNA and protein abundances due to maternal mRNA and protein loading. Therefore, proteome profiling is necessary to study gene expression dynamics during early *Xenopus* development. In contrast to mammals, single *Xenopus* eggs and embryos contain enough protein to allow identification and quantification of thousands of proteins using mass spectrometry-based proteomics. In addition to investigating developmental processes, single egg or blastomere proteomes can be used to study cell-to-cell variability at an unprecedented depth. In this protocol, we describe a mass spectrometry-based proteomics approach for the identification and absolute quantification of *Xenopus laevis* egg or embryo proteomes, including sample preparation, peptide fractionation and separation, and data analysis.

MATERIALS

It is essential that you consult the appropriate Material Safety Data Sheets and your institution's Environmental Health and Safety Office for proper handling of equipment and hazardous materials used in this protocol.

RECIPES: Please see the end of this protocol for recipes indicated by <R>. Additional recipes can be found online at http://cshprotocols.cshlp.org/site/recipes.

Reagents

ABC buffer (50 mM ammonium bicarbonate, freshly prepared)

Britton-Robinson (B&R) buffer (pH 11/pH 8/pH 2) <R>

Buffer A (0.1% formic acid)

Buffer B (80% acetonitrile, 0.1% formic acid)

C18 disks (Empore) and reagents for preparation of C18 StageTips (see Steps 11–14 of Swaney and Villen 2016)

Cysteine (3% in 0.25× MMR)

IAA buffer <R> (freshly prepared, kept in the dark)

MMR buffer (0.25×, pH 7.4) <R>

NaCl (0.5 M)

---

[4]These authors contributed equally to this work.

[5]Correspondence: michiel.vermeulen@science.ru.nl

NaOH (1 M)

Strong Anion eXchange (SAX) disk (Empore) and reagents for preparation of 20 plug SAX StageTips (Rappsilber et al. 2007)

Trifluoroacetic acid (TFA) (10%)

Trypsin (Promega)

UA buffer <R> (freshly prepared)

UPS2 standard solution (0.5 µg/µL) <R> (freshly prepared)

WCE-LS (whole cell extract-low salt) buffer <R> (freshly prepared)

Xenopus laevis eggs (fertilized in vitro)

Equipment

Centrifuge, tabletop (Eppendorf) at 4°C and 20°C

Combitips (Eppendorf)

Filters (30-kDa) (Microcon YM-30)

High-resolution mass spectrometer (e.g., time-of-flight or Orbitrap mass analyzer)

Incubator at 37°C

Nano-HPLC coupled to a C18 column of 25–30 cm

Pipette tips (Rainin P200) and equipment for preparation of StageTips (see Steps 11–14 of Swaney and Villen 2016)

Speed vacuum concentrator

Thermoshaker (Eppendorf) at 20°C

Tubes (0.2- and 1.5-mL)

METHOD

This method is based on the work of Smits et al. (2014) (Fig. 1). Yolk-deprived egg or embryo extracts are first digested with trypsin using the so-called Filter-Aided Sample Preparation (FASP) protocol (Wisniewski et al. 2009a) followed by SAX fractionation (Wisniewski et al. 2009b). A protein spike-in is used to facilitate quantification of the protein molecules (Schwanhausser et al. 2011). After sample preparation, peptides are measured in a mass spectrometer. Finally, data analysis is performed using the MaxQuant software suite (Cox and Mann 2008).

Embryo Collection and Lysis

1. Dejelly fertilized eggs by adding 3% cysteine in 0.25× Marc's modified Ringer's (MMR). Wash thoroughly four times with 0.25× MMR as soon as the embryos start touching each other.

2. Stage the embryos according to Nieuwkoop and Faber (1994), and collect single embryos in separate tubes. Remove any carryover MMR.

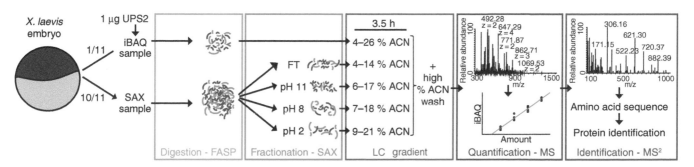

FIGURE 1. Schematic overview of the workflow. Reprinted from Smits et al. (2014) by permission of Oxford University Press.

3. Add 20 µL of WCE-LS buffer per sample and homogenize the embryos by gentle pipetting.

4. Centrifuge the samples at 3500*g* for 5 min at 4°C.

5. Transfer the supernatant to a fresh tube without touching the yolk/pigment pellet.

Filter-Aided Sample Preparation

All of the following steps are performed at 20°C.

6. Add UA buffer to the supernatant from Step 5 to a total volume of 220 µL. Incubate with gentle shaking for 5 min in a thermoshaker.

7. Transfer 20 µL of the lysate in UA buffer to a new tube.

   This is the FASP-iBAQ sample, which is not fractionated and is used for absolute quantification. The residual 200 µL is the FASP-SAX sample, which is subjected to fractionation to obtain deep proteome coverage.

8. Add 2 µL of UPS2 standard (0.5 µg/µL) and 178 µL of UA buffer to the FASP-iBAQ sample and mix.

9. Transfer both samples (FASP-iBAQ and FASP-SAX) to two 30-kDa filters. Centrifuge at 11,600*g* for 15 min. Discard the flowthrough (FT).

   See Troubleshooting.

10. Add 100 µL of IAA buffer to each filter. Incubate for 1 min in a thermoshaker at 600 rpm.

11. Incubate the filters for an additional 20 min protected from light and without shaking. Centrifuge at 11,600*g* for 15 min.

12. Wash the samples by adding 100 µL of UA buffer to each filter. Centrifuge at 11,600*g* for 15 min.

13. Repeat Step 12 two times. Discard the FT.

14. Wash the samples by adding 100 µL of ABC buffer to each filter. Centrifuge at 11,600*g* for 10 min.

15. Repeat Step 14 two times. Discard the FT.

16. Add trypsin to each filter as follows.

    i. To the FASP-iBAQ sample, add 0.1 µg of trypsin in ABC buffer to a total volume of 40 µL.

    ii. To the FASP-SAX sample, add 1.0 µg of trypsin in ABC buffer to a total volume of 40 µL.

17. Mix the samples at 600 rpm for 1 min and then incubate in a sealed plastic box containing moist paper towels overnight at 37°C.

18. Transfer each filter to a new tube. Centrifuge at 11,600*g* for 15 min. Retain the FT.

    Steps 18–20 are performed in the same tube. The FT from these steps are combined during centrifugation.

19. Process each filter as follows.

    i. To the FASP-iBAQ sample, add 50 µL of 0.5 M NaCl. Centrifuge at 11,600*g* for 10 min.

    ii. To the FASP-SAX sample, add 50 µL of ABC buffer. Centrifuge at 11,600*g* for 10 min.

20. Repeat Step 19.

    The FT of ∼140 µL from each filter represents the tryptic digest of each sample.

Strong Anion eXchange Fractionation (FASP-SAX Sample Only)

21. Prepare a 20 plug SAX StageTip according to Rappsilber et al. (2007).

22. Wash the StageTip by adding 200 µL of B&R buffer (pH 11). Centrifuge for 24 min at 1000*g*.

23. Add 140 µL of B&R buffer (pH 11) and 10 µL of 1 M NaOH to the FASP-SAX sample and mix.

    The pH should be between 11 and 12.

Cite this protocol as *Cold Spring Harb Protoc*; doi:10.1101/pdb.prot098376

24. Transfer the FASP-SAX sample to the StageTip.

25. Centrifuge the sample for 45 min at 500*g*. Retain the FT.

    The FT represents the FASP-SAX FT fraction.

    See Troubleshooting.

26. Transfer the StageTip to a new tube. Add 200 µL of B&R buffer (pH 11) to the StageTip. Centrifuge for 24 min at 1000*g*. Retain the FT.

    The FT represents the FASP-SAX (pH 11) fraction.

27. Repeat Step 26 with 200 µL of B&R buffer (pH 8) to obtain the FASP-SAX (pH 8) fraction.

28. Repeat Step 26 with 200 µL of B&R buffer (pH 2) to obtain the FASP-SAX (pH 2) fraction.

Peptide Desalting and Elution

Use all samples (1 FASP-iBAQ + 4 SAX fractions).

29. Prepare, wash and activate five C18 StageTips (or filtration tips) as described in Swaney and Villen (2016) (Steps 11–14).

30. Add 10 µL of 10% TFA to each sample. Apply each sample to a separate C18 StageTip.

31. Wash each StageTip by adding 30 µL of buffer A. Push the buffer slowly through the StageTips with an airtight, fitted Combitip.

32. Elute the peptides of each sample in a separate 0.2-mL tube by adding 30 µL of buffer B to each StageTip and pushing the buffer slowly through with a Combitip.

33. Concentrate the samples to ∼5 µL with a speed vacuum concentrator. Add 7 µL of buffer A to each sample (final volume ∼12 µL).

Mass Spectrometry

34. Inject 5 µL of each sample into the nano-HPLC coupled to a C18 column of 25–30 cm.

    The nano-HPLC acetonitrile gradient varies between the different fractions (Table 1).

35. Acquire mass spectra on a high-resolution mass spectrometer, e.g., time-of-flight or Orbitrap mass analyzer.

    For examples of optimized nano-HPLC and mass analyzer settings, see Richards et al. (2015).

Data Analysis

36. Perform data analysis using the MaxQuant software suite (www.maxquant.org). Specify for each FASP-SAX sample the same experiment name but different fraction numbers (assign uneven numbers only). Assign the FASP-iBAQ sample its own experiment name.

37. Upload the FASTA files with protein sequences of *Xenopus* and UPS2 proteins.

    Protein databases from Smits et al. (2014), Wuhr et al. (2014) or UniProt can also be used as a reference for the Xenopus proteome. A FASTA file containing the sequences of the UPS2 proteins can be downloaded from the supplier's webpage.

TABLE 1. Linear nano-HPLC acetonitrile (ACN) gradients for different samples

Sample	% ACN at 0 min	% ACN at 214 min	% ACN at 240 min
FASP-iBAQ	4	26	76
FASP-SAX FT fraction	4	14	76
FASP-SAX pH11 fraction	6	17	76
FASP-SAX pH8 fraction	7	18	76
FASP-SAX pH2 fraction	9	21	76

38. Empty the contents of the "ibaq.txt" file in the 1.X.X.X_MaxQuant\bin\conf\ folder.

39. Make sure to enable the "iBAQ" option to facilitate absolute quantification and enable the "match-between-run" option to share peptide identification information between runs.

40. After the MaxQuant analysis, find the quantified protein abundances in the proteinGroups.txt file.

> *The relative iBAQ quantification of the UPS2 proteins in the FASP-iBAQ sample can be used for absolute quantification of all identified proteins in the fractionated sample.*

- A linear regression between the supplied concentrations of the spike-in proteins and the measured iBAQ values can be used to calculate the abundance in femtomoles for all proteins identified in the FASP-iBAQ sample.

- A second linear regression between the absolute abundances of the quantified proteins in the FASP-iBAQ sample and the iBAQ values of same proteins in the FASP-SAX sample can then be used to extrapolate the absolute amounts of all identified proteins in the FASP-SAX sample.

> *See Troubleshooting.*

TROUBLESHOOTING

Problem (Steps 9 and 25): The 30-kDa filters or SAX StageTips are not completely cleared after centrifugation.
Solution: It is essential that the 30-kDa filters and StageTips are completely cleared before continuing to the next step. Increase the centrifugation time accordingly.

Problem (Step 40): The peptide identification rates are low after MaxQuant analysis.
Solution: To boost low peptide identification rates, keep temperature stable in Steps 6–13 to prevent peptide carbamylation by urea buffers.

DISCUSSION

While RNA-sequencing is often used to study genome-wide gene expression, it is not sufficient when studying early *Xenopus* development due to decoupled proteome and transcriptome dynamics (Smits et al. 2014; Peshkin et al. 2015). Sample preparation of early-stage *Xenopus* embryos for mass spectrometry requires proper removal of yolk and protein solubilization with strong detergents (Peuchen et al. 2016). Recent advances in the field of mass spectrometry-based proteomics allowed the study of early *Xenopus* proteome dynamics at an unprecedented depth (Sun et al. 2014; Wuhr et al. 2014). The relatively large protein content also enabled the first single egg and embryo deep-proteome studies (Smits et al. 2014; Lombard-Banek et al. 2016; Sun et al. 2016).

RECIPES

Britton-Robinson (B&R) Buffer (pH 11/pH 8/pH 2)

40 mM H_3PO_4
40 mM CH_3COOH
40 mM H_3BO_3

Titrate buffer to pH 11, pH 8, or pH 2 with 1 M NaOH. Store for up to 2 yr at room temperature.

Cite this protocol as *Cold Spring Harb Protoc*; doi:10.1101/pdb.prot098376

IAA Buffer

UA buffer <R> (freshly prepared)
50 mM iodoacetamide (IAA)

Prepare fresh before use and keep in the dark.

MMR Buffer (0.25×, pH 7.4)

22 mM NaCl
0.5 mM KCl
0.5 mM CaCl$_2$
0.25 mM MgCl$_2$
1.25 mM HEPES

Store for up to 2 yr at room temperature.

UA Buffer

8 M urea
0.1 M Tris-HCl (pH 8.5)
50 mM dithiothreitol

Prepare fresh before use.

UPS2 Standard Solution (0.5 µg/µL)

UA buffer <R> (freshly prepared)
UPS2 standard (Sigma-Aldrich) (0.5 µg/µL)
Dissolve the UPS2 standard (0.5 µg/µL) in UA buffer for >30 min with gentle shaking at 20°C.

Prepare fresh before use.

WCE-LS (Whole Cell Extract-Low Salt) Buffer

20 mM Tris-HCl (pH 8.0)
70 mM KCl
1 mM EDTA
10% glycerol
0.1% IGEPAL CA-630
5 mM dithiothreitol (DTT)
1× cOmplete EDTA-Free Protease Inhibitors (Roche)

Prepare fresh before use.

REFERENCES

Cox J, Mann M. 2008. MaxQuant enables high peptide identification rates, individualized p.p.b.-range mass accuracies and proteome-wide protein quantification. *Nat Biotechnol* **26:** 1367–1372.

Lombard-Banek C, Reddy S, Moody SA, Nemes P. 2016. Label-free quantification of proteins in single embryonic cells with neural fate in the cleavage-stage frog (*Xenopus laevis*) embryo using capillary electrophoresis electrospray ionization high-resolution mass spectrometry (CE-ESI-HRMS). *Mol Cell Proteomics* **15:** 2756–2768.

Nieuwkoop PD, Faber J. 1994. *Normal table of* Xenopus laevis *(Daudin): A systematical and chronological survey of the development from the fertilized egg till the end of metamorphosis.* Garland Pub, New York.

Peshkin L, Wuhr M, Pearl E, Haas W, Freeman RM Jr, Gerhart JC, Klein AM, Horb M, Gygi SP, Kirschner MW. 2015. On the relationship of protein and mRNA dynamics in vertebrate embryonic development. *Dev Cell* **35:** 383–394.

Peuchen EH, Sun L, Dovichi NJ. 2016. Optimization and comparison of bottom-up proteomic sample preparation for early-stage *Xenopus laevis* embryos. *Anal Bioanal Chem* **408:** 4743–4749.

Rappsilber J, Mann M, Ishihama Y. 2007. Protocol for micro-purification, enrichment, pre-fractionation and storage of peptides for proteomics using StageTips. *Nat Protoc* **2:** 1896–1906.

Richards AL, Hebert AS, Ulbrich A, Bailey DJ, Coughlin EE, Westphall MS, Coon JJ. 2015. One-hour proteome analysis in yeast. *Nat Protoc* **10:** 701–714.

Schwanhausser B, Busse D, Li N, Dittmar G, Schuchhardt J, Wolf J, Chen W, Selbach M. 2011. Global quantification of mammalian gene expression control. *Nature* **473:** 337–342.

Smits AH, Lindeboom RG, Perino M, van Heeringen SJ, Veenstra GJ, Vermeulen M. 2014. Global absolute quantification reveals tight regulation of protein expression in single *Xenopus* eggs. *Nucleic Acids Res* **42:** 9880–9891.

Sun L, Bertke MM, Champion MM, Zhu G, Huber PW, Dovichi NJ. 2014. Quantitative proteomics of *Xenopus laevis* embryos: Expression kinetics of nearly 4000 proteins during early development. *Sci Rep* **4:** 4365.

Sun L, Dubiak KM, Peuchen EH, Zhang Z, Zhu G, Huber PW, Dovichi NJ. 2016. Single cell proteomics using frog (*Xenopus laevis*) blastomeres isolated from early stage embryos, which form a geometric progression in protein content. *Anal Chem* **88:** 6653–6657.

Swaney DL, Villen J. 2016. Enrichment of phosphopeptides via immobilized metal affinity chromatography. *Cold Spring Harb Protoc* doi:10.1101/pdb.prot088005.

Wisniewski JR, Zougman A, Nagaraj N, Mann M. 2009a. Universal sample preparation method for proteome analysis. *Nat Methods* **6:** 359–362.

Wisniewski JR, Zougman A, Mann M. 2009b. Combination of FASP and StageTip-based fractionation allows in-depth analysis of the hippocampal membrane proteome. *J Proteome Res* **8:** 5674–5678.

Wuhr M, Freeman RM Jr, Presler M, Horb ME, Peshkin L, Gygi S, Kirschner MW. 2014. Deep proteomics of the *Xenopus laevis* egg using an mRNA-derived reference database. *Curr Biol* **24:** 1467–1475.

Cite this protocol as *Cold Spring Harb Protoc;* doi:10.1101/pdb.prot098376

INTACT Proteomics in *Xenopus*

Lauren Wasson,[1,2] Nirav M. Amin,[2,3] and Frank L. Conlon[1,2,3,4]

[1]*Department of Genetics, University of North Carolina-Chapel Hill, Chapel Hill, North Carolina 27599;*
[2]*University of North Carolina McAllister Heart Institute, University of North Carolina-Chapel Hill, Chapel Hill,*
North Carolina 27599; [3]*Department of Biology, University of North Carolina-Chapel Hill, Chapel Hill,*
North Carolina 27599

Analysis of the molecular mechanisms driving cell specification, differentiation, and other cellular processes can be difficult due to the heterogeneity of tissues and organs. Therefore, it is critical to isolate pure cell populations in order to properly assess the function of certain cell types in the context of a tissue. This protocol describes use of the INTACT (isolation of nuclei tagged in specific cell types) method in *Xenopus*, followed by proteomics analysis of nuclear protein complexes. The INTACT protocol utilizes two transgenes: (1) a three-part nuclear targeting fusion (NTF) consisting of a nuclear envelope protein (Nup35) that targets the NTF to the nuclear membrane, an enhanced green fluorescent protein (EGFP) cassette for NTF visualization in live animals, and a biotin ligase receptor protein (BLRP) that provides a substrate for the biotinylation of the NTF, and (2) the *E. coli* ligase BirA (which biotinylates the NTF) tagged to mCherry (for visualization). Either or both transgenes are driven by a tissue-specific promoter, making this protocol easily adaptable to proteomics analyses of immunoprecipitated complexes from INTACT-isolated nuclei of multiple tissue types to determine the composition of protein complexes in pure cell populations.

MATERIALS

It is essential that you consult the appropriate Material Safety Data Sheets and your institution's Environmental Health and Safety Office for proper handling of equipment and hazardous materials used in this protocol.

RECIPES: Please see the end of this protocol for recipes indicated by <R>. Additional recipes can be found online at http://cshprotocols.cshlp.org/site/recipes.

Reagents

Acetonitrile, HPLC grade (Fisher A998)

BCA Protein Assay Kit (Pierce) (Thermo Fisher Scientific 23225)

BirA-mCherry transgene (Schaffer et al. 2010)

> *Synthesize capped BirA-mCherry mRNA constructs using the mMessage mMachine Transcription Kit (Thermo-Fisher Scientific AM1344 for T7, AM1348 for T3, or AM1340 for SP6) according to the manufacturer's protocol. Note that the enzyme used for transcription will depend on the RNA transcription promoter present in the mRNA construct used (i.e., T7, T3, or SP6).*

Dulbecco's phosphate-buffered saline (DPBS) (1×, pH 7.4) (Thermo Fisher Scientific 14040117)

---

[4]Correspondence: Frank_Conlon@med.unc.edu

Cite this protocol as *Cold Spring Harb Protoc*; doi:10.1101/pdb.prot098384

Ethanol (100%)

Formic acid, LC-MS/MS grade (99%+) (Pierce 28905)

Methanol

Nuclear purification buffer (NPB) <R> (at 4°C)

In addition, prepare NPBb (NPB +0.5% BSA) and NPBt (NPB +0.1% Triton X-100).

Nuclear targeting fusion (NTF) transgene (Amin et al. 2014)

The NTF consists of three parts: a nuclear envelope protein (Nup35) that targets the NTF to the nuclear membrane, an EGFP cassette for NTF visualization in live animals, and a biotin ligase receptor protein (BLRP) that provides a substrate for the biotinylation of the NTF.

Clone the NTF transgene downstream of a tissue-specific promoter of choice (e.g., mlc2p or cardiac actin). For example, fusion downstream of the mlc2p regulatory element would drive NTF expression specifically in cardiomyocytes after stage 26 (Latinkic et al. 2004).

OptiPrep density gradient medium (Sigma-Aldrich D1556)

Prepare a solution of 30% OptiPrep in NPB.

Polyacrylamide gels and reagents for denaturing gel electrophoresis (Thermo Fisher Scientific)

NuPAGE 4%–12% Bis-Tris protein gels (NP0321)

NuPAGE antioxidant (NP0005)

NuPAGE LDS sample buffer (4×) (NP0008)

NuPAGE MOPS running buffer (20×) (NP0001)

NuPAGE sample reducing agent (10×) (NP0009)

RIPA buffer for *Xenopus* <R>

Streptavidin-conjugated magnetic Dynabeads (Invitrogen M-270)

Trypsin stock (0.5 µg/µL), sequencing grade (Promega V5111)

Store stock solution at −80°C and limit to <5 freeze/thaw cycles.

Immediately before use, prepare a working solution of 12.5 ng/µL trypsin in 20 µL of 50 mM ammonium bicarbonate.

Xenopus and reagents for transgenesis using REMI or method of choice (Amaya and Kroll 1999; Allen and Weeks 2005; Yergeau and Mead 2007)

Equipment

Cell strainer (100 µM) (Sigma-Aldrich CLS431752)

Centrifuge, refrigerated

Denaturing gel electrophoresis equipment

Embryo injector

Heat blocks at 37°C and 95°C

Liquid chromatography system and mass spectrometer

We recommend using a Dionex Ultimate 3000 nanoRSLC system coupled to an LTQ-Orbitrap Velos mass spectrometer.

Liquid nitrogen

Magnet (Thermo Fisher Scientific 12321D)

Mortar and pestle, ceramic (Thermo Fisher Scientific FB961A-3 and FB961K-M)

Polytetrafluoroethylene tissue grinder

Razor blades

Rotator

Sonicator (Bioruptor [Diagenode])

Stopcock, two way (Thermo Fisher Scientific 6460)

Styrofoam cooler

Syringe needle

Transfer pipettes, plastic (Thermo Fisher Scientific 242)

Cite this protocol as *Cold Spring Harb Protoc*; doi:10.1101/pdb.prot098384

Tubes, polypropylene (conical, 15- and 50-mL)
Tubes, polypropylene (rounded bottom, 2-mL)
Windex

METHOD

Generating INTACT Nuclei in *Xenopus*

This section describes generation of transgenic Xenopus *embryos harboring the NTF transgene under control of a tissue-specific promoter and injected with or without BirA-mCherry mRNA.*

1. Generate transgenic *Xenopus* containing the NTF transgene using REMI or the transgenesis method of choice (Amaya and Kroll 1999; Allen and Weeks 2005; Yergeau and Mead 2007).

   Stable transgenic lines assure uniform expression; however, this method has been used successfully with transient (mosaic) transgenic animals (Amin et al. 2014).

2. Inject 1 ng of *BirA-mCherry* mRNA into the transgenic *Xenopus* embryos at the 1-cell stage. As a negative control, generate NTF-transgenic embryos that do not receive *BirA-mCherry* injection.

   Injection at the 1-cell stage ensures distribution of BirA-mCherry throughout the entire developing embryo.

3. Culture the embryos to the desired stage of development.

Isolating INTACT Nuclei

This section describes the separation and isolation of nuclei from the NTF/BirA-mCherry transgenic embryos.

4. Harvest 100 embryos in 1× DPBS in a 2-mL round-bottom tube. Wash the embryos three times with 1× DPBS by gently resuspending them using a plastic transfer pipette and then letting them settle to the bottom of the tube.

   The required number of embryos should be determined empirically based on the percentage of cells in the embryos expressing the NTF. For most applications, 100 embryos should be sufficient; however, in instances where the NTF is only driven in a few cells (e.g., cardiac valve cells), more embryos will be required. Enough embryos should be harvested to recover 30 µg of nuclear protein as measured by the BCA assay (Step 30).

5. Using a syringe needle, poke four holes in the cap of a 50-mL conical tube. Remove the cap and secure the tube into a rack in a Styrofoam cooler. Fill the cooler and tube with liquid nitrogen.

6. Using a plastic transfer pipette, drop the embryos into the conical tube containing liquid nitrogen, minimizing the volume of transferred DPBS. Once the embryos have been transferred, screw the cap onto the tube tightly. Remove the tube from the cooler (using a paper towel or cryo-glove for protection) and invert to drain the liquid nitrogen. Store the frozen tissue at −80°C.

7. Wash a ceramic mortar and pestle once each with the following: Windex, ethanol, methanol, and ddH$_2$O. Wipe dry with a paper towel. Carefully pour liquid nitrogen into the mortar and let it evaporate to cool the ceramic.

8. Carefully pour liquid nitrogen into the ceramic mortar, add the embryos, and grind the frozen embryos to a powder using a cold pestle. Collect the frozen powder in a 15-mL conical tube.

9. Resuspend the frozen powder in 6 mL of NPB. Thaw the powder in NPB for 10 min on ice.

10. Transfer the lysate to a polytetrafluoroethylene tissue grinder and homogenize with 40 strokes.

11. Pass the lysate through a 100-µm cell strainer into a 50-mL conical tube.

12. Centrifuge the lysate at 1000g for 10 min at 4°C to collect crude nuclei.

13. Resuspend the nuclei in 6 mL of 30% Optiprep in NPB. Centrifuge at 1000g for 10 min at 4°C to harvest the enriched nuclei.

    A small portion of this suspension can be analyzed by phase-contrast or stained with DAPI to determine nuclear purity.

14. Wash the nuclei three times by gentle resuspension in 6 mL of NPB per wash.

Be sure to remove all traces of Optiprep before proceeding with subsequent steps.

Affinity-Isolating INTACT Nuclei

This section describes streptavidin-based isolation of nuclei. The addition of a large volume (9 mL) of NBPt ensures that contaminating cell types are washed away while the nuclei are preserved.

Perform Steps 15–26 at 4°C.

15. Resuspend the nuclei in 1 mL of NPB.
16. Incubate the nuclei with 50 µL of streptavidin-conjugated magnetic beads. Rotate for 30 min.

     Proceed to Steps 17–19 during this incubation.

17. At the start of the incubation in Step 16, preload P1000 tips with 1 mL of NPBb per tip. Lay the preloaded tips on their sides.
18. After 20 min of incubation, insert a pre-loaded tip vertically into a two-way stopcock and attach it to a magnet. Open the stopcock and drain the NPBb from the P1000 tip.
19. Add 1.2 mL of NPBt to the broad opening of the stopcock assembly.
20. After the 30-min incubation (Step 16), add 9 mL of NPBt to the nuclei/bead mixture.
21. Resuspend the mixture in a 10-mL pipette and insert it vertically into the broad opening of the stopcock assembly.
22. Slowly release the mixture through the stopcock assembly (~1 mL/30 sec).

     The bead-bound nuclei should be collected to the side of the P1000 tip by the magnet. The flow-through can be retained and processed through the remainder of the protocol (Steps 27–36) as a non-bead-bound control.

23. Remove the magnet and collect the bead-bound nuclei in a fresh tube in 1 mL of NPB.
24. Dilute the NPB/nuclei mixture (1 mL) to 10 mL with NPBt.
25. Add 1.2 mL of NPBt to the stopcock assembly.
26. Repeat Steps 21–23 with a new tip-magnet assembly.

Proteomic Profiling INTACT Nuclei

This section describes the lysis of nuclei, isolation and separation of protein complexes, digestion of proteins into peptides for mass spectrometry, and preliminary data analysis.

27. Resuspend the bead-bound nuclei in 100 µL of RIPA buffer. Incubate for 10 min on ice to lyse the nuclei.
28. Sonicate the lysate using a Bioruptor on high for 15 min (30 sec on/30 sec off) at 4°C.
29. Place the lysate on a magnet to elute the sample from the beads.
30. Measure the protein concentration using a BCA assay.
31. Add the appropriate amount of 4× LDS buffer and reducing agent to obtain a final concentration of 0.5× buffer to 30 µg of protein sample. Incubate the sample for 10 min at 95°C.
32. Load and run each sample on a 4%–12% Bis-Tris polyacrylamide gel. Add 500 µL of antioxidant to the middle chamber before running. Load empty 1× sample buffer in all empty lanes to prevent uneven running of the gel.

     We recommend 1× MOPS running buffer, since MOPS is ideally used to separate proteins between 14 kDa and >200 kDa, ensuring a thorough separation of the majority of the proteome.

33. Using a razor blade, slice each lane into 1-mm gel slices. Group 10 slices per sample.
34. Add trypsin (12.5 ng/µL) to the gel slices. Incubate overnight at 37°C to digest the proteins into peptides.
35. Extract the peptides in 0.5% formic acid/50% acetonitrile (ACN).

Cite this protocol as *Cold Spring Harb Protoc*; doi:10.1101/pdb.prot098384

36. Analyze the peptides by nanoliquid chromatography-tandom mass spectrometry.

Search the spectra using a Xenopus-specific protein sequence database; see Greco et al. (2012) for a detailed protocol.

DISCUSSION

Studies of cellular and molecular pathways that are required for the development and maintenance of multicellular organs and tissues are notoriously difficult due to the heterogeneity of these structures. Many strategies have been implemented to isolate pure cell populations to circumvent these issues, including physical isolation by laser microdissection (Golubeva et al. 2013) or isolation based on the expression of a tissue or cell specific marker (FACS) (Barker et al. 1975). However, these methods of isolation require that the cell type of interest expresses a unique marker or reporter construct in order to ensure isolation of a pure population of cells. In *Xenopus*, isolation of cell populations using antibody-based sorting is difficult due to the availability of antibodies against these unique markers. The INTACT (isolation of nuclei tagged in specific cell types) method has been utilized in plants, worms, and flies to isolate pure nuclear populations via in vivo biotinylation of the nuclear envelope (Deal and Henikoff 2010, 2011; Henry et al. 2012; Steiner et al. 2012). The biotinylated nuclei are then isolated using streptavidin beads, separating nuclei from the cell population of interest away from other contaminating cell types without requiring an antibody. These studies have led to analysis of gene expression and chromatin features in *Arabidopsis* root epidermis cells (Deal and Henikoff 2011), distinct cell types in the *Drosophila* brain (Henry et al. 2012), and muscle cells from *C. elegans* (Steiner et al. 2012). This protocol describes the adaptation of the INTACT method for use in *Xenopus*, and subsequent proteomics analysis of nuclear protein complexes (Amin et al. 2014). The INTACT protocol utilizes two transgenes, the NTF transgene and the *E. coli* ligase BirA tagged to mCherry. Either or both transgenes are driven under the control of a tissue-specific promoter (e.g., the *mlc2p* regulatory element fused upstream of the NTF to drive NTF expression specifically in cardiomyocytes after stage 26). The fact that the NTF can be expressed in a specific spatial and temporal manner makes this protocol easily adaptable to almost any cell or tissue type. Furthermore, proteomics analysis of immunoprecipitated complexes from INTACT-isolated nuclei of multiple tissue types can determine the composition of protein complexes in pure cell populations (Conlon et al. 2012).

A potential disadvantage of the INTACT method is that it requires the use of transgenic animals that express the NTF transgene. Ideally, stable transgenic lines would assure uniform expression of the transgene. Generation of stable transgenic lines can be technically challenging and time-consuming, however, and this method has been used successfully with transient (mosaic) transgenic animals (Amin et al. 2014).

RECIPES

Nuclear Purification Buffer (NPB)

10 mM Tris (pH 7.4)
40 mM NaCl
90 mM KCl
2 mM EDTA
0.5 mM EGTA
0.2 mM dithiothreitol (DTT)
0.5 mM phenylmethanesulfonyl fluoride (PMSF)
0.5 mM spermine (Sigma-Aldrich S3256)
0.25 mM spermidine (Sigma-Aldrich S2626)
1× Roche cOmplete Protease Inhibitor Cocktail (Sigma-Aldrich 11697498001)

Add DTT, PMSF, spermine, spermidine, and protease inhibitors to NPB immediately prior to use. Use at 4°C.

RIPA Buffer for Xenopus

50 mM Tris (pH 8.0)
150 mM NaCl
0.5% sodium deoxycholate
1% NP-40
0.1% SDS

Store for up to 1 mo at 4°C.

REFERENCES

Allen BG, Weeks DL. 2005. Transgenic *Xenopus laevis* embryos can be generated using φC31 integrase. *Nat Methods* **2:** 975–979.

Amaya E, Kroll KL. 1999. A method for generating transgenic frog embryos. *Methods Mol Biol* **97:** 393–414.

Amin NM, Greco TM, Kuchenbrod LM, Rigney MM, Chung MI, Wallingford JB, Cristea IM, Conlon FL. 2014. Proteomic profiling of cardiac tissue by isolation of nuclei tagged in specific cell types (INTACT). *Development* **141:** 962–973.

Barker CR, Worman CP, Smith JL. 1975. Purification and quantification of T and B lymphocytes by an affinity method. *Immunology* **29:** 765–777.

Conlon FL, Miteva Y, Kaltenbrun E, Waldron L, Greco TM, Cristea IM. 2012. Immunoisolation of protein complexes from *Xenopus. Methods Mol Biol* **917:** 369–390.

Deal RB, Henikoff S. 2010. A simple method for gene expression and chromatin profiling of individual cell types within a tissue. *Dev Cell* **18:** 1030–1040.

Deal RB, Henikoff S. 2011. The INTACT method for cell type–specific gene expression and chromatin profiling in *Arabidopsis thaliana. Nat Protoc* **6:** 56–68.

Golubeva Y, Salcedo R, Mueller C, Liotta LA, Espina V. 2013. Laser capture microdissection for protein and NanoString RNA analysis. *Methods Mol Biol* **931:** 213–257.

Greco TM, Miteva Y, Conlon FL, Cristea IM. 2012. Complementary proteomic analysis of protein complexes. *Methods Mol Biol* **917:** 391–407.

Henry GL, Davis FP, Picard S, Eddy SR. 2012. Cell type–specific genomics of *Drosophila* neurons. *Nucleic Acids Res* **40:** 9691–9704.

Latinkic BV, Cooper B, Smith S, Kotecha S, Towers N, Sparrow D, Mohun TJ. 2004. Transcriptional regulation of the cardiac-specific MLC2 gene during *Xenopus* embryonic development. *Development* **131:** 669–679.

Schaffer U, Schlosser A, Muller KM, Schafer A, Katava N, Baumeister R, Schulze E. 2010—SnAvi—A new tandem tag for high-affinity protein-complex purification. *Nucleic Acids Res* **38:** e91.

Steiner FA, Talbert PB, Kasinathan S, Deal RB, Henikoff S. 2012. Cell-type-specific nuclei purification from whole animals for genome-wide expression and chromatin profiling. *Genome Res* **22:** 766–777.

Yergeau DA, Mead PE. 2007. Manipulating the *Xenopus* genome with transposable elements. *Genome Biol* **8(Suppl 1):** S11.

Cite this protocol as *Cold Spring Harb Protoc*; doi:10.1101/pdb.prot098384

Methods for Investigating the Larval Period and Metamorphosis in *Xenopus*

Daniel R. Buchholz[1,3] and Yun-Bo Shi[2,3]

[1]*Department of Biological Sciences, University of Cincinnati, Cincinnati, Ohio 45244;* [2]*Section on Molecular Morphogenesis, Eunice Kennedy Shriver National Institute of Child Health and Human Development (NICHD), National Institutes of Health (NIH), Bethesda, Maryland 20892*

Anuran metamorphosis resembles postembryonic development in mammals, a period around birth when many organs/tissues mature into their adult form as circulating thyroid and stress hormone levels are high. Unlike uterus-enclosed mammalian embryos, tadpoles develop externally and undergo the dramatic changes of hormone-dependent development totally independent of maternal influence, making them a valuable model in which to study vertebrate postembryonic organ development and maturation. Various protocols have been developed and/or adapted for studying metamorphosis in *Xenopus laevis* and *X. tropicalis*, two highly related and well-studied frog species. Here, we introduce some of the methods for contemporary cell and molecular studies of gene function and regulation during metamorphosis.

INTRODUCTION

Amphibians undergo a biphasic development process (Dodd and Dodd 1976; Shi 1999). Their embryogenesis leads to the formation of free-feeding tadpoles, which after a finite period of growth undergo the second phase of development, metamorphosis. Metamorphosis in anurans (frogs and toads) is the most dramatic among amphibians, with essentially every organ/tissue changing drastically by the end of metamorphosis. While two dramatic phases of development exist only in amphibians among terrestrial vertebrates, all vertebrates undergo similar developmental processes (Tata 1993; Laudet 2011; Buchholz 2015). In mammals, the initial embryonic development leads to the formation of most organs. After a period of growth and enlargement, the organs then mature into the adult forms. This second period of development occurs around birth in mammals. While much has been learned about early embryonic development by using various vertebrate models, much less is known about postembryonic development, especially in mammals. This is largely because of the difficulty of manipulating and analyzing uterus-enclosed late-stage mammalian embryos and neonates, which are still dependent on the maternal supply of nutrients for survival.

A key aspect of postembryonic development is the peak in plasma thyroid hormone (TH) around birth, hatching and metamorphosis, which are equivalent periods in different animal species (Sachs and Buchholz 2017). Although birth still occurs in mammals in the absence of TH, many organ systems are abnormal, especially neural development (Hetzel 1989; Braverman and Utiger 2005).

---

[3]Correspondence: buchhodr@ucmail.uc.edu; Shi@helix.nih.gov

In frog metamorphosis, the role of TH is more significant, as TH signaling is necessary and sufficient for most developmental events that occur during metamorphosis (Dodd and Dodd 1976; Shi 1999). It is easy to block metamorphosis completely by inhibiting endogenous TH synthesis or to induce metamorphosis precociously by adding physiological levels of TH to the rearing water of premetamorphic tadpoles (Dodd and Dodd 1976; Shi 1999). Other factors (such as environmental rearing conditions or other hormones, including stress hormone) affect metamorphic events (Denver 2009; Denver et al. 2009); however, these factors influence metamorphosis mostly indirectly through their effects on TH signaling (Bonett et al. 2010). The dramatic TH-dependent changes and ease of experimental manipulation have made anuran metamorphosis a unique and valuable model for studying postembryonic development in vertebrates (Sachs and Buchholz 2017). Over the last few decades, two highly related anuran species, the allo-tetraploid *Xenopus laevis* and the diploid *X. tropicalis*, have been studied extensively for various aspects of metamorphosis (Furlow and Neff 2006; Brown and Cai 2007; Das et al. 2010; Sachs and Buchholz 2017). The methods introduced here focus on some of the techniques that are employed in contemporary cell and molecular studies of gene regulation and functions during metamorphosis. Many other protocols, including those for gene expression, transgenesis, and gene editing, can also be used for metamorphosis studies.

ANALYSIS OF GENE REGULATION AND FUNCTION DURING METAMORPHOSIS

All vertebrates experience a period of preadult development controlled by hormones; in mammals, this development occurs inside the uterus (Sachs and Buchholz 2017). Frog metamorphosis provides a valuable research model in which to study this developmental period because hormone-dependent developmental progression occurs in free-living tadpoles to a dramatic and easily manipulable degree. Frog metamorphosis has provided an in vivo testing ground for molecular mechanisms of gene regulation proposed based on cell culture studies, from thyroid hormone receptor function (Shi 2009) and natural and synthetic thyroid hormone receptor agonists and antagonists (Lim et al. 2002; Schriks et al. 2007) to chromatin remodeling (Wong et al. 1995), cofactor recruitment (Shi 2013) and long-range hormone-DNA interactions (Buisine et al. 2015).

The protocols introduced here represent fundamental and recent approaches to answering outstanding questions about the roles of genes in development in different tissues, the actions of thyroid hormone and its interactions with other hormones during development, and the molecular and genetic mechanisms underlying postembryonic developmental processes. These protocols range from general purpose procedures (tissue harvest, cell proliferation) to specific manipulations in specific tissues (transfections of brain and tail, intestine organ culture, and ChIA-PET from tail skin) (Fig. 1). Protocol 1: *Xenopus* Tadpole Tissue Harvest (Patmann et al. 2017) will apply to a great number of experiments performed on tadpoles. Tadpoles are large enough to harvest nearly all tissues from single individuals to assess morphological, histological, and transcriptional responses to thyroid hormone. Other experiments require a large volume of starting material. In either case, the descriptions and images provided in Protocol 1: *Xenopus* Tadpole Tissue Harvest (Patmann et al. 2017) represent, for researchers new to tadpole anatomy, a fundamental step in experimental procedures on frog metamorphosis. Protocol 5: Cell Proliferation Analysis During *Xenopus* Metamorphosis: Using 5-Ethynyl-2-Deoxyuridine (EdU) to Stain Proliferating Intestinal Cells (Okada and Shi 2017) is another widely applicable method, as cell proliferation plays a prominent role in many tissues during postembryonic developmental remodeling.

A valuable aspect of studying frog metamorphosis is that many organs, including tail, limb, lung, liver, and skin, can be cultured in vitro, thereby providing stringent experimental control over developmental changes that mimic what happens in the whole organism. Protocol 2: Organ Culture of the *Xenopus* Tadpole Intestine (Ishizuya-Oka 2017) describes procedures for studying the effects of hormones on intestinal remodeling and the developmental origin of stem cells in

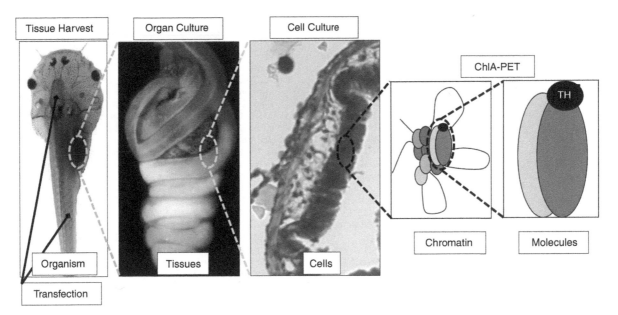

FIGURE 1. Levels of analysis of the *Xenopus* model to elucidate mechanisms of hormonal control of development. From left to right, the figure panels show decreasing biological levels from organism (tadpole) to tissues (whole intestine), cells (intestinal cross-section), chromatin structure (chromosomal looping) and molecules (receptor-hormone interaction, thyroid hormone [TH] bound to its nuclear receptor). The focal biological levels queried by the protocols introduced here are indicated. For tissue harvest, see Protocol 1: *Xenopus* Tadpole Tissue Harvest (Patmann et al. 2017); for transfection, see Protocol 3: Bulk Electroporation-Mediated Gene Transfer into *Xenopus* Tadpole Brain (Sáenz de Miera et al. 2018) and Protocol 4: In Vivo Transfection of Naked DNA into *Xenopus* Tadpole Tail Muscle (Marshall et al. 2017); for organ culture, see Protocol 2: Organ Culture of the *Xenopus* Tadpole Intestine (Ishizuya-Oka 2017); for cell culture, see Protocol 5: Cell Proliferation Analysis During *Xenopus* Metamorphosis: Using 5-Ethynyl-2-Deoxyuridine (EdU) to Stain Proliferating Intestinal Cells (Okada and Shi 2017); for ChIA-PET, see Protocol 6: Chromatin Immunoprecipitation for Chromatin Interaction Analysis Using Paired-End-Tag (ChIA-PET) Sequencing in Tadpole Tissues (Buisine et al. 2018a) and Protocol 7: Chromatin Interaction Analysis Using Paired-End-Tag (ChIA-PET) Sequencing in Tadpole Tissues (Buisine et al. 2018b).

vitro; the basic procedures detailed in that protocol will apply to other tadpole organs as well. Studying gene function in mammals often involves genetically modified lines. However, Protocol 3: Bulk Electroporation-Mediated Gene Transfer into *Xenopus* Tadpole Brain (Sáenz de Miera et al. 2018) and Protocol 4: In Vivo Transfection of Naked DNA into *Xenopus* Tadpole Tail Muscle (Marshall et al. 2017) open the door to rapid and targeted studies of gene function, thus bypassing the requirement for lengthy and laborious characterization of genetically modified lines. A central, contemporary question in gene regulation concerns chromatin conformation, transcription factor binding, and the location and role of enhancers in transcription initiation. Protocol 6: Chromatin Immunoprecipitation for Chromatin Interaction Analysis Using Paired-End-Tag (ChIA-PET) Sequencing in Tadpole Tissues (Buisine et al. 2018a) and Protocol 7: Chromatin Interaction Analysis Using Paired-End-Tag (ChIA-PET) Sequencing in Tadpole Tissues (Buisine et al. 2018b) describe methods central to addressing this question by detailing how to identify functional chromatin interactions genome-wide between binding sites of a transcription factor and the promoters of genes regulated by that transcription factor.

ACKNOWLEDGMENTS

Y.-B.S. was supported by the Intramural Research Program of the National Institute of Child Health and Human Development, National Institutes of Health.

REFERENCES

Bonett RM, Hoopfer ED, Denver RJ. 2010. Molecular mechanisms of corticosteroid synergy with thyroid hormone during tadpole metamorphosis. *Gen Comp Endocrinol* **168:** 209–219.

Braverman LE, Utiger RD (eds.). 2005. *Werner and Ingbar's the thyroid.* Lippincott, Williams, and Wilkins, Philadelphia.

Brown DD, Cai L. 2007. Amphibian metamorphosis. *Dev Biol* **306:** 20–33.

Buchholz DR. 2015. More similar than you think: Frog metamorphosis as a model of human perinatal endocrinology. *Dev Biol* **408:** 188–195.

Buisine N, Ruan X, Bilesimo P, Grimaldi A, Alfama G, Ariyaratne P, Mulawadi F, Chen J, Sung WK, Liu ET, et al. 2015. *Xenopus tropicalis* genome re-scaffolding and re-annotation reach the resolution required for in vivo ChIA-PET analysis. *PLoS One* **10:** e0137526.

Buisine N, Ruan X, Ruan Y, Sachs LM. 2018a. Chromatin immunoprecipitation for chromatin interaction analysis using paired-end-tag (ChIA-PET) sequencing in tadpole tissues. *Cold Spring Harb Protoc* doi: 10.1101/pdb.prot097725.

Buisine N, Ruan X, Ruan Y, Sachs LM. 2018b. Chromatin interaction analysis using paired-end-tag (ChIA-PET) sequencing in tadpole tissues. *Cold Spring Harb Protoc* doi:10.1101/pdb.prot104620.

Das B, Matsuda H, Fujimoto K, Sun G, Matsuura K, Shi YB. 2010. Molecular and genetic studies suggest that thyroid hormone receptor is both necessary and sufficient to mediate the developmental effects of thyroid hormone. *Gen Comp Endocrinol* **168:** 174–180.

Denver RJ. 2009. Stress hormones mediate environment-genotype interactions during amphibian development. *Gen Comp Endocrinol* **164:** 20–31.

Denver RJ, Glennemeier KA, Boorse GC. 2009. Endocrinology of complex life cycles: Amphibians. In *Hormones, brain and behavior,* 2nd ed. (ed. Pfaff DW, et al.), pp. 707–744. Academic Press, San Diego.

Dodd MHI, Dodd JM. 1976. The biology of metamorphosis. In *Physiology of the Amphibia* (ed. Lofts B), pp. 467–599. Academic Press, New York.

Furlow JD, Neff ES. 2006. A developmental switch induced by thyroid hormone: *Xenopus laevis* metamorphosis. *Trends Endocrinol Metab* **17:** 40–47.

Hetzel BS. 1989. *The story of iodine deficiency: An international challenge in nutrition.* Oxford University Press, Oxford.

Ishizuya-Oka A. 2017. Organ culture of the *Xenopus* tadpole intestine. *Cold Spring Harb Protoc* doi: 10.1101/pdb.prot097683.

Laudet V. 2011. The origins and evolution of vertebrate review metamorphosis. *Curr Biol* **21:** R726–R737.

Lim W, Nguyen NH, Yang HY, Scanlan TS, Furlow JD. 2002. A thyroid hormone antagonist that inhibits thyroid hormone action in vivo. *J Biol Chem* **277:** 35664–35670.

Marshall L, Girardot F, Demeneix BA, Coen L. 2017. In vivo transfection of naked DNA into *Xenopus* tadpole tail muscle. *Cold Spring Harb Protoc* doi: 10.1101/pdb.prot099366.

Okada M, Shi Y.-B. 2017. Cell proliferation analysis during *Xenopus* metamorphosis: Using 5-ethynyl-2-deoxyuridine (EdU) to stain proliferating intestinal cells. *Cold Spring Harb Protoc* doi: 10.1101/pdb.prot097717.

Patmann MD, Shewade LH, Schneider KA, Buchholz DR. 2017. *Xenopus* tadpole tissue harvest. *Cold Spring Harb Protoc* doi:10.1101/pdb.prot097675.

Sachs LM, Buchholz DR. 2017. Frogs model man: In-vivo thyroid hormone signaling during development. *Genesis* **55:** e23000.

Sáenz de Miera C, Parr E, Denver RJ. 2018. Bulk electroporation-mediated gene transfer into *Xenopus* tadpole brain. *Cold Spring Harb Protoc* doi: 10.1101/pdb.prot097014.

Schriks M, Roessig JM, Murk AJ, Furlow JD. 2007. Thyroid hormone receptor isoform selectivity of thyroid hormone disrupting compounds quantified with an in vitro reporter gene assay. *Environ Toxicol Pharmacol* **23:** 302–307.

Shi YB. 1999. *Amphibian metamorphosis: From morphology to molecular biology.* Wiley-Liss, New York.

Shi YB. 2009. Dual functions of thyroid hormone receptors in vertebrate development: The roles of histone-modifying cofactor complexes. *Thyroid* **19:** 987–999.

Shi YB. 2013. Unliganded thyroid hormone receptor regulates metamorphic timing via the recruitment of histone deacetylase complexes. *Curr Top Dev Biol* **105:** 275–297.

Tata JR. 1993. Gene expression during metamorphosis: An ideal model for postembryonic development. *Bioessays* **15:** 239–248.

Wong J, Shi YB, Wolffe AP. 1995. A role for nucleosome assembly in both silencing and activation of the *Xenopus* TR beta A gene by the thyroid hormone receptor. *Genes Dev* **9:** 2696–2711.

Xenopus Tadpole Tissue Harvest

Matthew D. Patmann,[1] Leena H. Shewade,[1] Katelin A. Schneider,[1] and Daniel R. Buchholz[1,2]

[1]*Department of Biological Sciences, University of Cincinnati, Cincinnati, Ohio 45221*

The procedures described here apply to *Xenopus* tadpoles from the beginning of feeding through the major changes of metamorphosis and are appropriate for downstream postoperative snap freezing for molecular analysis, fixation for histological analysis, and sterile organ culture. To the uninitiated, the most difficult aspects of tadpole tissue dissections are likely knowing the appearance and location of organs, and the difficulty manipulating and holding tadpoles in place to carry out the oftentimes fine and precise dissections. Therefore, images and stepwise instructions are given for the harvest of external organs (tail, head, eyes, tail skin, back skin, gills, thymus, hind limbs, forelimbs) and peritoneal organs (intestine, pancreas, liver, spleen, lungs, fat bodies, kidney/gonad complex), as well as brain, heart, and blood. Dissections are typically done under a dissection stereomicroscope, and two pairs of fine straight forceps, one pair of fine curved forceps, and one pair of microdissection scissors are sufficient for most tissue harvests.

MATERIALS

It is essential that you consult the appropriate Material Safety Data Sheets and your institution's Environmental Health and Safety Office for proper handling of equipment and hazardous materials used in this protocol.

RECIPES: Please see the end of this protocol for recipes indicated by <R>. Additional recipes can be found online at http://cshprotocols.cshlp.org/site/recipes.

Reagents

Ethanol (70%)

Frog water

Remove chlorine/chloramine from tap water by carbon filtration (Alternatively, reconstitute reverse-osmosis water to 800 µS using Instant Ocean or equivalent). Adjust the pH to 7.5 using sodium bicarbonate.

Ice

MS-222 solution (0.1% tricaine methanesulfonate [Sigma-Aldrich], in frog water)

Add sodium bicarbonate to 0.1% to adjust the pH to 7.0–7.5.

Phosphate-buffered saline (0.6×) <R>

Dilute 1× PBS 60:40 with reverse-osmosis or distilled water. Dilution is required because the osmolarity of amphibian blood is ~62% that of mammalian blood.

Xenopus tadpoles

[2]Correspondence: buchhodr@ucmail.uc.edu

Tadpoles at stages from the beginning of feeding through tail resorption (i.e., NF stages 45–66 [Nieuwkoop and Faber 1994]) can be obtained from suppliers listed at Xenbase (www.xenbase.org/other/obtain.do). Full-sized tadpoles around the stage of forelimb emergence (NF stages 57–58) are ideal for this protocol. Smaller tadpoles from the beginning of feeding (≤NF45) or tadpoles during metamorphosis (NF 61–66) will be more challenging for some tissues.

Equipment

Capillary tubes, micro-hematocrit, heparinized, 0.5-mm i.d., 75-mm length (Kimble-Chase, #40C505)
Dissection stereomicroscope
Finger bowl, 350-mL
Fish net (or tea strainer)
Forceps, curved (Dumont #7)
Forceps, straight (Dumont #5 or #55)
Kimwipes
Microdissection scissors, straight, w/ sharp points, 3.25-in.
Petri dishes, 100- and 150-mm diameter, plastic or glass
Razor blade
Syringe, 3-mL, fitted w/ 30-gauge needle

METHOD

Dissection tools are shown in Figure 1A. The forceps and scissors used in these procedures are delicate and damage easily. Care must be taken not to drop them, or even tap the tips into hard surfaces such as the surface of the Petri dish during dissection. Use straight forceps in the following procedures unless specifically instructed otherwise. When harvesting tissues from tadpoles that have undergone hormone or chemical treatments, avoid cross-contamination by rinsing the Petri dish, dissection tools, and tea strainer between treatment groups or samples.

Anesthetizing the Tadpole

1. Transfer tadpole from a large stock tank using a fish net (or from a small treatment tank with a tea strainer) into a 0.1% MS-222 solution in a finger bowl or 100-mm Petri dish.

2. After 1–2 min, verify that the tadpole is anesthetized by gently pinching the tadpole's tail with forceps.

   A properly anesthetized tadpole will not respond to a tail pinch but will still have a heartbeat.

3. Scoop the anesthetized tadpole with curved forceps. Place it in a 100-mm Petri dish under the dissection stereomicroscope to perform tissue harvest.

   Placing anesthetized tadpoles on a slightly damp Kimwipe tissue can help prevent them from sliding during dissections. Tissues for gene expression analysis need to be dissected on ice; place the 100-mm Petri dish containing the tadpole on top of a 150-mm Petri dish filled with ice. For genomic analyses, clean the Petri dish and dissection tools with Kimwipes and 70% ethanol between each individual.

Harvesting External Tissues

Tail and Head (Fig. 1B,C)

The tail and head are the easiest and most common tadpole organs to harvest. The tail has fewer tissues than the head, and these organs have different gene expression responses during metamorphosis. The size of the tail increases during growth and the ratio of tail skin to tail muscle changes through ontogeny; both of these features need to be taken into account when collecting tissue to maintain comparable results across treatments or developmental stages. Unlike the other tissues, a stereomicroscope is not needed to harvest the tail or head.

4. Use forceps to hold the tadpole in place while harvesting the tail and head.

   i. Using a razor blade, cut off the tail with a transverse cut posterior to the hind limbs.

   ii. Using a razor blade, cut off the head posterior to the eyes and anterior to the thymus glands (solid line in Fig. 1B).

Cite this protocol as *Cold Spring Harb Protoc*; doi:10.1101/pdb.prot097675

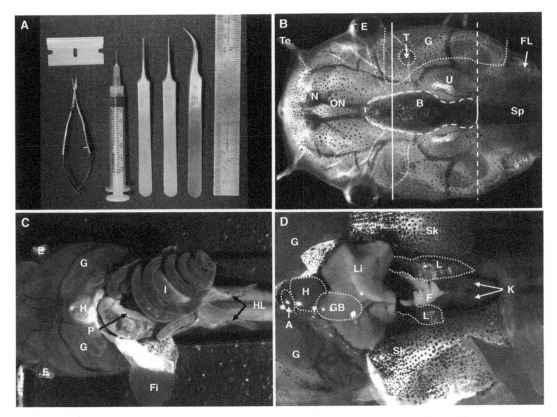

FIGURE 1. Dissection tools and tadpole organs in situ. (*A*) The dissection tools pictured are sufficient to harvest all tissues in this protocol except blood, which also requires heparinized capillary tubes (ruler shown for size). (*B*) In this dorsal view of the tadpole head, the organs are labeled only on the *right* side for clarity. The solid line is the cut site to harvest the head. The straight dashed line is the first cut site to harvest the brain, and the curved dashed line outlines the location of the brain. The dotted line is the cut site to harvest the gills, and the dotted circle outlines the thymus. (*C*) Ventral view of tadpole with skin and abdominal muscle reflected to the sides. Heart is still in pericardial cavity. (*D*) Ventral view of tadpole with intestine, pancreas, and spleen removed. The *right* lung is sharply curved upward and thus appears shorter than the left lung. The tadpole is Nieuwkoop and Faber (NF) stage 56 (Nieuwkoop and Faber 1994). Abbreviations: (A) aorta, (B) brain, (E) eye, (F) fat bodies, (Fi) fin on abdomen, (FL) forelimb, (G) gill, (GB) gallbladder, (H) heart, (HL) hind limb, (I) intestine, (K) kidney/gonad complex, (L) lung, (Li) liver, (N) nostril, (ON) olfactory nerve, (P) pancreas, (Sk) skin, (Sp) spine, (Te) tentacle, (T) thymus, (U) ultimobranchial body.

Eyes (Fig. 1B)

Eyes change during metamorphosis in position and in photopigment expression.

5. Secure the anesthetized tadpole by piercing the head along the midline with the two points of a pair of forceps.

6. Use curved forceps to partially pinch below the ball of the eye.

7. Slowly separate the eye from the body.

8. Carefully complete the pinching action below the eye as the eye is being removed.

   Take care while lifting the eye out to ensure that it separates cleanly from the optic nerve and connective tissue remains held by the curved forceps.

Tail Skin

Tail skin is more challenging to obtain than whole tail, but a highly uniform tissue (i.e., almost exclusively epidermis) is the result. The tail muscle (with spinal cord and notochord) can be harvested at the same time, if desired.

9. Use a razor blade to obtain the tail as described in Step 4.i.

10. Secure the tail by inserting open forceps into the muscle of the tail through the cut site, being careful not to pierce or break the skin.

Maintain a secure hold of the tail muscle with open forceps for the rest of the procedure.

11. Using curved forceps, pinch the tail skin at the cut end. Slowly pull a few millimeters of the skin toward the posterior end of the tail.

12. Turn the tail over (without releasing the muscle). Slowly pull the skin toward the posterior of the tail.

13. Repeat Steps 11–12, turning the tail over periodically to release the skin evenly down the length of the tail.

> *If done carefully, the skin will release from the muscle like pulling off a glove. The skin becomes difficult to separate from the muscle at the posterior tip of the tail and thus is not collected.*

Back Skin (Fig. 1B)

The back skin has a distinct developmental fate compared with the tail skin and skin covering the gills (Watanabe et al. 2002; Suzuki et al. 2009). The back skin will transform from the larval to adult skin by the end of metamorphosis rather than being resorbed like the skin covering the tail and gills.

14. Pinch up an edge of back skin to be harvested with forceps.

15. Use scissors to cut a small hole through the skin.

16. Reposition the forceps to hold the skin at the hole.

17. Use scissors to cut out the desired patch of skin, repositioning the forceps as necessary.

> *Make sure to avoid the skin covering the gills (Fig. 1B) because that skin remains larval in character and is resorbed during metamorphosis.*

Gills (Fig. 1B)

Gills—along with tail, pronephros, tentacles, and the overlying integument—are strictly larval structures that resorb completely at metamorphosis. The gills are comprised of a fine cartilaginous network covered with epithelium all enclosed within larval skin. The forelimbs and thymus are found in the posterior and central portion, respectively, within the gill compartment. The thymus is embedded in a space in the gill cartilage and will remain with the gills unless specifically removed.

18. Secure the anesthetized tadpole by piercing the head along the midline with the two points of a pair of forceps.

19. Using microdissection scissors, cut from the posterior side of the eye toward the brain (dotted line in Fig. 1B).

20. Hold the tadpole with forceps through the mouth. Cut posteriorly along the side of the brain to release the gills.

> *The ultimobranchial bodies can be easily pierced, releasing a milky substance. The skin and thymus (see below) can be removed to obtain a less complex tissue mixture.*

21. Repeat Steps 18–20 to harvest gills on the other side.

Thymus (Fig. 1B)

The thymus is particularly delicate and extremely easy to burst if not handled with caution. It decreases in size dramatically during metamorphosis, likely in response to high glucocorticoid levels. Repositioning of the thymus after gill resorption makes the thymus more difficult to locate in later stages of metamorphosis.

22. Secure the anesthetized tadpole by piercing the head along the midline with the two points of a pair of forceps.

23. Using microdissection scissors make two transverse cuts completely through the gill, one anterior and the other posterior to the thymus.

> *These cuts should begin at the lateral edge of the gill and extend to a point toward the midline beyond the thymus, but not into the brain.*

24. Using microdissection scissors, make a longitudinal incision connecting the end points of the first two transverse cuts, going straight down with the scissors through the dorsal surface.

25. Fold back the skin/cartilage flap created to expose the thymus, which is still encapsulated by internal gill tissue.

26. Use forceps to secure the opened gill area.

27. Using microdissection scissors, carefully trim away the jelly-like gill tissue from the surface of the thymus.

    It is very difficult to remove 100% of the gill tissue without causing the thymus to burst.

Hind and Forelimbs (Fig. 1B,C)

Limb buds form early in the larval period but require thyroid hormone produced during metamorphosis to grow and develop beyond the paddle stage. Larger limbs are less fragile and can be gripped anywhere along the length of the limb and cut off with scissors.

28. With two pairs of forceps positioned next to each other, pinch at the base of each hind limb to separate the limbs from where they join the posterior part of the abdomen. Expose the forelimbs from underneath the gill skin with forceps or scissors, then remove them as for hind limbs.

    i. Use the forceps to pinch at the base of the hind limb to separate it from where it joins to the posterior part of the abdomen.

    ii. Expose the forelimbs from underneath the gill skin using forceps or scissors.

    iii. Use the forceps to pinch at the base of the forelimb to separate it from the torso.

Harvesting the Brain, Heart, and Blood

Brain (Fig. 1B)

The brain is composed mostly of neurons and glia, but is highly heterogeneous in gene expression (Yao et al. 2008; Denver et al. 2009). Also, the pituitary is found in a pocket in the ventromedial region of the brain and is not readily separated by accident from the rest of the brain in this procedure. The spinal cord can also be harvested (Muñoz et al. 2015).

29. Secure the tadpole with forceps. Using a razor blade, make a single transverse cut at the base of the hindbrain, just anterior to the spinal cord (dashed line in Fig. 1B).

    For the next three steps, maintain hold of the head dorsally with a pair of forceps inserted through the anterior gills on both sides of the brain.

30. Using scissors, make straight cuts through the dorsal skin and cartilaginous braincase along the left and right sides of the brain.

31. Expose the brain by lifting the skin/cartilage flap up with forceps.

32. Insert closed forceps under the brain. Gently push the brain upward, taking care to allow the cranial nerves to break in the process.

    The ultimobranchial body can be pierced easily and release a milky substance.

Heart (Fig. 1C,D)

The heart is located along the midline, enclosed in a silver-colored pericardium just anterior to the peritoneal cavity. Very few studies have been performed on the heart larval-to-adult transition.

33. With the ventral side up, use two pairs of forceps to carefully tear and peel away the clear epidermis and silvery pericardium that covers the heart.

34. With the heart exposed, gently slide the forceps underneath it. Grasp it at the point where it connects to major arteries.

35. Pinch the forceps to cut the heart free of the arteries. Lift the heart out of the heart cavity, ensuring that it tears free cleanly.

    Assist the process by pinching any attached arteries with a second pair of forceps.

Blood

Because of the small blood volume of tadpoles, it is typical to obtain heparinized plasma rather than serum. This procedure should be done in <3 min to avoid altered stress hormone levels.

36. Expose the heart as described in Step 33.

    The heart should still be beating.

37. Use an absorbent tissue to blot dry the cardiac area.

38. With all interstitial fluid removed, position a heparinized capillary tube at a 45° angle to the aorta (Fig. 1D).

39. Nick the aorta with forceps to allow the blood to flow into the tube by capillary action.

    As the blood rises into the capillary, it might be necessary to decrease the angle to continue collection.

40. Transfer the blood into a fresh Eppendorf tube on ice by blowing it from the capillary tube with a P200 pipettor fitted with a yellow tip.

41. Centrifuge at 4000–6000 rpm for 20 min at 4°C.

42. Carefully remove the supernatant plasma using a P200 pipette. Store at −80°C.

    To collect a sufficient volume for measurements, blood from 2–3 tadpoles can be pooled before centrifugation.

Harvesting Peritoneal Tissues

Access the Peritoneum

The peritoneal cavity is accessed through two tissue layers: the ventral skin and the abdominal muscle wall. Both layers are thin and can be cut one at a time or simultaneously in the steps below. Sufficiently long cuts will allow the skin and muscle wall to be reflected back to expose the peritoneal organs, although only the large coiled intestine will be visible initially.

43. With forceps, pinch/grasp the ventral fin connected along the surface of the peritoneal cavity. Maintain this hold for the next step.

    The ventral fin regresses at later stages of metamorphosis. It is thus necessary to pinch the posterior abdominal skin directly at later stages.

44. Snip open a small hole into the posterior end of the peritoneal cavity.

    Care must be taken not to lacerate the intestines that lie just beneath the thin skin and muscle wall covering the peritoneal cavity.

45. While still holding the skin with forceps at the cut site, make lateral cuts with scissors through the skin and muscle wall along the posterior border of the peritoneum.

46. Maintaining hold with the forceps, make a posterior to anterior cut with scissors through the skin and muscle up the midline of the peritoneal cavity.

    Two pairs of good quality forceps can also serve to make several tears to similarly open the cavity.

Intestine (Fig. 1C)

The gastrointestinal tract consists of the esophagus, manicotto glandulare, intestine, and rectum. The manicotto is a tadpole-specific organ that is resorbed, and the stomach develops in that location during metamorphosis (Griffiths 1961). After intestinal remodeling that occurs during metamorphosis, a mesentery binds the loops of the greatly shortened intestine such that it cannot be straightened without many small cuts to the mesentery. For gene expression and histology, it is often important to flush the intestine of its contents.

47. Without damaging it, gently push but not uncoil the intestinal coil out of the cavity with forceps.

    The coil should roll out of the cavity while remaining attached at each end.

48. Clip the posterior end of the intestine with forceps where the intestine meets the rectum.

49. Clip the anterior end of the intestine around the level of the pancreas, where the pancreatic duct enters the gastrointestinal tract.

Cite this protocol as *Cold Spring Harb Protoc*; doi:10.1101/pdb.prot097675

50. To flush the intestine, gently grip the open end of the intestine with forceps without pinching it closed.

51. Holding the intestinal opening at a ~45° angle, insert a 30-gauge needle attached to a 3-mL syringe filled with 0.6× PBS into the opening.

> It is important to get the angle of needle and intestinal opening aligned to avoid piercing the side of the intestine with the needle's point.

52. Carefully secure the intestine onto the needle by gently squeezing the forceps. Use the syringe to slowly flush the gut contents from the intestine with 0.6× PBS.

> The intestine can be cut into two or more pieces to facilitate flushing the gut contents which will help avoid rupturing the intestinal wall.

Pancreas (Fig. 1C)

The white translucent pancreas is closely associated with the anterior loop of the intestine at the manicotto glandulare. The pancreas decreases greatly in size as it remodels to the adult version during metamorphosis. Like the liver, the pancreas breaks apart easily when pulled on with forceps.

53. Using two pairs of forceps, gently separate the pancreas from where it joins the intestine.

54. Once completely free of the intestine, gently grasp the pancreas with the forceps. Lift it out of the cavity.

> Do not pull on the pancreas directly with forceps as it will just tear and remain attached to the intestine.

Liver (Fig. 1D)

The tadpole liver is yellow/orange in color, has three lobes, and is located in the anterior portion of the peritoneal cavity. The liver is fairly small at early developmental stages but greatly increases in size toward the end of metamorphosis. Like the pancreas, it breaks apart easily when pulled with forceps. Embedded within the tadpole liver is the larval hematopoietic tissue, which is gradually replaced by adult hematopoietic stem cells during metamorphosis.

Because bile can interfere with procedures associated with gene expression analysis, care must be taken not to pierce the gallbladder while removing the liver.

55. Push the intestine out of the peritoneal cavity to reveal the anteriorly located liver.

56. Gently position curved forceps underneath the liver. Squeeze the curved forceps forcefully beneath, but not on, the liver.

57. While holding the curved forceps firmly, lift the liver out of the cavity by tearing the connections grasped by the forceps.

> If the liver does not come out in one piece, it can be removed in multiple pieces. Nevertheless, care should be taken not to shred the organ.

58. Use two pairs of forceps to clear away any extra tissue that might have been removed with the liver (e.g., the gallbladder, which is closely associated with the liver).

Spleen

The spleen is a tiny red ball found among the intestinal coils secured in place by two thin, black strands of connective tissue. It is not closely associated with any organ.

59. Carefully lift out coils of intestine until the spleen is located.

60. Using forceps, grasp the black strands of connective tissue that hold the spleen in place. Pull out the spleen.

> Do not use forceps to grasp the spleen directly, as it will rupture and release lymphocytes.

Lungs (Fig. 1D)

The lungs are small and inconspicuous in small tadpoles but grow in later staged tadpoles to extend the entire length of the peritoneal cavity and fill with air (Rose and James 2013).

61. Push the intestine out of the peritoneal cavity.

62. Remove the liver as described in Steps 56–57 to reveal the lungs in the anterior dorsolateral parts of the peritoneal cavity.

   The anterior end of the intestine will remain situated medially between the two lungs adjacent to the trachea.

63. To better access the lungs, use forceps to clip the anterior end of the intestine.

64. Secure the tadpole in place by piercing the head with the two points of the forceps.

65. Grasp the trachea with forceps. Gently pull upward, from anterior to posterior, to remove the lungs from the peritoneal cavity as a pair.

   The connective tissue of the lungs is fairly strong such that pulling the trachea can be done without damage to the lungs in nearly all cases.

66. Use forceps to trim any nonlung tissue away from the harvested lungs.

   Do not mistake the bronchial diverticula (Wassersug and Souza 1990) for nonlung tissue. These structures exist as small offshoots of the lung bronchii, which often contain small air bubbles in the anterior region of each lung.

Fat Bodies (Fig. 1D)

The abdominal fat bodies are yellow, finger-like structures that emanate from a pair of bilateral stalks at the anterior end of the kidney/gonad complex. When fat globules are released from the fat bodies, they spread quickly and glisten in the light. The size of the fat bodies generally increase during the larval period correlated with food intake.

67. Push the intestine out of the peritoneal cavity to reveal the fat bodies.

68. Grasp the stalks connecting the fat bodies to the kidney/gonad complex with forceps. Pull out to harvest.

   The kidneys might come out with the fat bodies, but they can be separated using forceps.

Kidney/Gonad Complex (Fig. 1D)

The kidney/gonad complex is located at the dorsal posterior region of the peritoneal cavity and is fairly well attached to the ventral side of the dorsal trunk muscles. The gonads are inconspicuous, thin translucent strands running nearly the length of the ventral side of kidney. The gonads are fragile and are not easily separated from the kidneys in the tadpole stages. Steroidogenic tissue producing glucocorticoids and mineralocorticoids (called "interrenal tissue" and homologous to the adrenal gland of mammals) is contained within the kidney capsule.

69. Push the intestine out of the peritoneal cavity. Remove the liver as described in Steps 56–57 to reveal the fat bodies and kidney/gonad complex.

70. Remove the fat bodies from the kidney/gonad complex by pinching and breaking the stalk of the fat bodies with two pairs of forceps.

71. Carefully separate the kidney/gonad complex by running a closed pair of forceps between the kidney and the back muscles.

   The kidney/gonad complex can break easily if pulled with forceps before this step.

72. Once separated from the back muscles, gently lift the kidneys from the peritoneal cavity.

DISCUSSION

The *Xenopus* tadpole is a well-established and valuable vertebrate model for hormone-dependent growth and development, molecular mechanisms of gene expression in vivo, and regeneration in tail, limbs, and spinal cord (Beck et al. 2009; Buchholz 2015; Sachs and Buchholz 2017). Most tadpole tissues have been examined to some extent to determine their function during tadpole life and to determine their hormone-dependent changes throughout metamorphosis (Pouyet and Beaumont 1975; Dodd and Dodd 1976; Fox 1983; Gilbert et al. 1996; Brown and Cai 2007). Many organs have been cultured in vitro, including tail, hind limbs, skin, intestine, liver, pancreas, lung, fat bodies, spinal cord, and gills (Derby 1968; Hanke 1978; Derby et al. 1979; Richmond and Pollack 1983; Mathisen and Miller 1989; Ishizuya-Oka and Shimozawa 1991; Tata et al. 1991; Helbing et al. 1992; Buchholz and Hayes 2005; Veldhoen et al. 2015). Supplementary anatomical aspects of tadpole skeletal and soft tissue structures are available (McDiarmid and Altig 1999).

Virtually all *Xenopus* tadpole tissues can be readily harvested. After harvest, tissues can be used in numerous types of analyses. For gene expression or hormone analysis, tissues must be harvested on ice and snap frozen immediately by placing the tissue with forceps into a prelabeled 1.5 mL Eppendorf tube that has been in direct contact with dry ice and then kept at −80°C for storage until assay. For histology and in situ hybridization, tissues can be fixed in ethanol, neutral buffered formalin, MEMFA, or other fixatives (Sive et al. 2000). Standard procedure for organs to be processed for immunohistochemistry is fixation in 4% paraformaldehyde at room temperature for 2 h (or overnight at 4°C) with rotation.

For organ culture, the sterility of internal organs such as liver and spleen must be maintained during the tissue harvest procedures. Before incision, treat the dissection tools and ventral integument with 70% ethanol. Care should be taken during dissection to avoid rupturing the intestine and thus releasing bacteria into the cavity. For external structures and the intestine, tadpoles can be pretreated overnight in sulfadiazine to inhibit bacterial growth, and harvested tissues are placed in sterile 0.6× L15 culture medium supplemented with antibiotics and antimycotics (Derby 1968; Buchholz and Hayes 2005).

RECIPE

Phosphate-Buffered Saline (PBS)

Reagent	Amount to add (for 1× solution)	Final concentration (1×)	Amount to add (for 10× stock)	Final concentration (10×)
NaCl	8 g	137 mM	80 g	1.37 M
KCl	0.2 g	2.7 mM	2 g	27 mM
Na_2HPO_4	1.44 g	10 mM	14.4 g	100 mM
KH_2PO_4	0.24 g	1.8 mM	2.4 g	18 mM
If necessary, PBS may be supplemented with the following:				
$CaCl_2 \cdot 2H_2O$	0.133 g	1 mM	1.33 g	10 mM
$MgCl_2 \cdot 6H_2O$	0.10 g	0.5 mM	1.0 g	5 mM

PBS can be made as a 1× solution or as a 10× stock. To prepare 1 L of either 1× or 10× PBS, dissolve the reagents listed above in 800 mL of H_2O. Adjust the pH to 7.4 (or 7.2, if required) with HCl, and then add H_2O to 1 L. Dispense the solution into aliquots and sterilize them by autoclaving for 20 min at 15 psi (1.05 kg/cm²) on liquid cycle or by filter sterilization. Store PBS at room temperature.

REFERENCES

Beck CW, Izpisúa Belmonte JC, Christen B. 2009. Beyond early development: *Xenopus* as an emerging model for the study of regenerative mechanisms. *Dev Dyn* **238:** 1226–1248.

Brown DD, Cai L. 2007. Amphibian metamorphosis. *Dev Biol* **306:** 20–33.

Buchholz DR. 2015. More similar than you think: Frog metamorphosis as a model of human perinatal endocrinology. *Dev Biol* **408:** 188–195.

Buchholz DR, Hayes TB. 2005. Variation in thyroid hormone action and tissue content underlies species differences in the timing of metamorphosis in desert frogs. *Evol Dev* **7:** 458–467.

Denver RJ, Hu F, Scanlan TS, Furlow JD. 2009. Thyroid hormone receptor subtype specificity for hormone-dependent neurogenesis in *Xenopus laevis*. *Dev Biol* **326:** 155–168.

Derby A. 1968. An in vitro quantitative analysis of the response of tadpole tissue to thyroxine. *J Exp Zool* **168:** 147–156.

Derby A, Jeffrey JJ, Eisen AZ. 1979. The induction of collagenase and acid phosphatase by thyroxine in resorbing tadpole gills in vitro. *J Exp Zool* **207:** 391–398.

Dodd MHI, Dodd JM. 1976. The biology of metamorphosis. In *Physiology of the amphibia* (ed. Lofts B), pp. 467–599. Academic Press, New York.

Fox H. 1983. *Amphibian morphogenesis*. Humana Press, Clifton, NJ.

Gilbert LI, Tata JR, Atkinson BG (eds.). 1996. *Metamorphosis: Postembryonic reprogramming of gene expression in amphibian and insect cells*. Academic Press, San Diego, CA.

Griffiths I. 1961. The form and function of the fore-gut in anuran larvae (Amphibia, Salientia) with particular reference to the *manicotto glandulare*. *Proc Zool Soc London* **137**: 249–283.

Hanke W. 1978. The adrenal cortex in amphibia. In *General, comparative, and clinical endocrinology of the adrenal cortex* (ed. Jones IC, Henderson IW), Vol. **2**, pp. 419–495, Academic Press, London.

Helbing C, Gergely G, Atkinson BG. 1992. Sequential up-regulation of thyroid hormone β receptor, ornithine transcarbamylase, and carbamyl phosphate synthetase mRNAs in the liver of *Rana catesbeiana* tadpoles during spontaneous and thyroid hormone-induced metamorphosis. *Dev Genet* **13**: 289–301.

Ishizuya-Oka A, Shimozawa A. 1991. Induction of metamorphosis by thyroid hormone in anuran small intestine cultured organotypically in vitro. *In Vitro Cell Dev Biol* **27A**: 853–857.

Mathisen PM, Miller L. 1989. Thyroid hormone induces constitutive keratin gene expression during *Xenopus laevis* development. *Mol Cell Biol* **9**: 1823–1831.

McDiarmid RW, Altig R (ed.). 1999. *Tadpoles: The biology of anuran larvae*. University of Chicago Press, Chicago, IL.

Muñoz R, Edwards-Faret G, Moreno M, Zuñiga N, Cline H, Larraín J. 2015. Regeneration of *Xenopus laevis* spinal cord requires Sox2/3 expressing cells. *Dev Biol* **408**: 229–243.

Nieuwkoop PD, Faber J (eds.). 1994. *Normal table of* Xenopus laevis *(Daudin): A systematical & chronological survey of the development from the fertilized egg till the end of metamorphosis*. Garland Publishing, New York.

Pouyet JC, Beaumont A. 1975. Ultrastructure of the larval pancreas of an anuran amphibian, Alytes obstetricans L, in organ culture. *CR Seances Soc Biol Paris* **169**: 846–850.

Richmond MJ, Pollack ED. 1983. Regulation of tadpole spinal nerve fiber growth by the regenerating limb blastema in tissue culture. *J Exp Zool* **225**: 233–242.

Rose CS, James B. 2013. Plasticity of lung development in the amphibian, *Xenopus laevis*. *Biology Open* **2**: 1324–1335.

Sachs LM, Buchholz DR. 2017. Frogs model man: *In vivo* thyroid hormone signaling during development. *Genesis* **55**: e23000.

Sive HL, Grainger RM, Harland RM (eds.). 2000. *Early development of Xenopus laevis: A laboratory manual*. Cold Spring Harbor Laboratory Press, Cold Spring Harbor, NY.

Suzuki K, Machiyama F, Nishino S, Watanabe Y, Kashiwagi K, Kashiwagi A, Yoshizato K. 2009. Molecular features of thyroid hormone-regulated skin remodeling in *Xenopus laevis* during metamorphosis. *Dev Growth Differ* **51**: 411–427.

Tata JR, Kawahara A, Baker BS. 1991. Prolactin inhibits both thyroid hormone-induced morphogenesis and cell death in cultured amphibian larval tissues. *Dev Biol* **146**: 72–80.

Veldhoen N, Stevenson MR, Helbing CC. 2015. Comparison of thyroid hormone-dependent gene responses in vivo and in organ culture of the American bullfrog (*Rana (Lithobates) catesbeiana*) lung. *Comp Biochem Physiol Part D: Genomics Proteomics* **16**: 99–105.

Wassersug RJ, Souza KA. 1990. The bronchial diverticula of *Xenopus* larvae: Are they essential for hydrostatic assessment? *Naturwissenschaften* **77**: 443–445.

Watanabe Y, Tanaka R, Kobayashi H, Utoh R, Suzuki K-I, Obara M, Yoshizato K. 2002. Metamorphosis-dependent transcriptional regulation of *xak-c*, a novel *Xenopus* type I keratin gene. *Dev Dyn* **225**: 561–570.

Yao M, Hu F, Denver RJ. 2008. Distribution and corticosteroid regulation of glucocorticoid receptor in the brain of *Xenopus laevis*. *J Comp Neurol* **508**: 967–982.

Cite this protocol as *Cold Spring Harb Protoc*; doi:10.1101/pdb.prot097675

Organ Culture of the *Xenopus* Tadpole Intestine

Atsuko Ishizuya-Oka[1]

Department of Biology, Nippon Medical School, Musashino, Tokyo 180-0023, Japan

During *Xenopus* metamorphosis, most tadpole organs remodel from the larval to adult form to prepare for adaptation to terrestrial life. Organ culture serves as an important tool for studying larval-to-adult organ remodeling independent of the effects of other parts of the body. Here, I introduce a protocol for organ culture in vitro using the *Xenopus laevis* tadpole intestine before metamorphic climax. During culture in the absence of exogenous 3,3′,5-triiodo-L-thyronine (T3), the most potent natural thyroid hormone, the intestine remains in its larval state without any metamorphic changes. In contrast, when T3 is added to the culture medium, the larval epithelium undergoes apoptosis, whereas adult stem cells appear, actively proliferate, and finally generate the differentiated adult epithelium within a week. At the same time, the surrounding nonepithelial tissues also develop. Thus, this culture model is useful for clarifying the control mechanisms of apoptosis in larval tissues, formation of adult stem cells, and cell proliferation and differentiation of adult tissues, all of which occur in harmony during natural metamorphosis. Moreover, a procedure for tissue recombination combined with organ culture provides a platform for investigating complex tissue interactions during organ remodeling. Such tissue recombination experiments will help to reveal the important role of nonepithelial tissues in larval epithelial apoptosis and/or adult stem cell development in the *X. laevis* intestine.

MATERIALS

It is essential that you consult the appropriate Material Safety Data Sheets and your institution's Environmental Health and Safety Office for proper handling of equipment and hazardous materials used in this protocol.

RECIPES: Please see the end of this protocol for recipes indicated by <R>. Additional recipes can be found online at http://cshprotocols.cshlp.org/site/recipes.

Reagents

CTS medium <R>
Dispase I (10,000 Protease Unit/ampule; Wako 386-02271)
Distilled water (DW)
> Sterilize before culture by autoclaving for 20 min at 120°C.

Hormone-containing medium <R>
Penicillin (10,000 U/mL)
Streptomycin (10,000 μg/mL)
Xenopus laevis tadpoles, reared in dechlorinated tap water

---

[1]Correspondence: a-oka@nms.ac.jp

Tadpoles at NF stage 57 (Nieuwkoop and Faber 1967), when the total length of the tadpole intestine reaches a maximum, are usually used, but those at earlier stages can also be used unless their intestine is too small to handle. However, do not use tadpoles at stage 58 or later, because their intestine occasionally undergoes remodeling without addition of T3.

Equipment

Autoclave

Centrifuge (benchtop)

Cleanroom

> *Clean bench or laminar flow hood can also be used instead of the cleanroom.*

Culture dishes (glass Petri dishes; 30-mm, 50-mm)

Dissecting instruments for small explants

Iris scissors, Wecker or Barraquer

Microspatula or sharp spoon for transferring the explants

Needle with metal handle for separating the tissues

Tweezers, Vigor #5 (antimagnetic)

Dry-heat sterilizer

Filter paper

Grids (stainless steel)

> *To make these reusable grids, cut the steel mesh in rectangles (1.5 cm × 2.5 cm) and bend it into a table-like shape where the top is 1.5 cm square and the sides are 0.5 cm high (see Fig. 1).*

Incubator preset to 26°C

Membrane filters (1.2-μm pores)

Microsyringes (1 mL)

Steel mesh

> *For making grids as above. Hole size is unimportant as long as the medium can pass through.*

Stereoscopic microscope

METHOD

To prevent contamination, before culture sterilize all glassware including culture dishes and dissecting instruments by autoclaving for 20 min at 120°C or dry-heating for 2 h at 120°C –160°C.

Preparation of Organ Culture Dishes

1. Place a stainless steel grid in a 30-mm culture dish, and place the 30-mm culture dish on top of a piece of filter paper laid in the bottom of a 50-mm culture dish with a lid (Fig. 1).

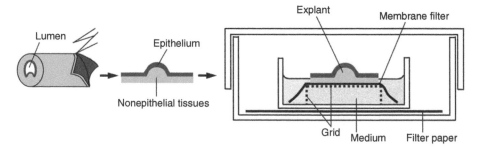

FIGURE 1. Preparation and organ culture of the *Xenopus* tadpole intestine.

Cite this protocol as *Cold Spring Harb Protoc*; doi:10.1101/pdb.prot097683

2. Cut a 1.2 µm-membrane filter to 1.5 cm × 2.5 cm and put it on top of the grid. Sterilize the whole setup including the dishes, the grid, and the filters by autoclaving for 20 min at 120°C.

> To keep the filter fixed on the grid, it is advisable to let the longer side of the filter hang over the grid to reach the bottom of the dish.

Preparation of Intestinal Explants

Steps 4–16 should be performed in the cleanroom, using scissors or tweezers under a stereoscopic microscope when required.

3. Immerse tadpoles in sterile DW containing 100 U/mL of penicillin and 100 µg/mL of streptomycin at room temperature overnight.

4. Cool a tadpole on ice and cut open its abdomen with scissors. Isolate the anterior part of the small intestine about 1.0–1.5 cm long from the opening of the bile duct, where a longitudinal fold is well developed and rich in connective tissue (Ishizuya-Oka and Shimozawa 1992). Place the intestine in a 30-mm dish containing coal-treated serum (CTS) medium at room temperature.

5. Cut the isolated anterior part of the small intestine into several pieces about 3 mm in length in the dish. If a significant amount of food remains in the intestinal lumen, remove it by gently streaming CTS medium into the lumen with a 1 mL-microsyringe to prevent contamination.

6. Slit open the intestine lengthwise in a dish containing CTS medium (Fig. 1) and wash it gently by stirring several times with a microspatula or sharp spoon.

> Opening of the tubular intestine helps to prevent contamination caused by apoptotic cells, which are sloughed off the epithelium and accumulate in the lumen of the tubular intestine after 3 d of T3-treatment (in hormone-containing medium).

7. Proceed to Step 8 for organ culture or Step 12 for tissue recombination.

Organ Culture of Intestinal Explants

8. Fill the 30-mm culture dish that was prepared and sterilized in Step 2 with about 3 mL of CTS medium (as a control) or the hormone-containing medium (to induce organ remodeling). Moisten the filter paper below the dish with sterile DW.

9. Put the intestinal explants prepared in Step 6 on the membrane filter in the culture dish (Fig. 1). Adjust the medium level with a micropipette so that it reaches the surface of the explants. Close the lid of the 50-mm dish.

10. Incubate the explants in an incubator at 26°C. Change the culture medium every other day. If the filter paper is dried, moisten it with sterile DW.

11. Transplant the explants to a new membrane filter set up in a new culture dish (prepared in Step 8) on the 5th day. Adjust the medium level to the surface of the explants as in Step 9.

> This transplantation is necessary to prevent contamination. Since the opened intestine becomes globular in shape with the epithelium outside and the other tissues inside, apoptotic cells sloughed off the epithelium are spread on the membrane filter after 3 d of T3-treatment (in hormone-containing medium).

Procedure for Tissue Recombination

12. Treat intestinal explants (prepared in Step 6) with 1000 U/mL of dispase in CTS medium in a dish for 30–45 min at 33°C–35°C. As the dispase degrades the basement membrane components, the epithelium gradually detaches from the connective tissue as a single-layer sheet. When the epithelium begins to be identified as a sheet-like structure, immediately separate it mechanically with a sharp needle or tweezers at room temperature.

> The best temperature and duration for dispase-treatment depend on the type and stage of the explant. As long as the epithelium is successfully separated, treatment with lower temperature and shorter duration is recommended for culture.

> See Troubleshooting.

13. Just after separation, transfer the epithelium and the nonepithelial tissues separately to new dishes containing CTS medium.

14. Put the nonepithelial tissues, whose epithelial side naturally faces the outside, on a membrane filter in a culture dish (prepared in Step 8), and then place the homologous (originating from the same organ) or heterologous (originating from a different [e.g., transgenic] organ) epithelium closely on top of the nonepithelial tissues in a single layer to recombine them.

15. Adjust the medium level to the surface of the recombinants and close the lid of the dish as in Step 9.

16. Incubate the recombinants in an incubator at 26°C as in Step 10 and transplant them as in Step 11.

> *When the epithelium successfully recombines with the nonepithelial tissue, the epithelium covers the entire outer surface of the nonepithelial tissues and firmly adheres to them. If the recombination fails, the epithelium easily falls off from the nonepithelial tissues during the first medium change.*

TROUBLESHOOTING

Problem (Step 12): After dispase-treatment, the epithelium is dissociated into small fragments or single cells and is hard to handle.

Solution: To separate the epithelium as a continuous tissue, reduce the temperature or duration of dispase-treatment as far as possible.

DISCUSSION

This organ culture protocol (Steps 1–11), which is a modified version of Trowell's liquid culture (Trowell 1959), can be used for other *Xenopus* organs. To do so, the best culture conditions must be determined for each organ to (1) prevent explants from contamination, (2) optimize the size of explants such that they are not too large to provide every cell with oxygen and nutrients, while retaining the tissue architecture, and (3) optimize concentrations of culture components to maintain explants or induce their remodeling. In *X. laevis* stage 57-tadpoles, the intestine remains as larval-type in CTS medium for at least 10 d (Ishizuya-Oka and Shimozawa 1991). Addition of 10 nM T3 to CTS medium induces larval epithelial apoptosis, but is not sufficient for adult epithelial development. In hormone-containing medium, which contains 5 µg/mL of insulin and 0.5 µg/mL of hydrocortisone in addition to T3, adult stem/progenitor cells actively proliferate and generate the fully differentiated adult epithelium by 7 d, following development of the connective tissue. Thereafter, the adult epithelium remains differentiated until at least the tenth day.

The issue recombination procedure (Steps 12–16) enables the analysis of epithelial–connective tissue interactions during *Xenopus* intestinal remodeling (Ishizuya-Oka and Shimozawa 1992). This protocol can also be performed on other organs once the best conditions to separate the tissues are determined. By taking advantage of transgenic tadpoles, this procedure should serve as a valuable tool to clarify the molecular mechanisms of the complex tissue interactions involved in organ remodeling in *Xenopus* (Hasebe et al. 2011).

Cite this protocol as *Cold Spring Harb Protoc*; doi:10.1101/pdb.prot097683

RECIPES

Charcoal-Treated Serum (CTS)

Inactivate fetal bovine serum (FBS) by incubating for 30 min at 56°C. To remove endogenous thyroid hormones in FBS, add 6.5 g of activated charcoal (powder, neutral) to FBS and stir gently overnight at 4°C with a stir bar (Yoshizato et al. 1980). Centrifuge FBS at 10,000 rpm for 1 h at 4°C in a benchtop centrifuge twice, filter sterilize the supernatant through a 0.22-μm filter, and store at −20°C.

CTS Medium

Reagent	Final concentration
Leibovitz's L-15 medium (GIBCO 11415-064)	60%
Charcoal-treated serum (CTS) <R>	10%
Penicillin (10,000 U/mL)	100 U/mL
Streptomycin (10,000 μg/mL)	100 μg/mL

Just before use, combine reagents to the desired volume in sterilized distilled water, and filter sterilize through a 0.22-μm filter.

Hormone-Containing Medium

Reagent	Final concentration
Leibovitz's L-15 medium	60%
Charcoal-treated serum (CTS) <R>	10%
3,3′,5-triiodo-L-thyronine (T3) (Sigma-Aldrich T6397; 404 μg/mL in 0.01 M NaOH)	10 nM
Insulin bovine (Sigma-Aldrich I664; 5 mg/mL in 0.005 N HCl)	5 μg/mL
Hydrocortisone (Sigma-Aldrich H0888; 0.5 mg/mL in distilled water)	0.5 μg/mL
Penicillin (10,000 U/mL)	100 U/mL
Streptomycin (10,000 μg/mL)	100 μg/mL

Just before use, combine reagents to the desired volume in sterilized distilled water, and filter sterilize through a 0.22-μm filter.

REFERENCES

Hasebe T, Buchholz DR, Shi Y-B, Ishizuya-Oka A. 2011. Epithelial-connective tissue interactions induced by thyroid hormone receptor are essential for adult stem cell development in the *Xenopus laevis* intestine. *Stem Cells* 29: 154–161.

Ishizuya-Oka A, Shimozawa A. 1991. Induction of metamorphosis by thyroid hormone in anuran small intestine cultured organotypically *in vitro*. *In Vitro Cell Dev Biol* 27A: 853–857.

Ishizuya-Oka A, Shimozawa A. 1992. Connective tissue is involved in adult epithelial development of the small intestine during anuran metamorphosis *in vitro*. *Roux's Arch Dev Biol* 201: 322–329.

Nieuwkoop PD, Faber J. 1967. *Normal table of* Xenopus laevis *(Daudin)*. North Holland Publishing, Amsterdam.

Trowell OA. 1959. The culture of mature organs in a synthetic medium. *Exp Cell Res* 16: 118–148.

Yoshizato K, Kikuyama S, Shioya N. 1980. Stimulation of glucose utilization and lactate production in cultured human fibroblasts by thyroid hormone. *Biochim Biophys Acta* 627: 23–29.

Bulk Electroporation-Mediated Gene Transfer into *Xenopus* Tadpole Brain

Cristina Sáenz de Miera,[1] Ethan Parr,[2] and Robert J. Denver[1,3]

[1]*Department of Molecular, Cellular and Developmental Biology, University of Michigan, Ann Arbor, Michigan 48109;* [2]*Undergraduate Research Opportunity Program, University of Michigan, Ann Arbor, Michigan 48109*

In vivo gene transfer is a powerful tool for investigating protein function and gene regulation in living organisms. Delivery of plasmid DNA to the brain of *Xenopus* tadpoles by bulk electroporation-mediated (EM) gene transfer can be used to study the effects of ectopic gene expression on development, physiology, and behavior. It can also be used to mark cells for lineage tracing, investigate the in vivo function of gene regulatory elements when linked to a reporter gene, and introduce mutations into the genome of transfected cells, among other applications. Bilateral EM gene transfer allows for transfection of both sides of the brain, whereas unilateral EM gene transfer enables analysis of the effects of forced gene expression on one side of the brain, with the other side serving as the control.

MATERIALS

It is essential that you consult the appropriate Material Safety Data Sheets and your institution's Environmental Health and Safety Office for proper handling of equipment and hazardous materials used in this protocol.

RECIPES: Please see the end of this protocol for recipes indicated by <R>. Additional recipes can be found online at http://cshprotocols.cshlp.org/site/recipes.

Reagents

Benzocaine (10% in ethanol)

To prepare the benzocaine stock solution, dissolve benzocaine at 10 g/100 mL in 100% ethanol.

Ethanol (75%)

Mineral oil (Fisher Scientific O121-4)

Plasmid DNA for transfection, including a plasmid that expresses a fluorescent protein (e.g., enhanced green fluorescent protein [EGFP]) to monitor transfection efficiency

Prepare plasmid DNA using an endotoxin-free plasmid purification kit (e.g., EndoFree Plasmid Maxi Kit [QIAgen 12362]). On the day of electroporation, dilute the purified plasmid DNA to a final concentration of 0.5–2 µg/µL in sterile 0.6% saline containing 0.01% Fast Green dye (Sigma-Aldrich F7258). Maintain the solution on wet ice until use (Step 11).

Powdered growth food (Sera Micron) or frog brittle (NASCO)

Steinberg's solution (100×) <R>

---

[3]Correspondence: rdenver@umich.edu

Supplemental material is available for this article at cshprotocols.cshlp.org.

Xenopus laevis tadpoles (premetamorphic stages Nieuwkoop and Faber (NF) 49–50 [Nieuwkoop and Faber 1956])

> *Healthy, actively growing tadpoles are essential to the success of this technique! Healthy X. laevis tadpoles reared at 23°C (water temperature) typically reach NF stage 50 from 10 to 12 d after hatching. In our experience, actively swimming tadpoles with a length of 1.8 cm or greater at NF stage 50 show higher rates of survival than smaller tadpoles that are less active.*
>
> *This method was developed for X. laevis; however, it also can be applied to X. tropicalis tadpoles.*

Equipment

Borosilicate glass capillary pipettes (3.5 inch length) (Drummond Scientific 3-000-203-G/X)

Dissection microscope and light source

Electrodes (3 × 5 mm gold-plated Genepaddles) (BTX Model 542)

Fluorescent stereoscope (for analyzing expression of EGFP in tadpole brain)

Hypodermic needle (30-gauge) and syringe (1-cc)

Insulated wires (Pomona Electronics)

Light stereoscope

Microinjector (Nanoject II Auto-Nanoliter Injector) (Drummond Scientific)

Micromanipulators (one for holding the microinjector and one for holding the electrodes)

Micropipette puller (e.g., Sutter Instrument P-97)

Microscope slides

Oscilloscope (e.g., Rigol DS1000E)

Paintbrush, small

Parafilm

Petri dishes, plastic (100-mm)

Rearing tanks

    10-L (for tadpole growth)

    2-L (one for anesthesia and one for recovery per treatment group)

> *Include an airstone in the recovery tanks. In our experience, oxygenation of the water provided by the airstone flow significantly improves tadpole survival after electroporation.*

SD-9 square pulse stimulator (Grass Instruments)

Stage or eyepiece micrometer

Tissue paper

Transfer pipettes, plastic (Fisher Scientific 13-711-7M)

METHOD

Tadpole Husbandry

1. After tadpoles hatch and become free swimming (NF stage 45), distribute animals into 10-L tanks at 2–3 tadpoles per liter.

2. Maintain tadpoles in dechlorinated tap water at 23°C on a 12L:12D photoperiod. Feed tadpoles twice per day with powdered growth food or frog brittle.

Electrical System Setup

3. Set the stimulator electric pulse to square wave, 25–30 V with a 15- to 25-msec pulse duration.

> *These settings are adapted for use with the specified electrodes and stimulator. Settings may differ for other instruments and should be determined empirically. Parameters to be monitored are (1) tadpole survival, and (2) transfection efficiency (monitored using EGFP).*

4. Connect the oscilloscope to the electrodes at the beginning of each experiment to monitor the square-wave pulses.

5. Connect the wires to the stimulator and the electrodes as shown in Figure 1A.

6. Clamp the electrodes in the micromanipulator and position over the base of the light stereoscope.

Injection Setup

7. Pull the capillary pipettes in advance of the experiment using a pipette puller.

   As a starting point, instructions for pulling pipettes for different purposes can be found in the Pipette Cookbook (Oesterle 2015), but the specific parameters must be determined empirically for each type of glass and the puller's filament. We use the following conditions with a Sutter Instrument P-97 pipette puller: Heat, Ramp; Pull, 100; Velocity, 100; Time, 150; Pressure, 500.

8. Open the pipettes by breaking the tips.

   i. Place a pulled pipette on a microscope slide, and using an upright light microscope, determine the point at which to break the pipette by measuring the tip diameter using an eyepiece or stage micrometer.

   ii. Break the tip of the pipette by sweeping it at the desired location with the back of a Pasteur pipette.

      We use pipettes with tips that are 30–40 µm in diameter and have a bevelled taper length of ∼8 mm.

9. Back-load the pipette with mineral oil using a 30-gauge needle and 1-cc syringe.

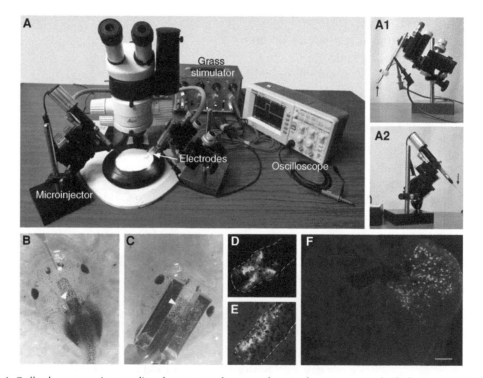

FIGURE 1. Bulk electroporation-mediated gene transfer procedure in the *Xenopus* tadpole brain. (*A*) Work station showing the location of equipment required for the procedure: Stimulator for square pulse generation, oscilloscope for recording the electric pulse delivery, microinjector, micromanipulator holding the electrodes (arrow), stereoscope and light source. (*A1*) Detail of the micromanipulator holding the electrodes (arrow). (*A2*) Detail of the micromanipulator holding the microinjector with the pulled pipette (arrow). (*B*) The location of the pipette over the head of the tadpole where it will be inserted into the brain (arrowhead) for injection of the plasmid solution. (*C*) Electrodes placed onto the head parallel to both sides of the tadpole brain. The shape of the brain ventricular system can be seen filled with the plasmid solution containing Fast Green dye (arrowhead). (*D*) EGFP expression in both sides of the tadpole brain 24 h after bilateral EM gene transfer. (*E*) EGFP expression in one side of the tadpole brain (*right* side) 24 h after unilateral EM gene transfer. (*F*) Cross-section showing EGFP fluorescence at the level of the optic tectum 5 d after unilateral EM gene transfer. Scale bar = 100 µm.

Cite this protocol as *Cold Spring Harb Protoc*; doi:10.1101/pdb.prot097691

10. Insert the pipette into the microinjector wire plunger and tighten the collet securely.

11. Fill the pipette with plasmid DNA solution (0.5–2 µg/µL with 0.01% Fast Green) as follows.

    i. Discharge approximately half of the mineral oil in the pipette onto a piece of tissue paper.

    ii. Pipette 5–15 µL of plasmid solution (according to the planned number of animals to be injected) onto a piece of Parafilm sterilized with 75% ethanol.

    iii. Fill the pipette with the plasmid solution, avoiding the introduction of air bubbles (Supplemental Movie S1).

    See Troubleshooting.

12. Position the microinjector over the base of the light stereoscope.

Electroporation

During the electroporation procedure, wipe off the electrodes frequently. Liquid and mucus from the tadpole accumulates on the electrode surface, and can interact with the transmission of electric pulses.

13. Anesthetize the tadpoles by placing them in a 2-L tank containing 1× Steinberg's Solution with 0.002%–0.005% benzocaine until unresponsive (3–5 min). Use a cut plastic transfer pipette to transfer the anesthetized tadpoles to the microscope base.

14. Place an anesthetized tadpole ventral-side-down on a stage made from a 100-mm Petri dish. Place a piece of folded paper tissue wetted with the anesthetic on top of the tadpole. Use a paintbrush to gently move the tadpole into the proper position.

15. Lower the microinjector to insert the pipette into the tadpole brain (Fig. 1B; Supplemental Movie S2). Insert the pipette to a depth that places it in the middle of the brain ventricle, but not so far in as to pierce through the ventral part of the brain.

    A pulled pipette, or a 30-gauge hypodermic needle attached to a 1-cc syringe, can be used to make a small puncture in the middle of the tadpole skull where the pipette will be inserted.

16. Inject 75–140 nL (adjusted based on tadpole size and desired plasmid amount to be delivered) at 23 nL per second.

    If the volume is too large, the solution will leak from the skull after injection, leading to poor transfection efficiency.

    The solution, observed by monitoring the Fast Green dye, diffuses in the cerebrospinal fluid, filling the entire ventricle (Fig. 1C; Supplemental Movie S2).

17. Wait 10 sec after the injection, and then slowly remove the pipette.

18. Lower the electrodes over the tadpole brain, placing the inner edges firmly on the tadpole head on either side of the brain.

    The distance between the electrodes will be determined by the size of the tadpole, but should not be <2 mm to prevent arcing (Fig. 1C; Supplemental Movie S3).

19. Manually deliver three to five single square wave pulses, 25–30 V, 15–25 msec pulse duration, using the stimulator (Supplemental Movie S3). To perform bulk electroporation on both sides of the brain, after delivering the first set of pulses, reverse the polarity on the stimulator and deliver a second set of pulses.

    During pulse delivery, the tadpole tail muscle and eyes should twitch. Small bubbles may appear along the edges of the electrodes. Large bubbles are a sign of electrocution.

20. Retract the electrodes and gently place the electroporated tadpole into a 2-L recovery tank containing dechlorinated tap water at 23°C.

    Tadpoles should start to move within 10 min. Some animals might take longer and can be checked for a heartbeat to verify that they are alive.

    See Troubleshooting.

In Vivo Screening for Fluorescent Expression

21. Twenty-four to 48 h after electroporation, check survival and monitor transfection efficiency by examining EGFP fluorescence using a fluorescent stereoscope (Fig. 1D,E).

    See Troubleshooting.

22. Select animals for further analysis based on the location and intensity of the fluorescence.

    See Hu et al. (2016) for a description, including figures, of how to select animals for further analysis. Note that strong autofluorescence of melanocytes in the head of X. tropicalis tadpoles may prevent analysis of fluorescence produced by the transfected plasmid. Accurate screening of the transfection may not be possible, and thus fluorescence may need to be determined a posteriori when the animals are killed and the brain dissected for analysis (Fig. 1F).

TROUBLESHOOTING

Problem (Step 11): The micropipette does not take up the plasmid solution.

Solution: The tip may be clogged or too thin. Discard the needle and prepare another one with a slightly wider tip and, perhaps, a greater bevel angle. In our experience, a good micropipette has a smooth tip that is 30–40 μm in diameter with a bevelled tip length of 30 μm. Check micropipette quality by verifying that it can be filled without resistance, and by first delivering several injections onto tissue paper before using it for animal injection.

Problem (Step 20): Tadpoles die following the electroporation procedure.

Solution: The animals may have been left too long in the anesthetic. Reduce the time of exposure to anesthesia. Alternatively, if bleeding is observed in the brain, it likely indicates that the voltage setting is too high, the number of pulses is too much, or the pulse duration is too long. Reduce the pulse voltage and/or duration.

Problem (Step 21): The survival rate the day after electroporation is low.

Solution: The tadpoles may have been in poor body condition. It is critical that tadpoles used for the procedure be healthy and actively growing. Check tadpole growth conditions; e.g., reduce tadpole density, increase frequency of feeding and check water quality. Sometimes poor growth may be related to crowding or other stressors present during embryonic development. The solution is to start with a new spawn.

Problem (Step 21): EGFP expression is low.

Solution: If low EGFP expression is observed in combination with a good survival rate, it may indicate that the electric pulse was not properly administered or was too low, or the plasmid concentration was too low. Revise the electrode positioning and increase the pulse voltage and/or duration. Plasmid quality can be checked by measuring plasmid concentration and purity (e.g., using a Nanodrop) and integrity (by agarose gel electrophoresis). Analysis of protein expression after transient transfection in tissue culture cells may also be used to verify plasmid quality and function. Repeat the electroporation with a new plasmid preparation.

DISCUSSION

Bulk electroporation of the *Xenopus* tadpole brain in vivo is a powerful approach for investigating the effects of ectopic gene expression on neural development and behavior. Native and mutant proteins (e.g., catalytically dead or dominant negative forms) can be expressed ectopically to modulate neural

Cite this protocol as *Cold Spring Harb Protoc*; doi:10.1101/pdb.prot097691

cell development and function, and morphological, physiological and behavioral outcomes can be analyzed. Unilateral EM gene transfer, in which the untransfected side of the brain serves as a control, can be paired with histochemical analysis to investigate the effects of ectopic gene expression on neural cell proliferation, migration, structure, and function. This technique is also useful for determining the functionality of gene regulatory elements in an in vivo context. For example, plasmids with native or mutant gene regulatory elements driving expression of reporter genes like firefly luciferase or EGFP can be transfected into tadpole brain, followed by biochemical or histochemical analysis of the reporter (Yao et al. 2007, 2008; Bagamasbad et al. 2015; Bender et al. 2017). The technique is also used to deliver antisense morpholino oligonucleotides (Bestman and Cline 2014), and may be used with RNAi constructs or CRISPR/Cas9 plasmids to knockdown or knockout genes in targeted cells.

RECIPE

Steinberg's Solution (100×)

0.5 g KCl
0.8 g Ca(NO$_3$)$_2$·4H$_2$O
2.1 g MgSO$_4$·7H$_2$O
34 g NaCl
119 g HEPES

Dissolve reagents in MilliQ-purified H$_2$O (18.2 Ω resistance, 23°C, pH 6.5–7.5). Bring the volume to 1 L and then adjust the pH to 7.4. Store at 4°C.

REFERENCES

Bagamasbad PD, Bonett RM, Sachs L, Buisine N, Raj S, Knoedler JR, Kyono Y, Ruan Y, Ruan X, Denver RJ. 2015. Deciphering the regulatory logic of an ancient, ultraconserved nuclear receptor enhancer module. *Mol Endocrinol* **29**: 856–872.

Bender MC, Sifuentes CJ, Denver RJ. 2017. Leptin induces mitosis and activates the canonical wnt/ß-catenin signaling pathway in neurogenic regions of *Xenopus* tadpole brain. *Front Endocrinol (Lausanne)* **8**: Art. 99.

Bestman JE, Cline HT. 2014. Morpholino studies in *Xenopus* brain development. In *Brain development. Methods in molecular biology (methods and protocols)* (ed. Sprecher SG), Vol. **1082**, pp. 155–171. Humana Press, Totowa, NJ.

Hu F, Knoedler JR, Denver RJ. 2016. A mechanism to enhance cellular responsivity to hormone action: Krüppel-like factor 9 promotes thyroid hormone receptor-β autoinduction during postembryonic brain development. *Endocrinology* **157**: 1683–1693.

Nieuwkoop PD, Faber J. 1956. *Normal table of* Xenopus laevis *(Daudin): A systematical and chronological survey of the development from the fertilized egg till the end of metamorphosis*. North-Holland Pub. Co., Amsterdam.

Oesterle A. 2015. *Pipette Cookbook 2015, P-97 and P-1000 micropipette pullers*. Rev. E Sutter Instruments Company, Novato, CA.

Yao M, Schulkin J, Denver RJ. 2008. Evolutionarily conserved glucocorticoid regulation of corticotropin-releasing factor expression. *Endocrinology* **149**: 2352–2360.

Yao M, Stenzel-Poore M, Denver RJ. 2007. Structural and functional conservation of vertebrate corticotropin-releasing factor genes: Evidence for a critical role for a conserved cyclic AMP response element. *Endocrinology* **148**: 2518–2531.

In Vivo Transfection of Naked DNA into *Xenopus* Tadpole Tail Muscle

Lindsey Marshall,[1] Fabrice Girardot,[1] Barbara A. Demeneix,[1] and Laurent Coen[1,2]

[1]*Evolution des Régulations Endocriniennes, Département RDDM, CNRS UMR 7221, MNHN, Sorbonne Université, Paris, France*

In vivo gene transfer systems are important to study foreign gene expression and promoter regulation in an organism, with the benefit of exploring this in an integrated environment. Direct injection of plasmids encoding exogenous promoters and genes into muscle has numerous advantages: the protocol is easy, efficient, and shows time-persistent plasmid expression in transfected muscular cells. After injecting naked-DNA plasmids into tadpole tail muscle, transgene expression is strong, reproducible, and correlates with the amount of DNA injected. Moreover, expression is stable as long as the tadpoles remain, or are maintained, in premetamorphic stages. By directly expressing genes and regulated promoters in *Xenopus* tadpole muscle in vivo, one can exploit the powerful experimental advantages of gene transfer systems in an intact, physiologically normal animal.

MATERIALS

It is essential that you consult the appropriate Material Safety Data Sheets and your institution's Environmental Health and Safety Office for proper handling of equipment and hazardous materials used in this protocol.

RECIPES: Please see the end of this protocol for recipes indicated by <R>. Additional recipes can be found online at http://cshprotocols.cshlp.org/site/recipes.

Reagents

DNA injection mixture

Dilute ultrapure stock solution(s) of plasmid(s) (obtained using a standard Maxiprep kit) to a final concentration of 0.1–0.5 g/L in 0.075 M NaCl with 1× Fast Green solution. Prepare the mixture on the day of injections, adapting the volume to the number of tadpoles to inject. Maintain at room temperature until use.

Fast Green stock solution (10×)

Dissolve 1 mg of Fast Green FCF dye (Sigma-Aldrich, F7252) in 1 mL of sterile, double-distilled water. Store at −20°C. Dilute to 1× before use.

MS222 solution (0.1%) <R>

Also known as tricaine methanesulfonate or ethyl 3-aminobenzoate methanesulfonate, MS222 is the main anesthetic used for aquatic models. MS222 can cause skin, eye, and even respiratory irritation. Use personal protective equipment as required.

Water, dechlorinated, 21°C

[2]Correspondence: laurent.coen@mnhn.fr

Water, sterile, double-distilled

Xenopus laevis tadpoles, premetamorphic stages NF55–NF57 (Nieuwkoop and Faber 1994)

> *Injections can be performed on younger tadpoles, but because of their smaller size and thinner tail muscles the success of the experiment is less assured. Likewise, older animals can be used, but the experiment must be designed such that it can be completed before the peak of metamorphosis when tail regression and muscle apoptosis occur.*

Equipment

Acupuncture needle, 0.25-mm diameter

Electrode puller (e.g., Narishige Group, Model Pb-7 or equivalent)

Gauze

Injection holder set for 1.0-mm microcapillary tubing (Narishige Group, Model IM-H1)

> *The set consists of an HI-7 injection holder, CT-1 PTFE tubing and Cl-1 tube connector.*

Microcapillary tubing, glass, 1.0 mm (Drummond Scientific, Cat. #1-000-0300)

Microdissection forceps

Microinjector (Narishige Group, Model IM-6)

Micromanipulator (Narishige Group, Model M-152)

> *This is used to precisely position the injection holder on three axes.*

P2 pipette (Gilson or equivalent)

Paraffin mold

Pipette tips, sterile, 10-μL (to fit a P2 pipette)

Steel sheet, galvanized, 35 cm × 45 cm × ≥ 2 mm

Stereomicroscope and light source

Syringe, 50-mL

Tanks

> *One tank is used for MS222; three for dechlorinated water (pre- and post-anesthesia, and post-injection).*

Teflon tubing, 1-mm i.d.

METHOD

Injection System Setup

1. Assemble the injection system (Fig. 1A):

    i. Attach the micromanipulator and the microinjector to the steel sheet.

    ii. Support the injection holder set with the micromanipulator.

    iii. Connect the microinjector to the injection holder set.

2. Using a syringe, fill the injection system with sterile double-distilled water through the inlet at the top of the microinjector (Fig. 1A).

3. Bleed any air bubbles from the injection system.

4. Use the electrode puller to prepare needles from glass microcapillary tubes.

    > *Settings for the Narishige model Pb-7 correspond to: No.1 Heater = 60; No. 2 Heater = 70.*

5. Use microdissection forceps to bevel the elongated needle extremity such that length of the elongated end is ~2–2.5 mm and the outer diameter of the needle tip is ~10–15 μm.

    > *This can be verified using a graduated stage micrometer measurement tool.*

6. Insert the needle in the injection holder containing a small piece of Teflon tube that will be tightened around the needle with the outer screw.

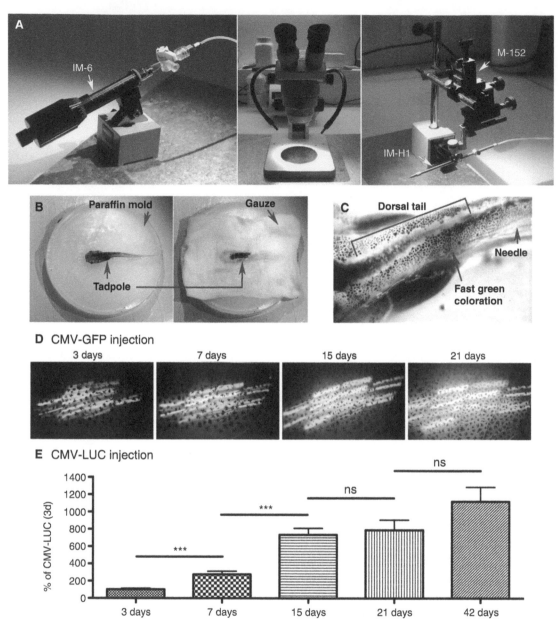

FIGURE 1. Workstation setup, tadpole positioning and injection, and potential analysis. Key steps in the protocol. (*A*) The workstation includes a microinjector (IM-6; *left*), microscope and light (*center*) and injection holder set (IM-H1) supported by a micromanipulator (M-152; *right*). (*B*) Tadpole positioned on a paraffin mold (*left*) and covered with a moistened gauze that has an opening exposing the tail (*right*). (*C*) Fast Green makes the mixture injected into the muscle visible. (*D*) Increasing GFP fluorescence observed on a live animal injected with a mixture containing a cytomegalovirus-green fluorescent protein (CMV-GFP) plasmid (200 ng/µL) at various times post-injection. (*E*) Quantification of luciferase (LUC) activity in extracts from muscles injected with a mixture containing a CMV-LUC plasmid (200 ng/µL, 10 animals per time point), showing a maximum activity detected after 15 d post-injection. (*D* and *E* adapted from Vivien et al. [2012].)

Tadpole Preparation

7. Anesthetize a NF55–NF57 stage tadpole by immersing it in freshly prepared MS222 solution.

   The tadpole should stop moving in ~15–30 sec.

8. Transfer the anesthetized tadpole to a tank of clean dechlorinated water to rinse off the MS222.

9. Gently place the tadpole ventral-side down on the paraffin mold. Cover with a piece of moist gauze into which an opening has been cut that allows access to the tail muscle (Fig. 1B).

Filling the Needle

10. Adjust the needle in a slightly oblique position with the tip visible under the microscope.

11. Aspirate 1 µL of DNA injection mixture into a 10-µL pipette tip. Gently remove the tip from the pipette. Using one's thumb, press the large end of the tip to force the liquid back toward the end.

12. While viewing under the microscope, approach the tip with the needle's end until it is in the liquid inside the pipette tip. Steadily holding that position, gradually move the injector dial counter-clockwise to load 1 µL of the injection mixture into the needle (see online Movie 1 at cshprotocols.cshlp.org).

    Control the flow carefully. Reserve an air space between the injection system water and the solution to prevent mixing.

Tadpole Injection (see Movie 2 at cshprotocols.cshlp.org)

13. Position the prepared tadpole under the microscope such that both the tadpole tail muscle and the filled needle are visible.

14. Select a suitable visual cue (e.g., a melanocyte) located 5–6 myomeres from the head and positioned at the top of the selected myomere. Pierce the skin at the selected point using the acupuncture needle.

    Wipe the needle clean with 70% ethanol between each injected animal.

15. Rotate the mold slightly. Position the micromanipulator to guide the needle into the tadpole muscle through the pierced hole.

16. With the needle's end now inside the myomere, rotate the mold back to its original position such that the tail muscle parallels the needle.

17. Drive the needle forward through three myomeres. Withdraw slightly.

    Resistance occurs each time the needle passes through a myomere, allowing them to be counted.

18. Slowly inject 1 µL of mixture by rotating the injector dial clockwise. Wait a few seconds.

    The mixture often diffuses into two to three myomeres.

    See Troubleshooting.

19. Slowly withdraw the needle from the tadpole, controlling the pressure to prevent injection of air bubbles.

    The mixture injected into the myomeres can be observed by the Fast Green coloration for a few minutes (Fig. 1C).

    See Troubleshooting.

20. Transfer the injected tadpole to a new tank of clean dechlorinated water and allow it to recover.

    It should start to swim within ~5 min.

    See Troubleshooting.

21. Maintain the tadpoles in dechlorinated water. Do not feed the tadpoles for 24 h following the injection, then feed every 2 d until the end of the experiment.

22. Depending on the reporter plasmid used, injected tadpoles can be monitored differently. For instance, fluorescent markers (e.g., GFP) can be monitored on living tadpoles (Fig. 1D), or luciferase activity can be quantified in extracted muscle tissue after tadpole euthanasia (Fig. 1E).

TROUBLESHOOTING

Problem (Step 18): The needle appears to be clogged.
Solution: Debris or muscle tissue can obstruct the needle tip. Slightly withdraw the needle and try again.

Problem (Steps 18 and 19): The liquid is released too fast in the muscle. Air bubbles can be seen in the tissue.

Solution: Air injected into the tadpole can lead to impaired muscle integrity and transfection efficiency. Repeat the injection with a new tadpole.

Problem (Step 19): No Fast Green coloration is seen in muscle immediately after injection.

Solution: No staining in muscle indicates that the mixture was not injected properly, possibly by reaching too far in the general cavity. Repeat the injection with a new tadpole.

Problem (Step 20): Tadpole does not recover following injection.

Solution: In general we do not observe mortality following this procedure. However, possibilities include animal left too long in MS222 or out the water. Repeat the injection with a new tadpole.

DISCUSSION

Among nonviral techniques, direct injection of plasmids encoding exogenous promoters and genes into muscle was first established in mice (Wolff et al. 1990). This method was adapted and applied to *Xenopus laevis* in our laboratory (de Luze et al. 1993). In vivo somatic gene transfer in *Xenopus* tadpole muscle is particularly relevant for studying biological processes in which the integrated context is especially important. For instance, this technique has been used to study physiological regulations by thyroid hormone during amphibian metamorphosis (Sachs et al. 1996, 1998; Nakajima and Yaoita 2003). Testing for endocrine-disrupting effects of increasing numbers of substances present in the environment also benefits from this technique, as the in vivo context integrates all possible targets of the compounds and their metabolites resulting in higher specificity (Turque et al. 2005; Punzon et al. 2013). We recently used this technique to show the possibility to induce cell reprogramming in vivo with a high efficiency. We showed that *Xenopus* differentiated skeletal muscle fibers could be reprogrammed to pluripotency after transfecting the Yamanaka factors (mouse Oct4, Sox2 and Klf4). This was the first time that induced reprogramming was reported to be feasible in vivo and that mammalian factors were shown to be able to reprogram nonmammalian cells (Vivien et al. 2012); others have confirmed these observations (Abad et al. 2013; Rosselló et al. 2013). This safe and inexpensive technique has been adapted to fish (Hansen et al. 1991) including zebrafish (L. Coen, unpublished data), *Xenopus tropicalis* (Rowe et al. 2002), and used in various species to develop immunization protocols (Donnelly et al. 1995; Lewis and Babiuk 1999; Pachuk et al. 2000).

RECIPE

MS222 Solution (0.1%)

MS222 (Sigma-Aldrich, A5040)	1 g
Sodium bicarbonate (Sigma-Aldrich, S5761)	1 g
Water, dechlorinated	1 L

Protect from light. Store for up to 2 wk at room temperature.

REFERENCES

Abad M, Mosteiro L, Pantoja C, Cañamero M, Rayon T, Ors I, Graña O, Megías D, Domínguez O, Martínez D, et al.. 2013. Reprogramming *in vivo* produces teratomas and iPS cells with totipotency features. *Nature* 502: 340–345.

de Luze A, Sachs L, Demeneix B. 1993. Thyroid hormone-dependent transcriptional regulation of exogenous genes transferred into *Xenopus* tadpole muscle *in vivo. Proc Natl Acad Sci* 90: 7322–7326.

Donnelly JJ, Ulmer JB, Liu MA. 1995. Protective efficacy of intramuscular immunization with naked DNA. *Ann N Y Acad Sci* 772: 40–46.

Hansen E, Fernandes K, Goldspink G, Butterworth P, Umeda PK, Chang KC. 1991. Strong expression of foreign genes following direct injection into fish muscle. *FEBS Lett* 290: 73–76.

Lewis PJ, Babiuk LA. 1999. DNA vaccines: A review. *Adv Virus Res* 54: 129–188.

Cite this protocol as *Cold Spring Harb Protoc*; doi:10.1101/pdb.prot099366

Nakajima K, Yaoita Y. 2003. Dual mechanisms governing muscle cell death in tadpole tail during amphibian metamorphosis. *Dev Dyn* **227**: 246–255.

Nieuwkoop PD, Faber J (eds.). 1994. *Normal table of* Xenopus laevis *(Daudin): A systematical & chronological survey of the development from the fertilized egg till the end of metamorphosis.* Garland Publishing, New York.

Pachuk CJ, McCallus DE, Weiner DB, Satishchandran C. 2000. DNA vaccines – Challenges in delivery. *Curr Opin Molec Ther* **2**: 188–198.

Punzon I, Latapie V, Le Mével S, Hagneau A, Jolivet P, Palmier K, Fini JB, Demeneix BA. 2013. Towards a humanized PPARγ reporter system for in vivo screening of obesogens. *Mol Cell Endocrinol* **374**: 1–9.

Rosselló RA, Chen C-C, Dai R, Howard JT, Hochgeschwender U, Jarvis ED. 2013. Mammalian genes induce partially reprogrammed pluripotent stem cells in non-mammalian vertebrate and invertebrate species. *Elife* **2**: e00036.

Rowe I, Coen L, Le Blay K, Le Mével S, Demeneix BA. 2002. Autonomous regulation of muscle fibre fate during metamorphosis in *Xenopus tropicalis*. *Dev Dyn* **224**: 381–390.

Sachs L, de Luze A, Lebrun JJ, Kelly PA, Demeneix BA. 1996. Use of heterologous DNA-based gene transfer to follow physiological, T3-dependent regulation of myosin heavy chain genes in *Xenopus* tadpoles. *Endocrinology* **137**: 2191–2194.

Sachs LM, de Luze A, Demeneix BA. 1998. Studying amphibian metamorphosis by *in vivo* gene transfer. *Ann NY Acad Sci* **839**: 152–156.

Turque N, Palmier K, Le Mével S, Alliot C, Demeneix BA. 2005. A rapid, physiologic protocol for testing transcriptional effects of thyroid-disrupting agents in premetamorphic *Xenopus* tadpoles. *Environ Health Perspect* **113**: 1588–1593.

Vivien C, Scerbo P, Girardot F, Le Blay K, Demeneix BA, Coen L. 2012. Non-viral expression of mouse Oct4, Sox2, and Klf4 transcription factors efficiently reprograms tadpole muscle fibers *in vivo*. *J Biol Chem* **287**: 7427–7435.

Wolff JA, Malone RW, Williams P, Chong W, Acsadi G, Jani A, Felgner PL. 1990. Direct gene transfer into mouse muscle in vivo. *Science* **247**: 1465–1468.

Cell Proliferation Analysis during *Xenopus* Metamorphosis: Using 5-Ethynyl-2′-Deoxyuridine (EdU) to Stain Proliferating Intestinal Cells

Morihiro Okada and Yun-Bo Shi[1]

Section on Molecular Morphogenesis, Eunice Kennedy Shriver National Institute of Child Health and Human Development (NICHD), National Institutes of Health (NIH), Bethesda, Maryland 20892

Proper cell proliferation is important for organ homeostasis and normal tissue development. Aberrations in cell proliferation, however, can give rise to degenerative diseases and cancer. Therefore, accurate and simple methods to evaluate cell proliferation are important and necessary to understand the pathways regulating cell proliferation and mechanisms underlying normal development and pathogenesis. The thymidine analog 5-ethynyl-2′-deoxyuridine (EdU), which is incorporated into DNA during active DNA synthesis (e.g., during S phase of the cell cycle), allows easy visualization of proliferating cells. Incorporated EdU can be detected without harsh chemical or enzymatic treatments and is fully compatible with a number of other staining methods, such as immunohistochemistry and in situ hybridization. This protocol describes how to detect proliferating cells using EdU staining in the intestines of *Xenopus* tadpoles (stages 54–66). Although this method was developed for studying intestinal metamorphosis, it should be applicable to other tissues/organs and other developmental stages as well.

MATERIALS

It is essential that you consult the appropriate Material Safety Data Sheets and your institution's Environmental Health and Safety Office for proper handling of equipment and hazardous materials used in this protocol.

RECIPES: Please see the end of this protocol for recipes indicated by <R>. Additional recipes can be found online at http://cshprotocols.cshlp.org/site/recipes.

Reagents

Click-iT Plus EdU Alexa Fluor 594 Imaging Kit (Thermo Fisher Scientific)
Ethanol (100%, 95%, 90%, 80%, and 70%)
Hoechst 33342 (10 mg/mL)
MEMFA fixative <R>
Paraffin wax
Phosphate-buffered saline (PBS, pH 7.2–7.6; 1×, 0.7×) <R>
Phosphate-buffered saline with 0.05% Tween-20 (PBST)
Reagents for situ hybridization, immunostaining, or a TUNEL assay (optional; see Step 7)
Vectashield antifade mounting medium (Vector Laboratories)

---

[1]Correspondence: Shi@helix.nih.gov

Xenopus tropicalis tadpoles and/or frogs (Nieuwkoop and Faber 1965)
Xylene

Equipment

Conical tube (5-mL)
Fluorescence microscope
Hamilton gas-tight syringe (25-µL; gauge: 33; needle length: 12 mm; point style 4; Hamilton Bonaduz)
Microscope slides and coverslips
Microtome
Needle (30-gauge)
Scalpel
Scissors
Slide warmer
Syringe
Water bath

METHOD

Preparation of EdU Solution and EdU Injection

1. Prepare a 10 mg/mL solution of EdU by adding PBS to EdU powder and mixing well in a water bath (55°C–65°C) until the EdU is fully dissolved.

    Although the manufacturer recommends dissolving EdU powder in dimethylsulfoxide (DMSO), in our experience most of the tadpoles died after injection of EdU dissolved in DMSO. Thus, we strongly recommend dissolving EdU in PBS. After use, store remaining stock solution at ≤−20°C.

2. Place tadpoles in ice water for 5–10 sec until they stop moving. Place tadpoles on a paper towel. With a Hamilton syringe, intraperitoneally inject 1.25, 10, and 10 µL of 10 mg/mL EdU into the abdominal cavity of stage 54, 62, and 66 *Xenopus tropicalis*, respectively, as shown in Figure 1. Place tadpoles back into the water in which they were reared.

    The volumes of injection were determined based on the relative sizes of the animals at different developmental stages.

Preparation of Intestinal Sections

3. Thirty to 60 min after EdU injection, anesthetize tadpoles by placing them on ice for 5–10 min. Euthanize tadpoles by decapitation with a scalpel or scissors. Isolate the intestine and flush the contents with 0.7× PBS using a 30-gauge needle and syringe.

    The flushing helps to reduce auto-fluorescence from the debris, which can make it difficult to visualize the EdU signal.

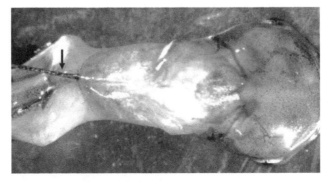

FIGURE 1. Photographic representation of intraperitoneal injections. The arrow points to the needle (gauge size 33) injecting into a stage 60 tadpole.

4. Fix the intestine in MEMFA fixative in a 5 mL tube for 2 h at room temperature or overnight at 4°C with shaking. To process the tissue for paraffin embedding, pass the tissue sequentially through graded ethanol solutions for dehydration: 70% ethanol, 80% ethanol, 90% ethanol, 95% ethanol, 100% ethanol, with a 30 min incubation for each ethanol concentration at room temperature. Soak the tissue with xylene three times for 30 min each at room temperature. Melt the paraffin by heating in a heat-proof vessel to 55°C–60°C. Soak the tissue in the paraffin three times for 1 h each. Embed the tissue into paraffin blocks.

5. Cut 5–7 μm thick intestinal sections with a microtome and mount sections on to slides. Dry overnight using a slide warmer (40°C–45°C).

6. Immerse the slides in xylene three times for 5 min each at room temperature, then sequentially in 100% ethanol, 90% ethanol, 80% ethanol, 70% ethanol, and PBST for 5 min each at room temperature.

7. If double labeling is desired, perform in situ hybridization, immunostaining, or a TUNEL assay as described (Okada et al. 2015).

   If double labeling will not be performed, proceed to Step 8.

   See Troubleshooting.

EdU Staining

8. Prepare Click-iT Plus reaction cocktail as per the manufacturer's instructions.

9. Incubate intestinal sections with 100–200 μL Click-iT Plus reaction cocktail per slide for 30 min at room temperature in the dark.

10. Remove the Click-iT Plus reaction cocktail and wash once with PBST for 5 min at room temperature.

11. Dilute the Hoechst 33342 solution 1:2000 in PBS to obtain a 1× Hoechst 33342 solution. Add 100–200 μL of 1× Hoechst 33342 solution per slide and incubate the sample for 30 min at room temperature in dark.

12. Remove the 1× Hoechst 33342 solution and wash once with PBST for 5 min.

13. Mount sections with coverslips using Vectashield antifade mounting medium.

14. View the slides immediately using a fluorescence microscope or store in the dark.

    See Troubleshooting.

TROUBLESHOOTING

Problem (Step 7): When carrying out double labeling with EdU and in situ hybridization, immuno-staining, or a TUNEL assay, the signal is weak or absent.
Solution: Consider the following.

- Antigen retrieval often interferes with EdU staining. Avoid the antigen retrieval by using a Decloaking Chamber (a pressure cooker which has been designed for heat-induced epitope retrieval) for immunostaining.

- Process sections for in situ hybridization, immunostaining, or TUNEL assays first, and then perform EdU staining.

- Ensure that the samples are protected from light during incubations.

Problem (Step 14): EdU signal is too low to detect.
Solution: Consider the following.

- Ensure that the ingredients in the Click-iT Plus reaction cocktail are mixed in the order listed in the manufacturer's instructions.

- Increase the labeling time. Sacrifice tadpoles 6–12 h after injection for tissues with poor proliferative capacity (see Step 3 above).

- Increase the amount of EdU for injection. Test a range of EdU concentrations to determine the optimal concentration (see Step 2 above).

DISCUSSION

Proper spatiotemporal regulation of cell proliferation is critical for the development of multicellular organisms. The measurement of cell proliferation is fundamental to the assessment of the processes underlying development and disease. Historically, measurement of the incorporation of radioactive nucleosides (e.g., ^{3}H-thymidine) was used to assess active DNA synthesis in the S phase of the cell cycle. Additionally, application of the nucleoside analog 5-bromo-2′-deoxyuridine (BrdU) later allowed antibody-based detection of proliferating cells (Gratzner et al. 1976). However, the denaturation step in the BrdU protocol can disrupt DNA integrity, thus affecting nuclear DNA counterstaining, and/or destroy cell morphology and antigen recognition sites. More recently, the thymidine analog EdU was developed and found to be a superior alternative to BrdU (Salic and Mitchison 2008). Incorporated EdU can be detected chemically without DNA denaturation and is fully compatible with a number of other staining methods.

EdU staining allows easy visualization of proliferating cells (Fig. 2). One of the major advantages of EdU staining is that it avoids harsh chemical or enzymatic treatments. Elimination of such treatments results in the preservation of cell morphology and antigen recognition sites, making the approach superior for double-labeling with in situ hybridization, immunohistochemistry, TUNEL assay, or other staining protocols including dye staining (e.g., Methyl green-pyronin Y staining) (Okada et al. 2015). Such double-labeling allows a direct view of the spatiotemporal localization of proliferating cells with other markers such as those for specific cell types or apoptotic cells and should find applications in both basic research and clinical development such as disease diagnosis.

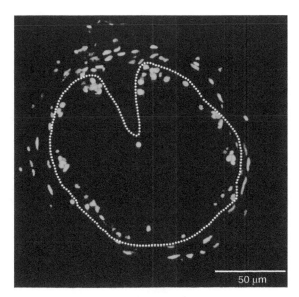

50 μm

FIGURE 2. EdU staining of *Xenopus* intestinal section. A cross-section of the intestine from premetamorphic stage 54 tadpole treated with 10 nM thyroid hormone for 6 d was used for EdU staining. The dotted lines depict the epithelium-mesenchyme boundary. See Okada et al. (2015) for more details.

RECIPES

MEM Salts (10×)

Reagent	Quantity (for 500 mL)	Final concentration
MOPS (pH 7.4)	104.65 g	1 M
EGTA	3.804 g	20 mM
MgSO$_4$	0.602 g	10 mM
H$_2$O	to 500 mL	

This solution can be stored for several months at room temperature, protected from light.

MEMFA Fixative

MEM salts (10×) <R>	0.4 mL
Formaldehyde (37%)	0.4 mL
H$_2$O	3.2 mL

Prepare fresh before use.

Phosphate-Buffered Saline (PBS)

Reagent	Amount to add (for 1× solution)	Final concentration (1×)	Amount to add (for 10× stock)	Final concentration (10×)
NaCl	8 g	137 mM	80 g	1.37 M
KCl	0.2 g	2.7 mM	2 g	27 mM
Na$_2$HPO$_4$	1.44 g	10 mM	14.4 g	100 mM
KH$_2$PO$_4$	0.24 g	1.8 mM	2.4 g	18 mM
If necessary, PBS may be supplemented with the following:				
CaCl$_2$·2H$_2$O	0.133 g	1 mM	1.33 g	10 mM
MgCl$_2$·6H$_2$O	0.10 g	0.5 mM	1.0 g	5 mM

PBS can be made as a 1× solution or as a 10× stock. To prepare 1 L of either 1× or 10× PBS, dissolve the reagents listed above in 800 mL of H$_2$O. Adjust the pH to 7.4 (or 7.2, if required) with HCl, and then add H$_2$O to 1 L. Dispense the solution into aliquots and sterilize them by autoclaving for 20 min at 15 psi (1.05 kg/cm^2) on liquid cycle or by filter sterilization. Store PBS at room temperature.

ACKNOWLEDGMENTS

Work in our laboratory is supported by the Intramural Research Program of National Institute of Child Health and Human Development, National Institutes of Health. M.O. was supported in part by a Japan Society for the Promotion of Science Research Fellowship for Japanese Biomedical and Behavioral Researchers at the National Institutes of Health.

REFERENCES

Gratzner HG, Pollack A, Ingram DJ, Leif RC. 1976. Deoxyribonucleic acid replication in single cells and chromosomes by immunologic techniques. *J Histochem Cytochem* 24: 34–39.

Okada M, Wen L, Miller TC, Su D, Shi YB. 2015. Molecular and cytological analyses reveal distinct transformations of intestinal epithelial cells during *Xenopus* metamorphosis. *Cell Biosci* 5: 74.

Salic A, Mitchison TJ. 2008. A chemical method for fast and sensitive detection of DNA synthesis in vivo. *Proc Natl Acad Sci* 105: 2415–2420.

Nieuwkoop PD, Faber J. 1965. *Normal table of* Xenopus laevis. 1st edn. North Holland Publishing, Amsterdam.

Chromatin Immunoprecipitation for Chromatin Interaction Analysis Using Paired-End-Tag (ChIA-PET) Sequencing in Tadpole Tissues

Nicolas Buisine,[1] Xiaoan Ruan,[2] Yijun Ruan,[2,3] and Laurent M. Sachs[1,4]

[1]Function and Mechanism of Action of Thyroid Hormone Receptor Group, UMR 7221 CNRS and Muséum National d'Histoire Naturelle, Sorbonne Universités, Paris 75005, France; [2]The Jackson Laboratory of Genomic Medicine, Farmington, Connecticut 06030; [3]The Department of Genetics and Developmental Biology, University of Connecticut, Farmington, Connecticut 06030

Proper gene expression involves communication between the regulatory elements and promoters of genes. Because regulatory elements can be located over a large range of genomic distances (from as close as a few hundred bp to as much as several Mb away), contact and communication between regulators and the core transcriptional machinery at promoters are mediated through DNA looping. Today, chromosome conformation capture (3C)-based methods efficiently probe chromosome folding in the nucleus and thus provide a molecular description of physical proximity between enhancer(s) and their target promoter(s). One such method, chromatin interaction analysis using paired-end-tag (ChIA-PET) sequencing, is a leading high-throughput method for detection of genome wide chromatin interactions. Briefly, the method involves cross-linkage of chromatin (-DNA) fibers in cells in situ, fragmentation of the fixed chromatin-DNA complexes by sonication, followed by enrichment of the chromatin complexes with a dedicated antibody through the process of immunoprecipitation (IP). Next, application of the ChIA-PET protocol followed by deep sequencing and mapping of reads to the reference genome reveals both binding sites and remote chromatin interactions mediated by the protein factors of interest. The method detailed here focuses on ChIP sample preparation and can be completed in ~5 d. The ChIA-PET method is detailed in an associated protocol. Because not all chromatin immunoprecipitation protocols are suitable for ChIA-PET, it is important to strictly follow this procedure before performing the ChIA-PET protocol.

MATERIALS

It is essential that you consult the appropriate Material Safety Data Sheets and your institution's Environmental Health and Safety Office for proper handling of equipment and hazardous materials used in this protocol.

RECIPES: Please see the end of this protocol for recipes indicated by <R>. Additional recipes can be found online at http://cshprotocols.cshlp.org/site/recipes.

Reagents

Prepare all stock solutions in nuclease-free water unless noted.

Absolute ethanol (Carlo Erba 308609)
Antibody to the protein of interest

---

[4]Correspondence: sachs@mnhn.fr

AquaPhenol (water-saturated, pH 8 for DNA extraction; MP Biomedicals 11AQUAPH01)
ChIP Assay Kit (Millipore 17-295)
Chromatin immunoprecipitation (ChIP) elution buffer <R>
Chloroform (99+% stabilized with amylenes; Sigma-Aldrich C2432)
Chromatin Immunoprecipitation Assay Kit (Millipore 17-295)
cOmplete, EDTA-free Proteinase Inhibitor Cocktail (Roche 11873580001)
DNA Ladder (100 bp; New England BioLabs N0467)
Formaldehyde (37%; Fluka 93482)
Glycerol (Sigma-Aldrich G5516)
Glycine (1.25 M; Sigma-Aldrich G7126)
Glycogen from *Mytilus edulis*, Blue mussel (Sigma-Aldrich G1767)
Isopropyl alcohol (ACS for analysis; Carlo Erba 415154)
Nuclease-free water (Ambion AM9937)
Nuclei extraction buffer <R>
Phenylmethylsulfonyl fluoride (PMSF; 0.1 M in isopropanol; Fluka 93482)
Power SYBR Green PCR Master Mix (Applied Biosystems 4367659)
Proteinase K solution (2 g/L; Sigma-Aldrich P2308)
Quant-iT dsDNA HS Assay Kit (Invitrogen Q32851)
Real-time quantitative polymerase chain reaction reagents (described in Bilesimo et al. 2011)
Sodium acetate (3 M, pH 5.5; Ambion AM9740)
Sodium chloride (5 M; Sigma-Aldrich S3014)
SDS lysis buffer <R>

> This buffer is also found in the ChIP Assay Kit (Millipore 17-295).

SYBR Green (10,000×; Invitrogen S33102)
Tadpoles tissues, stage NF-54 (see Step 1)
Tris-acetate-EDTA (TAE) (10× buffer, pH 8.0; Invitrogen 15558-026)
UltraPure Agarose-1000 (Invitrogen 16550100)

Equipment

Aerosol pipette tips with high recovery, filter and wide orifice (VWR Europe 732-054 North America 46620-642)
Bioruptor sonicator (Diagenode) associated with a thermoregulator (NESLAB RTE7)
Cell strainer (100 µm; BD 352360)
Centrifuge (benchtop; Sigma-Aldrich model 6K15 with rotor 11650 and swinging buckets 17677 for 50-mL tubes or 17659 for 15-mL tubes)
Centrifuge tubes (DNA LoBind, 1.5-mL, PCR clean; Eppendorf 0030 108.051)
Chemical hood
Dissecting microscope (Olympus, model SZ51)
Dounce homogenizer (Kontes Kimble, Wheaton)
Conical tubes (15 and 50 mL; BD 352096 & 352070)
Forceps (fine)
Freezer (−80°C)
Gel electrophoresis apparatus for 5 × 5 cm gel
Gel imager (Bio-Rad Chemidoc Touch Imaging System)
Incubator for cross-linking (AquaLytic M83649)
MicroAmp Optical 96-Well Reaction Plate with Barcode (Applied Biosystems 4306737)
MicroAmp Optical Adhesive Film (Applied Biosystems 4311971)
Phase-lock gel light 1.5-mL tubes (5prime 2302800)
qPCR machine (Applied Biotechnology QStudio6 4485692)
Qubit assay tubes (Invitrogen Q32856)

Cite this protocol as *Cold Spring Harb Protoc*; doi:10.1101/pdb.prot097725

Qubit fluorometer (Invitrogen Q32857)
Refrigerated microcentrifuge (Eppendorf 5418R)
Rotary agitator (Stuart SB3, speed 13 rpm)
Transfer pipettes with large opening (Dutscher Samco 043222S)
Transfer pipettes with small opening (Dutscher Samco 043233S)
Tube holder with metallic reflecting bar (Diagenode B01200013)
UV-visible spectrophotometer
Vortex mixer
Water bath

METHOD

Reaching high immunoprecipitation enrichment is crucial for ChIA-PET and may require optimization. It is therefore essential to know a priori a least one DNA locus target that will serve as a gold standard for optimization. Additionally, efficiency of chromatin fragmentation and immunoprecipitation are strongly dependent on cell type and antibodies. Therefore, we strongly advise running a pilot experiment by subjecting a sample to the whole chromatin preparation and immunoprecipitation procedure, to set optimized parameters for antibody dilution, cross-linking–decross-linking duration time and temperature, and amount of tissue. Also, this pilot experiment will help estimate the number of parallel samples to process to reach sufficient chromatin quantities for ChIA-PET. Insufficient amounts of chromatin lead to a poor ChIA-PET library and dramatically increase the risk of failure. High amounts of material are required for each ChIA-PET sample. To illustrate, four chromatin samples produced from eight tail fin skins each (Steps 1 to 28), provide enough raw material for 20 chromatin immunoprecipitations (Step 29 to 46) which, once pooled, should reach the required amount (>50–100 ng) of precipitated material to proceed to Protocol 7: Chromatin Interaction Analysis Using Paired-End-Tag (ChIA-PET) Sequencing in Tadpole Tissues (Buisine et al. 2018). When comparing physiological states and/or treatments, one has to consider control and treated/stage specific samples as well as biological replicates. When no biological replicates are available, experimental validation with additional methods is strongly encouraged.

This protocol is a long and sensitive procedure. Particular attention should be given to quality controls at key steps to decide whether or not to proceed to the next step.

Chromatin Isolation from Tissues

1. Place tissues in an ice-cold microcentrifuge tube.

   The following procedure is suited for eight tail fin skins, 12 hindlimb buds, eight livers, or eight brains of stage NF-54 tadpoles. To produce enough material for 20 chromatin immunoprecipitations, four chromatin samples should be prepared using, for example, 32 tail fin skins.

2. Add 1 mL of cold (4°C) nuclei extraction buffer and transfer tissues and buffer to an ice-cold Dounce homogenizer on ice.

   Nuclei Extraction Buffer without PMSF must be used within 5 h. Add PMSF in Nuclei Extraction Buffer just before use because the stability of PMSF is low.

3. Break the tissues by applying 10 strokes with the large clearance piston (type A) of the Dounce homogenizer.

   This dissociates the tissues. The homogenizer should be kept on ice.

4. Transfer the homogenate to a microcentrifuge tube using a transfer pipette with a large opening.

5. Add 28 μL of 37% formaldehyde.

   The timing and concentration of formaldehyde treatment should be adapted to the cell samples and protein factors under investigation. Overfixed chromatin is difficult to manipulate and increases background signal, while too short of a treatment leads to a poor capture of chromatin interactions. To determine the best fixation parameters, we advise performing several pilot experiments (going through the whole procedure) with different fixation parameters. The parameters indicated here have been shown to work well with a variety of different tissues (skin, limb bud, liver, and brain) and may be a good starting point, but they are by no means universal. Other tissues may require additional optimization. This step is crucial and emphasis on establishing appropriate fixation parameters should not be underestimated.

6. Incubate for 20 min on a rotary platform (13 rpm) in an incubator set at 20°C.

7. Add 114 μL of 1.25 M glycine and incubate as in Step 6 for 5 min.

8. Centrifuge at 2000g for 2 min at 4°C.

9. Discard the supernatant using a transfer pipette with narrow opening and place the tube on ice.

10. Resuspend the pellet in 1 mL of cold (4°C) nuclei extraction buffer using a P1000 pipette.

11. Vortex the tube for 5 sec and transfer the contents to a clean cold Dounce homogenizer using a transfer pipette with a large opening.

12. Break the cells by applying 10 strokes with the tight piston (type B) of the Dounce homogenizer.

    This leads to nucleus isolation. Keep the homogenizer on ice.

13. Transfer the homogenate with a narrow-opening transfer pipette to a 100 μm cell strainer positioned on the top of a 50 mL conical tube.

14. Centrifuge at 2000g for 2 min at 4°C in a centrifuge with swinging buckets.

15. Discard the supernatant using a P1000 pipette. Proceed immediately to Step 16.

Chromatin Fragmentation

16. Resuspend the nuclear pellet in 500 μL of SDS lysis buffer using a P1000 pipette.

    This step must be performed at room temperature to avoid SDS precipitation.

17. Transfer the suspension to a 15-mL conical tube.

    A hard plastic 15 mL tube (polystyrene or polyethylene-crystal clear tube) is preferred to increase sonication efficiency.

18. Close the tube with a cap attached to the metallic probe supplied by Diagenode.

    Check that the probe does not contact the tube wall and is positioned at the center. The bar reflects the ultrasound and thus optimizes sonication efficiency (Diagenode-patented system).

19. Place the tube in the Bioruptor sonicator water bath. Shear the chromatin with the following power setting: high for 30 min total time (30 sec "on"/30 sec "off").

    A refrigerated circulation bath set at 1°C is recommended to maintain constant temperature (<10°C) and to avoid overheating that would reverse cross-linking and degrade chromatin. Note that sonication prevents SDS precipitation at this low temperature. Vigorous sonication is recommended because it has the advantage of "shaking off" nonspecific interactions compared to other modes of chromatin fragmentation. Sonication strength should be adapted to fit samples and proteins under investigation. We recommend following the guidelines provided by the sonicator manufacturer to find optimum sonication parameters. Be aware that sonication quality is assessed by two parameters: the size of the fragmented DNA (tested from Steps 22 to 28), and immunoprecipitation capacity.

20. Transfer the sonicated chromatin to a new microcentrifuge tube and centrifuge at 16,000g for 10 min at 4°C.

    SDS precipitation has never been observed at this step.

21. Transfer the supernatant into a fresh microcentrifuge tube and place the tube on ice. Take three 10-μL aliquots, and store the rest of the sample at −80°C until use (Step 29).

Quality Control of Chromatin Fragmentation

22. Measure the absorbance (A_{260}) of two 10 μL aliquots from Step 21. Deduce the estimated DNA concentration from the average of the two measures.

    Use SDS lysis buffer as a blank. A concentration between 100 and 300 ng/μL is expected. Supernatants (Step 21) with a lower concentration should be discarded because the material is too diluted and thus not suitable for immunoprecipitation. Supernatants (Step 21) with a higher concentration must be diluted because of the increased viscosity of the solution, which will affect DNA quantification. Adjust the concentration of each chromatin sample (supernatant) as necessary (see Step 29).

Cite this protocol as *Cold Spring Harb Protoc*; doi:10.1101/pdb.prot097725

23. Dilute the third 10-µL aliquot of fragmented chromatin from Step 21 to a final DNA concentration between 100 and 300 ng/µL. Take 10 µL of the diluted sample, and add 0.8 µL of 0.5 M NaCl and 2 µL of 2 g/L proteinase K.

24. Incubate the sample from Step 23 overnight at 65°C.

    This procedure degrades proteins and reverses cross-link.

25. Take 5 µL of the digested sheared chromatin from Step 24 and add 5 µL of 50% glycerol solution.

    Do not use loading buffer with dye because after agarose gel electrophoresis, bromophenol blue has an apparent migration similar to that of chromatin fragments and may prevent their visualization.

26. Load 10 µL of the mix (digested sheared chromatin and glycerol from Step 25) and a DNA ladder on a 1× TAE, 1% agarose gel with SYBR Green.

27. Run the gel for 30 min at 50 V in 1× TAE buffer.

    Migration at low speed improves the resolution of DNA separation.

28. Visualize DNA with the gel imager.

    Good quality DNA usually appears as a smear with fragment size ranging from ~200 to 600 bp. Sometimes, migration can be affected by SDS precipitation. If this occurs, the remaining decross-linked DNA (Step 24) can be used to repeat the gel electrophoresis. If repeating the electrophoresis, first heat the tube at 37°C for a few seconds to completely dissolve the SDS.

    See Troubleshooting.

Chromatin Immunoprecipitation (ChIP)

29. Thaw the sample(s) from Step 21. Adjust the concentration of the sonicated chromatin to 100 ng/µL with SDS lysis buffer in a 15-mL conical tube. Mix, then aliquot the total amount of fragmented chromatin required for immunoprecipitation with the antibody of interest, for a negative control, and for evaluation of preimmunoprecipitation starting material to a new 15 mL conical tube (see below for details).

    As a starting point, the amount of sheared chromatin to use for each precipitation is 10 µg. Given that the ChIA-PET protocol requires 50 to 100 ng of purified immunoprecipitated DNA, it is necessary to adjust the number of samples to process accordingly (see note under the Method heading). We typically perform 20 immunoprecipitations. In addition, 30 µg of sheared DNA are required to control the quality of the immunoprecipitation. This includes an aliquot of the precipitated material, a negative control (no-antibody immunoprecipitation or an immunoprecipitation with preimmune serum) and an input/starting material aliquot (10 µg for each). Overall, in our hands, the typical amount of starting chromatin should be around 240 µg. The sonicated chromatin should be kept at room temperature to avoid SDS precipitation. Use a 50 mL tube if the amount of chromatin required is higher than 120 µg.

30. Dilute the sonicated chromatin 10-fold with ChIP dilution buffer with protease inhibitors. Vortex to mix.

    The ChIP Dilution Buffer is included in the ChIP Assay Kit from Millipore (0.01% SDS, 1.1% Triton X-100, 1.2 mM EDTA, 16.7 mM Tris-HCl pH 8.1, 167 mM NaCl). One tablet of cOmplete, EDTA-free Proteinase Inhibitor Cocktail and 70 µL of PMSF (0.1 M) are added to every 7 mL of ChIP dilution buffer before use.

31. Add 80 µL of Salmon Sperm DNA/Protein A agarose mix (from the ChIP Assay Kit) per 10 µg of sonicated DNA. Incubate for 30 min at 4°C with agitation on a rotary platform.

    This preclear step will reduce nonspecific background. Salmon Sperm DNA/Protein A agarose mix is included in the ChIP Assay Kit from Millipore (600 µg sonicated salmon sperm DNA, 1.5 mg BSA, 4.5 mg recombinant protein A), provided as a 50% gel suspension in 10 mM Tris-HCl pH 8, 1 mM EDTA, 0.05% sodium azide, in a final volume of 3 mL. Mix vigorously before use and pipette with large orifice pipette tips. Can be purchased independently (Millipore 16-157C). Appropriate protein A or G agarose should be used, according to the source of antibody.

32. Centrifuge at 2000g for 3 min at 4°C in a centrifuge with swinging buckets and collect the supernatant in 1 mL fractions in 23 microcentrifuge tubes (20 for samples and three for no antibody controls).

33. Add antibody at the desired final concentration to each 1 mL supernatant fraction, one of which will be used to estimate immunoprecipitation efficiency. Add preimmune serum (or no antibody) to one 1 mL supernatant fraction as a negative control. Reserve 100 µL of the supernatant fraction (~10 µg) to estimate the input/starting material. Store this sample at −20°C in a microcentrifuge tube.

> *The titer, batch, nature (mono/polyclonal) and source (mouse, sheep...) of antibodies strongly affect immunoprecipitation efficiency. This will have to be optimized in a pilot experiment.*

34. Incubate tubes overnight at 4°C with rotation.

35. Add 60 µL of Salmon Sperm DNA/Protein A agarose to each tube and incubate for 1 h at 4°C with rotation.

> *Use large orifice pipette tips.*

36. Centrifuge tubes at 2000*g* for 2 min at 4°C and discard the supernatants.

> *Carefully pipette the supernatant without disturbing the agarose pellet. The supernatant contains the unbound nonspecific sheared chromatin.*

37. Wash the pellets with 1 mL of Low Salt Immune Complex Wash Buffer and incubate for 5 min at 4°C with rotation.

> *The Low Salt Immune Complex Wash Buffer is included in the ChIP Assay Kit from Millipore (0.1% SDS, 1% Triton X-100, 2 mM EDTA, 20 mM Tris-HCl pH 8.1, 150 mM NaCl).*

38. Centrifuge tubes at 2000*g* for 2 min at 4°C and discard the supernatants.

> *Carefully pipette the supernatant without disturbing the agarose pellet.*

39. Wash the pellets with 1 mL of High Salt Immune Complex Wash Buffer and incubate for 5 min at 4°C with rotation.

> *The High Salt Immune Complex Wash Buffer is included in the ChIP Assay Kit from Millipore (0.1% SDS, 1% Triton X-100, 2 mM EDTA, 20 mM Tris-HCl pH 8.1, 500 mM NaCl).*

40. Centrifuge the tubes at 2000*g* for 2 min at 4°C and discard the supernatants.

> *Carefully pipette the supernatant without disturbing the agarose pellet.*

41. Wash the pellets with 1 mL of LiCl Immune Complex Wash Buffer and incubate for 5 min at 4°C with rotation.

> *The LiCl Immune Complex Wash Buffer is included in the ChIP Assay Kit from Millipore (0.25 M LiCl, 1% NP40, 1% deoxycholate, 1 mM EDTA, 10 mM Tris-HCl pH 8.1).*

42. Centrifuge the tubes at 2000*g* for 2 min at 4°C and discard the supernatant.

> *Carefully pipette the supernatant without disturbing the agarose pellet.*

43. Wash the pellets with 1 mL of TE Buffer and incubate for 5 min at 4°C with rotation.

> *The TE Buffer is included in the ChIP Assay Kit from Millipore (1 mM EDTA, 10 mM Tris-HCl pH 8).*

44. Centrifuge the tubes at 2000*g* for 2 min at 4°C and discard the supernatant.

> *Carefully pipette the supernatant without disturbing the agarose pellet.*

45. Repeat Step 43.

46. Combine all the tubes where antibody has been added into a single 15 mL conical tube or in a single 50 mL conical tube if the number of tubes to pool is higher than 12. Centrifuge at 2000*g* for 2 min at 4°C and discard the supernatant. Add 60 µL of TE for each pooled sample. Mix gently and take an aliquot for quality control (120 µL). This amount (120 µL) corresponds to the equivalent of one tube prepared at Step 33 and will be used to measure the quantity of ChIP-DNA fragments and the enrichment of the precipitation. Meanwhile, the remaining ChIP-DNA solution should be kept on ice before proceeding to the ChIA-PET protocol.

> *ChIP material is sensitive to protease and nuclease. Care should be taken to avoid contamination. ChIP-DNA fragments can be stored at 4°C for up to 2 wk before ChIA-PET library construction. Avoid long term storage to limit elution of DNA fragments from the beads and chromatin degradation.*

Cite this protocol as *Cold Spring Harb Protoc*; doi:10.1101/pdb.prot097725

47. Two tubes will be used to estimate immunoprecipitation efficiency: the 120 µL aliquot from Step 46, and the tube to which preimmune serum or no antibody was added at Step 33. Centrifuge these tubes at 2000*g* for 2 min at 4°C.

48. Discard the supernatants. Proceed to Step 49 to quantify immunoprecipitation efficiency.

    Carefully pipette the supernatant without disturbing the agarose pellet.

Quantification of ChIP-DNA Fragments and Measure of the Fold Enrichment

49. Add 100 µL of ChIP elution buffer to the two tubes from Step 48 (ChIP with and without antibody).

50. Incubate 30 min with rotation at room temperature.

51. Thaw the input sample from Step 33 on ice.

52. To ChIP (with and without antibody) and input samples (Step 33), add 4 µL 5 M NaCl, 2 µL 0.5 M EDTA, 4 µL 1 M Tris-HCl pH 6.1 and 2 µL 2 g/L Proteinase K. Mix vigorously.

    NaCl, EDTA, and Tris-HCl solutions are included in the ChIP Assay Kit from Millipore.

53. Incubate overnight at 65°C.

54. Centrifuge tubes briefly (20 sec) at 13,000*g* at room temperature to recover liquid from evaporation.

55. Add 300 µL of nuclease-free H_2O to each tube and centrifuge at 2000*g* for 2 min at room temperature.

56. Centrifuge 3 Phase lock gel tubes at 13,000*g* for 1 min at room temperature.

57. Transfer 400 µL from each tube of the eluted chromatin from Step 55 into a Phase Lock Gel tube, avoiding pipetting agarose.

58. Add 200 µL of chloroform and 200 µL of phenol and mix gently by inversion of the tube to create an emulsion.

    Do not vortex.

59. Centrifuge at 13,000*g* for 5 min at room temperature.

60. Transfer the aqueous phase from each tube to a new microcentrifuge tube.

    The solid interface prevents pipetting the organic phase. Usually, around 440 µL can be collected.

61. Add 44 µL of 3 M sodium acetate pH 5.5, 2 µL of 20 mg/mL glycogen, and 1 volume of cold isopropyl alcohol (kept at −20°C). Vortex to mix.

62. Incubate for 15 min at −80°C.

    Check that the liquid is not frozen.

63. Centrifuge at 16,000*g* for 30 min at 4°C.

    Ensure the presence of a small white pellet, which can detach from the tube wall. Be careful not to disturb the pellet in Steps 64–67.

64. Discard the supernatant.

65. Wash pellet twice with 1 mL of ice-cold 70% ethanol (kept at −20°C).

66. Centrifuge at 16,000*g* for 5 min at 4°C and remove supernatant after each wash.

67. Remove as much remaining ethanol as possible.

    To this end, centrifuge at 16,000g for 10 sec at room temperature and remove the liquid with a pipette.

68. Air dry the pellet by leaving the tube open for no more than 5 min at room temperature.

    Do not over dry the pellet, which makes it more difficult to dissolve in H_2O.

69. Resuspend the pellet in 22 µL of nuclease-free H_2O.

    Gently pipetting up and down can help. DNA can be stored at −20°C until the next step. Avoid freeze–thaw cycles.

70. Use two 2 µL aliquots to measure DNA concentration using a Qubit fluorometer with the Quant-iT dsDNA HS assay kit, as described by the manufacturer.

    Use 200 µL Qubit assay tubes.

71. Measure precipitation enrichment by real time quantitative polymerase chain reaction as previously described (Bilesimo et al. 2011). We use Prism 7300 system software (Applied Biosystems) to analyze the results in auto determination of cycle time (ct) mode. Polymerase chain reaction (PCR) efficiency is calculated for each primer set from the slope of a standard curve (with $R^2 >$ 0.99). The standard curve is a serial 10-fold dilution of each input in the range of utilization for the ChIP assays. The first dilution corresponds to 10% of input, the second corresponds to 1% of input, and the last one corresponds to 0.1% and was chosen as a conservative base line of detection because the primer efficiency with dilutions under 0.05% of input decreased and the variability increased. The values for the input and experimental samples are calculated from the standard curve. Results are expressed as percent of input calculated as followed: % of input = (AE) (Ct input − Ct ChIP) × Fd × 100%, with (AE) amplification efficiency of primer: 10–slope of standard curve (3.32), Ct input: mean of duplicates from dilution of input corresponding to 1%, Ct ChIP: mean of duplicates from ChIP sample and Fd: compensatory factor to balance the difference in amounts of ChIP and input DNA taken for qPCR.

    To measure the immunoprecipitation enrichment (ChIP signal normalized over input signal), compare the values between the samples with and without antibody (or preimmune serum), at a known DNA binding target of the protein of interest. In addition, values at a known DNA binding target should be highly contrasted compared to a negative control locus (no binding of the protein of interest). Primers must be well designed and tested.

    See Troubleshooting.

72. Proceed to Protocol 7: Chromatin Interaction Analysis Using Paired-End-Tag (ChIA-PET) Sequencing in Tadpole Tissues (Buisine et al. 2018) only if a total of 50 to 100 ng chromatin DNA is recovered. ChIA-PET also critically requires high immunoprecipitation enrichment. Successful experiments show an enrichment fold higher than 30 when comparing immunoprecipitation with and without antibody, and a binding contrast higher than 30-fold when comparing known binding locus versus no-binding locus (unpublished data). These threshold levels may be very specific to individual DNA-binding proteins.

TROUBLESHOOTING

Problem (Step 28): Incomplete fragmentation of chromatin by sonication and/or low amount of fragmented chromatin.

Solution: The expected size range of the fragmented chromatin DNA should be between 200 and 600 bp. In case of incomplete fragmentation (size range higher than 600 bp), avoid resonication and restart instead with a new batch of chromatin. The amount of sonicated material must be high enough for precipitation and library construction (see next problem).

Problem (Step 71): Low yield or low enrichment of chromatin immunoprecipitation products.

Solution: Yield of enriched chromatin-DNA on beads should be between 50 and 100 ng. Repeat immunoprecipitation and pool immunoprecipitation products until this yield is reached. Enrichment should be higher than 30-fold at a positive control DNA locus (a known DNA binding site for the protein of interest) after normalization using input DNA and negative control locus (at least 2000 bp away from the known binding site). If immunoprecipitation enrichment is too low, repeat immunoprecipitation with a new antibody (from a new batch or a new provider). Importantly, immunoprecipitation enrichment is batch dependent and should be evaluated for each antibody.

Cite this protocol as *Cold Spring Harb Protoc*; doi:10.1101/pdb.prot097725

DISCUSSION

In 2009, the Genome Institute of Singapore developed ChIA-PET to produce simultaneously, in a single experiment, a genome-wide binding profile, together with a map of the physical interactions between protein-bound DNA regions (Fullwood et al. 2009). The method was originally used for chromatin isolated from human cell lines (Fullwood et al. 2009; Handoko et al. 2011; Li et al. 2012; Zhang et al. 2013). In 2015, the method was used for the first time with chromatin isolated from *Xenopus tropicalis* tissues (Buisine et al. 2015). The objective was to better understand the molecular transcriptional mechanisms controlled by thyroid hormone and their receptors during metamorphosis. However, the ChIA-PET method is complex and thus requires high quality starting materials. Here, we describe a protocol for ChIP proven to provide appropriate products for ChIA-PET analysis. This protocol presents a powerful opportunity for *Xenopus* researchers to analyze gene regulation in the context of the 3D nucleus.

RECIPES

Chromatin Immunoprecipitation (ChIP) Elution Buffer

Freshly prepare a bicarbonate buffer by dissolving 84 mg of $NaHCO_3$ (Sigma-Aldrich S5761) in 9 mL of H_2O. Add 1 mL of 10% SDS (Promega V6553).

Nuclei Extraction Buffer

Reagent	Amount to add (for 10.5 mL)	Final concentration (1×)
Triton X-100 (Sigma-Aldrich T8787)	52.5 µL	0.5%
Tris-HCl (1 M, pH 7.5; ThermoFisher 15567-027)	105 µL	10 mM
$CaCl_2$ (1 M; Sigma-Aldrich C3306)	31.5 µL	3 mM
Sucrose (Sigma-Aldrich S0369)	882 mg	250 mM
DTT (0.1 M; Sigma-Aldrich D9779)	10.5 µL	0.1 mM
Complete Mini EDTA-free Proteinase Inhibitor Cocktail (Roche 11836170001)	1.5 tablets	1×
Phenylmethylsulfonyl fluoride (PMSF; 0.1 M in isopropanol; Fluka 93482)	105 µL	1 mM

Prepare the solution fresh before use. Add protease inhibitors (Complete Mini EDTA-free Proteinase Inhibitor Cocktail and PMSF) immediately before using the buffer.

SDS Lysis Buffer

Reagent	Amount to add (for 3.5 mL)	Final concentration (1×)
SDS (10%; Promega V6553)	350 µL	1%
Tris-HCl (1 M, pH 8.1)	175 µL	50 mM
EDTA (0.5 M, pH 8; Ambion AM9260G)	70 µL	10 mM
Complete Mini EDTA-free Proteinase Inhibitor Cocktail (Roche 11836170001)	0.5 tablet	1×
Phenylmethylsulfonyl fluoride (PMSF; 0.1 M in isopropanol; Fluka 93482)	35 µL	1 mM

Keep this buffer at room temperature to avoid SDS precipitation. Add protease inhibitors (Complete Mini EDTA-free Proteinase Inhibitor Cocktail and PMSF) immediately before using the buffer.

REFERENCES

Bilesimo P, Jolivet P, Alfama G, Le Mevel S, Havis E, Demeneix BA, Sachs LM. 2011. Specific histone lysine 4 methylation patterns define TR-binding capacity and differentiate direct T$_3$ responses. *Mol Endoc* **25:** 225–237.

Buisine N, Ruan X, Bilesimo P, Grimaldi A, Alfama G, Ariyaratne P, Mulawadi F, Chen J, Sung WK, Liu ET, et al. 2015. *Xenopus tropicalis* genome re-scaffolding and re-annotation reach the resolution for *in vivo* ChIA-PET analysis. *PLoS One* **10:** e0137526.

Buisine N, Ruan X, Ruan Y, Sachs LM. 2018. Chromatin interaction analysis using paired-end-tag (ChIA-PET) sequencing in tadpole tissues. *Cold Spring Harb Protoc* doi: 10.1101/pdb.prot104620.

Fullwood MJ, Liu MH, Pan YF, Liu J, Xu H, Mohamed YB, Orlov YL, Velkov S, Ho A, Mei PH, et al. 2009. An oestrogen-receptor-α-bound human chromatin interactome. *Nature* **462:** 58–64.

Handoko L, Xu H, Li G, Ngan CY, Chew E, Schnapp M, Lee CW, Ye C, Ping JL, Mulawadi F, et al. 2011. CTCF-mediated functional chromatin interactome in pluripotent cells. *Nat Genet* **43:** 630–638.

Li G, Ruan X, Auerbach RK, Sandhu KS, Zheng M, Wang P, Poh HM, Goh Y, Lim J, Zhang J, et al. 2012. Extensive promoter-centered chromatin interactions provide a topological basis for transcription regulation. *Cell* **148:** 84–98.

Zhang Y, Wong CH, Birnbaum RY, Li G, Favaro R, Ngan CY, Lim J, Tai E, Poh HM, Wong E, et al. 2013. Chromatin connectivity maps reveal dynamic promoter-enhancer long-range associations. *Nature* **504:** 306–310.

Cite this protocol as *Cold Spring Harb Protoc*; doi:10.1101/pdb.prot097725

Chromatin Interaction Analysis Using Paired-End-Tag (ChIA-PET) Sequencing in Tadpole Tissues

Nicolas Buisine,[1] Xiaoan Ruan,[2] Yijun Ruan,[2,3] and Laurent M. Sachs[1,4]

[1]*Function and Mechanism of Action of Thyroid Hormone Receptor Group, UMR 7221 CNRS and Muséum National d'Histoire Naturelle, Sorbonne Universités, Paris 75005, France;* [2]*The Jackson Laboratory of Genomic Medicine, Farmington, Connecticut 06030;* [3]*The Department of Genetics and Developmental Biology, University of Connecticut, Farmington, Connecticut 06030*

Proper gene expression involves communication between the regulatory elements and promoters of genes. Today, chromosome conformation capture (3C)-based methods efficiently probe chromosome folding in the nucleus and thus provide a molecular description of physical proximity through DNA looping between enhancer(s) and their target promoter(s). One such method, chromatin interaction analysis using paired-end-tag (ChIA-PET) sequencing is a powerful high-throughput method for detection of genome-wide chromatin interactions. Following enrichment of the chromatin complexes with a dedicated antibody, through a process of immunoprecipitation (IP), DNA fragments are end-joined with specifically designed DNA-linkers through proximity ligation. The DNA-linkers contain the binding site for the type II restriction enzyme *Mme*I, which cleaves 20 bp from each end of the ligated fragments, thus releasing a "paired end tag" (PET): [20 bp tag]-[linker]-[20 bp tag]. The PETs are then deep-sequenced and reads are mapped to the reference genome, revealing both binding sites, as well as remote chromatin interactions mediated by the protein factors of interest. The method detailed here focuses on ChIA-PET library construction and can be completed in 2 wk.

MATERIALS

It is essential that you consult the appropriate Material Safety Data Sheets and your institution's Environmental Health and Safety Office for proper handling of equipment and hazardous materials used in this protocol.

RECIPES: Please see the end of this protocol for recipes indicated by <R>. Additional recipes can be found online at http://cshprotocols.cshlp.org/site/recipes.

Reagents

Prepare all stock solutions in nuclease-free water unless noted.

Adaptor-A, top strand:

    5′-CCATCTCATCCCTGCGTGTCCCATCTGTTCCCTCCCTGTCTCAGNN-3′

    Order 250 nmole of each oligo listed. Oligos should be HPLC purified in desalted form.

Adaptor-A, bottom strand:

    5′-CTGAGACAGGGAGGGAACAGATGGGACACGCAGGGATGAGATGG-3′

Adaptor-B, top strand:

    5′-CTGAGACACGCAACAGGGGATAGGCAAGGCACACAGGGGATAGG-3′

[4]Correspondence: sachs@mnhn.fr

Adaptor-B, bottom strand:

    5′-CCTATCCCCTGTGTGCCTTGCCTATCCCCTGTTGCGTGTCTCAGNN-3′

Agilent DNA 1000 reagents (Agilent Technologies 5607-1505)

Binding and washing (B&W) buffer (2×) <R>

    Dilute to 1× as needed.

Buffer EB (QIAGEN 19086)

ChIA-PET elution buffer <R>

ChIA-PET wash buffer <R>

ChIP beads from Protocol 6: Chromatin Immunoprecipitation for Chromatin Interaction Analysis
    Using Paired-End-Tag (ChIA-PET) Sequencing in Tadpole Tissues (Buisine et al. 2018)

Circularization mix <R>

DNA ladder (25 bp; Invitrogen 10597-011)

Dynabeads M-280 streptavidin (10 mg/mL, 10 mL; Invitrogen 11206D)

E. coli DNA polymerase I (10 U/µL; New England Biolabs M0209L)

Ethanol

GlycoBlue (15 mg/mL; Ambion AM9516)

Half-Linker-A, **with one biotinylated nucleotide** (bold character), top strand
    5′-GGCCGCGATATCTTATCCAAC-3′

Half-Linker-A, bottom strand 5′-GTTGGATAAGATATCGC-3′

Half-Linker-A, nonbiotinylated, top strand 5′-GGCCGCGATATCTTATCCAAC-3′

Half-Linker-B, **with one biotinylated nucleotide**, top strand 5′-GGCCGCGATATACATTCCAAC-3′

Half-Linker-B, bottom strand 5′-GTTGGAATGTATATCGC-3′

Half-Linker-B, nonbiotinylated, top strand 5′-GGCCGCGATATACATTCCAAC-3′

Isopropanol

Loading dye (6×; Fermentas R0611)

MmeI (2 U/µL; New England Biolabs R0637L)

NEB buffer 4 (10×; New England Biolabs B7004S)

Novex 4%–20% TBE acrylamide gel (1.0 mm, 10 wells; Invitrogen EC6225BOX)

    Gels are fragile, and we recommend using a gel handler for their manipulation.

Novex 6% TBE acrylamide gel (1.0 mm, 5 wells; Invitrogen EC6264BOX)

Nuclease-free water (Ambion AM9937)

PCR master mix for ChIA-PET <R>

Phenol:chloroform:isoamyl alcohol (25:24:1, pH 7.9, 100 mL; Ambion AM9730)

Polymerase I master mix <R>

Polynucleotide kinase master mix <R>

Proteinase K solution (20 mg/mL; Sigma-Aldrich P2308)

S-adenosylmethionine (SAM; 32 mM; New England Biolabs B9003S)

Sodium acetate (3 M, pH 5.5; Ambion AM9740)

SYBR Green (10,000×; Invitrogen S33102)

T4 DNA ligase (30 U/µL; Fermentas EL0013)

T4 DNA ligase buffer (5× with PEG; Invitrogen 46300-018)

T4 DNA ligase buffer (10×; New England Biolabs B0202S)

T4 DNA polymerase (Promega M4215)

T4 DNA polymerase master mix <R>

T4 DNA polynucleotide kinase (10 U/µL; New England Biolabs M0201L)

TBE buffer (10×; Ambion AM9863)

TE Buffer (pH 8.0; Ambion AM9849)

TNE buffer for ChIA-PET <R>

Triton X-100 (Promega H5142)

Cite this protocol as *Cold Spring Harb Protoc*; doi:10.1101/pdb.prot104620

Equipment

Agilent DNA 1000 Kit (Agilent Technologies 5607-1504)

Bioanalyzer 2100 (Agilent)

Centrifuge tubes (DNA LoBind 1.5-mL PCR clean; Eppendorf 0030 108.051)

Chemical hood

Conical tubes (50 mL; BD 352070)

DarkReader Transilluminator (Clare Chemical Research DR-45M)

FEP centrifuge tubes (50 mL; Nalgene 3114-0050)

Freezer (−80°C)

Gel imager (Bio-Rad Chemidoc Touch Imaging System)

High-speed centrifuge (Sorvall RC 5C plus)

Incubator (Memmert INB500)

Intelli-Mixer rotator (Palico Biotech RM-2L; program used: F8, 30 rpm, U = 50, u = 60)

Magnetic Particle Collector (Invitrogen DynaMag-2)

MaXtract High Density tubes (25 × 50 mL; QIAGEN 129073)

NanoDrop spectrophotometer (Thermo Scientific)

Needle (21G; Becton Dickinson)

PAGE gel electrophoresis system (Invitrogen, model Novex Mini-Cell)

PCR tubes (0.2 mL; Axygen)

Refrigerated microcentrifuge (Eppendorf 5418R)

Screw-cap tubes (2 mL, presterilized; Axygen)

Single-edge blades (0.009)

Speed Vac (Tomy MV-100)

Spin-X Centrifuge Tube Filters (0.22 μm Pore CA Membrane, sterile; Costar 8160)

Tubes (0.6 mL; Axygen)

METHOD

This protocol requires biological material obtained from Protocol 6: Chromatin Immunoprecipitation for Chromatin Interaction Analysis Using Paired-End-Tag (ChIA-PET) Sequencing in Tadpole Tissues (Buisine et al. 2018). ChIP-DNA fragments should have been stored for no longer than 2 wk at 4°C to avoid elution of DNA fragments from the beads and chromatin degradation. Proceed to this protocol only if a total of 50 to 100 ng chromatin DNA is reached in the previous protocol.

The ChIA-PET analysis is a long and sensitive procedure (Zhang et al. 2012). Particular attention should be given to quality control steps to decide whether or not the procedure can proceed to the next step. A video tutorial highlights the most critical aspects of the protocol (Goh et al. 2012).

An overview of ChIA-PET analysis is illustrated in Figure 1.

Hybridizing Oligos to Make Double Stranded Half Linkers and Adaptors

1. Resuspend the following oligos in TNE buffer to make 100 μM solutions: Adaptor-A, top strand; Adaptor-A, bottom strand; Adaptor-B, top strand; Adaptor-B, bottom strand; Half-Linker-A, with one biotinylated nucleotide, top strand; Half-Linker-A, bottom strand; Half-Linker-B, with one biotinylated nucleotide, top strand; Half-Linker-B, bottom strand; Half-Linker-A, nonbiotinylated, top strand; Half-Linker-B, nonbiotinylated, top strand.

2. Vortex for 1 min and let tubes stand overnight at 4°C. Prepare five ratios of top oligos/bottom oligos (1:1, 1.5:1, 2:1, 1:1.5, 1:2) for the sets of oligos listed below for a total of 30 tubes, each with a volume of at least 12 μL (i.e., for a ratio of 1.5:1 mix together 7.5 μL of top oligonucleotide with 5 μL of bottom oligonucleotide):

   Adaptor-A top strand plus Adaptor-A bottom strand

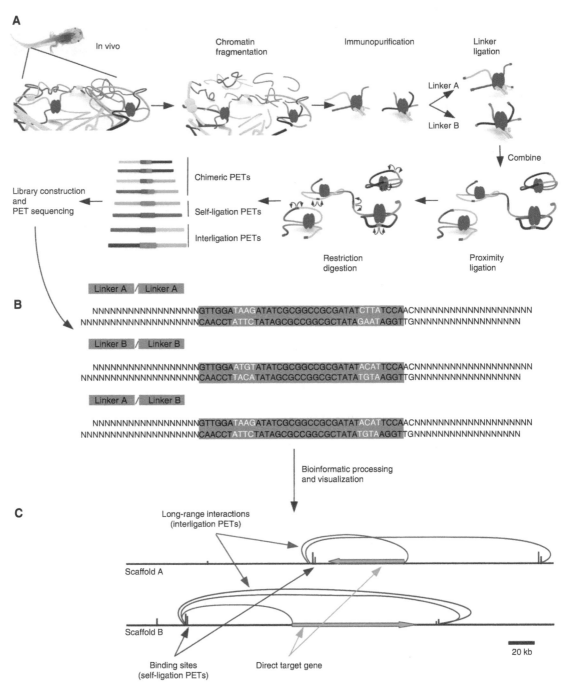

FIGURE 1. Overview of the ChIA-PET analysis. (*A*) Protein mediated chromatin loops in vivo are captured by ChIA-PET. After extraction, chromatin is fragmented by sonication and the protein of interest (together with bound DNA from a different locus (green and blue)) is captured with specific antibodies. Free DNA ends are then ligated to DNA linkers in two independent reactions (linker A (pink) and linker B (orange)). The two reactions are then combined and subject to "proximity ligation," where ligation between DNA fragments originating from different complexes are minimized in favor of independent ligation between DNA ends from the same complex. Restriction digest with the *Mme*l enzyme releases Paired End di-Tags (PETs) composed of ligated linkers together with 23 bp of flanking DNA. There are three kinds of PETs: self-ligation PETs capture the two ends of sonicated DNA fragments, interligation PETs capture the two ends of a DNA loop and chimeric PETs correspond to ligation products between complexes. The number of chimeric PETs is the key quality control parameter of the ChIA-PET protocol, because it estimates the level of cross-reaction products and indicates the quality of the "proximity ligation." Color gradient on DNA is related to genomic coordinates along chromosomes. (*B*) Linker PETs sequence readily distinguishes self-ligation and interligation from cross-ligation events. "N" corresponds to the flanking genomic DNA tags being sequenced. Highlighted residues correspond to the barcode used to differentiate Linker A from Linker B. Linkers colors are as in panel *A*. (*C*) Example of ChIA-PET output. Self-ligation PETs are used to derive a genome wide map of DNA binding, in a manner similar to ChIP-Seq. Interligation PETs describe physical interactions between binding sites resulting from chromatin loops in vivo. Combining both information identifies direct target genes regulated by transcription factors.

Cite this protocol as *Cold Spring Harb Protoc*; doi:10.1101/pdb.prot104620

Adaptor-B top strand plus Adaptor-B bottom strand
Half-Linker-A with one biotinylated nucleotide top strand plus Half-Linker-A bottom strand
Half-Linker-B with one biotinylated nucleotide top strand with Half-Linker-B bottom strand
Half-Linker-A nonbiotinylated top strand plus Half-Linker-A bottom strand
Half-Linker-B nonbiotinylated top strand plus Half-Linker-B bottom strand

3. Run on polymerase chain reaction (PCR) machine using the program:

i. Heat for 2 min at 95°C.

ii. Ramp from 95°C to 75°C at 0.1°C/sec, hold for 2 min at 75°C.

iii. Ramp from 75°C to 65°C at 0.1°C/sec, hold for 2 min at 65°C.

iv. Ramp from 65°C to 50°C at 0.1°C/sec, hold for 2 min at 50°C.

v. Ramp from 50°C to 37°C at 0.1°C/sec, hold for 2 min at 37°C.

vi. Ramp from 37°C to 20°C at 0.1°C/sec, hold for 2 min at 20°C.

vii. Ramp from 20°C to 4°C at 0.1°C/sec, hold at 4°C until ready to collect. Keep annealed oligos cold from this point on.

4. Measure the concentration of the annealed oligos using a NanoDrop spectrophotometer and dilute each tube of annealed oligos to 200 ng/µL.

5. Dilute 1 µL of each set of annealed oligos to 200 ng/10 µL. Run 200 ng of each single stranded oligo and 200 ng from each ratio of annealed oligos with 5 µL of 6× loading dye on the same 4%–20% TBE polyacrylamide gel. Load a 25 bp DNA ladder in one of the wells. Run the gel in 1× TBE at 180 V for 50 min.

6. Stain the gel with 8 µL of SYBR Green in 80 mL TBE 1× buffer for 7 min. Visualize the gel under the gel imager. The best ratio corresponds to the lane where no more top or bottom oligo are present.

End Blunting of ChIP-DNA Fragments

7. Centrifuge the ChIP beads from Step 46 of Protocol 6: Chromatin Immunoprecipitation for Chromatin Interaction Analysis Using Paired-End-Tag (ChIA-PET) Sequencing in Tadpole Tissues (Buisine et al. 2018) at 100g at 4°C for 1 min.

8. Discard the supernatant carefully without disturbing the bead pellet.

9. Resuspend the beads in 989.6 µL of T4 DNA polymerase master mix.

10. Add 10.4 µL of T4 DNA polymerase.

11. Mix and incubate in an incubator at 37°C for 40 min with rotation on the Intelli-Mixer.

12. Centrifuge tube at 100g for 1 min at 4°C.

13. Discard the end-blunting mix (supernatant) without disturbing the bead pellet.

14. Wash the beads three times with 1 mL of cold ChIA-PET wash buffer by centrifuging the tube at 100g for 1 min at 4°C and discarding the wash buffer carefully after each centrifugation. At the last resuspension in cold ChIA-PET wash buffer, split the beads into two microcentrifuge tubes as evenly as possible. The two tubes are named A and B.

15. Centrifuge both tubes at 100g for 1 min at 4°C and discard the supernatant.

Ligation of Biotinylated Half Linker A and B to Polished End

16. Prepare biotinylated Half-Linker-A and biotinylated Half-Linker B ligation mixes as follows: mix 1186.5 µL nuclease free H₂O with 300 µL 5× T4 DNA ligase buffer with PEG and 7.5 µL biotinylated annealed Half-Linker-A or -B (200 ng/µL) at the ratio determined to be ideal in Step 6.

17. To tube A from Step 15, add 996 µL of Half-Linker-A ligation mix and to tube B from Step 15, add 996 µL of Half-Linker-B ligation mix.

    Do not add ligase to the ligation mix. Ideally, in chromosome conformation capture procedures, all proximity ligation products are derived from DNA fragments bound together within a chromatin immunoprecipitation complex. To avoid chimeric ligation products between two chromatin immunoprecipitation complexes, proximity ligation is processed in large volume. However, to not confound data analysis if chimeric ligation occurs, linkers are barcoded. Adding Half-Linker-A and Half-Linker-B (4 bp differences) separately to the chromatin immunoprecipitated materials allows the production of three possible circularized products within the circularization mixture (Steps 35–37) with specific linker compositions (Half-Linker-A/Half-Linker-A, Half-Linker-B/Half-Linker-B and Half-Linker-A/Half-Linker-B). Among them, the hybrid Half-Linker-A/Half-Linker-B corresponds to the chimeric ligation product between two different immunoprecipitated chromatin complexes. The percentage of this hybrid is a quality control indicator to estimate nonspecific ligation.

18. Add 4 µL of T4 DNA ligase to tube A. Mix by inverting the tube. Wrap the tube in parafilm and place it on the Intelli-Mixer at room temperature.

19. Proceed identically with tube B (Step 18).

20. Incubate the two tubes overnight at 16°C (at least 16 h) with rotation on the Intelli-Mixer.

21. Centrifuge tubes A and B at 100*g* for 1 min at 4°C.

22. Wash the beads three times with 1 mL of cold ChIA-PET wash buffer by centrifuging at 100*g* for 1 min at 4°C and discarding the wash buffer carefully after each centrifugation. At the last resuspension, pool together the two tubes.

    For proper combination of the two tubes, remove the wash buffer from tube A. Mix the beads in tube B with wash buffer by pipetting up and down gently. Transfer the contents of tube B to tube A. Do not discard tube B. Centrifuge tube A at 100g for 1 min at 4°C and discard the supernatant. Add 1 mL of wash buffer to tube B to recover the remaining beads. Transfer the content of tube B to tube A. Discard tube B.

23. After combining the contents of the two tubes, centrifuge at 100*g* for 1 min at 4°C and discard the supernatant.

Phosphorylation of the "Linker-Added" DNA Fragments

24. Resuspend the beads in 980 µL of polynucleotide kinase master mix.

25. Add 20 µL of T4 DNA polynucleotide kinase.

26. Mix by inverting the tube and incubate for 50 min at 37°C with rotation on the Intelli-Mixer.

27. Centrifuge at 100*g* for 1 min at 4°C and discard the supernatant.

Elution of Chromatin-DNA Complex

28. Add 200 µL of ChIA-PET elution buffer and place the tube on the Intelli-Mixer at room temperature for 30 min.

    Do not place tube on ice because SDS crystallizes when cold.

29. Centrifuge at 100*g* for 1 min at 4°C and transfer the eluted chromatin DNA complex to a fresh microcentrifuge tube.

30. Wash the beads with 900 µL of QIAGEN Buffer EB and centrifuge at 100*g* for 1 min at 4°C.

31. Transfer the QIAGEN Buffer EB supernatant to the tube containing the first eluted chromatin DNA complex.

32. Transfer half of the collected eluate (550 µL) to a Spin-X column and centrifuge at 16,000*g* for 1 min at 4°C. Transfer the filtrate into a clean microcentrifuge tube and pass the remaining collected eluate through the same SpinX column as described above.

33. Combine the two filtrates into a 2 mL screw cap tube and add 90 µL of 20% Triton X-100 to sequester the SDS.

34. Incubate in a 37°C incubator for 1 h with rotation on the Intelli-Mixer.

Cite this protocol as *Cold Spring Harb Protoc*; doi:10.1101/pdb.prot104620

Circularization of Linker-Added DNA Fragment (Proximity Ligation)

35. Prepare 8776 µL of circularization mix on ice in a 50 mL conical tube and add 1190 µL of linker-added DNA complex from Step 34.

36. Mix by inverting the tube and add 33.33 µL of T4 DNA ligase.

37. Mix by inverting followed by a short centrifugation at 16,000g for 30 sec at room temperature and a light swirl. Incubate for 20 to 24 h at 16°C.

De-Crosslinking of Chromatin-DNA Complex (Removal of Protein)

38. Add 100 µL of Proteinase K, mix by flicking followed by a short spin at 16,000g for 30 sec at room temperature and a light swirl.

39. Incubate for 2 h at 50°C.

    It is critical that proteins are completely digested, or DNA will be lost in the following steps.

DNA Purification

40. Add 9 mL of nuclease free H$_2$O to the tube from Step 39.

    This is to make sure that the Nalgene FEP tube used Step 43 will be >80% full.

41. Transfer the solution to a MaXtract High Density 50 mL tube and add an equal volume of phenol-chloroform-isoamyl alcohol pH 7.9.

    Prior to use, centrifuge the MaXtrack High density tube at 3000g for 5 min at room temperature.

42. Mix vigorously and centrifuge at 3000g for 5 min at room temperature.

43. Transfer the aqueous phase (which stays above the gel) into a fresh 50 mL Nalgene tube.

44. Add 1.9 mL of sodium acetate 3 M, pH 5.5, 5 µL of GlycoBlue, and 19 mL of isopropanol.

45. Mix by inverting the tube and incubate overnight at −80°C to freeze the mixture.

46. Thaw the sample and centrifuge at 25,000g for 30 min at 4°C.

    Prechill the Sorvall high-speed centrifuge and the SS-34 rotor to 4°C.

47. Decant the supernatant by inverting the tube.

    Be careful as the pellet may detach from the tube.

48. Transfer the blue pellet to a 1.5 mL microcentifuge tube and centrifuge at 16,000g for 5 min at 4°C.

49. Wash the pellet twice with 1 mL of 75% ice-cold ethanol.

50. Remove all ethanol and dry pellet with a speed vacuum for 2 min.

51. Resuspend the pellet in 34 µL of QIAGEN Buffer EB.

MmeI Digestion to Release Captured Interaction PETs (iPETs)

52. Prepare the *MmeI* digestion master mix as follows: mix 7.5 µL 32 mM S-adenosylmethionine, 7.5 µL NEB buffer 4 (10×), and 7.5 µL nonbiotinylated annealed Half-Linker A and B mixed (200 ng/µL) at the ratio(s) determined to be ideal in Step 6.

    Nonbiotinylated Half-Linker is added to quench the excess of MmeI.

53. Add 15 µL of *MmeI* digestion master mix to the tube from Step 51 and pipette up and down to mix.

54. Add 1 µL of *MmeI* and mix by pipetting up and down.

55. Incubate for 2 h at 37°C.

Preparation of Dynabeads and Immobilization of iPET DNA

56. Transfer 50 µL of M280 Streptavidin Dynabeads suspension to a microcentrifuge tube.

    Mix the suspension well before transferring.

57. Centrifuge the beads at 100g for 1 min at 4°C and place the tube on the DynaMagTM-2 Magnet for 2 min to separate the beads from the buffers by magnetic force.

58. Discard the buffer and wash the Dynabeads with 150 µL of 2× binding and washing buffer (2× B&W Buffer) by inverting and flicking the tube. Centrifuge at 100g for 1 min at 4°C and place the tube on the DynaMagTM-2 Magnet for 2 min.

    Discard the buffer carefully. In all steps involving magnetic beads, manipulate beads carefully without disturbing the bead pellet. Do not let the beads dry out.

59. Repeat Step 58.

60. Resuspend the beads in 50 µL of 2× B&W buffer.

61. Transfer the 50 µL of *Mme*I digested mix from Step 55 to the 50 µL washed dynabeads.

62. Incubate for 45 min at room temperature on the Intelli-Mixer.

63. Centrifuge at 100g for 1 min at 4°C and place the tube on DynaMagTM-2 Magnet for 2 min.

64. Discard the *Mme*I digestion mix and wash beads three times with 150 µL of 1× B&W buffer as indicated in Step 58.

Ligation of Adaptors A and B to the Immobilized iPET-DNA (iPET Library Construction for High Throughput DNA Sequencing)

65. Prepare adaptor ligation master mix as follows: mix 54 µL nuclease free H_2O, 7.5 µL T4 DNA ligase buffer (10×), 6 µL annealed Adaptor A, and 6 µL annealed Adaptor B at the ratios determined to be ideal in Step 6.

    Adaptor A will anneal to the 5′end of iPET-DNA and adaptor B to its 3′end. Two types of adaptors are required for the polymerase chain reaction (PCR) amplification of the library and for the orientation of the DNA fragments at the sequencing step.

66. Discard the 1× B&W buffer from the tube from Step 64 and resuspend the beads with 49 µL of adaptor ligation master mix by pipetting up and down.

67. Add 1 µL of T4 DNA ligase and mix by pipetting up and down.

68. Incubate overnight (16 h) at 16°C with rotation on the Intelli-Mixer.

69. Centrifuge at 100g for 1 min at 4°C and place tube on DynaMagTM-2 Magnet for 2 min.

70. Discard the ligation mix and wash beads three times with 150 µL of 1× B&W buffer as indicated in Step 58.

Nick Translation of iPETs on Dynabeads

71. Discard the 1× B&W Buffer and resuspend the beads in 46 µL of polymerase I master mix by pipetting up and down.

    Nicks are formed during ligation of adaptors, as adaptors are nonphosphorylated to prevent self-concatenation. In addition, the manipulation of DNA (heat exposure, pipetting, or mixing) may introduce nicks. Thus, nick translation is performed to prevent these nicks from affecting subsequent PCR and sequencing steps.

72. Add 4 µL of *E. coli* DNA polymerase I and mix by pipetting up and down.

73. Incubate 2 h at room temperature with rotation on the Intelli-Mixer.

74. Centrifuge at 100g for 1 min at 4°C and put on DynaMagTM-2 Magnet for 2 min.

75. Discard the polymerase mix and wash three times with 150 µL of 1× B&W Buffer as indicated in Step 58.

Cite this protocol as *Cold Spring Harb Protoc*; doi:10.1101/pdb.prot104620

QC PCR Amplification of the iPET Library

For PCR optimization, it is essential to test different cycle conditions and volumes of beads. Use as few PCR cycles as possible to avoid overamplification that leads to PCR errors and reduces library complexity.

76. Discard the 1× B&W Buffer and resuspend the beads in 50 μL of QIAGEN Buffer EB.

77. Transfer 2 μL of resuspended beads to each of three 0.2 mL PCR tubes.

    Different volumes of beads can also be tested.

78. Add 48 μL of PCR master mix to each tube. Mix by pipetting up and down.

    Keep the tubes on ice as the PCR master mix is very sensitive to temperature.

79. Start three PCR thermocycler blocks and pause the temperature at 98°C before starting the following program: for 30 sec at 98°C (for 10 sec at 98°C, for 30 sec at 65°C, for 30 sec at 72°C) repeat 15 times, 17 times, or 19 times (each tube should be subjected to a different number of cycles), for 5 min at 72°C, end.

80. Briefly centrifuge the tubes at 16,000g for 30 sec at room temperature and transfer 25 μL from each PCR tube into a new tube.

81. Add 5 μL of 6× loading dye to each tube and mix well.

82. Load samples on a precast 4%–20% TBE polyacrylamide gel. Also load 250 ng of a 25 bp DNA ladder. Run the gel for 55 min at 180 V in 1× TBE buffer.

83. After migration, stain the gel with 8 μL of SYBR green in 80 mL TBE 1× buffer for 7 min at room temperature.

84. Visualize the 223 bp library band under the gel imager. The best PCR condition corresponds to the lane showing the highest level of only the 223 bp band after the lowest number of cycles.

    See Troubleshooting.

Large-Scale Amplification and Purification of the Library

85. Proceed to a large-scale amplification with the optimal PCR conditions identified in Step 84. Prepare 20 reaction tubes by adding 2 μL of resuspended beads to 48 μL of PCR master mix in each tube. Cycle as in Step 79, using the best number of cycles identified in Step 84.

86. Pool the 1200 μL of PCR products and split into two microcentrifuge tubes.

87. Add 60 μL of sodium acetate 3 M, pH 5.5, 2 μL of GlycoBlue, and 600 μL of isopropanol to each tube. Invert the tubes to mix.

88. Incubate tubes for 30 min at −80°C to freeze the mixture.

89. Centrifuge tubes at 16,000g for 30 min at 4°C.

90. Discard the supernatants and wash the pellets twice with 1 mL of ice-cold 75% ethanol each. After each wash, centrifuge at 16,000g for at least 5 min at 4°C to avoid sample loss.

91. Remove residual ethanol and dry pellets using a speed vacuum for 2 min.

92. Resuspend each pellet in 100 μL of TE buffer.

93. Add 5 μL of 6× loading dye to each tube and mix. Load 60 μL per well on a 1× TBE - 6% polyacrylamide gel. Also load 500 ng of a 25 bp ladder.

94. Run the gel in 1× TBE buffer at 200 V for 35 min.

95. Stain the gel for 7 min in 80 mL of 1× TBE with 8 μL of SYBR green.

96. Visualize the gel with a DarkReader transilluminator.

    The dark reader uses pure visible blue light. This excitation source prevents damage to DNA, which can occur with ultraviolet illumination. As previously (Step 84), a 223 bp band is expected.

    See Troubleshooting.

Elution of DNA from TBE Gel and Quantification of the ChIA-PET Library

97. Excise the library band with a single edge blade and place the pieces of cut DNA gel into 0.6 mL tubes that have been pierced at the bottom with a 21G needle.

    Make sure that the gel pieces are as small as possible and limited to the library band. Larger pieces will require multiple tubes at this step. If multiple tubes are used, pool the material at the precipitation step (Step 103) with a scaled-up amount of sodium acetate and isopropanol (do not use a higher amount of GlycoBlue).

98. Place the 0.6 mL tube inside a 2 mL screw cap tube and centrifuge at 16,000g for 5 min at 4°C.

    The gel is shredded by centrifugation through the tube and collected in the 2 mL tube.

99. Discard the 0.6 mL tube. Add 200 μL of TE buffer to the 2 mL tube and stir the gel with the pipette tip.

    Make sure that the gel pieces are immersed in the buffer.

100. Freeze for 2 h at −80°C followed by a 37°C incubation overnight.

101. Transfer the gel pieces and buffer to the top of a Spin-X column and centrifuge at 16,000g for 5 min at room temperature.

102. Rinse the 2 mL screw cap tube with 200 μL of TE buffer. Transfer to the same Spin-X column used in Step 101 and centrifuge at 16,000g for 30 sec at room temperature.

103. To the combined eluates, add 43 μL of 3 M sodium acetate pH 5.5, 2 μL of GlycoBlue and 430 μL of isopropanol.

104. Mix and incubate for at least 30 min at −80°C.

105. Centrifuge at 16,000g for 30 min at 4°C.

106. Wash the pellet twice with ice cold 75% ethanol and dry it with a speed-vac.

107. Resuspend the pellet in 15 μL of TE buffer.

108. Assess the quality and quantity of the library by using 1 μL of the sample in an Agilent DNA 1000 assay on a 2100 bioanalyzer.

    Proceed as indicated by the manufacturer. A typical electropherogram shows a peak at 223 bp corresponding to the library and two peaks at 15 and 1500 bp to molecular weight makers. The baseline should be flat. Be aware that Agilent has up to 10% error (size higher than expected). Estimation of the concentration is in the report.

109. The library is ready for sequencing on Illumina platform or another platform.

    To test the quality of the library, the generation of 15 to 20 million reads is ideal. If the quality is sufficient, proceed to complete sequencing. At least 80 million reads are required to saturate the library and generate PETs. Furthermore, a software package is required for processing of ChIA-PET sequence data (Li et al. 2010). The uniquely mapped PETs can be classified in two categories depending on the location of the two tags. First, if the two tags of the PET are on the same scaffold, same strand, and at a distance less than 3 Kb apart, this PET results from the self-ligation of a ChIP DNA fragment and is called a "self-ligation PET." When the two tags do not fit the previous criteria, the PETs are called "interligation PET." The two tags of an interligation PET could come from the same scaffold or different scaffolds and could be on different strands with different orientations. They are the result of ligation events between two different chromatin fragments coming from either two pieces of DNA brought together in the same chromatin complex by interacting proteins or two pieces of DNA coming from two different chromatin complexes. The latter case is clearly noise derived from chimeric ligation with hybrid Linker-A/Linker-B associated PETs. A successful library should contain high number of self-ligations and interligation PETs with little or no interaction from the chimeric PET within the same region. The noise is randomly distributed throughout the genome and corresponds to singletons or small clusters.

    See Troubleshooting.

TROUBLESHOOTING

Problem (Steps 76 to 84): Low and noisy amplification of the library.

Cite this protocol as *Cold Spring Harb Protoc*; doi:10.1101/pdb.prot104620

Solution: Optimize the PCR by changing cycle conditions and substrate amount. After gel electrophoresis, PCR products should show only a single and strong band at 223 bp. Avoid large numbers of amplification cycles, which introduce strong amplification biases in the library, and generate high redundancy of the sequenced reads, together with a low percent of the "uniquely mapped reads."

Problem (Step 96): Poor sequencing library construction (poor sequencing yield and poor read mapping).
Solution: Do not use UV illumination to visualize DNA.

Problem (Step 109): Weak or no interaction or binding sites after mapping of the sequencing reads.
Solution: Adjustment should be made to optimize chromatin immunoprecipitation conditions and construct of the library. Special care should be taken on chromatin immunoprecipitation steps.

DISCUSSION

In 2009, the Genome Institute of Singapore developed ChIA-PET to produce simultaneously, in a single experiment, a genome-wide binding profile, together with a map of the physical interactions between protein-bound DNA regions (Fullwood et al. 2009). Originally used for chromatin isolated from human cell lines, the method was used for the first time in 2015 with chromatin isolated from *Xenopus tropicalis* tissues (Buisine et al. 2015). The objective was to characterize the molecular transcriptional mechanisms controlled by thyroid hormone and their receptors during metamorphosis. ChIA-PET analysis of thyroid hormone receptor and transcriptome analysis of thyroid hormone response will document with unprecedented details the topology and the *cis* regulatory dynamic of direct target genes of thyroid hormone signaling. Preliminary data show that indeed, thyroid hormone receptors can act over large genomic distances. Note that ChIA-PET requires high-resolution genome annotation and assembly. Fortunately, a sufficiently precise quality of the *Xenopus tropicalis* genome is available to exploit ChIA-PET analysis (Buisine et al. 2015), a quality that is not so common compared to other sequenced genomes. Thus, the use and feasibility of ChIA-PET provide a powerful opportunity for *Xenopus* research to analyze gene regulation in the context of the 3D nucleus. There is no doubt that the ChIA-PET method is complex. However, new developments will undoubtedly make the ChIA-PET technology easier and accessible to a broader user base (Tang et al. 2015).

RECIPES

Binding and Washing (B&W) Buffer (2×)

Reagent	Amount (for 100 mL)	Final concentration (2×)
Tris-HCl (1 M, pH 7.5; ThermoFisher 15567-027)	1 mL	10 mM
EDTA (0.5 M, pH 8; Ambion AM9260G)	200 µL	1 mM
NaCl (5 M; Sigma-Aldrich S3014)	40 mL	2 M
Nuclease-free H₂O (Ambion AM9937)	58.8 mL	

Store the buffer at room temperature.

ChIA-PET Elution Buffer

Reagent	Amount to add (for 1 mL)	Final concentration (1×)
Tris-HCl (1 M, pH 7.5) (ThermoFisher 15567-027)	10 μL	10 mM
EDTA (0.5 M, pH 8) (Ambion AM9260G)	2 μL	1 mM
SDS (10%; Promega V6553)	100 μL	1%
Nuclease-free H$_2$O (Ambion AM9937)	888 μL	

Prepare the solution fresh before use.

ChIA-PET Wash Buffer

Reagent	Amount to add (for 500 mL)	Final concentration (1×)
Tris-HCl (1 M, pH 7.5; (ThermoFisher 15567-027)	5 mL	10 mM
EDTA (0.5 M, pH 8; Ambion AM9260G)	1 mL	1 mM
NaCl (5 M; Sigma-Aldrich S3014)	50 mL	500 mM
Nuclease-free H$_2$O (Ambion AM9937)	444 mL	

Store at room temperature for up to 1 mo.

Circularization Mix

Reagent	Amount (for 1 sample)
Nuclease-free H$_2$O (Ambion AM9937)	7776 μL
T4 DNA ligase buffer (10×; NEB B0202S)	1000 μL

Prepare the solution fresh before use. Use new ligase buffer if reagent is too old (it includes dithiothreitol, which oxidizes over time).

PCR Master Mix for ChIA-PET

Reagent	Amount (for 1 sample)
Nuclease-free H$_2$O (Ambion AM9937)	21 μL
Library construction primer 1 (10 μM; see below)	1 μL
Library construction primer 2 (10 μM; see below)	1 μL
Phusion High-Fidelity PCR Master Mix with HF buffer (Finnzymes F-531)	25 μL

Use the following primer sequences:

Library construction-primer 1: 5′-AATGATACGGCGACCACCGAGATCTA
CACCCTATCCCCTGTGTGCCTTG-3′

Library construction-primer 2:5′-CAAGCAGAAGACGGCATACGAGATCG
GTCCATCTCATCCCTGCGTGTC-3′

Prepare the solution fresh before use.

Cite this protocol as *Cold Spring Harb Protoc*; doi:10.1101/pdb.prot104620

Polymerase I Master Mix

Reagent	Amount (for 1.5 samples)
Nuclease-free H_2O (Ambion AM9937)	57.8 µL
NEB buffer 2 (10×; New England Biolabs B7002S)	7.5 µL
dNTP Mix (10 mM; Eppendorf 0032 003.109)	3.75 µL

Prepare the solution fresh before use.

Polynucleotide Kinase Master Mix

Reagent	Amount (for 1.5 samples)
Nuclease-free H_2O (Ambion AM9937)	1320 µL
T4 DNA ligase buffer (10×; NEB B0202S)	150 µL

Prepare the solution fresh before use. Use new ligase buffer if reagent is too old (it includes dithiothreitol, which oxidizes over time).

TNE Buffer for ChIA-PET

Reagent	Amount to add (for 500 mL)	Final concentration (1×)
Tris-HCl (1 M, pH 8; ThermoFisher 15568-025)	5 mL	10 mM
EDTA (0.5 M, pH 8; Ambion AM9260G)	0.1 mL	0.1 mM
NaCl (5 M; Sigma-Aldrich S3014)	5 mL	50 mM
Nuclease-free H_2O (Ambion AM9937)	489.9 mL	

Store at room temperature for up to 1 mo.

T4 DNA Polymerase Master Mix

Reagent	Amount (for 1.5 samples)
Nuclease free H_2O (Ambion AM9937)	1319.4 µL
10× buffer for T4 DNA polymerase (Promega M831A)	150 µL
dNTPs (10 mM; Eppendorf 0032 003.109)	15 µL

Prepare the solution fresh before use.

REFERENCES

Buisine N, Ruan X, Bilesimo P, Grimaldi A, Alfama G, Ariyaratne P, Mulawadi F, Chen J, Sung WK, Liu ET, et al. 2015. *Xenopus tropicalis* genome re-scaffolding and re-annotation reach the resolution required for in vivo ChIA-PET analysis. *PLoS One* 10: e0137526.

Buisine N, Ruan X, Ruan Y, Sachs LM. 2018. Chromatin immunoprecipitation for Chromatin Interaction Analysis using Paired-End-Tag (ChIA-PET) sequencing in tadpole tissues. *Cold Spring Harb Protoc* doi: 10.1101/pdb.prot097725.

Fullwood MJ, Liu MH, Pan YF, Liu J, Xu H, Mohamed YB, Orlov YL, Velkov S, Ho A, Mei PH, et al. 2009. An oestrogen-receptor-α-bound human chromatin interactome. *Nature* 462: 58–64.

Goh Y, Fullwood MJ, Poh HM, Peh SQ, Ong CT, Zhang J, Ruan X, Ruan Y. 2012. Chromatin Interaction Analysis with Paired-End Tag sequencing (ChIA-PET) for mapping chromatin interactions and understanding transcription regulation. *J Vis Exp* pii: 3770.

Li G, Fullwood MJ, Xu H, Mulawadi FH, Velkov S, Vega V, Ariyaratne PN, Mohamed YB, Ooi H-S, Tennakoon C, et al. 2010. ChIA-PET tool for comprehensive chromatin interaction analysis with paired-end tag sequencing. *Genome Biol* 11: R22.

Tang Z, Luo OJ, Li X, Zheng M, Zhu JJ, Szalaj P, Trzaskoma P, Magalska A, Wlodarczyk J, Ruszczycki B, et al. 2015. CTCF-mediated human 3D genome architecture reveals chromatin topology for transcription. *Cell* 163: 1611–1627.

Zhang J, Poh HM, Peh SQ, Sia YY, Li G, Mulawadi FH, Goh Y, Fullwood MJ, Sung WK, Ruan X, et al. 2012. ChIA-PET analysis of transcriptional chromatin interactions. *Methods* 58: 289–299.

Imaging Structural and Functional Dynamics in *Xenopus* Neurons

Hollis T. Cline[1]

Department of Neuroscience, Dorris Neuroscience Center, The Scripps Research Center, La Jolla, California 92039, USA

In vivo time-lapse imaging has been a fruitful approach to identify structural and functional changes in the *Xenopus* nervous system in tadpoles and adult frogs. Structural imaging studies have identified fundamental aspects of brain connectivity, development, plasticity, and disease and have been instrumental in elucidating mechanisms regulating these events in vivo. Similarly, assessment of nervous system function using dynamic changes in calcium signals as a proxy for neuronal activity has demonstrated principles of neuron and circuit function and principles of information organization and transfer within the brain of living animals. Because of its many advantages as an experimental system, use of *Xenopus* has often been at the forefront of developing these imaging methods for in vivo applications. Protocols for in vivo structural and functional imaging—including cellular labeling strategies, image collection, and image analysis—will expand the use of *Xenopus* to understand brain development, function, and plasticity.

BACKGROUND

Morphological analysis and time-lapse imaging in frog tadpoles and embryos have been used extensively to study cellular rearrangements during embryogenesis and development (Keller et al. 1989). A major advance occurred when O'Rourke and Fraser used laser scanning confocal microscopy to collect time-lapse images of DiI-labeled retinal ganglion cell axons in the optic tectum during initial stages of topographic map formation (O'Rourke and Fraser 1990). Complementing static images of retinotectal axons in previous studies, daily in vivo time-lapse imaging of single arbors over 4–5 d directly revealed dynamic changes in axon arbor structure that can only be inferred by comparing static images from different animals. Subsequent studies using a variety of time-lapse imaging protocols with interimaging intervals ranging from minutes to hours showed that new branches are constantly added and retracted in axon arbors and the net elaboration of the arbor over days occurs by the stabilization of a surprisingly small fraction of the total added branches (O'Rourke et al. 1994; Cohen-Cory and Fraser 1995; Witte et al. 1996; Rajan et al. 1999). Iontophoresis of DiI into single optic tectal neurons allowed high resolution in vivo time-lapse imaging of individual tectal cell dendritic arbors, revealing dynamic rearrangements of dendritic branches that generated complex dendritic arbors (Wu and Cline 1998, 2003; Wu et al. 1999). Whereas rapid acousto-optic deflector (AOD)-mediated confocal laser scanning, which minimized phototoxicity of the fluorescent dyes, was essential for the success of these initial in vivo imaging experiments, the combined use of less-phototoxic fluorescent proteins, such as green fluorescent protein (GFP) and custom-built two-photon microscopes (Denk and Svoboda

---

[1]Correspondence: cline@scripps.edu

1997), significantly improved imaging capacity, demonstrated in vivo in the *Xenopus* tadpole visual system (Sin et al. 2002; Ruthazer et al. 2003). Importantly, these advances permitted frequent images of neuronal structure to be collected without photodamage, increasing the temporal and spatial resolution of studies of structural dynamics.

The capacity to combine in vivo manipulations of visual experience, as well as molecular and cellular components in both presynaptic retinal axons and postsynaptic tectal neurons, with in vivo time-lapse structural imaging ushered in a heyday of mechanistic studies on neuronal development and connectivity. Studies demonstrated that the stabilization of newly added axon and dendritic branches is regulated by glutamate receptors and mediated by establishment of synaptic contacts (O'Rourke et al. 1994; Wu and Cline 1998; Sin et al. 2002; Haas et al. 2006; Ruthazer et al. 2006; Chen et al. 2010) and the participation of key signaling mechanisms—for instance, involving TrkB and BDNF (Cohen-Cory and Fraser 1995; Marshak et al. 2007), CaMKII (Wu et al. 1996; Zou and Cline 1996a,b, 1999; Wu and Cline 1998), and transcription factors (Chen et al. 2012). These studies also demonstrated fascinating retrograde regulation of axon arbor structure by manipulation of postsynaptic proteins. For instance, viral expression of constitutively active CaMKII in tectal neurons strengthened retinotectal synapses (Wu et al. 1996), acted as a stop-growing signal in tectal cell dendrites (Wu and Cline 1998) and retrogradely limited elaboration of presynaptic retinotectal axons (Zou and Cline 1996b, 1999). Coordinated control of retinotectal axon and tectal cell dendrite structure was also seen in experiments in which tectal cell expression of the visual activity regulated protein CPG15 increased tectal cell dendritic arbor growth, strengthened retinotectal synapses, and increased axon arbor complexity (Cantallops et al. 2000).

PROTOCOLS

Detailed protocols describing electroporation, morpholino-mediated knockdown, in vivo imaging, and accurate reconstruction of axonal and dendritic arbors (Haas et al. 2002; Bestman et al. 2006; Hewapathirane and Haas 2008; Bestman et al. 2012; Ruthazer et al. 2013a,b,c,d; Sáenz de Miera et al. 2018; Bestman and Cline 2020) have been instrumental in expanding the use of *Xenopus* tadpoles as an experimental system that is widely valued for studying mechanisms regulating neuronal development and plasticity (Ghiretti and Paradis 2011, 2014). For example, one of the Protocols in this collection, Protocol 4: Imaging the Dynamic Branching and Synaptic Differentiation of *Xenopus* Optic Axons In Vivo (Santos et al. 2020), describes in detail how to transfect individual retinal ganglion cells (RGCs) with different reagents and to simultaneously visualize optic axon arbor morphology and presynaptic sites in real time. Only recently has 4D analysis of arbor dynamics been automated (He and Cline 2011; Lee et al. 2013; He et al. 2016), facilitating the analysis of time-lapse structural plasticity data by speeding up analysis and increasing experiment reproducibility by decreasing interinvestigator influence on outcome measures. Protocol 1: In Vivo Time-Lapse Imaging and Analysis of Dendritic Structural Plasticity in *Xenopus laevis* Tadpoles (He et al. 2021) describes tectal neuronal electroporation, two-photon in vivo time-lapse imaging of complete dendritic arbors in isolated tectal neurons, reconstruction from the complete image stacks, and the 4D automated analysis and code. This flexible analysis facilitates quantitative comparisons between arbor branch dynamics over a range of time periods and a range of experimental conditions. The dynamic analysis of the 4D time-lapse data uses customized C++ software, 4DSPA, to analyze the branch dynamics, such as branch additions, branch growth, and retractions in reconstructed dendritic arbors across different time points. The protocol describes how arbor reconstruction data, using either Imaris or Neurolucida, are prepared and imported into 4DSPA, followed by the pairwise matching analysis of reconstructions from sequential time points and the serial dynamic analysis. Output files provide data on branch categories and can be used to generate images with branches color coded according to their branch dynamic categories.

Studies mentioned above in which pre- and postsynaptic neuronal compartments are visualized by expression of targeted proteins tagged with fluorescent reporters have been instrumental in dissecting

Cite this introduction as *Cold Spring Harb Protoc*; doi:10.1101/pdb.top106773

axonal and dendritic plasticity mechanisms related to synaptogenesis and synapse maturation. Tagging other organelles such as ribonucleoprotein particle (RNPs) (Bestman and Cline 2008), cytoskeletal proteins, and mitochondria (Plucinska et al. 2012) has also demonstrated fundamental elements regulating neuronal development and structural plasticity. Also included in this collection is Protocol 2: Imaging Mitochondrial Dynamics in the *Xenopus* Central Nervous System (CNS) (Feng and Bestman 2021). The protocol describes animal husbandry and bulk electroporation of *Xenopus* tadpoles to express cytosolic fluorescent protein (FP) to reveal cell morphology and a compatible FP targeted to mitochondria (Weber and Koster 2013). Multiple proteins can be coexpressed by separating the individual proteins within a single polypeptide with a self-cleaving peptide (Donnelly et al. 2001). The protocol also describes screening animals, in vivo confocal imaging of FP-labeled cells containing FP-tagged mitochondria, and image analysis using open source software Fiji/ImageJ (Schindelin et al. 2012) and Simple Neurite Tracer (Longair et al. 2011) to identify mitochondrial locations within individual cells. This protocol drives FP reporter expression in neural progenitor cells and their neural progeny by using a promotor that requires binding of endogenous Sox2 (Bestman et al. 2012), allowing comparative analysis of mitochondrial dynamics in neural progenitor cells and their neuronal progeny. Although neural progenitor cells do not use mitochondria for their primary energy source, they do contain many mitochondria, which recent studies suggest may be critical for neural progenitor cell fate and neurogenesis (Khacho et al. 2017; Khacho and Slack 2018; Bhaskar et al. 2020; Su et al. 2020). Recent studies have used time-lapse in vivo imaging to determine cell lineages of radial glial neural progenitor cells in the optic tectum (Bestman et al. 2012), to screen for genes that regulate tectal neural progenitor cell proliferation and differentiation (Bestman et al. 2015), and to examine interactions between radial glial endfeet and the vasculature (Lau et al. 2017). Imaging mitochondrial dynamics in neural progenitors in the *Xenopus* CNS, in which effects on cellular structure, cell proliferation, and cell fate can be easily assessed under a variety of experimental conditions, is likely to reveal elemental functions of mitochondria in neural progenitor cells.

Neuronal function and plasticity can be captured by imaging calcium dynamics in neuronal somata, dendrites, and axons, as well as subcellular compartments. Bulk or single-cell loading of chemical calcium indicators and expression of genetically encoded calcium indicators is widely used to monitor neuronal activity based on transient elevations of intracellular calcium levels (Garaschuk et al. 2006; Paredes et al. 2008; Grienberger and Konnerth 2012). Genetically encoded calcium indicators continue to be developed and improved (Chen et al. 2013; Inoue 2020; Salgado-Almario et al. 2020; Shemetov et al. 2020; Shen et al. 2020). Early studies in isolated *Rana pipiens* frog brains used membrane permeable fura-2 AM to examine and characterize calcium transients in retinotectal presynaptic compartments in response to optic nerve shock (Feller et al. 1996). Inspired by Feller et al. (1996), subsequent studies imaged visually evoked calcium dynamics in *Xenopus* tadpole retinotectal axons labeled by bulk delivery of calcium green 1 dextran (CaGD) into the retina (Edwards and Cline 1999). A key advantage of CaGD for this study was that CaGD delivered into the retina was transported into retinotectal axon arbors, which could then be imaged without interference from the labeling site. In addition, CaGD has a relatively large change in fluorescence emission in response to calcium as well as significant emission under resting conditions, which allowed imaging and reconstruction of axon arbor morphology under resting conditions and with visual stimulation of the eye. Visual stimulation in intact animals increased calcium levels in retinal axon arbors and even in fine filopodia and growth cones, previously seen only in studies in vitro. Correlative studies of visual stimulation induced calcium transients and axon structural dynamics imaged at 3-min intervals showed that new branches emerged from axons at sites of punctate calcium transients, inspiring later in vivo structural imaging studies investigating the relation between presynaptic sites and axon branch dynamics (Javaherian and Cline 2005; Ruthazer et al. 2006). More recently, calcium dynamics in populations of neurons in the optic tectum and olfactory system in *Xenopus* tadpoles have helped identify circuit-based computations and principles of information processing and plasticity in response to sensory inputs. This collection includes Protocol 5: Bulk Dye Loading for In Vivo Calcium Imaging of Visual Responses in Populations of *Xenopus* Tectal Neurons (Hogg and Haas 2021). The protocol includes micropipette preparation and injecting the tectum with chemical calcium indicator.

Bulk loading of the membrane permeable AM ester of Oregon Green 488 BAPTA (OGB-1) permits in vivo two-photon time-lapse imaging of large populations of neurons, here applied to the optic tectum and combined with in vivo presentations of visual stimuli. Protocols for two-photon imaging and image analysis are available in Sakaki et al. (2020). This strategy has been valuable for imaging neuronal network activity in healthy animals (Dunfield and Haas 2009; Podgorski et al. 2012) and in models of neurological disease (Hewapathirane et al. 2008). Alternate strategies to image calcium dynamics in individual neurons or populations include genetically encoded calcium indicators, such as the GCaMP family of indicators (Tian et al. 2009), which can be delivered by bulk or single electroporation and can be targeted to genetically defined cell types based on promoter driven expression (He et al. 2016).

Extensive studies in the Manzini and Schild laboratories have generated fundamental insights into the transformation and organization of odor representations in the vertebrate olfactory system. Here, Offner and colleagues from the Manzini laboratory provide Protocol 3: Whole-Brain Calcium Imaging in Larval *Xenopus* (Offner et al. 2020) for calcium imaging in the *Xenopus* tadpole olfactory system using local injection of calcium indicator dyes or a transgenic *Xenopus* line expressing GCaMP6s in neurons, which is available from the National *Xenopus* Resource (NXR) at the Woods Hole Marine Biological Lab. They describe fast volumetric multiphoton imaging strategies that reveal spatiotemporal dynamics of brain activity in different brain regions, indicative of stages of sensory information processing. The protocol describes preparation of the Fluo-4 AM calcium indicator, injecting the indicator into an experimental preparation consisting of the peripheral olfactory system and the brain, followed by 3D multiphoton calcium imaging of odor-evoked calcium responses and analysis of the imaging data using published algorithms and toolkits (Ramirez et al. 2014; Friedrich et al. 2017; Pnevmatikakis and Giovannucci 2017; Giovannucci et al. 2019).

These diverse protocols will enable further investigation into molecular, cellular, and circuit level mechanisms underlying brain development and function in *Xenopus*, extending the discovery power of this organism in both basic and disease-related research.

COMPETING INTEREST STATEMENT

The author declares no conflicts of interest with respect to the authorship or publication of this article.

REFERENCES

Bestman JE, Cline HT. 2008. The RNA binding protein CPEB regulates dendrite morphogenesis and neuronal circuit assembly in vivo. *Proc Natl Acad Sci* 105: 20494–20499. doi:10.1073/pnas.0806296105

Bestman JE, Cline HT. 2020. Morpholino studies in *Xenopus* brain development. *Methods Mol Biol* 2047: 377–395. doi:10.1007/978-1-4939-9732-9_21

Bestman JE, Ewald RC, Chiu SL, Cline HT. 2006. In vivo single-cell electroporation for transfer of DNA and macromolecules. *Nat Protoc* 1: 1267–1272. doi:10.1038/nprot.2006.186

Bestman JE, Lee-Osbourne J, Cline HT. 2012. In vivo time-lapse imaging of cell proliferation and differentiation in the optic tectum of *Xenopus laevis* tadpoles. *J Comp Neurol* 520: 401–433. doi:10.1002/cne.22795

Bestman JE, Huang LC, Lee-Osbourne J, Cheung P, Cline HT. 2015. An in vivo screen to identify candidate neurogenic genes in the developing *Xenopus* visual system. *Dev Biol* 408: 269–291. doi:10.1016/j.ydbio.2015.03.010

Bhaskar S, Sheshadri P, Joseph JP, Potdar C, Prasanna J, Kumar A. 2020. Mitochondrial superoxide dismutase specifies early neural commitment by modulating mitochondrial dynamics. *iScience* 23: 101564. doi:10.1016/j.isci.2020.101564

Cantallops I, Haas K, Cline HT. 2000. Postsynaptic CPG15 promotes synaptic maturation and presynaptic axon arbor elaboration in vivo. *Nat Neurosci* 3: 1004–1011. doi:10.1038/79823

Chen SX, Tari PK, She K, Haas K. 2010. Neurexin–neuroligin cell adhesion complexes contribute to synaptotropic dendritogenesis via growth stabilization mechanisms in vivo. *Neuron* 67: 967–983. doi:10.1016/j.neuron.2010.08.016

Chen SX, Cherry A, Tari PK, Podgorski K, Kwong YK, Haas K. 2012. The transcription factor MEF2 directs developmental visually driven functional and structural metaplasticity. *Cell* 151: 41–55. doi:10.1016/j.cell.2012.08.028

Chen TW, Wardill TJ, Sun Y, Pulver SR, Renninger SL, Baohan A, Schreiter ER, Kerr RA, Orger MB, Jayaraman V, et al. 2013. Ultrasensitive fluorescent proteins for imaging neuronal activity. *Nature* 499: 295–300. doi:10.1038/nature12354

Cohen-Cory S, Fraser SE. 1995. Effects of brain-derived neurotrophic factor on optic axon branching and remodelling in vivo. *Nature* 378: 192–196. doi:10.1038/378192a0

Denk W, Svoboda K. 1997. Photon upmanship: why multiphoton imaging is more than a gimmick. *Neuron* 18: 351–357. doi:10.1016/S0896-6273(00)81237-4

Donnelly MLL, Luke G, Mehrotra A, Li X, Hughes LE, Gani D, Ryan MD. 2001. Analysis of the aphthovirus 2A/2B polyprotein 'cleavage' mechanism indicates not a proteolytic reaction, but a novel translational effect: a putative ribosomal 'skip'. *J Gen Virol* 82: 1013–1025. doi:10.1099/0022-1317-82-5-1013

Dunfield D, Haas K. 2009. Metaplasticity governs natural experience-driven plasticity of nascent embryonic brain circuits. *Neuron* 64: 240–250. doi:10.1016/j.neuron.2009.08.034

Edwards JA, Cline HT. 1999. Light-induced calcium influx into retinal axons is regulated by presynaptic nicotinic acetylcholine receptor activity in vivo. *J Neurophysiol* 81: 895–907. doi:10.1152/jn.1999.81.2.895

Feller MB, Delaney KR, Tank DW. 1996. Presynaptic calcium dynamics at the frog retinotectal synapse. *J Neurophysiol* 76: 381–400. doi:10.1152/jn.1996.76.1.381

Feng MS, Bestman JE. 2021. Imaging mitochondrial dynamics in the *Xenopus* central nervous system (CNS). *Cold Spring Harb Protoc* doi:10.1101/pdb.prot106807

Friedrich J, Zhou P, Paninski L. 2017. Fast online deconvolution of calcium imaging data. *PLoS Comput Biol* 13: e1005423. doi:10.1371/journal.pcbi.1005423

Garaschuk O, Milos RI, Konnerth A. 2006. Targeted bulk-loading of fluorescent indicators for two-photon brain imaging in vivo. *Nat Protoc* 1: 380–386. doi:10.1038/nprot.2006.58

Ghiretti AE, Paradis S. 2011. The GTPase Rem2 regulates synapse development and dendritic morphology. *Dev Neurobiol* 71: 374–389. doi:10.1002/dneu.20868

Ghiretti AE, Paradis S. 2014. Molecular mechanisms of activity-dependent changes in dendritic morphology: role of RGK proteins. *Trends Neurosci* 37: 399–407. doi:10.1016/j.tins.2014.05.003

Giovannucci A, Friedrich J, Gunn P, Kalfon J, Brown BL, Koay SA, Taxidis J, Najafi F, Gauthier JL, Zhou P, et al. 2019. CaImAn an open source tool for scalable calcium imaging data analysis. *eLife* 8: e38173. doi:10.7554/eLife.38173

Grienberger C, Konnerth A. 2012. Imaging calcium in neurons. *Neuron* 73: 862–885. doi:10.1016/j.neuron.2012.02.011

Haas K, Jensen K, Sin WC, Foa L, Cline HT. 2002. Targeted electroporation in *Xenopus* tadpoles in vivo—from single cells to the entire brain. *Differentiation* 70: 148–154. doi:10.1046/j.1432-0436.2002.700404.x

Haas K, Li J, Cline HT. 2006. AMPA receptors regulate experience-dependent dendritic arbor growth in vivo. *Proc Natl Acad Sci* 103: 12127–12131. doi:10.1073/pnas.0602670103

He HY, Cline HT. 2011. Diadem X: automated 4 dimensional analysis of morphological data. *Neuroinformatics* 9: 107–112. doi:10.1007/s12021-011-9098-x

He HY, Shen W, Hiramoto M, Cline HT. 2016. Experience-dependent bimodal plasticity of inhibitory neurons in early development. *Neuron* 90: 1203–1214. doi:10.1016/j.neuron.2016.04.044

He HY, Lin CY, Cline HT. 2021. In vivo time-lapse imaging and analysis of dendritic structural plasticity in *Xenopus laevis* tadpoles. *Cold Spring Harb Protoc* doi:10.1101/pdb.prot106781

Hewapathirane DS, Haas K. 2008. Single cell electroporation in vivo within the intact developing brain. *J Vis Exp* 11: 705. doi:10.3791/705

Hewapathirane DS, Dunfield D, Yen W, Chen S, Haas K. 2008. In vivo imaging of seizure activity in a novel developmental seizure model. *Exp Neurol* 211: 480–488. doi:10.1016/j.expneurol.2008.02.012

Hogg PW, Haas K. 2021. Bulk dye loading for in vivo calcium imaging of visual responses in populations of *Xenopus* tectal neurons. *Cold Spring Harb Protoc* doi:10.1101/pdb.prot106831

Inoue M. 2020. Genetically encoded calcium indicators to probe complex brain circuit dynamics in vivo. *Neurosci Res* 169: 2–8. doi:10.1016/j.neures.2020.05.013

Javaherian A, Cline HT. 2005. Coordinated motor neuron axon growth and neuromuscular synaptogenesis are promoted by CPG15 in vivo. *Neuron* 45: 505–512. doi:10.1016/j.neuron.2004.12.051

Keller R, Cooper MS, Danilchik M, Tibbetts P, Wilson PA. 1989. Cell intercalation during notochord development in *Xenopus laevis*. *J Exp Zool* 251: 134–154. doi:10.1002/jez.1402510204

Khacho M, Slack RS. 2018. Mitochondrial dynamics in the regulation of neurogenesis: from development to the adult brain. *Dev Dyn* 247: 47–53. doi:10.1002/dvdy.24538

Khacho M, Clark A, Svoboda DS, MacLaurin JG, Lagace DC, Park DS, Slack RS. 2017. Mitochondrial dysfunction underlies cognitive defects as a result of neural stem cell depletion and impaired neurogenesis. *Hum Mol Genet* 26: 3327–3341. doi:10.1093/hmg/ddx217

Lau M, Li J, Cline HT. 2017. In vivo analysis of the neurovascular niche in the developing *Xenopus* brain. *eNeuro* 4: ENEURO.0030-17.2017. doi:10.1523/ENEURO.0030-17.2017

Lee PC, He HY, Lin CY, Ching YT, Cline HT. 2013. Computer aided alignment and quantitative 4D structural plasticity analysis of neurons. *Neuroinformatics* 11: 249–257. doi:10.1007/s12021-013-9179-0

Longair MH, Baker DA, Armstrong JD. 2011. Simple neurite tracer: open source software for reconstruction, visualization and analysis of neuronal processes. *Bioinformatics* 27: 2453–2454. doi:10.1093/bioinformatics/btr390

Marshak S, Nikolakopoulou AM, Dirks R, Martens GJ, Cohen-Cory S. 2007. Cell-autonomous TrkB signaling in presynaptic retinal ganglion cells mediates axon arbor growth and synapse maturation during the establishment of retinotectal synaptic connectivity. *J Neurosci* 27: 2444–2456. doi:10.1523/JNEUROSCI.4434-06.2007

Offner T, Daume D, Weiss L, Hassenklöver T, Manzini I. 2020. Whole-brain calcium imaging in larval *Xenopus*. *Cold Spring Harb Protoc* doi:10.1101/pdb.prot106815

O'Rourke NA, Fraser SE. 1990. Dynamic changes in optic fiber terminal arbors lead to retinotopic map formation: an in vivo confocal microscopic study. *Neuron* 5: 159–171. doi:10.1016/0896-6273(90)90306-Z

O'Rourke NA, Cline HT, Fraser SE. 1994. Rapid remodeling of retinal arbors in the tectum with and without blockade of synaptic transmission. *Neuron* 12: 921–934. doi:10.1016/0896-6273(94)90343-3

Paredes RM, Etzler JC, Watts LT, Zheng W, Lechleiter JD. 2008. Chemical calcium indicators. *Methods* 46: 143–151. doi:10.1016/j.ymeth.2008.09.025

Plucinska G, Paquet D, Hruscha A, Godinho L, Haass C, Schmid B, Misgeld T. 2012. In vivo imaging of disease-related mitochondrial dynamics in a vertebrate model system. *J Neurosci* 32: 16203–16212. doi:10.1523/JNEUROSCI.1327-12.2012

Pnevmatikakis EA, Giovannucci A. 2017. NoRMCorre: an online algorithm for piecewise rigid motion correction of calcium imaging data. *J Neurosci Methods* 291: 83–94. doi:10.1016/j.jneumeth.2017.07.031

Podgorski K, Dunfield D, Haas K. 2012. Functional clustering drives encoding improvement in a developing brain network during awake visual learning. *PLoS Biol* 10: e1001236. doi:10.1371/journal.pbio.1001236

Rajan I, Witte S, Cline HT. 1999. NMDA receptor activity stabilizes presynaptic retinotectal axons and postsynaptic optic tectal cell dendrites in vivo. *J Neurobiol* 38: 357–368. doi:10.1002/(SICI)1097-4695(19990215)38:3<357::AID-NEU5>3.0.CO;2-#

Ramirez A, Pnevmatikakis EA, Merel J, Paninski L, Miller KD, Bruno RM. 2014. Spatiotemporal receptive fields of barrel cortex revealed by reverse correlation of synaptic input. *Nat Neurosci* 17: 866–875. doi:10.1038/nn.3720

Ruthazer ES, Akerman CJ, Cline HT. 2003. Control of axon branch dynamics by correlated activity in vivo. *Science* 301: 66–70. doi:10.1126/science.1082545

Ruthazer ES, Li J, Cline HT. 2006. Stabilization of axon branch dynamics by synaptic maturation. *J Neurosci* 26: 3594–3603. doi:10.1523/JNEUROSCI.0069-06.2006

Ruthazer ES, Schohl A, Schwartz N, Tavakoli A, Tremblay M, Cline HT. 2013a. Bulk electroporation of retinal ganglion cells in live *Xenopus* tadpoles. *Cold Spring Harb Protoc* 2013: 771–775. doi:10.1101/pdb.prot076471

Ruthazer ES, Schohl A, Schwartz N, Tavakoli A, Tremblay M, Cline HT. 2013b. Dye labeling retinal ganglion cell axons in live *Xenopus* tadpoles. *Cold Spring Harb Protoc* 2013: 768–770. doi:10.1101/pdb.prot076463

Ruthazer ES, Schohl A, Schwartz N, Tavakoli A, Tremblay M, Cline HT. 2013c. In vivo time-lapse imaging of neuronal development in *Xenopus*. *Cold Spring Harb Protoc* 2013: 804–809.

Ruthazer ES, Schohl A, Schwartz N, Tavakoli A, Tremblay M, Cline HT. 2013d. Labeling individual neurons in the brains of live *Xenopus* tadpoles by electroporation of dyes or DNA. *Cold Spring Harb Protoc* 2013: 869–872. doi:10.1101/pdb.prot077149

Sáenz de Miera C, Parr E, Denver RJ. 2018. Bulk electroporation-mediated gene transfer into *Xenopus* tadpole brain. *Cold Spring Harb Protoc* 2018. doi:10.1101/pdb.prot097691

Sakaki KDR, Podgorski K, Dellazizzo Toth TA, Coleman P, Haas K. 2020. Comprehensive imaging of sensory-evoked activity of entire neurons within the awake developing brain using ultrafast AOD-based random-access two-photon microscopy. *Front Neural Circuits* 14: 33. doi:10.3389/fncir.2020.00033

Salgado-Almario J, Vicente M, Vincent P, Domingo B, Llopis J. 2020. Mapping calcium dynamics in the heart of zebrafish embryos with ratiometric genetically encoded calcium indicators. *Int J Mol Sci* **21**: 6610. doi:10.3390/ijms21186610

Santos RA, Rio RD Jr, Cohen-Cory S. 2020. Imaging the dynamic branching and synaptic differentiation of *Xenopus* optic axons in vivo. *Cold Spring Harb Protoc* doi: 10.1101/pdb.prot106823

Schindelin J, Arganda-Carreras I, Frise E, Kaynig V, Longair M, Pietzsch T, Preibisch S, Rueden C, Saalfeld S, Schmid B, et al. 2012. Fiji: an open-source platform for biological-image analysis. *Nat Methods* **9**: 676–682. doi:10.1038/nmeth.2019

Shemetov AA, Monakhov MV, Zhang Q, Canton-Josh JE, Kumar M, Chen M, Matlashov ME, Li X, Yang W, Nie L, et al. 2020. A near-infrared genetically encoded calcium indicator for in vivo imaging. *Nat Biotechnol* **39**: 368–377. doi:10.1038/s41587-020-0710-1

Shen Y, Nasu Y, Shkolnikov I, Kim A, Campbell RE. 2020. Engineering genetically encoded fluorescent indicators for imaging of neuronal activity: progress and prospects. *Neurosci Res* **152**: 3–14. doi:10.1016/j.neures.2020.01.011

Sin WC, Haas K, Ruthazer ES, Cline HT. 2002. Dendrite growth increased by visual activity requires NMDA receptor and Rho GTPases. *Nature* **419**: 475–480. doi:10.1038/nature00987

Su Y, Zhang W, Patro CPK, Zhao J, Mu T, Ma Z, Xu J, Ban K, Yi C, Zhou Y. 2020. STAT3 regulates mouse neural progenitor proliferation and differentiation by promoting mitochondrial metabolism. *Front Cell Dev Biol* **8**: 362. doi:10.3389/fcell.2020.00362

Tian L, Hires SA, Mao T, Huber D, Chiappe ME, Chalasani SH, Petreanu L, Akerboom J, McKinney SA, Schreiter ER, et al. 2009. Imaging neural activity in worms, flies and mice with improved GCaMP calcium indicators. *Nat Methods* **6**: 875–881. doi:10.1038/nmeth.1398

Weber T, Koster R. 2013. Genetic tools for multicolor imaging in zebrafish larvae. *Methods* **62**: 279–291. doi:10.1016/j.ymeth.2013.07.028

Witte S, Stier H, Cline HT. 1996. In vivo observations of timecourse and distribution of morphological dynamics in *Xenopus* retinotectal axon arbors. *J Neurobiol* **31**: 219–234. doi:10.1002/(SICI)1097-4695(199610)31:2<219::AID-NEU7>3.0.CO;2-E

Wu GY, Cline HT. 1998. Stabilization of dendritic arbor structure in vivo by CaMKII. *Science* **279**: 222–226. doi:10.1126/science.279.5348.222

Wu GY, Cline HT. 2003. Time-lapse in vivo imaging of the morphological development of *Xenopus* optic tectal interneurons. *J Comp Neurol* **459**: 392–406. doi:10.1002/cne.10618

Wu G, Malinow R, Cline HT. 1996. Maturation of a central glutamatergic synapse. *Science* **274**: 972–976. doi:10.1126/science.274.5289.972

Wu GY, Zou DJ, Rajan I, Cline H. 1999. Dendritic dynamics in vivo change during neuronal maturation. *J Neurosci* **19**: 4472–4483. doi:10.1523/JNEUROSCI.19-11-04472.1999

Zou DJ, Cline HT. 1996a. Control of retinotectal axon arbor growth by postsynaptic CaMKII. *Prog Brain Res* **108**: 303–312. doi:10.1016/S0079-6123(08)62548-0

Zou DJ, Cline HT. 1996b. Expression of constitutively active CaMKII in target tissue modifies presynaptic axon arbor growth. *Neuron* **16**: 529–539. doi:10.1016/S0896-6273(00)80072-0

Zou DJ, Cline HT. 1999. Postsynaptic calcium/calmodulin-dependent protein kinase II is required to limit elaboration of presynaptic and postsynaptic neuronal arbors. *J Neurosci* **19**: 8909–8918. doi:10.1523/JNEUROSCI.19-20-08909.1999

Cite this introduction as *Cold Spring Harb Protoc*; doi:10.1101/pdb.top106773

In Vivo Time-Lapse Imaging and Analysis of Dendritic Structural Plasticity in *Xenopus laevis* Tadpoles

Hai-yan He,[1,4] Chih-Yang Lin,[2] and Hollis T. Cline[3]

[1]*Department of Biology, Georgetown University, Washington, D.C. 20057, USA;* [2]*Department of Optoelectronics and Materials Engineering, Chung Hua University, Hsinchu 30012, Taiwan;* [3]*Neuroscience Department, Dorris Neuroscience Center, Scripps Research Institute, La Jolla, California 92037, USA*

In vivo time-lapse imaging of complete dendritic arbor structures in tectal neurons of *Xenopus laevis* tadpoles has served as a powerful in vivo model to study activity-dependent structural plasticity in the central nervous system during early development. In addition to quantitative analysis of gross arbor structure, dynamic analysis of the four-dimensional data offers particularly valuable insights into the structural changes occurring in subcellular domains over experience/development-driven structural plasticity events. Such analysis allows not only quantifiable characterization of branch additions and retractions with high temporal resolution but also identification of the loci of action. This allows for a better understanding of the spatiotemporal association of structural changes to functional relevance. Here we describe a protocol for in vivo time-lapse imaging of complete dendritic arbors from individual neurons in the brains of anesthetized tadpoles with two-photon microscopy and data analysis of the time series of 3D dendritic arbors. For data analysis, we focus on dynamic analysis of reconstructed neuronal filaments using a customized open source computer program we developed (4D SPA), which allows aligning and matching of 3D neuronal structures across different time points with greatly improved speed and reliability. File converters are provided to convert reconstructed filament files from commercial reconstruction software to be used in 4D SPA. The program and user manual are publicly accessible and operate through a graphical user interface on both Windows and Mac OSX.

MATERIALS

It is essential that you consult the appropriate Material Safety Data Sheets and your institution's Environmental Health and Safety Office for proper handling of equipment and hazardous materials used in this protocol.

RECIPES: Please see the end of this protocol for recipes indicated by <R>. Additional recipes can be found online at http://cshprotocols.cshlp.org/site/recipes.

Reagents

Albino *Xenopus laevis* embryos (stage 45–49; Nieuwkoop and Faber 1956)

Rear in 0.1× Steinberg's solution at 22°C with 12 h dark/12 h light cycles. Feed animals with tadpole food from stage 47 (Nieuwkoop and Faber 1956), when the yolk completely disappears (McKeown and Cline 2019). Starting from the first day of imaging, house animals individually in six-well plates to keep track of individual animals (and neurons) imaged.

[4]Correspondence: haiyan.he@georgetown.edu

Supplemental Material is available at cshprotocols.cshlp.org.

DNA plasmid (1–3 µg/µL in ddH$_2$O; mixed with 0.1% w/v Fast-green FCF)

For time-lapse structural imaging of the complete dendritic arbor, green fluorescent protein (GFP) or other fluorescent protein-expressing DNA constructs can be used. Amplification systems such as the Gal4-UAS binary system are recommended to ensure high expression of GFP. The Gal4 construct can be either subcloned into the same DNA construct or coelectroporated with the UAS-E1b-eGFP construct, which has a minimal e1b promoter under 14 repeats of the gal4 response element UAS (Bestman et al. 2012; He et al. 2016).

Fast-green FCF (Sigma-Aldrich; 0.1% in ddH$_2$O)

Filter-sterilize through a 0.2-µm filter and store for up to 1 y at room temperature.

MS-222 solution (3-aminobenzoic acid ethyl ester, Sigma-Aldrich; 0.02% in 0.1× Steinberg's solution)

Store for up to 1 mo at 4°C.

SeaPlaque low-melting temperature agarose (Lonza) (optional; see Step 4)
Steinberg's solution (0.1×) <R>
Sylgard 184 (Electron Microscopy Sciences)
Tadpole food (*Xenopus* Express)

Equipment

4D SPA and accessory programs (https://4dspa.web.nctu.edu.tw/)

Capacitor (3 µf)

Commercially available or open source 3D reconstructing software such as Imaris (Bitplane), Neurolucida (MBF Bioscience), or neuTube (Feng et al. 2015)

Computer (PC or Mac)

Epifluorescence microscope for screening

Glass coverslip

Grass SD9 stimulator

Incubator

Petri dish (35-mm)

Plates (six-well)

Two-photon microscope with a water-emersion objective (20×, NA 0.95)

We use a two-photon microscope custom modified from an Olympus (BX51) FV300 microscope (Sin et al. 2002). GFP signal is two-photon excited by a 910-nm laser generated by a Chameleon Ultra II laser (Coherent). Emitting fluorescence signal is detected by a R3896 photomultiplier tube (Hamamatsu), mounted to the side of a filter cube equipped with a 700LP dichroic mirror and a T560LPXR beam splitter and ET510/80M-2P; ET605/70M-2P filters (Chroma Technology) optimized for the detection of GFP and red fluorescent protein (RFP) signals. The objective is a 20× water immersion objective (Olympus XLUMPlanFL 0.95NA).

METHOD

The experimental protocols described here have been approved by the Institutional Animal Care and Use Committees of Georgetown University and Scripps Research Institute.

Tectal Cell Electroporation

1. Electroporate animals with GFP-expressing DNA constructs at stage 45–46.

   Two methods are commonly used to sparsely label individual tectal neurons with fluorescent protein for time-lapse imaging (Ruthazer et al. 2013): single cell electroporation (Bestman et al. 2006) and whole-brain electroporation (Haas et al. 2002; Bestman et al. 2012). To perform whole-brain electroporation, generate exponential decay voltage pulses through a Grass SD9 stimulator coupled to a capacitor (3 µf). To achieve sparse labeling using this technique, use a moderate concentration of DNA plasmids (1–3 µg/µL), lower voltage (34 V), and fewer (1–2) pulses with a 1.6-msec pulse duration (He et al. 2016).

2. After electroporation, return animals to 0.1× Steinberg's solution to recover. House in groups in the incubator until the day of screening at 22°C.

Cite this protocol as *Cold Spring Harb Protoc*; doi:10.1101/pdb.prot106781

3. Using an epifluorescence microscope, screen electroporated animals in a 35-mm Petri dish 1–7 d postelectroporation for well-isolated tectal neurons with high GFP expression that fills the whole neuron including all dendritic processes.

> *The best animals to be used for subsequent structural imaging are those with one brightly transfected GFP⁺ neuron in each tectal lobe.*

In Vivo Time-Lapse Imaging of Complete Dendritic Arbor

4. Anesthetize the animal by immersing it in 0.02% MS-222 solution for 1 min at room temperature. Move the anesthetized animal in 0.02% MS-222 to a Sylgard chamber that is hand-carved to snuggly fit the tadpole's body (Ruthazer et al. 2013). Make sure that the animal is submerged in the anesthetic solution and seal the chamber tightly with a glass coverslip to prevent movement of the animal during imaging.

> *Animals can stay healthy in the chamber for up to 10 min and should be removed from the chamber after each imaging session. For time-lapse imaging with shorter intervals, it is better to embed animals in 1% low melting temperature agarose and perfuse with 0.1× Steinberg's rearing solution (Ruthazer et al. 2013).*

5. Take z-stack images of the whole neuron under the two-photon microscope.

   i. Ensure that the complete dendritic arbor is encompassed. Collect time-lapse image stacks at regular intervals, which can range from minutes to 24 h.

   > *The shortest interval that can be achieved is limited by the amount of time it takes to acquire a full stack. It takes ∼5 min to finish a stack of size 512×512×300(z) on a galvanometer-based scanning system. On the other hand, the subsequent alignment can become increasingly difficult with longer intervals such as 24 h, because of large changes in the dendritic arbor structure as a result of rapid growth.*

   ii. Use the same zoom factor and z-step size (we use 1 μm) throughout the time series. Choose the zoom factor for the first time point wisely to leave room for potential neuronal growth.

   iii. Maintain the laser intensity constant.

   iv. Each time when fitting the animal in the Sylgard chamber for imaging, use care to position the animal the same way as in the previous time point to help with alignment of images in subsequent analysis.

Structural Data Reconstruction

6. Perform reconstruction of the complete dendritic arbor from raw image stacks with any one of multiple commercially available and open-source software.

> *We use the semimanual function in the Filament module of Imaris (Bitplane) to trace the dendritic arbor in the 3D image stack. Once the reconstruction is completed, the software generates measurements for basic filament characteristics, such as quantifications of arbor size and complexity (total branch tip number, total branch length, branch order, Sholl analysis, etc.). An example of a 3D image of a GFP-labeled neuron and the reconstructed dendritic arbor is shown in Supplemental Movie S1.*

Dynamic Analysis of the 4D Time-Lapse Structural Data

> *We use a customized C⁺⁺ software 4D SPA (Lee et al. 2013) to analyze the dynamic changes (growth and retraction of individual branches, branch tip dynamics, etc.) of the reconstructed dendritic arbors across different time points.*

7. Prepare the data as follows.

   i. Export the full filament data set as txt data file to be used as inputs for 4D SPA. In the txt filament file, the branches are represented as data points defined by their 3D location (coordinates) and are organized blockwise. Each block starts with a line defining the branch index (P #) and the number of nodes in that branch (N #). The first point is the starting (branching) point of the branch, and the last point in the block is the branch tip. The first point in the first block of the file represents the soma position.

ii. Use converting tools (HOCtoBCF, ASCtoBCF, and SWCtoBCF) available through supplemental online resources (https://4dspa.web.nctu.edu.tw/) to convert reconstructed filament files generated from the most commonly used reconstruction software programs (.hoc file in Imaris, .asc file in Neurolucida, and .swc file) to txt files that meet the requirement for 4D SPA. The converting tools run in the terminal of any operation system. In cases where the soma location is not the first point in the first block of the filament file, specify the coordinates of the actual soma location as input parameters in the converting command to allow reorganizing of the filament file structure to meet the requirements for 4D SPA. Reconstructed filament data files in txt format generated by other reconstruction software should also be usable as long as the data structure abides by the above-mentioned rules as in BCF format.

8. Perform pairwise matching analysis as follows.

i. Load the pair of reconstructed filament txt files of the same neuron into the 4D SPA software for alignment. Usually, these should be filaments traced from image stacks acquired at two consecutive time points (T1 and T2). The software will automatically highlight the branch to be aligned (matched) in T1 in the order of branch length (from longest to shortest).

ii. For each branch in T1, a list of candidate matching branches in the T2 filament will be provided, in the order of calculated matching score. Choose the best matching branch and confirm the match by clicking the "add" button. For branches in T1 that do not have a matching branch in T2, the software will still provide a candidate matching list. It is up to the user to designate the branch as retracted by clicking the 'retract' button.

iii. After every branch in T1 is designated either as "matched" or "retracted," all of the unmatched branches in T2 will be automatically designated as "newly added." If mistakes were made during the matching process, go back to the branch in question in T1 and reassign the proper matching branch (or designate as "retracted"). If the reassigned matching branch had been assigned to another T1 branch, the new operation will overwrite the previous matching result and render the other T1 branch "unmatched," which will be prompted for rematching. For an example of the matching process, see Supplemental Movie S2.

A tip to check for matching mistakes is to examine the matching results with the color coding on. If any branch appears at the same locations in T1 and T2 but is labeled as "retracted" in T1 and "newly added" in T2, it is likely a stable branch miscategorized.

iv. Once the alignment is completed for the pair, save the results by clicking "SaveFiles"—"Reports." This will generate a separate "T2_matched.txt" file for the T2 filament. Do this pairwise matching analysis for the next two time points (e.g., T2 and T3). Use the new "T2_matched.txt" file instead of the original T2.txt file for T2. Examples of the pairwise matching analysis results are shown in Figure 1 (top panel).

9. Perform serial dynamic analysis as follows:

i. Serial analysis of sequential images in a multitime-point time-lapse data set requires that branches have unique identifiers that are maintained through the image series. This is achieved through serial pair-wise matching analysis: T1 versus T2, T2 versus T3, and so on. Upon completion of each pairwise analysis, the 4D SPA program adjusts the indices so that matching (stable) branches are assigned with the same identifier that is consistent with the starting time point indices.

ii. To combine the multiple pairwise alignment results in data sets with more than two time points, run "MultiplePairAnalysis" in the terminal window with the report files generated by each pairwise matching analysis within the whole time series (*.rep.txt). A spreadsheet.txt file will be generated, in which branches are categorized as stable, newly added, retracted or transient. A stable branch is a branch that is present at all time

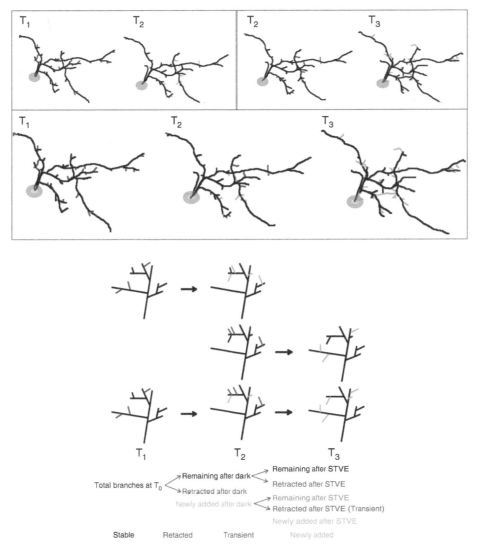

FIGURE 1. Example of dynamic analysis of reconstructed complete dendritic arbors of the same neuron in a time series. (*Bottom*) Illustration for the color-coding scheme. (*Top*) (*Top* panels) Pairwise matching analysis (T1 vs. T2; T2 vs. T3). (*Bottom* panels) Serial analysis. Dendritic branches are color-coded by their dynamic categories: black (stable branches), blue (retracted branches), green (newly added branches), magenta (transient branches). STVE: short-term visual enhancement. (Modified from He et al. 2016, with permission from Elsevier.)

points. A retracted branch is a branch that is present at the first time point but is lost at a later time point. A "newly added" branch is a branch that does not exist at the first time point but appears later and remains in place until the end time point. The "newly added" branch category includes those that appear at the last time point. A transient branch is a branch that appears at a time point after the first time point and is then retracted by the last time point (Fig. 1, bottom panel). The output file is a txt file (default "spreadsheet.txt").

iii. In addition to categorization of the branches, the dynamic changes (growth or retraction) of the stable branches can be analyzed subsequently. This spreadsheet file can also be loaded into 4D SPA together with the matched neuron files to generate filament images with branches color coded by their categories (as shown in Fig. 1, bottom panel). New neuron files (*_mat.txt) can then be saved with the branch category information to be used for further analysis in Matlab. An example of the format of the output spreadsheet file is shown below.

```
Total Time points <int>
    BranchID BranchType BaseBranchLength ΔBranchLength@T0, ..., Tn
    Example:
    3
    index type base T0 T1 T2
    0 Stable 131.658 0 -1.809 -8.333
    1 Stable 130.953 0 -2.664 -8.099
    ...
    27 Retracted 42.5477 0 -4.5266 #
    28 Retracted 41.2327 0 -5.3916 -2.7764
    ...
    65 Transient 39.9811 * 4.0722 -4.0722
    ...
    94 Newly_Added 12.8521 * * 3.8426
```

DISCUSSION

Quantitative analysis of time-lapse data requires particular care be taken in the initial 3D reconstructions of the neuronal structures, so errors in the identification of branch dynamic events do not arise from misidentification of branches throughout the imaging sequence. Recently, new reconstruction software has been made available that allows "serial reconstruction" by modifying the reconstructed filament file of the previous time point to reconstruct the filament file of the later time point (Feng et al. 2015). This significantly facilitates the matching and alignment process by having the reconstructed filaments of the same neuron done in the same primary structure framework. This potentially eliminates the necessity of alignment when it comes to 4D analysis, which is especially useful for reconstructing time series collected with short time intervals (minutes), with the major arbor structure remaining stable across time points. On the other hand, 4D SPA can be used for dynamic analysis on data not suitable for "serial reconstruction," as well as data already collected and reconstructed. Doing so avoids the time-consuming steps of redoing the reconstruction in the newer software. Further analysis of the 3D location of the dynamic branches is also possible by combining the MultiplePairAnalysis results with the matched txt files from the pairwise alignment.

RECIPE

Steinberg's Solution (0.1×)

58 mM NaCl
0.67 mM KCl
0.34 mM Ca(NO$_3$)$_2$
0.83 mM MgSO$_4$
3 mM HEPES

Adjust pH to 7.2. As an alternative to a 0.1× solution, prepare a 10× stock in advance and store it for up to 6 mo at 4°C.

REFERENCES

Bestman JE, Ewald RC, Chiu SL, Cline HT. 2006. In vivo single-cell electroporation for transfer of DNA and macromolecules. *Nat Protoc* **1:** 1267–1272. doi:10.1038/nprot.2006.186

Bestman JE, Lee-Osbourne J, Cline HT. 2012. In vivo time-lapse imaging of cell proliferation and differentiation in the optic tectum of *Xenopus laevis* tadpoles. *J Comp Neurol* **520:** 401–433. doi:10.1002/cne.22795

Cite this protocol as *Cold Spring Harb Protoc*; doi:10.1101/pdb.prot106781

Feng L, Zhao T, Kim J. 2015. neuTube 1.0: a new design for efficient neuron reconstruction software based on the SWC format. *eNeuro* **2:** ENEURO.0049-14.2014. doi:10.1523/ENEURO.0049-14.2014

Haas K, Jensen K, Sin WC, Foa L, Cline HT. 2002. Targeted electroporation in *Xenopus* tadpoles in vivo—from single cells to the entire brain. *Differentiation* **70:** 148–154. doi:10.1046/j.1432-0436.2002.700404.x

He HY, Shen W, Hiramoto M, Cline HT. 2016. Experience-dependent bimodal plasticity of inhibitory neurons in early development. *Neuron* **90:** 1203–1214. doi:10.1016/j.neuron.2016.04.044

Lee PC, He HY, Lin CY, Ching YT, Cline HT. 2013. Computer aided alignment and quantitative 4D structural plasticity analysis of neurons. *Neuroinformatics* **11:** 249–257. doi:10.1007/s12021-013-9179-0

McKeown CR, Cline HT. 2019. Nutrient restriction causes reversible G_2 arrest in *Xenopus* neural progenitors. *Development* **146:** dev178871. doi:10.1242/dev.178871

Nieuwkoop PD, Faber J. 1956. *Normal table of* Xenopus laevis *(Daudin); a systematical and chronological survey of the development from the fertilized egg till the end of metamorphosis.* North-Holland, Amsterdam.

Ruthazer ES, Schohl A, Schwartz N, Tavakoli A, Tremblay M, Cline HT. 2013. In vivo time-lapse imaging of neuronal development in *Xenopus*. *Cold Spring Harb Protoc* **2013:** 804–809.

Sin WC, Haas K, Ruthazer ES, Cline HT. 2002. Dendrite growth increased by visual activity requires NMDA receptor and Rho GTPases. *Nature* **419:** 475–480. doi:10.1038/nature00987

Imaging Mitochondrial Dynamics in the *Xenopus* Central Nervous System (CNS)

Martin Sihan Feng[1] and Jennifer E. Bestman[1,2]

[1]*Department of Biology and Neuroscience Program, William and Mary, Williamsburg, Virginia 23185, USA*

Notable for producing ATP via oxidative phosphorylation, mitochondria also control calcium homeostasis, lipogenesis, the regulation of reactive oxygen species, and apoptosis. Even within relatively simple cells, mitochondria are heterogeneous with regard to their shape, abundance, movement, and subcellular locations. They exist as interconnected, tubular networks and as motile organelles that are transported along the cytoskeleton for distribution throughout cells. These spatial and morphological features reflect variability in the organelle's capacity to synthesize ATP and support cells. Changes to mitochondria are believed to support cell function and fate, and mitochondrial dysfunction underlies disease in the nervous system. Here we describe an in vivo time-lapse imaging approach to monitor and measure the movement and position of the mitochondria in cells of the developing brain in albino *Xenopus laevis* tadpoles. The unparalleled benefit of using *Xenopus* for these experiments is that measurements of mitochondrial morphology and distribution in cells can be measured in vivo, where the surrounding neural circuitry and other inputs that influence these critical organelles remain intact. This protocol draws together techniques to label brain cells and capture the morphology of the cells and their mitochondria with 3D time-lapse confocal microscopy. We describe open-source methods to reconstruct cells in order to quantify the features of their mitochondria.

MATERIALS

It is essential that you consult the appropriate Material Safety Data Sheets and your institution's Environmental Health and Safety Office for proper handling of equipment and hazardous materials used in this protocol.

RECIPES: Please see the end of this protocol for recipes indicated by <R>. Additional recipes can be found online at http://cshprotocols.cshlp.org/site/recipes.

Reagents

Fast green solution (1%)

> *Dissolve fast green powder (Fisher Scientific F99-10) in endotoxin-free molecular grade water and filter through a 0.2-µm syringe filter.*

HEPES-buffered Steinberg's solution (100%) <R>

MS-222 (tricaine; 3-aminobenzoic acid ethyl ester)

> *Dissolve MS-222 powder in HEPES-buffered Steinberg's solution (100%) at a concentration of 0.02%.*

Plasmid DNA for cell transfection

[2]Correspondence: jebestman@wm.edu

TABLE 1. Useful Addgene (addgene.org) plasmids for generating expression constructs to label the cytosol and mitochondria using three-fragment MultiSite Gateway recombinational cloning

Addgene ID	Plasmid name	Gateway element or regulatory sequence	Localization	Fluorescent protein name ±2A self-cleaving peptide
Green fluorescence emission				
48348	pDONR P4-P1R-EGFP	5′ entry	Cytosol	EGFP
25899	pDONR221_EGFP	Middle entry	Cytosol	EGFP
80807	pME-GFP-P2A	Middle entry	Cytosol	EGFP_P2A
48349	pDONR P2R-P3-EGFP	3′ entry	Cytosol	EGFP
80826	p3E-P2A-GFP no-pA	3′ entry	Cytosol	P2A_EGFP
56529	Clover-Mito-7	CMV promoter	Mito	Clover
54160	mEmerald-Mito-7	CMV promoter	Mito	mEmerald
61373	4xnrUAS:GFP	UAS	Cytosol	EGFP
Red fluorescence emission				
48344	pDONR P4-P1R-mKate2	5′ entry	Cytosol	mKate2
25892	pDONR223_HcRed	Middle entry	Cytosol	HcRed
26031	543-p3E-2A-mCherrypA	3′ entry	Cytosol	mCherry
48345	pDONR P2R-P3-mKate2	3′ entry	Cytosol	mKate2
80828	p3E-P2A-mKate2 no-pA	3′ entry	Cytosol	P2A_mKate2
67707	p3E-p2a-tdTomato	3′ entry	Cytosol	P2A_tdTomato
61390	pCS2-TagRFPT.zf1	CMV promoter	Cytosol	tagRFP-T
55842	HcRed1-Mito-7	CMV promoter	Mito	hcRed1
55102	mCherry-Mito-7	CMV promoter	Mito	mCherry
55905	mRuby2-Mito-7	CMV promoter	Mito	mRuby2
58023	mTagRFP-T-Mito-7	CMV promoter	Mito	mTagRFP
58425	pclbw-mitoTagRFP-T	CMV promoter	Mito	tagRFP-T
Regulatory sequences				
61372	p5E-4xnrUAS (L4-R1)	5′ entry + UAS	-	-
34703	pSox2-bd::FP	Sox2 binding domain	-	-
80802	P5E-CMVmin	CMV promoter	-	-

Table 1 contains a selection of plasmids available from the plasmid repository Addgene (addgene.org) that target mitochondria and others that could be used for creating expression vectors using MultiSite Gateway cloning (Thermo-Fisher Scientific; Kwan et al. 2007; Protocol: Using Multisite LR Cloning to Generate a Destination Clone [Reece-Hoyes and Walhout 2018]). The expression of multiple proteins can be coregulated using gal4/UAS regulatory elements or by separating fluorescent protein sequences with the "self-cleaving" 2A peptide (Donnelly et al. 2001; Weber and Koster 2013; Fowler et al. 2016).

Tadpole food (TP-Tadpole Powder, *Xenopus* Express)

Xenopus laevis tadpoles, albino, stage 45 (Nieuwkoop and Faber 1994)

Equipment

Capillary glass

Capillary glass/micropipette puller (e.g., P97, Sutter Instruments)

Confocal microscope

Many suitable imaging system technologies can be used for in vivo time-lapse imaging; see Liebling (2011), Ruthazer et al. (2013), and Bayguinov et al. (2018). Bear in mind that the light sources and filters of the imaging system should be considered when selecting fluorescent reporters. A long-working-distance (~150-μm) objective is necessary to capture full 3D confocal stacks of cells in the Xenopus brain.

Containers for tadpoles (e.g., 2-cup food storage containers that hold 250 mL)

Course micromanipulators (e.g., MM33, Märzhäuser Wetzlar)

Coverslips

Custom electroporation electrode

The electroporation voltage pulses are delivered using an electrode that consists of a pair of ~1-mm platinum paddles (Bestman and Cline 2020).

Epifluorescence dissecting microscope

Fine brush

Image J package, Fiji (Schindelin et al. 2012)

Imaging chambers

Carve blocks of Sylgard elastomer so that the tadpoles are held in recesses just large enough that the top of their head is just at the surface of the Sylgard. Place the Sylgard blocks on a microscope slide so that they can be locked onto the stage of an upright microscope.

Laboratory incubator

Tadpoles are housed at 23°C with a 12 h light/12 h dark diurnal cycle.

Microcentrifuge

Microinjector/pressure injector and micropipette holder

Petri dish, 35-mm

Square pulse generator (e.g., Grass SD9 Stimulator)

Syringe filter (0.2-µm)

Tissue culture plate, six-well

Transfer pipette, plastic

METHOD

Tadpole Care

1. Rear groups of approximately 25 albino *Xenopus laevis* tadpoles in ∼250 mL HEPES-buffered Steinberg's solution (100%) in containers placed in a temperature- and light-controlled incubator. Tadpoles at stage 45 (Nieuwkoop and Faber 1994) are ready for electroporation and to be fed daily.

2. Anesthetize tadpoles by transferring them to a 35-mm Petri dish containing 0.02% MS222 in Steinberg's solution.

   They should be unresponsive after ∼90 sec. Tadpoles can be moved from dish to dish using a disposable plastic transfer pipette that has had its tip cut back so that the diameter is large enough to capture the tadpole.

Brain Cell Transfection with Targeted, In Vivo Electroporation

Consult Bestman and Cline (2020) for details and images of the capacitator circuit, electroporation electrode, and other equipment required for in vivo electroporation. See also Sec. 13, Protocol 3: Bulk Electroporation-Mediated Gene Transfer into Xenopus Tadpole Brain (Sáenz de Miera et al. 2018). Here we detail how to adapt this technique to target mitochondria in Xenopus tadpole brain cells.

3. Make an ∼20-µL aliquot of Fast Green and endotoxin-free plasmid DNA. For example, mix pSox2-bd::tagRFP-T (built from Addgene 34703 and 58023) (Day and Davidson 2009) and UAS::mitoEGFP (similar to Addgene 61373 with an amino-terminal Cox8 localization sequence) (Akitake et al. 2011), and dilute with water to achieve a final concentration of ∼0.05% dye and 0.5 µg/µL DNA. Store at 4°C and use within ∼1 mo.

4. Centrifuge the plasmid DNA mixture at ∼2000*g* for 15 sec at room temperature in a microfuge to remove debris. Fill ∼3 µL of the mixture into a pulled micropipette and secure it in the micropipette holder of the pressure injector. Use the recommendations for a Type E pipette as a guide (Osterele 2018) to form a long taper that can be broken back to a ∼50-µm tip diameter. (It should penetrate the skin without flexing or causing excess damage to the tadpole.)

5. Position anesthetized tadpoles under a dissecting microscope on a wet tissue.

6. Position the micropipette tip in the ventricle and with minimal pressure (<20 psi) and millisecond pulses, expel the DNA mixture. (The fast green dye should show that the DNA solution fills the ventricle with the edges of the brain visible in contrast.) Retract the micropipette.

 Cite this protocol as *Cold Spring Harb Protoc*; doi:10.1101/pdb.prot106807

7. Use the micromanipulator to position the electroporation electrode so that its paddles straddle the edges of the brain. Apply four 1.6-msec pulses of 35 V with the square pulse generator. Reverse the polarity and repeat. Retract the electrode.

8. Return tadpoles to the container of Steinberg's solution, and place the container back in the incubator. Wait ~24–36 h for fluorescent protein expression.

In Vivo Confocal Microscopy of Labeled Cells in *Xenopus* Brain

9. Anesthetize animals as in Step 2. Prioritize candidates by screening the anesthetized animals with a dissecting microscope equipped with epifluorescence. Aim for tadpoles containing isolated cells with strong fluorescent signals situated within 150 μm of the surface of the brain. Separate tadpoles into a six-well tissue culture plate containing Steinberg's solution in order to track their identity.

10. Position the tadpole with a fine brush in the Sylgard chamber, and place a coverslip over it and gently press. Blot the excess Steinberg's solution with a tissue and check that the coverslip has adhered to the Sylgard so that it is stable.

11. After the tadpole is positioned in its chamber, secure the chamber on the microscope stage and locate the cells that are fluorescently labeled in the brain.

12. Acquire full confocal *z*-stacks for each fluorescent protein signal (e.g., the cytosolic label of the cell morphology and the mitochondrially localized fluorescent signal). Obtain images that are sufficient to identify the mitochondrial label over background noise with minimal cytotoxicity. Settings will vary widely with imaging systems, but adjustments can be made to the laser power, detector sensitivity, or acquisition time, the dwell times and averaging levels, time-lapse interval, and duration of acquisition. Figure 1A is a maximum projection of a radial glial progenitor cell expressing cytosolic tagRFP-T and mitoEGFP acquired with a spinning disk confocal microscope. Figure 1B–D shows three time points of a 340-sec time lapse of the mito-EGFP with the following parameters: 100% laser, 100-msec exposure, 1-μm *z*-step interval.

13. Remove the chamber from the microscope stage, take off the coverslip, and gently flush out the tadpole from the chamber with Steinberg's solution into its well of the six-well plate.

> *The animals can be maintained for days in the incubator, removing them only for imaging sessions and changing the Steinberg's solution once a day.*

Image Analysis

14. Install the Image J package, Fiji (Schindelin et al. 2012), and load SNT (Simple Neurite Tracer) (Longair et al. 2011).

    i. Click on Help/Upgrade to open the ImageJ Updater.

    ii. Open Manage update site.

    iii. Check Neuroanatomy and apply changes.

    iv. Restart Fiji.

15. Proprietary image formats can often be imported using the Bio-Formats Importer plug-in (Linkert et al. 2010). Select "hyperstack" and "group files with similar names" to import each channel. The channels, *z*-stack, and time points can be assigned if needed using "swap dimensions."

16. Open Neuroanatomy Plugin > SNT and select the image file as instructed. The default view opens five windows: the image file, its two orthogonal views, the SNT window, and the path manager window where the reconstruction is stored. Generous and detailed instructions are available for SNT at imagej.net.

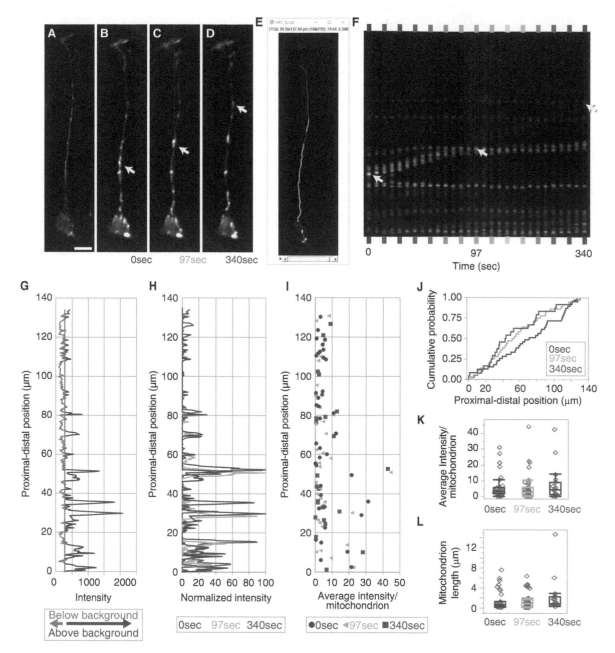

FIGURE 1. Rapid 3D time lapse of a neural progenitor cell in the stage 46 *Xenopus laevis* optic tectum. One day after electroporation of plasmid DNA to deliver cytosolic tagRFP-t and mitochondrially localized EGFP, the full 3D confocal stack (*z* = 1 μm) was acquired with a spinning disk confocal system (3i, Inc). A projection of the 44-μm stack is shown in *A*. (*B–D*) Projections of 3D confocal stacks acquired at 0, 175, 340 sec of the time lapse. (*E*) Example of the reconstruction generated with the SNT plug-in in Fiji. (*F*) The kymograph of the mito-EGFP signal within the radial process of the NPC at each time point (11.7-sec interval). The arrows in *B–D* and *F* indicate the movement of a single mitochondrion. (*G*) The intensity of mito-EGFP signal at 0 sec, showing the background level (275) that is subtracted to reveal the intensity along the radial process reconstruction in *E*. This calculation displays the relative positions of the mitochondrial start and end points. (*H*) The intensity (normalized to the maximum value of each time point) at the first, mid, and last time points shows that the majority of the mitochondria along the radial process hold stable positions over time. With the position of the mitochondria identified from the plot profile data in *H*, the average normalized intensity for each mitochondrion along the proximal–distal position of the cell can be calculated (*I*). (*J*) The cumulative probability plot of the position of the mitochondria suggests changes in the distribution along the process over time. With individual mitochondria identified from the Plot Profile results, the average intensity (*K*) and length (*L*) of the mitochondria can be calculated. Scale bar, 10 μm.

17. In the Option tab, select the channel that contains the mitochondrial label as the "Data Source" but view the cytosolic label so that the reconstruction will be completed using the channel that contains the cell morphology data.

18. Complete the reconstruction:

    i. Scroll within the z-stack and find the most proximal point of the cell's main neurite or major process and "start a new path" by clicking on this point.

    ii. Select a point ~20 μm more distal and select it. Choose "yes" if the quality of the reconstruction is suitable and save the segment. If the quality of the reconstruction is not suitable, choose "no" and click on a less remote point.

    iii. Continue to add segments until the process is reconstructed. Select "finish." This reconstruction appears in the Path Manager window where it can be named, filled, and refined. An example of a SNT path within a 3D stack file is shown in Figure 1E.

    iv. Add side branches by starting a new path while holding down shift-alt. A completed branch will appear as a "child" to the main path in the path manager window.

19. To save the intensity-position data, select the path in the Path Manager window, and click on Analyze > Plot Profile. A new window will appear with two columns of data: the position along the length of the 3D reconstruction and the intensity of the mitochondria channel at each position. Save (Data > Save Data) as a .csv spreadsheet to be processed and analyzed offline.

20. Produce a kymograph by creating a maximum intensity projection of the time-lapse file and drawing a segmented line along the process. Analyze > MultiKymograph produces a file (Fig. 1F) that reveals stationary and moving mitochondria (arrows in Fig. 1B–D,F). Analyze these data to reveal rate and direction of movement and levels of fusion and fission.

    In this example, the mitochondrion moves at 0.13 μm/sec and has a ~30% drop in fluorescence intensity as it passes by the stationary mitochondrion (Fig. 1F, 97 sec, arrow), suggesting a fission/fusion between the motile and stable mitochondria.

21. If necessary, process the images to aid in identifying the mitochondria. Use the following Fiji tools.

    • Thresholding to identify puncta by a cutoff intensity (Image > Adjust > Threshold).

    • WalkingAverage function within MultiKymograph (Paquet et al. 2014) to reduce noise in time-lapse data.

    • Median filter (Process > Filter) to reduce noise.

22. Identify and tabulate mitochondria from the image stacks or kymograph manually using the ROI Manager or Cell Counter tools.

23. Calculate the average maximum intensity background value for the 3D stack. Subtract this background value (275 for the 0 sec time point; Fig. 1G) from the Plot Profile data.

    The values that are greater than background indicate the presence of the mitochondria along the cell process.

24. To normalize intensity data so that comparisons can be made between images, divide the background-subtracted intensity of each image by its maximum value (Fig. 1H).

25. After identifying the mitochondria, determine the average normalized intensity of each organelle (Fig. 1I), and calculate features such as the distribution of their positions along the cell's process (Fig. 1J), average intensities (Fig. 1K), and lengths (Fig. 1L).

RECIPE

HEPES-Buffered Steinberg's Solution (100%)

Reagent	Quantity
KCl	0.05 g
$Ca(NO_3)_2\ 4H_2O$	0.08 g
$MgSO_4\ 7H_2O$	0.205 g
NaCl	3.4 g
HEPES	1.19 g

Dissolve in 1 L of dH_2O. Adjust the pH to 7.4 by adding NaOH. This working solution can be made from a 10× stock solution (stored at 4°C) by diluting with dH_2O. Check pH before using. The 100% working solution can be kept at room temperature for ~1 mo. (This recipe is from Zackson and Steinberg 1986.)

ACKNOWLEDGMENTS

We are grateful to the W&M Charles Center and Ferguson Awards for support of M.S.F. The generosity of the Grass Foundation and W&M Summer Faculty grants have supported J.E.B.

REFERENCES

Akitake CM, Macurak M, Halpern ME, Goll MG. 2011. Transgenerational analysis of transcriptional silencing in zebrafish. *Dev Biol* 352: 191–201. doi:10.1016/j.ydbio.2011.01.002

Bayguinov PO, Oakley DM, Shih CC, Geanon DJ, Joens MS, Fitzpatrick JAJ. 2018. Modern laser scanning confocal microscopy. *Curr Protoc Cytom* 85: e39. doi:10.1002/cpcy.39

Bestman JE, Cline HT. 2020. Morpholino studies in *Xenopus* brain development. *Methods Mol Biol* 2047: 377–395. doi:10.1007/978-1-4939-9732-9_21

Day RN, Davidson MW. 2009. The fluorescent protein palette: tools for cellular imaging. *Chem Soc Rev* 38: 2887. doi:10.1039/b901966a

Donnelly MLL, Luke G, Mehrotra A, Li XJ, Hughes LE, Gani D, Ryan MD. 2001. Analysis of the aphthovirus 2A/2B polyprotein 'cleavage' mechanism indicates not a proteolytic reaction, but a novel translational effect: a putative ribosomal 'skip'. *J Gen Virol* 82: 1013–1025. doi:10.1099/0022-1317-82-5-1013

Fowler DK, Stewart S, Seredick S, Eisen JS, Stankunas K, Washbourne P. 2016. A MultiSite Gateway toolkit for rapid cloning of vertebrate expression constructs with diverse research applications. *PLoS One* 11: e0159277. doi:10.1371/journal.pone.0159277

Kwan KM, Fujimoto E, Grabher C, Mangum BD, Hardy ME, Campbell DS, Parant JM, Yost HJ, Kanki JP, Chien CB. 2007. The Tol2kit: a multisite gateway-based construction kit for Tol2 transposon transgenesis constructs. *Dev Dyn* 236: 3088–3099. doi:10.1002/dvdy.21343

Liebling M. 2011. Imaging the dynamics of biological processes via fast confocal microscopy and image processing. *Cold Spring Harb Protoc* doi: 10.1101/pdb.top117

Linkert M, Rueden CT, Allan C, Burel JM, Moore W, Patterson A, Loranger B, Moore J, Neves C, MacDonald D, et al. 2010. Metadata matters: access to image data in the real world. *J Cell Biol* 189: 777–782. doi:10.1083/jcb.201004104

Longair MH, Baker DA, Armstrong JD. 2011. Simple Neurite Tracer: open source software for reconstruction, visualization and analysis of neuronal processes. *Bioinformatics* 27: 2453–2454. doi:10.1093/bioinformatics/btr390

Nieuwkoop PD, Faber J. 1994. *Normal table of* Xenopus laevis *(Daudin): a systematical & chronological survey of the development from the fertilized egg till the end of metamorphosis.* Garland Science, New York.

Osterele A. 2018. Pipette Cookbook 2018: P-97 & P-1000 Micropipette Pullers, Rev. F. Sutter Instruments, www.sutter.com/PDFs/pipette_cookbook.pdf

Paquet D, Plucinska G, Misgeld T. 2014. In vivo imaging of mitochondria in intact zebrafish larvae. *Methods Enzymol* 547: 151–164. doi:10.1016/B978-0-12-801415-8.00009-6

Reece-Hoyes JS, Walhout AJM. 2018. Using multisite LR cloning to generate a destination clone. *Cold Spring Harb Protoc* doi: 10.1101/pdb.prot094946

Ruthazer ES, Schohl A, Schwartz N, Tavakoli A, Tremblay M, Cline HT. 2013. In vivo time-lapse imaging of neuronal development in *Xenopus*. *Cold Spring Harb Protoc* doi: 10.1101/pdb.top077156

Sáenz de Miera C, Parr E, Denver RJ. 2018. Bulk electroporation-mediated gene transfer into *Xenopus* tadpole brain. *Cold Spring Harb Protoc* doi: 10.1101/pdb.prot097691

Schindelin J, Arganda-Carreras I, Frise E, Kaynig V, Longair M, Pietzsch T, Preibisch S, Rueden C, Saalfeld S, Schmid B, et al. 2012. Fiji: an open-source platform for biological-image analysis. *Nat Methods* 9: 676–682. doi:10.1038/nmeth.2019

Weber T, Koster R. 2013. Genetic tools for multicolor imaging in zebrafish larvae. *Methods* 62: 279–291. doi:10.1016/j.ymeth.2013.07.028

Zackson SL, Steinberg MS. 1986. Cranial neural crest cells exhibit directed migration on the pronephric duct pathway: further evidence for an in vivo adhesion gradient. *Dev Biol* 117: 342–353. doi:10.1016/0012-1606(86)90304-0

Whole-Brain Calcium Imaging in Larval *Xenopus*

Thomas Offner,[1] Daniela Daume, Lukas Weiss, Thomas Hassenklöver, and Ivan Manzini

Department of Animal Physiology and Molecular Biomedicine, University of Giessen, 35392 Giessen, Germany

Sensory systems detect environmental stimuli and transform them into electrical activity patterns interpretable by the central nervous system. En route to higher brain centers, the initial sensory input is successively transformed by interposed secondary processing centers. Mapping the neuronal activity patterns at all of those stages is essential to understand sensory information processing. Larval *Xenopus laevis* is very well-suited for whole-brain imaging of neuronal activity. This is mainly due to its small size, transparency, and the accessibility of both peripheral and central parts of sensory systems. Here we describe a protocol for calcium imaging at several levels of the olfactory system using focal injection of chemical calcium indicator dyes or a *Xenopus* transgenic line with neuronal GCaMP6s expression. In combination with fast volumetric multiphoton microscopy, the calcium imaging methods described can provide detailed insight into spatiotemporal activity of entire brain regions at different stages of sensory information processing. Although the methods are broadly applicable to the central nervous system, in this work we focus on protocols for calcium imaging of glomeruli in the olfactory bulb and odor-responsive neurons in the olfactory amygdala.

MATERIALS

It is essential that you consult the appropriate Material Safety Data Sheets and your institution's Environmental Health and Safety Office for proper handling of equipment and hazardous materials used in this protocol.

RECIPES: Please see the end of this protocol for recipes indicated by <R>. Additional recipes can be found online at http://cshprotocols.cshlp.org/site/recipes.

Reagents

Calcium indicator solution <R>

The addition of MK571 to the calcium indicator solution is optional. MK571 can block the extrusion of AM dyes from certain neurons by multidrug transporters (Manzini and Schild 2003), resulting in higher intracellular concentrations of calcium indicator.

Ethyl 3-aminobenzoate methanesulfonate (MS-222; Tokyo Chemical Industry T0941)
Frog ringer <R>
Larval *Xenopus laevis* albino (NASCO strain) or transgenic animals expressing the genetically encoded calcium indicator GCaMP6s under the *Xenopus* neuronal β-tubulin promoter (National *Xenopus* Resource NXR 0.0107)

We have performed these experiments on stage 52–54 Xenopus (Nieuwkoop and Faber 1956), but the techniques are most likely generally applicable to various stages.

---

[1]Correspondence: thomas.offner@physzool.bio.uni-giessen.de

Supplemental material is available at cshprotocols.cshlp.org.

Odorant solutions

For the olfactory system of aquatic animals, amino acids, amines, bile acids, and other water-soluble chemicals have proven to be potent stimuli. For a detailed description of components used in our system, see Gliem et al. (2013) and Friedrich and Korsching (1997). Dissolve odorants in frog ringer as 10 mM concentrated stock solutions and keep aliquots at −20°C. Prepare fresh odorant solutions of 100 µM concentration in frog ringer from stock for calcium imaging.

Equipment

Borosilicate glass capillaries

Draining system consisting of a bent syringe needle connected to a peristaltic pump (Cyclo II, Roth) via flexible silicone tubing

Fine forceps (e.g., Fine Science Tools, Dumont 5SF Forceps, Super fine 0.025 × 0.005-mm)

Fine spring scissors (e.g., Fine Science Tools, Vannas Spring Scissors, 2-mm blades)

Gravity-fed multichannel perfusion system (ALA-VM-8 Series + valve commander VC3-8xP Series; ALA Scientific) with common outflow funnel (Milli Manifold; 16 inlet ports; ALA Scientific) mounted to a micromanipulator (e.g., M-MT-AB2, Newport)

Imaging analysis tools (e.g., CaImAn; Fiji, etc.)

Microcentrifuge (capable of 16,100g)

Micropipette head stage with attached pressure application system composed of a syringe and a three-way valve connected via silicone tubing

The head stage should be mounted to a micromanipulator (e.g., Scientifica, PatchStar).

Nylon-stringed platinum grid (dimension of the recording chamber's recess)

Pinning needles (0.2-mm diameter)

Pipette puller (e.g., P 1000, Sutter Instruments)

Recessed recording chamber (compatible with microscope stages)

Silicone rubber-filled Petri dish (transparent)

Prepare by filling transparent plastic or glass Petri dishes of the desired diameter (we use 5.5-cm) with transparent silicone rubber compound. Leave to dry at room temperature until the rubber is completely hardened.

Upright fluorescent stereomicroscope for dye injection (bright-field and fluorescent light source including appropriate filters for calcium indicator and dextran tracer; e.g., Olympus BX51WI; light source AMH-200-F6S; Andor Technology-Oxford Instruments)

Upright multiphoton microscope with fast scanning mirrors and piezo z-drive for volumetric calcium imaging (e.g., Nikon A1R MP)

METHOD

All scientific procedures on living animals used in this protocol were approved by the regional board (RP Giessen Az: V54-19c2015h01 GI 15/7; Niedersächsisches Landesamt für Verbraucherschutz und Lebensmittelsicherheit, Oldenburg, Germany, Az: 16/2136) in accordance with the German animal welfare law and the European legislation for the protection of animals used for scientific purposes (2010/63/EU).

Whole-Mount Preparation

An instructional video of the preparation can be found at https://www.jove.com/video/54108/recording-temperature-induced-neuronal-activity-through-monitoring (2:06–4:06; Brinkmann et al. 2016).

1. Anesthetize larvae in 0.02% MS-222 (pH 7.6 in tap water) until complete irresponsiveness.

   Tail movements usually cease after 2–3 min, and the animal is generally irresponsive after 3–5 min.

2. Assess proper anesthesia and kill the animal by severing the brainstem using fine scissors.

   Anesthesia is confirmed by the absence of any motoric responses of the animal when using a small metal spatula to gently touch it or to create shock waves by tapping on the bottom of the beaker.

Cite this protocol as *Cold Spring Harb Protoc*; doi:10.1101/pdb.prot106815

3. Remove excess tissue rostral of the nostril to facilitate odorant application during the experiment.

4. Cut out a rectangular block including the entire peripheral olfactory system and the central nervous system.

5. Pin the explant to the silicone-filled Petri dish with fine needles and submerge it in a drop of frog ringer to prevent the tissue from drying out.

6. Expose the central nervous system by removing the superincumbent palatial tissue with forceps and scissors without injuring or stretching essential nerve bundles (here olfactory nerves).

Calcium Indicator Dye Injection

7. Fix the preparation in a recessed plastic recording chamber using a nylon-stringed platinum grid. Fill the recess with additional frog ringer to keep the explant submerged. Position the recording chamber under the upright stereomicroscope (see Fig. 1).

8. Pull micropipettes from borosilicate glass capillaries with pipette resistances in the range of 8–10 MΩ.

9. Fill a micropipette with up to 3 µL of calcium indicator solution. Avoid air bubble formation in the pipette tip.

10. Mount the micropipette to the head stage and pierce the pipette tip into the brain region of interest via the micromanipulator under bright-field or fluorescent illumination.

11. Build up pressure by compressing air in the syringe connected to the tubing system with the valve closed.

12. Open the valve to inject the dye, and adjust the manually applied pressure in relation to the observed dye extrusion from the micropipette under fluorescent illumination (cascade blue dextran fluorescence).

13. Relieve the pressure after 5 sec at the maximum. If the pipette gets clogged or the injected volume is not sufficient, withdraw the pipette and repeat pressure application once more at a slightly distinct injection site. Otherwise use a new micropipette.

14. Determine the number of injection sites depending on the size of the target region. To maximize dye load for the olfactory bulb neurons and keep mechanical damage minimal, inject at three adjacent sites.

15. Incubate the preparation for 30 min at room temperature before imaging. A slight increase in the faint green fluorescence of Fluo-4 is observed in the target region after successful uptake of the AM dye into neurons.

3D-Multiphoton Calcium Imaging of Odor-Evoked Neuronal Activity

16. Position the recording chamber on the stage of a multiphoton microscope.

17. Add odorant solutions to the additional syringes of the perfusion system. Adjust the positioning of the perfusion system outflow tube using the micromanipulator so that applied odorants reach the olfactory organ.

> The valve commander and a common outlet funnel enable switching between frog ringer perfusion and individual odorants without interrupting the continuous flow in the recording chamber.

18. Create a stable flow of frog ringer between the outflow tube and the suction syringe positioned caudal to the preparation.

> A flow rate of ~50 µL/sec was used in our experiments, and therefore the estimated volume of odorant solution applied was 250 µL for 5 sec of stimulation.

19. Record volumetric time series experiments in the fast scanning mode of the multiphoton microscope (see Supplemental Movie S1 and Supplemental Movie S1 Legend). For our goal of

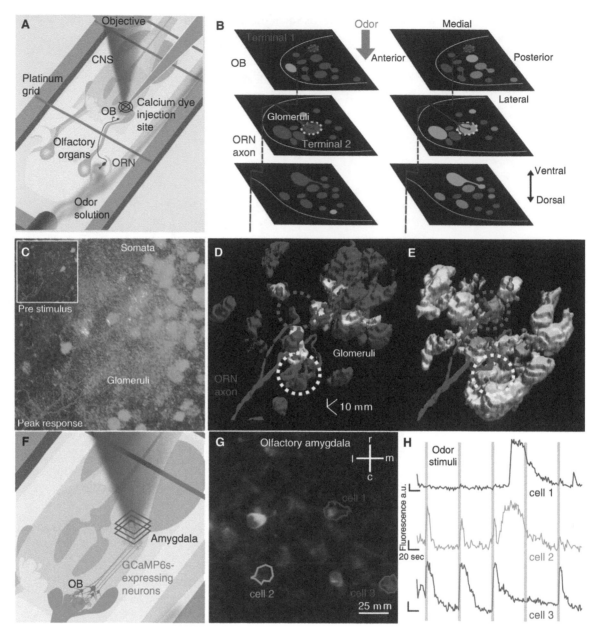

FIGURE 1. Volumetric multiphoton calcium imaging of odor representations in the olfactory bulb and olfactory amygdala of larval *Xenopus laevis*. (*A*) Larval *Xenopus* whole-mount preparation including the peripheral olfactory system and the central nervous system. Site of calcium indicator injection in the olfactory bulb (OB) is indicated. Single olfactory receptor neurons (ORNs) and their axonal projections in the OB (red) can be labeled before calcium imaging experiments via single-cell electroporation of dextran-coupled fluorophores (Hassenklöver and Manzini 2014). (*B*) Volumetric multiphoton calcium imaging of multiple image planes before (*left*) and after (*right*) odor stimulation. Increase in calcium-dependent fluorescence (dark to lighter green) can be observed in neurons and olfactory glomeruli of the OB (ovoid shapes) in proximity to individual ORN axons (red) and their terminals (magenta and yellow). (*C*) Maximum intensity projection of an imaged OB volume including ORN axons (red) and the calcium indicator signal of dye-loaded glomeruli and neurons (green). The peak fluorescence response of the OB neuronal network after odorant application (light green) is shown. (*Inset*) Calcium indicator fluorescence of the same region before stimulation. (*D,E*) Activity-based 3D reconstruction of the OB glomerular map activated by individual odorants (arginine, light blue; lysine, dark blue; tryptophan, dark red; tryptophan and histidine, orange) in proximity to the axonal terminals of two ORNs (yellow/magenta outlines) of a single ORN (light red). (*F–H*) Multiphoton imaging of odor responses in the olfactory amygdala of larval *Xenopus* with neuronal expression of the genetically encoded calcium indicator GCaMP6s (*F*, light and dark green). Maximum fluorescence intensity projection of measured olfactory amygdala volume including GCaMP6s-expressing neurons (*G*). (*H*) Fluorescent signal time course of selected GCaMP6s-expressing neurons from *G* (cell 1–3). Gray bars indicate stimulations with different odorant groups (amino acids, amines, bile acids, etc.). (m) Medial, (l) lateral, (r) rostral, (c) caudal.

Cite this protocol as *Cold Spring Harb Protoc*; doi:10.1101/pdb.prot106815

mapping long-lasting odor-evoked calcium transients, we scan brain volumes of 25–35 image planes (512 × 512 px; resolution: 0.33–0.66 µm/px; interplane distance of 3–6 µm) at frequencies between 0.5 and 1 Hz.

20. When applying multiple odor stimuli, record at least 10 sec of baseline activity before exposure and wait at least 60 sec between stimuli to cover the entire calcium response and avoid desensitization effects.

21. Repeat every stimulus at least twice and aim for recordings with little lateral or axial drift of the sample throughout the recording. *Xenopus* brain explants can be imaged for 1–2 h at room temperature under ideal conditions.

Processing and Evaluation of Calcium Imaging Data

22. Process and evaluate stimulus-induced neuronal activity in the different brain regions using published calcium imaging analysis toolkits such as CaImAn (Giovannucci et al. 2019) based on algorithms described in detail in Pnevmatikakis et al. (2014, 2017) and Friedrich et al. (2017).

23. Compensate for motion artifacts of the sample along the x- or y-axis by running a piecewise-rigid motion correction algorithm (NoRMCorre; Pnevmatikakis and Giovannucci 2017).

24. Improve data quality for subsequent analysis steps by denoising, deconvolution, and demixing of the motion-corrected image data by CaImAn (Giovannucci et al. 2019).

25. Derive fluorescent time traces and spatial footprints of stimulus-responsive regions of interest from constrained nonnegative matrix factorization (Pnevmatikakis et al. 2014), and select by choosing additional desired quality criteria for the reactive regions of interest in CaImAn (Giovannucci et al. 2019).

26. (Optional) Visualize odor representations by creating fluorescence intensity difference maps from the 4D image data (e.g., using CaImAn denoised movies). Subtract the peak fluorescence intensity of stimulus correlated responses from the baseline mean fluorescence before the stimulus for each pixel of the image stack.

DISCUSSION

Focal bulk loading of chemical calcium indicators (Garaschuk et al. 2006; Paredes et al. 2009) and the employment of animals that genetically encode calcium indicators such as GCaMP6s (Chen et al. 2013) are established methods to monitor neuronal activity based on transient elevations of intracellular calcium levels (Grienberger and Konnerth 2012). Previous work from our laboratory and others includes the development of protocols for calcium imaging of neuronal activity in the olfactory bulb of aquatic vertebrates on multiple levels (Friedrich and Korsching 1997; Yaksi and Friedrich 2006; Manzini et al. 2007; Yaksi et al. 2007; Chen et al. 2009; Junek et al. 2010; Gliem et al. 2013; Brinkmann et al. 2016). Our protocol illustrates how to record neuronal activity in several brain regions of larval *Xenopus laevis*, thus complementing and extending previous protocols (Manzini et al. 2002; Brinkmann et al. 2016). Injection of calcium dye solution via manual pressure application often comes at the cost of exact knowledge of the injected volume. Nevertheless, adjustments of pressure based on the observed dye extrusion are helpful to compensate varying tissue resistances in different brain regions or in animals of different developmental stages. By using (over)saturated calcium indicator solution for focal bulk loading we achieved an elevated level of dye load in neurons. The resulting high signal-to-noise ratio despite the short pixel dwell times of fast resonant scanning is crucial for the 3D reconstruction of entire neuronal networks from the calcium-dependent fluorescence increase. Additional injection of calcium dye in larvae expressing genetically encoded calcium indicators (e.g., a red calcium dye complementing the green GCaMP6s) allows simultaneous recording of different types of neurons or neuronal modules across the larval central nervous system. Although we focus on odor-evoked neuronal activity patterns in the olfactory bulb and olfactory amygdala, the size, easy preparation, and translucency of *Xenopus* larvae provide an excellent framework for using our calcium imaging protocol in other neural subsystems of their central nervous system.

RECIPES

Calcium Indicator Solution

Dissolve 50 µg of Fluo-4 AM dye (Thermo Fisher F14201) in 5 µL of dimethyl sulfoxide (DMSO; Sigma-Aldrich D8418) before adding 10 µL of Pluronic F-127 (Biotium 59005) and 35 µL of frog ringer <R> to the solution. Prepare this solution on the day of the experiment with all ingredients at room temperature; store for the day at room temperature.

Centrifuge solution at 16,100*g* for 60 sec at room temperature. Transfer the supernatant to a new tube and discard the pellet. Add 0.3 µL of the fluorescent tracer cascade-blue dextran (Thermo Fisher D1976; 3 mM in frog ringer) to the solution to facilitate the observation of dye extrusion into the tissue.

(Optional) Add 3 µL of MK571 (50 µM in frog ringer; Alexis Biochemicals) to the calcium indicator solution.

Frog Ringer

Reagent	Amount (per L)	Final concentration
NaCl	5.727 g	98 mmol/L
KCl	0.149 g	2 mmol/L
CaCl$_2$ dihydrate	0.147 g	1 mmol/L
MgCl$_2$ hexahydrate	0.407 g	2 mmol/L
D-glucose	0.900 g	5 mmol/L
Sodium pyruvate	0.550 g	5 mmol/L
Hydroxyethyl piperazineethanesulfonic acid (HEPES)	2.383 g	10 mmol/L

Prepare in distilled H$_2$O. Adjust the pH to 7.8 and confirm that the osmolarity is 230 mOsmol/L. Store for 1–2 wk at 4°C.

ACKNOWLEDGMENTS

Work in our laboratory is supported by Deutsche Forschungsgemeinschaft (DFG) Grants 4113/3-1 and 4113/4-1.

REFERENCES

Brinkmann A, Okom C, Kludt E, Schild D. 2016. Recording temperature-induced neuronal activity through monitoring calcium changes in the olfactory bulb of *Xenopus laevis*. *J Vis Exp* 112: e54108. doi:10.3791/54108

Chen T-W, Lin B-J, Schild D. 2009. Odor coding by modules of coherent mitral/tufted cells in the vertebrate olfactory bulb. *Proc Natl Acad Sci* 106: 2401–2406. doi:10.1073/pnas.0810151106

Chen T-W, Wardill TJ, Sun Y, Pulver SR, Renninger SL, Baohan A, Schreiter ER, Kerr RA, Orger MB, Jayaraman V, et al. 2013. Ultrasensitive fluo-rescent proteins for imaging neuronal activity. *Nature* 499: 295–300. doi:10.1038/nature12354

Friedrich RW, Korsching SI. 1997. Combinatorial and chemotopic odorant coding in the zebrafish olfactory bulb visualized by optical imaging. *Neuron* 18: 737–752. doi:10.1016/S0896-6273(00)80314-1

Friedrich J, Zhou P, Paninski L. 2017. Fast online deconvolution of calcium imaging data. *PLoS Comput Biol* 13: e1005423. doi:10.1371/journal.pcbi .1005423

Cite this protocol as *Cold Spring Harb Protoc*; doi:10.1101/pdb.prot106815

Garaschuk O, Milos R-I, Konnerth A. 2006. Targeted bulk-loading of fluorescent indicators for two-photon brain imaging *in vivo*. *Nat Protoc* **1**: 380. doi:10.1038/nprot.2006.58

Giovannucci A, Friedrich J, Gunn P, Kalfon J, Koay A, Taxidis J, Naja F, Gauthier JL, Tank DW, Chklovskii D, et al. 2019. CaImAn: an open source tool for scalable calcium imaging data analysis. *Elife* **8**: e38173. doi:10.7554/eLife.38173

Gliem S, Syed AS, Sansone A, Kludt E, Tantalaki E, Hassenlöver T, Korsching SI, Manzini I. 2013. Bimodal processing of olfactory information in an amphibian nose: odor responses segregate into a medial and a lateral stream. *Cell Mol Life Sci* **70**: 1965–1984. doi:10.1007/s00018-012-1226-8

Grienberger C, Konnerth A. 2012. Imaging calcium in neurons. *Neuron* **73**: 862–885. doi:10.1016/j.neuron.2012.02.011

Hassenlöver T, Manzini I. 2014. The olfactory system as a model to study axonal growth patterns and morphology *in vivo*. *J Vis Exp* 52143. doi:10.3791/52143

Junek S, Kludt E, Wolf F, Schild D. 2010. Olfactory coding with patterns of response latencies. *Neuron* **67**: 872–884. doi:10.1016/j.neuron.2010.08.005

Manzini I, Schild D. 2003. Multidrug resistance transporters in the olfactory receptor neurons of *Xenopus laevis* tadpoles. *J Physiol* **546**: 375–385. doi:10.1113/jphysiol.2002.033175

Manzini I, Rössler W, Schild D. 2002. cAMP-independent responses of olfactory neurons in *Xenopus laevis* tadpoles and their projection onto olfactory bulb neurons. *J Physiol* **545**: 475–484. doi:10.1113/jphysiol.2002.031914

Manzini I, Brase C, Chen T-W, Schild D. 2007. Response profiles to amino acid odorants of olfactory glomeruli in larval *Xenopus laevis*. *J Physiol* **581**: 567–579. doi:10.1113/jphysiol.2007.130518

Nieuwkoop PD, Faber J. 1956. *Normal table of* Xenopus laevis *(Daudin)*. North Holland, Amsterdam.

Paredes RM, Etzler JC, Watts LT, Lechleiter JD. 2009. Chemical calcium indicators. *Methods* **46**: 143–151. doi:10.1016/j.ymeth.2008.09.025

Pnevmatikakis EA, Giovannucci A. 2017. NoRMCorre: an online algorithm for piecewise rigid motion correction of calcium imaging data. *J Neurosci Methods* **291**: 83–94. doi:10.1016/j.jneumeth.2017.07.031

Pnevmatikakis EA, Gao Y, Soudry D, Pfau D, Lacefield C, Poskanzer K, Bruno R, Yuste R, and Paninski L. 2014. A structured matrix factorization framework for large scale calcium imaging data analysis. ArXiv doi:ArXiv1409.2903

Pnevmatikakis EA, Soudry D, Gao Y, Machado TA, Merel J, Pfau D, Reardon T, Mu Y, Lacefield C, Yang W, et al. 2017. Simultaneous denoising, deconvolution, and demixing of calcium imaging data. *Neuron* **89**: 285–299. doi:10.1016/j.neuron.2015.11.037

Yaksi E, Friedrich RW. 2006. Reconstruction of firing rate changes across neuronal populations by temporally deconvolved Ca^{2+} imaging. *Nat Methods* **3**: 377–383. doi:10.1038/nmeth874

Yaksi E, Judkewitz B, Friedrich RW. 2007. Topological reorganization of odor representations in the olfactory bulb. *PLoS Biol* **5**: e178. doi:10.1371/journal.pbio.0050178

Imaging the Dynamic Branching and Synaptic Differentiation of *Xenopus* Optic Axons In Vivo

Rommel Andrew Santos, Rodrigo Del Rio, Jr., and Susana Cohen-Cory[1]

Department of Neurobiology and Behavior, University of California Irvine, Irvine, California 92697, USA

In the developing *Xenopus* tadpole visual system, the targeting and branching of optic axons in the brain is a dynamic process that is closely intertwined with the morphological differentiation and maturation of their postsynaptic neurons and with the formation, stabilization, and elimination of functional synapses. The coordinated addition and retraction of axonal and dendritic branches guides the gradual recognition between pre- and postsynaptic neuronal partners, which subsequently allows synaptic connections to be formed. Axon and dendrite branching and selective synapse formation and stabilization are developmental mechanisms largely orchestrated by an array of signaling molecules that interact in vivo for the proper formation of functional visual circuits. In vivo real-time imaging of individual fluorophore-labeled neurons in living *Xenopus* tadpoles has allowed investigation of molecular and cellular mechanisms mediating circuit assembly at a cellular level in the intact organism. In this protocol, we describe the use of bulk and single-cell electroporation to rapidly and efficiently transfect individual retinal ganglion cells (RGCs) with different reagents and to simultaneously visualize optic axon arbor morphology and presynaptic sites in real time. Similar techniques for labeling and visualizing RGC axons can be combined with the use of morpholino antisense oligonucleotides, as we describe here, to alter gene expression cell autonomously.

MATERIALS

It is essential that you consult the appropriate Material Safety Data Sheets and your institution's Environmental Health and Safety Office for proper handling of equipment and hazardous materials used in this protocol.

RECIPES: Please see the end of this protocol for recipes indicated by <R>. Additional recipes can be found online at http://cshprotocols.cshlp.org/site/recipes.

Reagents

Agar (Fisher BioReagents BP1423)
Agarose, low melt (IBI Molecular Biology Grade IB70056) (optional; see Step 17)
Dextran, Alexa Fluor 488 (3000 MW, 6 mM; Invitrogen)
Endofree Maxi Prep Kit (QIAGEN)
Methylcellulose (2%, prepared in 1× MR) (optional; see Step 17)
Modified rearing solution (1× MR)<R>
Morpholino antisense oligonucleotide (Gene Tools)

> *Custom-made morpholino antisense oligonucleotides of about 25 short-chain subunits complementary to the RNA of interest are designed to block translation of a protein of interest. These oligos contain a fluorophore tag (lissamine) for visualization of neurons.*

---

[1]Correspondence: scohenco@uci.edu

Plasmid constructs coding for GFP-synaptobrevin (Alsina et al. 2001) and tdTomato (Clontech), both controlled by CMV promoters

TE buffer for plasmids <R>

Tricaine methanesulfonate (Finquel MS-222)

Xenopus tadpoles, stages 30–45 (Nieuwkoop and Faber 1956)

This protocol has been optimized for Xenopus laevis *tadpoles but could be tried with* Xenopus tropicalis *tadpoles as well.*

Equipment

Aluminosilicate glass capillaries (1.00 mm outer diameter, 0.64 mm inner diameter, 10 cm length; Sutter Instrument AF100-64-10)

Anode and cathode copper electrodes, custom-made (see Falk et al. 2007)

Build a custom-made copper electrode by soldering copper alloy (0.5 mm in width) onto the end of an electrical lead wire. Mount two individual copper electrodes in parallel onto a rod (electrode holder, H-13, Narishige). Connect lead wires of the electrodes to an electronic stimulator via a banana plug.

Cell culture dish, six-well

Dissection microscope equipped with a mercury arc lamp

E-Series Electrode Holder with wire and handle (glass, 1.0 mm outer diameter; Warner Instruments E45P-M10NH)

Fine forceps (Electron Microscopy Sciences 72700-D) or a microelectrode beveler

Imaging analysis software (Neuromantic from OmicX or MetaMorph from Molecular Devices, Inc.)

Laser-scanning confocal microscope (Zeiss LSM780)

Any laser-scanning confocal microscope equipped with argon (488 nm excitation, 10% neutral density filter) and HeNe (543 nm excitation) lasers and 515/30 nm (barrier) and 605/32 nm (band-pass) filters is suitable for these studies. We use an LSM780 confocal microscope also equipped with a MaiTai Ti: Sapphire two-photon laser system for two-photon excitation.

Micropipette puller (Sutter Instrument)

Petri dishes (100 × 20 mm; Falcon)

Picospritzer injector (Picospritzer III, General Valve)

Square pulse electronic stimulator (Grass SD9 or similar, A-M Systems)

Standard size harp slice grid (ALA Scientific Instruments)

Tadpole trenches, custom made (two sizes)

Carve one trench the size of stage 30–34 tadpoles and another the size of stage 43–45 tadpoles from Sylgard (Silicone Elastomer Kit) molded in small Petri dishes (35 × 10 mm, Falcon).

Three-axis manual micromanipulators (M-333, Narishige)

Two are needed: one to hold the custom-made anode and cathode copper electrodes and another to hold the Picospritzer III injector.

METHOD

RGCs can be transfected with either plasmids (Steps 1–7) or morpholinos (Steps 8–13) followed by in vivo imaging and analysis of axon arbors (Steps 14–23).

Transfection of RGCs with DNA Plasmids by Electroporation to Visualize Axon Arbors and Presynaptic Sites

1. Prepare plasmid constructs using an Endofree Maxi Prep Kit (QIAGEN). Reconstitute plasmids in TE buffer at a stock concentration of 5–8 µg/µL.

2. Prepare a 2–5 µL mix of GFP-synaptobrevin and tdTomato plasmid DNA at equimolar amounts of both plasmids (1 µg/µL) in TE buffer. Load the plasmid solution into an aluminosilicate glass capillary pulled to a relatively long tapered tip. Using fine forceps or a microelectrode beveler, snip open the tip of the capillary to ~5–10 µm.

3. Anesthetize tadpoles at stage 30–34 by immersing in diluted tricaine methanesulfonate at a final concentration of 0.05% (w/v) in 1× MR, for 5 min at room temperature.

4. Transfer a single, anesthetized tadpole into a Petri dish containing a 0.05% tricaine methanesulfonate anesthetic-saturated Sylgard cushion with a custom-made trench carved to hold tadpoles of that size and stage. Place the tadpole on its side and gently place a standard size harp slice grid to hold the tadpole in place such that the tadpole's right eye is positioned and made available for pressure injection of DNA reagents and for electroporation (Fig. 1A). Ensure that the tadpole is fully immersed in the anesthetic solution.

5. Place the glass capillary containing the plasmid mix inside a Picospritzer III pipette holder and mount onto a three-axis manual micromanipulator. Have another three-axis manual micromanipulator holding the custom-made anode and cathode copper electrodes connected to a Grass SD9 stimulator.

6. Using a dissecting microscope, with one of the micromanipulators holding the glass capillary pipette as a guide, puncture the tadpole's eye with the micropipette containing the DNA mix to inject the solution into the anterior chamber, near the lens (Fig. 1A). Using the second micromanipulator holding the pair of copper electrodes 0.1 mm apart to span the diameter of the eye, place the electrodes gently on top of the two edges of the tadpole's eye (Fig. 1A). Using the Picospritzer's pedal, pressure-inject about ∼2–4 nL of DNA mix using repeated pulses of 20 psi, 15 msec duration. Simultaneously, use the electronic stimulator to deliver single currents of 40 V, 200 Hz, 2 msec delay, 2 msec duration.

   A total of 15 concurrent pressure pulses and single current pulses can be delivered simultaneously into the eye, reversing polarity after every five pulses.

   See Troubleshooting.

7. After transfection, transfer the tadpole to a clean Petri dish with fresh rearing solution (1× MR). Allow tadpoles to recover and develop in fresh 1× MR at room temperature to stage 43–45 for in vivo imaging (Steps 14–17).

   Pharmacologic treatment can be made right after transfection or at any stage (drug supplemented in 1× MR); alternatively, the drugs can be directly injected into tectum of anesthetized tadpoles.

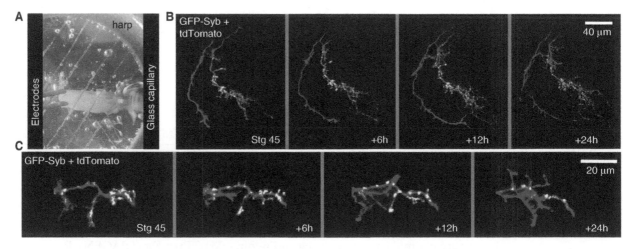

FIGURE 1. (*A*) A magnified view of the eye electroporation setup with the stage 34 tadpole held in place by a harp slice grid and a pair of anode-cathode copper electrodes approaching the eye from the *left* and a glass capillary needle containing the plasmid DNA mix approaching the lens from the *right*. (*B*) Confocal reconstructions of individual RGC axons show the localization of GFP-synaptobrevin puncta (yellow) within specific regions of the arborizing tdTomato-labeled axons (red) imaged in a stage 45 tadpole. (*C*) Magnified portions of an RGC axon arbor imaged at stage 45, and 6, 12, and 24 h later better illustrate the dynamic changes in individual axon branches and GFP-synaptobrevin clusters.

 Cite this protocol as *Cold Spring Harb Protoc*; doi:10.1101/pdb.prot106823

Single-Cell Coelectroporation of Morpholino Oligos and Fluorescent Dyes to Knock Down Gene Expression and Visualize RGC Axon Morphology

8. Mix 1 μL of custom-designed morpholino oligos (300 nM; tagged with a negatively charged fluorophore—lissamine, Gene Tools) with 1 μL 6 mM Alexa Fluor 488 Dextran.

    Coelectroporation of fluorescently tagged dextran and morpholino oligos facilitates visualization and imaging of axonal arbors. The fluorescent dextran acts as a cell-filling dye that fills the entire extent of the axon arbor, with its fluorescence signal remaining stable for repeated imaging when using two-photon excitation.

9. Load the reagent mixture into an aluminosilicate glass capillary pulled to have a stubby tapered tip with a small opening of 0.5–1.0 μm. Attach the glass capillary to the silver wire cathode electrode holder (E-series holder with wire and handle) and connect to the stimulator. Mount the cathode electrode onto a three-axis manual micromanipulator.

10. Anesthetize a stage 43 tadpole by immersion in diluted tricaine methanesulfonate at a final concentration of 0.05% (w/v) in 1× MR. Transfer the anesthetized tadpole into a 35 × 10 mm Petri dish containing an anesthetic-saturated Sylgard cushion with a custom-made trench carved to hold tadpoles of that size and stage. Position the tadpole in the trench on its side so that the tadpole's right eye faces upward and is accessible for injection. Gently place a standard size harp slice grid on top of the Sylgard bed to hold the tadpole in place.

11. Fully immerse the tadpole in anesthetic by adding 0.05% tricaine methanesulfonate solution and place a silver wire anode electrode, connected to the electronic stimulator, in the liquid to serve as a ground wire.

12. Place the glass capillary tip near the area of the retina to be injected and gently puncture through the skin of the embryo. Using the stimulator pedal, deliver five to six electrical currents at 20 V, 200 Hz, 2 msec delay, and 2 msec duration to transfect individual retinal neurons.

    It is recommended to perform the electroporation with a microscope equipped with a mercury arc lamp to confirm that cells are transfected with the fluorophore-tagged morpholino and fluorescent dextran cell-filling dye.

13. After transfection, transfer the tadpoles to fresh rearing solution (1× MR) and raise to stage 45 for in vivo imaging (Steps 14–17) and/or drug manipulation (optional).

In Vivo Real-Time Imaging

14. Prior to imaging, anesthetize tadpoles at stage 43–45 by immersing in diluted tricaine methanesulfonate at a final concentration of 0.05% (w/v) in 1× MR. Place tadpoles in a Petri dish with an agar cushion (2.5% w/v agar gel in 1× MR) saturated with anesthetic. Use an epifluorescence microscope equipped with a low-magnification objective to screen tadpoles for the presence of fluorescently labeled retinal axons innervating the tectum contralateral to the transfected eye.

    For all screening and imaging procedures, it is preferable to use upright microscopes equipped with water-dipping objectives.

    See Troubleshooting.

15. Transfer the selected, anesthetized tadpole into a 35 × 10-mm Petri dish containing an anesthetic-saturated Sylgard cushion with a custom-made trench carved to hold tadpoles of that stage. Position the tadpole with its dorsal side up and place a standard size harp slice grid gently on top of the tadpole to hold it in place.

16. Use a laser scanning confocal microscope equipped with a water dipping objective (40× or 60× magnification) to focus on the tadpole's midbrain contralateral to the eye injected and collect thin optical z-sections through the entire extent of the axon arbor.

    For dual-wavelength imaging (GFP-synaptobrevin and tdTomato, Steps 1–7), collect image stacks at 1.0 μm intervals simultaneously at two wavelengths (488 and 543 nm), below saturation levels, with minimal gain and contrast enhancements. For single-wavelength imaging (Alexa 488 Dextran, Steps 8–13), obtain image stacks at 5- to 10-μm intervals at a wavelength of 760–780 nm with two-photon excitation.

See Troubleshooting.

17. Immediately after imaging, transfer the tadpole to fresh 1× MR and place in a labeled and identified well in a six-well plate. Remove rearing solution and again replace with fresh 1× MR to eliminate any trace of anesthetic. Let the tadpole recover and swim freely until the next imaging time point.

    We recommend that imaging be performed at consistent time intervals as determined by the investigator/ protocol. For GFP-synaptobrevin and tdTomato dual-label imaging (Steps 1–7), depending on the intensity of the label, the microscope settings, and the strength of the laser used, it is possible to remount the tadpole in the plate and image RGC axons in live tadpoles at multiple time intervals. Tadpoles can be imaged every 2 h for 2 to 3 d, and up to 3 wk after initial imaging when imaging either once per day or every few days and when left to recover in fresh rearing solution between imaging sessions. Alternatively, tadpoles can be imaged at short intervals (every 15 min for a total of 12 h) when immobilized in low-melting agarose. To immobilize tadpoles in agarose, place tadpoles head-side up on an agar-cushioned Petri dish and then embed in molten 1.5% w/v low-melting agarose in 1× MR. Alternatively, use a viscous 2% methylcellulose solution to immobilize tadpoles for in vivo imaging as in Paredes et al. (2015). Note that fluorescent dextran-labeled axons (Steps 8–13) are more susceptible to damage during short-interval repetitive imaging due to dye phototoxicity.

Analysis of Axon Arbors

18. Use the microscopy analysis software of preference (i.e., Neuromantic, MetaMorph) to create three-dimensional reconstructions of the fluorescently labeled axon arbors from the confocal *z*-stacks (Fig. 1B). Using the software's tracing function, carefully trace, plane-by-plane, the three-dimensional image of each individual axon from its confocal *z*-stacks.

    It is preferable to select files from tadpoles with only a single axon arbor labeled for the analysis of data or with axon arbors that are clearly distinguishable from neighboring axons when multiple axons are labeled.

19. Obtain a quantitative measure of the number of branches per axon arbor by manually counting the number of branch tips from each axon's three-dimensionally reconstructed tracings (Meta-Morph) or by using the automated function of a tracing software (i.e., Neuromantic).

    Tracings of each individual axon can be compared to identify individual branches that remain stable from one imaging time point to the next, are newly added, or are eliminated between observation intervals to evaluate branch dynamics at different imaging periods (Manitt et al. 2009; Shirkey et al. 2012).

20. Obtain a quantitative measure of the total length of the axon arbor using the automated function of the software (i.e., Neuromantic or Metamorph) and using the first branch or a common landmark most-proximal to the first branch within each axon as a starting mark. Convert pixel values to microns to obtain a cumulative length value for all axon branches.

    Measurements from each individual axon obtained at different time points can be compared to quantify changes in growth over time.

21. For axons with GFP-synaptobrevin expression (Steps 1–7), identify the location of each clearly distinguishable GFP fluorescence puncta (green channel) on the three-dimensional reconstruction of the axon arbor. Use the software to move plane-by-plane through the dual-color confocal image *z*-stacks and carefully mark each puncta location.

22. Compare the location of each GFP-labeled puncta among the individual reconstructions and tracings of the imaged axon (i.e., 0, 6, 12, 24 h; see Fig. 1C) to characterize the distribution of GFP-labeled presynaptic sites on the individual RGC axons (Steps 1–7).

    Distinct morphological axon landmarks (bends in the axon and puncta pattern) can be used to identify puncta and evaluate changes in puncta distribution between observation time points. Similar to branch dynamics, puncta that remain stable, are added, or are eliminated from one imaging time point to the next can be quantified and expressed as percent changes between observation intervals (Alsina et al. 2001; Hu et al. 2005).

23. Count the total number of GFP-labeled puncta per axon arbor traced to obtain a measure of total puncta number and puncta density (puncta per total arbor length) for each axon and each imaging time point (Steps 1–7).

Cite this protocol as *Cold Spring Harb Protoc*; doi:10.1101/pdb.prot106823

TROUBLESHOOTING

Problem (Step 6): Mixture containing DNA plasmids does not come out of injection needle.

Solution: The DNA mixture is a very viscous solution so encountering a clogged injection needle is common. One would need to soak the tip of the injection needle in rearing solution and increase the pressure settings of the Picospritzer III to change the volume of DNA delivered. Further breaking the needle taper is also an option.

Problem (Steps 6 and 16): There are several fluorescently labeled axons in a GFP-synaptobrevin and tdTomato coelectroporated tadpole, but not all have the dual label.

Solution: The efficiency of cotransfection of GFP-synaptobrevin and tdTomato plasmid DNA into RGCs may vary. The majority of neurons are cotransfected and coexpress both GFP-synaptobrevin and tdTomato, but in few cases neurons may delay the expression and accumulation of one of the two proteins at the time of screening (see Fig. 1B), or express tdTomato or GFP-synaptobrevin protein only. By coelectroporating multiple embryos for each experiment and screening and selecting for tadpoles with single, isolated axons with strong GFP-synaptobrevin and tdTomato colabeling (Step 14) and analyzing the individual axon through careful tracing of the confocal z-stacks (Step 18) one can distinguish a dual-labeled axon from neighboring axons that are singly labeled.

Problem (Step 14): Transfected tadpoles exhibit overlabeling of retinal axons, which makes it difficult to analyze individual axon arbors and synapses.

Solution: The screening step is used to optimize imaging of axons exhibiting dual fluorescent labels (GFP-synaptobrevin and tdTomato, Steps 1–7; lissamine-tagged morpholino and Alexa 488 Dextran, Steps 8–13) and to select for tadpoles with only one or few dual-labeled axon arbors that can easily be distinguishable from each other during image acquisition and data analysis. We recommend decreasing the voltage setting and adjusting the time duration of the electrical current and/or the number of electrical currents delivered during electroporation at Step 6 or 12 (depending on the reagent). For example, for plasmid electroporation, settings can be adjusted to deliver single currents at a lower voltage (15–20 V).

Problem (Step 16): There is high-background GFP-synaptobrevin signal in the cytoplasm.

Solution: During image acquisition, we recommend optimizing imaging acquisition parameters using minimal laser intensity and laser gain to avoid oversaturated GFP background signal in the cytoplasm. Setting the image capture parameters so that distinct GFP fluorescence puncta, measuring between 0.5 and 1.0 μm^2 in size (size of smallest puncta observed), have pixel intensity values below saturation levels and above background (150–245 intensity value) permits clear differentiation of single GFP-synaptobrevin synaptic clusters in the reconstructed axonal arbors.

DISCUSSION

Methods optimized in our laboratory that use plasmid DNAs coding for fluorescent chimeric proteins to visualize RGC axons and their presynaptic sites in living tadpoles and fluorophore labeled dextran and morpholino antisense oligonucleotide to label and down-regulate protein expression cell-autonomously are presented here. These methods were adapted from our publications that first characterized synaptic sites in RGC axons in living tadpoles (Alsina et al. 2001; Hu et al. 2005) and from protocols that optimized electroporation methods developed by Holt and colleagues and Cline and colleagues (Falk et al. 2007; Haas et al. 2002; see Santos et al. 2018). These methods can be easily modified and adapted to perform similar manipulations in other neurons or circuits within the developing *Xenopus* nervous system and/or to label or manipulate neurons at different developmental stages. In vivo imaging of optic axons allows for an increased depth in understanding of mechanisms mediating axon pathfinding and axon arborization, as well as synaptic connectivity and developmental plasticity in the retinotectal system. In our laboratory, these methods have been applied alone and in combination with complementary methods that label postsynaptic midbrain neurons (optic tectal neurons) to visualize coincident changes in morphology and synaptic connectivity during the development of pre- and postsynaptic neurons in the retinotectal circuit (Nagel et al. 2015). The accessibility of the *Xenopus* visual system at multiple developmental stages combined with the flexibility of adapting these approaches to transgenic animal lines or to tadpoles with cell-type-specific alterations in gene expression allows for a deeper understanding of molecules and factors implicated in multiple aspects of optic axon development (Marshak et al. 2007; Paredes et al. 2015). DNA constructs coding for wild-type or mutant proteins can be used in combination with those highlighted here to simultaneously visualize axon morphology and presynaptic connectivity while also overexpressing those genes to increase protein function and/or alter specific signaling pathways (Marshak et al. 2007). Using dual-imaging approaches together with pharmacologic treatments or brain-targeted drug injections has also proven useful in differentiating effects of molecules implicated in axon pathfinding, targeting, branching, and presynaptic differentiation (Manitt et al. 2009; Shirkey et al. 2012). Moreover, by combining real-time imaging over a period of hours or days using morpholino oligos as highlighted here, it is possible to further differentiate signaling pathways that affect multiple axon developmental processes through single-cell or cell-population gene knockdown approaches (Santos et al. 2018).

RECIPES

Modified Rearing Solution (1× MR)

60 mM NaCl
0.67 mM KCl
0.34 mM $Ca(NO_3)_2$
0.83 mM $MgSO_4$
10 mM HEPES (pH 7.4)
40 mg/L gentamycin

Supplement with 0.001% phenylthiocarbamide to prevent melanocyte pigmentation. Store for up to 1 wk at 4°C.

TE Buffer for Plasmids

Reagent	Quantity (for 1 L)	Final concentration
Tris (1 M, pH 7.5)	10 mL	10 mM
EDTA (0.5 M)	2 mL	1 mM
H_2O	to 1 L	

Store at 4°C.

REFERENCES

Alsina B, Vu T, Cohen-Cory S. 2001. Visualizing synapse formation in arborizing optic axons in vivo: dynamics and modulation by BDNF. *Nat Neurosci* 4: 1093–1101. doi:10.1038/nn735

Falk J, Drinjakovic J, Leung KM, Dwivedy A, Regan AG, Piper M, Holt CE. 2007. Electroporation of cDNA/Morpholinos to targeted areas of em-

bryonic CNS in *Xenopus*. *BMC Dev Biol* 7: 107. doi:10.1186/1471-213X-7-107

Haas K, Jensen K, Sin WC, Foa L, Cline HT. 2002. Targeted electroporation in *Xenopus* tadpoles in vivo—from single cells to the entire brain. *Differentiation* 70: 148–154. doi:10.1046/j.1432-0436.2002.700404.x

Cite this protocol as *Cold Spring Harb Protoc*; doi:10.1101/pdb.prot106823

Hu B, Nikolakopoulou AM, Cohen-Cory S. 2005. BDNF stabilizes synapses and maintains the structural complexity of optic axons in vivo. *Development* 132: 4285–4298. doi:10.1242/dev.02017

Manitt C, Nikolakopoulou AM, Almario D, Nguyen S, Cohen-Cory S. 2009. Netrin participates in the development of retinotectal synaptic connectivity by modulating axon arborization and synapse formation in the developing brain. *J Neurosci* 29: 11065–11077. doi:10.1523/JNEUROSCI.0947-09.2009

Marshak S, Nikolakopoulou AM, Dirks R, Martens GJM, Cohen-Cory S. 2007. Cell autonomous TrkB signaling in presynaptic retinal ganglion cells mediates axon arbor growth and synapse maturation during the establishment of retinotectal synaptic connectivity. *J Neurosci* 27: 2444–2456. doi:10.1523/JNEUROSCI.4434-06.2007

Nagel AN, Marshak S, Manitt C, Santos RA, Piercy MA, Mortero SD, Shirkey-Son NJ, Cohen-Cory S. 2015. Netrin-1 directs dendritic growth and connectivity of vertebrate central neurons in vivo. *Neural Dev* 10: 14. doi:10.1186/s13064-015-0041-y.

Nieuwkoop PD, Faber J. 1956. *Normal table of* Xenopus laevis *(Daudin)*. North Holland, Amsterdam.

Paredes R, Ishibashi S, Borrill R, Robert J, Amaya E. 2015. *Xenopus*: an in vivo model for imaging the inflammatory response following injury and bacterial infection. *Dev Biol* 408: 213–228. doi:10.1016/j.ydbio.2015.03.008

Santos RA, Fuertes AJC, Short G, Donohue KC, Shao H, Quintanilla J, Malakzadeh P, Cohen-Cory S. 2018. DSCAM differentially modulates pre- and postsynaptic structural and functional central connectivity during visual system wiring. *Neural Dev* 13: 22. doi:10.1186/s13064-018-0118-5

Shirkey N, Manitt C, Zuniga L, Cohen-Cory S. 2012. Dynamic responses of *Xenopus* retinal ganglion cell axon growth cones to netrin-1 as they innervate their *in vivo* target. *Dev Neurobiol* 72: 628–648. doi:10.1002/dneu.20967

Bulk Dye Loading for In Vivo Calcium Imaging of Visual Responses in Populations of *Xenopus* Tectal Neurons

Peter W. Hogg and Kurt Haas[1]

Cellular and Physiological Sciences, University of British Columbia, Vancouver, British Columbia V6T2B5, Canada

Bulk loading of neurons with fluorescent calcium indicators in transparent albino *Xenopus* tadpoles offers a rapid and easy method for tracking sensory-evoked activity in large numbers of neurons within an awake developing brain circuit. In vivo two-photon time-lapse imaging of an image plane through the optic tectum allows defining receptive field properties from visual-evoked responses for studies of single-neuron and network-level encoding and plasticity. Here, we describe loading the *Xenopus* tadpole optic tectum with the membrane-permeable AM ester of Oregon Green 488 BAPTA-1 (OGB-1 AM) for in vivo imaging experiments.

MATERIALS

It is essential that you consult the appropriate Material Safety Data Sheets and your institution's Environmental Health and Safety Office for proper handling of equipment and hazardous materials used in this protocol.

RECIPES: Please see the end of this protocol for recipes indicated by <R>. Additional recipes can be found online at http://cshprotocols.cshlp.org/site/recipes.

Reagents

Albino *Xenopus laevis* tadpoles, stage 47–50 (reared in 10% Steinberg's solution)
Calcium-free Ringer's solution for *Xenopus* neurons <R>
Clear casting silicone (optional; see Step 12)
MS-222 solution (0.02% w/v in 10% Steinberg's solution, with pH adjusted to 7.4 with NaOH)
Nitrogen (compressed gas)
Oregon Green 488 BAPTA-1 AM stock solution (OGB-1 AM; 10 mM) <R>
Oxygen (gas)
Pancuronium dibromide (PCD; 2 mM)
Steinberg's solution (10%)

> *Dilute 100% stock solution 1/10 with Milli-Q H_2O and adjust pH to 7.4 with HCl.*

Steinberg's solution (100%) <R>

---

[1]Correspondence: kurt.haas@ubc.ca

Supplemental Material is available at cshprotocols.cshlp.org.

Equipment

Aluminum foil
Borosilicate glass capillaries with filament (O.D., 1.5 mm; I.D., 0.75 mm)
Centrifuge (benchtop)
Craft knife
Dissection scope (optionally, a fluorescent stereoscope)
Flaming/Brown micropipette puller (P-97 Sutter Instrument)
Forceps
Glass tadpole dish
Imaging chamber

> Critical requirements for implementing calcium imaging in Xenopus tadpoles are appropriate imaging chambers capable of sample stabilization, perfusion with oxygenated 10% Steinberg's solution, and means to provide controlled visual stimuli. For an example imaging chamber meeting these requirements, see Sakaki et al. (2020).

Kimwipes
Micromanipulator
Needle tips (26-gauge)
Paintbrush
Petri dish (5-cm)
Picospritzer
Plastic transfer pipettes
Polymerase chain reaction (PCR) tubes
Scissors
Silicone-cast tadpole bed (optional; see Step 12)
Two-photon or confocal microscope
Water bath

METHOD

Preparation of Working OGB-1 AM Solution

1. Preheat a water bath to 50°C.

2. Thaw OGB-1 AM stock solution for 15 min in the 50°C water bath.

3. Vortex and centrifuge the tube of stock solution at 6000 RPM for 2 min at room temperature every 5 min to homogenize the solution.

4. Dilute the thawed stock at a ratio of 1:10 with calcium-free Ringer's solution for *Xenopus* neurons.

   > Only 4–6 µL of diluted dye is needed and can be stored in a PCR tube. The diluted dye should be a translucent neon yellow–green color.

5. Wrap the outside of the PCR tube with aluminum foil to minimize dye photobleaching from the light exposure.

   > The dye can remain at room temperature and should be used immediately.

Micropipette Preparation

6. Pull thin micropipettes from borosilicate glass capillaries using a micropipette puller.

   > The micropipettes should have a gradual taper and be 10–12 mm in length.

7. Place the micropipette upside-down into the PCR tube of diluted OGB-1 AM, allowing the pipette to backfill.

8. Once the pipette is adequately filled, with dye reaching past the taper of the micropipette, connect the pipette to a Picospritzer, allowing for the precise control of pressurized nitrogen gas through the capillary.

9. Mount the micropipette onto a micromanipulator mounted on a dissection microscope.

10. Using forceps, break the tip of the micropipette; check the flow of dye when the Picospritzer is set between 12 and 20 PSI.

Injection of OGB-1 AM

11. Transfer a tadpole from 10% Steinberg's solution to MS-222 solution to induce anesthesia (~1 min). Use plastic transfer pipettes that have been cut with scissors such that the diameter of the tip is large enough for a tadpole to enter to move tadpoles.

12. Place the tadpole under a dissection microscope. Use a paint brush to manipulate the orientation of the tadpole such that the dorsal side is facing upward. Use either a silicone-cast tadpole bed or a moist Kimwipe in a 5-cm Petri dish to hold the tadpole in place. Ensure the tectum is centered in the microscope field of view.

    Tadpole beds can be created by pouring clear casting silicone into a 5-cm Petri dish and carving a tadpole shaped cavity into the cured silicone with a craft knife.

    A fluorescent stereoscope is ideal for visualizing injection and filling brain tissue while learning the technique but is not necessary.

13. After the tadpole is appropriately orientated, use a 26-gauge needle to make a small incision in the skin next to the tectum contralateral to the eye to which visual stimuli will be presented (e.g., if visual stimuli will be presented to the left eye, make the incision laterally to the right tectum).

    This premade incision will allow the thin micropipette tip to enter the brain without breaking (Fig. 1). When preparing multiple tadpoles, replace the needle frequently, approximately every three tadpoles, so that only sharp needles are used.

14. Guide the micropipette tip through the incision and into the optic tectum. Once positioned, set the Picospritzer to 15 PSI or less (exact values depend on pipette tip geometry and extent of breaking the tip with tweezers). Slowly fill the tectum with dye over 45–60 sec with continuous pressure.

    Do not allow the brain to swell. If the brain swells, the diameter of the micropipette is too large. Either replace the micropipette or reduce the PSI such that the flow of dye does not cause tissue swelling.

15. Remove the micropipette from the tectum and turn off the Picospritzer. Make note of the rate of flow of the dye as the micropipette is removed. Close the valve on the Picospritzer. If dye is no

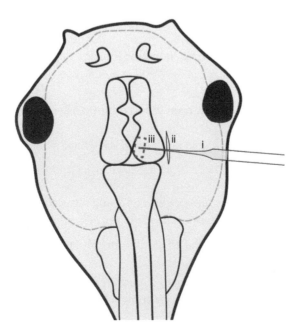

FIGURE 1. Schematic diagram for tectal pressure injections of the right tectum. The filled micropipette (i) should enter the premade incision (Step 13) (ii). The downward angle of the micropipette should be <15° so that the micropipette enters the tectum laterally and not from the dorsal surface. The tip of the micropipette should be directed to the target area (iii), the caudal portion of the optic tectum near the midline. This region of the tectum has a high density of visually responsive cells.

Cite this protocol as *Cold Spring Harb Protoc*; doi:10.1101/pdb.prot106831

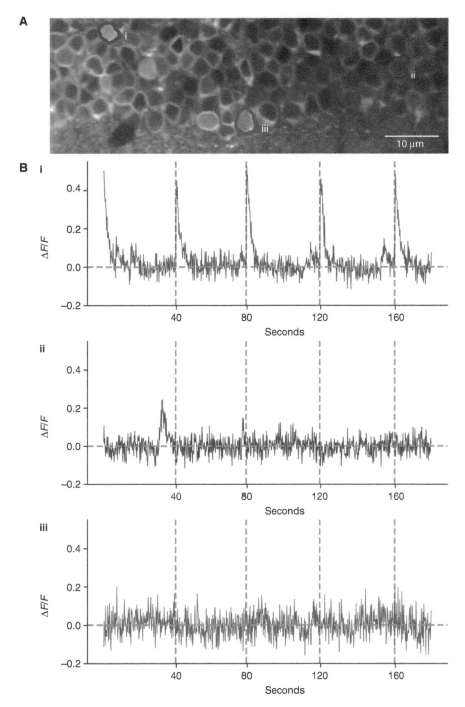

FIGURE 2. In vivo two-photon imaging of visually invoked responses. (*A*) A representative frame from a 5-min-long time series acquired by two-photon microscopy (Supplemental Movie S1). The optic tectum of a stage 48 tadpole was prepared using the above protocol. A single plane of the optic tectum was imaged at 5 Hz while the tadpole was presented with a 50-msec full-field OFF stimulus every 40 sec. Somas were located in the region outlined in Figure 1. Calcium traces for three ROI are shown in *B*. (*B*) 200 sec of calcium traces from three separate somas are shown. Vertical lines indicate when the 50-msec OFF stimuli were presented. The raw time series was denoised and stabilized and region of interest (ROI) and Δ*F/F* were exected from the time series using CaImAn (Giovannucci et al. 2019). The tectal neurons displayed both invoked and noninvoked responses.

longer flowing, this is an indication that the tip is clogged, and an inadequate amount of dye was injected into the tectum. Either break the tip of the micropipette further or replace the micropipette.

In cases where an inadequate amount of dye is injected, it is best practice to start over with a new tadpole to minimize chances of the brain swelling.

16. Remove the tadpole from the stage of the microscope and transfer it to glass tadpole dish containing 10% Steinberg's solution. Allow the tadpole to recover from the procedure and anesthetic for 20–60 min before use in imaging experiments.

If using a fluorescent stereoscope, tadpoles may be screened using epifluorescence before removal.

In Vivo Calcium Imaging

17. After the tadpole has fully recovered from the anesthetic, paralyze the tadpole by transferring it into a room temperature bath containing 2 mM of PCD for 5 min.

This reversible muscle relaxant will immobilize the tadpole for ∼3 h.

18. Transfer the tadpole to an imaging chamber that is accessible to a microscope.

19. Present the tadpole with the visual stimulus of choice.

Tadpoles can be presented with various visual stimuli. For example, full-field OFF and ON stimuli trigger responses in larger populations of tectal neurons. Moving bar stimuli trigger responses in smaller populations of tectal neurons.

20. Acquire images with a microscope for rapid time-lapse imaging. Use a confocal microscope for surface responses or a two-photon microscope for imaging of deep optical sections.

This procedure has been used successfully with a laser scanning two-photon microscope using a 910-nm excitation wavelength (Fig 2; Supplemental Movie S1). Set the detector sensitivity such so there are no saturated pixels and the full dynamic range of the dye is recorded. Image acquisition should occur at a minimum of 5 Hz.

For detailed protocols for in vivo imaging and image analysis, see Sakaki et al. (2020).

DISCUSSION

Pressure delivery of fluorescent calcium indicators has been useful for studying populations of neurons in both fish and mice (Brustein et al. 2003; Stosiek et al. 2003). This method for labeling *Xenopus* tectal neurons with OGB-1 AM provides a very versatile experimental preparation to investigate receptive field properties and experience-driven plasticity at the single-neuron and population network levels (Dunfield and Haas 2009, 2010; Podgorski et al. 2012). This preparation can also be used to model disease states, such as allowing direct in vivo imaging of brain neuronal populations during early-life seizures as a model of epileptogenesis (Hewapathirane et al. 2008).

RECIPES

Calcium-Free Ringer's Solution for Xenopus Neurons

NaCl	100 mM
KCl	2 mM
EGTA	1 mM
MgCl$_2$	1 mM
HEPES	5 mM

Adjust the pH to 7.4 with HCl. Autoclave before use. Store at room temperature.

Oregon Green 488 BAPTA-1 AM Stock Solution (OGB-1 AM; 10 mM)

Prepare a stock solution of 10 mM Oregon Green 488 BAPTA-1 AM by preheating Pluronic F-127 (20% w/v in DMSO) in a 50°C water bath. Pipette 40 µL of heated Pluronic F-127 into a fresh vial of Oregon Green 488 BAPTA-1 AM containing 50 µg of dye. Heat this 10 mM stock solution for an additional 15 min, vortexing every 5 min. Check the color of the solution; it should be a yellow–orange. Store the solution for up to 4–6 mo at −20°C. Optional: Store with Drierite to minimize water moisture contamination.

Steinberg's Solution (100%)

Reagent	Amount to add (for 1 L)
NaCl	3.4 g
KCl	0.05 g
$Ca(NO_3)_2 \cdot 4H_2O$	0.08 g
$MgSO_4 \cdot 7H_2O$	0.205 g
Tris	0.56 g
H_2O	to 1 L

Adjust pH to 7.4 with HCl.

REFERENCES

Brustein E, Marandi N, Kovalchuk Y, Drapeau P, Konnerth A. 2003. "In vivo" monitoring of neuronal network activity in zebrafish by two-photon Ca^{2+} imaging. *Pflugers Arch Eur J Physiol* **446:** 766–773. doi:10.1007/s00424-003-1138-4

Dunfield D, Haas K. 2009. Metaplasticity governs natural experience-driven plasticity of nascent embryonic brain circuits. *Neuron* **64:** 240–250. doi:10.1016/j.neuron.2009.08.034.

Dunfield D, Haas K. 2010. In vivo single-cell excitability probing of neuronal ensembles in the intact and awake developing *Xenopus* brain. *Nat Protoc* **5:** 841–848. doi:10.1038/nprot.2010.10.

Giovannucci A, Friedrich J, Gunn P, Kalfon J, Brown BL, Koay SA, Taxidis J, Najafi F, Gauthier JL, Zhou P, et al. 2019. CaImAn an open source tool for scalable calcium imaging data analysis. *Elife* **8:** e38173. doi:10.7554/eLife.38173

Hewapathirane DS, Dunfield D, Yen W, Chen S, Haas K. 2008. In vivo imaging of seizure activity in a novel developmental seizure model. *Exp Neurol* **211:** 480–488. doi:10.1016/j.expneurol.2008.02.012

Podgorski K, Dunfield D, Haas K. 2012. Functional clustering drives encoding improvement in a developing brain network during awake visual learning. *PLoS Biol* **10:** 17–21. doi:10.1371/journal.pbio.1001236

Sakaki KDR, Podgorski K, Dellazizzo Toth TA, Coleman P, Haas K. 2020. Comprehensive imaging of sensory-evoked activity of entire neurons within the awake developing brain using ultrafast AOD-based random-access two-photon microscopy. *Front Neural Circuits* **14:** 1–18. doi:10.3389/fncir.2020.00001

Stosiek C, Garaschuk O, Holthoff K, Konnerth A. 2003. In vivo two-photon calcium imaging of neuronal networks. *Proc Natl Acad Sci* **100:** 7319–7324. doi:10.1073/pnas.1232232100

Neurophysiological and Behavioral Analysis in *Xenopus*

Ben G. Szaro[1]

Department of Biological Sciences, University at Albany, State University of New York, Albany, New York 12222, USA

Because of its resilience to hypoxia and trauma, the frog has long been a favored preparation of neurophysiologists. Its use has led to the discovery of many fundamental properties of neurons and neural circuits. Neurophysiologists were originally attracted to *Xenopus* embryos, tadpoles, and frogs because of their ready availability, their external development, and the anatomical accessibility and relatively simple neural circuitry of the *Xenopus* visual, locomotory, and vocalization systems. Nowadays, the sequencing of *Xenopus* genomes and the panoply of tools for manipulating gene expression have created new opportunities for neurophysiologists to address the molecular underpinnings of how neurons generate behaviors in a vertebrate. Here, we introduce protocols for harnessing the power of *Xenopus* for performing electrophysiological studies of neural circuitry in the developing optic tectum and spinal cord, as well as in vocalization, and for studying the ontogeny of locomotory behavior.

THE FROG AT THE FOUNDATIONS OF NEUROPHYSIOLOGY

The use of frogs as an animal model for electrophysiology is as old as the science itself. In the eighteenth century, Galvani showed that frog spinal cord uses electricity to stimulate leg movements (Galvani 1791; Bernardi 2000). As electrophysiology burgeoned in the twentieth century, neurophysiologists took advantage of the resilience of frog tissue and its ability to function in hypoxic environments at room temperature to make fundamental discoveries in neurotransmission. For example, using vagal nerve and frog heart, Otto Loewi (Nobel Prize in Physiology or Medicine, 1936) showed the chemical nature of neural modulations of heart rate, laying the foundations for the discovery of acetylcholine and other neurotransmitters (Loewi 1921; Loewi and Navaratil 1926). Then, by teasing out single nerve fibers from breast muscle, Edgar Adrian (Nobel Prize in Physiology or Medicine, 1932) showed that the intensity of a stimulus is reflected in the frequency of firing of nerve impulses (Adrian and Zotterman 1926). Joseph Erlanger and Herbert Spencer Gasser (Nobel Prize in Physiology or Medicine, 1944) further characterized the physiological role of axons, using frog sciatic nerve to establish the relationship between axon caliber and impulse conduction velocity (reviewed in Perl 1994). Also using frog sciatic nerve, Alan Hodgkin (Nobel Prize in Physiology or Medicine, 1963) was one of the first to show that electrical currents are causal agents in neural transmission (Hodgkin 1937a,b), and by studying the end plate potential at the frog neuromuscular junction, Paul Fatt and Bernard Katz (Nobel Prize in Physiology or Medicine, 1970) showed the quantal nature of synaptic transmission (Fatt and Katz 1951).

In addition to these seminal studies on the nature of neurotransmission between neurons and their target cells, the frog also figured prominently in early studies of information processing and neural

---

[1]Correspondence: bszaro@albany.edu

Cite this introduction as *Cold Spring Harb Protoc*; doi:10.1101/pdb.top106849

plasticity in the central nervous system. Characterizing the response properties of the frog retina and optic nerve, Ragnar Granit (Nobel Prize in Physiology or Medicine, 1967) established the visual properties of receptive fields in the vertebrate eye, complementing Keffer Hartline's work in limulus (Granit and Therman 1935). In the 1950s, working at MIT on frog retina and optic nerve, two neurophysiologists, Jerome Lettvin and Humberto Maturana, collaborating with two computational neuroscientists, Walter Pitts and Warren McCulloch, provided among the first convincing evidence that image processing in the central nervous system begins in the retina (Lettvin et al. 1959). Using extracellular recordings from optic arbors in the tectum, this same group also provided electrophysiological confirmation that the return of vision after optic nerve injury in frog involved regeneration of cut axons to their proper places in the tectum (Maturana et al. 1959). This had been shown previously in behavioral studies in frog by Roger Sperry (Sperry 1944). Michael Gaze, working at the University of Edinburgh, published similar results the same year in *Xenopus laevis* and added additional proof of the locus specificity of the retinotectal projection by rotating tadpole eyes at the time of injury (Gaze 1959). A year earlier, Gaze had published his electrophysiological studies of the normal retinotectal projection in *Xenopus* (Gaze 1958). Together, these two papers marked the beginnings of *Xenopus* as a model system for neurophysiological studies and heralded its advantages for studying neural development and plasticity.

THE DEVELOPING *XENOPUS* VISUAL SYSTEM

Prior to Gaze's studies in *Xenopus laevis*, neurophysiologists had primarily used ranids. The advantages of *Xenopus* over ranids, particularly for studies of neural development, first became apparent in the late 1930s when *Xenopus* were imported in large numbers from Africa into European hospitals for use in pregnancy tests (Elkan 1938). Whereas adult ranids must be kept in modified terrariums containing both dry land and pools of water and will eat only live insects, adult *Xenopus* remain aquatic throughout life and will feed on carrion. Ranid tadpoles graze on vegetation, whereas *Xenopus* tadpoles filter-feed on algae and suspended plant matter. These features greatly simplify *Xenopus* husbandry in the laboratory. Moreover, injecting *Xenopus* breeding pairs with human chorionic gonadotropin yields hundreds of embryos on demand at any time of year, thereby greatly facilitating studies of development. *Xenopus* embryos also develop much faster than ranids, becoming swimming tadpoles in a matter of days and metamorphosing into frogs in ∼2 mo versus 2 yr for ranids. These features, along with the accessibility of the *Xenopus* visual system and its resilience to surgical manipulations throughout development, attracted many investigators interested in development of the vertebrate visual system to Gaze's laboratory. As they established their own research groups, they introduced *Xenopus* as a model for studying the development of the vertebrate visual system around the world. Consequently, since the 1960s, the *Xenopus* retinotectal system has featured prominently in studies of how neural maps form (reviewed extensively in Jacobson 1991; Schmidt 2020).

These extensive studies make *Xenopus* an attractive system for relating neuroanatomical development to developmental changes in the properties of neural circuitry. Development of the *X. laevis* retina begins in late neural plate stages (Nieuwkoop and Faber [NF] stage [st.] 15), when bilaterally symmetric regions of the anterior neural plate become specified (Szaro et al. 1985; Eagleson et al. 1995). They evaginate soon thereafter to form readily visible external eye cups by NF st. 26. Morphogenetic movements from the ventral diencephalon and optic stalk continue to contribute to ventral retina through tail bud stages (Jacobson and Hirose 1978; Holt 1980). At day 2 (NF st. 28), the retinal ganglion cells, which project axons from retina to contralateral optic tectum, begin to withdraw from the mitotic cycle (Jacobson 1968; Beach and Jacobson 1979). Soon thereafter, they then extend their axons into the nascent optic nerve to arrive at the contralateral optic tectum by NF st. 35/36 (Gaze et al. 1974; Grant et al. 1980; Szaro et al. 1989). This happens even before the remaining cell types of the retina become fully differentiated (NF st. 41), making it possible to study tectal circuit properties while visually driven behaviors emerge. Development of the visual system continues through metamorphosis and into adulthood as new retinal cells are continually added from stem

Cite this introduction as *Cold Spring Harb Protoc*; doi:10.1101/pdb.top106849

cells located in the ciliary margin. At these later stages, the disparate growth patterns of the retina and tectum require that the retinotectal projection remains plastic as synaptic connections shift to keep the retina and tectum in register with the visual world (Gaze et al. 1979; Fraser 1983). In addition, as tadpoles progress through prometamorphic stages into metamorphic climax, the eyes shift from the sides of the animal to point forward, giving the froglet binocular vision. This process accompanies the formation of an indirect ipsilateral projection to the optic tectum, which is mediated through the isthmus optic nucleus and is kept in register with the direct contralateral projection through NMDA-receptor mediated, activity-dependent shifts in synaptic connections (Udin and Sherer 1990; Udin and Grant 1999; Udin 2008). Also during metamorphosis, growth of the ventral retina outpaces that of dorsal retina, forming new, direct ipsilateral projections to thalamic nuclei (Khalil and Székely 1976; Hoskins and Grobstein 1985). Electrophysiologically, thalamic projections have been studied less than the retinotectal projection has, creating new opportunities for investigating their involvement in the more complex behaviors that emerge at these later stages.

THE DEVELOPING *XENOPUS* LOCOMOTORY SYSTEM

Anatomically, the development of the brainstem and spinal cord in *Xenopus* is as well documented as the visual system, providing the context for physiologists to study the ontogeny of locomotion (Roberts and Clarke 1982; van Mier and ten Donkelaar 1984; Nordlander 1984). The anatomical simplicity of tadpole locomotory circuits further enables physiologists to address neural circuitry in a system having relatively fewer neurons than other vertebrate neural circuits (Roberts et al. 2010). Large sensory, hindbrain, and spinal neurons begin to withdraw from the mitotic cycle at the end of gastrulation, just as the neural plate is being induced by underlying mesoderm (Vargas-Lizardi and Lyser 1974; Lamborghini 1980). Neurite outgrowth commences a few hours later, in neural tube stages (NF st. 17–19), and the earliest reflexes, which manifest as simple C-flexures, can be elicited in day-old neurula at NF st. 22–24 (Jacobson and Huang 1985; Roberts et al. 2010). Coculturing dissociated neural tube and somites at these stages has enabled physiologists to study in detail the emerging electrical properties of neurons and myoblasts, as well as the formation of neuromuscular synapses at first contact (Tabti and Poo 1991). The accessibility of these cells in the intact embryo also make this system amenable to intracellular and whole cell recordings in situ. Collectively, these studies have provided essential insights into such processes as the ontogeny of ion channels and the action potential (Spitzer and Lamborghini 1976; Spitzer and Baccaglini 1976; Ribera and Spitzer 1987; Ribera and Spitzer 1989; Ribera and Spitzer 1990; Desarmenien et al. 1993; Gu et al. 1994), the activity dependence of neurotransmitter choice (Bixby and Spitzer 1984; Borodinsky et al. 2004), the development of neurotransmitter receptor properties (Rohrbough and Spitzer 1996; Prime et al. 1999), and the earliest interactions between neurons and muscle cells during synapse formation (Brehm et al. 1984; Xie and Poo 1986; Evers et al. 1989).

The ontogeny of locomotory behaviors in *Xenopus* follows a stereotypic progression, beginning with simple reflexes emerging at the beginning of the second day of development, at NF st. 22–24. Over the next few days, these simple reflexes progress to tactile-induced short bursts of swimming, and then on to the full swimming behaviors needed for foraging and feeding (Roberts et al. 2010). Within a week, these swimming behaviors become integrated with other sensory modalities, including vision, to become more complex. Throughout these stages, the tadpole hindbrain and spinal cord are fully accessible for electrophysiological studies (Li et al. 2002), and the advent of inexpensive digital video cameras and microcomputers now enables one to document and analyze these behaviors at greater resolution than ever before (Dong et al. 2009).

EX VIVO PREPARATIONS FOR STUDYING COMPLEX NEURAL CIRCUITRY AND BEHAVIOR IN *XENOPUS*

As frogs become sexually mature after metamorphosis, they develop complex mating behaviors. These behaviors include hormonally driven, sexually dimorphic mating calls, which vary among *Xenopus* species. The resilience of the frog brain to shock and its ability to function in a hypoxic environment have once again brought *Xenopus* to the fore for studies of the neural circuitry involved in vocalization. Just as Otto Loewi studied the neural inputs of the heart by placing it into organ culture, investigators now make organ cultures of the intact brain and spinal cord, along with the attached peripheral nerve roots, to conduct physiological and anatomical studies (Luksch et al. 1996). These ex vivo organ cultures can be kept alive for several days in vitro, allowing neurophysiologists to perform pharmacological manipulations, anatomical studies, and electrophysiological recordings of neural circuits under highly controlled conditions, providing them with advantages similar to those of mammalian brain slice preparations. These *Xenopus* ex vivo cultures have the added advantage of encompassing an entire neural circuit, with measurable behavioral outputs in the form of trains of action potentials, which travel along the nerve roots innervating laryngeal muscles. Such preparations are now being used to study the neural drivers of sexually dimorphic mating calls (Rhodes et al. 2007; Zornik and Kelley 2008; Nasipak and Kelley 2012).

PROTOCOL SUMMARIES

Coupling these preparations with the highly cost-effective techniques for manipulating gene expression that are now available in *Xenopus* creates new prospects for gaining unprecedented insights into the molecular underpinnings of neural functions. Additionally, the now-completed sequencing of the *Xenopus laevis* and *Xenopus tropicalis* genomes has revealed extraordinarily high synteny with the human genome, making the identification of orthologous genes more straightforward than is the case for other anamniotes and the invertebrates (Hellsten et al. 2010; Session et al. 2016). Thus, based on past performance and these new resources, *Xenopus* arguably offers excellent prospects for future discoveries of novel, clinically relevant properties of the nervous system.

The accompanying protocols are designed to introduce researchers to using *Xenopus* for such electrophysiological and behavioral studies. The former can be readily implemented by most neurophysiology laboratories, and the latter use relatively simple, cost-effective digital imaging and microcomputers for analysis, making them broadly accessible even to laboratories not set up for electrophysiology. Yuhao Luo, Wanhua Shen, and Hollis Cline provide instructions for performing in vivo electrophysiological recordings from individual tectal neurons to study retinotectal circuitry in the developing (NF st. 47–49) *Xenopus* tadpole tectum (see Protocol 1: Electrophysiological Recording for Study of Retinotectal Circuitry [Luo et al. 2020]). Their protocol provides an effective and affordable method to measure spiking activity and to map excitatory and inhibitory receptive fields in individual neurons, soon after swimming behavior commences in the intact animal. Their custom-built recording chamber with a back-projector screen mimics the visual stimulus from the natural environment, with the advantage that multiple types of recordings can be performed in one cell without having to change the stimulus equipment. Masaki Hiramoto and Hollis Cline describe a method for making continuous long-term extracellular recordings from neurons in intact tadpoles for up to 2 d (see Protocol 2: Tetrode Recording in the *Xenopus laevis* Visual System Using Multichannel Glass Electrodes [Hiramoto and Cline 2020]). Recordings can be made through the four channels of the electrodes from several cells at once, permitting one to observe how local circuit properties and visual fields of individual neurons evolve. Using an ex vivo tadpole brain preparation and whole-cell electrophysiological recordings from individual tectal cells, a protocol by Kara Pratt describes electrophysiological assays for characterizing effects of manipulating expression of macromolecules in the visual system on tectal cells, during the time the retinotectal projection becomes visually responsive (see Protocol 3: Electrophysiological Approaches to Studying Normal and Abnormal Retinotectal Circuit Development in the *Xenopus* Tadpole [Pratt 2020]). The parameters measured in this protocol include the total amount of retinal ganglion cell input received by a tectal neuron, the strength of individual retinal ganglion cell inputs onto a given tectal neuron, AMPA/NMDA receptor ratios, the probability of presynaptic transmitter release, excitatory/inhibitory ratios, the overall amount of synaptic drive received by individual tectal neurons, and individual synapse strength. Wen-Chang

Cite this introduction as *Cold Spring Harb Protoc*; doi:10.1101/pdb.top106849

Li describes comparable procedures for making in situ whole-cell recordings in NF st. 37/38 tadpole to study the early development of motor centers in the brainstem and spinal cord (see Protocol 4: Making In Situ Whole-Cell Patch-Clamp Recordings from *Xenopus laevis* Tadpole Neurons [Li 2020]). Arseny Khakhalin describes a visual avoidance assay and how to use it to monitor sensory processing and sensorimotor transformations in the developing vertebrate brain. His setup assesses the integration of visual cues with motor behaviors in tadpoles at any stage after vision and swimming commence (see Protocol 5: Analysis of Visual Collision Avoidance in *Xenopus* Tadpoles [Khakhalin 2020]). Virgilio Lopez, Arseny Khakhalin, and Carlos Aizenman describe a relatively inexpensive, automated method for delivering timed vibratory stimuli to a group of tadpoles to induce swimming and schooling behavior (see Protocol 6: Schooling in *Xenopus laevis* Tadpoles as a Way to Assess Their Neural Development [Lopez et al. 2020]). Their method controls a digital camera to document tadpole positions, and analysis is done by microcomputer using ImageJ and custom Python scripts, which are provided in their protocol. Their method can be used to judge the effects of reverse genetic manipulations on the emergence of integrated swimming behavior during development and could be readily adapted to observe the reemergence of swimming behaviors in tadpoles after central nervous system (CNS) injury. Last, for studying the neural circuitry of vocalization, Ayako Yamaguchi provides a protocol for removing the intact central nervous system from postmetamorphic *Xenopus* frogs. This ex vivo preparation is useful for performing a variety of techniques, ranging from extracellular and intracellular electrophysiological recordings and calcium imaging to surgical and pharmacological manipulation of neurons (see Protocol 7: Ex Vivo Brain Preparation to Analyze Vocal Pathways of *Xenopus* Frogs [Yamaguchi 2021]).

CONCLUSION

Using the visual, locomotory, and vocalization systems of *Xenopus*, the protocols presented here provide cost-effective and powerful means of studying the fundamental properties of neural circuits and how they develop to generate behaviors in a vertebrate.

REFERENCES

Adrian ED, Zotterman Y. 1926. The impulses produced by sensory nerve-endings. Part II. The response of a single end organ. *J Physiol* **61:** 151–171. doi:10.1113/jphysiol.1926.sp002281

Beach DG, Jacobson M. 1979. Patterns of cell proliferation in the retina of the clawed frog during development. *J Comp Neurol* **189:** 671–698. doi:10.1002/cne.901830308

Bernardi W. 2000. The controversy on animal electricity in eighteenth-century Italy: Galvani, Volta, and others. In *Nuova Voltianas, studies on Volta and his times* (eds. Bevilacqua F, Fregonese L), pp. 101–114. Hoepli, Milan.

Bixby JL, Spitzer NC. 1984. The appearance and development of neurotransmitter sensitivity in *Xenopus* embryonic spinal neurons *in vitro*. *J Physiol* **353:** 143–155. doi:10.1113/jphysiol.1984.sp015328

Borodinsky LN, Root CM, Cronin JA, Sann SB, Gu X, Spitzer NC. 2004. Activity-dependent homeostatic specification of transmitter expression in embryonic neurons. *Nature* **429:** 523–530. doi:10.1038/nature02518

Brehm P, Kidokoro Y, Moody-Corbett F. 1984. Acetylcholine receptor channel properties of *Xenopus* muscle cells in culture. *J Physiol* **357:** 203–217. doi:10.1113/jphysiol.1984.sp015497

Desarmenien MG, Clendening B, Spitzer NC. 1993. *In vivo* development of voltage-dependent ionic currents in embryonic *Xenopus* spinal neurons. *J Neurosci* **13:** 2575–2581. doi:10.1523/JNEUROSCI.13-06-02575.1993

Dong W, Lee RH, Xu H, Yang S, Pratt KG, Cao V, Song YK, Nurmikko A, Aizenman CD. 2009. Visual avoidance in *Xenopus* tadpoles is correlated with the maturation of visual responses in the optic tectum. *J Neurophysiol* **101:** 803–815. doi:10.1152/jn.90848.2008

Eagleson G, Ferreiro B, Harris WA. 1995. Fate of the anterior neural ridge and the morphogenesis of the *Xenopus* forebrain. *J Neurobiol* **28:** 146–158. doi:10.1002/neu.480280203

Elkan ER. 1938. The *Xenopus* pregnancy test. *Br Med J* **2:** 1253–1274. doi:10.1136/bmj.2.4067.1253

Evers J, Laser M, Sun YA, Xie ZP, Poo MM. 1989. Studies of nerve-muscle interaction in *Xenopus* cell culture: analysis of early synaptic events. *J Neurosci* **9:** 1523–1539. doi:10.1523/JNEUROSCI.09-05-01523.1989

Fatt P, Katz B. 1951. An analysis of the end-plate potential recorded with an intracellular electrode. *J Physiol* **115:** 320–370. doi:10.1113/jphysiol.1951.sp004675

Fraser SE. 1983. Fiber optic mapping of the *Xenopus* visual system: shift in the retinotectal projection during development. *Dev Biol* **95:** 505–511. doi:10.1016/0012-1606(83)90053-2

Galvani L. 1791. De viribus electricitatis in motu musculari Commentarius. *De bononiensi scientiarum et artium instituto atque academia Commentarii (Bononiae)* **7:** 363–418.

Gaze RM. 1958. The representation of the retina on the optic lobe of the frog. *Q J Exp Physiol Cogn Med Sci* **43:** 209–214. doi:10.1113/expphysiol.1958.sp001318

Gaze RM. 1959. Regeneration of the optic nerve in *Xenopus laevis*. *Q J Exp Physiol Cogn Med Sci* **44:** 290–308.

Gaze RM, Keating MJ, Chung SH. 1974. The evolution of the retinotectal map during development in *Xenopus*. *Proc R Soc Lond B* **185:** 301–330. doi:10.1098/rspb.1974.0021

Gaze RM, Ostberg A, Chung SH. 1979. The relationship between retinal and tectal growth in larval *Xenopus*: implications for the development of the retinotectal projection. *J Embryol Exp Morphol* **53:** 103–143.

Granit R, Therman PO. 1935. Excitation and inhibition in the retina and optic nerve. *J Physiol* **83**: 359–381. doi:10.1113/jphysiol.1935.sp003235

Grant P, Rubin E, Cima C. 1980. Ontogeny of the retina and optic nerve in *Xenopus laevis*. I. Stages in the early development of the retina. *J Comp Neurol* **189**: 593–616. doi:10.1002/cne.901890403

Gu X, Olson EC, Spitzer NC. 1994. Spontaneous neuronal calcium spikes and waves during early differentiation. *J Neurosci* **14**: 6325–6335. doi:10.1523/JNEUROSCI.14-11-06325.1994

Hellsten U, Harland RM, Gilchrist MJ, Hendrix D, Jurka J, Kapitonov V, Ovcharenko I, Putnam NH, Shu S, Taher L, et al. 2010. The genome of the Western clawed frog *Xenopus tropicalis*. *Science* **328**: 633–636. doi:10.1126/science.1183670

Hiramoto M, Cline HT. 2020. Tetrode recording in the *Xenopus laevis* visual system using multichannel glass electrodes. *Cold Spring Harb Protoc* doi:10.1101/pdb.prot107086

Hodgkin AL. 1937a. Evidence for electrical transmission in nerve. Part I. *J Physiol* **90**: 183–210. doi:10.1113/jphysiol.1937.sp003507

Hodgkin AL. 1937b. Evidence for electrical transmission in nerve. Part II. *J Physiol* **90**: 211–232. doi:10.1113/jphysiol.1937.sp003508

Holt CE. 1980. Cell movements in *Xenopus* eye development. *Nature* **287**: 850–852. doi:10.1038/287850a0

Hoskins SG, Grobstein P. 1985. Development of the ipsilateral retinothalamic projection in the frog *Xenopus laevis*. II. Ingrowth of optic nerve fibers and production of ipsilaterally projecting retinal ganglion cells. *J Neurosci* **5**: 920–929. doi:10.1523/JNEUROSCI.05-04-00920.1985

Jacobson M. 1968. Cessation of DNA synthesis in retinal ganglion cells correlated with the time of specification of their central connections. *Dev Biol* **17**: 202–218. doi:10.1016/0012-1606(68)90061-4

Jacobson M. 1991. Neuronal specificity and development of neuronal projection maps. In *Developmental neurobiology*, pp. 453–537. Plenum, New York.

Jacobson M, Hirose G. 1978. Origin of the retina from both sides of the embryonic brain: a contribution to the problem of crossing at the optic chiasma. *Science* **202**: 637–639. doi:10.1126/science.705349

Jacobson M, Huang S. 1985. Neurite outgrowth traced by means of horseradish peroxidase inherited from neuronal ancestral cells in frog embryos. *Dev Biol* **110**: 102–113. doi:10.1016/0012-1606(85)90068-5

Khakhalin A. 2020. Analysis of visual collision avoidance in *Xenopus* tadpoles. *Cold Spring Harb Protoc* doi:10.1101/pdb.prot106914

Khalil SH, Székely G. 1976. The development of the ipsilateral retinothalamic projections in the *Xenopus* toad. *Acta Biol Acad Sci Hung* **27**: 253–260.

Lamborghini JE. 1980. Rohon–Beard cells and other large neurons in *Xenopus* embryos originate during gastrulation. *J Comp Neurol* **189**: 323–333. doi:10.1002/cne.901890208

Lettvin JY, Maturana HR, McCulloch WS, Pitts WH. 1959. What the frog's eye tells the frog's brain. *Proc Inst Radio Engrs* **47**: 1940–1951. doi:10.1109/JRPROC.1959.287207

Li WC. 2020. Making in situ whole-cell patch-clamp recordings from *Xenopus laevis* tadpole neurons. *Cold Spring Harb Protoc* doi:10.1101/pdb.prot106856

Li WC, Soffe SR, Roberts A. 2002. Spinal inhibitory neurons that modulate cutaneous sensory pathways during locomotion in a simple vertebrate. *J Neurosci* **22**: 10924–10934. doi:10.1523/JNEUROSCI.22-24-10924.2002

Loewi O. 1921. On the humoral propagation of cardiac nerve action. Communication I. *Pflügers Arch* **189**: 239–242. doi:10.1007/BF01738910

Loewi O, Navaratil E. 1926. On the humoral propagation of cardiac nerve action. Communication X. *Pflügers Arch* **214**: 678–688. doi:10.1007/BF01741946

Lopez V III, Khakhalin A, Aizenman C. 2020. Schooling in *Xenopus laevis* tadpoles as a way to assess their neural development. *Cold Spring Harb Protoc* doi:10.1101/pdb.prot106906

Luksch H, Walkowiak W, Muñoz A, ten Donkelaar HJ. 1996. The use of in vitro preparations of the isolated amphibian central nervous system in neuroanatomy and electrophysiology. *J Neurosci Methods* **70**: 91–102. doi:10.1016/S0165-0270(96)00107-0

Luo Y, Shen W, Cline HT. 2020. Electrophysiological recording for study of retinotectal circuitry. *Cold Spring Harb Protoc* doi:10.1101/pdb.prot106880

Maturana HR, Lettvin JY, McCulloch WS, Pitts WH. 1959. Evidence that cut optic fibers in a frog regenerate to their proper places in the tectum. *Science* **130**: 1709–1710. doi:10.1126/science.130.3390.1709

Nasipak B, Kelley DB. 2012. Developing laryngeal muscle of *Xenopus laevis* as a model system: androgen-driven myogenesis controls fiber type transformation. *Dev Neurobiol* **72**: 664–675. doi:10.1002/dneu.20983

Nordlander RH. 1984. Developing descending neurons of the early *Xenopus* tail spinal cord in the caudal spinal cord of early *Xenopus*. *J Comp Neurol* **228**: 117–128. doi:10.1002/cne.902280111

Perl E. 1994. The nobel prize to erlanger and gasser. *FASEB J* **8**: 782–783. doi:10.1096/fasebj.8.10.8050679

Pratt KG. 2020. Electrophysiological approaches to studying normal and abnormal retinotectal circuit development in the *Xenopus* tadpole. *Cold Spring Harb Protoc* doi:10.1101/pdb.prot106898

Prime L, Pichon Y, Moore LE. 1999. N-Methyl-D-aspartate-induced oscillations in whole cell clamped neurons from the isolated spinal cord of *Xenopous laevis* embryos. *J Neurophysiol* **82**: 1069–1073. doi:10.1152/jn.1999.82.2.1069

Rhodes HJ, Heather JY, Yamaguchi A. 2007. *Xenopus* vocalizations are controlled by a sexually differentiated hindbrain central pattern generator. *J Neurosci* **27**: 1485–1497. doi:10.1523/JNEUROSCI.4720-06.2007

Ribera AB, Spitzer NC. 1987. Both barium and calcium activate neuronal potassium currents. *Proc Natl Acad Sci* **84**: 6577–6581. doi:10.1073/pnas.84.18.6577

Ribera AB, Spitzer NC. 1989. A critical period of transcription required for differentiation of the action potential of spinal neurons. *Neuron* **2**: 1055–1062. doi:10.1016/0896-6273(89)90229-8

Ribera AB, Spitzer NC. 1990. Differentiation of I_{KA} in amphibian spinal neurons. *J Neurosci* **10**: 1886–1891. doi:10.1523/JNEUROSCI.10-06-01886.1990

Roberts A, Clarke JDW. 1982. The neuroanatomy of an amphibian embryo spinal cord. *Phil Trans R Soc Lond B* **296**: 195–212. doi:10.1098/rstb.1982.0002

Roberts A, Li WC, Soffe SR. 2010. How neurons generate behavior in a hatchling amphibian tadpole: an outline. *Front Behav Neurosci* **4**: 16.

Rohrbough J, Spitzer NC. 1996. Regulation of intracellular Cl⁻ levels by Na⁺-dependent Cl⁻ cotransport distinguishes depolarizing from hyperpolarizing GABA_A receptor-mediated responses in spinal neurons. *J Neurosci* **16**: 82–91. doi:10.1523/JNEUROSCI.16-01-00082.1996

Schmidt JT. 2020. *Self organizing neural maps: the retinotectal map and mechanisms of neural development from retina to tectum*, pp. 1–453. Academic Press, Cambridge, MA.

Session AM, Uno Y, Kwon T, Chapman JA, Toyoda A, Takahashi S, Fukui A, Hikosaka A, Suzuki A, Kondo M, et al. 2016. Genome evolution in the allotetraploid frog *Xenopus laevis*. *Nature* **538**: 336–343. doi:10.1038/nature19840

Sperry RW. 1944. Optic nerve regeneration with return of vision in anurans. *J Neurophysiol* **7**: 351–361. doi:10.1152/jn.1944.7.1.57

Spitzer NC, Baccaglini PI. 1976. Development of the action potential in embryo amphibian neurons *in vivo*. *Brain Res* **107**: 610–616. doi:10.1016/0006-8993(76)90148-7

Spitzer NC, Lamborghini JE. 1976. The development of the action potential mechanism of amphibian neurons in culture. *Proc Natl Acad Sci* **73**: 1641–1645. doi:10.1073/pnas.73.5.1641

Szaro B, Ide C, Kaye C, Tompkins R. 1985. Regulation in the neural plate of *Xenopus laevis* demonstrated by genetic markers. *J Exp Zool* **234**: 117–129. doi:10.1002/jez.1402340114

Szaro BG, Lee VM, Gainer H. 1989. Spatial and temporal expression of phosphorylated and non-phosphorylated forms of neurofilament proteins in the developing nervous system of *Xenopus laevis*. *Brain Res Dev Brain Res* **48**: 87–103. doi:10.1016/0165-3806(89)90095-3

Tabti N, Poo MM. 1991. Culturing spinal neurons and muscle cells from *Xenopus* embryos. In *Culturing nerve cells* (ed. Banker G, Goslin K), pp. 137–154. MIT Press, Cambridge.

Udin SB. 2008. Isthmotectal axons maintain normal arbor size but fail to support normal branch numbers in dark-reared *Xenopus laevis*. *J Comp Neurol* **507**: 1559–1570. doi:10.1002/cne.21633

Udin SB, Grant S. 1999. Plasticity in the tectum of *Xenopus laevis* binocular maps. *Prog Neurobiol* **59**: 81–106. doi:10.1016/S0301-0082(98)00096-3

Udin SB, Sherer WJ. 1990. Restoration of the plasticity of binocular maps by NMDA after the critical period in *Xenopus*. *Science* **249**: 669–672. doi:10.1126/science.2166343

Cite this introduction as *Cold Spring Harb Protoc*; doi:10.1101/pdb.top106849

van Mier P, ten Donkelaar HJ. 1984. Early development of descending pathways from the brain stem to the spinal cord in *Xenopus laevis*. *Anat Embryol* **170:** 295–306. doi:10.1007/BF00318733

Vargas-Lizardi P, Lyser KM. 1974. Time of origin of Mauthner's neuron in *Xenopus laevis* embryos. *Dev Biol* **38:** 220–228. doi:10.1016/0012-1606(74)90002-5

Xie ZP, Poo MM. 1986. Initial events in the formation of neuromuscular synapse: rapid induction of acetylcholine release from embryonic neuron. *Proc Natl Acad Sci* **83:** 7069–7073. doi:10.1073/pnas.83.18.7069

Yamaguchi A. 2021. *Ex vivo* brain preparation to analyze vocal pathways of *Xenopus* frogs. *Cold Spring Harb Protoc* doi: 10.1101/pdb.prot106872

Zornik E, Kelley DB. 2008. Regulation of respiratory and vocal motor pools in the isolated brain of *Xenopus laevis*. *J Neuroci* **28:** 612–621. doi:10.1523/JNEUROSCI.4754-07.2008

Electrophysiological Recording for Study of *Xenopus* Retinotectal Circuitry

Yuhao Luo,[1] Wanhua Shen,[1,3] and Hollis T. Cline[2]

[1]*Zhejiang Key Laboratory of Organ Development and Regeneration, College of Life and Environmental Sciences, Hangzhou Normal University, Hangzhou, Zhejiang, 311121, China;* [2]*The Dorris Neuroscience Center, Department of Neuroscience, The Scripps Research Institute, La Jolla, California 92037, USA*

The innervation of the optic tectum of *Xenopus* by retinal ganglion cells controls visual information processing and behavioral output. Several indicators can be used to evaluate the functional inputs/outputs of tectal neurons, such as spontaneous activity, visually evoked currents, temporal receptive fields, and spatial receptive fields. Analysis of multiple functional properties in the same neurons allows increased understanding of mechanisms underlying visual system function and plasticity. Patch-clamp recordings combined with gene expression or morpholino-mediated knockdown techniques have been especially powerful in the study of specific genes during development and circuit function. The protocol described here provides instructions for performing in vivo electrophysiological recordings from individual tectal neurons to study retinotectal circuitry in the developing *Xenopus* tectum.

MATERIALS

It is essential that you consult the appropriate Material Safety Data Sheets and your institution's Environmental Health and Safety Office for proper handling of equipment and hazardous material used in this protocol.

RECIPES: Please see the end of this protocol for recipes indicated by <R>. Additional recipes can be found online at http://cshprotocols.cshlp.org/site/recipes.

Reagents

Bleach
Deionized (DI) H_2O
Extracellular bath saline <R>
Internal recording saline <R>
Steinberg's solution <R>
Tricane methanesulfonate (MS-222; 0.02% in 1× Steinberg's solution)
Tubocurarine (50 mM)
Xenopus tadpoles from stage 42 to 49 (Nieuwkoop and Faber 1994)

Equipment

Ag/AgCl electrode and pipette holder
Digidata 1440A data acquisition system (Molecular Devices)

---

[3]Correspondence: shen@idrbio.org

Disposable sterile plastic syringe

Dissecting stereomicroscope (Nikon model SMZ 1500 or similar)

Epifluorescence microscope (equipped with 10× or 20× objective, Nikon Instruments)

Faraday cage

Forceps (Dumont no. 5; Fine Science Tools)

Gravity perfusion system

High vacuum grease (Dow Corning)

Insect pins

Kodak neutral density filters

Light-tight materials to seal Faraday cage

MATLAB (The MathWorks)

Microforge (Narishige MF-830)

Micromanipulator controller (e.g., MP-225, Sutter Instrument)

Micropipette puller (P-97, Sutter Instrument)

Micropipettes with filaments (O.D. 1.5 mm, I.D. 0.86 mm; borosilicate glass capillaries, Warner Instruments)

Patch-clamp amplifier (Multiclamp 700B amplifier, Molecular Devices)

pClamp and Clampfit software (10.3, Molecular Devices)

Petri dish

Plastic syringe (1-mL)

Projector (Samsung SP-P310ME LED)

Recording chamber with a back-projector screen

This protocol uses a custom-made recording chamber (see Fig. 1 and Shen et al. 2011) in which a stimulus is presented to a Xenopus retina from a projector to one side of the chamber with a 4.5 × 4.5 cm² viewing area. A rotatable bar with a cube of silicone polymer (Sylgard) is inserted across the chamber. Alternatively, the visual stimuli can be presented using the multicore optical image fiber (FIGH-30–650S, Myriad Fiber) (Dong et al. 2009; Van Horn et al. 2017); see Discussion.

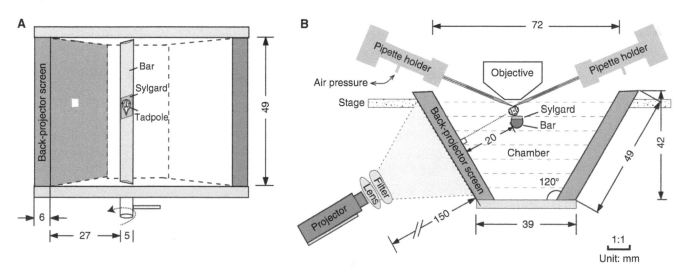

FIGURE 1. Schematic diagram of the recording chamber. (*A*) Overview of the recording chamber. The gray screen is made of a back-projector screen, which presents the stimulus to the eye while the animal is in the chamber. The bar and white walls are made of acrylic glass. Tadpole is secured on a Sylgard cushion (5 × 8 mm) with insect pins and stuck to the rotatable bar with grease. The angle of the bar is adjusted such that the retina faces vertically to the back-projector screen. The white square represents one stimulus grid on the black background. (*B*) Front view of the recording chamber. A convex lens is placed in front of the projector to control the projected image size. The filters are used to adjust the luminance. The condenser is removed before placing the chamber on the stage. Scale: mm.

Screen luminance meter (SM208, M&A Instruments Inc)
Silicone tubing
Sylgard-184 (Electron Microscopy Sciences 24236-10)
Three-way stopcock valve
Transfer pipette

METHOD

All animal protocols were approved by the Institutional Animal Care and Use Committee of the Scripps Research Institute and the local ethics committee of the Hangzhou Normal University.

Recording Chamber Preparation

1. Remove the specimen holder and the light condenser from the epifluorescence microscope so that the recording chamber fits.

2. Mount the recording chamber on the stage of the inverted microscope.

3. Adjust the distance between the projector and screen until all of the stimulus images of the 8×8 grid (see Step 18) fit within the animal's field of view.

Patch-Clamp System Preparation

4. Seal the Faraday cage with light-tight materials.

5. Silver new electrodes by soaking in fresh bleach overnight at room temperature and mount the Ag/AgCl electrode to the pipette holder.

    This will deposit a layer of AgCl on the recording and ground electrodes.

6. Pull recording micropipettes with glass capillaries using a pipette puller according to the manufacturer's instructions. Inspect the shape of the micropipettes under a microforge.

7. Custom-make a plastic syringe as follows: Heat and melt the plastic tubing of a disposable sterile plastic syringe and manually pull a disposable sterile needle-like injector. Control the speed used to pull the syringe while heating to obtain a suitable size of the plastic tubing to fit into the micropipette.

8. Use the custom-made plastic syringe to add internal saline to the micropipettes. Add sufficient internal solution so that the electrode is submerged. Place the micropipettes in the pipette holder.

9. Turn on the amplifier, digitizer, and Clampex software in the computer.

10. Move micropipettes to the chamber with a micromanipulator and check the resistances.

    The resistance of recording pipette should be in the range of 7 to 9 MΩ. This indicates a suitable tip size required to form a giga-ohm seal and maintain a long-term whole-cell recording without the cell membrane resealing.

Tadpole Preparation for Patch Clamp

11. Set and maintain the room temperature at 20°C.

12. Using a plastic transfer pipette, move tadpoles into a Petri dish containing MS-222 solution. Anesthetize tadpoles in this solution for ∼5 min.

13. Gently cut the dorsal skin with a sharp needle and peel the skin back with forceps to expose the tectum under a dissecting microscope (He et al. 2018).

14. Rinse the tadpole with extracellular saline to wash out remaining MS-222. Secure the tadpole on a Sylgard cushion with pins.

Cite this protocol as *Cold Spring Harb Protoc*; doi:10.1101/pdb.prot106880

15. Fill the recording chamber with extracellular saline with 0.05 mM tubocurarine (Fig. 1B). Attach the Sylgard to the rotatable bar with vacuum grease in the recording chamber and submerge the tadpole in the extracellular saline during recordings.

    Keep the tadpole submerged in the extracellular saline during all recordings. The continuous perfusion of extracellular saline with a gravity perfusion system is recommended to keep the saline fresh. Start recording cells when tadpole is immobilized by tubocurarine after 5 min. Each tadpole can typically be maintained in a healthy state for 2–4 h after preparation, and we routinely record three to four cells from each tectum.

16. Adjust the angle of the bar so that the tadpole's eyes are vertically facing the back-projector screen.
    The computer-generated image is projected to the retina via the screen.

Mapping Receptive Fields

17. Adjust the brightness of the projector with Kodak neutral density filters and control the image size with a convex lens.

    The luminance determines the strength of visually evoked currents. Measure the luminance every time with a luminance meter before the recording. The range of the luminance should be 3 to 55 cd/m². Place a photodiode beside the recording chamber and connect it to the analog input on the front of the Digidata with a BNC cable. Set up a new channel in the Clampex and monitor the stimulus of light ON/OFF.

18. Generate the stimulus image via MATLAB using a separate computer.

    The stimulus consists of an 8×8 grid of 0.5×0.5 cm² nonoverlapping squares covering the visual field. White squares on a dark background or dark squares on a white background are presented for 1.5 sec in random order, 64 times with 5-sec intervals, until the entire visual field is mapped.

19. Move the stage to find a clear image of the optic tectum under the microscope.

    Record the neurons from the same area of the optic tectum for each recording to decrease the variation of responses due to differential maturation stages along the rostrocaudal tectal axis.

20. Gently break a micropipette (made in Step 6) tip with a Kimwipe and move it toward the tectum. Apply a gentle negative pressure in the broken micropipette, and then use it to remove a piece of the membranes (dura mater and pia mater) overlying the tectum to expose the tectal cells. Keep the major tectum intact, and choose healthy cells to record.

21. Make a pressure control device by connecting a 1-mL syringe with one of the female luer locks of a three-way valve. Connect the other female luer lock to silicone tubing, the other end of which is connected to the pipette holder. Use the syringe to maintain a gentle positive pressure in the micropipette.

22. Use the micromanipulator to move the patch pipette close to the cell.

23. Slowly push the micropipette against the cell until the measured resistance rises.

24. Remove positive pressure and apply negative pressure to obtain a giga-ohm seal.

25. Run the stimulation protocol generated by pClamp software and collect cell-attached recordings for three repetitions per cell.

    Hold the cell membrane current at 0 pA. Record cell spikes for 20 min. Repeat the recording to eliminate possible contamination by spontaneous activity.

    See Troubleshooting.

26. Apply a brief strong negative pressure to break the cell membrane and continue whole-cell voltage clamp recording.

    See Troubleshooting.

27. Run the same stimuli to record excitatory and inhibitory currents by holding the membrane potential at −60 mV and 0 mV, respectively.

    Voltage clamp the membrane potential at −60 mV to record AMPAR-mediated currents or at 0 mV to record GABA$_A$R-mediated currents. Monitor the series resistance live and collect data from those cells with changes of series resistance within 10%. Filter signals at 2 kHz sampled at 10 kHz.

    See Troubleshooting.

Data Analysis

28. For cell-attached recording, determine the average number of total spikes per stimulus and map the spiking receptive field (sRF) or temporal receptive field (tRF) using MATLAB.

29. For whole-cell recording, compute and normalize the total synaptic charge transfer over 600 msec from the onset of stimulus to map the excitatory receptive field (eRF) and inhibitory receptive field (iRF) using MATLAB.

    Include all values that are larger than three times the standard deviation of spontaneous activity in the analysis of spatial receptive fields. Include only neurons from which both eRF and iRF are successfully recorded in the correlation analysis (Shen et al. 2011).

    See Troubleshooting.

30. Represent spatial receptive fields for synaptic currents on a gray-scale map, according to the normalized total charge transfer in response to the visual stimulus.

    See Troubleshooting.

TROUBLESHOOTING

Problem (Step 25–27): Baseline shifts and recordings (currents or spikes) are unstable over time.
Solution: Wait 10 min and allow the cell to reach a stable condition after the formation of cell-attached or whole-cell mode.

Problem (Steps 27, 29, and 30): Each stimulus presented on different grids evokes similar values of currents.
Solution: Reduce the luminance intensity of white squares by applying neutral density filters to avoid the saturation of light sources. Measure the luminance intensity with a luminance meter every time before performing the visual stimulation experiment.

DISCUSSION

This protocol is based on our study showing that the inhibition to excitation ratio regulates visual system responses and behavior in developing *Xenopus* in vivo (Shen et al. 2011; He et al. 2018). This protocol provides an effective method to map spiking, excitatory, and inhibitory receptive fields in individual neurons without physically removing the lens from the retina as shown before (Engert et al. 2002; Tao and Poo 2005). The custom-built recording chamber with a back-projector screen mimics the visual stimulus from the natural environment. The advantage of using the chamber is that multiple recordings (spontaneous activity, receptive field mapping, and visually evoked responses) can be performed in one cell without changing the stimulus equipment. An alternative way to present visual stimuli is to use a rather expensive piece of medical equipment, such as the multicore optical image fiber (FIGH-30–650S, Myriad Fiber) (Dong et al. 2009; Van Horn et al. 2017).

RECIPES

Extracellular Bath Saline

Reagent	Amount (for 1 L)	Final concentration
NaCl	6.72 g	115 mM
KCl	0.30 g	4 mM
$CaCl_2 \cdot 2H_2O$	0.44 g	3 mM
$MgCl_2 \cdot 6H_2O$	0.61 g	3 mM
HEPES	1.19 g	5 mM
Glycine	0.00075 g	0.01 mM
Glucose	1.80 g	10 mM
Tubocurarine	0.068 g	0.1 mM

To prepare 1 L of extracellular saline, dissolve the reagents listed above in 800 mL of ddH_2O. Adjust the pH to 7.2 with NaOH, and then add ddH_2O to 1 L. The final osmolarity is 255 mOsm. Store at 4°C.

Internal Recording Saline

Reagent	Amount (for 20 mL)	Final concentration
Potassium gluconate	515.4 mg	110 mM
KCl	11.9 mg	8 mM
NaCl	5.8 mg	5 mM
$MgCl_2 \cdot 6H_2O$	6.1 mg	1.5 mM
HEPES	95.3 mg	20 mM
EGTA	3.8 mg	0.5 mM
ATP	44.1 mg	4 mM
GTP	3.1 mg	0.3 mM

To prepare 20 mL of internal saline, dissolve the reagents listed above in 15 mL of ddH_2O. Adjust the pH to 7.2 with KOH, and then add ddH_2O to 20 mL. The final osmolarity is 255 mOsm. Store at −20°C, and thaw on ice just before the experiment.

Steinberg's Solution

Reagent	Amount (for 100× stock)	Final concentration (for the 100× solution)
NaCl	34 g	582 mM
KCl	0.5 g	6.7 mM
$Ca(NO_3)_2 \cdot 4H_2O$	0.8 g	3.4 mM
$MgSO_4 \cdot 7H_2O$	2.06 g	8.3 mM
HEPES	119 g	500 mM

To prepare 1 L of 100× Steinberg's solution, dissolve the reagents listed above in 800 mL of ddH_2O. Adjust the pH to 7.5 with NaOH, and then add ddH_2O to 1 L. Store the solution at 4°C. To prepare 10 L of 1× Steinberg's solution, pour 100 mL of stock and add ddH_2O to 10 L. Store at room temperature.

COMPETING INTEREST STATEMENT

The authors declare no conflicts of interest with respect to the authorship and/or publication of this article.

ACKNOWLEDGMENTS

This work was supported by the National Nature Sciences Foundation of China (NSFC 31871041 to W.S.) and the United States National Institutes of Health (EY011261 and EY027437 to H.T.C.).

Author contributions: All authors listed have made a substantial, direct, and intellectual contribution to the work and approved it for publication.

REFERENCES

Dong W, Lee RH, Xu H, Yang S, Pratt KG, Cao V, Song YK, Nurmikko A, Aizenman CD. 2009. Visual avoidance in *Xenopus* tadpoles is correlated with the maturation of visual responses in the optic tectum. *J Neurophysiol* 101: 803–815. doi:10.1152/jn.90848 .2008

Engert F, Tao HW, Zhang LI, Poo MM. 2002. Moving visual stimuli rapidly induce direction sensitivity of developing tectal neurons. *Nature* 419: 470–475. doi:10.1038/nature00988

He HY, Shen WH, Zheng LJ, Guo X, Cline HT. 2018. Excitatory synaptic dysfunction cell-autonomously decreases inhibitory inputs and disrupts structural and functional plasticity. *Nat Commun* 9: 2893. doi:10.1038/s41467-018-05125-4

Nieuwkoop P, Faber J. 1994. *Normal table of* Xenopus laevis *(Daudin)*. Garland, New York.

Shen WH, McKeown CR, Demas JA, Cline HT. 2011. Inhibition to excitation ratio regulates visual system responses and behavior in vivo. *J Neurophysiol* 106: 2285–2302. doi:10.1152/jn.00641.2011

Tao HW, Poo MM. 2005. Activity-dependent matching of excitatory and inhibitory inputs during refinement of visual receptive fields. *Neuron* 45: 829–836. doi:10.1016/j.neuron.2005.01.046

Van Horn MR, Strasser A, Miraucourt LS, Pollegioni L, Ruthazer ES. 2017. The gliotransmitter D-serine promotes synapse maturation and axonal stabilization in vivo. *J Neurosci* 37: 6277–6288. doi:10.1523/JNEURO SCI.3158-16.2017

Cite this protocol as *Cold Spring Harb Protoc*; doi:10.1101/pdb.prot106880

Tetrode Recording in the *Xenopus laevis* Visual System Using Multichannel Glass Electrodes

Masaki Hiramoto[1] and Hollis T. Cline

The Dorris Neuroscience Center, Department of Neuroscience, The Scripps Research Institute, La Jolla, California 92037, USA

The *Xenopus* tadpole visual system shows an extraordinary extent of developmental and visual experience–dependent plasticity, establishing sophisticated neuronal response properties that guide essential survival behaviors. The external development and access to the developing visual circuit of *Xenopus* tadpoles make them an excellent experimental system in which to elucidate plastic changes in neuronal properties and their capacity to encode information about the visual scene. The temporal structure of neural activity encodes a significant amount of information, access to which requires recording methods with high temporal resolution. Conversely, elucidating changes in the temporal structure of neural activity requires recording over extended periods. It is challenging to maintain patch-clamp recordings over extended periods and Ca^{2+} imaging has limited temporal resolution. Extracellular recordings have been used in other systems for extended recording; however, spike amplitudes in the developing *Xenopus* visual circuit are not large enough to be captured by distant electrodes. Here we describe a juxtacellular tetrode recording method for continuous long-term recordings from neurons in intact tadpoles, which can also be exposed to diverse visual stimulation protocols. Electrode position in the tectum is stabilized by the large contact area in the tissue. Contamination of the signal from neighboring neurons is minimized by the tight contact between the glass capillaries and the dense arrangement of neurons in the tectum. This recording method enables analysis of developmental and visual experience–dependent plastic changes in neuronal response properties at higher temporal resolution and over longer periods than current methods.

MATERIALS

It is essential that you consult the appropriate Material Safety Data Sheets and your institution's Environmental Health and Safety Office for proper handling of equipment and hazardous materials used in this protocol.

RECIPES: Please see the end of this protocol for recipes indicated by <R>. Additional recipes can be found online at http://cshprotocols.cshlp.org/site/recipes.

Reagents

3-aminobenzoic acid ethyl ester (Tricane or MS222; Sigma-Aldrich; 0.01% w/v, pH 7.4, in Steinberg's solution)

Albino or wild-type *Xenopus laevis* tadpoles (obtained by in-house fertilization or from XEN EXPRESS; stage 46–48 [Nieuwkoop and Faber 1956])

---

[1]Correspondence: hiramoto@scripps.edu

Copyright © 2023 Cold Spring Harbor Laboratory Press; all rights reserved
Cite this protocol as *Cold Spring Harb Protoc*; doi:10.1101/pdb.prot107086

Bleach

External and internal recording solutions <R>

The composition of the internal and external solutions is the same.

Pancuronium dibromide (10 μM in external solution)

Steinberg's solution (100%) <R>

Equipment

Electrical tester

Glass pipette (tip diameter ∼20–30 μm)

Needle, 30-gauge

Materials for the glass tetrode electrode holder

Ag–AgCl Electrode Pellet, 2 × 4 mm (Warner Instruments, E201, 64-1305)

Ceramic tile (Sutter CTS)

Clear acrylic plates (4-mm thickness)

Coaxial cables

Compression cap polycarb for 2-mm glass holder (Warner Instruments, QC-20, 64-1296)

Cross-theta capillaries made of borosilicate glass (outer diameter = 2 mm, inner diameter = 1.4 mm, septum = 0.0088 mm, length = 100 mm; Hilgenberg GmbH 1401731)

Gold-plated 1-mm pin (four; Warner Instruments, WC1-10, 64-1325)

Gold-plated 2-mm pin (four; Warner Instruments, WC2-5, 64-1326)

Jack assembly, 2-mm (four; Warner Instruments, HC-21, 64-1286)

M3 long nut, 2-cm

This is part of the handle for the electrode holder.

M3 threaded rod, 2-cm (five)

This is part of the handle for the electrode holder.

Pipette seal (four, Warner Instruments, PS-20, 64-1292)

Silicone tube (ID 1 mm, OD 3 mm) × 3 m

Silver wire

Syringe, 1-mL

Wire seal (four, Warner Instruments, WS-1, #64-1298)

Materials for recording setup

Acrylic round rod (ϕ = 7 mm)

Air pump and air stone

Digitizer (Molecular Device, DIGIDATA 1440A)

Epoxy glue or O-rings (ID 6 mm, OD 8 mm)

Faraday cage with front doors

Forceps

Head stage (four; Molecular Device CV-7B)

Insect pins

Patch-clamp amplifier (two; Molecular Device, Multiclamp 700A/700B)

Perfusion system

Peristaltic pump

Petri dish

Photodiode

Rear projection screen (Alternative Screen Solutions, Sticky Screen, https://store.gooscreen.com/stickyscreen-rear-projection-screen-window-film)

Shielded wires (ϕ0.2–0.4 mm)

Short acrylic pipe (OD 2 mm × 5 mm)

Short acrylic pipe (OD 3 mm × 5 mm)

Cite this protocol as *Cold Spring Harb Protoc*; doi:10.1101/pdb.prot107086

Sutter P-97 pipette puller with a 3.0 mm × 3.0 mm square box filament (Sutter FB-330B)

Sylgard

Syringe, 60-mL

Vacuum grease

Visual stimulation equipment: miniprojector, etc.

METHOD

This technique employs a pulled four-channel glass capillary pipette based on the electrode design used in industry and patch-clamp recording instrumentation available in many laboratories. A flat blunt end on the glass capillary creates an electrically shielded space that can enclose from one to several neurons for recording. This excludes convergence of weak signals from distant cells that increase uncertainty in the spike-sorting process. Contact with the shaft of the glass electrode stabilizes the electrode in the tectum over extended recording sessions. Aeration and perfusion extend survival up to 1 or 2 d, which has not been achieved with previous recording methods. The four-channel electrodes collect signals from several cells, which enables the study of components in a local circuit.

This protocol has been approved by The Scripps Research Institute IACUC (Protocol# 08-0083).

Make a Glass Tetrode Electrode Holder

1. Prepare the holder for the four-channel glass electrode (Fig. 1A–C), which is a key component of this setup. This holds the cross-theta glass capillary (Hilgenberg-GmbH), originally designed for industrial use. The diameter of the glass electrode tip is ~40 µm and covers approximately three to five neurons in the tectum (Fig. 1D,E).

    Our custom-made holder (Fig. 1A) is modified from the basic design of the theta-capillary holder, THS-M20PH, from Warner Instruments, which is commonly used for focal stimulation of neural tissues. We increased the number of channels from two to four. If the length of L3 is too short, the four Ag–AgCl wires are difficult to insert.

2. Set an O-ring around the glass capillary and the gaskets at the electric terminals to make an airtight seal.

3. Insert a 2-mm pin into a 2-mm jack assembly (Fig. 1A, four circles). Attach a silicone tube (internal diameter = 1 mm) to the pipe protruding from the holder (Fig. 1A, left). Connect the other end of the tube to a 1-mL syringe. Control the pressure in the canal through the tube.

4. Set the silver wires in the holder. Cut the length of the four wires to different lengths corresponding to the different channels. This helps in inserting the wires into the four channels of the capillary. The longest wire should reach the shoulder of the pipette. Soak the silver wires in bleach for 5 min.

Make the Recording Setup

5. Make a recording chamber with clear acrylic plates (4-mm-thickness). Paste the rear projection screens on both sides of the chamber with a screen size of 60 mm (width) × 55 mm (height). Glue these screen parts to two trapezoid acrylic side panels with a height of 60 mm and sides of 64 mm and 35 mm (Fig. 2A). The size of the bottom panel is 35 mm × 60 mm.

    The distance from the tadpole's eye to the screen is ~20 mm.

6. To make a stage for tadpoles, penetrate an acrylic round rod (ϕ = 7 mm) through the two trapezoid plates at 18 mm from the upper side. Flatten the upper side of the round rod. Then seal the holes with epoxy glue or O-rings. Make a 3-mm hole on the bottom panel, and glue a short pipe (OD 3 mm × 5 mm) into the hole. Connect the recording chamber and 60-mL syringe with a silicone tube (ID 2 mm × 1 m). Make a 2-mm hole in the trapezoid panel, and glue a short pipe (OD2 mm × 5 mm) into the hole. Connect a silicone tube (ID 1 mm) with an outlet from a peristaltic pump. Connect the inlet to the peristaltic pump to a silicone tube (ID 1–2 mm × 20

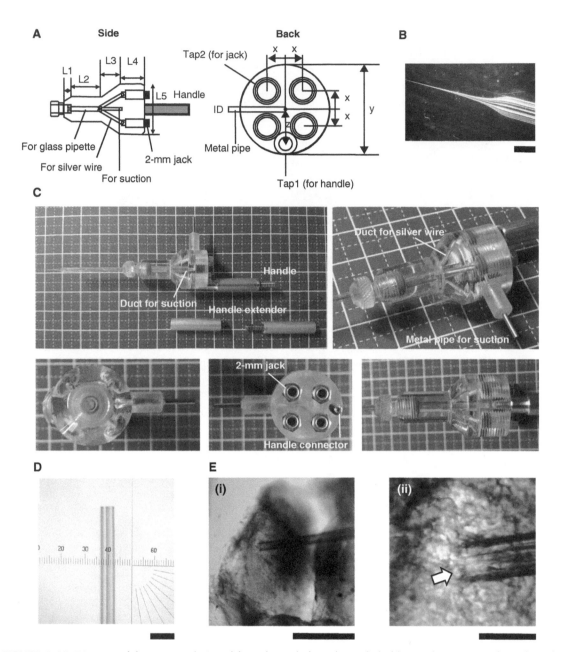

FIGURE 1. (*A*) Diagram of the custom-designed four channel glass electrode holder. (*Left*) projection from the side view, (*right*) projection from the connector side. The design is modified from a Warner Instruments theta glass electrode holder (https://www.warneronline.com/theta-glass-holders). It has a duct for a glass pipette and five branches (*left*; three branches are visible). The four branches are for silver wires and the other duct at the center (*left*) is for a suction line, as labeled in the figure. Although the four branches for the wires merge, the wires do not touch each other unless they are twisted. The suction line is connected to a metal pipe (*right*). Each silver wire passes through a φ2-mm compression cap (Warner QC-20) and contacts a 2-mm jack (Warner HC-21), which is screwed into the top side of the capillary holder. A 2-mm pin connector (Warner WC2-5) is connected to a 1-mm pin connector (Warner, 1–10) through a shielded cable. The 1-mm pin connector is inserted into the 1-mm jack on the patch-clamp head stage (Molecular Device CV-7B). The other end of the 2-mm pin connector is inserted into the 2-mm jack on the capillary holder. The ground terminals on the four head stages are connected to a single wire. The wire is soldered to a ground pellet (Warner E201). Size (mm) x: 3.8, y: 20, z: 5.4, ID: 1, tap1: 4–40 × 4.7, tap2: 10–32 × 6, L1: 3, L2: 13, L3: 8, L4: 9 L5: 19. (*B*) Photo of a four-channel glass electrode pulled on a Sutter P-97 micropipette puller (shank = 13 mm; the setting on Sutter P-97: 3-mm box heater, heat = RAMP, pull = 60, velocity = 83, delay = 135, *P* = 200). (*C*) Photos of the electrode holder. (*D*) Glass electrodes with a blunt end. (*E*) Electrode positioned on the tectum. Arrow points to a tectal cell. Note the diameter of the neuron compared with the pipette tip. The pipette tip has four openings corresponding to the four channels. Scale bars, (*B*) 2 mm, (*D*) 100 µm, (*E*) (*i*) 400 µm, (*ii*) 40 µm.

Cite this protocol as *Cold Spring Harb Protoc*; doi:10.1101/pdb.prot107086

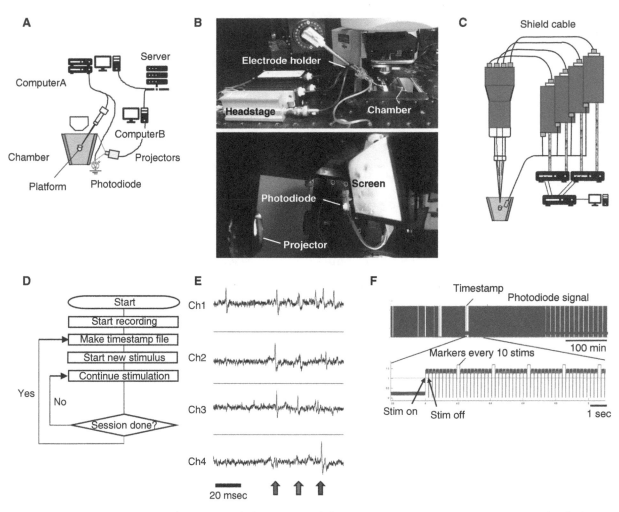

FIGURE 2. (*A*) Diagram of setup. Tetrode data are recorded in computer A. A rear-projection screen and a platform to hold the animal are attached to a recording chamber. Images are generated in computer B and projected onto the rear-projection screen. Stimulus timing is monitored by a photodiode. (*B*) (*Top*) Picture of the mounted electrode and head stage on the manipulator. (*Bottom*) Picture of the miniprojector, photodiode, and screen on the side of the recording chamber. (*C*) Wiring of the four-channel glass electrode with the head stage assembly and amplifier (MultiClamp 700A/B). (*Bottom*) Picture of the recording chamber and the projector. The photodiode is placed to the *left* of the screen. (*D*) A flowchart of continuous recording. Neural activity is recorded continuously. Temporal boundaries of recording sessions are marked with a timestamp file generated by the image presentation computer (computer B). (*E*) Spike waveforms recorded by individual channels. Spike signals from the same cells are detected by multiple channels (colored arrows). (*F*) An example of the photodiode signal. The red lines identify session start times marked by the timestamp file. Timing of each stimulus is identified by the rising edge of the photodiode signal.

cm). Set the silicone tube inside of the 60-mL syringe. Set an air stone connected to an air pump in the 60-mL syringe.

7. For the tetrode recording system, use the same amplifiers and digitizer as used for patch-clamp recording (Fig. 2A,B). Use four channels for amplification (Fig. 2C). Connect the electrodes to the input terminals of the head stages using shielded wires, which should be as short as possible.

8. Connect the ground lines to a single Ag–AgCl pellet. To record visually evoked activity, place the animal on the stage in the recording chamber as described in Step 15 (Fig. 2C). Project images for visual stimulation from the miniprojector to the rear-projecting screen. Image details depend on the purpose of the experiment, and images can be generated using a variety of software, including MATLAB. To identify the time of stimulus presentation, a small dot is presented on a photodiode located outside of the rear-projecting screen when the stimulus is presented.

Recording Procedure

9. Pull the cross-theta glass electrodes with a Sutter P-97 puller with a 3.0 mm × 3.0 mm square box filament using the following parameters: temp: RAMP + 20, pull = 30, Vel = 140, time = 20.

10. Insert the Ag–AgCl wires into the cross-theta capillary. Make sure the wires in the holder are dry. Insert the longest wire into the first canal, then set the second longest one into the neighboring canal. Insert the other wires this way. If the wires become bent many times during this process, replace the wires.

    See Troubleshooting.

11. When the wires are set, check the isolation between electrodes by measuring electric resistance between the gold-plated 2-mm jacks using an electrical tester. If the resistance is <1 megaohm, the wires are contacting each other in the holder. Remove the pipette and insert the wires again.

12. Break the glass electrode. Follow these steps to generate a flat cutting plane (Fig. 1D) at the electrode tip for tight contact with the tissue:

    i. Break the very end of the tip with forceps.

    ii. Dip the tip into the external solution to fill the end of the capillary to 5 mm.

    iii. Make a scratch at the $\phi = 40$ μm position on the capillary using a ceramic tile. The tip should break when the tile scratches the capillary. If the tip does not break, the ceramic tile is not sharp enough. Ceramic tiles dull easily.

    iv. Attach the electrode to the head-stage assembly, and dip the tip into the external solution applying negative pressure to fill the capillary. Monitor voltage in current-clamp mode ($I = 0$). The voltage changes when the external solution comes into the capillary. Wait until the baseline stabilizes.

    See Troubleshooting.

13. Prepare the animal for recording. Anesthetize a tadpole in a Petri dish containing 0.01% MS222 for <1 min. Transfer the animal to a strip of Sylgard in a Petri dish filled with external solution containing 10 μM pancuronium dibromide. Pin the animal to the Sylgard using insect pins.

14. To expose the tectal neurons for recording, cut the dorsal midline of the tectum through the skin using a 30-gauge needle attached to a syringe. Then, flip one of the tectal lobes. Remove the membrane covering the ventricular surface of the tectum using suction through a glass pipette (tip diameter ∼20–30 μm). This exposes the neurons. Recordings will be collected from the exposed neurons.

    The tectal neurons are located on the ventral side of the tectal lobe.

15. Transfer the tadpole on the Sylgard strip to the recording chamber made of clear acrylic plates as described in Figure 2A. Fill the chamber with external solution. Attach the Sylgard to the platform bar in the recording chamber using vacuum grease. Place the electrode tip on the exposed surface of the tectum.

16. Apply weak negative pressure (−0.73 psi) through the syringe connected to the pipe on the electrode holder (Fig. 1A,C). Use the scale on the syringe to control the pressure. Close the front door of the Faraday cage. Wait until the baseline and the spiking are stabilized (∼10 min). If clear spikes are not observed, start the process over with a new glass pipette. Contact between the glass electrodes and tissue will be stable for a long time, from 8 h to 1–2 d.

    See Troubleshooting.

17. Turn on the peristaltic pump and the air pump for aeration. The peristaltic pump makes a spike-like periodic noise. Turn on the pump only between recording sessions or use a gravity feed system.

Cite this protocol as *Cold Spring Harb Protoc*; doi:10.1101/pdb.prot107086

It is likely that tadpole tissues can tolerate less oxygen in the external solution compared to mammalian tissue because the animals live in poorly oxygenated water. We observed visual stimulation–evoked neuronal activity over 15–24 h without aeration when the chamber volume was 100 cc. Nevertheless, aeration ensures a longer recording time.

18. Start recording. Perform recording in current-clamp mode ($I = 0$). Select the gap-free mode on the recording software. Spikes are usually detected in the first several trials. Failure to detect spikes may be due to poor positioning of the tetrode. If this occurs, reposition the tetrode. In successful trials, the amplitude of spikes becomes larger over the first ~10 min of recording because the contact between the pipette and neurons stabilizes and the electric shield between the pipette and tissue increases.

19. Carry out a desired protocol. For instance, use the following protocol to test whether tectal cell visual response properties change in response to exposure to particular stimuli (Fig. 2D):

    i. Present image stimuli to characterize response properties of recorded neurons. Signals from the same cells are detected through the four channels (Fig. 2E).

    ii. The computer for image presentation makes a setting data file in which stimulus information is described. The timestamp of the file also indicates the approximate start time of the stimulus.

    iii. Identify the exact start time of the visual stimulus based on the photodiode signal and the timestamp of the setting file data (Fig. 2F).

       See Troubleshooting.

20. Perform spike sorting with wave_clus (https://github.com/csn-le/wave_clus) (Fig. 3; Quiroga 2012; Chaure et al. 2018). We modified the code to use the same waveform in sorting spikes in different sessions. In our experiment, the stringency (temperature) of the clustering condition was set so that sizes of major clusters were at the local maximum (Fig. 3). Change the stringency to match the purpose of the analysis.

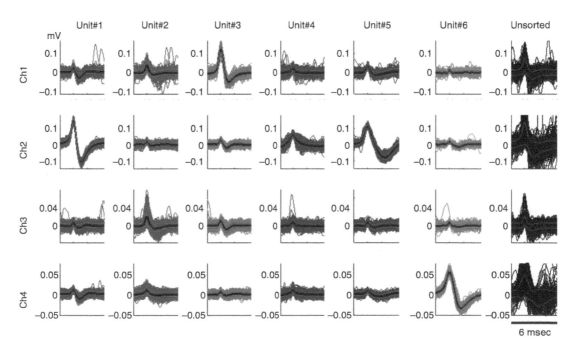

FIGURE 3. (*A*) Waveforms of sorted spikes. Unit #5 and Unit #1 are likely the same cells. The *rightmost* column shows units that are not grouped into the clusters. The wave data are digitized at 10 kHz.

TROUBLESHOOTING

Problem (Step 10): The Ag wires cannot be inserted during electrode preparation.
Solution: This is likely due to metal fatigue of wires. Use new wires.

Problem (Step 12): The electrode tip is not flat after breaking.
Solution: This can occur if the ceramic tile is old or dirty. Either use a new tile or clean the existing tile with 70% ethanol.

Problem (Step 12): The baseline voltage reading is unstable during electrode preparation.
Solution: An unstable baseline can occur as a result of incomplete soldering or lost Cl coating on the wire because of multiple attempts to insert Ag wires. Check the connection with a tester.

Problem (Step 16): No spikes were observed during recording.
Solution: This can occur because the electrode tip size is too small. Replace with an electrode with a larger tip. It can also occur because of poor recording position. Reposition the electrode in the tectum.

Problem (Step 16): Only small spikes were observed during recording.
Solution: This can occur because the electrode tip size is too large. Replace with an electrode with a smaller tip.

Problem (Step 16): A large local field potential is observed during recording.
Solution: This can occur when the pipette tip is not flat. Remake the electrode with a flat tip. To make the tip flat, make a deeper scratch or use a new and clean ceramic tile.

Problem (Step 19): Spikes were only recorded for a short time.
Solution: This can occur because of unstable contact between neurons and the pipette. If this is the case, use a clean pipette. It can also occur as a result of animal movement. If the tadpole is young (<stage 44), use a higher concentration (20 μM) of pancuronium dibromide or use other immobilization methods.

Problem (Step 19): Animals do not live long during recording.
Solution: Animals may die during recording sessions because they were damaged during preparation for recording. In future preparations, be more careful in animal preparation, taking care to not damage the tail. Another possible explanation for animals dying during the recording session is infection. Sterilize the chamber with 70% ethanol to prevent infection. A third explanation is that the animals experienced a lack of oxygen. For future experiments, increase intervals between recording sessions for longer aeration periods.

DISCUSSION

The *Xenopus* visual circuit is an excellent system in which to study fundamental principles of plasticity (Constantine-Paton and Law 1978; Meyer 1998; Zhang et al. 1998; Engert et al. 2002; Sin et al. 2002). Currently, neural activity in the *Xenopus* tectum is mainly studied by patch-clamp recording (Wu et al.

1996; Aizenman et al. 2003; Pratt et al. 2008; Ciarleglio et al. 2015) and Ca^{2+} imaging (Dunfield and Haas 2009; Hiramoto and Cline 2009; Xu et al. 2011; Podgorski et al. 2012). These techniques have drawbacks: The duration of patch-clamp recordings is limited and Ca^{2+} imaging has poor temporal resolution, restricting the study of circuit organization and plasticity. Methods that are amenable to longer recording duration would allow investigations of forms of plasticity that occur over longer timescales (Ruthazer et al. 2003; Lim et al. 2010; Hiramoto and Cline 2014). Extracellular recording is such a method. Single unit extracellular recording with metal electrodes has been used in adult *Xenopus* retinotectal mapping studies (Gaze 1958). Juxtacellular recording with metal-filled glass with platinum-plated tips electrodes has also been used in *Xenopus* (Udin and Scherer 1990); however, these metal electrodes are less sensitive than glass electrodes. Multichannel glass electrode recording therefore offers multiple advantages over other recording methods. Because tetrodes are typically used in systems with relatively large-amplitude spikes, we modified the tetrode design to detect the small-amplitude spikes in developing *Xenopus* optic tectal neurons. This modification is particularly valuable for investigating development and plasticity of neuronal and circuit properties, because in many cases, neurons with rich plasticity are immature and do not generate large spikes. The large contact area between the shaft of the glass electrode and the tissue stabilizes the electrode within the tissue over extended periods of time. The tight seal between the neurons and the multiple openings of the pipette increase the sensitivity of the recording.

We wished to develop multiunit recording capacity for longer timescales, with the following parameters in mind: the ability to collect stable recordings over time; the ability to perform reliable spike sorting by limiting contaminating signals from distant cells; and the capacity to collect spikes from multiple neighboring cells. This protocol is optimized to collect long-term recordings to study changes in visual response properties of individual neighboring neurons.

The success of the glass tetrode recordings depends on high-quality data, which are essential for accurate spike sorting and data analysis. Although recordings with metal electrodes detect signals from many neurons, the glass pipette electrode records activity from relatively few neurons, based on several factors. First, the glass pipette records cells in a limited area in the tissue because the pipette and the tissue make an electric shield, excluding signals outside of the pipette. Excluding signals outside of the pipette contact area makes spike sorting easy. In addition, neurons within the cell-dense layers of the *Xenopus* tectum are closely packed. Usually, only three to five units are sorted from the tectal recording, roughly matching the expected number of the tectal cells in the shielded area (cell body = ~10 μm, each pipette tip = ~15 μm). Because only three to five cells are in the recording area, the waveforms across the channels are usually distinct between units. The spike waveform can change gradually over a long recording session. This is partly due to movement of the pipette location or changes in the property of the neurons. This recording method reduces the likelihood that changes in waveform over time are due to pipette movement because of the stable contact of the glass electrode shaft with the tissue. Sometimes changes in the waveform may be caused by changes in the biophysical properties of the neurons. To track these units over the recording sessions, methods by Dhawale et al. (2017), in which they track the changes of waveforms and identify units that change waveforms over time, may be useful.

Because this electrode shields signals from other regions, local field potentials (LFPs) are not typically detected. To record LFPs from larger groups of neurons, pipettes with larger diameters (>70-μm) can be used. This will decrease signal amplitudes from single neurons.

In summary, we present a method for multichannel juxtacellular glass electrode recording from neurons in *Xenopus* optic tectum. The advantages of this method are its sensitivity to detect small spikes and the ability to maintain stable recordings in an intact animal over extended periods of time. Tetrode recording using the four-channel capillaries are useful to study developmental plasticity in the *Xenopus* visual system and may also be useful to study plasticity in other circuits.

RECIPES

External and Internal Recording Solutions

Reagent	Final concentration
NaCl	115 mM
KCl	4 mM
CaCl$_2$	3 mM
MgCl$_2$	3 mM
HEPES	5 mM
Glucose	10 mM
Glycine	10 μM

Adjust the pH to 7.2 with NaOH. Adjust the osmolarity to 255. Filter-sterilize and store for up to 1 mo at 4°C.

Steinberg's Solution (100%)

Reagent	Amount to add (for 1 L)
NaCl	3.4 g
KCl	0.05 g
Ca(NO$_3$)$_2 \cdot$4H$_2$O	0.08 g
MgSO$_4 \cdot$7H$_2$O	0.205 g
Tris	0.56 g
H$_2$O	to 1 L

Adjust pH to 7.4 with HCl.

REFERENCES

Aizenman CD, Akerman CJ, Jensen KR, Cline HT. 2003. Visually driven regulation of intrinsic neuronal excitability improves stimulus detection in vivo. *Neuron* **39:** 831–842. doi:10.1016/S0896-6273(03)00527-0

Chaure FJ, Rey HG, Quian Quiroga R. 2018. A novel and fully automatic spike-sorting implementation with variable number of features. *J Neurophysiol* **120:** 1859–1871. doi:10.1152/jn.00339.2018

Ciarleglio CM, Khakhalin AS, Wang AF, Constantino AC, Yip SP, Aizenman CD. 2015. Multivariate analysis of electrophysiological diversity of *Xenopus* visual neurons during development and plasticity. *Elife* **4:** e11351. doi:10.7554/eLife.11351

Constantine-Paton M, Law MI. 1978. Eye-specific termination bands in tecta of three-eyed frogs. *Science* **202:** 639–641. doi:10.1126/science.309179

Dhawale AK, Poddar R, Wolff SB, Normand VA, Kopelowitz E, Olveczky BP. 2017. Automated long-term recording and analysis of neural activity in behaving animals. *Elife* **6:** e27702. doi:10.7554/eLife.27702

Dunfield D, Haas K. 2009. Metaplasticity governs natural experience-driven plasticity of nascent embryonic brain circuits. *Neuron* **64:** 240–250. doi:10.1016/j.neuron.2009.08.034

Engert F, Tao HW, Zhang LI, Poo MM. 2002. Moving visual stimuli rapidly induce direction sensitivity of developing tectal neurons. *Nature* **419:** 470–475. doi:10.1038/nature00988

Gaze RM. 1958. The representation of the retina on the optic lobe of the frog. *Q J Exp Physiol Cogn Med Sci* **43:** 209–214. doi:10.1113/expphysiol.1958.sp001318

Hiramoto M, Cline HT. 2009. Convergence of multisensory inputs in *Xenopus* tadpole tectum. *Dev Neurobiol* **69:** 959–971. doi:10.1002/dneu.20754

Hiramoto M, Cline HT. 2014. Optic flow instructs retinotopic map formation through a spatial to temporal to spatial transformation of visual information. *Proc Natl Acad Sci* **111:** E5105–E5113. doi:10.1073/pnas.1416953111

Lim BK, Cho SJ, Sumbre G, Poo MM. 2010. Region-specific contribution of ephrin-B and Wnt signaling to receptive field plasticity in developing optic tectum. *Neuron* **65:** 899–911. doi:10.1016/j.neuron.2010.03.008

Meyer RL. 1998. Roger Sperry and his chemoaffinity hypothesis. *Neuropsychologia* **36:** 957–980. doi:10.1016/S0028-3932(98)00052-9

Nieuwkoop PD, Faber J. 1956. *Normal table of* Xenopus laevis *(Daudin)*. Elsevier-North Holland, Amsterdam.

Podgorski K, Dunfield D, Haas K. 2012. Functional clustering drives encoding improvement in a developing brain network during awake visual learning. *PLoS Biol* **10:** e1001236. doi:10.1371/journal.pbio.1001236

Pratt KG, Dong W, Aizenman CD. 2008. Development and spike timing–dependent plasticity of recurrent excitation in the *Xenopus* optic tectum. *Nat Neurosci* **11:** 467–475. doi:10.1038/nn2076

Quiroga RQ. 2012. Spike sorting. *Curr Biol* **22:** R45–R46. doi:10.1016/j.cub.2011.11.005

Ruthazer ES, Akerman CJ, Cline HT. 2003. Control of axon branch dynamics by correlated activity in vivo. *Science* **301:** 66–70. doi:10.1126/science.1082545

Sin WC, Haas K, Ruthazer ES, Cline HT. 2002. Dendrite growth increased by visual activity requires NMDA receptor and Rho GTPases. *Nature* **419:** 475–480. doi:10.1038/nature00987

Udin SB, Scherer WJ. 1990. Restoration of the plasticity of binocular maps by NMDA after the critical period in *Xenopus*. *Science* **249:** 669–672. doi:10.1126/science.2166343

Cite this protocol as *Cold Spring Harb Protoc*; doi:10.1101/pdb.prot107086

Wu G, Malinow R, Cline HT. 1996. Maturation of a central gluta-matergic synapse. *Science* **274:** 972–976. doi:10.1126/science.274.5289.972

Xu H, Khakhalin AS, Nurmikko AV, Aizenman CD. 2011. Visual experi-ence–dependent maturation of correlated neuronal activity patterns in a developing visual system. *J Neurosci* **31:** 8025–8036. doi:10.1523/JNEUROSCI.5802-10.2011

Zhang LI, Tao HW, Holt CE, Harris WA, Poo M. 1998. A critical window for cooperation and competition among developing retinotectal synapses. *Nature* **395:** 37–44. doi:10.1038/25665

Protocol 3

Electrophysiological Approaches to Studying Normal and Abnormal Retinotectal Circuit Development in the *Xenopus* Tadpole

Kara G. Pratt

Department of Zoology and Physiology, and Program in Neuroscience, University of Wyoming, Laramie, Wyoming 82071, USA

The *Xenopus* tadpole retinotectal projection is the main component of the amphibian visual system. It comprises the retinal ganglion cells (RGCs) in the eye, which project an axon to synapse onto tectal neurons in the optic tectum. There are many attributes of this relatively simple projection that render it uniquely well-suited for studying the functional development of neural circuits. One major experimental advantage of this circuit is that it can be genetically or pharmacologically altered and then assessed at high resolution via whole-cell electrophysiological recordings using an ex vivo isolated brain preparation. This protocol provides instructions for performing such electrophysiological investigations using the ex-vivo-isolated brain preparation. It allows one to measure many different aspects of synaptic transmission between the RGC axons and individual postsynaptic tectal neurons, including AMPA (α-amino-3-hydroxy-5-methyl-4-isoxazolepropionic acid) to NMDA (*N*-methyl-D-aspartate) ratios, strength of individual RGC axons, paired pulse facilitation, and strength of individual synapses.

MATERIALS

It is essential that you consult the appropriate Material Safety Data Sheets and your institution's Environmental Health and Safety Office for proper handling of equipment and hazardous materials used in this protocol.

RECIPES: Please see the end of this protocol for recipes indicated by <R>. Additional recipes can be found online at http://cshprotocols.cshlp.org/site/recipes.

Reagents

2,3-dioxo-6-nitro-sulfamoyl-benzo[f]quinoxaline (NBQX; 20 mM stock)
External recording solution for whole-cell electrophysiological recordings <R>
Internal recording solution for whole-cell electrophysiological recordings <R>
Steinberg's solution (100× stock) <R>
> This solution is the tadpole-rearing solution.

Tetrodotoxin (TTX; 2 mM stock)
Tricaine methanesulfonate (MS222; 0.02% in 1× Steinberg's solution)
Xenopus laevis tadpole larvae between developmental stages 39/40 and 48/49 (staging according to Nieuwkoop and Faber 1994)

Correspondence: kpratt4@uwyo.edu

Equipment

Dissection microscope

Electrophysiology rig outfitted with the following components:

   Amplifier with voltage-clamp capability (Molecular Devices Multiclamp 700B)

   Bipolar stimulation electrode (FHC cluster electrode CE2C75)

   Digitizer (Digidata 1440A)

   Fixed-stage upright microscope designed for electrophysiology experiments (Zeiss Axio Examiner A1)

   Micromanipulator (Sutter)

   Stimulus isolator (A.M.P.I. Iso-flex)

Forceps, #5

Glass recording pipettes with a resistance between 10 MΩ and 12 MΩ

Minutien insect pins (F.S.T. 26002-10)

Needle, 25-gauge

Petri dish

Recording dish with a block of Sylgard 184 silicone elastomer (approximate dimensions of the block: 10 mm [length] × 10 mm [width] × 5 mm [height]) attached to the floor via medical-grade silicone grease (FST 29051-35)

Transfer pipette

METHOD

All procedures described here have been approved by the University of Wyoming's Institutional Animal Care and Use Committee (IACUC). Carry out all steps, including recording, at room temperature.

Whole-Brain Dissection and Preparing to Record

1. To anesthetize the tadpole, use a transfer pipette to place the tadpole into a small Petri dish containing 0.02% tricaine methanesulfonate (MS222) in 1× Steinberg's solution.

   Tadpoles will become fully anesthetized, which is characterized by cessation of swimming, within ~5 min.

2. Transfer the anesthetized tadpole to the recording dish containing external recording solution for whole-cell electrophysiological recordings.

   Carry out Steps 3–7 under the dissection microscope.

3. Using insect pins, secure the tadpole to the Sylgard attached to the floor of the recording dish.

4. Using a sterile 25-gauge needle, gently peel away the skin overlying the brain and hindbrain.

5. Using the same 25-gauge needle, fillet open the brain along the dorsal midline of the rostrocaudal axis (i.e., a midsagittal transection), from the hindbrain to the olfactory bulbs (OBs), such that the roof of the neural tube is severed while the floor plate remains intact.

   For a video and detailed description of this dissection see Liu et al. (2018a). A properly filleted preparation is shown in Figure 1B.

6. Next, remove the filleted whole-brain preparation from the tadpole by first making a transverse cut through the caudal end of the hindbrain such that it is completely detached from the spinal cord, and then running the needle gently underneath the brain, in a caudal-to-rostral direction, to sever all the lateral and ventral nerve fibers.

7. Secure the isolated brain to the block of Sylgard by placing one pin in an OB and one in the hindbrain. Discard the remainder of the tadpole.

8. Move the recording dish with the brain preparation from the dissecting microscope and position it on the microscope stage of the electrophysiology rig.

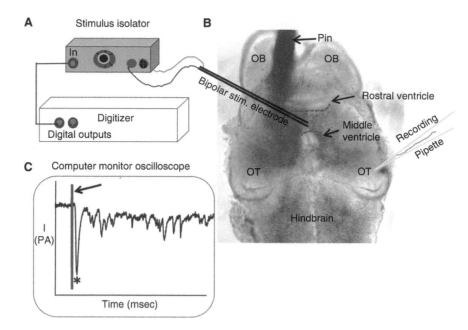

FIGURE 1. Whole-brain preparation with bipolar stimulation electrode for measuring RGC-evoked responses. (*A*) The bipolar stimulation electrode is placed on the (*B*) optic chiasm of the whole brain preparation where the RGCs from the two eyes cross. The optic chiasm is not visible, but it resides in between the rostral, horizontal-shaped ventricle and the round middle ventricle. Red dashed lines indicate the caudal side of rostral ventricle and the rostral side of the middle ventricle. The strength of bipolar electrode stimulation is controlled by the stimulus isolator, which is triggered (ON/OFF) by the computer via the digitizer. (*C*) On the computer monitor, the black arrow is pointing to the stimulus artifact created by activation of the bipolar electrode. The asterisk is lined up in time with the peak of the monosynaptic portion of RGC-evoked response. (OB) Olfactory bulb, (OT) optic tectum.

9. To access tectal neuron somata, remove the overlying ventricular membrane using a broken glass recording pipette. Break the pipette tip by lightly dragging it across a delicate task wiper such that the tip breaks somewhere along its taper, preferably at the narrower end. Control the broken pipette with the electrode micromanipulator. Viewing through a 60× or 40× objective, move the pipette tip over the surface of the somata until it catches onto the ventricular membrane, and then move the pipette tip such that the membrane is pulled away. Only remove a small portion of the membrane, sufficient to uncover 20 somata or so.

10. To prepare to activate retinal ganglion cell (RGC) axons to measure RGC-evoked responses, place a bipolar electrode on the optic chiasm, where the axons from the two eyes cross (Fig. 1A,B).

    The bipolar electrode is driven by a stimulus isolator. Connecting the stimulus isolator to the digitizer's digital output allows for the activation of the RGC axons to be electronically triggered and shows on the recorded trace precisely when the RGC axons were activated (Fig. 1C, computer monitor).

11. To carry out whole-cell recordings, visualize tectal neuron somata under a 60× (optimal) or 40× objective. If applicable, identify neurons electroporated with fluorescently tagged plasmid DNAs and/or morpholinos via appropriate filters. Record individual neurons using a 10 MΩ–12 MΩ glass recording pipette containing internal recording solution for whole-cell electrophysiological recordings.

    Extended activation of the fluorophore in living neurons can cause phototoxicity. Once the electroporated neuron is identified, subsequent steps should be performed under bright-field illumination.

Electrophysiology

Steps 12–18 describe different types of electrophysiological recordings used to characterize different aspects of synaptic transmission between the retinal ganglion cell axons and the postsynaptic tectal neurons. These recordings can be performed in any order.

12. To determine the total amount of RGC input received by a given tectal neuron, voltage-clamp the neuron at −60 mV. Increase the strength of the stimulation electrode until the amplitude of the

evoked synaptic response no longer increases. This amplitude reflects the total amount of direct monosynaptic RGC input received by the postsynaptic neuron (Liu et al. 2018b).

RGC-evoked responses are composed of a fast and synchronous monosynaptic component, which is purely glutamatergic, followed by a slower, less synchronous component due to activation of local recurrent excitatory and inhibitory circuits (as shown on the computer monitor in Fig. 1C).

The AMPA component of the monosynaptic response is measured with the neuron clamped at −60 mV. Because the AMPA component is temporally distinct from the recurrent portion of the evoked response, it is not necessary to block inhibitory postsynaptic currents (IPSCs) by adding picrotoxin to the external recording solution. However, it is also fine if picrotoxin is present so that, for example, AMPA:NMDA ratios (see Step 14), which require that picrotoxin be present to block inhibitory events, can also be measured.

See Troubleshooting.

13. To measure the strength of individual RGC axons onto a given tectal neuron, voltage clamp the neuron at −60 mV. Adjust the strength of the stimulation electrode such that an evoked synaptic response is observed in no more than 50% of the trials. This is referred to as the "minimal stimulation" protocol. At this stimulation strength, the responses are likely due to activation of a single RGC axon (Pratt and Aizenman 2007). Estimate the number of different axons received by the neuron by dividing the average maximum response amplitude by the average minimal response amplitude.

    See Troubleshooting.

14. To measure AMPA:NMDA ratios, prepare external recording solution containing 100 µM picrotoxin to block local inhibitory currents that may overlap in time with the relatively slower NMDA receptor (NMDAR)-mediated current. Adjust the strength of the bipolar electrode to evoke a reliable response that is between 40% and 60% of the maximum response. Measure RGC-evoked AMPA receptor (AMPAR)-mediated currents with the neuron voltage-clamped at −60 mV. Record NMDAR-mediated currents at +55 mV to relieve the voltage-dependent Mg^{2+} block. The peak NMDAR-mediated current is typically reported as the averaged amplitude between 20 and 30 msec after the response onset (Wu et al. 1996; Aizenman and Cline 2007).

    The ratio of AMPAR-mediated current amplitude to NMDAR-mediated current amplitude increases as synapses mature. Deviation in this ratio reflects abnormal synapse formation.

    See Troubleshooting.

15. To determine the probability of presynaptic transmitter release by measuring paired pulse facilitation, adjust the bipolar electrode to evoke a reliable response that is between 40% and 60% of maximum. Activate RGC axons twice (paired pulse) with an interstimulus interval of 50 msec. Record responses with the voltage clamped at −60 mV. Calculate the amount of facilitation as the amplitude of the second excitatory postsynaptic current (EPSC) divided by the first, EPSC2/EPSC1 (Aizenman and Cline 2007; Liu et al. 2018b). The amount of facilitation is negatively correlated with release probability.

    See Troubleshooting.

16. To measure excitatory and inhibitory recurrent activity, record maximum evoked responses at −40 mV and at +5 mV, respectively, and then calculate excitatory:inhibitory (E:I) ratios. Because inhibitory network is measured here, picrotoxin should not be included in the external recording solution for these recordings. Because this activity is asynchronous and more spread out over time, it is quantified as the response area (i.e., charge) within a 250-msec time window beginning immediately after the peak of monosynaptic response (Truszkowski et al. 2016).

    RGC stimulation activates local excitatory and inhibitory network activity in the optic tectum. Deviations in the E:I ratio are known to impair visually guided behaviors.

    See Troubleshooting.

17. To determine the overall amount of synaptic drive received by individual tectal neurons, record spontaneous synaptic currents with the neuron clamped at −60 mV. At this voltage, both excit-

atory and inhibitory synaptic currents are inward. Isolate excitatory currents by holding the cell at −40 mV, the reversal potential for IPSCs, or by blocking GABAergic transmission by adding 100 μM picrotoxin to the external recording solution (Pratt and Aizenman 2007; Hamodi and Pratt 2014). Isolate spontaneous inhibitory events by holding the cell at ∼+5 mV, the reversal potential for excitatory (AMPA) mixed ion currents (Truszkowski et al. 2016).

18. To measure individual synapse strength, add 1 μM TTX to the external recording solution. TTX will block action potential–driven synaptic events. The action potential–independent synaptic currents (aka "minis") represent the spontaneous release of transmitter from single release sites. Record mini excitatory postsynaptic currents (mEPSCs) with the neuron voltage clamped at −60 mV and with 100 μM picrotoxin added to the external recording solution to block inhibitory events. Record mini inhibitory postsynaptic currents (mIPSCs) with the neuron voltage clamped at −80 mV and with 20 μM NBQX added to the external recording solution to block mEPSCs (Liu et al. 2018b).

TROUBLESHOOTING

Problem (Steps 12–16): No RGC-evoked response is observed upon activation of the bipolar stimulation electrode.

Solution: This outcome is most likely due to improper placement of the bipolar stimulation electrode such that it is not touching down on the optic chiasm. This is a common problem given that the optic chiasm is not actually visible. It can be readily corrected by repositioning the bipolar electrode as shown in Figure 1.

Problem (Step 15): Recurrent activity interferes with visualizing the second EPSC in the paired pulse protocol.

Solution: Recurrent activity that can sometimes interfere with the second EPSC can be minimized by blocking IPSCs, which constitute a large portion of this local recurrent activity. IPSCs can be minimized by recording responses with the neuron voltage clamped at −40 mV, the reversal potential for the inhibitory (chloride) currents. Also, decreasing the stimulation strength of the bipolar electrode a tad will lessen recurrent activity.

DISCUSSION

This set of whole-cell electrophysiological recordings was derived from several published works (namely, Wu et al. 1996; Aizenman and Cline 2007; Pratt and Aizenman 2007; Truszkowski et al. 2016; Liu et al. 2018b). These references showcase the different recording configurations at work and the types of data they generate.

These recordings are routinely used to characterize retinotectal synaptic transmission from developmental stage 42 (5–6 dpf), when the RGCs have just begun to form synapses onto tectal neuron dendrites, through developmental stage 49 (∼18–21 dpf), by which time the (normal) projection is more refined, supporting robust visually guided behaviors (Pratt and Aizenman 2007).

In addition to studying normal circuit development, these recordings are often used to characterize how altering a specific molecule of interest in postsynaptic tectal neurons may impact the functional development of the circuit. The function of a given molecule can be altered pharmacologically by adding a drug or chemical to the rearing medium. Pharmacological manipulations are global, meaning both presynaptic RGCs and postsynaptic tectal neurons are exposed to the drug/chemical. They are also fast-acting. Genetic manipulations, such as morpholinos or plasmid DNA electroporated into tectal neurons, are also routinely used (Haas et al. 2002; Falk et al. 2007; Bestman and Cline 2014). Electroporation allows morpholinos or plasmid DNA to be targeted specifically to postsynaptic tectal neurons. Because plasmid DNA must be transcribed and translated, it requires relatively more time to begin to elicit effects. Transfection rates for electroporation of plasmid DNAs tend to be low, allowing for cell-autonomous effects to be identified. The advantage inherent to both of these modes of manipulation is that they can be applied specifically during retinotectal synapse formation and maturation, allowing all earlier stages of development to proceed as normal. This is especially advantageous given that many molecules play different roles throughout development.

RECIPES

External Recording Solution for Whole-Cell Electrophysiological Recordings

Reagent	Final concentration
NaCl	115 mM
KCl	2 mM
CaCl$_2$	3 mM
MgCl$_2$	3 mM
HEPES	5 mM
Glycine	10 mM
Glucose	10 mM
Picrotoxin (optional[a])	0.1 mM

Prepare solution with deionized H$_2$O. Adjust pH to 7.25 using 1 N NaOH, and adjust osmolarity to 255 mOsm using sucrose (if prepared properly, osmolarity should not need to be adjusted). Prepare fresh on the day of the experiment.
[a]Do not include when recording inhibitory network activity.

Internal Recording Solution for Whole-Cell Electrophysiological Recordings

Reagent	Final concentration
K-gluconate	100 mM
KCl	8 mM
NaCl	5 mM
MgCl$_2$	1.5 mM
HEPES	20 mM
EGTA	10 mM
ATP	2 mM
GTP	0.3 mM

Prepare the solution with deionized H$_2$O. Adjust the pH to 7.2 using 1 N KOH, and adjust osmolarity to 255 mOsm using sucrose. Store at −80°C.

Steinberg's Solution (100× Stock)

Reagent	Final concentration
KCl	6.7 mM
Ca(NO$_3$)$_2$ 4H$_2$O	3.4 mM
MgSO$_4$ 7H$_2$O	8.3 mM
NaCl	580 mM
HEPES	499.37 mM

Prepare the 100× stock solution with deionized H$_2$O. Adjust pH to 7.5 using 10 N NaOH. Store. at 4°C. To prepare a 1× working solution, dilute 1/100 in deionized H$_2$O (i.e., 1 part 100× stock + 99 parts H$_2$O).

REFERENCES

Aizenman CD, Cline HT. 2007. Enhanced visual activity in vivo forms nascent synapses in the developing retinotectal projection. *J Neurophysiol* **97**: 2949–2957. doi:10.1152/jn.00452.2006.

Bestman JE, Cline HT. 2014. Morpholino studies in *Xenopus* brain development. *Methods Mol Biol* **1082**: 155–171. doi:10.1007/978-1-62703-655-9_11

Falk J, Drinjakovic J, Leung KM, Dwivedy A, Regan AG, Piper M, Holt C. 2007. Electroporation of cDNA/Morpholinos to targeted areas of embryonic CNS in *Xenopus*. *BMC Dev Biol* **7**: 107. doi:10.1186/1471-213X-7-107.

Haas K, Jensen K, Sin WC, Foa L, Cline HT. 2002. Targeted electroporation in *Xenopus* tadpoles in vivo: from single cells to the entire brain. *Differentiation* **70**: 148–154. doi:10.1046/j.1432-0436.2002.700404.x

Hamodi AS, Pratt KG. 2014. Region-specific regulation of voltage-gated intrinsic currents in the developing optic tectum of the *Xenopus* tadpole. *J Neurophysiol* **112**: 1644–1655. doi:10.1152/jn.00068.2014

Liu Z, Donnelly KB, Pratt KG. 2018a. Preparations and protocols for whole cell patch clamp recording of *Xenopus laevis* tectal neurons. *J Vis Exp* **133**: 57465. doi:10.3791/57465

Liu A, Thakar A, Santoro SW, Pratt KG. 2018b. Presenilin regulates retinotectal synapse formation through EphB2 receptor processing. *Dev Neurobiol* **78**: 1171–1190. doi:10.1002/dneu.22638.

Nieuwkoop PD, Faber J. 1994. *Normal table of* Xenopus laevis *(Daudin)*. Routledge, Abingdon, UK.

Pratt KG, Aizenman CD. 2007. *Homeostatic regulation of intrinsic excitability and synaptic transmission in a developing visual circuit. J Neurosci* **27**: 8268–8277. doi:10.1523/JNEUROSCI.1738-07.2007

Truszkowski TL, James EJ, Hasan M, Wishard TJ, Liu Z, Pratt KG, Cline HT, Aizenman CD. 2016. Fragile X mental retardation protein knockdown in the developing *Xenopus* tadpole optic tectum results in enhanced feedforward inhibition and behavioral deficits. *Neural Dev* **11**: 14. doi:10.1186/s13064-016-0069-7.

Wu G, Malinow R, Cline HT. 1996. Maturation of a central glutamatergic synapse. *Science* **274**: 972–976. doi:10.1126/science.274.5289.972

Cite this protocol as *Cold Spring Harb Protoc*; doi:10.1101/pdb.prot106898

Making In Situ Whole-Cell Patch-Clamp Recordings from *Xenopus laevis* Tadpole Neurons

Wen-Chang Li

School of Psychology and Neuroscience, St Andrews, Fife, KY16 9TS, United Kingdom

Xenopus laevis tadpoles have been an excellent, simple vertebrate model for studying the basic organization and physiology of the spinal cord and motor centers in the brainstem. In the past, intracellular recordings from the spinal and brainstem neurons were primarily made using sharp electrodes, although whole-cell patch-clamp technology has been around since the early 1980s. In this protocol, I describe the dissections and procedures needed for in situ whole-cell patch-clamp recording, which has become routine in tadpole neurophysiology since the early 2000s. The critical step in the dissections is to delicately remove some ependymal cells lining the tadpole neurocoele in order to expose clean neuronal somata without severing axon tracts. Whole-cell recordings can then be made from the somata in either current- or voltage-clamp mode.

MATERIALS

It is essential that you consult the appropriate Material Safety Data Sheets and your institution's Environmental Health and Safety Office for proper handling of equipment and hazardous materials used in this protocol.

RECIPES: Please see the end of this protocol for recipes indicated by <R>. Additional recipes can be found online at http://cshprotocols.cshlp.org/site/recipes.

Reagents

3-aminobenzoic acid ester (MS-222; 0.1% in saline)

Store for up to a couple of weeks in a sealed container at room temperature.

α-bungarotoxin (100 μM in distilled water; store at −20°C in 100-μL aliquots)

For a working solution, dilute 100 μL of stock to 1 mL with saline. Store for up to a couple of weeks between uses at 4°C.

Intracellular pipette solution for *Xenopus* <R>
Saline <R>
Xenopus laevis tadpoles, stage 37/38 (Nieuwkoop and Faber 1956)

Equipment

Equipment for dissections (Steps 1–12):
  Blunt and fine tungsten needles

Correspondence: wl21@st-andrews.ac.uk

Prepare by dipping the end of a short piece of 200-μm tungsten wire glued onto a handle into the NaOH etching solution (Li et al. 2014). Incubate longer in NaOH to prepare the fine needle.

Cold-light source with light guides

Dissection bath with rotatable Sylgard stage (Fig. 1A; Li et al. 2002)

Dissection microscope with magnification up to 80–120 times

Electrical etching kit (24-V, 1500-mA AC adaptor, connected to a plastic dish filled with 5 M NaOH solution with a pair of forceps to hold needles and pins at one end of the electrical circuit)

This kit is used to make tungsten dissection needles and pins.

Fine forceps (Dumont Tweezers, WPI)

Forceps

These are used for moving pins.

Microcentrifuge tubes

Petri dish

Pins

Prepare by dipping the end of a 50-μm tungsten wire a few mm long into the NaOH etching solution (Li et al. 2014).

Transfer pipette

Equipment for recordings (Steps 13–28; see Fig. 1C for details on how the equipment is connected):

Coarse manipulators (MX10; SD Instruments)

Coverslip (50 mm × 33 mm, 0.17-mm thick)

Digitizer (CED power 1401 from Cambridge Electronic Design)

Electrical stimulator (DS3; Digitimer)

Extracellular amplifier (four-channel; A-M Systems)

This is used to amplify ventral root signals.

Filters, 0.2-μm (Whatman Puradisc 13 syringe filters, Sigma-Aldrich)

Motor-driven fine control manipulator (MX763OR manipulators; SD Instruments)

Nikon E600FN microscope with 4× and water immersion 40× or 60× objectives and a translation table

Patch-clamp amplifier (Multiclamp 700B; Molecular Devices)

Peristaltic pump

Pipette puller (P-97; Sutter Instrument)

Recording bath with rotatable Sylgard stage (Fig. 1B; Li et al. 2002)

Use Vaseline to attach a thin piece of coverslip to the bottom of the recording chamber to allow light transmission from below on the Nikon E600FN microscope. Leave a gap in the stage to avoid light distortion by Sylgard.

Sampling software (Signal version 5 from Cambridge Electronic Design)

Single barrel capillary glass with microfilament for patch clamp (A-M systems)

Syringes, 1-mL

Three-way connectors

Vaseline

Workstation PC

METHOD

Dissections and Tadpole Preparation

1. Anesthetize a stage 37/38 tadpole (Nieuwkoop and Faber 1956) with 0.1% MS-222 for ∼20 sec in a Petri dish at room temperature.

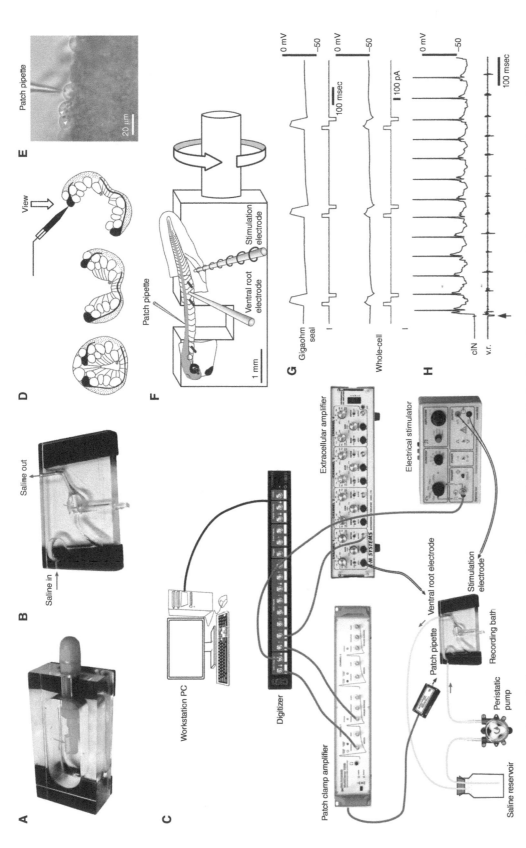

FIGURE 1. Recording tadpole spinal neurons. (A) A dissection bath made from a block of polycarbonate 20-mm-thick, 40-mm-wide, and 80-mm-long. (B) A polycarbonate recording bath block 6-mm-deep, 40-mm-wide, and 70-mm-long with a 10-mm through hole in the middle, the bottom of which is sealed with a piece of coverslip glass and Vaseline. Blue arrows in A and B point at the Sylgard blocks glued onto the rotatable plastic rods, and the black magnetic strips at the bottom are for fixing the baths onto base plates. (C) The connections between the main equipment. The thick line to the chassis means cable sockets are in the back panels. Electrodes, microscope, and manipulators are omitted for simplicity. Red arrows in B and C indicate the direction of saline flow for the recording bath. (D) A diagram of tadpole spinal cord cross sections showing neurons (ovals, two examples filled), ependymal cells (rectangles), and axon tracts (dotted area). The spinal cord is split open (middle diagram) and tilted to allow view of exposed neurons on the E600FN microscope after removal of ependymal cells (right diagram). (E) A photo of a segment of the spinal cord after dissections in D. The patch pipette is placed on a potential sensory Rohon–Beard neuron (dorsal is up, rostral is to the left). White arrowheads point at the yolk platelets inside neurons. (F) Arranging electrodes around a dissected tadpole (central nervous system shown in gray) pinned at the edge of a rotatable Sylgard-lined platform over the gap, to evoke and record fictive swimming and make whole-cell recordings. Solid short lines are pins through the notochord and myotomes. (G) Using brief current pulses (I) to monitor the patch pipette seal with the cellular membrane and progression to whole-cell recordings. (H) The activity of a commissural interneuron (cIN) recorded in whole-cell current-clamp mode at the beginning of a fictive swimming episode initiated by a brief electrical stimulus (0.5 msec, arrow). (v.r.) Ventral root. (D and F, Adapted from Li et al. 2002. E, Reprinted, with permission, from Winlove and Roberts 2012, © John Wiley and Sons.)

2. Pin the tadpole through its notochord at two locations onto the rotatable Sylgard stage in the dissection bath filled with saline illuminated by the cold light (Li et al. 2014).

3. Make a vertical cut in the tadpole dorsal fin near the hindbrain by crushing the skin against the Sylgard using the blunt tungsten needle. Extend the cut by sliding the needle tip inside the dorsal fin along the body axis.

4. Free the tadpole and transfer it to a microcentrifuge tube containing 1 mL saline with 10 μM α-bungarotoxin with a small transfer pipette. Incubate for 20–30 min for immobilization at room temperature.

5. After immobilization, repin the tadpole onto the Sylgard stage in the dissection bath.

6. Remove the majority of the yolk belly in the trunk using a blunt dissection needle (Fig. 1F).

7. Cut free the middle strip of skin connecting the head and tail skin using the fine dissection needle.

8. Peel off the skin strip using a pair of fine forceps to expose the swimming myotomes.

9. Push a blunt dissection needle between the spinal cord and the swimming muscles on the exposed side along the body axis. Cut one end of the loosened muscles. Remove the contracted lump of muscles using a pair of fine forceps. Remove the loose tissue using a pair of fine forceps or by rubbing the side of the dissection needle gently against the spinal surface.

10. Pierce the dorsal roof of the hindbrain and open up the half-exposed spinal cord by gently pushing the dissection needle inside the neurocoele (Fig. 1D).

11. To expose more ventral spinal neurons, remove some ependymal cells lining the neurocoele (Fig. 1D). Clear loose tissue by gently rubbing the side of a dissection needle against the exposed spinal somata.

12. Free the tadpole. Transfer it to the recording bath filled with saline (Fig. 1B) and repin it onto the small rotatable Sylgard stage at its edge (Fig. 1F).

Recordings

See Figure 1C for recording setup.

13. Tilt the Sylgard stage and tadpole to an optimal angle for clear visualization of exposed somata in the spinal cord under the Nikon E600FN microscope (Fig. 1D–F).

14. Circulate saline in the recording chamber with a peristaltic pump at ∼2 mL per min.

15. Fill the stimulating and ventral root suction electrodes with saline by pulling their connected syringes. Place the electrodes on tadpole tail skin to start fictive swimming and muscle cleft to monitor fictive swimming using the coarse manipulators (Fig. 1F).

16. Pull patch pipettes from the capillary glass using the p-97 puller to the desired shape (Fig. 1E) and DC resistance 5–10 MΩ.

17. Pull a filler tube from a plastic pipette tip over a small flame. Attach the filler tube to a small syringe with a 0.2-μm filter to back-fill a patch pipette with the intracellular solution.

18. Mount the patch pipette on the patch-clamp amplifier head stage. Make sure the tubing connecting to the pipette is airtight. Apply gentle pressure (7 cm H$_2$O) by compressing ∼2 mL air trapped in the tubing by 10 μL using another 1-mL syringe. Use a three-way connector to lock the pressure before the patch electrode touches the cell membrane.

19. Lower the pipette into the saline using the motorized manipulator, and move it close to the exposed spinal cord with the 4× objective.

20. In the Multiclamp 700B controller software, choose the recording mode as current-clamp (IC) for the channel connected to the head stage holding the patch pipette.

   Set up test step current injection pulses in Signal software outputs (2 msec with a gap of 3 msec and alternating polarities at 50 Hz, 100 pA; Fig. 1G) for monitoring the progression of the gigaohm seal and whole-cell recording.

Cite this protocol as *Cold Spring Harb Protoc*; doi:10.1101/pdb.prot106856

21. Under the 40× water immersion lens, use the motorized manipulator to advance the patch pipette to contact an exposed neuron soma.

22. Neutralize electrode capacitance (stray capacitance) by turning up its reading in the Multiclamp 700B controller until the recording trace thickens but does not oscillate.

23. Once the pipette touches the soma, release the pressure and apply gentle suction to the pipette (14–35 cm H_2O, expand 2 mL of air by 20–50 µL). The voltage responses to test current injections should change from that resembling the step currents to monophasic with increased amplitude (Fig. 1G). Very often the gigaohm seal forms within seconds.

24. Once a gigaohm seal is achieved, apply large zapping currents configured in Signal software to rupture the membrane inside the pipette tip (2–5 nA, two to five pulses at 100 Hz and 2 msec in duration).

25. Determine if the membrane is ruptured as follows: If the membrane is ruptured and whole-cell recording achieved, test pulse responses will change instantly to smaller monophasic shapes (Fig. 1G).

    See Troubleshooting.

26. Balance bridge by turning up the series resistance value in the Multiclamp 700B controller.

27. Stimulate tadpole skin electrically to start fictive swimming (0.5-msec to 1-msec pulses, can be triggered using Signal software output), and record intracellular activity in current-clamp mode (Fig. 1H).

    See Troubleshooting.

TROUBLESHOOTING

Problem (Step 25): Difficulty is encountered in rupturing the cellular membrane.

Solution: It can be difficult to rupture the cellular membrane if the osmolarity of the pipette solution is high relative to that of saline. Saline osmolarity should be 15–20 mOsm/L H_2O higher than that of the pipette solution. If it is not, adjust the pipette solution osmolarity by adding sucrose or the saline osmolarity by changing the NaCl concentration. Membrane rupture can also be difficult if the pipette has formed a false gigaohm seal with loose tissue or ependymal cells. Be sure to clean the exposed somata thoroughly in Step 11. Replace the pipette that has formed a false seal and avoid contacting loose tissue before touching the soma membrane.

Problem (Step 27): A small action potential overshoot is observed.

Solution: Make sure the capacitance neutralization is properly adjusted in Step 22 before whole-cell recordings. Series resistance can rise during recordings because of the movement of yolk platelets (Fig. 1E, arrowheads). If this occurs, sudden suction can be applied to unclog the pipette by quickly pulling out the cork of the connected 1-mL syringe. If this fails, use a new patch pipette and repatch the neuron.

DISCUSSION

Whole-cell recordings from tadpole neurons were first achieved in cultured spinal neurons (Ribera and Spitzer 1987; Dale 1991), but success with in situ recordings remained limited (Rohrbough and Spitzer 1996; Prime et al. 1999) until 2002 (Li et al. 2002). Because any membranous tissue can seal up the pipette, difficulty exists in generating a genuine gigaohm seal with a neuronal membrane. Fine dissections are required to expose the somata to improve seal success. The need for surgical exposure of the soma membrane has prevented its application to neurons deep in the central nervous system. It is also more difficult to see neurons in the brainstem where transmitted light is obscured by thicker tissue and Sylgard. Once established, a successful whole-cell recording can be sustained for several hours.

The conventional whole-cell recording can be adapted to expand its applications. For example, perforated patch-clamp recordings can be achieved by including amphotericin B or gramicidin (Kyrozis and Reichling 1995) in the pipette solution. Tadpole neuronal dendrites are short at stage 37/38. This allows good space clamp to analyze voltage-dependent ionic currents and synaptic currents in voltage-clamp mode (Li 2015). Neurobiotin or fluorescent dyes like Alexa Fluor 488 at 0.1% can also be added to the pipette solution to observe neuronal anatomy after fixation and processing or in situ (Li et al. 2002). New features in Signal 5 and 6 sampling software have recently been developed to enable dynamic clamp recordings to remove or add different "ion channels" and "synaptic connections" to the recorded neurons.

RECIPES

Intracellular Pipette Solution for Xenopus

Reagent	Final concentration
K-gluconate	100 mM
MgCl$_2$	2 mM
EGTA	10 mM
HEPES	10 mM
Na$_2$ATP	3 mM
NaGTP	0.5 mM

Prepare 50 mL of solution in deionized water. Adjust pH to 7.3 using 1 M KOH. Divide into 100-μL aliquots (or 500-μL aliquots if subsequent filtering is to be carried out). Freeze at −20°C. Store for up to a few months.

Saline

Reagent	Final concentration
NaCl	115 mM
KCl	3 mM
CaCl$_2$	2 mM
NaHCO$_3$	2.4 mM
MgCl$_2$	1 mM
HEPES	10 mM

Prepare saline solution in distilled water. Adjust pH to 7.4 with 5 M NaOH. Store for up to several months at room temperature.

REFERENCES

Dale N. 1991. The isolation and identification of spinal neurons that control movement in the *Xenopus* embryo. *Eur J Neurosci* **3:** 1025–1035. doi:10.1111/j.1460-9568.1991.tb00039.x

Kyrozis A, Reichling DB. 1995. Perforated-patch recording with gramicidin avoids artifactual changes in intracellular chloride concentration. *J Neurosci Methods* **57:** 27–35. doi:10.1016/0165-0270(94)00116-X

Li WC. 2015. Selective gating of neuronal activity by intrinsic properties in distinct motor rhythms. *J Neurosci* **35:** 9799–9810. doi:10.1523/JNEUROSCI.0323-15.2015

Li WC, Soffe SR, Roberts A. 2002. Spinal inhibitory neurons that modulate cutaneous sensory pathways during locomotion in a simple vertebrate. *J Neurosci* **22:** 10924–10934. doi:10.1523/JNEUROSCI.22-24-10924.2002

Li WC, Wagner M, Porter NJ. 2014. Behavioral observation of *Xenopus* tadpole swimming for neuroscience labs. *J Undergrad Neurosci Educ* **12:** A107–A113.

Nieuwkoop PD, Faber J. 1956. *Normal tables of* Xenopus laevis *(Daudin)*. North Holland, Amsterdam.

Prime L, Pichon Y, Moore LE. 1999. *N*-methyl-D-aspartate–induced oscillations in whole cell clamped neurons from the isolated spinal cord of *Xenopus laevis* embryos. *J Neurophysiol* **82:** 1069–1073. doi:10.1152/jn .1999.82.2.1069

Ribera AB, Spitzer NC. 1987. Both barium and calcium activate neuronal potassium currents. *Proc Natl Acad Sci* **84:** 6577–6581. doi:10.1073/pnas .84.18.6577

Rohrbough J, Spitzer NC. 1996. Regulation of intracellular Cl$^-$ levels by Na$^+$-dependent Cl$^-$ cotransport distinguishes depolarizing from hyperpolarizing GABA$_A$ receptor–mediated responses in spinal neurons. *J Neurosci* **16:** 82–91. doi:10.1523/JNEUROSCI.16-01-00082 .1996

Winlove CI, Roberts A. 2012. The firing patterns of spinal neurons: in situ patch-clamp recordings reveal a key role for potassium currents. *Eur J Neurosci* **36:** 2926–2940. doi:10.1111/j.1460-9568.2012.08208.x

Protocol 5

Analysis of Visual Collision Avoidance in *Xenopus* Tadpoles

Arseny S. Khakhalin

Biology Program, Bard College, Annandale-on-Hudson, New York 12504, USA

In teaching, the best exam questions are those that seem simple at first but can lead to deep and nuanced conversations. Similarly, to probe brain development, we should look for behaviors that are easy to evoke and quantify, but that are demanding, malleable, and inherently variable. Visual collision avoidance is an example of such a behavior; it is ecologically relevant, robust, and easy to record, but also nuanced and shaped by the sensory history of the animal. Here we describe how to set up a visual avoidance assay and how to use it to test sensory processing and sensorimotor transformations in the vertebrate brain.

MATERIALS

It is essential that you consult the appropriate Material Safety Data Sheets and your institution's Environmental Health and Safety Office for proper handling of equipment and hazardous materials used in this protocol.

RECIPE: Please see the end of this protocol for recipes indicated by <R>. Additional recipes can be found online at http://cshprotocols.cshlp.org/site/recipes.

Reagents

Tadpole-rearing medium <R>
> *Other rearing media, such as the popular 0.1× Steinberg solution, are also acceptable.*

Xenopus laevis tadpoles (typically, Nieuwkoop–Faber stages 45–49 [Nieuwkoop and Faber 1994])

Equipment

Acrylic light filters (optional; see Step 10)
Behavioral protocols (https://github.com/khakhalin/Xenopus-Behavior)
Computer with Internet access
IR light source (e.g., Univivi U48R LED wide angle source for CCTV security) (optional; see Step 10)
Plastic Petri dish (8.5 cm in diameter)
Plastic transfer pipette
Projection table materials
> *We use 1100 cm (36′) of 20 × 40 mm (1″ × 2″) pine beam, 32 wood screws (e.g., #8 × 1″ flat head), four furniture felt pads, two pieces of clean acrylic 31 × 31 cm (1′ × 1′), and a white disposable plastic apron (see Step 1 and Fig. 1A).*

Projector

Correspondence: khakhalin@bard.edu

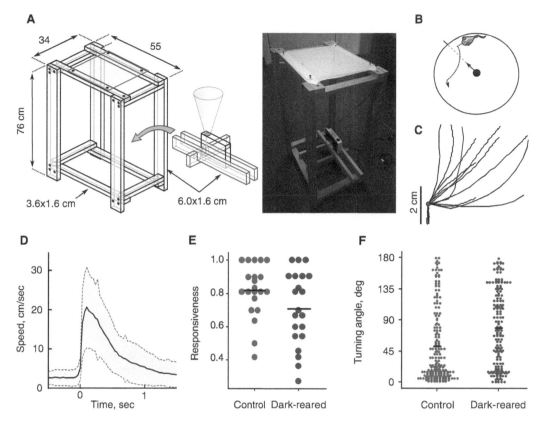

FIGURE 1. (*A*) The projection table. (*B*) Schematics of the experiment. (*C*) Representative avoidance trajectories. (*D*) Average (solid) and standard deviation (dashed) values of swimming speed for 500 fast avoidance responses. (*E*) Response probabilities for normal and dark-reared tadpoles. (*F*) Turning angles for normal and dark-reared tadpoles. These data were previously presented in Khakhalin et al. (2014) and Ramirez-Vizcarrondo et al. (2015) but were analyzed in a new way for this protocol.

The visual stimulus will be projected on the floor of the chamber, and the tadpole should be able to see the image even at sharp angles. Therefore, a standard LCD monitor cannot be used for visual stimulation. Instead, use either a projector, an old CRT monitor placed horizontally, or an iPad-like tablet. We recommend using a short-throw USB pico projector (i.e., one capable of producing a focused image at a distance of ~1 m; e.g., AAXA P300). The downsides of CRT monitors are that the surface is never quite flat, they heat the water, and video recording is compromised by a refresh beam artifact. This protocol also works with high-end tablets, but we have not explored this approach systematically.

Tracking software, such as http://ctrax.sourceforge.net or http://sourceforge.net/projects/buridan/ (optional; see Step 10)

Web camera

METHOD

Projecting Device

1. Set up the projecting device.

   We recommend using a short-throw USB pico projector and a screen made of a piece of white disposable polyethylene apron, fixed between two pieces of clean acrylic (Fig. 1A). The visual stimulus will be projected on the floor of the chamber. The tadpole should be able to see the image even at sharp angles. As a test, if you cannot see the image on the screen when looking along the surface, the tadpole will also not see it.

Recording

2. In a browser, open the repository of *Xenopus* behavioral protocols: https://github.com/khakhalin/Xenopus-Behavior. Navigate to the current version of the Collision Avoidance Stimulator (Fig. 1B). Project the image.

   The stimulator program consists of an outline for the arena, a dark circle in the middle that can be controlled with the keyboard (see Step 6), and parameter controls on the left.

3. Adjust the background lightness to keep the contrast high without blinding the tadpole.

   To assess the optimal lightness for your setup, run a series of experiments with different contrasts, and pick the contrast with the highest response rate (the ratio of the number of responses to the number of stimuli).

4. Place a Petri dish on top of the screen and fill it 1–1.5 cm deep with tadpole-rearing medium. Make the arena of the program match the position of the Petri dish; adjust the radius if needed.

5. Set the stimulation parameters. Wait for at least 20 sec or preferably 30 sec between the stimuli to prevent habituation. Make the black dot ~5 mm in diameter (comparable to the size of a tadpole), traveling at a speed of 3–5 cm/sec (comparable to the speed of a tadpole).

   Making the circle faster or smaller (e.g., $v = 4$ cm/sec, $d = 4$ mm) increases the chances of triggering a "fast" poorly coordinated escape response, whereas keeping the circle slower or larger (e.g., $v = 4$ cm/sec, $d = 12$ mm) allows tadpoles to implement more spatially informed course corrections (Khakhalin et al. 2014; Khakhalin 2019).

6. Before running actual experiments, practice targeting. Transfer one tadpole into the Petri dish arena using a standard plastic transfer pipette. Target the circle using LEFT/RIGHT keys, then send it toward the tadpole by pressing the UP key.

   As healthy tadpoles tend to travel around the edges of the arena, the best strategy is to keep the target fixed, and time the pressing of the UP key such that the stimulus, traveling to the side of a dish, and the tadpole, traveling along the side of the dish, will collide. Sending the circle toward the tadpole sets the timer in the left bottom corner of the screen. The timer controls the color of the circle, making it pale during the interstimulus interval, to prevent habituation. Try not to press the LEFT/RIGHT keys while the circle is in motion, as adding lateral motion will increase the overall speed of the stimulus. You can return the circle to the center by hitting the DOWN key.

7. Before the experiment, place tadpoles for each experimental condition in a separate marked bowl, and keep the experimenter blinded to the identity of each group. Prepare extra bowls for tadpoles that have already been tested.

8. Transfer one tadpole into the Petri dish arena using a standard plastic transfer pipette. Run the protocol (with ~2 min for the transfer, 3 min acclimation, 30 sec between the stimuli, and 20 stimuli, the experiment will take 15 min per tadpole). Repeat for all tadpoles, alternating between experimental groups.

9. Record a video for postprocessing, but also quantify each trial as you go, as either a "success "(if the avoidance response was triggered; Fig. 1C,D) or a "failure."

Analysis

The stimulation program presented here can also be used for multisensory experiments, as it can deliver a sound either when the circle hits the wall, or immediately before, or immediately after that. These multisensory experiments, however, fall outside the scope of the current protocol.

10. Select one of the following approaches for analyzing the data.

    - Compare rates of avoidance responses across different treatment groups (Fig. 1E).

      This is the simplest experimental design. It requires minimal equipment, can be done by eye during the experiment, and can later be verified from the video (scored by another person). This design is also appropriate for teaching labs.

      To better probe brain development, one can sweep across a range of values for circle speeds, sizes, or contrast, and compare the results between groups (Henriet et al. 2017).

- Track the tadpole position at every frame manually (with ImageJ), or automatically, using commercial (Noldus EthoVision; CleverSys AquaScan) or open source (CeTrAn, C-trax, etc.; Colomb et al. 2012; Chao et al. 2015) software.

  This is a more sophisticated analysis. Tracking of both the stimulus and the tadpole gives access to several more variables, such as the peak swimming speed, the angle of the turn during the avoidance maneuver (Fig. 1F), and the distance to the circle when the response is triggered. It also allows automated classification of responses into "successes" and "failures."

- Add a near-infrared (IR) light source below the projection screen, and turn the camera to IR-only, by replacing a glass IR filter inside it with a stack of three colored (red, green, and blue) acrylic filters (Truszkowski et al. 2017).

  In this configuration, the camera will only record the tadpole, but not the stimulus, which still allows for some analysis of trajectories. This approach is used because not all software solutions can track both the circle and the tadpole.

TROUBLESHOOTING

Problem: Tadpoles are sluggish and do not swim, staying in one place; yet the moment they are returned to the tank, they start swimming again.

Solution: Many possible solutions exist. Consider the following.

- Give the tadpole a minute to acclimate, then startle it with a gentle tap. Ideally, a tadpole should always be swimming, with a speed of 1–3 cm/sec.

- Make sure the medium is cold enough. *X. laevis* tadpoles prefer temperatures slightly colder than room temperature (18°C–21°C). If the projector heats the dish, replace the media regularly.

- Add ambient light to the chamber. Use a cardboard box to screen the arena, and make the chamber (cardboard box) white from the inside. Tadpoles stop swimming if the environment is too dark.

- Run the experiments earlier in the morning: On a 12/12 light cycle with "dawn" at 7 a.m., tadpoles seem most active between 9 a.m. and 12 p.m., and often stop responding in the afternoon.

- Do not feed tadpoles before the experiment.

- Adjust the brightness down to reduce retinal bleaching or increase the contrast.

- If working without a screening chamber (cardboard box), reduce movement to avoid visual habituation.

DISCUSSION

A typical fast avoidance maneuver involves a sharp turn of $\sim90° \pm 40°$ performed at a distance of ~1 cm from the circle, followed by a brief acceleration to 12 ± 4 cm/sec. A slow course correction is produced in response to larger, slower circles and involves a shallower turn of $\sim70° \pm 50°$ and acceleration to $\sim7 \pm 4$ cm/sec (Khakhalin et al. 2014).

A typical responsiveness for this protocol is 80% for control Nieuwkoop–Faber stage 49 tadpoles (15–30 d postfertilization, if raised at 20°C; Khakhalin et al. 2014), but responsiveness drops to 70% in dark-reared tadpoles (Ramirez-Vizcarrondo et al. 2015), 50% for tadpoles adjusted to strong visual stimulation (Jang et al. 2016), 40% for animals with mild neurodevelopmental abnormalities (James et al. 2015), and 20% in nutritionally restricted tadpoles raised in isolated wells (Khakhalin 2019). A power analysis shows that with 20 stimuli per tadpole (10 min of recording), base response probability of 70%, 20 animals in each group, and $\alpha = 0.05$, one can detect a drop of responsiveness to 60% with 80% power. Together, this makes visual collision avoidance a powerful tool to dissect abnormalities in the sensorimotor system of *Xenopus* tadpoles, from retinal dysfunction, through sensorimotor transformation proper, and down to locomotor deficits.

RECIPE

Tadpole-Rearing Medium

15 mM NaCl
0.5 mM KCl
1.0 mM MgSO$_4$
150 µM KH$_2$PO$_4$
50 µM NaHPO$_4$
1.0 mM CaCl$_2$
0.7 mM NaHCO$_3$
0.5 mg/L methylene blue

Prepare as needed in deionized H$_2$O, and check that the pH is 7.2–7.6 before use.

REFERENCES

Chao R, Macía-Vázquez G, Zalama E, Gómez-García-Bermejo J, Perán JR. 2015. Automated tracking of *Drosophila* specimens. *Sensors* **15**: 19369–19392. doi:10.3390/s150819369

Colomb J, Reiter L, Blaszkiewicz J, Wessnitzer J, Brembs B. 2012. Open source tracking and analysis of adult *Drosophila* locomotion in Buridan's paradigm with and without visual targets. *PLoS One* **7**: e42247. doi:10.1371/journal.pone.0042247

Henriet E, Mannioui A, Khakhalin A, Zalc B. 2017. A behavioral test to evaluate the functional consequences in a *Xenopus laevis* model of inducible-demyelination and myelin repair. *Multiple Sclerosis J* **23**: 996.

James EJ, Gu J, Ramirez-Vizcarrondo CM, Hasan M, Truszkowski TL, Tan Y, Oupravanh PM, Khakhalin AS, Aizenman CD. 2015. Valproate-induced neurodevelopmental deficits in *Xenopus laevis* tadpoles. *J Neurosci* **35**: 3218–3229. doi:10.1523/JNEUROSCI.4050-14.2015

Jang EV, Ramirez-Vizcarrondo C, Aizenman CD, Khakhalin AS. 2016. Emergence of selectivity to looming stimuli in a spiking network model of the optic tectum. *Front Neural Circuits* **10**: 95. doi:10.3389/fncir.2016.00095

Khakhalin A. 2019. Graph analysis of looming-selective networks in the tectum, and its replication in a simple computational model. bioRxiv doi:10.1101/589887

Khakhalin AS, Koren D, Gu J, Xu H, Aizenman CD. 2014. Excitation and inhibition in recurrent networks mediate collision avoidance in *Xenopus* tadpoles. *Eur J Neurosci* **40**: 2948–2962. doi:10.1111/ejn.12664

Nieuwkoop PD, Faber J. 1994. *Normal table of* Xenopus laevis *(Daudin): a systematical and chronological survey of the development from the fertilized egg till the end of metamorphosis.* Garland, New York.

Ramirez-Vizcarrondo C, Hasan M, Gu J, Khakhalin A, Aizenman C. 2015. Novel behavioral assays to model neurodevelopmental disorders in the *Xenopus laevis* tadpole. *FASEB J* **29**: 657.1.

Truszkowski TL, Carrillo OA, Bleier J, Ramirez-Vizcarrondo CM, Felch DL, McQuillan M, Truszkowski CP, Khakhalin AS, Aizenman CD. 2017. A cellular mechanism for inverse effectiveness in multisensory integration. *Elife* **6**: e25392. doi:10.7554/eLife.25392

Cite this protocol as *Cold Spring Harb Protoc*; doi:10.1101/pdb.prot106914

Schooling in *Xenopus laevis* Tadpoles as a Way to Assess Their Neural Development

Virgilio Lopez III,[1] Arseny S. Khakhalin,[2] and Carlos Aizenman[1]

[1]*Department of Neuroscience, Brown University, Providence, Rhode Island 02912, USA;* [2]*Biology Program, Bard College, Annandale-on-Hudson, New York 12504, USA*

Escape behaviors, orienting reflexes, and social behaviors in *Xenopus laevis* tadpoles have been well-documented in the literature. Schooling behavior experiments allow for the observation of tadpole social interactions and in the past have been used to characterize behavioral deficits in models of neurodevelopmental disorders. Unlike other species of frogs, *Xenopus* tadpoles show polarized schooling. Not only do tadpoles aggregate, they also swim in the same direction. Quantifying both aggregation and relative swim angle can give us an important measure of social behavior and sensory integration. Past iterations of these experiments have required the continued presence of an experimenter throughout the duration of each trial and relied on expensive software for subsequent data analysis. The instrument configuration and analysis protocol outlined here provide an automated method to assess schooling by delivering a series of timed vibratory stimuli to a group of tadpoles to induce swimming behavior and then controlling a camera to document their positions via still images. Both stimulus delivery and image acquisition are automated using the Python programming language. Analysis is done using ImageJ and custom Python scripts, which are provided in this protocol. The specific equipment configuration and scripts shown here provide one solution, but other equipment and custom scripts can be substituted.

MATERIALS

It is essential that you consult the appropriate Material Safety Data Sheets and your institution's Environmental Health and Safety Office for proper handling of equipment and hazardous materials used in this protocol.

RECIPES: Please see the end of this protocol for recipes indicated by <R>. Additional recipes can be found online at http://cshprotocols.cshlp.org/site/recipes.

Reagents

10× Steinberg's rearing solution <R>

Dilute 1:100 to create 10% Steinberg's solution before rearing tadpoles in solution.

The recipe describes how to prepare a 10× stock Steinberg's solution. In practice, Xenopus tadpoles are reared in 10% Steinberg's solution, which means that the recipe yields a solution that must be diluted 100 times to produce an appropriate rearing medium. The reasons for this confusing naming system seem to be purely historical.

Xenopus laevis tadpoles, stage 46–49 (Nieuwkoop and Faber 1994)

Correspondence: Carlos_Aizenman@brown.edu

Equipment

Double-sided tape or Velcro
Flat bottom glass bowl (17-cm-diameter)
GoPro Hero 7 with any flexible mount of choice
IOT Controllable Power Relay
Jintai Dental Laboratory Vibrator
LabJack U3-HV

This is an inexpensive analog-digital data acquisition device for controlling vibration delivery by a computer.

LED light tracing tablet (e.g., Picture/Perfect light pad)
Microsoft Excel
Plastic transfer pipette
Ruler (optional; see Step 13)
Software (available in github repository listed below, or by provided links)

Custom scripts for this project: https://github.com/khakhalin/Xenopus-Behavior
Fiji: https://imagej.net/Fiji/Downloads
Jupyter: https://jupyter.org/install.html
LabJack drivers and software: https://labjack.com/support/software/installers
LabJack Python library: https://labjack.com/support/software/examples/ud/labjackpython
Python 3.X: https://www.python.org/

METHOD

All of these procedures have passed IACUC review at Brown University.

LabJack Software Installation and Directory Creation

1. Install Python 3 and Jupyter notebooks.

   We recommend using the Anaconda distribution package, but installing Python and Jupyter separately will also work. Step-by-step instructions can be found here: https://jupyter.readthedocs.io/en/latest/install/ notebook-classic.html.

2. Download or clone the GitHub repository located at: https://github.com/khakhalin/Xenopus-Behavior.

   For the protocol described here, you will only need the files in the "01_Schooling" folder.

3. Follow the instructions in the "instructions.md" file to download and set up all necessary software and drivers to run the LabJack-U3 and GoPro.

Equipment Setup

See Figure 1A.

4. Connect one wire from the FIO4 pin on the LabJack-U3 to the positive terminal on the IOT power relay's removable screw terminal block and one wire from the adjacent GND pin on the LabJack-U3 to the negative terminal.

5. Plug the LabJack-U3 into your computer via USB. The green light on the device should illuminate.

6. Plug the dental vibrator into one of the "normally off" outlets on the Power Relay and ensure that the vibrator is turned on high.

7. Connect the power relay to an outlet and switch on.

   A red light should appear.

Cite this protocol as *Cold Spring Harb Protoc*; doi:10.1101/pdb.prot106906

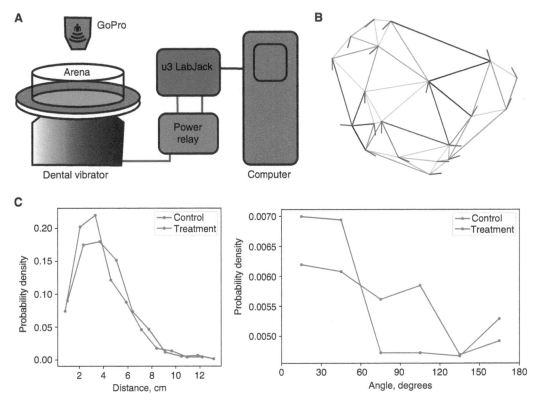

FIGURE 1. Schematic of equipment setup and analysis. (*A*) Diagram outlining experimental setup using the U3 LabJack, power relay, dental vibrator, and GoPro camera. (*B*) Result of triangulation function showing tadpole positions and swimming angle. (*C*) Output of analysis function showing distribution of inter-tadpole distances and angles, comparing a control and treatment group.

8. Open a Command prompt/Terminal window, ensure that you are in the directory with the supplied Python programs, and run the following code:

```
$ Python u3vibrate.py
```

> *This should trigger a series of four vibrations, which indicates that you have successfully installed the appropriate drivers to connect to the LabJack-U3.*

9. Set up the camera above the arena using a flexible mount. Once the camera has been successfully connected to the computer via WiFi, open a second Command prompt/Terminal window and run the following command:

```
$ Python GoProStream.py
```

> *A live feed of your GoPro camera should open. You are now ready to run your experiment.*

10. Secure the LED light tracing tablet to the dental vibrator platform with double-sided tape or Velcro to provide illumination from the bottom.

11. Add 350 mL of 10% Steinberg's solution to a large round dish, 17 cm in diameter. Place the dish on the dental vibrator, directly beneath the GoPro.

> *It is helpful to include some barriers to hold the bowl in place, should the vibrations cause it to move throughout the duration of the experiment.*

12. Use the screen on the GoPro to adjust the placement of the vibrator and dish to ensure it is in the camera's frame.

> *The distance between the GoPro and the arena will vary depending on the specifics of your setup. Typically it is ~50 cm above the dish.*

13. Add 15 to 20 stage 46–49 *Xenopus laevis* tadpoles to the dish using a plastic transfer pipette. Leave an object of known length somewhere in-frame, such as a small ruler or some marked distance of known length.

    See Troubleshooting.

14. Run the following command in the first Command Prompt window:

    ```
    $ Python schooling_experiment.py
    ```

    The entire experiment is executed using a programmed loop. The loop consists of an image acquisition command, a 150-sec wait time, a vibratory stimulus command, and another 150-sec wait time. Therefore, each image and stimulus event is separated by a 300-sec wait period, but the two events are offset from each other by 150 sec. This loop is set to execute for 4000 sec, which in the end will result in 12 acquired images for each experiment. Images acquired by the GoPro are saved as automatically named .JPG files, making the sequence clear. The timing parameters for the experiment can be easily altered by the experimenter by changing the values for the timer (line 29) and wait times (lines 46 & 50) in the script.

    This experiment can be customized a number of ways in order to suit the needs of the experimenter. The schooling_experiment.py script is written to run on a set timer, which can be shortened or lengthened in order to yield more or fewer images for analysis. To do this, simply change the value given on line 29 of the code, which is currently set to 4000 sec by default.

Data Analysis

Fiji (Recording Position and Orientation of Tadpoles in Each Frame)

15. Open the first image in Fiji.

16. Use the line tool to measure the known distance included at the beginning of the experiment.

17. Select Analyze > Set scale and include the measured distance in the "known distance" box. Be sure to change the unit of length to the unit used for the measured distance (e.g., cm) and select the "global" box to apply this scale to all subsequent images.

18. Use the multipoint tool to mark the head and gut of each individual tadpole such that each tadpole has a sequential odd number and even number marking for its head and gut, respectively. Take care that the two spots create a vector that aligns with the direction in which the tadpole's body is pointing. If necessary, adjust the contrast in the image files to facilitate marking.

19. Select Analyze > Measure and copy the results into an Excel sheet. Repeat this process for every image, adding the measured data to the same Excel sheet.

    When finished, the Excel sheet will have 12 blocks of data, representing x–y coordinates for each tadpole, in all 12 images.

    Alternatively, use the ROI Manager tool in Fiji to select the points, which will allow the user to save the ROIs with the image.

20. Save the Excel sheet as a CSV file "input_control.csv."

Jupyter Notebook (Calculating Intertadpole Distances and Relative Swimming Angle, Generating Data Output Files)

21. Run the Jupyter notebooks environment, and navigate to your repository. Open the "schooling_analysis" notebook.

22. Look through the notebook (perhaps using the "Find" command), and make sure the names of all input and output CSV files are correct.

    By default, the notebook attempts to read files "input_control.csv" and "input_treatment.csv" from the "data" subfolder, and then saves the analysis results as "output_processed_control.csv," and "output_processed_treatment.csv" in the same subfolder.

23. Run the notebook, which will process the data as follows.

Cite this protocol as *Cold Spring Harb Protoc*; doi:10.1101/pdb.prot106906

i. The program will perform a Delaunay triangulation (a standard way to identify a set of "neighboring" points for every point on a plane) and find intertadpole distances and angles between orientations of neighboring tadpoles (Fig. 1B).

ii. The results, including figures and Kolmogorov–Smirnov test P-values for distance and angle comparisons will appear under respective sections (Fig. 1C). In schooling tadpoles, one would observe more short distances and fewer medium distances, compared to nonschooling (randomly distributed) tadpoles. The distribution of angles will be uniform (flat) for randomly oriented tadpoles, and declining for tadpoles that co-orient.

iii. The last few sections of the code compare data from a single set of experiments (control experiments by default) to reshuffled data, as a way to estimate P-values for a null hypothesis of "no schooling." The data are shuffled by taking all observed tadpoles and assigning them to random frames, which preserves the spatial and angular distributions of the data but removes any patterns and any coordination between neighboring tadpoles.

TROUBLESHOOTING

Problem (Step 13): We have found that tadpole activity decreases as the day goes on, with the highest amount of swimming activity occurring within 1–3 h into the start of their diurnal cycle (tadpoles are housed in a 12:12 light/dark cycle).

Solution: Take care with regard to the timing of schooling experiments. Ideally, duplicate experiments should be run around the same point of their L/D cycle. Carefully set up specific criteria for inclusion of control and experimental groups in the analysis. These criteria could involve a baseline swimming threshold or accurate schooling in the control group. Criteria may vary by age or nature of the experiment, but experimenters should take care to set these a priori criteria for inclusion into their experimental groups.

DISCUSSION

Experiments in tadpole schooling, as a method to characterize sensory integration in the brain, were previously described in Wassersug et al. (1981) and James et al. (2015). Schooling behavior occupies an interesting niche at the overlap between orienting responses and social behaviors, making it a productive "litmus test" for neurodevelopmental deficits (Katz et al. 1981; Roberts et al. 2000; Simmons et al. 2004; Lee et al. 2010; Villinger and Waldman 2012). Here we describe a logistic improvement on the "manual" schooling protocol by automating camera control (Iturbe 2020) and streamlining analysis of photos with ImageJ (Schindelin et al. 2012).

In our experimental setup, tadpoles organize themselves into small groups within a dish swimming together in one direction. Formation of these schools takes ~30–60 sec after a vibration stimulus is provided to induce swimming behavior. Thus, we deliver this stimulus with a given time interval so we can sample several schooling configurations from a single group of tadpoles. The included analysis script provides an output file that contains the result of a triangulation used to calculate distances between neighboring tadpoles in each image. It also calculates the relative angles between the orientations of neighboring tadpoles. For normal schooling, the distribution of intertadpole distances will be nonrandom, with more values at shorter distances and longer distances and fewer "intermediate distances," representing the tadpole–tadpole distances within a school and between schools.

Our general approach to experiment design is to have a treatment group and a control group of 20 tadpoles each from the same clutch to account for variations within clutches. Each experiment is run with at least five control groups and five treatment groups. Distributions of controls and treatment groups can be compared directly using nonparametric statistics (e.g., Kolmogorov–Smirnov test), or each group can be compared to a random distribution to test for presence or absence of schooling (see a simulation notebook on GitHub for details and a power analysis for these tests). It is important to remember that treatments that result in higher or lower levels of overall swimming activity will almost always show up as differences in schooling behavior. Therefore, it is helpful to assess general swimming activity separately, by using video tracking to compare baseline swimming speed between groups.

RECIPE

10× Steinberg's Rearing Solution

Reagent	Amount
NaCl	34.0 g
HEPES	119.0 g
$Ca(NO_3)_2 \cdot 4H_2O$	0.8 g
$MgSO_4 \cdot 7H_2O$	2.06 g
KCl	0.5 g

Completely dissolve all reagents in 850 mL of dH_2O. Adjust the pH to 7.5 with 10 N NaOH. Add dH_2O to reach a final volume of 1 L, and store at 4°C.

REFERENCES

Iturbe K. 2020. Unofficial GoPro API Library for Python—connect to GoPro via WiFi, GitHub Repository https://github.com/KonradIT/gopro-py-api

James EJ, Gu J, Ramirez-Vizcarrondo CM, Hasan M, Truszkowski TL, Tan Y, Oupravanh PM, Khakhalin AS, Aizenman CD. 2015. Valproate-induced neurodevelopmental deficits in *Xenopus laevis* tadpoles. *J Neurosci* **35**: 3218–3229. doi:10.1523/jneurosci.4050-14.2015

Katz LC, Potel MJ, Wassersug RJ. 1981. Structure and mechanisms of schooling in tadpoles of the clawed frog, *Xenopus laevis. Anim Behav* **29**: 20–33. doi:10.1016/S0003-3472(81)80148-0

Lee RH, Mills EA, Schwartz N, Bell MR, Deeg KE, Ruthazer ES, Marsh-Armstrong N, Aizenman CD. 2010. Neurodevelopmental effects of chronic exposure to elevated levels of pro-inflammatory cytokines in a developing visual system. *Neural Dev* **5**: 2. doi:10.1186/1749-8104-5-2

Nieuwkoop PD, Faber J. 1994. *Normal table of* Xenopus laevis *(Daudin): a systematical and chronological survey of the development from the fertilized egg till the end of metamorphosis.* Garland, New York.

Roberts A, Hill NA, Hicks R. 2000. Simple mechanisms organise orientation of escape swimming in embryos and hatchling tadpoles of *Xenopus laevis. J Exp Biol* **203**: 1869–1885.

Schindelin J, Arganda-Carreras I, Frise E, Kaynig V, Longair M, Pietzsch T, Preibisch S, Rueden C, Saalfeld S, Schmid B, et al. 2012. Fiji: an open-source platform for biological-image analysis. *Nat Methods* **9**: 676–682. doi:10.1038/nmeth.2019

Simmons AM, Costa LM, Gerstein HB. 2004. Lateral line-mediated rheotactic behavior in tadpoles of the African clawed frog (*Xenopus laevis*). *J Comp Physiol A* **190**: 747–758. doi:10.1007/s00359-004-0534-3

Villinger J, Waldman B. 2012. Social discrimination by quantitative assessment of immunogenetic similarity. *Proc Biol Sci* **279**: 4368–4374. doi:10.1098/rspb.2012.1279

Wassersug RJ, Andrew ML, Michael JP. 1981. An analysis of school structure for tadpoles (Anura: Amphibia). *Behav Ecol Sociobiol* **9**: 15–22.

Cite this protocol as *Cold Spring Harb Protoc*; doi:10.1101/pdb.prot106906

Ex Vivo Brain Preparation to Analyze Vocal Pathways of *Xenopus* Frogs

Ayako Yamaguchi[1]

School of Biological Sciences, University of Utah, Salt Lake City, Utah 84112-0840, USA

Understanding the neural basis of behavior is a challenging task for technical reasons. Most methods of recording neural activity require animals to be immobilized, but neural activity associated with most behavior cannot be recorded from an anesthetized, immobilized animal. Using amphibians, however, there has been some success in developing in vitro brain preparations that can be used for electrophysiological and anatomical studies. Here, we describe an ex vivo frog brain preparation from which fictive vocalizations (the neural activity that would have produced vocalizations had the brain been attached to the muscle) can be elicited repeatedly. When serotonin is applied to the isolated brains of male and female African clawed frogs, *Xenopus laevis*, laryngeal nerve activity that is a facsimile of those that underlie sex-specific vocalizations in vivo can be readily recorded. Recently, this preparation was successfully used in other species within the genus including *Xenopus tropicalis* and *Xenopus victorianus*. This preparation allows a variety of techniques to be applied including extracellular and intracellular electrophysiological recordings and calcium imaging during vocal production, surgical and pharmacological manipulation of neurons to evaluate their impact on motor output, and tract tracing of the neural circuitry. Thus, the preparation is a powerful tool with which to understand the basic principles that govern the production of coherent and robust motor programs in vertebrates.

MATERIALS

It is essential that you consult the appropriate Material Safety Data Sheets and your institution's Environmental Health and Safety Office for proper handling of equipment and hazardous materials used in this protocol.

RECIPES: Please see the end of this protocol for recipes indicated by <R>. Additional recipes can be found online at http://cshprotocols.cshlp.org/site/recipes.

Reagents

Ethyl 3-aminobenzoate methanesulfonate (MS-222; 1.3% w/v in dH$_2$O)
Frog saline <R>
Ice
Serotonin hydrochloride (28 mM in frog saline)

> *Prepare fresh before use and keep on ice.*

Xenopus frogs, sexually mature males and females

---

[1]Correspondence: a.yamaguchi@utah.edu

Equipment

Aspirator bottle with a bottom hose connection
Compressed oxygen (99% O_2, 1% CO_2) with a regulator (special order from most vendors)
Corneoscleral punch
Data acquisition board
Differential amplifier
Differential suction electrodes
Fine forceps
Flow regulator
Gas dispersion tubes, glass
Iridectomy scissors
Light source for stereomicroscope
Microelectrode amplifier
Micromanipulators
Minutien pins (0.1-mm-diameter stainless steel insect pins)
Needle (27-gauge)
Operating scissors (4.5″ straight sharp/sharp) (two pairs; see Steps 4 and 6)
PC with a data acquisition software
Petri dish (15-cm) coated with 1-mm-thick Sylgard (two; see Steps 5 and 10)

> *Pour mixed Sylgard into the Petri dish and let it dry for 24 h.*

Petri dish (6-cm) coated with 1-mm-thick Sylgard

> *Pour mixed Sylgard into the Petri dish and let it dry for 24 h.*

Silver chloride ground wire
Stereomicroscope
Syringe (1-mL)
Thumb-dressing forceps
Tray for ice
Tungsten electrodes (1-MΩ)
Vacuum suction pump

METHOD

This procedure was reviewed and approved by the IACUC committee at the University of Utah.

Brain Isolation Procedure

1. Anesthetize the frog by injecting males with 0.3 mL or females with 1.0 mL of 1.3% (w/v) MS-222 into its dorsal lymph sac on the dorsal torso via a 27-gauge needle. Confirm the depth of anesthesia after 10–30 min by lack of response to a toe pinch.

2. While waiting for the frog to be anesthetized, move the frog saline from the refrigerator to ice and oxygenate via gas dispersion tube.

3. Once the frog is deeply anesthetized, wrap it in a paper towel and place it on a tray filled with ice for 5 min.

4. Place the frog on the ice tray dorsal side up. Remove the skin off of the dorsal surface of the torso and the head using operating scissors and dressing forceps.

   > *The scissors and forceps used to handle skin should be set aside, and not be used for the rest of the dissection to prevent skin secretions from contaminating the isolated brain.*

5. Isolate the skull and the rostral end of the spinal cord from the rest of the body. Place the skull (Fig. 1A) into a 15-cm Petri dish coated with Sylgard filled with oxygenated ice-cold saline (from Step 2) placed on ice.

Cite this protocol as *Cold Spring Harb Protoc*; doi:10.1101/pdb.prot106872

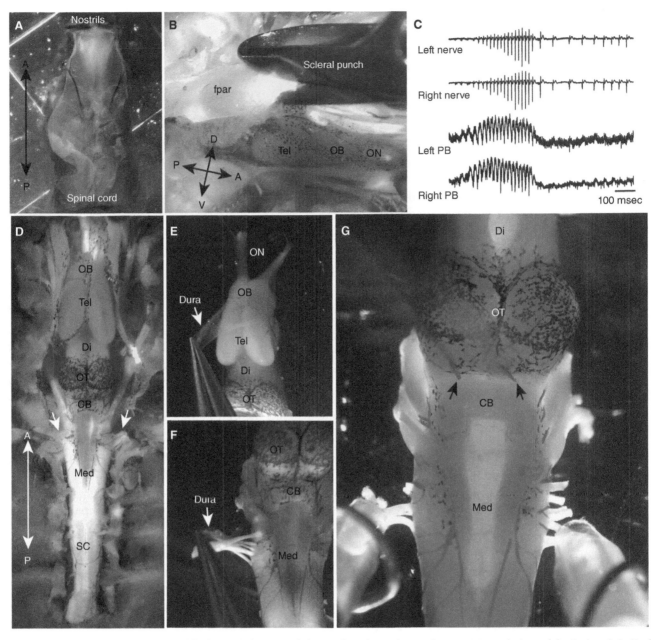

FIGURE 1. Procedure for brain isolation and electrophysiological recordings. (*A*) Dorsal view of the isolated skull of *Xenopus laevis*. The arrow shows the orientation of the brain. (*B*) The frontoparietal bone (fpar, the rostral portion of the dorsal skull) being removed using a scleral punch. Telencephalon (Tel) and olfactory nerve (ON) are visible under the dura mater. Arrows show the orientation of the brain. (*C*) Electrophysiological recordings obtained from the *left* and *right* laryngeal nerves (*top* two traces) and *left* and *right* amphibian parabrachial (PB) nuclei (*bottom* two traces) in response to serotonin applied to the whole brain. (*D*) Dorsal view of the frog brain after the dorsal skull is removed. Arrows point to the laryngeal nerves. (*E*) The dura is removed from the dorsal surface of the telencephalon of an isolated brain. (*F*) Five rootlets of the laryngeal nerves. (*G*) Positions of the *left* and *right* suction electrodes placed over the laryngeal nerves (*bottom* of the image), and *left* and *right* tungsten electrodes (arrows) placed in the amphibian parabrachial nuclei at the rostral end of the cerebellum (CB). (A) Anterior, (P) posterior, (D) dorsal, (V) ventral, (OB) olfactory bulb, (Di) diencephalon, (OT) optic tectum, (Med) medulla, (SC) spinal chord.

When removing the skull from the body, avoid puncturing internal organs to prevent digestive enzymes to contaminate your brain preparation.

6. From this step forward, use a stereomicroscope with illumination. Remove muscles from the dorsal surface of the skull and spinal cord using iridectomy and operating scissors.

7. Expose the dorsal surface of the brain by removing bones using a corneoscleral punch, starting from the rostral end of the skull. Insert the lower jaw of the punch between the skull and brain, close the punch gently to secure the end of the bone, and twist the punch to break and remove fragments of the dorsal skull (Fig. 1B).

8. Once the entire dorsal surface of the brain is exposed, remove the bones surrounding the laryngeal nerves. It is best to have >1-mm laryngeal nerves attached to the brain to obtain nerve recordings using suction electrodes. Crush the prootic bone and work your way posteriorly to uncover the nerve.

9. Once the dorsal surface of the brain is exposed and the laryngeal nerves are freed from the skull (Fig. 1D, marked with arrows), lift up the brain by holding onto the olfactory nerves using a fine forceps and cut the ventral cranial nerves (optic nerves, etc.) using iridectomy scissors.

10. Move the isolated brain into a new 15-cm Petri dish coated with Sylgard filled with cold oxygenated saline, this time without the ice tray, and pin the brain down to the dish using minutien pins.

11. Remove the dura mater from the dorsal surface of the brain (Fig. 1E). Remove the pia over the cerebellum to facilitate the penetration of the tungsten electrodes to obtain local field potential recordings from the amphibian parabrachial nucleus (formerly known as dorsal tegmental area of the medulla [DTAM]; see, e.g., Zornik and Yamaguchi 2012). Remove the pia by lifting the edge of the pia along the ventricle and inserting the iridectomy scissors between the pia and the brain.

12. Identify the laryngeal nerves, which are made of four to six rootlets (individual variation). The most caudal rootlets contain axons for laryngeal motor neurons (Fig. 1F). Trim the rostral two rootlets to obtain activity from the axons of laryngeal motoneurons.

13. Leave the brain in the Petri dish while oxygenating the saline for 1 h via tubing placed in the saline as the temperature of the dish and the brain slowly increase to room temperature.

Procedure to Obtain Nerve and Local Field Potential Recordings

14. After 1 h, transfer the brain into a recording chamber (a 6-cm Petri dish coated with 1-mm-thick Sylgard) using fine forceps and pin the brain to the bottom of the dish using minutien pins.

15. Superfuse the brain with oxygenated saline (at room temperature, in an aspirator bottle, oxygenated with a gas dispersion tube) by supplying 100 mL/h of saline via flow regulator while the overflow of oxygenated saline is removed via a vacuum suction pump to maintain the total volume of the recording chamber to be 20 mL.

16. Place the silver chloride ground wire into the bath.

17. Place suction electrodes onto both laryngeal nerves by slowly drawing the nerve into the electrode by applying gentle suction, being careful not to pull the nerve. The position of the suction electrode along the nerve is not critical.

18. Place tungsten electrodes into both parabrachial nuclei (PB; Fig. 1G) to obtain local field potential (LFP) recordings using micromanipulators. The best LFP recordings with the largest signal amplitude are obtained when the tungsten electrodes are placed in a location ~650-μm deep from the surface of the cerebellum, 650-μm left or right of the midline along the rostral edge of the cerebellum (see Fig. 1G).

19. Check to see if all four channels record activities. Most brains show sporadic breathing-related tonic activity (~0.5-sec-long) that can be easily distinguished from a very brief (~10-msec) vocal compound action potential (see Fig. 1C).

Cite this protocol as *Cold Spring Harb Protoc*; doi:10.1101/pdb.prot106872

20. Stop the superfusion and apply 40 µL of the serotonin hydrochloride stock solution directly to the brain in the recording chamber (20 mL volume) manually using a 1-mL pipette to achieve the final concentration quickly.

21. Record fictive vocalizations from the two suction electrodes, and premotor activity from the tungsten electrodes (Fig. 1C). After 5 min of observation, reinstate the superfusion at a high rate (250 mL/h) for 5 min, followed by slower superfusion (100 mL/h) for 50 min.

    See Troubleshooting.

22. In most isolated brains, serotonin can be reapplied to elicit fictive vocalizations five or six times throughout the day.

TROUBLESHOOTING

Problem (Step 21): A common problem with this procedure, especially when the experimenter is new to the procedure, is that the isolated brain does not generate fictive vocalizations. The *Xenopus* brain is very soft compared to mammalian brains, and it is easy to damage the tissue during dissection.

Solution: If the preparation does not generate fictive vocalizations, the brain can still be used for an anatomical project. Alternatively, the brain can be used to analyze the respiratory activity, a tonic burst of activity that lasts for ∼400 msec (often repeated every ∼10 sec) that typically persists even if the brain does not generate fictive vocalizations.

DISCUSSION

Until recently, amphibian in vitro brain preparations have primarily been used for electrophysiological and anatomical studies (Luksch et al. 1996; Christensen-Dalsgaard and Walkowiak 1999). This protocol describes an ex vivo frog brain preparation from which fictive vocalizations can be elicited repeatedly (Rhodes et al. 2007), which has recently been used in *Xenopus tropicalis* and *Xenopus victorianus* (Barkan et al. 2018).

The ex vivo preparation described here is a powerful preparation that allows analyses of neuronal activity in the context of the expression of behavior. For example, electrophysiological recordings from vocal neurons (whole-cell patch-clamp recordings or local field potential recordings) can be obtained while the fictive vocalizations are generated. Doing so allows easy interpretation of how the neuronal activity contributes to the generation of vocal motor programs by analyzing the timing of neural activity and the compound action potentials recorded from the laryngeal nerve (Rhodes et al. 2007; Zornik and Yamaguchi 2012; Lawton et al. 2017; Barkan et al. 2017, 2018). Furthermore, pharmacological or surgical manipulation can be applied to a specific part of the central vocal pathways and the causal relations between the activity of the manipulated vocal nuclei or the projections between the vocal nuclei in generating coordinated motor programs can be evaluated (Rhodes et al. 2007; Zornik and Yamaguchi 2012; Barkan et al. 2017, 2018; Lawton et al. 2017; Yamaguchi et al. 2017). Although the ability to selectively target a population of neurons using these manipulation techniques is currently poor, the development of viral vectors (Yamaguchi et al. 2018) to deliver optogenetic actuators to a selective population of neurons will allow us to activate or inhibit specific neurons. Similarly, the identification of unique receptors and ion channels expressed by vocal neurons (Inagaki et al. 2020) will enhance our ability to selectively manipulate target vocal neurons in the future.

To date, we have used the preparation to study central vocal pathways of *Xenopus laevis*, but the preparation can also be used to examine the function of the central respiratory pathways, because the glottal motoneurons contained in the laryngeal nerve also remain active in this ex vivo preparation. Furthermore, by stimulating the sensory nerves, the preparation can be used to analyze the central sensory pathways.

RECIPE

Frog Saline

Reagent	Amount to be added to 1 L	Concentration
NaCl	6.02 g	103 mM
HEPES	2.38 g	1 mM
Dextrose	1.98 g	11 mM
CaCl$_2$ (1 M)	2 mL	2 mM
KCl (1 M)	2 mL	2 mM
MgCl$_2$ (1 M)	0.5 mL	0.5 mM
NaHCO$_3$	1.09 g	13 mM

Mix the top six reagents in dH$_2$O. Adjust the pH to 7.8 by adding 10 N NaOH before NaHCO$_3$ is added. Store for up to 1 mo at 4°C.

REFERENCES

Barkan CL, Zornik E, Kelley DB. 2017. Evolution of vocal patterns: tuning hindbrain circuits during species divergence. *J Exp Biol* **220:** 856–867. doi:10.1242/jeb.146845

Barkan CL, Kelley DB, Zornik E. 2018. Premotor neuron divergence reflects vocal evolution. *J Neurosci* **38:** 5325–5337. doi:10.1523/JNEUROSCI .0089-18.2018

Christensen-Dalsgaard J, Walkowiak W. 1999. *In vitro* and *in vivo* responses of saccular and caudal nucleus neurons in the grassfrog (*Rana temporaria*). *Eur J Morphol* **37:** 206–210. doi:10.1076/ejom.37.2 .206.4745

Inagaki RT, Raghuraman S, Chase K, Steele T, Zornik E, Olivera B, Yamaguchi A. 2020. Molecular characterization of frog vocal neurons using constellation pharmacology. *J Neurophysiol* **123:** 2297–2310.

Lawton KJ, Perry WM, Yamaguchi A, Zornik E. 2017. Motor neurons tune premotor activity in a vertebrate central pattern generator. *J Neurosci* **37:** 3264–3275. doi:10.1523/JNEUROSCI.2755-16.2017

Luksch H, Walkowiak W, Muñoz A, ten Donkelaar HJ. 1996. The use of in vitro preparations of the isolated amphibian central nervous system in neuroanatomy and electrophysiology. *J Neurosci Methods* **70:** 91–102. doi:10.1016/S0165-0270(96)00107-0

Rhodes HJ, Yu HJ, Yamaguchi A. 2007. *Xenopus* vocalizations are controlled by a sexually differentiated hindbrain central pattern generator. *J Neurosci* **27:** 1485–1497. doi:10.1523/JNEUROSCI.4720-06.2007

Yamaguchi A, Barnes JC, Appleby T. 2017. Rhythm generation, coordination, and initiation in the vocal pathways of male African clawed frogs. *J Neurphysiol* **117:** 178–194. doi:10.1152/jn.00628.2016

Yamaguchi A, Woller DJ, Rodrigues P. 2018. Development of an acute method to deliver transgenes into the brains of adult *Xenopus laevis*. *Front Neural Circuits* **12:** 92. doi:10.3389/fncir.2018.00092.

Zornik E, Yamaguchi A. 2012. Coding rate and duration of vocalizations of the frog, *Xenopus laevis*. *J Neurosci* **32:** 12102–12114. doi:10.1523/ JNEUROSCI.2450-12.2012

Cite this protocol as *Cold Spring Harb Protoc*; doi:10.1101/pdb.prot106872

Experimental Platform Using the Amphibian *Xenopus laevis* for Research in Fundamental and Medical Immunology

Jacques Robert[1]

Department of Microbiology and Immunology, University of Rochester Medical Center, Rochester, New York 14620

The amphibian *Xenopus* constitutes a powerful, versatile, and cost-effective nonmammalian model with which to investigate important contemporary issues of immunity relevant to human health such as ontogeny of immunity, self-tolerance, wound healing, autoimmunity, cancer immunity, immunotoxicology, and adaptation of host immune defenses to emerging pathogens. This model system presents several attractive features: an external developmental environment free of maternal influence that allows for easy experimental access from early life stages; an immune system that is remarkably similar to that of mammals; the availability of large-scale genetic and genomic resources; invaluable major histocompatibility complex (MHC)-defined inbred strains of frogs; and useful tools such as lymphoid tumor cell lines, monoclonal antibodies, and MHC tetramers. Modern reverse genetic loss-of-function and genome-editing technologies applied to immune function further empower this model. Finally, the evolutionary distance between *Xenopus* and mammals permits distinguishing species-specific adaptation from more conserved features of the immune system. In this introduction, the advantages and features of *Xenopus* for immunological research are outlined, as are existing tools, resources, and methods for using this model system.

INTRODUCTION

Xenopus has firmly been established as the model of choice for fundamental and medical immunological research. Its immune system is one of the most comprehensively studied besides those of mammals and chickens, and it shows remarkable similarity to that of mammals (for reviews, see Robert and Ohta 2009; Robert and Cohen 2011; Robert 2016). Indeed, the fully sequenced and annotated genomes of two *Xenopus* species, *X. tropicalis* and *X. laevis*, have provided compelling evidence that there is a high degree of similarity between the gene repertoires of *Xenopus* species and those of mammals, including humans (Hellsten et al. 2010; Session et al. 2016).

Over a century of studies have helped to characterize the *Xenopus* immune system. It contains primary lymphoid organs, as in mammals, including the thymus (the site of T-cell differentiation) and the spleen (a reservoir of antibody-producing B cells, T cells, and other leukocytes). As in mammals, *Xenopus* innate immune cells include neutrophils, basophils, eosinophils, monocytes, macrophages, and natural killer (NK) cells, and the adaptive T and B lymphocytes express a wide repertoire of somatically generated receptors that require recombination-activating genes (RAGs) 1 and 2. The development, activation, and restriction of T cells are directed by polymorphic molecules of the major histocompatibility complex (MHC). MHC class I molecules (for cytotoxic responses) and MHC class II molecules (for helper T cell responses) have been identified (for reviews, see Du Pasquier et al. 1989;

[1]Correspondence: Jacques_Robert@URMC.Rochester.edu

Robert and Ohta 2009), as have nonpolymorphic MHC class I-like molecules important for innate-like T (iT) cells, which express a very limited repertoire of T cell receptors compared to conventional T cells (Robert and Edholm 2014). iT cells are also found in humans (e.g., iNKT cells and mucosal-associated T [MAIT] cells) and are of growing medical interest (for reviews, see Robert and Edholm 2014; Castro et al. 2015).

XENOPUS AS AN IMMUNOLOGICAL MODEL

Because of its mammalian-like immunity and the fact that it is easy to propagate and experimentally manipulate in the laboratory, *Xenopus* is ideally suited to serve as a powerful model system for multiple areas of immunological research, some of which are listed below.

- *Model for developmental immunology:* In contrast to humans and mice, anuran amphibians such as *Xenopus* embryos develop externally, thus free of maternal influences. This allows easy access for experimental manipulation and visualization at all stages of development and provides a unique model for perinatal immunology. Comparison of the immune system at two distinct life stages, larval and adult, affords intriguing insights into immune function specialization (as summarized in Fig. 1). In addition, drastic developmental remodeling occurs during metamorphosis in *Xenopus*, which also affects the immune system. The larval thymus undergoes histolysis, losing most of its thymocytes, and becomes repopulated at metamorphosis completion by new stem cells that differentiate into "adult-type" T cells (for review, see Rollins-Smith 1998). To prevent auto-immunity, T-cell education needs to establish tolerance against the many new adult-type proteins. T-cell-deficient animals can easily be generated by sublethal γ-irradiation or by thymectomy within the first week after fertilization, before the immigration of stem cells. In addition, an athymic strain has been generated by genome editing in *X. tropicalis* (Nakai et al. 2016). Therefore, *Xenopus* is a valuable experimental model to study T-cell ontogeny. With advances in genomic, genetic, and genome-editing technologies, *Xenopus* will offer new approaches to reveal the functions of immunologically relevant genes and immune cells.

- *Model for iT cell immunity:* To date *Xenopus* is the only species besides mammals where an immune surveillance system based on nonpolymorphic MHC class I-like molecules directing the development and function of iT cells has been characterized. Concomitant with a naturally

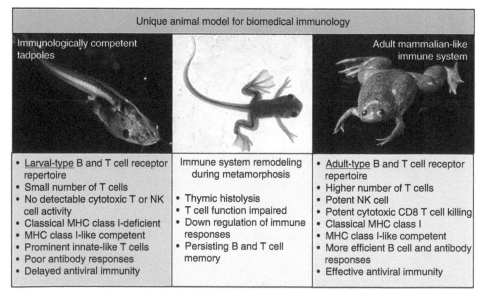

FIGURE 1. Schematic overview of the major developmental steps of the *Xenopus laevis* immune system.

Cite this introduction as *Cold Spring Harb Protoc*; doi:10.1101/pdb.top106625

deficient classical MHC class I function and a diversification of MHC class I-like genes, tadpoles rely on an adaptive immune system dominated by iT cells. In contrast, mainly conventional T cells with a minor iT cell fraction are present in adult frogs. This provides a unique model system (and an alternative to mice) to investigate in vivo the role of these important T-cell types in antimicrobial and tumor immunity as well as in immune homeostasis and autoimmunity (for reviews, see Robert and Edholm 2014; Castro et al. 2015; Banach and Robert 2017a).

- *Model for immune tolerance: Xenopus* is an attractive comparative model to explore self-tolerance because allotolerance to minor H antigens on transplanted adult skin can be induced just before or during metamorphosis when immune regulation is temporarily altered (Flajnik et al. 1987). During this developmental period, it is possible to establish specific tolerance to minor H antigens that persists after metamorphosis in adult frogs and that relies on a nondeletional ("split") anergic-like process (for review, see Houssaint and Flajnik 1990). Genes encoding MHC molecules are also differentially regulated in tadpoles and adult frogs (Rollins-Smith 1998). The change in MHC gene expression during metamorphosis coincides with new T-cell differentiation in the thymus. Thus, the amenability of *Xenopus* to experimental manipulation (e.g., thymectomy, blocking, or accelerating metamorphosis) from early development is convenient for addressing questions about MHC restriction, autoimmunity, and the development of self-tolerance that are challenging to study in mammalian models.

- *Model for tumor immunity and tumorigenesis: X. laevis* and more recently *X. tropicalis* have proven instrumental for exploring novel and innovative approaches for cancer biology and immunotherapy (for reviews, see Banach and Robert 2017a; Hardwick and Philpott 2018). Thymic lymphoid tumor lines, which are tumorigenic after transplantation into compatible MHC-defined inbred strains, or isogenic clones can be used to investigate tumor immunity (Banach and Robert 2017a). Tadpole transparency, temperature tolerance, and accessibility are ideal for studying the tumor microenvironment by intravital microscopy in real time (Haynes-Gimore et al. 2015). The implementation of genome-editing technology in *Xenopus*, especially *X. tropicalis*, is another promising avenue to identify and characterize mutations in human oncogenes, those relevant for cancer development (for review, see Dimitrakopoulou et al. 2019). As such, the comparative tumor immunity model in *Xenopus* can significantly contribute to designing and testing novel immunotherapeutic approaches against cancer.

- *Model for immunity to important infectious diseases: Xenopus* plays a key role in understanding infectious diseases that plague amphibians worldwide and contribute to their decline. Notably, *X. laevis* is the leading experimental model to study host defense and pathogenesis during ranavirus infection and to evaluate the contribution of immunocompromised animals in disease dissemination (for reviews, see Chen and Robert 2011; Chinchar and Waltzek 2014). Comparative studies (e.g., of tadpole vs. adult resistance to ranavirus or mycobacterial infections [for review, see Hyoe and Robert 2019] or of *X. tropicalis* vs. *X. laevis* resistance to chytrid fungal infections) can provide novel fundamental insights into virulence and immune escape mechanisms (for review, see Rollins-Smith et al. 2009; Grogan et al. 2018). The *Xenopus* skin secretes multiple antimicrobial peptides that are highly active against human immunodeficiency virus (HIV) as well as against many human gram negative and positive bacteria, which is of high biomedical relevance (for reviews, see Zasloff 1987; Rollins-Smith et al. 2009).

- *Model for wound healing, inflammation, and regeneration:* The *Xenopus* tadpole life stage offers an advantageous system for regeneration research (for reviews, see Slack et al. 2008; Chen et al. 2014; Li et al. 2016). The amputation of the tadpole's limb or tail results in the regeneration of a completely new tail that is fully functional. Immune cells (e.g., macrophages) and an inflammatory immune response are critically involved in this process as well as during wound healing. The transparency of tadpoles is ideal for intravital studies (Haynes-Gimore et al. 2015; Paredes et al. 2015), and they are amenable to single-cell transcriptomics (Aztekin et al. 2019).

- *Model for immunotoxicology:* Because of its external development free of maternal influence, *Xenopus* is a sensitive and cost-effective model system to investigate both acute and chronic effects resulting from early life exposure to water pollutants (e.g., endocrine disruptors of immune homeostasis and of immune responses to infectious diseases), which is gaining interest and attention from the public and the scientific community (Robert et al. 2018; Nagel et al. 2020).

TOOLS AND RESOURCES FOR *XENOPUS* IMMUNOLOGY RESEARCH

Importantly, tools (antibodies and cell lines) as well as genetically defined inbred clones and transgenic strains that are indispensable for immunological research are available in *Xenopus* and can be obtained at the *Xenopus laevis* Research Resource for Immunobiology (https://www.urmc.rochester.edu/microbiology-immunology/xenopus-laevis.aspx). These include monoclonal antibodies (mAbs) specific for *X. laevis* NK, B, and T cells and immune-related molecules such as CD8 and MHC (Robert and Ohta 2009). It is noteworthy that some of these mAbs, including the anti-IgY (11D5) and anti-CD8 (AM22), also cross-react with *X. tropicalis*. Furthermore, given the support of the genome and inbred strains developed in *X. tropicalis* (and available in the Japanese and U.S. stock centers), information obtained in *X. laevis* about gene expression and gene loss-of-function can be easily transposed.

The accompanying protocols include convenient immune assays adapted to *Xenopus* such as flow cytometry analysis (Protocol 4: Flow Cytometric Analysis of *Xenopus* Immune Cells [Edholm 2018]) and antibody response analysis by enzyme-linked immunosorbent assay (ELISA) (Protocol 9: Assessing Antibody Responses to Pathogens or Model Antigens in *Xenopus* by Enzyme-Linked Immunosorbent Assay (ELISA) [De Jesús Andino and Robert 2019]). These are followed by the use of invaluable MHC-defined inbred strains and cloned animals for in vivo immune assays such as skin grafting (Protocol 1: Skin Grafting in *Xenopus laevis*: A Technique for Assessing Development and Immunological Disparity [Izutsu 2019]), tumor transplantation (Protocol 2: Collagen-Embedded Tumor Transplantations in *Xenopus laevis* Tadpoles [Banach and Robert 2017b]), elicitation of peritoneal leukocytes (Protocol 8: Elicitation of *Xenopus laevis* Tadpole and Adult Frog Peritoneal Leukocytes [Grayfer 2018]), and adoptive cell transfer (Protocol 3: Adoptive Transfer of Fluorescently Labeled Immune Cells in *Xenopus* [Rhoo and Robert 2019]). Notably, adoptive cell transfer and tumor transplantation can also use MHC defined isogenic clones (e.g., LG-6 and LG-15). These clones are produced by gynogenesis from diploid eggs of *X. laevis* x *X. gilli* interspecies hybrids (Kobel and Du Pasquier 1975; Du Pasquier et al. 1977). Since irradiated sperm is used, only genetically identical females are produced. Finally, assays not easily achieved in mammalian models are described, including the convenient generation of immune-deficient animals by sublethal irradiation (Protocol 6: Lymphocyte Deficiency Induced by Sublethal Irradiation in *Xenopus* [Rollins-Smith and Robert 2019]), by thymectomy at an early developmental stage (Protocol 5: Larval Thymectomy of *Xenopus laevis* [Mashoof et al. 2018]), and by RNAi-mediated loss of function by transgenesis using the I-SceI meganuclease (Protocol 7: RNAi-Mediated Loss of Function of *Xenopus* Immune Genes by Transgenesis [Edholm and Robert 2018]).

CONCLUSION

In summary, the combination of attractive and unconventional features with conserved mammalian-like immunity empowers *Xenopus* as an influential multifaceted experimental platform for exploring fundamental (e.g., evolution and ontogeny) as well as medical (e.g., cancer, inflammation, and infectious disease) aspects of immunity.

REFERENCES

Aztekin C, Hiscock TW, Marioni JC, Gurdon JB, Simons BD, Jullien J. 2019. Identification of a regeneration-organizing cell in the *Xenopus* tail. *Science* **364**: 653–658.

Aztekin C, Hiscock TW, Butler R, De Jesús Andino F, Robert J, Gurdon JB, Jullien J. 2020. The myeloid lineage is required for the emergence of a regeneration permissive environment following

Xenopus tail amputation. *Development* 147: dev185496. doi:10.1242/dev.185496

Banach M, Robert J. 2017a. Tumor immunology viewed from alternative animal models—the *Xenopus* story. *Curr Pathobiol Rep* 5: 49–56.

Banach M, Robert J. 2017b. Collagen-embedded tumor transplantations in *Xenopus laevis* tadpoles. *Cold Spring Harb Protoc* doi:10.1101/pdb.prot097584

Castro CD, Luoma AM, Adams EJ. 2015. Coevolution of T-cell receptors with MHC and non-MHC ligands. *Immunol Rev* 267: 30–55.

Chen G, Robert J. 2011. Antiviral immunity in amphibians. *Viruses* 3: 2065–2086.

Chen Y, Love NR, Amaya E. 2014. Tadpole tail regeneration in *Xenopus*. *Biochem Soc Trans* 42: 617–623.

Chinchar VG, Waltzek TB. 2014. Ranaviruses: not just for frogs. *PLoS Pathog* 10: e1003850.

De Jesús Andino F, Robert J. 2019. Assessing antibody responses to pathogens or model antigens in *Xenopus* by enzyme-linked immunosorbent assay (ELISA). *Cold Spring Harb Protoc* doi:10.1101/pdb.prot099234

Dimitrakopoulou D, Tulkens D, Van Vlierberghe P, Vleminckx K. 2019. *Xenopus tropicalis*: joining the armada in the fight against blood cancer. *Front Physiol* 10: 48.

Du Pasquier L, Miggiano V, Kobel HR, Fischbert M. 1977. The genetic control of histocompatibility reactions in natural and laboratory-made polyploid individuals of the clawed toad *Xenopus*. *Imunogenetics* 5: 129–141.

Du Pasquier L, Schwager J, Flajnik MF. 1989. The immune system of *Xenopus*. *Annu Rev Immunol* 7: 251–275.

Edholm E-S. 2018. Flow cytometric analysis of *Xenopus* immune cells. *Cold Spring Harb Protoc* doi:10.1101/pdb.prot097600

Edholm E-S, Robert J. 2018. RNAi-mediated loss of function of *Xenopus* immune genes by transgenesis. *Cold Spring Harb Protoc* doi:10.1101/pdb.prot101519

Flajnik MF, Hsu E, Kaufman JF, Pasquier LD. 1987. Changes in the immune system during metamorphosis of *Xenopus*. *Immunol Today* 8: 58–64.

Grayfer L. 2018. Elicitation of *Xenopus laevis* tadpole and adult frog peritoneal leukocytes. *Cold Spring Harb Protoc* doi:10.1101/pdb.prot097642

Grogan LF, Robert J, Berger L, Skerratt LF, Scheele BC, Castley JG, Newell DA, McCallum HI. 2018. Review of the amphibian immune response to chytridiomycosis, and future directions. *Front Immunol* 9: 2536.

Hardwick LJA, Philpott A. 2018. *Xenopus* models of cancer: expanding the oncologist's toolbox. *Front Physiol* 9: 1660.

Haynes-Gimore N, Banach M, Brown E, Dawes R, Edholm ES, Kim M, Robert J. 2015. Semi-solid tumor model in *Xenopus laevis*/gilli cloned tadpoles for intravital study of neovascularization, immune cells and melanophore infiltration. *Dev Biol* 408: 205–212.

Hellsten U, Harland RM, Gilchrist MJ, Hendrix D, Jurka J, Kapitonov V, Ovcharenko I, Putnam NH, Shu S, Taher L, et al. 2010. The genome of the Western clawed frog *Xenopus tropicalis*. *Science* 328: 633–636.

Houssaint E, Flajnik M. 1990. The role of thymic epithelium in the acquisition of tolerance. *Immunol Today* 11: 357–360.

Hyoe RK, Robert J. 2019. A *Xenopus* tadpole alternative model to study innate-like T cell-mediated anti-mycobacterial immunity. *Dev Comp Immunol* 92: 253–259.

Izutsu Y. 2019. Skin grafting in *Xenopus laevis*: a technique for assessing development and immunological disparity. *Cold Spring Harb Protoc* doi:10.1101/pdb.prot099788

Kobel HR, Du Pasquier L. 1975. Production of large clones of histocompatible, fully identical clawed toads (*Xenopus*). *Immunogenetics* 2: 87–91.

Li J, Zhang S, Amaya E. 2016. The cellular and molecular mechanisms of tissue repair and regeneration as revealed by studies in *Xenopus*. *Regeneration (Oxford, England)* 3: 198–208.

Mashoof S, Breaux B, Criscitiello MF. 2018. Larval thymectomy of *Xenopus laevis*. *Cold Spring Harb Protoc* doi:10.1101/pdb.prot099192

Nagel SC, Kassotis CD, Vandenberg LN, Lawrence BP, Robert J, Balise VD. 2020. Developmental exposure to a mixture of unconventional oil and gas chemicals: a review of effects on adult health, behavior, and disease. *Mol Cell Endocrinol* doi:10.1016/j.mce.2020.110722

Nakai Y, Nakajima K, Robert J, Yaoita Y. 2016. Ouro proteins are not essential to tail regression during *Xenopus tropicalis* metamorphosis. *Genes Cells* 21: 275–286.

Paredes R, Ishibashi S, Borrill R, Robert J, Amaya E. 2015. *Xenopus*: an in vivo model for imaging the inflammatory response following injury and bacterial infection. *Dev Biol* 408: 213–228.

Rhoo KH, Robert J. 2019. Adoptive transfer of fluorescently labeled immune cells in *Xenopus*. *Cold Spring Harb Protoc* doi:10.1101/pdb.prot097592

Robert J. 2016. The immune system of amphibians. In *Encyclopedia of immunobiology* (ed. Ratcliffe MJH), pp. 486–492. Elsevier, Amsterdam.

Robert J, Cohen N. 2011. The genus *Xenopus* as a multispecies model for evolutionary and comparative immunobiology of the 21st century. *Dev Comp Immunol* 35: 916–923.

Robert J, Edholm ES. 2014. A prominent role for invariant T cells in the amphibian *Xenopus laevis* tadpoles. *Immunogenetics* 66: 513–523.

Robert J, Ohta Y. 2009. Comparative and developmental study of the immune system in *Xenopus*. *Dev Dyn* 238: 1249–1270.

Robert J, McGuire CC, Kim F, Nagel SC, Price SJ, Lawrence BP, De Jesús Andino F. 2018. Water contaminants associated with unconventional oil and gas extraction cause immunotoxicity to amphibian tadpoles. *Toxicol Sci* 166: 39–50.

Rollins-Smith LA. 1998. Metamorphosis and the amphibian immune system. *Immunol Rev* 166: 221–230.

Rollins-Smith LA, Robert J. 2019. Lymphocyte deficiency induced by sublethal irradiation in *Xenopus*. *Cold Spring Harb Protoc* doi:10.1101/pdb.prot097626

Rollins-Smith LA, Ramsey JP, Reinert LK, Woodhams DC, Livo LJ, Carey C. 2009. Immune defenses of *Xenopus laevis* against *Batrachochytrium dendrobatidis*. *Front Biosci* 1: 68–91.

Session AM, Uno Y, Kwon T, Chapman JA, Toyoda A, Takahashi S, Fukui A, Hikosaka A, Suzuki A, Kondo M, et al. 2016. Genome evolution in the allotetraploid frog *Xenopus laevis*. *Nature* 538: 336–343.

Slack JM, Lin G, Chen Y. 2008. The *Xenopus* tadpole: a new model for regeneration research. *Cell Mol Life Sci* 65: 54–63.

Zasloff M. 1987. Magainins, a class of antimicrobial peptides from *Xenopus* skin: isolation, characterization of two active forms, and partial cDNA sequence of a precursor. *Proc Natl Acad Sci* 84: 5449–5453.

Skin Grafting in *Xenopus laevis*: A Technique for Assessing Development and Immunological Disparity

Yumi Izutsu[1]

Department of Biology, Faculty of Science, Niigata University, Niigata 950-2181, Japan

Skin grafting in the amphibian *Xenopus laevis* has been used to detect not only allogeneic antigens that differ by minor H antigens or by one MHC haplotype, but also to detect ontogeny-specific antigens (including both emerging adult- and disappearing larval-specific) during metamorphosis. To understand the mechanisms underlying allogeneic tolerance or immune responses against larval- and/or adult-specific antigens, a complete MHC homozygous, inbred strain is the most appropriate experimental model. The inbred J strain established in Japan is used here. Owing to complete histocompatibility, the inbred J strain shows no grafted skin rejection among the same strain of adult frogs, and its genuine homozygosity was reconfirmed by genomic sequence analysis in 2016. Therefore, the J strain enables immunologists and embryologists to understand evolutionary processes as well as immunological events and tissue remodeling mechanisms present during development. Furthermore, an F1 hybrid between the J strain and a GFP-labeled transgenic line is available from our laboratory and can be used as a model for long-term cell tracking. This protocol explains the methodology for skin grafting in *X. laevis* to determine immunological discrepancies between the host and donor. It is also possible to trace cell and tissue fates in the hosts during early embryogenesis and during complete development from larvae to adults, which is extremely difficult to perform using other species.

MATERIALS

It is essential that you consult the appropriate Material Safety Data Sheets and your institution's Environmental Health and Safety Office for proper handling of equipment and hazardous materials used in this protocol.

RECIPES: Please see the end of this protocol for recipes indicated by <R>. Additional recipes can be found online at http://cshprotocols.cshlp.org/site/recipes.

Reagents

MS222 (3-aminobenzoic acid ethyl ester) (Sigma-Aldrich A5040)

A 1% stock solution of MS222 is stable for several months when stored in a dark glass bottle at 4°C. For anesthetization of adult frogs, prepare 0.05% MS222 solution in 1× Steinberg's solution containing 50 µg/mL gentamicin.

OCT compound (Sakura 4583)
Paraformaldehyde (4% in phosphate-buffered saline)

---

[1]Correspondence: izutsu@gs.niigata-u.ac.jp

Steinberg's solution (1×) <R>

To prepare a working Steinberg's solution containing 50 μg/mL gentamicin, add 1/1000 volume of gentamicin stock solution (50 mg/mL [Sigma-Aldrich G1397]) plus an equal volume of 0.2 N NaOH to 1× Steinberg's solution. Check that the pH is neutral and adjust as needed. Store at 4°C.

Xenopus laevis (inbred J strain and/or gfp-Tg hybrid line) in rearing containers.

The X. laevis inbred J strain was established by Tochinai and Katagiri (1975). In 1948, four pairs of founder frogs were inbred by gonadotropic hormone-induced mating. In 1973, it was found that these inbred frogs showed no short-term skin graft rejection among siblings of the same strain, indicating immunologically high homogeneity. After repeated single-pair mating for over 21 generations, this inbred line showed no long-term skin rejection (Izutsu and Yoshizato 1993). The detailed history of the J strain is described by Session et al. (2016). The J strain is available from a domestic animal vendor in Japan, Watanabe Zoushoku (http://www5d.biglobe.ne.jp/~zoushoku/top.htm).

The gfp transgenic line (gfp-Tg) was generated in our laboratory by nuclear transplantation using wild-type X. laevis carrying the gfp gene under the control of the Xenopus heat shock promoter (hsp70) (Mukaigasa et al. 2009). The partially inbred F4 gfp-Tg line was obtained by hormone-induced mating; the strongest GFP-expressing individuals were selected from among the siblings of the F3 gfp-Tg line (Otsuka-Yamaguchi et al. 2017). The F1 gfp-Tg hybrid J line can be used as a host for skin grafting experiments, as the J strain donor cells grafted to the hybrid recipients can be easily distinguished after induction of GFP expression by heat shock.

Equipment

Cotton sticks (Johnson and Johnson) (optional; see Step 8)

Digital camera (AxioCam HRc; ZEISS)

Filters (0.2-μm pore size) (Millipore GSWP04700)

Fine forceps (110-mm total length) (Natsume Seisakusho [Tokyo] MA-55) (sterilized with 70% ethanol)

Glass beakers (200-mL) (Iwaki Glass; Asahi Techno Glass, Tokyo)

Kimwipes

Liquid nitrogen

Microscissors (105-mm total length, 7-mm blade length) (Natsume Seisakusho MB-50-7) (sterilized with 70% ethanol)

Microscope (Leica M60)

Petri dishes (10-cm) (Corning 353803)

Plastic scale (3×3 mm^2)

Graph paper can be copied to an overhead projector sheet and cut with scissors to create a plastic scale.

Suture needle with thread (6-0 blue nylon polyamide) (Nescosuture ET0806NA45-KF2)

Water aspirator pump (Sigma-Aldrich)

METHOD

All treatments should be done aseptically to avoid infection.

1. One to two days before the operation, transfer the frogs to a rearing container of filter-sterilized (0.22 μm) or autoclaved tap water. Keep them without feeding until the day of the operation.

2. On the day of the operation, gently wash the frogs in a glass beaker 10 times using sterilized water.

3. Anesthetize the animals by immersing them in a glass beaker containing 0.05% MS222 solution.

   The duration of anesthesia should not exceed 30 min.

   See Troubleshooting.

4. Place a donor animal on its back on a sheet of Kimwipe of suitable size in a 10-cm Petri dish. Pour ice-cold 1× Steinberg's solution with gentamicin over the animal until the top of the body is almost immersed in solution.

   The Kimwipe sheet prevents the slipping of the donor animal during the operation.

Sufficient solution is necessary because the skin can easily dry and curl up without contact with water once excised from the donor. Once the skin is curled up, it is difficult to use for grafting.

5. Place a plastic scale (3×3 mm^2) on the ventral skin of the donor for reference (Fig. 1A). Excise a 1.5- to 3.0-mm^2 area of skin from the white-colored donor belly (distinguishable from the dark brown-colored back skin of the host) using fine forceps and microscissors. Leave the piece of skin on the ventral side of the donor until it is transplanted (Fig. 1B).

6. Transfer a host animal to the same dish. Trim the same-size area from the dorsal skin of the host to fit the donor skin graft (Fig. 1C, cyan-dotted area). Remove the trimmed skin from the host immediately and arrange the donor and host adjacent to each other.

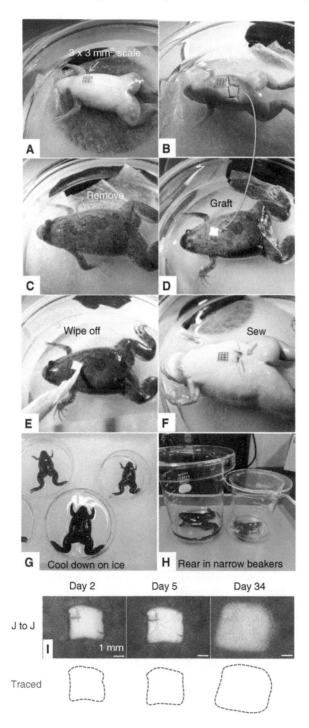

FIGURE 1. Schematic representation of the skin grafting procedure. (A) Anesthetize the J strain donor. Place a plastic scale (3×3 mm^2) on the ventral skin. (B) Cut grafts corresponding to the scale. Leave the grafts on the ventral skin. (C) Remove an area of dorsal skin (cyan-dotted area) from the J strain host to create a hole. (D) After making a hole, immediately slide the graft underwater into the hole. Otherwise, the graft will dry and curl up, making it difficult to graft. (E) Carefully wipe the grafted skin to keep the graft dry just after the operation. This treatment raises the success rate dramatically. (F) Use a sutures needle to sew the ventral skin of the donor. (Note that this individual is different from the frog shown in A and B.) (G) Keep the hosts on ice to prevent movement. Otherwise, the grafts may become displaced easily. (H) Rear the hosts in narrow glass beakers for 2–3 d to prevent jumping or overturning. (I) After transplantation, once the grafted skin from the white-colored donor belly is accepted by the host frog, it grows as part of the host skin (see Day 34). Blood vessels are observable on Day 5. The grafted skin area is indicated by the traced area (blue-dotted lines) in the images. Tracing is used for image analysis. Note that since the grafted donor skin tissue is completely accepted as self-tissue by the host, the graft continues to develop with the growth of the young host and the boundary between the skin of the white donor and that of the brown host appears blurred and blended.

Cite this protocol as *Cold Spring Harb Protoc*; doi:10.1101/pdb.prot099788

7. Add ice-cold 1× Steinberg's solution with gentamicin to the dish until the tops of both animals are immersed. Transfer the skin graft by gently slipping the donor skin underwater into the graft bed (the hole on the back of the host) using forceps (Fig. 1D).

    Care must be taken to not expose the skin graft to air during the operation.

8. After grafting, immediately remove the Steinberg's solution using a water aspirator pump. Wipe off the solution around the graft >5 times, particularly at the junction between the graft and the host skin, using a Kimwipe or a cotton stick (Fig. 1E).

    This quick wipe-off helps the donor skin adhere firmly to the host skin.

9. If necessary, close the back slit of the donor with sutures (Fig. 1F).

10. Post-operation, move the host frog to a cold 200-mL glass dish or beaker containing a small amount of ice-cold 1× Steinberg's solution with gentamicin to prevent drying of the animal. Place the dish on ice for 1 h (Fig. 1G).

    The temperature and time required for cooling can be changed depending on the size or stage of the animal.

11. Transfer the dish and incubate for 2–4 h in the dark at 4°C.

    This prevents the animal from moving and jumping and prevents the grafted skin from flaking off.

    See Troubleshooting.

12. Transfer the dish and incubate overnight in the dark at 16°C.

    See Troubleshooting.

13. Rear the host animal in a narrow glass beaker for 2–3 d at normal temperature (23°C–24°C) (Fig. 1H) with a small amount of 1× Steinberg's solution with gentamicin.

    The narrow glass beaker prevents inversion or jumping.

    The Steinberg's solution, which prevents the animal from completely drying up, should be replaced often to prevent infection.

14. Observe the grafted skin using a microscope (Fig. 1I). Confirm that the graft is successfully attached and that its original size (e.g., 3×3 mm^2) is maintained.

    With successful attachment to the host, thin blood vessels extending into the graft should be visible around 2–3 d after grafting (Fig. 1I, Day 5).

15. Maintain host frogs at normal temperature (23°C–24°C) in rearing containers with a suitable volume of dechlorinated tap water, at a maximum density of five frogs per 2 L. Feed frogs and change the rearing water every day.

16. Photograph the host animal grafts every 2–3 d using a digital camera. Trace the areas of the skin grafts (Fig. 1I, traced).

    The traced areas can be measured and compared to their original sizes at Day 1 using the public domain image processing and analysis program "ImageJ" (https://imagej.nih.gov/ij/).

17. To determine the cell and tissue fates of the grafted cells, prepare the tissues for immunohistochemistry.

    i. Dissect the grafts, including some of the host skin, using fine forceps and microscissors.

    ii. Place the skin samples on 0.22-μm filter membranes (e.g., 3×3 mm^2) to keep the skin flat.

    iii. Embed the tissues in OCT and then dip in liquid nitrogen.

    iv. Fix the tissue sections with 4% paraformaldehyde for 10 min.

    Tissue samples can be prepared any time after proper graft attachment (Step 14), up to 6 mo after grafting. To look for features of skin graft rejection, sections should be prepared when blood vessel numbers increase or inflammation is observed. Migration of effector T cells and degeneration of tissue may be visible.

    Sometimes GFP fluorescence, derived from the gfp-Tg line, quenches during immunostaining even after fixation. Therefore, GFP visualization must be performed before immunostaining. Two different colors, e.g., Alexa488 (green)- and Cy3 (red)-conjugated antibodies, may be used. GFP fluorescence can be detected using anti-GFP antibodies.

TROUBLESHOOTING

Problem (Steps 3, 11, and 12): Animals vomit during the course of grafting.
Solution: If the animals are fully fed, they sometimes vomit during and just after the operation. At least 1 d before the operation, the animals should be fasted.

DISCUSSION

Because genomic information is now available for the *X. laevis* J strain (Session et al. 2016), it is possible to produce genome-edited J strain frogs. Cell and tissue transplantation can be used in animals with the J strain genetic background as an innovative strategy to address unresolved questions about mechanisms of development. These experiments also enable both genetic labeling for cell tracking and ectopic gene induction in the grafted cells. Although the J strain has been inbred for decades, it does not possess any problematic characteristics that might affect experimental planning or development and growth when compared to wild-type frogs. While the body size of the J strain is smaller than that of the wild-type frog at the same age, and while J strain frogs produce a smaller number of eggs, a sexually mature J strain female produces approximately 1000–2000 eggs. During hormone-induced mating, 80%–90% of shed eggs are usually fertilized; therefore, it is easy to produce more than 1000 metamorphosed frogs from one mating. Since the J strain has perfect histocompatibility, it can be used for experiments without the immunological barrier, which is not possible in many other cold-blooded animals.

RECIPE

Steinberg's Solution (1×)

3.4 g/L NaCl
0.05 g/L KCl
0.08 g/L $Ca(NO_3)_2 \cdot 4H_2O$
0.1025 g/L $MgSO_4$
0.56 g/L Tris-HCl (pH 7.4)
10 mg/L phenol red

For convenience, prepare a 10× stock solution. Store the 10× stock solution for no longer than a few months at room temperature.

ACKNOWLEDGMENTS

I thank Dr. Shin Tochinai for the critical reading. I also thank Haruka Kobayashi and Akira Nakamura for technical assistance. Work in my laboratory is supported by Grant-in-Aid from Nagase Science Technology Foundation (to Y.I.) and by Grant-in-Aid for Scientific Research (C) from the Ministry of Education, Culture, Sports, Science, and Technology (15K06992 to Y.I.).

REFERENCES

Izutsu Y, Yoshizato K. 1993. Metamorphosis-dependent recognition of larval skin as non-self by inbred adult frog (*Xenopus laevis*). *J Exp Zool* **266:** 163–167.

Mukaigasa K, Hanasaki A, Maéno M, Fujii H, Hayashida S, Itoh M, Kobayashi M, Tochinai S, Hatta M, Iwabuchi K, et al. 2009. The keratin-related Ouroboros proteins function as immune antigens mediating tail regression in *Xenopus* metamorphosis. *Proc Natl Acad Sci* **106:** 18309–18314.

Otsuka-Yamaguchi R, Aiko Kawasumi-Kita A, Kudo N, Izutsu Y, Tamura K, Yokoyama H. 2017. Cells from subcutaneous tissues contribute to scarless skin regeneration in *Xenopus laevis* froglets. *Dev Dyn* **248:** 586–597.

Session AM, Uno Y, Kwon T, Chapman JA, Toyoda A, Takahashi S, Fukui A, Hikosaka A, Suzuki A, Kondo M, et al. 2016. Genome evolution in the allotetraploid frog *Xenopus laevis*. *Nature* **538:** 336–343.

Tochinai S, Katagiri C. 1975. Complete abrogation of immune response to skin allografts and rabbit erythrocytes in the early thymectomized *Xenopus*. *Dev Growth Differ* **17:** 383–394.

Collagen-Embedded Tumor Transplantations in *Xenopus laevis* Tadpoles

Maureen Banach[1,2] and Jacques Robert[1]

[1]*Department of Microbiology and Immunology, University of Rochester Medical Center, 14-642 Rochester, New York 14642*

The *Xenopus laevis* tadpole provides a valuable model for studying tumorigenesis and tumor immunity by intravital real-time microscopy. Using well-characterized thymic lymphoid tumor lines (15/0 and ff-2) that are transplantable into their compatible hosts (LG-15 isogenic clones and the F inbred strain, respectively), a system of semisolid tumor engraftment has been designed. Because these lymphoid tumor cell lines are not adherent and grow in suspension, they are first immobilized in a matrix of type I rat tail collagen before transplantation as a semisolid tumor graft under the transparent dorsal skin in the head region of a tadpole. This semisolid tumor engraftment is amenable to manipulation and permits real-time visualization of tumor growth, neovascularization, collagen rearrangements, immune cell infiltration, and formation of the tumor microenvironment.

MATERIALS

It is essential that you consult the appropriate Material Safety Data Sheets and your institution's Environmental Health and Safety Office for proper handling of equipment and hazardous materials used in this protocol.

RECIPES: Please see the end of this protocol for recipes indicated by <R>. Additional recipes can be found online at http://cshprotocols.cshlp.org/site/recipes.

Reagents

Amphibian PBS (APBS) <R>

Earle's balanced salt solution (EBSS, 10×)

Iscove-derived tumor media (Robert et al. 1994) (e.g., MSF tumor medium; see Step 8)

Rat tail collagen I (Corning 354249; 100 mg supplied as a liquid in 0.02 N acetic acid)

Tricaine methanesulfonate (TMS, MS-222, Western Chemical, Inc.)

Tumor cell line of interest (1×10^6 cells) and compatible animal recipient

    15/0 tumor cells and LG-15 or LG-6 cloned tadpoles

    ff-2 tumor cells and MHC homozygous F inbred strain tadpoles

15/0 tumor cells are derived from isogenic X. laevis/gilli LG-15 clones, and are transplantable into LG-15 or LG-6 cloned tadpoles. ff-2 tumor cells are derived from and transplantable into X. laevis tadpoles of the F inbred strain. Culture media and conditions have been described by Du Pasquier and Robert (1992) and Robert et al. (1994).

---

[2]Correspondence: maureen_banach@urmc.rochester.edu

Equipment

Centrifuge (benchtop)
Compressed air tank with blood gas mixture (5% CO_2, 21% O_2, 74% N)
Culture plates (six-well, sterile)
Forceps, stainless (small, well-sharpened; e.g., #5 forceps from A. Dumont and Fils, Switzerland)
Incubator at 27°C, with modular incubator chamber
Microcentrifuge tubes (1.5-mL Eppendorf)

METHOD

Preparing the Collagen Setting Solution

Keep all solutions, reagents, and tubes chilled on ice throughout the following section. Higher temperatures will polymerize the collagen.

1. In a chilled microcentrifuge tube, slowly mix 10× EBSS into the collagen at a ratio of 1:5 (v:v). Pipette up and down until the solution is thoroughly mixed. Avoid air bubbles.

2. In 10 µL increments, add more 10× EBSS until the color of the mixture changes to light orange with a ring of light pink on the top.

   This color indicates a neutral pH, which allows the collagen to polymerize at room temperature.

3. Once the required color is achieved, centrifuge the tube briefly to collect the contents. Keep the collagen setting solution on ice.

   See Troubleshooting.

Embedding Tumor Cells into the Collagen Matrix

4. Wash 1×10^6 cells (15/0 or ff-2) in amphibian phosphate-buffered saline (APBS) three times. Centrifuge the cells at 1000 rpm for 10 min at 4°C. Aspirate the APBS.

5. To prepare ten 10-µL grafts of 100,000 cells/graft, add 100 µL of collagen setting solution to the cells. Slowly pipette up and down to evenly mix the cells.

6. For each graft, pipette 10 µL of the cell mixture into one well of a sterile six-well culture plate, forming a single drop in each well. Leave the plates at room temperature for 10 min to allow the collagen to polymerize.

   See Troubleshooting.

7. Transfer the plates into a modular incubator chamber, fill the chamber with blood gas mixture, and incubate for 1 h at 27°C.

8. If the collagen-embedded tumors will not be transplanted immediately, add 2 mL of MSF tumor medium to each well so that the samples are completely covered. Keep samples at 27°C in the incubator.

   To achieve the best results, the grafts should be transplanted within 24 h.

Transplanting the Collagen-Embedded Tumors

Use F inbred tadpoles for transplantations with ff-2 tumors, and LG-15 or LG-6 tadpoles for transplantations with 15/0 tumors.

9. Anesthetize a tadpole with 0.1 g/L of TMS solution.

10. Make a small subcutaneous incision on the tadpole's head, posterior to the eyes. Broaden the incision to form a pocket. Using forceps, insert a collagen-embedded tumor into the pocket.

    Cite this protocol as *Cold Spring Harb Protoc*; doi:10.1101/pdb.prot097584

11. Allow the tadpole to recover from the procedure.

See Troubleshooting.

TROUBLESHOOTING

Problem (Step 3): The collagen setting solution polymerizes before mixing with the tumor cells.
Solution: The reagents were not sufficiently chilled. Keep all reagents and cells on ice before preparing grafts in a six-well plate.

Problem (Step 6): The collagen setting solution does not polymerize.
Solution: The pH of the collagen setting solution was not optimized to neutral. The color should be light orange with a ring of light pink on the top.

Problem (Step 6): Air bubbles are present in the grafts.
Solution: Air bubbles can be introduced into the grafts while mixing the collagen setting solution. Collagen is very viscous, so the collagen setting solution should be pipetted very slowly to avoid air bubbles.

Problem (Step 6): Tumor cells are not dispersed evenly in the graft.
Solution: While mixing the collagen setting solution with the tumor cells, make sure the cells are evenly dispersed in the mixture.

Problem (Step 11): The collagen-embedded tumor transplant does not stay attached.
Solution: For transplantation of collagen-embedded cells, the incision and the pocket under the tadpole skin should be as small as possible.

DISCUSSION

With the aid of intravital microscopy (confocal, two photon), the collagen-embedded tumor graft model offers an opportunity to study the in vivo and real-time processes involved in tumor estab-

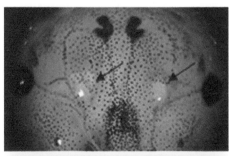
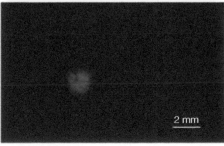

FIGURE 1. Dorsal view of a tadpole of the inbred MHC homozygous F strain grafted with semisolid ff-2 tumors on both sides of the head viewed by a fluorescent dissecting microscope. ff-2 tumor cells (1×10^5) are unlabeled (*right* side) or labeled with PHK-26 (*left* side). Bar size, 2 mm.

lishment and growth, neovascularization, collagen remodeling, infiltration of immune cells, migration of melanophores, and perhaps metastasis (Haynes-Gilmore et al. 2015).

Collagen-embedded tumor grafts are easily manipulated, for example, using tumor cells fluorescently labeled with CFSE or PKH-26 (Fig.1). The tumors can be harvested at different times post-engraftment, and different techniques (such as microscopy, flow cytometry, or real-time polymerase chain reaction [RT-PCR]) can be applied to study the various infiltrating host cells. Furthermore, different experimental setups can be performed on the same tadpole, for example, by transplanting a wild-type tumor on one side and a tumor mutant on the other side of the tadpole's head.

RECIPES

APBS

25 mL H$_2$O (sterile; Merck)
100 mL PBS (phosphate-buffered saline; Invitrogen)

In this recipe, PBS is adjusted to amphibian osmolarity (225 ± 5 mOsm/L) by the addition of H$_2$O.

Earle's Balanced Salt Solution (EBSS, 10×)

100 mL Earl's Balanced Salt Solution (EBSS) (10×) (Sigma-Aldrich E7510)
0.2 M NaHCO$_3$
0.15 M NaOH
42.5 mL APBS (amphibian PBS)

To prepare 150 mL of 10× EBSS, combine the above reagents and sterilize through 0.22-micron filter. Store aliquots at −20°C.

REFERENCES

Du Pasquier L, Robert J. 1992. In vitro growth of thymic tumor cell lines from *Xenopus*. *Dev Immunol* 2: 295–307.

Haynes-Gilmore N, Banach M, Brown E, Dawes R, Edholm ES, Kim M, Robert J. 2015. Semi-solid tumor model in *Xenopus laevis/gilli* cloned tadpoles for intravital study of neovascularization, immune cells and melanophore infiltration. *Dev Biol* 408: 205–212.

Robert J, Guiet C, Du Pasquier L. 1994. Lymphoid tumors of *Xenopus laevis* with different capacities for growth in larvae and adults. *Dev Immunol* 3: 297–307.

Cite this protocol as *Cold Spring Harb Protoc*; doi:10.1101/pdb.prot097584

Adoptive Transfer of Fluorescently Labeled Immune Cells in *Xenopus*

Kun Hyoe Rhoo[1,2] and Jacques Robert[1]

[1]*Department of Immunology Microbiology and Virology, University of Rochester, Medical Center, New York 14620*

Adoptive cell transfer from inbred adult *Xenopus* to inbred tadpoles is a useful way to study the dissemination of immune cells or pathogen-infected immune cells in tadpoles. For example, *Xenopus* peritoneal leukocytes (PLs) can be readily infected by pathogens such as Frog virus 3 (FV3) and *Mycobacterium marinum* (*M. marinum*). By transferring fluorescently labeled, FV3-infected PLs into tadpoles, we observed infiltration of these cells into the tadpole's brain, which indicates that FV3-infected PLs can cross blood brain barrier. Taking advantage of tadpoles' transparency, fluorescently labeled immune cells can be tracked in real time using fluorescence microscopy.

MATERIALS

It is essential that you consult the appropriate Material Safety Data Sheets and your institution's Environmental Health and Safety Office for proper handling of equipment and hazardous materials used in this protocol.

RECIPES: Please see the end of this protocol for recipes indicated by <R>. Additional recipes can be found online at http://cshprotocols.cshlp.org/site/recipes.

Reagents

Amikacin (25 mg/mL) (optional; see Step 2)

Amphibian phosphate-buffered saline (APBS; sterile) <R>

Amphibian serum free (ASF) medium supplemented with fetal bovine serum (FBS) <R>

Cells obtained from Protocol 4: Flow Cytometric Analysis of *Xenopus* Immune Cells (Edholm 2018)

Cells can be isolated from the animals' blood, peritoneal leukocytes (PLs), lymphoid tissues (spleen, thymus, and bone marrow) or nonlymphoid tissues such as kidney, intestine, or liver. Follow Protocol 4: Flow Cytometric Analysis of Xenopus *Immune Cells (Edholm 2018) to isolate cells for transfer. Use MHC homozygous inbred J(j/j) adult frogs or a different inbred strain such as F(f/f) for the procedure. Cell yield will be different depending on what tissue is used. For example, we can isolate at least 1×10^6 cells from the peritoneal lavage of a single adult animal.*

More specific immune cell subsets can be used for transfer (e.g., B or T cells) by further purification or enrichment using either magnetic cell sorting or fluorescence-activated cell sorting (FACS). Prepare cells under sterile conditions.

Red fluorescent membrane dye PKH26 (Sigma-Aldrich) or green fluorescent CFSE tracer (Invitrogen)

Tadpoles, J inbred strain, stage 55

---

[2]Correspondence: Kunhyoe_rhoo@urmc.rochester.edu

Texas red dextran (Sigma-Aldrich)
Trypan blue solution (Thermo Fisher Scientific)
Trypsin-EDTA (0.25%; Gibco)
Tween 80 (optional; see Step 2)
Viral (FV3) or bacterial (*M. marinum*) pathogens (optional; see Step 2)

Equipment

Beckman Coulter Allegra 21R Centrifuge
Conical tubes (15-mL)
Glass capillaries
Hemocytometer
Incubator (27°C)
Inverted fluorescence microscope (Axiovert 200)
Microcentrifuge tubes (1.5-mL)
Micropipette puller (Sutter Instrument Co.)
Microscope slides and coverslips
Six-well plate

METHOD

1. Seed between 3×10^5 to 1×10^6 isolated cells per well in a six-well plate. Incubate cells in each well with 2 mL of ASF medium with FBS in a standard incubator for 1 d at 27°C to rest the activated cells before infection.

2. If using macrophages infected with pathogens for adoptive transfer, infect these cells with viral (FV3) or bacterial (*M. marinum*) pathogens of interest with a multiplicity of infection (MOI) between 0.1 and 1 for a desired time in culture.

Infect Cells with FV3

i. Follow the in vitro infection protocol described in De Jesus Andino et al. (2016).

Infect Cells with M. marinum

ii. Enrich PLs for macrophages using heat-killed *E. coli* stimulation (Robert et al. 2014).

iii. Co-incubate PLs with 0.1 MOI *M. marinum* in APBS containing 0.05% Tween 80 for 2 h in culture at 27°C.

   The volume depends on the number of cells that you desire to infect. For example, if you have approximately 1 million cells to infect, you may seed them in one well of six-well plate and use total volume of 3–5 mL. The number of cells also depends on the yield (i.e., how many PLs or liver cells were isolated).

iv. Remove the supernatant, and replace it with ASF medium with FBS containing amikacin (100 µg/mL). Incubate for 2 h to remove unphagocytosed *M. marinum*.

   Use the same volume of ASF with FBS containing amikacin as you used for M. marinum *infection. For example, if* M. marinum *infection occurs in the six-well plate with total volume will be 3–5 mL, the new medium that replaces it will also be 3–5 mL.*

   Proceed to Step 3.

3. Remove supernatant. Detach the cells from the six-well plate by adding 1 mL of 0.25% trypsin-EDTA to each well. Incubate for 5 min at 27°C. Place collected cells in 15-mL conical tubes. (Cells from multiple wells may be pooled.) Centrifuge cells for 10 min at 400*g* at 4°C. Remove supernatant.

4. Resuspend cells in 1 mL of APBS, and transfer the cells to a microcentrifuge tube. Count the cells using a hemocytometer. Pellet the cells as in Step 3.

Cite this protocol as *Cold Spring Harb Protoc*; doi:10.1101/pdb.prot097592

5. Fluorescently label the cells with either CFSE or PKH26.

Label the Cells with PKH26 (Red Fluorescence)

i. Resuspend up to 2 × 107 cells in each 1 mL of Diluent C (a component of the PKH labeling kit). Then mix cells with an equal volume of 4 μM PKH26. Incubate the mixture for 15 min at 27°C.

   The final cell concentration will be 1 × 10^7 mL, and the final concentration of PKH26 will be 2 μM.

ii. Stop the reaction by adding 2 mL of ASF with FBS.

iii. Pellet the cells by centrifugation at 9000g for 30 sec at room temperature. Decant the supernatant, and wash with 1 mL of APBS.

iv. Repeat Step 5.iii two more times.

v. Resuspend the pellet in 1 mL of APBS. To determine if the cells are properly labeled with red fluorescence, place an aliquot of cells on a glass microscope slide and look for red signal under the fluorescence microscope.

Label the Cells with CFSE (Green Fluorescence)

vi. Resuspend up to 2.5 × 10^6 cells in 50 μL of 40 μM CFSE (diluted in DMSO; included in supplier package) in a microcentrifuge tube. Incubate for 15 min at 27°C.

vii. Quench CFSE by adding 250 μL of ASF with FBS, and incubate for 7 min at 27°C. Flick with fingers every 2 min.

viii. Add 1 mL of APBS and pellet the cells by centrifugation at 9000g for 30 sec at room temperature. Decant the supernatant.

ix. Repeat Step 5.viii two more times.

x. Resuspend the pellet in 1 mL of APBS. To determine if the cells are properly labeled with green fluorescence, place an aliquot of cells on a glass microscope slide and look for green signal under the fluorescence microscope.

6. Count a sample of the cells using a hemocytometer and determine cell death by Trypan blue exclusion test. Note the percent of cell death as you proceed further.

7. Inject 100,000 labeled cells intraperitoneally into J inbred tadpoles of stage 55 (3 to 4 wk of age) as follows.

i. Centrifuge the fluorescently labeled cells at 9000g for 30 sec at room temperature, and resuspend the cells in 10 μL of sterile APBS.

ii. Use a micropipette puller to generate microinjecting needles from standard glass capillaries (see Protocol 8: Elicitation of *Xenopus laevis* Tadpole and Adult Frog Peritoneal Leukocytes [Grayfer 2018]).

iii. Load the cells in the back opening of the glass capillary using a thin micropipette tip. Inject the cells directly into the tadpole's peritoneum using the microinjection technique described in Protocol 8: Elicitation of *Xenopus laevis* Tadpole and Adult Frog Peritoneal Leukocytes (Grayfer 2018).

   Alternatively, the labeled cells can be injected into the blood circulation through the aorta using the same microinjection technique above.

8. Visualize adoptively transferred cells using intravital fluorescence microscopy in the whole tadpoles or the harvested organs.

   The cells can be visualized within a few hours to a day. See Troubleshooting.

   Alternatively, cells can be isolated from organs or recovered from the peritoneal fluid for flow cytometry analysis (see Protocol 4: Flow Cytometric Analysis of Xenopus *Immune Cells [Edholm 2018]; Paredes et al. 2015).*

TROUBLESHOOTING

Problem (Step 8): Poor fluorescence signal of the labeled cells is observed.

Solution: When CFSE is purchased, it comes in several aliquots. Because CFSE is sensitive to humidity, reusing aliquots is not recommended. Check the fluorescence before the adoptive transfer to make sure that the fluorescence can be detected in culture.

Problem (Step 8): Adoptively transferred cells not found.

Solution: Ensure that the donor and recipient are both inbred J strains by skin grafting or MHC class Ia gene typing (Flajnik and Du Pasquier 1990). Since some cells are fragile and die rapidly, it is a good idea to save some cells in culture as a backup and measure cell death after a few days. If necessary, it is also possible to transfer more cells (e.g., 1×10^6 cells).

DISCUSSION

This technique can be used to study homing or infiltration of adoptively transferred cells into particular organs or tissues. For example, we used this adoptive cell transfer system to study the ability of FV3-infected PLs to cross the blood brain barrier. We have successfully visualized the donor's peritoneal macrophages infected with FV3 infiltrating into the recipient tadpole's brain at stage 55 when the BBB is fully functional (De Jesus Andino et al. 2016). In another example, adoptive cell transfer in *Xenopus* has been used to study the immunization role of heat shock proteins (hsp). When PLs pulsed with gp96 chaperoning minor H-Ag were adoptively transferred from minor H-Ag-mismatched donors to recipients, the transferred cells were successfully detected in the recipient's spleen and the appearance of tumors were delayed in these immunized recipients (Maniero and Robert 2004).

Using confocal microscopy instead of conventional fluorescence microscopy is an attractive approach to trace fluorescently labeled cells in real time. Confocal microscopy can locate the adoptive transferred cells in three-dimensional structure with high resolution. Recently, we successfully used confocal microscopy to study the dynamic structure of cell aggregation (e.g., granulomas) following *M. marinum* infection. In parallel, it is also possible to label blood vessels by injecting Texas Red Dextran into the heart (25 mg/mL; 10 µL of volume) to provide a better localization of infiltrated cells.

RECIPES

Amphibian Phosphate-Buffered Saline (APBS)

Sodium chloride (NaCl)	6.6 g/L
Sodium phosphate (Na_2HPO_2)	1.15 g/L
Potassium phosphate (KH_2PO_4)	0.2 g/L

Adjust pH to 7.7 with 10 N NaOH, and filter-sterilize through a 0.2-µm filter. Store at room temperature for up to 6 mo.

Cite this protocol as *Cold Spring Harb Protoc*; doi:10.1101/pdb.prot097592

Amphibian Serum-Free (ASF) Medium Supplemented with Fetal Bovine Serum (FBS)

Reagent	Quantity
Iscove's DMEM basal medium, powdered (Gibco 12200-036)	1 package
Insulin (Sigma-Aldrich 19278-5mL)	10 mL
Nonessential amino acids (Gibco 11140-050)	10 mL
Penicillin-streptomycin (Gibco 15070-063)	10 mL
Primatone (Sheffield Products Division)	3 mL
2-mercaptoethanol	1 mL
$NaHCO_3$	3.02 g

Combine the above ingredients in water. Adjust the pH to 7.0 with 10 N NaOH, and bring the final volume to 1 L. Filter through a 0.2-µm filter, and store at 4°C. On the day of use, prepare the desired volume of ASF by adding 30% double-distilled water, 10% fetal bovine serum (FBS), and 10 µg/mL kanamycin to the appropriate volume of solution.

This recipe was adapted from Robert et al. (2004).

REFERENCES

De Jesus Andino F, Jones L, Maggirwar SB, Robert J. 2016. Frog Virus 3 dissemination in the brain of tadpoles, but not in adult *Xenopus*, involves blood brain barrier dysfunction. *Sci Rep* **6**: 22508.

Edholm E-S. 2018. Flow cytometric analysis of *Xenopus* immune cells. *Cold Spring Harb Protoc* doi: 10.1101/pdb.prot097600.

Flajnik MF, Du Pasquier L. 1990. The major histocompatibility complex of frogs. *Immunol Rev* **113**: 47–63.

Grayfer L. 2018. Elicitation of *Xenopus laevis* tadpole and adult frog peritoneal leukocytes. *Cold Spring Harb Protoc* doi: 10.1101/pdb.prot097642.

Maniero GD, Robert J. 2004. Phylogenetic conservation of gp96-mediated antigen-specific cellular immunity: New evidence from adoptive cell transfer in *Xenopus*. *Transplantation* **78**: 1415–1421.

Paredes R, Ishibashi S, Borrill R, Robert J, Amaya E. 2015. *Xenopus*: An in vivo model for imaging the inflammatory response following injury and bacterial infection. *Dev Biol* **408**: 213–228.

Robert J, Gantress J, Cohen N, Maniero GD. 2004. *Xenopus* as an experimental model for studying evolution of hsp-immune system interactions. *Methods* **32**: 42–53.

Robert J, Grayfer L, Edholm ES, Ward B, De Jesus Andino F. 2014. Inflammation-induced reactivation of the ranavirus Frog Virus 3 in asymptomatic *Xenopus laevis*. *PLoS One* **9**: e112904.

Protocol 4

Flow Cytometric Analysis of *Xenopus* Immune Cells

Eva-Stina Edholm[1]

Department of Microbiology and Immunology, University of Rochester Medical Center, Rochester, New York 14620

Flow cytometry is a versatile analytical platform capable of multiparameter analysis of more than a thousand individual cells per second. This technique is used to measure the physical and chemical characteristics of individual cells in a heterogeneous cell suspension as they pass through one or multiple lasers. Physical properties, such as size and internal complexity, are recorded as light scattering at different angles and are expressed as forward- and side-scatter, respectively. Following light excitation, fluorochromes conjugated to antibodies or intercalated with different cellular components reemit light at distinct wavelengths. This can identify a broad array of cell specific antigens, further defining distinct cell subsets based on activation, lineage, and developmental stage. The combination of labels that can be used depends on the laser used to excite the fluorochromes and on the detector and available antibodies. With the growing number of *Xenopus*-specific antibodies, flow cytometry can be used to identify, isolate, and characterize distinct immune cell subsets. In this protocol, different methods to obtain single-cell suspensions from various *X. laevis* tissues are described. A standard three-parameter procedure defining viability and two cell-surface markers is then described.

MATERIALS

It is essential that you consult the appropriate Material Safety Data Sheets and your institution's Environmental Health and Safety Office for proper handling of equipment and hazardous materials used in this protocol.

RECIPES: Please see the end of this protocol for recipes indicated by <R>. Additional recipes can be found online at http://cshprotocols.cshlp.org/site/recipes.

Reagents

Amphibian phosphate-buffered saline (APBS) <R>
Anti-*Xenopus* monoclonal antibodies

> Anti-Xenopus *antibodies can be found at the* Xenopus laevis *Research Resource for Immunobiology: https://www.urmc.rochester.edu/microbiology-immunology/xenopus-laevis.aspx.*

Enzyme(s) for tissue disruption (e.g., collagenase) (optional; see Step 16)
Flow cytometry staining buffer (FCSB) <R>
Fluorochrome-conjugated secondary antibodies
Fluorochrome-conjugated streptavidin
Ice
Percoll (1.130 ± 0.005 g/mL in sterile APBS) (optional; see Step 4)
Propidium iodide (1 mg/mL in molecular grade water)

---

[1]Correspondence: eva-stina.i.edholm@uit.no

Sodium heparin (1000 U/mL) (optional; see Step 3)
Trypan blue
Xenopus laevis adults

Equipment

Cell strainer (optional; see Step 16)
Centrifuge (benchtop, refrigerated, with swinging bucket rotors for 5- and 15-mL tubes)
Compound microscope
Dissection kit (optional; see Steps 10, 14)
FACS tubes (5 mL, round-bottom, polystyrene)
Flow cytometer

> *A number of different flow cytometry instruments are available, ranging in capability from a single to >18 different detectors. Due to the limited number of frog-specific antibodies currently available we find that a personal benchtop flow cytometer with two lasers (typically blue and red), two light scatter detectors and four fluorescence detectors with optical filters optimized for the detection of popular fluorochromes is sufficient.*

Hemocytometer
Ice bucket
Microcentrifuge
Microcentrifuge tubes (1.5-mL, flip-top)
Petri dish (35 × 10 mm) (optional; see Step 11)
Polyester mesh (70-μm; 28% open area) (optional; see Step 11)
Scalpel blade (optional; see Step 15)
Tubes (15-mL, conical)
Tweezers (small, micro-mashing with ¼ inch × ½ inch pads; part # MASMIC from Arrow Springs) (optional; see Step 11)

METHOD

Preparation of Cells from Adult *X. laevis*

Isolation of Peritoneal Leukocytes

An example of a scatter profile for peritoneal leucocytes is shown in Figure 1B.

1. Collect peritoneal leukocytes as described in Protocol 8: Elicitation of *Xenopus laevis* Tadpole and Adult Frog Peritoneal Leukocytes (Grayfer 2018).

2. Perform a cell count and viability analysis using a hemocytometer and trypan blue exclusion: Prepare a 0.4% solution of trypan blue in APBS (pH 7.3). Add 10 μL of trypan blue solution to a 10-μL cell suspension. Load onto a hemocytometer and count the cells.

   > *Cells that have taken up trypan blue are nonviable.*

Isolation of Blood Leukocytes

3. Collect 1–2 mL of blood in APBS containing 50 U/mL sodium heparin as anticoagulant. Adjust to a final volume of 4 mL.

4. Add 4 mL of a 51% (v/v) solution of 1.130 ± 0.005 g/mL Percoll stock diluted in APBS to a 15-mL centrifuge tube. Carefully layer 4 mL of blood/heparin/APBS mixture on top without disturbing the Percoll layer.

5. Centrifuge at 300*g* for 25 min at 4°C.

   > *Centrifugation over the Percoll gradient should be performed with slow acceleration and a slow or no break to prevent disruption of the interphase.*

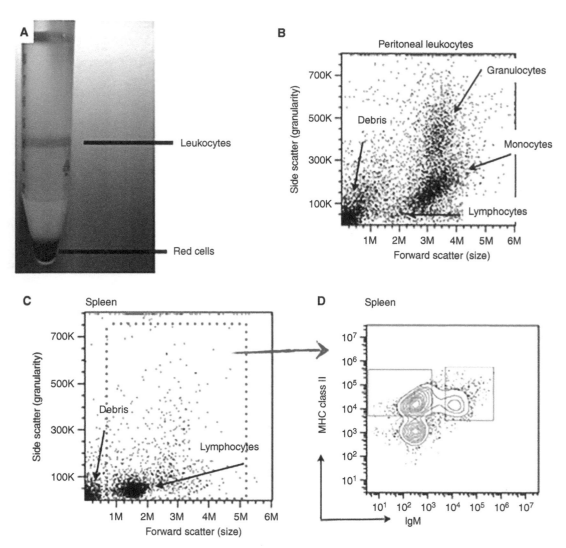

FIGURE 1. (*A*) Isolation of leukocytes from blood samples as described in Step 6. (*B,C*) Identification of different cell populations based on forward- and side-scatter characteristics. (*B*), peritoneal leucocytes (Steps 1–2), (*C*) spleen cells (Steps 10–13). (*D*) Identification of specific cell populations based on fluorescence. Live (propidium iodine negative) spleen cells from (*C*) stained with anti-IgM followed by goat-anti-mouse FITC and biotinylated anti-MHC class II primary antibody and APC conjugated streptavidin as described in Steps 19–36.

6. Collect the cloudy interface, which contains the leucocytes, using a 1-mL pipette and transfer to a 1.5-mL microcentrifuge tube (see Fig. 1A).

7. Pellet cells at 300*g* for 5 min at 4°C. Resuspend the pellet in 1 mL of APBS. Mix gently.

8. Repeat Step 7 twice.

9. Perform Step 2.

Isolation of Leukocytes from Lymphoid Tissue

An example of a scatter profile for spleen cells is shown in Figure 1C.

10. Harvest tissue.

> *Frogs, unlike mammals, do not have lymph nodes; the main lymphoid organ is the spleen. Large fractions of T cells are also found in the thymus.*

11. Place a 1 inch square-sized mesh into a 35 × 10 mm Petri dish. Soak the mesh with APBS. Place the tissue in the middle of the mesh and place a second 1 inch square-sized mesh on top of the

Cite this protocol as *Cold Spring Harb Protoc*; doi:10.1101/pdb.prot097600

tissue. Using micro mashing tweezers push the cells through the mesh using repetitive horizontal movements. Once the tissue is disrupted tilt the dish 45° and rinse with APBS. Collect the cell suspension and transfer to 1.5 mL microcentrifuge tubes.

12. Pellet the cells by centrifugation at 300g for 5 min at 4°C. Discard the supernatants.

    Resuspend the pellets in APBS and combine in a final volume of 1 mL.

13. Perform Step 2.

    To remove red blood cells from the sample, Percoll gradient separation (Steps 4–9) can be performed.

Isolation of Leukocytes from Nonlymphoid Tissue

14. Harvest the tissue of interest and rinse with APBS.

15. Cut the tissue into 1–2 mm pieces using a scalpel blade and place in a 15-mL centrifuge tube with 5 mL of APBS.

16. Add an appropriate amount of enzyme(s), such as collagenase, diluted in APBS and incubate under conditions recommended by the enzyme manufacturer. Disperse cells by gently pipetting and filter through a cell strainer to remove clumps.

17. Pellet the cells by centrifugation at 300g for 5 min at 4°C. Discard the supernatant. Add 4 mL of APBS.

18. Perform Steps 4–9.

Staining Procedure

Prepare the following controls: (1) unstained cells; (2) cells incubated with an isotype control secondary antibody alone; (3) cells incubated with fluorochrome-conjugated streptavidin alone; and (4) fluorochromes minus one control, i.e., prepare samples that have a combination of all fluorochromes used in the experiment except one. Depending on the number of fluorochromes used, the number of fluorochromes minus one control will vary. For example, in a three color staining protocol with anti-CD8-FITC, anti-CD4-PE, and anti-CD3-APC, three fluorochromes minus one control are needed (1. anti-CD8-FITC and anti-CD4-PE; 2. anti-CD8-FITC and anti-CD3-APC; and 3. anti-CD4-PE and anti-CD3-APC).

A schematic showing different staining procedures (i.e., indirect and direct staining) is shown in Figure 2.

19. Dilute cells in FCSB to 1×10^6–1×10^7 cells/mL. Aliquot 100 µL of cell suspension into each test and control FACS tube.

20. Add 100 µL of hybridoma supernatant containing 50–100 µg/mL of the primary antibody. Mix gently.

21. Incubate on ice for 30 min.

22. Pellet the cells by centrifugation at 300g for 5 min at 4°C. Discard the supernatant.

    Resuspend the cells gently in 1 mL of FCSB.

23. Repeat Step 22 but resuspend the cells in 100 µL of FCSB.

24. Dilute the appropriate fluorochrome-labeled secondary antibody 1:100 (0.1–1 µg/ mL) in 100 µL of FCSB. Add to the cell suspension. Incubate for 30 min on ice. Protect from light.

    Titrate the secondary antibody to determine which dilution allows for the strongest specific signal with the least background.

25. Pellet the cells by centrifugation at 300g for 5 min at 4°C. Discard the supernatant. Resuspend the cells gently in 1 mL of FCSB.

26. Repeat Step 25 but resuspend cells in 100 µL of FCSB.

    If single staining is performed, proceed to Step 36.

27. Biotinylate a primary antibody according to Mao (2010) and dilute to 1 µg/mL in FCSB. Add 100 µL of the diluted biotinylated antibody to the cells.

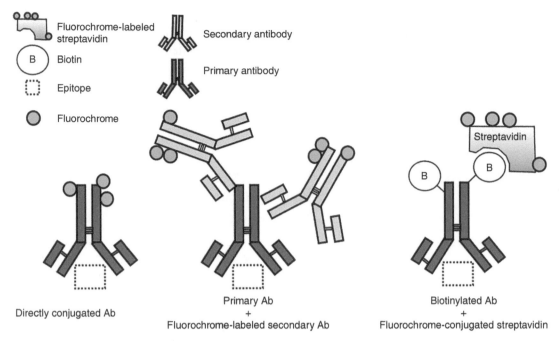

FIGURE 2. Schematic showing the different staining procedures described; indirect staining using primary and secondary antibodies, and direct staining using fluorochrome-conjugated primary antibodies and biotinylated primary antibodies.

Perform a titration curve to determine which dilution of the biotinylated primary antibody allows for the strongest specific signal with the least background.

28. Incubate on ice for 30 min.

29. Pellet the cells by centrifugation at 300*g* for 5 min at 4°C. Discard the supernatant.
    Resuspend the cells gently in 1 mL of FCSB.

30. Repeat Step 29 but resuspend the cells in 100 μL of FCSB.

31. Dilute fluorochrome-conjugated streptavidin 1/100 in FCSB and add 100 μL to the cells.

32. Incubate on ice for 30 min.

33. Pellet the cells by centrifugation at 300*g* for 5 min at 4°C. Discard the supernatant.
    Resuspend the cells gently in 1 mL of FCSB.

34. Repeat Step 33 but resuspend the cells in 400 μL of FCSB.

35. Just before analysis, add 0.5 μL of propidium iodide to each sample.

36. Analyze samples by flow cytometry according to the instructions of the FACS instrument manufacturer.
    An example showing MHC class II and IgM staining in spleen cells is shown in Figure 1D.

    See Troubleshooting.

TROUBLESHOOTING

Problem (Step 36): Background caused by autofluorescence.
Solution: Use a fluorescent probe with high fluorescence intensity that emits light in the detection channel that is least effected by autofluorescence.

Problem (Step 36): Spectral overlaps.
Solution: Choose a combination of fluorochromes that have little to no spectral overlap.

 Cite this protocol as *Cold Spring Harb Protoc*; doi:10.1101/pdb.prot097600

Problem (Step 36): Background caused by nonspecific antibody binding.

Solution: Increase the amount of BSA in the FCSB to 2%–5%. Alternatively, incubate cells with 0.5% *X. laevis* serum for 30 min on ice before staining.

DISCUSSION

These procedures give a rapid and informative assessment of a large number of cells within a heterogeneous cell population. The combination of light scatter, which provides information about cell size and granularity, with specific cell-surface antigens is one of the most widely used applications of flow cytometry and provides valuable insight into the composition of cell populations. In addition, flow cytometry can be used to measure a long and constantly expanding list of parameters, including expression of intracellular antigens, total DNA content (for analysis of the cell cycle, cell kinetics, ploidy, etc.), phagocytosis (Grayfer and Robert 2014), apoptosis (De Jesús Andino et al. 2015), proliferation (Morales and Robert 2007), cytotoxicity assays (Haynes-Gilmore et al. 2014), and enzyme activity (Grayfer and Robert 2015). In addition, based on the same principles, fluorescent activated cell sorting is used to physically sort specific cell populations to purify populations of interest.

RECIPES

APBS

25 mL H$_2$O (sterile; Merck)
100 mL PBS (phosphate-buffered saline; Invitrogen)

In this recipe, PBS is adjusted to amphibian osmolarity (225 ± 5 mOsm/L) by the addition of H$_2$O.

Flow Cytometry Staining Buffer (FCSB)

Add 10 g of bovine serum albumin to 700 mL of amphibian phosphate-buffered saline (APBS) <R>. Stir well until dissolved. Add 50 µL of 1 M sodium azide. (Sodium azide inhibits metabolic activity, so do not add sodium azide if recovering cell function is required—i.e., if cells are to be collected for cell culture or functional assays.) Adjust the volume to 1000 mL by adding additional APBS. Filter-sterilize through a 0.2-µm filter and store at 4°C.

REFERENCES

De Jesús Andino F, Grayfer L, Chen G, Chinchar VG, Edholm ES, Robert J. 2015. Characterization of Frog Virus 3 knockout mutants lacking putative virulence genes. *Virology* **485:** 162–170.

Grayfer L. 2018. Elicitation of *Xenopus laevis* tadpole and adult frog peritoneal leukocytes. *Cold Spring Harb Protoc* doi:10.1101/pdb.prot097642.

Grayfer L, Robert J. 2014. Divergent antiviral roles of amphibian (*Xenopus laevis*) macrophages elicited by colony-stimulating factor-1 and interleukin-34. *J Leukoc Biol* **96:** 1143–1153.

Grayfer L, Robert J. 2015. Distinct functional roles of amphibian (*Xenopus laevis*) colony-stimulating factor-1- and interleukin-34-derived macrophages. *J Leukoc Biol* **98:** 641–649.

Haynes-Gilmore N, Banach M, Edholm ES, Lord E, Robert J. 2014. A critical role of non-classical MHC in tumor immune evasion in the amphibian *Xenopus* model. *Carcinogenesis* **35:** 1807–1813.

Mao SY. 2010. Biotinylation of antibodies. *Methods Mol Biol* **588:** 49–52.

Morales HD, Robert J. 2007. Characterization of primary and memory CD8 T-cell responses against ranavirus (FV3) in *Xenopus laevis*. *J Virol* **81:** 2240–2248.

Protocol 5

Larval Thymectomy of *Xenopus laevis*

Sara Mashoof,[1] Breanna Breaux,[1] and Michael F. Criscitiello[1,2,3]

[1]*Comparative Immunogenetics Laboratory, Department of Veterinary Pathobiology, College of Veterinary Medicine and Biomedical Sciences, Texas A&M University, College Station, Texas 77843;* [2]*Department of Microbial Pathogenesis and Immunology, College of Medicine, Texas A&M Health Science Center, Texas A&M University, College Station, Texas 77843*

In jawed vertebrates from sharks to mammals, the thymus is the primary (or central) lymphoid tissue where T cells develop and mature. The particular stromal cell types, cytokine environment, and tissue organization in the thymus are essential for V(D)J recombination, positive selection for major histocompatibility complex recognition, and negative selection against self-peptide recognition of most αβ T cells. The thymectomy operation on *Xenopus* tadpole larva described here creates a T-cell–deficient model suitable for many immunology studies.

MATERIALS

It is essential that you consult the appropriate Material Safety Data Sheets and your institution's Environmental Health and Safety Office for proper handling of equipment and hazardous materials used in this protocol.

RECIPES: Please see the end of this protocol for recipes indicated by <R>. Additional recipes can be found online at http://cshprotocols.cshlp.org/site/recipes.

Reagents

Carbenicillin

Human chorionic gonadotropin (hCG; 200 U/mL in amphibian phosphate-buffered saline [PBS] <R>)

NaCl, saturated aqueous solution

Tricaine methanesulfonate (MS222)

Xenopus laevis, males and females, breeding age

Equipment

Beakers, plastic

Cheesecloth

Dissection microscope

Metal plate stage, grounded

Microcautery device (e.g., an aphid Zapper, a device designed for stylectomy)

---

[3]Correspondence: mcriscitiello@cvm.tamu.edu

Our device delivers a VHF pulse of 10 msec-duration and a power of 10 watts to the target tissue via an abraded tungsten wire.

Needles, tungsten

Sharpen the tungsten needles by electrolysis (Brady 1965). The same Zapper instrument used for microcauterization can be used to fashion desired needle points by bobbing the wire tip up and down in a saturated sodium chloride solution.

Pipettes, plastic
Water aerator

METHOD

There are differing views as to the ideal developmental window for thymectomies to be performed. Although thymus visualization is more difficult, earlier thymectomies—e.g., from day 4 to day 7 post-spawn (Nieuwkoop and Faber stages 45–48 [Nieuwkoop and Faber 1994])—are recommended. This is the period where the thymus is developing, but lymphoid lineage cells have not yet migrated to the thymus (Horton and Manning 1972; Arnall and Horton 1987). Past this time, T cells can mature and escape into the periphery, resulting in incomplete T cell ablation.

Spawning

Methods for spawning Xenopus laevis vary. We use hCG injections into the dorsal lymph sacs to induce amplexus.

1. Prime a male *Xenopus* with 40 U and a female *Xenopus* with 20 U of hCG. Leave in separate tanks overnight.

2. Administer an additional 250 U of hCG to the female the following morning.

3. Four hours later, add the male and female to a shared tank containing spawning mops, nylon mesh, or marbles. Cover the tank to keep it dark and avoid disturbances.

Anesthesia

Individuals vary in susceptibility to MS222 toxicity. Overexposure to MS222 reduces the chance of tadpole recovery. Do not leave tadpoles in the MS222 bath for longer than 15 min.

4. Dilute 30 mg of MS222 in 100 mL of buffered vivarium system H_2O in a plastic beaker.

5. Construct a tadpole "net" by looping a plastic pipette back on itself and covering it with a piece of cheesecloth. Secure the cheesecloth to the pipette with tape (Fig. 1D).

6. Using the net, capture a tadpole. Place it in the MS222 bath.

7. Monitor the tadpole for 1–3 min until it has stopped moving.

8. Retrieve the tadpole from the anesthesia using the net.

    If it is still moving, return it to the MS222 bath.

Thymectomy

9. Wet a piece of cheesecloth with dechlorinated tank water. Place it on the dissection microscope stage. Position the cloth such that a section overlaps with the grounded metal plate stage, illuminating the cloth from below in the field of view.

10. Using the net, transfer the anesthetized tadpole to the stage cheesecloth so that the light shines through it. Position the tadpole on its ventral side with either the eyes or tail facing forward (Fig. 1F).

11. Locate the thymus using the microscope.

    The thymus is located bilaterally caudal and medial to the eyes and lateral to the dark central nervous system (Fig. 2). Note that the thymus shown in Figure 2 is at a later stage of development where it is melanized and easily visualized; at the present stage, the thymus will not be melanized and will appear as a small, clear oval or circular sac (Fig. 1F).

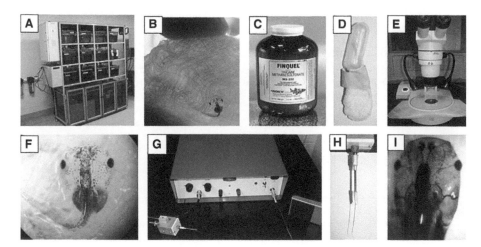

FIGURE 1. Procedural progression. (*A*) The recirculating system housing the frogs. (*B*) *Xenopus* tadpole at eight days of age. (*C*) MS222 used for tadpole anesthesia. (*D*) Tadpole net formed from cheesecloth over bent transfer pipette. (*E*) The dissection microscope. (*F*) Same tadpole as shown in (*B*), as viewed under the dissection microscope. (*G*) The microcautery apparatus used for the *Xenopus* thymectomy. (*H*) Close-up of tungsten needle. (*I*) A fully thymectomized tadpole with no visible thymus.

12. Optionally, dab the surface of the tadpole covering the developing thymus with a small piece of tissue paper to dry the skin.

    If thymectomizing later in the optimal window (e.g., day 7 post-spawn), using the dissection needle to rupture the skin and provide better exposure of the thymus before administering the electrical pulse can increase accuracy and improve efficiency of ablation.

13. Use the microcautery apparatus needle to burn the bilateral thymus organ away on one side of the tadpole:

    i. While looking through the microscope, lower the needle until it is visible.

    ii. Place the needle on the tadpole near the thymus.

    iii. Press the foot pedal to administer an electrical pulse.

       Ideally, the thymus should be burned with only one quick pulse on the thymic lobe (Fig. 1). If the thymus is still visible, a second pulse can be administered; however, this decreases the chance of survival and should be avoided when possible.

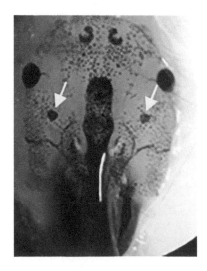

FIGURE 2. Localization of thymus. In *Xenopus* tadpole the thymus is located (yellow arrows) bilaterally caudal and medial to the eyes and lateral to the dark central nervous system.

Cite this protocol as *Cold Spring Harb Protoc*; doi:10.1101/pdb.prot099192

14. Repeat Steps 13.i–13.iii on the other half of the bilateral organ on the other side of the animal.

   Rotating the cheesecloth allows use of one's dominant hand for both procedures, but takes precious time to reposition. With practice, both sides of the tadpole can be ablated with a single positioning.

Recovery

15. Immediately after thymic ablation use the net to transfer the tadpole to an ice bath. Cool the animal for 3 sec.

16. Transfer the tadpole to a plastic beaker of aerated regular tank water containing carbenicillin at a concentration of 80 mg/L. Allow the tadpole to recover for 1 h.

   See Troubleshooting.

17. Transfer the tadpole to a designated tank in the primary recirculating *Xenopus* husbandry system.

18. Repeat Steps 4–17 with the remaining tadpoles.

   Screen thymectomized tadpoles for regrowth of the thymus at 1 and 2 wk after the surgery (see Steps 19–22). See Troubleshooting.

Screening

19. Dilute 30 mg of MS222 in 100 mL of system H_2O in a plastic beaker.

20. Wet a piece of cheesecloth with regular tank water. Place it on the dissection microscope stage.

21. Use the net to transfer a thymectomized tadpole into the MS222 bath. Remove the tadpole from the anesthesia after 30 sec.

   Complete immobilization is not necessary for screening.

22. Transfer the tadpole onto the stage cheesecloth. Carefully inspect it through the microscope.

   If a melanized thymus can be detected on either side of the tadpole (see, e.g., Fig. 2 versus Fig. 1I), the thymectomy was unsuccessful. Tadpoles with thymic regrowth should be euthanized by exposure to an overdosed bath of MS222 (>1000 mg/L) for at least 1 h.

TROUBLESHOOTING

Problem (Step 16): The tadpole does not recover from anesthesia.

Solution: MS222 can be toxic at high doses or with slower processing times. Variable size and pharmacogenetics will cause some tadpoles not to recover in the aerated post-surgery tank. Lower levels of MS222 (e.g., 10 mg/100 mL [Rollins-Smith et al. 1996] or 20 µg/mL [Tochinai 1975]) have been used successfully and might increase survival.

Problem (Step 18): The tadpoles die in the days following the thymectomy.

Solution: This surgery can have a low survival rate because of the traumatic nature of the procedure. However, if you experience very low rates, several variables can be altered. The ice bath post-thymectomy is optional: Although lowering the body temperature lowers the metabolic rate and can therefore increase survival, it comes at the cost of increased handling of the tadpole, which disrupts the outer mucous membrane. Similarly, some have performed the thymectomy procedure itself on a chilled agar plate. The use of antibiotics in the recovery tank is also optional, and could disrupt mutualistic bacterial skin and gastrointestinal flora.

DISCUSSION

The thymus is the primary lymphoid tissue for T cell development (Cooper et al. 1966; Criscitiello et al. 2010). Additionally, the frog undergoes class switch recombination of the immunoglobulin heavy chain locus in a translocon organization and somatic hypermutation driven by specialized follicular antigen presentation (Neely et al. 2018). The frog thymectomy model has been used for decades (Horton and Manning 1972) and is now an established model of adaptive immunity absent the effect of T cells in a metamorphosing tetrapod vertebrate (see, e.g., Sakuraoka and Tochinai 1993; Ono and Tochinai 1995; Kinney et al. 1996; Robert et al. 1997; Horton et al. 2003; Mashoof et al. 2013).

The relatively superficial position of the bilateral organ in developing anuran amphibians makes the frog thymectomy model straightforward. Alternatively, in the Northern leopard frog *Rana pipiens*, it can even be aspirated using a fine glass micropipette (Rollins-Smith and Cohen 1982). A late larval thymectomy model has also been used in *Xenopus laevis* to explore the persistence of larval T cells post-metamorphosis (Rollins-Smith et al. 1996).

RECIPE

Amphibian Phosphate-Buffered Saline (APBS)

Sodium chloride (NaCl)	6.6 g/L
Sodium phosphate (Na_2HPO_2)	1.15 g/L
Potassium phosphate (KH_2PO_4)	0.2 g/L

Adjust pH to 7.7 with 10 N NaOH, and filter-sterilize through a 0.2-μm filter. Store at room temperature for up to 6 mo.

REFERENCES

Arnall JC, Horton JD. 1987. In vivo studies on allotolerance perimetamorphically induced in control and thymectomized *Xenopus*. *Immunology* **62:** 315–319.

Brady J. 1965. A simple technique for making very fine, durable dissecting needles by sharpening tungsten wire electrolytically. *Bull World Health Organ* **32:** 143–144.

Cooper MD, Peterson RDA, South MA, Good RA. 1966. The functions of the thymus system and the bursa system in the chicken. *J Exp Med* **123:** 75–102.

Criscitiello MF, Ohta Y, Saltis M, McKinney EC, Flajnik MF. 2010. Evolutionarily conserved TCR binding sites, identification of T cells in primary lymphoid tissues, and surprising trans-rearrangements in nurse shark. *J Immunol* **184:** 6950–6960.

Horton JD, Manning MJ. 1972. Response to skin allografts in *Xenopus laevis* following thymectomy at early stages of lymphoid organ maturation. *Transplantation* **14:** 141–154.

Horton TL, Stewart R, Cohen N, Rau L, Ritchie P, Watson MD, Robert J, Horton JD. 2003. Ontogeny of *Xenopus* NK cells in the absence of MHC class I antigens. *Dev Comp Immunol* **27:** 715–726.

Kinney KS, Felten SY, Cohen N. 1996. Sympathetic innervation of the amphibian spleen: Developmental studies in *Xenopus laevis*. *Dev Comp Immunol* **20:** 51–59.

Mashoof S, Goodroe A, Du CC, Eubanks JO, Jacobs N, Steiner JM, Tizard I, Suchodolski JS, Criscitiello MF. 2013. Ancient T-independence of mucosal IgX/A: Gut microbiota unaffected by larval thymectomy in *Xenopus laevis*. *Mucosal Immunol* **6:** 358–368.

Neely HR, Guo J, Flowers EM, Criscitiello MF, Flajnik MF. 2018. "Double-duty" conventional dendritic cells in the amphibian *Xenopus* as the prototype for antigen presentation to B cells. *Eur J Immunol* **48:** 430–440.

Nieuwkoop PD, Faber J (eds.). 1994. *Normal table of* Xenopus laevis *(Daudin): A systematical and chronological survey of the development from the fertilized egg till the end of metamorphosis*. Garland Publishing, New York.

Ono M, Tochinai S. 1995. Demonstration of cells possessing tolerance-inducing activity in *Xenopus laevis* rendered tolerant perimetamorphically. *Transplantation* **60:** 66–70.

Robert J, Guiet C, Cohen N, Du Pasquier L. 1997. Effects of thymectomy and tolerance induction on tumor immunity in adult *Xenopus laevis*. *Int J Cancer* **70:** 330–334.

Rollins-Smith LA, Cohen N. 1982. Effects of early larval thymectomy on mitogen responses in leopard frog (*Rana pipiens*) tadpoles. *Dev Comp Immunol* **6:** 303–309.

Rollins-Smith LA, Needham DA, Davis AT, Blair PJ. 1996. Late thymectomy in *Xenopus* tadpoles reveals a population of T cells that persists through metamorphosis. *Dev Comp Immunol* **20:** 165–174.

Sakuraoka J, Tochinai S. 1993. Demonstration of cells involved in rejection of tolerogenic grafts in tolerant *Xenopus*. *Dev Comp Immunol* **17:** 439–447.

Tochinai S. 1975. Success of thymectomy from larval *Xenopus* at the earliest stage of its histogenesis. *Zool Mag* **84:** 138–141.

Cite this protocol as *Cold Spring Harb Protoc*; doi:10.1101/pdb.prot099192

Protocol 6

Lymphocyte Deficiency Induced by Sublethal Irradiation in *Xenopus*

Louise A. Rollins-Smith[1,2,4] and Jacques Robert[3]

[1]*Departments of Pathology, Microbiology, and Immunology and of Pediatrics, Vanderbilt University School of Medicine, Nashville, Tennessee 37232;* [2]*Department of Biological Sciences, Vanderbilt University, Nashville, Tennessee 37235;* [3]*Department of Microbiology and Immunology, University of Rochester Medical Center, Rochester, New York 14642*

In many studies of diseases affecting amphibians, it is important to determine to what extent lymphocyte-mediated defenses are involved. For example, in studies of the nature of the immune response of *Xenopus laevis* to the amphibian chytrid fungus, *Batrachochytrium dendrobatidis*, it was essential to determine if mucosal antimicrobial peptides or lymphocyte-mediated immunity was most important for resistance to this skin pathogen. In this protocol, we describe a method for sublethal irradiation to reduce lymphocyte numbers. Briefly, *X. laevis* adults or tadpoles are exposed to 9 Gy (900 rads) of irradiation applied by exposure to a cesium source or gamma irradiator to reduce lymphocyte populations in the spleen.

MATERIALS

It is essential that you consult the appropriate Material Safety Data Sheets and your institution's Environmental Health and Safety Office for proper handling of equipment and hazardous materials used in this protocol.

Reagents

Dechlorinated tap water for frogs (filter-sterilized)
Penicillin stock solution (10,000 IU/mL)
Streptomycin stock solution (10,000 µg/mL)
Xenopus laevis adults (small; 4–5 g in weight) or tadpoles

Equipment

Clean opaque plastic containers (12 cm × 8.5 cm × 5 cm) with vented lids to accommodate small
   Xenopus laevis adults (4–5 g in weight) or tadpoles
Sterile, plastic 500-mL container to dilute antibiotics
X-ray source (alternative sources such as a Cesium-137 source or a Cobalt-60 source could be
   substituted)

> *For studies at Vanderbilt, the X-ray source was a 300 kVp/10 mA tube manufactured by Pantak (East Haven, Connecticut, USA). The half value layer was 0.73 mm Cu. In Rochester, we are now using a Gammacell 40 Exactor Low Dose device that uses Cesium-137 sources and is designed for low dose irradiations of small animals such as mice.*

---

[4]Correspondence: louise.rollins-smith@vanderbilt.edu

Cite this protocol as *Cold Spring Harb Protoc*; doi:10.1101/pdb.prot097626

METHOD

1. Place small *X. laevis* frogs at ~1–2 yr of age (4–5 g) individually in plastic containers in a small volume of dechlorinated tap water (~25 mL) to keep the frogs moist. Cover the vented containers with a lid to prevent the frogs from escaping. If using tadpoles, place them individually in similar containers with less water (10 mL).

2. Expose the frogs or tadpoles to the irradiation source for a sufficient time to absorb 9 Gy (900 rads) of irradiation.

   In the Ramsey et al. (2010) study, the frogs in their containers were placed 80 cm from the source, and the dose was applied at a rate of 0.8 Gy/min for exactly 11.25 min.

3. At the end of the irradiation period, replace the original water with a small volume of fresh dechlorinated water sufficient to completely immerse the frogs (~150 mL). Dilute stock solution antibiotics such that the water contains 0.005 IU of penicillin and 0.005 µg of streptomycin as a precaution to limit possible bacterial infections due to the irradiation.

   See Troubleshooting.

TROUBLESHOOTING

Problem (Step 3): Because the treatment compromises the immune system, any asymptomatic infection may be revealed.

Solution: It is recommended to include a control group that is only sublethally irradiated and not undergoing any other treatment. Notably, when we investigated adaptive immunity against ranavirus pathogens, several of the controls that had only been irradiated (not infected experimentally) died from ranaviral infection, which led to the discovery of persistent asymptomatic ranaviral infections in a significant fraction of *X. laevis* obtained from different suppliers (Robert et al. 2007).

DISCUSSION

Xenopus laevis is a species frequently used as a model for studies of amphibian immunity to disease. There is a rich literature describing the complexity of the amphibian immune system, based largely on studies using *X. laevis*. (See Robert and Ohta 2009; Flajnik 2018 for recent reviews.)

Compared to mice for which 10 Gy is a lethal dose, *X. laevis* adults and tadpoles are relatively more resistant to radiation and tolerate 9 Gy of irradiation without significant mortality. The dose of sublethal irradiation recommended here is sufficient to reduce spleen cell numbers (Ramsey et al. 2010), impair tumor rejection (Rau et al. 2001), impair the ability to resist infection and reemergence of disease caused by ranaviruses (Robert et al. 2005, 2007), and impair the ability to resist infection caused by *B. dendrobatidis* (Ramsey et al. 2010). Analysis of the spleens of irradiated frogs showed that the numbers of leukocytes were markedly reduced by the irradiation (Ramsey et al. 2010). Irradiated frogs maintained their weight for greater than 45 d, suggesting that the irradiation did not impair the functions of the gastrointestinal tract (Ramsey et al. 2010). Irradiated frogs showed no evidence of skin damage due to the procedure (Rau et al. 2001; Ramsey et al. 2010). It is, however, possible to use a lower irradiation dose (6 Gy) to further minimize side effects on skin and intestine epithelia. These lower doses have been found to be effective for depletion of thymocytes in *X. laevis* adults and tadpoles, but they did not markedly affect the thymic stroma, a more radio-resistant compartment (Goyos et al. 2009). Similarly, a 10 Gy dose was also effective to inhibit survival of thymocytes of *X. tropicalis* (Goyos et al. 2011) and would likely be applicable to other nonmodel amphibians to help characterize adaptive immune responses to pathogens.

Cite this protocol as *Cold Spring Harb Protoc*; doi:10.1101/pdb.prot097626

REFERENCES

Flajnik MF. 2018. A cold-blooded view of adaptive immunity. *Nat Rev Immunol* **18:** 438–453.

Goyos A, Ohta Y, Guselnikov S, Robert J. 2009. Novel nonclassical MHC class Ib genes associated with CD8 T cell development and thymic tumors. *Mol Immunol* **46:** 1775–1786.

Goyos A, Sowa J, Ohta Y, Robert J. 2011. Remarkable conservation of distinct nonclassical MHC class I lineages in divergent amphibian species. *J Immunol* **186:** 372–381.

Ramsey JP, Reinert LK, Harper LK, Woodhams DC, Rollins-Smith LA. 2010. Immune defenses against a fungus linked to global amphibian declines in the South African clawed frog, *Xenopus laevis. Infect Immun* **78:** 3981–3992.

Rau L, Cohen N, Robert J. 2001. MHC-restricted and -unrestricted CD8 T cells. *Transplantation* **72:** 1830–1835.

Robert J, Ohta Y. 2009. Comparative and developmental study of the immune system in *Xenopus. Dev Dyn* **238:** 1249–1270.

Robert J, Morales H, Buck W, Cohen N, Marr S, Gantress J. 2005. Adaptive immunity and histopathology in frog virus 3-infected *Xenopus. Virology* **332:** 667–675.

Robert J, Abramowitz L, Gantress J, Morales HD. 2007. *Xenopus laevis*: A possible vector of ranavirus infection? *J Wildl Dis* **43:** 645–652.

RNAi-Mediated Loss of Function of *Xenopus* Immune Genes by Transgenesis

Eva-Stina Edholm[1] and Jacques Robert

Department of Microbiology and Immunology, University of Rochester Medical Center, Rochester, New York 14620

Generation of transgenic frogs through the stable integration of foreign DNA into the genome is well established in *Xenopus*. This protocol describes the combination of transgenesis with stable RNA interference as an efficient reverse genetic approach to study gene function in *Xenopus*. Initially developed in the fish medaka and later adapted to *Xenopus*, this transgenic method uses the I-SceI meganuclease, a "rare-cutter" endonuclease with an 18 bp recognition sequence. In this protocol, transgenic *X. laevis* with knocked down expression of a specific gene are generated using a double promoter expression cassette. This cassette, which is flanked by I-SceI recognition sites, contains the shRNA of choice under the control of the human U6 promoter and a green fluorescent protein (GFP) reporter gene under the control of the human EF-1α promoter. Prior to microinjection the plasmid is linearized by digestion with I-SceI and the entire reaction is then microinjected into one-cell stage eggs. The highly stringent recognition sequence of I-SceI is thought to maintain the linearized plasmid in a nonconcatamerized state, which promotes random integration of the plasmid transgene in the genome. The injected embryos are reared until larval stage 56 and then screened for GFP expression by fluorescence microscopy and assessed for effective knockdown by quantitative RT-PCR using a tail biopsy. Typically, the I-SceI meganuclease transgenesis technique results in 35%–50% transgenesis efficiency, a high survival rate (>35%) and bright nonmosaic GFP expression. A key advantage of this technique is that the high efficiency and nonmosaic transgene expression permit the direct use of F_0 animals.

MATERIALS

It is essential that you consult the appropriate Material Safety Data Sheets and your institution's Environmental Health and Safety Office for proper handling of equipment and hazardous materials used in this protocol.

RECIPES: Please see the end of this protocol for recipes indicated by <R>. Additional recipes can be found online at http://cshprotocols.cshlp.org/site/recipes.

Reagents

Agarose gel DNA extraction kit
Annealing buffer for shRNA (10×) <R>
Competent bacteria
DNA sequencing reagents
DNeasy Blood and Tissue Kit (QIAGEN) (optional; Step 15)

[1]Correspondence: Eva-Stina_Edholm@urmc.rochester.edu

Ethanol (optional; see Step 10)

Ficoll (lyophilized powder, Type 400-DL)

Gentamicin (50 mg/mL)

GFP-specific PCR primers (optional; Step 15)

I-SceI-GFP-huU6-I-SceI plasmid (available upon request from the *Xenopus laevis* Research Resource for Immunobiology, https://www.urmc.rochester.edu/microbiology-immunology/xenopus-laevis.aspx)

I-SceI meganuclease (New England Biolabs R06943)

Luria Bertani (LB) ampicillin agar plates

Modified Barth's saline (MBS) <R>

PCR reagents for standard and qRT-PCR and target gene-specific primers

Phenol:chloroform:isoamyl alcohol (25:24:1) (optional; see Step 10)

Plasmid DNA isolation kits (mini and midi or maxi scale)

Restriction endonucleases appropriate for experimental design

shRNA oligonucleotides (designed according to Steps 1–2)

T4 DNA ligase

U6 promoter sequencing primer

Water (DNase/RNase-free) (optional; see Step 10)

Equipment

Agarose gel electrophoresis apparatus

Beaker (3 L minimum)

DNA sequencing apparatus

Fluorescence microscope

Heat block

Hot plate

Incubator

Microcentrifuge tubes (1.5 mL)

Microinjector (PLI-100, Harvard Apparatus or equivalent)

PCR apparatus (for standard and quantitative PCR)

Petri dishes (60 × 15 mm and 35 × 12 mm)

Scalpel (optional; Step 15)

METHOD

shRNA Design and I-SceI-GFP-huU6-shRNA-I-SceI Plasmid Preparation

1. Design shRNA oligonucleotides according to experimental goals. Include restriction sites at the 5′ ends of each oligonucleotide to facilitate shRNA cloning between the huU6 promoter and the I-SecI restriction site of I-SceI-GFP-huU6-I-SceI. (See Fig. 1 for potential restriction sites, which include BbsI, HindIII, ClaI, SalI, XhoI, and KpnI.) Use the following guidelines to design effective shRNAs.

   - Determine whether the gene of interest has one or multiple splice variants and decide whether to target all splice forms of the gene or specific variants. Select exons for targeting accordingly.

   - Design the shRNA so that the first nucleotide is a purine.

   - Choose a sequence with low (30%–50%) GC content.

   - Avoid the first 50–100 nt downstream from the start codon and the 100 nt upstream of the stop codon. Also avoid 5′- and 3′-UTRs. These regions contain binding sequences for regulatory proteins that may affect the accessibility of the RNA target sequence.

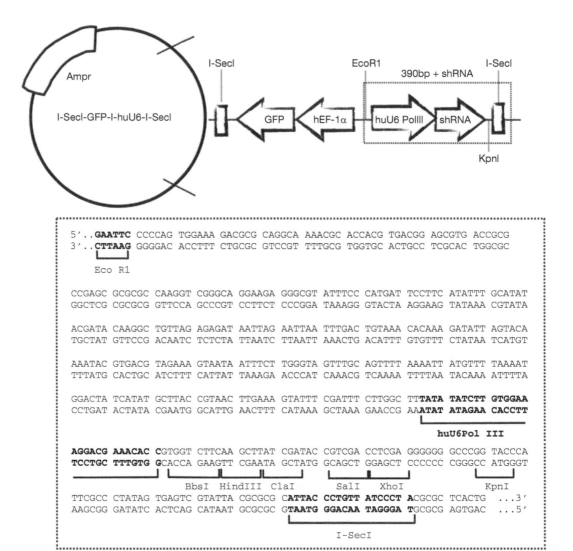

FIGURE 1. I-SceI-GFP-huU6-shRNA-I-SceI vector map and multiple-cloning site sequence. The sequence covering the shRNA cloning site includes the locations of potential restriction sites used to insert the shRNA into the vector.

- Design more than one shRNA per gene target.
- Ensure that the target sequence is specific for the gene of interest. Potential off targets can be identified by blast searching the shRNA sequence against the *X. laevis* genome available at http://www.xenbase.org/entry/.

  There are several commercial and noncommercial websites available to assist in designing knockdown shRNAs and appropriate controls, including http://www.invivogen.com/sirnawizard/, http://biodev. extra.cea.fr/DSIR/DSIR.html, and http://www.genelink.com/sirna/shRNAi.asp.

  shRNA oligonucleotide design is an empirical process and the gene silencing efficiency of each shRNA needs to be determined experimentally.

2. Design a control shRNA by randomly scrambling the target sequence.

   It is important to check the control shRNA for potential off target effects by blast searching the shRNA sequence against the X. laevis *genome.*

3. Synthesize or commercially obtain 40 nmol of each shRNA oligonucleotide in the 5′-phosphor-ylated form.

4. Resuspend 4 nmol of each shRNA in 10 µL 2× Annealing buffer and then combine the two oligonucleotides in a 1.5 mL microcentrifuge tube. Mix gently.

Cite this protocol as *Cold Spring Harb Protoc*; doi:10.1101/pdb.prot101519

5. Boil 2 L of water in a beaker. When water reaches boiling point turn off the hot plate and add the 1.5 mL microcentrifuge tube containing the oligonucleotides. Leave for a minimum of 6 h (or overnight) to allow oligonucleotides to anneal.

6. Double digest 1 μg of the I-SceI-GFP-huU6-I-SceI vector using restriction enzymes for the sites in the shRNA oligonucleotides. Digest for 1 h at 37°C. Heat inactivate digestion for 20 min at 65°C. Run the digest on a 1% agarose gel and perform gel purification of the digested vector using an agarose gel DNA extraction kit.

7. Ligate the annealed oligonucleotides from Step 5 into the digested I-SceI-GFP-huU6-I-SceI vector from Step 6 using T4 ligase. Combine 3.3 μL annealed shRNA, 3.3 μL digested I-SceI-GFP-huU6-I-SceI, 1 μL T4 ligase and 2 μL T4 10× buffer. Adjust volume to 20 μL with molecular grade H_2O. Ligate overnight at 16°C.

8. Transform 2.5 μL of the ligation reaction into competent bacteria using standard protocols. Select transformants on LB ampicillin agar plates.

9. Purify plasmid DNA from transformants using standard protocols and verify correct shRNA insertion by sequencing using U6 promoter primers.

10. Prepare plasmid DNA for digestion and microinjection using standard "midi- or maxi-prep" scale kits for isolation of pure plasmid DNA.

    Plasmid DNA can be further purified by phenol:chloroform:isoamyl alcohol (25:24:1) extraction, ethanol precipitation, and resuspension in DNase/RNase-free water.

11. Digest 1 μg I-SceI-GFP-huU6-shRNA-I-SceI with 10 units I-SceI meganuclease in a 25 μL volume for 40 min at 37°C.

    Store the I-SceI meganuclease at −80°C at all times (storage at −20°C leads to inefficient or no transgene incorporation into the genome). Freeze thawing of both the enzyme and the buffer significantly reduces transgenesis efficiency; therefore use single aliquots.

Microinjection of *Xenopus* Eggs

The microinjection of fertilized Xenopus eggs is well documented and the precise methodology used will depend on the type of equipment available.

12. Fertilize and dejelly eggs according to standard protocols and place eggs in a 35 × 12 mm injection Petri dish containing 0.3× MBS solution with 4% Ficoll. Ensure the eggs are completely submerged in the solution such that they are not subject to surface tension. Inject each egg with 10 nL of the digest from Step 11 containing 80 pg DNA and 1×10^{-3} U I-SceI.

    Use the I-SceI-GFP-huU6-shRNA-I-SceI digest within 15–30 min. If more eggs are to be injected, set up fresh digests. It is crucial that injection is performed within 30 min after fertilization and within 30 min after plasmid digestion to increase the likelihood of early integration. To avoid mosaicism it is advised that eggs are not injected once the first signs of cell division are observed.

13. Transfer injected eggs to a 60 × 15 mm Petri dish containing 0.3× MBS solution with 4% Ficoll and incubate for 4 h at 13°C.

    This delays cell division to extend the time for transgene integration.

14. Transfer the embryos to Petri dishes containing 0.3× MBS with 50 μg/mL gentamicin and rear the embryos at 18°C until hatching. Ensure that the embryos are completely submerged and not too crowded; between 50–100 embryos per dish is suitable.

    To ensure optimal survival of the embryos it is important to change the medium daily and to remove any dead embryos, which might otherwise compromise embryo survival.

    See Troubleshooting.

15. At developmental stage 58, screen tadpoles for GFP fluorescence using a fluorescence stereomicroscope.

    Alternatively, amputate the distal third of the tail with a scalpel and extract genomic DNA. Perform polymerase chain reaction (PCR) for the GFP transgene with GFP specific primers.

16. Select tadpoles with successful insertion of the GFP-containing vector and perform quantitative PCR to determine expression of the target gene. Verify target gene knockdown by comparing expression levels in shRNA-injected tadpoles with scrambled shRNA-injected age-matched controls.

See Troubleshooting.

TROUBLESHOOTING

Problem (Steps 14–15): Knockdown of a specific gene results in embryonic lethality.
Solution: Clone the shRNA of choice in a vector containing an H1 RNA polymerase III promoter repressed by the Tet-repressor. The expression of the shRNA can then be induced by inhibiting the tetracycline-element-specific repressor (TetR) from binding and blocking transcription with the tetracyclin analogue, doxycycline.

Problem (Step 16): Low to no knockdown of the gene of interest.
Solution: Design multiple shRNAs targeting the same gene and assess efficiency.

DISCUSSION

Different strategies to generate transgenic *Xenopus* frogs have been developed with specific molecular and technical features. The meganuclease approach is straightforward and results in high transgenesis efficiency with uniform expression of the transgene as well as stable germline transmission (Ogino et al. 2006a,b). One key advantage of this technique is that, because substantial numbers of nonmosaic embryos with high levels of knockdown are produced, it allows F_0 animals to be used for experimentation within a month after microinjection (Edholm et al. 2013). Another advantage of this knockdown technique is that different levels of silencing can be obtained in different animals, which may result in variable phenotypes. In addition, I-SceI meganuclease-mediated transgenesis has also been optimized for use with isogenic clones (LG 15 and LG 6 [Nedelkovska and Robert 2012; Nedelkovska et al. 2013]).

RELATED INFORMATION

We recommend Jensen et al. (2012) and Bofill-De Ros and Gu (2016) for in-depth discussion of shRNA design. These articles highlight important points to consider and potential pitfalls. In addition, see Ishibashi et al. (2012) for a detailed protocol describing I-SecI meganuclease-mediated transgenesis in *X. tropicalis*.

RECIPES

Annealing Buffer for shRNA (10×)

Reagent	Concentration
Tris-HCl (1 M, pH 7.5)	5 mL
NaCl (5 M)	10 mL
EDTA (0.5 M)	1 mL
H$_2$O	34 mL

Sterilize by autoclaving, and store at 4°C.

Cite this protocol as *Cold Spring Harb Protoc*; doi:10.1101/pdb.prot101519

Modified Barth's Saline (MBS)

$CaCl_2$ (0.1 M)

MBS salts (10×)

For a 1× solution of MBS, mix 100 mL of 10× MBS salts with 7 mL of 0.1 M $CaCl_2$. Adjust the volume up to 1 liter with H_2O. Store at room temperature.

MBS Salts (10×)

NaCl (880 mM)

KCl (10 mM)

$MgSO_4$ (10 mM)

HEPES (50 mM, pH 7.8)

Omit HEPES if MBS is to be used for oocyte maturation.

$NaHCO_3$ (25 mM)

Adjust pH to 7.8 with NaOH. Autoclave. Store at room temperature.

REFERENCES

Bofill-De Ros X, Gu S. 2016. Guidelines for the optimal design of miRNA-based shRNAs. *Methods* **103:** 157–166.

Edholm ES, Albertorio Saez LM, Gill AL, Gill SR, Grayfer L, Haynes N, Myers JR, Robert J. 2013. Nonclassical MHC class I-dependent invariant T cells are evolutionarily conserved and prominent from early development in amphibians. *Proc Natl Acad Sci* **110:** 14342–14347.

Ishibashi S, Love NR, Amaya E. 2012. A simple method of transgenesis using I-SceI meganuclease in *Xenopus*. *Methods Mol Biol* **917:** 205–218.

Jensen SM, Schmitz A, Pedersen FS, Kjems J, Bramsen JB. 2012. Functional selection of shRNA loops from randomized retroviral libraries. *PLoS One* **7:** e43095.

Nedelkovska H, Robert J. 2012. Optimized transgenesis in *Xenopus laevis/gilli* isogenetic clones for immunological studies. *Genesis* **50:** 300–306.

Nedelkovska H, Edholm ES, Haynes N, Robert J. 2013. Effective RNAi-mediated β2-microglobulin loss of function by transgenesis in *Xenopus laevis*. *Biol Open* **2:** 335–342.

Ogino H, McConnell WB, Grainger RM. 2006a. High-throughput transgenesis in *Xenopus* using I-SceI meganuclease. *Nat Protoc* **1:** 1703–1710.

Ogino H, McConnell WB, Grainger RM. 2006b. Highly efficient transgenesis in *Xenopus tropicalis* using I-SceI meganuclease. *Mech Dev* **123:** 103–113.

Protocol 8

Elicitation of *Xenopus laevis* Tadpole and Adult Frog Peritoneal Leukocytes

Leon Grayfer[1]

Department of Biological Sciences, George Washington University, Washington, D.C. 20052

Peritoneal lavage of *Xenopus laevis* tadpoles and adult frogs is a reliable way of isolating resident and/or recruited innate immune populations. This protocol details the isolation of tadpole and adult amphibian (*Xenopus laevis*) peritoneal leukocytes. The isolated cells are comprised predominantly of innate immune populations and chiefly of mononuclear and polymorphonuclear granulocytes. As described here, these cells are typically elicited by peritoneal injections of animals with heat-killed *Escherichia coli*, causing peritoneal accumulation of inflammatory cell populations, which are then isolated from the stimulated animals by lavage. *E. coli*-mediated elicitation of tadpole and adult peritoneal leukocytes greatly enhances the total numbers of recovered cells, at the cost of their inflammatory activation. Conversely, lavage may be performed on naïve, unstimulated animals to isolate nonactivated cells with much lower yield. This protocol represents a reliable means of deriving tadpole and adult frog innate immune cell populations, and the conditions of the stimulation may be amended to suit the specifics of a given experimental design.

MATERIALS

It is essential that you consult the appropriate Material Safety Data Sheets and your institution's Environmental Health and Safety Office for proper handling of equipment and hazardous materials used in this protocol.

RECIPES: Please see the end of this protocol for recipes indicated by <R>. Additional recipes can be found online at http://cshprotocols.cshlp.org/site/recipes.

Reagents

Amphibian phosphate-buffered saline (APBS) <R>
Complete-amphibian medium (C-AM) <R>
E. coli (DH5α or alternative laboratory strain)
LB medium <R>
Tricaine mesylate (0.1%) buffered with sodium bicarbonate (0.5 g/L)
Tricaine mesylate (1%) buffered with sodium bicarbonate (0.5 g/L)
Trypan blue
Xenopus laevis tadpoles (stage of choosing) or adult frogs (Nieuwkoop and Faber 1967)

Equipment

Aquarium filter floss or sponge
Bunsen burner

[1]Correspondence: leon_grayfer@gwu.edu

Centrifuge (preferably refrigerated)
Centrifuge tubes (1.5, 15, and 50 mL)
Hemocytometer
Needles (25G, 1 inch; 18G, 1.5 inch)
Parafilm
Pasteur pipettes
Plastic spoon
Rubber bulb (small)
Shaking incubator (37°C)
Syringe (1 mL and 10 mL)

METHOD

Preparation of Heat-Killed *E. coli*

1. Inoculate 40 mL of LB medium with DH5α (or alternative laboratory strain) *E. coli* and grow while shaking overnight at 37°C.

2. The following day collect the bacteria by centrifugation at 6000*g* for 10 min at 4°C, resuspend the pellet in 4 mL of APBS, and boil for 1 h at 100°C to heat-kill the bacteria.

3. Cool the heat-killed *E. coli* on ice and make 0.5 mL aliquots. Use immediately or store at −20°C.

4. Proceed to Step 5 to inject tadpoles and Step 21 to inject adult frogs.

Injection of *Xenopus laevis* Tadpoles with Heat-Killed *E. coli* and Peritoneal Leukocyte Collection

5. The day before injections, isolate the tadpoles to be used in the experiment from general housing and house separately overnight.

6. Prepare glass injection needles by pulling Pasteur pipettes to a fine point over a flame of a Bunsen burner.

7. The following day, anesthetize the tadpoles by immersion in 0.1% tricaine mesylate solution buffered with sodium bicarbonate (0.5 g/L).

   Tadpoles will cease movement after 2–5 min. Wait an additional 1–2 min after cessation of movement to ensure that animals are fully sedated.

8. Pipette the heat-killed *E. coli* in 5 (for stage NF 50 tadpoles) to 10 µL (for stage NF 54+ tadpoles) droplets onto a wide section of parafilm on a sturdy surface.

9. Gently place anesthetized tadpoles on their side on top of moist aquarium filter floss or a moist sponge.

10. Attach a finely pulled needle to a small rubber bulb and carefully take up the heat-killed *E. coli* droplet into the tip of the needle, taking care not to take up the suspension past the tip of the needle.

    Make sure that you are comfortable taking up and expelling small volumes with a glass needle and a rubber bulb before you begin this procedure.

11. Using the thumb and forefinger of your nondominant hand, gently distend the tadpole's abdomen (Fig. 1A).

12. Using your dominant hand, carefully insert the *E. coli*-containing needle 1–2 mm into the tadpole's abdomen and carefully expel 95% of the volume into the animal by gently squeezing the bulb.

    Take care not to inject the entire volume into the tadpole, thus avoiding possible introduction of air into the animal.

13. Using a plastic spoon, pick up the injected tadpole and place into water. Monitor until the animal regains movement.

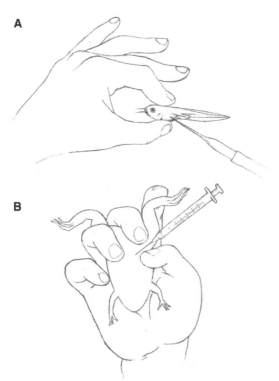

FIGURE 1. (*A*) Peritoneal injection and lavage of a *X. laevis* tadpole. (*B*) Peritoneal injection and lavage of adult *X. laevis*.

14. Three days after *E. coli* injection, anesthetize the tadpole(s) by immersion in 0.1% tricaine mesylate solution buffered with sodium bicarbonate (0.5 g/L).

    Tadpoles will cease movement after 2–5 min. Wait an additional 1–2 min after cessation of movement to ensure that animals are fully sedated.

    Note that peritoneal leukocytes begin to accumulate within the first 24 h of injection (primarily granulocytes) and isolating cells 3 d after injection is a suggested time.

15. Pipette APBS in 50 μL droplets onto a wide section of parafilm.

16. Gently place anesthetized tadpole(s) on their side on top of moist aquarium filter floss or a moist sponge.

17. Following the procedures described in Steps 10–12, take up approximately one-third of the APBS droplet into the tip of the needle (using a small rubber bulb) and inject it into the tadpole, preferably into the same needle entry site.

18. Gently pull the liquid back into the needle and eject it into a 1.5 mL microfuge tube. Repeat twice, with fresh APBS volumes and collect the total 50 μL of lavaged fluid into the same tube over ice.

19. Collect the isolated leukocytes by centrifugation at 600*g* for 10 min at 4°C. Wash the cells with 1 mL APBS and collect by centrifugation at 600*g* for 10 min at 4°C. Discard supernatants.

20. Resuspend the cells in a small volume (100–200 μL) of complete amphibian medium and enumerate by hemocytometer counts using trypan blue exclusion viability stain.

    Using stage NF 54 tadpoles, the yield should be ~1–5 × 10⁶ cells.

Injection of Adult *Xenopus laevis* with Heat-Killed *E. coli* and Peritoneal Leukocyte Collection

21. The day before injections, isolate the animals to be used in the experiment from general housing and house separately overnight.

22. The following day, anesthetize the adult *X. laevis* by immersion in 1% tricaine mesylate solution buffered with sodium bicarbonate (0.5 g/L).

Cite this protocol as *Cold Spring Harb Protoc*; doi:10.1101/pdb.prot097642

Frogs will cease movement after 2–5 min. Wait an additional 1–2 min after cessation of movement to ensure that animals are fully sedated.

If you are comfortable with handling the animals, you may inoculate them with E. coli without sedation, if injections are performed quickly.

23. Take up an appropriate amount of the prepared heat-killed *E. coli* suspension into a 1 mL syringe fitted with a 25G 1 inch needle and place within reach.

24. Pick up individual anesthetized frogs with your nondominant hand so that the top of their head is resting in the palm of your hand and their legs are dangling on either side of your middle finger (Fig. 1B).

25. Distend the frog's abdomen by gently putting pressure on their sides with your thumb and pinky finger (Fig. 1B).

26. With your dominant hand, pick up the syringe with the heat-killed *E. coli*, gently pierce the frog's lower abdomen (2–3 mm), and make sure not to insert the needle too far into the animal. Inject 100 µL of the *E. coli* suspension into the animal.

27. Immediately place the animal in a shallow container of water, so that its nostrils are exposed to air but the majority of its body is submerged underwater.

28. Monitor the animal(s) as they recover from sedation, making sure that the exposed areas of their bodies are moist (to prevent desiccation) and that their nostrils are above water at all times. Alternatively, you may keep the sedated frogs covered with wet filter floss, to ensure that they do not dry out.

29. Once the animals recover movement, place them back into separate housing and monitor.

30. Three days after *E. coli* injection, anesthetize the adult frogs by immersion in 1% tricaine mesylate solution buffered with sodium bicarbonate (0.5 g/L).

    Frogs will cease movement after 2–5 min. Wait an additional 1–2 min after cessation of movement to ensure that animals are fully sedated.

    Note that peritoneal leukocytes begin to accumulate within the first 24 h of injection (primarily granulocytes) and isolating cells 3 d after injection is a suggested time.

31. Fill a 10 mL syringe fitted with an 18G 1.5 inch needle with APBS and place within reach.

32. Repeat Steps 24–26, injecting 5 mL of APBS (rather than *E. coli*) into the frog's peritoneum, preferably into the same needle entry site.

33. Carefully unscrew/unhinge the needle from the syringe while making sure that the needle remains lodged in the frog's abdomen.

    When detaching the syringe, take care not to move the needle further into the animal.

34. Hold the animal above an opened 15 mL conical tube (on ice) so that the wide part of the needle is directly above the tube. While firmly holding the animal, gently massage its sides, forcing the injected APBS to drip out of the needle. If necessary, gently twist and adjust the positioning of the needle within the incision to promote the leukocyte-containing APBS recovery.

35. When 4.5–5 mL of APBS has been collected from the animal, gently reattach the 10 mL syringe containing the other 5 mL of APBS to the needle while it is still in the animal.

36. Inject the frog with the remainder of the APBS and recover the peritoneal leukocytes as above. Collect both 5 mL fractions into the same 15 mL tube.

37. Monitor the animal as in Steps 27–29.

38. Collect the isolated leukocytes by centrifugation at 600*g* for 10 min at 4°C, wash with 1 mL APBS and again collect by centrifugation at 600*g* for 10 min at 4°C. Discard supernatant.

39. Resuspend the cells in complete amphibian medium (300–500 µL) and count using a hemocytometer and trypan blue exclusion viability stain.

    Using a 2–2.5 inch adult X. laevis, the yield should be $1–2 \times 10^7$ leukocytes.

DISCUSSION

The use of peritoneal lavage of *Xenopus laevis* tadpoles and adult frogs to isolate resident and/or recruited innate immune populations is described further in De Jesús Andino et al. (2012), Fites et al. (2013), and De Jesús Andino et al. (2016). Both tadpole and adult frog resident cells appear to be comprised predominantly of mononuclear phagocytes. Elicitation of peritoneal cells with heat-killed *E. coli* accumulates both macrophage- and granulocyte-lineage cells in animal peritonea, so these are the populations that are retrieved upon peritoneal lavage of these animals (Du Pasquier et al. 1985; Morales et al. 2010). Elicitation with heat-killed *E. coli* before lavage results in significantly greater cell yields than achieved when performing peritoneal lavage on unstimulated animals. However, it should be noted that the derived cells are activated and no longer represent the resident peritoneal leukocyte populations but rather inflammatory infiltrates, with elevated expression of inflammatory immune genes. These cells may be returned to relatively more baseline states by incubating them in complete amphibian medium for 24 h at 27°C with 5% CO_2 before commencing subsequent experimentation. Alternatively, animals can be lavaged without the heat-killed *E. coli* elicitation. While the total peritoneal leukocyte yields will be substantially lower ($1–5 \times 10^5$ cells for tadpoles and $1–5 \times 10^6$ cells from adult frogs, respectively), these cells will be representative of respective tadpole and adult frog resident peritoneal populations and not activation-biased by exposure to an inflammatory stimulus (heat-killed *E. coli*). Finally, animals may be injected with putative leukocyte chemo-attractants and/or growth factors. In this case injections and lavage would be performed using the methods and volumes specified above.

RELATED INFORMATION

Related reagents and resources are available via the *Xenopus laevis* research resource for immunology: https://www.urmc.rochester.edu/microbiology-immunology/xenopus-laevis/protocols.

RECIPES

Amphibian Phosphate-Buffered Saline (APBS)

Sodium chloride (NaCl)	6.6 g/L
Sodium phosphate (Na_2HPO_2)	1.15 g/L
Potassium phosphate (KH_2PO_4)	0.2 g/L

Adjust pH to 7.7 with 10 N NaOH, and filter-sterilize through a 0.2-µm filter. Store at room temperature for up to 6 mo.

Complete-Amphibian Medium (C-AM)

Mammalian serum-free medium <R>	200 mL
Triple-distilled H_2O (purchased or from filtering system)	60 mL
Fetal bovine serum	5 mL
Penicillin-streptomycin (10,000 units/mL; Gibco, 15140122)	5 mL
Gentamycin (50 mg/mL)	80 µL

Combine all reagents, filter-sterilize through a 0.2-µm filter, and store at 4°C.

Cite this protocol as *Cold Spring Harb Protoc*; doi:10.1101/pdb.prot097642

LB (Luria-Bertani) Liquid Medium

Reagent	Amount to add
H_2O	950 mL
Tryptone	10 g
NaCl	10 g
Yeast extract	5 g

Combine the reagents and shake until the solutes have dissolved. Adjust the pH to 7.0 with 5 N NaOH (~0.2 mL). Adjust the final volume of the solution to 1 L with H_2O. Sterilize by autoclaving for 20 min at 15 psi (1.05 kg/cm^2) on liquid cycle.

For solid medium, see the recipe entitled "Media containing agar or agarose."

Mammalian Serum-Free Medium (MSF)

Reagent	Quantity (for 1 L)
Iscove's modified Dulbecco's medium, powder	1 pkg
Penicillin-streptomycin (10,000 units/mL; Gibco, 15140122)	10 mL
Non-essential amino acids solution (10 mM; Gibco, 11140050)	10 mL
Insulin from bovine pancreas (5 mg/mL)	10 mL
2-mercaptoethanol	3 µL
Peptone Primatone RL (10% in H_2O)	3 mL
NaHCO$_3$	3.02 g

Adjust pH to 7.7 with 10 N NaOH, and filter-sterilize through a 0.2-µm filter. Store at 4°C.

REFERENCES

De Jesús Andino F, Chen G, Li Z, Grayfer L, Robert J. 2012. Susceptibility of *Xenopus laevis* tadpoles to infection by the ranavirus Frog-Virus 3 correlates with a reduced and delayed innate immune response in comparison with adult frogs. *Virology* **432:** 435–443.

De Jesús Andino F, Jones L, Maggirwar SB, Robert J. 2016. Frog Virus 3 dissemination in the brain of tadpoles, but not in adult *Xenopus*, involves blood brain barrier dysfunction. *Sci Rep* **6:** 22508.

Du Pasquier L, Flajnik MF, Guiet C, Hsu E. 1985. Methods used to study the immune system of *Xenopus* (Amphibia, Anura). *Immunol Methods* **3:** 425–465.

Fites JS, Ramsey JP, Holden WM, Collier SP, Sutherland DM, Reinert LK, Gayek AS, Dermody TS, Aune TM, Oswald-Richter K, et al. 2013. The invasive chytrid fungus of amphibians paralyzes lymphocyte responses. *Science* **342:** 366–369.

Morales HD, Abramowitz L, Gertz J, Sowa J, Vogel A, Robert J. 2010. Innate immune responses and permissiveness to ranavirus infection of peritoneal leukocytes in the frog *Xenopus laevis*. *J Virol* **84:** 4912–4922.

Nieuwkoop PD, Faber J. 1967. *Normal table of* Xenopus laevis *(Daudin)*. North Holland Publishing, Amsterdam.

Assessing Antibody Responses to Pathogens or Model Antigens in *Xenopus* by Enzyme-Linked Immunosorbent Assay (ELISA)

Francisco De Jesús Andino[1] and Jacques Robert[1]

Department of Microbiology and Immunology, University of Rochester Medical Center, Rochester, New York 14642

Xenopus laevis-specific monoclonal antibodies recognize IgM and IgY antibodies not only from *X. laevis* but also *X. tropicalis* as well as a variety of amphibian species including *Ranidae*, *Bufonidae*, and even some salamanders. These reagents are very useful to assess antibody responses from the serum or other animal secretions (e.g., peritoneal fluid). We present here an enzyme-linked immunosorbent assay (ELISA) optimized for amphibians that permits users to detect and titrate the presence of each type of antibody (IgM and IgY) produced against particular pathogens (e.g., virus, bacteria, or fungus) or antigens (e.g., DNP-KLH).

MATERIALS

It is essential that you consult the appropriate Material Safety Data Sheets and your institution's Environmental Health and Safety Office for proper handling of equipment and hazardous materials used in this protocol.

Reagents

Amphibian PBS (1× PBS diluted by adding 30% v/v distilled water)
This is used only for injection in the amphibian recipient.

Blocking buffer (1% BSA in 1× PBS)
ELISA substrate (3,3′,5,5′-tetramethylbenzidine [TMB])
Freund's adjuvant (optional; see Step 1)
H_2SO_4 (1 M)
LB broth (optional; see Step 6)
Pathogen or antigen of choice (see preparation instructions in Steps 1 and 6):

Chytrid fungus (*Batrachochytrium dendrobatidis*) (Ramsey et al. 2010)
Dinitrophenylated keyhole limpet hemocyanin (DNP-KLH) (Du Pasquier et al. 1985)
Heat-killed *E. coli* bacteria (e.g., Stratagene XL-blue) (Robert et al. 2014)
Mycobacterium (*Mycobacterium marinum*) (Shirtcliffe et al. 2004)
Ranavirus (frog virus 3 [FV3]) (De Jesús-Andino et al. 2016)
PBS (10×, mammalian; e.g., OmniPur 10× PBS, premixed powder)
Dilute in ddH$_2$O, filter through 0.2-µm sterile filter flask, and store at 4°C.

PenStrep (Gibco 15070-063) or ethacridine lactate (10 µg/mL)
Secondary antibody (goat anti-mouse IgG-horseradish peroxidase conjugated [IgG-HRP])

---

[1]Correspondence: francisco_dejesus@urmc.rochester.edu; jacques_robert@urmc.rochester.edu

Copyright © 2023 Cold Spring Harbor Laboratory Press; all rights reserved
Cite this protocol as *Cold Spring Harb Protoc*; doi:10.1101/pdb.prot099234

Tricaine methane sulfonate (TMS; 0.1% or 1 g/L; Western Chemicals MS-222) buffered with 0.5 g/L sodium bicarbonate (Fisher Scientific S-2333)

We use sodium bicarbonate to keep the pH of the final solution near the ambient environmental pH for Xenopus laevis.

Wash buffer (1× PBS containing 0.05% Tween 20)

To increase stringency, 1% BSA as well as 1% NaCl can be added.

Xenopus laevis adults

Xenopus laevis-specific antibodies (mouse monoclonal antibodies 11D5 (IgY) and 10A9 (IgM), which are available upon request from the *Xenopus laevis* Research Resource for Immunobiology [https://www.urmc.rochester.edu/microbiology-immunology/xenopus-laevis.aspx])

Equipment

96-well plates (sterile, flat-bottomed, polystyrene)
Benchtop centrifuge (Beckman Coulter Allegra X-30R centrifuge)
Conical centrifuge tubes (10-mL)

Glass tubes are preferable for optimal blood coagulation.

ELISA reader instrument (SpectraMax M5 with Softmax Pro 6.4 software, optical system monochromator, and xenon flash lamp, with the fluorescence intensity at 420 nm and 25°C)
Needles (22G, 1½ inch)
Pulled glass needle, sterile (Du Pasquier et al. 1985; Nedelkovska et al. 2010)
Syringes, sterile (1-mL)

METHOD

Immunization

1. Immunize frogs by intraperitoneal injection in the abdominal region using a 1 mL sterile syringe with a 22 G, 1½ inch needle of the following pathogens or antigens diluted in amphibian PBS.

   - Ranavirus FV3: 1×10^6 PFU in 100 µL volume of amphibian PBS per adult frog; (Maniero et al. 2006).

   - *Mycobacterium marinum:* 1×10^5 CFU in 100 µL volume of amphibian PBS per adult frog.

   - Heat-killed *E. coli:* 100 µL of 10^8 bacteria/mL in amphibian PBS; (Robert et al. 2014) or *Batrachochytrium dendrobatidis:* 10 µL/g body weight of heat-killed *B. dendrobatidis* (mixed zoospores and maturing sporangia) at a concentration of 5×10^7 cells/mL (Ramsey et al. 2010).

   - DNP-KLH: doses can range from 2–10 µg/g body weight emulsified in complete Freund's adjuvant (Du Pasquier et al. 1985)

2. After immunization with the pathogen or antigen, anesthetize frogs by immersion in a 0.1% tricaine methane sulfonate buffered with 0.5 g/L sodium bicarbonate. Leave frogs in the TMS for up to 5 min or until all movement ceases.

3. Collect blood from the dorsal tarsus vein of adult *Xenopus laevis* using a sterile pulled glass needle as described (Du Pasquier et al. 1985; Nedelkovska et al. 2010). One to two mL of blood can be obtained from one average sized (~200 g) frog. Collect the blood in a 10 mL conical glass centrifuge tube placed on ice. After bleeding, place the frog(s) in water containing an antiseptic (2.5 mL/L PenStrep or ethacridine lactate) for a day (Du Pasquier et al. 1985; Nedelkovska et al. 2010).

4. Let the blood coagulate overnight at 4°C in a tube closed with a cap.

5. In a benchtop centrifuge, centrifuge the blood for 15 min at 1000g at 4°C. Collect the serum (supernatant), and store at −20°C until use.

Antigen or Pathogen Absorption

6. Dilute pathogen or antigen used to immunize frogs in 1× PBS pH 8.0 unless otherwise noted (see below). Place 100 μL of diluted antigen per well of a 96-well plate and incubate overnight at 4°C.

   The number of wells will depend of the number of assays you wish to perform. See Step 8 to plan appropriate controls.

   - Ranavirus FV3: Grow and purify live FV3 from baby hamster kidney (BHK-21) cell lines incubated at 30°C for 5–6 d as previously described (De Jesús-Andino et al. 2016). Dilute virus (0.5 to 1×10^7 PFU per well) in 100 μL 1× PBS (Maniero et al. 2006).
   - *Mycobacterium marinum* (1×10^2 to 1×10^7 CFU): Boil *M. marinum* for an hour and sonicate for 2 min (Shirtcliffe et al. 2004). Pellet *M. marinum* by centrifugation for 15 min at 3500 rpm and resuspend in 1× PBS + 0.05% Tween 80.
   - *E. coli*: prepare an overnight culture in 25 mL LB broth and incubate at 37°C. Boil the culture for 1 h and centrifuge for 15 min at 3500 rpm at 4°C. Resuspend in 2.5 mL of 1× PBS (∼10^8 bacteria/mL) (Robert et al. 2014).
   - *Batrachochytrium dendrobatidis* (JEL 197): heat-kill for 20 min at 60°C. Dilute fungus in 1× PBS with ∼5×10^4 cells per well (Ramsey et al. 2010).
   - DNP-KLH: dilute in 1× PBS (1–10 μg/mL; Du Pasquier et al. 1985).

Blocking, Antibody Incubations, and Development

7. Remove solution from Step 6 from all wells and wash each well by adding 200 μL of blocking buffer per well. Incubate for 10 min at room temperature and discard the supernatant. Perform this wash three times as described.

   Unbound antigen/pathogen will be removed at this step.

8. Add 100 μL/well of *Xenopus* (or other species) serum dilutions (2 to 3 dilutions between 1:50–1:1000) from immunized (Step 1) and naïve animals. Use the following dilutions: viral infection: 1:50 to 1:200 dilutions; bacterial immunization: 1:100 to 1:1000 dilutions; fungal immunization: 1:100 to 1:1000 dilutions; or DNP-KLH immunization: 1:100 to 1:1000 dilutions in blocking buffer. Each sample should be tested in triplicate. It also very important to always coat with negative and positive controls (note that all wells will be coated with antigen from Step 6).

   - Negative controls can include: (i) normal nonimmunized *Xenopus* serum; and/or (ii) PBS containing 1% bovine serum albumin (BSA).
   - Positive controls can include: (i) an antiserum known to contain antibodies, serving as a control for the binding of the *Xenopus*-specific secondary monoclonal antibody; (ii) the secondary mAbs (10A9 or 11D5) serving as control for the binding and signal of tertiary HRP-conjugated goat anti-mouse antibody.
   - Specificity controls coated with the antigen (pathogen or DNP-KLH) but without antiserum (i) or without secondary mAbs (ii).

9. Incubate 1–3 h at room temperature or overnight at 4°C.

10. Remove the extra serum. Wash each well 5× for 10 min each wash with 100 μL wash buffer at room temperature.

11. Add 100 μL/well of 11D5 (to detect antigen-specific IgY) supernatant or 10A9 (to detect antigen-specific IgM) supernatant containing 50–100 μg/mL of the primary antibody diluted 2× in blocking buffer. Incubate for 2 h at room temperature or overnight at 4°C.

Cite this protocol as *Cold Spring Harb Protoc*; doi:10.1101/pdb.prot099234

12. Remove the unbound or extra supernatant. Wash each well 5× for 10 min each wash with 100 µL wash buffer at room temperature.

13. Add 100 µL/well of the goat anti-mouse IgG-HRP diluted 1:5000 in blocking buffer. Incubate for 1 h at room temperature.

14. Wash each well 6 to 8× with 100 µL wash buffer for 10 min each wash at room temperature.

15. Incubate each well with 100 µL of 1 Step Ultra TMB for 30–60 min at room temperature.

16. Block and incubate reaction with 100 µL of 1 M H_2SO_4 for ~5 min. Read plates at 420 nm. (The different reagents used and steps are depicted in Fig. 1.)

    See Troubleshooting.

TROUBLESHOOTING

Problem (Step 16): There is too much background (e.g., signal detected in negative control).
Solution: Increase the stringency of the washes by increasing the molarity of the NaCl and/or by increasing the percent of detergent.

Problem (Step 16): No signal upon plate reading.
Solution: The different controls should permit users to determine whether secondary or tertiary antibody is working.

DISCUSSION

ELISA provides a reliable and very sensitive method to monitor an immune response. Since blood samples can be collected multiple times from the same animal, it is an ideal technique to determine the kinetics of an immune response. When planning to incorporate an ELISA in any experiments there are some aspects important to consider.

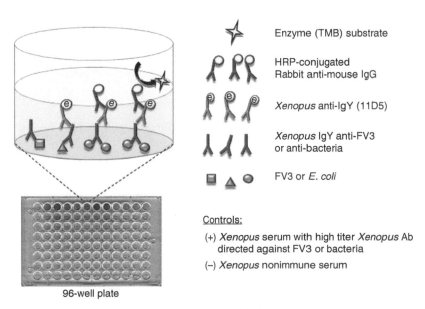

Enzyme (TMB) substrate

HRP-conjugated
Rabbit anti-mouse IgG

Xenopus anti-IgY (11D5)

Xenopus IgY anti-FV3
or anti-bacteria

FV3 or *E. coli*

Controls:

(+) *Xenopus* serum with high titer *Xenopus* Ab
directed against FV3 or bacteria

(–) *Xenopus* nonimmune serum

96-well plate

FIGURE 1. Schematic of ELISA depicting the different reagents added sequentially into a culture well of a 96-well flat-bottom plates. The sequential addition of reagents is listed from bottom to top on the right.

Type of ELISA, Plates and Consistency

They are four general types of ELISAs: direct (using labeled primary antibody), indirect (involving two binding processes of primary antibody and labeled secondary antibody), sandwich (quantifying antigens between two layers of antibodies, a capture and detection antibody), and competitive (using a second antigen to compete and determine the binding specificity). The type of ELISA used for a particular experiment will depend on multiple factors including the complexity of the experimental samples, the reagents or antibodies available (e.g., availability of a secondary antibody directed against the primary antibody) and the level of sensitivity required (e.g., indirect and sandwich ELISA are more sensitive than direct ELISA because the binding of secondary and tertiary antibodies to additional targets will enhance the signal.

One of the most common plates used for ELISA assays is the flat-bottomed, 96-well polystyrene plate based on its consistency, minimizing edge effects (e.g., overlap toward the outer edges of the well during the analysis, more uniform cell layer compared to the round-bottomed plates) and giving optimal optical conditions for the data collection.

Using multiwell plates, multichannel pipettes, and reservoirs (plate washers) will provide a high consistency and faster results. Improperly calibrated or dirty pipettes will cause cross contamination in your samples, resulting in variation in your results. Also, it is important to make sure that the levels of the samples in your multichannel pipette match, as sometimes tips are not well-attached to the pipette, which will affect your results.

Antisera

Blood coagulation can be influenced by the type of tube used. Conical glass tubes are the most suitable for blood coagulation. However, plastic tubes can be used.

Secondary Antibodies

They are usually two types of secondary antibodies used for ELISA: monoclonal and polyclonal antibodies. Because a monoclonal antibody recognizes a single epitope it is more specific but if it cross-reacts nonspecifically with proteins in the assay, it can be challenging to reduce this nonspecific binding. Monoclonal antibodies are more likely than polyclonal antibodies to contain a fraction of nonspecific cross-reacting antibodies. However, this nonspecific background can usually be reduced by diluting the antibody. It also possible to absorb the antibody on cells (e.g., erythrocytes, splenocytes) from the species from which the sample tested originates before using it for this assay. It is also important to test samples in duplicate or triplicate including known standards (positive and negative controls) and to test several dilutions for optimal results and quantitation.

Coating, Washing, and Blocking Buffers

For coating the antigen, buffer controlling the pH such as PBS or Tris-HCl containing 1% NaCl are preferred. A slightly basic pH (8.0) is usually optimal for the maximal binding of antigens. The blocking buffer is important to prevent nonspecific binding of the antibodies used, especially against the plastic wall of the multiwell plate itself. This can affect the sensitivity of the assay (nonspecific signal) detected by the ELISA instrument. In addition to blocking, it is very important to wash extensively between each step of the ELISA. Wash buffer should be the same PBS or Tris-HCl saline type containing a nonionic detergent such as Tween 20 to prevent binding by the negative electric charges of the antibodies. The wash buffer is used to remove nonbound reagents, decreasing background and enhancing specific signal. It is recommended to wash at least 3–5 times between each step, but more wash cycles can be added to decrease the background. The small amounts of the retained buffer in the multiwell plate should be removed to prevent dilution of the reagents used. This can be done by careful aspiration or tapping the plate upside down on an absorbent paper. Insufficient or excessive washing may lead to high background signal or decreased sensitivity, respectively.

Cite this protocol as *Cold Spring Harb Protoc;* doi:10.1101/pdb.prot099234

REFERENCES

Du Pasquier L, Flajnik M, Guiet C, Hsu E. 1985. Methods used to study the immune system of *Xenopus*. *Immunol Methods* **3**: 425–464.

De Jesús-Andino F, Letitia L, Maggirwar S, Robert J. 2016. Frog Virus 3 dissemination in the brain of tadpoles, but not in adult *Xenopus*, involves blood brain barrier dysfunction. *Sci Rep* **6**: 22508.

Maniero G, Morales H, Gantress J, Robert J. 2006. Generation of a long-lasting, protective, and neutralizing antibody response to the ranavirus FV3 by the frog *Xenopus*. *Dev Comp Immunol* **30**: 649–657.

Nedelkovska H, Cruz-Luna T, McPherson P, Robert J. 2010. Comparative *in vivo* study of gp96 adjuvanticity in the frog *Xenopus laevis*. *J Vis* **16**: 2026.

Ramsey JP, Reinert LK, Harper LK, Woodhams DC, Rollins-Smith LA. 2010. Immune defenses against *Batrachochytrium dendrobatidis*, a fungus linked to global amphibian declines, in the South African clawed frog, *Xenopus laevis*. *Infect Immunol* **78**: 3981–3992.

Robert J, Grayfer L, Edholm ES, Ward B, De De Jesús-Andino F. 2014. Inflammation-induced reactivation of the ranavirus Frog Virus 3 in asymptomatic *Xenopus laevis*. *PLoS One* **9**: e112904.

Shirtcliffe PM, Easthope SE, Weatherall M, Beasley R. 2004. Effect of repeated intradermal injections of heat-inactivated *Mycrobacterium bovis* Bacillus Calmette-Guerin in adult asthma. *Clin Exp Allergy* **34**: 207–212.

Xenopus, a Model to Study Wound Healing and Regeneration: Experimental Approaches

Paula G. Slater, Miriam Palacios, and Juan Larraín[1]

Center for Aging and Regeneration, Facultad de Ciencias Biológicas, P. Universidad Católica de Chile, Santiago de Chile, Chile 7820436

Xenopus has been widely used as a model organism to study wound healing and regeneration. During early development and at tadpole stages, *Xenopus* is a quick healer and is able to regenerate multiple complex organs—abilities that decrease with the progression of metamorphosis. This unique capacity leads us to question which mechanisms allow and direct regeneration at stages before the beginning of metamorphosis and which ones are responsible for the loss of regenerative capacities during later stages. *Xenopus* is an ideal model to study regeneration and has contributed to the understanding of morphological, cellular, and molecular mechanisms involved in these processes. Nevertheless, there is still much to learn. Here we provide an overview on using *Xenopus* as a model organism to study regeneration and introduce protocols that can be used for studying wound healing and regeneration at multiple levels, thus enhancing our understanding of these phenomena.

XENOPUS AS A MODEL ORGANISM TO STUDY WOUND HEALING AND REGENERATION

Understanding how regeneration processes occur has been an ancient question. Considering *Stedman's Medical Dictionary* definition of regeneration, "reproduction or reconstitution of a lost or injured part," both the study of organ regeneration per se and wound healing are crucial to understand the cellular, molecular, and physiological mechanisms underlying regeneration. This knowledge can then be used in the pursuit of treatments for different diseases, wounds, or injuries. Herein we recapitulate the advantages of using *Xenopus* as a model organism to study wound healing and regeneration and highlight different experimental approaches and methods to study such processes.

Many animal models have been historically used to study regeneration, from nonmammals to mammals, each of them presenting specific advantages and disadvantages (Table 1). *Xenopus* has been positioned as an ideal model. To start, they can be induced to lay hundreds to thousands of eggs (Cline and Kelly 2012), and their maintenance and breeding is simple and of low cost (Sive et al. 2000; Edwards-Faret et al. 2017), leading to experiments with large sampling and high statistical power. Additionally, performing surgeries and manipulating their embryos is easy because of their size and external development; likewise, they can resist and overcome surgeries of different complexities with simple surgery care (Harland and Grainger 2011). These characteristics, in addition to a well-annotated genome (Hellsten et al. 2010; Session et al. 2016), make *Xenopus* an outstanding animal model to perform high-throughput experiments, such as RNA sequencing and proteomics (Amin et al. 2014;

[1]Correspondence: jlarrain@bio.puc.cl

TABLE 1. Comparison of some model organisms widely used for regeneration studies

Model organism	Advantages	Disadvantages	References
Xenopus	Regenerative and nonregenerative stages Large offspring External development Genome available Molecular and genetic techniques Transgenic lines available High synteny with mammals	Fewer antibodies available Injuries not closely analogous to those in mammals	Session et al. 2016; Ishibashi et al. 2008; Hellsten et al. 2010; Edwards-Faret et al. 2017; Slack et al. 2004; James-Zorn et al. 2015; Phipps et al. 2020
Axolotl	Excellent regenerative capacities Large offspring External development Genome available Molecular and genetic techniques High synteny with mammals	Only regenerative stages (does not allow comparison with nonregenerative stages) Fewer antibodies available Improved annotation and genomic assembly are needed More extensive transgenic line repository is needed. Injuries not closely analogous to those in mammals	Joven et al. 2019; Joven and Simon 2018; Gerber et al. 2018; Nowoshilow et al. 2018; Smith et al. 2019; Khattak et al. 2014
Zebrafish	Excellent regenerative capacities Large offspring External development Genetics Genome available Molecular and genetic techniques Extensive transgenic lines available	Only regenerative stages (does not allow comparison with nonregenerative stages) Fewer antibodies available Injuries not closely analogous to those in mammals Evolutionarily further from mammals	Azevedo et al. 2011; Itou et al. 2012; Marques et al. 2019; Varshney et al. 2015; Farah et al. 2016
Spiny mouse	Mammal Expectedly high protein homology with *Mus musculus* (many antibodies available) Genetic information available	Only regenerative stages (does not allow comparison with nonregenerative stages) Lack of an annotated genome Fewer tools and techniques (relatively new model) Small offspring, long gestation More complex maintenance and postsurgery care	Gawriluk et al. 2019; Gawriluk et al. 2016; Haughton et al. 2016; Maden and Varholick 2020; Brant et al. 2019; Seifert et al. 2012

Collart et al. 2014; Lee-Liu et al. 2014, 2018; Sun et al. 2014; Peshkin et al. 2015). Furthermore, transgenic lines can be generated in large numbers and at low cost (Ishibashi et al. 2008). Techniques using transcription activator-like effective endonucleases (TALENs) and clustered regularly interspaced short palindromic repeat (CRISPR)–Cas9 (Ken-ichi et al. 2013; Guo et al. 2014; Sakane et al. 2014; Nakajima and Yaoita 2015) or injection or electroporation of DNA, mRNA, or morpholinos (Eide et al. 2000; Gómez et al. 2003; Bestman et al. 2006; Blum et al. 2015) are effective in elucidating gene functions. Even though *Xenopus laevis* has a relatively long generation time (7–12 mo) and an allotetraploid genome, making genetic experiments harder to perform as a result of more complex genome organization and gene content, the sister species *Xenopus tropicalis* is an alternative for genetics experimentation because of its diploid genome and shorter generation time (3–4 mo). There is extreme similarity between both species, which allows some interchange of experimental results (Karpinka et al. 2015).

Moreover, *Xenopus* tadpole stages are able to regenerate many tissues and organs, including the spinal cord, lens, tail, and limbs; this ability decreases through metamorphosis progression and is almost completely lost after metamorphosis (Filoni et al. 1995, 1997; Slack et al. 2004; Gaete et al. 2012). This transition offers the ability to compare the responses to damage in regenerative versus nonregenerative stages, in order to untangle the mechanisms responsible for regeneration competence (Gaete et al. 2012; Lee-Liu et al. 2014, 2018; Muñoz et al. 2015). Furthermore, functional recovery can be evaluated through simple behavioral tests: Classification of swimming behavior (Gaete et al. 2012) and measurement of free swimming distances (Muñoz et al. 2015; Edwards-Faret et al. 2017) can be used for spinal cord and tail regeneration evaluation, and optomotor response or visual avoidance behavior can be used for visual system regeneration assessment (McKeown et al. 2013). This positions *Xenopus* as a good model organism for performing preclinical trials.

Cite this introduction as *Cold Spring Harb Protoc*; doi:10.1101/pdb.top100966

Even though some specific antibodies and reagents are still lacking, the *Xenopus* community has an established database resource, Xenbase (www.xenbase.org), that works as a repository for information about the genome, genes, expression profiles, gene function, and useful reagents and biological data obtained from *Xenopus* research (James-Zorn et al. 2015). Likewise, there are community resources, such as the National *Xenopus* Resource (Marine Biological Laboratory), the European *Xenopus* Resource Centre (University of Portsmouth), the Biological Resource Center *Xenopus* (University of Rennes 1), and the National BioResource Project (Hiroshima University), that rear and distribute transgenic lines.

EXPERIMENTAL SYSTEMS FOR THE STUDY OF WOUND HEALING AND REGENERATION IN *XENOPUS*

Regeneration can be addressed at many levels: (1) epimorphic regeneration, involving the formation of a blastema that guides the regeneration of a complex structure (e.g., limb regeneration); (2) tissue regeneration, considering regeneration without blastema formation (i.e., spinal cord, lens, and embryonic regeneration); (3) cellular regeneration, which encompasses reconstruction of a damaged cell (e.g., axon regeneration); and (4) wound healing, implying scar-free or scar-based repair of a tissue (e.g., epidermis) (Carlson 2007). Here we provide an overview of some of the latest *Xenopus* experimental approaches that allow the study of regeneration at different levels (Fig. 1), many of which are described in detail in the accompanying protocols.

Limb regeneration capacities have been studied during development (Komala 1957; Suzuki et al. 2006; Keenan and Beck 2016); nonetheless, what hinders these studies is variation in the degree of regeneration among tadpoles. For decreasing this variation, special attention should be given to larval maintenance as well as to precision and consistency of limb amputation; see Protocol 1: Studies of Limb Regeneration in Larval *Xenopus* (Mescher and Neff 2019) and Figure 1, Limb amputation. As limb regeneration involves different tissues, it is a great model for studying the effect of different compounds on successful regeneration and patterning of these tissues (King et al. 2012; Mescher et al. 2013).

The inability of mammals to regenerate spinal cord is determined by cellular intrinsic and extrinsic factors (Kaplan et al. 2015). On one hand, mammalian spinal cord axons are able to grow in a regenerative permissive environment (Richardson et al. 1980; David and Aguayo 1981), providing evidence that extrinsic factors, present in the spinal cord environment, restrict or favor axon regeneration. On the other hand, stem cells, grafted into an injured mammalian spinal cord, are able to differentiate into neurons despite the nonpermissive environment (Lu et al. 2012), proving that there are some intrinsic factors, within the neurons, hampering spinal cord regeneration in mammals. Even though specific protocols for generating different types of spinal cord injuries, including their pros and cons, have been published (Polezhaev and Carlson 1972; Lee-Liu et al. 2013; Edwards-Faret et al. 2017; Phipps et al. 2020), they cannot discriminate between intrinsic and extrinsic factors. Transplantation experiments, involving regenerative stages as donors and nonregenerative stages as hosts, allow one to study intrinsic factors of regenerative cells in a nonregenerative environment (Méndez-Olivos et al. 2017). Cell transplantation experiments can be performed in the spinal cord; see Protocol 2: Cell Transplantation as a Method to Investigate Spinal Cord Regeneration in Regenerative and Nonregenerative *Xenopus* Stages (Méndez-Olivos and Larraín 2018) and Figure 1, Transplant.

Stages of lens regeneration are well-characterized (Freeman 1963; Henry 2003; Henry and Tsonis 2010). In *Xenopus*, different stages of lens regeneration can be studied in whole animals after lentectomy; see Protocol 3: Methods for Examining Lens Regeneration in *Xenopus* (Henry et al. 2019b) and Figure 1, Lentectomy. In addition, one can prepare ex vivo eye tissue cultures to examine specific eye tissue interactions; see Protocol 4: Ex Vivo Eye Tissue Culture Methods for *Xenopus* (Henry et al. 2019a) and Figure 1, Ex vivo eye culture.

Finally, to understand the involvement of specific genes, proteins, or signaling pathways in wound healing, oocytes and embryos can be studied. If a quick assay is needed, mechanical wounding can be performed. If greater consistency of wounding is needed or if the interest is on studying the early steps or

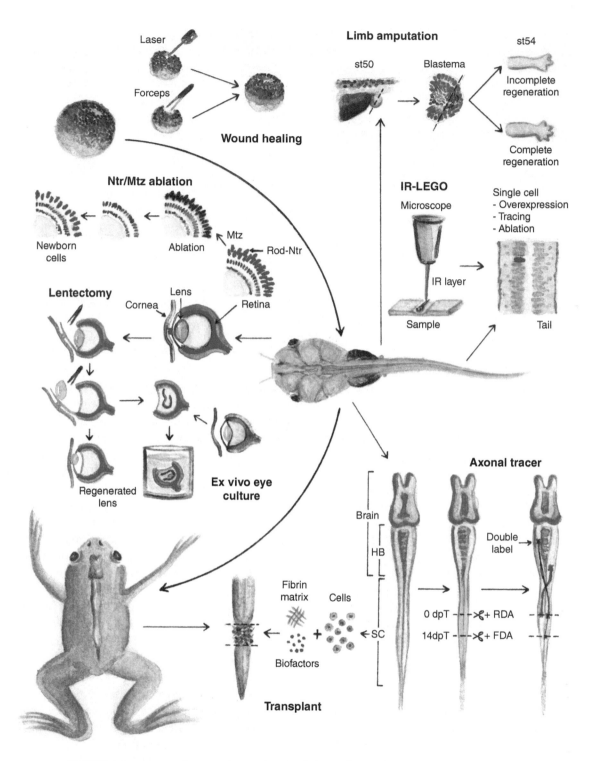

FIGURE 1. Experimental approaches and methods for studying wound healing and regeneration in *Xenopus*. Shown are some of the *Xenopus* life cycle stages and methods (introduced in the text) that are used to study different phenomena: oocytes for wound healing, tadpoles for limb and lens regeneration, and tadpoles in combination with froglets for transplantation studies.

the dynamics of wound healing, laser wounding is preferred. The involvement of different signaling pathways in wound healing can be achieved by using glutathione-S-transferase (GST) pull-down assays of signaling molecules. See Protocol 5: Investigating the Cellular and Molecular Mechanisms of Wound Healing in *Xenopus* Oocytes and Embryos (Li and Amaya 2019) and Figure 1, Wound healing.

CELLULAR AND MOLECULAR METHODS FOR THE STUDY OF REGENERATION MECHANISMS

Many groups have shed light on the mechanisms involved in regeneration (for reviews, see Lee-Liu et al. 2017; Phipps et al. 2020). These include cellular migration (Yoshii et al. 2007; Aztekin et al. 2019), proliferation and differentiation (Yoshino and Tochinai 2004; Gaete et al. 2012; McKeown et al. 2013; Muñoz et al. 2015), and inflammation and the immune response (Mescher et al. 2017), among others.

Considering the cases of tail, spinal cord, limb, and retina injury, the regeneration of nerve connections is crucial (Gaze 1959; Filoni and Paglialunga 1990; Zhao and Szaro 1994; Taniguchi et al. 2008). Thus, relevant questions arise: Which neuronal connections are recovered? Which neuronal nuclei are involved in the regenerative process? Are the axons regenerating and regrowing (e.g., sprouting), or are new neurons being generated? These questions can be addressed by using double axonal tracing in an injured spinal cord; see Protocol 6: Tracing Central Nervous System Axon Regeneration in *Xenopus* (Gibbs and Szaro 2018) and Figure 1, Axonal tracer. This sequential retrograde double-labeling approach uses dextran amines that are incorporated only in terminals or cut axons, allowing the labeling of regenerated axons.

Additionally, understanding the contribution of specific cell types on cell death, proliferation, and regeneration is another recurring question in the regeneration field. This can be addressed through cellular ablation, and the specificity of the observation relies on precise spatial and temporal control of the ablation of a specific cell type. The **nitroreductase/metronidazole** (NTR/Mtz) system addresses these requirements; see Protocol 7: Rod-Specific Ablation Using the Nitroreductase/Metronidazole System to Investigate Regeneration in *Xenopus* (Martinez-De Luna and Zuber 2018) and Figure 1, Ntr/Mtz ablation. The NTR/Mtz system expresses NTR under the control of a cell-type-specific promoter, and, following treatment with Mtz, a cytotoxic product is generated in the NTR-expressing cells, resulting in cell death.

Finally, if the interest focuses on single cells or a cluster of cells for determining (1) the cell fate of a specific cell, (2) the role of a particular gene in a specific cell, or (3) the role of a particular cell during the spinal cord regeneration process, the infrared laser-evoked gene operator (IR-LEGO) system can be used; see Protocol 8: Infrared Laser-Mediated Gene Induction at the Single-Cell Level in the Regenerating Tail of *Xenopus laevis* Tadpoles (Hasugata et al. 2018) and Figure 1, IR-LEGO. IR-LEGO allows gene induction at the single-cell level by generating local heat shock by laser irradiation or ablation of a specific cell by using higher-laser-power irradiation. The IR-LEGO system is highly robust and generates focused cellular damage (Kamei et al. 2009).

In summary, *Xenopus* is clearly an outstanding model organism to study wound healing and regeneration, and the *Xenopus* community has constantly been developing approaches and methods to improve research in this area. Continued advances in understanding regenerative mechanisms in *Xenopus* may provide novel insights to improve regeneration in humans.

COMPETING INTEREST STATEMENT

The authors declare no conflicts of interest.

ACKNOWLEDGMENTS

Our work is supported by FONDECYT 1180429 (for J.L.) and 3190820 (for P.G.S.) Figure 1 was created by @mimipalacios_art.

REFERENCES

Amin NM, Tandon P, Osborne Nishimura E, Conlon FL. 2014. RNA-seq in the tetraploid *Xenopus laevis* enables genome-wide insight in a classic developmental biology model organism. *Methods* **66:** 398–409. doi:10.1016/j.ymeth.2013.06.009

Azevedo AS, Grotek B, Jacinto A, Weidinger G, Saude L. 2011. The regenerative capacity of the zebrafish caudal fin is not affected by repeated amputations. *PLoS One* **6:** e22820. doi:10.1371/journal.pone.0022820

Aztekin C, Hiscock T, Marioni J, Gurdon J, Simons B, Jullien J. 2019. Identification of a regeneration-organizing cell in the *Xenopus* tail. *Science* **364:** 653–658. doi:10.1126/science.aav9996

Bestman JE, Ewald RC, Chiu S-L, Cline HT. 2006. In vivo single-cell electroporation for transfer of DNA and macromolecules. *Nat Protoc* **1:** 1267. doi:10.1038/nprot.2006.186

Blum M, De Robertis EM, Wallingford JB, Niehrs C. 2015. Morpholinos: antisense and sensibility. *Dev Cell* **35:** 145–149. doi:10.1016/j.devcel.2015.09.017

Brant JO, Boatwright JL, Davenport R, Sandoval AGW, Maden M, Barbazuk WB. 2019. Comparative transcriptomic analysis of dermal wound healing reveals de novo skeletal muscle regeneration in *Acomys cahirinus*. *PLoS One* **14:** e0216228. doi:10.1371/journal.pone.0216228

Carlson B. 2007. *Principles of regenerative biology*, 1st ed, pp. 1–23. Academic, San Diego.

Cline HT, Kelly D. 2012. *Xenopus* as an experimental system for developmental neuroscience: introduction to a special issue. *Dev Neurobiol* **72:** 463–464. doi:10.1002/dneu.22012

Collart C, Owens ND, Bhaw-Rosun L, Cooper B, De Domenico E, Patrushev I, Sesay AK, Smith JN, Smith JC, Gilchrist MJ. 2014. High-resolution analysis of gene activity during the *Xenopus* mid-blastula transition. *Development* **141:** 1927–1939. doi:10.1242/dev.102012

David S, Aguayo AJ. 1981. Axonal elongation into peripheral nervous system "bridges" after central nervous system injury in adult rats. *Science* **214:** 931–933. doi:10.1126/science.6171034

Edwards-Faret G, Muñoz R, Méndez-Olivos EE, Lee-Liu D, Tapia VS, Larraín J. 2017. Spinal cord regeneration in *Xenopus laevis*. *Nat Protoc* **12:** 372–389. doi:10.1038/nprot.2016.177

Eide FF, Eisenberg SR, Sanders TA. 2000. Electroporation-mediated gene transfer in free-swimming embryonic *Xenopus laevis*. *FEBS Lett* **486:** 29–32. doi:10.1016/S0014-5793(00)02124-4

Farah Z, Fan H, Liu Z, He JQ. 2016. A concise review of common animal models for the study of limb regeneration. *Organogenesis* **12:** 109–118. doi:10.1080/15476278.2016.1205775

Filoni S, Paglialunga L. 1990. Effect of denervation on hindlimb regeneration in *Xenopus laevis* larvae. *Differentiation* **43:** 10–19. doi:10.1111/j.1432-0436.1990.tb00425.x

Filoni S, Bernardini S, Cannata SM. 1995. Differences in the decrease in regenerative capacity of various brain regions of *Xenopus laevis* are related to differences in the undifferentiated cell populations. *J Hirnforsch* **36:** 523–529.

Filoni S, Bernardini S, Cannata SM, D'Alessio A. 1997. Lens regeneration in larval *Xenopus laevis*: experimental analysis of the decline in the regenerative capacity during development. *Dev Biol* **187:** 13–24. doi:10.1006/dbio.1997.8598

Freeman G. 1963. Lens regeneration from the cornea in *Xenopus laevis*. *J Exp Zool* **154:** 39–65. doi:10.1002/jez.1401540105

Gaete M, Muñoz R, Sánchez N, Tampe R, Moreno M, Contreras EG, Lee-Liu D, Larraín J. 2012. Spinal cord regeneration in *Xenopus* tadpoles proceeds through activation of Sox2-positive cells. *Neural Dev* **7:** 13. doi:10.1186/1749-8104-7-13

Gawriluk TR, Simkin J, Thompson KL, Biswas SK, Clare-Salzler Z, Kimani JM, Kiama SG, Smith JJ, Ezenwa VO, Seifert AW. 2016. Comparative analysis of ear-hole closure identifies epimorphic regeneration as a discrete trait in mammals. *Nat Commun* **7:** 11164. doi:10.1038/ncomms11164

Gawriluk TR, Simkin J, Hacker CK, Kimani JM, Kiama SG, Ezenwa VO, Seifert AW. 2019. Mammalian musculoskeletal regeneration is associated with reduced inflammatory cytokines and an influx of T cells. bioRxiv doi:10.1101/723783

Gaze R. 1959. Regeneration of the optic nerve in *Xenopus laevis*. *Q J Exp Physiol Cogn Med Sci* **44:** 290–308.

Gerber T, Murawala P, Knapp D, Masselink W, Schuez M, Hermann S, Gac-Santel M, Nowoshilow S, Kageyama J, Khattak S, et al. 2018. Single-cell analysis uncovers convergence of cell identities during axolotl limb regeneration. *Science* **362:** eaaq0681. doi:10.1126/science.aaq0681

Gibbs KM, Szaro BG. 2018. Tracing central nervous system axon regeneration in *Xenopus*. *Cold Spring Harb Protoc* doi:10.1101/pdb.prot101030

Gómez TM, Harrigan D, Henley J, Robles E. 2003. Working with *Xenopus* spinal neurons in live cell culture. *Methods Cell Biol* **71:** 129–156. doi:10.1016/S0091-679X(03)01008-2

Guo X, Zhang T, Hu Z, Zhang Y, Shi Z, Wang Q, Cui Y, Wang F, Zhao H, Chen Y. 2014. Efficient RNA/Cas9-mediated genome editing in *Xenopus tropicalis*. *Development* **141:** 707–714. doi:10.1242/dev.099853

Harland RM, Grainger RM. 2011. *Xenopus* research: metamorphosed by genetics and genomics. *Trends Genet* **27:** 507–515. doi:10.1016/j.tig.2011.08.003

Hasugata R, Hayashi S, Kawasumi-Kita A, Sakamoto J, Kamei Y, Yokoyama H. 2018. Infrared laser-mediated gene induction at the single-cell level in the regenerating tail of *Xenopus laevis* tadpoles. *Cold Spring Harb Protoc* doi:10.1101/pdb.prot101014

Haughton CL, Gawriluk TR, Seifert AW. 2016. The biology and husbandry of the African spiny mouse (*Acomys cahirinus*) and the research uses of a laboratory colony. *J Am Assoc Lab Anim Sci* **55:** 9–17.

Hellsten U, Harland RM, Gilchrist MJ, Hendrix D, Jurka J, Kapitonov V, Ovcharenko I, Putnam NH, Shu S, Taher L. 2010. The genome of the Western clawed frog *Xenopus tropicalis*. *Science* **328:** 633–636. doi:10.1126/science.1183670

Henry JJ. 2003. Cell and molecular biology of lens regeneration. *Int Rev Cytol* **228:** 195–264. doi:10.1016/S0074-7696(03)28005-0

Henry JJ, Tsonis PA. 2010. Molecular and cellular aspects of amphibian lens regeneration. *Prog Retin Eye Res* **29:** 543–555. doi:10.1016/j.preteyeres.2010.07.002

Henry JJ, Perry KJ, Hamilton PW. 2019a. Ex vivo eye tissue culture methods for *Xenopus*. *Cold Spring Harb Protoc* doi:10.1101/pdb.prot101535

Henry JJ, Perry KJ, Hamilton PW. 2019b. Methods for examining lens regeneration in *Xenopus*. *Cold Spring Harb Protoc* doi:10.1101/pdb.prot101527

Ishibashi S, Kroll KL, Amaya E. 2008. A method for generating transgenic frog embryos. *Methods Mol Biol* **461:** 447–466. doi:10.1007/978-1-60327-483-8_31

Itou J, Oishi I, Kawakami H, Glass TJ, Richter J, Johnson A, Lund TC, Kawakami Y. 2012. Migration of cardiomyocytes is essential for heart regeneration in zebrafish. *Development* **139:** 4133–4142. doi:10.1242/dev.079756

James-Zorn C, Ponferrada VG, Burns KA, Fortriede JD, Lotay VS, Liu Y, Brad Karpinka J, Karimi K, Zorn AM, Vize PD. 2015. Xenbase: core features, data acquisition, and data processing. *Genesis* **53:** 486–497. doi:10.1002/dvg.22873

Joven A, Simon A. 2018. Homeostatic and regenerative neurogenesis in salamanders. *Prog Neurobiol* **170:** 81–98. doi:10.1016/j.pneurobio.2018.04.006

Joven A, Elewa A, Simon A. 2019. Model systems for regeneration: salamanders. *Development* **146:** dev167700. doi:10.1242/dev.167700

Kamei Y, Suzuki M, Watanabe K, Fujimori K, Kawasaki T, Deguchi T, Yoneda Y, Todo T, Takagi S, Funatsu T. 2009. Infrared laser–mediated gene induction in targeted single cells in vivo. *Nat Methods* **6:** 79–81. doi:10.1038/nmeth.1278

Kaplan A, Ong Tone S, Fournier AE. 2015. Extrinsic and intrinsic regulation of axon regeneration at a crossroads. *Front Mol Neurosci* **8:** 27. doi:10.3389/fnmol.2015.00027

Karpinka JB, Fortriede JD, Burns KA, James-Zorn C, Ponferrada VG, Lee J, Karimi K, Zorn AM, Vize PD. 2015. Xenbase, the *Xenopus* model organism database; new virtualized system, data types and genomes. *Nucleic Acids Res* **43:** D756–D763. doi:10.1093/nar/gku956

Keenan SR, Beck CW. 2016. *Xenopus* limb bud morphogenesis. *Dev Dyn* **245:** 233–243. doi:10.1002/dvdy.24351

Ken-ichi TS, Isoyama Y, Kashiwagi K, Sakuma T, Ochiai H, Sakamoto N, Furuno N, Kashiwagi A, Yamamoto T. 2013. High efficiency TALENs enable F0 functional analysis by targeted gene disruption in *Xenopus laevis* embryos. *Biol Open* **2:** 448–452. doi:10.1242/bio.20133855

Khattak S, Murawala P, Andreas H, Kappert V, Schuez M, Sandoval-Guzman T, Crawford K, Tanaka EM. 2014. Optimized axolotl (*Ambystoma mexicanum*) husbandry, breeding, metamorphosis, transgenesis and tamoxifen-mediated recombination. *Nat Protoc* 9: 529–540. doi:10.1038/nprot.2014.040

King MW, Neff AW, Mescher AL. 2012. The developing *Xenopus* limb as a model for studies on the balance between inflammation and regeneration. *Anat Rec* 295: 1552–1561. doi:10.1002/ar.22443

Komala Z. 1957. Comparative investigations on the course of ontogenesis and regeneration of the limbs in *Xenopus laevis* tadpoles in various stages of development. *Folia Biol* 5: 1–51.

Lee-Liu D, Edwards-Faret G, Tapia VS, Larraín J. 2013. Spinal cord regeneration: lessons for mammals from non-mammalian vertebrates. *Genesis* 51: 529–544. doi:10.1002/dvg.22406

Lee-Liu D, Moreno M, Almonacid LI, Tapia VS, Muñoz R, von Marées J, Gaete M, Melo F, Larraín J. 2014. Genome-wide expression profile of the response to spinal cord injury in *Xenopus laevis* reveals extensive differences between regenerative and non-regenerative stages. *Neural Dev* 9: 12. doi:10.1186/1749-8104-9-12

Lee-Liu D, Méndez-Olivos EE, Muñoz R, Larraín J. 2017. The African clawed frog *Xenopus laevis*: a model organism to study regeneration of the central nervous system. *Neuroscience letters* 652: 82–93. doi:10.1016/j.neulet.2016.09.054

Lee-Liu D, Sun L, Dovichi NJ, Larraín J. 2018. Quantitative proteomics after spinal cord injury (SCI) in a regenerative and a nonregenerative stage in the frog *Xenopus laevis*. *Mol Cell Proteomics* 17: 592–606. doi:10.1074/mcp.RA117.000215

Li J, Amaya E. 2019. Investigating the cellular and molecular mechanisms of wound healing in *Xenopus* oocytes and embryos. *Cold Spring Harb Protoc* doi:10.1101/pdb.prot100982

Lu P, Wang Y, Graham L, McHale K, Gao M, Wu D, Brock J, Blesch A, Rosenzweig ES, Havton LA. 2012. Long-distance growth and connectivity of neural stem cells after severe spinal cord injury. *Cell* 150: 1264–1273. doi:10.1016/j.cell.2012.08.020

Maden M, Varholick JA. 2020. Model systems for regeneration: the spiny mouse, *Acomys cahirinus*. *Development* 147: dev167718. doi:10.1242/dev.167718

Marques IJ, Lupi E, Mercader N. 2019. Model systems for regeneration: zebrafish. *Development* 146: dev167692. doi:10.1242/dev.167692

Martinez-De Luna RI, Zuber ME. 2018. Rod-specific ablation using the nitroreductase/metronidazole system to investigate regeneration in *Xenopus*. *Cold Spring Harb Protoc* doi:10.1101/pdb.prot100974

McKeown CR, Sharma P, Sharipov HE, Shen W, Cline HT. 2013. Neurogenesis is required for behavioral recovery after injury in the visual system of *Xenopus laevis*. *J Comp Neurol* 521: 2262–2278. doi:10.1002/cne.23283

Méndez-Olivos EE, Larraín J. 2018. Cell transplantation as a method to investigate spinal cord regeneration in regenerative and nonregenerative *Xenopus* stages. *Cold Spring Harb Protoc* doi:10.1101/pdb.prot101006

Méndez-Olivos EE, Muñoz R, Larraín J. 2017. Spinal cord cells from premetamorphic stages differentiate into neurons and promote axon growth and regeneration after transplantation into the injured spinal cord of non-regenerative *Xenopus laevis* froglets. *Front Cell Neurosci* 11: 398. doi:10.3389/fncel.2017.00398

Mescher AL, Neff AW. 2019. Studies of limb regeneration in larval *Xenopus*. *Cold Spring Harb Protoc* doi:10.1101/pdb.prot100990

Mescher AL, Neff AW, King MW. 2013. Changes in the inflammatory response to injury and its resolution during the loss of regenerative capacity in developing *Xenopus* limbs. *PLoS One* 8: e80477. doi:10.1371/journal.pone.0080477

Mescher AL, Neff AW, King MW. 2017. Inflammation and immunity in organ regeneration. *Dev Comp Immunol* 66: 98–110. doi:10.1016/j.dci.2016.02.015

Muñoz R, Edwards-Faret G, Moreno M, Zuñiga N, Cline H, Larraín J. 2015. Regeneration of *Xenopus laevis* spinal cord requires Sox2/3 expressing cells. *Dev Biol* 408: 229–243. doi:10.1016/j.ydbio.2015.03.009

Nakajima K, Yaoita Y. 2015. Highly efficient gene knockout by injection of TALEN mRNAs into oocytes and host transfer in *Xenopus laevis*. *Biology Open* 4: 180–185. doi:10.1242/bio.201410009

Nowoshilow S, Schloissnig S, Fei JF, Dahl A, Pang AWC, Pippel M, Winkler S, Hastie AR, Young G, Roscito JG, et al. 2018. The axolotl genome and the evolution of key tissue formation regulators. *Nature* 554: 50–55. doi:10.1038/nature25458

Peshkin L, Wuhr M, Pearl E, Haas W, Freeman RM Jr, Gerhart JC, Klein AM, Horb M, Gygi SP, Kirschner MW. 2015. On the relationship of protein and mRNA dynamics in vertebrate embryonic development. *Dev Cell* 35: 383–394. doi:10.1016/j.devcel.2015.10.010

Phipps LS, Marshall L, Dorey K, Amaya E. 2020. Model systems for regeneration: *Xenopus*. *Development* 147: dev180844. doi:10.1242/dev.180844

Polezhaev LV, Carlson BM. 1972. *Loss and restoration of regenerative capacity in tissues and organs of animals*. Harvard University Press, Boston.

Richardson P, McGuinness U, Aguayo A. 1980. Axons from CNS neurones regenerate into PNS grafts. *Nature* 284: 264–265. doi:10.1038/284264a0

Sakane Y, Sakuma T, Kashiwagi K, Kashiwagi A, Yamamoto T, Suzuki KT. 2014. Targeted mutagenesis of multiple and paralogous genes in *Xenopus laevis* using two pairs of transcription activator-like effector nucleases. *Dev Growth Differ* 56: 108–114. doi:10.1111/dgd.12105

Seifert AW, Kiama SG, Seifert MG, Goheen JR, Palmer TM, Maden M. 2012. Skin shedding and tissue regeneration in African spiny mice (*Acomys*). *Nature* 489: 561–565. doi:10.1038/nature11499

Session AM, Uno Y, Kwon T, Chapman JA, Toyoda A, Takahashi S, Fukui A, Hikosaka A, Suzuki A, Kondo M. 2016. Genome evolution in the allotetraploid frog *Xenopus laevis*. *Nature* 538: 336–343. doi:10.1038/nature19840

Sive H, Grainger R, Harland R. 2000. *Early development of Xenopus laevis: a laboratory manual*. Cold Spring Harbor Laboratory Press, Cold Spring Harbor, New York.

Slack J, Beck C, Gargioli C, Christen B. 2004. Cellular and molecular mechanisms of regeneration in *Xenopus*. *Philos Trans R Soc Lond B Biol Sci* 359: 745–751. doi:10.1098/rstb.2004.1463

Smith JJ, Timoshevskaya N, Timoshevskiy VA, Keinath MC, Hardy D, Voss SR. 2019. A chromosome-scale assembly of the axolotl genome. *Genome Res* 29: 317–324. doi:10.1101/gr.241901.118

Sun L, Bertke MM, Champion MM, Zhu G, Huber PW, Dovichi NJ. 2014. Quantitative proteomics of *Xenopus laevis* embryos: expression kinetics of nearly 4000 proteins during early development. *Sci Rep* 4: 4365. doi:10.1038/srep04365

Suzuki M, Yakushiji N, Nakada Y, Satoh A, Ide H, Tamura K. 2006. Limb regeneration in *Xenopus laevis* froglet. *Sci World J* 6: 26–37. doi:10.1100/tsw.2006.325

Taniguchi Y, Sugiura T, Tazaki A, Watanabe K, Mochii M. 2008. Spinal cord is required for proper regeneration of the tail in *Xenopus* tadpoles. *Dev Growth Differ* 50: 109–120. doi:10.1111/j.1440-169X.2007.00981.x

Varshney GK, Sood R, Burgess SM. 2015. Understanding and editing the zebrafish genome. *Adv Genet* 92: 1–52. doi:10.1016/bs.adgen.2015.09.002

Yoshii C, Ueda Y, Okamoto M, Araki M. 2007. Neural retinal regeneration in the anuran amphibian *Xenopus laevis* post-metamorphosis: transdifferentiation of retinal pigmented epithelium regenerates the neural retina. *Dev Biol* 303: 45–56. doi:10.1016/j.ydbio.2006.11.024

Yoshino J, Tochinai S. 2004. Successful reconstitution of the non-regenerating adult telencephalon by cell transplantation in *Xenopus laevis*. *Dev Growth Differ* 46: 523–534. doi:10.1111/j.1440-169x.2004.00767.x

Zhao Y, Szaro BG. 1994. The return of phosphorylated and nonphosphorylated epitopes of neurofilament proteins to the regenerating optic nerve of *Xenopus laevis*. *J Comp Neurol* 343: 158–172. doi:10.1002/cne.903430112

Studies of Limb Regeneration in Larval *Xenopus*

Anthony L. Mescher[1,2,3] and Anton W. Neff[1,2]

[1]*Center for Developmental and Regenerative Biology, Indiana University, Bloomington, Indiana 47405;*
[2]*Department f Anatomy and Cell Biology, Indiana University School of Medicine at Bloomington, Bloomington, Indiana 47405*

A basic protocol is given for animal maintenance and surgery in studies of hindlimb regeneration in larval *Xenopus laevis*. Unlike urodele limbs, those of larval frogs typically show much more variation in the extent of regeneration after amputation. Such variation can be reduced by optimizing the conditions of larval maintenance to regulate the rates of growth and development, by selecting only larvae with normal rates of growth and morphological development for experimental use, and by attention to precision and consistency in the proximo–distal level of surgical amputation.

MATERIALS

It is essential that you consult the appropriate Material Safety Data Sheets and your institution's Environmental Health and Safety Office for proper handling of equipment and hazardous materials used in this protocol.

RECIPES: Please see the end of this protocol for recipes indicated by <R>. Additional recipes can be found online at http://cshprotocols.cshlp.org/site/recipes.

Reagents

Anesthetic solution

> *Dissolve 0.1 g of benzocaine (ethyl 4-aminobenzoate; Sigma-Aldrich) or MS222 (ethyl 3-aminobenzoate methanesulfonate; Sigma-Aldrich) in 10 mL of absolute ethanol. Dilute to 1 L with artificial pond water. If necessary, buffer to pH 7. Prepare fresh weekly, or as needed.*

Artificial pond water <R>
Fixative (e.g., 4% formalin or 100% methanol) (optional; see Step 17)
Tadpole Powder (*Xenopus* express)
Xenopus embryos

> *Obtain embryos as described by Sive et al. (2000).*

Equipment

Aquarium aerator
Containers for larvae, glass or plastic, 3-L
Culture plate, 10-cm diameter
Gauze
Spring scissors, 10- to 15-mm cutting edge, Castroviejo or Noyes (Roboz or Fine Science Tools)
Stereomicroscope

---

[3]Correspondence: mescher@indiana.edu

METHOD

Husbandry of *Xenopus laevis* Larvae

1. At approximately developmental Stage 45 (as defined by Nieuwkoop and Faber 1994), separate the larvae into containers containing 2.5 L artificial pond water, at a population density not exceeding 1 larva per 500 mL.

2. Maintain larvae at $23 \pm 2°C$ in natural light or a 12/12 dark/light cycle. Feed once a day with tadpole powder in an amount that allows all to be eaten before the next feeding.

3. Change artificial pond water daily or every second day:

    i. Remove larvae carefully with a soft net.

    ii. Wipe biofilm from the container.

    iii. Add fresh pond water.

    > *Ensure that the fresh water is at the same temperature as that removed.*

    iv. Replace the larvae.

4. Examine the larvae daily. Eliminate those with abnormalities or significantly reduced rates of growth or external development.

Anesthesia and Hindlimb Amputation of Larvae

Excellent fate maps of developing Xenopus laevis *hindlimbs were published by Tschumi (1957) and recently reprinted by Keenan and Beck (2016).*

5. Select an appropriate number of larvae for the specific experiment (10–15 animals is recommended), ensuring that all are at the desired developmental stage.

    > *Under ideal environmental conditions (Nieuwkoop and Faber 1994), hindlimb buds appear at Stage 46 and become longer than they are broad at Stage 50, ~15 d after fertilization.*

6. Place 1–2 larvae in a small bowl or 10-cm tissue culture plate containing 40–50 mL of freshly prepared anesthetic solution.

7. Place an immobilized larva on gauze moistened with artificial pond water. Position the larva on its side to allow a good view of the developing hindlimb when placed on the stage of the stereomicroscope.

8. Blot excess water from the exposed larval flank and hindlimb. Focus the microscope on the latter.

9. Using spring scissors, amputate the hindlimb at a 90° angle to the proximal–distal axis, taking care to do so at the precise proximal–distal level desired, and at the exact same level in each successive limb.

    > *See Troubleshooting.*

10. If bilateral amputation is desired, use a plastic spoon to carefully turn the larva over. Amputate the other hindlimb as above.

11. Place each amputated larva in ~1 L of artificial pond water aerated continuously with an aquarium aerator until fully awake and swimming normally.

12. Repeat Steps 6–11 until all the animals have been processed.

13. When all experimental larvae are active and swimming normally, place in containers of artificial pond water. Maintain husbandry at a density similar to that before surgery.

14. Visually monitor hindlimb regeneration and animal development daily. Eliminate larvae that develop abnormalities or show significantly reduced rates of overall growth or external development.

    > *See Discussion for the importance of this procedural step.*

15. At the desired time point or regenerative stage postamputation, place the larvae individually again in the anesthetic. Use spring scissors to remove the regenerating limbs at the pelvis (or other level, as desired).

16. Immediately place each limb in an appropriate fixative or extraction buffer as required for histological or molecular analysis, respectively.

17. After removal of the experimental tissues, euthanize each larva by decapitation or by reimmersion in anesthetic, followed by placement in 4% formalin or 100% methanol.

TROUBLESHOOTING

Problem (Step 9): In hindlimbs at Stage 57 or later, the cut end(s) of skeletal elements can protrude slightly after amputation.
Solution: Cut the protruding skeletal tissue again flush with the soft tissues of the limb stump.

DISCUSSION

Anuran tadpoles and various urodele species such as newts (*Notophthalmus viridescens*, *Pleurodeles waltl*, *Cynops pyrrhogaster*, and *Cynops orientalis*) or axolotls (*Ambystoma mexicanum*) are the only tetrapod research models capable of complete limb regeneration after amputation (Song et al. 2010). Regeneration in both anurans and urodeles appears similar in terms of the tissue interactions required for limb regrowth and development, but the process in anuran larvae differs from that of urodeles in at least two respects. Importantly, in anurans the capacity to successfully pattern a new limb declines during larval development and is gradually lost, first in proximal portions of the limb and then in progressively more distal levels, during the onset of metamorphosis (Schotté and Harland 1943). The gradual decline of regenerative ability in developing *Xenopus laevis* hindlimbs—showed by both Komala (1957) and Dent (1962)—is indicated by the reduced number of articulated digits, with the first digits lost being those formed last during larval development (Muneoka et al. 1986). The developmental decline of regenerative capacity allows important processes that follow amputation, such as localized gene expression changes (Grow et al. 2006) and inflammation (Mescher et al. 2013), to be compared at stages when regeneration is expected to be complete or incomplete (King et al. 2012).

The second difference between anuran and urodele limb regeneration is that although patterning in the latter is almost always perfect (or nearly so), morphogenesis in regenerating hindlimbs of anuran tadpoles is often quite variable even in limbs amputated well before metamorphosis, which in *Xenopus* occurs during Stages 57–66 (Nieuwkoop and Faber 1994). A summary figure of Dent (1962) suggests that limbs amputated through the presumptive region of the distal thigh at Stages 53 and 55 are expected to regenerate four- and three-digit limbs, respectively. However his quantitative data show that more than one-third of the hindlimbs amputated at those stages failed to achieve those results. Similar variability in digit patterning with this system has also been reported by other investigators (e.g., Shimizu-Nishikawa et al. 2003; Slack et al. 2004; King et al. 2012).

The variability of regeneration in *Xenopus* hindlimbs was investigated by Nye and Cameron (2005) and discussed in a review by Beck et al. (2009). Nye and Cameron (2005) showed that such variation is reduced by attention to precision and consistency in the amputation plane, and by ensuring normal larval growth and development. They found that even under "optimal" growth conditions laboratory-raised *Xenopus* larvae go through metamorphosis ∼1 wk later than the larvae described by Nieuwkoop and Faber (1994), which were collected in nature and "…had developed under the *most natural* conditions [their italics]… the water was green with algae…" (Nieuwkoop and Faber 1994). Even with well-fed sibling larvae raised at the low population density used in this protocol, growth and development often slow and can eventually stop in a significant percentage of animals. When hindlimbs

Cite this protocol as *Cold Spring Harb Protoc*; doi:10.1101/pdb.prot100990

of these abnormal larvae are amputated, regeneration occurs very slowly and with poor patterning compared to that in normally developing, younger larvae amputated at the same Nieuwkoop–Faber stage (Nye and Cameron 2005; Beck et al. 2009). Nye and Cameron (2005) recommended that each laboratory establish that the growth curve of their tadpoles approximates that of Nieuwkoop and Faber (1994) and that slow-growing or developmentally abnormal tadpoles be eliminated from experimental analyses of limb regeneration to avoid excessively variable results.

As discussed by Mescher (2017), it is of interest that maturation of the immune system continues in *Xenopus* larvae whose external morphological development has slowed (Ruben et al. 1972; Rollins-Smith et al. 1997). Those studies showed that slow-growing larvae at an early morphological stage, when complete hindlimb regeneration is expected, can have immunological capabilities typical of later stages, when regeneration is normally defective. We have reported initial studies on the localization of antigen-presenting cells in the skin of developing *Xenopus* hindlimbs (Mescher et al. 2007). The data are consistent with the view that this maturation of cutaneous immunity could elicit qualitatively different inflammatory responses to amputation that lead to the developmental decline of regenerative capacity, similar to the process by which a tendency for postinjury scarring increases during skin maturation in other vertebrates.

RECIPE

Artificial Pond Water

1. Prepare a stock solution containing 175 g of NaCl and 35 g of $CaCl_2$ in 2 L of distilled/deionized water.

2. Prepare a second stock solution containing 5 g of $NaHCO_3$ in 2 L of distilled/deionized water.

3. To 70 L (44 gal) of distilled/deionized water in an HDPE plastic barrel on casters (e.g., Rubbermaid Commercial Products) add 280 mL of each stock solution and 25 mL of NovAqua water conditioner (Kordon).

   This "water conditioner" buffers the pH, adds other electrolytes, and removes chlorine, chloramine, and heavy metals.

4. Keep the barrel lidded and the contents aerated.

Prepare fresh every 3–4 d.

REFERENCES

Beck CW, Izpisúa Belmonte JC, Christen B. 2009. Beyond early development: *Xenopus* as an emerging model for the study of regenerative mechanisms. *Dev Dyn* 238: 1226–1248.

Dent JN. 1962. Limb regeneration in larvae and metamorphosing individuals of the South African clawed toad. *J Morph* 110: 61–77.

Grow M, Neff AW, Mescher AL, King MW. 2006. Global analysis of gene expression in *Xenopus* hindlimbs during stage-dependent complete and incomplete regeneration. *Dev Dyn* 235: 2667–2685.

Keenan SR, Beck CW. 2016. *Xenopus* limb bud morphogenesis. *Dev Dyn* 245: 233–243.

King MW, Neff AW, Mescher AL. 2012. The developing *Xenopus* limb as a model for studies on the balance between inflammation and regeneration. *Anat Rec* 295: 1552–1561.

Komala Z. 1957. Comparative investigations on the course of ontogenesis and regeneration of the limbs in *Xenopus laevis* tadpoles in various stages of development. *Folia Biol* 5: 1–51. [in Polish].

Mescher AL. 2017. Macrophages and fibroblasts during inflammation and tissue repair in models of organ regeneration. *Regeneration* 4: 39–53.

Mescher AL, Wolf WL, Moseman EA, Hartman B, Harrison C, Nguyen E, Neff AW. 2007. Cells of cutaneous immunity in *Xenopus*: Studies during larval development and limb regeneration. *Dev Comp Immunol* 31: 383–393.

Mescher AL, Neff AW, King MW. 2013. Changes in the inflammatory response to injury and its resolution during the loss of regenerative capacity in developing *Xenopus* limbs. *PLoS One* 8: e80477.

Muneoka K, Holler-Dinsmore G, Bryant SV. 1986. Intrinsic control of regenerative loss in *Xenopus laevis* limbs. *J Exp Zool* 240: 47–54.

Nieuwkoop PD, Faber J (eds.). 1994. *Normal table of* Xenopus laevis *(Daudin): A systematical & chronological survey of the development from the fertilized egg till the end of metamorphosis.* Garland Publishing, New York.

Nye HLD, Cameron JA. 2005. Strategies to reduce variation in *Xenopus* regeneration studies. *Dev Dyn* 234: 151–158.

Rollins-Smith LA, Flajnik MF, Blair PJ, Davis AT, Green WF. 1997. Involvement of thyroid hormones in the expression of MHC class I antigens during ontogeny in *Xenopus*. *Dev Immunol* 5: 133–144.

Ruben LN, Stevens JM, Kidder GM. 1972. Suppression of the allograft response by implants of mature lymphoid tissues in larval *Xenopus laevis*. *J Morph* 138: 457–465.

Schotté OE, Harland M. 1943. Amputation level and regeneration in limbs of late *Rana clamitans* tadpoles. *J Morphol* **73:** 329–362.

Shimizu-Nishikawa K, Takahashi J, Nishikawa A. 2003. Intercalary and supernumerary regeneration in the limbs of the frog, *Xenopus laevis*. *Dev Dyn* **227:** 563–572.

Sive HL, Grainger RM, Harland RM. 2000. *Early development of* Xenopus laevis: *A laboratory manual.* Cold Spring Harbor Laboratory Press, Cold Spring Harbor, NY.

Slack JMW, Beck CW, Gargioli C, Christen B. 2004. Cellular and molecular mechanisms of regeneration in *Xenopus*. *Philos Trans R Soc Lond B Biol Sci* **359:** 745–751.

Song F, Li B, Stocum DL. 2010. Amphibians are research models for regenerative medicine. *Organogenesis* **6:** 141–150.

Tschumi PA. 1957. The growth of the hindlimb bud of *Xenopus laevis* and its dependence upon the epidermis. *J Anat* **91:** 149–173.

Cite this protocol as *Cold Spring Harb Protoc*; doi:10.1101/pdb.prot100990

Cell Transplantation as a Method to Investigate Spinal Cord Regeneration in Regenerative and Nonregenerative *Xenopus* Stages

Emilio E. Méndez-Olivos[1] and Juan Larraín[1,2]

[1]*Center for Aging and Regeneration, Faculty of Biological Sciences, Pontificia Universidad Católica de Chile, Alameda 340, Santiago, Chile*

Mammals are not capable of regenerating their central nervous system (CNS); anamniotes, however, can regenerate in response to injury. The mechanisms that explain the different regenerative capabilities include: (i) extrinsic mechanisms that consider the cellular environment and extracellular matrix composition, (ii) intrinsic factors implicating the presence or absence of genetic programs that promote axon regeneration, and (iii) the presence or absence of neural stem and progenitors cells (NSPCs) that allow neurogenesis. *Xenopus laevis* is able to regenerate its CNS during larval stages (i.e., the regenerative stage [R-stage]). However, concomitant with metamorphosis this capacity decreases and is lost completely in juvenile froglets (i.e., nonregenerative stages [NR-stages]). The loss of the regenerative ability correlates with a reduction in the percentage of Sox2$^+$ cells, which are putative NSPCs. This protocol shows the effect of transplantation of spinal cord cells from R-stage *Xenopus* larvae into NR-stage froglets. Using this procedure, it is possible to study axon regeneration and stem cell biology in vivo.

MATERIALS

It is essential that you consult the appropriate Material Safety Data Sheets and your institution's Environmental Health and Safety Office for proper handling of equipment and hazardous materials used in this protocol.

RECIPES: Please see the end of this protocol for recipes indicated by <R>. Additional recipes can be found online at http://cshprotocols.cshlp.org/site/recipes.

Reagents

Bovine serum albumin (BSA; 0.5 mg/mL in L-15 Medium)

Ethyl 3-aminobenzoate methanesulfonate salt (0.02% [w/v] in 0.1× MBS [MS-222; Sigma-Aldrich A5040])

The solution can be stored for 1 mo at 4°C.

Fibrinogen (100 mg/mL in phosphate-buffered saline [PBS]; Sigma-Aldrich F6755)

Final medium <R>

L-15 Medium (Sigma-Aldrich L4386)

[2]Correspondence: jlarrain@bio.puc.cl

Modified Barth's solution (MBS) (1×) <R>

Penicillin

StemPro Accutase cell dissociation reagent (Gibco A1110501)

This enzyme has protease and collagenase activity to allow tissue and cell dissociation.

Streptomycin

Thrombin (Sigma-Aldrich T5772; 100 U/mL in PBS)

Thrombin cleaves fibrinogen into fibrin to form a clot.

Trypan Blue solution (Sigma-Aldrich T8154)

Trypan Blue is a vital stain that allows discrimination between dead and living cells because it is only incorporated by cells with a damaged membrane.

Xenopus laevis animals (stages NF50 and NF66; Nieuwkoop and Faber 1994)

Transgenic animals make the identification of donor and host cells easier.

Equipment

Forceps (e.g., FST, Student Dumont #5)

Hemocytometer

Light source (e.g., Fiber-Lite Mi-150 Illuminators [Dolan-Jenner])

Microcentrifuge

Microscissors, straight (e.g., Vannas-Tübingen spring scissors [FST 15003-08])

Needle (23-g) (optional; see Step 13)

Parafilm M (Sigma-Aldrich P7793)

Petri dish (glass, 100-mm)

Stereomicroscope (e.g., Nikon, SMZ745T)

Tissue paper (e.g., Kimwipes)

Tubes (microcentrifuge, 1.5-mL)

Vortex mixer (e.g., V1 Plus [Boeco])

METHOD

Spinal Cord Isolation and Tissue Dissociation

Use three to four tadpoles to obtain sufficient cells per each froglet to be transplanted.

1. Place a larva at NF Stage 50 into a Petri dish with 50 mL of MS-222. After 3 min, check for an escape reflex.

   The absence of a response indicates that the animal is anesthetized.

2. Transfer the larva to a dissection surface under the stereomicroscope.

3. Remove and discard the head and the posterior third of the tail.

4. Holding the remaining part of the trunk, use tweezers to pull the spinal cord out of the body.

   See Troubleshooting.

5. Transfer the spinal cord to a 1.5-mL microcentrifuge tube containing 500 µL of Accutase. Seal the tube with Parafilm.

   i. Incubate at room temperature with agitation (e.g., using a vortex mixer) for 1 h.

   ii. To avoid having the spinal cords adhere to the walls of the tube, mix the tubes by manual inversion every 15 min.

6. Add 1 mL of 0.5 mg/mL BSA in L-15 Medium to inactivate the Accutase.

7. Pellet the cells by centrifugation at 200*g* for 5 min. Resuspend the pellet in 1 mL of final medium.

8. Stain a 10-μL aliquot with trypan blue. Count the cells using a hemocytometer.

    See Troubleshooting.

9. Pellet the cells again by centrifugation at 200*g* for 5 min. Carefully remove and discard the supernatant.

10. Resuspend the cells in final medium to attain a concentration of 1×10^5 cells/μL.

    There is no need for further supplementation with growth factors.

Spinal Cord Resection

11. Place a froglet at NF Stage 66 into a Petri dish containing 50 mL MS-222 for 10 min. Check for the absence of an escape reflex.

    See Troubleshooting.

12. Place the froglet under a stereomicroscope.

13. Using microscissors (or a 23-g needle) as a scalpel, open the dorsal skin with a longitudinal cut above the spinal cord between the fifth and sixth vertebrae.

14. Carefully lift the dorsal muscles to expose the vertebrae.

15. To expose the spinal cord, use a pair of tweezers to perform a dorsal laminectomy and remove the fifth and sixth vertebrae.

16. Using microscissors, perform a 1-mm resection of the spinal cord.

    Make sure that no uninjured nerve tracts are left.

17. Blot the area using tissue paper until bleeding stops. Leave a cavity between both stumps to place the cells.

Transplantation Procedure

18. Mix 1 μL of the cell suspension (1×10^5 cells/μL) obtained from the NF Stage 50 spinal cord (Step 10) with 0.5 μL of fibrinogen and 0.5 μL of thrombin.

    Clot formation will begin as soon as the thrombin and fibrinogen mix. The growth factors in the suspension are diluted to a final concentration of 10 ng/mL.

19. Immediately transfer the mixture to the resected site using a micropipette.

    See Troubleshooting.

20. Two minutes after cell transplantation, reappose the dorsal muscle and skin. Hold closed with tweezers for a few seconds.

    Avoid disturbing the clot in the spinal cord.

21. Carefully transfer the froglet to a recipient Petri dish containing 0.1× MBS supplemented with penicillin (100 μg/mL) and streptomycin (100 μg/mL). Use a volume just sufficient to cover the animal.

    See Troubleshooting.

TROUBLESHOOTING

Problem (Step 4): The spinal cord breaks during the dissection.
Solution: Try to remove the surrounding muscle to expose the spinal cord before pulling it out.

Problem (Step 8): Low amount of living cells or clusters of cells
Solution: We use 60 spinal cords per mL of the enzyme. With higher numbers, there is no guarantee of a good yield.

Problem (Step 11): Froglets are still awake after 10 min.
Solution: Froglets weighing 1 g or less should be anesthetized in 10 min. Larger animals take longer.

Problem (Step 19): The quantity of liquid overflows the injury site.
Solution: The cavity should hold up to 2 μL. If the liquid overflows the site, check the size of resection. Larger volumes require larger pieces to be resected.

Problem (Step 21): The froglets die soon after the surgery.
Solution: Allow the animals to recover in a volume just barely sufficient to cover, as they will be unable to swim to reach the surface to breathe.

DISCUSSION

This procedure was adapted from transplantation experiments performed in rats (Lu et al. 2012). Here we use froglets at NF Stage 66 (Nieuwkoop and Faber 1994) as host, but we have obtained very similar results using NF Stage 56. It takes twenty to thirty days for the transplanted cells to populate the lesion site; their progression can be studied using classic hematoxylin and eosin staining of paraffin or cryosections. Although the histological organization is sufficiently different between donor and host tissue, the use of transgenic animals expressing fluorescent proteins in a tissue-specific manner makes the distinction between the two easier. Transplanted animals can be maintained up to 60 d without problems. The efficiency of this procedure can be improved by testing other growth factors (as was demonstrated on rodents [Lu et al. 2012]), performing transplantations at different days after resection, making the lesion in an anterior part of the spinal cord, or using animals at different stages as donors or hosts. The effects of immunosuppressing the host froglets can also be tested by thymectomy, as previously performed in limb transplantation experiments (Lin et al. 2013).

This procedure is compatible with anterograde tracing to study axon regeneration, gain- and loss-of-function experiments, pharmacological studies, and functional recovery analysis. Given that there is no regeneration after transection in froglets (Beattie et al. 1990; Gaete et al. 2012; Muñoz et al. 2015), this procedure provides an opportunity to understand the mechanisms that allow NSPCs to improve spinal cord regeneration, a process that has been poorly studied (Assinck et al. 2017).

RECIPES

Final Medium

Reagent	Final concentration
BDNF (R&D Systems 248-BD-025/CF)	20 ng/mL
Bovine serum albumin (Sigma-Aldrich)	0.5 mg/mL
EGF (R&D Systems 2028-EG-200)	20 ng/mL
FGF (R&D Systems 3139-FB-025/CF)	20 ng/mL

Prepare fresh in L-15 Medium (Sigma-Aldrich L4386).

Modified Barth's Solution (MBS) (1×)

Reagent	Final concentration (1×)
$CaCl_2$	0.41 mM
$Ca(NO_3)_2$	0.33 mM
HEPES	10 mM
KCl	1.0 mM
$MgSO_4$	0.82 mM
NaCl	88 mM
$NaHCO_3$	2.4 mM

Adjust the pH to 7.4. Prepare as a 10× solution and then autoclave. Store at 4°C for up to 1 mo. Dilute to 0.1× and add antibiotics before use, and store for up to 3 d.

ACKNOWLEDGMENTS

This work was supported by funds from CARE Chile UC-Centro de Envejecimiento y Regeneración (PFB 12/2007), Fondo Nacional de Desarrollo Científico y Tecnologíco (FONDECYT) (1141162) to J.L. E.E.M.-O. is a Comisión Nacional de Investigación Científica y Tecnológica (CONICYT) Ph.D. fellow 21120400.

REFERENCES

Assinck P, Duncan GJ, Hilton BJ, Plemel JR, Tetzlaff W. 2017. Cell transplantation therapy for spinal cord injury. *Nat Neurosci* 20: 637–647.

Beattie MS, Bresnahan JC, Lopate G. 1990. Metamorphosis alters the response to spinal cord transection in *Xenopus laevis* frogs. *J Neurobiol* 21: 1108–1122.

Gaete M, Muñoz R, Sánchez N, Tampe R, Moreno M, Contreras EG, Lee-Liu D, Larraín J. 2012. Spinal cord regeneration in *Xenopus* tadpoles proceeds through activation of Sox2-positive cells. *Neural Dev* 7: 13.

Lin G, Chen Y, Slack JMW. 2013. Imparting regenerative capacity to limbs by progenitor cell transplantation. *Dev Cell* 24: 41–51.

Lu P, Wang Y, Graham L, McHale K, Gao M, Wu D, Brock J, Blesch A, Rosenzweig ES, Havton LA, et al. 2012. Long-distance growth and connectivity of neural stem cells after severe spinal cord injury. *Cell* 150: 1264–1273.

Muñoz R, Edwards-Faret G, Moreno M, Zuñiga N, Cline HT, Larraín J. 2015. Regeneration of *Xenopus laevis* spinal cord requires Sox2/3 expressing cells. *Dev Biol* 408: 229–243.

Nieuwkoop PD, Faber J (eds.). 1994. *Normal table of* Xenopus laevis *(Daudin): A systematical & chronological survey of the development from the fertilized egg till the end of metamorphosis.* Garland Publishing, New York.

Methods for Examining Lens Regeneration in *Xenopus*

Jonathan J. Henry,[1,3] Kimberly J. Perry,[1] and Paul W. Hamilton[2]

[1]*Department of Cell and Developmental Biology, University of Illinois, Urbana, Illinois 61801;* [2]*Department of Biology, Illinois College, Jacksonville, Illinois 62650*

Some vertebrates are able to regenerate the lens following its removal. This includes species in the genus *Xenopus* (i.e., *X. laevis*, *X. tropicalis*, and *X. borealis*), the only anurans known to undergo lens regeneration. In *Xenopus* the regenerated lens is derived de novo from cells located within the basal-most layer of the larval corneal epithelium, and is triggered by factors provided by the neural retina. In larval frogs the corneal epithelium is underlain by an endothelium separated from the corneal epithelium except for a small central attachment (i.e., the "stromal-attracting center"). This connection grows larger as the stroma forms and the frogs approach metamorphosis. Here we provide instructions for performing lentectomies (removal of the original lens) to study lens regeneration.

MATERIALS

It is essential that you consult the appropriate Material Safety Data Sheets and your institution's Environmental Health and Safety Office for proper handling of equipment and hazardous materials used in this protocol.

RECIPES: Please see the end of this protocol for recipes indicated by <R>. Additional recipes can be found online at http://cshprotocols.cshlp.org/site/recipes.

Reagents

1/20× NAM (normal amphibian medium) <R>

Frog culture water

> Use autoclaved dechlorinated tap water (or salt-balanced reverse-osmosis water).

Tadpole anesthetic/antibiotic solution <R>

Tadpole antibiotic solution <R>

Xenopus embryos

> The capacity for regeneration is high in larval X. laevis *and weaker in some other species because of the more rapid healing of the corneal endothelium, which prematurely cuts off the inducing factors (Freeman 1963; Henry and Elkins 2001; Filoni et al. 1997, 2006). Stages 48–52 or even younger larvae are typically used for these experiments, as both the rate of regeneration (i.e., the growth of the new lens) and the number of animals that can undergo regeneration gradually declines as frogs approach metamorphosis; animals can no longer regenerate a lens after metamorphosis (Stage 66; Freeman [1963]).*

---

[3]Correspondence: j-henry4@illinois.edu

Cite this protocol as *Cold Spring Harb Protoc*; doi:10.1101/pdb.prot101527

Equipment

Clay, plastalina, dark green (Van Aken International 10119)

Forceps, #5 Dumont INOX (Polysciences Inc. 07379)

Sharp, perfectly matched tips are essential to grasp the lens capsule. Standard INOX forceps must be sharpened. Use a small rubber band to hold the forceps tips together in alignment. Sharpen using an EZE-LAP Pocket Diamond Fish Hook Sharpener, Model S. The cutting action of diamonds is very aggressive, so proceed carefully to avoid shortening the forceps and making them too blunt (see "before" and "after" images in the inset of Fig. 1O).

Glass rod tool

Use a gas burner to melt the tip of a short glass Pasteur pipette. Heat only the tip, rotating the glass for even heating. The open end should seal completely and a small rounded ball will form at the tip, ~1–2 mm in diameter.

Hood, sterile

Microscissors (Oban BioScissors, Oban Precision Instruments)

Standard scissors tend to push tissues out and away from the cutting edges, making these operations more difficult. Oban BioScissors are ideal for these surgeries because they have a pincer-type action that grips the corneal tissue just before the cut is made. Similar scissors can be made from #5 forceps.

Microscope, dissection

Plastic dish, 60-mm diameter

Line the bottom of the dish with clay, spreading it evenly with one's fingertips. The clay layer should be ~7–8 mm thick.

Transfer scoop

A small stainless steel or plastic spoon, used to handle the animals.

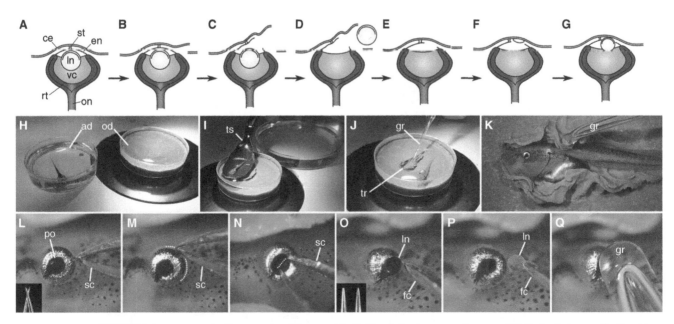

FIGURE 1. Lentectomy in *Xenopus laevis* larvae. (*A–G*) Simple lentectomy diagram. (*H*) Tadpole is anesthetized in a dish containing anesthetic/antibiotic solution, transferred to the operating dish (*I*) following sedation, and immobilized on its side in a clay trough (*J*). Flaps of clay are used to secure the tadpole (*K*), and lentectomy begins with an incision on the posterior side of the eye (*L–M*; microscissors are shown in the inset, dotted line indicates the incision). (*N*) An incision is made across the pupillary opening in the corneal endothelium and the intact lens capsule (containing the lens cells) is removed (*O–P*). *Inset* in *O* shows forceps before (*left*) and after (*right*) sharpening. The corneal epithelium is smoothed over the eye with a glass rod tool (*Q*). Abbreviations: (ad) anesthetic dish, (ce) corneal epithelium, (en) corneal endothelium, (fc) forceps, (gr) glass rod, (ln) lens, (od) operating dish, (on) optic nerve, (po) pupillary opening, (rt) retina, (sc) scissors, (st) connecting stalk, (ts) transfer spoon, (tr) trough, (vc) vitreous chamber.

METHOD

Lentectomies must be carried out using a dissection microscope, ideally located in a clean hood. Use sterile disposable plasticware. Sanitize all work surfaces, tools, and operating dishes with 70% ethanol before and after use.

The procedure has been described by Freeman (1963), Henry and Mittleman (1995) and Beck (2012) and is shown schematically in Figure 1A–G.

Lentectomy Operation

1. Anesthetize the animal in a small Petri dish containing anesthetic/antibiotic solution (Fig. 1H) for 2–3 min until it is unresponsive to touch and has lost the righting reflex.

   To limit the animal's time of exposure, no more then two animals should be anesthetized together in this dish. Typically, it takes only 1–3 min to complete each surgery. Ideally, anesthesia should be limited to no longer than 10 min (Hamilton and Henry 2014).

2. Transfer the anesthetized animal to the clay-lined operating dish using the transfer scoop (Fig. 1I).

   Animals should remain covered with anesthetic/antibiotic solution.

3. Using the glass rod tool, form a small trough in the clay the same size and depth as the tadpole (Fig. 1J). Raise up a few folds of clay along the sides of the trough to use as flaps to restrain the animal.

   This recess can be used for subsequent operations.

4. Using the glass rod tool, position the animal in the trough on its side with one eye upward for surgery (Fig. 1K). Gently push the flaps of clay down along the sides of the abdomen to secure the animal.

   Avoid compressing the heart region, which will continue to beat under anesthesia.

5. Use microscissors to cut along one edge of the corneal epithelium (Fig. 1B,L).

6. Make a series of continuous cuts around one-third of the circumference of the eye along the posterior edge (Fig. 1M).

   If the cut is too big the cornea can flap open and the wound might not heal properly.

7. Reach under the cut epithelium with microscissors to pierce the deeper corneal endothelium (Fig. 1C,N):

   This is the more difficult step, as the eyeball tends to move.

   i. To steady the eyeball, approach the eye from straight above, at a high angle, with the tips of the scissors initially straddling the pupillary opening.

   If you approach the eye from a lower angle the scissors will not cut the tissue, and will tend to slide off the slippery surface.

   ii. Make a straight cut through the cornea endothelium across the center of the pupillary opening.

   Some force is required, but not enough to damage the deeper lens tissues. Care must be taken not to cut the iris or retinal tissue at the edge of the pupillary opening, as there are large blood vessels located there.

8. Taking care to grasp only the exposed outer surface (i.e., the lens capsule), use sharpened forceps to grab the lens capsule and remove it (Fig. 1D,O–P). Hold the lens by only the very tips of the forceps.

   The lens should be removed intact and undamaged. An undamaged lens will appear uniformly spherical and completely transparent. Wipe the forceps clean before and after each use to remove any lens debris from previous surgeries.

   See Troubleshooting.

9. Smooth the cornea back into place with the glass rod tool to help close the wound (Fig. 1E,Q).

10. Release the clay flaps with the glass rod tool. Remove the animal from the recess.

 Cite this protocol as *Cold Spring Harb Protoc*; doi:10.1101/pdb.prot101527

11. Over the course of 10 min, transfer the animal through three washes of antibiotic solution. Transfer the animal to a larger recovery bowl containing antibiotic solution.

    Gentle agitation on a rocker will hasten recovery from the anesthetic (Hamilton and Henry 2014). The cornea heals rapidly within 12–24 h; continue antibiotic use throughout this period. After 24 h, frog culture water can be used instead of 1/20× NAM. Antibiotics are not required once the cornea has healed; lens regeneration will occur whether or not antibiotics are used.

12. At the conclusion of the experiments (or in cases where the lens cannot be completely removed) euthanize animals following accepted standards (e.g., 2–3 g/L of MS222 for larvae; 250 mg/L benzocaine HCl for adult animals).

TROUBLESHOOTING

Problem (Step 8): The lens is irregular in shape or cloudy.

Solution: The lens has been damaged. It is likely that lens cells were left inside the vitreous chamber; any remaining lens epithelial tissues can proliferate and reform a lens (i.e., an alternative form of lens regeneration). One should therefore assume that if the lens has not been removed intact, lens cells are likely to be present; those animals should be discarded.

DISCUSSION

The stages of lens regeneration have been thoroughly described by Freeman (1963; see also Henry 2003; Henry et al. 2008; Henry and Tsonis 2010). Distinct cellular changes occur within the first 24 h following lens removal: cells of the basal corneal epithelium become cuboidal and the nuclei in a significant percentage of these cells will possess only one nucleolus, rather than two. A small lentoid (placode) that expresses lens crystallin proteins generally appears within 5 d, which can be recognized in histological sections (Fig. 1F). At that time the new lens cells begin to express lens crystallins that can be detected by antibodies (Henry and Mittleman 1995). These lentoids will continue to grow larger over time to form spherical lenses within one to two weeks. These lenses will become visible through the pupillary opening in live animals (Fig. 1G). One can also visualize lens regeneration using live transgenic animals that express GFP driven by the γ-crystallin enhancer/promoter (Henry and Elkins 2001). The regenerated lens will eventually grow to reach the original size in a few weeks to a month.

RECIPES

1/20× NAM (Normal Amphibian Medium)

Reagent	Volume
10× NAM salts solution <R>	5 mL
0.1 M NaPO$_4$, pH 7.5 <R>	20 mL
0.1 M NaHCO$_3$ <R>	0.5 mL

Bring to 1 L with autoclaved dH$_2$O. Prepare using sterile technique; do not resterilize this working solution. (This recipe was adapted from Slack 1984.)

0.1 M NaHCO$_3$

Dissolve 0.42 g NaHCO$_3$ in 50 mL of dH$_2$O. Filter-sterilize using a 0.2-μm surfactant-free cellulose acetate membrane filter unit. Store frozen as 0.5-mL aliquots.

0.1 M NaPO₄, pH 7.5

Reagent	Quantity	Concentration
$Na_2HPO_4 \cdot 7H_2O$	22.5 g	84 mM
$NaH_2PO_4 \cdot H_2O$	2.2 g	16 mM

Bring up to 1 L with dH_2O. Adjust pH to 7.5. Autoclave.

10× NAM Salts Solution

Reagent	Quantity	Concentration
NaCl	65.0 g	110 mM
KCl	1.5 g	2 mM
$Ca(NO_3)_2 \cdot 4H_2O$	2.4 g	1 mM
$MgSO_4 \cdot 7H_2O$	2.5 g	1 mM
$Na_2EDTA \cdot 2H_2O$	0.4 g	1 mM

Bring up to 1 L with dH_2O. Autoclave.

Modified Leibovitz's L-15 Medium

1. Add 13.8 g of L-15 powdered medium (Leibovitz's L-15 medium, with glutamine; Sigma-Aldrich L4386) to 1300 mL of dH_2O. Mix well.
2. Adjust pH to 7.6 with 1 N NaOH or 1 N HCl.
3. Add 149 mL of fetal bovine serum (FBS, certified, heat-inactivated; Gibco 10082147).
4. Adjust the final volume to 1.493 L.
5. Filter-sterilize with a 0.2-µm surfactant-free cellulose acetate membrane filter unit.
6. Freeze the medium or maintain it at 4°C for up to 4 wk.

Tadpole Anesthetic/Antibiotic Solution

Reagent	Concentration
Penicillin G (Sigma-Aldrich P3032)	100 U/mL
Streptomycin (Sigma-Aldrich S1277)	100 µg/mL
Tricaine methanesulfonate (MS-222; Sigma-Aldrich A5040)	1:2000 (w:v)

Prepare fresh, just before use, in 1/20× NAM (Normal Amphibian Medium) <R>.

Tadpole Antibiotic Solution

Reagent	Concentration
Penicillin (Sigma-Aldrich P3032)	100 U/mL
Streptomycin (Sigma-Aldrich S1277)	100 µg/mL

Prepare fresh, just before use, in modified Leibovitz's L-15 medium <R>.

ACKNOWLEDGMENTS

J.J.H. is funded by National Institutes of Health/National Eye Institute (NIH/NEI) grant EYO23979.

REFERENCES

Beck CW. 2012. Studying regeneration in *Xenopus*. In Xenopus *protocols* (eds. Hoppler S, Vise P), pp. 525–539. Humana Press, Totowa, NJ.

Filoni S, Bernardini S, Cannata SM, D'Alessio A. 1997. Lens regeneration in larval *Xenopus laevis*: Experimental analysis of the decline in the regenerative capacity during development. *Dev Biol* **187:** 13–24.

Filoni S, Bernardini S, Cannata SM. 2006. Experimental analysis of lens-forming capacity in *Xenopus borealis* larvae. *J Exp Zoolog A Comp Exp Biol* **305:** 538–550.

Freeman G. 1963. Lens regeneration from the cornea in *Xenopus laevis*. *J Exp Zool* **154:** 39–65.

Hamilton PW, Henry JJ. 2014. Prolonged in vivo imaging of *Xenopus laevis*. *Dev Dyn* **243:** 1011–1019.

Henry JJ. 2003. The cellular and molecular bases of vertebrate lens regeneration. *Int Rev Cytol* **228:** 195–264.

Henry JJ, Elkins ME. 2001. Cornea-lens transdifferentiation in the anuran, *Xenopus tropicalis*. *Dev Genes Evol* **211:** 377–387.

Henry JJ, Mittleman J. 1995. The matured eye of *Xenopus laevis* tadpoles produces factors that elicit a lens-forming response in embryonic ectoderm. *Dev Biol* **171:** 39–50.

Henry JJ, Tsonis PA. 2010. Molecular and cellular aspects of amphibian lens regeneration. *Prog Retin Eye Res* **29:** 543–555.

Henry JJ, Wever JM, Veragara MN, Fukui L. 2008. *Xenopus*, an ideal vertebrate system for studies of eye development and regeneration. In *Animal models for eye research* (ed. Tsonis PA), pp. 57–92. Academic Press, Amsterdam.

Slack JMW. 1984. Regional biosynthetic markers in the early amphibian embryo. *J Embryol Exp Morphol* **80:** 289–319.

Protocol 4

Ex Vivo Eye Tissue Culture Methods for *Xenopus*

Jonathan J. Henry,[1,3] Kimberly J. Perry,[1] and Paul W. Hamilton[2]

[1]*Department of Cell and Developmental Biology, University of Illinois, Urbana, Illinois 61801;* [2]*Department of Biology, Illinois College, Jacksonville, Illinois 62650*

Lens regeneration can be studied in whole animals following removal of the original lens (lentectomy). However, culturing a whole animal can be impractical for assays involving small molecule inhibitors or proteins. Ex vivo eye tissue culture is an alternative approach for examining lens regeneration. The ex vivo culture system offers certain advantages when compared to the in vivo regeneration assay, as the percentage of cases showing lens differentiation can exceed that seen in whole animals. This culture system also allows for the treatment of eye tissues in small volumes, which helps ensure reproducibility and reduces the amount (and cost) of small-molecule inhibitors or exogenous proteins, etc., necessary to conduct an experiment. Additionally, different eye tissues can be combined, such as nontransgenic and transgenic tissues (e.g., eyecup and cornea) that carry reporters or inducible transgenes. This approach represents a very useful tool in the analysis of lens regeneration or for simply culturing specific eye tissues, and can be used to culture either *Xenopus laevis* or *Xenopus tropicalis* eye tissues.

MATERIALS

It is essential that you consult the appropriate Material Safety Data Sheets and your institution's Environmental Health and Safety Office for proper handling of equipment and hazardous materials used in this protocol.

RECIPES: Please see the end of this protocol for recipes indicated by <R>. Additional recipes can be found online at http://cshprotocols.cshlp.org/site/recipes.

Reagents

1/20× NAM (normal amphibian medium) <R>
Anesthetic solution

> *Dilute tricaine methanesulfonate (MS222; Sigma-Aldrich A5040) 1:2000 (w:v) in 1/20× NAM.*

Frog culture water

> *Use autoclaved dechlorinated tap water (or salt-balanced reverse-osmosis water).*

Xenopus eye culture medium <R>
Xenopus tadpoles

> *Stages 48–52 or even younger larvae are typically used for these experiments.*

Equipment

Clay, plastalina, dark green (Van Aken International 10119)

[3]Correspondence: j-henry4@illinois.edu

Forceps, #5 Dumont INOX, super-thin tips (Polysciences Inc. 07379)

Sharp, perfectly matched tips are essential to grasp the lens capsule. Standard INOX forceps must be sharpened. Use a small rubber band to hold the forceps tips together in alignment. Sharpen using an EZE-LAP Pocket Diamond Fish Hook Sharpener, Model S. The cutting action of diamonds is very aggressive, so proceed carefully to avoid shortening the forceps and making them too blunt.

Glass rod tool

Use a gas burner to melt the tip of a short glass Pasteur pipette. Heat only the tip, rotating the glass for even heating. The open end should seal completely and a small rounded ball will form at the tip, ~1–2 mm in diameter.

Hood, sterile

Microscissors (e.g., Oban BioScissors, Oban Precision Instruments)

Standard scissors tend to push tissues out and away from the cutting edges, making these operations more difficult. BioScissors are ideal for these surgeries because they have a pincer-type action that grip the cornea tissue just before the cut is made. Similar scissors can be made from #5 forceps.

Microscope, dissection

Petri dish, plastic, 60-mm diameter

Line the bottom of the dish with clay, spreading it evenly with one's fingertips. The clay layer should be ~7–8 mm thick.

Petri dishes, plastic, 35-mm diameter

Pipettor, automatic

Tissue culture dish (e.g., 24-well)

Transfer scoop

A small stainless steel or plastic spoon, used to handle the animals.

Transfer tips for pipettes

Cut aerosol-free plastic pipette tips with a razor blade to an opening of ~2-mm diameter. Autoclave before use.

METHOD

Lentectomies must be carried out using a dissection microscope, ideally located in a clean hood. Use sterile disposable plasticware. Sanitize all work surfaces, tools, and operating dishes with 70% ethanol before and after use.

1. Perform a lentectomy on the *Xenopus* tadpole as described in Steps 1–10 of Protocol 3: Methods for Examining Lens Regeneration in *Xenopus* (Henry et al. 2018) on the host eye's vitreous chamber (Fig. 1D–G), except that during Step 6 of that procedure, cut the cornea completely around the circumference of the eye (Fig. 1E,K–L). The cut cornea will remain attached to the eye by the central stalk (i.e., the stromal attracting center, Fig. 1E,L).

   Rotate the operating dish during the cutting procedure so that the cutting hand does not have to be relocated. To aid in this process, a small turntable made of magnetic stainless steel that fits into the base of the dissection scope can be used. Self-adhesive magnetic tape can be affixed to the bottom of the operating dish to help it adhere to the turntable for greater stability. Construction of this turntable is beyond the scope of this article.

2. Use the microscissors to cut the central stalk to detach the cornea (Fig. 1F). Retain or discard the cornea, as needed.

3. In the same operating dish, have a second animal ready as the cornea donor. Use the microscissors to cut the corneal epithelium around the entire periphery of the eye (Fig. 1A–B).

4. Once the donor cornea is cut free along the periphery, use the microscissors to cut the central stalk (Fig. 1C). Hold onto the cornea with forceps, as it is transparent and easy to lose in the dish.

   As a variation of this procedure, one can use the host's own cornea. In this case, there is no need to cut the central attachment. The persistent attachment can help to ensure that the cornea remains tucked in the vitreous chamber.

5. Using the sharpened forceps, grab the edge of the freed donor cornea and tuck the cornea inside the host's vitreous chamber (Fig. 1H,M).

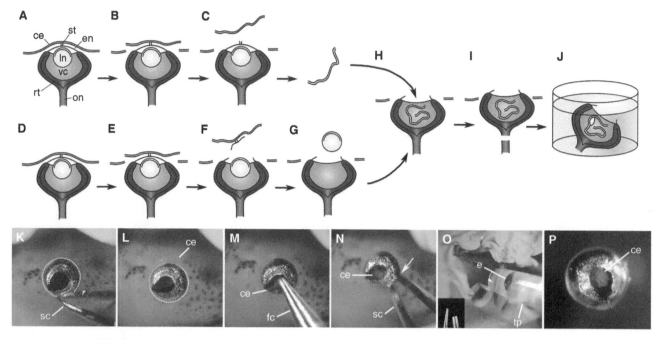

FIGURE 1. Ex vivo eye culture protocol to combine excised host and donor eye tissues (*A–J*). The donor corneal epithelium (*A–C*) is removed (*K–L*) and implanted inside the vitreous chamber of the host eye (*D–H,M*). The eye is then excised (*I,N*) and transferred to a culture well (*J,O–P*). *Arrow* in *N* shows where the attachments are severed to remove the eye. *Inset* in *O* shows pipette tip before (*left*) and after (*right*) trimming. Abbreviations: (ce) corneal epithelium, (e) eye, (en) endothelium, (fc) forceps, (ln) lens, (on) optic nerve, (rt) retina, (sc) scissors, (st) connecting stalk, (tp) transfer pipette, (vc) vitreous chamber.

> *The corneal tissue tends to stick to the forceps. If necessary, use the glass ball tool to hold the corneal tissue inside the eye while removing the forceps from the eyecup.*

6. Excise the eyecup (with the cornea tucked inside) by using microscissors to cut all of the eye muscle attachments, as well as the optic nerve and blood vessels that lie deeper to the eye. Make the cuts carefully and do not squeeze the eyeball, as the cornea can come out during this process (Fig. 1I,N).

   > *See Troubleshooting.*

7. Use an automatic pipettor fitted with a sterilized pipette tip to transfer the freshly enucleated eyecup to a 35-mm dish containing *Xenopus* eye culture medium. Over the course of a few minutes, serially transfer the eyecup to three dishes of this culture medium (i.e., three washes).

   > *Use one pipette tip for the initial transfer and a fresh tip for subsequent transfers.*

8. Transfer the eyecup to the well of a tissue culture dish containing 500–1000 µL of *Xenopus* eye culture medium.

   > *If necessary, up to five eyes can be placed in a single well (Fig. 1J,P). However, the eyes must not touch, otherwise they might fuse together. Once an eye has been placed in the culture well, verify that the cornea has remained tucked inside the eye.*

9. Change culture media daily or every other day. Monitor for bacterial or fungal contamination. Immediately discard cultures that show any bacterial or fungal growth.

   > *The medium will appear cloudy and the pH indicator will change to yellow if there is contamination.*

10. Maintain cultures as needed for the experimental purposes.

    > *We typically culture eyes for up to 14 d, although the cultures can be maintained longer under these conditions. Lens formation can be observed in as little as 7 d; longer times will result in larger, more well-developed lenses (Fukui and Henry 2011).*

    > *See Troubleshooting.*

Cite this protocol as *Cold Spring Harb Protoc*; doi:10.1101/pdb.prot101535

TROUBLESHOOTING

Problem (Step 6): The cornea starts to protrude from the eyecup.
Solution: Tuck the cornea back in with the forceps.

Problem (Step 10): Corneas are expelled from the eyecup during the culturing process.
Solution: This can occur because the vitreous chamber tends to contract following lens removal. Therefore, the cornea diameter might need to be trimmed to make it fit inside the vitreous chamber. Trimming the cornea is easier to accomplish before the central stalk is cut (Step 4).

DISCUSSION

The procedure for establishing eye cultures in *Xenopus laevis* was described by Fukui and Henry (2011). These methods can also be used to isolate and culture any eye tissue, such as the retina, cornea, or lens, as well as whole eyes. Because these tissues can be cultured in a relatively small volume (0.5–1 mL), one is able to treat these tissues with reduced amounts of costly small molecule inhibitors, activators, or exogenous proteins, including growth factors, etc., as might be necessary to undertake certain experiments. Additionally, different eye tissues can be combined, such as nontransgenic and transgenic tissues (e.g., eyecup and cornea) that carry reporters or inducible transgenes (Fukui and Henry 2011; Hamilton and Henry 2014).

RECIPES

1/20× NAM (Normal Amphibian Medium)

Reagent	Volume
10× NAM salts solution <R>	5 mL
0.1 M NaPO$_4$, pH 7.5 <R>	20 mL
0.1 M NaHCO$_3$ <R>	0.5 mL

Bring to 1 L with autoclaved dH$_2$O. Prepare using sterile technique; do not resterilize this working solution. (This recipe was adapted from Slack 1984.)

0.1 M NaHCO$_3$

Dissolve 0.42 g NaHCO$_3$ in 50 mL of dH$_2$O. Filter-sterilize using a 0.2-µm surfactant-free cellulose acetate membrane filter unit. Store frozen as 0.5-mL aliquots.

0.1 M NaPO$_4$, pH 7.5

Reagent	Quantity	Concentration
Na$_2$HPO$_4$·7H$_2$O	22.5 g	84 mM
NaH$_2$PO$_4$·H$_2$O	2.2 g	16 mM

Bring up to 1 L with dH$_2$O. Adjust pH to 7.5. Autoclave.

10× NAM Salts Solution

Reagent	Quantity	Concentration
NaCl	65.0 g	110 mM
KCl	1.5 g	2 mM
$Ca(NO_3)_2 \cdot 4H_2O$	2.4 g	1 mM
$MgSO_4 \cdot 7H_2O$	2.5 g	1 mM
$Na_2EDTA \cdot 2H_2O$	0.4 g	1 mM

Bring up to 1 L with dH_2O. Autoclave.

Modified Leibovitz's L-15 Medium

1. Add 13.8 g of L-15 powdered medium (Leibovitz's L-15 medium, with glutamine; Sigma-Aldrich L4386) to 1300 mL of dH_2O. Mix well.
2. Adjust pH to 7.6 with 1 N NaOH or 1 N HCl.
3. Add 149 mL of fetal bovine serum (FBS, certified, heat-inactivated; Gibco 10082147).
4. Adjust the final volume to 1.493 L.
5. Filter-sterilize with a 0.2-μm surfactant-free cellulose acetate membrane filter unit.
6. Freeze the medium or maintain it at 4°C for up to 4 wk.

Xenopus Eye Culture Medium

Reagent	Concentration
Amphotericin B (Sigma-Aldrich A9528)	2.5 μg/mL
Penicillin (Sigma-Aldrich P3032)	100 U/mL
Marbofloxacin (Sigma-Aldrich 34039)	4 μg/mL
Streptomycin (Sigma-Aldrich S1277)	100 μg/mL

Prepare fresh, just before use, in modified Leibovitz's L-15 medium <R>. (Note: Marbofloxacin is added to prevent Pseudomonas contamination.)

ACKNOWLEDGMENTS

J.J.H. is funded by National Institutes of Health/National Eye Institute (NIH/NEI) grant EYO23979.

REFERENCES

Fukui L, Henry JJ. 2011. FGF signaling is required for lens regeneration in *Xenopus laevis*. *Biol Bull* **221**: 137–145.

Hamilton PW, Henry JJ. 2014. Prolonged in vivo imaging of *Xenopus laevis*. *Dev Dyn* **243**: 1011–1019.

Henry JJ, Perry KJ, Hamilton PW. 2018. Methods for examining lens regeneration in *Xenopus*. *Cold Spring Harb Protoc* doi: 10.1101/pdb.prot101527.

Slack JMW. 1984. Regional biosynthetic markers in the early amphibian embryo. *J Embryol Exp Morphol* **80**: 289–319.

Cite this protocol as *Cold Spring Harb Protoc*; doi:10.1101/pdb.prot101535

Investigating the Cellular and Molecular Mechanisms of Wound Healing in *Xenopus* Oocytes and Embryos

Jingjing Li[1] and Enrique Amaya[2,3]

[1]*Department of Craniofacial Development and Stem Cell Biology, Dental Institute, King's College London, London SE1 9RT, United Kingdom;* [2]*Division of Cell Matrix Biology & Regenerative Medicine, Faculty of Life Sciences, The University of Manchester, Manchester M13 9PT, United Kingdom*

The African clawed frog *Xenopus* has remarkable capacities to heal wounds rapidly and to regenerate complex tissues. Because of its experimental tractability, studies using *Xenopus* oocytes, embryos, and larvae have contributed extensively to our understanding of the molecular and cellular mechanisms underpinning wound healing and tissue regeneration. In this protocol, we describe wound-healing assays following mechanical or laser injuries of oocytes and multicellular epithelia in *Xenopus laevis* embryos. We also explain how to perform assays aimed at investigating the cellular and molecular events during wound healing, including gene knockdown and overexpression experiments. In the latter assays, we explore the use of biochemical pull-down assays to investigate the activity of Rho GTPases, as well as the injection of mRNAs encoding fluorescent proteins or probes, followed by quantitative confocal image analyses to assay the dynamics of cytoskeletal components and their regulators.

MATERIALS

It is essential that you consult the appropriate Material Safety Data Sheets and your institution's Environmental Health and Safety Office for proper handling of equipment and hazardous materials used in this protocol.

RECIPES: Please see the end of this protocol for recipes indicated by <R>. Additional recipes can be found online at http://cshprotocols.cshlp.org/site/recipes.

Reagents

Agarose

Antibodies, anti-Cdc42 (Cell Signaling Technology 2462) (for GST pull-down assays only)

Antibodies, anti-Rac1/2/3 (Cell Signaling Technology 2465) (for GST pull-down assays only)

Antibodies, anti-RhoA (Santa Cruz Biotechnology SC-179) (for GST pull-down assays only)

Bovine serum albumin (BSA)

Constructs (optional; see Steps 1, 7.ii, 7.iv, 11)

> *Embryos can be injected with moesin-gfp to observe actin dynamics (Li et al. 2013), gfp-α-tubulin to observe microtubule dynamics (Woolner and Papalopulu 2012), GEM-GECO or C2-mrfp to observe calcium dynamics (Clark et al. 2009; Soto et al. 2013), or gfp/mcherry-caax to label the plasma membrane and thus outline the cell*

---

[3]Correspondence: enrique.amaya@manchester.ac.uk

Cite this protocol as *Cold Spring Harb Protoc*; doi:10.1101/pdb.prot100982

boundaries (Li et al. 2013). Use pak1-gst mRNA to detect Rac1 and Cdc42 activities; to detect RhoA activity, coinject mRNA for egfp-rhotekinGBD and rhoa (Li et al. 2013). Details of a suggested selection of plasmids— including pCS2-GFP/mCherry-moesin, pCS107-RhoA, pCS107-Pak1-GST, and pCS107-rGBD-GST—can be found in Li et al. (2013) and Soto et al. (2013). All published plasmids have been deposited in the Amaya laboratory repository and are available on request.

Freon-113 (CAS 76-13-1) (for GST pull-down assays only)

Carbon tetrachloride (Millipore Sigma-Aldrich 319961, CAS 56-23-5) or 1,1,2,2-tetrachloroethane (Millipore Sigma-Aldrich 185434, CAS 79-34-5) are acceptable substitutes.

Glutathione Sepharose 4B beads (GE Healthcare 17-0756-01) (for GST pull-down assays only)
GST binding buffer (for GST pull-down assays only) <R>
GST wash buffer (for GST pull-down assays only) <R>
Marc's modified Ringer's (MMR) (10×) <R>

Dilute to 0.1× before use.

Normal amphibian medium (NAM) (10×) <R>
Xenopus embryo lysis buffer (for GST pull-down assays only) <R>
Xenopus oocytes and/or embryos
Additional reagents for western blotting (for GST pull-down assays only)

Equipment

Centrifuge, refrigerated (for GST pull-down assays only)
Coverslips
Forceps, fine (Dumont #5)
Imaging chamber, steel (optional; for laser wounding experiments [Fig. 1C,D])
Microscope, confocal, equipped with a MicroPoint pulsed nitrogen-pumped ablation dye laser (e.g., Laser Science, Inc.)
Microscope, dissection, equipped with LED illumination and a 10×–75× zoom range
Nutator (for GST pull-down assays only)
Petri dishes, with or without glass bottom
Tissue homogenizer (for GST pull-down assays only)
Tubes, microcentrifuge, 1.5-mL (for GST pull-down assays only)
Vortexer (for GST pull-down assays only)
Additional reagents and equipment for western blotting (for GST pull-down assays only)

METHOD

Mechanical Wounding Assays Using *Xenopus* Embryonic Epithelia

1. If gain- or loss-of-function experiments are desired, inject 1–2-cell stage embryos with antisense morpholino oligonucleotides or in-vitro–transcribed mRNAs targeting or encoding the gene products of interest, respectively, before raising the embryos to the blastula or tailbud stages.

2. Culture embryos in 0.1× MMR at either room temperature or 16°C until the blastula or tailbud stages.

3. Transfer embryos to 75% (v/v) NAM containing 0.2% BSA in an agarose dish.

   For drug treatment assays, preincubate embryos for 30 min before wounding experiments.

4. Perform epithelial wounding experiments at the desired stage of development:

   When regions of the epithelium are removed, it is common to remove some of the yolk cells under the skin. This should have minimal effect on the experiment. However, the size of the wound in Steps 4.ii and 4.iv should be kept as consistent as possible to ensure consistent results between replicates and reliable comparisons between different treatments.

Cite this protocol as *Cold Spring Harb Protoc*; doi:10.1101/pdb.prot100982

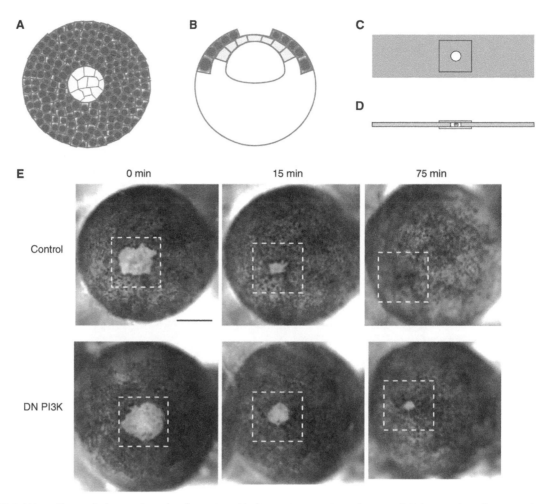

FIGURE 1. Wounding and observation of embryonic epithelium. (*A*) *Top* view of a superficial wound on the animal side of a blastula embryo. Pigmented cells are of the superficial layer, and nonpigmented cells are of the deep layer, which is kept intact in the experiment. (*B*) Transverse side view of a superficial wound on a blastula embryo. (*C*) Self-made imaging chamber for laser wounding and observation. The big (gray) slide is made of steel, with a hole in the middle. (*D*) *Side* view of the chamber, with an embryo mounted inside the hole and sealed from both sides with coverslips. (*E*) Sample pictures of wound healing in control and dominant-negative (DN) PI3K-overexpressing blastula stage embryos. Note that, although control wounds heal completely by 75 min postinjury, DN PI3K-expressing wounds maintain incompletely healed wounds at 75 min postinjury. Wounded areas are highlighted in dashed squares. Scale bar: 200 µm.

For Blastula Stage Epithelial Wounding Experiments

i. Use clean fine forceps to remove the vitelline membranes of a Stage 8 embryo. Approach from the vegetal (i.e., unpigmented) side of the embryo to limit damage to the animal (i.e., pigmented) side.

ii. Carefully remove a desired area of the superficial outer (pigmented) layer of the animal side (e.g., an 8–10 cell diameter area), leaving the deep (unpigmented or lightly pigmented) layer intact (Fig. 1A,B).

For Tailbud Stage Epithelial Wounding Experiments

iii. Use clean fine forceps to remove the vitelline membrane from the embryo.

iv. Remove, pinch or puncture a desired area of the epithelium in the flank of the embryo (e.g., one-third the length of the trunk of the embryo).

5. Leave embryos in 75% NAM throughout the experiment.

6. Image the embryos:

To Assess the Overall Effect of a Gene or a Drug on Wound Healing

i. Observe wounded embryos using either a dissection microscope, or low magnification (e.g., 10×–20×) using a confocal microscope. Image every 20 min.

Completion of wound closure takes a few hours (examples of successful and incomplete healing are shown in Fig. 1E). Wound closure under a stereoscope can be measured by outlining the leading edge of the epithelium (Fig. 1E).

See Troubleshooting.

To Observe the More Rapid Dynamics of the Cytoskeleton or Signaling Events

ii. Image every 2 min using a confocal microscope equipped with a 60× objective.

Depending on the process under investigation, images might need to be taken more or less frequently to capture its dynamic nature.

Laser Wounding Assays in *Xenopus* Oocytes and Embryonic Epithelia

7. Prepare the experimental tissues:

For Oocyte Laser Wounding Assays

i. Prepare oocytes as described in Sive et al. (2000).

ii. Inject morpholino or mRNAs as desired.

iii. Culture the oocytes at 16°C until needed.

For Embryonic Epithelial Laser Wounding Assays

iv. Inject embryos with morpholinos or mRNAs at the 1–2 cell stage.

v. Culture until Stage 9 in 0.1× MMR.

8. Mount oocytes or embryos in a glass bottom dish or coverslip-sealed imaging chamber (Fig. 1C, D) in 75% NAM. Fit the dish or imaging chamber on a confocal stage.

9. Make laser wounds on the surface of the oocyte or the embryo.

A possible combination of pulsed laser settings using the MicroPoint laser to ablate the tissue is: 561 nm wavelength, power 40 and pulse 10. The user is advised to adjust the settings and equipment to fulfill their specific experimental requirements.

10. Observe and record the wound healing process immediately postwounding.

Take images as frequently as possible, being conscious to ensure minimal photo-damage while capturing the dynamics of the proteins or signals under investigation.

GST Pull-Down to Quantify Activation of Small Rho GTPases: Rac, Cdc42, and RhoA

11. Inject *Xenopus* embryos at the 1–2-cell stage with desired mRNAs:

To Detect Rac1 and Cdc42 Activities

i. Inject 500 pg mRNA of pak1-gst.

To Detect RhoA Activity

ii. Coinject 500 pg mRNA of egfp-rhotekinGBD and 125 pg rhoa mRNA.

12. Grow embryos to Stage 8 or 9 in 0.1× MMR.

13. Resuspend and transfer 300 µL Glutathione Sepharose beads into 1.5 mL microcentrifuge tubes.

14. Centrifuge at 1000g for 30 sec at 4°C. Remove the upper aqueous phase without disturbing the bead pellet.

15. Wash the beads with 300 µL GST binding buffer three times at the same centrifugation speed and temperature.

16. Resuspend the beads in 150 µL of GST binding buffer. Place on ice until use.

17. Wound Stage 8 or 9 embryos with forceps in 75% NAM.

18. Collect 30–50 embryos per condition or time point in 1.5 mL microcentrifuge tubes. Remove as much liquid as possible. Place the tubes on ice.

19. Homogenize each tube of embryos in 500 µL freshly prepared and prechilled embryo lysis buffer.

20. Centrifuge at maximum speed (e.g., >16,000g) for 15 min at 4°C.

21. Transfer the supernatant to another prechilled microcentrifuge tube. Add 200 µL prechilled Freon 113 to each tube. Vortex for 15 sec.

22. Centrifuge at maximum speed for 10 min at 4°C.

23. Transfer the upper aqueous phase to a new microcentrifuge tube (∼450 µL total volume).

    Withdraw and reserve 5% of this volume (i.e., ∼9 µL) as input.

24. Add an equal volume (450 µL) of GST binding buffer and 50 µL of prewashed beads (from Step 16) to each tube. Incubate on an end-to-end nutator for 30 min (for Rac1 and Cdc42) or 1 h (for RhoA) at 4°C.

25. Centrifuge the tubes at 1000g for 30 sec at 4°C. Remove the buffer.

26. Wash the beads once with 500 µL precooled GST binding buffer, and twice with precooled GST wash buffer. To mix, gently invert the tubes 3–4 times; do *not* pipette. Centrifuge at 1000g for 30 sec at 4°C between each wash.

27. Detect and measure levels of active Rac, Cdc42, and RhoA using Western blot as described in Li et al. (2013).

TROUBLESHOOTING

Problem (Step 6.i): The leading edge of the wound is difficult to locate.

Solution: Normally, the leading edge should be clear to see because of the color difference between the two epithelial layers. If it is difficult to locate the edge, embryos can be fixed at the end of experiments and stained using phalloidin. For confocal imaging, embryos can be injected with mRNAs of proteins that either localize to the cell membrane or bind to F-actin; both outline the leading edge during wound healing.

DISCUSSION

We present three protocols to assay wound healing in *Xenopus* oocytes or embryos: mechanical wounding in macroscale, and laser wounding and GST pull-down assays at the cellular and molecular level. Of the three, mechanical wounding and observation (Steps 1–6) is the most accessible. When combined with gene knockdown/knockout or chemical treatment, it provides a rapid and robust assessment whether a certain gene, protein, or signaling pathway is involved in wound healing (Li et al. 2013, 2016; Soto et al. 2013). Nonetheless, it does require some practice to make the size and depth of wounds consistent, which can affect the speed and quality of healing.

Laser wounding (Steps 7–10) has two major advantages. First, because the wounding laser is often mounted on (or a part of) a confocal or multiphoton system, it is easy to injure and then immediately image using the same system. Second, the observation of the cellular and molecular events can be started seconds or even milliseconds postwounding, which cannot be achieved after mechanical wounding. For this reason, laser wounding has become the method of choice in a variety of experiments to assess the molecular and cellular bases of *Xenopus* wound healing (Clark et al. 2009; Burkel et al. 2012; Soto et al. 2013; Davenport et al. 2016). Specifically, membrane and cytoskeletal dynamics—and how they are regulated by small Rho GTPases and their effectors near the wound edge—were studied using a single-cell laser-wounding model (Burkel et al. 2012; Davenport et al. 2016). Multicellular cytoskeletal networks and signal propagation to mobilize a sheet of epithelium in multicellular wound healing were studied using an embryonic wounding model (Clark et al. 2009; Soto et al. 2013). On the other hand, the challenge of this protocol is to maintain normal tissue growth or vitality over time, particularly in an experiment lasting hours. Therefore, tests and calibrations of tissue health should always be performed before an experiment.

In the past decade, fluorescence resonance energy transfer (FRET)-based tools to measure activation of small Rho GTPases have been well developed (Fritz and Pertz 2016), especially in cultured cells (Santiago-Medina et al. 2012). However, the use of FRET tools in *Xenopus* embryo is still limited (Yamashita et al. 2016), making biochemical approaches such as the active Rho GST pull-down assays (Steps 11–27) valid and powerful methods to examine the dynamics of these signaling molecules. Also, because the readout of this assay is based on a collection of embryos, individual effects are reduced to a minimum, giving a more robust measurement of the activity of the detected molecules. A current limitation of this protocol is the limited availability of antibodies in *Xenopus*; thus, detection of endogenous proteins pulled down is not always possible. In such cases, overexpression of tagged target proteins can be used as an alternative for the measurements.

RECIPES

GST Binding Buffer

25 mM Tris-HCl, pH 7.5
30 mM MgCl$_2$
40 mM NaCl
1 mM dithiothreitol
0.5% Nonidet P-40
cOmplete, Mini Protease Inhibitor (1 tablet/10 mL; Roche 04693124001)
PhosStop Phosphatase Inhibitor (1 tablet/10 mL; Roche PHOSS-RO)

Prepare fresh. Keep on ice until use.

GST Wash Buffer

25 mM Tris-HCl, pH 7.5
30 mM MgCl$_2$
40 mM NaCl

Prepare fresh. Keep on ice until use.

Cite this protocol as *Cold Spring Harb Protoc;* doi:10.1101/pdb.prot100982

Marc's Modified Ringer's (MMR) (10×)

20 mM $CaCl_2$
50 mM HEPES (pH 7.5)
20 mM KCl
10 mM $MgCl_2$
1 M NaCl

Adjust pH with NaOH to 7.5. Sterilize by autoclaving. Store at room temperature.

Normal Amphibian Medium (NAM) (10×)

1.1 M NaCl
20 mM sodium phosphate (pH 7.4)
20 mM KCl
10 mM $Ca(NO_3)_2$
10 mM $MgSO_4$
10 mM $NaHCO_3$
1 mM EDTA

Store at room temperature.

Xenopus Embryo Lysis Buffer

50 mM Tris-HCl (pH 7.5)
100 mM NaCl
10 mM $MgCl_2$
1 mM dithiothreitol
5% glycerol
1% Nonidet P-40
cOmplete, Mini Protease Inhibitor (1 tablet/10 mL; Roche 4693124001)
PhosStop Phosphatase Inhibitor (1 tablet/10 mL; Roche PHOSS-RO)

Prepare fresh. Keep on ice until use.

REFERENCES

Burkel BM, Benink HA, Vaughan EM, Dassow von G, Bement WM. 2012. A Rho GTPase signal treadmill backs a contractile array. *Dev Cell* 23: 384–396.

Clark AG, Miller AL, Vaughan E, Yu H-YE, Penkert R, Bement WM. 2009. Integration of single and multicellular wound responses. *Curr Biol* 19: 1389–1395.

Davenport NR, Sonnemann KJ, Eliceiri KW, Bement WM. 2016. Membrane dynamics during cellular wound repair. *Mol Biol Cell* 27: 2272–2285.

Fritz RD, Pertz O. 2016. The dynamics of spatio-temporal Rho GTPase signaling: Formation of signaling patterns. *F1000Res* 5: 749.

Li J, Zhang S, Soto X, Woolner S, Amaya E. 2013. ERK and phosphoinositide 3-kinase temporally coordinate different modes of actin-based motility during embryonic wound healing. *J Cell Sci* 126: 5005–5017.

Li J, Zhang S, Amaya E. 2016. The cellular and molecular mechanisms of tissue repair and regeneration as revealed by studies in *Xenopus*. *Regeneration (Oxf)* 3: 198–208.

Santiago-Medina M, Myers JP, Gomez TM. 2012. Imaging adhesion and signaling dynamics in *Xenopus laevis* growth cones. *Dev Neurobiol* 72: 585–599.

Sive HL, Grainger RM, Harland RM. 2000. *Early development of* Xenopus laevis*: A laboratory manual.* Cold Spring Harbor Laboratory Press, Cold Spring Harbor, NY.

Soto X, Li J, Lea R, Dubaissi E, Papalopulu N, Amaya E. 2013. Inositol kinase and its product accelerate wound healing by modulating calcium levels, Rho GTPases, and F-actin assembly. *Proc Natl Acad Sci* 110: 11029–11034.

Woolner S, Papalopulu N. 2012. Spindle position in symmetric cell divisions during epiboly is controlled by opposing and dynamic apicobasal forces. *Dev Cell* 22: 775–787.

Yamashita S, Tsuboi T, Ishinabe N, Kitaguchi T, Michiue T. 2016. Wide and high resolution tension measurement using FRET in embryo. *Sci Rep* 6: 28535.

Tracing Central Nervous System Axon Regeneration in *Xenopus*

Kurt M. Gibbs[1] and Ben G. Szaro[2,3]

[1]*Department of Biology & Chemistry, Morehead State University, Morehead, Kentucky 40351;* [2]*Department of Biological Sciences, State University of New York at Albany, Albany, New York 12222*

Axonal tracing allows visualizing connectivity between neurons, providing useful information about structure, neuronal location, and function of the nervous system. Identifying regenerating axons and their neuron cell bodies present the particular challenges of labeling the projections of interest while unambiguously demonstrating regrowth of those axons that have been damaged. In the developing brain, an additional labeling challenge arises, as new connections are being made throughout the duration of an experiment. Various strategies have been used to label regenerating axons, including transgenic animals expressing neuron-specific fluorescent proteins, and application of a single labeling molecule after axotomy and regeneration. However, the single label approach is limited in its application to the developing brain, primarily because it leads to the conclusion that every axon that is labeled has regenerated. Double-labeling overcomes these obstacles by identifying regenerating cells as those that are labeled with two different tracing molecules. Moreover, the use of dextran amines, which are only taken up by injured axons and transported retrogradely, provides further confidence of labeling regenerating axons and neuron cell bodies. The procedure described herein provides a straightforward method for using fluorescently labeled dextran amines to identify regenerating supraspinal neurons in *Xenopus*, but can be applied to other areas of the central and peripheral nervous system as well.

MATERIALS

It is essential that you consult the appropriate Material Safety Data Sheets and your institution's Environmental Health and Safety Office for proper handling of equipment and hazardous materials used in this protocol.

Reagents

Aqueous mounting medium (e.g., Fluoromount-G with DAPI [SouthernBiotech])

Dextran amine (fluorescent, 10 kDa, lysine-fixable; e.g., Fluorescein [D1820] or Tetramethylrhodamine [D1817] from Molecular Probes)

> *Dissolve 25 mg of fluorescent dextran amine in 200 µL of deionized, sterile water. Store 10-µL aliquots in 0.2-mL tubes in the dark at −20°C.*

Dry ice (optional; see Step 22)

Ethyl 3-aminobenzoate methanesulfonate (MS-222, Sigma-Aldrich E10521; 0.02% [w/v], buffered with 0.02% sodium bicarbonates pH 7.0–7.5, in frog rearing water)

Freezing medium (e.g., TFM [Triangle Biomedical Sciences])

---

[3]Correspondence: bszaro@albany.edu

Gentamicin (50 mg/mL; Sigma-Aldrich G1397)

Paraformaldehyde (3.7% in 0.1 M sodium phosphate buffer)

Sodium phosphate buffer (0.1 M, pH 7.4)

Sucrose (30% in 0.1 M sodium phosphate buffer)

Xenopus laevis albino tadpoles (Nieuwkoop-Faber Stage 53–54)

The central nervous system (CNS) and vertebral column are visible through the skin of albino tadpoles, allowing easy identification of the thoracic spinal cord.

Equipment

Beeswax (Fisher Scientific S25192A)

Cover glass (#1)

Cryotome

Embedding molds (disposable, $15 \times 15 \times 5$ mm; Fisher Scientific)

Forceps (fine, Dumont #5)

Freezer (−80°C)

Kimwipes

Micro Knife with 15° cutting angle (Fine Science Tools 10315-12)

Microscope (epifluorescence or confocal)

Microscope slides (positively charged; e.g., Fisherbrand Superfrost Plus)

Minutien pins (0.20-mm; Fine Science Tools 26002-20)

Parafilm

Petri dish

Pin holder (Fine Science Tools 26016-12)

Rubber bands

Scalpel

Spring scissors (Vannas; Fine Science Tools 15000-00)

Tubes (0.2-mL; e.g., ThermoFisher Scientific AB-0620)

Tubes (conical, 50-mL; Falcon)

Vortexer

METHOD

Label Preparation

Use the dehydrated drops within 1 wk for the most vibrant labeling results.

1. Thaw a single tube of fluorescently labeled dextran amine.

   A single tube will produce 10 drops, which is enough to label 40 animals.

2. Pipette 1-µL drops onto a strip of Parafilm. Allow them to air-dry in the dark.

Spinal Cord Transection and Labeling

3. Anesthetize tadpoles by immersion in 0.02% MS-222 until reflexive movement elicited with a tail pinch is absent.

4. Transfer the tadpole to a Petri dish lined with beeswax that is molded to the tadpole's body shape; allowing for water to be added to cover the gills. Stabilize with rubber bands and insect pins (Fig. 1A).

5. Using Vannas scissors and fine forceps, incise the dorsal skin and axial musculature to reveal the cartilaginous vertebral column.

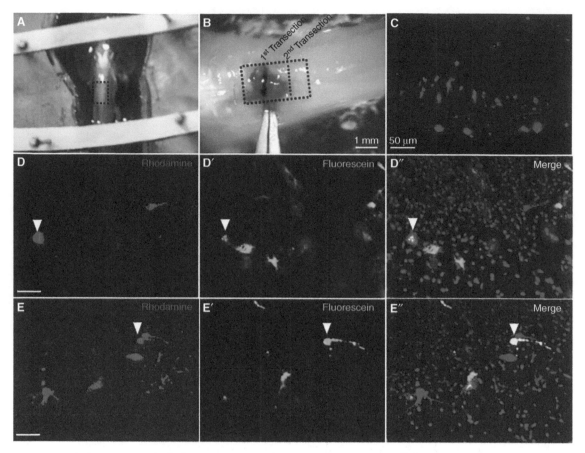

FIGURE 1. Labeling transected axons in the spinal cord with 10 kDa fluorescent dextran amines. Images were captured using a Zeiss Axioimager epifluorescence microscope and 20× objective. (*A*) Tadpole restrained in a beeswax-lined Petri dish. The dashed rectangle surrounds the site of spinal cord transection. (*B*) Rhodamine dextran amine label is applied to the newly transected tadpole spinal cord. The dashed line indicates the future site of the second transection and application of fluorescein dextran amine. (*C*) Parasagittally sectioned spinal cord and hindbrain showing labeled axons and neurons in the ventral region following rhodamine dextran amine application. Dorsal is up, rostral is left. (*D,E*) At higher magnification in transverse section, double-labeling with fluorescein and rhodamine in the regenerating tadpole spinal cord reveals regenerating neurons (arrowheads) in the medial (*D,D′,D″*) and inferior (*E,E′,E″*) hindbrain reticular formation, as indicated by the yellow color (green and red overlay in *D′* and *E′*). DAPI is used to stain nuclei blue. Dorsal is up. Scale bars in *C, D,* and *E* = 50 μm.

6. Remove the vertebral lamina using fine forceps and Vannas scissors to cut the cartilaginous lamina bilaterally, being careful not to injure the spinal cord.

7. Use a second pair of Vannas scissors (reserved for soft tissue only) to carefully transect the spinal cord while avoiding the dorsal spinal vein.

8. Confirm complete transection by inserting a Minutien pin into the transection site with a dorsal-to-ventral motion.

9. After transection, blot the blood and cerebral spinal fluid with a Kimwipe until bleeding has stopped.

10. Under a stereomicroscope, cut the 1-μL dextran amine drops (from Step 2) into four equal pieces using a micro knife.

11. Use fine forceps to transfer a single piece over the lesion site. Use a Minutien pin mounted in a pin holder to insert it into the transection site (Fig. 1B).

    The crystal will rehydrate and eventually diffuse from the injury site. Alternatively, when labeling smaller nerves, Minutien pins can be dipped in fluorescent dextran amine, allowed to dry in the dark, and used in place of the crystallized form.

Cite this protocol as *Cold Spring Harb Protoc*; doi:10.1101/pdb.prot101030

12. Allow the tadpole to recover in the dish for ∼2 min.

13. Use fine forceps to close the skin over the lesion by apposition.

14. Transfer the tadpole to an aerated 2-L recovery tank containing 1 mg/L gentamicin.

15. Monitor the recovery of the tadpole for the following 2 wk.

> *The following day, the tadpole will be lying on the bottom of the tank, completely paralyzed below the level of the lesion. If caudal tail movement is observed, complete transection was not achieved. After ∼1 wk, the tadpole will float on the surface and begin to show signs of occasional voluntary movement. By 2 wk, the tadpole should be able to perform short bouts of voluntary swimming, indicating regenerated supraspinal axons.*

16. Repeat Steps 3–14 using a dextran amine labeled with a fluorophore sufficiently spectrally separated from the first.

> *The second spinal cord transection and labeling should be performed 1-mm caudal to the first (Fig. 1B).*

17. Allow tadpoles to recover for 2 d to ensure that the second label has been transported to their cells of origin in the mid- and hindbrain before euthanizing them for analysis.

CNS Harvesting and Embedding

18. Euthanize tadpoles using an overdose of MS-222.

19. Rapidly dissect the CNS. Immerse in 3.7% paraformaldehyde in 0.1 M sodium phosphate buffer (15 mL per brain). Fix overnight at 4°C.

20. The following day, incubate the CNS in 0.1 M sodium phosphate buffer for 2 h at room temperature to remove excess paraformaldehyde from the tissue.

21. Infiltrate the CNS overnight with 30% sucrose in 0.1 M sodium phosphate buffer (as a cryoprotectant).

> *The tissue is fully infiltrated when it sinks to the bottom of the tube.*

22. Embed the CNS in a disposable mold using TFM. Freeze on dry ice or in a −80°C freezer.

> *Frozen specimens can be stored in an airtight container at −80°C for several weeks.*

Sectioning and Visualization

23. Cut 20-μm sections from frozen specimens using a cryotome set at −16°C.

> *To visualize cell bodies of supraspinal neurons whose axons regenerate, section the mid- and hindbrain transversely (Fig. 1D,E). To visualize labeled axons, section the spinal cord parasagittally (Fig. 1C).*

24. Mount sections on Superfrost Plus microscope slides.

> *To prevent fading of the fluorescent signal, place slides on a tray covered with aluminum foil until sectioning is complete.*

25. Dry slides for 2 h at room temperature.

26. Rehydrate by washing 3 times with 0.1 M sodium phosphate buffer, 5 min each wash.

27. Mount the slides using aqueous mounting medium. Apply a cover glass immediately. Image as soon as possible.

> *Store the slides in the dark at 4°C to prevent the fluorescent signal from fading.*

DISCUSSION

Like several anuran amphibians, *Xenopus laevis* offers the opportunity to study the underlying factors enabling successful CNS regeneration (Forehand and Farel 1982; Beattie et al. 1990; Gibbs and Szaro 2006; Gaete et al. 2012). *Xenopus laevis* tadpoles can regenerate descending supraspinal axons after complete spinal cord transection but begin to lose regenerative ability as they enter the prometamorphic stages (Gibbs et al. 2011; Tapia et al. 2017). Identifying neurons capable of regenerating their axons is the first step toward understanding why a specific type of neuron regenerates whereas neighboring neurons do not.

Various studies have used dextran amines to identify tracts, their cells of origin, and terminations in the peripheral and central nervous system of *Xenopus*. Fluorescent dextran amines also allow the versatility of incorporating immunofluorescence into the analysis, which can be used to identify further, specific populations of cells by neurotransmitter subtype (Sánchez-Camacho et al. 2002a,b). Unlike the carbocyanine dyes (e.g., DiO, DiI) that diffuse through cell membranes, dextran amines are aqueous, noncytotoxic, cell-impermeable tracers that are only taken up by terminals or cut axons (Fritzsch and Sonntag 1991). Using the method detailed above, dextran amines label both the cut axons and the cytoplasm. As the dextran amine is taken up, it is packaged into vesicles that are retrogradely transported to the cell body, giving the cell body a combined filled and vesicular staining pattern. The time needed to label neuronal perikarya by retrograde movement of dextran amines varies with molecular weight (Fritzsch 1993); we have found that 10 kDa molecules yield the best result. By using sequential retrograde double-labeling, we have found that neuronal perikarya containing both fluorophores represent those cells whose axons were cut and labeled twice, indicating true regeneration.

Although dextran amines are also used as anterograde tracers, when labeling transected axons the anterograde portion of severed axons undergo Wallerian degeneration and do not contribute to the overall analysis. When performing double-labeling with dextran amines, care should be taken in selecting the emission spectra of the fluorophores and filter profiles of the microscope to ensure minimal spectral overlap, which can lead to erroneous interpretation of the results.

REFERENCES

Beattie MS, Bresnahan JC, Lopate G. 1990. Metamorphosis alters the response to spinal cord transection in *Xenopus laevis* frogs. *J Neurobiol* **21**: 1108–1122.

Forehand CJ, Farel PB. 1982. Anatomical and behavioral recovery from the effects of spinal cord transection: Dependence on metamorphosis in anuran larvae. *J Neurosci* **2**: 654–662.

Fritzsch B. 1993. Fast axonal diffusion of 3000 molecular weight dextran amines. *J Neurosci Methods* **50**: 95–103.

Fritzsch B, Sonntag R. 1991. Sequential double labelling with different fluorescent dyes coupled to dextran amines as a tool to estimate the accuracy of tracer application and of regeneration. *J Neurosci Methods* **39**: 9–17.

Gaete M, Muñoz R, Sánchez N, Tampe R, Moreno M, Contreras EG, Lee-Liu D, Larraín J. 2012. Spinal cord regeneration in *Xenopus* tadpoles proceeds through activation of Sox2-positive cells. *Neural Dev* **7**: 13.

Gibbs KM, Szaro BG. 2006. Regeneration of descending projections in *Xenopus laevis* tadpole spinal cord demonstrated by retrograde double labeling. *Brain Res* **1088**: 68–72.

Gibbs KM, Chittur SV, Szaro BG. 2011. Metamorphosis and the regenerative capacity of spinal cord axons in *Xenopus laevis*. *Eur J Neurosci* **33**: 9–25.

Sánchez-Camacho C, Marín O, López JM, Moreno N, Smeets WJAJ, ten Donkelaar HJ, González A. 2002a. Origin and development of descending catecholaminergic pathways to the spinal cord in amphibians. *Brain Res Bull* **57**: 325–330.

Sánchez-Camacho C, Marín O, ten Donkelaar HJ, González A. 2002b. Descending supraspinal pathways in amphibians: III. Development of descending projections to the spinal cord in *Xenopus laevis* with emphasis on the catecholaminergic inputs. *J Comp Neurol* **446**: 11–24.

Tapia VS, Herrera-Rojas M, Larraín J. 2017. JAK-STAT pathway activation in response to spinal cord injury in regenerative and non-regenerative stages of *Xenopus laevis*. *Regeneration (Oxf)* **4**: 21–35.

Cite this protocol as *Cold Spring Harb Protoc*; doi:10.1101/pdb.prot101030

Rod-Specific Ablation Using the Nitroreductase/Metronidazole System to Investigate Regeneration in *Xenopus*

Reyna I. Martinez-De Luna[1,4] and Michael E. Zuber[1,2,3]

[1]*Department of Ophthalmology and the Center for Vision Research;* [2]*Department of Biochemistry and Molecular Biology;* [3]*Department of Neuroscience and Physiology, SUNY Upstate Medical University, Syracuse, New York 13210*

Genetically controlled cell type–specific ablation provides a reproducible method to induce regeneration that can be temporally and spatially controlled. Until recently, regeneration studies in *Xenopus* have relied on surgical methods to stimulate regeneration. These methods are labor intensive and not as reproducible as a genetically controlled approach. In this protocol we describe selective ablation of rod photoreceptors in the premetamorphic *Xenopus laevis* retina using the nitroreductase/metronidazole (NTR/Mtz) system. We use the XOPNTR transgenic line in which the *Xenopus* Rhodopsin promoter drives rod photoreceptor-specific expression of the bacterial enzyme, NTR. Exposure of transgenic tadpoles to Mtz for 2 d completely ablates rods by 7 d after initial Mtz exposure. Removal of Mtz allows rods to regenerate and makes rod-specific ablation reversible and amenable for regeneration studies. The protocol presented here is applicable to the selective ablation of any cell type with the use of appropriate cell type–specific promoters.

MATERIALS

It is essential that you consult the appropriate Material Safety Data Sheets and your institution's Environmental Health and Safety Office for proper handling of equipment and hazardous materials used in this protocol.

RECIPES: Please see the end of this protocol for recipes indicated by <R>. Additional recipes can be found online at http://cshprotocols.cshlp.org/site/recipes.

Reagents

5′-bromo-2′-deoxyuridine (BrdU) (Sigma-Aldrich B5002; 10 mM in frog water)
Dissolve 3.07 g of BrdU in 1 L of frog water by stirring for ∼30 min. Freeze aliquots at −20°C.

Agarose gel (1%) electrophoresis reagents
DMSO-only control solution <R>
Ethanol (95%)
Frog water <R>
Hydrochloric acid (1 M)
Immunohistochemistry reagents
Metronidazole (Mtz) treatment solution <R>

[4]Correspondence: martiner@upstate.edu

Mouse anti-Transducin antibody (Santa Cruz, sc-136143, diluted 1:500)

Mouse anti-XAP-2 antibody (DSHB, Clone 5B9, diluted 1:25)

MS-222 (Tricaine) (1% in frog water)

> *Dissolve 0.5 g of MS-222 in 50 mL of frog water. Store at 4°C.*

Optimal cutting temperature (OCT) compound

Paraformaldehyde (PFA; 4% in phosphate-buffered saline [PBS])

> *Add 4 g of PFA in 75 mL of ddH$_2$O. Add 10 mL of phosphate buffered saline (PBS; 10×, pH 7.4, without CaCl$_2$ or MgCl$_2$) <R> plus three to four drops of 10 M NaOH. Stir gently at 60°C until the PFA is dissolved. Cool. Readjust pH to 7.4. Filter-sterilize to remove any particulate matter. Use immediately or aliquot and store at −20°C for up to 1 yr. Avoid repeated freeze–thawing.*

PCR primers for NTR (forward: 5′-CGCTAAATCCTTTGTTGCTGACGC-3′; reverse: 5′-GTTGAACACGTAATTACCGGCAGC-3′)

PCR reagents

Rabbit anti-BrdU antibody (Thermo Fisher PA5-32256, diluted 1:200)

Reagents for crude genomic DNA extraction (see Showell and Conlon 2009) or genomic DNA extraction kit

Sucrose (20% in phosphate-buffered saline [PBS])

> *Dissolve 20 g of sucrose in 80 mL of phosphate-buffered saline (PBS; 1×, pH 7.4) <R>. Make up to 100 mL with PBS. Filter sterilize using a 0.2 µm membrane filter. Store at 4°C.*

XOPNTR transgenic frogs Xla.Tg(rho:Eco.nfsA)Zuber (National *Xenopus* Resource) or Xla.Tg(-508/+41XOP:NTR) Line #49 (European *Xenopus* Resource Centre)

Equipment

Agarose gel electrophoresis equipment

Aluminum foil

Immunohistochemistry equipment

Incubator

Jars (0.5 L, Nalgene polypropylene straight-sided wide mouth; [Thermo Fisher, 2118-0016PK] or 0.5 gallon Flex Tanks [Nasco, SB19271J])

Microcentrifuge tubes (1.5 mL)

Net or soft plastic hair catcher

PCR thermocycler

PCR tubes

Stir plate

Watchmaker's forceps (No. 5) or scalpel blade

METHOD

Generation of Embryos for Metronidazole Treatment

1. Generate embryos from a natural mating of F$_1$ XOPNTR animals or by in vitro fertilization of eggs from an XOPNTR female with sperm from a wild-type male and culture to stage 50 (Viczian and Zuber 2010).

2. At stage 50, transfer tadpoles to treatment containers containing frog water using a soft plastic hair catcher. Distribute equal numbers of tadpoles into experimental (Mtz) and control (DMSO-only) treatment groups. Protect containers from light and keep tadpoles at a density of no more than one tadpole per 20 mL.

> *The total number of tadpoles required for treatment and control groups depends on the percentage of transgenic tadpoles obtained from the XOPNTR mating pair used and the number of experimental condi-*

tions and time points to be tested. Our F₁ XOPNTR mating pairs generate ~50% XOPNTR+ tadpoles, so we collect 20–30 embryos per time point. Rod ablation is observed in nearly 100% of XOPNTR+ tadpoles permitting statistically significant results with 6–8 tadpoles analyzed per condition.

We use either 500 mL Nalgene Polypropylene Straight-Sided Wide Mouth Jars or 0.5 gallon Flex Tanks covered in foil. Overcrowding tadpoles at a density higher than 1 tadpole per 10 mL results in poor health with spontaneous rod degeneration in wild-type and transgenic tadpoles.

3. Before treatment, collect the desired number of XOPNTR tadpoles for this time point (treatment day 0 [T0]) from treatment and control groups.

   i. Collect tails for NTR genotyping. Anesthetize tadpoles in 1% MS-222 in frog water for 10 min. Amputate the tail using a pair of watchmaker's forceps or scalpel blade and transfer to a 1.5 mL microcentrifuge tube. Store at −20°C. Between each tadpole, wash forceps or scalpel blade in 1 M HCl, 95% ethanol, and lastly two washes of ddH₂O.

   ii. Fix the head in 4% PFA for 45 min at room temperature. Cryoprotect in 20% sucrose in PBS at 4°C overnight. Embed in OCT (see Uribe and Gross 2007).

Mtz Treatment and 5-Bromo-2′-Deoxyuridine (BrdU) Labeling

4. About 40 min before the time chosen for the start of ablation (T0), prepare the necessary amounts of Mtz treatment and DMSO-only control solutions.

   Prepare sufficient solutions for 20 mL per tadpole. The BrdU in the Mtz treatment and DMSO-only control solutions is included to label new rods born after ablation. We have found BrdU+ rods after BrdU treatment during the first week of incubation [from T0 to recovery day 5 (R5) (day 1 to day 7)] or during the third week (from R21 to R28) (Fig. 1A). BrdU labeling combined with rod-specific markers allows the generation of these cells to be traced.

5. Transfer the predistributed treatment and control tadpoles into Mtz treatment and DMSO-only control solutions.

   Perform the incubation in either Nalgene Jars or Flex Tanks depending on the volume. If treating more than 25 tadpoles (total volume <500 mL) use Flex Tanks.

6. Incubate tadpoles in the dark at 22°C from T0 (day 0) to T2 (day 2) (Fig. 1A).

7. At T2, collect the desired number of control and experimental tadpoles and process as described in Step 3.

Recovery and BrdU Labeling

8. Transfer remaining control and experimental tadpoles into fresh frog water containing 1 mM BrdU at a density of no more than one tadpole per 20 mL.

9. Incubate in the dark at 22°C until R5 (day 7) (Fig. 1A). Change BrdU frog water and feed tadpoles daily.

10. Collect the desired number of BrdU-treated tadpoles at R5 and process as described in Step 3.

11. Grow half of the remaining tadpoles in frog water without BrdU at 22°C in the dark at a density of no more than one tadpole per 20 mL until R28 and the other half until R58, processing tadpoles as described in Step 3 at each time point.

    BrdU labeling is performed for 1 wk periods. Longer incubation negatively affects the health of the tadpoles. If desired, tadpoles can be incubated with BrdU in any week during the recovery period (i.e., other than from T0 to R5) to label any newborn rods.

NTR Genotyping

12. Extract genomic DNA from all tail samples collected at the various time points.

    Crude genomic DNA can be prepared (see Showell and Conlon 2009) or genomic DNA can be extracted using a commercially available kit. The selected method should be tested first to ensure the genomic DNA yield and quality are sufficient for detection of the transgene by polymerase chain reaction (PCR).

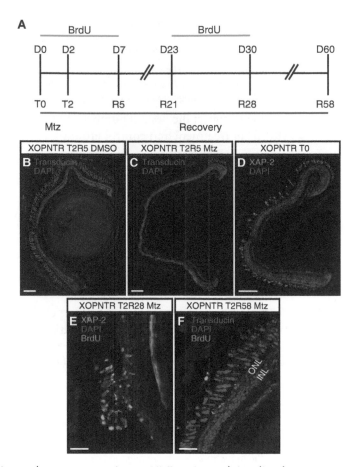

FIGURE 1. Rod ablation and recovery experiment. (*A*) Experimental time line for treatment with Mtz (red line) and BrdU (green lines), and recovery (blue line). (D) day, (T) Mtz treatment day, (R) recovery day. (*B*) Control retina (DMSO-treated) with healthy rods at T2R5. (*C*) Mtz-treated retina at T2R5; (*D*) Untreated retina (T0) with spontaneous rod degeneration (possibly resulting from high XOPNTR transgene copy number); (*E*) BrdU-labeled newborn rods in an Mtz-treated T2R28 retina labeled with BrdU from T0 to R5 (*F*) BrdU-labeled newborn rods in an Mtz-treated T2R58 retina labeled with BrdU from R21 to R28. Scale bars: B–D, 50 μM; E and F, 25 μM.

13. Using genomic DNA as template, set up the following PCR reaction for each sample:

Genomic DNA (from Step 12)	2.5 μL
NRT forward primer (100 μM)	0.1 μL
NTR reverse primer (100 μM)	0.1 μL
dNTPs (10 mM)	0.25 μL
10× PCR buffer	1.25 μL
Taq DNA polymerase	1 unit
Nuclease free H_2O	to 12.5 μL

14. Run PCRs using the following cycling parameters:

Number of cycles	Denaturation	Annealing	Polymerization
1	2 min at 94°C		
34	30 sec at 94°C	30 sec at 56°C	30 sec at 72°C
1			5 min at 72°C

Cite this protocol as *Cold Spring Harb Protoc*; doi:10.1101/pdb.prot100974

15. Analyze 8 μL of PCR product on a 1% agarose gel.

> *A 355 bp band indicates the presence of the XOPNTR transgene. Samples that lack an amplicon are wild-type tadpoles.*
>
> *See Troubleshooting*

Analysis of Rod Ablation and Regeneration

16. Examine rod ablation and regeneration in PFA-fixed heads from all time points using immunohistochemistry (IHC) against rod specific markers and BrdU (Martinez-De Luna et al. 2013).

> *The extent of rod ablation can be detected using transducin (rod inner and outer segments) or XAP-2 (rod outer segments) IHC. In tadpoles obtained from our F_1 mating pair, rods are completely ablated by R5 (day 7). The extent of regeneration is best assessed using transducin IHC.*
>
> *Newborn rods can be detected by BrdU IHC. In our F_2 tadpoles, generation of new rods begins during the first week and continues into the third week. We have not treated tadpoles with BrdU beyond the third week. Rod regeneration is nearly complete by R58 (day 60).*
>
> *See Troubleshooting.*

TROUBLESHOOTING

Problem (Step 15): There is an insufficient number of NTR-positive tadpoles.

Solution: Empirically determine the number of NTR positive tadpoles obtained from each breeding pair to determine a suitable sample size. We obtain 50% NTR-positive tadpoles from our F_1 breeding pair. With 50% NTR transgenic tadpoles, 20–30 tadpoles collected per time point are sufficient to analyze three to six different antibodies by double or triple immunolabeling.

Problem (Step 16): The rods are not ablated following Mtz treatment.

Solution: Mtz treatment must be done at 22°C and in the dark. Enzyme activity of NTR is significantly less efficient at temperatures below 22°C.

Problem (Step 16): There is no BrdU incorporation.

Solution: Confirm that BrdU was administered at a final concentration of 1 mм. Also confirm rod ablation; however, BrdU incorporation should be observed at the ciliary marginal zone regardless of rod ablation. We have observed the generation of new rods during the first five recovery days after rod ablation and a smaller number of rods generated during recovery days 21–28 (days 23–30). We have not investigated rod generation at other time points.

Problem (Step 16): Rods degenerate in both wild-type and XOPNTR animals in the absence of Mtz treatment (T0 samples).

Solution: Rod degeneration in controls indicates that the animals are unhealthy. This could be a consequence of overcrowding, insufficient food or unclean rearing conditions.

Problem (Step 16): Rods degenerate in XOPNTR animals in the absence of Mtz treatment (T0 samples).

Solution: Excess transgene expression can compromise rod health. We have observed spontaneous rod degeneration in tadpoles expressing high levels of NTR. Empirically determine the status of rods (by transducin IHC) in untreated stage 50 tadpoles in preliminary experiments with each breeding pair to be used.

Problem (Step 16): Regeneration is not observed.

Solution: Empirically determine the Mtz treatment time necessary to ablate rods for each mating pair. We have found that the time needed to ablate rods is similar between generations, but that prolonged treatment prevents regeneration in F_1 and F_2 generation tadpoles. We have tested different Mtz treatment times ranging from a half day to 35 d on tadpoles obtained from the F_0 founder and F_1 mating pairs. For tadpoles obtained from the F_0 founder, a treatment of 10 d ablated rods in the central retina between day 6 and 10, and nearly complete regeneration was observed by recovery day 30 (Choi et al. 2011). Last, treatment time should be initiated at stage 50 or later because regeneration was observed to be stage dependent (Langhe et al. 2017).

DISCUSSION

Here, we described a protocol for rod-specific ablation using the NTR/Mtz system in *X. laevis*. In this method, cell type–specific ablation was accomplished using XOPNTR transgenic frogs in which the *Escherichia coli* enzyme, NTR, is exclusively expressed in rods under control of the *Xenopus* Rhodopsin promoter (Choi et al. 2011). Although not yet tested, the NTR/Mtz system should be amenable to application in *X. tropicalis*.

Mtz Treatment Time, Rod-Specific Ablation, and Regeneration

The duration of Mtz treatment required to ablate rods in the central retina is similar across different generations of our XOPNTR line. In the F_1 generation, a 10 d treatment completely ablated rods between days 6 and 10, whereas in the F_2 generation, rods were completely ablated by day 7. The slight decrease in the number of days needed for rod ablation may be caused by an increase in the copy number of the transgene in the F_2 versus the F_1 generation. We have not determined the number of transgene copies present in the parents of each generation. Similarly the differences in the time required for rod regeneration between the F_1 and F_2 generations may also result from an increase in the transgene copy number.

Advantages and Disadvantages of Cell Type–Specific Ablation

Genetically controlled cell type–specific ablation using the NTR/Mtz method is a powerful tool for studying regenerative mechanisms because it is reversible and provides both spatial and temporal control of cell ablation. The NTR/Mtz system is also a useful tool to study the molecular mechanisms of degeneration. The XOPNTR transgenic line is a model of retinitis pigmentosa (RP) because persistent Mtz treatment results in the secondary death of cone photoreceptors, an event observed in RP (Choi et al. 2011). Müller cell gliosis, a secondary change observed in human retinal degeneration, was also observed in the XOPNTR retina with prolonged Mtz treatment (Choi et al. 2011). Another advantage of this system is that cell type–specific ablation can be combined with TUNEL staining and BrdU/5-ethynyl-2′-deoxyuridine (EdU) labeling to track the rate of cell death, and the location of regenerated tissue. In addition BrdU/EdU pulse-chase experiments allow determination of when and which new cell types are produced during the regenerative response. Lastly, the NTR/Mtz system is not limited to studies of retinal regeneration and degeneration. In zebrafish, promoters such as cardiac myosin light chain 2 (cmlc2), liver-type fatty acid binding protein (l-fabp), and insulin (ins) have been used to ablate cardiomyocytes, hepatocytes, and pancreatic B-cells, respectively (Curado et al. 2007). In *X. laevis*, the NTR/Mtz system has also been used to ablate oligodendrocytes of the optic nerve using the mbp promoter (Kaya et al. 2012).

Although a powerful tool, one disadvantage of this type of genetic ablation is that its specificity depends on the specificity of the promoter, making prevalidation of the intended driver essential. It is also advisable to use available transgene constructs that have NTR fused to a fluorescent protein (for examples see, Sekizar et al. 2015; Langhe et al. 2017). This allows easy identification of transgenic and nontransgenic animals without the need for PCR genotyping. An increasing number of *Xenopus*

Cite this protocol as *Cold Spring Harb Protoc*; doi:10.1101/pdb.prot100974

transgenic lines are available from the *Xenopus* stock centers, facilitating the testing for promoter specificity and targeted cell ablation.

RECIPES

DMSO-Only Control Solution

Reagent	Amount (1 L)	Final concentration
DMSO	0.2 mL	0.2%
BrdU (10 mM in frog water)	0.1 L	1 mM
Frog water <R>		

Prepare fresh on the day of use in glass bottles covered in foil. Add DMSO to frog water. Add 10 mM BrdU stock solution to a final concentration of 1 mM. Stir to mix.

Frog Water

Reagent	Amount (200 L)	Final concentration
Na_2HPO_4 (0.4 M)	1 L	2 mM
Instant Ocean Sea Salt solution	1 L	0.5 g/L
Reverse osmosis (RO) water	198 L	-

Prepare frog water in a 200-L carboy. Frog water can be stored long-term at room temperature. To prevent the growth of green algae, periodically scrub the carboy with a brush without detergent, and rinse with water.

Mtz Treatment Solution

Reagent	Amount (1 L)	Final concentration
Metronidazole (Mtz; Sigma-Aldrich M1547)	1.71 g	10 mM
DMSO	0.2 mL	0.2%
BrdU (10 mM in frog water)	0.1 L	1 mM
Frog water <R>		

Prepare fresh on the day of use in glass bottles covered in foil. Add Mtz powder plus DMSO to frog water and stir for 30 min. Add 10 mM BrdU stock solution to a final concentration of 1 mM. (Note that it is essential to use Mtz from Sigma-Aldrich [M1547]. Other sources of Mtz, and even different Mtz products from Sigma-Aldrich, are not reliable.)

Phosphate-Buffered Saline (PBS)

Reagent	Amount to add (for 1× solution)	Final concentration (1×)	Amount to add (for 10× stock)	Final concentration (10×)
NaCl	8 g	137 mM	80 g	1.37 M
KCl	0.2 g	2.7 mM	2 g	27 mM
Na_2HPO_4	1.44 g	10 mM	14.4 g	100 mM
KH_2PO_4	0.24 g	1.8 mM	2.4 g	18 mM
If necessary, PBS may be supplemented with the following:				
$CaCl_2 \cdot 2H_2O$	0.133 g	1 mM	1.33 g	10 mM
$MgCl_2 \cdot 6H_2O$	0.10 g	0.5 mM	1.0 g	5 mM

PBS can be made as a 1× solution or as a 10× stock. To prepare 1 L of either 1× or 10× PBS, dissolve the reagents listed above in 800 mL of H_2O. Adjust the pH to 7.4 (or 7.2, if required) with HCl, and then add H_2O to 1 L. Dispense the solution into aliquots and sterilize them by autoclaving for 20 min at 15 psi (1.05 kg/cm^2) on liquid cycle or by filter sterilization. Store PBS at room temperature.

ACKNOWLEDGMENTS

This work was supported by a Hendricks Bridge Grant Award (to M.E.Z.) and the Lions Club District 20-Y1 of Central New York.

REFERENCES

Choi RY, Engbretson GA, Solessio EC, Jones GA, Coughlin A, Aleksic I, Zuber ME. 2011. Cone degeneration following rod ablation in a reversible model of retinal degeneration. *Invest Ophthalmol Vis Sci* **52:** 364–373.

Curado S, Anderson RM, Jungblut B, Mumm J, Schroeter E, Stainier DY. 2007. Conditional targeted cell ablation in zebrafish: A new tool for regeneration studies. *Dev Dyn* **236:** 1025–1035.

Kaya F, Mannioui A, Chesneau A, Sekizar S, Maillard E, Ballagny C, Houel-Renault L, Dupasquier D, Bronchain O, Holtzmann I, et al. 2012. Live imaging of targeted cell ablation in *Xenopus*: A new model to study demyelination and repair. *J Neurosci* **32:** 12885–12895.

Langhe R, Chesneau A, Colozza G, Hidalgo M, Ail D, Locker M, Perron M. 2017. Müller glial cell reactivation in *Xenopus* models of retinal degeneration. *Glia* **65:** 1333–1349.

Martinez-De Luna RI, Ku RY, Lyou Y, Zuber ME. 2013. Maturin is a novel protein required for differentiation during primary neurogenesis. *Dev Biol* **384:** 26–40.

Sekizar S, Mannioui A, Azoyan L, Colin C, Thomas JL, Du Pasquier D, Mallat M, Zalc B. 2015. Remyelination by resident oligodendrocyte precursor cells in a *Xenopus laevis* inducible model of demyelination. *Dev Neurosci* **37:** 232–242.

Showell C, Conlon FL. 2009. Tissue sampling and genomic DNA purification from the western clawed frog *Xenopus tropicalis*. *Cold Spring Harb Protoc* doi: 10.1101/pdb.prot5294.

Uribe RA, Gross JM. 2007. Immunohistochemistry on cryosections from embryonic and adult zebrafish eyes. *Cold Spring Harb Protoc* doi: 10.1101/pdb.prot4779.

Viczian AS, Zuber ME. 2010. Tissue determination using the animal cap transplant (ACT) assay in *Xenopus laevis*. *J Vis Exp* **39:** e1932.

Protocol 8

Infrared Laser-Mediated Gene Induction at the Single-Cell Level in the Regenerating Tail of *Xenopus laevis* Tadpoles

Riho Hasugata,[1] Shinichi Hayashi,[2] Aiko Kawasumi-Kita,[3] Joe Sakamoto,[4] Yasuhiro Kamei,[4,5] and Hitoshi Yokoyama[1,6]

[1]*Department of Biochemistry and Molecular Biology, Faculty of Agriculture and Life Science, Hirosaki University, Aomori 036-8561, Japan;* [2]*Fujii Memorial Institute of Medical Sciences, Tokushima University, Tokushima 770-8503, Japan;* [3]*Laboratory for Developmental Morphogeometry, RIKEN Center for Biosystems Dynamics Research, Kobe, Hyogo 650-0047, Japan;* [4]*Spectrography and Bioimaging Facility, National Institute for Basic Biology, Myodaiji, Okazaki, Aichi 444-8585, Japan;* [5]*Department of Basic Biology in the School of Life Science of the Graduate University for Advanced Studies (SOKENDAI), Okazaki, Aichi 444-8585, Japan*

We describe a precise and reproducible gene-induction method in the amphibian, *Xenopus laevis*. Tetrapod amphibians are excellent models for studying the mechanisms of three-dimensional organ regeneration because they have an exceptionally high regenerative ability. However, spatial and temporal manipulation of gene expression has been difficult in amphibians, hindering studies on the molecular mechanisms of organ regeneration. Recently, however, development of a *Xenopus* transgenic system with a heat-shock-inducible gene has enabled the manipulation of specific genes. Here, we applied an infrared laser-evoked gene operator (IR-LEGO) system to the regenerating tail of *Xenopus* tadpoles. In this method, a local heat shock by laser irradiation induces gene expression at the single-cell level. After amputation, *Xenopus* tadpoles regenerate a functional tail, including spinal cord. The regenerating tail is flat and transparent enabling the targeting of individual cells by laser irradiation. In this protocol, a single neural progenitor cell in the spinal cord of the regenerating tail is labeled with heat-shock-inducible green fluorescent protein (GFP). Gene induction at the single-cell level provides a method for rigorous cell-lineage tracing and for analyzing gene function in both cell-autonomous and noncell-autonomous contexts. The method can be modified to study the regeneration of limbs or organs in other amphibians, including *Xenopus tropicalis*, newts, and salamanders.

MATERIALS

It is essential that you consult the appropriate Material Safety Data Sheets and your institution's Environmental Health and Safety Office for proper handling of equipment and hazardous materials used in this protocol.

RECIPES: Please see the end of this protocol for recipes indicated by <R>. Additional recipes can be found online at http://cshprotocols.cshlp.org/site/recipes.

Reagents

Dechlorinated water
Tricaine (5×) in Holtfreter's solution <R>

Prepare 1× tricaine by diluting 5× tricaine in dechlorinated water.

---

[6]Correspondence: yokoyoko@hirosaki-u.ac.jp

Supplemental Material is available for this article at cshprotocols.cshlp.org.

Xenopus laevis adults (sexually mature, transgenic containing a gene of interest that is directly tagged with GFP [or another fluorescent protein such as mCherry], under control of the heat shock [*hsp70*] promoter)

Xenopus laevis adults (sexually mature, wild-type)

Equipment

Digital camera system (mounted on the inverted microscope)
Dish (glass-bottom, e.g., D1113OH, Matsunami Glass)
Dish (plastic, 35 mm)
Incubator
Inverted microscope (e.g., IX73, Olympus)
IR-LEGO-200 system (Sigma Koki, or equivalent)
Laser power meter (e.g., Ophir Vega with a sensor head [10A-V1.RoHS], Ophir Optronics Solutions)
Marker pen (black oil-ink, e.g., K-177N, Shachihata)
Pipette (disposable, plastic)
Surgical blade (sterile, e.g., 2976#10, Feather Safety Razor)

METHOD

Tail Amputation of *Xenopus laevis* Tadpoles

1. Obtain transgenic *Xenopus laevis* embryos by crossing transgenic with wild-type adult frogs.

2. Rear embryos at 16°C until the first tadpole stage, which is ~4–6 d after fertilization. Then rear tadpoles at 22°C–23°C.

   Embryos will be transgenic and wild-type. Transgenic embryos are most reliably selected after irradiation and observation of GFP fluorescence.

   Rearing at 16°C minimizes undesirable (so-called leaky) expression from the hsp70 promoter (Wheeler et al. 2000).

   See Troubleshooting.

3. Select stage 41–42 tadpoles according to Nieuwkoop and Faber (1994).

   Intestinal looping is a good indicator for selection at these stages.

   Remaining tadpoles after stage selection (i.e., tadpoles not at stage 41/42) can be examined to determine sufficient expression of the transgene after heat shock and to estimate the percentage of tadpoles expressing the gene. Tadpoles can be whole-body heat shocked for 30 min at 34°C (Beck et al. 2003). Robust GFP fluorescence should be seen by one day after heat shock (see Supplemental Movie S1).

4. Anesthetize a tadpole in 1× tricaine under careful observation until the tadpole stops moving.

   Anesthetize for as short a time as possible because tadpoles can die if exposed to tricaine for too long.

5. Transfer the tadpole into dechlorinated water for a few seconds to remove excess tricaine from the tadpole.

6. Place the tadpole in a 35 mm plastic dish. Amputate the tadpole tail at the appropriate level with a surgical blade (Fig. 1A).

7. Transfer the tadpole to dechlorinated water.

8. Rear the tadpole in dechlorinated water with appropriate feeding.

   Avoid overfeeding because it may decrease the survival rate of tadpoles at these stages.

Cite this protocol as *Cold Spring Harb Protoc*; doi:10.1101/pdb.prot101014

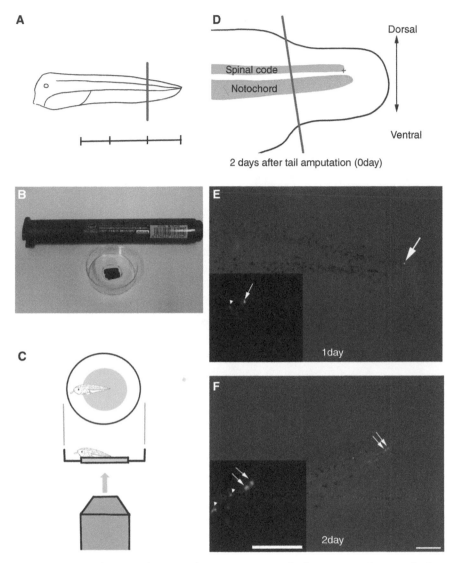

FIGURE 1. Single-cell-level induction of GFP in the regenerating tail of a *Xenopus laevis* tadpole. F6 transgenic *Xenopus laevis* tadpoles containing a *gfp* gene under the control of a heat-shock promoter (*hsp70*) were prepared by mating the F5 transgenic line (Mukaigasa et al. 2009) with wild-type frogs. (*A*) At stage 41/42, the tadpole tail was amputated at one third of the distance from the tail tip to the cloaca. The ruler indicates the position of amputation. The red line indicates the amputation plane. (*B*) Black marker pen (*top*) and glass-bottom dish painted with the marker pen (*bottom*). (*C*) Illustration of infrared laser irradiation applied to the regenerating tail of a tadpole. The IR laser was applied through the 20× objective lens of an inverted microscope (laser power at the target point was 10 mW). A tadpole tail was tightly adhered to a glass-bottom dish on its *left* or *right* lateral side. (*D*) Illustration of a regenerating tadpole tail 2 d after amputation. The orange region represents the notochord. The green region represents the spinal cord. The red line indicates the amputation plane. The red cross indicates the point targeted by the infrared laser. Neural progenitor cells (target cells) were located around the distal edge of the regenerating spinal cord. (*E,F*) A single neural progenitor cell was induced to express GFP by IR-LEGO. The orientation of the sample is same as that in (*D*). A cell expressed GFP 1 d after irradiation (3 d after tail amputation) (*E*), and then two GFP-labeled cells were observed in the same sample 2 d after irradiation (*F*). The sample in (*E,F*) was photographed at the same magnification and under the same camera conditions. White arrows indicate the irradiated cell and its descendants. White arrowheads indicate cells with nonspecific (so-called leaky) faint GFP fluorescence. Insets are high-power views. These observations indicate that 2 d after irradiation a single irradiated cell had divided into two cells. Scale bar = 200 μm.

Infrared-Laser-Evoked Gene Operator (IR-LEGO) Method

The IR-LEGO microscope system is a simple optical construction. We do not provide here a description of the optical adjustment and laser power measurement steps; please refer to the manufacturer's instructions for this. Alternatively, please contact Y. Kamei (ykamei@nibb.ac.jp) for advice on self-build IR-LEGO systems.

9. Confirm the laser focal point. Paint the inner surface of a glass-bottom dish with a black oil-ink marker pen (Fig. 1B). Place the dish onto an inverted microscope stage and adjust the focus to the painted surface. The laser irradiation will ablate the black ink at the focus point if the focus is correctly adjusted to the glass bottom. Confirm the laser focal point, and mark it (e.g., with crosshairs) on the preview screen of the digital camera system.

   Note that some brands of marker pens cannot be used for this purpose.

10. Select a tadpole whose tail has been amputated and anesthetize in 1× tricaine under careful observation until the tadpole stops moving.

    Anesthetize for as short a time as possible because tadpoles can die if exposed to tricaine for too long. Process one tadpole at a time because the time it takes to irradiate a tadpole is unpredictable.

    Move the tadpole to a glass-bottom dish and rinse with dechlorinated water. Remove water as completely as possible with a disposable plastic pipette.

11. Observe the tadpole under the inverted microscope and locate its spinal cord. Identify a target neural progenitor cell and move the glass-bottom dish so that the target cell is at the center of the target-sight crosshairs (i.e., at the laser focal point). Set the laser power and shutter speed to the appropriate values and apply IR laser to the target locus (Fig. 1C,D). Move the tadpole to dechlorinated water.

    Laser power and shutter speed should be empirically determined (see Troubleshooting points for Step 13). A shutter speed of 1 sec is typical.

12. Twenty-four hours after irradiation, observe the tadpoles under the inverted microscope and select those with a single GFP-labeled neural progenitor cell.

    It takes more than half a day for irradiated cells to emit bright GFP fluorescence; therefore, an irradiated cell may proliferate to two or more GFP fluorescence-emitting cells. If observations need to start from a single cell, select only those tadpoles with a single GFP-labeled cell at this time. Cells with faint GFP are often observed in tail tissues that were not subjected to the heat shock (arrowheads, Fig. 1E,F). Although this phenomenon is especially observed in individuals receiving efficient gene transfer, a high level of transgene expression can be induced by heat shock in these animals. The emission of GFP fluorescence from irradiated cell(s) (arrows, Fig. 1E,F) is stronger than that of the nonspecific GFP of nonirradiated cells (arrowheads, Fig. 1E,F), and can be easily distinguished.

    See Troubleshooting.

TROUBLESHOOTING

Problem (Step 2): Undesirable (leaky) gene expression and/or developmental abnormalities are observed in transgenic embryos in the absence of heat shock.

Solution: Undesirable gene expression from the *hsp70* promoter is often observed and causes abnormal embryonic development when embryos are reared at room temperature. To prevent undesirable gene expression, it is important to keep embryos at 16°C (Wheeler et al. 2000). If transgenic embryos are obtained by natural mating, put the mating pair of adult frogs in a container in an incubator at 16°C to allow fertilized eggs to develop at the appropriate temperature.

Problem (Step 12): No or very few GFP-labeled cells are observed after irradiation.

Solution: Check whether the tadpole tail is tightly adhered to the glass-bottom dish at irradiation. Then, check the distance from the surface of the glass bottom to the focal point of laser irradiation. You can estimate this distance by counting the revolutions of the focus knob of the microscope. Ideally, the distance should be shorter than 100 μm. Confirm the laser power at a sample position using a laser power meter. Then, increase the laser power gradually. Alternatively, you can try using another objective lens with a different magnification or numerical aperture for the irradiation. Usually, a cellular response cannot be observed after irradiation at very low laser power; therefore, the laser power should be coarsely increased until induction of GFP fluorescence is observed. Then the optimal irradiation conditions can be determined by finely decreasing the laser power (see also the next problem and solution).

Problem (Step 12): Irradiated cell(s) seem to be dying.

Solution: If irradiation is too strong, cells change to a yellowish color and ring-shaped fluorescence is often observed in the irradiated area (Kawasumi-Kita et al. 2015). In this case, cells in the central area of the irradiated spot, which receive the strongest irradiation, die and those in the peripheral area, which receive weaker irradiation, survive and express the transgene. If the sample was strongly irradiated, cell survival can be assessed by staining cell nuclei of the irradiated sample with a dye (e.g., TO-PRO-3, Thermo Fisher). If cell death is induced by irradiation, degradation of cell nuclei can be observed within several hours (Kawasumi-Kita et al. 2015). In this case, the laser power should be decreased so as not to induce such cell death.

DISCUSSION

The IR-LEGO system using an IR laser (wavelength 1480 nm) provides highly efficient and reproducible gene induction with minimal cellular damage compared with other gene expression-inducible systems that use a heat-shock promoter (Kamei et al. 2009). IR-LEGO can be used to label cells in various multicellular organisms (Deguchi et al. 2009) as long as the targeted cell(s) can be focused on by the IR-laser. IR-LEGO has already been applied to multiple organs in anuran and urodele amphibians (Kawasumi-Kita et al. 2015). While it is still difficult to target an inner layer of cells that is distant from the bottom of the glass dish, the laser irradiation can be spatially manipulated in a precise manner. In *Xenopus laevis* and the newt, *Pleurodele waltl*, genes can be induced in different sized areas after a single irradiation, from a single cell to a cell cluster <500 µm in diameter (skeletal muscle) or 200 µm in diameter (limb bud and blastema) (Hayashi et al. 2014; Kawasumi-Kita et al. 2015). Thus, by modifying the laser power and the objective lens used for irradiation, IR-LEGO can induce gene expression in targets ranging from a single cell to a cluster of cells a few hundred micrometers in diameter. The irradiation can be applied precisely in the micrometer range, especially if the microscope has a precision motorized stage system.

The application of IR-LEGO will facilitate regeneration studies in amphibians. The blastema, which plays a major role in organ regeneration of amphibians and fish, is not thought to be a population of identical cells but a heterogeneous population of cells with different differentiation potentials (Tanaka and Reddien 2011). Therefore, the ability to label a single cell is crucial for tracking blastema cell fates and elucidating organ regeneration mechanisms. The *Xenopus laevis* tadpole can regenerate a functional tail consisting of spinal cord, notochord, and muscles after amputation (e.g., Beck et al. 2003; Gaete et al. 2012) and is, therefore, a valuable model system for studying new therapies for spinal cord injury in mammals, including humans. In the tail-regeneration process of early-stage *Xenopus* tadpoles, highly proliferative Sox2-positive neural progenitor cells play an important role in spinal cord regeneration (Gaete et al. 2012). Neural progenitor cells during *Xenopus* tail regeneration are a convenient target for single-cell gene induction by IR-LEGO (Hayashi et al. 2014), because they are easily recognized in the flat and transparent tissues of the regenerating tail. This technique can be used to trace cell fate for several days, and for the functional analysis of a particular gene at the single-cell level. Such single-cell-level analyses are indispensable for distinguishing cell-autonomous from noncell-autonomous gene functions in regeneration phenomena, as previously described by Hayashi et al. (2014).

Once proficient in IR-LEGO at the single-cell-level for *Xenopus laevis* neural progenitor cells, a researcher can be confident to optimize the irradiation conditions for other tissues, such as limb blastema cells or similar progenitor cells in the regenerating tail of other amphibians (e.g., *Xenopus tropicalis*, newts, and salamanders). The gene induction by IR-LEGO described here only persists for a few days; however, combining the IR-LEGO system with the Cre-loxP system could enable longer-term cell labeling for lineage tracing, as shown in medaka (Shimada et al. 2013). IR-laser irradiation using a higher laser power can also be used in cell ablation experiments to eliminate specific cells in a developing or regenerating organ (Zeng et al. 2016). IR-LEGO at the single-cell level is a precise and reproducible method for gene manipulation and cell-fate analysis in amphibians.

RECIPE

Tricaine (5×) in Holtfreter's Solution

Reagent	Quantity for 500 mL of 5× solution	Final concentration (5×)
NaCl	8.75 g	300 mM
KCl	0.125 g	3 mM
CaCl$_2$	0.25 g	4.5 mM
NaHCO$_3$	0.500 g	12.5 mM
Tricaine (ethyl 3-aminobenzonate methanesulfonate)	1.25 g	2.5 mg/mL (0.25%)

To prepare 500 mL, dissolve all reagents in 500 mL deionized H$_2$O. Sterilization is not required. Store at 4°C.

ACKNOWLEDGMENTS

We thank Dr. Yumi Izutsu for providing the *hsp70*-GFP Tg *Xenopus laevis* frogs. We thank all staff members of the Spectrography and Bioimaging Facility and the National Institute for Basic Biology (NIBB) Core Research Facilities for their kind support in setting up the IR-LEGO experiments in *Xenopus laevis*. This work was supported by the NIBB Collaborative Research Program to H.Y. (16-509, 17-505) and to A.K.-K. (Representative investigator: Yoshihiro Morishita; 16-521, 17-512); Japan Society for Promotion of Science (JSPS) KAKENHI Grant Number 16H04790 to H.Y.; JSPS KAKENHI Grant Number 17H06258 to Y.K.; Asahi Glass Foundation to H.Y.; Takeda Science Foundation to H.Y.; Uehara Memorial Foundation to H.Y.; Hirosaki University Grant for Exploratory Research by Young Scientists and Newly-appointed Scientists to H.Y.; Hirosaki University Institutional Research Grant for Young Investigators to H.Y.

REFERENCES

Beck CW, Christen B, Slack JM. 2003. Molecular pathways needed for regeneration of spinal cord and muscle in a vertebrate. *Dev Cell* 5: 429–439.

Deguchi T, Itoh M, Urawa H, Matsumoto T, Nakayama S, Kawasaki T, Kitano T, Oda S, Mitani H, Takahashi T, et al. 2009. Infrared laser-mediated local gene induction in medaka, zebrafish and *Arabidopsis thaliana. Dev Growth Differ* 51: 769–775.

Gaete M, Muñoz R, Sánchez N, Tampe R, Moreno M, Contreras EG, Lee-Liu D, Larraín J. 2012. Spinal cord regeneration in *Xenopus* tadpoles proceeds through activation of Sox2-positive cells. *Neural Dev* 7: 13.

Hayashi S, Ochi H, Ogino H, Kawasumi A, Kamei Y, Tamura K, Yokoyama H. 2014. Transcriptional regulators in the Hippo signaling pathway control organ growth in *Xenopus* tadpole tail regeneration. *Dev Biol* 396: 31–41.

Kamei Y, Suzuki M, Watanabe K, Fujimori K, Kawasaki T, Deguchi T, Yoneda Y, Todo T, Takagi S, Funatsu T, et al. 2009. Infrared laser-mediated gene induction in targeted single cells in vivo. *Nat Methods* 6: 79–81.

Kawasumi-Kita A, Hayashi T, Kobayashi T, Nagayama C, Hayashi S, Kamei Y, Morishita Y, Takeuchi T, Tamura K, Yokoyama H. 2015. Application of local gene induction by infrared laser-mediated microscope and temperature stimulator to amphibian regeneration study. *Dev Growth Differ* 57: 601–613.

Mukaigasa K, Hanasaki A, Maéno M, Fujii H, Hayashida S, Itoh M, Kobayashi M, Tochinai S, Hatta M, Iwabuchi K, et al. 2009. The keratin-related Ouroboros proteins function as immune antigens mediating tail regression in *Xenopus* metamorphosis. *Proc Natl Acad Sci* 106: 18309–18314.

Nieuwkoop PD, Faber J. 1994. *Normal table of* Xenopus laevis *(Daudin)*. Garland Publishing, New York.

Shimada A, Kawanishi T, Kaneko T, Yoshihara H, Yano T, Inohaya K, Kinoshita M, Kamei Y, Tamura K, Takeda H. 2013. Trunk exoskeleton in teleosts is mesodermal in origin. *Nat Commun* 4: 1639.

Tanaka EM, Reddien PW. 2011. The cellular basis for animal regeneration. *Dev Cell* 21: 172–185.

Wheeler GN, Hamilton FS, Hoppler S. 2000. Inducible gene expression in transgenic *Xenopus* embryos. *Curr Biol* 10: 849–852.

Zeng CW, Kamei Y, Wang CT, Tsai HJ. 2016. Subtypes of hypoxia-responsive cells differentiate into neurons in spinal cord of zebrafish embryos after hypoxic stress. *Biol Cell* 108: 357–377.

Chemical Screening and Toxicity Testing

André W. Brändli[1,2]

[1]*Walter-Brendel-Center of Experimental Medicine, University Hospital Munich, Ludwig-Maximilians-University Munich, 81377 Munich, Germany*

The chemical space is vast, encompassing potentially billions of natural and synthetic molecules, which are for the most part uncharted with regard to their pharmaceutical, therapeutic, or toxicological potential. Determining the biological efficacy or harm of these chemicals presents both an enormous opportunity and a challenge to society. Chemical screening is the first step in development of novel therapeutical agents. The process typically involves searching chemical libraries for small organic molecules that have biological activities that might be useful in addressing pathological conditions for which there are unmet medical needs. Toxicology, in contrast, investigates effects of chemicals that are harmful to human or animal health or the environment in general. *Xenopus* is an exceptionally effective animal model system for assaying both potential therapeutic and toxicological effects. Here I introduce protocols that detail how *Xenopus* extracts, embryos, and tadpoles can be used in chemical screening and toxicity testing.

WHY *XENOPUS?*

Xenopus is the most accessible vertebrate animal system at all stages of its life cycle. Unlike any mammalian system, *Xenopus* eggs and embryos are readily available in huge numbers and easy to observe, while being amenable to every state-of-the-art molecular technique. Many of these virtues are shared by zebrafish, but *Xenopus* has a significant advantage over fish models in that it shares at least a hundred million years longer of common evolutionary descent with humans. Decades of research have shown the extensive similarities of amphibians such as *Xenopus* to humans both at the molecular level (genomes, signaling molecules, and biochemical pathways) and the cellular level (histology and organ function). Many key discoveries about the mechanisms and molecules directing mammalian development were made on the basis of findings from studies performed in *Xenopus*. It has also contributed important insights into human biology and disease pathology. *Xenopus* embryos and tadpoles should thus remain important animal models in biomedical research, including as a system for discovery of novel drug candidates and for studies of the positive and negative effects of different chemical compounds.

CHEMICAL SCREENING IN VITRO AND IN VIVO USING *XENOPUS*

The identification of biologically active compounds and characterization of their effects in biological systems can involve in vitro or in vivo systems. In vitro test systems can range from purified proteins to

---

[2]Correspondence: abrandli@med.lmu.de

cell extracts to cell and organ culture systems, whereas in vivo testing requires the use of whole organisms. *Xenopus* offers attractive possibilities for both in vitro and in vivo screening and characterization of the effects of such compounds.

Powerful in vitro assay systems based on extracts derived from *Xenopus* eggs or early embryos have been developed over the last three decades. For example, *Xenopus* egg extracts provide for a more complex environment than purified proteins as they represent the concentrated cytosol with intact organelles and protein complexes. They lack the plasma membrane but retain activities that organize the mitotic spindle and other cytosolic processes, such as cell cycle control and microtubule assembly. The protocol by Broadus and Lee demonstrates that egg extracts can be used in a multiwell format to identify small organic molecules that target important cellular pathways (see Protocol 1: Chemical Screening Using Cell-Free *Xenopus* Egg Extract [Broadus and Lee 2018]).

Xenopus embryos and tadpoles represent intact multicellular vertebrate organisms harboring tissues and organs that show complex biochemical and physiological interactions that maintain homeostasis and growth. Unlike mammalian model systems, they are ideal whole organisms for chemical library screening, given their small size (1–4 mm) and their ability to be raised externally in simple saline solutions. Thousands of embryos can be generated and arrayed each day in 48- or 96-well plates for chemical screening and testing as outlined by Gull and colleagues (see Protocol 2: Screening of Chemical Libraries Using *Xenopus* Embryos and Tadpoles for Phenotypic Drug Discovery [Gull et al. 2022]). Chemicals may readily penetrate into the embryo via the skin or be taken up by ingestion and gill filtration in older animals. Sullivan and Levin (2018) have developed a clever chemical screening assay using *Xenopus* tadpoles to identify the bioelectric signaling and neurotransmitter machinery underlying tail regeneration. Their comprehensive protocol uses known chemical compounds with defined pharmacology. In a hierarchical approach, tadpoles are probed with tiers of compounds progressively narrowing down possible candidate pathways and target classes (see Protocol 3: Inverse Drug Screening of Bioelectric Signaling and Neurotransmitter Roles: Illustrated Using a *Xenopus* Tail Regeneration Assay [Sullivan and Levin 2018]).

XENOPUS MODEL SYSTEMS FOR TOXICOLOGY

Man-made chemicals released into the environment can have devastating effects in delicate ecosystems as well as interfere with the well-being of domestic animals and the human population. The threat is ever-increasing given the continuous development of novel chemicals for industrial purposes or applications in agriculture. In addition, hundreds of novel drug candidates are identified annually with the hope of improving animal and human health, and their effects must also be assessed.

The utility of *Xenopus* model systems for toxicology and early assessment of adverse side effects of drug candidates is shown by three protocols. Synthetic endocrine disruptors can be a source of declining health of animals. Spirhanzlova et al. (2019) present a high-throughput protocol to identify compounds with endocrine-disrupting potential. The protocol uses *Xenopus* tadpoles to detect endocrine-disrupting activities by monitoring drug-induced changes in thyroid gene expression or, alternatively, expression of a green fluorescent protein (GFP) reporter gene in transgenic tadpoles (see Protocol 4: Following Endocrine-Disrupting Effects on Gene Expression in *Xenopus laevis* [Spirhanzlova et al. 2019]). *Xenopus* embryos and tadpoles can also be used as highly sensitive and convenient whole-organism test systems to detect potential adverse effects of drug candidates early in the drug development process. Saide and Wheeler (2020) detail two hepatoxicity protocols designed to pick out harmful compounds in *Xenopus* embryos, which replicate similar protocols carried out in mammals. One uses a liver-specific microRNA as an indicator of hepatocellular damage in vivo; the other measures the concentration of glutathione as an indicator of paracetamol-induced liver injury (see Protocol 5: In Vivo Assessment of Drug-Induced Hepatotoxicity Using *Xenopus* Embryos [Saide and Wheeler 2020]).

Teratogenic compounds perturb embryonic and fetal development with devastating effects for animal and human health that include birth defects, cancer, and reduced fertility. Hence, the early identification of drug candidates with teratogenic potential and their elimination from further preclinical evaluation is essential in drug development. Fort and colleagues have been pioneers in developing the standardized Frog Embryo Teratogenesis Assay (FETAX) as a cost-effective alternative to mammalian model systems for assessment of the teratogenicity of chemical compounds (see Protocol 6: Frog Embryo Teratogenesis Assay—*Xenopus* (FETAX): Use in Alternative Preclinical Safety Assessment [Fort and Mathis 2018]).

CONCLUDING REMARKS

Screening and identification of chemical compounds with novel biological activities is an essential step in the development of novel therapeutics to treat human disease. In addition, assessment of how a drug candidate perturbs molecular signaling, cell biological functions, embryonic development, and organ physiology is important in preclinical drug development and can allow early discovery of potential adverse effects before entering clinical testing in humans. *Xenopus* models, either in vitro or in vivo, represent predictive and cost-effective alternatives to mammalian test systems for drug discovery and development.

ACKNOWLEDGMENTS

I am grateful to Hazel Sive for her critical input. Work in my laboratory was supported by the University Hospital Munich and in part by a grant of the European Commission (EU FP7 Program, EuRenOmics Grant Agreement 305608).

REFERENCES

Broadus MR, Lee E. 2018. Chemical screening using cell-free *Xenopus* egg extract. *Cold Spring Harb Protoc* doi:10.1101/pdb.prot098277

Fort DJ, Mathis M. 2018. Frog embryo teratogenesis assay—*Xenopus* (FETAX): use in alternative preclinical safety assessment. *Cold Spring Harb Protoc* doi:10.1101/pdb.prot098319

Gull M, Schmitt SM, Kälin RE, Brändli AW. 2022. Screening of chemical libraries using *Xenopus* embryos and tadpoles for phenotypic drug discovery. *Cold Spring Harb Protoc* doi:10.1101/pdb.prot098269

Saide K, Wheeler GN. 2020. In vivo assessment of drug-induced hepatotoxicity using *Xenopus* embryos. *Cold Spring Harb Protoc* doi:10.1101/pdb.prot106096

Spirhanzlova P, Leemans M, Demeneix BA, Fini JB. 2019. Following endocrine-disrupting effects on gene expression in *Xenopus laevis*. *Cold Spring Harb Protoc* doi:10.1101/pdb.prot098301

Sullivan KG, Levin M. 2018. Inverse drug screening of bioelectric signaling and neurotransmitter roles: illustrated using a *Xenopus* tail regeneration assay. *Cold Spring Harb Protoc* doi:10.1101/pdb.prot099937

Protocol 1

Chemical Screening Using Cell-Free *Xenopus* Egg Extract

Matthew R. Broadus[1] and Ethan Lee[2,3]

[1]*Department of Cell Biology, Harvard Medical School, Boston, Massachusetts 02115;* [2]*Department of Cell and Developmental Biology, Vanderbilt Ingram Cancer Center, Vanderbilt, University Medical Center, Nashville, Tennessee 37232*

Most drug screening methods use purified proteins, cultured cells, and/or small model organisms such as *Xenopus*, zebrafish, flies, or nematodes. These systems have proven successes in drug discovery, but they also have weaknesses. Although purified cellular components allow for identification of compounds with activity against specific targets, such systems lack the complex biological interactions present in cellular and organismal screens. In vivo systems overcome these weaknesses, but the lack of cellular permeability, efflux by cellular pumps, and/or toxicity can be major limitations. *Xenopus laevis* egg extract, a concentrated and biologically active cytosol, can potentially overcome these weaknesses. Drug interactions occur in a near-physiological milieu, thereby functioning in a "truer" endogenous manner than purified components. Also, *Xenopus* egg extract is a cell-free system that lacks intact plasma membranes that could restrict drug access to potential targets. Finally, *Xenopus* egg extract is readily manipulated at the protein level: Proteins are easily depleted or added to the system, an important feature for analyzing drug effects in disease states. Thus, *Xenopus* egg extract offers an attractive media for screening drugs that merges strengths of both in vitro and in vivo systems.

MATERIALS

It is essential that you consult the appropriate Material Safety Data Sheets and your institution's Environmental Health and Safety Office for proper handling of equipment and hazardous materials used in this protocol.

RECIPES: Please see the end of this protocol for recipes indicated by <R>. Additional recipes can be found online at http://cshprotocols.cshlp.org/site/recipes.

Reagents

Chemical library

Select a library based on the specific screen.

Energy reaction mix (ER mix) <R>

Reagents for assay detection

Kits are commercially available and should be chosen based on the specific application. Light-based readouts (e.g., luminescence, fluorescence, etc.) are favored in chemical screening for their robust signals, ease in execution, quantifiability, and rapidity of setup. Other types of readout (e.g., immunoblotting or radioactive-based assays) require considerable time and effort to perform and are cumbersome when large libraries are used.

Recombinant proteins of interest

---

[3]Correspondence: Ethan.Lee@vanderbilt.edu

Cite this protocol as *Cold Spring Harb Protoc*; doi:10.1101/pdb.prot098277

Commercially available kits for in vitro transcription/translation represent rapid methods to produce recombinant components and allow for incorporation of labeled amino acids (e.g., [^{35}S]-methionine and -cysteine). If large amounts of recombinant protein are required, protein from bacteria, Sf9/baculovirus, or other systems might be necessary.

Xenopus egg extract (XE)

Many methods exist for preparing XE (e.g., Cross and Powers 2008a,b; see also Sec. 7, Protocol 1: Preparation of Cellular Extracts from Xenopus Eggs and Embryos [Good and Heald 2018]) and often it can be useful to tailor the extract preparation to the specific screen design. For example, to monitor β-catenin turnover in the Wnt pathway, a medium-speed extract (Chen et al. 2014) is preferable to the type of low-speed extract (e.g., Murray and Kirschner 1989; Murray 1991; Maresca and Heald 2006) often used in many cell cycle assays. Other specific XE preparation protocols are available for the study of cytoskeletal dynamics (Lebensohn et al. 2006; Maresca and Heald 2006), nucleocytoplasmic transport (Chan and Forbes 2006), DNA replication (Walter et al. 1998), nuclear assembly (Newport 1987), nuclear size regulation (Edens and Levy 2016), DNA damage (Willis et al. 2012), and apoptosis (Kornbluth and Evans 2001; Deming and Kornbluth 2006). XE can also be used to study signal transduction (see Bermudez et al. [2017] and references therein), DNA replication (e.g., Raspelli et al. 2017), ubiquitin metabolism (Verma et al. 2004), and protein turnover (Shennan 2006). In particular, XE screens have successfully identified small molecule inhibitors of actin polymerization (Peterson et al. 2004), proteasomal degradation (Verma et al. 2004), cyclin turnover (Salic and King 2005), DNA damage-repair (Landais et al. 2009), and Wnt signaling (Thorne et al. 2011).

Equipment

Detector (e.g., luminometer, fluorescence plate reader, ELISA plate reader, etc.)

Foil covers for multiwell plates

Multiwell plate (e.g., 96-well)

The type of plate will depend on the screen readout and the brand of detector.

Multichannel pipette or automated dispenser

Vortex mixer

Additional equipment (e.g., incubator, centrifuge) might also be required for specific screens.

METHOD

Before screening, it is important to optimize reaction conditions and perform a pilot screen. Parameters that can affect the assay (e.g., time, component concentration, temperature, etc.) should be carefully assessed to obtain greatest signal-to-noise ratio, thereby enhancing the statistical performance of the screen. Z-factor scoring, an indicator of assay quality, reflects the dynamic range and data variation of an assay. A favorable Z-factor is often used as a way to determine the quality of an assay to increase the chances of performing a successful chemical screen (Zhang et al. 1999; Broadus et al. 2015). Also, compounds should be tested with appropriate controls. For example, if a luciferase activity serves as the readout, compounds should be tested against the luciferase enzyme itself. The following method uses a 96-well plate but can be adjusted for other formats (e.g., 384-well plates).

1. Chill 96-well plates on ice.

   This prevents spurious reactions once XE and other components are added.

2. If starting with frozen XE, quickly hand-thaw aliquots (or thaw in a 30°C water bath). Place the tubes on ice before thawing is complete (i.e., small amounts of frozen XE should still be visible).

   The use of frozen versus freshly prepared XE depends on the particular assay. Certain cellular activities (e.g., translation) can be severely inhibited by freeze/thawing.

3. Thaw 20× ER Mix and other reagents necessary for the assay (e.g., recombinant proteins, luciferase reagents, etc.). Place them on ice.

4. Add ER Mix to a final concentration of 1× to the thawed XE on ice.

5. Add predetermined concentrations of any additional components required for the assay to the XE mix.

   As a reference/example, recombinant luciferase-fusions are typically added with initial starting activity of at least 10,000 relative luminescence units per μL.

6. Aliquot 10 μL of the reaction mix into each well of the plate on ice.

Each plate screened should have at least six wells dedicated to running controls in triplicate (i.e., three wells for a positive control and three wells for a negative control).

7. Using a multichannel pipette or automated dispenser, dispense the drug library and controls into their respective wells.

   For efficiency, use a stock drug library previously arrayed into 96-well plates. Because the protein concentration of XE is high (~50 mg/mL), free drug concentrations can be severely underestimated. A much higher compound concentration will therefore be required relative to purified or cell-based assays; an initial small molecule concentration of ~500 μM is recommended. Note that high concentrations of dimethyl sulfoxide (often used to solubilize drugs) can inhibit biochemical reactions; its concentration should therefore be limited and its effects carefully tested during assay development.

8. Seal plates with a foil cover. By hand (or by low-speed vortexing), gently mix plates without splashing the contents along the sides of the well or lid. Centrifuge plates briefly after mixing to ensure coalescence.

9. Allow reactions to proceed by incubating plates at the appropriate temperature.

   This step is dependent on the particular reaction used and should be carefully optimized before screening. For many reactions, incubation at room temperature works well and a special incubator is not necessary. For reactions requiring long incubation times (hours), plates should be incubated inside a humidified chamber to minimize evaporation.

10. Terminate reactions in a manner consistent with the readout method. For example, if assaying for luciferase activity, the reaction is terminated with at least 10× dilution of Luciferin reagent.

11. Analyze reactions according to the assay readout.

    For luciferase-based assays, luminescence levels are measured using a luminometer.

DISCUSSION

Typically, compounds that increase or decrease signals by more than three standard deviations are considered significant. The usefulness of this value depends on the number of "hits" that can be handled in a secondary screen. Each screen protocol should be individually assessed using high-quality controls. Ideally, samples should be assayed in triplicate. Compounds that meet the significance cutoff in all three replicates are considered "candidate leads" and represent potential modulators of the biological process. In summary, XE recapitulates a large number of complex biological processes and shows strengths of both in vitro and in vivo systems. The advantages offered by XE makes it an attractive medium for identifying small molecules that target important and complex cellular pathways.

RECIPE

Energy Reaction Mix (ER Mix)

Reagent	Amount to add	Final concentration (20×)
Adenosine triphosphate (Sigma-Aldrich A2383)	10.1 mg	20 mM
MgCl$_2$ (Fisher BP214)	1.7 mg	20 mM
Creatine phosphate (Sigma-Aldrich 2380)	31.7 mg	150 mM
Creatine phosphokinase (Sigma-Aldrich C3755)	600 μg	600 μg/mL

Dissolve in 1 mL of deionized water. Mix by vortexing. Store 50-μL aliquots frozen at −20°C for several months; for longer-term storage, maintain samples at −80°C.

Cite this protocol as *Cold Spring Harb Protoc*; doi:10.1101/pdb.prot098277

REFERENCES

Bermudez JG, Chen H, Einstein LC, Good MC. 2017. Probing the biology of cell boundary conditions through confinement of *Xenopus* cell-free cytoplasmic extracts. *Genesis* **55**: e23013.

Broadus MR, Yew PR, Hann SR, Lee E. 2015. Small-molecule high-throughput screening utilizing *Xenopus* egg extract. *Methods Mol Biol* **1263**: 63–73.

Chan RC, Forbes DI. 2006. In vitro study of nuclear assembly and nuclear import using *Xenopus* egg extracts. *Methods Mol Biol* **322**: 289–300.

Chen TW, Broadus MR, Huppert SS, Lee E. 2014. Reconstitution of β-catenin degradation in *Xenopus* egg extract. *J Vis Exp* **88**: 51425.

Cross MK, Powers M. 2008a. Obtaining eggs from *Xenopus laevis* females. *J Vis Exp* **18**: 890.

Cross MK, Powers M. 2008b. Preparation and fractionation of *Xenopus laevis* egg extracts. *J Vis Exp* **18**: 891.

Deming P, Kornbluth S. 2006. Study of apoptosis in vitro using the *Xenopus* egg extract reconstitution system. *Methods Mol Biol* **322**: 379–393.

Edens LJ, Levy DL. 2016. A cell-free assay using *Xenopus laevis* embryo extracts to study mechanisms of nuclear size regulation. *J Vis Exp* **114**: 54173.

Good MC, Heald R. 2018. Preparation of cellular extracts from *Xenopus* eggs and embryos. *Cold Spring Harb Protoc* doi: 10.1101/pdb.prot097055.

Kornbluth S, Evans EK. 2001. Analysis of apoptosis using *Xenopus* egg extracts. *Curr Protoc Cell Biol* **11**: Unit 11.12.

Landais I, Sobeck A, Stone S, LaChapelle A, Hoatlin ME. 2009. A novel cell-free screen identifies a potent inhibitor of the Fanconi anemia pathway. *Int J Cancer* **124**: 783–792.

Lebensohn AM, Ma L, Ho H-YH, Kirschner MW. 2006. Cdc42 and PI(4,5) P$_2$-induced actin assembly in *Xenopus* egg extracts. *Methods Enzymol* **406**: 156–173.

Maresca TJ, Heald R. 2006. Methods for studying spindle assembly and chromosome condensation in *Xenopus* egg extracts. *Methods Mol Biol* **322**: 459–474.

Murray AW. 1991. Cell cycle extracts. *Methods Cell Biol* **36**: 581–605.

Murray AW, Kirschner MW. 1989. Cyclin synthesis drives the early embryonic cell cycle. *Nature* **339**: 275–280.

Newport J. 1987. Nuclear reconstitution in vitro: Stages of assembly around protein-free DNA. *Cell* **48**: 205–217.

Peterson JR, Bickford LC, Morgan D, Kim AS, Ouerfelli O, Kirschner MW, Rosen MK. 2004. Chemical inhibition of N-WASP by stabilization of a native autoinhibited conformation. *Nat Struct Mol Biol* **11**: 747–755.

Raspelli E, Falbo L, Costanzo V. 2017. *Xenopus* egg extract to study regulation of genome-wide and locus-specific DNA replication. *Genesis* **55**: e22996.

Salic A, King RW. 2005. Identifying small molecule inhibitors of the ubiquitin-proteasome pathway in *Xenopus* egg extracts. *Methods Enzymol* **399**: 567–585.

Shennan KIJ. 2006. *Xenopus* egg extracts: A model system to study proprotein convertases. *Methods Mol Biol* **322**: 199–212.

Thorne CA, Lalfeur B, Lewis M, Hanson AJ, Jernigan KK, Weaver DC, Huppert KA, Chen TW, Wichaidit C, Cselenyi CS, et al. 2011. A biochemical screen for identification of small-molecule regulators of the Wnt pathway using *Xenopus* egg extracts. *J Biomol Screen* **16**: 995–1006.

Verma R, Peters NR, D'Onofrio M, Tochtrop GP, Sakamoto KM, Varadan R, Zhang M, Coffino P, Fushman D, Deshaies RJ, et al. 2004. Ubistatins inhibit proteasome-dependent degradation by binding the ubiquitin chain. *Science* **306**: 117–120.

Walter J, Sun L, Newport J. 1998. Regulated chromosomal DNA replication in the absence of a nucleus. *Mol Cell* **1**: 519–529.

Willis J, DeStephanis D, Patel Y, Gowda V, Yan S. 2012. Study of the DNA damage checkpoint using *Xenopus* egg extracts. *J Vis Exp* **5**: e4449.

Zhang J-H, Chung TDY, Oldenburg KR. 1999. A simple statistical parameter for use in evaluation and validation of high throughput screening assays. *J Biomol Screen* **4**: 67–73.

Screening of Chemical Libraries Using *Xenopus* Embryos and Tadpoles for Phenotypic Drug Discovery

Mazhar Gull,[1,2,5] Stefan M. Schmitt,[1,3,5] Roland E. Kälin,[1,4] and André W. Brändli[1,6]

[1]*Walter-Brendel-Center of Experimental Medicine, University Hospital, Ludwig-Maximilians-University Munich, 81377 Munich, Germany*

Phenotypic drug discovery assesses the effect of small molecules on the phenotype of cells, tissues, or whole organisms without a priori knowledge of the target or pathway. Using vertebrate embryos instead of cell-based assays has the advantage that the screening of small molecules occurs in the context of the complex biology and physiology of the whole organism. Fish and amphibians are the only classes of vertebrates with free-living larvae amenable to high-throughput drug screening in multiwell dishes. For both animal classes, particularly zebrafish and *Xenopus*, husbandry requirements are straightforward, embryos can be obtained in large numbers, and they develop ex utero so their development can be monitored easily with a dissecting microscope. At 350 million years, the evolutionary distance between amphibians and humans is significantly shorter than that between fish and humans, which is estimated at 450 million years. This increases the likelihood that drugs discovered by screening in amphibian embryos will be active in humans. Here, we describe the basic protocol for the medium- to high-throughput screening of chemical libraries using embryos of the African clawed frog *Xenopus laevis*. Bioactive compounds are identified by observing phenotypic changes in whole embryos and tadpoles. In addition to the discovery of compounds with novel bioactivities, the phenotypic screening protocol also allows for the identification of compounds with in vivo toxicity, eliminating early hits that are poor drug candidates. We also highlight important considerations for designing chemical screens, choosing chemical libraries, and performing secondary screens using whole mount in situ hybridization or immunostaining.

MATERIALS

It is essential that you consult the appropriate Material Safety Data Sheets and your institution's Environmental Health and Safety Office for proper handling of equipment and hazardous materials used in this protocol.

RECIPES: Please see the end of this protocol for recipes indicated by <R>. Additional recipes can be found online at http://cshprotocols.cshlp.org/site/recipes.

Reagents

Prepare all solutions in sterile, deionized, milliQ-filtered H₂O.

Chemical library collections (see Table 1 for examples)
Compound E stock solution (2 mM) <R>

[2]Present address: Sartorius CellGenix GmbH, Freiburg, Germany

[3]Present address: Merck KGaA, Darmstadt, Germany

[4]Present address: Neurosurgical Research, Department of Neurosurgery, University Hospital, Ludwig-Maximilians-University Munich, Munich, Germany

[5]These authors contributed equally to this work.

[6]Correspondence: abrandli@med.lmu.de

TABLE 1. Selection of commercial chemical libraries

Library	Source	Compounds	Format	Description
Libraries of approved drugs, drug candidates, and drug-like compounds				
Phenotypic screening library	Selleck Chemicals	3751	10 or 2 mM in DMSO, 10 or 2 mM in H_2O	Bioactive compounds with identified targets and FDA-approved drugs
FDA-approved and Passed Phase I Drug Library	Selleck Chemicals	3355	10 or 2 mM in DMSO, 10 or 2 mM in H_2O	Drugs that are marketed around the world or have passed clinical phase I
Preclinical/Clinical Compound Library	Selleck Chemicals	3081	10 or 2 mM in DMSO, 10 or 2 mM in H_2O	Preclinical and clinical compounds that are structurally diverse, medicinally active, and cell-permeable
Express-Pick Library	Selleck Chemicals	3010	10 mM in DMSO	Compounds featuring different core structures and structural diversity
The Spectrum Collection	MicroSource Discovery Systems	2560	10 mM in DMSO	U.S. and international drugs, drug candidates, and natural products
The Pharmakon Collection	MicroSource Discovery Systems	1760	10 mM in DMSO	Compounds approved in the United States, Europe, and/or Asia
Prestwick Chemical Library	Prestwick Chemical Libraries	1520	10 mM in DMSO	Diverse small molecules, 98% of which are approved drugs (FDA, EMA, and other agencies)
U.S. Drug Collection	MicroSource Discovery Systems	1360	10 mM in DMSO	Compounds that have FDA-approved or reached clinical trials in the United States
LOPAC[1280]	MilliporeSigma/ Sigma-Aldrich	1280	10 mM in DMSO	A library of pharmacologically active compounds
Tocriscreen 2.0	Tocris Biosciences	1280	10 mM in DMSO	Biologically active compounds covering a wide range of pharmacological targets
SCREEN-WELL FDA Approved Drug Library	Enzo	775	10 mM in DMSO	FDA-approved drug compounds
SCREEN-WELL ICCB Known Bioactives Library	Enzo	472	10 mM in DMSO	Biologically active small organic molecules developed in cooperation with the Harvard Institute of Chemistry and Cell Biology (ICCB)
International Drug Collection	MicroSource Discovery Systems	400	10 mM in DMSO	Compounds that are or have been marketed in Europe and/or Asia but not in the United States
Prestwick Original Molecules Library	Prestwick Chemical Libraries	344	10 mM in DMSO	Original and exclusive drug-like compounds
Tocris FDA-Approved Drugs Library	Tocris Biosciences	190	10 mM in DMSO	Selection of FDA-approved drugs
Libraries of inhibitors and target class–specific compounds				
Kinase Inhibitor Library	Selleck Chemicals	1766	10 or 2 mM in DMSO, 10 or 2 mM in H_2O	Kinase inhibitors, mostly ATP competitive, some are FDA-approved
GPCR Compounds Library	Selleck Chemicals	1273	10 or 2 mM in DMSO, 10 or 2 mM in H_2O	Compounds targeting G protein–coupled receptors (GPCRs)
Tyrosine Kinase Inhibitor Library	Selleck Chemicals	533	10 or 2 mM in DMSO, 10 mM in H_2O	Tyrosine kinase inhibitors, some are FDA-approved
Cytokine Inhibitor Library	Selleck Chemicals	462	10 or 2 mM in DMSO, 10 or 2 mM in H_2O	Kinase inhibitors mostly ATP-competitive, some are FDA-approved
Prestwick GPCR Drug Library	Prestwick Chemical Libraries	265	10 mM in DMSO	Approved drugs known to interact primarily with GPCRs
Ion Channel Ligand Library	Selleck Chemicals	231	10 or 2 mM in DMSO, 10 or 2 mM in H_2O	Small-molecule modulators targeting diverse ion channels
Tocrisscreen Kinase Inhibitor Library	Tocris Biosciences	160	10 mM in DMSO	Compounds targeting >60 different kinases
Inhibitor Library I	MilliporeSigma/ Sigma-Aldrich	80	10 mM in DMSO	Inhibitors against mainly tyrosine, AGC, and atypical families of kinases
Inhibitor Library II	MilliporeSigma/ Sigma-Aldrich	80	10 mM in DMSO	Inhibitors against mainly CMGC and CaMK kinases
Inhibitor Library III	MilliporeSigma/ Sigma-Aldrich	84	10 mM in DMSO	Inhibitors against mainly CMGC, CaMK, AGC, and STE kinases
Inhibitor Library IV	MilliporeSigma/ Sigma-Aldrich	83	0.5–100 mM in DMSO	Inhibitors against mainly tyrosine kinases and tyrosine phosphatases

(continued)

TABLE 1. *Continued*

Library	Source	Compounds	Format	Description
SCREEN-WELL Kinase Inhibitor Library	Enzo	80	10 mM in DMSO	Known kinase inhibitors of well-defined activity
SCREEN-WELL Nuclear Receptor Ligand Library	Enzo	74	10 mM in DMSO	Compounds with defined, putative, and potential activity at nuclear receptors, receptor agonists, and antagonists

Dimethyl sulfoxide (DMSO) (Sigma-Aldrich 154938; purity 99%)

L-cysteine solution (2.0% w/v) <R>

Marc's Modified Ringer solution (MMR) (10× stock) <R>

> *Dilute to 0.1× with H_2O before use.*

MEMFA fixative for *Xenopus* <R>

Phosphate-buffered saline (PBS) <R>

> *Adjust pH to 7.4.*

Screening medium <R>

Tricaine (0.5 mg/mL in 0.1× MMR)

Xenopus laevis embryos

> *Obtain wild-type or albino embryos by in vitro fertilization using sexually mature* Xenopus laevis *males and females. Alternatively, embryos may also be obtained from natural matings (see Sec. 2, Protocol 2: Obtaining* Xenopus laevis *Embryos [Shaidani et al. 2021]). Where appropriate, embryos from genetically modified* Xenopus *strains or embryonic disease models generated by antisense-morpholino oligonucleotide knockdown can be used. Embryos and tadpoles are staged according to Nieuwkoop and Faber (NF) (Nieuwkoop and Faber 1994).*

Equipment

Aluminum foil

Cell culture incubator (18°C–23°C; humidified)

Cell culture plates, 48-well clear flat bottom, TC-treated, polystyrene (Corning Falcon 353078)

Dumont #5 forceps (stainless steel; Sigma-Aldrich)

Fume hood

Microcentrifuge tubes (1.5-mL)

Micropipettes (P20, P200, P1000) with filter tips of appropriate sizes

Multichannel pipette or liquid handling robot (e.g., Aquarius, Tecan) (see Step 1)

Nutator (Clay Adams BD 421125)

Parafilm M wrapping film (Fisher Scientific)

Petri dishes (6-cm, 10-cm, uncoated plastic)

Plastic plates, 96-well, clear round-bottom, polystyrene (Corning 3788)

Plastic wrap (e.g., Saran Wrap)

Stereomicroscope (e.g., Carl Zeiss SV6) with two-armed fiber optic illumination

> *This microscope is used for embryo handling and phenotype scoring.*

Stereomicroscope (e.g., Leica M205 FA) equipped with a digital camera

> *This microscope is used for imaging of embryos.*

Transfer pipettes (5-mL, plastic, disposable)

METHOD

An overview of the general strategy for phenotypic drug discovery using Xenopus *embryos is shown in Figure 1. The method described below is a protocol for the primary screen. Options for secondary screens are described subsequently.*

Adhere to local IACUC regulations regarding animal handling.

Cite this protocol as *Cold Spring Harb Protoc*; doi:10.1101/pdb.prot098269

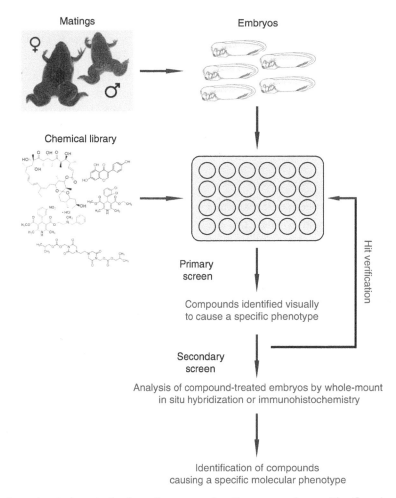

FIGURE 1. High-throughput phenotypic drug discovery using *Xenopus* embryos. The flowchart shows the steps involved in phenotypic chemical screening. Once embryos have reached the desired embryonic stage, they are arrayed in 48-well plates and treated with single compounds taken from the chemical library of choice. In the primary screen, embryos are scored visually by bright-field microscopy for the desired developmental phenotype(s) or morphological change(s). Positive hits are verified by repeating the primary chemical screen. In the secondary screen, embryos are treated with verified compounds and subsequently analyzed by whole-mount in situ hybridization or immunohistochemistry. Figure adapted from Kälin et al. (2009).

Chemical Library Handling

1. Dilute stock compounds of the chemical library to a concentration of 2 mM in DMSO (50 µL volume). Transfer aliquots of the chemicals from the mother plates to the daughter plates containing DMSO using a liquid handling robot or a multichannel pipette.

   Most compounds of commercial chemical libraries are predissolved in DMSO at a concentration of 10 mM or occasionally at 2 mM. Note that depending on the library, some compounds might be dissolved in H₂O instead of DMSO.

2. Store diluted stock plates at −20°C (2 yr) or −80°C (5 yr).

Embryo Staging, Selection, and Distribution

3. Once in vitro fertilized *Xenopus* eggs have reached the two- or four-cell stages, remove the jelly coats of the embryos by gentle swirling in 2% L-cysteine solution at room temperature for 3–4 min using a nutator until the jelly coats are dissolved. Rinse embryos with 0.1× MMR solution at room temperature until no cysteine odor can be detected.

4. Divide the embryos using Dumont #5 forceps into batches of comparable embryonic stages. Use plastic pipettes to transfer them into several large 10 cm Petri dishes containing 0.1× MMR. Avoid overcrowding by pooling fewer than 100 embryos per dish.

5. Culture embryos at 23°C in a humidified incubator until they reach the desired embryonic stage to initiate compound testing. Pool embryos of the same embryonic stage.

   The culture temperature can range between 18°C and 23°C. The lower the temperature, the slower embryonic development proceeds.

   See the subsection Timing and Duration of Exposure to Chemical Compounds below for guidance regarding timing.

   See Troubleshooting.

6. Inspect embryos with a stereomicroscope. Use a plastic transfer pipette to remove and dispose of any abnormal (i.e., unfertilized, necrotic, or misshapen) embryos. Repeat the selection process again later, if necessary.

   This step guarantees that only embryos of highest quality will be used for chemical screening.

Chemical Library Screening

7. Aliquot 1 mL screening medium to each well of a plastic 48-well dish.

8. Use a plastic transfer pipette to place five embryos of the desired embryonic stage into each well.

   For consistent results, the embryos should not be younger than late blastula or early gastrula stages. With multiple embryos per well, the penetrance of a compound-induced phenotype can be assessed. Five embryos per well are optimal.

9. To start compound screening, add 10 µL aliquots of the diluted chemicals to the wells using a P20 micropipette with filter tips, resulting in a final concentration of 20 µM for each compound. Consider the following controls: embryos treated with screening solution only (negative control) and embryos treated with a reference compound that is known to induce a specific phenotype (positive control). To guarantee equal screening conditions, include controls in each 48-well dish.

   Compounds that elicit a phenotype will be retested at concentrations that may range between 1 µM and 50 µM. If you are screening larger chemical libraries (more than 100 compounds), performing the initial screen at 20 µM is a good starting concentration. The γ-secretase inhibitor compound E (20 µM) can serve as a positive, edema-inducing reference compound (Kälin et al. 2009).

   See Troubleshooting.

10. Incubate embryos in a humidified incubator at 23°C until they have reached the desired end point. Typically, the time frame is between 1 and 5 d.

    Add H_2O to each well if fluid evaporation is observed. Do not grow embryos beyond stage 45 in 48-well dishes as they become too large (>10 mm).

    See Troubleshooting.

Scoring of Phenotypes

11. Score any visible phenotypes of the embryos every 24 h using a stereomicroscope. Record embryonic mortality in each well. Remove dead embryos using a plastic transfer pipette and discard. Use a new transfer pipette each time to avoid contaminating wells with compounds.

    See the subsection Types of Phenotypes below for details on phenotypes.

    Chemicals are considered to be active when at least 80% (four out of five embryos) display the same phenotype (i.e., edema, lethality, or other phenotypes). Independent phenotype scoring by multiple investigators is useful to reduce false positives or to avoid missing subtle phenotypes.

    See Troubleshooting.

Cite this protocol as *Cold Spring Harb Protoc*; doi:10.1101/pdb.prot098269

Imaging

12. At the end of the drug screening assay, anesthetize *Xenopus* embryos in tricaine (0.5 mg/mL) in 0.1× MMR for at least 5 min at room temperature.

13. Capture images of anesthetized embryos in solution using a stereomicroscope equipped with a high-resolution digital camera. Anesthetized embryos can be imaged either in the 48-well dishes or after transfer to 6-cm Petri dishes.

Fixing Embryos for Storage

14. After imaging, remove 0.1× MMR solution and fix *Xenopus* embryos by adding 1 mL of 1× MEMFA to each well of the 48-well dish. Incubate embryos in a fume hood for 2 h at room temperature.

    Make sure the embryos are completely covered with MEMFA.

15. Remove the fixative. Wash the embryos with 1× PBS three times for 15 min each at room temperature using a nutator.

16. Store embryos in 1× PBS at 4°C in the 48-well dishes. Seal dishes with plastic wrap to prevent evaporation.

    Fixed embryos may be used for further analysis by in situ hybridization or immunohistochemical staining.

Confirmatory Testing

17. Retest all bioactive compounds that have elicited specific phenotypes to eliminate "false positive" hits. A bioactive compound should modulate a biological process in a dose-dependent manner. Therefore, validate the top hits at multiple concentrations. For example, if the screen was performed at 20 μM, perform a dose curve starting at 80 μM with twofold serial dilution down to 1.25 μM.

    It is also advisable that all hit compounds of interest are repurchased or resynthesized and retested in the original screening assay. Besides weeding out false positives, this will confirm the identity of the compound in the chemical library stock plate.

Detection of Phenotypes

18. Observe compound-induced morphological phenotypes by direct observation with a stereomicroscope.

    The human eye is a superb tool for detecting even subtle morphological changes or altered behavior of treated embryos. An experienced researcher will screen several dozen 48-well plates per day, which amounts to well over 1000 compounds per day. This enables moderate phenotypic screening campaigns using chemical libraries comprised of a few thousand compounds to be performed in a matter of weeks.

    See the Discussion for information on types of phenotypes and automated imaging systems.

TROUBLESHOOTING

Problem (Step 5): Embryos are heterogeneous in age.

Solution: Heterogenous embryos can be caused by overcrowding of collected embryos, sick embryos, or poor clutch quality. These problems can be potentially remedied by placing <100 embryos per 10-cm Petri dish, removing all dying or misshaped embryos before plating, and using alternative parents to generate embryos.

Problem (Step 9): The chemical library is too large to be screened by a single investigator.

Solution: An investigator can usually handle the screening of up to 200 compounds per day. If the library is too large to screen in a single day, screen embryos with pools of five compounds and then repeat the screen with single compounds for pools testing positive.

Problem (Step 10): Embryos appear to be drying out.

Solution: This can occur because of excess evaporation of the screening medium. To prevent evaporation, check embryos regularly and add H_2O to wells to compensate for evaporation. Alternatively, seal a 48-well plate with Parafilm M wrapping film.

Problem (Step 10): Embryos die during incubation with compounds.

Solution: Embryos may die for any of the following reasons: toxicity of the test compound, the pH of the screening solution deviating from pH 7.4–7.5, treatment of embryos with compounds too early in embryogenesis, poor embryo quality leading to premature death, lack of food, or large size of embryos. Potential solutions are:

1. The chemical in question may be known to cause embryonic lethality; try reducing compound concentrations.

2. Adjust the pH of screening solution to 7.5.

3. Use older embryos, if possible.

4. Remove dying embryos immediately to protect the remaining embryos in the well; if untreated control embryos are dying, repeat the chemical screen with embryos obtained from alternative parents.

5. The 48-well format for drug screening is optimal for embryos up to about stage 45. If your end point requires older embryos, adapt the screening protocol for 24-well plates. Keep in mind that older embryos will require feeding, which will complicate chemical screening.

Problem (Step 11): The phenotype is weak and affects only a minority of embryos.

Solution: A weak phenotype may occur because the compound concentration is too low to detect a fully penetrant phenotype (dosing effect). Another possible reason is that embryos were treated either too early or too late in embryogenesis. Solutions to these problems are to retest the compounds at higher concentrations and/or to start exposing embryos either earlier or later to the compound.

DISCUSSION

The use of *Xenopus* embryos as an alternative to zebrafish larvae for whole organism–based phenotypic drug discovery screening was first proposed in 2004 (Brändli 2004) followed by a small-scale study of 14 compounds demonstrating that they induced a range of morphological phenotypes in *Xenopus* embryos (Tomlinson et al. 2005). Subsequent chemical screens were designed to identify compounds inducing a specific phenotype, altering morphology, or modifying organ function(s) (Kälin et al. 2009; Tomlinson et al. 2009; Dush et al. 2011; Tanaka et al. 2016; Willsey et al. 2021). Although chemical screens vary in their phenotypic readouts, certain steps are shared across different strategies. Here, we would like to highlight some general considerations when designing chemical screens using *Xenopus* embryos and tadpoles.

Design of a Chemical Screen

Central to any chemical screening strategy is that the anticipated phenotypes caused by bioactive compounds should be specifically linked to the biology or physiology of the targeted tissue or organ of the *Xenopus* embryo or tadpole. Hence, the aims and readouts of a chemical screen using whole organisms must be considered carefully. What is the developmental process that should be modulated or altered by treatment with a small molecule? Which biochemical pathway(s) should be inhibited or modulated? Is the anticipated morphological readout predictive of the underlying biochemical pathway? Is the screen supposed to be broad, targeting development of an entire organ, or more narrowly aimed at specific signaling molecules or pathways modulating morphogenesis or organ physiology? It is also important to consider whether the end point of the chemical screen leads to

Cite this protocol as *Cold Spring Harb Protoc*; doi:10.1101/pdb.prot098269

(1) a phenotype that can be easily scored by bright-field imaging, (2) a phenotype that requires the detection of a specific marker (i.e., mRNA or protein by in situ hybridization or immunohistochemistry), or (3) a behavioral phenotype. In general, the ideal readout of the chemical screen should be rapid, sensitive, accurate, and reliable to detect the desired modulation of a biological, developmental, or physiological process in vivo. Every phenotypic readout has its advantages and disadvantages. Phenotype detection by visual inspection is rapid, can be performed multiple time over several days, and does not require time-consuming tissue processing. In addition, the human eye is superb at picking up both subtle and complex phenotypes. However, internal morphological alterations are usually not detected. Whole-mount in situ hybridization and immunohistochemical staining can reveal changes in the spatial expression of marker genes at high resolution and can be considered semiquantitative. However, these procedures require fixing embryos and thus terminating the chemical screen. Furthermore, they are time-consuming, taking at least 3 d to complete. Finally, transgenic *Xenopus* tissue- or pathway-specific reporter lines are convenient for fluorescence imaging–based screening, particularly for assaying alteration of internal organs. However, generating a stable transgenic line requires access to suitable promoters to drive reporter gene expression and can easily take one to two years.

Selection of a Chemical Library

The ideal chemical library possesses as much chemical diversity as possible and a high proportion of compounds with drug-like physicochemical parameters. A large selection of chemical libraries is available from a variety of commercial vendors and the majority of chemical screens performed in zebrafish and *Xenopus* have used these libraries. They range from large libraries consisting of thousands of compounds to small collections of compounds consisting of inhibitors or targeting specific pharmacological classes or biochemical pathways (Table 1). When choosing a chemical library for screening with *Xenopus* embryos, it is important to consider the level of characterization of compounds in the library and whether it is possible to purchase single chemicals for retesting and validation. Using libraries containing FDA-approved drugs or known bioactive compounds has the advantage that toxicological and mechanistic information is readily available. However, using these libraries is biased toward less novel mechanisms and may yield compounds which are patent-protected, limiting their development as potential therapeutics. Commercial libraries have an important advantage over in-house synthetic libraries. The former typically consist of compounds that are drug-like, adhering to Lipinski's rules (Lipinski et al. 2001). Critical factors include the compound's molecular weight, hydrophobicity, and number of hydrogen bond donors and acceptors (Lipinski et al. 2001). A further critical parameter is the logarithm of the partition ratio between octanol and water (log P). The log P value of a compound correlates well with its membrane permeability. Generally, compounds with log P values greater than +1 are absorbed well by zebrafish and *Xenopus* embryos (Wheeler and Brändli 2009). Low molecular weights, typically 200–700 Da, and favorable partition coefficient values afford good drug penetration, pharmacokinetics, and dynamics. On the other hand, synthetic libraries have the advantage that they may contain small molecules with unique properties and novel pharmacological activities. These libraries are therefore suitable for identifying new functional chemicals and biological targets. However, once discovered, these novel compounds may require significant efforts to determine their mechanisms of action. Unfortunately, many commercial synthetic libraries consist of compounds that are easy to synthesize or are biased toward a few simple core structures rather than a broad array of diverse compounds. Public repositories such as those of the U.S. National Institutes of Health (https://ncats.nih.gov/preclinical/core/compound) or the Drug Repurposing Hub of the Broad Institute (https://clue.io/repurposing) offer attractive alternatives to commercial chemical libraries. Similar to commercial sources, public repositories offer access to curated and annotated collections of U.S. Food and Drug Administration (FDA)-approved drugs, clinical trial drugs, and preclinical tool compounds with companion information resources. Prior to embarking on a screening campaign, it is, however, important to inquire about the availability of individual compounds for validation and retesting purposes.

Targeted chemical screens with small libraries of compounds of known functionality are also powerful. These libraries are composed of compounds targeting a specific type of biochemical activity, such as kinases, neurotransmitters, or G-protein-coupled receptors. Using this screening approach can uncover previously unknown biochemical pathways regulating a developmental process in *Xenopus* embryos or tadpoles. Ultimately, the choice of the optimal chemical library is highly dependent on the type of screen and the aim of the chemical screen.

Choice of Animals to Be Treated

Most chemical screens are performed using wild-type embryos or tadpoles, which are obtained from different clutches to ensure consistent phenotype penetrance and reduce the impact of genetic variability. Wild-type screens are straightforward given that wild-type frogs are readily available and the embryos do not require genotyping before chemical screening. In phenotypic screens that could be obstructed by the development of pigment cells, the use of unpigmented embryos from the mating of albino frogs should be considered. If chemically induced alterations in internal organs are the aim of the investigation, it is best to use embryos obtained from transgenic *Xenopus* lines expressing fluorescent reporter proteins. The availability of suitable *Xenopus* reporter lines presently represents a major limitation for these types of chemical screens. Finally, *Xenopus* models of human diseases generated either by knocking down gene functions using morpholino antisense oligonucleotides or by CRISPR–Cas-mediated gene editing represent attractive starting points for chemical screening and phenotypic drug discovery (Schmitt et al. 2014). In these cases, the aim of the chemical screening would be to identify compounds able to suppress or ameliorate the disease phenotype in vivo using mutant embryos or tadpoles. Paramount for this approach to work is that the *Xenopus* disease model has to recapitulate the key pathophysiological hallmarks of the human disease.

Timing and Duration of Exposure to Chemical Compounds

Appropriate timing of compound addition depends on the biological process to be modulated. Generally, exposure of embryos or tadpoles to the compounds should start before and continue until the process of interest is completed. Nieuwkoop and Faber's *Normal Table of* Xenopus laevis provides detailed information on the timing and development of various tissues and organs during *Xenopus* development (Nieuwkoop and Faber 1994). In addition, Xenbase is a valuable online resource that integrates diverse biological, gene expression, and phenotypic data and should be consulted before settling on the timing parameters of chemical screens. Toxicity is a further factor to keep in mind. Extended exposures to chemicals increase the death rates of embryos compared to the negative control embryos. Furthermore, younger embryos are more vulnerable to the harmful effects of some compounds. In our experience, treatment of embryos before they have reached the gastrulation stage significantly increases the death rates of embryos compared to the negative controls. Additionally, prolonged compound exposure can reveal teratogenicity or cytotoxicity of test compounds. It is therefore advisable to go for extended compound exposure times in the initial screening campaign, which is typically performed at one fixed final concentration. Compounds found to be toxic can subsequently be retested at lower doses. A washout step to remove the chemicals can be considered if there is a possibility that the drug could interfere with a later essential developmental step that is not the aim of the chemical screen or if the readout of the screen (e.g., an enzymatic reporter gene assay) could be affected by certain compounds present in the chemical library.

Anticipated Results

Toxicity

The general toxicity of chemical library compounds depends on (1) final compound concentration, and (2) the embryonic stage used at first exposure. Ideally, each compound would be screened using a range of concentrations (e.g., from 0.1 to 100 μM). However, this is not practical when larger libraries containing thousands of compounds are used for chemical screening. For the initial screen, a single

Cite this protocol as *Cold Spring Harb Protoc*; doi:10.1101/pdb.prot098269

final compound concentration of 10 or 20 μM has been found by many laboratories to be a good compromise. Regarding general compound toxicities, younger embryonic stages, such as pregastrula stages, are more sensitive than older ones. The toxicity rate (i.e., the percentage of compounds causing embryonic lethality) is also dependent on the type of chemical library used for screening. Chemical libraries comprised of FDA-approved drugs or known bioactive compounds, such as the LOPAC[1280] library (see Table 1), contain compounds that have been subjected to toxicity assessments. These libraries usually have lower toxicity rates than synthetic compound libraries. We screened the LOPAC[1280] library for bioactive compounds using *Xenopus* late tailbud stage embryos (stage 31, 37 h postfertilization) and found that 4% of the compounds caused either embryonic or larval lethality (Kälin et al. 2009). Similar toxicity rates were reported in chemical screens using zebrafish (Kaufman et al. 2009).

Hit Rates

Hit rates are the percentage of compounds that elicit the predefined "desired" phenotype, and they are highly variable, ranging from the lower single digits down to 0%. Multiple parameters, including dosing, treatment time point, and length of compound exposure, influence the observed hit rates. In general, lower hit rates are a more promising outcome of a chemical screen than high rates, which could indicate a general problem in the chemical screening strategy or a lack of specificity of the compounds at the dose(s) tested. Furthermore, hit rates vary significantly between different chemical libraries. Screens using libraries of known bioactive compounds usually result in higher hit rates than those relying on libraries of synthetic compounds with unknown pharmacology. Generally, screening concentrations of 10 to 30 μM produces acceptable hit rates. With a compound concentration of 20 μM, we observed a hit rate of 3.7% in a screen of the LOPAC[1280] library for compounds capable of inducing edema formation in *Xenopus* tadpoles (Kälin et al. 2009). Others reported hit rates of 1.4% and 2.2% depending on the chemical library used (Tomlinson et al. 2009). Generally, a hit rate of 1%–3% is reasonable to expect.

Detection of Phenotypes with Automated Systems

Various automated imaging and motion-tracking systems are commercially available. They are, however, costly and time-consuming to implement. Automated morphological screens are also complicated by the fact that *Xenopus* embryos change orientation once they start moving. Movements develop from simple local twitching of a few swimming muscles starting at stage 37/38 to free forward-moving, upright swimming by stage 45 (Muntz 1975). Customized imaging systems can track phenotypes in moving embryos and tadpoles; however, we only recommended them in the following cases: (a) when chemical library screening becomes a core activity of the laboratory; (b) when the goal is to screen chemical libraries larger than 10,000 compounds; or (c) when performing complex behavioral screens (see the subsection Behavioral Readouts below).

Types of Phenotypes

There are many phenotypes that can be scored by observation of wild-type *Xenopus* embryos and tadpoles after compound treatment. Two general types of phenotypes, morphological and behavioral alterations, can be scored by eye. Chemical screens can be designed to identify compounds that induce a specific change in morphology. Multiple traits, such as alterations in body axis, eye shape, skin pigmentation, or heartbeat, can be easily examined. Additionally, an increasing number of transgenic lines expressing tissue-specific fluorescent reporter proteins are being developed in the *Xenopus* community (see https://www.xenbase.org). These can be used to conveniently examine compound-induced changes in internal organs, such as the heart, pronephric kidney, and blood vessels. The aim of such experiments is to generate a hypothesis about biochemical pathways underlying a developmental process. The discovery of such chemical modifiers could also be useful if they can be correlated with particular disease pathways.

Morphological and Developmental Phenotypes

A variety of morphological and developmental phenotypes caused by chemical perturbation have been characterized. For example, alterations in pigment cell development and migration are easily scored by visual inspection. High-throughput screens with *Xenopus* embryos led to the identification of several compounds that disrupt pigment cell development (Tomlinson et al. 2009). Interestingly, one of these compounds is structurally related to lefunomide, which is now being developed as potent inhibitor of melanoma growth (White et al. 2011). Similarly, anticancer drug candidates were identified by screening for compounds that perturb gastrulation or neural crest migration in vivo (Tanaka et al. 2016). Heterotaxia (reversal organ laterality) is a further morphological feature that can be easily scored by examining the looping of the heart and gut in compound-treated embryos. A small molecule screen for left–right asymmetry in *Xenopus* embryos identified a pyridine analog named heterotaxin. This compound disrupts both heart and digestive organ laterality and acts by inhibiting TGF-ß signaling (Dush et al. 2011). Pathophysiological phenotypes, such as edema formation, can also serve as a starting point for a chemical screen with *Xenopus* embryos. Edema or fluid-filled swellings can be used as convenient indicators of impaired cardiovascular, lymphatic, and/or excretory system functions. Using edema formation as the primary end point in a small-molecule screen, we identified several compounds regulating lymphatic and blood vascular development (Kälin et al. 2009). Importantly, an adenosine A1 receptor antagonist was also able to block neovascularization in adult mice. In a final example, an oncology drug screen using *Xenopus tropicalis* embryos identified 17 compounds that enhance or suppress neural progenitor cell proliferation in vivo (Willsey et al. 2021). Three of these compounds were involved in estrogen signaling. Given that alterations of cortical neurogenesis have been implicated in the vulnerability to autism spectrum disorders, estrogen might mitigate the effects of disparate autism gene mutations.

Behavioral Readouts

Beyond morphology, behavioral readouts can be also used for chemical screens in whole organisms. There are numerous physiological and behavioral phenotypes that can be assessed. For example, changes in the movement or swimming pattern can be scored in response to compound treatment. Compounds altering behavior in *Xenopus* embryos or tadpoles frequently elicit their activity by acting on the central nervous system. This in turn offers the opportunity for the discovery of novel neuro-active drugs, which is impossible using in vitro systems as they cannot recapitulate the biology and neurophysiology of an entire organism. Large-scale behavioral screens using *Xenopus* embryos have to date not been reported. In zebrafish, screens were performed to identify small molecules that alter acute photometer response or interfere with rest and wakefulness patterns (Kokel et al. 2010; Rihel et al. 2010; Chiu et al. 2016). Others have screened for compounds interfering with habituation to acoustic startle (Wolman et al. 2011) or convulsive and electrographic seizures (Baraban et al. 2013). Similar screens are also possible with *Xenopus* embryos. Given that the evolutionary distance to humans is significantly shorter for amphibians than fish, therapeutic translation of bioactive compounds might be more straightforward (Wheeler and Brandli 2009). Seizures characterized by uncontrolled tail bends and excessive turning can be induced in tadpoles by bath application of several common chemoconvulsants including pentylenetetrazole (PTZ), a $GABA_A$ antagonist (Hewapathirane et al. 2008). Using a *Xenopus* tadpole model of PTZ-induced epilepsy, a previously unknown neuroprotective role of polyamines was discovered (Bell et al. 2011). *Xenopus* tadpoles have also long been used in anesthesiology research (Downes and Courogen 1996). Two forms of behavioral assessments, loss of spontaneous swimming movements and the loss of elicited movement, are consistent and simple end points to assess anesthetic potency with *Xenopus* tadpoles. Importantly, the relative EC_{50} values for most anesthetics are strikingly similar between *Xenopus* and mammals, including humans (Woll and Eckenhoff 2018). Hence, tadpole bioassays are a cost-effective way to determine relative drug tolerance and cross-tolerance before entering more expensive testing in mammalian animal models. In addition, high-throughput screening methods for nonvolatile and volatile anesthetics using *Xenopus* tadpoles have been recently described (Woll

Cite this protocol as *Cold Spring Harb Protoc*; doi:10.1101/pdb.prot098269

and Eckenhoff 2018). Irrespective of the type of behavioral assay used, behavioral screens are ideally performed using automated imaging and motor activity tracking, because complex phenotypes are difficult to follow by eye. In addition, measuring, quantitation and statistical analysis can be automated using commercial motion-tracking systems. Overall, the potential of employing *Xenopus* embryos and tadpoles in behavioral assays to identify neuroactive and psychotropic drugs has to date not been fully tapped.

Secondary Screening

Depending on the nature of the primary phenotypic screen, a secondary screen may be necessary to determine those compounds affecting the process of interest or to further confirm the bioactivity of the compounds. Examples include subtle morphological phenotypes, particularly those affecting internal organs, changes in mRNA levels, or protein modifications, such as phosphorylation or degradation, which are not evident to the naked eye. Secondary assay methods rely either on in situ hybridization, immunohistochemistry, or reporter protein expression using transgenic lines. In situ hybridization uses an antisense mRNA probe to determine the spatial expression of a specific mRNA in fixed embryos or tadpoles. A color reaction linked to the antisense mRNA probe localizes expressed transcripts to specific tissues and organs. Immunohistochemistry is used to identify the expression levels of a specific protein or post-translational modification via specific antibodies. Finally, it is possible to screen transgenic lines expressing fluorescent reporters, such as the green fluorescent protein, in a tissue-specific manner. Detailed protocols for whole mount in situ hybridization and immunostaining of *Xenopus* embryos and tadpoles have been published elsewhere (see Sec. 9, Protocol 1: Whole-Mount RNA In Situ Hybridization and Immunofluorescence of *Xenopus* Embryos and Tadpoles [Willsey 2021], Sec. 4, Protocol 5: Whole-Mount Immunocytochemistry in *Xenopus* [Klymkowsky 2018], and Sec. 4, Protocol 4: Whole-Mount In Situ Hybridization of *Xenopus* Embryos [Saint-Jeannet 2017]). An example of a two-step chemical screening strategy involving whole-mount hybridization was reported by us in the past (Kälin et al. 2009). The aim of the screen was to identify novel compounds modulating angiogenesis and/or lymphangiogenesis in *Xenopus* tadpoles. The first step involved a simple phenotypic read-out (edema formation or larval lethality) to select 66 bioactive compounds from a 1280 compound library. This was followed by semiautomated in situ hybridization analysis using vascular and lymphatic marker genes to identify 32 compounds that could interfere with blood vascular or lymphatic development, respectively. In situ hybridization allowed for a detailed classification of bioactive compounds by the type of vascular and lymphatic phenotypes detected, including defective vasculogenesis or angiogenesis, ectopic angiogenic sprouting, and defective lymph angiogenesis. Other examples of *Xenopus* chemical screens that rely on secondary screens with whole mount in situ hybridization and/or immunohistochemistry can be found here (Dush et al. 2011; Tanaka et al. 2016).

Concluding Remarks

Chemical screening using *Xenopus* embryos and tadpoles has become a very useful approach for the discovery of bioactive compounds and drug-like molecules with the potential of becoming therapeutic agents in the future. This is illustrated best by the discovery of novel compounds that interfere with angiogenesis or lymph vessel development in vivo (Kälin et al. 2009) and drug candidates with anticancer activities (White et al. 2011). Phenotypic drug discovery using *Xenopus* embryos can be performed in a medium- to high-throughput manner by screening chemical libraries consisting of thousands of compounds. In vivo testing allows for the study of the effects of compounds in a complex biological system. In addition, the toxicity of compounds is simultaneously assessed and compounds with adverse effects can be eliminated early from the drug discovery process. Finally, the emergence of gene editing technologies paves the way for the development of more sophisticated genetic *Xenopus* disease models in the coming years. It therefore is anticipated that chemical screening using *Xenopus* embryos and tadpoles will become an indispensable tool in therapeutic drug discovery.

RECIPES

Compound E Stock Solution (2 mM)

Dissolve 0.5 mg of the γ-secretase inhibitor compound E (Merck 565790) in 543 μL DMSO (Sigma-Aldrich 154938). Aliquot and store for up to 1 yr at −20°C.

L-Cysteine Solution (2% w/v)

Dissolve 2.0 g L-cysteine (Sigma-Aldrich C7352; purity 98.0%) in 100 mL deionized H_2O. No sterilization is required. Freshly prepare before use.

Marc's Modified Ringer Solution (MMR) (10× stock)

1 M NaCl
20 mM KCl
10 mM $MgSO_4$
20 mM $CaCl_2$
50 mM HEPES (pH 7.5)

Prepare a 10× stock in deionized H_2O, adjust the pH to 7.8 with NaOH, and sterilize by autoclaving. Store solution indefinitely at room temperature.

MEM Salts (10×; pH 7.4)

209.3 g MOPS
76.08 g EGTA
2.46 g $MgSO_4$

Add to a 2-L beaker. Make up to 1 L with deionized H_2O. Adjust pH to 7.4 using NaOH and sterilize by autoclaving. Store at 4°C protected from light (wrapped in foil).

MEMFA Fixative for Xenopus

1 mL MEM salts (10×; pH 7.4) <R>
1 mL 37% formaldehyde
8 mL H_2O

Freshly prepare on the day and store at 4°C.

Phosphate-Buffered Saline (PBS)

Reagent	Amount to add (for 1× solution)	Final concentration (1×)	Amount to add (for 10× stock)	Final concentration (10×)
NaCl	8 g	137 mM	80 g	1.37 M
KCl	0.2 g	2.7 mM	2 g	27 mM
Na_2HPO_4	1.44 g	10 mM	14.4 g	100 mM
KH_2PO_4	0.24 g	1.8 mM	2.4 g	18 mM

If necessary, PBS may be supplemented with the following:

$CaCl_2 \cdot 2H_2O$	0.133 g	1 mM	1.33 g	10 mM
$MgCl_2 \cdot 6H_2O$	0.10 g	0.5 mM	1.0 g	5 mM

PBS can be made as a 1× solution or as a 10× stock. To prepare 1 L of either 1× or 10× PBS, dissolve the reagents listed above in 800 mL of H_2O. Adjust the pH to 7.4 (or 7.2, if required) with HCl, and then add H_2O to 1 L. Dispense the solution into aliquots and sterilize them by autoclaving for 20 min at 15 psi (1.05 kg/cm²) on liquid cycle or by filter sterilization. Store PBS at room temperature.

Screening Medium

0.1× MMR (dilute from Marc's Modified Ringer Solution (MMR) (10× stock) <R>)
1% DMSO (Sigma-Aldrich 154938)

Freshly prepare.

ACKNOWLEDGMENTS

We are grateful to Sabine D'Avis for excellent technical assistance. The work was supported by funds from the University Hospital Munich and in part by a grant of the European Commission (EU FP7 Program, EuRenOmics Grant Agreement 305608) to A.W.B.

Competing interests: A.W.B. and R.E.K. hold U.S. and European patents related to chemical screening in amphibians. The other authors declare no competing interests.

REFERENCES

Baraban SC, Dinday MT, Hortopan GA. 2013. Drug screening in *Scn1a* zebrafish mutant identifies clemizole as a potential Dravet syndrome treatment. *Nat Commun* **4**: 2410. doi:10.1038/ncomms3410

Bell MR, Belarde JA, Johnson HF, Aizenman CD. 2011. A neuroprotective role for polyamines in a *Xenopus* tadpole model of epilepsy. *Nat Neurosci* **14**: 505–512. doi:10.1038/nn.2777

Brändli AW. 2004. Prospects for the *Xenopus* embryo model in therapeutics technologies. *Chimia* **58**: 695–702. doi:10.2533/0009429047 77677443

Chiu CN, Rihel J, Lee DA, Singh C, Mosser EA, Chen S, Sapin V, Pham U, Engle J, Niles BJ, et al. 2016. A zebrafish genetic screen identifies neuromedin U as a regulator of sleep/wake states. *Neuron* **89**: 842–856. doi:10.1016/j.neuron.2016.01.007

Downes H, Courogen PM. 1996. Contrasting effects of anesthetics in tadpole bioassays. *J Pharmacol Exp Ther* **278**: 284–296.

Dush MK, McIver AL, Parr MA, Young DD, Fisher J, Newman DR, Sannes PL, Hauck ML, Deiters A, Nascone-Yoder N. 2011. Heterotaxin: a TGF-β signaling inhibitor identified in a multi-phenotype profiling screen in *Xenopus* embryos. *Chem Biol* **18**: 252–263. doi:10.1016/j.chembiol.2010.12.008

Hewapathirane DS, Dunfield D, Yen W, Chen S, Haas K. 2008. In vivo imaging of seizure activity in a novel developmental seizure model. *Exp Neurol* **211**: 480–488. doi:10.1016/j.expneurol.2008.02.012

Kälin RE, Bänziger-Tobler NE, Detmar M, Brändli AW. 2009. An *in vivo* chemical library screen in *Xenopus* tadpoles reveals novel pathways involved in angiogenesis and lymphangiogenesis. *Blood* **114**: 1110–1122. doi:10.1182/blood-2009-03-211771

Kaufman CK, White RM, Zon L. 2009. Chemical genetic screening in the zebrafish embryo. *Nat Protoc* **4**: 1422–1432. doi:10.1038/nprot.2009.144

Klymkowsky MW. 2018. Whole-mount immunocytochemistry in *Xenopus*. *Cold Spring Harb Protoc* doi:10.1101/pdb.prot097295

Kokel D, Bryan J, Laggner C, White R, Cheung CY, Mateus R, Healey D, Kim S, Werdich AA, Haggarty SJ, et al. 2010. Rapid behavior-based identification of neuroactive small molecules in the zebrafish. *Nat Chem Biol* **6**: 231–237. doi:10.1038/nchembio.307

Lipinski CA, Lombardo F, Dominy BW, Feeney PJ. 2001. Experimental and computational approaches to estimate solubility and permeability in drug discovery and development settings. *Adv Drug Deliv Rev* **46**: 3–26. doi:S0169-409X(00)00129-0

Muntz L. 1975. Myogenesis in the trunk and leg during development of the tadpole of *Xenopus laevis* (Daudin 1802). *J Embryol Exp Morphol* **33**: 757–774. doi:10.1242/dev.33.3.757

Nieuwkoop PD, Faber J. 1994. *Normal table of* Xenopus laevis *(Daudin): a systematical and chronological survey of the development from the fertilized egg till the end of metamorphosis*. Garland Publishing, New York.

Rihel J, Prober DA, Arvanites A, Lam K, Zimmerman S, Jang S, Haggarty SJ, Kokel D, Rubin LL, Peterson RT, et al. 2010. Zebrafish behavioral profiling links drugs to biological targets and rest/wake regulation. *Science* **327**: 348–351. doi:10.1126/science.1183090

Saint-Jeannet JP. 2017. Whole-mount in situ hybridization of *Xenopus* embryos. *Cold Spring Harb Protoc* doi:10.1101/pdb.prot097287

Schmitt SM, Gull M, Brändli AW. 2014. Engineering *Xenopus* embryos for phenotypic drug discovery screening. *Adv Drug Deliv Rev* **69–70**: 225–246. doi:10.1016/j.addr.2014.02.004

Shaidani N-I, McNamara S, Wlizia M, Horb ME. 2021. Obtaining *Xenopus laevis* embryos. *Cold Spring Harb Protoc* doi:10.1101/pdb.prot106211

Tanaka M, Kuriyama S, Itoh G, Kohyama A, Iwabuchi Y, Shibata H, Yashiro M, Aiba N. 2016. Identification of anti-cancer chemical compounds using *Xenopus* embryos. *Cancer Sci* **107**: 803–811. doi:10.1111/cas.12940

Tomlinson ML, Field RA, Wheeler GN. 2005. *Xenopus* as a model organism in developmental chemical genetic screens. *Mol Biosyst* **1**: 223–228. doi:10.1039/b506103b

Tomlinson ML, Rejzek M, Fidock M, Field RA, Wheeler GN. 2009. Chemical genomics identifies compounds affecting *Xenopus laevis* pigment cell development. *Mol BioSyst* **5**: 376–384. doi:10.1039/b818695b

Wheeler GN, Brändli AW. 2009. Simple vertebrate models for chemical genetics and drug discovery screens: lessons from zebrafish and *Xenopus*. *Dev Dyn* **238**: 1287–1308. doi:10.1002/dvdy.21967

White RM, Cech J, Ratanasirintrawoot S, Lin CY, Rahl PB, Burke CJ, Langdon E, Tomlinson ML, Mosher J, Kaufman C, et al. 2011. DHODH modulates transcriptional elongation in the neural crest and melanoma. *Nature* **471**: 518–522. doi:10.1038/nature09882

Willsey HR. 2021. Whole-mount RNA in situ hybridization and immunofluorescence of *Xenopus* embryos and tadpoles. *Cold Spring Harb Protoc* doi:10.1101/pdb.prot105635

Willsey HR, Exner CRT, Xu Y, Everitt A, Sun N, Wang B, Dea J, Schmunk G, Zaltsman Y, Teerikorpi N, et al. 2021. Parallel in vivo analysis of large-effect autism genes implicates cortical neurogenesis and estrogen in risk and resilience. *Neuron* **109**: 788–804.e788. doi:10.1016/j.neuron.2021.01.002

Woll KA, Eckenhoff RG. 2018. High-throughput screening to identify anesthetic ligands using *Xenopus laevis* tadpoles. *Methods Enzymol* **602**: 177–187. doi:10.1016/bs.mie.2018.01.007

Wolman MA, Jain RA, Liss L, Granato M. 2011. Chemical modulation of memory formation in larval zebrafish. *Proc Natl Acad Sci* **108**: 15468–15473. doi:10.1073/pnas.1107156108

Inverse Drug Screening of Bioelectric Signaling and Neurotransmitter Roles: Illustrated Using a *Xenopus* Tail Regeneration Assay

Kelly G. Sullivan[1,2] and Michael Levin[1,3]

[1]*Biology Department, and Allen Discovery Center at Tufts University, Medford, Massachusetts 02155*

Xenopus embryos and larvae are an ideal model system in which to study the interplay between genetics, physiology, and anatomy in the control of structure and function. An important emerging field is the study of bioelectric signaling, the exchange of ion- and neurotransmitter-mediated messages among all types of cells (not just nerve and muscle cells), in the regulation of growth and form during embryogenesis, regeneration, and cancer. To facilitate the mechanistic investigation of bioelectric events in vivo, it is necessary to identify the endogenous signaling machinery involved in any patterning process of interest. This protocol uses the tail regeneration assay in *Xenopus* to perform an inverse drug screen; tiers of known compounds are used to probe the involvement of increasingly specific classes of bioelectric and neurotransmitter machinery. By using a hierarchical approach, large classes of targets are ruled out in early rounds, focusing attention on progressively narrower sets of proteins. Such a screen avoids many of the limitations of a molecular-genetic targeting approach and provides a rapid and efficient way to focus on specific targets. Usually, <10 experiments are needed to determine whether bioelectrics and/or neurotransmitter signaling are involved in the process of interest. This protocol describes the strategy in the context of a semiquantitative analysis of tail regeneration but can be applied to any assay in *Xenopus* or other small aquatic model system (e.g., zebrafish). Given the ever-increasing toolkit of chemical genetics, such screens represent a powerful and versatile methodology for probing the physiological circuits underlying pattern regulation.

MATERIALS

It is essential that you consult the appropriate Material Safety Data Sheets and your institution's Environmental Health and Safety Office for proper handling of equipment and hazardous materials used in this protocol.

RECIPES: Please see the end of this protocol for recipes indicated by <R>. Additional recipes can be found online at http://cshprotocols.cshlp.org/site/recipes.

Reagents

DMSO
Drugs (selected according to experimental goals)

---

[2]Present address: Perelman School of Medicine, University of Pennsylvania, Philadelphia, PA
[3]Correspondence: michael.levin@tufts.edu

Copyright © 2023 Cold Spring Harbor Laboratory Press; all rights reserved
Cite this protocol as *Cold Spring Harb Protoc*; doi:10.1101/pdb.prot099937

It is not possible to list here all of the compounds that might be used in a screen; however, a typical set of compounds is given in the tables in Adams and Levin (2006a,b).

Marc's modified Ringer's (MMR) (1×) <R>

Dilute with water to 0.1× before use.

Tricaine (MS-222)

Water (18 MΩ pure)

Equipment

Dissection microscope (such as the Zeiss SV6) with fiberlight illumination

Incubator (18°C)

Microcentrifuge tubes (0.5 and 1.5 mL)

Micropipettes (P20, P200, and P1000) and filter tips of appropriate sizes

Negative pressure fume hood

Petri dishes (6 and 10 cm, uncoated plastic)

Probe or tweezers (#5 tweezers; Fine Science Tools)

Scalpel handle and #10 scalpel blades

Transfer pipette (5 mL, plastic, disposable)

Vortex mixer

METHOD

This hierarchical tree-like pharmacological screen is designed to implicate a specific pathway in a patterning assay such as tail regeneration. As an example, we focus on components relevant to developmental bioelectricity (Adams 2008; Adams and Levin 2013; Levin 2013, 2014a,b; Tseng and Levin 2013; Bates 2015). Ion channels and pumps produce endogenous voltage gradients, which redistribute neurotransmitters and other small molecules that serve as second-messengers linking bioelectric signals to downstream transcriptional targets. The goal of this protocol is to use a hierarchical representation of the main pathway components in a loss-of-function screen to implicate specific targets in a tail regeneration assay. In this tiered screen, less-specific reagents are used first, to rule out or implicate larger classes of targets. Each outcome allows one to focus on certain subfamilies of each node of the tree, using progressively more specific reagents (Adams and Levin 2006a,b). The power of the screen derives from the negative results obtained with a broad-targeting reagent, which enables all of its child nodes to be discarded; this enables a logarithmic decrease in the number of targets to be considered. Typical screens require <10 Experiments to zero in on a set of targets of high importance. These can then be validated using several chemically distinct compounds with the same candidate target, and by gene-specific knockdown. The Xenopus model offers both the ability to characterize endogenous regenerative mechanisms and a background on which to screen for regeneration-augmenting reagents (Slack et al. 2008; Tseng and Levin 2008).

Screen Design and Drug Preparation

1. Decide between a loss-of-function screen (for identification of an endogenous component necessary for tail regeneration) and a gain-of-function screen (for identification of a pathway that, when activated, induces regeneration in a nonregenerative background, or alters regenerative morphology). In the former case, use st. 41 larvae, which normally regenerate, to discover which reagents cause a specific inhibition of regeneration. In the latter case, use approximately st. 47 larvae, which are in the refractory stage (Beck et al. 2003; Slack et al. 2004), to discover pathways that can overcome the regenerative block.

   Figures 1 and 2 show the logic of the screen.

2. Select drugs that target the top tier of the set of pathways under investigation. Selection of drugs for subsequent rounds of screening are chosen depending on the results of the previous round.

   Figure 2 shows some sample drug trees. For example, if one is interested in ion channels (Fig. 2A), one would start by testing BaCl₂ (a nonspecific potassium channel blocker) and general blockers of chloride, sodium, and calcium channels. If only BaCl₂ had an effect on regeneration, the rest of the tree can be skipped. In the next

(A)

(B)

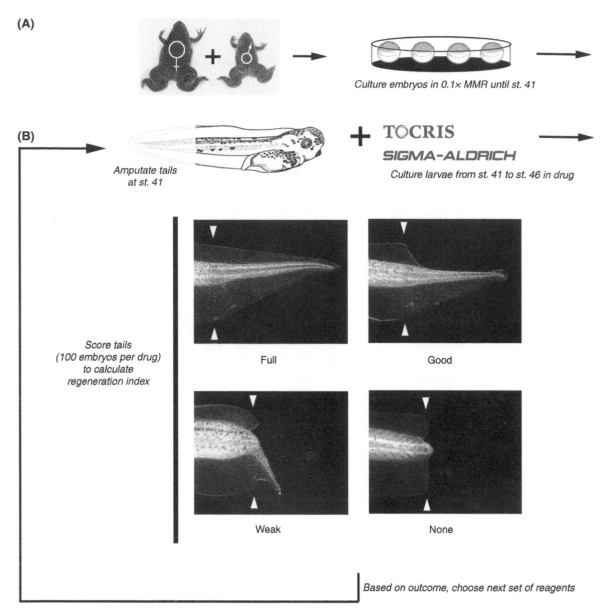

Culture embryos in 0.1× MMR until st. 41

Amputate tails
at st. 41

TOCRIS
SIGMA-ALDRICH

Culture larvae from st. 41 to st. 46 in drug

Score tails
(100 embryos per drug)
to calculate
regeneration index

Full

Good

Weak

None

Based on outcome, choose next set of reagents

FIGURE 1. The tail amputation assay. (*A*) *Xenopus* embryos are grown to st. 41, and the distal third of their tails amputated with a scalpel. They are then placed into solutions containing a reagent targeting a family of pathway components, and grown for 1 wk. (*B*) They are then scored under a dissecting microscope and each is assigned to one of four categories based on the quality of regeneration. Based on the combined regeneration index (calculated from the proportions of animals in the full, good, weak, or none categories), the next round of the screen selects more-specific reagents to probe targets belonging to those families implicated to have an effect on regeneration.

round more-specific blockers of voltage-gated, inward-rectifier, and two-pore channels can be tested. Lists of reagents can be obtained from our website: http://ase.tufts.edu/biology/labs/levin/resources/protocols.htm

3. Calculate the amount of each drug needed for the day's experiments; each batch of larvae needs 10–15 mL of final solution.

   The working concentration of a drug may need to be determined. Consult published mammalian studies to determine an appropriate starting dose for a compound. Drugs can be tested in Xenopus at doses of 10×, 1×, and 0.1× of the typical mammalian concentration in an initial toxicity test; expose larvae at the desired stage to the drug for several days and ensure normal development and lack of edema or ill health. Duration of drug exposure can also be varied.

   See Troubleshooting.

Cite this protocol as *Cold Spring Harb Protoc*; doi:10.1101/pdb.prot099937

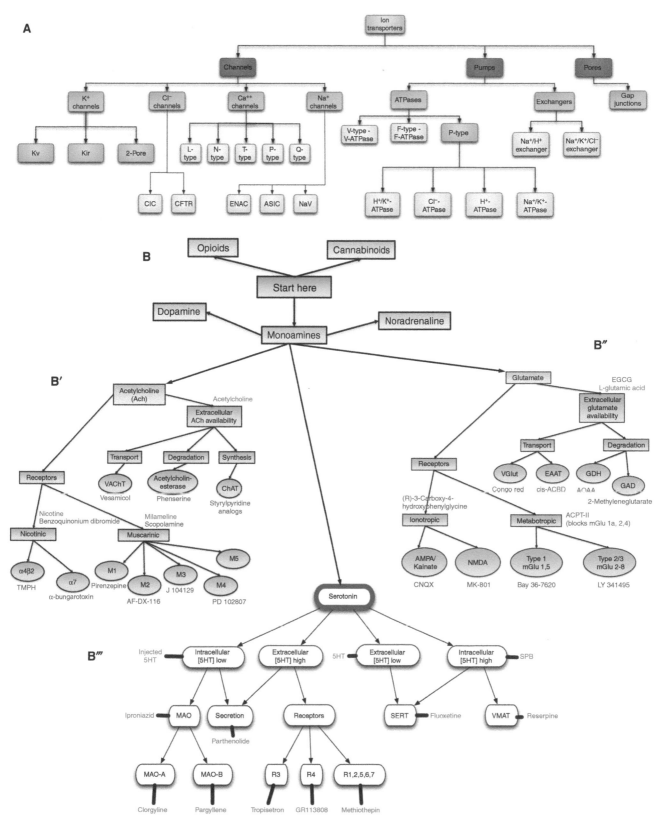

FIGURE 2. Sample drug screen trees. (*A*) A sample hierarchical tree for ion transport, indicating families and subfamilies of targets typically involved in bioelectric signaling processes. (*B*) A sample hierarchical tree for neurotransmitter signaling including activating (in green) and blocking (in red) reagents for nodes. Three specific subtrees are illustrated here in some detail, Acetylcholine (*B'*), Glutamate (*B''*), and Serotonin (*B'''*).

4. Working in a negative pressure fume hood, make stocks of each drug (in water whenever possible, or in DMSO when necessary), usually at a 1000× working concentration. Make all drug stocks fresh on the day of use, and keep protected from bright light. Freeze aliquots but avoid freeze–thaw cycles.

> *Many drugs are light-sensitive and degrade over time at room temperature. Many channel and neurotransmitter blockers are toxic if absorbed; therefore, ensure appropriate safety precautions are used at all times.*

> *See Troubleshooting.*

Amputation Assay

5. Prepare 6-cm Petri dishes containing 12–15 mL of 0.1× MMR and appropriate concentrations of drugs. Also prepare control dishes containing matching concentrations of drug vehicle.

6. Put 50–100 larvae of the appropriate age into 10-cm Petri dishes containing 12–15 mL of 0.1× MMR.

> *At st. 41, larvae do not need to be anesthetized. However, for refractory screening of st. 47 larvae, it is impossible to perform the amputation without anesthetic due to larval movement. A brief exposure to tricaine (MS-222 at 0.02% for 5 min) can be used; however, tricaine anesthesia should be minimized because anesthetics (including tricaine) target ion channels and inhibit regeneration, and can thus confound results of such screens. If tricaine is used, larvae must be washed out twice after amputation before placing in their final solutions.*

7. Working quickly under a dissecting microscope, use a scalpel to slice off the posterior third of the tail. At st. 41, the larvae will be lying immobile on their sides (as will st. 47 larvae after anesthesia). While holding the scalpel in a comfortable position, operate on larvae that are lying perpendicular to the scalpel; do not try to move the scalpel to accommodate larvae lying at awkward angles. Rather, rotate the dish to bring new larvae toward and perpendicular to the scalpel. When all of the larvae lying conveniently relative to the scalpel have been cut, swirl the dish (or use a disposable pipette to swirl the MMR) to bring new ones into position. Continue until almost all larvae have been cut. Discard any that are not perfectly normal in morphology. A typical cut bites down into the plastic of the dish (there is no need to saw across the tail), so change the scalpel blade after cutting approximately 20 tails.

8. Swirl the dish to redistribute the animals and randomize for any differences in cutting level during the amputation session. Using a plastic transfer pipette (the end of which has been cut to make the opening big enough to accommodate larvae, but not so big as to transfer excess medium), pick up 10 larvae at a time and deposit them into a drug dish. Do not drop them; instead, deposit them under the liquid, as larvae can be damaged by the air–water interface. Use a new pipette to fill each dish.

9. Fill all dishes with the same number of individuals. Do not overcrowd (place no more than 30 larvae in a 10-cm diameter Petri dish).

10. Keep the dishes away from bright light and place in an incubator at 18°C.

11. Check dishes daily and remove dead/decaying larvae. Record phenotypes that might be remodeled (disappear) over time.

> *See Troubleshooting.*

Scoring Regeneration

> *To quantify and compare regeneration efficiency of larvae under different conditions we devised an ordinal measure: the "regeneration index" (RI) (Adams et al. 2007).*

12. One week after amputation, anesthetize all larvae by adding tricaine to a final concentration of 0.05%. Leave the larvae in the drug dish plus tricaine while scoring. Swirl the dish so that the immobilized animals collect in the center. Using a dissecting microscope, assign each animal to one of the four following categories (illustrated in Fig. 1):

Cite this protocol as *Cold Spring Harb Protoc*; doi:10.1101/pdb.prot099937

Full	complete regeneration (regenerated tail, indistinguishable from uncut controls);
Good	robust regeneration with minor defects (e.g., missing fin, curved axis);
Weak	poor regeneration (hypomorphic regenerates);
None	no regeneration.

Use a probe (or closed tweezers) to move each animal to the side when it has been scored.

See Troubleshooting.

13. For each dish, calculate the percentage of larvae in each category and multiply by 3, 2, 1, or 0 for Full, Good, Weak, or None, respectively, and sum the results.

The regeneration index (RI) for any given dish will range from 0 to 300, with the extreme values corresponding to no regeneration and full regeneration, respectively, in all larvae. The RI evaluates the efficiency of regeneration at the single-dish level and allows ready comparison of the effect of treatments with controls. This is a more sensitive metric than length because it takes into account both outgrowth and dorsoventral patterning. However, subsequent studies can use more sophisticated morphometric approaches for detailed detection and characterization of specific patterning effects of selected treatments. Typically, a regeneration index below 100 indicates abrogation of regenerative ability. For statistical analysis, the results of multiple treatments can be compared using a Kruskal–Wallis test on the percentage of larvae in each category (or a similar test appropriate for unpaired nonparametric data). If the test indicates significant differences, Dunn's multiple comparisons corrected for tied ranks can be used for pairwise analysis between the treatment conditions.

TROUBLESHOOTING

Problem (Step 3): Determination of working drug concentration

Solution: A prescreen using uncut larvae will quickly reveal the highest dose of each drug that can be used without generalized toxicity. It is almost always possible to find a dose of a compound that gives informative results in a specific assay but does not cause overall ill health. In cases where this is not possible (sometimes, a drug will cause interesting regenerative effects but also unwanted developmental defects), alternative drugs targeting the same family can almost always be found. If a drug fails to cause any effects even at 10-fold the concentration used in the mammalian literature, it can be considered a negative result and its child nodes can be safely ignored in subsequent rounds.

Problem (Step 3): When testing different durations of drug exposure, it is difficult to know if a drug remains in tissues and is still active upon removal of larvae from drug-containing media.

Solution: Do not rely on wash-out. Drug exposures should be tested according to the initial time point of exposure because it cannot be assumed that once a drug-containing medium is removed from larvae that the drug is no longer active (many blockers are irreversible and bind tightly).

Problem (Step 4): When preparing drug aliquots, the drug comes from the supplier in very small (mg) quantities.

Solution: Do not try to weigh out the powder, which may be invisible in the vial and is often subject to static electricity, which can make the powder disperse. Carefully open the cap to the minimum extent necessary and use a Pipetteman to add the desired amount of solvent directly to the tube. Close the tube, vortex for 30 sec, pulse centrifuge, and then aliquot into 0.5–1.5 mL microcentrifuge tubes. This ensures that wherever the powder is located in the tube, it will end up in the solution. If using DMSO, use the drug's solubility data to prepare final concentrations of drug that contain <0.1% DMSO because excess DMSO can affect biological processes. Typical concentrations can be ~10 mg/mL.

Problem (Step 11): Degradation of drug in medium over time.

Solution: Most drugs do not need to be refreshed during the period of the experiment, but some may. Refer to a drug's material safety data sheet or to studies using the drug in mammalian cells for information regarding the drug's volatility and half-life. In some cases, a fresh solution needs to be provided each day. Transfer the larvae gently to prevent damage to the delicate regeneration buds.

Problem (Step 12): Responses vary among individual animals.
Solution: Perform each round of the screen with offspring from the same parents (combine their eggs) and repeat at least once on a different day with offspring from different parents.

Problem (Step 12): For gain-of-function tests, control larvae in refractory period stages still regenerate their tails.
Solution: The refractory period can be elusive and may depend on the time of year, batch of frogs, density of animals in the dish, and other unknown conditions; the stage timing of its beginning and end can vary considerably. It is essential to have no-treatment controls; if these do not regenerate, it is safe to conclude the refractory period has been identified. If regeneration occurs, try again with another batch and vary the timing of the experiment. For a given season of experiments, several weeks of testing are often required to determine the ideal time period at which controls do not regenerate. When this time period has been identified, it is recommended to perform as many experiments as possible, because the time period may shift within a few weeks, or disappear entirely for months at a time.

DISCUSSION

Growth and form are regulated by many physiological processes (Oates et al. 2012; Sullivan et al. 2016). Among these, bioelectric signaling using ion translocators and neurotransmitters provides an important and highly tractable set of control nodes for the manipulation of patterning and regenerative responses (Levin 2014b; Sullivan et al. 2016). Genetic screens in model systems (Perathoner et al. 2014) and human patients (Masotti et al. 2015; Simons et al. 2015; Adams et al. 2016) have identified a number of channelopathies. However, bioelectrical networks have some unique features that require different approaches compared with typical screens for chemical signaling factors because necessary-and-sufficient factors for downstream steps are physiological states, not gene products. For example, resting potential is the sum of the activities of many different ion translocators. Thus, single gene knockouts do not usually reveal the entirety of a bioelectric phenotype because of extensive compensation from other channels.

The advantage of drug screens is that most compounds simultaneously hit multiple family members, allowing the rapid identification of interesting channel types that would be too expensive to perform via combinatorial genetic knockouts. A huge number of reagents exist via suppliers such as Sigma-Aldrich and Tocris, and the literature describes numerous ion channel- and neurotransmitter-targeting compounds, many of which are teratogenic (Hernandez-Diaz and Levin 2014) and can be used to probe development and regeneration (Blackiston et al. 2017). Even highly nonspecific drugs can be useful because if they do *not* cause a phenotype, one can immediately rule out all of their various targets from further consideration. Inverse drug screens are a subset of the chemical genetics field (Wheeler and Brandli 2009; Wheeler and Liu 2012); they are an ideal tool to rapidly focus attention on specific targets for further molecular investigation, and to quickly determine whether bioelectrics or similar pathways are involved in a process of interest. The increasing availability of pharmacological agents means that this general strategy becomes more powerful over time. It has been successfully used to identify ion transporters and neurotransmitters involved in left–right patterning, tail regeneration, neural outgrowth, and anterior–posterior polarity during regeneration (Levin et al. 2002; Fukumoto et al. 2005; Adams et al. 2006a,b; Hibino et al. 2006; Tseng et al. 2010; Beane et al. 2011; Blackiston et al. 2015, 2017; Lobikin et al. 2015a).

It is important, however, to remember that pharmacological agents may have off-target effects, particularly at high concentrations. This can often be mitigated by cross-checking the effects of

Cite this protocol as *Cold Spring Harb Protoc*; doi:10.1101/pdb.prot099937

multiple drugs with the same putative target, but the possibility cannot be eliminated entirely. Moreover, information from a supplier about target specificity and selectivity may be incomplete; a thorough analysis of the published literature on each reagent that returns a positive hit in the assay is advised. Finally, the function of certain drugs may vary in different model systems because of differences in enzymatic clearance of the drug, temperature-dependency (drugs used in mammals are usually deployed at higher temperatures compared with their use in *Xenopus*), membrane permeability, and abundance of a drug's target protein in the organism. Thus, bracketing doses for drugs characterized in other model systems may be necessary when used in *Xenopus*.

After a screen is complete, and a small set of targets (usually <5) has been identified, it is most common to: (1) use the relevant drugs at different time points after regeneration to narrow down the window of their involvement in the regrowth process; (2) characterize the effects of the drug using markers of proliferation (H3P), apoptosis (Caspase3), regeneration (MSX1), and neural patterning (acetylated tubulin); (3) obtain native clones of relevant genes and study their expression patterns; (4) construct dominant negative proteins or morpholinos/siRNAs to target specific genes; (5) use voltage reporter dyes (Oviedo et al. 2008; Adams and Levin 2012b; Adams and Levin 2012a; Kruger and Bohrmann 2015) and neurotransmitter sensors (Balaconis and Clark 2012; Cash and Clark 2012, 2013; Awqatty et al. 2014; Ruckh and Clark 2014) to characterize the relevant endogenous physiological events in vivo and, ultimately, (6) build physiological models of the process (Blackiston et al. 2015; Lobikin et al. 2015a; Ferreira et al. 2016; Pietak and Levin 2016). Subsequently, misexpression approaches using light- (Adams et al. 2014) or ligand-gated (Lobikin et al. 2015b) channels can be used to spatially and temporally control relevant signals and test model hypotheses. The development of machine learning tools is also being integrated with the data from such screens to aid the design of interventions with biomedical application (Lobo and Levin 2015; Lobo et al. 2017).

This protocol focuses on bioelectric and neurotransmitter pathways; however, the tail amputation protocol can be used as an assay for any class of treatment: for example, physical (heat, radiation, vibration, light, magnetic), chemical (pharmacological targeting of any cell or developmental pathway), genetic (mutagenesis, transgenesis, mRNA injection/electroporation), or surgical (transplantation). The process of tail regeneration offers the experimenter opportunities to study the effects of any treatment within a coordinated process of growth and morphogenesis in a flat, transparent (optically tractable) appendage containing numerous cell and tissue types. Due to the stereotypical anatomy of the tail and the many markers available, it is an excellent in vivo model patterning system with which to probe the roles of any desired pathway or signaling modality. At the same time, the tiered strategy can be extended to other assays. By exploiting pharmacological reagents that have varying degrees of specificity, chemical genetics can be significantly accelerated through the use of hierarchical screens to study any process of interest in *Xenopus* development, regeneration, behavior, or physiology.

RECIPE

Marc's Modified Ringer's (MMR) (1×)

0.1 M NaCl
2 mM KCl
1 mM MgSO$_4$
2 mM CaCl$_2$
5 mM HEPES (pH 7.8)
0.1 mM EDTA

Sterilize by autoclaving, and store at room temperature.
Common alternative formulations of MMR omit EDTA and are adjusted to pH 7.4.

ACKNOWLEDGMENTS

We thank Dany S. Adams, the members of the Levin laboratory, and many others in the field of chemical genetics for their helpful suggestions in the development of the inverse drug screen methodology. We gratefully acknowledge support of the Templeton World Charity Foundation (TWCF0089/AB55), the W. M. Keck Foundation (5903), and the Allen Discovery Center program through The Paul G. Allen Frontiers Group (12171).

REFERENCES

Adams DS. 2008. A new tool for tissue engineers: Ions as regulators of morphogenesis during development and regeneration. *Tissue Eng* 14: 1461–1468.

Adams DS, Levin M. 2006a. Inverse drug screens: A rapid and inexpensive method for implicating molecular targets. *Genesis* 44: 530–540.

Adams DS, Levin M. 2006b. Strategies and techniques for investigation of biophysical signals in patterning. In *Analysis of growth factor signaling in embryos* (ed. Whitman M, Sater AK), pp. 177–262. Taylor and Francis Books.

Adams DS, Levin M. 2012a. General principles for measuring resting membrane potential and ion concentration using fluorescent bioelectricity reporters. *Cold Spring Harb Protoc* 2012: 385–397.

Adams DS, Levin M. 2012b. Measuring resting membrane potential using the fluorescent voltage reporters DiBAC4(3) and CC2-DMPE. *Cold Spring Harb Protoc* 2012: 459–464.

Adams DS, Levin M. 2013. Endogenous voltage gradients as mediators of cell–cell communication: Strategies for investigating bioelectrical signals during pattern formation. *Cell Tissue Res* 352: 95–122.

Adams D, Masi A, Levin M. 2006a. *Xenopus* tadpole tail regeneration requires the activity of the proton pump V-ATPase, and proton pumping is sufficient to partially rescue the loss-of-function phenotype. *Dev Biol* 295: 355–356.

Adams DS, Robinson KR, Fukumoto T, Yuan S, Albertson RC, Yelick P, Kuo L, McSweeney M, Levin M. 2006b. Early, H+-V-ATPase-dependent proton flux is necessary for consistent left-right patterning of non-mammalian vertebrates. *Development* 133: 1657–1671.

Adams DS, Masi A, Levin M. 2007. H+ pump-dependent changes in membrane voltage are an early mechanism necessary and sufficient to induce *Xenopus* tail regeneration. *Development* 134: 1323–1335.

Adams DS, Lemire JM, Kramer RH, Levin M. 2014. Optogenetics in Developmental Biology: Using light to control ion flux-dependent signals in *Xenopus* embryos. *Int J Dev Biol* 58: 851–861.

Adams DS, Uzel SG, Akagi J, Wlodkowic D, Andreeva V, Yelick PC, Devitt-Lee A, Pare JF, Levin M. 2016. Bioelectric signalling via potassium channels: A mechanism for craniofacial dysmorphogenesis in KCNJ2-associated Andersen-Tawil Syndrome. *J Physiol* 594: 3245–3270.

Awqatty B, Samaddar S, Cash KJ, Clark HA, Dubach JM. 2014. Fluorescent sensors for the basic metabolic panel enable measurement with a smart phone device over the physiological range. *Analyst* 139: 5230–5238.

Balaconis MK, Clark HA. 2012. Biodegradable optode-based nanosensors for in vivo monitoring. *Anal Chem* 84: 5787–5793.

Bates E. 2015. Ion channels in development and cancer. *Annu Rev Cell Dev Biol* 31: 231–247.

Beane WS, Morokuma J, Adams DS, Levin M. 2011. A chemical genetics approach reveals H,K-ATPase-mediated membrane voltage is required for planarian head regeneration. *Chem Biol* 18: 77–89.

Beck CW, Christen B, Slack JM. 2003. Molecular pathways needed for regeneration of spinal cord and muscle in a vertebrate. *Dev Cell* 5: 429–439.

Blackiston DJ, Anderson GM, Rahman N, Bieck C, Levin M. 2015. A novel method for inducing nerve growth via modulation of host resting potential: Gap junction-mediated and serotonergic signaling mechanisms. *Neurotherapeutics* 12: 170–184.

Blackiston DJ, Vien K, Levin M. 2017. Serotonergic stimulation induces nerve growth and promotes visual learning via posterior eye grafts in a vertebrate model of induced sensory plasticity. *npj Regen Med* 2: 8.

Cash KJ, Clark HA. 2012. In vivo histamine optical nanosensors. *Sensors (Basel)* 12: 11922–11932.

Cash KJ, Clark HA. 2013. Phosphorescent nanosensors for in vivo tracking of histamine levels. *Anal Chem* 85: 6312–6318.

Ferreira F, Luxardi G, Reid B, Zhao M. 2016. Early bioelectric activities mediate redox-modulated regeneration. *Development* 143: 4582–4594.

Fukumoto T, Kema IP, Levin M. 2005. Serotonin signaling is a very early step in patterning of the left-right axis in chick and frog embryos. *Curr Biol* 15: 794–803.

Hernandez-Diaz S, Levin M. 2014. Alteration of bioelectrically-controlled processes in the embryo: A teratogenic mechanism for anticonvulsants. *Reprod Toxicol* 47: 111–114.

Hibino T, Ishii Y, Levin M, Nishino A. 2006. Ion flow regulates left–right asymmetry in sea urchin development. *Dev Genes Evol* 216: 265–276.

Kruger J, Bohrmann J. 2015. Bioelectric patterning during oogenesis: Stage-specific distribution of membrane potentials, intracellular pH and ion-transport mechanisms in *Drosophila* ovarian follicles. *BMC Dev Biol* 15: 1.

Levin M. 2013. Reprogramming cells and tissue patterning via bioelectrical pathways: Molecular mechanisms and biomedical opportunities. *Wiley Interdiscip Rev Syst Biol Med* 5: 657–676.

Levin M. 2014a. Endogenous bioelectrical networks store non-genetic patterning information during development and regeneration. *J Physiol* 592: 2295–2305.

Levin M. 2014b. Molecular bioelectricity: How endogenous voltage potentials control cell behavior and instruct pattern regulation in vivo. *Mol Biol Cell* 25: 3835–3850.

Levin M, Thorlin T, Robinson KR, Nogi T, Mercola M. 2002. Asymmetries in H+/K+-ATPase and cell membrane potentials comprise a very early step in left-right patterning. *Cell* 111: 77–89.

Lobikin M, Lobo D, Blackiston DJ, Martyniuk CJ, Tkachenko E, Levin M. 2015a. Serotonergic regulation of melanocyte conversion: A bioelectrically regulated network for stochastic all-or-none hyperpigmentation. *Sci Signal* 8: ra99.

Lobikin M, Pare JF, Kaplan DL, Levin M. 2015b. Selective depolarization of transmembrane potential alters muscle patterning and muscle cell localization in *Xenopus laevis* embryos. *Int J Dev Biol* 59: 303–311.

Lobo D, Levin M. 2015. Inferring regulatory networks from experimental morphological phenotypes: A computational method reverse-engineers planarian regeneration. *PLoS Comput Biol* 11: e1004295.

Lobo D, Lobikin M, Levin M. 2017. Discovering novel phenotypes with automatically inferred dynamic models: A partial melanocyte conversion in *Xenopus*. *Sci Rep* 7: 41339.

Masotti A, Uva P, Davis-Keppen L, Basel-Vanagaite L, Cohen L, Pisaneschi E, Celluzzi A, Bencivenga P, Fang M, Tian M, et al. 2015. Keppen-Lubinsky syndrome is caused by mutations in the inwardly rectifying K+ channel encoded by *KCNJ6*. *Am J Hum Genet* 96: 295–300.

Oates AC, Morelli LG, Ares S. 2012. Patterning embryos with oscillations: Structure, function and dynamics of the vertebrate segmentation clock. *Development* 139: 625–639.

Oviedo NJ, Nicolas CL, Adams DS, Levin M. 2008. Live imaging of planarian membrane potential using DiBAC4(3). *Cold Spring Harb Protoc* 2008: pdb.prot5055.

Perathoner S, Daane JM, Henrion U, Seebohm G, Higdon CW, Johnson SL, Nusslein-Volhard C, Harris MP. 2014. Bioelectric signaling regulates size in zebrafish fins. *PLoS Genet* 10: e1004080.

Pietak A, Levin M. 2016. Exploring instructive physiological signaling with the Bioelectric Tissue Simulation Engine (BETSE). *Front Bioeng Biotechnol* 4: 55.

Ruckh TT, Clark HA. 2014. Implantable nanosensors: Toward continuous physiologic monitoring. *Anal Chem* 86: 1314–1323.

Simons C, Rash LD, Crawford J, Ma L, Cristofori-Armstrong B, Miller D, Ru K, Baillie GJ, Alanay Y, Jacquinet A, et al. 2015. Mutations in the voltage-gated potassium channel gene KCNH1 cause Temple–Baraitser syndrome and epilepsy. *Nat Genet* **47**: 73–77.

Slack JM, Beck CW, Gargioli C, Christen B. 2004. Cellular and molecular mechanisms of regeneration in *Xenopus*. *Philos Trans R Soc Lond B Biol Sci* **359**: 745–751.

Slack JM, Lin G, Chen Y. 2008. The *Xenopus* tadpole: A new model for regeneration research. *Cell Mol Life Sci* **65**: 54–63.

Sullivan KG, Emmons-Bell M, Levin M. 2016. Physiological inputs regulate species-specific anatomy during embryogenesis and regeneration. *Commun Integr Biol* **9**: e1192733.

Tseng AS, Levin M. 2008. Tail regeneration in *Xenopus laevis* as a model for understanding tissue repair. *J Dent Res* **87**: 806–816.

Tseng A, Levin M. 2013. Cracking the bioelectric code: Probing endogenous ionic controls of pattern formation. *Commun Integr Biol* **6**: 1–8.

Tseng AS, Beane WS, Lemire JM, Masi A, Levin M. 2010. Induction of vertebrate regeneration by a transient sodium current. *J Neurosci* **30**: 13192–13200.

Wheeler GN, Brandli AW. 2009. Simple vertebrate models for chemical genetics and drug discovery screens: Lessons from zebrafish and *Xenopus*. *Dev Dyn* **238**: 1287–1308.

Wheeler GN, Liu KJ. 2012. *Xenopus*: An ideal system for chemical genetics. *Genesis* **50**: 207–218.

Following Endocrine-Disrupting Effects on Gene Expression in *Xenopus laevis*

Petra Spirhanzlova,[1] Michelle Leemans,[1] Barbara A. Demeneix,[1,2] and Jean-Baptiste Fini[1,2]

[1]*Evolution des Régulations Endocriniennes, Département "Adaptation du Vivant," UMR 7221 MNHN/CNRS, Sorbonne Université, Paris 75006, France*

Endocrine-disrupting chemicals (EDCs), found in all categories of chemicals, are suspected to be a cause of declining well-being and human health, both as single molecules and as mixtures. It is therefore necessary to develop high throughput methods to assess the endocrine-disrupting potential of multiple chemicals currently on the market that are as yet untested. An advantage of in vivo chemical screening is that it provides a full spectrum of physiological impacts exerted by a given chemical. *Xenopus laevis* is an ideal model organism to test thyroid axis disruption in vivo as thyroid hormones (THs) are highly conserved across vertebrates and orchestrate tadpole metamorphosis. In particular, NF stage 45 *Xenopus laevis* are most apt for in vivo screening as at this stage the tadpoles possess all the main elements of thyroid hormone signaling (thyroid receptors, deiodinases transporters) and are metabolically competent, while fitting into multiple well plates, allowing the use of small amounts of test chemicals. One way to assess the endocrine-disrupting potential of chemicals or mixtures thereof is to analyze gene expression in organisms after a short time exposure to the chemical(s). Here we describe a protocol using *Xenopus laevis* embryos to detect endocrine disruption of the thyroid axis by analysis of gene expression and an alternative protocol for fluorescence read-out using a transgenic GFP-expressing *Xenopus laevis* line. Taken together, these methods allow detection of subtle changes in TH signaling by EDCs that either activate or inhibit TH signaling in vivo.

MATERIALS

It is essential that you consult the appropriate Material Safety Data Sheets and your institution's Environmental Health and Safety Office for proper handling of equipment and hazardous materials used in this protocol.

RECIPES: Please see the end of this protocol for recipes indicated by <R>. Additional recipes can be found online at http://cshprotocols.cshlp.org/site/recipes.

Reagents

Agilent RNA 6000 Pico Kit

Aquatic system water, chlorine-free

> *Filter through an aquatic water system (e.g., Veolia custom system).*

Dimethyl sulfoxide (DMSO; CAS 67-68-5)

[2]Correspondence: fini@mnhn.fr; bdem@mnhn.fr

Endocrine-disrupting compound(s) of interest dissolved in DMSO to a concentration 10,000× higher than the final test concentration

Ethyl 3-aminobenzoate methanesulfonate salt (MS222; CAS 886-86-2) solution (1 g/L) <R>
   Adjust pH between 7.2 and 8.

Evian water in 75 cL glass bottle or another mineral water with reproducible quality

High capacity cDNA RT kit (Life technologies 4368813) or Reverse Transcription Master Mix (Fluidigm 100-6299)

Human chorionic gonadotropin (hCG; CAS 9002-61-3; Sigma-Aldrich CG5-1VL)

Ice

Liquid nitrogen

Milli-Q reference water

Nuclease-free water (ThermoFisher Scientific AM9937)

Power SYBR Master Mix (Life Technologies 4368708)

Primers, 10 pM (Eurofins; see Table 1)

RNAse-free 0.8% agarose gel (optional; see Step 22)

RNAqueous–micro kit (Life Technologies AM1931)

RnaseZap (ThermoFisher Scientific AM9780; optional; see Step 17)

Sodium hydroxide (NaOH 0.1 N; CAS 1310-73-2; optional; see Step 17)

Transgenic *Xenopus laevis* line expressing GFP (e.g., Tg(*thibz*:GFP) [Fini et al. 2007]) (optional; see Step 36)

3,3′,5-Triiodo-L-thyronine sodium salt (T$_3$) solution <R>

Xenopus laevis mature male and female (WT) or *Xenopus laevis* NF stage 45 tadpoles

Equipment

96-well black, conical well plate (Greiner Bio-one 651209; optional; see Step 38)

384-well clear hardshell plate (ThermoFisher Scientific 4309849)

Air pump and stone

Adhesive covers for 384 well plates (ThermoFisher Scientific 4311971)

BioAnalyzer (Agilent)

Black foil to cover breeding chamber

TABLE 1. Primers for genes involved in thyroid signaling in *Xenopus*

thra F	CGCCTTGGTCTCTTCGGAT
thra R	CCCATACATTGGCTGTTCTTTCTTT
thrb F	AAGAGTGGTTGATTTTGCCAAAA
thrb R	AGGGACATGATCTCCATACAACAG
dio1 F	CAGCAGATGAATGGGGATTGA
dio1 R	TGTCTAACACTACTGGGCAAGAAGGT
dio2 F	AGGCTGAGTGTGGACTTGCT
dio2 R	TGACCTGCTTGTAGGCATCCA
dio3 F	CACAAAAAGTGCGACCAAACG
dio3 R	GCCTTGTTGCAGTTTACT
thibz F	ACCTCCACAGAATCAGCAGC
thibz R	GCAGAGAACGAGCAAGGAGT
klf9 F	TGTGGCAAAGTTTATGGGAAGTCT
klf9 R	GGCGTTCACCTGTATGGACTCT
lat2 F	CAAGAGATGCACTAAAGCTGCC
lat2 R	CCTTGCTTCCAACACCCGAT
odc F	TGAAAACATGGGTGCCTACA
odc R	AAGTTCCATTCCGCTCTCCT
ef1a F	TGGATATGCCCCTGTGTTGGATT
ef1a R	TCCACGCACATTGGCTTTCCT

Microcentrifuge (e.g., Eppendorf 5417 R)

Dissecting tools (forceps, etc.)

Fluorescent microscope equipped with 25× objective and long pass GFP filter (optional; see Step 40)

Freezer (−20°C)

Glass aquarium (25 L)

GraphPad Prism 6

Greiner polypropylene capped tubes (50 mL; Greiner Bio-one 227261)

> *Throughout the protocol, polypropylene tubes should be washed with dechlorinated aquatic system water before use.*

ImageJ (optional; Step 42)

Incubator

Microcentrifuge tubes 0.2–1.7 mL, flip top, 100% polypropylene

MS Excel

Multidistribution pipette

NanoDrop (ThermoScientific)

Needles (0.4 × 20 mm; Terumo NN2719R)

PCR machine (Bio-Rad)

PCR tube strips (0.2 mL; Eppendorf)

Pipette tips (0.2–5000 µL)

Plastic grid (optional; see Step 1)

Plastic transfer bucket

QC Capture pro (QImaging; optional; see Step 40)

QuantStudio 6 flex QPCR machine (Life Technologies)

QuantStudio Real-Time PCR Software

Syringes 1 mL (Terumo SS+01H1)

Transfer pipettes with large bulb (1 mL; Samco Scientific 222-1s)

Transfer pipettes with extended fine tip (1 mL; Samco Scientific 223-1S)

Transparent flat six-well plates (TPP, Switzerland, 1613470)

Vortex mixer

METHOD

Breeding

1. Prepare a breeding chamber by filling a glass tank with dechlorinated water. The temperature of the water should be equilibrated to 21°C. Place the breeding chamber in a quiet place.

   Placing a grid at the bottom of the tank will prevent adults from eating the eggs (optional).

2. Place one *Xenopus laevis* female and one *Xenopus laevis* male in plastic transport buckets filled with dechlorinated water equilibrated to 21°C.

   The probability of successful mating is increased by selecting a female with a prominent cloaca and pear-shaped body and a male with dark nuptial pads on the forelimbs.

3. Prepare a solution of hCG by transferring all liquid provided in the kit into the vial containing the hCG powder using a syringe. This will create a solution with a final concentration of 1000 U/mL. Vortex to ensure that the powder is fully dissolved.

   Once prepared, the chorion gonadotropin solution can be stored at −20°C until the next usage. Do not thaw it more than 3 times.

4. Inject the female with a starting dose of 500 units and the male with 50 units of hCG solution into the dorsal lymph sac located at the lower back of the animal between the stitch marks using 0.4× 20 mM needles. Gently massage the area after injection.

Cite this protocol as *Cold Spring Harb Protoc*; doi:10.1101/pdb.prot098301

5. Place the male and female together into the previously prepared breeding chamber. Wrap the tank with black foil to ensure that the couple won't be disturbed and place a weight (e.g., brick) on top of the lid.

6. Leave the couple overnight. On the next day verify amplexus (mating position in which the male clasps the female with his front limbs around the waist to align cloacae) and egg-laying.

   See Troubleshooting.

7. Return the frogs to their normal housing condition at the end of the next day.

   Xenopus laevis males can next be used for breeding after 2 wk, whereas Xenopus laevis females should not be used for breeding for at least 4–6 mo.

8. Leave the eggs in the breeding tank with an air pump and air stone for a week until the tadpoles reach stage 45 (Nieuwkoop and Faber 1994). Renew the water in the breeding chamber if it is turbid and maintain a stable temperature of 21°C.

   See Troubleshooting.

Chemical Exposure

In this protocol, tadpoles are exposed to chemicals in the absence or presence of T_3 (5 nM). The T_3 addition stimulates production of TRß (Thyroid Hormone Receptor Beta) from stage 37 onwards (Turque et al. 2005; Fini et al. 2012, 2017), thereby amplifying responses.

A thyroid hormone agonist is a good positive control when considering experiments in relation to the thyroid axis. Most of the time 5 nM T_3 solution is used as it is the actual hormone acting at the receptor level. An indirect induction could be studied using T_4 instead of T_3. As a control of nonthyroid hormone related action, we usually use NH-3 (Lim et al. 2002), the only TR antagonist (in presence of T_3) that has been described.

All steps are performed at room temperature unless specified.

9. Dilute the stock solutions of compound(s) of interest in DMSO to a concentration 10,000× higher than the final test concentration. Usually 0.2–1.7 mL 100% polypropylene microcentrifuge tubes are used for this purpose. If you are testing a range of concentrations of a chemical, prepare a cascade dilution in DMSO starting from the highest concentration of stock solution (illustrated in Fig. 1A).

   The 10,000× stock solution can be prepared fresh daily or aliquoted and stored at −20°C to avoid repeated thawing and freezing. Use a new aliquot each day.

   Not all chemicals are soluble in DMSO. Some dissolve more easily in water, ethanol, or methanol. The appropriate solvent should be checked using information available on the product. When a solvent other than DMSO is used, adjust the solutions of control groups accordingly, for example by using 0.01% ethanol or methanol when solutions are prepared in ethanol or methanol, respectively.

10. Rinse two 50 mL polypropylene tubes with dechlorinated aquatic system water. Add 10 mL of Evian water to one of these tubes and 10 mL 5 nM T_3 solution to the other tube. Add 1 µL of 10,000× concentrated test solution using a pipette with a 10 µL tip to create a 1× test solution. Vortex the solutions thoroughly.

11. Rinse all wells of a six-well plate with dechlorinated aquatic system water. Add 8 mL of Evian water to each well. Using a 1 mL transfer pipette with large bulb (polyethylene) with the open section cut (widen the tip of the transfer pipette by scalpel or scissors to diameter of at least 4 mM), transfer 15 st.45 tadpoles per well of the six-well plate (containing Evian water).

    Throughout the protocol, transfer tadpoles as described in this step.

    In this standard assay, 15 tadpoles are used per exposure group per replicate. In this protocol one chemical of interest is tested under two conditions: in the absence and presence of 5 nM T_3. A standard test plate therefore contains one well corresponding to a control (Evian water + 0.01% DMSO), one well containing the chemical of interest diluted in Evian water, one well containing 5 nM T_3 + 0.01% DMSO, and one well corresponding to the chemical of interest diluted in 5 nM T_3. If more than one concentration of the chemical of interest is tested, one control and one 5 nM T_3 group for the experiment are sufficient.

12. Remove the liquid from all six wells using a transfer pipette. To one well, add 8 mL of 1× chemical of interest in Evian water prepared in Step 10 using either a P5000 pipette (twice) or a 10 mL

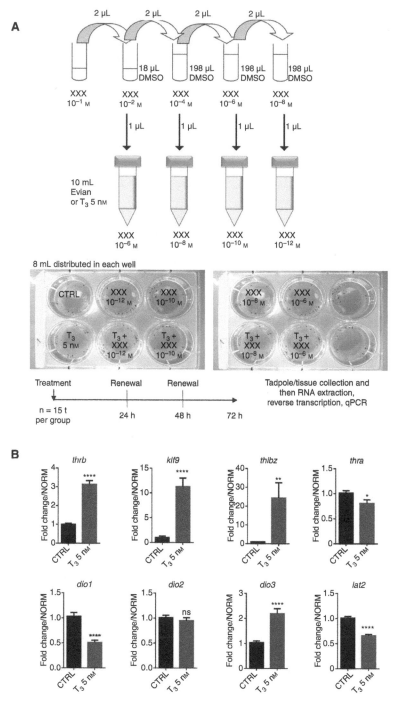

FIGURE 1. (*A*) Graphical description of treatment. An example is given of a substance to be tested in a starting solution at 10^{-1} M in DMSO. Targeted dilutions are 10^{-12}, 10^{-10}, 10^{-8}, and 10^{-6} M. First, a cascade dilution in DMSO is done to obtain a 10,000 times more concentrated solution than the targeted dilution. Then 1 µL of each "mother dilution" is diluted in either 10 mL of Evian water or 5 nM T_3 (prepared in Evian water). CTRL or T_3 treatments are, respectively, 1 µL DMSO in 10 mL water and 1 µL DMSO in 10 mL 5nM T_3. For each concentration tested, 8 mL of the prepared solution (10 mL) are pipetted into the wells and the plates are incubated at 23°C. Daily renewal is done at a given time. After 72 h, tissues/tadpoles are collected. (*B*) Quantification of expression of eight typical thyroid hormone-related genes are given as examples obtained in brain tissue after 3-d exposure with T_3 5 nM. Genes encoding transcription factors up-regulated in brain tissue after a 3-d exposure with T_3 are: thyroid hormone receptor beta (*thrb*) with two- to threefold induction, kruppel like factor 9 (*klf9*) with a 10-fold induction, and thyroid hormone induced bzip factor (*thibz*) with a 20- to 30-fold induction. *dio3* (encoding the D3 enzyme which inactivates T_3 into T_2) is documented to be up-regulated (two- to threefold induction) by T_3 while *dio1* (encoding D1) is observed to be down-regulated (twofold reduction) and *dio2* (encoding the activating enzyme D2 which converts T_4 into T_3) remains unchanged. Two other genes encoding the nuclear receptor alpha (*thra*) and the membrane transporter lat2 (*lat2*) are documented to be down-regulated. These results are the pool of three independent experiments with $n = 4$ replicates in each experiment. Total RNA extraction for each replicate was obtained from two brains. 500 ng of total RNA was used for reverse transcription and 1:20 dilution cDNA was used for qPCR. See Table 1 for primers. Fold changes are presented as mean + SEM using histograms. Statistics were done using pairwise nonparametric Mann–Whitney test with (****) $P < 0.0001$, (**) $P < 0.01$, (*) $P < 0.05$.

Cite this protocol as *Cold Spring Harb Protoc*; doi:10.1101/pdb.prot098301

pipette. In a second well add 8 mL of the chemical of interest diluted in 5 nM T_3 in the same manner. In the third and fourth wells, add 8 mL of 0.01% DMSO (add 1 µL of 100% DMSO to 10 mL of Evian or T_3 solution) to control groups of tadpoles (control or T_3).

13. Place the six-well plate into a dark incubator for the next 72 h at 23°C.

14. Renew the solutions every day at approximately the same time (± 1 h). Remove any dead tadpoles. If the mortality exceeds 20% in a well (three tadpoles out of the 15) do not include the well in the analysis. If this happens in a control group, stop and repeat the experiment.

    See Troubleshooting.

15. Proceed to Step 16 for RT-qPCR analysis or Step 36 for imaging analysis.

RT-qPCR

16. After 72 h of exposure, rinse the tadpoles by transferring them to a 50 mL plastic tube containing 40 mL of room temperature Evian water.

    Gene expression can also be assessed after a shorter exposure period e.g., 24 h exposure.

17. Transfer the tadpoles into a 50 mL plastic tube containing 100 mg/L MS-222.

    Tadpoles should stop moving after 30 sec.

    If interested in gene expression in specific tissues, dissect tissues for RT-qPCR analysis using nuclease-free tools prepared by washing in detergent (RNAseZap or SDS) and 0.1 N NaOH, followed by at least three washes in MilliQ water and overnight sterilization at 180°C. Clean tools and the dissection surface frequently with RNAseZap during dissections.

    See Troubleshooting.

18. Rinse the tadpoles by transferring them to a 50 mL plastic tube containing 40 mL of room temperature Evian water.

19. Place one tadpole per nuclease free microcentrifuge tube containing 100 µL of lysis solution from a total RNA extraction kit (e.g., Ambion RNAqueous). From one replicate at least 5 tadpoles (or tubes containing tissues such as two tadpole brains) of each group should be used for the gene expression analysis.

    If you intend to perform RT-qPCR on stage 45–47 Xenopus laevis brains, two brains per tube are sufficient to provide enough RNA.

20. Immediately extract RNA using an RNA extraction kit or snap-freeze samples in lysis buffer using liquid nitrogen to process them later. To extract RNA, thaw the samples on ice and continue RNA extraction.

    The DNAse I treatment step is important for RT-qPCR, even though it is often listed as optional in the RNA kit manual.

21. Use a Nanodrop spectrophotometer to estimate the obtained RNA concentration of each sample (at this point there should be 20 samples—5 replicates each for chemical of interest in Evian water, chemical of interest in T_3, and Evian water and T_3 controls). Quality can be checked at this step by looking at the 260/280 and 260/230 ratios.

    The 260/280 ratio should be over 1.8 and is expected to be close to 2.0. The 260/230 ratio should be above 2.0. A lower 260/230 ratio may indicate solvent contamination. This contamination can be removed by using specific purification columns (e.g., RNeasy MinElute Cleanup Kit (QIAGEN ref. 74204)).

    Note that at this step, transfer RNA, ribosomal RNA, and digested genomic DNA interfere with the quantification of targeted mRNA. A method based on total RNA specific dye (QUBIT) is available for accurate concentration estimation (in samples processed without DNAse treatment, for example, because of usage of TaqMan probes).

22. Determine the integrity and quality of the extracted RNA by estimating the 18S/28S ratios. We recommend the use of an Agilent Bioanalyzer with Agilent RNA 6000 Pico Kit. Only samples with RIN (RNA integrity number) higher than 7.5 should be used for further RT-qPCR.

A standard RNAse free 0.8% agarose gel could be used if RNA quantity is not limiting. On a gel, in the intact RNA sample the 18S and 28S ribosomal RNA bands should be clearly visible, while degraded RNA should appear as low molecular weight smear. Moreover, the 28S band should be bigger than 18S band.

23. Perform real time-polymerase chain reaction (RT-PCR) using an RT-PCR kit of your choice (e.g., High-Capacity cDNA Reverse Transcription Kit—Applied Biosystems or Reverse Transcription Master Mix—Fluidigm). We usually perform reverse transcription on 125 ng (Fluidigm) to 500 ng (Applied Biosystems) RNA.

    Note that the Fluidigm kit precludes doing an RT minus control. While performing RT-qPCR, an RT minus control should be included on the plate. This control is usually prepared during the reverse transcription step by replacing the enzyme reverse transcriptase with water and serves to verify that samples are not contaminated by genomic DNA, which might be amplified during the qPCR reaction. The RT minus control is expected to result in no amplified product. Some kits do not allow omission of reverse transcriptase; in that case RNA at the appropriate concentration should be added directly to the qPCR plate. A water control should be also included on the qPCR plate by replacing the cDNA sample with water. This control serves to verify that primers do not form primer dimers and that the primer mix is not contaminated by DNA.

24. Create your plate schema trying to put all your target and reference genes on the same plate (384-well clear hardshell plate). (See Fig. 2 for schema example.) Perform at least duplicates of each sample (triplicates are recommended). Follow the minimum information for publication of quantitative real-time PCR experiments (MiQE) (Bustin et al. 2009).

    A list of primers for genes involved in thyroid signaling (e.g., thrα, thrβ, dio1, dio2, dio3, thibz, klf9, lat2) is accessible in Table 1. In our example schema, every sample is tested with every primer pair. Do not forget to include at least two reference genes (i.e., those not affected by endocrine disrupters) for normalization. For more information about Xenopus reference genes consult (Mughal et al. 2018).

25. Prepare 1:20 dilutions of each cDNA with nuclease free water (e.g., 5 µL of cDNA in 95 µL of nuclease free water).

26. Prepare a series of qPCR mixes for each primer pair in Table 1 (genes involved in thyroid signaling in *Xenopus*). Each mix (target volume 5 µL of mix per well) will contain 0.15 µL of reverse primer (10 pM), 0.15 µL of forward primer (10 pM), 1.7 µL of nuclease free water and 3 µL of Power SYBR master mix per reaction. To ensure that sufficient amount of mix is prepared, increase the total volume, corresponding to the number of wells where you use this primer pair (including controls), by 10%. Work on ice.

    Volumes have been adapted for minimal reagent consumption with 384-well plates. If a 96-well plate is used, the minimal target volume per well is 13 µL. In that case the mix contains 0.3 µL of forward primer (10 pM), 0.3 µL of reverse primer (10 pM), 7.5 µL of Power SYBR master mix and 4.9 µL of nuclease free water per reaction.

27. In each well of the 384-well plate add 1 µL of diluted cDNA prepared in Step 25 and 5 µL of mix prepared in Step 26. Include a water control (1 µL of nuclease free water and 5 µL of qPCR mix per well). Work on ice.

    If you use a 96-well plate machine, pipette 2 µL of diluted cDNA and 13 µL of mix prepared in Step 26.

    At this step, the use of a multidistribution pipette is recommended.

28. Run qPCR using a dedicated machine and program to provide comparative C_t measurements including melting curve. An example is the QuantStudio 6 flex QPCR machine (Life technologies) with Quant Studio program.

qPCR Analysis

A brief description of qPCR analysis is provided below. For a more detailed explanation and other options of analysis refer to Bustin et al. 2009; Livak and Schmittgen 2001; Nolan et al. 2006; or Pfaffl 2001.

29. Export C_t values of your samples. If you worked with sample duplicates or triplicates as recommended, calculate the mean C_t values for each sample.

30. In MS Excel calculate the geometric mean of the C_t values of reference genes for each sample (e.g., geometric mean of the C_t values of *ef1α* and *odc* for each individual sample).

Cite this protocol as *Cold Spring Harb Protoc*; doi:10.1101/pdb.prot098301

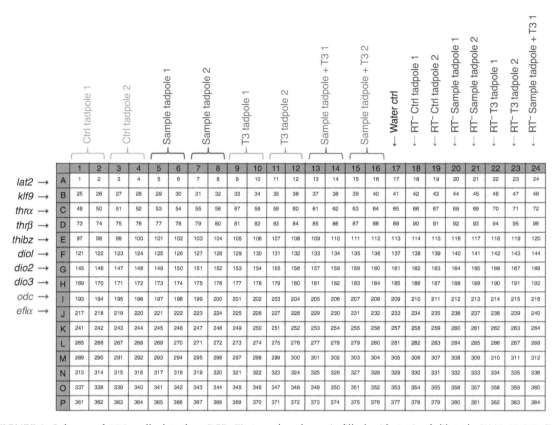

FIGURE 2. Schema of 384-well plate for qPCR. First, each column is filled with 1 μL of diluted cDNA (1:20). For example, column 1 is filled with 1 μL of 1:20 cDNA synthetized from RNA extracted from control tadpole #1. Column 2 is filled with diluted cDNA corresponding to control tadpole #2, etc. Note, that only 2 samples per group are depicted on this schema. Preferably, create a plate schema which includes all groups present in the experiment (in this case Evian control, sample in Evian, T_3 control, and sample in T_3). We aim to analyze gene expression of at least five samples from the same group per experimental replicate. The gene expression study described in this protocol would require three qPCR plates in total. In addition to this example plate, a second plate would include groups ctrl tadpoles 3 and 4 (columns 1–4), sample tadpoles 3 and 4 (columns 5–8), T_3 tadpoles 3 and 4 (columns 9–12), and sample + T_3 tadpoles 3 and 4 (columns 15–16). The last plate would contain the remaining samples (number 5) of each group. Water control wells (column 17) contain 1 μL of nuclease free water. RT- wells (columns 17–24) contain 1 μL of RT reaction run without the presence of reverse transcriptase if the RT kit allows that. If the reverse transcriptase is not provided separately in the kit, RNA of the corresponding sample can be used instead (1.25 ng per well). In the second step, 5 μL of qPCR mixes with primers specific for each gene is added to corresponding rows including water and RT⁻ rows. For example, mix containing reverse and forward primers for the gene *lat2* is distributed to the entire row A. It is always better to avoid placing your reference genes (in red) at the border rows of your plate (row A and P) to avoid possible drying of the wells.

31. Subtract the geometric mean of the C_t values of reference genes from the C_t values of your sample in order to calculate the ΔC_t values (= C_t of the gene of interest – geometric mean of reference genes of the same sample).

32. Calculate a median value of the ΔC_t values of all genes in the control group. This is a normalization step, if you wish to normalize to another group (e.g., T_3), calculate a median of this group.

33. Subtract the median value of the control group from the ΔC_t value of your sample to create a $\Delta\Delta C_t$.

34. Calculate a fold change using the formula = $2^{-\Delta\Delta Ct}$.

35. Use a program to perform a statistical analysis of the data (e.g., GraphPad Prism).

    Several examples of typical gene expression patterns of 72 h exposure to 5 nM T_3 solution are shown in Fig. 1B.

    See Troubleshooting.

Imaging (Optional)

Today many laboratories use in vivo bioassays based on transgenic organisms featuring genetic constructs driving expression of fluorescent proteins to link gene expression with physiological effects. If using a transgenic Xenopus laevis line expressing GFP (e.g., Tg(thibz:GFP) (Fini et al. 2007)), fluorescence can be quantified as described below. Changes in fluorescence are a result of potential disruption at multiple levels; synthesis, transport and/or metabolism of thyroid hormones in response to endocrine disruption.

36. After 72 h of exposure, rinse the tadpoles with Evian water in a 50 mL plastic tube as in Step 16.

37. Transfer the tadpoles into a 50 mL plastic tube containing ~40 mL of 100 mg/L MS-222.

38. Use a 2 mL transfer pipette with cut tip to place the tadpoles (one tadpole per well) in a black conical well 96-well plate. Leave an empty well between each group (for example, we test four groups: a control Evian group, a control T_3 group, a sample diluted in Evian group, and a sample diluted in T_3 group. There are 15 tadpoles in each group. These tadpoles should be plated consecutively with one empty well between groups).

39. Using a fine transfer pipette, position each tadpole in its well. The appropriate position of the tadpole will depend on which organ expresses GFP in the transgenic line you are using. For instance, if you wish to photograph the brain, place the tadpole dorsally so the ventral surface is uppermost. Make sure that the head of the tadpole is positioned in the middle of the well and the tail is curled around the body. Remove the majority of liquid.

40. Take color images with a 25× objective and 3 sec exposure time using a fluorescence microscope equipped with GFP long pass filters. QC Capture pro (QImaging) software can be used for image acquisitions.

    See Troubleshooting.

41. At the end of the experiment, fill each well with 1 g/L MS-222 to euthanize the tadpoles and dispose of the carcasses appropriately.

Image Analysis and Statistics

42. Open all images of one exposure group in ImageJ.

43. Exclude nonspecific signals by splitting the image into three layers (red, blue, and green channels) followed by subtraction of the red and blue channel from the green one.

44. Calculate the integrated density of the green channel. The data are expressed in relative units of fluorescence (RFU).

45. Use an appropriate program for statistical analysis of the data (e.g., GraphPad Prism).

    See Troubleshooting.

TROUBLESHOOTING

Problem (Step 6): No amplexus occurred; female doesn't produce any eggs; eggs are not fertilized.
Solution: Respect the recommended intervals between breedings (2 wk and 4–6 mo for males and females, respectively). Select females with prominent cloaca and pear-shaped bodies and males with dark nuptial pads at the inner side of forelimbs. Increase the amount of hCG injected in the females—never exceed 700 units—and preprime them in the morning with 50–100 units.

Problem (Step 8): High mortality of the tadpoles before NF stage 45.
Solution: Verify that the water temperature doesn't drop below 21°C. Change the water every time it becomes turbid.

Problem (Step 8): Tadpoles develop slowly or are small.

Solution: Verify that the water temperature doesn't drop below 21°C. An optimal solution is to place the tank into 23°C incubator.

Problem (Step 14): High mortality of the tadpoles during exposure not related to the toxicity of test compound.

Solution: Put the tadpoles into six-well plates 24 h before start of exposure and keep them overnight in a 23°C incubator. If the mortality exceeds 10% the experiment should not be considered valid and should be stopped.

Problem (Step 17): Tadpoles keep moving after the anaesthesia with MS222.

Solution: Verify that the MS222 solution is not older than 3 wk and that it was stored in the dark (old MS222 or MS222 exposed to light turns yellow).

Problem (Steps 35 and 45): T_3 induction is not detected.

Solution: Make sure that the correct dilution was used. Do not use stock stored at −20°C after 1 yr of storage. Prevent usage of plastic during the process (we have validated that the use of low binding tips, 5000 μL tips, and six-well plate (TPP) is okay); for example, do not use a manual dispenser (e.g., Eppendorf Varispenser).

Problem (Step 40): Tadpoles have strong yellow color under the fluorescence microscope.

Solution: Yellow color of tadpoles may interfere with image analysis and create false positive results. If the yellow signal seen under the fluorescent microscope is in the head region of the tadpoles, they are probably dead and possibly dry, potentially due to slow manipulation. Plate only as many tadpoles at the same time as you can position and photograph within 30–45 min. A yellow signal at the level of the intestines/yolk is frequent in tadpoles younger stage than stage 46. The gut area can be cropped away using the program Image J before further image analysis.

DISCUSSION

Amphibian metamorphosis is a thyroid hormone-dependent process, hence its use in the AMA (amphibian metamorphosis assay) OECD test guideline (OECD TG 231; Opitz et al. 2005). However, given the unknown endocrine-disrupting potential of the large numbers of chemicals currently on the market or to be tested before marketing, it is crucial to develop more rapid in vivo bioassays. Our group developed a 3-d test using a transgenic *Xenopus* line Tg(*thibz*:GFP) (Fini et al. 2007). This line harbors a genetic construct composed of the thyroid response elements upstream of the coding sequence for the transcription factor TH/bZIP that drives GFP expression with a thyroid hormone stimulus. This test is currently undergoing final validation at the level of the OECD as the *Xenopus* Embryonic Thyroid Assay (XETA). Using this principle of a fluorescent reporter upstream of a promoter specific for an endocrine response, a number of transgenic teleost lines have been developed to detect steroidal disruption (Brion et al. 2012; Sébillot et al. 2014; Spirhanzlova et al. 2015). Here, we describe a more general exposure protocol using wild type *Xenopus laevis* larvae to address effects of endocrine disruptors on gene expression. Besides the evolutionary conservation of TH signaling and brain development, the amphibian model provides experimental access to early critical windows of development that are essential for optimal later development, including that of the nervous system. This protocol has been used mainly for detecting chemicals with thyroid hormone disrupting capacities but could also be used with adjustments for critical windows appropriate for other endocrine axes. Aquatic embryo models have the additional advantages of having external development and large numbers of larvae produced by the female at each mating. In addition, early developmental stages of fish and tadpoles, which still feed from their reserves of yolk, are not classed as laboratory animals under European Union regulations (Directive 2010/63/EU).

RECIPES

3,3′,5-Triiodo-L-Thyronine Sodium Salt Solution (T₃; 5 nM)

1. Make a 0.01 M stock solution of T_3 by adding 100 mg of T_3 powder (Sigma-Aldrich; CAS 55-06-1) to 14.86 mL of 30% NaOH (prepared in Milli-Q [ultrapure] water). Vortex the 0.01 M stock solution well, aliquot, and store at −20°C.

2. Make a 0.0001 M intermediate solution of T_3 fresh every day by adding 5 µL of 0.01 M stock solution of T_3 to 495 µL of Evian water (use Evian water in a 75 cL glass bottle or another mineral water with reproducible quality).

   Be careful that neither precipitate nor crystals are present in the T_3 stock solution.

3. Add 5 µL of 0.0001 M T_3 to 100 mL of Evian water to create a final 5 nM T_3 solution.

MS222 Solution (0.1%)

MS222 (Sigma-Aldrich, A5040)	1 g
Sodium bicarbonate (Sigma-Aldrich, S5761)	1 g
Water, dechlorinated	1 L

Protect from light. Store for up to 2 wk at room temperature.

ACKNOWLEDGMENTS

We thank Sébastien Le Mével for his input on the methods that are routinely used. We thank Gérard Benisti, Philippe Durand, and Jean-Paul Chaumeil for animal care. This protocol has been refined thanks to work supported by grants from Centre National de la Recherche Scientifique (CNRS), Muséum National d'Histoire Naturelle (MNHN), and from French Ministry of Environment MEDD, PNRPE_THYDIS 2013-N°CHORUS-2101207963, PNREST THYPEST EST-2014-122, and European Union EDC MIX RISK_GA N°634880.

REFERENCES

Brion F, Le Page Y, Piccini B, Cardoso O, Tong SK, Chung B, Kah O. 2012. Screening estrogenic activities of chemicals or mixtures in vivo using transgenic (cyp19a1b-GFP) zebrafish embryos. *PLoS One* 7: e36069.

Bustin SA, Benes V, Garson JA, Hellemans J, Huggett J, Kubista M, Mueller R, Nolan T, Pfaffl MW, Shipley GL, et al. 2009. The MIQE guidelines: Minimum information for publication of quantitative real-time PCR experiments. *Clin Chem* 55: 611–622.

Fini JB, Le Mevel S, Turque N, Palmier K, Zalko D, Cravedi JP, Demeneix BA. 2007. An in vivo multiwell-based fluorescent screen for monitoring vertebrate thyroid hormone disruption. *Environ Sci Technol* 41: 5908–5914.

Fini JB, Le Mével S, Palmier K, Darras VM, Punzon I, Richardson SJ, Clerget-Froidevaux MS, Demeneix BA. 2012. Thyroid hormone signaling in the *Xenopus laevis* embryo is functional and susceptible to endocrine disruption. *Endocrinology* 153: 5068–5081.

Fini JB, Mughal BB, Mével SL, Leemans M, Lettmann M, Spirhanzlova P, Affaticati P, Jenett A, Demeneix BA. 2017. Human amniotic fluid contaminants alter thyroid hormone signalling and early brain development in *Xenopus* embryos. *Sci Rep* 7: 43786.

Lim W, Nguyen NH, Yang HY, Scanlan TS, Furlow JD. 2002. A thyroid hormone antagonist that inhibits thyroid hormone action in vivo. *J Biol Chem* 277: 35664–35670.

Livak KJ, Schmittgen TD. 2001. Analysis of relative gene expression data using real-time quantitative PCR and the $2^{-\Delta\Delta C_T}$ Method. *Methods San Diego Calif* 25: 402–408.

Mughal BB, Leemans M, Spirhanzlova P, Demeneix B, Fini JB. 2018. Reference gene identification and validation for quantitative real-time PCR studies in developing *Xenopus laevis*. *Sci Rep* 8: 496.

Nieuwkoop PD, Faber J. 1994. *Normal table of* Xenopus laevis *(Daudin): A systematical and chronological survey of the development from the fertilized egg till the end of metamorphosis*. Garland Publishing, New York.

Nolan T, Hands RE, Bustin SA. 2006. Quantification of mRNA using real-time RT-PCR. *Nat Protoc* 1: 1559–1582.

OECD. 2009. *Test No. 231: Amphibian metamorphosis assay*, OECD Publishing, Paris.

Opitz R, Braunbeck T, Bögi C, Pickford DB, Nentwig G, Oehlmann J, Tooi O, Lutz I, Kloas W. 2005. Description and initial evaluation of a *Xenopus* metamorphosis assay for detection of thyroid system-disrupt-

Cite this protocol as *Cold Spring Harb Protoc*; doi:10.1101/pdb.prot098301

ing activities of environmental compounds. *Environ Toxicol Chem* **24:** 653–664.

Pfaffl MW. 2001. A new mathematical model for relative quantification in real-time RT-PCR. *Nucleic Acids Res* **29:** e45.

Sébillot A, Damdimopoulou P, Ogino Y, Spirhanzlova P, Miyagawa S, Du Pasquier D, Mouatassim N, Iguchi T, Lemkine GF, Demeneix BA, et al. 2014. Rapid fluorescent detection of (anti)androgens with spiggin-gfp medaka. *Environ Sci Technol* **48:** 10919–10928.

Spirhanzlova P, Leleu M, Sébillot A, Lemkine GF, Iguchi T, Demeneix BA, Tindall AJ. 2015. Oestrogen reporter transgenic medaka for non-invasive evaluation of aromatase activity. *Comp Biochem Physiol C Toxicol Pharmacol* **179:** 64–71.

Turque N, Palmier K, Le Mével S, Alliot C, Demeneix BA. 2005. A rapid, physiologic protocol for testing transcriptional effects of thyroid-disrupting agents in premetamorphic *Xenopus* tadpoles. *Environ Health Perspect* **113:** 1588–1593.

In Vivo Assessment of Drug-Induced Hepatotoxicity Using *Xenopus* Embryos

Katy Saide and Grant N. Wheeler[1]

School of Biological Sciences, University of East Anglia, Norwich Research Park, Norwich NR4 7TJ, United Kingdom

Failure to predict drug-induced toxicity reactions is a major problem contributing to a high attrition rate and tremendous cost in drug development. Drug screening in *X. laevis* embryos is high-throughput relative to screening in rodents, potentially making them ideal for this use. *Xenopus* embryos have been used as a toxicity model in the frog embryo teratogenesis assay on *Xenopus* (FETAX) for the early stages of drug safety evaluation. We previously developed compound-screening methods using *Xenopus* embryos and believe they could be used for in vitro drug-induced toxicity safety assessment before expensive preclinical trials in mammals. Specifically, *Xenopus* embryos could help predict drug-induced hepatotoxicity and consequently aid lead candidate prioritization. Here we present methods, which we have modified for use on *Xenopus* embryos, to help measure the potential for a drug to induce liver toxicity. One such method examines the release of the liver-specific microRNA (miRNA) miR-122 from the liver into the vasculature as a result of hepatocellular damage, which could be due to drug-induced acute liver injury. Paracetamol, a known hepatotoxin at high doses, can be used as a positive control. We previously showed that some of the phenotypes of mammalian paracetamol overdose are reflected in *Xenopus* embryos. Consequently, we have also included here a method that measures the concentration of free glutathione (GSH), which is an indicator of paracetamol-induced liver injury. These methods can be used as part of a panel of protocols to help predict the hepatotoxicity of a drug at an early stage in drug development.

MATERIALS

It is essential that you consult the appropriate Material Safety Data Sheets and your institution's Environmental Health and Safety Office for proper handling of equipment and hazardous materials used in this protocol.

RECIPES: Please see the end of this protocol for recipes indicated by <R>. Additional recipes can be found online at http://cshprotocols.cshlp.org/site/recipes.

Reagents

5-sulfosalicylic acid hydrate (SSA; 6.5% [w/v])
Agarose gel (2% [w/v]; Sigma-Aldrich)
Bradford assay kit (BIO-RAD)
GSH (0.1 mM; Sigma)
GSH assay reagent <R>
GSH buffer <R>

---

[1]Correspondence: grant.wheeler@uea.ac.uk

GSH reductase (13 U/mL in GSH buffer; Sigma-Aldrich)

HCl (10 mM)

Liquid nitrogen

MEM salts (10×; pH 7.4) <R>

MEMFA fixative for *Xenopus* <R>

miRCURY LNA PCR Primer mix for *Xenopus* miR-103 (Exiqon)

miRCURY LNA PCR Primer mix for *Xenopus* miR-122 (Exiqon)

MMR (10×; pH 7.5) <R>

Paracetamol and/or drug to be tested (2× final desired concentration)

PBS(P) <R>

PBS containing 0.1% Tween 20 (PBST)

RNA isolation and cDNA synthesis kit

RNase-free water

SYBR Green PCR Master Mix (Applied Biosystems)

Tricaine (0.5 mg/mL in 0.1× MMR)

X. laevis embryos (Nieuwkoop and Faber (NF) stage 1; Nieuwkoop and Faber 1994)

Equipment

Applied Biosystems 7500 Fast Real-Time PCR system

Breathe-Easy sealing membranes (Sigma)

Culture incubator (set to appropriate temperature)

Culture plate (96- or 24-well; non-cell-culture-grade; Fisher Scientific)

Dumont #5 forceps (stainless steel; Sigma-Aldrich)

These are ultrafine for careful manipulation of embryos.

Eppendorf pestle

Freezer (−20°C)

Glass vials with screw caps (3.5 mL; SGL)

Light microscope with charge-coupled device (CCD) digital camera for whole-mount imaging of embryos

Long-handled scalpel (10A blades)

MicroAmp optical 96-well plate (Applied Biosystems)

Microcentrifuge

Microcentrifuge tubes

Parafilm M wrapping film (Fisher Scientific)

Pasteur pipette

To use plastic Pasteur pipettes, cut the end off with scissors. For glass pipettes, mark the end with a diamond pen, break off cleanly and fire the end briefly to melt any sharp edges.

Petri dish (3 cm^2 and 10 cm^2; Fisher Scientific)

Razor

Spectrophotometer such as GloMax Explorer System (Promega)

Stereomicroscope with two-armed fiber optic illuminator to allow easy adjustment of the angle of illumination

Vortex

METHOD

Before examining hepatoxicity, the overall toxicity of a small molecule or compound needs to be determined, including the maximum dose that embryos can be exposed to before they all consistently die (Saide et al. 2019). This will determine a range of concentrations to use to test for hepatoxicity below the maximum dose.

Assays for Liver Toxicity

These protocols were developed using paracetamol, which is associated with acute liver injury in humans after overdose administration, as a model drug. This drug can be used as a positive control in these methods and for comparisons with drugs/compounds being tested.

1. Harvest NF stage 1 *X. laevis* embryos as in Al-Yousuf et al. (2017) and incubate in 0.1× MMR in Petri dishes at 13°C–23°C until stage 38. Regularly check the embryos (at least twice daily or more at early stages) and remove any dead embryos until the correct NF stage has been reached.

   It should take roughly 4 d to reach stage 38.

2. In a 96-well plate, add 125 µL of a 2× final concentration of paracetamol or the drug to be tested dissolved in 0.1× MMR. For the negative control add 125 µL of 0.1× MMR to the well. Add one stage 38 embryo (Nieuwkoop and Faber 1994) in 125 µL of 0.1× MMR to each well using a cut-off pipette tip.

   The final volume of each well is 250 µL.

   When testing paracetamol, a range of 0–5 mM generated a paracetamol-induced liver injury phenotype that was similar to that observed in other animal models and humans (Vliegenthart et al. 2014; Verstraelen et al. 2016; Saide et al. 2019).

3. For one biological replicate (*n* = 1), use up to 10 embryos per drug concentration. Consequently for five biological replicates (*n* = 5), use a total of 50 embryos.

   One biological replicate is defined as the embryos from one female.

4. Seal plates with Breathe-Easy sealing membranes to prevent evaporation, but allow exchange of gases.

5. Incubate embryos from stage 38 to stage 45 at 23°C.

   This takes ~2–3 d.

   Stage 38 corresponds to late organogenesis when the heart, vascular system, and other organs such as the liver are well developed. During stages 38–45 the mouth opens and the gills and gut start to function.

   Proceed to Step 6 to measure differences in gene expression due to drug treatment, or Step 13 to measure changes in free glutathione due to drug treatment.

Expression of miR-122 in Different Tissues of Embryos

6. Following drug treatment until stage 45, place the embryos into a clean Petri dish of tricaine (0.5 mg/mL in 0.1× MMR) and incubate for 1 h at 23°C. Transfer the embryos to a Petri dish coated with 2% agarose and dissect the embryos into tail and gut tissue using a simple razor as shown in Figure 1.

   Following this step, samples can be stored at −80°C. Place the samples in a 1.5-mL microcentrifuge tube, remove all fluid, and snap-freeze in liquid nitrogen for at least 2 min. We usually freeze 10 embryos per 1.5 mL microcentrifuge tube.
   The gut section contains the liver. The tail section does not contain any part of the liver and so liver markers observed are assumed to be in the extensive vasculature of the tail.

7. Pool the guts and tails separately for 10 embryos. Using a commercial kit, isolate the RNA and synthesize cDNA.

   We routinely use around 10 ng of RNA for cDNA synthesis.

qRT-PCR

8. Dilute the cDNA 1:80 in RNase-free water.

   We empirically determined that this was the optimum dilution factor.

9. Prepare reactions in a MicroAmp optical 96-well plate. Add 5 µL of the diluted cDNA, 0.5 µL of miRCURY LNA PCR Primer mix for *Xenopus* miR-122 (Exiqon), 7.5 µL of 2× SYBR Green PCR Master Mix (Applied Biosystems), and 2 µL of RNase-free water for a final volume of 15 µL.

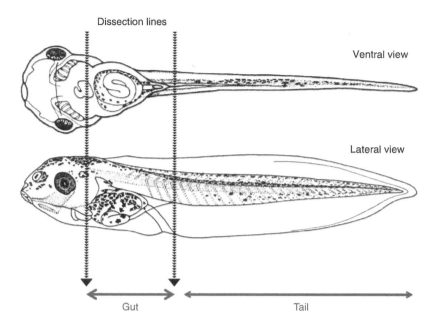

FIGURE 1. Dissection for qRT-PCR miR-122 detection. After drug exposure, stage 45 *Xenopus* embryos were anesthetized and dissected along the lines indicated here (black arrows) to obtain gut (purple arrow) and tail (red arrow) regions. These regions were then processed to detect miR-122 expression by qRT-PCR. (Adapted from Saide et al. 2019.)

We used miR-103 as a quantitative control for miR-122 expression of treated embryos using 0.5 μL of miRCURY LNA PCR Primer mix for Xenopus miR-103 (Exiqon).

Samples derived from embryos produced from different mothers are biological replicates. We test biological replicates three times (technical replicates).

10. Carry out PCR using an Applied Biosystems 7500 Fast Real-Time PCR system under the following conditions: 10 min at 95°C, 40 cycles for 10 sec at 95°C, and 1 min at 60°C.

qRT-PCR Statistical Analysis

11. Analyze gene expression using the Livak method (Livak and Schmittgen 2001). Take an average C_T value from the technical replicates and normalize that value to the C_T value for miR-103 expression and also to the untreated samples (embryos that received no drug treatment).

12. Determine the fold change in gene expression using the formula: $2^{-\Delta\Delta C_T}$ and convert this number into a logarithmic ($\log_{10}$). Perform Mann–Whitney tests between gut and tail tissues from embryos treated with the same concentration of drug to determine statistical significance.

Measuring Free GSH

Many compounds are metabolized in the liver to facilitate their excretion. In many cases, including paracetamol, this involves the metabolic addition of GSH to a molecule. Here we provide a method to measure the depletion of free GSH in response to a drug.

13. At the end of the incubation period described in Step 5, transfer 10 embryos that were treated with the same conditions into one microcentrifuge tube. Place the tube on ice.

14. Remove as much of the incubation solution as possible. Add 125 μL of 10 mM HCl.

15. Homogenize embryos using an Eppendorf pestle and vortexing.

16. Centrifuge at 14,000*g* for 5 min at 4°C.

17. Transfer 25 µL of the supernatant to another tube, flash freeze in liquid nitrogen, and store at −80°C for protein quantification using a Bradford assay (see Steps 25–26). Replace the 25 µL with 25 µL of 6.5% (w/v) SSA.

18. Remove all of the supernatant including the SSA and flash-freeze in liquid nitrogen. Store at −80°C until ready to perform the GSH assay.

GSH Assay

19. Prepare the following standard GSH concentrations: 0, 1, 2, 5, 10, 20, 30, and 40 nmol/mL using GSH buffer and 0.1 mM GSH. Keep on ice.

20. Mix 20 µL of the standard or sample with 200 µL of GSH assay reagent in one well of a 96-well plate and incubate for 5 min at room temperature.

21. To each well add 50 µL of 13 U/mL GSH reductase and gently shake before placing on the plate reader.

22. Immediately measure the absorption of the plate at 405 nm using a spectrophotometer.

23. Take 11 readings at 15-sec intervals.

24. Use the reading with the best standard curve to calculate the sample results (nmol/mL).

Bradford Assay

25. Prepare standard concentrations of BSA dissolved in 10 mM HCl: 0, 0.1, 0.2, 0.4, 0.8, 1.2, 1.6, 2, and 4 mg/mL.

26. Perform the assay according to the manufacturer's instructions as follows.

    i. Add 5 µL of each sample from Step 17 to one well of a 96-well plate with 25 µL of reagent A + S (1 mL reagent A + 20 µL reagent S) and 200 µL of reagent B.

    ii. Incubate for 15 min at room temperature.

    iii. Read the plate absorption at 595 nm using a spectrophotometer.

    iv. Divide the GSH results (nmol/mL) by the Bradford result (mg/mL) for each sample to give the final result (nmol/mg).

DISCUSSION

Predicting drug-induced toxicity is a big problem for pharma in the 21st century. *X. laevis* embryos could be ideal for screening new drugs and compounds because screening in *Xenopus* embryos is comparatively high-throughput relative to screening in rodents (Tomlinson et al. 2009). We have previously developed methods to both screen compounds for novel phenotypes and test toxicity (Tomlinson et al. 2009; Tomlinson et al. 2012; Wheeler and Liu 2012; Webster et al. 2016; Al-Yousuf et al. 2017; Saide et al. 2019). In addition, *Xenopus* embryos have been used as a toxicity model for drugs in their early stages of drug safety evaluation in FETAX (Leconte and Mouche 2013).

Xenopus embryos could assist in vitro drug-induced toxicity safety assessment in the early phases of drug development before moving on to expensive preclinical trials in mammals (Fig. 2). The procedures outlined here can provide an assessment of the potential for a novel compound to cause hepatotoxicity in the tadpole model. Previous work in *Xenopus* suggests this could be translatable to the incidence of toxicity in higher vertebrates such as mice and humans (Saide et al. 2019). Many drugs such as paracetamol are metabolized in the liver, and it is the metabolites that are often toxic. Preliminary work has shown that using mass spectroscopy of tadpole tissue to identify drug metabolites, such as those generated from paracetamol, is feasible (G. N. Wheeler, unpubl.). The tail/ vascular assay described here is also amenable to protein assays such as ELISAs. Known biomarkers

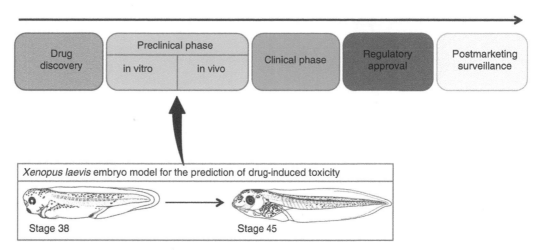

FIGURE 2. Drug development with *Xenopus laevis*. The proposed use of *Xenopus laevis* embryos in assessing hepatoxicity would be in the early preclinical phase for the prediction of drug-induced toxicity. *Xenopus* embryos could bridge the gap between in vitro and in vivo safety studies. (From Saide 2018.)

that are released into the blood in hepatotoxic situations that could be tested in this way include alanine aminotransferase, aspartate aminotransferase, alkaline phosphatase, glutamate dehydrogenase, cytochrome C, and keratin 18.

RECIPES

GSH Assay Reagent

0.28 mg/mL NADPH (Sigma)
0.4 mg/mL DTNB (5,5-dithio-bis-(2-nitrobenzoic acid); Sigma)

Prepare in GSH Buffer <R>. Store indefinitely at −20°C.

GSH Buffer

143 mM NaH_2PO_4
6.3 mM EDTA

Prepare in dH_2O. Store for up to 1 year at 4°C.

MEMFA Fixative for Xenopus

1 mL MEM salts (10×; pH 7.4) <R>
1 mL 37% formaldehyde
8 mL H_2O

Freshly prepare on the day and store at 4°C.

MEM Salts (10×; pH 7.4)

209.3 g MOPS
76.08 g EGTA
2.46 g $MgSO_4$

Add to a 2-L beaker. Make up to 1 L with deionized H_2O. Adjust pH to 7.4 using NaOH and sterilize by autoclaving. Store at 4°C protected from light (wrapped in foil).

MMR (10×; pH 7.5)

58.45 g NaCl
1.5 g KCl
2 g MgCl$_2$
2.7 g CaCl$_2$
11.9 g HEPES

Add to a 2-L beaker. Make up to 1 L with deionized H$_2$O. Adjust pH to 7.5 using NaOH and sterilize by autoclaving. Store indefinitely at room temperature.

PBS(P)

Reagent	Quantity (for 1 L)	Final concentration (10×)
NaH$_2$PO$_4$	2.56 g	18.6 mM
Na$_2$HPO$_4$	11.94 g	84.1 mM
NaCl	102.2 g	1.75 M

Adjust the pH to 7.4 using NaOH or HCl as necessary. This recipe produces a 10× stock solution; prepare 1× PBS(P) by diluting with H$_2$O. Both 1× and 10× PBS(P) can be kept indefinitely at room temperature.

REFERENCES

Al-Yousuf K, Webster CA, Wheeler GN, Bombelli FB, Sherwood V. 2017. Combining cytotoxicity assessment and *Xenopus laevis* phenotypic abnormality assay as a predictor of nanomaterial safety. *Curr Protoc Toxicol* 73: 20.13.21–20.13.33. doi:10.1002/cptx.25

Leconte I, Mouche I. 2013. Frog embryo teratogenesis assay on *Xenopus* and predictivity compared with in vivo mammalian studies. *Methods Mol Biol* 947: 403–421. doi:10.1007/978-1-62703-131-8_29

Livak KJ, Schmittgen TD. 2001. Analysis of relative gene expression data using real-time quantitative PCR and the 2$^{-\Delta\Delta C_T}$ method. *Methods* 25: 402–408. doi:10.1006/meth.2001.1262

Nieuwkoop PD, Faber J. 1994. *Normal table of* Xenopus laevis *(Daudin)*. Garland Publishing Inc, New-York.

Saide K. 2018. "Development of a non-mammalian, pre-clinical screening tool for the predictive analysis of drug toxicity." Doctoral thesis, University of East Anglia. Retrieved from https://ueaeprints.uea.ac.uk/id/eprint/67269/

Saide K, Sherwood V, Wheeler GN. 2019. Paracetamol-induced liver injury modelled in *Xenopus laevis* embryos. *Toxicol Lett* 302: 83–91. doi:10.1016/j.toxlet.2018.09.016

Tomlinson ML, Rejzek M, Fidock M, Field RA, Wheeler GN. 2009. Chemical genomics identifies compounds affecting *Xenopus laevis* pigment cell development. *Mol Biosyst* 5: 376–384. doi:10.1039/b818695b

Tomlinson ML, Hendry AE, Wheeler GN. 2012. Chemical genetics and drug discovery in *Xenopus*. *Methods Mol Biol* 917: 155–166. doi:10.1007/978-1-61779-992-1_9

Verstraelen S, Peers B, Maho W, Hollanders K, Remy S, Berckmans P, Covaci A, Witters H. 2016. Phenotypic and biomarker evaluation of zebrafish larvae as an alternative model to predict mammalian hepatotoxicity. *J Appl Toxicol* 36: 1194–1206. doi:10.1002/jat.3288

Vliegenthart AD, Tucker CS, Del Pozo J, Dear JW. 2014. Zebrafish as model organisms for studying drug-induced liver injury. *Br J Clin Pharmacol* 78: 1217–1227. doi:10.1111/bcp.12408

Webster CA, Di Silvio D, Devarajan A, Bigini P, Micotti E, Giudice C, Salmona M, Wheeler GN, Sherwood V, Bombelli FB. 2016. An early developmental vertebrate model for nanomaterial safety: bridging cell-based and mammalian toxicity assessment. *Nanomedicine (Lond)* 11: 643–656. doi:10.2217/nnm.15.219

Wheeler GN, Liu KJ. 2012. *Xenopus*: an ideal system for chemical genetics. *Genesis* 50: 207–218. doi:10.1002/dvg.22009

Cite this protocol as *Cold Spring Harb Protoc*; doi:10.1101/pdb.prot106096

Frog Embryo Teratogenesis Assay—*Xenopus* (FETAX): Use in Alternative Preclinical Safety Assessment

Douglas J. Fort[1] and Michael Mathis

Fort Environmental Laboratories, Inc., Stillwater, Oklahoma 74074

The primary objective of this protocol is to provide alternative developmental toxicity data using the frog embryo teratogenesis assay—*Xenopus* (FETAX) model for preclinical safety assessment. FETAX is most useful in the prioritization of developmental toxicity hazard for sets of discovery-level compounds. Assessment of teratogenic potential is based on teratogenic indices (TI), which measure the teratogenic potential of a given test material. The relative hazard ranking is based on a set of weighted endpoints that include developmental toxicity potency, teratogenic potential, and growth inhibition. Because of the importance of potency and teratogenic potential in determining relative hazard, these endpoints are weighted 2×. Growth inhibition is weighted 1×. These data determine a generic ranking. The generic hazard rank is based on numerical endpoint data derived from an assessment of potency, teratogenic potential as determined by the TI value, and concentration at which growth inhibition is detected when expressed as a proportion of the 4-d LC50 value. With the generic rank, the lower the generic hazard ranking the greater the developmental toxicity hazard relative to the other materials evaluated. However, the relative severity of the malformation syndromes induced is not incorporated into the generic ranking process. An objective evaluation of the severity of the deformities induced is an important process in the evaluation of FETAX results; therefore, a final definitive ranking process is used. A final definitive ranking is determined using the generic rank and the severity of the malformations induced (weighted 5×).

MATERIALS

It is essential that you consult the appropriate Material Safety Data Sheets and your institution's Environmental Health and Safety Office for proper handling of equipment and hazardous materials used in this protocol.

RECIPES: Please see the end of this protocol for recipes indicated by <R>. Additional recipes can be found online at http://cshprotocols.cshlp.org/site/recipes.

Reagents

3-Aminobenzoic acid ethyl ester (MS-222) stock solution (200 mg/L) <R>
6-Aminonicotinamide (6-AN) stock solutions (2500 and 5.5 mg/L) <R>
Dimethyl sulfoxide (DMSO) (Sigma-Aldrich, 154938; CAS No. 67-68-5; purity 99.0%)
FETAX solution <R>

> *Alternatively, dechlorinated tap water may be used as a substitute for FETAX solution.*

---

[1]Correspondence: djfort@fortlabs.com

Supplemental material is available at cshprotocols.cshlp.org

Hydrochloric acid (6 N) (optional; see Step 8)

L-cysteine solution (2.0%)

> *Dissolve 2.0 g L-cysteine (Sigma-Aldrich, C7352; CAS No. 52-90-4; purity 98.0%) in 100 mL FETAX solution.*

Neutral buffered formalin (NBF) (10.0% v/v) (Thermo-Scientific, 5705; CAS No. 50-00-0; purity 99.0%)

Sodium hydroxide (6 N) (optional; see Step 8)

Test compounds dissolved in FETAX solution (aqueous) or in DMSO (solvent), as appropriate

Equipment

Analytical balance (Ohaus E10640; Cole-Palmer EW-11012-40)

Digital camera (Sony Cyber-Shot; B&H Photo DCS-H300)

Excel (Microsoft)

ICp program (ToxCalc, Tidepool Scientific Software)

Incubator (307C; Fisher 11-679-25C)

Microscope (Stereomaster Zoom; Fisher 12-562-1)

Petri dishes (glass) (optional; see Step 6)

Petri dishes (plastic, 60 mm; Fisher FB0875713A)

Petri dishes (plastic, 100 mm; Fisher FB0875712)

pH meter (Oakton 700; Fisher 13-620-748)

SigmaScan (SPSS)

Syringes (5 mL; Fisher 14-823-35)

Topload balance (Ohaus Adventurer; Fischer AX8201)

Transfer pipettes (3 mL; Fisher 13-711-7)

Trays

METHOD

FETAX Embryo Staging and Selection

1. Remove the jelly coat from *Xenopus* embryos by gentle swirling in 2% (w/v) L-cysteine solution for 3–4 min. Rinse embryos with FETAX solution until no L-cysteine odor exists.

2. Divide embryos into several large Petri dishes (100 mm) and add FETAX solution.

3. Level one selection (first sorting). With the aid of a dissecting microscope and transfer or Pasteur pipettes, remove and dispose of all abnormal embryos, i.e., those that are necrotic, mottling, unfertilized or misshapen, or have a bleeding yolk.

4. Level two selection (second sorting). After 1–2 h at 23°C ± 1°C, during which cleavage continues, sort embryos a second time, again removing any abnormal embryos.

   This selection guarantees the highest quality embryos for testing.

5. Hold selected embryos at 23°C ± 1°C until ready to start the FETAX assay.

   The embryos should be at mid-blastula stage 8 to early gastrula stage 11.

FETAX Assay

6. Count out enough small Petri dishes (60 mm) to test the following: each sample concentration in duplicate, four 100% FETAX solution control replicates, four solvent control replicates (only if required for sample solubility, ≤1% (v/v) DMSO is recommended), and two replicates for each of two concentrations (5.5 mg/L and 2500 mg/L) of the reference toxicant, 6-AN.

   Disposable plastic Petri dishes may be used unless the chemical is known to react with plastic. In this case, glass Petri dishes should be used.

7. Place Petri dishes on trays and label dish lids for identification.

8. Record the pH of each sample and control to be tested. For chemical stock solutions, the pH must be between 6.5 and 9.0. If the pH is outside this range, adjust with a weak solution of sodium hydroxide or hydrochloric acid and record the adjusted pH.

9. To each dish, add 10 mL of the appropriate concentration of test material or control solution, corresponding to the labels on the Petri dish lids.

10. Randomize the Petri dishes on each tray.

11. To each Petri dish, add 20 embryos selected at random.

   It is desirable for all embryos used in a test to be from the same clutch of eggs. If this is not possible, new controls must be used for each different clutch tested.

12. Place each tray into an incubator with the temperature maintained at $23°C \pm 1°C$. The assay start time is referred to as study day (SD) 0.

13. Take mortality counts and renew test or control solution in each Petri dish every 24 h for 3 d (SD 1, 2, and 3). Using a dissecting microscope, observe and record larval mortality in each dish. Remove dead larvae from the Petri dishes and discard. Using a plastic transfer pipette, remove most of the 10 mL volume from a Petri dish (do not desiccate embryos completely) and replace with 10 mL of fresh sample or control solution. Repeat this process for each dish, using a new transfer pipette for each different set of samples.

14. After 96 h (SD 4), check the stage of the embryos in the FETAX solution control dishes using a dissecting microscope. If at least 90% of the embryos are at stage 46, the test is ready to take down. If less than 90% are at stage 46, extend the test for up to 3 h or until 90% of embryos have reached stage 46.

15. Using a transfer pipette, remove most of the 10 mL volume from each Petri dish and replace with 200 mg/L MS-222 until the larvae are sedated. Remove the MS-222 and add 8–10 mL of 10% (v/v) NBF to preserve the embryos.

16. Place fixed larvae in a designated holding area for scoring (Step 17), digital measurement of body length (Steps 18 and 19), and archiving (Step 20). Store larvae in 10% NBF until archived (Step 20).

   A designated holding area is a location where the Petri dishes will not be disturbed, such as a wire rack, archive closet, or isolated laboratory bench.

17. Score larvae using a dissecting microscope. List specific malformations (if any), survival, and total specimen counts. Record scoring information and findings using the scoring sheet shown Figure 1.

   See also Bantle et al. 1998 for additional examples of scoring and data sheets. Electronic data scoring sheets are available as Supplemental Material online at cshprotocols.cshlp.org.

18. Take digital photos of each Petri dish ensuring that each larva is clearly represented in the photo. Take additional close-up photos as warranted detailing specific abnormalities. Ensure photos are correctly labeled to guarantee accurate photo identification later.

19. Upload the digital photographs of the test larvae into SigmaScan and measure whole body length, from nose to tip of tail. Transfer the measurement data to Microsoft Excel for statistical evaluation. Determine the percent growth relative to the control before further statistical evaluation.

20. Archive larvae from a given replicate of the control or treatment in labeled 7 mL glass vials with fresh 10% NBF.

Data Analyses and Interpretation

21. Estimate the 4-d LC50 and 4-d EC50 (malformation) using probit, trimmed Spearman-Karber analysis (Hamilton et al. 1977, or an equivalent method. Estimate the IC25 (growth) using the ICp program or equivalent.

SCORE SHEET / MALFORMATION DATA

Investigator:		Client/Project-WO No.:		Test Start Date:	
Test Material:	Controls	Test No.:		Test End Date:	
Test Type:				Test Score Date:	

Malformation:	Sample ID/Concentration (unit)													5.5mg/L 6AN	2500mg/L 6AN
Stunted															
Edema															
Blister															
Head/Facial															
Jaw															
Mouth															
Eye															
Brain															
Cardiac															
Gut															
Hemorrhage															
Tail/Myotome															
Notochord															
Fin															
Other:															
Other:															
Other:															
Other:															
Other:															
Total Malformed															
Total Survived															
Total Tested															
% Mortality															
% Malformed															
Avg % Mortality															
SEM (Mortality)															
Avg % Malformed															
SEM (Malformed)															

FIGURE 1. Examples of scoring sheets for controls and treatments. Electronic data scoring sheets are available as Supplemental Material online at cshprotocols.cshlp.org.

The ICp program will generate a linear interpolation (IC25 estimate), a bootstrap mean, and 95% confidence limits, when appropriate.

22. Calculate teratogenic indices (TI), minimum concentration to inhibit growth (MCIG), generic rank, and definitive rank, as required according to experimental goals.

See the Discussion section for a description of the equations for these values and their application.

DISCUSSION

Hazard Assessment and Discovery Prioritization

FETAX is most useful for the prioritization of developmental toxicity hazard for sets of discovery-level compounds. At least five discovery chemicals can be easily assessed concurrently, and often 20 or more are assessed in a high throughput approach with the goal being to determine which behave most favorably and least favorably in the group. Assessment of teratogenic potential is based on teratogenic indices (TI), which measure the teratogenic potential of a given test material. Ultimately, the TI value represents the area between the curve representing embryo lethality and the curve representing embryonic malformation. By convention, compounds with TI values >1.5 have at least some terato-

Cite this protocol as *Cold Spring Harb Protoc*; doi:10.1101/pdb.prot098319

	SCORE SHEET / MALFORMATION DATA											

Investigator:			Client/Project-WO No.:					Test Start Date:	
Test Material:			Test No.:					Test End Date:	
Test Type:								Test Score Date:	

Malformation:	Sample No./Sample ID/Concentration (*unit*)											
Stunted												
Edema												
Blister												
Head/Facial												
Jaw												
Mouth												
Eye												
Brain												
Cardiac												
Gut												
Hemorrhage												
Tail/Myotome												
Notochord												
Fin												
Other:												
Other:												
Other:												
Other:												
Other:												
Total Malformed												
Total Survived												
Total Tested												
% Mortality												
% Malformed												
Avg % Mortality												
SEM (Mortality)												
Avg % Malformed												
SEM (Malformed)												

FIGURE 1. *Continued.*

genic potential; a larger TI value indicates a greater teratogenic potential. The TI is expressed by the following equation:

$$TI = 4 - day\ LC50/4 - day\ EC50\ (malformation).$$

The relative hazard ranking is based on a set of weighted endpoints that include developmental toxicity potency, teratogenic potential, and growth inhibition. Because of the importance of potency and teratogenic potential in determining relative hazard, these endpoints are weighted 2×. Growth inhibition is weighted 1×. These data determine generic ranking of the test compounds. The generic hazard rank is based on numerical endpoint data derived from an assessment of potency, teratogenic potential as determined by the TI value, and concentration at which growth inhibition is detected when expressed as a proportion of the 4-d LC50 value (see equation below). With the generic rank, the lower the generic hazard ranking the greater the developmental toxicity hazard relative to the other materials evaluated. However, the relative severity of the malformation syndromes induced is not incorporated into the generic ranking process. An objective evaluation of the severity of the deformities induced is an important process in evaluating results of FETAX, or of any other teratological test; therefore, a final definitive ranking process is used. A final definitive ranking is determined using

TABLE 1. Generic and definitive hazard rankings of example test chemicals based on FETAX testing[a]

Test material	Potency (EC50 [malformation]	Teratogenic potential (TI)	Growth inhibition potential (MCIG/LC50)[b]	Generic rank[c]	Severity factor	Definitive rank[d]
1	3 (2)	3 (2)	1	13 (3)	3 (5)	28 (3)
2	4 (2)	4 (2)	4	20 (4)	6 (5)	50 (5)
3	5 (2)	5 (2)	3	23 (5)	4 (5)	43 (4)
4	2 (2)	2 (2)	4	8 (2)	2 (5)	18 (2)
5	6 (2)	6 (2)	6	30 (6)	5 (5)	55 (6)
6-AN[e]	1 (2)	1 (2)	2	6 (1)	1 (5)	11 (1)

[a]In this assessment model, 1 represents the most potent (lowest EC50 [malformation value], greatest TI value, and lowest MCIG value), and 6 the least potent (lowest TI value, and greatest MCIG value) of the set evaluated. The number in parenthesis represents the weighting factor applied. Based on the final definitive ranking scores, the lower the score, the greater the developmental toxicity hazard.

[b]Minimum Concentration to Inhibit Growth/LC50.

[c]Generic Rank = (ranking for potency [EC50, malformation] × 2 [weighting factor]) + (ranking for teratogenic potential [TI] × 2 [weighting factor]) + (ranking for growth inhibition). Presented as total score with rank in parenthesis.

[d]Definitive Rank = (generic ranking [see previous equation]) + (ranking for severity of malformation syndromes × 5 [weighting factor]). Presented as total score with rank in parenthesis.

[e]6-aminonicotinamide reference teratogen provided for comparison.

the generic rank and the severity of the malformations induced (weighted 5×). The fact that the assessment of the severity of malformation is weighted 5× is indicative of the importance of including observational data specific for each material that is not directly reflected in the generic assessment process. As with the generic rank, a lower definitive ranking is indicative of greater potential hazard. This ranking process is based on large FETAX databases accumulated over many years.

Minimum concentration to inhibit growth (MCIG)

= IC25 or Lowest Observed Effect Concentration (LOEC) [growth]/4 − d LC50.

A lower MCIG value indicates greater a growth inhibition potential in the absence of lethality.

The concentrations selected for testing are based on the LC50 and EC50 (malformation), and not on those that inhibit growth; therefore, growth data are not as valuable in evaluating relative developmental toxicity hazard as potency and teratogenic potential. The rationale in using both an assessment of overt teratogenic potential (TI) and relative hazard (generic and definitive hazard rank) is that the combination of these endpoints provides the greatest opportunity to obtain useful data sets using requisite test chemical sample sizes.

Generic Rank

= (ranking for teratogenic potency based on EC50 [malformation] × 2 [weighting factor])

+ (ranking for teratogenic potential[TI] × 2 [weighting factor])

+ ranking for growth inhibition

Definitive Rank = (generic ranking [see previous equation])

+ (ranking for severity of malformation syndromes × 5 [weighting factor])

As an example, a study with five fictitious discovery chemicals is considered in Table 1. Evaluation of the five discovery chemicals is relative to the positive control teratogen, 6-AN. For the determination of generic rank, 6-AN was the most potent in terms of EC50 (malformation warranting an assignment of 1, whereas, test material 5 had the greatest EC50 (least potent) warranting an assignment of six. The same criteria were applied to teratogenic potential as determined by the TI value. Again, 6-AN with the greatest TI value and test material 5 with the lowest TI value were assigned the

Cite this protocol as *Cold Spring Harb Protoc*; doi:10.1101/pdb.prot098319

rank of 1 and 6, respectively. With respect to growth inhibition potential (MCIG/LC50), test material 1 had the lowest MCIG relative to the MCIG values warranting a rank of 1, whereas test material 6 had the greatest value warranting a rank of 6. Multiplying each rank by the rank factor yields the generic rank. Not surprisingly in this set of test materials, 6-AN had the lowest total and test material 5 had the greatest, indicating that 6-AN posed the greatest hazard and test material 5 the least. The severity of the malformation induced is an important factor in defining the ranking of the materials. In this case, the severity of the malformations induced by 6-AN was greater than that of any of the test materials evaluated, with malformations induced by test material 5 being the least severe. The overall values determined indicated that the definitive hazard ranking was 6-AN > 4 > 1 >> 3 > 2 >> 5.

Ecotoxicological Assessment

FETAX data may be used to assess ecological risk through ecological risk assessment. To characterize risk, additional data to estimate ecological responses to chemicals in the environment under given exposure conditions is required. Relevant data include: hazard quotients (HQs = chemical concentration to a selected screening benchmark) to quantify hazard; and estimations of exposure (USEPA 2016). FETAX data may be readily used to determine HQs by supplying benchmark data. Determination of the developmental toxicity of a given chemical, including endpoint sensitivity may be performed using FETAX. Field assessment of relevant environmental concentrations of the given chemical may then be used to assign HQs. It may be helpful to use data from native amphibian species, if such data exist. However, in many cases such data will not be available and *Xenopus* will serve as a useful surrogate. Estimating exposure may include an assessment of the area of concern; addressing the proportion of the area that includes amphibian habitat and the home range of the amphibian species of concern. Assessment of routes of exposure is necessary to determine likelihood. For amphibians, oral (consumption of sediment or food), and dermal (aerial spray or exposure to contaminated soil, sediment, or water) are the most likely routes. Assessment of contaminated food sources including insects and small fish will be necessary. Determination of bioavailability of the chemical of concern and bioaccumulation rates will also be needed in assessing exposure. Finally, assessment of life stage sensitivity to a chemical and its presence in the area of concern is critically important with amphibians. Although fish have been used historically as a surrogate for early life stage amphibians, the process of advanced development, including limb development, metamorphosis, and terrestrial life, cannot be extrapolated from other species. In summary, FETAX data are useful in establishing HQs for ecological risk assessments, including for amphibian species, and such data are needed to provide adequate hazard data.

RELATED INFORMATION

The methods presented herein are based on those of ASTM E1439-98 (2014), Nieuwkoop and Faber (1994), Bantle et al. (1998), Dawson and Bantle (1987), and Fort and Rogers (2005).

RECIPES

3-Aminobenzoic Acid Ethyl Ester (MS-222) Stock Solution (200 mg/L)

Dissolve 0.2 g of MS-222 (Sigma-Aldrich A5040; CAS No. 886-86-2) in 800 mL of FETAX solution <R>. Add 200 mg of sodium bicarbonate (NaHCO$_3$) and dilute to 1 L with FETAX solution. Sterilization is not required. Store at 4°C.

6-Aminonicotinamide (6-AN) Stock Solutions (2500 mg/L and 5.5 mg/L)

For a 2500 mg/L stock solution, dissolve 2.5 g of 6-AN in 1 L of FETAX solution <R>. Store at 1°C–9°C. If crystals form, warm on a heating stir plate until dissolved. Ensure temperature has returned to 23°C ± 1°C before use.

For a 5.5 mg/L stock solution, add 2.2 mL of 2500 mg/L 6-AN stock solution to 1 L FETAX solution <R>. Store at 1°C–9°C.

FETAX Solution

NaCl (Sigma-Aldrich, S9888; CAS No. 7647-14-5; purity 99.0%)	0.625 g
NaHCO$_3$ (Sigma-Aldrich, S8875; 144-55-8; purity 99.0%)	0.096 g
MgSO$_4$ (Sigma-Aldrich, M7506; CAS No. 7487-88-9; purity 99.0%)	0.075 g
CaSO$_4$ • 2 H$_2$O (Sigma-Aldrich, C3771; CAS No. 10101-41-4; purity 99.0%)	0.06 g
KCl (Sigma-Aldrich, P3911; CAS No. 7447-40-7; purity 99.0%)	0.03 g
CaCl$_2$ • 2 H$_2$O (Sigma-Aldrich, C5080; CAS No. 10035-04-8; purity 99.0%)	0.02 g

Dissolve in 1 L of deionized water. Store at 4°C. The final product is a reconstituted water medium (Dawson and Bantle 1987).

ACKNOWLEDGMENTS

Kevin Todhunter, Jennifer Staines, and Elisabeth Alder assisted with the preparation of this protocol.

REFERENCES

American Society for Testing and Materials. 2014. *Standard Guide for Conducting the Frog Embryo Teratogenesis Assay—Xenopus (FETAX), ASTM E1439-98: Annual Book of ASTM Standards*, v. **11.05**, p. 826–836. ASTM, Philadelphia, PA.

Bantle JA, Dumont JN, Finch RA, Linder G, Fort DJ. 1998. *Atlas of abnormalities: a guide for the performance of FETAX*, 2nd ed, pp. 72. Oklahoma State Publications Department.

Dawson DA, Bantle JA. 1987. Development of a reconstituted water medium and initial validation of FETAX. *J Appl Toxicol* **7**: 237–244.

Fort DJ, Rogers R. 2005. Enhanced frog embryo teratogenesis assay: *Xenopus* model using *Xenopus tropicalis*. In *Techniques in aquatic toxicology*, Vol. 2, p. 39–54. CRC Press, Taylor and Francis Group, New York.

Hamilton MA, Russo RC, Thurston RV. 1977. Trimmed Spearman-Karber method of estimated median lethal concentrations in toxicity bioassay. *Environ Sci Technol* **11**: 714–719.

Nieuwkoop PD, Faber J. 1994. *Normal tables of* Xenopus laevis *(Daudin)*, pp. 252. Garland Publishing, London.

USEPA. Conducting and Ecological Risk Assessment: Phase 2 – Characterization, https://www.epa.gov/risk/conducting-ecological-risk-assessment, updated on September 23, 2016, accessed November 22, 2017.

Index